Joseph M. Moran

WEATHER STUDIES
INTRODUCTION TO ATMOSPHERIC SCIENCE

FIFTH EDITION

Education Program

American Meteorological Society

The American Meteorological Society
Education Program

The American Meteorological Society (AMS), founded in 1919, is a **scientific and professional** society. Interdisciplinary in its scope, the Society actively promotes the development and dissemination of information on the atmospheric and related oceanic and hydrologic sciences. AMS has more than 14,000 professional members from more than 100 countries and over 175 corporate and institutional members representing 40 countries.

The Education Program is the initiative of the American Meteorological Society fostering the teaching of the atmospheric and related oceanic and hydrologic sciences at the precollege level and in community college, college and university programs. It is a unique partnership between scientists and educators at all levels with the ultimate goals of (1) attracting young people to further studies in science, mathematics and technology, and (2) promoting public scientific literacy. This is done via the development and dissemination of scientifically authentic, up-to-date, and instructionally sound learning and resource materials for teachers and students.

AMS Weather Studies, a component of the AMS education initiative since 1999, is an introductory undergraduate meteorology course offered partially via the Internet in partnership with college and university faculty. *AMS Weather Studies* provides students with a comprehensive study of the principles of meteorology while simultaneously providing classroom and laboratory applications focused on current weather situations. It provides real experiences demonstrating the value of computers and electronic access to time-sensitive data and information.

Developmental work for *AMS Weather Studies* was supported by the Division of Undergraduate Education of the National Science Foundation under Grant No. DUE - 9752416.

This project was supported, in part,
by the
National Science Foundation
Opinions expressed are those of the authors and
not necessarily those of the Foundation

Weather Studies: Introduction to Atmospheric Science/ Joseph M. Moran. — 5th edition
ISBN-10: 1-935704-95-8
ISBN-13: 978-1-935704-95-9
Copyright © 2012 by the American Meteorological Society

Published by the American Meteorological Society
45 Beacon Street, Boston, MA 02108

The 1st and 2nd editions of this book were published under the title *Online Weather Studies*.

Cover photograph: Iliamna Volcano in the Aleutian Chain, Alaska, by Michael Melford © National Geographic Stock.

BRIEF CONTENTS

Preface

Acknowledgements

Chapter 1	Monitoring the Weather	1
Chapter 2	Atmosphere: Origin, Composition & Structure	27
Chapter 3	Solar & Terrestrial Radiation	61
Chapter 4	Heat, Temperature & Atmospheric Circulation	105
Chapter 5	Air Pressure	141
Chapter 6	Humidity, Saturation & Stability	165
Chapter 7	Clouds, Precipitation & Weather Radar	203
Chapter 8	Wind & Weather	247
Chapter 9	Atmosphere's Planetary Circulation	277
Chapter 10	Weather Systems of Middle Latitudes	321
Chapter 11	Thunderstorms & Tornadoes	361
Chapter 12	Tropical Weather Systems	403
Chapter 13	Weather Analysis & Forecasting	439
Chapter 14	Light & Sound in the Atmosphere	471
Chapter 15	Climate & Climate Change	493
Appendix I	Conversion Factors	541
Appendix II	Milestones in the History of Atmospheric Science	543
Appendix III	Climate Classification	553
Glossary		561
Index		595

BRIEF CONTENTS

CONTENTS

CHAPTER 1: MONITORING THE WEATHER   1

Case-in-Point
    Blizzards Then and Now 2
Weather and Climate 4
Accessing Weather Information 5
Time Keeping 6
Weather Systems and Weather Maps 7
Describing the State of the Atmosphere 12
Weather Satellite Imagery 13
Weather Radar 16
Sky Watching 18
Conclusions 19
Basic Understandings/Enduring Ideas 19
Key Terms/Review/Critical Thinking 21
For Further Explanation
    Weather Maps, Historical Perspective 22
    Global Positioning System 24

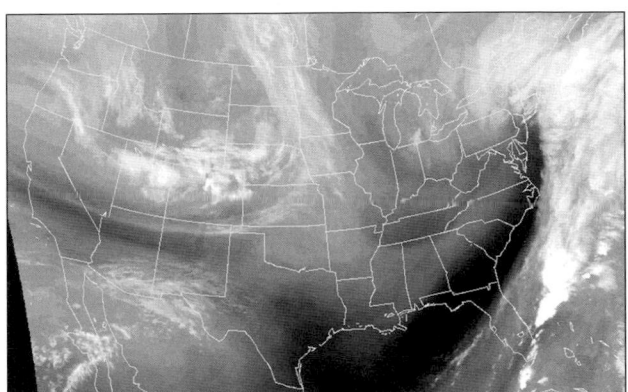

CHAPTER 2: ATMOSPHERE: ORIGIN, COMPOSITION & STRUCTURE   27

Case-in-Point
    African Origins of Wind-Borne Dust in the Americas 28
Evolution of the Atmosphere 31
    Primeval Phase 31
    Modern Phase 34
    Air Pollution 35
Investigating the Atmosphere 37
    The Scientific Method 37
    Atmospheric Models 38
Monitoring the Atmosphere 39
    Surface Observations 39
    Historical Perspective 40
    Upper-Air Observations 42
    Historical Perspective 43
    Remote Sensing 45
Temperature Profile of the Atmosphere 46
The Ionosphere and the Aurora 47
Conclusions 49
Basic Understandings/Enduring Ideas 49
Key Terms/Review/Critical Thinking 51
For Further Exploration
    The Atmosphere of Mars 52

Radio Transmission and the Ionosphere 55
Space Weather Prediction 57

CHAPTER 3: SOLAR & TERRESTRIAL RADIATION 61

Case-in-Point
    Ancient Astronomical Calendars 62
Electromagnetic Spectrum 64
Radiation Laws 66
Input of Solar Radiation 68
    Solar Altitude 69
    Earth's Motions in Space and the Seasons 70
    The Solar Constant 74
Solar Radiation and the Atmosphere 76
Stratospheric Ozone Shield 78
Solar Radiation and Earth's Surface 83
Global Solar Radiation Budget 84
Outgoing Infrared Radiation 85
    The Greenhouse Effect 85
    Greenhouse Gases 86
    The Callendar Effect 87
    Possible Impacts of Global Warming 90
Monitoring Radiation 91
Conclusions 93
Basic Understandings/Enduring Ideas 93
Key Terms/Review/Critical Thinking 95
For Further Exploration
    Hazards of Solar Ultraviolet Radiation 96
    Solar Power 100

CHAPTER 4: HEAT, TEMPERATURE & ATMOSPHERIC CIRCULATION 105

Case-in-Point
    Extreme Heat of Death Valley, CA 106
Distinguishing Temperature and Heat 108
    Temperature Scales and Heat Units 108
    Measuring Air Temperature 110
Heat Transfer Processes 112
    Radiation 112
    Conduction and Convection 112
    Phase Changes of Water 114
Thermal Response and Specific Heat 115
    Thermal Inertia 116
    Maritime and Continental Climates 116
Heat Imbalance: Atmosphere versus Earth's Surface 117
    Latent Heating 118
    Sensible Heating 119
    Bowen Ratio 120

Heat Imbalance: Tropics versus Middle and High Latitudes 121
    Heat Transport by Air Mass Exchange 122
    Heat Transport by Storms 122
    Heat Transport by Ocean Circulation 122
Why Weather? 123
Variation of Air Temperature 124
    Local Radiation Budget 124
    Cold and Warm Air Advection 125
    Urban Heat Island Effect 126
Conclusions 129
Basic Understandings/Enduring Ideas 129
Key Terms/Review/Critical Thinking 131
For Further Exploration
    Managing Weather Risk 133
    Heating and Cooling Degree-Days 135
    Wind Chill 137

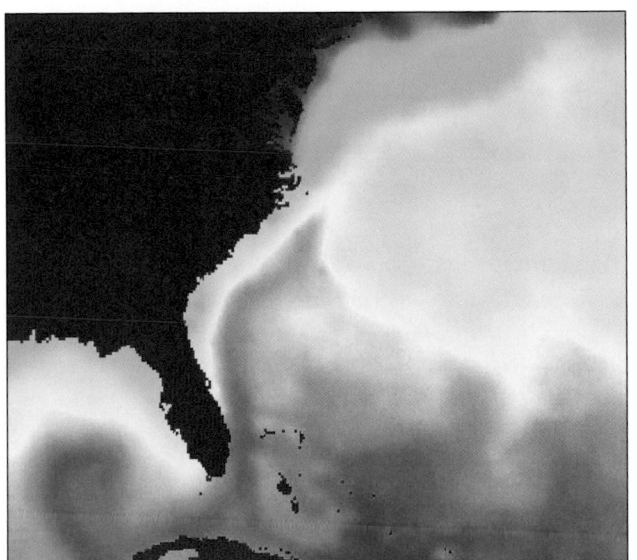

CHAPTER 5: AIR PRESSURE   141

Case-in-Point
    Air Pressures on Mount Everest 142
Defining Air Pressure 143
Air Pressure Measurement 143
Air Pressure Units 145
Variation in Air Pressure with Altitude 145
Horizontal Variations in Air Pressure 148
    Influence of Temperature and Humidity 149
    Influence of Diverging and Converging Winds 151
Highs and Lows 151
The Gas Law 151
Expansional Cooling and Compressional Warming 152
    Conservation of Energy 153
    Adiabatic Process 154
Conclusions 155
Basic Understandings/Enduring Ideas 155
Key Terms/Review/Critical Thinking 157
For Further Exploration
    Human Responses to Changes in Air Pressure 158
    Comparing Air and Water Pressure 161
    Determining Altitude from Air Pressure 162

CHAPTER 6: HUMIDITY, SATURATION & STABILITY   165

Case-in-Point
    Atmospheric Rivers 166
Global Water Cycle 168
    Transfer Processes 170
    Global Water Budget 171

How Humid Is It? 172
    Vapor Pressure 172
    Mixing Ratio, Specific Humidity, and Absolute Humidity 172
    Saturated Air 173
    Relative Humidity 175
    Dewpoint 175
    Precipitable Water 176
Monitoring Water Vapor 178
    Humidity Instruments 178
    Water Vapor Satellite Imagery 181
How Air Becomes Saturated 182
Atmospheric Stability 183
    Soundings 184
    Stüve Diagram 185
Lifting Processes 187
Conclusions 189
Basic Understandings/Enduring Ideas 189
Key Terms/Review/Critical Thinking 191
For Further Exploration
    Heat, Humidity and Human Comfort 193
    Measuring Evaporation 197
    Atmospheric Stability and Air Quality 198

CHAPTER 7: CLOUDS, PRECIPITATION & WEATHER RADAR   **203**

Case-in-Point
    Aircraft Contrails and Cloud Cover 204
Cloud Formation 206
    The Curvature Effect 206
    Role of Nuclei 206
    Supercooled Water 208
Cloud Classification 208
    High Clouds 209
    Middle Clouds 210
    Low Clouds 211
    Clouds Having Vertical Development 212
    Unusual Clouds 213
Fog 216
Precipitation Processes 219
    Terminal Velocity 219
    Warm-Cloud Precipitation 220
    Cold-Cloud Precipitation 221
    Holes and Canals in Clouds 222
Forms of Precipitation 223
Acid Deposition 227
Weather Radar: Locating Precipitation 229
    Reflectivity Mode 229
    Velocity (Doppler) Mode 230
    Dual-Polarization Weather Radar 232

Phased Array Weather Radar 233
Measuring Precipitation 233
    Rain and Snow Gauges 233
    Remote Sensing of Precipitation 235
Conclusions 236
Basic Understandings/Enduring Ideas 237
Key Terms/Review/Critical Thinking 239
For Further Exploration
    Clouds by Mixing 241
    Rainmaking 242
    When Is It Too Cold or Too Warm to Snow? 245

CHAPTER 8: WIND & WEATHER   247

Case-in-Point
    Sinking of the Edmund Fitzgerald 248
Forces Governing the Wind 249
    Pressure Gradient Force 249
    Centripetal Force 250
    Coriolis Effect 251
    Friction 253
    Gravity 254
    Summary 254
Wind: Joining Forces 255
    Hydrostatic Equilibrium 255
    Geostrophic Wind 255
    Gradient Wind 256
    Surface Winds in Highs and Lows 257
Continuity of Wind 260
Monitoring Wind Speed and Direction 261
Scales of Atmospheric Circulation 265
Conclusions 266
Basic Understandings/Enduring Ideas 266
Key Terms/Review/Critical Thinking 268
For Further Exploration
    Lake-Effect Snow 269
    Wind Profilers 272
    Wind Power 273

CHAPTER 9: ATMOSPHERE'S PLANETARY CIRCULATION   277

Case-in-Point
    Drought in the Sahel of West Africa 278
Idealized Circulation Pattern 280
Features of the Planetary-Scale Circulation 281
    Pressure Systems and Wind Belts 282
    Winds Aloft 283
    Trade Wind Inversion 284

Seasonal Shifts 285
Ocean Surface Currents 286
Monsoon Circulation 287
Long Waves in the Westerlies 290
Zonal and Meridional Flow Patterns 290
Blocking Systems and Weather Extremes 291
Jet Streams 294
Cyclone Development 295
El Niño and La Niña 296
Historical Perspective 297
Ekman Transport 299
Neutral Conditions in the Tropical Pacific 301
Warm Phase 302
Cold Phase 304
Predicting and Monitoring ENSO 306
Frequency of El Niño and La Niña 309
North Atlantic Oscillation 310
Arctic Oscillation 311
Pacific Decadal Oscillation 312
Conclusions 313
Basic Understandings/Enduring Ideas 313
Key Terms/Review/Critical Thinking 315
For Further Exploration
Defining Drought 317
Historic El Niño Episodes of 1982-83 and 1997-98 319

CHAPTER 10: WEATHER SYSTEMS IN MIDDLE LATITUDES 321

Case-in-Point
Early Observations on East Coast Cyclones 322
Air Masses 323
North American Types and Source Regions 323
Air Mass Modification 324
Frontal Weather 325
Stationary Front 326
Warm Front 327
Cold Front 328
Occluded Fronts 329
Summary 330
Extratropical Cyclones 331
Life Cycle 331
Conveyor Belt Model 333
Cyclone Weather 335
Principal Cyclone Tracks 336
Cold Side/Warm Side 337
Winter Storms 339
Cold- and Warm-Core Systems 340
Anticyclones 341
Arctic and Polar Highs 341

       Warm Highs 342
       Anticyclone Weather 342
Local and Regional Circulation Systems 343
       Sea (or Lake) Breeze and Land Breeze 344
       Mountain Breeze and Valley Breeze 344
       Chinook Wind 345
       Santa Ana Wind 346
       Katabatic Wind 347
       Desert Winds 348
       Heat Burst 349
Conclusions 350
Basic Understandings/Enduring Ideas 350
Key Terms/Review/Critical Thinking 352
For Further Exploration
       The Case of the Missing Storm 354
       Nor'easters 355
       Santa Ana Winds and California Fire Weather 358

CHAPTER 11: THUNDERSTORMS & TORNADOES   361

Case-in-Point
       Super Tornado Outbreak of 2011 362
Thunderstorm Life Cycle 364
       Towering Cumulus Stage 364
       Mature Stage 365
       Dissipating Stage 367
Thunderstorm Classification 367
Where and When 369
Severe Thunderstorms 371
Thunderstorm Hazards 373
       Lightning 373
       Downbursts 376
       Derecho 378
       Flash Floods 379
       Hail 381
Tornadoes 384
Tornado Characteristics 384
Where and When 385
Tornado Hazards and the EF-Scale 387
The Tornado-Thunderstorm Connection 389
Monitoring Tornadic Thunderstorms 391
Conclusions 393
Basic Understandings/Enduring Ideas 394
Key Terms/Review/Critical Thinking 396
For Further Exploration
       Lightning Safety 397
       Hail Suppression 399
       Tornado Look-Alikes 401

CHAPTER 12: TROPICAL WEATHER SYSTEMS 403

Case-in-Point
    Lessons Learned from Hurricane Katrina 404
Weather in the Tropics 406
Hurricane Characteristics 406
Where and When 408
Hurricane Life Cycle 411
Hurricane Hazards 413
    Inland Flooding 413
    Wind 417
    Storm Surge 417
    Saffir-Simpson Hurricane Wind Scale 420
Trends in Hurricane Frequency 423
Hurricane Threat to the Southeast United States 424
    Barrier Islands 425
    Evacuation 426
Long Range Forecasting of Atlantic Hurricanes 428
Hurricane Modification 429
Conclusions 429
Basic Understandings/Enduring Ideas 429
Key Terms/Review/Critical Thinking 431
For Further Exploration
    Polar Lows with Hurricane Characteristics 433
    Naming Hurricanes 434
    Variability in Atlantic Hurricane Activity 437

CHAPTER 13: WEATHER ANALYSIS & FORECASTING 439

Case-in-Point
    Evolution of Tornado Forecasting 440
International Cooperation 441
Acquisition of Weather Data 442
    Surface Weather Observations 442
    Upper-Air Weather Observations 444
Weather Data Assimilation, Depiction and Analysis 445
    Surface Weather Maps 446
    Upper-Air Weather Maps 446
Weather Prediction 449
    Numerical Weather Forecasting 449
    Forecasting Tropical Cyclones 451
    Forecasting for Aviation 454
    Forecasting Severe Storms 455
    River and Flood Forecasting 456
    Marine Forecasting 457
    Space Weather Forecasting 457
    Forecast Skill 457
    Long-Range Forecasting 458
    Single-Station Forecasting 460

       Private Sector Forecasting 461
Communication and Dissemination 461
       Weather-Ready Nation 462
Conclusions 462
Basic Understandings/Enduring Ideas 462
Key Terms/Review/Critical Thinking 464
For Further Exploration
       Marine Weather Statements 466
       Chaos and the Limits to Forecasting 468

CHAPTER 14: LIGHT & SOUND IN THE ATMOSPHERE   471

Case-in-Point
       Fata Morgana 472
Atmospheric Optics 473
       Visible Light and Color Perception 473
       Red Sun, White Clouds, and Blue Sky 473
       Halo 475
       Rainbow 478
       Corona 480
       Glory 480
       Mirage 481
       Sunrise-Sunset and Twinkling Stars 482
       Twilight 483
Atmospheric Acoustics 484
       Sound Waves 484
       Thunder 485
       Sonic Boom 486
       Aeolian Sounds 486
Conclusions 486
Basic Understandings/Enduring Ideas 486
Key Terms/Review/Critical Thinking 489
For Further Exploration
       Blue Haze 490

CHAPTER 15: CLIMATE & CLIMATE CHANGE   493

Case-in-Point
       Global Warming and Sea Level Rise 494
Earth's Climate System 495
       The Climatic Norm 495
       Climatic Anomalies 496
Climate Boundary Conditions 497
Global Climate Patterns 498
       Temperature 498
       Precipitation 499
       Climate Classification 503
The Climate Record 503

Geologic Time 503
Past Two Million Years 506
Instrument-Based Climate Trends 509
Causes of Climate Variability and Change 510
Solar Variability 512
Earth's Orbit 514
Volcanoes 516
Earth's Surface Properties 518
Human Activity 518
Anthropogenic versus Natural Forcing of Climate 520
The Climate Future 521
Global Climate Models 521
Search for Cycles and Analogs 521
Enhanced Greenhouse Effect and Global Warming 522
Potential Impacts of Global Climate Change 523
Shrinking Glaciers and Rising Sea Level 523
Arctic Environment 527
Other Impacts 530
Conclusions 530
Basic Understandings/Enduring Ideas 530
Key Terms/Review/Critical Thinking 533
For Further Exploration
Sources of Proxy Climate Data 535
Lessons of the Climate Past 540

Grinnell Glacier from Mt. Gould

APPENDIX I: CONVERSION FACTORS   **541**

APPENDIX II: MILESTONES IN THE HISTORY OF ATMOSPHERIC SCIENCE   **543**

APPENDIX III: CLIMATE CLASSIFICATION   **553**

GLOSSARY   **561**

INDEX   **595**

PREFACE

Welcome to *Weather Studies*! You are embarking on an exciting study of the science of the atmosphere. The purpose of this book is to provide you with background information on the properties of the atmosphere, the scientific principles that govern weather and climate, how the atmosphere interacts with the other components of the Earth system, and the implications of those interactions for humankind. *Weather Studies* was developed by the Education Program of the American Meteorological Society (AMS) with support from the National Science Foundation (NSF) and the National Oceanic and Atmospheric Administration (NOAA).

Weather Studies may serve as a stand-alone textbook in a traditional undergraduate college course on atmospheric science, meteorology, or weather and climate. *Weather Studies* also serves as the textbook for *AMS Weather Studies*, a turnkey course package developed, licensed, and nationally implemented by AMS. A companion *Investigations Manual* plus course website provide students with twice-weekly investigations, partially delivered online, that use real-world data to complement chapter goals.

AMS Weather Studies is guided by new findings of learning science redefining what it means to be proficient in science. According to the National Research Council of the National Academies, Board on Science Education (2007), students who are proficient in science "(1) know, use, and interpret scientific explanations of the natural world, (2) generate and evaluate scientific evidence and explanations, (3) understand the nature and development of scientific knowledge, and (4) participate productively in scientific practices and discourse." These strands of proficiency are learning goals in *AMS Weather Studies* that address the knowledge and reasoning skills essential for students to become proficient in the basic understandings of meteorology/climatology so that they may participate as atmospheric science literate citizens.

The course package, *AMS Weather Studies*, follows learning science in providing strategically designed "student encounters with science that take place in real time and over a period of months and years (e.g., *learning progressions*)." *AMS Weather Studies* seeks to engage learners in exploring their world by investigating meaningful questions. Investigations, tied directly to each chapter, have printed and online components that make use of atmospheric information/data available via the *AMS Weather Studies course website*. Investigations engage students in observation, prediction, data analysis, inference, and critical thinking. The course presents opportunities for students to collaborate with their instructor and fellow students as together they negotiate understanding. Application of information-age technology provides the student with experience in retrieving and analyzing real-world data (much in real-time) and sharing interpretations. Throughout the course, students assemble learning materials for assessment purposes.

AMS Weather Studies engages the reader in scientific inquiry and the scientific method. Atmospheric science is particularly suited to these aims because it is an applied science that readily lends itself to familiar everyday illustrations. The inquiry-based approach is designed to promote development of critical thinking skills. The emphasis on scientific methodology provides a perspective on the contributions of many atmospheric scientists and the challenges still facing the field. Early on, the reader realizes that weather is not an arbitrary act of nature, weather forecasting has its limits, and the climate future is uncertain. Integrated throughout the book are topics of contemporary interest including, for example, threats to the stratospheric ozone shield, severe weather, and the potential implications of global climate change.

This book is divided into 15 chapters. Each of the first 12 chapters corresponds to one week of the *AMS Weather Studies* course. Chapters are organized so that concepts build logically one upon the other so that Earth's atmosphere emerges as an interactive system subject to physical laws. Topics covered include sources of weather information (Chapter 1), origin, composition and structure of the atmosphere (Chapter 2), Earth's radiation budget and controls of temperature (Chapters 3 and 4), air pressure (Chapter 5), humidity, clouds and precipitation (Chapter 6 and 7), forces governing atmospheric circulation (Chapter 8), planetary-scale circulation (Chapter 9), weather systems of middle latitudes (Chapter 10), thunderstorms and tornadoes (Chapter 11), and tropical storms and hurricanes (Chapter 12). The final three chapters cover weather analysis and forecasting (Chapter 13), atmospheric optics and acoustics (Chapter 14), and climate and climate change (Chapter 15). All chapters have accompanying investigations in the *Investigations Manual*.

Each chapter opens with a driving question, a broad-based query that links chapter concepts and provides a central focus for that week's study, a list of learning objectives, and a *Case-in-Point*, an authentic, relevant, and real-life event or issue that highlights or applies one or more of the main concepts covered in the chapter. The Case-in-Point previews the chapter and seeks to engage reader interest early on. Chapter 12 (Tropical Weather Systems), for example, opens with a discussion of lessons learned from Hurricane Katrina (2005). Chapter content is science-rich and features some of the most recent results of meteorological research. Each chapter closes with a list of *Basic Understandings* and *Enduring Ideas*, *Key Terms* as well as questions for *Review* and *Critical Thinking*. One or more essays ("*For Further Exploration*") that appear at the end of each chapter address in some depth specific topics that complement or supplement a concept covered in the narrative. Examples include *Space Weather Prediction, Extreme Heat of Death Valley, CA, Lake-Effect Snow*, and *Lessons of the Climate Past*. *Historical Perspectives*, included in many chapters, provide an in depth story of how some of the major innovations in the science and technology of meteorology developed. All terms bold-faced in the narrative are defined in the *Glossary* at the back of the book. Appendixes cover unit conversions, milestones in the history of atmospheric science, and climate classification.

In this fifth edition, we have updated the science and case studies, added many new photographs and line drawings, and introduced QR codes enabling access to animations. New, expanded or significantly revised topics include the evolution of Earth's atmosphere (Chapter 2), record low levels of ozone in the Arctic stratosphere (Chapter 3), atmospheric rivers (Chapter 6), acid deposition (Chapter 7), upgrade of weather radar (Chapter 7), La Niña of 2011 and associated weather extremes (Chapter 9), local and regional circulation systems (Chapter 10), record-breaking 2011 tornado season (Chapter 11), derecho and bow echo (Chapter 11), Hurricane Irene (Chapter 12), and the role of feedback in climate variability and change (Chapter 15).

In summary, *AMS Weather Studies* is pedagogically guided by a teaching approach where students explore their world by investigating meaningful questions. *Weather Studies* plus the investigations inspire students to explore many of their own questions. Investigations have printed and online components that make use of weather and climate data available via the course website. Investigations engage students in observation, prediction, data analysis, inference, and critical thinking (processes of science). The course presents opportunities for collaboration among students as together they negotiate understanding. Application of information-age technology helps students to develop their ability to retrieve and analyze real-world data and share interpretations. Throughout the course, students assemble a variety of artifacts which may be used for authentic assessment. They develop core enduring ideas in atmospheric science, to carry with them throughout the course and following course completion.

James A. Brey
Director, AMS Education Program

ACKNOWLEDGEMENTS

AMS Weather Studies learning materials are the products of collaboration among many individuals with extensive scientific backgrounds and teaching experience. This book is primarily the work of Joseph M. Moran of the American Meteorological Society's Education Program and Professor Emeritus of Earth Science at the University of Wisconsin-Green Bay.

Elizabeth W. Mills, Associate Director of the AMS Education Program, had a major editorial role in researching topics, interpreting and synthesizing the suggestions of reviewers, and assembling the manuscript for internal consistency and publication. Kira A. Nugnes, Managing Editor/Content Specialist, AMS Education Program, oversaw the editorial process of the textbook from beginning to end. She played a major role in reviewing chapters, finding appropriate figures and researching recent weather events. Bernard A. Blair, Education Publications & Instructional Technology Manager, AMS Education Program, met the numerous challenges in formatting the final manuscript into this book with his exceptional skill and attention to detail, and oversaw overall book production.

The fifth edition of *Weather Studies* was a team effort that benefited greatly from critical content and editorial reviews provided by the professional staff of the AMS Education Program: James A. Brey, Director, Ira W. Geer, Senior Education Fellow, Katie L. O'Neill, Content Specialist, and Robert S. Weinbeck (also of SUNY College at Brockport). AMS staff members Maureen N. Moses and Lindsey A. Johnson played a major role in designing the textbook cover.

We acknowledge with much appreciation the advice, constructive criticisms and other contributions to *Weather Studies* provided by the following individuals: Edward J. Hopkins of the University of Wisconsin-Madison, Steven J. Meyer of the University of Wisconsin-Green Bay, and Margaret F. Boorstein of the C.W. Post Campus of Long Island University. We especially thank Louis W. Uccellini, Director of the NOAA National Centers for Environmental Prediction and Margaret Anne (Peggy) LeMone, Senior Scientist Emerita, National Center for Atmospheric Research (NCAR) and author of *The Stories Clouds Tell* for their ongoing advice and support. We thank all past instructors of *AMS Weather Studies* for their very helpful comments.

We acknowledge with great appreciation the assistance of the following persons who answered questions about supplied images, or reviewed sections for this fifth edition: Gregory Dalyai, Melody Magnus, Donald K. Rinker, and Paul Stokols of NOAA Headquarters; Jason Dunion of the NOAA/AOML/Hurricane Research Division; Edward J. Olenic and David Unger of the NOAA Climate Prediction Center; John Derber, Geoff DiMego, Mary Hart, Dingchen Hou, Shrinivas Moorthi, and Yuejian Zhu of the NOAA Environmental Modeling Center; Eric Blake, James L. Franklin, and Christopher W. Landsea of the NOAA National Hurricane Center; Adam Allgood, Michael J. Brewer, Jake Crouch, Richard Heim, and David Miskus of the NOAA National Climatic Data Center; Melissa Buras of the NOAA National Data Buoy Center; Barry W. Eakins of the NOAA National Geophysical Data Center; Susan Cobb of the NOAA National Severe Storms Laboratory; Douglas A. Biesecker of the NOAA Space Weather Prediction Center; Greg Carbin, and Andy Dean of the NOAA Storm Prediction Center; Jerry Griffin of the NOAA/NWS Training Center; Antonia Hedrick of the Bureau of Land Management; William Hawthorne and John Whelan of the Federal Aviation Administration; Julie Jones of the National Renewable Energy Laboratory; Lin Chambers of the NASA Langley Research Center; Jerome Lafeuille and Sophie Schlingemann of the World Meteorological Organization; Clark Christensen and Hyrum K. Wright of Brigham Young University; Phil Klotzbach of Colorado State University; Peter J. Webster of the Georgia Institute of Technology; Alex DeCaria and the entire Millersville University Meteorology Program; Alan Robock of Rutgers University; Brian Fuchs of the University of Nebraska-Lincoln; Jim Kurdzo of the University of Oklahoma; William R. Iseminger of the Cahokia Mounds State Historic Site; Keith S. Barr of Lockheed Martin Coherent Technologies; Lindsay Nyce of the Raytheon Company; Charles Whisenant of The Arab Tribune; Lauren Newberry of the Weather Risk Management Association; and the National Snow and Ice Data Center.

We continue to acknowledge the assistance of persons whose help was instrumental in previous editions of this text: William Blackmore, Joel W. Cline, Laura A. Cook, Ronald S. Gird, and Douglas C. Young of NOAA Headquarters; David J. Schwab of the NOAA Great Lakes Environmental Research Laboratory; Paul J. Kocin of the NOAA Hydrometeorological Prediction Center; Lauren L. Morone of NOAA NCEP Central Operations; Colin J. McAdie, Robert J. Berg, and William L. Read of the NOAA National Hurricane Center; Robert A. Luke of the NOAA National Data Buoy Center; Timothy D. Crum of the NOAA Radar Operations Center; Joseph T. Schaefer and Roger Edwards of the NOAA Storm Prediction Center; Joan O'Bannon of the NOAA National Severe Storms Laboratory; Jennifer M. Dover of the NOAA Sterling Field Support Center; Brian C. Ciemnecki and Jeffrey S. Tongue of NOAA/NWS New York City Forecast Office; Ronald J. Miller of NOAA/NWS Spokane Forecast Office; Kent R. Baxter, Brandon Bolinski, Daniel Catlett, Robin Danforth, Matthew Green, and Marshall L. Mabry of FEMA; Barbara A. Cohen of the NASA Marshall Space Flight Center; Lois Sumey of the National Park Service Rocky Mountain National Park; Christopher F. Keohan of the International Civil Aviation Organization; Miroslav Ondras and Judith C. C. Torres of the World Meteorological Organization; Daniel Reinert of Johannes Gutenberg Universität Mainz; Fred M. Phillips of New Mexico Tech University; Paul G. Knight of Penn State University; Paul E. Roundy of SUNY at Albany; Murray Cameron Peel of The University of Melbourne; Robert A. Norheim of the University of Washington; Roscoe Clark of the Casualty Research Associates; Greg Payne of the Franklin County, OH Engineer's Office; John A. Dutton of Storm Exchange; Valerie Cooper of the Weather Risk Management Association; and the NOAA Missouri Basin River Forecast Center.

Norman J. Frisch of Brockport, NY did an excellent job of turning line drawings into final art. Bernard A. Blair and Elizabeth W. Mills of the AMS Education Program met the numerous technical challenges of editing and formatting the book with their exceptional attention to detail, dedication, and enthusiasm.

A special note concerns the use of units in *Weather Studies*. Generally the International System of Units (abbreviated SI, for Systéme Internationale d'Unitès) is employed with equivalent English or other units following in parentheses. Exceptions are units used by convention or convenience in meteorology or the user community (e.g., knots, calories, millibars). Also, the equivalence between units is given in context; that is, where general estimates are used, approximate values are shown in all units. Conversion factors are given in Appendix I.

Development work culminating in *AMS Weather Studies* was initially supported by the National Science Foundation under Grant No. DUE-9752416. Subsequent development of *AMS Weather Studies* was funded by the National Science Foundation through its *Opportunities for Enhancing Diversity in the Geosciences (OEDG) Program*, Grant No. GEO-0119740, and its *Course, Curriculum and Laboratory Improvement–National Dissemination (CCLI-ND) Program*, Grant No. DUE-0126032.

WEATHER STUDIES
INTRODUCTION TO ATMOSPHERIC SCIENCE

A National Oceanic and Atmospheric Administration (NASA) scientist launching an ozonesonde weather balloon at Summit, a National Science Foundation (NSF) supported research station on the Greenland Ice Sheet. [Courtesy NASA]

MONITORING THE WEATHER

Chapter Highlights

Case-in-Point
   Blizzards Then and Now
Weather and Climate
Accessing Weather Information
Time Keeping
Weather Systems and Weather Maps
Describing the State of the Atmosphere
Weather Satellite Imagery
Weather Radar
Sky Watching
Conclusions
Basic Understandings/Enduring Ideas
Key Terms/Review/Critical Thinking
For Further Exploration
   Weather Maps, Historical Perspective
   Global Positioning System

Learning Objectives

Distinguish between weather and climate, meteorology and climatology.

List some of the readily available sources of regional and national weather information.

Explain why the drawing of a weather map requires weather observations that are simultaneous.

Identify the principal weather systems that are plotted on a surface weather map.

Distinguish between the types of weather usually associated with high pressure systems and low pressure systems.

Characterize the relationship between air masses and fronts.

Distinguish between geostationary and polar-orbiting weather satellites.

Identify the advantages of infrared satellite images versus visible satellite images in monitoring the Earth-atmosphere system.

Present the fundamental principles of weather radar.

Give some examples of how the habit of observing the sky and changing cloud cover might provide indications of future changes in the weather.

What are some basic characteristics of the atmosphere and weather?

Case-in-Point

Blizzards Then and Now

Across the Dakota Territory and Nebraska, 12 January 1888 began mild, almost spring-like, a welcome respite from several weeks of persistent bitter cold. Farmers took advantage of the pleasant weather, attending to chores which took them into pastures some distance from the safety of their homes. Neither the farmers nor the other residents of the isolated settlements that dotted the Northern Prairie could know that a ferocious blizzard was sweeping southeastward over eastern Montana and bearing down on the Dakota Territory and Nebraska with tragic consequences. The blizzard struck suddenly with winds strengthening to gale-force, plunging temperatures, and swirling snow that cut visibility to near zero. In some places the wind-chill plummeted to near −40 °C (−40 °F). Many farmers and others caught out in the open became disoriented and were unable to find their way to shelter.

In Nebraska, the storm hit just as school was letting out for the day. Because the day started mild, many students were inadequately dressed for the rapidly deteriorating conditions as they trudged home from country schools through blinding snow and cold winds. More than a hundred children never made it to their destinations, victims of exposure (hypothermia). All told, perhaps as many as 500 people perished from the storm, the deadliest blizzard on record in the American prairie. Because so many school children perished, the storm is referred to as the *Children's Blizzard*.

About two months later, on 12-14 March 1888, a powerful coastal storm accompanied by heavy snow and strong winds blasted the Eastern Seaboard from Washington, DC, northward into southern New England. This *Blizzard of '88* brought to a standstill virtually all activity. Particularly hard hit was New York City (Figure 1.1) which reported a three-day snowfall total of 53 cm (21 in.). Winds up to 65 km per hr (40 mph) whipped snow into drifts 4.5 to 6.0 m (15 to 20 ft) deep. Much greater snowfalls (generally 100 to 125 cm or 40 to 50 in.) and deeper snowdrifts (to 12 m or 40 ft) were reported over southeastern New York, western Connecticut, and western Massachusetts. On land, more than 300 people lost their lives to exposure, accidents, or overexertion, while at sea almost 200 ships were sunk or damaged with nearly 100 lives lost. Hundreds of trains stalled in deep snow drifts, marooning passengers for days; and telephone and telegraph lines were severed all over the Northeast.

The *Children's Blizzard* and the *Blizzard of '88* were total surprises for the people impacted. At the time, the U.S. Army Signal Corps had operated the nation's weather service for more than 17 years, and observers at some 154 weather stations nationwide telegraphed weather observations three times daily to headquarters in Washington, DC. *Indications* (later called weather forecasts) were issued based on surface weather observations alone. In St. Paul, MN, an experimental office was charged with forecasting cold waves and blizzards for the Northern Plains. The technology of the day and the limited understanding of the workings of the atmosphere put meteorologists at a distinct disadvantage in forecasting storms, especially storms that tracked offshore, parallel to the Atlantic coast (called *nor'easters*). There was no means of monitoring the upper atmosphere, no satellites, no radar, no wireless communications with ships at sea, and no computer models for forecasting the weather.

Efforts to communicate weather warnings to the public were often unsuccessful. The public received most (if not all) of their weather information from newspapers that may have

FIGURE 1.1
Scene of the aftermath of the blizzard of 12-14 March 1888 in New York City. [NOAA Photo Library, Historic NWS Collection]

been a day or two old. The officer in charge of the St. Paul office knew that a blizzard was heading toward the Dakota Territory and Nebraska and sent warnings via telegraph. However, the telegraph lines ran along the railroad tracks, too far away from most of the region's residents for them to receive any advance warning.

The historic blizzards of 1888 were rivaled in ferocity by four nor'easters that hit major metropolitan areas of the mid-Atlantic during the winter of 2009-2010. Each storm brought widespread 25-50 cm (10-20 in.) swaths of wind-driven snowfall, with greater amounts locally. Three of the 2010 storms impacted the 15 million residents of the Washington, DC-Philadelphia, PA, urban corridor (Figure 1.2). Two arrived back-to-back (4-7 and 9-11 February), causing widespread power outages and shutting down commercial aviation, ground transportation, shipping, and the local economy. School systems closed for at least a week and offices of the Federal Government closed for four successive days. These storms were primarily responsible for establishing new seasonal snowfall records in Richmond, VA (71.1 cm or 28.0 in.), Washington, DC (142.2 cm or 56.0 in.), Baltimore, MD (195.6 cm or 77.0 in.), and Philadelphia, PA (199.9 cm or 78.7 in.).

FIGURE 1.2
The Capitol Building in Washington, DC, on 6 February 2010, following the first of two back-to-back nor'easters. [Photo by Carrie Smith, NOAA Central Library]

Weather forecasters were much better prepared in 2009-2010 because the science of meteorology, including our understanding of coastal winter storms, matured considerably since 1888. Today, Earth-orbiting weather satellites continuously monitor a storm's developing cloud shield and track its movements, radar locates snow bands sweeping onshore, and observational data from the surface and upper atmosphere feed into sophisticated numerical weather forecast models running on supercomputers. Well in advance of a storm's arrival, the National Weather Service (NWS) issues winter storm watches and warnings that are rapidly communicated to the public and public service agencies. Some school districts and businesses announce closings more than 12 hours before the first snowflake is expected to fall giving people time to stock up on food and other supplies in case they became marooned at home as many were in the mid-Atlantic area during the winter of 2009-2010.

While the historic winter storms of 1888 and 2009-2010 clearly demonstrate how weather analysis and forecasting have benefited from advances in scientific knowledge and technology, there is much more to be learned regarding the Earth's atmosphere. It is now evident, for example, that atmospheric conditions that favor development of nor'easters are linked to complex changes in the planetary-scale circulation. We have more on this topic in Chapter 10.

For more, refer to:
David Laskin, 2004. The Children's Blizzard. New York, NY: Harper Collins Publishers, 307 p.

We are about to embark on a systematic study of our continually changing atmospheric environment. We will observe weather as it happens while becoming familiar with some of the basic scientific principles and understandings that govern the atmospheric environment. As our study progresses, we will combine our observational experiences with what we have learned in exploring a wide variety of atmospheric phenomena ranging from cold waves to hurricanes. Long before the end of the course, you will come to realize that the behavior of the atmosphere is not arbitrary; its workings are explainable in scientific terms, drawn from many scientific disciplines, such as physics and chemistry. At the same time, questions are likely to persist regarding past, present, and future weather and climate.

If you are counting on studying storms, tornadoes, fronts, hailstorms, and the like right away, you will not be disappointed. Significant weather events will be observed and discussed as they happen via information delivered by the course website, and they will be explored in greater detail after relevant scientific foundations have been established. You can begin some activities now—today in fact—that will involve you in weather events, and enrich and enliven your study of the atmosphere. Our objective in this first chapter is to introduce and describe some of the tools and basic understandings that will guide your investigation.

Weather and Climate

Everyone has considerable experience with (and understanding of) the subject matter of this book. After all, each of us has been living with weather all our lives. Regardless of where we live or what we do, we are well aware of the far-reaching influence of weather. To a large extent, weather dictates how we dress, how we drive, and even our choice of recreational activities. Before setting out in the morning, most of us check the latest weather forecast on the Internet, TV, or radio, or glance out the window to look at the sky or read the thermometer. Every day we gather information on weather through our senses, the media, and perhaps our own weather instruments. And from that experience, we develop a basic understanding of how the atmosphere functions.

Weather is defined as the state of the atmosphere at a particular place and time, described in terms of the current value of such quantitative variables as temperature, humidity, cloudiness, precipitation, and wind speed and direction. Thousands of weather stations around the world monitor these weather variables at Earth's surface at least

hourly every day. A place and time must be specified when describing weather because the atmosphere is dynamic and its state changes from one place to another and with time. At the same time that the weather is cold and snowy in New York City, it might be warm and humid in Miami, and hot and dry in Phoenix. From personal experience, we know that tomorrow's weather may differ markedly from today's weather. *If you don't like the weather, wait a minute* is an old saying that is not far from the truth in many areas of the nation. **Meteorology** is the study of the atmosphere, processes that cause weather, and the life cycle of weather systems.

While weather often varies from one day to the next, we are aware that the weather of a particular location tends to follow reasonably consistent seasonal variations, with temperatures higher in summer and lower in winter. In addition, we associate the tropics with warmer weather and seasonal temperature contrasts that are less than at higher latitudes. Because arithmetic averages of weather elements (e.g., temperature, precipitation) taken over a span of years are often the easiest way of describing typical weather conditions, **climate** is popularly defined as weather conditions at some locality averaged over a specified time interval. By international convention, average values of weather elements are computed for a 30-year period beginning with the first year of a decade. At the close of a decade, the averaging period is moved forward ten years. Current climatic summaries are based on weather records from 1981-2010. Thirty-year average monthly and annual temperatures and precipitation totals are commonly used to describe the climate of some locality. Other useful climatic parameters include average seasonal snowfall, length of growing season, percent of possible sunshine, and number of days with dense fog. Climate is the ultimate environmental control in that it governs, for example, what crops can be cultivated, the fresh water supply, and the average heating and cooling requirements for homes.

In addition to average values of weather elements, the description of a locale's climate encompasses extremes in weather (e.g., highest and lowest temperature, greatest 24-hr snowfall and rainfall). Tabulation of extreme values usually covers the entire period of record (or at least the period when observations were made at the same location). Records of weather extremes provide information on the variability of climate at a particular place and give a more complete and useful description of climate. Farmers, for example, are interested in knowing not only the average summer rainfall, but also the frequency of extremely wet or dry summers. **Climatology** is the study of climate, its controls, and spatial and temporal variability.

Accessing Weather Information

Until now, for most of us, keeping track of the weather has been a casual part of daily life. From now on, or at least for the duration of this course, weather observation will be a more formal and regular activity. As a key part of this course, you are encouraged to access current weather data frequently. Today, this is relatively simple to do. Whether through the Internet, televsion, radio, or software applications, weather information is available practically everywhere!

We suggest that you tune to a televised weathercast at least once a day. Weathercasts are routine segments of the local morning, noon, and evening news reports. All-news channels, such as CNN and MSNBC, also feature weather segments throughout the day. In addition, *The Weather Channel* broadcasts national and local weather reports and forecasts 24 hours a day.

Another rich source of weather information is the Internet. The Internet provides real-time access to national and regional weather maps, satellite and radar images depicting large-scale cloud, precipitation, and atmospheric circulation patterns, weather forecasts, plus updates on environmental issues, such as trends in global climate, stratospheric ozone, and air quality. All NWS Forecast Offices maintain websites that provide a variety of current meteorological, climatological, and hydrological information. This course's website also brings you a host of real-time weather data via the Internet (Figure 1.3). There are hundreds of additional websites that offer other types of weather services, such as live webcams to view local weather conditions and email or text alerts for severe weather events.

Weather information may be also obtained by reading reports in local or national newspapers or by listening to the radio. Most newspapers include a weather column or page featuring maps and statistical summaries. Radio stations provide the latest local weather conditions and forecasts, but may not include a summary of weather conditions across the nation unless some newsworthy event has occurred such as a tornado outbreak or hurricane.

Since the dawn of the computer era, a more efficient and much more popular way to retrieve weather information is through weather software applications for desktops, laptops, tablets, and mobile devices. There are thousands of different weather apps that disseminate a variety of weather data. From a local beach forecast to

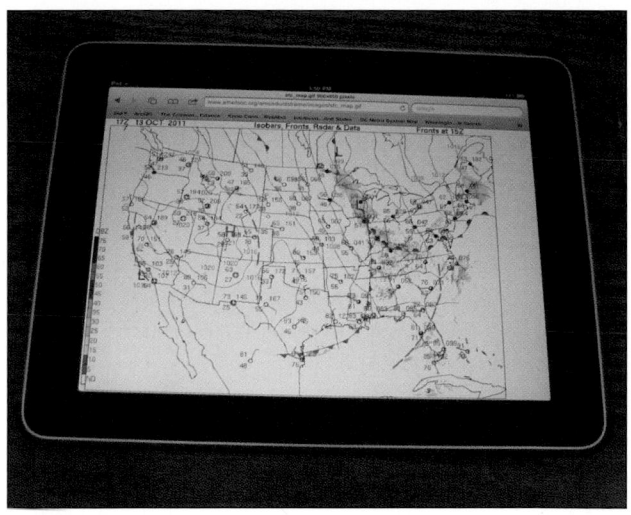

FIGURE 1.3
A tablet PC displaying a surface weather map linked from the course website. This is just one of about 50 customized weather maps provided by the NOAA National Centers for Environmental Prediction for this course.

tracking real-time radar, apps have revolutionized the way people get weather information.

A valuable source of local weather and climate information is the **NOAA Weather Radio (NWR)**. As a public service, the *National Oceanic and Atmospheric Administration (NOAA)*, the parent organization of the *National Weather Service (NWS)*, operates low-power, VHF high band FM radio transmitters that broadcast continuous weather information (e.g., regional conditions, local forecasts, marine warnings, and local climatological statistics) directly from NWS Forecast Offices 24 hours a day, 7 days a week. A series of messages is repeated every 4 to 6 minutes and some messages are updated hourly. When weather-related hazards threaten, NWS forecasters interrupt routine reports and issue the appropriate watches, warnings, and advisories.

The NWR is an "all-hazards" radio network. Working with federal, state, and local emergency managers and the Federal Communications Commission's Emergency Alert System, the NWR provides the public with warnings and post-event information for all types of hazards: weather, natural (e.g., earthquakes, tsunamis, volcanic eruptions), technological (e.g., oil spills, chemical leaks), and other non-weather emergencies (e.g., terrorist attacks, Amber alerts or 911 Telephone outages).

A special *weather radio* is required to receive NWR transmissions because the seven broadcast frequencies (from 162.40 to 162.55 MHz) are outside the range of standard AM/FM radios. Some weather radios are designed to sound an alarm or switch on automatically

FIGURE 1.4
At the push of a button, this weather radio broadcasts weather reports, forecasts, watches and warnings issued by the National Weather Service. [Photo courtesy of Bruce Allen, Midland Radio]

when a weather watch or warning or other emergency information is issued (Figure 1.4). The latest generation weather radio is equipped with the *Specific Area Message Encoding (S.A.M.E.)* feature that sounds an alarm only if a weather watch or warning is issued for a specific county or limited local area programmed (selected) by the user. Depending on terrain, the maximum range of NOAA weather radio broadcasts is approximately 65 km (40 mi). As of this writing, about 1000 transmitters are operating in all 50 states, the District of Columbia, Puerto Rico, the U.S. Virgin Islands, and the U.S. Pacific Territories. Expansion of this service is expected to eventually bring 95% of the U.S. population within range of NOAA weather radio broadcasts. In many communities, NOAA weather radio broadcasts are also available through audio programming channels on TV. Environment Canada's Meteorological Service has transmitters at 199 sites in its *Weatheradio network*, operating on the same seven frequencies as the NWR network and broadcasting in the nation's two official languages.

Time Keeping

Weather observations are made simultaneously at weather stations around the world. Simultaneous observations are necessary if the state of the atmosphere over a large area is to be portrayed accurately on a weather map at a specific time. This portrayal is also the basis for weather forecasting. Weather maps used in this course are given in *Z time*. The meaning of Z time and the conventional basis for time keeping are subjects of this section.

For millennia, humans have kept track of their activities by the daily motions of the Sun. Local noon was a convenient reference, marking the time of day when the Sun reached its highest point in the observer's sky. However, locations only a few tens of kilometers to the east or west would have different *Sun times*. Beginning in the mid-1800s, with advances in transportation and communication made possible by railroads and telegraphy, travel east or west could be confusing and even potentially hazardous because of local time differences. To eliminate confusion, the railroads argued for the standardized time keeping scheme we use today. Civil time zones were instituted in the United States and Canada on 18 November 1883 to standardize time keeping in North America. The concept of international time zones was officially adopted on 22 October 1884 at the International Meridian Conference, held in Washington, DC.

Time zones (Figure 1.5) were established based on longitude, which is measured as so many degrees east and west of the *prime meridian*, that is, zero degrees longitude. Because some of the best early astronomical determinations of time had been made at The Old Royal Observatory in Greenwich, England, the 1884 Conference designated the meridian of longitude passing through the observatory as the prime meridian. For more than 50 years, Greenwich Mean Time (GMT) was used for essentially all meteorological observations. GMT is based upon the daily rotation of Earth with respect to a "mean Sun." Often the single letter Z (phonetically pronounced "zulu") designated the time within the Greenwich Time zone (centered on the prime meridian). Today, the preferred time system is the more precise Coordinated Universal Time or Universel Temps Coordinné (UTC), based on an atomic clock and reckoned according to the stars. For practical purposes, Z, GMT, and UTC are equivalent.

Earth makes one complete rotation (360 degrees) on its axis with respect to the Sun once every 24 hrs. Hence, Earth rotates through 15 degrees of longitude every hour (360 degrees divided by 24 hrs equals 15 degrees per hr). Ideally, Earth should be divided into 24 civil time zones each having a width of 15 degrees of longitude. The central meridian of each time zone is defined as a longitude that is evenly divisible by 15. For example, the central meridian of the Central Time Zone in North America is 90 degrees W longitude. Ninety degrees divided by 15 equals 6 so that Central Standard Time (CST) is six hours different from the time at Greenwich. Earth rotates eastward so that Greenwich

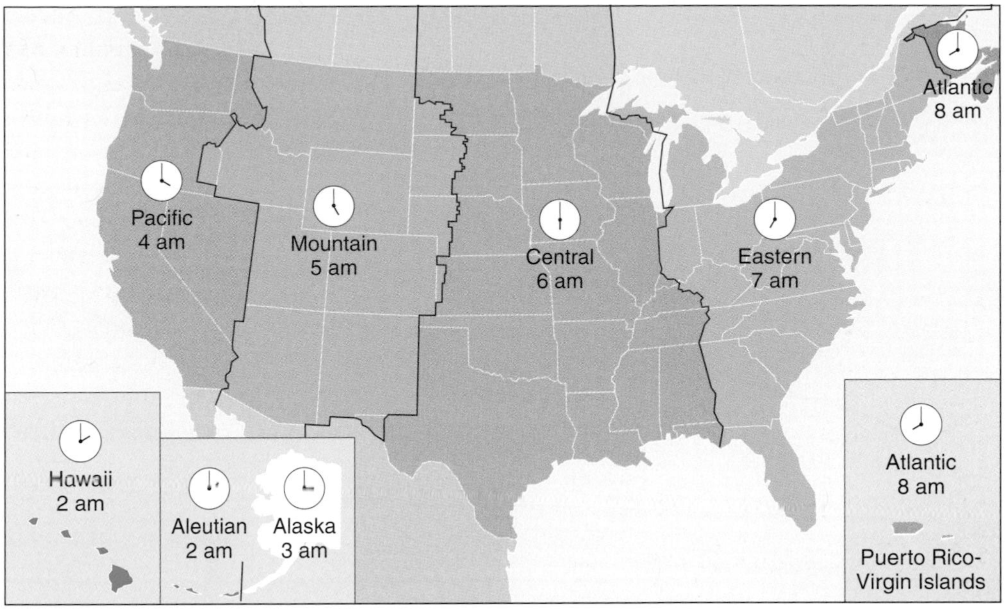

FIGURE 1.5
Times zones in the United States, Puerto Rico and the Virgin Islands, southern Canada, and northern Mexico showing Standard Times equivalent to 1200 Z.

time is ahead of CST by 6 hrs. When it is noon at Greenwich, it is 7 a.m. Eastern Standard Time (EST), 6 a.m. CST, 5 a.m. Mountain Standard Time (MST), and 4 a.m. Pacific Standard Time (PST).

Boundaries between most time zones have been adjusted to accommodate political boundaries of various nations. In a few cases, nations adhere to a local civil time that may differ by one half hour from the central meridian of that time zone. To head off potential confusion, time is expressed according to a 24-hr clock so that, for example, 7:45 a.m. is 0745 and 2:20 p.m. is 1420. While most of the United States observes Daylight Saving Time, UTC is fixed and does not shift to a summer schedule. Hence, you will have to adjust the time by one hour where and when Daylight Saving Time is observed. For example, during summer, residents of the U.S. Central Time Zone lag Greenwich time by 5 instead of 6 hrs. Hence, 0700 Central Daylight Time (CDT) = 0600 Central Standard Time (CST) = 1200 Z. The Energy Policy Act of 2005 mandated a change in dates for Daylight Saving Time in the U.S. beginning in 2007. Daylight Saving Time now begins on the second Sunday of March (when time "springs forward" one hour) and ends the first Sunday of November (when time "falls back" one hour).

By international agreement, surface weather observations are taken at least four times per 24 hrs, that is, at 0000 Z, 0600 Z, 1200 Z, and 1800 Z, with upper-air measurements via balloon-borne instruments (radiosondes) made at 0000 Z and 1200 Z. In the United States, surface observations are taken hourly (at the top of the hour), composite radar charts are also issued hourly at 35 minutes past the hour, and fronts are analyzed on weather maps every 3 hrs beginning at 0000 Z.

Weather Systems and Weather Maps

So that your weather watching is more meaningful and useful from the beginning of this course, the remainder of this chapter is devoted to a description of what to watch for, beginning with features plotted on the national weather map.

Temperature, dewpoint, wind, and air pressure are among the many atmospheric variables that are routinely monitored by instruments at surface weather stations. These data are transmitted to a central facility where they are assembled and plotted on surface weather maps. Special symbols are used on national weather maps to identify and locate the principal weather-makers, that is, pressure systems and fronts (Figure 1.6). For an historical sketch on the origins of weather maps, see the first *For Further Exploration* discussion at the end of this chapter.

Pressure systems are of two main types, *highs* (or anticyclones) and *lows* (or cyclones). The high and low designations refer to air pressure. We can think of

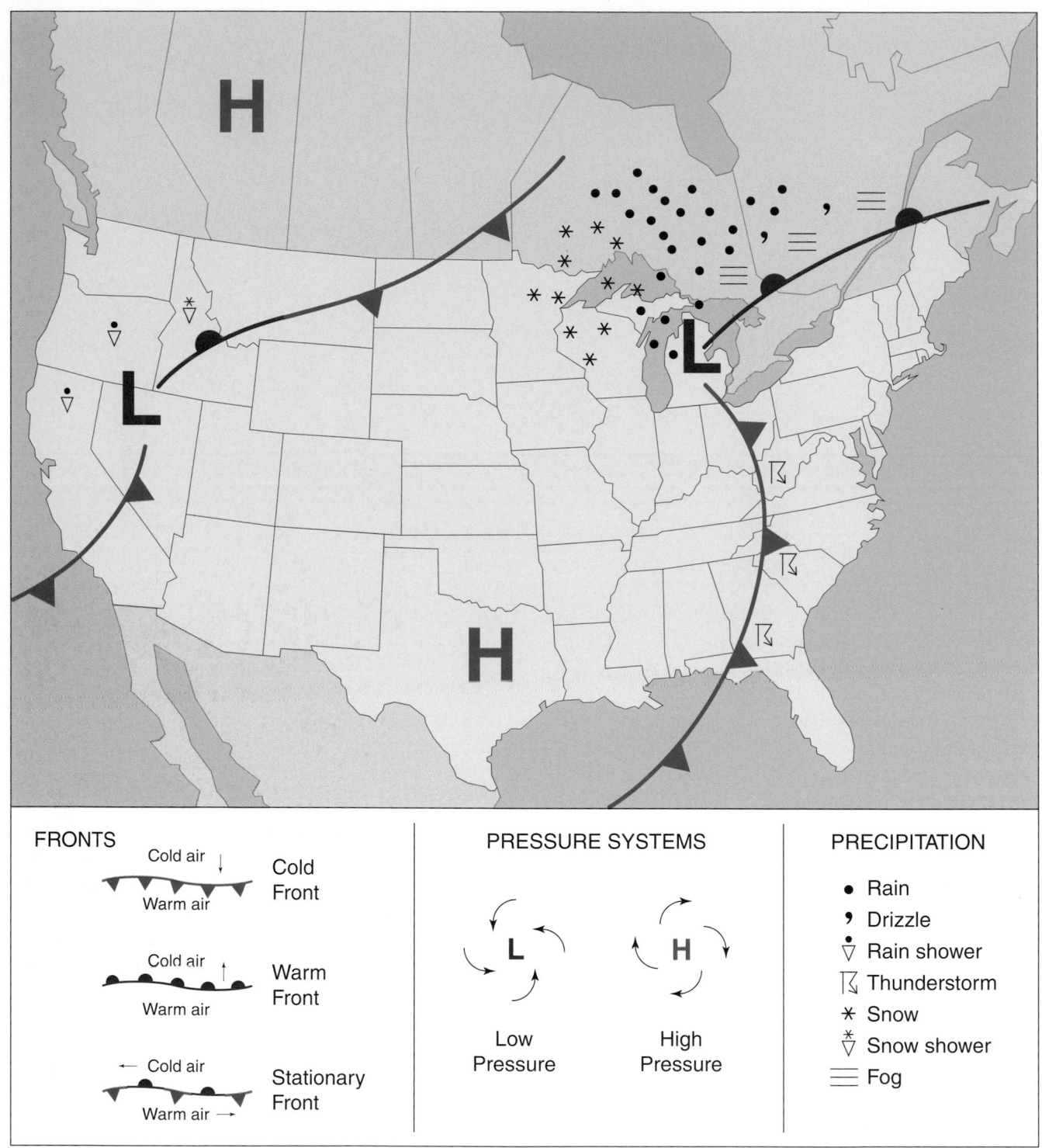

FIGURE 1.6
On a weather map, special symbols represent the state of the atmosphere over a broad geographical area at a specific time.

air pressure as the weight per unit area of a column of air that extends upward from Earth's surface (or any altitude within the atmosphere) to the top of the atmosphere. At any specified time, air pressure at the Earth's surface varies from one place to another across the continent.

On a weather map, *H* or *High* symbolizes the center of regions where the air pressure is relatively high compared to surrounding areas. Conversely, *L* or *Low* signifies the center of regions where the air pressure is relatively low compared to surrounding areas.

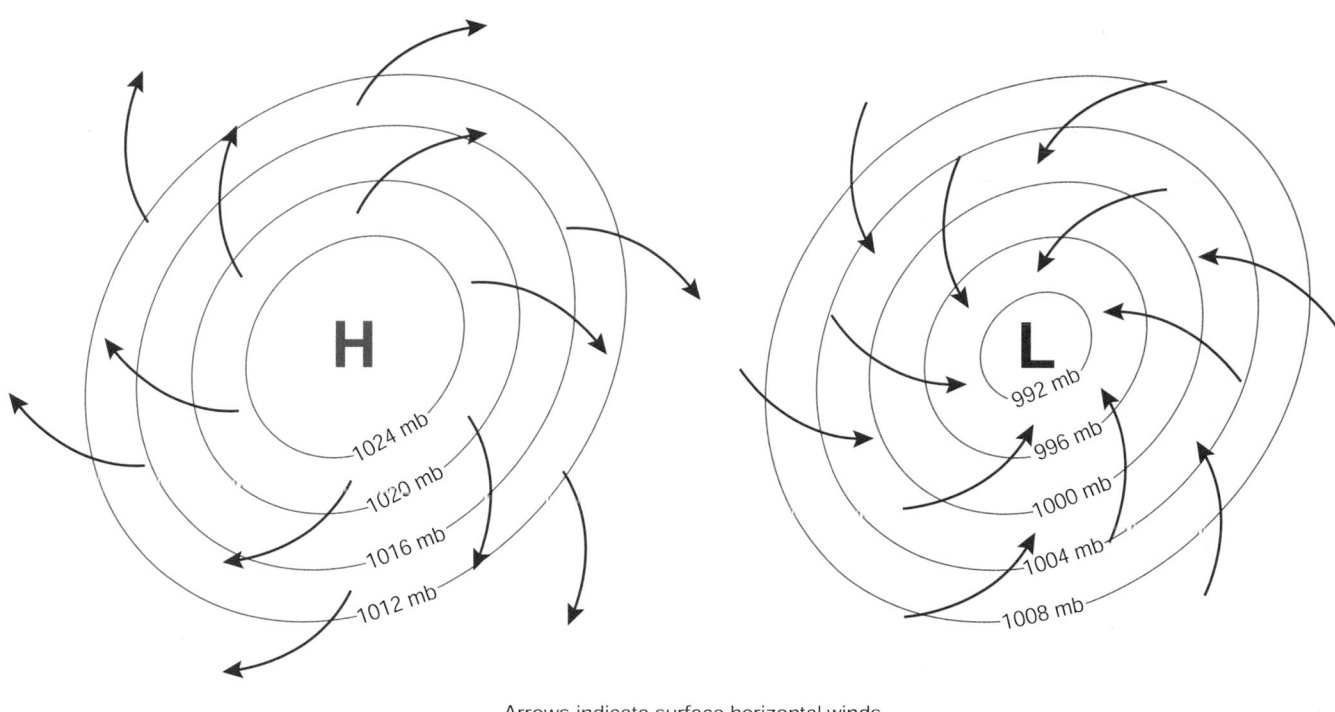

Arrows indicate surface horizontal winds

A   B

FIGURE 1.7
Viewed from above in the Northern Hemisphere, winds near the Earth's surface blow (A) clockwise and outward from a high-pressure system, and (B) counterclockwise and inward into a low-pressure system. Blue lines are isobars, passing through places having the same air pressure in millibars (mb). Isobars are drawn at intervals of 4 millibars.

As you examine weather maps, note the following about pressure systems:

- Highs (anticyclones) feature descending air and are usually accompanied by fair weather; hence they are described as *fair-weather systems*. Highs that originate in northwestern Canada bring cold, dry weather in winter and cool, dry weather in summer to much of the coterminous United States. Highs that develop further south over land may bring hot, dry weather in summer and mild, dry weather in winter. Furthermore, summer winds around a warm high centered over the southeastern states transport warm, humid air from the Gulf of Mexico over a broad swath of the eastern U.S. into southeastern Canada.
- Viewed from above in the Northern Hemisphere, surface winds in a high-pressure system blow in a clockwise and outward spiral as shown in Figure 1.7A. Calm conditions or light winds are typical over a broad area at the center of a high.
- Lows (cyclones) feature ascending air and typically produce cloudy, rainy or snowy weather; hence they are described as *stormy-*

weather systems. An exception may be lows that develop over broad regions of arid or semiarid terrain, such as the Desert Southwest, especially in summer. In such areas, intense solar heating of the ground raises the air temperature and lowers the surface air pressure, producing a low that remains stationary over the hot ground and is not accompanied by stormy weather.
- Viewed from above in the Northern Hemisphere, surface winds in a low-pressure system blow in a counterclockwise and inward spiral as shown in Figure 1.7B.
- Both highs and lows move with the prevailing wind several kilometers above the surface, generally eastward across North America, and as they do, the weather changes at places in their paths. Highs follow lows and lows follow highs causing the weather to shift from stormy to fair (that can be monitored as changes in air pressure). As a general rule, highs track toward the east and southeast whereas lows track toward the east and northeast. An important exception is tropical cyclones (e.g., hurricanes, tropical storms) that often move from east to west over

the tropical North Atlantic Ocean before turning north and then eastward if they reach the middle latitudes.

- Lows that track across the northern United States or southern Canada are more distant from sources of moisture and usually produce less rain or snowfall than lows that track further south (such as lows that travel out of eastern Colorado and move along the Gulf Coast or up the Eastern Seaboard).

- Weather on the left side (west and north) of an extratropical cyclone's track (path) tends to be relatively cold, whereas weather on the right (east and south) of the cyclone's track tends to be relatively warm. For this reason, winter snows are most likely to the west and north of the path of a low-pressure system.

Air masses and fronts are also important weather-makers. An **air mass** is a huge volume of air covering hundreds of thousands of square kilometers that has relatively uniform temperature and humidity properties in its horizontal extent. An air mass is typically associated with a large high pressure system and its specific characteristics depend on the types of surfaces over which the air mass forms (its *source region*) and travels. Cold air masses form at higher latitudes over surfaces that are often snow or ice covered, whereas warm air masses form in the lower latitudes where the Earth's surface is relatively warm. Humid air masses form over moist maritime surfaces (e.g., Pacific Ocean, Gulf of Mexico), whereas dry air masses develop over dry continental surfaces (e.g., Desert Southwest, northwestern Canada). The four basic types of air masses are cold and dry, cold and humid, warm and dry, and warm and humid.

A **front** is a narrow transition zone between air masses that differ in temperature, humidity, or both. Fronts form where contrasting air masses meet, and the associated air movements often give rise to cloudiness and precipitation. The most common fronts are stationary, cold, and warm; weather map symbols for all three are shown in Figure 1.6. As the name implies, a *stationary front* is just that, stationary (or nearly so). On both sides of a stationary front, winds blow roughly parallel to the front but in opposite directions. A shift in wind direction may cause a portion of a stationary front to advance northward (becoming a warm front) or southward (becoming a cold front).

At the same pressure, warm air is less dense than cold air so that a warm air mass advances by gliding up

and over a retreating cold air mass. The cold air forms a wedge under the warm air. The leading edge of warm air at the Earth's surface is plotted on a weather map as a *warm front* (Figure 1.8A). On the other hand, cold air advances by sliding under and pushing up the less dense warm air and the leading edge of cold air at the Earth's surface is plotted on a weather map as a *cold front* (Figure 1.8B). Consequently, the boundary between warm and cold air associated with a warm front slopes more gently with altitude than does the boundary between warm and cold air associated with a cold front.

As you examine surface weather maps, note the following about air masses and fronts:

- In response to regular seasonal changes in the duration and intensity of sunlight, polar air masses are much colder in winter and milder in summer. By contrast, in the tropics, sunlight is nearly uniform in duration and intensity throughout the year so that tropical air masses are warm and exhibit less seasonal variation in temperature.

- An air mass modifies (becomes warmer, colder, more humid, drier) as it moves away from its source region. The degree of modification depends on the properties of the surface over which the air mass travels. For example, a cold air mass warms more if it travels over ground that is bare rather than snow-covered.

- Fronts are three-dimensional and the map symbol for a front is plotted where the frontal surface intersects Earth's surface.

- Most cloudiness and precipitation associated with a warm front occur over a broad band, often hundreds of kilometers wide, in advance of where the front intersects Earth's surface. Widespread precipitation ahead of a warm front generally is light to moderate in intensity and may persist at a particular location from 12 to 24 hrs or longer.

- Most cloudiness and precipitation associated with a cold front occur as a narrow band along or just ahead of where the front intersects Earth's surface. Although precipitation often is showery and may last from a few minutes to a few hours, it can be very intense.

- Wind direction (and at times wind speed) differs on opposite sides of a front.

- Some fronts are not accompanied by cloudiness or precipitation. A shift in wind direction, together with changes in air pressure, tempera-

FIGURE 1.8
The two most common types of fronts are (A) a warm front that marks the boundary between advancing relatively warm (less dense) air and retreating cold (more dense) air, and (B) a cold front that marks the boundary between advancing cold air and retreating warm air. In both diagrams the fronts are shown in vertical cross-section with the vertical scale greatly exaggerated.

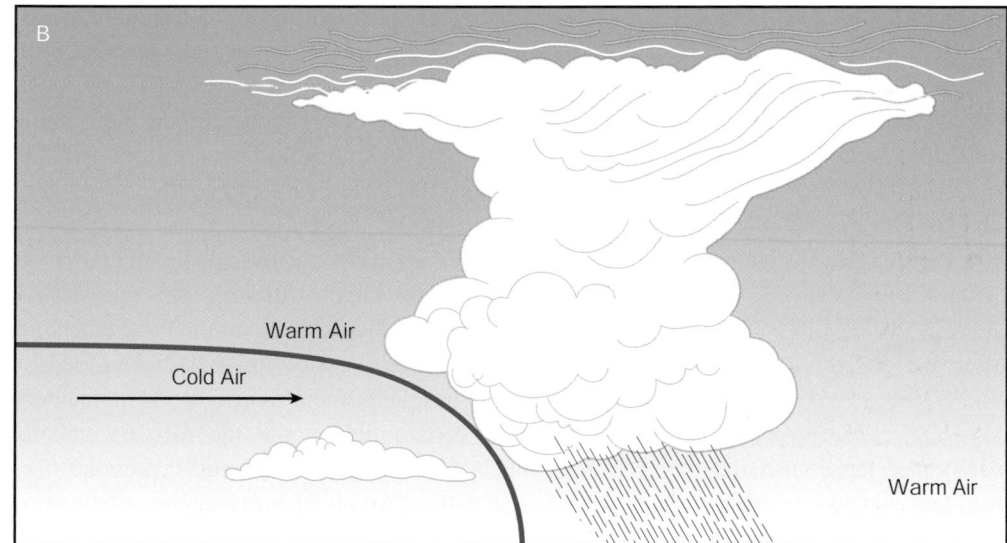

ture and/or humidity typically accompany the passage of the front.

- In summer, air temperatures can be nearly the same ahead of and behind a cold front. In that case, the air masses on opposite sides of the front differ primarily in humidity; that is, the air mass ahead of the advancing front is more humid (and therefore less dense) and the air mass behind the front is less humid (denser). With passage of the front, refreshingly drier air replaces uncomfortably humid air.

- Cold and warm fronts are plotted on a weather map as heavy lines that are usually anchored at the center of a low-pressure system. The counterclockwise and inward circulation about a low brings contrasting air masses together to form fronts. By contrast, the outward flow of relatively homogeneous air from a high precludes development of fronts.

- A low-pressure system may develop along a stationary front and travel rapidly like a large ripple from west to east along the front.

- Thunderstorms and associated severe weather (e.g., tornadoes, hail) most often develop south and southeast of the center of a low-pressure system, that is, in the warm, humid air mass located between the cold front and the warm front.

As you monitor national and regional weather maps, also watch for the following:

- Cool sea breezes or lake breezes may push inland perhaps 10 to 50 km (6 to 30 mi), lowering summer afternoon temperatures in coastal areas.
- Beginning in late fall and continuing through much of the winter, lake-effect snows fall in narrow bands on the downwind (eastern and southern) shores of the Great Lakes and Utah's Great Salt Lake. Such snowfalls are highly localized and can be very heavy.
- Severe thunderstorms and tornadoes are most common in spring and early summer across the central United States, especially from east Texas northward to Nebraska and from Iowa eastward to central Indiana.
- Thunderstorms are most frequent in Florida, on the western High Plains and eastern slopes of the Rockies, but are relatively rare along the Pacific coast and on the Hawaiian Islands.
- Tropical storms and hurricanes occasionally impact the Atlantic and Gulf coasts, primarily from August through October. However, these tropical weather systems are rare on the West Coast.

Describing the State of the Atmosphere

Internet, television, and newspaper weather reports usually include statistical summaries of present and past weather conditions. These data are compiled to form a climatology (description of the climate) of a locale. It is useful to briefly comment on the meaning of the most common weather parameters:

- *Maximum temperature.* The highest air temperature recorded over a 24-hr period, usually between midnight of one day and midnight of the next day. Typically, but not always, the day's maximum temperature occurs in the early to mid-afternoon. In the United States, surface air temperatures are reported in degrees Fahrenheit (°F) and in most other countries in degrees Celsius (°C).
- *Minimum temperature.* The lowest air temperature recorded over a 24-hr period, usually between midnight of one day and midnight of the next day. Typically, but not always, the minimum temperature occurs around sunrise.
- *Dewpoint* (or *frost point*). The temperature to which air must be cooled at constant pressure to become saturated with water vapor and for dew (or frost) to begin forming on relatively cold surfaces.
- *Relative humidity.* A measure of the actual concentration of the water vapor component of air compared to the concentration the air would have if saturated with water vapor. Relative humidity is always expressed as a percentage. Because the saturation concentration varies with air temperature so too does the relative humidity. On most days, the relative humidity is highest during the coldest time of day (around sunrise) and lowest during the warmest time of day (early to mid-afternoon).
- *Precipitation amount.* Depth of rainfall or melted snowfall over a 24-hr period, usually from midnight of one day to midnight of the next, measured to one-hundredth of an inch in the United States and in millimeters elsewhere. On average, 10 inches of freshly fallen snow melt down to about 1 inch of water.
- *Air pressure.* The weight of a column of air over a unit area of Earth's surface. With a mercury barometer, the traditional instrument for measuring air pressure, the pressure exerted by the atmosphere supports a column of mercury to a certain height and the mercury column fluctuates up and down as the air pressure rises and falls. This is the reason air pressure is commonly reported in units of length, that is, inches or millimeters of mercury. In the United States, meteorologists express air pressure in millibars (mb), a unit of pressure equal to a hectopascal (hPa). The average air pressure at sea-level is 1013.25 mb, corresponding to the pressure exerted by a 29.92 in. (760 mm) column of mercury. Falling air pressure over a span of several hours often signals an approaching low-pressure system and a turn to stormy weather. Rising air pressure, on the other hand, indicates an approaching high-pressure system, with clearing skies or continued fair weather.
- *Wind direction* and *wind speed.* According to meteorological tradition, wind direction is the compass direction *from which* the wind blows

(Figure 1.9). For example, a southeast wind blows from the southeast toward the northwest and a west wind blows from the west toward the east. As a general rule, at middle latitudes a wind shift from east to northeast to north to northwest is accompanied by falling air temperatures. On the other hand, a wind shift from east to southeast to south usually brings warmer weather. Over a broad area about the center of a high-pressure system calm air or light winds prevail. Wind speed tends to increase markedly as a cold front passes a location and winds are particularly strong and gusty in the vicinity of thunderstorms.

- *Sky cover.* Based on the fraction of the sky that is cloud covered, the sky is described as clear (no clouds), a few clouds (1/8 to 2/8 cloud cover), scattered clouds (3/8 to 4/8), broken clouds (5/8 to 7/8), and overcast (completely cloud-covered). All other factors being equal, nights are coldest when the sky is clear and the air is relatively dry. An overcast sky prevents the day's minimum temperature from falling and the day's maximum temperature from rising more than otherwise.
- *Weather watch.* Issued by the National Weather Service when hazardous weather (e.g., tornadoes, heavy snowfall) is considered possible based on current or anticipated atmospheric conditions.
- *Weather warning.* Issued by the National Weather Service when hazardous weather is imminent or actually taking place.

Weather Satellite Imagery

Satellite images and video loops (composed of successive images) are routine components of many televised and Internet-delivered weather reports. Some newspaper weather pages also feature satellite images. Sensors on weather satellites orbiting Earth provide a unique and valuable perspective on the state of the atmosphere over broad areas and enable meteorologists to remotely measure temperature and humidity as well as to locate and track weather systems.

The weather satellite images most familiar to us are those obtained by sensors onboard a **Geostationary Operational Environmental Satellite (GOES)** orbiting Earth about 36,000 km (22,300 mi) above the planet's equator. A GOES satellite revolves around Earth at the same rate and in the same direction as the planet rotates so that the satellite is always positioned over the same spot

FIGURE 1.9
Wind vane atop the Smithsonian Institution Building in Washington, DC. The arrow of the wind vane points in the direction from which the wind is blowing, as confirmed by the flags that are stretched out in the downwind direction. [Photo by R.S. Weinbeck]

on Earth's surface and its sensors have a consistent field of view (Figure 1.10). The *sub-satellite point*, the location on Earth's surface directly below the satellite, is essentially on the equator. Two geostationary satellites, one over South America (near 75 degrees W longitude) and the other over the eastern Pacific (approximately 135 degrees W longitude), provide a complete view of much of North America and adjacent portions of the Pacific and Atlantic Oceans to latitudes of about 60 degrees. Considerable distortion sets in near the edge of this field of view so that polar-orbiting satellites complement geostationary satellites in monitoring the planet.

A **Polar-orbiting Operational Environmental Satellite (POES)** travels in a relatively low altitude (800 to 1000 km, 500 to 600 mi) nearly north-south orbit passing close to the poles (Figure 1.11). The satellite's orbit traces out a plane in space while the planet continually rotates on its axis through the plane of the satellite's orbit. With each orbit, points on Earth's surface (except near the poles) move eastward so that onboard sensors sweep out successive overlapping north/south strips. A polar-orbiting satellite that follows the Sun (sun-synchronous) passes over the same area twice each 24-hr day. Other polar-orbiting satellites are positioned so that they require several days before passing over the same point on Earth's surface.

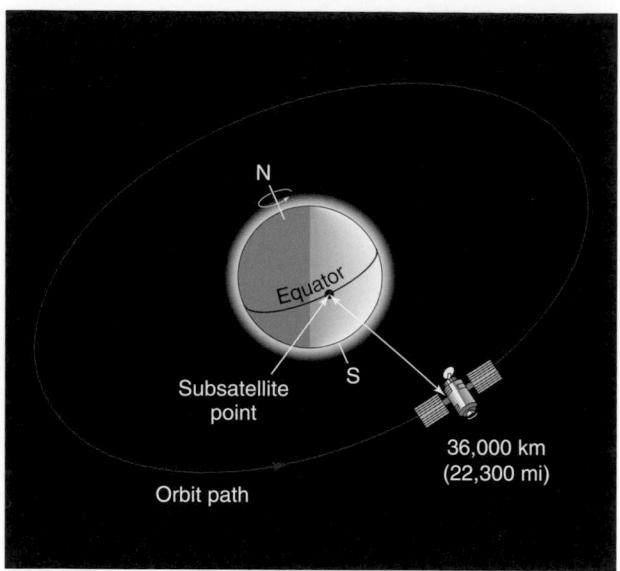

FIGURE 1.10
A geostationary weather satellite (GOES) orbits Earth so that it is always positioned over the same equatorial spot on Earth's surface.

Satellite-borne sensors are either passive or active. Passive sensors measure radiation coming from the Earth-atmosphere system, that is, visible solar radiation reflected by the planet and invisible infrared radiation emitted by the planet. Sunlight reflected by Earth's surface and atmosphere produces images that are essentially black and white photographs of the planet.

FIGURE 1.11
A polar-orbiting weather satellite (POES) orbits Earth in a nearly north-south direction passing near the poles.

Sun-lit, highly reflective surfaces such as cloud tops and snow-covered ground appear bright white whereas less reflective surfaces such as evergreen forests and the ocean appear much darker. Cloud patterns on a **visible satellite image** are of particular interest to atmospheric scientists (Figure 1.12). From analysis of cloud patterns displayed on the image, they can identify not only a specific type of weather system (such as a hurricane), but also the stage of its life cycle and its direction of movement when a sequence of images is animated.

A second type of sensor onboard a weather satellite detects infrared radiation (IR). IR is an invisible form of radiation emitted by all objects continually, both day and night. Hence, an **infrared satellite image** of the planet can provide useful information at any time whereas visible weather satellite images are available only during daylight hours. (Because of around-the-clock availability, infrared satellite images are usually shown on television weathercasts.) IR signals are routinely calibrated to give the temperature of objects in the sensor's field of view. This is possible because the intensity of IR emitted by an object depends on its surface temperature; that is, relatively warm objects emit more intense IR than do relatively cold objects. IR-derived temperatures are calibrated on a gray scale such that the brightest white indicates the lowest temperature and dark gray indicates the highest temperature. Alternately, a color scale is used for enhancement so that, for example, reds and oranges may represent high temperatures and blues and violets signal low temperatures.

The temperature dependency of IR emission makes it possible, for example, to distinguish low clouds from high clouds on an IR satellite image. In the part of the atmosphere where most clouds occur (the lowest 10 km or 6 mi or so), air temperature usually decreases with increasing altitude. The tops of high clouds are colder than the tops of low clouds and emit less intense IR. In the sample IR satellite image in Figure 1.13, taken at the same time as Figure 1.12, high clouds appear bright white whereas low clouds are darker. IR sensors are also tuned to detect those wavelength bands where, in cloud-free areas, some of the radiation emitted from Earth's surface passes directly out to space. In this way, we can determine the surface temperature of the ground, snow cover, fog or ocean and lake where the sky is cloud-free.

Water vapor satellite images enable meteorologists to identify and track the movements of moisture plumes within the atmosphere over distances of thousands of kilometers. Water vapor is invisible and not detectable on visible or conventional infrared satellite

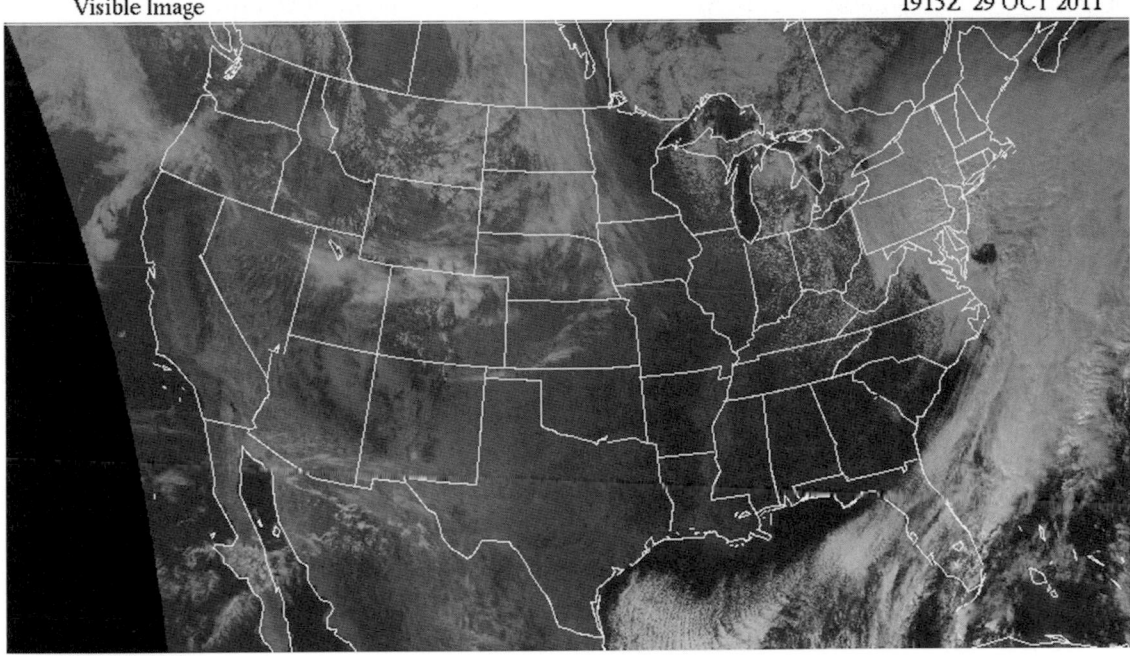

Visible Image   1915Z 29 OCT 2011

NCEP/NWS/NOAA

FIGURE 1.12
A sample visible satellite image from sensors onboard a geostationary weather satellite. The cloud swirl over the northeastern U.S. is associated with a historic early season nor'easter that dumped more than 20 inches of snow in parts of interior Maine, Massachusetts, New Hampshire, and New York with a few locations receiving more than 30 inches. For the most current visible satellite image, go to the course website and in the "Satellite" section - select "Visible-Latest", or scan the QR code.

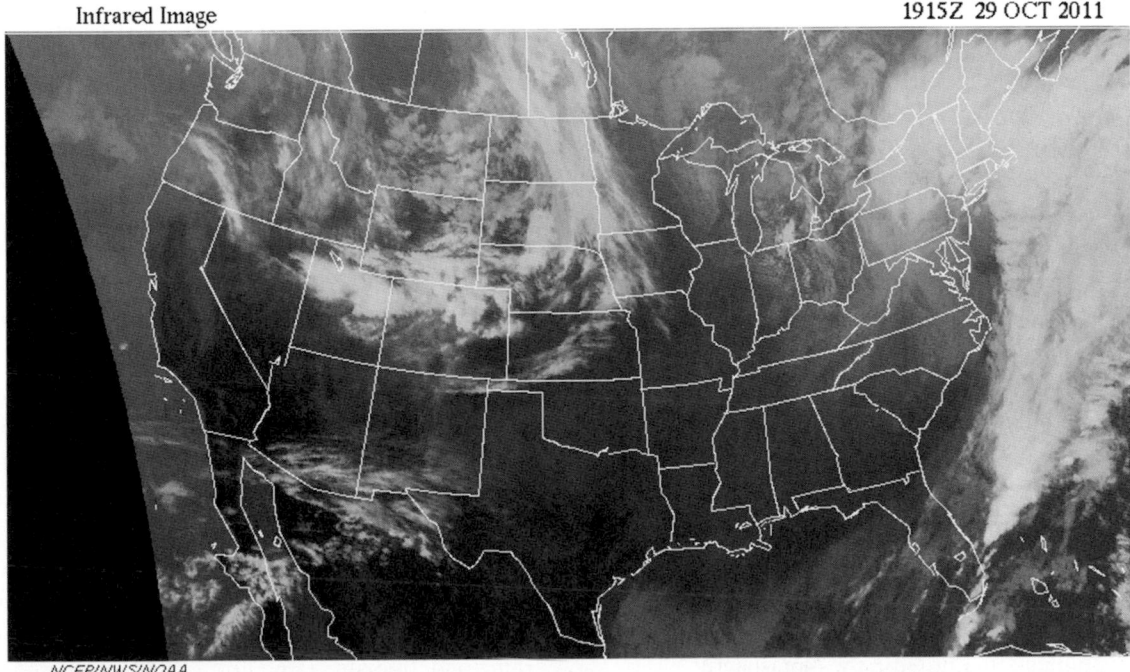

Infrared Image   1915Z 29 OCT 2011

NCEP/NWS/NOAA

FIGURE 1.13
A sample infrared satellite image from sensors onboard a geostationary satellite, taken at the same time as Figure 1.12. For the most current infared satellite image, go to the course website and in the "Satellite" section - select "Infrared - Latest." This website also includes a current 24-hr infrared animation which you can view now by scanning the QR code.

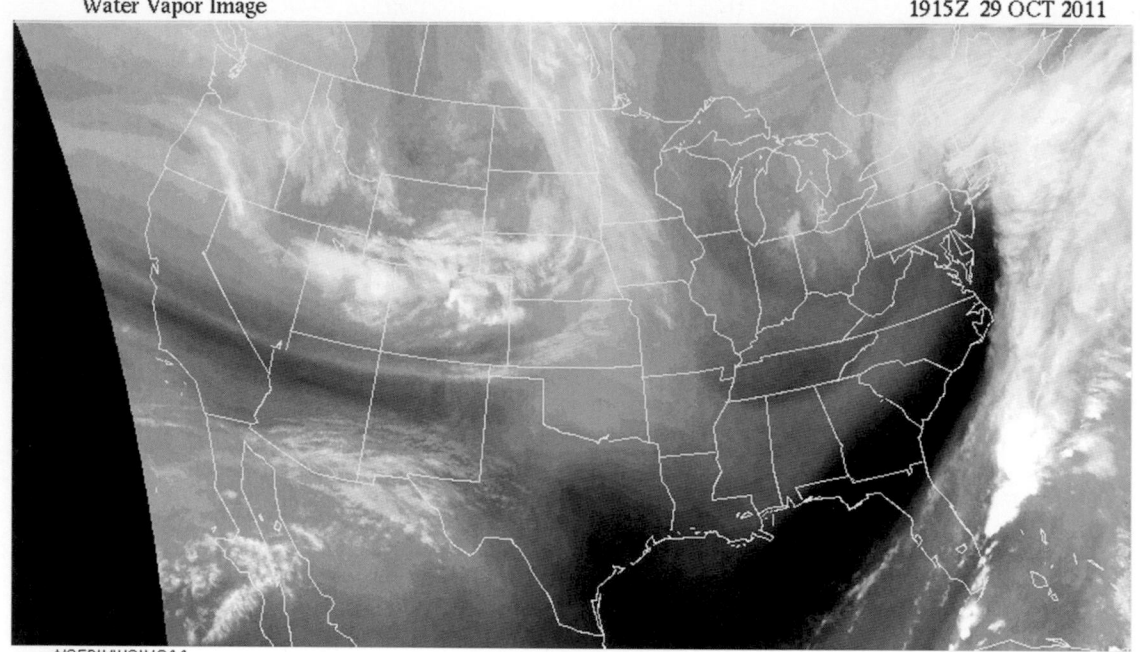

Water Vapor Image   1915Z 29 OCT 2011

NCEP/NWS/NOAA

FIGURE 1.14
A sample water vapor satellite image from sensors onboard a geostationary satellite, taken at the same time as Figure 1.12 and Figure 1.13. For the most current water vapor satellite image, go to the course website and in the "Satellite" section - select "Water Vapor - Latest." This website also includes a current 24-hr water vapor animation which you can view now by scanning the QR code.

images. But water vapor efficiently absorbs and emits certain wavelength bands of IR so that sensors onboard weather satellites that are sensitive to these wavelengths can detect water vapor. Water vapor imagery displays the water vapor concentration at altitudes between about 5000 and 12,000 m (16,000 and 40,000 ft) on a gray scale (Figure 1.14). At one extreme, dark gray indicates almost no water vapor whereas at the other extreme, milky white signals a relatively high concentration of water vapor. Upper-atmospheric clouds appear milky to bright white on water vapor images, masking water vapor concentrations in the image. Water vapor animation, like the one linked from this course's website, can reveal the air circulation pattern in weather systems prior to widespread cloud formation.

In addition to monitoring the weather, satellites have countless applications crucial to today's society, including the *Global Positioning System (GPS)*, which allows for precise location and navigation. Meteorologists use GPS to track instrumentation probing the atmosphere and slight variations in GPS signals can provide insight into atmospheric variables such as water vapor. To learn how GPS works, see this chapter's second *For Further Exploration*.

Weather Radar

Weather radar complements satellite surveillance of the atmosphere by locating and tracking the movement of areas of precipitation and monitoring the circulation within small-scale weather systems such as thunderstorms. Weather radar continually emits pulses of microwave energy that are reflected by atmospheric targets such as raindrops, snowflakes, or hailstones. The reflected signal is displayed as blotches (*radar echoes*) on computer displays with a superimposed map of the region surrounding the radar. The heavier the precipitation, the more intense is the echo. Echo intensity is calibrated on a color scale so that light green indicates light rain whereas dark red signals heavy rain or hail.

Some radar images appearing on weather broadcasts or the Internet are color coded according to precipitation type, so the public can distinguish between snow, rain, and freezing rain. From analysis of radar echoes, meteorologists can determine the intensity of thunderstorms, track the movement of areas of precipitation, and predict when precipitation is likely to begin or end at a particular place. Composite maps of

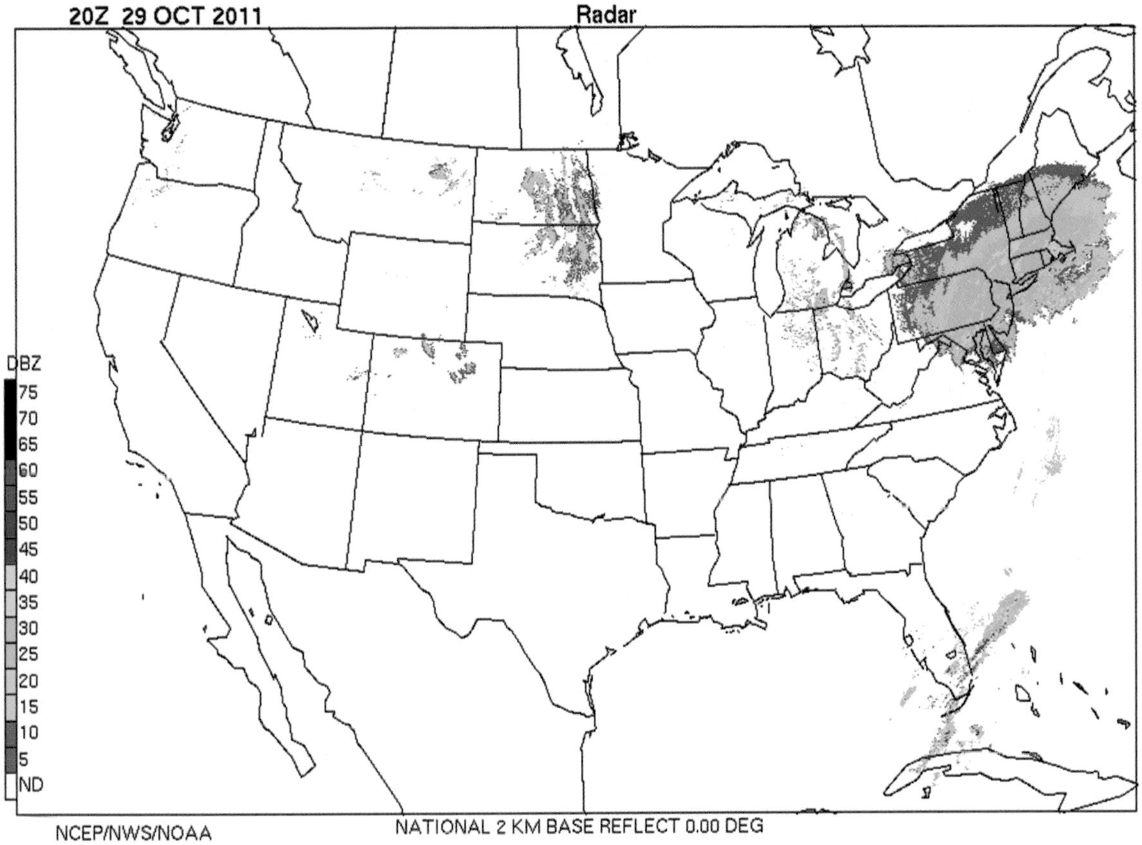

20Z 29 OCT 2011   Radar

DBZ
75
70
65
60
55
50
45
40
35
30
25
20
15
10
5
ND

NCEP/NWS/NOAA   NATIONAL 2 KM BASE REFLECT 0.00 DEG

FIGURE 1.15
A sample composite national radar image (approximately the same time as Figures 1.12 through 1.14). For the most current national radar image, go to the course website and in the "Radar" section - select "Radar." This website also includes a current 24-hr radar animation which you can view now by scanning the QR code.

radar echoes from a number of weather radars around the nation are useful in following the progress of large-scale weather systems (Figure 1.15). By overlaying radar echoes on satellite images, areas of precipitation and clouds can be displayed simultaneously, thereby helping to locate the most active portion of a weather system (Figure 1.16). These composites are often animated.

Weather radar also monitors the movement of raindrops, snowflakes, and hailstones within a storm system. Using the same operating principle as the device that measures the speed of a pitched baseball (*Doppler effect*), weather radar detects the motion of precipitation particles within the storm system, which then can provide an indication of the system's circulation. Early identification of the circulation within a thunderstorm cloud that is a precursor of a developing tornado is a potentially life-saving application of Doppler weather radar. With advance warning of a tornadic circulation before a funnel cloud touches the ground, people have time to seek shelter or move out of its path. Radar is discussed in greater detail in Chapter 7.

FIGURE 1.16
A sample infrared satellite image of the eastern United States, overlaid with radar echoes. This type of image allows areas of precipitation and clouds to be displayed simultaneously.

Sky Watching

At this beginning stage in our study of the atmosphere and weather, it is also a good idea to develop the habit of observing the sky, watching for changes in clouds and cloud cover. As the ancients before us found, sky watching makes us more aware of the dynamic nature of the atmosphere and may reveal clues to future weather. Here are some things to watch for:

- Clouds are aggregates of tiny water droplets, ice crystals, or some combination of both. Ice-crystal clouds such as cirrus clouds, occur at high altitudes where air temperatures are relatively low; they have a fibrous or wispy appearance (Figure 1.17). Water-droplet clouds such as cumulus clouds, occur at lower altitudes where temperatures are higher; their edges tend to be more sharply defined (Figure 1.18).
- A cloud that is very near or actually in contact with Earth's surface is fog (Figure 1.19). By convention, *fog* is a suspension of tiny water droplets or ice crystals in air that reduces horizontal visibility to less than 1000 m (0.621 mi).
- Some clouds form in horizontal layers (*stratiform clouds*) whereas others appear puffy (*cumuliform*

FIGURE 1.18
These relatively low clouds are composed of tiny water droplets and have more sharply defined edges than ice-crystal clouds.

clouds). Stratiform clouds develop where air ascends gently over a broad region whereas cumuliform clouds are produced by more vigorous ascent of air over a much smaller area. Often stratiform clouds develop ahead of a warm front while cumuliform clouds, especially those having great vertical development, form along or just ahead of a cold front.

- Arrival of high, thin clouds in the western sky at a middle latitude location is often the first sign of an approaching warm front. In time, clouds gradually lower and thicken so that eventually they block out the Sun during the day or the Moon at night.
- The day may begin with clear skies, but after several hours of bright sunshine, small white clouds appear, resembling puffs of cotton floating in the sky (Figure 1.20). These fair-weather *cumulus clouds* usually vaporize rapidly near sunset.
- During certain atmospheric conditions, cumulus clouds build vertically and merge laterally, eventually forming a thunderstorm cloud, called a *cumulonimbus cloud* (Figure 1.21). All thunderstorms are accompanied by lightning and thunder. Severe thunderstorms can produce torrential rains, large hail, strong and gusty winds, and even tornadoes.
- Clouds situated at different altitudes sometimes move horizontally in different directions and perhaps at different speeds (Figure 1.22). Because most clouds move with the wind, such motion indicates that the horizontal wind shifts direction (and speed) with increasing altitude.

FIGURE 1.17
These high thin cirrus clouds appear fibrous because they are composed of mostly tiny ice crystals. [Photo by Captain Albert E. Theberge Jr., NOAA Corps (ret.), NOAA NWS Collection]

FIGURE 1.19
Fog reduces visibility along a highway. [Photo by NOAA]

FIGURE 1.20
Fair weather cumulus clouds are most common during the warmest time of day. [Photo by Ralph F. Kresge, NOAA NWS Collection]

FIGURE 1.21
Cumulonimbus (thunderstorm) clouds always produce lightning and sometimes heavy rain, hail, or strong and gusty surface winds.

FIGURE 1.22
Clouds at different altitudes may move horizontally in different directions indicating a change in wind direction with altitude. [Photo from the NASA S'COOL Project]

Conclusions

We can learn much about the atmosphere and weather by keeping track of local, regional, and national weather patterns via the Internet, television, radio, and newspapers. Weather maps, satellite images, and radar displays are particularly valuable in obtaining the "big picture" when following the development and movement of weather systems across a broad region. In addition, we are well advised to develop the habit of watching the sky for changing conditions locally and to monitor weather instruments if available. In this way, we are able to get involved with the subject matter of this course from the beginning and what we learn becomes more meaningful and practical. Our study of the atmosphere, weather, and climate continues in the next chapter with an examination of the evolution, composition, and structure of the atmosphere.

Basic Understandings

- From the experiences of everyday life, all of us have acquired basic understandings of the workings of the atmosphere, weather, and climate.
- Weather is the state of the atmosphere at some place and time, described quantitatively in terms of such variables as temperature, humidity, and wind speed. Climate encompasses weather conditions at some locality averaged over a specified time period plus extremes in weather.
- We can access weather information via the Internet, software applications, television,

newspapers, and radio, including the NOAA Weather Radio.

- A weather map portrays graphically the state of the atmosphere over a broad geographical region at a specified time. On a weather map, H or High symbolizes a locale where air pressure is relatively high compared to the surrounding area. Viewed from above in the Northern Hemisphere, surface winds blow clockwise and outward about the center of a high. A high (or anticyclone) features generally fair weather.

- On a weather map, L or Low symbolizes a locale where air pressure is relatively low compared to the surrounding area. Viewed from above in the Northern Hemisphere, surface winds blow counterclockwise and inward about the center of a low. A low (or cyclone) often features stormy weather.

- An air mass is a huge volume of air covering hundreds of thousands of square kilometers that is relatively uniform horizontally in temperature and humidity. The temperature and humidity of an air mass depend on the properties of its source region and the nature of the surface over which the air mass travels.

- A front is a narrow transition zone between air masses that contrast in temperature and/or humidity. Fronts form where air masses meet and associated air movements often give rise to cloudiness and precipitation that have characteristics dependent on the type of front. The most common types of fronts are stationary, warm, and cold.

- Sensors onboard weather satellites orbiting the planet provide a unique and valuable perspective on the state of the atmosphere and enable meteorologists to remotely measure temperature and humidity and track weather systems. These sensors measure two types of radiation: sunlight that is reflected or scattered by Earth and infrared radiation that is emitted by Earth. Weather satellites are in either geostationary or polar orbits and generate data that are processed into visible and infrared images.

- Weather radar continually emits pulses of microwave radiation that are reflected by raindrops, snowflakes, or hailstones. The reflected signal appears as blotches, known as radar echoes, on a television-type screen. From analysis of radar echoes, meteorologists determine the intensity and track the movement of areas of precipitation. Using the Doppler effect, weather radar can also monitor the circulation of air within a storm system and provide advance warning of severe weather.

- Observing the development, type, and movement of clouds may provide clues as to future weather.

Enduring Ideas

- Weather is described in terms of variables of state, including temperature, dewpoint, wind, and air pressure. Near real-time information about weather around the world is readily available via the Internet, television, and AM/FM radio. The NOAA Weather Radio is an important medium for receiving immediate watches, warnings, and advisories regarding extreme weather and other hazards proximate to your locality.

- Weather maps commonly plot air pressure and/or temperature as these variables help illustrate the principal weather-makers, pressure systems and fronts. High pressure systems, representing specific air mass types, are separated by low pressure systems and their associated fronts. The atmosphere is in continuous motion with air sinking and flowing clockwise and outward near the Earth's surface in Highs and flowing counterclockwise and inward near the surface and rising in Lows. Rising air often leads to cloud and precipitation development.

- Satellite and radar imagery are valuable tools in weather analysis and forecasting. Visible, infrared, and water vapor imagery monitor the ever-changing state of the atmosphere and weather systems moving across the globe. Radar imagery offers a detailed look into the structure, movement, and circulation within precipitation systems.

- Sky watching helps us discern the current state of the atmosphere and often reveals clues to future weather. Clouds are defined by their height (low, middle, or high), nature of appearance (stratiform or cumuliform), and whether they are producing precipitation.

Key Terms

weather

climate

meteorology

climatology

NOAA Weather Radio (NWR)

pressure systems

air pressure

air mass

front

GOES

POES

visible satellite image

infrared satellite image

water vapor satellite image

weather radar

Review

1. Distinguish between weather and climate.
2. What is the advantage of describing the climate of a locality in terms of both average weather plus extremes in weather?
3. Identify the various sources of weather information that are available to the public.
4. Describe the type of weather that usually accompanies a high (anticyclone) and a low (cyclone) in middle latitudes.
5. What is an air mass? What governs the temperature and humidity of an air mass?
6. Explain why clouds and precipitation are often associated with fronts.
7. Distinguish between GOES and POES weather satellites.
8. What advantages does an infrared satellite image offer over a visible satellite image for monitoring the state of the atmosphere?
9. Describe how weather radar detects the location and movement of areas of precipitation.
10. What causes high thin cirrus clouds to appear fibrous or wispy?

Critical Thinking

1. Identify several of the technological advances of the 20th century that significantly improved weather observation and forecasting.
2. Why are near-simultaneous weather observations essential for drawing weather maps?
3. How does the local air pressure change as an anticyclone approaches your location?
4. Describe the general characteristics of a cloud that may produce lightning, thunder, and heavy rain.
5. Explain how it is possible for stratiform clouds at different altitudes to be moving in different directions.
6. Speculate on why the highest temperature of the day usually occurs in mid-afternoon rather than noon.
7. Describe how the composition and appearance of a cloud change with increasing altitude.
8. Why does snow-covered ground appear bright white on a visible satellite image?
9. In forecasting the weather, what are some of the advantages of overlying a composite radar image on an infrared satellite image?
10. An intense low pressure system tracks northeastward just offshore from Cape Hatteras, NC, to just east of Cape Cod, MA. Describe the general direction of surface winds over New England as the center of the storm moves toward Nova Scotia.

Weather Maps, Historical Perspective

Weather maps shown on television, the Internet, or in newspapers are graphical models that display weather observations and analyses. The purpose of a weather map is to represent the state of the atmosphere over a broad geographical area at a specific time thereby permitting the identification of weather systems such as fronts, cyclones and anticyclones. Analyzing a chronological sequence of weather maps enables meteorologists to determine the movement of weather systems, a key component of weather forecasting.

In meteorology, *weather analysis* refers to the sequence of steps involved in the interpretation of a graphical (e.g., weather map) display of the distribution of one or more weather elements (e.g., temperature, air pressure). Typically, an important part of the analysis consists of drawing isopleths. An *isopleth* is a contour line that joins places having the same value of some weather element. Examples include *isotherms*, connecting locales with the same air temperature, and *isobars*, joining places with the same air pressure. Isopleths plotted on a weather map form patterns that are interpreted by meteorologists.

During the 18th century, European and North American scientists made weather observations because of their interest in weather for its own sake and weather's influence on agriculture and human health. By the early 19th century, some weather observers had assembled reasonably continuous weather records, but typically they were not part of an organized reporting network. Their observations were primarily for the purpose of describing the climate. Lacking any means for rapid communication of weather observations, these data were not useful for weather analysis and forecasting.

The first weather maps appeared in the early 19th century but took years to assemble because of sparse data, non-standard observation practices, and poor communications systems. Heinrich Wilhelm Brandes (1777-1834), a professor at the University of Breslau (in modern Poland), was one of the first to recognize the value of the spatial display of weather data in depicting the state of the atmosphere. Credit goes to Brandes for drawing the first weather map in 1819 showing an intense storm centered over the English Channel on 6 March 1783. However, the 36-year time delay meant that his analysis was not useful for weather forecasting. In 1819, Brandes also plotted 365 daily weather maps for the year 1785.

The source of data for Brandes' weather maps was the network of 39 volunteer weather observers in 18 nations (mostly European) operated by the Meteorological Society of the Palatinate in Mannheim. This first attempt at organized international weather observation lasted from 1781 to 1792. The Society published the data (i.e., wind, temperature, pressure, humidity, precipitation) but no maps in a series of a dozen volumes from 1783 to 1795. Brandes plotted departures (anomalies) from normal air pressure and drew isopleths joining locations having the same departure value. He also plotted winds on the same map and observed how air rotated around a storm and converged inward toward the storm center. By viewing a sequence of weather maps based on simultaneous observations, Brandes concluded that reduced air pressure was associated with a storm system.

In the United States, the first weather maps appeared in the fourth and fifth decades of the 19th century. In 1837, James P. Espy (1785-1860) published the first U.S. weather map based on observations from widely spaced locales. Espy's map had no isobars and used arrows to represent wind directions associated with a storm system centered near Silver Lake, NY, on 20 June 1836. Elias Loomis (1811-1889), professor of mathematics at Western Reserve College in Cleveland, OH, employed a similar approach as Brandes in his 1843 study of a storm that occurred on 16 February 1842. Loomis analyzed a series of maps on which were plotted simultaneous observations from 131 locations to determine the circulation and track of the storm. In so doing, Loomis produced the first *synoptic weather map* with pressure, temperature, wind, sky condition, and precipitation plotted on the same map.

Until the mid-1800s, the utility of weather maps for forecasting purposes was stymied by a lack of a rapid means of communicating observational data. That changed with the invention of the electric telegraph in 1844. Now weather observations could be transmitted rapidly over long distances making possible up-to-date weather maps and the first practical system of weather analysis and forecasting (Figure 1). The first telegraph-linked weather observation networks in the U.S. were operated by Cleveland Abbe (1838-1916) at Cincinnati's Mitchell Astronomical Observatory, Joseph Henry (1797-1878) at the Smithsonian Institution, and the U.S. Army Signal Corps (predecessor to the U.S.

Weather Bureau and National Weather Service). Through the years with technological advances in instrumentation and communications, weather maps became more accurate in their portrayal of the state of the atmosphere and more useful for weather analysis and forecasting (Chapter 2).

FIGURE 1
A U.S. Army Signal Corps weather map dated 1 September 1872 based on observations telegraphed to Washington, DC. The Signal Corps weather network was the predecessor to the U.S. Weather Bureau and National Weather Service. (NOAA Photo Library)

Global Positioning System

Satellite navigation was the brainchild of the U.S. Department of Defense. Similar to 18th century governments offering prize money to whoever solved the longitude problem (creating a reliable means of determining the longitude of a ship at sea), the U.S. government spent over $12 billion to develop a satellite-based system that could specify a location to near pinpoint accuracy at all times and in all types of weather. Originally intended for U.S. Armed Forces, with continuous positioning and navigation data available worldwide, the *Global Positioning System (GPS)* dates to the 1978 launch of the first Navstar satellite. In the 1980s, the government made the system available for civilian use, which began mostly on large ships, and completed the original satellite network in 1994. The U.S. government owns and operates GPS as a national resource.

Not to be outdone by its Cold War adversary, the Soviet Union developed a similar system known as the GLObal NAvigation Satellite System (GLONASS), which Russia made publically available in 2007. Other satellite navigation systems in operation or development include the European Union's Galileo (expected to be operational by 2014), The People's Republic of China's COMPASS (2020), and regional systems operated by India (IRNSS) covering India and the Northern Indian Ocean (2012), as well as one by Japan (QZSS) encompassing Asia and Oceania (2013).

At least 24 Navstar satellites are needed to provide continuous service for GPS. Once this was achieved in the early 1990s, inexpensive, small portable receivers were developed and civilian use soared. GPS is widely used for navigation by ships, planes, delivery trucks, automobiles, and, increasingly, cell phones. Today non-military sectors (consumers and commercial customers) account for 92% of GPS equipment sales. Sales of GPS equipment worldwide grew from $3.5 billion in 2003 to $9.9 billion in 2010.

The principle behind GPS is *trilateration* (related to triangulation) from satellites that are in unobstructed view of the receiver (Figure 1). At least four satellites are needed for maximum precision, with each satellite in orbit at 19,000 km (12,000 mi) and completing two orbits in less than 24 hrs. Ranging signals travel from satellites to GPS receivers by line of sight through the atmosphere, clouds, and materials such as glass, but not most solids such as buildings and mountains. Hence, GPS units do not function indoors or underground, or underwater. The time it takes a ranging signal to travel from satellite to receiver is used to determine distance to the location. Atomic clock times, taken from the satellite, are embedded in the code of the ranging signal and compared with a code generated by the clock of the receiver. Radio signals travel at a finite speed (300,000 km per sec, or 186,000 mi per sec, the speed of light) so that elapsed time is easily converted to distance. All GPS satellites are identifiable by their coded signals and their orbits are regular, allowing refinement of triangulated distances. Using an inexpensive, widely available receiver, anyone can readily locate his/her position within a few meters. More sophisticated military receivers can achieve location accuracy in centimeters.

With a single satellite, a location could be at any point on a sphere, centered about the satellite, with a radius equal to the distance between the location and the satellite. With a second satellite, the location must be where the spheres overlap. The two spheres have different radii so their intersection is a circle. A third satellite describes yet another sphere, intercepting the other two spheres, and narrowing the location to the two points where all three spheres intersect. Usually one point is in space or Earth's interior while the other is on Earth's surface. Computer algorithims in the GPS receiver can distinguish between correct and spurious locations.

The fourth (and any additional) satellite synchronizes the slight differences in timing between the precise atomic clocks onboard GPS satellites and the less accurate quartz clocks in GPS receivers. In practice, most receivers select the best signals from many satellites so that assumptions about the correct point of intersection are unnecessary. Also, the more satellites in the line of sight of a GPS receiver, the greater the accuracy.

Plans are currently underway to increase the accuracy of GPS by broadcasting new signals from the navigation satellites. One of the anticipated benefits of these new signals is a reduction in interference caused by Earth's *ionosphere* (a region of the upper atmosphere containing a relatively high concentration of electrically charged particles). GPS satellite radio signals slow as they pass first through the ionosphere and then the electrically neutral atmosphere, delaying the expected arrival of the ranging signal. It is possible to correct for ionospheric delay, as it is frequency dependent, by using dual-frequency GPS receivers. The delay due to the neutral atmosphere is not frequency dependent but depends on the components of the atmosphere (a mixture of dry gases and water vapor).

Interestingly, measurement of atmospheric delays in GPS signals allows atmospheric water vapor to be monitored. As discussed in detail elsewhere in this book, water vapor is one of the most important components of the atmosphere. Through the energy involved in the phase changes of water, heat is transferred within the Earth-atmosphere system and contributes to the weather. Water vapor is also a greenhouse gas that plays a critical role in Earth's climate system, not only absorbing and emitting infrared (heat) radiation, but also affecting cloud and aerosol formation and the chemistry of the lower atmosphere. Monitoring atmospheric water vapor is therefore a significant meteorological application of GPS technology.

GPS signals can also arrive at receivers from nearby reflecting surfaces, a phenomenon known as *multipath signals*. Normally, multipath signals interfere with those received directly from the satellites, reducing positioning accuracy. Researchers from the National Oceanic and Atmospheric Administration/Earth System Research Laboratory (NOAA/ESRL) Physical Sciences Division (PSD), however, use GPS multipath signals reflecting off ocean or land to learn about the sea state, soil moisture, snowpack, and sea ice. Land surface reflections of the radio wavelength emitted by GPS transmitters are especially sensitive to these environmental parameters.

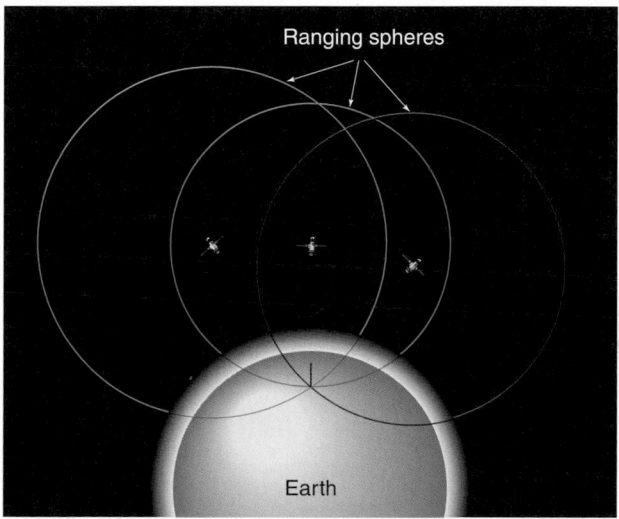

FIGURE 1
To determine its location on Earth's surface, a GPS receiver measures the travel times of radio signals sent by three satellites of known orbital location. Travel times are converted to distances. This trilateration technique is more precise when a fourth satellite is used to synchronize the clocks on the GPS satellites and receiver.

This view of Earth's horizon as the sun sets over the Pacific Ocean was taken by an Expedition 7 crewmember onboard the International Space Station (ISS). Anvil tops of thunderclouds are also visible. [Courtesy NASA]

ATMOSPHERE: ORIGIN, COMPOSITION, & STRUCTURE

Chapter Highlights

Case-in-Point
  *African Origins of Wind-Borne Dust in the
    Americas*
Evolution of the Atmosphere
  Primeval Phase
  Modern Phase
  Air Pollution
Investigating the Atmosphere
  The Scientific Method
  Atmospheric Models
Monitoring the Atmosphere
  Surface Observations
  Upper-Air Observations
  Remote Sensing
Temperature Profile of the Atmosphere
The Ionosphere and the Aurora
Conclusions
Basic Understandings/Enduring Ideas
Key Terms/Review/Critical Thinking
For Further Exploration
  The Atmosphere of Mars
  Radio Transmission and the Ionosphere
  Space Weather Prediction

Learning Objectives

Describe the principal events and processes involved in the evolution of Earth's atmosphere.

Explain the role played by the ocean in the evolution of the atmosphere.

Distinguish between layers of the atmosphere.

Describe how human activity plays a role in the evolution of the atmosphere.

Present some examples of how minor components (gases and aerosols) of the atmosphere are essential for life on Earth.

Explain how scientists employed the scientific method in analyzing the possible causes of the Antarctic ozone hole.

Demonstrate the role of models in scientific inquiry.

Provide examples of the various types of scientific models used in meteorology and climatology.

Summarize how meteorologists monitor surface and upper-atmospheric conditions such as temperature, humidity, and winds.

Sketch the average vertical temperature profile of the atmosphere.

Distinguish between the properties of the troposphere and stratosphere.

Describe the cause and significance of the ionosphere.

Explain the connection between solar activity and the occurrence of the aurora.

What is the composition and structure of the atmosphere?

African Origins of Wind-Borne Dust in the Americas

For many years, scientists suspected that significant amounts of red iron- and clay-rich soils on islands throughout the Caribbean and in Bermuda owed their origin partially to dust transported by northeast trade winds from the arid lands of North Africa. They also proposed that wind-borne North African dust supplies nutrients to the Amazon rain forest of South America. Remote sensing by satellite now confirms transport of dust from North Africa to the western North Atlantic, Caribbean Sea, and Gulf of Mexico (Figure 2.1). In recent years, scientists have found evidence that a pool of warm dusty air originating over the Saharan Desert may suppress tropical cyclone activity (e.g., hurricanes) in the North Atlantic, adversely affect air quality over the southeastern United States, and when the dust settles into the ocean, may contribute to toxic red tides in the Gulf of Mexico and threaten coral reefs in the Caribbean.

As weather systems sweep across North Africa (i.e., the Sahara and Sub-Saharan Africa), their strong turbulent winds lift dust particles from the dry topsoil and can carry them to altitudes of 5000 m (16,500 ft) or higher. The prevailing northeast trade winds then transport plumes of the smallest dust particles across the Atlantic and over the Caribbean, Central America, and the southeastern United States. (Florida receives more than half of all North African dust delivered to the United States.) This transoceanic journey takes about 1 to 2 weeks and dust plumes are observed primarily from June to October, peaking in July. Persistent drought in sub-Saharan Africa (from the mid-1960s through the early 1990s) may explain an increase in the volume of dust transported to the western North Atlantic.

In 2004, scientists reported on a new satellite-based technique that enables them to track and analyze the westward movement of the *Saharan Air Layer (SAL)*. They found that the SAL is uniformly warm and dusty, extends to an altitude approaching 5500 m (18,000 ft), and covers an area somewhat larger than the 48 contiguous United States. Over the Atlantic Ocean, the SAL is underlain by a shallow (900-1800 m or 3000-5900 ft) layer of cool, humid air (the *marine layer*). Remarkably, the SAL retains its warm dusty characteristics as it travels many thousands of kilometers across the Atlantic. Absorption of solar radiation by the suspended dust particles is one reason why the SAL remains warm. In 2006, scientists confirmed the strong inverse correlation between mean dust coverage and tropical cyclone activity over the North Atlantic. For reasons given in Chapter 12, conditions in the SAL tend to

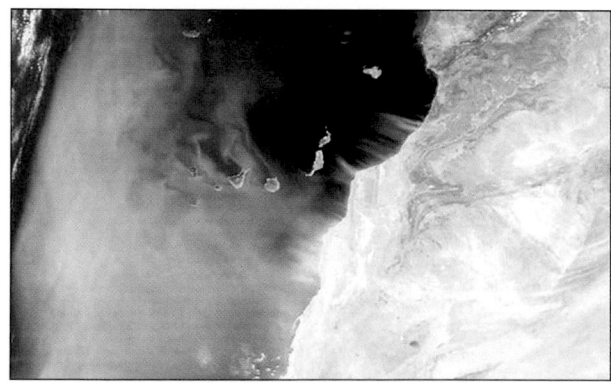

A

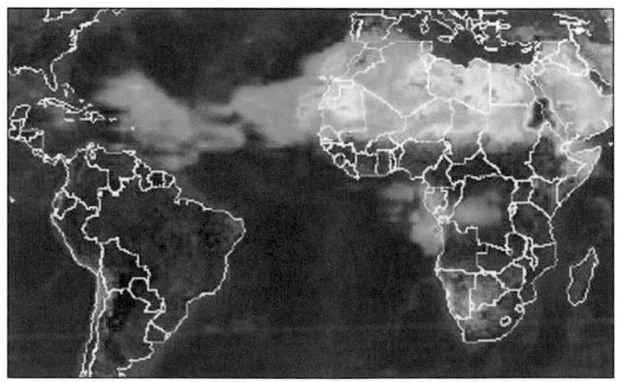

B

FIGURE 2.1

(A) A massive dust storm over northwest Africa and the Sahara Desert. This satellite image was acquired on 11 February 2001 by SeaWiFS. [Courtesy of the SeaWiFS Project, NASA/Goddard Space Flight Center and ORBIMAGE] (B) This satellite image taken by NASA's Total Ozone Mapping Spectrometer (TOMS) instrument shows dust and smoke blowing off land sources in North Africa and moving westward across the Atlantic as a dust plume on 2 July 1999. [Courtesy of NASA]

inhibit the formation of tropical cyclones. When the SAL engulfs a hurricane (or a tropical precursor to a hurricane), the system weakens.

North African dust contributes to a reddish haze and colorful sunsets in the Caribbean (Figure 2.2) and over much of the southeastern United States. The dust particles also can harbor in their microscopic cracks and crevices, bacteria and fungi, potential disease-causing organisms that may pose a health risk for people suffering from respiratory illnesses or having weakened immune systems. Even without these pathogens, the dust is known to trigger allergic and respiratory reactions. Furthermore, the North African dust arrives over the southeastern United States at the time of year when photochemical air pollutants are at their highest levels, further exacerbating regional air quality problems.

In the summer of 2001, scientists reported that the iron in North African dust particles fertilizes the Gulf of Mexico waters, increasing the probability of harmful algal blooms, commonly known as *red tides*. Red tides have been implicated in the die-off of fish, marine mammals, and birds as well as respiratory problems and skin irritations in humans. Enhanced levels of iron enable specialized bacteria to convert nitrogen gas dissolved in seawater to an organic form that triggers an explosive growth in populations of potentially toxic algae. North African dust may also harm coral reefs in the Caribbean through nutrient enrichment by similarly spurring the growth of populations of algae and phytoplankton that colonize the same environment as coral and interfere with its growth.

The story of long-range transport of North African dust illustrates how our investigation of the atmosphere is key to a better understanding not only of weather and climate but also environmental issues that involve the atmosphere and atmospheric processes. Hence, our study focuses on the atmosphere's interactions with the other components of the Earth system, including land, ocean, and living organisms. The significance of these interactions will become evident as we develop basic understandings in atmospheric science.

FIGURE 2.2
Photo of the Saharan Air Layer taken from the NOAA P-3 Orion airplane near Barbados. Saharan dust gives the sky an orange glow during this late afternoon sunset in the eastern Caribbean. Small cumulus clouds can be seen poking through the top of the dust layer. [Photo by Jason Dunion, NOAA/HRD]

Almost everyone seems interested in the weather, probably because it affects virtually every aspect of daily life—our clothing, the price of orange juice and coffee in the grocery store, and sometimes even the outcome of a football game. Tranquil, pleasant weather allows us to enjoy a variety of outdoor activities. A turn to stormy weather can bring mixed blessings: heavy rains wash out picnics but also benefit crops wilting under the searing summer Sun. Occasionally, the weather is severe, and the impact may range from mere inconvenience to a disaster that is costly in human lives and property (Figure 2.3). Thick fog causes

flight delays and cancellations, a night of subfreezing temperatures takes its toll on Florida citrus, and heavy snowfall snarls commuter traffic. But these impacts pale in comparison to the death, injury, and property damage that a tornado or hurricane can cause.

Regardless of where we live, each of us is well aware from personal experience that weather is variable. This variability prompted Mark Twain, not one to shy from exaggeration, to quip of spring weather in New England: "In the spring I have counted one hundred and thirty-six different kinds of weather inside of four-and-

A

B

C

D

FIGURE 2.3

Every season has distinct types of severe weather possible. (A) In fall, heavy rains from tropical systems or strong extratropical cyclones often cause flooding in the Northeast. In this photo, rainfall of more than an inch affected attendance at a festival in the Binghampton, NY, area. [NOAA/NWS] (B) Heavy winter snows can make driving hazardous. A plow works to keep the street passable as a blizzard in December 2006 hit Denver, CO, with up to 28 inches of snow. [FEMA/Michael Rieger] (C) The EF-5 tornado on 22 May 2011 tore a path through Joplin, MO, seven miles long and half a mile wide. This photo, taken on 1 August 2011, shows there was still much work to be done to recover from the extreme damage. [FEMA/Elissa Jun] (D) In summer, storm surge from tropical storms and hurricanes causes beach erosion. This photo taken at Rodanthe, NC, on 2 September 2011 shows waves pounding the supports of raised homes. These homes previously had yards of beach in front of them that was eroded by Hurricane Irene. [FEMA/Tim Burkitt]

twenty hours."[1] Of course, weather is not equally variable everywhere. For example, the temperature contrast between winter and summer is much more pronounced in the Dakotas, where summers are very warm and winters are bitterly cold, than in south Florida, where the weather is usually warm year round.

This chapter covers the evolution of the atmosphere, the composition of air, how meteorologists monitor surface and upper-air properties of the atmosphere, the average temperature profile of the atmosphere, the thermal subdivisions of the atmosphere, and the special electromagnetic characteristics of the upper atmosphere.

Evolution of the Atmosphere

The atmosphere is the site of weather and climate, the main subjects of this book. Earth's **atmosphere** is a relatively thin layer of gases and tiny, suspended particles that envelopes the globe. The atmosphere is one component of the *Earth system* that also consists of the *hydrosphere* (ocean, glaciers, rivers, lakes, groundwater), *geosphere* (rocks, minerals, soil), and *biosphere* (plants, animals). All components of the Earth system interact with one another in ways that maintain a habitable environment.

Compared to the planet's diameter (12,740 km or 7918 mi), the atmosphere is like the thin skin of an apple (Figure 2.4). About half the atmosphere's mass is concentrated within about 5500 m (18,000 ft) of Earth's surface and 99% of its mass is below an altitude of 32 km (20 mi).[2] Yet, this thin atmospheric skin is essential for life and the orderly functioning of physical and biological processes operating on Earth. The atmosphere shields organisms from exposure to hazardous levels of solar ultraviolet radiation; it has the gases necessary for life-sustaining processes of photosynthesis and cellular respiration; and it is involved in supplying the water required by all forms of life.

The ancient Greeks thought that air was one of the four elements (earth, air, water, and fire). However, by the beginning of the 19th century, chemists such as Joseph Black, Daniel Rutherford and Joseph Priestley had isolated various constituent gases from the air, demonstrating that air is a mixture of gases that are either elements or compounds. Chemists have identified a chemical *element*, such as hydrogen or oxygen, to consist of atoms that

FIGURE 2.4
Planet Earth, viewed from space, appears as a "Blue Marble" with its surface mostly ocean water and partially obscured by swirling cloud masses. The atmosphere, seen most clearly in Northern Hemisphere polar latitudes as a thin light blue arc on the edge of the image, is like the thin skin of an apple compared to the planet's diameter. [NASA image by Reto Stöckli, based on data from NASA and NOAA]

cannot be broken down farther by ordinary means whereas a *compound* contains molecules consisting of chemically bonded atoms of various elements in fixed proportions, such as water (H_2O) or carbon dioxide (CO_2). Mixtures, like Earth's atmosphere, contain a collection of elements and/or compounds that can be separated by ordinary physical means.

Earth's atmosphere is the product of a lengthy evolutionary process that began at the planet's birth more than 4.5 billion years ago. Astronomers scanning the solar system and geologists analyzing evidence obtained from meteorites, rocks, and fossils have given us a reasonable, albeit incomplete, account of the origins of the atmosphere. At least one thing is known with a high degree of confidence: The evolution of the atmosphere was closely linked to the formation of the ocean and the coming of life on the planet.

PRIMEVAL PHASE

The atmosphere and other components of the Earth system (hydrosphere, geosphere, and biosphere) co-evolved during the vast expanse of Earth history. More than 4.5 billion years ago, Earth, the Sun, and the entire solar system evolved from an immense rotating cloud of

---

[1] "Address of Mr. Samuel L. Clemens: The Weather in New England," New England Society in the City of New York, Annual Report, 1876, p. 59.

[2] For unit conversions, see Appendix I.

FIGURE 2.5
The leftmost "pillar" of interstellar hydrogen gas and dust in M16, the Eagle Nebula. [Courtesy of NASA/NSSDC Photo Gallery]

cosmic dust, ice, and gases, called a **nebula** (Figure 2.5). Temperature, density, and pressure were highest at the center of the nebula and gradually decreased outward. With temperatures exceeding 400 °C (750 °F) at the nebula's center, ice and the lighter elements vaporized, streaming toward the outer reaches of the nebula. Consequently, residual dry rocky masses formed the inner planets, including Earth. Further out, meteorites and the less-dense giant planets Saturn and Jupiter formed.

Although it began as a dry rocky mass, today, with about 71% of the surface covered by ocean water and 10% of the land area overlain by glacial ice, Earth is the water planet. The ocean is the immediate source of water that is a key component of the atmosphere, occurring in all three phases, and composing clouds (ice crystals and/or water droplets) and precipitation. Water vapor and clouds slow the escape of Earth's heat to space, elevating the temperature of the lower atmosphere (the *greenhouse effect*). Where did the water originally come from?

In 2011, a team of European scientists presented new evidence for the explanation that comets and asteroids delivered most water to Earth. The Herschel Space Observatory satellite detected a reservoir of very cold water vapor associated with a star, 175 light-years from Earth, in the early stages of planet formation. Similar reservoirs located in the outer reaches of the evolving solar system would have made comets and asteroids water-rich. Vaporized and dispersed water condensed within comets

beyond Jupiter and Saturn. Composed of approximately equal amounts of meteoritic dust and ice, a **comet** is a relatively small mass that travels in a highly elliptical orbit around the Sun. Jupiter's strengthening gravitational attraction may have drawn a multitude of comets from the outer toward the inner reaches of the solar system, on a collision course with Earth. Boosting confidence in the comet origin of water was the recent discovery that the ratio of deuterium to ordinary hydrogen in the comet Hartley 2 was the same as in the ocean. (*Deuterium* is an isotope of hydrogen with one proton and one neutron in its nucleus, and is very rare on Earth; a normal hydrogen atom has only a single proton.)

Asteroids, large bodies of rock many kilometers across, also formed in reservoirs of very cold water vapor just inside Jupiter's orbit during the latter stages of Earth's formation. At the time, the asteroid belt consisted of rocks ranging in size from dust to small planets, and these scattered inward, including towards Earth. Asteroids are 10% ice by mass. A period of heavy bombardment of Earth's surface by comets and asteroids began about 4.1 billion years ago and spread a veneer of water over Earth's surface. The bombardment ended about 3.8 billion years ago. By then most of Earth's water was in place.

Another possibility is that Earth's water is indigenous. If the center of the nebula was cooler than previously assumed, the materials present in the inner solar system, which formed the Earth, could have been water-rich.

As Earth's atmosphere developed, it was mostly hydrogen (H) and helium (He) with a few hydrogen compounds, including methane (CH_4) and ammonia (NH_3). Because these atoms and molecules are relatively light and have high molecular speeds, Earth's weak gravitational field and high temperatures allowed the early atmosphere to escape to space. By 4.4 billion years ago, the planet's accumulating mass produced a gravitational pull sufficiently strong enough to retain a thin gaseous envelope of volcanic origin, Earth's *primeval atmosphere*.

The principal source of Earth's atmosphere is **outgassing**, the release of gases from rock through volcanic eruptions and the planet's rocky surface when meteorites strike. Perhaps as much as 85% of all outgassing took place within a million years of the planet's formation, though it continues today at a much slower pace (Figure 2.6). Earth's primeval atmosphere was mostly carbon dioxide (CO_2), with some nitrogen (N_2) and water vapor (H_2O), and trace amounts of methane, ammonia, sulfur dioxide (SO_2), hydrogen sulfide (H_2S), and hydrochloric acid (HCl). Radioactive decay of the

FIGURE 2.6
Gas plume at Kilauea's Puʻu ʻŌʻō crater, Hawai'i. [Courtesy of Hawaiian Vulcano Observatory/United States Geological Survey]

potassium-40 isotope (^{40}K) in the planet's bedrock added argon (Ar), an inert (chemically non-reactive) gas, to the atmospheric mix. Dissociation of water vapor into its constituent atoms, hydrogen and oxygen, by high-energy solar ultraviolet (UV) radiation, contributed a small quantity of free oxygen while the lighter, faster moving hydrogen escaped to space. Oxygen also combined with other elements in various chemical compounds.

From 4.5 to 2.5 billion years ago, the Sun was 30% dimmer than today but Earth was still warm because of the abundance of CO_2. (Earth's CO_2-rich atmosphere was 10 to 20 times denser than today.) Carbon dioxide slows the escape of Earth's heat to space, and raised the average surface temperature to as high as 85 °C to 110 °C (185 °F to 230 °F), significantly higher than the current average surface temperature of 15° C (59 °F).

By 4 billion years ago, the planet began cooling and the Earth system changed. Cooling caused atmospheric water vapor to condense into clouds, producing rain. Precipitation and runoff from landmasses gave rise to the ocean, eventually covering 95% of the planet's surface. The *global water cycle*, which cooled Earth's surface even more through evaporation, and its largest reservoir, the ocean, were in place. Rains also caused a substantial decline in the concentration of atmospheric CO_2, which dissolves in rainwater and produces weak carbonic acid that then chemically reacts with bedrock, locking carbon in rocks and minerals. Diminishing amounts of atmospheric CO_2 further lowered surface temperatures. On land, the physical and chemical breakdown of rock (*weathering*) and erosion delivered carbon-containing sediment to the ocean. Runoff washed dissolved CO_2 directly into the sea and some atmospheric CO_2 dissolved in ocean surface waters as their temperatures fell. (CO_2 is more soluble in cold water.)

Although carbon dioxide has been only a minor component of the atmosphere for at least 3.5 billion years, its concentration fluctuated greatly during the geologic past and altered global climate and life on Earth. All other factors being equal, more CO_2 in the atmosphere means higher temperatures at the Earth's surface. Since peaking at 5000 ppmv about 550 million years ago, the concentration of atmospheric CO_2 generally has declined. However, this downward trend was punctuated by many episodes of large-scale volcanic activity responsible for temporary upturns in atmospheric CO_2 and a considerably warmer global climate. An example is the *Middle Eocene Climatic Optimum (MECO)*, about 40 million years ago, one of the hottest intervals in Earth history. According to temperature reconstructions based on deep-sea sediment cores, sea-surface temperatures in the southwest Pacific Ocean rose 3 to 6 Celsius degrees (5.4 to 11 Fahrenheit degrees). Warming has been attributed to a million years of active volcanism along Asia's southern border, increasing atmospheric CO_2 to more than 1000 ppmv.

Atmospheric CO_2 levels also fluctuated during the Pleistocene Ice Age, 1.8 million to 10,500 years ago, decreasing during episodes of glacial expansion and increasing during episodes of glacial recession.

Living organisms have also played an important role in Earth's evolving atmosphere, primarily through **photosynthesis**, in which green plants use sunlight, water, and carbon dioxide to produce sugars and oxygen. Although vegetation is a *sink* for carbon dioxide, geochemical processes were more important than photosynthesis in removing CO_2 from the atmosphere. Based on the fossil record, photosynthesis dates to about 2.7 billion years ago when cyanobacteria first appeared in the ocean. However, it was not until 2.3 billion years ago that significant amounts of oxygen began accumulating in the atmosphere. Hundreds of millions of years passed before atmospheric oxygen levels spiked in the *Great Oxidation Event*. Why the lengthy delay?

Initially, whatever oxygen was available in the ocean combined with marine sediments thereby preventing the escape of the oxygen to the atmosphere. Eventually, however, the sediments were completely oxidized. The available oxygen then dissolved in ocean water and once that was saturated, began escaping into the atmosphere. In 2007, scientists reported that 2.5 billion years ago, there was also a global shift in dominant volcanism from underwater to terrestrial accompanied by an important change in the composition of eruptive gases. The eruptive gases from submarine volcanism consisted of molecules that react with oxygen. With submarine volcanism reduced, this submarine sink for oxygen ceased to operate. On the other hand, the eruptive gases from terrestrial volcanism consisted of molecules that are inert toward oxygen. Hence, the change in dominant volcanism favored a buildup in oxygen in both the ocean and atmosphere.

As oxygen emerged as a major component of Earth's atmosphere, the ozone shield formed. Within the *stratosphere*, incoming solar ultraviolet (UV) radiation drives chemical reactions that convert oxygen (O_2) to ozone (O_3) and back (Chapter 3). Absorption of UV radiation in these reactions prevents potentially lethal intensities of UV radiation from reaching Earth's surface. UV radiation penetrates ocean water only to relatively shallow depths so that marine life was able to exist at intermediate and lower depths. However, by 440 million years ago, formation of the *stratospheric ozone shield* made it possible for organisms to thrive in surface waters and, eventually, on land.

The concentration of atmospheric oxygen has fluctuated significantly over the past 550 million years, varying between 13% and 31%. These fluctuations were linked to imbalances in the rates of weathering of organic carbon and pyrite (FeS_2), which decreases atmospheric oxygen, and the deposition of these materials with other sediments which increases atmospheric oxygen. Oxygen today composes 21% of the air we breathe.

Nitrogen (N_2), also a product of outgassing, is the most abundant atmospheric gas (78%) mostly because it is relatively inert and its molecular speeds are too slow to escape Earth's gravitational pull. Furthermore, nitrogen is less soluble in water than other atmospheric gases. Therefore, nitrogen cycles out of the atmosphere only very slowly. While nitrogen continues to be generated as a minor component of volcanic eruptions, today the principal process whereby nitrogen enters the atmosphere is denitrification, which accompanies the bacterial decay of plants and animals. Nitrogen is removed from the atmosphere by *biological fixation*, direct nitrogen uptake by leguminous plants, such as clover and soybeans, and *atmospheric fixation*, when the intense heat of lightning causes nitrogen to combine with oxygen to form nitrates (Figure 2.7).

FIGURE 2.7
Nitrogen is removed from the atmosphere during the process of atmosphere fixation, caused by the intense heat of lightning. [NOAA]

MODERN PHASE

Ultimately, these gradual evolutionary processes produced the modern atmosphere. The lower atmosphere continually circulates so that the principal atmospheric gases are well mixed and occur almost everywhere in the same relative proportions up to an altitude of about 80 km (50 mi). That portion of the atmosphere is called the **homosphere**. Above 80 km, gases are stratified such that concentrations of the heavier gases decrease more rapidly with altitude than do concentrations of the lighter gases. The region of the atmosphere above 80 km is the **heterosphere**.

Nitrogen and oxygen are the chief gaseous components of the homosphere. Not counting water vapor (which has a highly variable concentration), nitrogen (N_2) occupies 78.08% by volume of the homosphere, and oxygen (O_2) is 20.95% by volume. Henry Cavendish (1731-1810) discovered this basic gaseous composition of air in 1781. The next most abundant gases are argon (0.93%) and carbon dioxide (0.0389%). As shown in Table 2.1, the atmosphere also has small quantities of helium (He), methane (CH_4), hydrogen (H), ozone (O_3), and many other gases. Unlike the atmosphere's principal gases, the percent volume of some of these trace gases varies with time and location within the homosphere.

In the heterosphere, above about 150 km (95 mi), oxygen is the chief atmospheric gas but occurs primarily in the atomic (O) rather than diatomic (O_2) form. Solar ultraviolet radiation dissociates O_2 into its constituent atoms. Two oxygen atoms can recombine to form a molecule only by colliding with another atom or molecule, but the number of molecules per unit volume (the *number density*) at these high altitudes is so low that such collisions are infrequent. At lower altitudes the intensity of incoming

TABLE 2.1
Gases Composing Dry Air in the Lower
Atmosphere (below 80 km)

Gas	% by Volume	Parts per Million
Nitrogen (N_2)	78.08	780,840.0
Oxygen (O_2)	20.95	209,460.0
Argon (Ar)	0.93	9,340.0
Carbon dioxide (CO_2)	0.03890	389.0
Neon (Ne)	0.00180	18.0
Helium (He)	0.00052	5.2
Methane (CH_4)	0.00014	1.4
Krypton (Kr)	0.00010	1.0
Nitrous oxide (N_2O)	0.00005	0.5
Hydrogen (H)	0.00005	0.5
Xenon (Xe)	0.000009	0.09
Ozone (O_3)	0.000007	0.07

solar ultraviolet radiation is less, thereby reducing the rate of dissociation of O_2. Also, with more molecules per unit volume at lower altitudes, molecular collisions are more frequent. Below about 100 km (60 mi), oxygen atoms recombine at a faster rate than oxygen molecules dissociate, so that oxygen occurs mostly as O_2.

Earth's nitrogen/oxygen-dominated atmosphere contrasts strikingly with the CO_2-rich atmospheres of our neighboring planets Venus and Mars. The atmosphere of Venus is almost 100 times denser than Earth's atmosphere and features an average surface temperature of about 460 °C (860 °F). The Martian atmosphere, on the other hand, is much thinner than Earth's atmosphere and has an average surface temperature ranging from about −60 °C (−76 °F) at the equator to as low as −123 °C (−189 °F) at the poles. These differences exist even though all three planets likely began with chemically similar atmospheres. The atmospheres of Earth, Mars, and Venus evidently followed different evolutionary paths; for more on the Martian atmosphere, see this chapter's first *For Further Exploration*.

In addition to gases, Earth's atmosphere contains minute liquid and solid particles, collectively called **aerosols**. A flashlight beam in a darkened room reveals an abundance of tiny dust particles floating in the air. Individual aerosols are too small to be visible but in aggregates, such as the multitude of water droplets and ice crystals composing clouds, they may be visible. Most

aerosols occur in the lower atmosphere, near their sources on Earth's surface; they derive from wind erosion of soil (e.g., Saharan dust), ocean spray, forest fires, volcanic eruptions, and industrial, agricultural, and other human activities. Also, some aerosols, such as meteoric dust, enter the atmosphere from above.

We may be tempted to dismiss as unimportant those substances that make up only a small fraction of the atmosphere, but the significance of an atmospheric gas or aerosol is not necessarily related to its concentration. Water vapor, carbon dioxide, and ozone (O_3) occur in very low concentrations, yet they are essential for life. Most water vapor is confined to the lowest kilometer or so of the atmosphere and is never more than about 4% by volume even in the most humid places on Earth (e.g., over tropical rainforests and seas). But without water vapor, the planet would have no water cycle, no rain or snow, no ocean, and no fresh water. Although comprising only 0.0389% of the lower atmosphere, carbon dioxide (CO_2) is essential for photosynthesis. Furthermore, water vapor and carbon dioxide absorb and emit infrared radiation, elevating the temperature of the lower atmosphere to a range that makes life possible on Earth. Although the volume percentage of ozone is minute, the chemical reactions responsible for its formation (from oxygen) and dissociation (to oxygen) in the stratosphere (mostly at altitudes between 30 and 50 km) shield organisms on Earth's surface from potentially lethal intensities of solar UV radiation.

The aerosol concentration of the atmosphere is also relatively small, yet these suspended particles participate in important processes. As demonstrated in Chapter 7, some aerosols function as nuclei that promote the development of clouds. Furthermore, some aerosols (e.g., volcanic dust, sulfurous particles) affect the climate by interacting with incoming solar radiation (Chapter 15).

AIR POLLUTION

Human activity also plays a significant role in the composition of air primarily by contributing air pollutants. An **air pollutant** is a gas or aerosol in the atmosphere that occurs at a concentration that threatens the wellbeing of living organisms (especially humans) or disrupts the orderly functioning of the environment. Many of these substances occur naturally in the atmosphere. Sulfur dioxide (SO_2) and carbon monoxide (CO), for example, are normal minor gaseous components of the atmosphere that become pollutants when their concentrations approach or exceed the tolerance limits of organisms. At sufficiently high concentrations, sulfur dioxide irritates the throat and lungs and impairs the respiratory system's defenses against foreign particles and bacteria. Carbon monoxide is a colorless, odorless, and tasteless asphyxiating agent (reducing the blood's oxygen-carrying ability). Certain air pollutants, however, are not natural constituents of the atmosphere and some of these are hazardous to human health even at very low concentrations. An example is benzene, which is a known carcinogen, a cancer-causing agent.

A substance that is harmful immediately upon emission into the atmosphere is designated a *primary air pollutant*. Carbon monoxide in automobile exhaust is an example. In addition, within the atmosphere, chemical reactions involving primary air pollutants, both gases and aerosols, produce *secondary air pollutants*. An example is *photochemical smog*, a mixture of aerosols, ozone, nitrogen oxides, and hydrocarbons, generated by the action of sunlight on motor vehicle exhaust, some industrial emissions, and volatile organic compounds emitted by certain vegetation (Figure 2.8).

Air pollutants are products of both natural events and human activities. Natural sources of air pollutants include forest fires, dispersal of pollen, wind erosion of soil, decay of dead plants and animals, and volcanic eruptions. The single most important human-related (anthropogenic) source of air pollutants is the internal combustion engine that propels most motor vehicles. Many industrial sources also contribute to air quality problems. Unless emissions are controlled, pulp and paper mills, zinc and lead smelters, oil refineries, and chemical plants can be major sources of air pollutants. Additional pollutants come from fuel combustion for space heating and generation of electricity (Figure 2.9),

FIGURE 2.8
Aerial photo showing a layer of smog over Ontario, CA, near Los Angeles. [Photo courtesy of Dan Lack/NOAA]

FIGURE 2.9
Conventional coal-fired electric power plant in Green Bay, WI. Note the pile of coal (to the left) and the high tension wires (to the right) that deliver electricity to northeastern Wisconsin. The plant's capacity is 410 Megawatts (MW). While the primary fuel is sub-bituminous coal, it can be substituted by natural gas. Carbon dioxide and water vapor are emitted from the tall stack but scrubbers remove sulfur oxide, and an electrostatic precipitator captures particulate emissions.

refuse burning, and various agricultural activities such as crop dusting.

The U.S. Environmental Protection Agency (EPA) has set ambient air quality standards for six criteria air pollutants: carbon monoxide, lead, nitrogen oxides, ozone, particulate matter, and sulfur dioxide. *Ambient air* is the outside air that we breathe. Standards for the six criteria air pollutants are based on scientific studies that demonstrate that adverse effects are likely only after concentrations exceed a specific threshold value (hence, the *criteria* designation). Standards are of two types, primary and secondary. *Primary air quality standards* are defined as the maximum exposure levels that can be tolerated by humans without ill effects. *Secondary air quality standards* are defined as the maximum levels that are allowable to minimize the impact on crops, visibility, personal comfort, and climate. Emission standards for stationary sources (e.g., coal-fired electric power plants) and mobile sources (e.g., motor vehicles) are set to ensure, theoretically at least, that once pollutants enter the atmosphere, ambient air quality standards are not exceeded.

A geographical region where the ambient air meets or does better than primary air quality standards is designated an *attainment area*. Compliance is established by direct sampling of air over specified time periods. A region where the ambient air does not meet the primary standard for one or more criteria air pollutants is designated a *nonattainment area* for those pollutants. In that case, remedial action (i.e., reduction of emissions) is required.

Investigating the Atmosphere

Our understanding of the atmosphere, weather, and climate is the culmination of centuries of painstaking inquiry by scientists representing many disciplines. Physicists, chemists, astronomers, and others have applied basic scientific principles in unlocking the mysteries of the atmosphere. The roots of modern meteorology, in fact, go back to around BCE 350-340 and Aristotle's *Meteorologica*, the first treatise on atmospheric science.[3] Although much progress has been made over the years, many questions remain regarding the workings of the atmosphere, weather, and climate. Today's atmospheric scientists (meteorologists and climatologists) continue the efforts of their predecessors, and although armed with more sophisticated tools such as satellites, radar, and supercomputers, they still rely on the scientific method of inquiry.

THE SCIENTIFIC METHOD

The **scientific method** is a systematic form of inquiry involving observation, interpretation, speculation, and reasoning. The Antarctic ozone hole illustrates how scientists applied the scientific method. In 1985, the British Antarctic Survey team reported a drastic decline in the amount of stratospheric ozone over Antarctica during the Southern Hemisphere spring (mainly September and October). The region where ozone was depleted encompassed an area almost as large as North America and was later dubbed the *Antarctic Ozone Hole*. In checking records from ground-based instruments, scientists found evidence of massive ozone depletion during each of the eight prior Antarctic springs.

At first, scientists dismissed the Antarctic ozone hole as a product of instrument error—not unexpected considering the brutal environmental conditions in Antarctica. Other scientists argued that the Antarctic ozone hole was real but the product of natural seasonal changes in the polar atmospheric circulation. A third hypothesis attributed the Antarctic ozone hole to pollution. In an effort to solve the mystery, an intensive field study was launched during the Antarctic spring of 1987 involving monitoring of the Antarctic stratosphere by satellites, specially outfitted aircraft, and instrumented balloon probes. That field investigation not only confirmed the existence of the Antarctic ozone hole, but also detected exceptionally high levels of chlorine monoxide (ClO), a gas known from laboratory studies to be a product of chemical reactions

that destroy ozone. For many scientists, discovery of ClO in the Antarctic stratosphere was the "smoking gun" linking ozone depletion to a group of chemicals known as chlorofluorocarbons (CFCs). We have more to say about CFCs and the threat to stratospheric ozone in Chapter 3.

From our Antarctic hole ozone example, it is evident that the scientific method involves a sequence of steps in which scientists (1) identify specific questions related to the problem at hand; (2) propose an answer to one of these questions in the form of an educated guess; (3) state the educated guess in such a way that it can be tested, that is, formulate a **hypothesis**; (4) predict the outcome of the test if the hypothesis were correct; (5) test the hypothesis by checking to see if the prediction is correct; and (6) reject or revise the hypothesis if the prediction is wrong. A hypothesis that has stood the test of time and is generally accepted by the scientific community is a **scientific theory**.

In our Antarctic ozone illustration we divided the scientific method into a sequence of steps but in actual practice, scientists do not always follow this scheme cookbook style, and discrete steps often combine into a seamless avenue of inquiry. Furthermore, the scientific method is not a formula for creativity because it does not provide the key idea, hunch, or educated guess that spontaneously springs to mind and forms the basis of a hypothesis. Rather, it is a technique for assessing the validity of a creative idea regardless of its origins.

A hypothesis is a tool that suggests new field studies, experiments or observations, or opens new avenues of scientific inquiry. Hence, even a rejected hypothesis may be fruitful. Above all, scientists must keep in mind that a hypothesis is merely a working assumption that may be accepted, modified, or rejected. They must be objective in evaluating a hypothesis and not allow personal biases or expectations to cloud that evaluation. In fact, scientists search for observations or information that could disprove their hypothesis. If they find such evidence, they revise their hypothesis to account for disparate observations or information. Inquiry, creative thinking, and imagination are stifled when a hypothesis is considered immutable.

A new hypothesis (or an old, resurrected one) is sometimes hotly debated within the scientific community. History attests to the natural human reluctance to accept new ideas that threaten to displace long-held notions. Also, disagreements among scientists on a particularly controversial issue sometimes receive much media attention, which may confuse the general public. Although considerable scientific consensus prevails within the atmospheric science community concerning the impact of burning of fossil fuels (e.g., coal, oil) on global climate, some issues of disagreement persist

---

[3] For a timeline of historical events in atmospheric science, see Appendix II.

such as the magnitude of future climate change. Throughout history, debate among scientists has been an essential step in the process of reaching understanding. Disagreements generate useful suggestions, stimulate new thinking, and uncover errors. In fact, debate and skepticism buffer the scientific community from too hastily accepting new ideas. If a hypothesis survives the scrutiny and skepticism of scientists, it is probably correct.

ATMOSPHERIC MODELS

In applying the scientific method, scientists usually find that models aid inquiry and this is certainly the case in the atmospheric sciences. Because we use models throughout this book, we consider at the outset the general objectives and limitations of scientific models, especially as they apply to meteorology and climatology.

A **scientific model** is an approximate representation or simulation of a real system. A **system** consists of a set of components that interact in an orderly way according to the laws of nature. The *Earth-atmosphere system*, for example, encompasses the Earth's highly variable surface (e.g., grasslands, forests, urban areas, lakes, oceans, glaciers et al.) plus the atmosphere. A model includes only what are considered to be the essential variables or characteristics of a system while omitting details assumed to be non-essential. For example, to learn how to improve the fuel efficiency of automobiles, engineers might examine a scale-model automobile in a wind tunnel. An automobile can be designed to minimize its air resistance, thereby increasing its fuel efficiency (miles per gallon). The shape of the model automobile is the critical variable and is the focus of study. Other variables, such as the weight of the model automobile, are important, but not relevant to the wind tunnel experiment and can be ignored. In constructing a model, trial and error often determine which variables are or are not essential.

Because models are not cluttered with extraneous and distracting details, they may provide important insights as to how things interact, or they may facilitate critical (or creative) thinking about complex phenomena. Models also are used to predict how a real system might respond to internal or external forcing (e.g., climate change). Depending on their particular function, scientific models are classified as conceptual, graphical, physical, or numerical.

A *conceptual model* is a statement of a fundamental law or relationship. The geostrophic wind (described in Chapter 8) is a conceptual model that relates the interaction of certain forces operating in the atmosphere to straight, horizontal winds blowing at altitudes of 1000 m (3300 ft) or higher. A *graphical*

model compiles and displays data in a format that readily conveys meaning. For example, a weather map integrates weather observations taken simultaneously at hundreds of locations into a coherent representation of the state of the atmosphere at a specific time (Chapter 1).

A *physical model* is a small-scale (miniaturized) version of a system. Almost half a century ago, scientists began studying the patterns exhibited by a fluid in a flat rotating pan in experiments designed to simulate the planetary scale atmospheric circulation. More recently, Purdue University scientists simulated a tornadic circulation in a specially designed Tornado Vortex Chamber, measuring 5 m wide, 5 m deep, and 8 m high (Figure 2.10). During an experiment, precision instruments monitored air motions within small tornado-like vortices (swirls) that developed. From this model, scientists improved their understanding of the internal characteristics of real tornadoes.

Today, atmospheric scientists depend on *numerical models* rather than physical models for investigating the atmosphere, weather, and climate. A numerical model consists of one or more mathematical equations that describe the relationship among variables of a system. Numerical models of the Earth-atmosphere system are usually programmed on a computer that can accommodate enormous quantities of data and perform numerous and often repetitive calculations with lightning speed. Variables in the numerical model (e.g., temperature, winds, cloud cover) are manipulated, individually or in groups, to assess their influence on the behavior of the system.

Numerical models have been used to forecast weather since the 1950s. More recently, other types of numerical models have been used to predict the potential impact on the global climate of rising levels of atmospheric carbon dioxide and other infrared-absorbing *(greenhouse)* gases. As discussed in more detail in Chapters 3 and 15, the atmospheric carbon dioxide concentration has been rising for more than a century, primarily as a byproduct of fossil fuel burning. Global warming is the likely consequence because, as noted earlier, carbon dioxide slows the escape of Earth's heat to space. Atmospheric scientists employ numerical global climate models in experiments to compute the magnitude of warming that might accompany a continued increase in atmospheric carbon dioxide concentration. Essentially they follow three steps: (1) Design a global-scale model that simulates Earth's climate system to accurately depict the long-term average pattern of air temperature worldwide. (2) Holding all other climate-control variables constant, elevate the carbon dioxide concentration (typically, it is doubled),

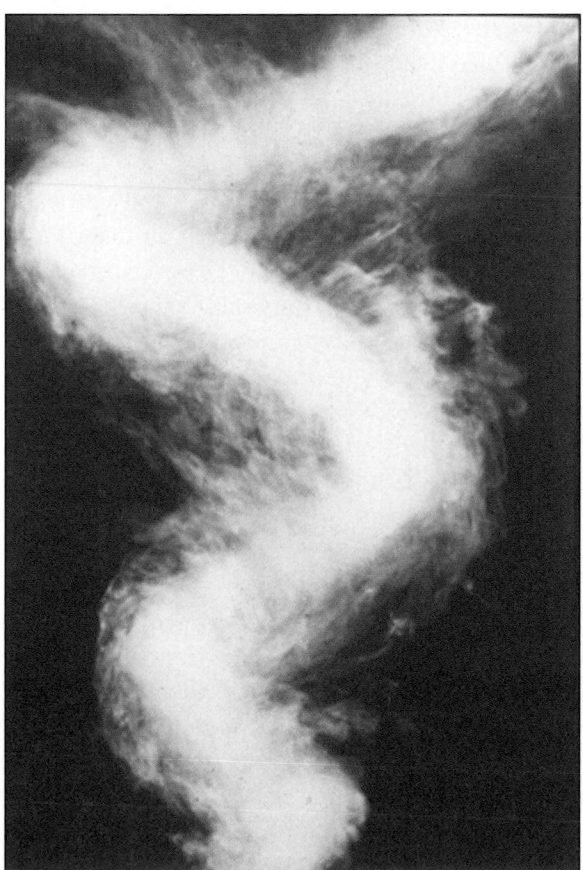

FIGURE 2.10
(A) External view of Purdue University's Tornado Vortex Chamber, used to simulate the circulation within tornadoes (B). [Courtesy of John T. Snow, University of Oklahoma, and C.R. Church, Miami University of Ohio]

and the global climate model computes a new worldwide average air temperature pattern. (3) Subtract the present temperature pattern from the temperature pattern predicted for the CO_2-enriched atmosphere to obtain the net temperature change.

All models *simulate* reality and as such are subject to error. A weather map portrays the state of the atmosphere at a particular time based on weather observations at discrete locations (weather stations), which may be hundreds of kilometers apart. (Weather satellite imagery provides a much more continuous field of view but the spatial resolution of satellite sensors is still limited.) This state of the atmosphere is used as a starting point for numerical models that are used to forecast the weather. Missing or erroneous observational data may mean that numerical models are not properly initialized, reducing the accuracy of a forecast. Another potential difficulty with numerical models concerns the accuracy of their component equations. Individually or together, equations only approximate the way a system really works in nature, and may not adequately account for all relevant variables. This is one reason why the skill of long-range weather forecasting, based on numerical models of the Earth-atmosphere system, degrades in accuracy as the forecast period lengthens (Chapter 13). Despite these potential vulnerabilities, models are becoming more accurate due to enhanced observational data from satellites, a more complete understanding of the workings of the atmosphere, and rapid increases in computing capacity. Hence, numerical model forecasts are generated faster and at higher resolution.

Monitoring the Atmosphere

Much of what we know about Earth's atmosphere, weather, and climate is derived from direct (*in situ*) and remote sensing of the Earth-atmosphere system. Since the invention of the first weather instruments in the 17th century, monitoring of the atmosphere has undergone considerable refinement. Denser monitoring networks, more sophisticated instruments and communications systems, and better trained observers have produced an increasingly detailed, reliable, and representative record of weather and climate. Monitoring has progressed on two fronts: surface and upper-air observations.

SURFACE OBSERVATIONS

On 1 July 1891, the nation's weather network was transferred from military to civilian hands in a new Weather Bureau within the U.S. Department of Agriculture, with a special mandate to provide weather and climate guidance

Historical Perspective

SURFACE OBSERVATIONS

Early observers kept records of weather conditions using primitive instruments or qualitative descriptions, jotting them down in journals or diaries. In North America, the first systematic weather observations were made in 1644-1645 at Old Swedes Fort (now Wilmington, DE) by Reverend John Campanius (1601-1683), chaplain of the Swedish military expedition. Campanius had no weather instruments, but he wrote in his diary qualitative descriptions of temperature, humidity, wind, and weather. Campanius returned to Sweden in 1648 but some fifty years would pass before his grandson published his weather observations.

Long-term instrument-based temperature records began in Philadelphia, PA, in 1731; Charleston, SC, in 1738; and Cambridge, MA, in 1753. The New Haven, CT, temperature record began in 1781 and continues uninterrupted today. Thomas Jefferson (third President of the United States) and the Reverend James Madison (president of the College of William and Mary) are credited with taking the nation's first simultaneous weather observations. Over a six-week period in 1778, they recorded temperature, air pressure, and wind at their respective homes in Monticello and Williamsburg, VA.

On 2 May 1814, James Tilton, M.D., U.S. Surgeon General, issued an order that directed the Army Medical Corps to begin a diary of weather conditions at army posts. His objective was to assess the relationship between weather and the health of the troops, for it was widely believed at the time that weather and its seasonal changes were important factors in the onset of disease. Even well into the 20th century, more troops lost their lives to disease than combat. Tilton also wanted to learn more about the climate of the then sparsely populated interior of the continent.

In 1818, Joseph Lovell, M.D., succeeded Tilton as Surgeon General and issued formal instructions for taking weather observations. In 1826, Lovell began summarizing and publishing the data and, for this reason, Lovell, rather than Tilton, is sometimes credited with founding the federal government's system of weather and climate observations. By 1838, 16 Army posts had recorded at least 10 complete (although not always successive) years of weather observations. By the close of the American Civil War, weather records had been tabulated for varying periods at 143 Army posts.

In the mid-1800s, Joseph Henry (1797-1878), first secretary of the Smithsonian Institution in Washington, DC, established a national network of observers who mailed monthly weather reports to the Smithsonian. The number of observers (mostly farmers, educators, or public servants) peaked at nearly 600 just prior to the American Civil War. Henry understood the value of rapid communication of weather data and quickly realized the potential of the recently invented electric telegraph in achieving this goal. In 1849, Henry persuaded the heads of several telegraph companies to direct their telegraphers in major cities to take weather observations at the opening of each business day and to transmit these data free of charge. With the availability of simultaneous weather observations, in 1850 Henry was able to prepare the first current national weather map. Beginning in 1856, he regularly displayed the daily weather map for public viewing in the Great Hall of the Smithsonian building. By 1860, 42 telegraph stations, most east of the Mississippi River, were participating in the Smithsonian network.

The success of Henry's Smithsonian network and another telegraphic-based network operated by Cleveland Abbe (1838-1916) in Cincinnati, OH, persuaded the U.S. Congress to establish a telegraph-based storm warning system for the Great Lakes. In the 1860s, surprise storms sweeping across the Great Lakes were responsible for a great loss of life and property from shipwrecks. President Ulysses S. Grant (1822-1885) signed the Congressional resolution into law on 9 February 1870 and the network, initially composed of 24 stations, began operating on 1 November 1870 under the auspices of the U.S. Army Signal Corps. Although originally authorized for the Great Lakes, in 1872, Congress appropriated funds expanding the storm-warning network to the entire nation. The new weather observation network soon encompassed stations previously operated by the Army Medical Department, Smithsonian Institution, U.S. Army Corps of Engineers, and Cleveland Abbe. With the expansion of telegraph service nationwide, the number of Signal Corps stations regularly reporting daily weather observations reached 110 by 1880.

for farmers. Later, aviation's growing need for weather information spurred the shift of the Weather Bureau to the Commerce Department on 1 July 1940. Many cities saw their Weather Bureau offices relocated from downtown to an airport, usually in a rural area well outside the city. In 1965, the Weather Bureau was reorganized as the National Weather Service (NWS) within the Environmental Science Services Administration (ESSA), which became the National Oceanic and Atmospheric Administration (NOAA) in 1971.

FIGURE 2.11
The National Weather Service's Automated Surface Observing System (ASOS) consists of electronic meteorological sensors, computers, and communication ports that record and transmit atmospheric conditions automatically 24 hours a day. The small image insert shows instrumentation that is atop the red and white pole and is not displayed on the larger image. [Photo by K.A. Nugnes]

During the early and mid-1990s, facilities of the National Weather Service underwent an extensive $4.5 billion modernization with the goal of upgrading the quality and reliability of weather observations and forecasts. Today, NWS Forecast Offices operate at 123 locations nationwide. The NWS and the Federal Aviation Administration (FAA) operate 884 automated weather stations, many at airports, which have replaced the old system of manual hourly observations. This **Automated Surface Observing System (ASOS)** consists of electronic sensors, computers, and fully automated communications ports (Figure 2.11). Twenty-four hours a day, ASOS feeds data to NWS Forecast Offices and airport control towers. Nearly 1100 additional automatic weather stations, which are funded by other federal and state agencies, supply hourly weather data from smaller airports.

In addition to the numerous weather stations that provide observational data mostly for weather forecasting and aviation, another 10,400 cooperative weather stations are scattered across the nation. These stations are cooperative in that volunteers supply their time and labor to monitor instruments and the National Weather Service provides instruments and data management. The principal mission of member stations of the **NWS Cooperative Observer Network** is to record data for climatic, hydrologic, and agricultural purposes. Observers report 24-hr precipitation totals and maximum/minimum temperatures based on observations made daily by 8 a.m. local time (Figure 2.12); some cooperative observers also report river levels. In the past, observers telephoned their reports to the local NWS

FIGURE 2.12
A National Weather Service cooperative observer station at Charleston, SC. Inside the wooden instrument shelter are sensors for measuring air temperature and humidity. In front of the shelter is a standard rain gauge.

Weather Forecast Office. Now a new program enables them to enter data into a computer that formats and transmits data to computer workstations in the NWS *Advanced Weather Interactive Processing System (AWIPS)*.

UPPER-AIR OBSERVATIONS

Today, radiosondes are launched simultaneously at 12-hr intervals (at 0000Z and 1200Z) from hundreds of ground stations around the world (Figure 2.13). The balloon bursts at an altitude of about 30,000 m (100,000 ft), and the instrument package descends to the surface under a parachute. In the United States, about 20% of radiosondes are recovered, refurbished, and reused. Each radiosonde contains a prepaid mailbag and instructions to its finder for returning the instrument to the National Weather Service.

As of this writing, the National Weather Service is implementing the *Radiosonde Replacement System (RRS)* to upgrade the current generation of radiosondes. The RRS consists of a global positioning system (GPS) tracking antenna, GPS radiosondes that operate at a radio frequency of 1680 MHz, plus a NT-based computer workstation. The new system provides more detailed and accurate upper-air data. Readings will be available at 1-second intervals, corresponding to altitude levels about 5 m (16 ft) apart.

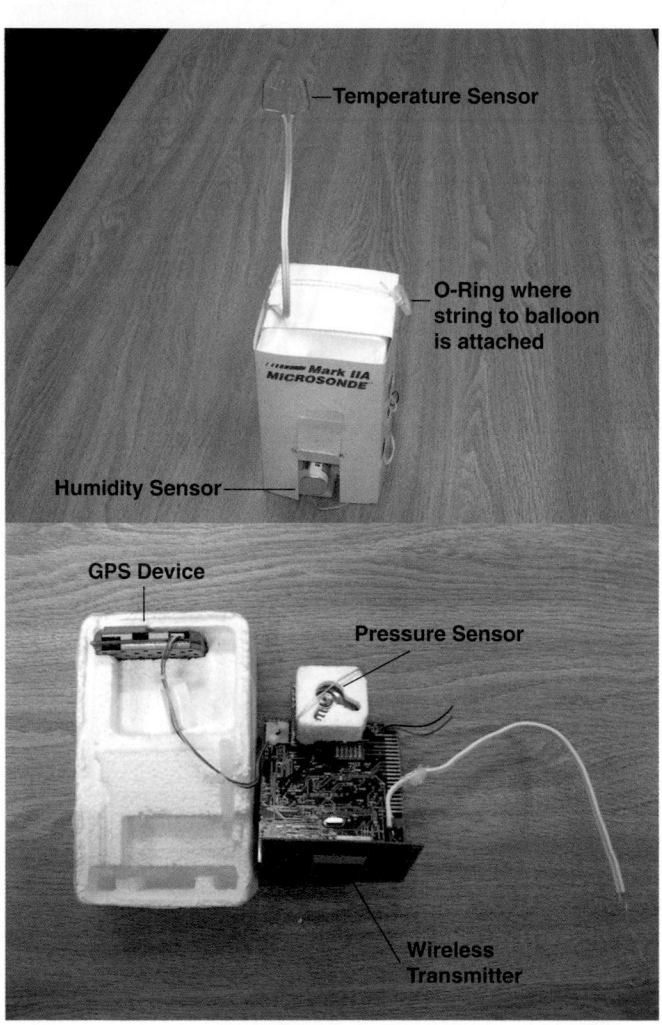

A

B

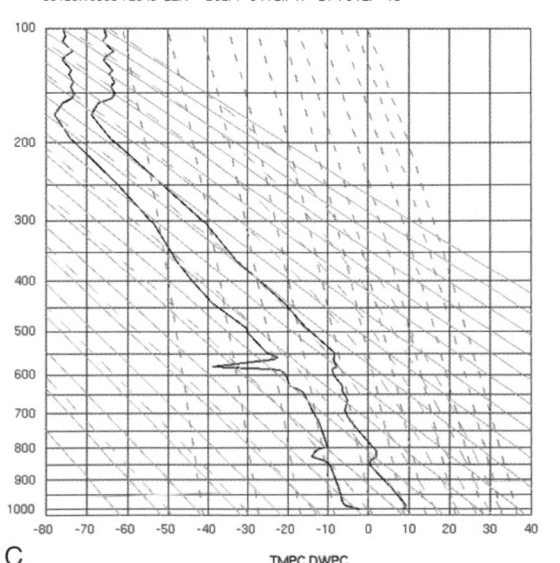

C

FIGURE 2.13
A radiosonde (A) [Courtesy Jerry Griffin/NOAA] is a small instrument package equipped with a radio transmitter. Borne by a hydrogen or helium-filled balloon (B), a radiosonde measures vertical profiles of air temperature, pressure, and dewpoint (C) up to an altitude of about 30,000 m (100,000 ft). Tracking horizontal movements of the radiosonde gives a profile of wind speed and direction (symbols on right).

UPPER-AIR OBSERVATIONS

Early meteorologists faced a challenge in taking weather measurements above Earth's surface. Some weather stations were established on mountaintops, but the influence of the mountain itself meant that observations were not representative of conditions in the free atmosphere. Kites were an early means of investigating the upper atmosphere. In July 1749 in Glasgow, Scotland, Alexander Wilson attached several thermometers to six paper kites and flew them in tandem. He designed the apparatus so that the thermometers would fall to the ground unbroken at predetermined intervals. In this way, Wilson was the first to obtain a free-air temperature profile of the lower atmosphere (up to an altitude of perhaps 60 m or 200 ft).

On 27 August 1804, the French scientists J. L. Gay-Lussac (1778-1850) and Jean Biot (1774-1862) ushered in the age of manned balloon exploration. They took air samples, measured temperature and humidity, and on one ascent reached an altitude of 7000 m (23,000 ft). On 5 September 1862, the British scientist James Glaisher (1822-1911) and his fellow aeronaut Henry Coxwell took weather instruments aloft in a balloon ascent over Wolverhampton, England. In less than an hour, they ascended to about 7600 m (25,000 ft.). They nearly perished from severe cold and oxygen deprivation in reaching an altitude of 9000 m (29,500 ft) setting a new manned balloon altitude record.

Through the early part of the 20th century, weather instruments borne by kites, aircraft, and balloons provided data chiefly on the lowest 5000 m (16,000 ft) of the atmosphere. On 4 August 1894 at Harvard's Blue Hill Observatory near Boston, MA, kites were used for the first time to carry aloft a self-recording thermometer (*thermograph*). This instrument provided a vertical profile of air temperature to an altitude of 427 m (1400 ft) above the ground. The next year at Mount Weather Observatory, VA, scientists used kites equipped with a recording instrument, called a *meteorograph*, to profile altitude variations in air pressure, temperature, humidity, and wind speed. Wind direction was determined from the ground using a theodolite, an instrument that measures azimuth and elevation angles. On 3 October 1907, a box kite equipped with a meteorograph set a new world record in climbing to an altitude of 7044 m (23,111 ft). On 5 May 1910, a train of 10 instrumented kites reached an altitude of 7265 m (23,835 ft).

The success of kite experiments convinced the U.S. Weather Bureau in 1907 to begin operating a network of weather kite stations at several locations, mostly in the central part of the nation (Figure 2.14).

FIGURE 2.14
In the early 20th century, scientists probed the atmosphere up to altitudes of about 3000 m (10,000 ft) using recording weather instruments (top) attached to box kites (bottom). [NOAA Photo Library]

Box kites equipped with meteorographs profiled the atmosphere up to a maximum altitude of about 3000 m (10,000 ft). The longest and last operating station, at Ellendale, ND, was closed in 1933. Among reasons cited for discontinuing the kite network was the relatively low maximum altitude reached by kites and the interruption in observations during episodes of light winds or calm air. For several years afterward, the U.S. Weather Bureau relied mostly on regular 5 a.m. (EST) aircraft observations originating at as many as 30 locations to probe the atmosphere to altitudes up to 4900 m (16,000 ft), but this practice proved too hazardous and costly.

A leap forward in monitoring higher altitudes came in the late 1920s with invention of the first radiosonde (Figure 2.15). A radiosonde is a small instrument package equipped with a radio transmitter that is carried aloft by a helium- (or hydrogen-) filled balloon. This device transmits altitude readings, called a sounding, of temperature, air pressure, and dewpoint (a measure of humidity) to a ground station. Radiosonde data are received immediately; no recovery of a recording instrument is required. The first official U.S. Weather Bureau radiosonde was launched at East Boston, MA, in 1937. By World War II, meteorologists were tracking radiosonde movements from ground stations using radio direction-finding antennas, thereby monitoring variations in wind direction and speed with altitude. A radiosonde used in this way is called a rawinsonde.

FIGURE 2.15
Launch of an early version of the radiosonde, developed by the U.S. Bureau of Standards, on 7 May 1936 at the Washington Airport blimp hangar. This slighly preceeded the first official U.S. Weather Bureau radiosonde launch in 1937. [NOAA's National Weather Service Collection]

The U.S. operates almost 90 radiosonde observation (RAOB) stations, a quarter of which use special high-altitude balloons that obtain data from altitudes above 30,000 m (100,000 ft). About 70 RAOB stations are located in the continental U.S., 14 in Alaska, 2 in Hawaii, and one each in Guam and Puerto Rico. Occasionally, meteorological rockets reach much higher altitudes (up to 100 km or 62 mi) but data from these probes are primarily for research purposes. In some urban areas, low-level soundings (up to 3000 m or 9800 ft) monitor atmospheric conditions to assess air pollution potential.

A dropwindsonde is similar to a rawinsonde except that instead of being launched by a balloon from a surface station, it is dropped from an aircraft. The instrument package descends on a parachute at about 18 km per hr (11 mph) and along the way radios data back to the aircraft every few seconds. The dropwindsonde was developed at the *National Center for Atmospheric Research (NCAR)* in Boulder, CO, to obtain soundings over the ocean where conventional rawinsonde stations are few and far between. Dropwindsondes provide vertical profiles of air temperature, pressure, dewpoint, and wind. They are often deployed by "hurricane hunter" aircraft that fly out to investigate offshore tropical storms and hurricanes that are approaching the Atlantic or Gulf Coasts.

REMOTE SENSING

Remote sensing refers to the measurement of environmental conditions by processing signals that are either emitted by an object or reflected back to a signal source. Remote sensing instruments are never in direct contact with whatever is being measured. In meteorology and climatology, radar and Earth-orbiting satellites are used extensively in remote sensing of the Earth-atmosphere system. **Radar** (acronym for *ra*dio *d*etection *a*nd *r*anging) is a valuable tool for determining the location, movement, and intensity of areas of precipitation. Weather radar emits microwave signals and receives reflected signals from targets as it continually scans the lower atmosphere. The first reported use of radar for meteorological purposes was on 20 February 1941, when a radar unit on the south coast of England tracked a thunderstorm a distance of 11 km (7 mi). More widespread application of radar for weather analysis began shortly after World War II.

Early weather radars were short-range surplus military units. Not until the mid-1950s, following major tornado and hurricane disasters, did the U.S. Congress authorize funds for purchase and installation of new long-range radar units designed specifically for meteorological purposes. In 1959, these units, called *WSR-57* radars, went into service for hurricane detection along the Atlantic and Gulf Coasts and for tornado and severe thunderstorm surveillance over the central United States. By 1964, *WSR-57* radars were operating at 32 sites and by the late 1960s were a routine feature of televised weathercasts in many parts of the nation. Weather radar was upgraded in 1974 with installation of *WSR-74* radars. During the early and mid-1990s, as a key component in the modernization of the National Weather Service, the obsolete *WSR-57* and *WSR-74* radars were replaced by the more technologically sophisticated *WSR-88D* radars (Figure 2.16). (WSR stands for Weather Surveillance Radar; the number refers to the year of development; and, D is for Doppler.) As of this writing, a network of 159 *WSR-88D* weather radar units monitors atmospheric conditions over much of the nation. *Phased array weather radar* is currently being tested as a

FIGURE 2.16
This radome for a *WSR-88D* radar unit houses a rotating dish antenna that sends out pulses of microwave energy and receives microwave energy reflected by precipitation particles.

possible replacement for the *WSR-88D* units. Weather radar is discussed in greater detail in Chapter 7.

Robert H. Goddard (1882-1945) is credited with conducting the first rocket probe of the atmosphere in 1929. The payload of Goddard's primitive rocket included a thermometer and a barometer. World War II spurred advances in rocketry so that by the late 1940s, instrumented rockets were investigating the composition of the middle and upper atmosphere. *Rocketsondes* have been used routinely during the past five decades to monitor winds, density, and ozone concentration above levels attained by radiosondes. A small rocket is launched carrying an instrument package similar to a radiosonde that parachutes back to Earth's surface.

In March 1947, a vertically fired V2 rocket, equipped with a camera, took the first successful photographs of Earth's cloud cover from altitudes of 110-165 km (70-100 mi). This and subsequent rocket probes of the upper atmosphere convinced scientists of the value of cloud pattern photography in monitoring weather systems and inspired the first serious proposals for orbiting a weather satellite. In the mid- to late 1950s, the United States' fledgling space program was directed at developing a launch vehicle (rocket) capable of putting a satellite in Earth orbit. With the former Soviet Union's successful orbiting of *Sputnik I* on 4 October 1957, the age of remote sensing by satellite had begun.

Surveillance of the Earth-atmosphere system by satellite began on 13 October 1959 with the U.S. launch of *Explorer VII*. Among other instruments onboard the Juno rocket was a flat-plate radiometer, developed by Verner E. Suomi (1915-1995) and Robert Parent of the University of Wisconsin-Madison. The radiometer provided the first measurements of Earth's radiation budget from space and established the important role of clouds in absorbing solar radiation. *Explorer VII* transmitted data continuously until February 1961. On 1 April 1960 the United States orbited the world's first weather satellite, *TIROS-1 (Television and Infrared Observation Satellite)*. Since then, numerous weather satellites equipped with increasingly sophisticated sensors have been launched into orbit.

Weather satellites are invaluable tools in weather observation and storm surveillance. These *eyes in the sky* offer distinct advantages over the network of surface weather stations by providing a broad and nearly continuous field of view. Surface weather stations are discrete and often widely spaced data sources, and weather observations are sparse or absent over vast areas of Earth's surface, especially the ocean. Five geostationary satellites provide nearly complete coverage of the globe poleward to about 60 degrees latitude. Onboard weather satellites are sensors that monitor cloud patterns, temperature, water vapor concentration, upper-air winds, rainfall, and the life cycles of severe storms. By the early 1990s, vertical profiling of the atmosphere's temperature and humidity by satellite became routine practice.

Today, application of various technologies provides detailed information on the state of the atmosphere. The remainder of this chapter considers two important properties of the atmosphere: its average vertical temperature profile and electromagnetic characteristics of the upper atmosphere.

Temperature Profile of the Atmosphere

Earlier in this chapter, we described the vertical structure of the atmosphere in terms of composition (*homosphere* versus *heterosphere*). However, for convenience of study, the atmosphere typically is subdivided into concentric layers based upon the vertical profile of the average air temperature (Figure 2.17). Almost all weather occurs

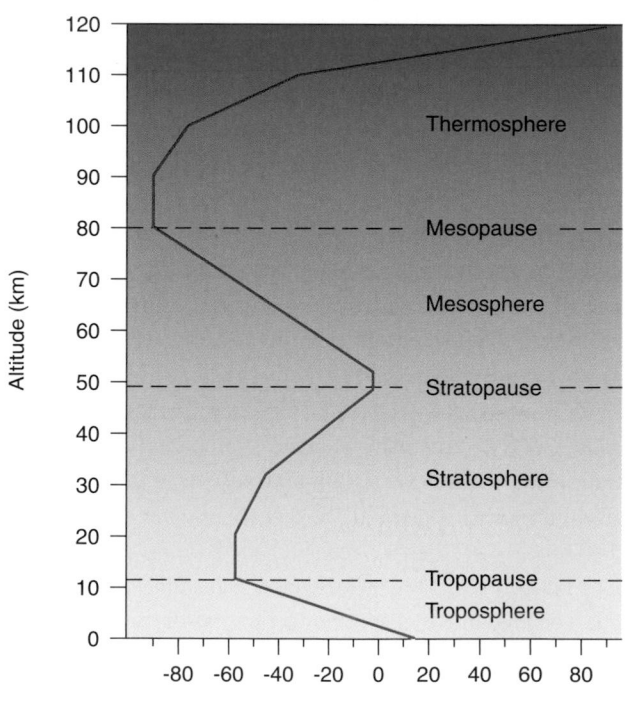

FIGURE 2.17

Variation in average air temperature with altitude within the atmosphere. Based on this vertical temperature profile, the atmosphere is subdivided vertically into the troposphere, stratosphere, mesosphere, and thermosphere.

FIGURE 2.18
Within the troposphere, average air temperature falls with increasing altitude so that it is generally colder on mountain peaks than in lowlands. [Photo by A.A. Nugnes]

within the lowest layer, the **troposphere**, which extends from Earth's surface to an average altitude ranging from about 6 km (3.7 mi) at the poles to about 20 km (12 mi) at the equator. Normally, but not always, the temperature within the troposphere falls with increasing altitude. Hence, the air temperature is usually lower on mountaintops than in surrounding lowlands (Figure 2.18). On average, within the troposphere, the air temperature drops 6.5 Celsius degrees for every 1000 m increase in altitude (3.5 Fahrenheit degrees per 1000 ft). The upper boundary of the troposphere, called the **tropopause**, is a transition zone between the troposphere and the next higher layer, the stratosphere.

The **stratosphere** extends from the tropopause up to about 50 km (31 mi). On average, the temperature does not change with increasing altitude through the lower portion of the stratosphere. A constant temperature condition is described as *isothermal*. Above about 20 km (12 mi), the temperature rises with increasing altitude up to the top of the stratosphere, the **stratopause**. At the stratopause, the temperature is not much lower than it is at sea level. The stratosphere is ideal for jet aircraft travel because it is above the weather, offering excellent visibility and generally smooth flying conditions. But because little air is exchanged between the troposphere and the stratosphere, pollutants that reach the lower stratosphere may persist there for lengthy periods. Gases and aerosols thrown into the stratosphere during violent volcanic eruptions, for example, can reside there for many months to years and perhaps trigger short-term variations in climate (Chapter 15). Other pollutants (e.g., CFCs)

threaten the protective ozone layer within the stratosphere (Chapter 3).

The stratopause is the transition zone between the stratosphere and the next higher layer, the **mesosphere** (or "middle" layer). Within this portion of the atmosphere, the temperature once again falls with increasing altitude. The mesosphere extends up to the **mesopause**, which is about 80 km (50 mi) above Earth's surface and features the lowest average temperature in the atmosphere (about −95 °C or −139 °F). Above this layer is the **thermosphere**, where temperatures at first are isothermal and then rise rapidly with increasing altitude. Within the thermosphere, air temperature is particularly sensitive to incoming solar radiation and is more variable with time than in any other region of the atmosphere.

The Ionosphere and the Aurora

A third way of describing the vertical structure of the atmosphere is in terms of concentration of ions, especially in the upper atmosphere. An **ion** is an atom or group of atoms that carries an electrical charge. The **ionosphere**, located primarily within the thermosphere, from a base of 70 to 80 km (43 to 50 mi) to an indefinite altitude, is a region of the atmosphere defined by a relatively high concentration of ions and electrons. High-energy solar radiation entering the upper atmosphere strips electrons from oxygen and nitrogen atoms and molecules, converting them to positively charged ions. The highest concentration of ions is in the lower portion of the thermosphere. Changes in the concentration of the principal constituents of the ionosphere with altitude are plotted in Figure 2.19.

Although conditions in the upper atmosphere apparently do not influence day-to-day weather in the troposphere, the ionosphere reflects radio waves and is important for conventional long-distance radio transmission. Radio signals travel in straight lines and bounce back and forth between Earth's surface and the ionosphere. By repeated reflections, a radio signal may travel completely around the globe. For more on this phenomenon, see this chapter's second *For Further Exploration*.

The ionosphere is also the site of the spectacular **aurora** (Latin for "dawn")—*aurora borealis* (northern lights) in the Northern Hemisphere and *aurora australis* (southern lights) in the Southern Hemisphere. Auroras often appear in the night sky as a variety of visual displays including overlapping

Principal constituents of the ionosphere (45° N, equinox)

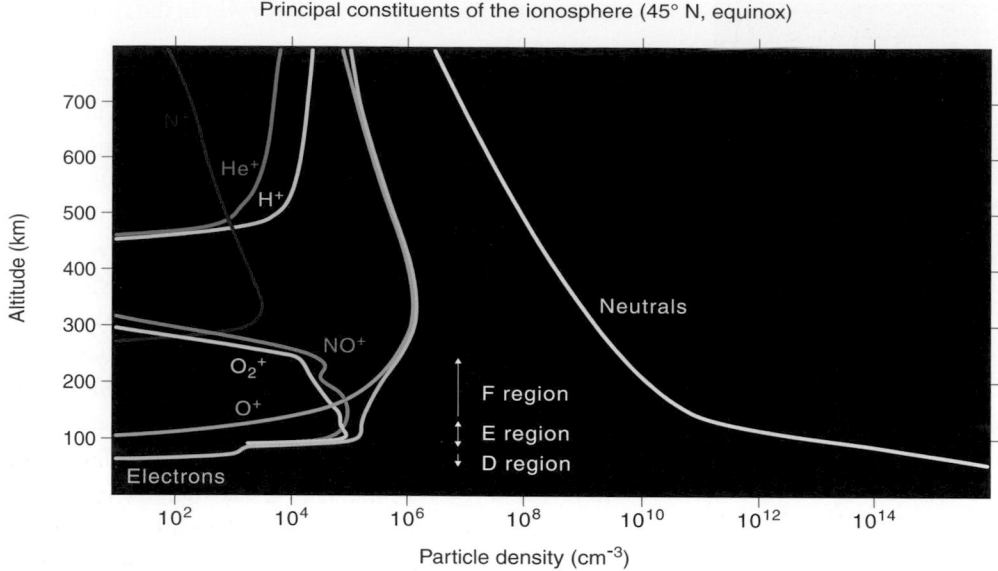

FIGURE 2.19
Average variation with altitude of the particle density of the principal constituents of the ionosphere at 45 degrees N on the equinox. [National Center for Atmospheric Research High Altitude Observatory and The COMET Program]

curtains of greenish-white light, occasionally fringed with pink (Figure 2.20). The bottom of the curtains is at an altitude of about 100 km (62 mi) and the top is at 400 km (250 mi) or higher.

An aurora is triggered by the **solar wind**, a stream of high-speed electrically charged subatomic particles (protons and electrons) that continually emanates from the Sun and travels through space at speeds of 400 to 500 km (250 to 300 mi) per second. Like a rock in a mountain stream, Earth's magnetic field deflects the solar wind and, in so doing, is deformed into a shape seen in cross-section

in Figure 2.21, producing Earth's **magnetosphere**. The solar wind compresses the front end of the magnetosphere and elongates the back end. A complex interaction between the solar wind and the magnetosphere generates beams of electrons that collide with atoms and molecules within the ionosphere, ripping apart molecules, exciting atoms, and increasing ion and election densities. As atoms shift down from their excited (energized) state and ions combine with free electrons, they emit radiation, part of which is visible as the aurora. Excited nitrogen molecules emit pinkish or magenta light, whereas excited oxygen atoms emit greenish light.

The aurora is usually visible only at higher latitudes. Earth's magnetic field channels some solar wind particles into two doughnut-shaped belts that are centered on the planet's north and south geomagnetic poles. These belts of more or less continuous auroral activity, known as *auroral ovals*, are located between 20 and 30 degrees of latitude from the geomagnetic poles and bulge equatorward on the dark (night) side of the planet.

FIGURE 2.20
Aurora borealis (northern lights) nearly directly overhead, photographed at Menasha, WI, at 9:30 p.m. CST on 5 November 2001. The concentric arcs of orange light are reflections from a light at a nearby parking lot. [Courtesy of John Beaver, University of Wisconsin, Fox Valley]

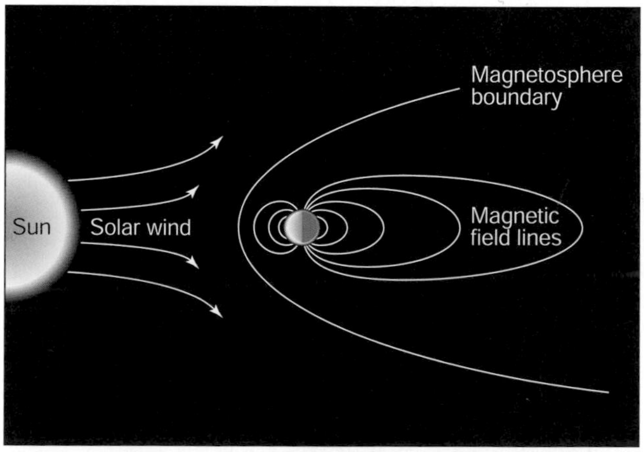

FIGURE 2.21
Earth's magnetic field deflects the solar wind and is thereby deformed into the teardrop-shaped magnetosphere that surrounds the planet.

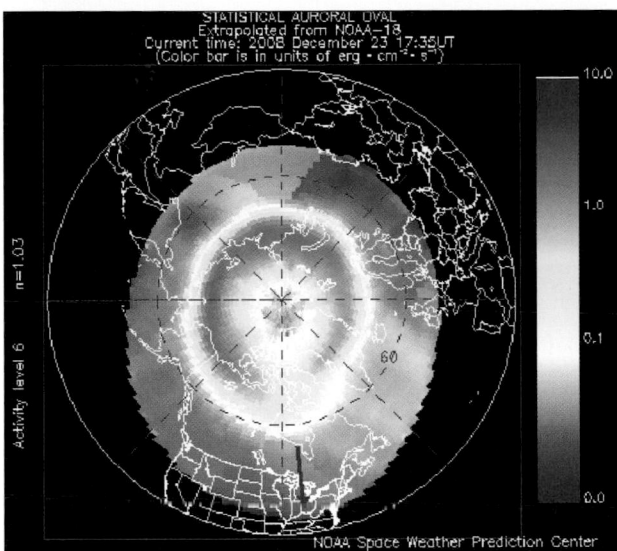

FIGURE 2.22
The Northern Hemisphere auroral oval, extrapolated from measurements taken by the NOAA POES satellite. [Courtesy of NOAA Space Weather Prediction Center]

The Northern Hemisphere auroral oval (Figure 2.22) is centered on the northwest tip of Greenland at latitude 78.5 degrees N and longitude 69 degrees W.

Auroral activity varies directly with solar activity. When the Sun is quiet, the auroral oval shrinks, but when the Sun is active, the auroral oval expands toward the equator, and the aurora may be visible across southern Canada and the northern United States or, rarely, further south. A rare, spectacular display occurred on 24 October 2011 and was viewed in more than half of all U.S. states, as far south as California, New Mexico, Arkansas, and Tennessee. Solar activity follows a roughly 11-year cycle, with the most recent solar maximum in 2000 (Chapter 15). As of this writing, solar cycle 24 is underway and is expected to peak in 2013.

In 1989, during a particularly active solar phase, the aurora was visible as far south as Mexico's Yucatán Peninsula. Gigantic explosions, called *solar flares*, characterize an active Sun. A solar flare is a brief event (lasting perhaps an hour) that produces a shock wave that propagates rapidly (500 to 1000 km per sec) through the solar wind. Collision of the shock wave with the magnetosphere causes the auroral oval to expand equatorward. During some active solar episodes, the phenomenon that produces auroras also disrupts the operation of electric power grids, telecommunications systems, and satellites. For more on space weather impacts and prediction, see this chapter's third *For Further Exploration*.

Conclusions

In this chapter we covered the origin, evolution, composition, and structure of Earth's atmosphere. We emphasized the importance of minor gases in the functioning of the atmosphere, and surveyed some of the various technologies used to monitor the Earth-atmosphere system. We also saw how the vertical profile of average air temperature is a convenient basis for subdividing the atmosphere into four layers. Because the primary focus of this course is weather and climate, we will be concerned primarily with atmospheric processes operating within the troposphere.

Our next major objective is to examine the driving force behind weather. To do so, we require an understanding of energy input and energy conversions within the Earth-atmosphere system. In the next chapter we learn how the Sun supplies the energy that drives the atmosphere's circulation. As we will see, circulation of the atmosphere ultimately is responsible for variations in weather from one place to another and with time.

Basic Understandings

- Earth's atmosphere covers the planet as a thin envelope of gases and suspended solid and liquid particles (aerosols).
- The modern atmosphere is the product of a lengthy evolutionary process that began about 4.5 billion years ago. Outgassing (the release of gases from rock through volcanic eruptions and meteorite impacts) played a key role in the evolution of the atmosphere by contributing water vapor, carbon dioxide, and nitrogen. Photosynthesis is the principal source of free oxygen in the atmosphere.
- Within the homosphere, the lowest 80 km (50 mi) of the atmosphere, the principal atmospheric gases, nitrogen (N_2) and oxygen (O_2), occur everywhere in the same relative proportion (about 4 to 1).
- The significance of an atmospheric gas or aerosol is not necessarily related to its relative concentration. Water vapor, carbon dioxide, and ozone are minor in concentration but extremely important in the life-sustaining roles they play on Earth.
- An air pollutant is a gas or aerosol occurring in concentrations that adversely affect the wellbeing of organisms (especially humans) or disrupt the orderly functioning of the environment. Human

activities and natural processes are sources of air pollutants.

- The scientific method is a systematic form of inquiry that requires the formulation and testing of hypotheses. A hypothesis is a working assumption that may be accepted, modified, or rejected. Debate and disagreement among scientists on controversial issues are usual and important elements of the scientific method.

- A scientific model is a simulation of a real system and may be conceptual, graphical, physical, or numerical. Examples of models commonly used in meteorology and climatology include weather maps, weather forecast models, and global climate models.

- The Army Medical Corps operated the nation's first weather observation network in the 1800s. Invention of the telegraph made possible rapid communication of simultaneous weather observations and the first near real-time weather maps.

- In the 1860s, growing public concern over the great loss of life and property from surprise storms sweeping the Great Lakes spurred the federal government to create a telegraph-linked national weather observation network in 1870 under the auspices of the U.S. Army Signal Corps, forerunner of today's National Weather Service.

- Through the years, tools for investigating the upper atmosphere advanced from instrumented kites and manned balloons to radiosondes, radar, rockets, and satellites.

- A radiosonde consists of an instrument package and radio transmitter carried aloft by balloon to altitudes up to 30,000 m (100,000 ft); it provides vertical profiles (called soundings) of air temperature, pressure, and dewpoint (a measure of humidity).

- Remote sensing refers to the measurement of environmental conditions by processing signals that are either emitted by an object or reflected back to a signal source. In meteorology and climatology, radar and Earth-orbiting satellites are used in remote sensing.

- The atmosphere is subdivided into four concentric layers (troposphere, stratosphere, mesosphere, and thermosphere) based on the average vertical temperature profile. Almost all weather takes place in the troposphere, the lowest subdivision of the atmosphere.

- The ionosphere is a region of electrically charged particles situated primarily within the thermosphere and consists of a relatively high concentration of charged particles (ions and electrons).

- An aurora (northern or southern lights) is a spectacular visual display in the night sky at higher latitudes, sometimes appearing as curtains of color. It develops in the ionosphere when the solar wind interacts with Earth's magnetic field and varies with solar activity.

Enduring Ideas

- The Earth has undergone a series of significant changes since coming into existence about 4.5 billion years ago, ultimately resulting in a surface covered by 71% ocean water and an atmosphere composed predominantly of nitrogen and oxygen.

- Most weather occurs in the troposphere, due to its favorable temperature structure and the presence of most atmospheric water vapor. Water vapor, carbon dioxide, and ozone occur in minute concentrations in the atmosphere, but they are necessary for life.

- Atmospheric models are essential to the understanding and prediction of weather. Forecasters rely on complex numerical models and a network of surface and upper air observations to generate predictions on various time scales.

- The ionosphere, located in the upper atmosphere, is defined by a high concentration of ions and electrons. Changes in the Sun's surface and atmosphere can greatly influence activity in the ionosphere.

Key Terms

atmosphere	scientific model	troposphere
nebula	system	tropopause
comet	Automated Surface Observing	stratosphere
asteroid	System (ASOS)	stratopause
outgassing	NWS Cooperative Observer	mesosphere
photosynthesis	Network	mesopause
homosphere	radiosonde	thermosphere
heterosphere	sounding	ionosphere
air pollutant	rawinsonde	ion
scientific method	dropwindsonde	aurora
hypothesis	remote sensing	solar wind
scientific theory	radar	magnetosphere

Review

1. What is thought to be the principal source of water on Earth?
2. The chief source of Earth's early atmosphere was outgassing. Explain what is meant by outgassing. Does outgassing still operate today?
3. Explain how living organisms played an important role in the evolution of Earth's atmosphere.
4. Distinguish between the homosphere and the heterosphere.
5. The significance of an atmospheric gas or aerosol is not necessarily related to its concentration. Explain this statement and provide a few examples.
6. Distinguish among the various types of scientific models used in meteorology and climatology.
7. What was the significance of the invention of the electric telegraph in weather observation and forecasting?
8. What is the principal mission of the National Weather Service Cooperative Observer Network?
9. What is a radiosonde? Describe the function served by a radiosonde launch.
10. Compare the properties of the troposphere with those of the stratosphere.

Critical Thinking

1. Describe some ways whereby the atmosphere interacts with Earth's hydrosphere, geosphere, and biosphere.
2. Identify some natural processes and human activities that can alter the concentration of carbon dioxide (CO_2) in the atmosphere. How might variations in the atmospheric concentration of CO_2 affect global climate?
3. Explain how auroral activity varies with changes in solar activity.
4. A mountaintop is closer to the Sun than the surrounding lowlands and yet mountaintops are colder than lowlands. Explain why.
5. During the evolution of Earth's atmosphere, how did formation of the global water cycle affect the concentration of carbon dioxide in the atmosphere?
6. What is the significance of stratospheric ozone for life on Earth?
7. Under what circumstances is a natural component of the atmosphere considered an air pollutant?
8. A radiosonde balloon typically bursts at altitudes above about 30,000 m (100,000 ft). Explain why.
9. The altitude of the tropopause generally decreases with increasing latitude. Explain why.
10. What circumstances led to the establishment of a national weather service in 1870?

For Further Exploration

The Atmosphere of Mars

In 1976, sensors onboard the National Aeronautics and Space Administration's (NASA's) *Viking* spacecraft confirmed speculation that the Martian atmosphere differs considerably from Earth's. The Martian atmosphere is 95% carbon dioxide, 2% to 3% nitrogen, 1% to 2% argon, and 0.1% to 0.4% oxygen. Also, the Martian atmosphere is much thinner, with an average surface air pressure only 0.6% of Earth's. More than 4 billion years ago, however, in the early stages of the planets' formation, their atmospheres were probably quite similar in composition.

Outgassing produced the primeval atmospheres of both Earth and Mars. Gases were released from ancient planetary rock through volcanic eruptions and the impact of meteorites on the rocky surfaces of both planets. Because the source rocks on both planets were chemically equivalent, the gases were likely the same. The primeval atmosphere of the two planets would have been mostly carbon dioxide, some nitrogen and water vapor, and trace amounts of other gases. As noted earlier in this chapter, geochemical processes and photosynthesis altered Earth's primeval atmosphere until nitrogen and oxygen were the principal gases and carbon dioxide only a minor one. In the Martian atmosphere, carbon dioxide has always been the chief gaseous component despite a significant decline in its density.

Contrasts in the volcanic histories of Earth and Mars may help to explain why the atmospheres of the two planets evolved differently. On Mars, the bulk of volcanic activity apparently took place during the planet's first 2 billion years, whereas on Earth volcanism has been nearly continuous. The decline in volcanism on Mars cut the supply of nitrogen even as the original nitrogen escaped Mars' relatively weak gravitational field. Gravity, the force that holds an atmosphere to a planet, is about 38% weaker on Mars than on Earth because Mars is only 11% of Earth's mass.

Initially, the CO_2-rich Martian atmosphere produced a surface pressure perhaps three times that of Earth's modern atmosphere. However, as volcanic activity diminished, less CO_2 was released to the atmosphere even as carbon dioxide left the Martian atmosphere. Enormous amounts escaped the weak gravity into space, some adhered to the dust that blankets the planet's surface, and some may have been locked in carbonate rocks. Without carbon dioxide to slow the loss of the planet's heat to space, surface temperatures on Mars plunged.

FIGURE 1
The surface of Mars viewed from space is marked by numerous craters produced by meteorite impacts. Running water apparently cut canyons up to 8 km (5 mi) deep and flowed into a large basin in the upper right of the image. Volcanoes (dark spots) are visible on the left side of the image. [Courtesy of NASA and the U.S. Geological Survey]

Today, the mean temperature on the Martian surface (Figure 1) ranges from −60 °C (−76 °F) at the equator to −123 °C (−189 °F) at the poles, much too low for liquid water. Sensors on Mars-orbiting spacecraft have detected frozen water buried under layers of surface dust and rocky debris. The amount of water ice detected within the upper few kilometers of the planet, if spread as liquid over the entire surface of Mars, would cover the planet to a depth of about 11 m (36 ft). In just the upper few meters, there is about twice the volume of Lake Superior, the largest of the North American Great Lakes. Mars' perennial polar ice caps are mostly water ice with a covering of dry ice (frozen carbon dioxide). Severe cold on Mars means no common life forms, no photosynthesis, and hence, only trace amounts of free oxygen and no ozone.

Planetary scientists believe that during at least the first billion years of its existence, Mars was sufficiently warm enough for water to flow on the planet's surface. Sensors onboard Mars-orbiting satellites and rovers (robots) placed on the planet analyzed its surface for signs of past liquid water. The *Mariner* satellite missions of the 1960s and *Viking* orbiter missions of the 1970s photographed what were described as vast flood channels and networks of valleys on the Martian surface. Scientists proposed that these features were eroded by running water.

About two decades later, NASA's Pathfinder mission to Mars confirmed this hypothesis. On 4 July 1997, the Pathfinder spacecraft landed on an ancient floodplain that had been scoured by catastrophic flooding billions of years ago. Also, Pathfinder's tiny rover, *Sojourner*, photographed pebbles and rocks on the Martian surface having physical properties indicative of transport by running water. In 2005, images taken by sensors onboard the European Space Agency's *Mars Express* spacecraft revealed a frozen body of water in a flat area near the equator that at one time was the size of the North Sea. A layer of ash or dust about 1 to 20 m (3 to 66 ft) thick mantles the surface thereby preventing the ice from vaporizing (sublimating).

In January 2004, twin Mars Exploration Rovers (MERs), *Opportunity* and *Spirit*, landed at sites on opposite sides of Mars. Both uncovered conclusive evidence that liquid water once flowed on Mars. *Spirit* located a rock showing signs of chemical alteration by water and *Opportunity* detected minerals (i.e., salts, evaporites) and rock features (e.g., mud cracks, ripple marks) indicating that water not only altered rock but also rock formed from sediment that settled out of a shallow saline sea. In 2011, *Opportunity* completed a 3-year, 21.5 km (13.3 mi) journey to a very large crater known as Endeavour that contains clay minerals indicative of wet conditions.

On 25 May 2008, NASA's *Phoenix Mars Lander* set down on Mars' arctic region. Using a robotic arm, the *Phoenix* uncovered lumps of ice in a shallow trench (Figure 2). Water ice was detected in a soil sample only 5 cm (2 in.) beneath the surface. Chemical analysis identified several water-soluble elements and inorganic salt compounds (e.g., sodium and potassium chloride) in the soil samples, which points to past interaction with liquid water.

Scientists have also gained a wealth of information regarding water on Mars via the *Mars Reconnaissance Orbiter (MRO)* (Figure 3). Launched in 2005 to investigate the hydrologic history of Mars, the MRO is equipped with a radar sounder, known as *Shallow Subsurface Radar (SHARAD)*, capable of detecting subsurface ice deposits. The MRO's *High Resolution Imaging Science Experiment (HiRISE)* can detect boulders as small as 0.5 m (20 in.). The Orbiter detected extensive water glaciers buried under about 10 m (33 ft) of dust and rocky debris in portions of Mars' southern and northern mid-latitudes. The volume of buried glaciers is estimated at 10% of the volume of frozen water

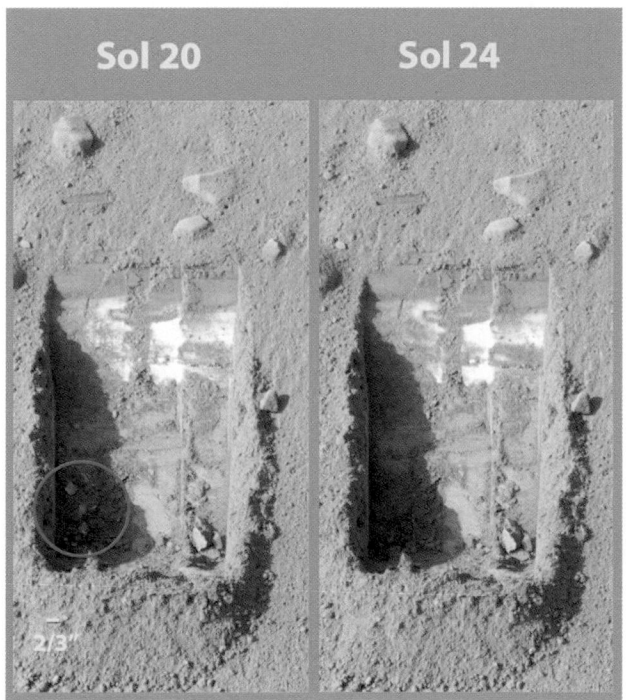

FIGURE 2
These images were acquired by the NASA Phoenix Mars Lander on the 21st and 25th days of the mission, or Sols 20 and 24 (15 June 2008 and 19 June 2008). The left image shows lumps of ice (encircled) uncovered in a shallow trench. In the right image, four days later, the ice sublimated and no longer appears. [Courtesy of NASA/JPL-Caltech/University of Arizona/Texas A&M University]

in Mars polar ice caps. In December 2008, the Mars Reconnaissance Orbiter detected carbonate minerals, indicating that Mars had neutral to alkaline water more than 3.6 billion years ago and that different types of watery environments were later present. A greater variety of watery environments increases the chances that one or more of them may have supported life.

Mounting data from rovers and orbiters (especially the high resolution information provided by MRO) is prompting scientists to question earlier geological interpretations that pointed to free flowing water on the Martian surface. According to HiRISE, great conical fans on the flanks of large craters may have been generated by impactors striking a water-rich crust and triggering muddy flows downslope. The journeys of the *Opportunity* and *Spirit* rovers reveal that the rates of erosion on the Martian surface were extremely slow so that running water could not have been a major player 3 billion years ago.

Today Mars is a frigid and foreboding place swept by strong winds and massive dust storms that sometimes encircle the entire planet. Although the composition of Mars' primeval atmosphere was probably similar to that of Earth, a reduction in volcanic outgassing and a weaker gravitational field caused the Martian atmosphere to follow a distinctly different path. On Mars, carbon dioxide has remained the principal atmospheric gas despite its significant decline in concentration, which caused surface temperatures to plunge. Indications that water now frozen under a dusty mantle once flowed as liquid on the Martian surface are consistent with this planetary-scale climate change. But as the evidence mounts, the early phases of Mars look to be drier than previously assumed.

FIGURE 3
The Mars Reconnaissance Orbiter using its SHARAD radar (top view). This image is an artist's concept of a view looking down on the MRO. The spacecraft is pictured using its SHARAD radar to "look" under the surface of Mars. The SHARAD instrument will seek liquid or frozen water within the first few hundred feet (up to a kilometer) under the martian surface. [NASA]

Radio Transmission and the Ionosphere

Reception of radio signals (waves) from distant locations at night is not at all unusual. Late-night radio listeners in Illinois can, for example, routinely pick up WBZ, a Boston radio station (1030 on the AM dial) even though the station's transmitter is more than 1500 km (930 mi) away and its signals receive no help from orbiting satellites.

Arrival of distant radio waves at night and their disappearance during sunlit hours is due to interactions of those waves with the ionosphere. A radio wave is a form of *electromagnetic radiation* that travels in essentially straight paths in all directions away from its source transmitter (Chapter 3). Earth is essentially a sphere so its surface gradually curves under and away from direct radio waves. A *quiet zone*, where direct radio waves are not received, begins about 160 km (100 mi) from a transmitter tower of average height. At night, beyond the distant edge of the quiet zone, reception resumes because radio waves are reflected back to Earth's surface from the upper ionosphere.

Recall from elsewhere in this chapter that the ionosphere is a region of the upper atmosphere containing a relatively high concentration of ions (electrically charged particles) and free electrons. Highly energetic solar radiation splits molecular nitrogen (N_2) and molecular oxygen (O_2) into atoms, positively charged ions, and free electrons. The production rate of ions and electrons depends on two factors, both of which vary with altitude: (1) the *number density* of atoms and molecules available for ionization, which decreases rapidly with altitude, and (2) the intensity of solar radiation, which increases with altitude. Combined, these two factors maximize the concentration of ions and free electrons in the ionosphere.

By convention, the ionosphere is subdivided vertically into several layers. From lowest to highest, layers are designated D (upper mesosphere below 100 km), E (100 to 120 km), F_1 (150 to 190 km), and F_2 (above 200 km). The original basis for this subdivision was the belief that each layer is a distinct zone of maximum electron density. Measurements by rockets and satellites, however, show that the ionosphere is not made up of discrete layers; rather, electron density increases nearly continuously with altitude to a maximum at an average altitude close to 300 km (186 mi). The D, E, F_1, and F_2 labels in this discussion refer to regions within the ionosphere.

Radio waves that enter the ionosphere interact with free electrons in the D, E, and F regions and are either absorbed or reflected back toward Earth's surface. At night, in the absence of ionizing radiation, the D region virtually disappears as ions and electrons recombine into neutral particles. The recombination rate depends on air density; that is, the higher the density, the greater the likelihood of collision of particles and capture of electrons by positive ions. The F_1 and F_2 regions slowly recombine at night and separate again during the day. In the E and F regions, air is so rarefied that collisions are infrequent, and although the E region weakens, these regions persist through the night. Radio waves that reach the F region are reflected back toward Earth's surface. (Exceptions are radio waves that enter the F region nearly vertically; these waves pass on into space.) At night, because of F-region reflection, radio waves propagate many hundreds of kilometers from their point of origin (Figure 1).

With the return of the Sun's ionizing radiation during the day, the D region redevelops. Most radio waves that reach the

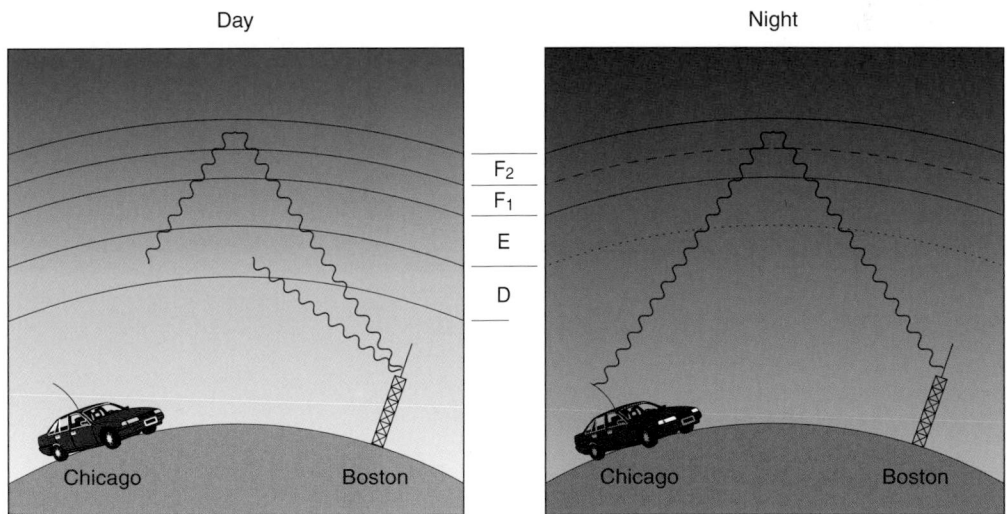

Day	Night

F_2
F_1
E
D

Chicago   Boston   Chicago   Boston

FIGURE 1
During the day (left image), radio waves are either directly absorbed by the D region or reflected back to the D region where they are absorbed. At night (right image), the D region virtually disappears and radio waves can propagate hundreds of kilometers from their origin.

D region are absorbed instead of propagating into the E and F regions. The F region reflects waves that penetrate the D region back to the D region where they are absorbed. Consequently, during sunlit hours, radio-wave propagation is not aided by F-region reflection. In summary, radio waves travel greater distances at night because the upper ionosphere reflects them back to Earth's surface.

Ionizing radiation from the Sun generates the ionosphere so that any solar activity that disturbs the flow of this radiation may affect ion density and consequently, radio communication on Earth. *Sudden ionospheric disturbances (SIDs)*, typically lasting 15 to 30 minutes, are caused by bursts of ultraviolet radiation emitted from the Sun. Ionization temporarily increases, D-region absorption strengthens, and radio transmissions fade. The same solar activity responsible for the aurora also increases ionization and causes radio fadeout.

For Further Exploration

Space Weather Prediction

We are all familiar with the effects of weather on daily life, but did you know that changes in the Sun's atmosphere can also dramatically affect many of the technologies we rely upon? Space weather can significantly impact the availability of electricity, cell phone performance, the GPS system in our car, our safety during a trans-polar airline flight, satellite images we view on TV and the Internet, and even our ability to make a credit card purchase. In fact, economists estimate that the adverse impacts of space weather can cost $200 to $400 million per year, with the potential for much greater losses. As society becomes increasingly dependent on technologies vulnerable to space weather, the need for accurate monitoring and prediction grows. For this reason, NOAA's Space Weather Prediction Center (SWPC) in Boulder, CO, continually monitors solar changes and forecasts Earth's space environment.

As described in the American Meteorological Society's 2008 Policy Statement on Space Weather, *space weather refers to the variable conditions on the Sun and in the space environment that can influence the performance and reliability of spaceborne and groundbased technological systems, as well as endanger life or health. Most space weather occurs because emissions from the Sun influence the space environment around Earth, as well as other planets.*

Just as the Earth-atmosphere system is in a constant state of flux, so too is the Sun. Several solar features that can dramatically affect the near-Earth environment include sunspots, solar flares, coronal mass ejections, and coronal holes. *Sunspots* are relatively cool, dark areas that dot the surface of the portion of the Sun's atmosphere called the photosphere, and are generated where an intense magnetic field suppresses the flow of gases transporting heat from the Sun's interior (Figure 1). In cross-section, a medium-sized sunspot is about as large as Earth and has an average lifetime of days to weeks. The average number of sunspots varies systematically over an approximately 11-year cycle and during solar maximum (when sunspots are most numerous), solar activity, and effects on Earth's space environment are relatively high.

A *solar flare* is a major disturbance of the Sun's chromosphere that emits high-energy radiation and high-velocity streams of electrically charged subatomic particles (Figure 2). A solar flare likely results from the tearing and reconnection of strong magnetic fields. Flares have a lifetime of minutes to hours, and are the Sun's largest explosive events.

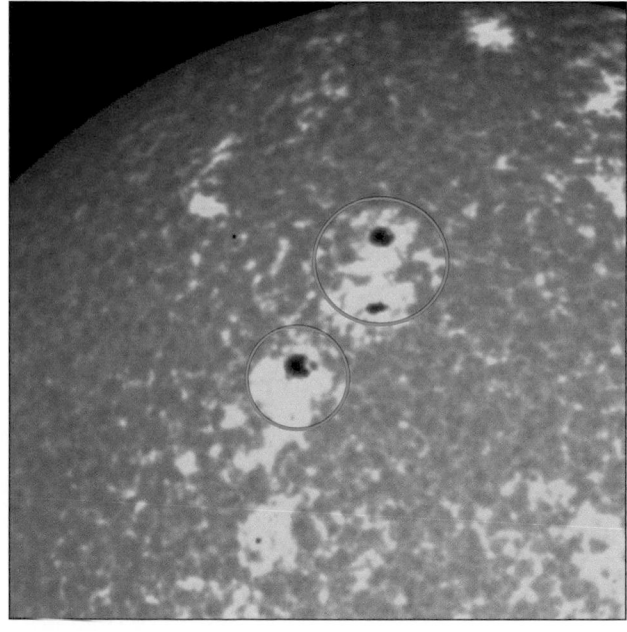

FIGURE 1
UV image of a portion of the Sun with sunspot groups encircled. In this image, the dark sunspots are accompanied by bright white blotches called faculae. [NASA/Goddard Space Flight Center]

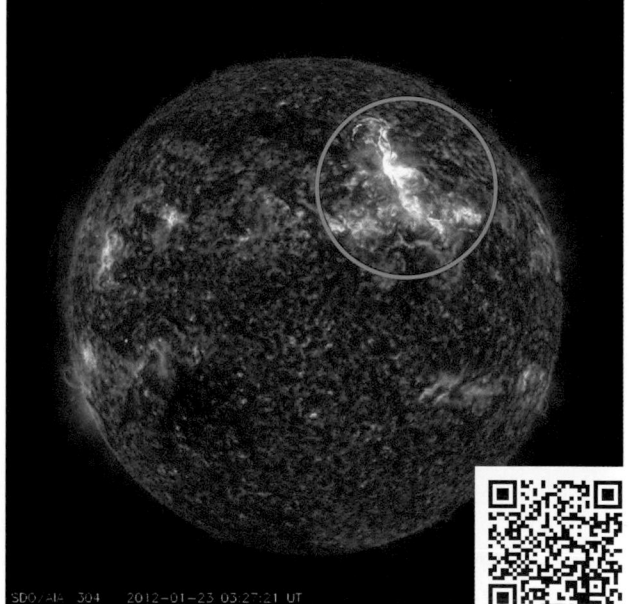

FIGURE 2
A massive solar flare erupts on the surface of the Sun on 23 January 2012. Flares have a lifetime from minutes to hours, and are the Sun's largest explosive events. [Courtesy of NASA/SDO]

A *Coronal mass ejection* is a sudden and violent release of bubbles or tongues of gas and magnetic fields from the Sun's outer atmosphere known as the corona (Figure 3). *Coronal holes* are large areas of the Sun's corona that are cooler and darker than surrounding areas (Figure 4). Their open magnetic field structure allows for a strong, continuous outflow of the *solar wind*, which can disturb Earth's geomagnetic field.

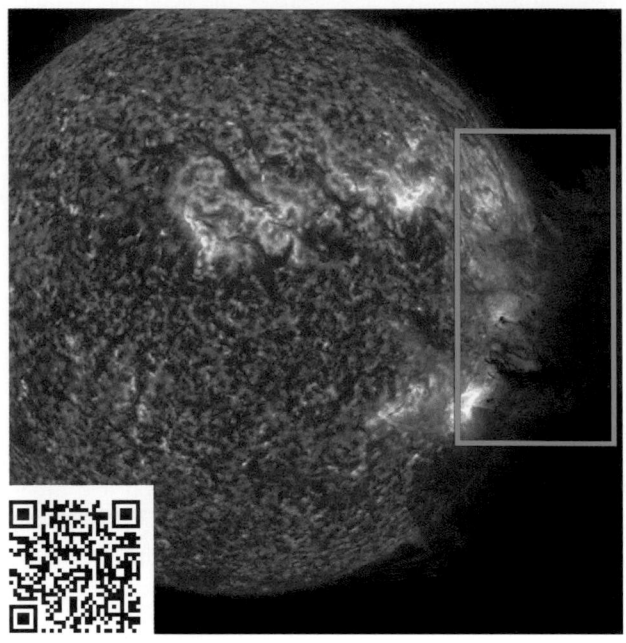

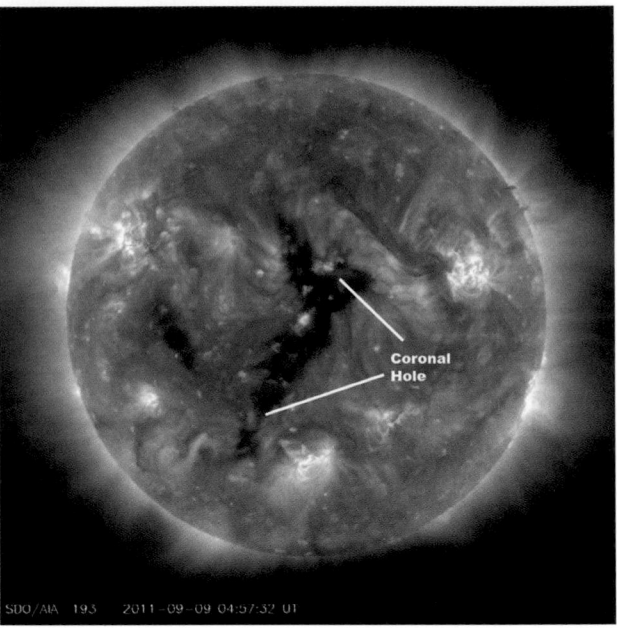

FIGURE 3
A coronal mass ejection as viewed by the Solar Dynamics Observatory (SDO) on 7 June 2011. [Courtesy of NASA/SDO]

FIGURE 4
The Sun as viewed by the SDO on 9 September 2011. The elongated black area near the center is a coronal hole. [NASA/SDO]

Solar flares are closely related to sunspots as the magnetic field in sunspots stores energy that is explosively released in solar flares. Solar flares and coronal mass ejections are sometimes related, but usually occur independently. The direction in which flares, ejections, and gusts in the solar wind are projected from the Sun into space determines the impact on the near-Earth environment. If these violent events are steered in Earth's direction, effects can begin to show up within an hour, making monitoring and prediction crucial.

The NOAA Space Weather Prediction Center (SWPC) is charged with the monitoring, prediction, and communication of space weather phenomena to the public. The SWPC recently developed scales, akin to the Richter scale for earthquakes, to communicate critical information to the rapidly growing number of people who rely upon technology vulnerable to space weather. For example, government and privately-owned satellites are particularly susceptible, due to their location above Earth's atmosphere and their sensitive electronics. Satellites receive and transmit a wide variety of data, such as weather information, military surveillance, communications signals, GPS data, radio broadcasts (e.g., SiriusXM), cell phone conversations, and credit card information.

SWPC scales relate space weather events to their probable effects on various technological systems; each scale has five intensity levels ranging from minor (1) to extreme (5). Scales cover three categories: geomagnetic storms, solar radiation storms, and radio blackouts. Geomagnetic storms are disturbances in the geomagnetic field caused by gusts in the solar wind. Solar radiation storms refer to elevated radiation levels due to an increase in the number of energetic particles from the Sun. Radio blackouts are ionospheric disturbances caused by solar X-ray emissions. Figure 5 describes a severe solar radiation storm (S4).

Following a minimum in solar activity over the past few years, the Sun has entered a new period of activity called Solar Cycle 24, which is expected to peak in 2013. During that period, powerful solar flares and coronal mass ejections are much more likely to occur.

Category		Effect	Physical measure	Average frequency (1 cycle = 11 years)
Scale	Descriptor	Duration of event will influence severity of effects		
Solar radiation storms			Flux level of > = 10 MeV particles (ions)	Number of events when flux level was met (number of storm days)
S 4	Severe	Biological: unavoidable radiation hazard to astronauts on EVA; passengers and crew in high-flying aircraft at high latitudes may be exposed to radiation risk. Satellite operations: may experience memory device problems and noise on imaging systems; star-tracker problems may cause orientation problems, and solar panel efficiency can be degraded. Other systems: blackout of HF radio communications through the polar regions and increased navigation errors over several days are likely.	10^4	3 per cycle

FIGURE 5
Description of a severe solar radiation storm (S4). [Image adapted from the NOAA Space Weather Prediction Center table at http://www.swpc.noaa.gov/NOAAscales/index.html#SolarRadiationStorms]

On Saturday, Aug. 27, 2011, International Space Station astronaut Ron Garan used a high definition camera to film one of the sixteen sunrises astronauts see each day. [Courtesy NASA]

SOLAR & TERRESTRIAL RADIATION

Chapter Highlights

Case-in-Point
  Ancient Astronomical Calendars
Electromagnetic Spectrum
Radiation Laws
Input of Solar Radiation
  Solar Altitude
  Earth's Motions in Space and the Seasons
  The Solar Constant
Solar Radiation and the Atmosphere
Stratospheric Ozone Shield
Solar Radiation and Earth's Surface
Global Solar Radiation Budget
Outgoing Infrared Radiation
  The Greenhouse Effect
  Greenhouse Gases
  The Callendar Effect
  Possible Impacts of Global Warming
Monitoring Radiation
Conclusions
Basic Understandings/Enduring Ideas
Key Terms/Review/Critical Thinking
For Further Exploration
  Hazards of Solar Ultraviolet Radiation
  Solar Power

Learning Objectives

Identify the principal characteristics of electromagnetic radiation and the electromagnetic spectrum.

Distinguish among the various forms of electromagnetic radiation.

Explain how solar altitude influences the intensity of solar radiation received at Earth's surface.

Describe the causes of the astronomical seasons.

Define the solar constant.

Describe the several types of interactions that take place as solar radiation travels through the atmosphere, encountering gases and aerosols.

Explain the significance of the stratospheric ozone layer for life on Earth.

Identify the principal threat to the stratospheric ozone shield.

Describe the interactions that take place when solar radiation strikes Earth's surface.

Explain the role of the ocean in the global solar radiation budget.

Contrast solar radiation with terrestrial infrared radiation.

Identify the gases responsible for the greenhouse effect.

Explain how the continued buildup of greenhouse gases in the atmosphere is likely to cause further global warming.

List some of the possible societal implications of global warming.

How does energy flow into and out of the Earth-atmosphere system maintain Earth as a habitable planet?

Ancient Astronomical Calendars

Knowing when the seasons begin and end is critical to the timing of planting and harvesting of crops (e.g., to reduce the likelihood of being exposed to a killing frost). In monsoon climates, the timing of the rainy and dry seasons is closely tied to the solar cycle.

Ancient people were well aware of the annual solar cycle and the march of the seasons. While these people did not possess a calendar in the modern sense, their knowledge of the Sun's path through the sky and other regular astronomical events inspired them to construct astronomical calculators in the form of elaborate megaliths.

Probably the best known of the ancient astronomical calculators is Stonehenge located in Southern England, the earliest portion of which dates to about BCE 2950 (Figure 3.1). As early as the 18th century, scientists noticed that the horseshoe arrangement of these great stones opened in the direction of sunrise on the summer solstice. They also point in the direction of the mid-winter sunset. In more recent years, scientists discovered that the arrangement of stones and other features at Stonehenge could be used to predict solar and lunar eclipses. People living at Cahokia (a community just east of modern St. Louis, MO) erected similar solar calendars consisting of wooden posts arranged in circles (Figure 3.2). These Woodhenge calendars date from about CE 900 to 1100.

Predating Stonehenge by some two thousand years, the oldest astronomical calendar discovered so far consists of megaliths and a stone circle, located near Nabta in the Nubian Desert

FIGURE 3.1

Stonehenge, located in Southern England, is perhaps the best known of the ancient astronomical calendars. It is composed of a horseshoe arrangement of large standing stones. [Photo by Matthew Brennan]

of southern Egypt (Figure 3.3). The global positioning system confirms the stone alignment with the position of the rising Sun on the summer solstice (as it would have been 6000 years ago). For the people of that time and place, this was a significant date because monsoon rains typically began shortly after the summer solstice.

Chankillo is a 2300 year-old Peruvian ruin located in a coastal desert about 400 km (250 mi) north of Lima. Until recently, archaeologists were unsure of Chankillo's function. Three thick concentric walls on its hill-top location suggested a fortress, but there was no water supply. Relics suggested rituals and possibly a

FIGURE 3.2

A reconstruction of Woodhenge on the original site near modern day St. Louis, MO. [Photo courtesy of William R. Iseminger, Cahokia Mounds State Historic Site]

temple. In 2007, researchers Iván Ghezzi of the Pontificia Universidad Católica del Peru and Clive Ruggles of the University of Leicester, UK presented convincing evidence that at least part of Chankillo served as a solar observatory, the oldest one in the Americas.

While visiting Chankillo in 2001, Ghezzi observed a string of 13 nearly evenly spaced towers on a nearby ridge (Figure 3.4), oriented in a roughly north-south direction, and having a total length of about 300 m (985 ft). The towers were about 5 m (16 ft) apart and had a height of 2 to 6 m (7 to 20 ft), with flat tops that measured 11 to 13 m (36 to 43 ft) by 6 to 9 m (20 to 30 ft). Ghezzi suspected that the array of towers was an astronomical tool, pre-dating by 1800 years the solar calendars that reputedly were used by the Incas of Peru.

FIGURE 3.3

A model of the Nabta calendar in the Aswan Nubia museum. [Photo by Raymbetz/Creative Commons Attribution-Share Alike 3.0 Unported]

Returning to the site in 2004, Ghezzi discovered two observation sites, one located about 200 m (660 ft) east of the towers and the other about 200 m west of the towers. Ghezzi proposed that the ancients followed the progress of the seasons from day to day by viewing the towers and noting the location of the rising Sun (from the western observation site) and setting Sun (from the eastern observation site). Over the course of a solar year, the location of sunrise and sunset shifted from the June solstice at the northern end of the tower array to the December solstice at the southern end, and then back again to complete the annual solar cycle. The Chankillo solar calendar had an accuracy of 2 to 3 days per year. Furthermore, the dual observation sites offered the ancients an opportunity for a second daily observation in the event that a thick cloud cover obscured their view of the Sun during the other observation.

FIGURE 3.4

Sun rise on the June solstice at the ancient solar observatory at Chankillo, Peru erected 2300 years ago. [Courtesy of Iván Ghezzi]

Radiant energy from the Sun drives the circulation of the atmosphere and powers winds and storms. **Energy** is defined as the capacity for doing work and occurs in many different forms such as radiation and heat. The circulation of the atmosphere plays a major role in determining weather and its temporal and spatial variability.

The Sun ceaselessly emits energy to space in the form of electromagnetic radiation. A very small portion of that energy is intercepted by the Earth-atmosphere system and converted to other forms of energy including, for example, heat and the kinetic energy of the winds. Energy cannot be created nor destroyed although it can be converted from one form to another. This is the **law of energy conservation** (also known as the *first law of thermodynamics*).

In this chapter, we examine the basic properties of electromagnetic radiation, laws governing electromagnetic radiation, how solar radiation interacts with the components of the Earth-atmosphere system, and the conversion of solar radiation to heat. The Earth-atmosphere system responds to solar heating by emitting infrared radiation. Some of this infrared radiation is absorbed by certain atmospheric gases. In turn, those gases emit infrared radiation downward, elevating the temperature of Earth's surface and lower atmosphere to levels that make life possible (the *greenhouse effect*). We begin with the nature of electromagnetic radiation and some of the properties of its various forms.

Electromagnetic Spectrum

Planet Earth is bathed continuously in **electromagnetic radiation**, which has both electrical and magnetic properties. Essentially all objects absorb and emit electromagnetic radiation. Forms of electromagnetic radiation include radio waves, microwaves, infrared radiation, visible light, ultraviolet radiation, X-rays, and gamma radiation. Together, they make up the **electromagnetic spectrum**, illustrated in Figure 3.5.

All types of electromagnetic radiation travel as waves that are differentiated by wavelength or frequency. **Wavelength** is the distance between successive wave crests (or equivalently, wave troughs), as shown in Figure 3.6. **Wave frequency** is the number of crests (or troughs) passing a given point in a specified period of time, usually one second. Passage of one complete wave is called a *cycle*, and a frequency of one cycle per

FIGURE 3.5

The electromagnetic spectrum. The various forms of electromagnetic radiation are distinguished by wavelength in micrometers (μm), millimeters (mm), meters (m), and kilometers (km).

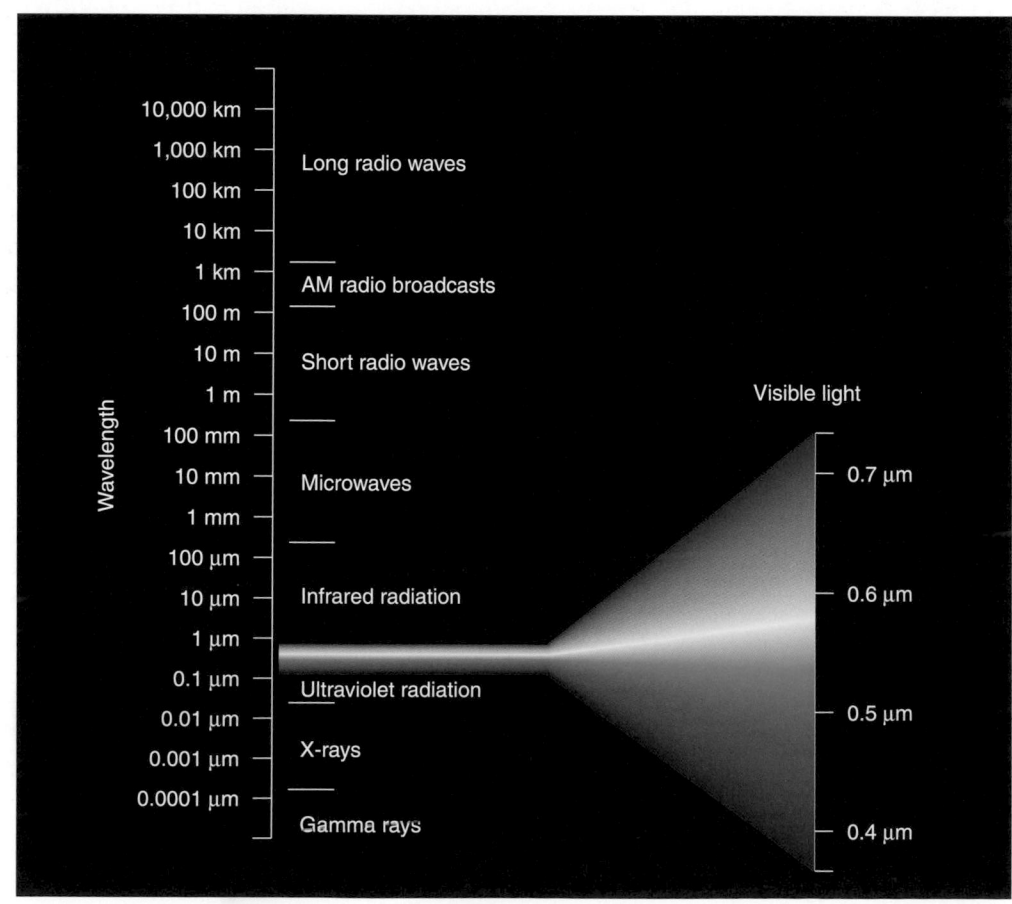

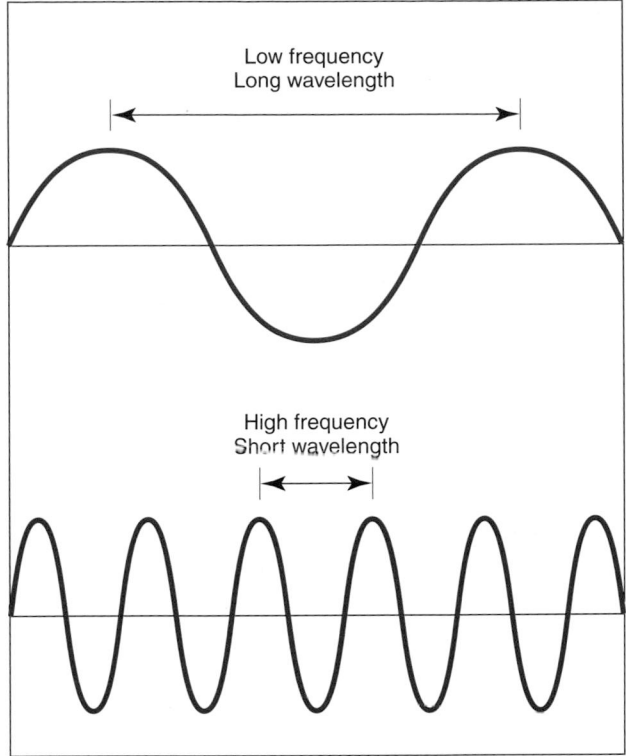

**Low frequency
Long wavelength**

**High frequency
Short wavelength**

FIGURE 3.6

Wavelength is the distance between two successive crests or, equivalently, between two successive troughs. Wavelength is inversely related to wave frequency.

second is defined as 1.0 hertz (Hz). Wave frequency is inversely proportional to wavelength; that is, the higher the frequency, the shorter is the wavelength. Radio waves have frequencies in the millions of hertz with wavelengths up to hundreds of kilometers. By contrast, at the other end of the electromagnetic spectrum, gamma rays have frequencies as high as 10^{24} (a trillion trillion) Hz and wavelengths as short as 10^{-14} (a hundred trillionth) m. Furthermore, higher frequency radiation with shorter wavelengths has higher energy levels than lower frequency (longer wavelength) radiation.

Electromagnetic waves travel through space and may pass through gases, liquids, and solids. In a vacuum, all electromagnetic waves travel at the maximum possible speed of 300,000 km (186,000 mi) per sec, commonly called the *speed of light*. All forms of electromagnetic radiation slow when passing through materials, their speed varying with wavelength and type of material. As electromagnetic radiation passes from one medium into another, it may be reflected or refracted (i.e., bent) at the interface. This happens, for example, when solar radiation strikes the ocean surface at an oblique angle.

Some radiation is reflected upward into the atmosphere and some is bent downward (refracted) as it penetrates the water. Electromagnetic radiation is also absorbed, that is, converted to heat energy.

Although the electromagnetic spectrum is continuous, different names are assigned to different segments because we detect, measure, generate, and use those segments in different ways. Furthermore, the various types of electromagnetic radiation do not begin or end at precise points along the spectrum. For example, red light shades into invisible infrared radiation (infrared, meaning *below red*). At the other end of the visible portion of the electromagnetic spectrum, violet light shades into invisible ultraviolet radiation (ultraviolet, meaning *beyond violet*).

Beyond visible light on the electromagnetic spectrum and in order of increasing energy level, increasing frequency, and decreasing wavelength are ultraviolet radiation (UV), X-rays, and gamma radiation. All three forms of radiation occur naturally, but can also be produced. All have medical uses: ultraviolet radiation is a potent germicide; X-rays are used as a powerful diagnostic tool; and both X-rays and gamma radiation are used to treat cancer patients. These three highly energetic types of radiation are dangerous as well as useful. Ultraviolet radiation can cause irreparable damage to the light-sensitive cells of the eye. Staring at the Sun (for instance, during a partial solar eclipse) can permanently blind a person, unless a filter is used to block out ultraviolet radiation. Also, overexposure to UV, X-rays, or gamma radiation can cause sterilization, cancer, mutations, or damage to a fetus. Fortunately, Earth's atmosphere blocks most incoming UV and virtually all X-rays and gamma radiation through absorption by the rarefied gases in the ionosphere and stratosphere. Without this protective atmospheric shield, life as we know it would not exist.

At lower frequencies and longer wavelengths, UV radiation shades into **visible radiation**, that is, radiation that is perceptible by the human eye. White light is *polychromatic*, composed of multiple colors associated with individual wavelength bands. The wavelength of visible light ranges from about 0.40 μm at the violet end to approximately 0.70 μm at the red end. (One *μm* is a millionth of a meter, one-tenth the thickness of a human hair.) Visible light is essential for many activities of plants and animals. In plants, light provides the energy needed for photosynthesis; it also regulates the opening of buds and flowers in spring and the dropping of leaves in autumn. For animals, light governs the timing of reproduction, hibernation, and migration and for many species, makes vision possible.

Between red light and microwave radiation on the electromagnetic spectrum is **infrared radiation (IR)**. IR is not visible, but we can feel the heat it generates when it is intense, as it is, for example, when emitted by a hot stove. Actually, every known object, including you and this book, emit small amounts of infrared radiation. As demonstrated later in this chapter, absorption and emission of IR by certain atmospheric gases is responsible for significant warming of the lower atmosphere (the *greenhouse effect*), resulting in a habitable planet.

At longer wavelengths is the microwave portion of the electromagnetic spectrum, which spans wavelengths of about 0.1 to 1000 millimeters. Microwave frequencies are used for radio communication, in microwave ovens, and in radar. At the low energy, low frequency, long wavelength end of the electromagnetic spectrum are radio waves. Wavelengths range from a fraction of a centimeter up to hundreds of kilometers, and frequencies can extend to a billion hertz. FM (frequency modulation) radio waves, for example, span 88 million to 108 million Hz; hence, the familiar 88 and 108 at opposite ends of the FM radio band.

Radiation Laws

Several physical laws describe the properties of electromagnetic radiation emitted by a perfect radiator, known as a blackbody. By definition, a **blackbody** at a constant temperature absorbs all radiation that is incident on it and emits all the radiant energy it absorbs. A blackbody is both a perfect absorber and perfect emitter of radiation. The range and intensity of the wavelengths of emitted radiation are related to the temperature of the blackbody. Surfaces of real objects may approximate blackbodies for certain wavelengths of radiation but not for others. Fresh fallen snow, for example, is very nearly a blackbody for infrared radiation but not for visible light. (Note that *blackbody* does not refer to color.) Although neither the Sun nor Earth is a blackbody, their absorption and emission of radiation are close enough to that of a blackbody so we can apply blackbody radiation laws to them with very useful results. Here we apply two blackbody radiation laws: Wien's displacement law and the Stefan-Boltzmann law.

Whereas all known objects emit and absorb all forms of electromagnetic radiation, the wavelength of most intense radiation (λ_{max}) emitted by a blackbody is inversely proportional to the absolute temperature (T) of the object. That is,

$$\lambda_{max} = C/T$$

where C, a constant of proportionality, has the value of 2897 if λ_{max} is expressed in micrometers (µm), and T is in kelvin (K). *Absolute temperature* is the number of kelvins above absolute zero (−273.15 °C or −459.67 °F). This is a statement of **Wien's displacement law**. As shown in Figure 3.7, a significant portion of the radiation emitted (or absorbed) by an object is found around this peak maximum wavelength.

According to Wien's displacement law, hot objects (such as the Sun) emit radiation that peaks at relatively short wavelengths, whereas relatively cold objects (such as the Earth-atmosphere system) emit peak radiation at longer wavelengths (and lower frequencies). The top heating elements of an electric stove provide an illustration of Wien's displacement law. After switching on a heating element, its metal coils warm and we readily feel the heat. We are actually feeling invisible infrared radiation. As the coil temperature continues to rise, the coils emit more intense infrared radiation with peak emission at progressively shorter wavelengths. That is, as the coil temperature rises, the coil's peak radiation is displaced from the infrared toward the near-infrared portion of the

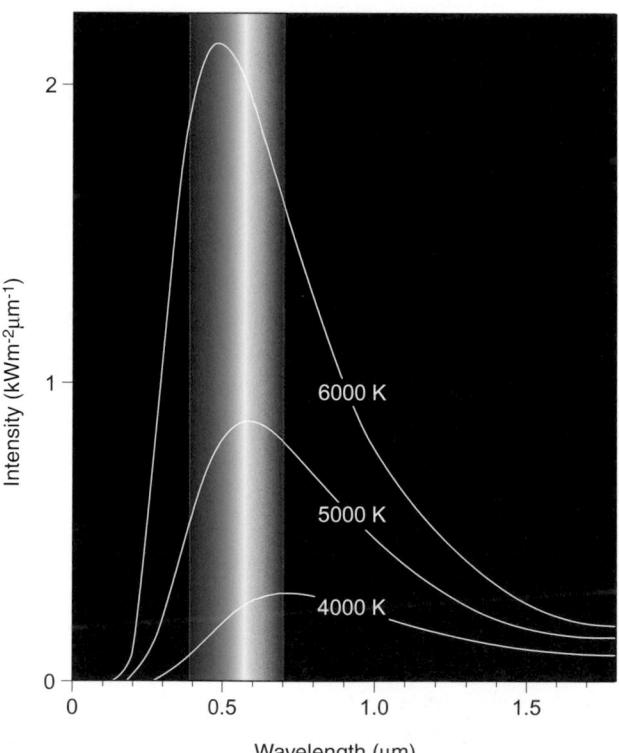

FIGURE 3.7
With increasing absolute temperature (K), the wavelength of maximum radiation emitted or absorbed by an object decreases. This is known as Wien's displacement law. The total amount of energy emitted or absorbed by the object, as indicated by the area under the individual curves, increases by the fourth power of the temperature, as specified by the Stefan-Boltzmann law.

electromagnetic spectrum. The highest setting on the stove will cause the heating element to glow dull red as some of the radiation is in the visible light spectrum.

Radiation emitted by the Sun is similar to that emitted by a blackbody at a temperature of about 6000 °C (11,000 °F). Figure 3.8 shows the flux (energy transfer per unit time per unit area) of solar radiation received at the top of the atmosphere as a function of wavelength. The Sun emits a band of radiation (at wavelengths mostly between 0.25 and 2.5 µm) that is most intense at a wavelength of about 0.5 µm (in the green of visible light, as seen in Figure 3.7). The flux of radiation emitted by Earth's surface is similar to that emitted by a blackbody at a temperature of about 15 °C (59 °F) peaking in the infrared (Figure 3.9). Earth's surface emits a broad band of infrared radiation (at wavelengths mostly between 4 and 24 µm) with peak intensity at a wavelength of about 10 µm.

The area under the blackbody curves in Figures 3.8 (for the Sun) and 3.9 (for Earth's surface) represent the total radiation energy emitted per unit time per unit surface area at all wavelengths. Note that the vertical scales in the two figures are much different. According to the **Stefan-Boltzmann law**, the total energy flux emitted by a blackbody across all wavelengths (E) is proportional to the fourth power of the absolute temperature (T⁴) of the object. That is,

$$E = \sigma T^4$$

where σ is the Stefan-Boltzmann constant of proportionality. This relationship implies that a small change in the temperature of a blackbody results in a much greater change in the total amount of radiational energy emitted by the blackbody during a given time interval. The Sun radiates at a much higher temperature than does Earth's surface, so that the Stefan-Boltzmann law predicts that the rate of the Sun's energy output per square meter is almost 190,000 times that of the Earth-atmosphere system to space.

Electromagnetic radiation at shorter wavelengths is more energetic than radiation at longer wavelengths. Hence, X-rays are described as high energy radiation whereas radio waves are relatively low energy.

As solar radiation travels away from the Sun and spreads outward into space in all directions, its intensity (energy per unit area) diminishes rapidly, as the inverse square of the distance traveled. The radiation emanating in all directions from a nearly point source like the Sun spreads out as ever larger spheres, with the intensity being reduced as a function of the surface area of the sphere

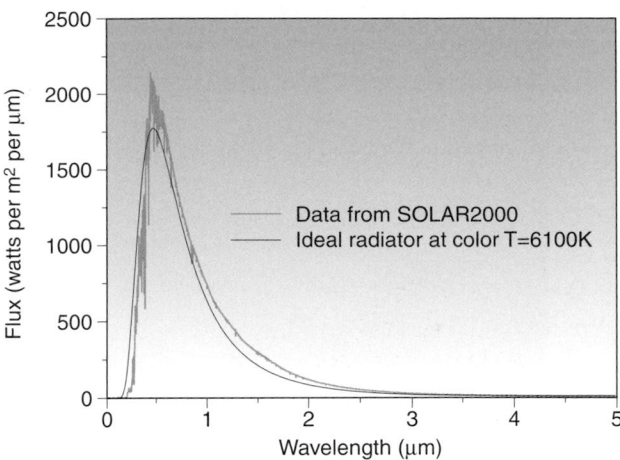

FIGURE 3.8
The flux of solar radiation incident at the top of the atmosphere as a function of wavelength. Radiation emitted by the Sun is similar to that emitted by a blackbody (ideal radiator) at a temperature of about 6000 °C (about 6100 K). [NOAA, Space Environment Center and Space Environment Technologies/SpaceWx]

(where the area is a function of the square of the sphere's radius). According to this **inverse square law**, doubling the distance traversed by radiation reduces its intensity to (1/2)² or 1/4 of its initial value.

Earth orbits the Sun at a distance such that it receives an intensity of solar radiation, resulting in a temperature range that supports life as we know it (Figure 3.10). Venus is closer to the Sun and receives nearly twice as much solar radiation per unit intercepted area as Earth. Venus is too hot for life, with surface temperatures averaging about 460 °C (860 °F). Mars, which is farther away from the Sun, receives solar

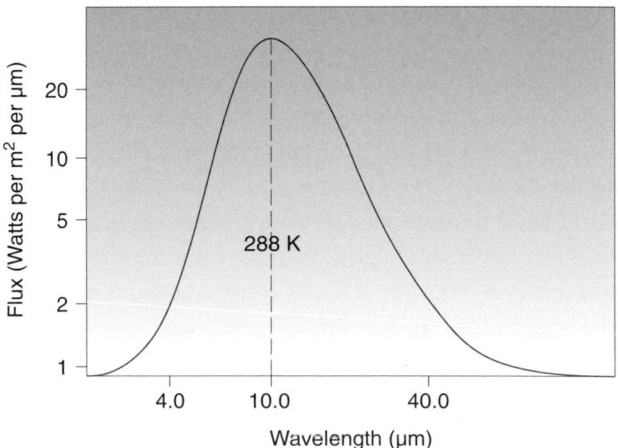

FIGURE 3.9
Flux of radiation as a function of wavelength emitted by a blackbody radiating at about the same average temperature as Earth's surface, that is, 15 °C (288 K).

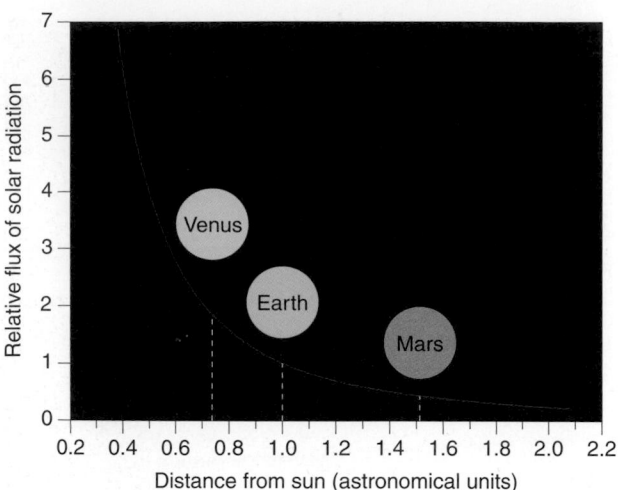

FIGURE 3.10
As radiation flows from the Sun into space, its intensity decreases as the inverse square of the distance traveled. Venus, closer to the Sun than Earth, receives about twice as much solar radiation as Earth. Mars, on the other hand, is farther from the Sun than Earth and receives roughly half as much solar radiation as Earth. [Courtesy of E.J. Hopkins]

radiation less than one-half as intense as Earth and is too cold for life, with surface temperatures averaging about −53 °C (−63 °F).

Because Earth rotates on its axis relatively rapidly, making one rotation with respect to the Sun on a daily basis, the incident solar radiation is effectively spread over the planet's surface. This more even distribution of sunlight makes the planet more habitable than if only one side of the planet were exposed to sunlight for a longer time. Our Moon rotates on its axis on a slower basis (one rotation period being approximately the same as one lunar revolution of Earth). This slow lunar rotation results in a large temperature variation on the lunar surface over the 29.5 days of a lunar synodic day (one complete Moon rotation relative to the Sun). According to NASA's World Book, the temperature on the lunar equator varies between −173 °C (−280 °F) at night to 127 °C (260 °F) in the daytime. The absence of a lunar atmosphere also contributes to this considerable temperature range.

Several decades of satellite measurements as well as a relatively constant planetary temperature indicates that the total energy (in the form of solar radiation) absorbed by planet Earth is essentially equal to the total energy (in the form of infrared radiation) emitted by the Earth-atmosphere system to space. A balance between energy input and energy output is known as **global radiative equilibrium** and is an example of the *law of energy conservation*. We examine this important concept in more detail in the next chapter.

Input of Solar Radiation

The Sun, the star closest to Earth, is a huge gaseous body composed almost entirely of hydrogen (about 80% by mass) and helium with internal temperatures that may exceed 20 million °C. The ultimate source of solar energy is a continuous nuclear fusion reaction in the Sun's interior. Simply put, in this reaction, four hydrogen nuclei (protons) fuse to form one helium nucleus (an alpha particle). However, the mass of one helium nucleus is about 0.7% less than the mass of the four hydrogen nuclei. This mass lost in the fusion of hydrogen to helium is converted to energy as described by Albert Einstein's mass-energy equivalence principle:

$$E = mc^2$$

where mass, m (in kg), is related to energy, E (in joules), with c being the speed of light (300,000 km per sec or 186,000 mi per sec). Note that c^2 is a huge number so that even a very small mass converts to an enormous quantity of energy. Some of the energy produced by nuclear fusion in the Sun is used to bind the helium nucleus together; the rest of the energy is radiated and transferred by convection to the Sun's surface, and then radiated in all directions to space.

At radiating temperatures near 6100 K (11,000 °F), the visible surface of the Sun, known as the **photosphere**, is much cooler than the Sun's interior. A network of huge, irregularly shaped convective cells, called *granules*, is responsible for the photosphere's honeycomb appearance. A typical granule is about 1000 km (600 mi) across, although some *supergranules* may be 30,000 to 50,000 km (18,500 to 31,000 mi) in diameter. Most granules have a life expectancy of only a few minutes and consist of a broad central area of rising hot gas surrounded by a thin layer of cooler gas sinking back toward the center of the Sun. This zone of convective activity encompasses the outer 200,000 km (125,000 mi) of the Sun.

Relatively dark, cool areas, called **sunspots**, usually occurring in pairs, dot the surface of the photosphere. Sunspot temperatures may be 400 to 1800 Celsius degrees (720 to 3240 Fahrenheit degrees) lower than the photosphere's average temperature. Bright areas, known as *faculae*, usually occur near sunspots. Changes in the number of sunspots and faculae accompany changes in solar energy output and may influence Earth's climate (Chapter 15).

Outward from the photosphere is the **chromosphere**, consisting of ions of hydrogen and helium at 4000 °C to 40,000 °C (7200 °F to 72,000 °F). Beyond the chromosphere is the outermost portion of the Sun's atmosphere, the **solar corona**, a region of extremely hot (1 to 4 million °C) and highly rarefied ionized gases (predominantly hydrogen and helium) that extends millions of kilometers into space. The solar wind originates in the corona and is intensified by solar flares that erupt from the photosphere into the corona (Chapter 2). In its orbit about the Sun, planet Earth intercepts only about one two-billionth of the enormous quantity of energy continually radiated by the Sun to space.

SOLAR ALTITUDE

At middle latitudes, the intensity of solar radiation striking Earth's surface varies significantly with the time of day and season. At local solar noon, the summer Sun is higher in the sky than the winter Sun, causing midday solar rays striking Earth's surface to be more concentrated in summer than in winter. Even over the course of a single day, regular changes occur in incoming solar radiation striking the Earth's surface; that is, the solar beam is more concentrated at noon than at sunrise or sunset. Hence, the angle of the Sun above the horizon, called the **solar altitude**, influences the intensity of solar radiation received at Earth's surface. At the place on Earth where the Sun is directly overhead, the local solar altitude has its maximum value of 90 degrees and solar rays are most concentrated (Figure 3.11). Whenever the Sun is positioned lower in the sky, solar radiation spreads over a larger area of Earth's horizontal surface and thus is less intense. *Local solar noon* is the time of day at essentially any location on Earth when the Sun appears at its highest point in its daily path across the local sky. At that time, the Sun would be on your local meridian (north/south line passing through your location and cutting the horizon at the north and south points).

Earth is so distant from the Sun (mean distance of about 150 million km or 93 million mi) that solar radiation reaches the planet essentially as parallel beams of uniform intensity. But the nearly spherical Earth presents a curved surface to incoming solar radiation so that the noon solar altitude always varies with latitude (Figure 3.12). The intensity of solar radiation actually striking Earth's atmosphere is greatest at the latitude where the noon Sun is in the zenith (solar altitude of 90 degrees) and decreases with distance north and south of that latitude.

Solar altitude also influences the interaction between incoming solar radiation and the atmosphere.

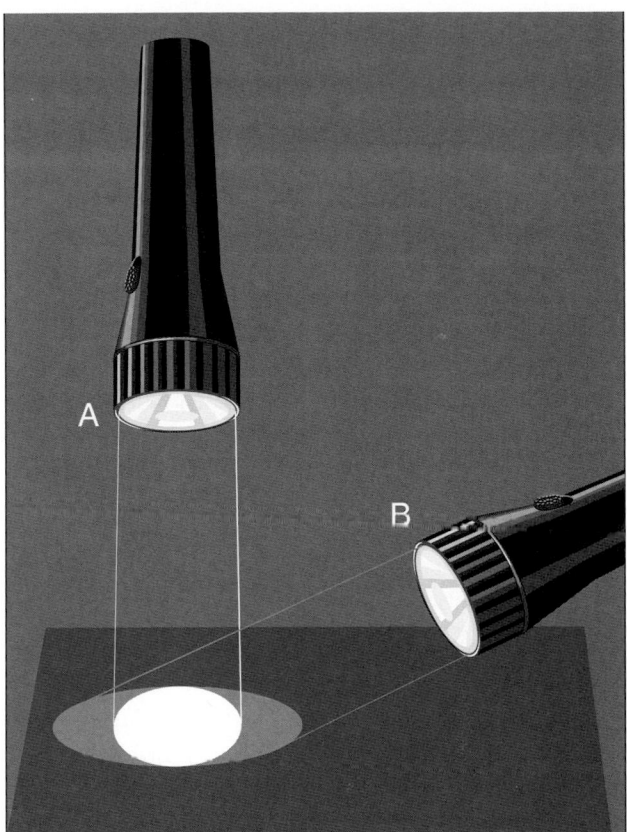

FIGURE 3.11
The intensity of solar radiation striking Earth's surface per unit area varies with the solar altitude. Consider this analogous situation: (A) A flashlight beam shines on a horizontal surface most intensely when the flashlight shines from directly overhead (analogous to a solar altitude of 90 degrees). (B) At an angle decreasing from 90 degrees, the flashlight beam spreads over an increasing area of the horizontal surface so that the light is less concentrated (less radiational energy received per unit area).

Decreasing solar altitude lengthens the path of the Sun's rays through the atmosphere (Figure 3.13). As the path lengthens, the greater interaction of solar radiation with clouds, gases and aerosols reduces its intensity. Even if the sky were cloud-free, the longer the path of solar radiation through the atmosphere, the less intense is the radiation striking Earth's surface. The interactions between incoming solar radiation and atmosphere (scattering, reflection, and absorption) are described later in this chapter.

While solar altitude influences the intensity of solar radiation striking Earth's horizontal surface per unit area, the length of daylight affects the total amount of solar radiational energy that is received each day. For example, in summer the altitude of the noon Sun is lower at high latitudes than at middle latitudes. Nonetheless, the greater length of daylight at high latitudes may translate

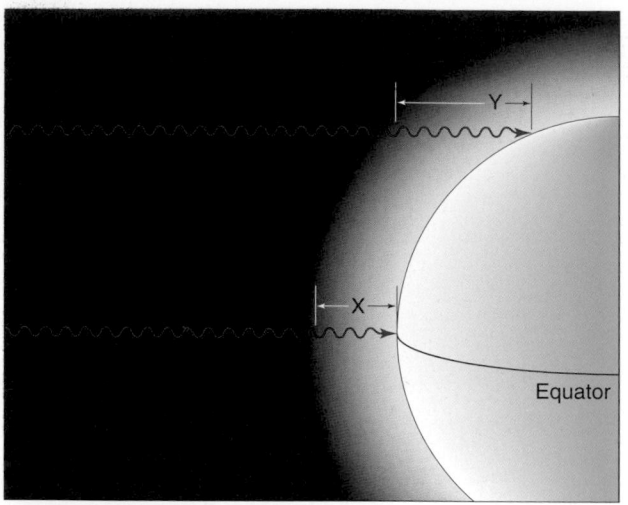

FIGURE 3.12
Solar radiation's path through the atmosphere lengthens with decreasing solar altitude, that is, as the Sun moves lower in the sky. *X* is the path length at high solar altitude and *Y* is the path length at low solar altitude. Path *Y* may approach being 30 times longer than *X*.

into more total radiation striking Earth's surface during a 24-hr day (Figure 3.14). Variations in both solar altitude and length of daylight accompany the annual march of the seasons. Before examining these relationships, first consider the fundamental motions of Earth in space: rotation of the planet on its spin axis and the planet's revolution about the Sun.

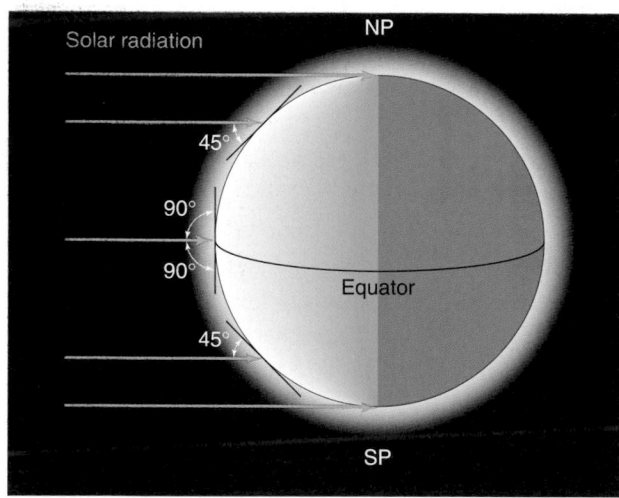

FIGURE 3.13
On any day of the year, the noon solar altitude always varies with latitude because Earth presents a curved surface to the incoming solar beam. In this example for an equinox, the solar altitude is 90 degrees at the equator and decreases with latitude (toward the poles). Hence, solar radiation striking horizontal surfaces per unit area is most intense at the equator and least intense at the poles.

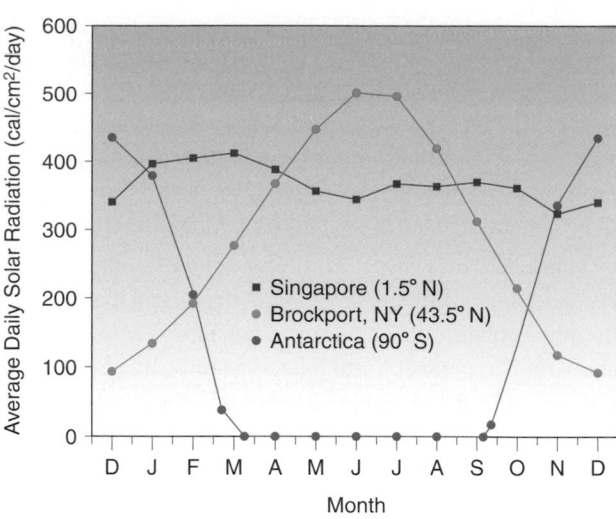

FIGURE 3.14
Average daily solar radiation (in calories per cm² per day) by month at Singapore, Brockport, NY, and Antarctica. During part of the year, the polar (Antarctic) and mid-latitude (Brockport) locations receive more solar radiation daily than the near-equator location (Singapore).

EARTH'S MOTIONS IN SPACE AND THE SEASONS

Rotation of Earth on its axis accounts for day and night. Once every 24 hrs, Earth completes one rotation on its spin axis. At any instant, half the planet is illuminated by visible solar radiation (day), while the other half is in darkness (night).

Over the course of one year, which is actually 365.2422 days, Earth makes one complete revolution about the Sun in a slightly elliptical orbit (Figure 3.15). Earth's orbital eccentricity, that is, its deviation from a circular orbit, is so slight that the Earth-to-Sun distance varies by only about 3.3% through the year. Earth is closest to the Sun (147 million km or 91 million mi) on about 3 January and farthest from the Sun (152 million km or 94 million mi) on about 4 July. These are the current dates of **perihelion** and **aphelion**, respectively. In the Northern Hemisphere, Earth is closest to the Sun in winter and farthest from the Sun in summer. Because of the spreading of radiation described by the inverse square law, Earth intercepts about 6.7% more solar radiation at perihelion than at aphelion. The eccentricity of Earth's orbit about the Sun cannot explain the seasons. What does account for seasons?

The answer is found in the tilt of Earth's spin axis relative to the plane of its orbit. Because of this tilt, Earth's equatorial plane at the present time is inclined 23 degrees 27 minutes to the plane defined by the planet's annual orbit about the Sun (Figure 3.16). Thus, Earth's spin

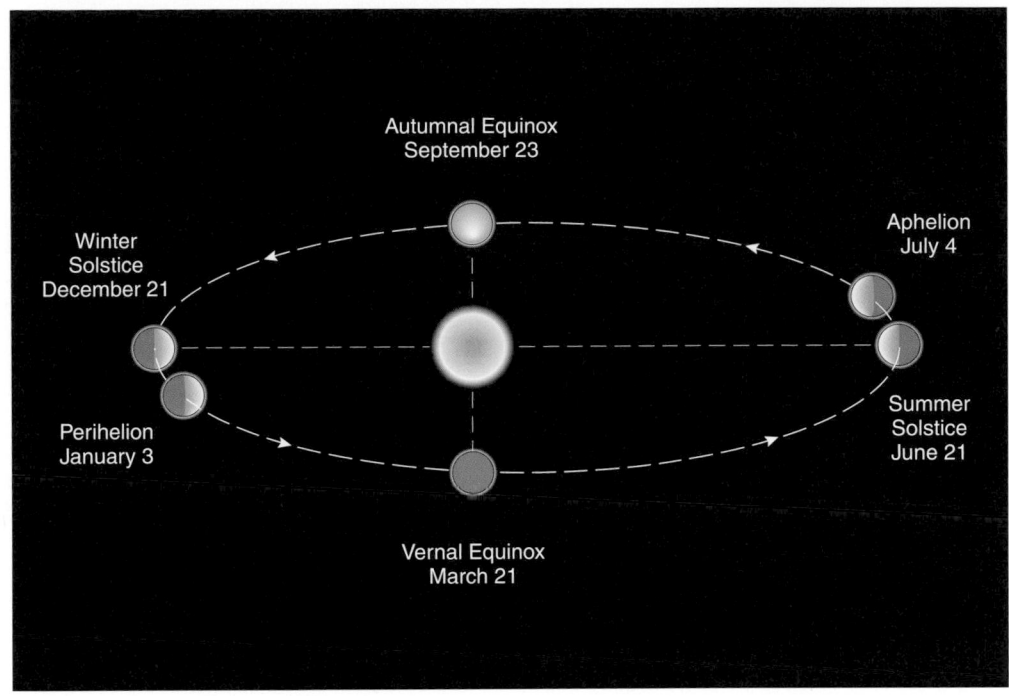

FIGURE 3.15
Earth's orbit is an ellipse with the Sun located at one focus. Earth is closest to the Sun at *perihelion* (about 3 January) and farthest from the Sun at *aphelion* (about 4 July). Note that the eccentricity of Earth's orbit is greatly exaggerated in this drawing.

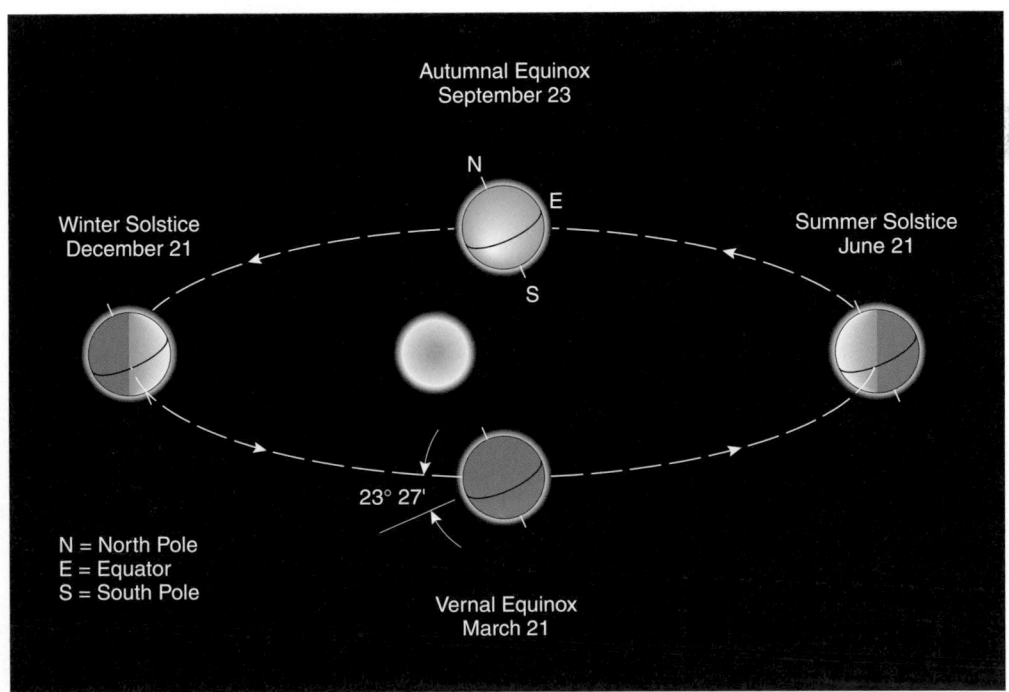

FIGURE 3.16
The seasons change because Earth's equatorial plane is inclined (at 23 degrees, 27 minutes) to its orbital plane. The seasons given are for the Northern Hemisphere. Note that the eccentricity of Earth's orbit is greatly exaggerated in this drawing.

axis is tilted 23 degrees 27 minutes from a line oriented perpendicular to Earth's orbital plane. During Earth's annual revolution about the Sun, its spin axis remains in the same alignment with respect to the background stars (the North Pole always points toward *Polaris*, the North Star during the current millennium) while its orientation to the Sun changes continually. Accompanying these changes are regular variations in solar altitude and length of daylight, which in turn affect the intensity and total amount of solar radiation received at different latitudes on Earth's surface. If Earth's rotational axis were perpendicular to its orbital plane (no tilt), Earth's axis would always have the same orientation to the Sun. Without the axial tilt, only changes in Earth-Sun distance between aphelion and perihelion would produce a seasonal contrast and it would be slight.

How does Earth's orientation to the Sun change over the course of a year? Viewed from Earth's surface, the latitude where the Sun's rays are most intense (overhead with a solar altitude of 90 degrees) shifts from 23 degrees 27 minutes south of the equator to 23 degrees 27 minutes north of the equator, and then back to 23 degrees 27 minutes south. On about 21 March and again on about 23 September, the Sun's noon position is directly over the equator. At these times, day and night are approximately equal in length (12 hrs) everywhere, except at the poles (Figure 3.17). For this reason, these dates are the **equinoxes** (from the Latin for "equal nights").

Following the equinoxes, the Sun continues its apparent journey toward its maximum poleward locations. On or about 21 June, the Sun's noon rays are vertical at

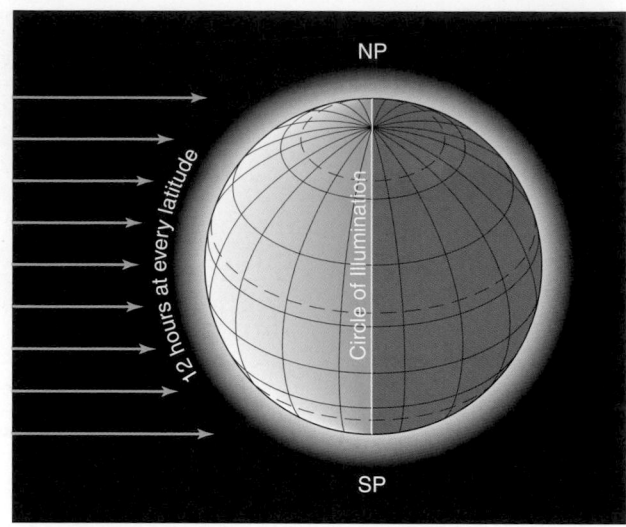

FIGURE 3.17
At the autumnal and spring equinoxes, the noon solar altitude is greatest (90 degrees) at the equator. Day and night are about equal in length except at the poles, where the Sun appears to travel along the local horizon for the entire day.

23 degrees 27 minutes N, the latitude circle known as the **Tropic of Cancer**. As shown in Figure 3.18, daylight is continuous north of the **Arctic Circle** (66 degrees 33 minutes N) and absent south of the **Antarctic Circle** (66 degrees 33 minutes S). Elsewhere, days are longer than nights in the Northern Hemisphere, where it is the first day of astronomical summer, and days are shorter than nights in the Southern Hemisphere, where it is the first day of astronomical winter. Hence, this is a **solstice**

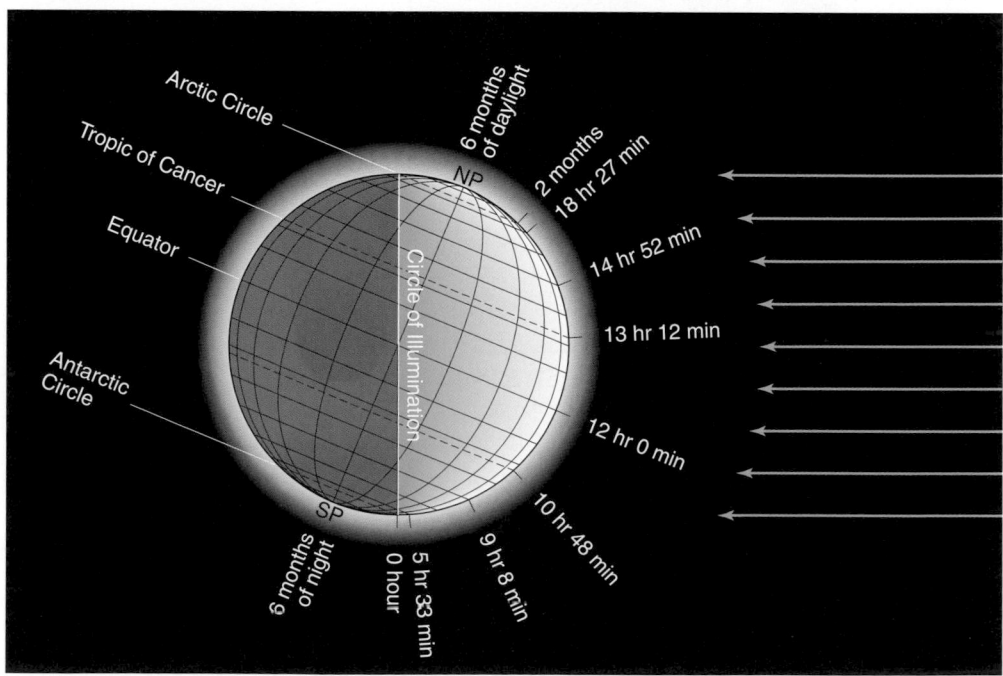

FIGURE 3.18
On the Northern Hemisphere summer solstice (about 21 June), the noon solar altitude is greatest (90 degrees) at 23 degrees 27 minutes N, and days are longer than nights everywhere north of the equator. Duration of daylight is given for every 20 degrees of latitude.

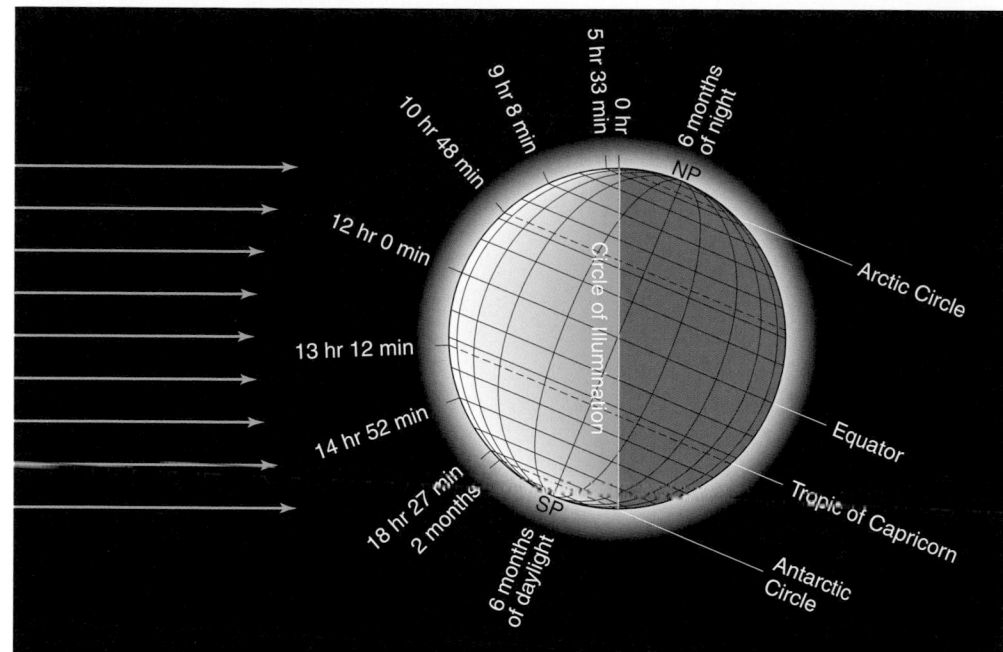

FIGURE 3.19
On the Northern Hemisphere winter solstice (about 21 December), the noon solar altitude is greatest (90 degrees) at 23 degrees 27 minutes S, and days are shorter than nights everywhere north of the equator. Duration of daylight is given for every 20 degrees of latitude.

date—the summer solstice in the Northern Hemisphere and winter solstice in the Southern Hemisphere. Solstice is from the Latin, *solstitium*, referring to the Sun standing still as its latitudinal position changes very little from day to day at this time of year.

On or about 21 December, the noon Sun is directly over 23 degrees 27 minutes S, the latitude circle known as the **Tropic of Capricorn** (Figure 3.19). Daylight is continuous south of the Antarctic Circle and absent north of the Arctic Circle. Elsewhere, nights are longer than days in the Northern Hemisphere, where it is the first day of astronomical winter, and days are longer than nights in the Southern Hemisphere, where it is the first day of astronomical summer. Thus, this is the date of the winter solstice in the Northern Hemisphere and the summer solstice in the Southern Hemisphere.

As Earth's orientation to the Sun changes through the course of a year, so does the daily path of the Sun through the local sky. Figure 3.20 portrays the path of the Sun through the sky from sunrise to sunset for solstices and equinoxes at the equator, a middle latitude Northern Hemisphere location, and the North Pole. The high point of each path is the Sun's position at local solar noon (except at the polar location). (Because of one's location within a civil time zone and the time of the year, along with daylight saving time, local solar noon may differ by as much as one hour from noon as indicated by a clock.) At the middle latitude location and North Pole, the altitude of the noon Sun is greatest on the summer solstice, but at the equator, the noon solar altitude is greatest on the equinoxes (when the Sun is directly overhead). Note that on the summer solstice at the North Pole, the Sun circles the sky at a constant solar altitude (about 23.5 degrees) over the 24-hr day.

Ignoring atmospheric effects, solar radiation incident on a horizontal Earth surface is at maximum intensity where the noon Sun is directly overhead. North and south of that latitude, the intensity of noontime solar radiation diminishes because the solar altitude decreases. At the equinoxes, solar rays are most intense at the equator at noon and decrease with latitude toward the poles. On the Northern Hemisphere summer solstice, solar rays are most intense at local noon along the Tropic of Cancer and decrease to zero at the Antarctic Circle. On that day, the noon Sun has the same altitude at 47 degrees N as at the equator, although the Sun appears 23.5 degrees on opposite sides of each latitude's respective zenith. On the Northern Hemisphere winter solstice, solar rays are most intense along the Tropic of Capricorn and decrease to zero at the Arctic Circle.

As noted earlier, days and nights are approximately equal in length (12 hrs) everywhere on Earth (except at the poles) on only two days of the year, the spring and autumnal equinoxes. The length of daylight on the equinoxes is not precisely 12 hrs because of the optical effects of the atmosphere on the solar beam. Times of sunrise and sunset refer to the hour and minute when the upper edge of the Sun (not its center) is on the horizon. The atmosphere refracts

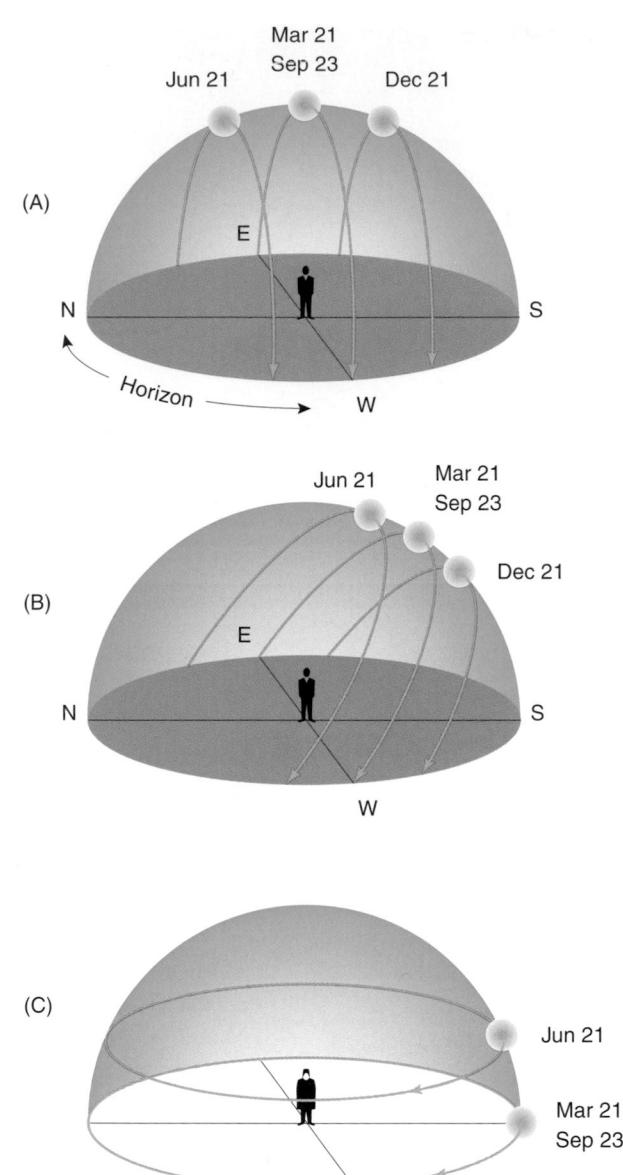

FIGURE 3.20
Path of the Sun through the sky on the solstices and equinoxes at (A) the equator, (B) middle latitudes of the Northern Hemisphere, and (C) the North Pole.

(bends) the incoming solar beam downward so that the Sun appears to be higher than it actually is, lengthening the period of daylight. For example, at Washington, DC the day length is 12 hrs and 6 minutes on the autumnal equinox (on or about 23 September).

Ignoring the atmospheric effects, between the March and September equinoxes, days are longer than nights in the Northern Hemisphere and days are shorter than nights in the Southern Hemisphere. Between the September and March equinoxes, days are shorter than nights in the Northern Hemisphere and days are longer than nights in the Southern Hemisphere. Furthermore, the seasonal (winter-to-summer) contrast in length of daylight increases with increasing latitude (Figure 3.21). At the equator, the period of daylight is essentially the same year round. At 60 degrees N, the length of daylight varies from 5 hrs 53 minutes on 22 December to 18 hrs 52 minutes on 21 June, while approaching the North Pole, daylight length varies from continuous winter darkness to 24-hr summer daylight.

THE SOLAR CONSTANT

For convenience of study, the solar energy input into the Earth-atmosphere system is often based

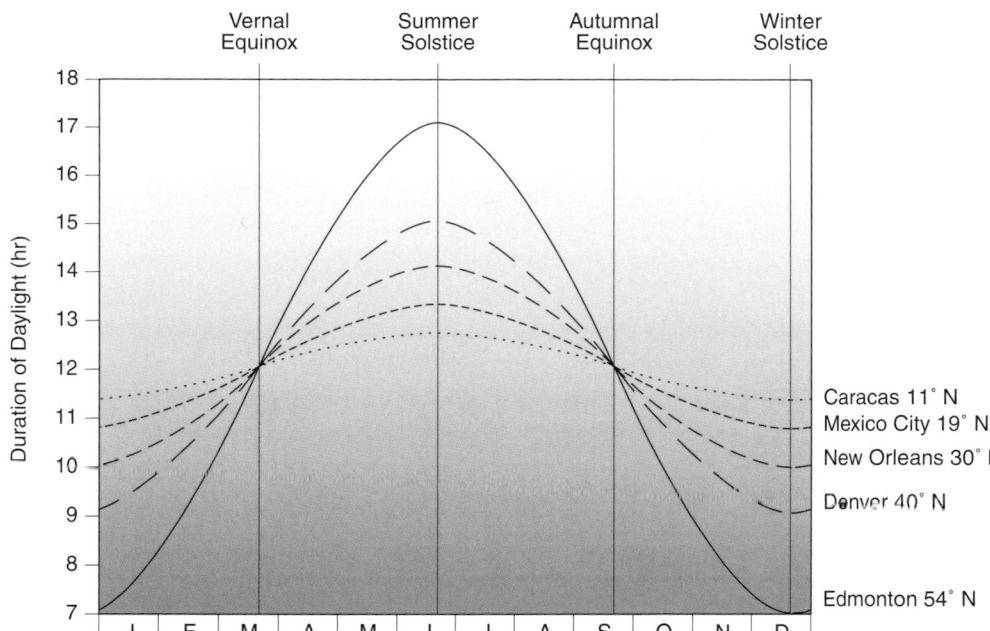

FIGURE 3.21
Variation in the length of daylight through the year increases with increasing latitude.

on what is called the solar constant. The **solar constant** is defined as the rate at which solar radiation falls on a unit area of a flat surface located at the outer edge of the atmosphere and oriented perpendicular to the incoming solar beam when Earth is at its mean distance from the Sun. The *constant* designation is misleading because solar energy output actually fluctuates by a very small fraction of a percent over a year and exhibits longer-term climatically significant variations (Chapter 15) while the distance between Sun and Earth changes during the year. The solar constant averages about 1.97 calories per square centimeter per minute (cal/cm²/min), or 1368 watts per square meter (W/m²).

Suppose that the incoming solar radiation were spread evenly over the rapidly rotating planet. What would be the intensity of solar radiation per unit area of Earth's surface (neglecting atmospheric effects)? Visualize Earth as intercepting a continuous stream of solar energy passing through a disk with the same radius as Earth (Figure 3.22). The radius, R, of the disk is the same as that of the planet and the surface area of a disk is given by

$$\pi R^2$$

where π is about 3.1416. At any point in time, a disk of energy is spread uniformly over the nearly spherical Earth having a surface area given by

$$4\pi R^2$$

Since the surface area of Earth is four times the area of the disk, the solar constant value is reduced to ($\pi R^2/4\pi R^2$) or ¼ of its original magnitude, that is, about 0.5 cal/cm²/min or 342 W/m² when spread over the entire Earth.

The rate of total solar energy input for the planet varies through the course of a year, from a maximum when Earth is closest to the Sun (perihelion), to a minimum when Earth is farthest from the Sun (aphelion). At perihelion, Earth is about 3.3% closer to the Sun than at aphelion. Applying the inverse square law, the planet intercepts about 6.7% more radiation at perihelion (2.04 cal/cm²/min or 1417 W/m²) than at aphelion (1.91 cal/cm²/min or 1326 W/m²).

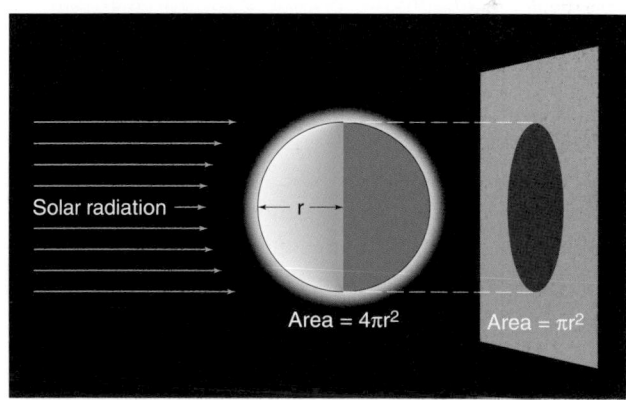

FIGURE 3.22
The area of a disk is ¼ of the area of a sphere. Hence, if incoming solar radiation is spread uniformly over Earth, the flux of solar radiation is reduced to ¼ of the solar constant.

FIGURE 3.23
Distribution of solar radiation received at the top of the atmosphere by latitude and day of the year in watts per square m. Solid white lines in polar latitudes mark the equatorward limits of the region experiencing 24 hours of uninterrupted daylight. [Courtesy of E.J. Hopkins]

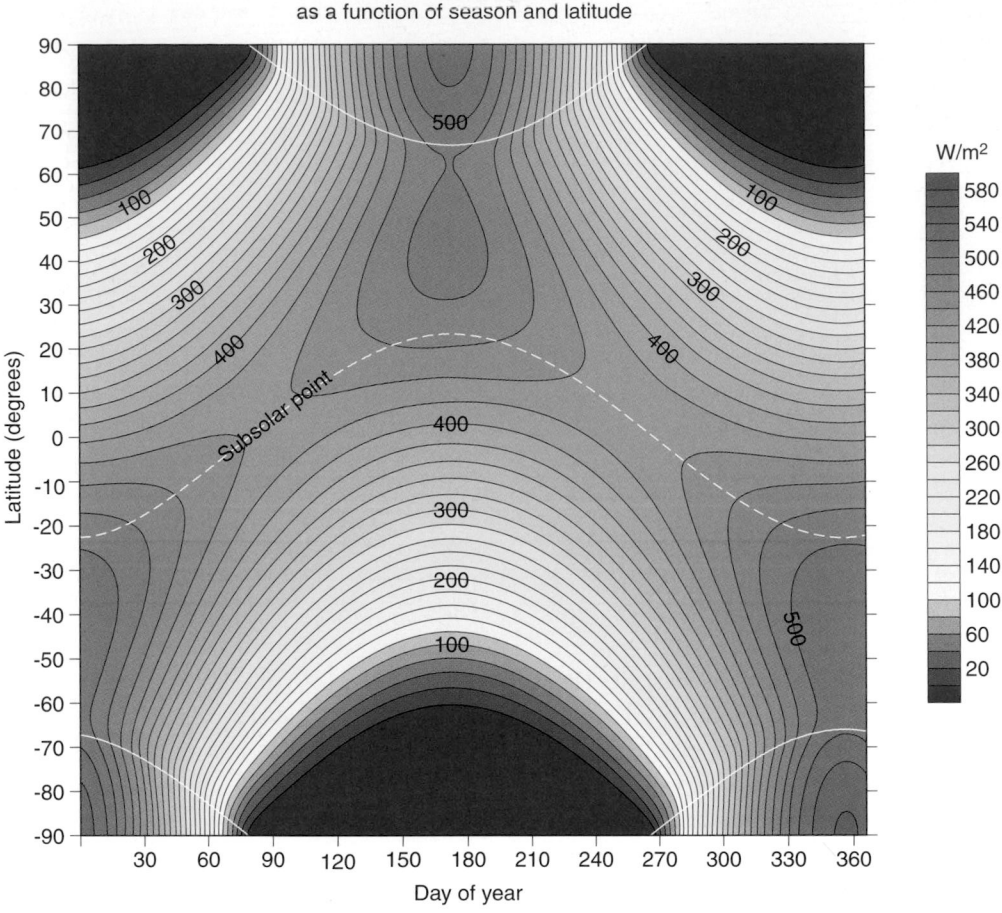

Extra-atmospheric irradiance in W/m² as a function of season and latitude

The perihelion/aphelion contrast in solar energy input coupled with the seasonal variation in radiation has implications for global climate. The Southern Hemisphere receives more radiation during summer and less radiation during winter than does the Northern Hemisphere. Figure 3.23 summarizes the distribution of solar radiation received at the top of the atmosphere by latitude and day of the year in watts per square m. Consequently, all other factors being equal, we might expect a greater winter-to-summer temperature contrast in the Southern Hemisphere. However, the ocean covers more surface area in the Southern Hemisphere than in the Northern Hemisphere. The Southern Hemisphere is 19.1% land and 80.9% ocean whereas the Northern Hemisphere is 39.3% land and 60.7% ocean. In the Southern Hemisphere the larger percentage of ocean surface area coupled with the relatively great thermal inertia of ocean water moderates seasonal temperature differences and largely offsets the greater seasonal contrast in incoming solar radiation (Chapter 4).

Solar Radiation and the Atmosphere

Solar radiation interacts with gases and aerosols as it travels through the atmosphere. These interactions consist of scattering, reflection, and absorption. Solar radiation that is not scattered or reflected back to space, or absorbed by gases and aerosols, reaches Earth's surface, where further interactions take place. Within the atmosphere, the percentage of solar radiation that is absorbed (*absorptivity*) plus the percentage scattered or reflected (*albedo*) plus the percentage transmitted to Earth's surface (*transmissivity*) must equal 100%. This relationship is another example of the *law of energy conservation*.

In the process of **scattering**, a particle disperses solar radiation in different directions—forward, backward, and sideways. Within the atmosphere, both gas molecules and aerosols (including the tiny water droplets and ice crystals composing clouds) scatter solar radiation, with important differences. Scattering

FIGURE 3.24
Clouds scatter solar radiation equally at all wavelengths, with the strongest scattering in the forward and backward directions.

by molecules and other particles that are less than one-tenth the wavelength of light is wavelength dependent with the preferential scattering of blue-violet light by nitrogen and oxygen molecules being the principal reason for the color of the clear daytime sky. This type of scattering, known as Rayleigh Scattering, is further described in Chapter 14. On the other hand, the tiny water droplets and ice crystals that compose clouds (but are much larger than molecules) scatter visible solar radiation equally at all wavelengths so that clouds appear white (Figure 3.24). This type of scattering, called Mie Scattering, occurs when particles are about the same size as the wavelength of light (Chapter 14). Scattering of a solar ray by cloud droplets and ice crystals is strongest in the forward and backward directions. Airborne particles (aerosols) suspended in the atmosphere, having diameters similar to the wavelength of light, also scatter the sunlight giving a milky appearance to the sky.

Reflection, a special case of scattering, takes place at the interface between two different media, such as air and cloud, when some of the radiation striking that interface is redirected (backscattered). The fraction of incident radiation that is backscattered by airborne particles or reflected by a surface (or interface) is the **albedo** of that surface, that is,

albedo = (reflected radiation)/(incident radiation)

where albedo is expressed either as a percentage or a fraction. Surfaces having a high albedo reflect a relatively large fraction of incident solar radiation and appear light in color. Surfaces having a low albedo reflect a relatively small fraction of incident solar radiation and appear dark in color. Because low-albedo surfaces absorb more sunlight, they typically are warmer than those that are lighter (having a higher albedo).

Within the atmosphere, much of the incident sunlight is backscattered, making cloud tops appear as the most important reflectors of solar radiation (Figure 3.25). Cloud top albedo depends primarily on cloud thickness and varies from under 40% for thin clouds (less than 50 m, or 165 ft, thick) to 80% or more for thick clouds (more than 5000 m, or 16,500 ft, thick). For this reason, during daytime, a high thin veil of cirrus clouds appears much brighter than the underside of a much thicker thunderstorm (cumulonimbus) cloud. The average albedo for all cloud types and thickness is about 55%, and, at any point in time, clouds cover about 60% of the planet. All other factors being constant, solar radiation reaching Earth's surface is more intense and daytime surface temperatures are higher when the sky is clear rather than cloudy.

Scattering and reflection within the atmosphere alter only the direction of incoming solar radiation. **Absorption**, however, is an energy conversion process whereby some of the radiation striking the surface of an object is converted to heat energy. Oxygen, ozone, water vapor, and various aerosols (including cloud particles) absorb a portion of the incoming solar radiation. Absorption by atmospheric gases varies by wavelength; that is, a specific gas may absorb strongly in some wavelengths, but weakly or not at all in other wavelengths. As a consequence of absorption, radiation

FIGURE 3.25
Cloud tops strongly backscatter visible solar radiation and appear very bright, as in this view from an airplane above Michigan. [Photo by Kristina Rebelo]

in that wavelength band is removed before the Sun's rays reach Earth's surface.

Within the stratosphere, oxygen (O_2) and ozone (O_3) strongly absorb solar ultraviolet radiation at wavelengths shorter than 0.3 µm. Oxygen absorbs UV at very short wavelengths (less than 0.2 µm), and ozone absorbs longer UV (0.22 to 0.29 µm). The net effect of this absorption is twofold: (1) a significant reduction in the intensity of UV that reaches Earth's surface and (2) a marked warming of the upper stratosphere (Figure 2.17). The clear atmosphere is essentially transparent to solar radiation in the wavelength range between about 0.3 and 0.8 µm (mostly visible radiation). Water vapor absorbs solar infrared radiation in certain wavelength bands greater than 0.8 µm. Clouds are relatively poor absorbers of solar radiation, typically absorbing less than 10% of the solar radiation that strikes the cloud top, although exceptionally thick clouds such as thunderclouds absorb somewhat more.

Stratospheric Ozone Shield

Since the late 1970s, scientists have measured a steady decline in the total volume of ozone (O_3) in the stratosphere, averaging a few percent per decade. Much greater depletion of stratospheric ozone takes place each spring over Antarctica. The increase in intensity of incoming solar ultraviolet radiation, as there is less stratospheric ozone to absorb it, has serious implications for human health and well-being.

Ozone is a relatively unstable molecule made up of three atoms of oxygen, distinct from oxygen molecules (O_2). The peak concentration of ozone, about 10 parts per million by volume (ppmv), occurs in the middle stratosphere (Figure 3.26). Total atmospheric ozone is measured in *Dobson units (DU)*, where 1 DU equals 2.69×10^{16} ozone molecules. If all the ozone were removed from an average column of air, from Earth's surface to space, and brought to standard temperature (0°C) and pressure (1013.25 mb), it would form a layer only about 3 mm deep. On average, the amount of columnar ozone increases with latitude from about 250 DU at the equator to more than 300 DU at middle latitudes, and more than 400 DU near the poles.

Depending on where it is located within the atmosphere, ozone has either a negative or positive impact on life. Ozone in the lower troposphere is a serious air pollutant and one of the chief constituents of photochemical smog (Chapter 2). In the stratosphere, the formation and photodissociation of ozone shields organisms at Earth's surface from exposure to lethal intensities of solar ultraviolet (UV) radiation. Despite that it is such a minor component of the atmosphere, without the **stratospheric ozone shield**, life as we know it could not exist.

Within the stratosphere, two sets of competing chemical reactions, both powered by solar ultraviolet radiation, continually generate and destroy ozone (Figure 3.27). During ozone production, UV strikes an oxygen molecule (O_2) causing it to split into two free oxygen atoms (O), which collide with molecules of oxygen (O_2) to form ozone molecules (O_3). Simultaneously, ozone is destroyed. When ozone absorbs ultraviolet radiation, the molecule splits into one free oxygen atom (O) and one molecule of oxygen (O_2). The free oxygen atom collides with an ozone molecule to form two molecules of oxygen. These opposing chemical reactions achieve equilibrium such that 1 gigaton of ozone is produced each year and 1 gigaton is destroyed. (One gigaton equals 1 billion tons.)

When ultraviolet radiation (at different wavelengths) passes through the atmosphere, it is absorbed and prevented from reaching Earth's surface; only a small amount of UV radiation passes through the stratospheric ozone shield. Therefore, it takes prolonged exposure to sunlight, which contains small quantities of UV, to cause serious health problems. A thinner ozone shield would mean more intense UV radiation reaching Earth's surface and, for humans, a greater risk of skin cancer, cataracts of the eye, and immune deficiencies. As a general rule, every 1% decline in stratospheric ozone concentration creates a 2% increase in the intensity of UV that passes through the ozone shield. Various studies suggest that a 2.5% thinning

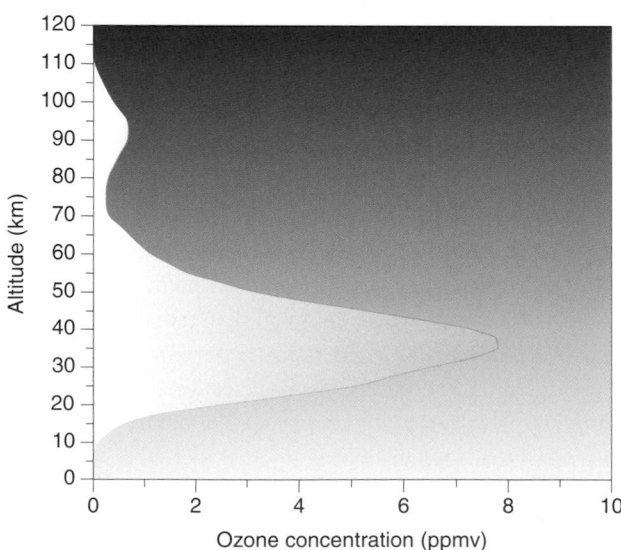

FIGURE 3.26
The concentration of ozone (O_3) peaks in the stratosphere. [Source: U.S. Standard Atmosphere, 1976]

OZONE PRODUCTION

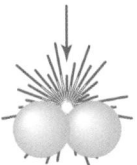

High energy ultraviolet radiation strikes an oxygen molecule...

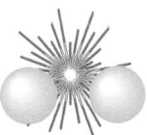

...and causes it to split into two free oxygen atoms.

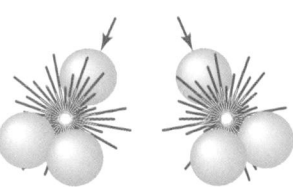

The free oxygen atoms collide with molecules of oxygen...

To form ozone molecules.

OZONE DESTRUCTION

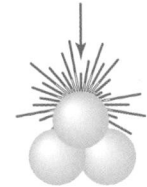

Ozone absorbs a range of ultraviolet radiation...

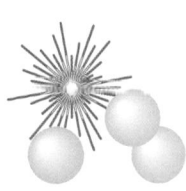

...splitting the molecule into one free oxygen atom and one molecule of ordinary oxygen.

The free oxygen atom then can collide with an ozone molecule...

To form two molecules of oxygen.

FIGURE 3.27
Within the stratosphere, two sets of competing chemical reactions continually produce and destroy ozone (O_3). [Adapted from "Ozone: What is it and why do we care about it?" *NASA Facts*, NASA Goddard Space Flight Center, Greenbelt, MD, 1993.]

of the ozone shield, with a 5% increase in UV, could boost the rate of human skin cancer by 10%.

The most dangerous portion of UV radiation that reaches Earth's surface is *UVB*, which spans the wavelength band from 0.29 to 0.32 μm. The amount of UVB that reaches Earth's surface depends on the cloudiness and dustiness of the atmosphere. For more on the human health hazards of overexposure to solar UV radiation and protection, refer to this chapter's first *For Further Exploration*.

As illustrated in Figure 3.28, chemicals that threaten to destroy stratospheric ozone have natural and industrial origins. Since the 1970s, a group of chemicals known as *chlorofluorocarbons (CFCs)* have received considerable attention for their negative impact on the ozone shield. First synthesized in 1928 by the American chemist Thomas Midgley, Jr., (1889-1944), CFCs were widely used as chilling (heat-transfer) agents in refrigerators and air conditioners as well as aerosol spray propellants (CFC-11 and CFC-12). Others were used as a solvent and cleaner of electronic circuit boards (CFC-13). CFC-11 is responsible for about half of observed ozone depletion. Another threat to the ozone shield comes from halons, compounds chemically related to CFCs, which are used in

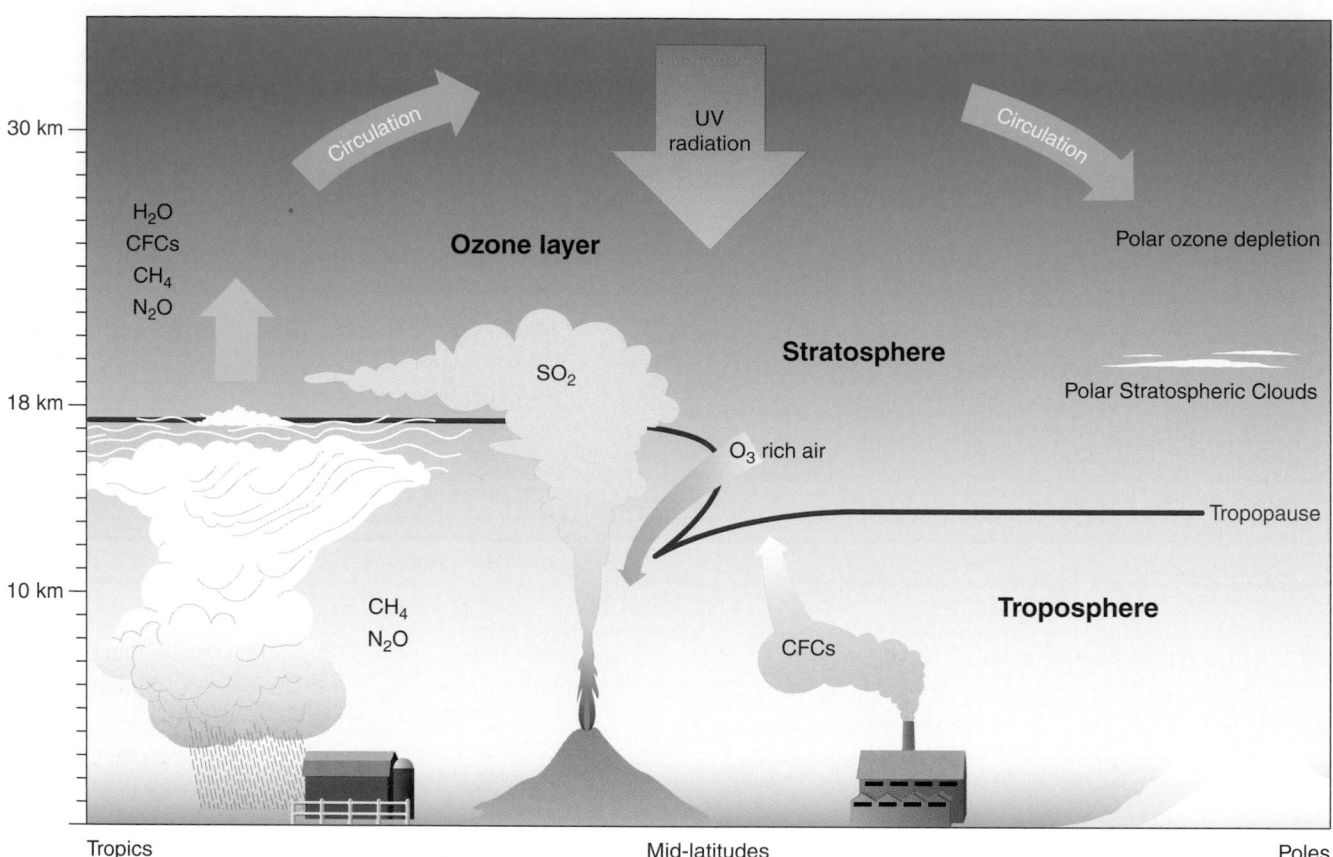

FIGURE 3.28
Chemicals that threaten to destroy stratospheric ozone have both industrial and natural sources. Some are injected into the stratosphere during violent volcanic eruptions and in the exhaust of high flying aircraft. Additionally, ozone-destroying chemicals are transported from sources on Earth's surface by strong convection currents in the tropics that build thunderstorm clouds (cumulonimbus) that surge upward through the tropopause and into the stratosphere. These chemicals include chlorofluorocarbons (CFCs), methane (CH_4), nitrous oxide (N_2O), and water vapor. Within the stratosphere, intense ultraviolet (UV) radiation breaks down these compounds with byproducts to form *radicals*, including nitrogen dioxide (NO_2) and chlorine monoxide (ClO), that readily react with and destroy even more stratospheric ozone. Furthermore, cloud particles such as those found in polar stratospheric clouds (PSC) indirectly contribute to stratospheric ozone loss. [NASA Earth Observatory, adapted from Barbara Summey, SSAI]

fire extinguishers. Halons release bromine (Br) that reacts with and destroys ozone. In fact, halons are ten times more effective than CFCs in thinning the ozone shield.

Certain CFCs are inert (chemically non-reactive) in the troposphere, where they accumulated for decades in concentrations measured in parts per billion by volume (ppbv). Atmospheric circulation transports CFCs into the stratosphere where, at altitudes above about 25 km (15 mi), intense UV radiation breaks down CFCs, releasing chlorine (Cl), which readily reacts with, and destroys, ozone. Products of this reaction are chlorine monoxide (ClO) and molecular oxygen (O_2). Chlorine (Cl) is a catalyst in chemical reactions, readily giving up the oxygen it bonded with to break apart another ozone molecule. In this way, a single chlorine atom destroys tens of thousands of ozone molecules before cycling out of the atmosphere.

Frank Sherwood Rowland (1927-2012) and Mario J. Molina of the University of California at Irvine first warned of the threat of CFCs to the stratospheric ozone shield in 1974. Five years later, the use of CFCs as propellants in common household aerosol sprays such as deodorants, hairsprays, and furniture polish was banned in the United States, Canada, Norway, and Sweden. For their pioneering research on the depletion of stratospheric ozone, Rowland, Molina (now at the Massachusetts Institute of Technology), and P.J. Crutzen (of the Max Planck Institute for Chemistry in Mainz, Germany) received the 1995 Nobel Prize in Chemistry.

As noted in Chapter 2, the first sign of thinning of the stratospheric ozone shield came from Antarctica. The British Antarctic Survey conducted ground-based measurements of atmospheric ozone using an instrument

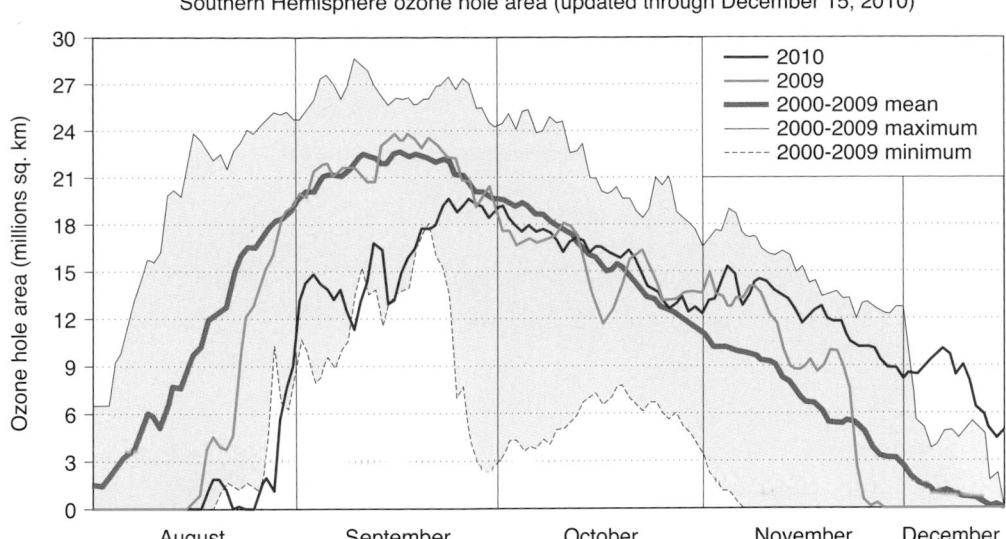

Southern Hemisphere ozone hole area (updated through December 15, 2010)

FIGURE 3.29
The Antarctic ozone hole is a widespread area of strato-spheric ozone depletion that develops over Antarctica during the Southern Hemi-sphere spring. The mean area encompassed by the Antarctic ozone hole dur-ing 2000-2009 increased from near zero at the begin-ning of August, peaked in September, and declined to zero in December. [NOAA and NASA Goddard Space Flight Center]

known as a *spectrophotometer*. Although first detected in 1981, discovery of the ozone hole over Antarctic was not publicly announced until 1985. For about six weeks during the Southern Hemisphere spring (September and October), the ozone layer in the Antarctic stratosphere at altitudes from 11 to 23 km (7 to 14 mi) thinned drastically then recovered in November (Figure 3.29). The **Antarctic ozone hole** is defined as the thinning of the ozone layer significantly below pre-1979 levels, that is, the Antarctic ozone hole is the area where the total columnar ozone is less than 220 DU. (Prior to formation of the ozone hole, ozone levels in Antarctica were normally lowest in the Southern Hemisphere in autumn.) Research conducted during the National Ozone Expeditions to the U.S. McMurdo Station in 1986-87, and NASA aircraft flights into the Antarctic stratosphere in 1987, measured relatively high concentrations of chlorine monoxide (ClO). This finding established a convincing link between the Antarctic ozone hole and CFCs.

The main space-borne observations of stratospheric ozone levels are from the *Total Ozone Mapping Spectrometer (TOMS)* flown aboard NASA's NIMBUS-7 satellite (1978-93), the Soviet's Meteor-3 (1991-94), and NASA's Earth Probe (1996-present) and Aura (2004-present). In addition, balloon-borne instruments and rocketsondes monitor the vertical structure of the ozone hole. Measurements indicate that the Antarctic ozone hole has steadily deepened (greater depletion of ozone) since the late 1970s. From 21-26 September 2006, the average size of the ozone hole was 27.5 million km² (10.6 million mi²), an area larger than North America and the largest on record. On 24 September 2006, the hole covered 29.5 million km²

(11.4 million mi²) (Figure 3.30), matching the record single-day extent originally set on 9 September 2000. When the ozone hole is at its maximum, southern portions of Argentina and Chile occasionally are exposed to elevated levels of UV.

How does the Antarctic ozone hole form and why does it fill in by November? During the long, dark Antarctic winter, extreme radiational cooling causes temperatures in the stratosphere to plunge below −88 °C (−126 °F). At these frigid temperatures, **polar stratospheric clouds (PSCs)**, composed of tiny particles of water ice, nitric acid, and sulfuric acid, form. They provide the small surfaces for chlorine and bromine compounds, which are inert toward ozone, to convert to active forms. Once the Sun reappears in the spring, the active compounds begin destroying ozone.

Ozone depletion takes place in the winter and spring when the Antarctic is cut off from the planetary-scale atmospheric circulation by the **circumpolar vortex**, a belt of strong winds that encircles the outer margin of the Antarctic continent. A month or so into spring, however, the circumpolar vortex begins to weaken, allowing warmer ozone-rich air from lower latitudes to invade the Antarctic stratosphere. As the air mixes, polar stratospheric clouds vaporize and the stratospheric ozone concentration returns to normal levels; thus the Antarctic ozone hole fills. Year to year fluctuations in the magnitude of ozone depletion and the area of the ozone hole are likely due to inter-annual variations in atmospheric circulation at high latitudes.

Might it be possible for an ozone hole to form over the Arctic, a region not far from human population centers? Scientists investigating stratospheric chemistry in the Arctic in early 1989 discovered ozone-destroying chlorine compounds and a slight thinning of ozone. Although some ozone depletion takes place in the Arctic during the Northern Hemisphere spring (March to May), an ozone hole comparable to the one over Antarctica was considered unlikely for two reasons. For one, in winter the Arctic stratosphere averages about 10 Celsius degrees (18 Fahrenheit degrees) warmer than the Antarctic stratosphere, making formation of polar stratospheric clouds unlikely. Secondly, the circumpolar vortex that surrounds the Arctic weakens earlier than its Antarctic counterpart. However, an exceptionally cold Arctic winter coupled with an unusually persistent circumpolar vortex could translate into significant ozone depletion in the Arctic.

For one week near the end of March 2011, the thickness of the ozone column over the Arctic fell to minimum values of 220 to 230 DU. Although much higher than minimum values routinely observed during the Antarctic spring (100 DU), the Arctic minimum was among the lowest ever measured in that region. For almost a full month, ozone column values were less than 250 DU. Key to the ozone decline in the Arctic was an unusual episode of exceptionally low stratospheric temperatures that persisted from December 2010 through March 2011. Temperatures were sufficiently low that polar stratospheric clouds formed. Because the stratospheric ozone loss was not as severe in the Arctic compared to the Antarctic, some scientists do not consider what happened in the Arctic in March 2011 to be an ozone hole. Other scientists disagree, pointing out the similarity in stratospheric chemistry between the Antarctic and Arctic (e.g., record high levels of ClO in the Arctic stratosphere during the spring of 2011).

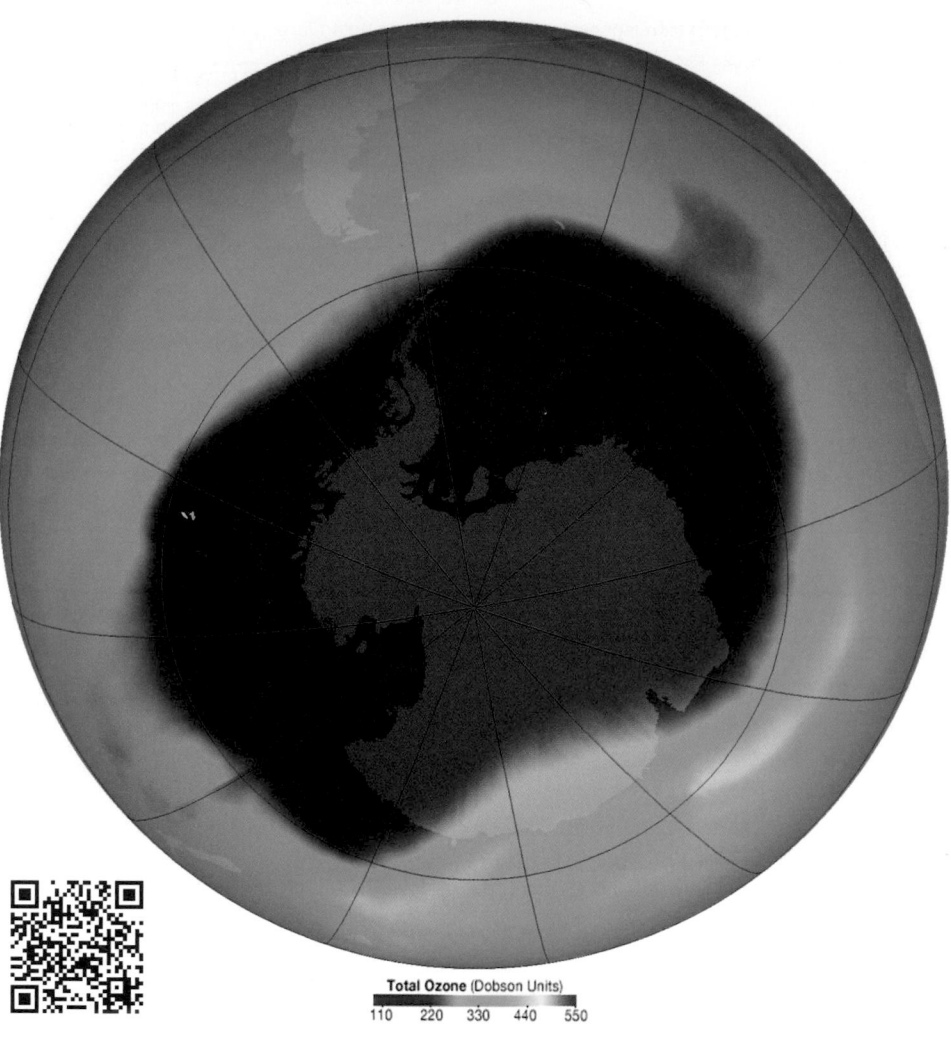

Total Ozone (Dobson Units)
110 220 330 440 550

FIGURE 3.30
From 21st to 30th September 2006, the average area of the Antarctic ozone hole was the largest ever observed at 27.5 million square km (10.6 million square mi). This image, created from measurements generated by the Ozone Monitoring Instrument on NASA's Aura satellite, shows the ozone hole on 24 September 2006. The ozone hole on this day tied the single-day largest area of 29.5 million square km (11.4 million square mi), which was previously reached on 9 September 2000. Blue and purple colors indicate lower amounts of ozone, and greens and yellows display higher levels. [NASA]

World-wide acceptance of the threat of CFCs, and other *ozone-depleting substances (ODSs)*, to the stratospheric ozone shield prompted the United Nations Environmental Programme (UNEP) in 1987 to draft the *Montreal Protocol on Substances That Deplete the Ozone Layer*. This international treaty entered into full force on 1 January 1989 and by 16 September 2009, all 192 UN member states had ratified the original Montreal Protocol. The initial goal of the treaty was to cut CFC production in half by 1992 (compared to 1986 levels). However, the seriousness of the problem led to seven subsequent amendments including expanding the list of regulated substances (e.g., to include bromine-containing halons) and requiring the worldwide phase out of the manufacture and use of CFCs beginning in January 1996.

CFCs and halons have long atmospheric lifetimes of, at least, a century. Hence, these substances will continue to threaten the stratospheric ozone shield. The 2006 UNEP *Scientific Assessment of Ozone Depletion Concentration*s reported that ozone-depleting chlorine levels in the stratosphere were on the decline but recovery of stratospheric ozone to pre-1980 levels is not likely until the middle of the 21st century in middle latitudes and a decade or two later in polar regions.

Solar Radiation and Earth's Surface

Incoming solar radiation has both direct and diffuse components. Solar radiation that passes directly through the atmosphere to Earth's surface is the direct component that provides bright sunshine and casts shadows. Solar radiation that is scattered and/or reflected to Earth's surface is the diffuse component, such as found during daytime under overcast skies. Direct plus diffuse solar radiation that strikes Earth's surface is either reflected or absorbed depending on the surface albedo. The fraction that is not reflected at the surface is absorbed (that is, converted to heat).

As noted earlier, light surfaces reflect more solar radiation than dark surfaces. Skiers and snow boarders who have been sunburned or snow-blinded on the slopes on a sunny day are well aware of the high reflectivity of a snow cover. The albedo of fresh-fallen snow typically ranges between 75% and 95%; that is, 75% to 95% of the solar radiation striking a fresh snow cover is reflected, and the rest (5% to 25%) is absorbed (converted to heat). A fresh snow cover is highly reflective because the snow surface consists of a multitude of randomly oriented crystals, each having many reflecting surfaces. As snow ages, the surface albedo declines because snow crystals convert to spherical ice particles having fewer reflecting surfaces. The albedo of old snow typically ranges between 40% and 60%. At the other extreme, the albedo of a dark surface, such as a black topped road or a spruce forest, may be as low as 5%. Differences in albedo explain why light-colored clothing is usually a more comfortable choice than dark-colored clothing on a sunny, hot day. Albedo values of some common surfaces types are listed in Table 3.1.

In the visible satellite image in Figure 3.31, the surface of the Great Lakes appears dark because of

TABLE 3.1
Average Albedo (Reflectivity) of Some Common Surface Types for Visible Solar Radiation

Surface	Albedo (% reflected)
Deciduous forest	15-18
Coniferous forest	9-15
Tropical rainforest	7-15
Tundra	15-35
Grasslands	18-25
Desert	25-30
Sand	30-35
Soil	5-30
Green crops	15-25
Sea ice	30-40
Fresh snow	75-95
Old snow	40-60
Glacial ice	20-40
Water body (high solar altitude)	3-10
Water body (low solar altitude)	10-100
Asphalt road	5-10
Urban area	14-18
Cumulonimbus cloud	90
Stratocumulus cloud	60
Cirrus cloud	40-50

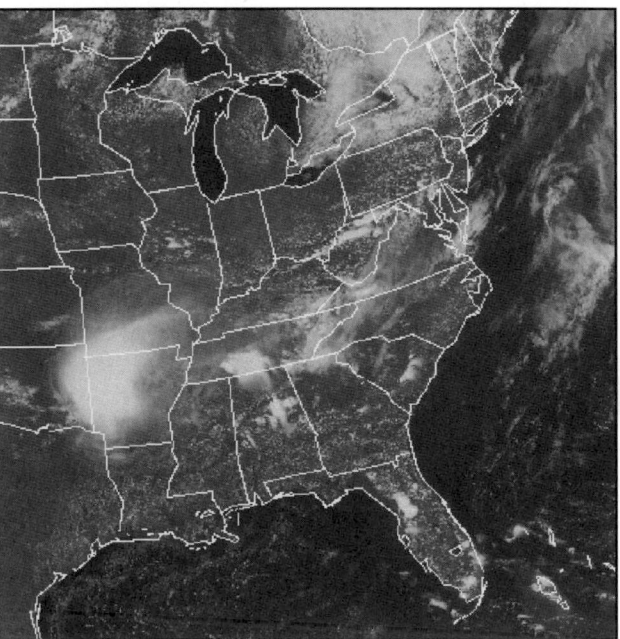

FIGURE 3.31
In this visible satellite image, water surfaces appear black because of their low albedo. Clusters of high-albedo clouds (appearing white) are visible across the view. [NOAA/NCEP]

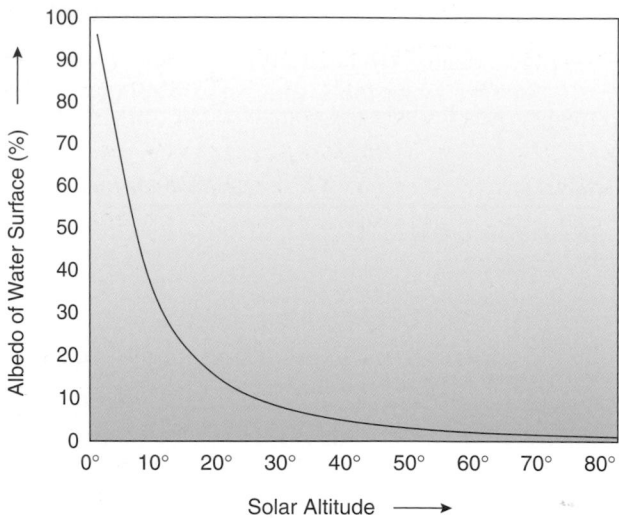

FIGURE 3.32
The albedo of a flat and undisturbed water surface under clear skies decreases with increasing solar altitude. A wave-covered water surface has a slightly higher albedo at high solar altitudes and a slightly lower albedo at low solar altitudes.

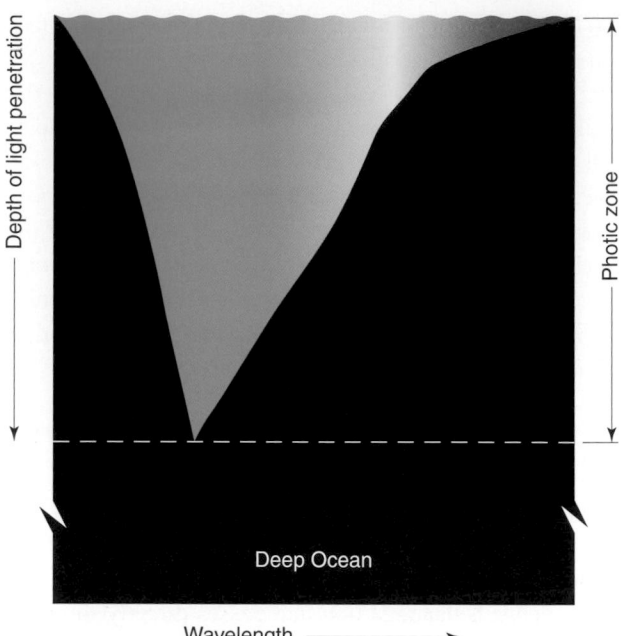

FIGURE 3.33
Visible solar radiation is selectively absorbed by wavelength as it penetrates the ocean's surface waters. Green and blue light penetrate to greater depths. The scale break lines in the deep ocean indicate that this portion of the ocean is on average considerably deeper than the photic zone.

its low albedo and strong absorption of solar radiation. The albedo of the surface of a body of water, such as the ocean or lake, varies with the angle of the Sun above the horizon (*solar altitude*). Under clear skies, the albedo of a flat, tranquil water surface decreases with increasing solar altitude (Figure 3.32). The albedo approaches a mirror-like 100% near sunrise and sunset (when the solar altitude is near 0 degrees) but declines sharply as the solar altitude approaches 20 degrees. With overcast skies, only diffuse solar radiation strikes the water surface and the albedo varies little with solar altitude and is uniformly less than 10%. On a global basis, the albedo of the ocean surface averages only 8%; that is, the ocean absorbs on average 92% of incident solar radiation.

Whereas the clear atmosphere is relatively transparent to solar radiation, the ocean absorbs most solar radiation within relatively shallow depths. As shown in Figure 3.33, the ocean's absorption of the visible portion of solar radiation is selective by wavelength. Water absorbs the longer wavelength light (i.e., reds and yellows) more efficiently than the shorter wavelength light (i.e., greens and blues) so that green and blue light penetrate to greater depths. Within clear, clean water, red light is completely absorbed within about 15 m (50 ft) of the surface, whereas green and blue-violet light may penetrate to depths approaching 250 m (800 ft), thereby determining the lower boundary of the photic zone. More green and blue light is scattered to our eyes, explaining the blue/green color of the open ocean. Particles suspended in

the water significantly boost absorption so that sunlight often is completely absorbed at shallower depths. In fact, some near-shore coastal waters are so turbid (murky) that little if any sunlight reaches much below 10 m (35 ft). Suspended particles preferentially scatter yellow and green light giving these waters their characteristic color.

Significant changes in surface albedo occur seasonally and affect the fraction of incident solar radiation that is converted to heat. In autumn, loss of leaves from deciduous trees raises the surface albedo of forested areas. At middle and high latitudes, significant increases in surface albedo accompany the winter freeze-over of lakes, formation of sea ice, and development of a persistent snow cover.

Global Solar Radiation Budget

Measurements by sensors onboard Earth-orbiting satellites indicate that the Earth-atmosphere system scatters or reflects back to space on average about 30% of the solar radiation intercepted by the planet. This is Earth's **planetary albedo**. The atmosphere (i.e., a mixture of gases, aerosols, and clouds) absorbs only about 23% of the total solar radiation intercepted by the Earth-atmosphere system. In other words, the atmosphere is relatively

TABLE 3.2
Earth's Solar Radiation Budget

Reflected by the Earth-atmosphere system	30%
Absorbed by the atmosphere	23%
Absorbed by Earth's surface	47%
Total	100%

transparent to solar radiation. The remaining 47% of solar radiation is absorbed by Earth's surface, chiefly because of the low average albedo of ocean water covering about 71% of the surface of the globe. The global annual solar radiation budget is summarized in Table 3.2.

Earth's surface is the principal recipient of solar heating, and heat is transferred from Earth's surface to the atmosphere, which eventually radiates this energy to space. Earth's surface is thus the main source of heat for the atmosphere; that is, the atmosphere is heated from below. This is evident in the vertical temperature profile of the troposphere. Normally, air is warmest close to Earth's surface, and air temperature drops with increasing altitude within the troposphere, that is, away from the main source of heat (Figure 2.17). A secondary heat source is found at the top of the stratosphere, where absorption of ultraviolet radiation by oxygen and ozone molecules at altitudes below 50 km (31 mi) results in the relatively warm stratopause.

Outgoing Infrared Radiation

If solar radiation were continually absorbed by the Earth-atmosphere system without any compensating flow of heat out of the system to space, Earth's surface temperature would rise steadily. Eventually, life would be extinguished and the ocean would boil away. Global radiative equilibrium keeps the planet's temperature in check; that is, the outgoing emission of heat to space in the form of infrared radiation balances the incoming solar radiational heating of the Earth-atmosphere system. However, long-term global climate change can modify the Earth's IR emission, and shift the Earth-atmosphere system to a new equilibrium. We will explore this balance later in the chapter. Although solar radiation is supplied only to the illuminated half of the planet at any instant, infrared radiation is emitted to space ceaselessly, day and night, from all over the planet. This explains why nights

are usually colder than days and why air temperatures typically drop throughout the night.

While the clear atmosphere is relatively transparent to solar radiation, certain gases in the atmosphere impede the escape of infrared radiation to space, thereby elevating the temperature of the lower atmosphere. This important climate control is the *greenhouse effect*.

THE GREENHOUSE EFFECT

The **greenhouse effect** refers to the heating of Earth's surface and lower atmosphere caused by strong absorption and emission of infrared radiation by certain atmospheric gases, known as **greenhouse gases**. Solar radiation and terrestrial infrared radiation peak in different portions of the electromagnetic spectrum, their properties differ, and they interact differently with the atmosphere. As noted earlier, the atmosphere directly absorbs only about 23% of the solar radiation intercepted by the planet. The atmosphere absorbs a greater percentage of the infrared radiation emitted by Earth's surface, and the atmosphere, in turn, radiates some of the absorbed radiation as IR to space and some back to Earth's surface. Hence, Earth's surface is heated by absorption of both solar radiation and atmosphere-emitted infrared radiation.

The similarity in radiational properties between infrared-absorbing atmospheric gases and the glass or plastic glazing of a greenhouse is the origin of the term greenhouse effect. Greenhouse glazing, like the atmosphere, is relatively transparent to visible solar radiation but strongly absorbs infrared radiation. A greenhouse, where plants are grown, takes advantage of the radiational properties of glazing (Figure 3.34). Sunlight readily

FIGURE 3.34
The glazing of a greenhouse behaves similarly to certain gases (e.g., water vapor, carbon dioxide) in the atmosphere that are transparent to sunlight but strongly absorb and emit infrared radiation.

penetrates greenhouse glazing and much of it is absorbed (converted to heat) within the greenhouse. Objects in the greenhouse emit infrared radiation that is strongly absorbed by glazing. The glazing, in turn, emits IR out to both the atmosphere and back to the greenhouse interior, thereby raising the temperature within the greenhouse.

Although the greenhouse analogy is widely cited, absorption and emission of IR radiation by glazing is only part of the reason why the interior of a greenhouse is relatively warm. A greenhouse is also a shelter from the wind and reduces heat loss to the external environment by conduction and convection (Chapter 4). As a rule, the thinner the greenhouse glazing and the higher the external wind speed, the more important is the shelter effect compared to the radiational effect. For this reason, some atmospheric scientists argue that the greenhouse analogy is inappropriate and the phenomenon should be renamed the *atmospheric effect*. Nonetheless, *greenhouse effect* is such a commonly used term (especially by the media) that we continue use of the term in this book.

The greenhouse effect is responsible for considerable warming of Earth's surface and lower atmosphere. Viewed from space, the planet radiates at about −18 °C (0 °F), whereas the average temperature at the Earth's surface is about 15 °C (59 °F). The temperature difference is due to the greenhouse effect and amounts to

$$[15\ °C − (−18\ °C)] = 33\ \textbf{Celsius degrees}$$
or
$$[59\ °F − 0\ °F] = 59\ \textbf{Fahrenheit degrees.}$$

Without the greenhouse effect, Earth would be too cold to support most forms of plant and animal life.

GREENHOUSE GASES

Water vapor is the principal greenhouse gas, having a clear-sky contribution of 60% to the greenhouse effect. Other gases contributing to the greenhouse effect are carbon dioxide (26%), ozone (8%), methane, nitrous oxide (N_2O), and halocarbons. As shown in Figure 3.35, the percentage of infrared radiation absorbed by each gas varies with wavelength. An **atmospheric window** is a range of wavelengths over which little or no radiation is absorbed. A *visible atmospheric window* extends from about 0.3 to 0.7 μm and the major *infrared atmospheric window* is from about 8 to 13 μm. Notably, this latter window includes the wavelength of the planet's peak infrared emission (about 10 μm). Through this window,

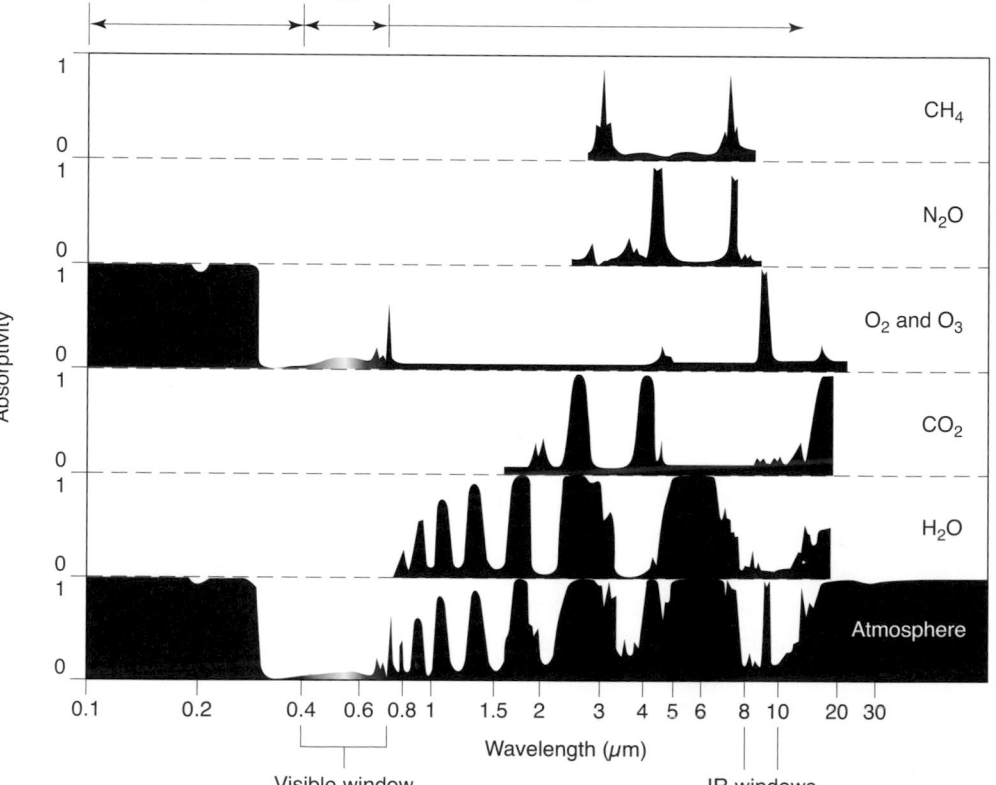

FIGURE 3.35
Absorption of radiation by selected gaseous components of the atmosphere and the atmosphere in total as a function of wavelength. *Absorptivity* is the fraction of radiation absorbed and ranges from 0 to 1 (0% to 100% absorption). Absorptivity is very low or near zero in *atmospheric windows*. Note the infrared windows near 8 and 10 micrometers.

most heat from the Earth-atmosphere system escapes to space as infrared radiation. IR sensors onboard Earth-orbiting satellites monitor this upwelling radiation, which is calibrated in terms of the surface temperature of the radiating object: the higher the temperature, the more intense is the emission of IR radiation (Chapter 1).

Warming caused by atmospheric water vapor is evident even at the scale of mesoclimates or local climates. Consider an example. Locations in the Desert Southwest and along the Gulf Coast are at about the same latitude and receive essentially the same input of solar radiation on clear days. In both regions, summer afternoon high temperatures commonly top 32 °C (90 °F). At night, however, air temperatures often differ markedly. Air is relatively dry (low humidity) in the Southwest so that infrared radiation readily escapes to space and air temperatures near Earth's surface may drop well under 15 °C (59 °F) by dawn. People who camp in the desert are well aware of the dramatic fluctuations in temperature between day and night. Infrared radiation does not escape to space as readily through the Gulf Coast atmosphere, where the air is more humid. Water vapor strongly absorbs outgoing IR and emits IR back towards Earth's surface so that early morning low temperatures may dip no lower than the 20s Celsius (70s Fahrenheit). The smaller diurnal (day to night) temperature contrast along the Gulf Coast is due to more water vapor and a stronger greenhouse effect.

Clouds are composed of IR-absorbing water droplets and/or ice crystals and also contribute to the greenhouse effect. All other factors being equal, nights usually are warmer when the sky is cloud-covered than when the sky is clear. Even high, thin cirrus clouds through which the Moon is visible can reduce the nighttime temperature drop at Earth's surface by several Celsius degrees. Clouds thus affect climate in two opposing ways. By absorbing and emitting IR, clouds warm Earth's surface and by reflecting solar radiation, clouds cool Earth's surface. On a global scale, which one of these two opposing effects is more important? Analysis of measurements by satellite of incoming and outgoing radiation shows that clouds have a net cooling effect on global climate. That is, all other factors being equal, a more extensive cloud cover would cool the planet whereas a less extensive cloud cover would warm the planet.

THE CALLENDAR EFFECT

The **Callendar effect** is the theory that global climate change can be brought about by enhancement of Earth's natural greenhouse effect through elevated levels of atmospheric CO_2 from anthropogenic sources, principally the burning of fossil fuels. The theory is named for engineer Guy Stewart Callendar (1898-1964) who investigated the link between global warming and fossil fuel combustion beginning in the late 1930s.

Systematic monitoring of atmospheric carbon dioxide began in 1957 at NOAA's Mauna Loa Observatory in Hawaii under the direction of Charles D. Keeling (1928-2005) of Scripps Institution of Oceanography. Situated on the northern slope of Earth's largest volcano, 3397 m (11,140 ft) above sea level in the middle of the Pacific Ocean, the observatory is sufficiently distant from major sources of air pollution so carbon dioxide levels are considered representative of, at least, the Northern Hemisphere. Since the same year, atmospheric CO_2 has been monitored at the South Pole station of the U.S. Antarctic Program, and that record closely parallels the one at Mauna Loa. The Mauna Loa record (the *Keeling curve*) shows a sustained increase in average annual atmospheric carbon dioxide concentration from about 316 ppmv (parts per million by volume) in 1959 to 389 ppmv near the end of 2011 (Figure 3.36). Superimposed on this upward trend is an annual carbon dioxide cycle caused by seasonal changes in photosynthesis from Northern Hemisphere vegetation, which shows how the level of carbon dioxide falls during the growing season to a minimum in October, recovers over winter, and reaches a maximum in May. Northern Hemisphere vegetation dominates because that hemisphere contains more than twice the land mass of the Southern Hemisphere.

The upward trend in atmospheric carbon dioxide was underway long before Keeling's monitoring. It may have begun thousands of years ago, when land was cleared for farmland and settlements via the burning of vegetation, as well as the decay of wood residue and reduced photosynthetic removal of carbon dioxide from the atmosphere. By the middle of the 19th century, the Industrial Revolution spurred a more rapid rise in atmospheric CO_2, as a byproduct of the burning of coal and other fossil fuels. Today, fossil fuel combustion accounts for roughly 75% of the increase in atmospheric carbon dioxide while deforestation (and other land clearing) is responsible for the balance. Based on chemical analysis of ancient air bubbles trapped in glacial ice cores, the concentration of atmospheric CO_2 is now 35% higher than in the pre-industrial era. With continued growth in fossil fuel combustion, atmospheric carbon dioxide concentration could top 550 ppmv (double its pre-industrial level) by the year 2100.

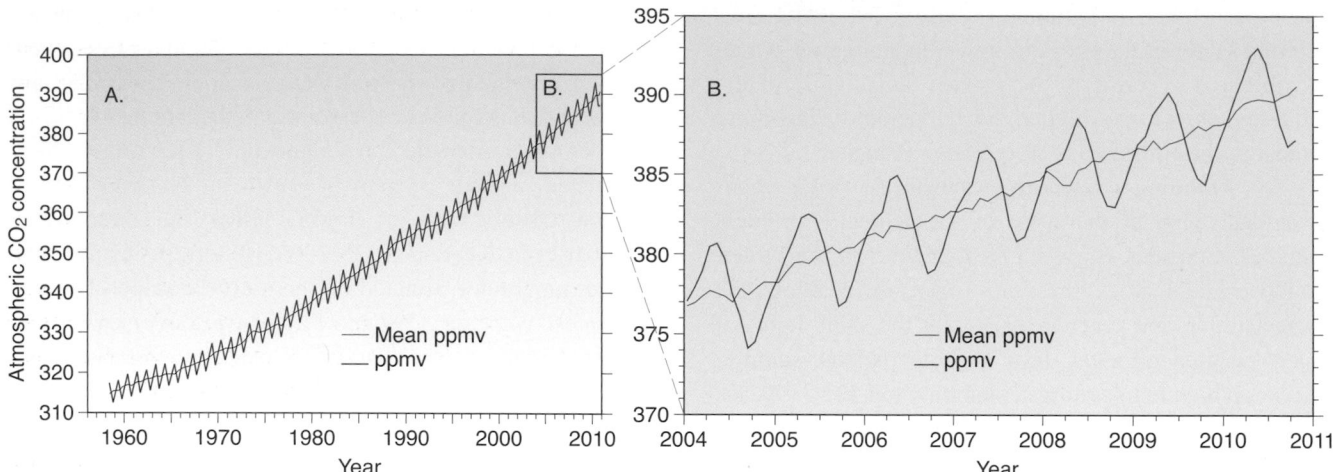

FIGURE 3.36

Concentration of atmospheric carbon dioxide (CO_2) as measured at the Mauna Loa Observatory, Hawaii (the Keeling curve). (A) The entire Mauna Loa record to date. (B) Recent monthly mean values centered on the middle of each month. Red curves include the influence of annual cycles in photosynthesis and cellular respiration. Black curves have been corrected for the average seasonal cycle. [Source: C.D. Keeling et al., Scripps Institution of Oceanography, University of California, La Jolla, CA; Dr. Pieter Tans, NOAA/ESRL]

Besides carbon dioxide, rising levels of other infrared-absorbing gases (e.g., methane, nitrous oxide, and chlorofluorocarbons) enhance the greenhouse effect. Glacial ice core records indicate that the concentration of methane (CH_4) in the atmosphere is now greater than at any time in the past 400,000 years. From 700 ppbv (parts per billion by volume) in 1750 to 1660 ppbv in 1985, the global average atmospheric concentration of methane increased by 237% since the industrial revolution. As shown in Figure 3.37, the growth rate slowed over the past three decades, generally trending downward from the early 1980s until 2005, and trending mostly upward since 2005. Methane, the principal component of natural gas, is a product of the decay of organic matter in the absence of oxygen (*anaerobic decay*). Anthropogenic sources (60% of the total) include rice cultivation, raising cattle, landfills, and termites. (As people build more homes and structures made of wood, the termite population explodes due to increased food sources, and more methane is emitted when the termites eventually decay.) Additional human-influenced sources include coal mining, gas leaks, wastewater treatment, petroleum systems, and biomass burning. The largest natural source of methane is decomposition of organic carbon in wetlands. Methane is removed from the system in the troposphere, where photochemical oxidation with hydroxyl radicals (OH^+) produces methyl radical CH_3 and H_2O. The atmospheric lifetime of methane is about 12 years.

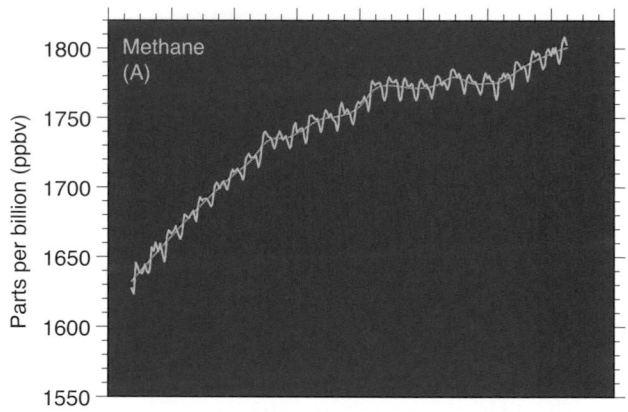

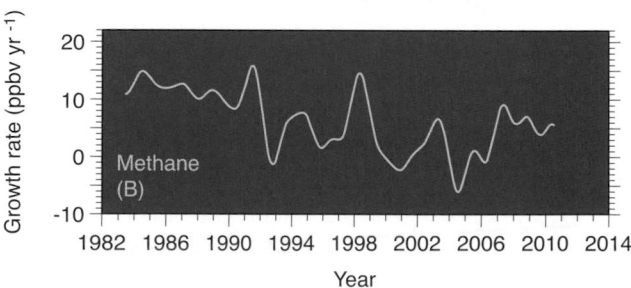

FIGURE 3.37

(A) The globally averaged atmospheric methane concentration increased from about 1620 ppbv in 1983 to a near steady state of about 1760 ppbv from 1999 to 2005, then rising to about 1818 ppbv by 2011. (B) The globally averaged growth rate of atmospheric methane (in ppbv per year) was more variable. The growth rate is the product of a delicate balance between global sources and sinks of methane. [Adapted from *Enigma of the Recent Methane Budget*, Martin Heimann, Nature, 11 August 2011, Vol. 476: 7359, pg. 157-158, Original data and graphics from E. Dlugokencky/NOAA-GMD]

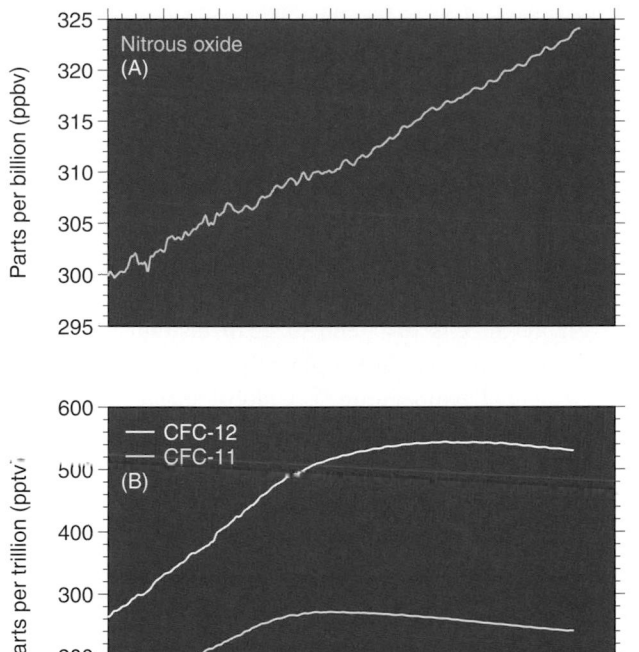

FIGURE 3.38
Average atmospheric concentration of the greenhouse gases (A) nitrous oxide in ppbv and (B) CFC-11 and CFC-12 in pptv beginning in 1978. [NOAA ESRL Global Monitoring Division]

The global average atmospheric concentration of nitrous oxide (N_2O) has increased from 270 ppbv in 1750 to recent levels of 324 ppbv (Figure 3.38A). The principal anthropogenic sources of N_2O in the U.S. include fertilizer application, fossil fuel combustion, and nitric acid production, while natural sources (bacterial breakdown of nitrogen in soils and in the ocean) account for more than 60% of N_2O emissions. *Photolysis* (breakdown by sunlight) in the stratosphere is the main N_2O removal mechanism. The atmospheric lifetime of nitrous oxide is about 110 years.

Spurred by international efforts to protect the stratospheric ozone shield (discussed earlier in this chapter), atmospheric concentrations of the chlorofluorocarbons CFC-11 and CFC-12 began leveling off in the early 1990s (Figure 3.38B).

Because they occur in extremely low concentrations, methane and nitrous oxide are very efficient absorbers of infrared radiation. This is because the efficiency of an IR-absorbing gas to absorb IR decreases as its concentration increases. To provide perspective on the relative importance of these gases, the **global warming potential (GWP) index** was developed. According to the U.S. Environmental Protection Agency (EPA), the "GWP for a particular greenhouse gas is the ratio of heat trapped by one unit mass of the greenhouse gas to that of one unit mass of CO_2 over a specified time period." Over a 100 year period, methane, for example, is 21 times more effective at trapping heat in the atmosphere when compared to CO_2, and nitrous oxide is about 310 times more effective.

Unless compensated for, the enhancement of Earth's greenhouse effect is likely to cause further global warming, which has spurred interest in reducing dependency on fossil fuels in favor of alternative energy sources such as solar power. The second *For Further Exploration* at the end of this chapter deals with solar power, while wind power is discussed in a Chapter 8 *For Further Exploration*.

As noted in Chapter 2, atmospheric scientists use numerical global climate models to predict the magnitude of global warming that could accompany the upswing in atmospheric carbon dioxide. Based on global climate models, in 2007 the **Intergovernmental Panel on Climate Change (IPCC)**, a standing international committee set up by the World Meteorological Organization and UNEP, concluded that global warming since the mid-20th century *very likely* (estimated probability of greater than 90%) was caused mostly by human activities. In a report issued six years earlier, the IPCC described the human role in global warming as *likely* (estimated probability of higher than 66%). The November 2007 *Synthesis Report of the IPCC Fourth Assessment Report* concluded that: "Warming of the climate system is unequivocal as is now evident from observations of increases in global average air and ocean temperatures, widespread melting of snow and ice, and rising global average sea level." The Report goes on to state that: "Most of the increase in globally-averaged temperatures since the mid-20th century is very likely due to the observed increase in anthropogenic greenhouse gas concentrations."

According to the 2007 IPCC Assessment Report, climate models predict that over the subsequent 20 years, the global mean annual temperature will increase at an average rate of 0.2 Celsius degree per decade. Depending on future greenhouse gas emissions, climate models project that the globally averaged surface temperature will rise by 1.8 to 4.0 Celsius degrees (3.2 to 7.2 Fahrenheit degrees) through the present century. As discussed in more detail in Chapter 15, climate change is geographically non-uniform (in both magnitude and direction) so that this projected rise in global mean annual temperature does not represent what might happen everywhere. For example, *polar amplification* suggests that the amount of warming will be greater at higher latitudes. Across the planet, enhancement of the greenhouse effect could cause a change in climate in greater magnitude than any over the past 10,000 years.

POSSIBLE IMPACTS OF GLOBAL WARMING

What are the potential societal impacts of an enhanced greenhouse effect and global warming? According to estimates by the IPCC in 2002, North American climatic zones could shift northward by as much as 550 km (350 mi), affecting all sectors of society. In areas where heat and moisture stress would cut crop yields, current farming practices would have to change. Across the North American grain belt, for example, higher temperatures and more frequent and prolonged drought might necessitate a switch from corn to wheat and still require increased irrigation.

Continued global warming would cause mean sea level (msl) to rise in response to melting of land-based polar ice sheets and mountain glaciers, along with thermal expansion of seawater. (The melting of floating sea ice does not raise sea level.) Amplification of the global warming trend at higher latitudes threatens the ice sheets of West Antarctica and Greenland, which have recently shown signs of accelerating mass loss. Also showing accelerating shrinkage since the middle of the 20th century, especially since the mid-1970s, are the world's mountain glaciers. Tide gauging stations and, since 1993, microwave altimeters on Earth-orbiting satellites have documented an 8 cm (7.1 in.) rise in msl during the 20th century. The 21st century will see a rise in msl ranging from 20 to 60 cm (8 to 24 in.) due to global warming, according to the 2007 IPCC Assessment. Thermal expansion of ocean waters, portions of which are already warming, would account for more than 60% of the rise while melting glaciers are responsible for the other 40%. Higher mean sea level would accelerate coastal erosion by wave action, allow seawater to inundate wetlands, estuaries and islands, and make low-lying coastal plains more vulnerable to storm surges. For people now living on low-lying islands, such as the Maldives, abandonment may be the only option as sea level rises.

Another major impact of global warming is the shrinkage of Arctic sea ice cover, a trend that has been underway over the past three decades. If the trend continues, the Arctic Ocean will be seasonally ice-free within two to three decades. While the melting of floating sea ice does not raise sea level, it can significantly alter the climate of the Arctic by triggering an ice-albedo feedback loop, further accelerating the melting of sea ice and amplifying warming. There is more on this phenomenon in Chapter 15.

On the positive side, global warming would lengthen the growing season at high latitudes. Warmer winters would also lengthen the navigation season on lakes, rivers, and harbors where seasonal ice cover is a hindrance. For most people living in Canada or the northern U.S., energy savings due to less space heating in winter would more than compensate for the greater energy requirements anticipated with more air conditioning needed with warmer summers.

Recent trends in climate are consistent with changes predicted by global climate models. Plotted in Figure 3.39 is the 1880 to 2010 instrument-derived record of variations in (1) global (land and sea-surface) mean annual temperature, (2) global mean sea-surface temperature, and (3) global mean land surface temperature. In all three cases, the temperature is expressed as a departure (in Celsius degrees and Fahrenheit degrees) from the long-term (1901 to 2010) average. For reasons given in Chapter 4, these temperature series assembled by NOAA's *National Climatic Data Center (NCDC)* indicate greater year-to-year variability over land than ocean. While the global mean temperature (land and ocean) trends downward during the late 1800s, it begins to rise from the early 1910s through 1940. Until about 1970, the temperature is steady or falling, and then rises again through the 1980s to 2010. The overall variation in temperature is mostly in the range of ±0.5 Celsius degree (±0.9 Fahrenheit degree) about the century average. This record is a global average; regionally, temperatures could be greater or less than is presented here.

In 2011, NOAA reported that 2010 was the 34th consecutive year in which global temperatures were above the 1901 to 2000 average. Combined global land and ocean annual surface temperatures in 2010 tied with 2005 as the warmest, at 0.62 ± 0.07 Celsius degrees above the 20th century average, since the standardized instrumental record began in 1880. As shown in Figure 3.39, most of the warming in the 20th century took place after 1980 (0.65 ± 0.15 Celsius degrees). NASA data shows that the global climate has warmed by about 0.2 of a Celsius degree per decade from the late 1970s to 2010. Climate reconstructions (mostly in the Northern Hemisphere) indicate that the 20th century was the warmest in 1000 years (Chapter 15).

Some critics question the integrity of hemispheric or global temperature records. They cite improvements in sophistication and reliability of current weather instruments compared to those from decades past, changes in location and exposure for those instruments at long-term weather stations, huge gaps in monitoring networks, especially over the ocean, and the warming influence of urbanization (*urban heat islands*) as potential sources

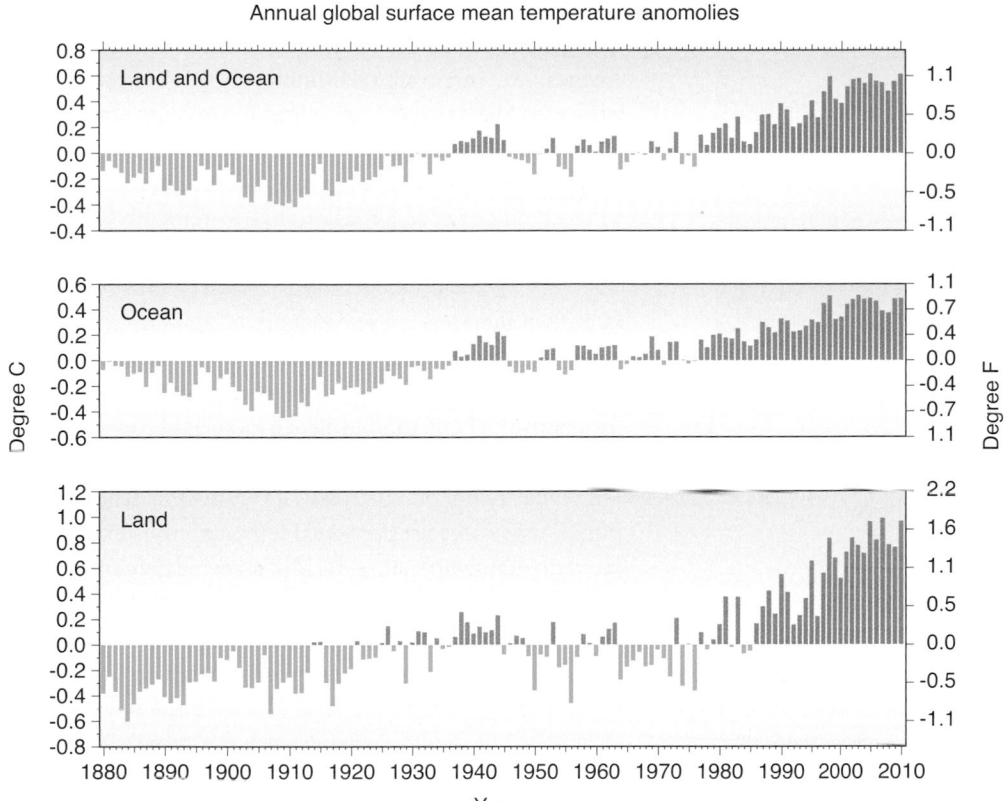

Annual global surface mean temperature anomolies

FIGURE 3.39
Instrument-derived trends in mean annual global (land plus ocean), sea-surface, and land temperatures; expressed as departures in Celsius degrees and Fahrenheit degrees from the long-term (1901-2010) period average. [NOAA, National Climatic Data Center]

of error. However, even when careful statistical analysis accounts for these concerns, a clear warming trend is evident over the past century.

The general consensus in the scientific community holds that a global-scale warming trend has prevailed since the end of the Little Ice Age. The most reasonable scientific explanation for the observed warming trend is the steady build-up of carbon dioxide in the atmosphere (primarily from the combustion of fossil fuels) and the consequent enhancement of the natural greenhouse effect. Continuation of the upward trend in the concentration of atmospheric CO_2 and other infrared-absorbing gases inevitably will lead to global warming throughout this century and probably beyond.

Many experts argue that society has so much at stake that action should be taken immediately to head off enhanced greenhouse warming. They call for sharp reductions in coal and oil consumption, greater reliance on non-carbon energy sources, higher energy efficiencies (e.g., more vehicle miles per gallon), and a halt to deforestation along with massive reforestation. Even if greenhouse warming is less than predicted, such actions will alleviate other serious environmental problems. For example, reducing reliance upon fossil fuels also cuts air pollution. Coal and oil burning produces not only CO_2 but also

SO_2 and particulate matter that are serious air pollutants. Chapter 15 has more on climate and climate change.

Monitoring Radiation

An understanding of Earth's energy budget requires accurate measurement of incoming solar radiation and outgoing infrared radiation emitted by the Earth-atmosphere system. The **pyranometer** is the standard instrument for measuring the intensity of solar radiation striking a horizontal surface. The instrument consists of a sensor enclosed in a transparent hemisphere that transmits total (direct plus diffuse) solar short-wave (less than 3.0 μm) radiation (Figure 3.40). In one design, the sensor is a disk made up of alternating black and white wedge-shaped segments in a star-like pattern. Black wedges are highly absorptive whereas white wedges are highly reflective. Differences in absorptivity and albedo mean that the temperatures of the black and white portions of the sensor respond differently to the same intensity of solar radiation. The temperature contrast between the black and white segments is calibrated in terms of radiation flux (e.g., W/m^2) or flux per unit time. The instrument should be situated where it will not be affected by shadows, any nearby highly reflective surface,

FIGURE 3.40
Solar pyranometer, the standard instrument for monitoring solar radiation in the wavelength band from 0.3 to 3.0 micrometers. [Photo by E.J. Hopkins]

or other sources of radiation which could interfere with measurement of solar radiation.

A pyranometer may be linked electronically to a pen recorder or computer that traces a continuous record of incident solar radiation (Figure 3.41). In this graph, the upper curve displays the rate of incoming solar radiation on the winter solstice (December 22) at Brockport, NY. The sky was partly cloudy in the morning and clear in the afternoon. Incident solar radiation during partly cloudy conditions may occasionally be higher than if the sky was clear due to scattering from the edges of clouds. The lower curve shows a cloudy day on the winter solstice, where the rate of incoming radiation at Earth's surface is significantly less.

As discussed in Chapter 1, sensors onboard weather satellites measure two types of radiation emanating from the Earth-atmosphere system: scattered or reflected visible solar radiation and emitted infrared radiation. The first signal is processed into visible satellite images and the second is processed into infrared satellite images. An **infrared radiometer** is an instrument that measures the intensity of infrared radiation emitted by the surface of some object such as land, cloud, or ocean. The temperature dependency of IR emission means that infrared sensors can be calibrated to remotely sense the temperature of Earth's surface or cloud tops.

As noted earlier in this chapter, for a *blackbody* (a perfect radiator) the wavelength of maximum radiation emission varies inversely with the radiation temperature (*Wien's displacement law*). Hence, a sensor that measures the intensity of radiation emitted by a surface can determine the radiation temperature of that object. For a blackbody, the actual temperature equals the radiation temperature. For all other objects, the actual temperature is higher than the radiation temperature, the difference depending on the object's **emissivity**, a measure of how closely an object approximates a blackbody. The emissivity of an ideal blackbody is 1.0 (or 100%) and less than 1.0 for all other objects. As a general rule, every 1% change in emissivity is associated with a change in the object's surface temperature of slightly less than 1.0 Celsius degree.

With adjustments made for differences in emissivity of objects within its field of view, an IR sensor (radiometer) onboard a weather satellite measures the temperature of land and sea surfaces and the tops of clouds. Based on measured differences in IR emission, sensors can distinguish low clouds, which are relatively warm, from high clouds, which are relatively cold. These measurements are calibrated on a gray scale (in which low clouds appear dark gray and high clouds appear bright white) or a color scale. One advantage of IR satellite images over visible satellite images is their availability both day and night.

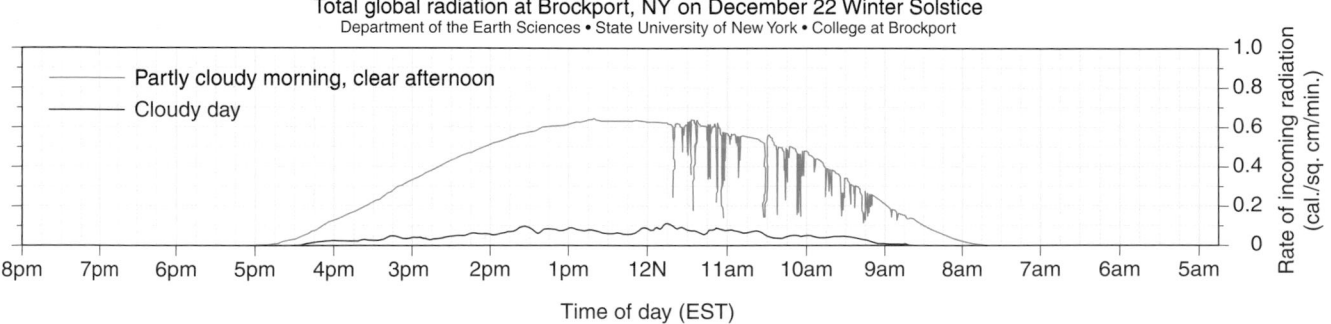

FIGURE 3.41
Pyranometer measurements of the rate of incoming radiation on two winter solstice days at Brockport, NY (note that time increases toward the left). The upper curve shows a partly cloudy morning and clear afternoon. The lower curve shows a cloudy day. [Courtesy of I.W. Geer and R.S. Weinbeck]

Conclusions

Earth intercepts only a tiny fraction of the total energy output of the Sun. Regular motions of the spherical planet in space (rotation on its spin axis and revolution about the Sun) distribute this energy unequally within the Earth-atmosphere system. As solar radiation passes through the atmosphere, a portion is scattered and reflected back to space, a portion is absorbed (i.e., converted to heat), while the rest is transmitted to Earth's surface either as direct or diffuse radiation. Overall, however, the atmosphere is relatively transparent to solar radiation and that which reaches Earth's surface is either reflected or absorbed. The Earth-atmosphere system responds to solar heating by emitting infrared radiation to space. Greenhouse gases and clouds absorb and emit IR in all directions including toward Earth's surface, thereby significantly elevating the average temperature of the lower troposphere (the greenhouse effect).

Over the long term, net incoming solar radiation is balanced by the infrared radiation emitted to space by the Earth-atmosphere system. Absorption of solar radiation causes system warming whereas emission of infrared radiation to space causes system cooling. Within the Earth-atmosphere system, however, the rates of radiational heating and radiational cooling are not the same everywhere. In the next chapter, we examine the reasons for these energy imbalances and their implications for atmospheric circulation and weather. But first we distinguish between heat and temperature, describe heat transfer processes, examine the significance of thermal response and heat capacity, and identify the various controls of air temperature.

Basic Understandings

- All objects emit and absorb energy in the form of electromagnetic radiation. The many forms of electromagnetic radiation make up the electromagnetic spectrum and are distinguished by wavelength, frequency, and energy level.
- Earth and the Sun closely approximate perfect radiators (blackbodies) so that blackbody radiation laws may be applied to them to describe and predict their radiational characteristics.
- Wien's displacement law predicts that the wavelength of most intense radiation is inversely proportional to the absolute temperature of a radiating object. Hence, solar radiation peaks in the visible portion of the electromagnetic spectrum whereas the Earth-atmosphere (terrestrial) radiation peaks in the infrared (IR).
- According to the Stefan-Boltzmann law, the total energy radiated by an object at all wavelengths is directly proportional to the fourth power of the object's absolute temperature. Hence, the much hotter Sun emits exponentially more radiational energy per unit area per unit time than does the much cooler Earth-atmosphere system.
- The intensity of radiation emitted by the Sun, so far away to be considered a point source, decreases as the inverse square of the distance traversed by the radiation. Thus, the intensity of the average solar radiation reaching Earth's orbit, referred to as the solar constant, is sufficient to help maintain Earth as a habitable planet.
- The total energy (in the form of solar radiation) that is absorbed by the Earth-atmosphere system is offset by the total energy (in the form of infrared radiation) emitted by Earth to space. A balance of energy input and output is known as global radiative equilibrium and is essentially the condition on planet Earth.
- Solar altitude, the angle of the Sun above the local horizon, influences the intensity of solar radiation that strikes Earth's surface at that location. All other factors being equal, as the solar altitude increases, the intensity of solar radiation received at Earth's surface (energy per unit area) also increases.
- As a consequence of Earth's elliptical orbit about the Sun, the planet's nearly spherical shape, rotation, and tilted spin axis, solar radiation is distributed unevenly over Earth's surface and changes through the course of a year.
- In the middle and high latitudes of the hemisphere experiencing spring or summer, maximum local solar altitudes are higher and daylight is longer, resulting in the receipt of more solar radiation, and ultimately higher air temperatures. In middle and high latitudes of the hemisphere experiencing fall or winter, maximum local solar altitudes are lower, daylight is shorter, there is less solar radiation, and air temperatures are lower.
- Solar radiation that is not absorbed by the atmosphere (converted to heat) or scattered or reflected to space reaches Earth's surface. With scattering, a particle disperses solar radiation

in all directions. Reflection, a special case of scattering, takes place at the interface between two media when some of the radiation striking the surface is backscattered.

- Absorption of ultraviolet radiation during the natural formation and destruction of ozone within the stratosphere shields organisms at and near Earth's surface from exposure to potentially lethal levels of UV radiation.

- Solar radiation that reaches Earth's surface is either reflected or absorbed depending on the surface albedo. Light-colored surfaces have a relatively high albedo for visible radiation whereas dark-colored surfaces have a relatively low albedo and typically have a higher surface temperature than light-colored surfaces when exposed to sunlight.

- The atmosphere is heated primarily from below; that is, on average heat flows from Earth's surface to the overlying air. This is evident in the average temperature profile of the troposphere as air temperature drops with increasing altitude.

- Water vapor, carbon dioxide, and several other atmospheric trace gases absorb outgoing IR and emit infrared radiation in all directions including downward, thereby significantly elevating the average temperature of Earth's surface and the lower atmosphere. Clouds also contribute to this greenhouse effect. Water vapor is the principal greenhouse gas.

- Clouds affect climate in two opposing ways. By absorbing and emitting IR, clouds warm Earth's surface and by reflecting incoming solar radiation, clouds cool Earth's surface. On a global scale, clouds have a net cooling effect on climate.

- The Callendar effect is the theory that global climate change can be brought about by enhancement of Earth's natural greenhouse effect by elevating levels of atmospheric CO_2 from anthropogenic sources, principally the burning of fossil fuels.

- The pyranometer is the standard instrument for monitoring the total flux of solar radiation whereas the infrared radiometer selectively monitors IR.

Enduring Ideas

- All objects emit electromagnetic radiation with the wavelength of most intense radiation dependent upon the object's temperature. Hot objects emit their most intense radiation at shorter wavelengths (higher frequencies) than cooler objects. The total energy flux emitted by a blackbody is proportional to the fourth power of the temperature. Intensity of radiation received diminishes rapidly with distance from the source, as the inverse square of distance traveled.

- Solar altitude influences the intensity of solar radiation received at Earth's surface and the length of daylight affects the total amount of solar radiation received each day. Variations in solar altitude and length of daylight are due to the tilt of Earth's spin axis relative to the plane of Earth's orbit.

- Solar radiation interacts with gases and aerosols composing the atmosphere, resulting in scattering, reflection, and absorption. That which reaches Earth's surface is either reflected or absorbed (i.e., converted to heat). The atmosphere is relatively transparent to solar radiation so that Earth's surface (mostly the ocean) is the main source of heat for the atmosphere. Global radiative equilibrium states that the outgoing emission of heat to space in the form of infrared radiation balances the incoming solar radiational heating of the Earth-atmosphere system. Absorption and emission of IR by certain atmospheric gases (chiefly water vapor) is responsible for the greenhouse effect and key to a habitable planet.

- Continual formation of ozone (from oxygen) and decomposition of ozone (to oxygen) in the stratosphere are natural processes that absorb potentially lethal intensities of incoming ultraviolet radiation. Threats to the ozone shield by certain chemicals (e.g., CFCs) led to a worldwide ban on the manufacture and use of ozone depleting substances.

- Infrared-absorbing atmospheric gases (i.e., water vapor, carbon dioxide, methane) are responsible for the greenhouse effect and make for a habitable planet. The latest IPCC report concluded that global warming is occurring and is very likely the consequence of mostly human activities, especially the combustion of fossil fuels and the release of carbon dioxide into the atmosphere.

Key Terms

energy
law of energy conservation
electromagnetic radiation
electromagnetic spectrum
wavelength
wave frequency
visible radiation
infrared radiation
blackbody
Wien's displacement law
Stefan-Boltzmann law
inverse square law
global radiative equilibrium
photosphere
sunspots

chromosphere
solar corona
solar altitude
perihelion
aphelion
equinoxes
Tropic of Cancer
Arctic Circle
Antarctic Circle
Tropic of Capricorn
solar constant
scattering
reflection
albedo
absorption

stratospheric ozone shield
Antarctic ozone hole
polar stratospheric clouds (PSCs)
circumpolar vortex
planetary albedo
greenhouse effect
greenhouse gases
atmospheric window
Callendar effect
global warming potential
Intergovernmental Panel on Climate
    Change (IPCC)
pyranometer
infrared radiometer

Review

1. What is the basis for subdividing the electromagnetic spectrum into various forms of electromagnetic radiation?
2. What is a blackbody? Is the Earth-atmosphere system a blackbody?
3. Apply Wien's displacement law in comparing the radiational properties of the Sun versus those of the Earth-atmosphere system.
4. How does solar altitude affect the length of the path of solar radiation through Earth's atmosphere?
5. In the Northern Hemisphere, Earth is closer to the Sun during the winter than during the summer and yet winter is colder than summer. Please explain.
6. What is the significance of the Tropic of Cancer and the Tropic of Capricorn relative to incident solar radiation?
7. Describe the interactions of incoming solar radiation with the components of the atmosphere.
8. Why does the Antarctic ozone hole appear in the Southern Hemisphere spring?
9. All other factors being the same, how does cloud cover affect the day's minimum air temperature?
10. Provide a convincing argument that water vapor is the principal greenhouse gas.

Critical Thinking

1. How is global radiative equilibrium an example of the law of energy conservation?
2. Speculate on some means whereby Earth's planetary albedo might increase. What would be the climate implications?
3. Explain how the Antarctic ozone hole is caused mainly by catalytic destruction of ozone by chlorine.
4. Why does atmospheric carbon dioxide concentration exhibit an annual cycle?
5. Although the Keeling curve and the trend in global mean annual temperature show increasing trends, the two curves are not parallel. Explain why.
6. The concentration of methane in the atmosphere is so low that it must be measured in parts per billion by volume (ppbv). Why then is methane considered an important greenhouse gas?
7. Suggest some strategies that promise to reduce the amount of carbon dioxide vented to the atmosphere.
8. What is the significance of the Stefan-Boltzmann law when applied to the Sun versus the Earth-atmosphere system?
9. According to the inverse square law, if the distance between Earth and the Sun were triple its present magnitude, the solar constant would be reduced to what fraction of its present value?
10. On the equinox, the length of daylight is not precisely 12 hours. Explain why.

Hazards of Solar Ultraviolet Radiation

Bright, sunny skies are welcomed as an opportunity for outdoor activities and gatherings. Especially during warm weather, people will spend as much time as possible in the Sun, some sunbathing for hours in pursuit of a perfect tan (Figure 1). While sunlight plays an important role in the body's production of vitamin D, overexposure can cause serious health problems. In both cases, it is the ultraviolet (UV) spectrum of solar radiation that interacts with our skin.

There are three types of ultraviolet radiation, categorized by wavelength: UVA (0.32-0.40 μm), UVB (0.29-0.32 μm), and UVC (0.20-0.29 μm). UVC rays, the smallest and most energetic of the three, readily kill exposed cells. Fortunately for life on Earth, the *stratospheric ozone shield* prevents UVC from reaching Earth's surface. UVB, however, can pass through the ozone shield. Although less energetic than UVC, UVB still can cause damage to cells, resulting in sunburn, skin cancer, and other health problems. While UVA, the weakest of the three, poses the least health hazard, it indirectly accelerates aging of the skin and exacerbates the damaging effects of UVB exposure.

FIGURE 1
Sunbathers run the risk of excessive exposure to solar ultraviolet (UV) radiation that can cause serious health problems.
[Photo by K.A. Nugnes]

Some consequences of overexposure to solar ultraviolet radiation appear within hours (e.g., sunburn) whereas others are the accumulation of years of overexposure (e.g., skin cancer and cataracts). Probably the most noticeable impact of long-term overexposure is *photoaging*, premature aging of the skin due to the Sun. UV rays that penetrate the skin damage structural proteins (collagen and elastin) responsible for skin's strength and resilience. As damage progresses, the skin develops a leathery texture, wrinkles, and irregular pigmentation. People who have had long-term exposure to the Sun, perhaps because of an outdoor occupation or recreation, may exhibit symptoms of photoaging before their 30s while people with minimal exposure begin to show signs of aging in their 40s.

Ultraviolet radiation also contributes to certain types of cataracts, a major cause of vision impairment worldwide. A *cataract* is a cloudy (opaque) region that develops in the lens of the eye where light focuses to produce sharp, clear images. If untreated, cataracts can cause partial or total blindness. Research suggests that 20% of all cataracts are related to UV exposure. Additionally, excess exposure to UV rays suppresses the body's immune system.

Ultraviolet radiation also plays a key role in the development of skin cancer, the most common form of cancer in the U.S. Researchers continue to accumulate evidence that links UVB radiation overexposure to the development of both melanoma and non-melanoma skin cancer (basal cell carcinomas and squamous cell carcinomas). UVB damages the DNA in skin cells, resulting in uncontrolled cell growth and subsequent formation of tumors.

The American Cancer Society estimates that each year more than 2 million cases of non-melanoma skin cancer are diagnosed nationwide. About 1 in 5 Americans will contract this disease, perhaps 90% will stem from overexposure to the Sun's UV rays. Responsible for 75% of skin cancer deaths annually (about 8700 of 11,790), melanoma is the most virulent form of skin cancer and the number of cases has more than doubled in the U.S. over the past decade; about 68,000 new cases were reported in 2010. Fortunately, all types can be successfully treated if detected in their early stage and skin cancer is one of the most preventable forms of cancer.

Protection from solar ultraviolet radiation begins with monitoring the level of exposure. In 1994, NOAA's National Weather Service (NWS) and the U.S. Environmental Protection Agency (EPA) introduced a special index so that the general public could avoid overexposure while outdoors. In 2004, the *Ultraviolet (UV) Index* was revised based on recommendations of the World Health Organization (WHO), the World Meteorological Organization (WMO), the United Nations Environment Programme (UNEP), and the International Commission on Non-Ionizing Radiation Protection. Within the index, the higher the rating, the greater the expected exposure to UV and the quicker a person is to sunburn. As shown in Figure 2, values of the *UV Index* range from < 2 (low risk) to 11 or higher (extremely high risk).

The National Weather Service uses a computer model to relate the ground-level strength of solar UV radiation (in the range of 0.29 to 0.40 μm) from forecasted stratospheric ozone levels and cloud cover, elevation of the ground, and solar altitude (which varies with latitude, day of year, and time of day). Satellite sensor measurements of current ozone levels are also used to forecast ozone levels for the following day. At present the UV Index does not take into account the effects of air pollution, haze, or variations in surface properties, such as snow-cover. As of this writing, index values are issued for 58 major metropolitan areas across the nation. A sample UV Index forecast for the nation is shown in Figure 3.

The safe level of exposure to UVB depends upon skin phototype. People who burn instead of tan (most-sensitive phototype) can tolerate considerably less exposure to UV than people who tan and rarely burn (least-sensitive phototype).

Exposure Category	UVI Range	Sun protection messages
Low	< 2	• Wear sunglasses on bright days. In winter, reflection off snow can nearly double UV strength. • If you burn easily, cover up and use sunscreen.
Moderate	3 to 5	• Take precautions, such as covering up and using sunscreen, if you will be outside. • Stay in shade near midday when the sun is strongest.
High	6 to 7	• Protection against sunburn is needed. • Reduce time in sun between 11 a.m. and 4 p.m. • Cover up, wear a hat and sunglasses, and use sunscreen.
Very high	8 to 10	• Take extra precautions. Unprotected skin will be damaged and can burn quickly. • Try to avoid the sun between 11 a.m. and 4 p.m. Otherwise, seek shade, cover up, wear a hat and sunglasses, and use sunscreen.
Extreme	11 +	• Take all precautions. Unprotected skin can burn in minutes. Beachgoers should know that white sand and other bright surfaces reflect UV and will increase UV exposure. • Avoid the sun between 11 a.m. and 4 p.m. • Seek shade, cover up, wear a hat and sunglasses, and use sunscreen.

FIGURE 2
The *Ultraviolet (UV) Index*, designed by the National Weather Service and the U.S. Environmental Protection Agency, to advise the public of the potential health hazards of exposure to solar ultraviolet radiation.

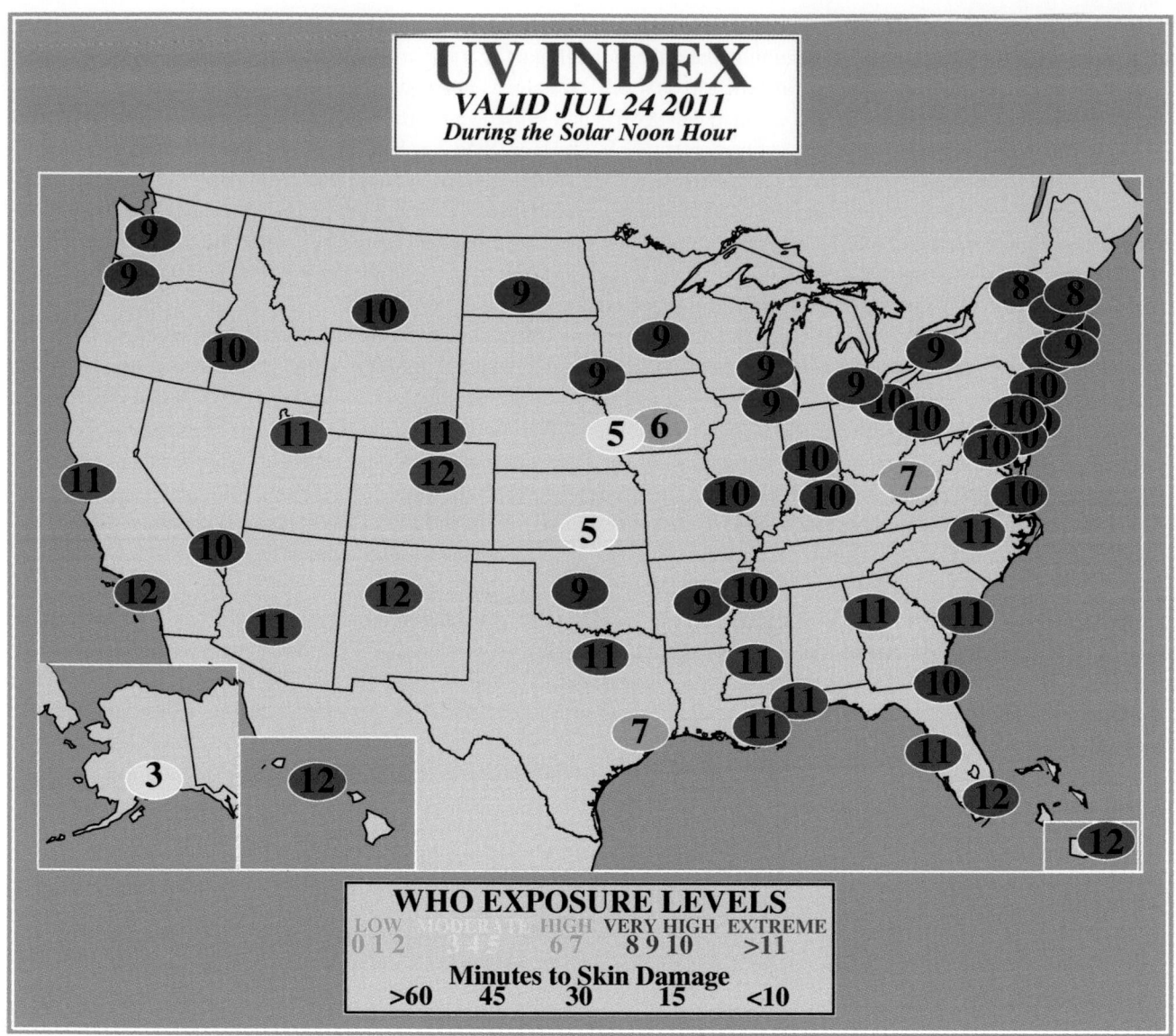

FIGURE 3
A sample UV Index forecast. Note the many major cities with extreme levels of UV radiation. [Image courtesy of Craig Long, NOAA/National Weather Service]

However, the UV Index is intended only as a general guide. Skin phototype varies considerably within the human population so some will require more protection than indicated by the UV Index.

If possible, people should avoid the Sun when its rays are most intense between about 10 a.m. and 2 p.m. local solar time. This general rule fails to account for the width of time zones (typically 15 degrees of longitude), daylight saving time, or seasonal changes in solar altitude. Leith Holloway (formerly of the Geophysical Fluid Dynamics Laboratory in Princeton, NJ) developed a simple and reliable guide for gauging UV exposure. On a clear midsummer day when the solar altitude increases through the morning, so does the intensity of UV striking Earth's surface – until the Sun is about 45 degrees above the horizon. At this point, the intensity of UV radiation striking Earth's surface increases much more rapidly because the solar beam passes through less and less atmosphere. In fact, total UV exposure at solar altitudes greater than 45 degrees is about 5 times greater than the total exposure during the remaining daylight hours.

As the Sun passes 45 degrees, the shadows on a horizontal surface become shorter than the objects casting them. Thus Holloway created the *shadow rule*: when a person's shadow is shorter than he or she is tall, then either sunscreen should be applied or the Sun avoided. While a person's shadow is longer, then the health risk is significantly less. Simply stated: *"Short shadow? Seek shade!"*

Solar altitudes are greatest in late spring and early summer (May through August in the Northern Hemisphere) and in tropical latitudes, so UV intensity is also greatest then. Unfortunately, in both that season and area people spend more time outdoors and wear less protective clothing. Since skin cancers develop on exposed skin surfaces, they are most common on the forearms, face, and neck. Hence, protective clothing over skin that frequently burns, such as a wide-brimmed hat, long-sleeved shirt or tightly woven fabric, is the best protection. Sunglasses that are UV-rated worn along with a wide brimmed hat provide adequate eye protection.

Sunscreens have ingredients that selectively absorb UV wavelengths and should be applied to all exposed skin approximately 20 minutes before venturing outdoors. Medical experts strongly recommend a broad-spectrum sunscreen that provides protection against both UVB and UVA, which is labeled on the bottle.

All sunscreen bottles also are labeled with their *Sun Protection Factor* (*SPF*), an estimation of how long treated skin can be safely exposed to the Sun versus unprotected skin. For example, if a person's skin becomes pink after 20 minutes in the Sun, applying a sunscreen with SPF 15 should provide 5 hours of protection (20 minutes × 15 = 300 minutes). The higher the SPF, the longer the protection lasts. Health professionals recommend a minimum SPF of 15, but a higher SPF protects better for those who are fair-skinned, spend significant time outdoors, or live at a high elevation. (UV intensity increases by 2% for every 300-m (1000-ft) increase in elevation.) Optimal protection only comes when sunscreen is liberally applied. Perspiration and water remove sunscreen (even so-called waterproof sunscreen), so it should be reapplied every two hours, even on cloudy days.

Solar Power

Solar power is a renewable energy source that will last as long as the Sun, billions of years. According to N.S. Lewis of the California Institute of Technology, "more energy from the Sun hits the Earth in 1 hour than all of the energy consumed by humans in an entire year." If all the solar radiation that is transmitted through the day-lit atmosphere were distributed uniformly over Earth's surface, about 180 watts would illuminate every square meter. Solar power is an environmentally attractive alternative to conventional energy sources, especially coal, oil, and nuclear fuels. Furthermore, the cost of solar power is becoming more competitive with the price of conventional fuels in a number of applications and environments.

    The greatest challenge for solar technology is the variability of solar radiation. Only half the Earth faces the Sun at any instant, while the intensity of solar radiation varies by day, hour, location and weather. Seasonal fluctuations are especially pronounced at middle and high latitudes. Figure 1 shows the average daily solar radiation incident on a horizontal Earth surface per month (in kWh per m² per day) for the contiguous United States, derived from the 1961-1990 National Solar Radiation Data Base (NSRDB) for January and July. Figure 2 is a global map of the annual average daily solar radiation received at Earth's surface for the entire globe from July 1983 through June 2005.

    In relatively sunny, eastern Washington State the average annual solar radiation striking Earth's surface is 194 W per m² but monthly mean values range from 50 to 343 W per m², a sevenfold difference through the course of the year. At the same time, minimal cloud cover and seasonal variability make the Desert Southwest the most promising

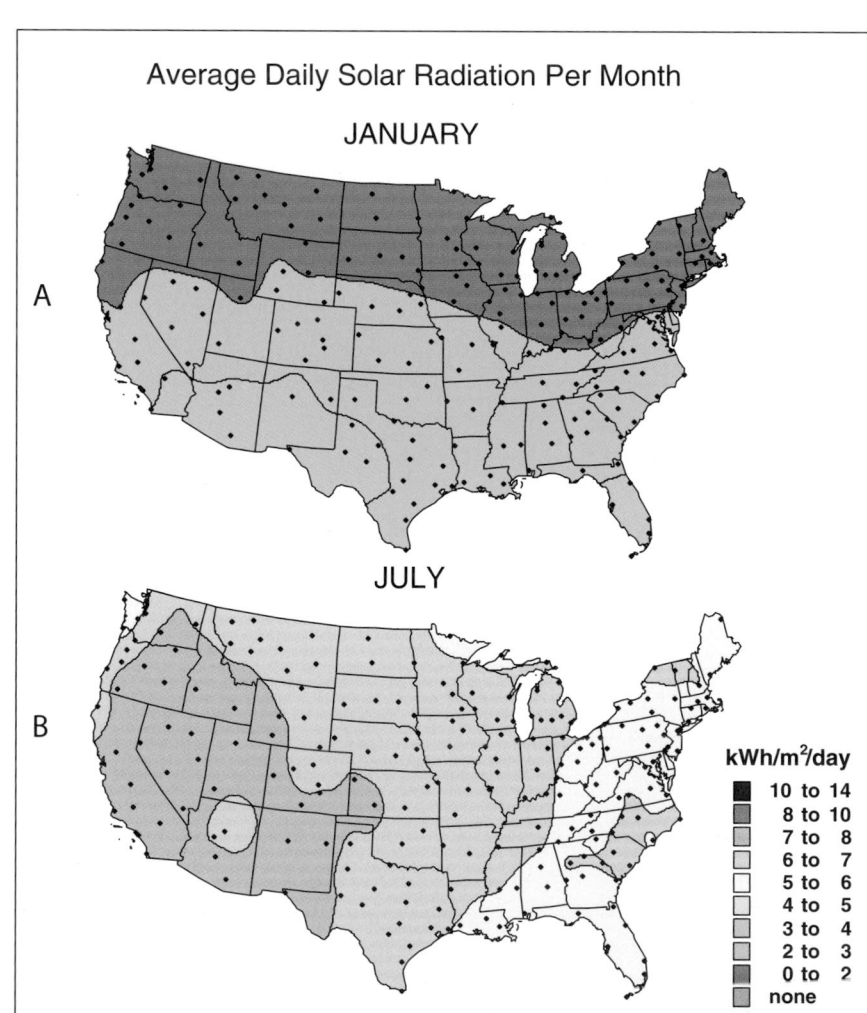

FIGURE 1

The average daily solar radiation received on a flat-plate collector facing south on a horizontal surface (in kWh per m² per day) for (A) January and (B) July. [National Renewable Energy Laboratory, Resource Assessment Program]

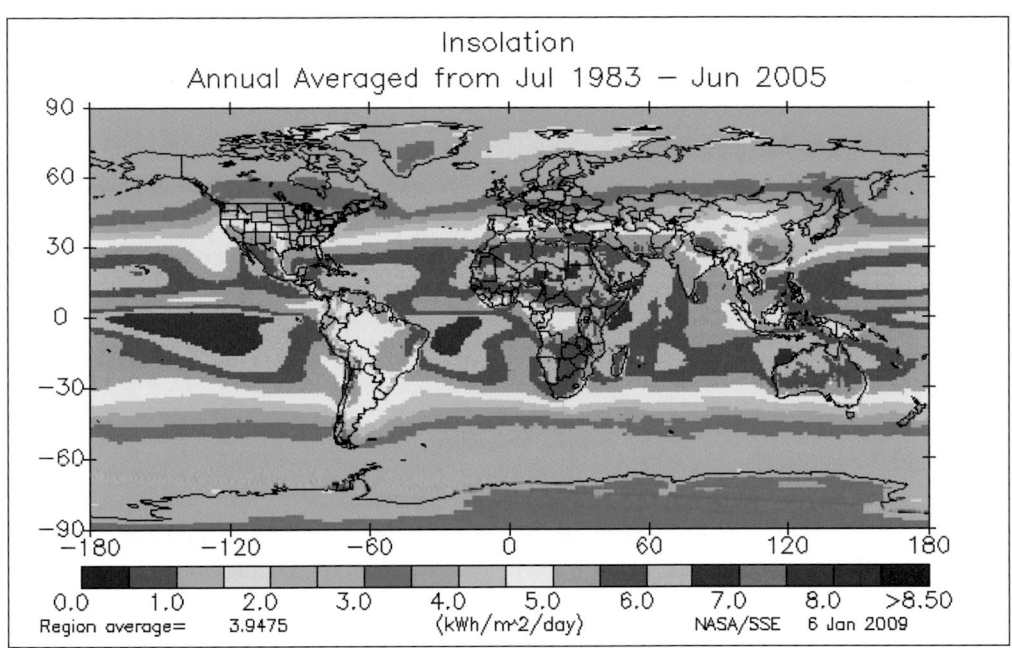

FIGURE 2
Annual average incoming solar radiation at Earth's surface (in kWh per m² per day) based on the period July 1983 through June 2005.

area for solar power in the U.S. Also, fluctuations in consumer demand do not always match the supply of solar energy, requiring some form of energy storage. Household solar collectors often store excess heat in insulated water tanks or compartments filled with rocks. At middle and high latitudes, a conventional heating system is needed as a backup, particularly during extended spells of cloudy or very cold weather.

The traditional *flat plate solar collector* is a framed panel of glass designed to capture solar power. Sunlight passes through two layers of glass before being absorbed by a blackened (low albedo) metal plate. Heat is then conducted from the absorbing plate to either air or liquid, which conveys the heat to its destination via a fan or pump. Solar collectors typically capture 30% to 50% of the solar energy that strikes them and are most commonly used for space or water heating in small buildings such as homes, schools, and apartment houses. Flat plate solar collectors do not produce temperatures high enough to convert water to steam, so these devices are not suitable for most industrial purposes.

Tilting solar collectors toward the equator partially compensates for the solar radiation variability due to changes in solar altitude. The advantage of tilted versus horizontal collectors depends on average cloud cover and latitude of the site. An optimal situation occurs in winter at middle latitude sites favored by clear skies, where a tilted solar collector can double the amount of solar radiation absorbed. While some solar collectors can also track the Sun so they are always perpendicular to the solar beam (Figure 3), their cost usually exceeds the value of the energy obtained.

Scientists and engineers are developing ways to efficiently convert solar energy to electricity on a large scale. In one conversion method, a *power tower system*, an array of computer-controlled flat mirrors, called *heliostats*, track the Sun and focus its radiation on a single heat collection point at the top of a tower. Concentrated sunlight in these systems can produce temperatures up to 480 ºC (900 ºF), high enough to convert water to high-pressure steam (just as the burning of coal does) for driving turbines that generate electricity. Other designs utilize a large parabolic dish or trough as a collection device.

An alternative to solar-driven turbines for generating electricity is the *photovoltaic cell* (*PV cell*), also known as a *solar cell.* Solar cells convert solar radiation directly into electricity and routinely power many devices, including handheld calculators, parking lot lights, applications in remote locations, and space vehicles. Solar cells use sunlight to create a *voltage*, that is, a difference in electrical potential, in a diode. When sunlight strikes the diode, an electric current flows through the circuit to which it is connected. Only special materials develop the necessary voltage to produce a direct current when exposed to sunlight. These materials, called *semiconductors*, are composed of highly purified silicon to which tiny amounts of certain impurities have been added. Semiconductors are manufactured in the form of tiny wafers or sheets. Electricity travels through metal contact wires on the front and back of the wafer. Groups of wafers are wired together to form photovoltaic modules, and these are interconnected to form a photovoltaic panel.

A limitation to PV cells is their relatively low *energy conversion efficiency*, that is, the percentage of solar radiation striking a photovoltaic cell that is converted to electrical energy. The efficiency of a handheld solar-powered calculator, for example, is 3% or less. At the other extreme, solar cells that power Earth-orbiting satellites have the highest efficiencies, approaching 50%. The best commercial crystalline silicon cells typically have an efficiency of a little more than 25%. Mass-produced photovoltaic panels have conversion efficiencies of 15% to 18%. (The energy conversion efficiency of conventional coal electricity varies between 40% and 60%.) Among the many factors that contribute to low solar-cell efficiency are cell reflectivity, conversion of radiation to heat, and the sensitivity of cells to only a portion of the solar spectrum.

Worldwide, the total accumulated installed PV capacity is about 15 gigawatts (GW), only 0.375% of the 4000 GW of all installed electric generation capacity. PV industry representatives predict that PV electric power generation will become a major contributor in 20 years. Over the past two decades, the PV power industry has benefited from technological advances and market growth; PV manufacturing output grew by a factor of 200 (according to the *Sunpower Corporation* in San Jose, CA). Technological advances have lowered the silicon requirement from 15 grams per watt in 2000 to 5.6 grams per watt in 2009 and today's $1.40 per watt cost of manufacture is projected to decrease to $1.00 per watt within the next 5 years.

FIGURE 3
The solar power array at Nellis Air Force Base uses tracking devices to keep the solar panels pointed toward the Sun. Tilted toward the south, each set of solar panels rotates around a central bar to track the sun from east to west. [Credit: Nellis AFB]

As a general rule, the cost of production increases with the more efficient solar cells, so future cost reductions are unlikely to be achieved through boosting the efficiency of traditional solar cells. Instead, the development of solar cells that are less efficient, but also less expensive to manufacture, is likely. An example is a *thin-film solar cell* consisting of a film of silicon or other light-sensitive substance deposited on a base material (whereas conventional solar cells consist of individual silicon crystals).

In the future, multi-megawatt, solar cell power plants are expected to feed electricity into regional grids at costs competitive with conventional power plants. The world's largest PV plant, completed in September 2010, is Canada's Sarnia Photovoltaic Power Plant. Consisting of 1.3 million thin-film PV panels that cover 966,000 square m, it generates 80 megawatts of peak power, though its capacity factor (ratio of the actual output of a power plant to its potential output if it had operated at full capacity over a specified time period) is expected to be only 17%. The outlook for more multimegawatt PV power plants is based on current trends that are developing efficient thin-film solar cells, as well as declining manufacturing costs made possible by mass production and economy of scale. In addition, international efforts to shift to a carbon-free electric grid, to offset anthropogenic contributions to global climate change and the inevitable decline in supplies of fossil fuels, should spur greater demand for solar power in the future.

Cumulus clouds developing over a wheat field in eastern Colorado, signaling that atmospheric conditions are favorable for convective currents to lift humid air and produce clouds via condensation. [Photo by Carlye Calvin. Courtesy of University Corporation for Atmospheric Research]

HEAT, TEMPERATURE, & ATMOSPHERIC CIRCULATION

Chapter Highlights

Case-in-Point
  Extreme Heat of Death Valley, CA
Distinguishing Temperature and Heat
  Temperature Scales and Heat Units
  Measuring Air Temperature
Heat Transfer Processes
  Radiation
  Conduction and Convection
  Phase Changes of Water
Thermal Response and Specific Heat
  Thermal Inertia
  Maritime and Continental Climates
Heat Imbalance: Atmosphere versus Earth's
  Surface
  Latent Heating
  Sensible Heating
  Bowen Ratio
Heat Imbalance: Tropics versus Middle and
  High Latitudes
  Heat Transport by Air Mass Exchange
  Heat Transport by Storms
  Heat Transport by Ocean Circulation
Why Weather?
Variation of Air Temperature
  Local Radiation Budget
  Cold and Warm Air Advection
  Urban Heat Island Effect
Conclusions
Basic Understandings/Enduring Ideas
Key Terms/Review/Critical Thinking
For Further Exploration
  Managing Weather Risk
  Heating and Cooling Degree-Days
  Wind Chill

Learning Objectives

Distinguish between temperature and heat.
Compare and contrast the various
  temperature scales.
Describe the instrumentation used to monitor
  temperature.
Summarize how the second law of
  thermodynamics applies to temperature
  gradients.
Explain how heat is transferred via radiation,
  conduction and convection.
Explain how heat is transferred via phase
  changes of water.
Demonstrate how heat and temperature are
  related through specific heat.
Contrast a continental climate with a
  maritime climate.
Describe the imbalances in radiational
  heating and radiational cooling within the
  Earth-atmosphere system.
Distinguish between sensible heating and
  latent heating.
Explain why latent heating is more important
  than sensible heating on a global scale.
Identify the various processes responsible for
  poleward heat transport within the Earth-
  atmosphere system.
Distinguish between radiational controls and
  air mass controls of air temperature.
Explain the seasonal and diurnal lag between
  radiation and temperature.
Provide some examples of how properties of
  Earth's surface influence air temperature.
Identify several factors that contribute to the
  formation of an urban heat island.

*What are the causes and consequences of heat transfer within
the Earth-atmosphere system?*

Case-in-Point

Extreme Heat of Death Valley, CA

Located in central eastern California and at the edge of Nevada, within the Mojave Desert, Death Valley endures the hottest and driest climate in North America. Designated a National Monument in 1933 and a National Park in 1994, the sand dunes, salt-flats, badlands, canyons, and mountains encompass 5270 mi^2 (13,649 km^2).

Although weather observations in Death Valley date to 1861, official record keeping began in 1911 when a permanent U.S. Weather Bureau cooperative observing station was established at Greenland Ranch, now known as Furnace Creek Ranch (Figure 4.1). With verifiable observations from 1911 to 2011, Death Valley's average daily high temperature ranges from 18 °C (65 °F) in December and January to 46 °C (115 °F) in July. The average daily low temperature similarly ranges from December and January at 4 °C (39 °F) to July at 31 °C (88 °F).

FIGURE 4.1

The cooperative weather observing station at Furnace Creek Ranch in Death Valley.

Death Valley holds many temperature records. On 10 July 1913 at Greenland Ranch, the temperature reached 56.6 °C (134 °F), the highest temperature ever recorded in the Western Hemisphere and, at the time, the highest recorded on Earth. The record stood until 13 September 1922 when El Azizia, Libya reported a temperature of 58 °C (136 °F). Today, the Death Valley temperature of 56.6 °C still stands as the highest recorded temperature in the Western Hemisphere. Since 1913, Death Valley has reported a temperature of 53.9 °C (129 °F) on four occasions, most recently on 6 July 2007.

Death Valley is also known for extended episodes of extreme summer heat. In the summer of 2001, the temperature at Furnace Creek reached or exceeded 37.8 °C (100 °F) for 154 consecutive days—a record. During the summer of 1996, the hottest summer on record at Death Valley, the temperature reached 48.9 °C (120 °F) on 40 successive days and 43.3 °C (110 °F) for 105 days in a row. The record high temperature of 56.6 °C (134 °F) was set during a hot episode from 8 to 14 July 1913, when the maximum temperature each day was at least 52.8 °C (127 °F).

Since 1911, air temperatures were recorded by standard thermometers mounted within a regulation wooden instrument shelter with louvered sides. A shelter is designed to provide ventilation and shield instruments from exposure to direct sunshine, the night sky, and precipitation. Usually the shelter is about 1.5 m (4.9 ft) above ground level (although the one in use at the time of the record high temperature at Greenland Ranch in 1913 was 1.1 m or 3.5 ft above ground). This is an important consideration when analyzing the 1913 record as daytime temperatures at ground level are often considerably higher than shelter temperatures. For example, on 15 July 1972 the ground temperature at Furnace Creek reached a record high of 93.9 °C (201 °F) while the shelter temperature was 53.3 °C (128 °F).

A combination of topography, the prevailing atmospheric circulation, and intense solar radiation is responsible for the extreme heat and desert conditions of Death Valley (Figure 4.2). At the lowest point, Badwater Basin, the valley floor is 86 m (282 ft) below sea level. (Furnace Creek Ranch is 24 km or 15 mi north of Badwater Basin and 54 m or 178 ft below sea level.) The Valley itself is narrow, 210 km (130 mi) long but only 10 to 23 km (6 to 14 mi) wide, and surrounded by high steep mountains. Prior to entering Death Valley, moist prevailing winds blowing inland from the Pacific Ocean encounter five mountain ranges, including the

Sierra Nevada and Panamint Range, with the highest peak of 3368 m (11,049 ft). As will be discussed in greater detail in Chapter 6, clouds and precipitation develop where winds blow upslope (windward slopes) and dry conditions prevail where winds blow downslope (leeward slopes). Having lost considerable moisture on the windward slopes of the mountain ranges, winds are exceptionally dry when they reach Death Valley, so clear skies are frequent.

This rain shadow effect causes the average annual precipitation in Death Valley to be only 4.9 cm (1.92 in.). The greatest total rainfall recorded in a calendar year was 11.53 cm (4.54 in.) in 1913, and again in 1983, while the greatest rainfall in a 12-month stretch amounted to 15.47 cm (6.09 in.) in 1997 to 1998. During

FIGURE 4.2
A combination of many environmental factors is responsible for Death Valley's extreme climate; it is the hottest and driest place in North America.

the driest episode on record, 1.63 cm (0.64 in.) of rain fell during a 40-month period, from 1931 to 1934. Furnace Creek reported a run of 385 days without measurable precipitation from 30 December 1952 to 18 January 1954 and from 29 December 1988 through 17 January 1990. High temperatures and meager moisture mean that only those plants and animals adaptable to exceptionally stressful conditions survive, yet Death Valley is home to a surprisingly great diversity of plants and animals, some indigenous to the area.

The low rainfall, due to the rain shadow, also assists in the buildup of heat. Solar radiation heats the dry desert floor of rock, dirt, and sparse vegetation and, because there is scant moisture to absorb heat energy via evaporation, it directly raises the temperature of the ground. The hot surface heats the overlying air via conduction and convection (confined to the valley by the surrounding mountain rim). Typically the air is too dry for the ascending branch of convection currents to produce clouds while the descending branch of convection currents is compressed, heated, and cloud-free. Details on ascending and descending air are found in Chapter 5.

With little protective vegetation to anchor the soil, on the rare occasions when rain does fall in Death Valley, the results can seem surprising. Even a brief downpour can cause considerable erosion and property damage, and locally heavy rainfall brings dangerous flash flooding (Figure 4.3). For example, on 15 August 2004, rainfall estimated from 2.5 to 5 cm (1 to 2 in.) washed out large sections of roadways and eroded trails in Death Valley, with two reported fatalities. Yet any significant rain also brings a spectacular bloom of colorful wild flowers across the valley floor.

FIGURE 4.3
Road sign alerts travelers to the possibility of flash flooding due to brief periods of heavy rainfall in Death Valley National Park, CA.

Temperature is one of the most important variables used to describe the state of the atmosphere; it is a usual element in weather observations and forecasts. From everyday experience, we know that air temperature varies with time: from one season to another, between day and night, and even from one hour to the next. Air temperature also varies from one place to another: highlands and higher latitudes are usually colder than lowlands and lower latitudes.

We also know that temperature and heat are closely related concepts. Heating a pan of soup on the stove raises the temperature of the soup whereas dropping an ice cube into a warm beverage lowers the temperature of the beverage. Granted that the two concepts are related, what is the precise distinction between heat and temperature? This is one of the questions we consider in this chapter. We also compare temperature scales, describe how temperature is measured, explain how heat is transferred in response to a temperature gradient, and describe how the temperature of a substance responds to an addition or loss of heat.

This chapter then examines heat imbalances that develop within the Earth-atmosphere system and the various processes whereby heat is transferred from Earth's surface to the troposphere and from the tropics to middle and high latitudes. All this enables us to better understand atmospheric circulation, energy conversions within the Earth-atmosphere system, and the controls of air temperature both globally and locally.

Distinguishing Temperature and Heat

All matter is composed of atoms or molecules that are in continual vibrational, rotational, and/or translational motion. The energy represented by this motion is called kinetic molecular energy or just *kinetic energy*, the energy of motion. In any substance, atoms or molecules actually exhibit a range of kinetic energies, as they move randomly at various speeds. **Temperature** is directly proportional to the *average* kinetic energy of the atoms or molecules composing a substance. At the same temperature, the average kinetic molecular energy of water is the same regardless of the volume of the water.

Internal energy encompasses all the energy in a substance, that is, the kinetic energy of atoms and molecules plus the potential energy arising from forces between atoms or molecules. If two objects have different temperatures (different average kinetic molecular energies) and are brought into contact, molecular collisions will transfer energy between the two objects; we call this energy in transit **heat**. Heat transferred from an object reduces the internal energy of that object whereas heat absorbed by an object increases its internal energy.

Differences in temperature rather than differences in internal energy govern the direction of heat transfer. For example, a cup of coffee at 21 °C has more internal energy than a teaspoon of coffee at 21 °C, but no heat is transferred between them. Heat energy is always transferred from a warmer object to a colder object. Heat is not necessarily transferred from an object having greater internal energy to an object with less internal energy. Consider, for example, a hot marble (at 40 °C) that is dropped into 5 liters of cold water (at 5 °C). The water has much more internal energy than the marble; nonetheless, heat is transferred from the warmer marble to the cooler water.

The following clarifies the distinction between temperature and heat. A cup of water at 60 °C (140 °F) is much hotter than a bathtub of water at 30 °C (86 °F); that is, the average kinetic energy of water molecules is greater at 60 °C than at 30 °C. Although lower in temperature, the greater volume of water in the bathtub means that it contains more total kinetic molecular energy than does the cup of water. If in both cases, the water is warmer than its environment, heat is transferred from water to its surroundings. However, much more heat energy must be removed from the bathtub water than from the cup of water for both to cool to the same temperature, for instance a room temperature of 20 °C (68 °F).

From the above discussion, we might assume that any substance, including air, always changes temperature whenever it gains or loses heat. However, this is not necessarily the case. Water is a component of air and occurs in all three phases (ice crystals, droplets, and vapor), and, as we will see later in this chapter, heat must be either absorbed from or released to the environment, for water to change phase. Furthermore, air is a compressible mixture of gases; that is, an air sample can change volume. As discussed in Chapter 5, heat energy is required for the work of expansion or compression of air. Hence, as a sample of air gains or loses heat, that heat may be involved in some combination of temperature change, phase change of water, or volume change.

TEMPERATURE SCALES AND HEAT UNITS

When measuring temperature, a temperature scale is needed that is based upon recognizable and reproducible reference points. For most scientific purposes, temperature is measured on the Celsius scale. First popularized in 1742 by the Swedish astronomer Anders Celsius (1701-1744), the Celsius temperature scale has the numerical convenience of a 100-degree interval between the readily reproducible freezing point and boiling point of pure water at sea level. A German physicist, Gabriel Daniel Fahrenheit (1686-1736), had introduced another scale in 1714 that was also based on the phase changes of water. Today, the United States is one of only a few nations still making public reports of surface weather conditions using the Fahrenheit temperature scale. If a thermometer graduated in both scales is immersed in a beaker containing an equilibrium mixture of ice and water, the temperature will be 0 degrees on the Celsius scale (0 °C) and 32 degrees on the Fahrenheit scale (32 °F). In boiling water at average sea level air pressure, an equilibrium between water and water vapor will have temperature readings of 100 °C and 212 °F.

Average kinetic molecular energy decreases with falling temperature. The theoretical temperature at which all molecular motion ceases, called **absolute zero**, corresponds to −273.15 °C (−459.67 °F). Actually, some atomic-level activity likely occurs at absolute zero, but an object at that temperature emits no electromagnetic radiation. On the Kelvin scale, temperature is the number of *kelvins* above absolute zero; hence, the Kelvin scale is a more direct measure of average kinetic molecular activity than either the Celsius or Fahrenheit temperature scales. The Kelvin scale is named for Lord Kelvin (William Thomson, 1824-1907) who developed it in the mid 19th century. Whereas units of temperature are expressed in degrees Celsius (°C) on the Celsius scale and degrees Fahrenheit (°F) on the Fahrenheit scale, temperature is expressed simply in kelvins (K) on the Kelvin scale. Nothing can be colder than absolute zero so the Kelvin scale has no negative values and a one-kelvin increment corresponds precisely to a one-degree increment on the Celsius scale. The three temperature scales are contrasted in Figure 4.4, and conversion formulas are presented in Table 4.1.

Most people would hardly notice a half-degree Celsius change in air temperature. But for operators of a Midwestern natural gas utility, a half-degree change in the average winter temperature can mean a significant change in consumer energy demand for space heating that also translates into a major fluctuation in revenue stream for the utility. Energy utilities are one among many different weather-sensitive businesses. According to the 2008 report by the U.S. Bureau of Economic

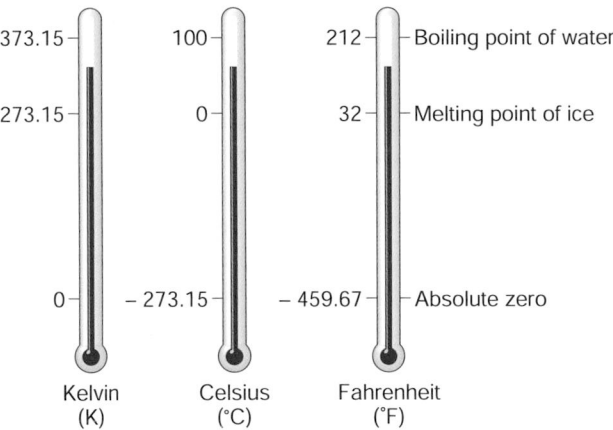

FIGURE 4.4
A comparison of three temperature scales: Kelvin, Celsius, and Fahrenheit.

TABLE 4.1
Temperature Conversion Formulas

$$°F = 9/5 \ °C + 32°$$
$$°C = 5/9 \ (°F − 32°)$$
$$K = 5/9 \ (°F + 459.67)$$
$$K = °C + 273.15$$

Analysis, an estimated one-third of the $14.28 trillion U.S. Gross Domestic Product (GDP) is weather sensitive, representing about $4.76 trillion of the nation's economic activity. Besides utilities, weather-sensitive businesses include transportation, agriculture, retail, tourism, recreation, and insurance. For example, a mild winter may mean fewer weather delays for aircraft and cross-country truckers, less risk of freezing temperatures in the Florida citrus groves, fewer northerners vacationing in southern Arizona, and reduced consumer demand for snow shovels, skis, and winter coats. According to a report published in the June 2011 issue of the *Bulletin of the American Meteorological Society*, the year to year dollar variation in U.S. economic activity caused by variability in the weather could be 3.4% or $485 billion of the 2008 GDP. See this chapter's first *For Further Exploration* for how weather-sensitive businesses can manage their weather risk.

As described in detail in this chapter's second *For Further Exploration*, heating degree-days are a measure of household energy consumption for space heating. Heating degree-days are computed for days when the average daily outdoor air temperature is less than 65 °F. For example, if the average temperature for the day is 35 °F, there are 65 − 35 = 30 heating degree-days for the day. The cumulative total of degree-days is tabulated throughout the heating season, defined as running from 1 July through the following 30 June.

Although temperature is a convenient way of describing the degree of hotness or coldness of an object, we can quantify heat energy directly. Until recently, atmospheric scientists commonly measured heat energy in units called calories, where one **calorie (cal)** is defined as the quantity of heat needed to raise the temperature of 1 gram of water 1 Celsius degree (technically, from 14.5 °C to 15.5 °C). (The term *calorie* has two definitions, which can cause confusion. The calorie unit is also used to measure the energy content of food and is actually 1000 heat calories or 1.0 kilocalorie. Here we refer to the "small" heat calorie.) Today, the more usual unit used in scientific research for energy of any form, including heat, is the *joule (J)*. One calorie equals 4.1868 J. In the English system, heat is quantified as British thermal units. One **British thermal unit (Btu)** is defined as the amount of heat required to raise the temperature of 1 pound (lb) of water 1 Fahrenheit degree (technically, from 62 °F to 63 °F). One Btu is equivalent to 252 cal and to 1055 J.

MEASURING AIR TEMPERATURE

A **thermometer** is the usual instrument for measuring air temperature. Galileo Galilei (1564-1642) is credited with its invention in 1592. (Actually, Galileo invented a *gas thermoscope*, a simple instrument that provided relative temperature measurements without reference to a scale. Air in a glass bulb expands or contracts as the air temperature changes, thus changing the level of colored water in the neck of the instrument.) A common type of thermometer consists of a liquid-in-glass tube attached to a graduated scale (Figure 4.5). Typically, the liquid is alcohol (which freezes at −117 °C or −179 °F) but mercury (which freezes at −39 °C or −38 °F) has been used in the past. Both the glass and the liquid (alcohol or mercury) expand when heated and contract when cooled but the liquid much more so than the glass. As air warms, heat is transferred to the thermometer, and the expanding liquid rises in the glass tube. The temperature can be read from the scale etched on the glass or printed on the surface of the plate on which the glass tube is mounted. As air cools, heat is transferred from the thermometer, and the liquid contracts, dropping in the tube. While this type of thermometer is relatively inexpensive and accurate, it must be read directly and not remotely.

During the 19th century, self-recording thermometers, a variation of the common liquid-in-glass thermometers, were perfected to record the daily maximum and minimum temperatures that often did not occur at regu-

larly scheduled observation times. A maximum recording thermometer, similar to a traditional non-electric clinical thermometer, contains a constriction in the bore of the glass tube between the bulb and the etched portion of the stem, designed to break the mercury thread as the temperature begins to fall from the daily maximum reading. A minimum recording thermometer uses alcohol and contains a small glass index located inside the alcohol column in the bore of the thermometer. Surface tension of the liquid keeps the index located just below the meniscus of the liquid column as temperatures drop. The index is left behind if temperatures rise. This index remains at the minimum temperature position until reset.

Another type of thermometer employs an electrical conductor whose resistance changes with fluctuations in the temperature, permitting a calibration between electrical resistance and temperature. Figure 4.6 shows the change of temperature during passage of a cold front as determined by an electronic thermometer. A radiosonde is equipped with this type of thermometer. Digital read-out thermometers, widely available in a variety of consumer products, are also of this type. Some electronic thermometers give remote temperature readings by using wireless transmission or mounting the sensor at the end of a long cable joined to the instrument. The latter system has replaced standard liquid-in-glass thermometers at National Weather Service facilities nationwide. The liquid-in-glass maximum/minimum thermometers used in the *NWS Cooperative Observing Network* for more than a century are now being replaced by electronic thermometers

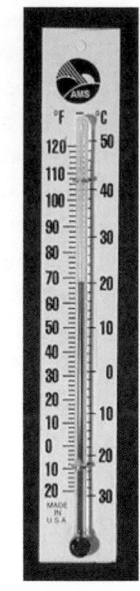

FIGURE 4.5
Liquid-in-glass thermometer. In this instrument, the liquid is alcohol.

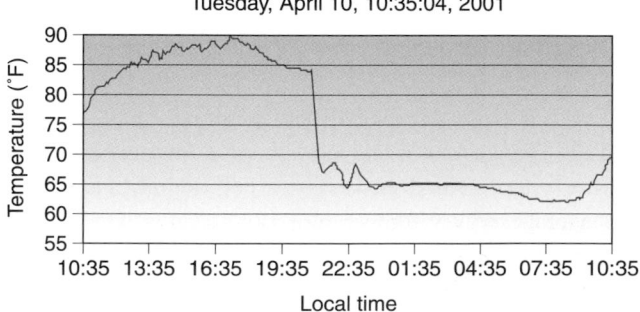

FIGURE 4.6
Record of air temperature over a 24-hr period showing the passage of a cold front.

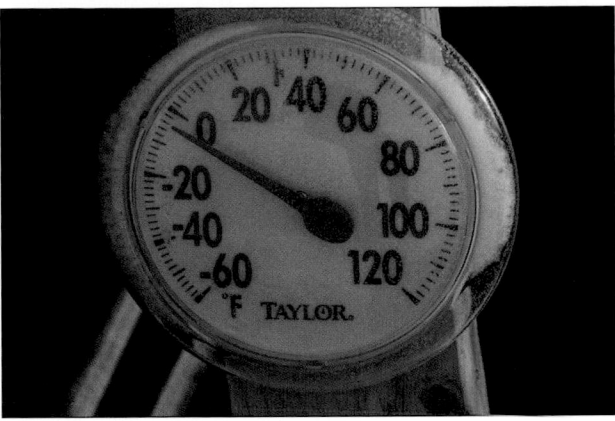

FIGURE 4.7
A bimetallic thermometer measuring air temperature at the Applied Physics Laboratory Ice Station, 180 nautical miles off the north coast of Alaska. [U.S. Navy photo by Chief Mass Communication Specialist Shawn P. Eklund]

and a remote digital readout display, called the *Maximum-Minimum Temperature System (MMTS)*.

Another common type of thermometer uses a bimetallic sensing element to take advantage of the expansion and contraction that accompany the heating and cooling of metals (Figure 4.7). A bimetallic sensing element consists of strips of two different metals welded together back to back. The two metals have different rates of thermal expansion; that is, one metal expands more than the other in response to the same amount of heating. Because the two metals are bonded together, heating causes the bimetallic strip to bend; the greater the heating, the greater the bending. For example, the rate of thermal expansion of brass is about twice that of iron so that a bimetallic strip composed of those two metals will bend in the direction of the iron when heated. Gears or levers translate the response of the bimetallic strip to a pointer and a dial calibrated to read in °C or °F. Alternatively, this sensor may be connected to a pen and a clock-driven drum to produce a continuous trace of temperature with time; this instrument is called a **thermograph**. Also, as described in Chapter 3, satellite sensors (radiometers) monitor surface temperatures remotely by measuring the intensity of emitted infrared radiation.

Regardless of the type of thermometer used, two important properties of the instrument are accuracy and response time. For most weather and climate purposes, a thermometer that is accurate to within 0.3 Celsius degree (0.5 Fahrenheit degree) will suffice. *Response time* refers to the rapidity at which an instrument resolves changes in temperature. Electrical resistance thermometers have rapid response times, liquid-in-glass somewhat less,

whereas bimetallic thermometers tend to be more sluggish. Because of differences in response time, a switch from a liquid-in-glass thermometer to an electronic thermometer may produce a discrepancy between the before and after temperature readings, thus altering the long-term climate record of a station.

For in situ measurements of air temperatures to be representative of the local atmospheric environment in which it is located, ideally a thermometer should be adequately ventilated, in thermal equilibrium with its surroundings, and shielded from precipitation, direct sunlight (to reduce unwanted heating of the instrument), and the night sky (preventing excessive radiative heat loss). Shielding prevents heat exchanges with the temperature sensor due to solar radiation, emission of infrared radiation, and phase changes of water. Enclosing temperature sensors (and other weather instruments) in a white, louvered wooden shelter had been standard practice for official temperature measurements for more than a century (Figure 2.12). Currently though, the sensor for the National Weather Service electronic thermometer is mounted inside a ventilated shield made of white plastic and the digital read-out box is located indoors (Figure 4.8). So that temperature readings acquired at different locations are comparable, the shelter should be located in an open grassy area well away from trees, buildings, or other obstacles, and at a standard height (1.5 m or about 5 ft) above the ground. As a general rule, the shelter should be no closer than four times the height of the nearest obstacle. If a shelter is not available, mounting a thermometer outside a window on the shady north side of a building is usually sufficient for general purposes.

In addition to using standard thermometers, air temperature can sometimes be estimated in unconventional

FIGURE 4.8
Enclosure for the National Weather Service electronic temperature sensor shields the instrument from direct sunshine and precipitation. This shield also protects a humidity sensor.

ways. One interesting and surprisingly accurate method is to count cricket chirps, assuming, of course, that you have a reliable supply of crickets. Crickets are cold-blooded organisms so their activity (including frequency of chirping) depends on air temperature. For air temperatures above about 12 °C (54 °F), the number of cricket chirps heard in an 8-second period plus 4 approximates the air temperature in °C.

Heat Transfer Processes

Air temperature varies from one place to another in large part because of imbalances in rates of radiational heating and radiational cooling within the Earth-atmosphere system. A change in temperature with distance is known as a **temperature gradient**. A global-scale air temperature gradient prevails between the hot equator and the cold poles (a horizontal temperature gradient). Another is the vertical temperature gradient between the relatively mild Earth's surface and the relatively cold tropopause (Figure 2.17).

Heat flows in response to a temperature gradient, according to the **second law of thermodynamics**. Simply put, this law states that all systems tend toward a state of disorder. You probably have personal experience with some implications of the second law, exemplified by uniform temperatures. If you were to open your car window on a cold day, temperatures inside and out would soon be equal (disordered). The presence of a gradient of any kind within a system signals relative states of order within that system. Hence, as a system tends toward disorder, gradients decrease. The second law predicts that where a temperature gradient exists, heat is transferred in a direction so as to eliminate the gradient; that is, heat flows from where the temperature is higher toward where the temperature is lower. In addition, the greater the temperature difference (i.e., the steeper the temperature gradient), the more rapid is the rate of heat transfer. Within the Earth-atmosphere system, heat is transferred via radiation, conduction, and convection. Also, when water changes phase, heat is either absorbed from or released to the environment.

RADIATION

Radiation describes both a form of energy and a means of energy transfer (Chapter 3). Electromagnetic radiation consists of a spectrum of waves traveling at the speed of light. Unlike conduction and convection, radiation requires no intervening physical medium; that is, it can travel through a vacuum. Although not actually a vacuum, interplanetary space is so rarefied that conduction and convection play essentially no role in transporting

energy from the Sun to Earth (or any other planet). Rather, radiation is the principal means whereby the Earth-atmosphere system gains energy from the Sun. Radiation is also the principal means whereby heat escapes from the planet to space thus maintaining a habitable environment on Earth (Chapter 3).

Absorption of radiation consists of the conversion of electromagnetic energy to heat. By contrast, emission of electromagnetic energy involves a loss of heat from the radiating object to the environment. All objects both absorb and emit electromagnetic radiation. If an object absorbs radiation at a greater rate than it emits radiation, the temperature of the object will rise. This type of imbalance in the flux of radiation is known as **radiational heating**. If an object emits radiation at a greater rate than it absorbs radiation, the temperature of the object will fall. This type of imbalance in the flux of radiation is called **radiational cooling**. At radiative equilibrium, when absorption and emission of radiation are equal, the object's temperature remains constant.

Radiative equilibrium does not necessarily mean that the temperature stays constant among all components of a system. Over the long-term, the entire Earth-atmosphere system is approximately in radiative equilibrium with surrounding space so that the effective planetary temperature remains relatively constant from year to year. Nonetheless, heat may be redistributed among the various components of the Earth-atmosphere system (for example, among the ocean, land, and glaciers) so that air temperature at a specified location may undergo significant short- and long-term variations. Hence, global radiative equilibrium does not preclude climate variability or climate change (Chapter 15).

CONDUCTION AND CONVECTION

Heat can be conducted within a substance or between substances that are in direct physical contact. **Conduction** (of heat) refers to the transfer of kinetic energy of atoms or molecules via collisions between neighboring atoms or molecules. This is why the temperature of a metal spoon rises when placed in a steaming cup of coffee. As the more energetic molecules of the hot coffee collide with the less energetic atoms of the cooler spoon, some kinetic energy is transferred to the atoms of the spoon. These atoms then transmit some of their heat energy, via collisions, to neighboring atoms, so that heat is conducted up the handle of the spoon and eventually the handle becomes warm to the touch.

Some substances are better conductors of heat than others. **Heat conductivity** is defined as the ratio of the rate of heat transfer across an area to the temperature gradient. Hence, in response to a specified temperature gradient, substances with higher heat conductivities have greater rates of heat transfer. As a rule, solids are better conductors of heat than are liquids, and liquids are better heat conductors than gases. At one extreme, metals are excellent conductors of heat whereas at the other extreme, still air is a very poor conductor of heat. Heat conductivities of some common substances are listed in Table 4.2, ranked from most to least conductive.

Differences in heat conductivity can be the reason one object feels colder than another, even though both objects have the same temperature. For example, at the same room temperature, a metallic object feels colder than a wooden object. The metallic object conducts heat away from your warmer hand more rapidly than the wooden object does, due to the much greater heat conductivity of metal. Consequently, you have the sensation that the metal is colder than the wood.

The relatively low heat conductivity of air makes it a good heat insulator. Heat conductivity is lower for still air than for air in motion so to avoid additional heat loss by convection, air must be confined. For example, in a thick fiberglass blanket used as attic insulation, the motionless air trapped between individual fiberglass filaments is primarily responsible for inhibiting heat loss. Similarly, the heat conductivity of a fresh snow cover is low because of the air trapped between

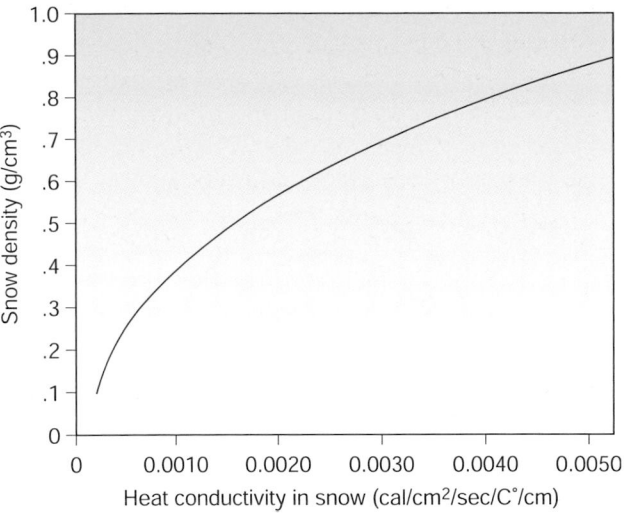

FIGURE 4.9
A thick layer of fresh snow is a good heat insulator primarily because of air trapped between the individual snowflakes. (Air is a poor conductor of heat.) But in time, the snow settles, air escapes, snow density increases, and the snow cover's insulating property diminishes.

individual snowflakes. A thick snow cover (20 to 30 cm or 8 to 12 in.) can thus inhibit or prevent freezing of the underlying soil, even though the temperature of the overlying air drops well below freezing. In time, however, a snow cover loses some of its insulating property as air escapes and the snow settles, snow density increases, and heat conductivity increases (Figure 4.9).

During the day, heat is conducted from Sun-warmed ground to cooler overlying air, but because air is a poor conductor of heat, conduction is significant only in a very thin layer of air in immediate contact with Earth's surface. Much more important than conduction in transporting heat vertically within the troposphere is convection. **Convection** is the transport of heat within a fluid via motions of the fluid itself. Although conduction takes place in solids, liquids, or gases, convection generally occurs only in liquids or gases. (An important exception is the convection currents in Earth's interior under conditions of tremendous confining pressure.)

Convection in the atmosphere is the consequence of differences in air density. At the same pressure, cold air is denser than warm air. As heat is conducted from the relatively warm ground to cooler overlying air, that air is heated, becoming less dense and thus more buoyant compared to the surrounding air. Cool, denser air from above sinks and replaces the warmer, less dense air at the ground forcing it to rise (buoyancy). (This also happens when cold tap water flows into a tub of hot water; that

TABLE 4.2	
Heat Conductivity of Some Familiar Substances[a]	
Copper	0.92
Aluminum	0.50
Iron	0.16
Ice (at 0 °C)	0.0054
Limestone	0.0048
Concrete	0.0022
Water (at 10 °C)	0.0014
Dry sand	0.0013
Air (at 20 °C)	0.000061
Air (at 0 °C)	0.000058

[a]*Heat conductivity* is defined as the quantity of heat (in calories) that would flow through a unit area of a substance (cm^2) in one second in response to a temperature gradient of one Celsius degree per centimeter. Hence, heat conductivity has units of calories per cm^2 per sec per C° per cm.

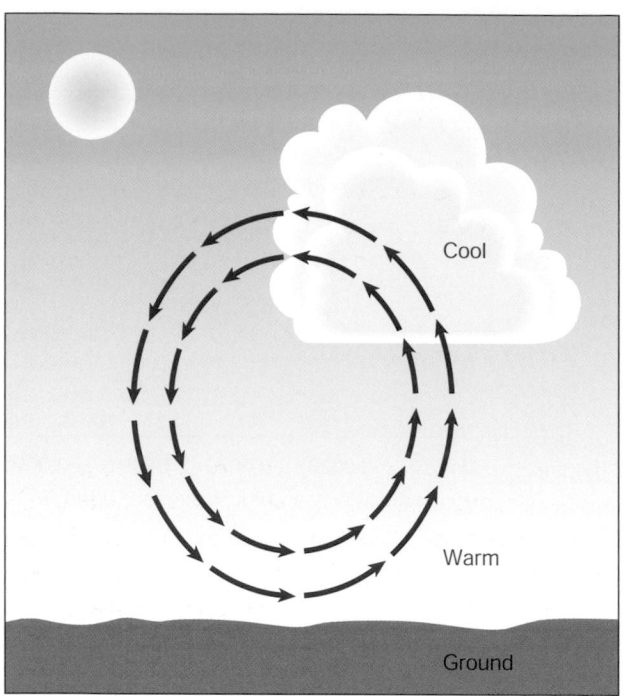

FIGURE 4.10
Convection currents transport heat from Earth's surface into the troposphere.

is, the denser cold water sinks and forces the less dense hot water to rise.) Ascending warm air expands, cools and eventually sinks back to the ground. Meanwhile, air now in contact with the warm ground is heated and rises—displaced by the sinking cooler denser air. In this way, as illustrated in Figure 4.10, convection transports heat upward from Earth's surface sometimes reaching thousands of meters into the troposphere.

We can readily observe convection currents in a pan of water on a hot stove. By adding a drop or two of food coloring to the water, we can see the circulating water redistributing heat that is conducted from the bottom of the pan into the water. In this example, as well as in the troposphere, conduction and convection work together in transferring heat. The combination of conduction and convection is known as **sensible heating**, as we can *sense* a change in temperature in response to such heating.

PHASE CHANGES OF WATER

Water is one of the very few substances that can occur naturally in all three phases within the temperature and pressure ranges found at and near Earth's surface. Water occurs as a crystalline solid (ice or snow), liquid, and gas (water vapor), and is continually changing phase as environmental conditions vary (Figure 4.11).

Depending on the type of phase change, water either absorbs heat from its environment or releases heat to its environment. The quantity of heat that is involved in phase changes of water is known as **latent heat**, where the term "latent" refers to heat that is "hidden" until released. Heat is absorbed or released during phase changes because of differences in molecular activity represented by the three physical phases of water, which for water, is of unusually great magnitude.

In the solid phase (ice), water molecules are relatively inactive and vibrate about a fixed location in an ice crystal lattice. Hence, an ice cube or any other piece of ice tends to maintain its shape. In the liquid phase, molecules are less strongly bonded and move about with greater freedom, so that liquid water takes the shape of its container. In the vapor phase, water molecules exhibit maximum activity and diffuse readily throughout the entire volume of a container. A change in phase is thus linked to a change in level of molecular activity, which is brought about by either an addition or loss of heat (Figure 4.12). Heat is absorbed from the environment during those changes to higher energy states: *melting* (phase change from solid to liquid), *evaporation* (phase change from liquid to vapor), and *sublimation* (phase change directly from solid to vapor). Heat is released to the environment during those changes to lower

FIGURE 4.11
At and near Earth's surface, water occurs in all three phases: solid (ice and snow), liquid, and vapor. Clouds are composed of tiny water droplets and/or ice crystals.

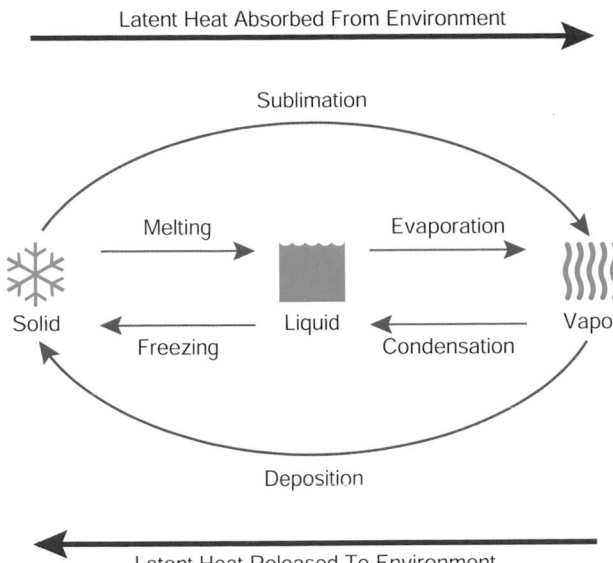

FIGURE 4.12
When water changes phase, heat energy is either absorbed from or released to the environment.

energy states: *freezing* (phase change from liquid to solid), *condensation* (phase change from vapor to liquid), and *deposition* (phase change directly from vapor to solid). During any phase change, heat is exchanged between water and its environment. Although the temperature of the environment changes in response, the temperature of the water undergoing the phase change remains constant until the phase change is complete (i.e., ice at 0 °C melting to 0 °C water). That is, the available heat (latent heat) is involved exclusively in changing the phase of water and not its temperature.

The transfer of heat via conduction, convection, and radiation oftentimes involves changes in the phase of water. The latent heat that is supplied to change liquid water to vapor in one place (e.g., evaporation at the ocean surface) is released to the environment when that water vapor condenses at some other place (e.g., cloud formation in the atmosphere). Prior to condensation, winds can transport water vapor thousands of kilometers horizontally and to altitudes of thousands of meters so that atmospheric circulation can bring about long-range transport of latent heat, helping to make the planet habitable.

Heat transfer between the human body and its surroundings (via radiation, conduction and convection) plus phase changes of water are important controls of human comfort and well being. For example, the combination of strong winds and low air temperature may accelerate heat loss from the body to the extent that a person risks frostbite or hypothermia (a dangerous

lowering of the body's core temperature). Hence, in winter in cold climates, weather forecasts usually include the *wind chill equivalent temperature* (or simply *wind chill*) along with the actual air temperature. The wind chill is an air temperature index that takes into account heat loss from exposed skin caused by the combined effect of wind and low air temperature. More information on wind chill is contained in this chapter's third *For Further Exploration*.

Thermal Response and Specific Heat

Whether by radiation, conduction, convection, or latent heating, transfer of heat from one place to another within the Earth-atmosphere system is accompanied by changes in temperature. A heat gain causes a rise in temperature whereas a heat loss causes a drop in air temperature.

The temperature change accompanying an input (or output) of a specified quantity of heat varies from one substance to another. The amount of heat that will raise the temperature of 1 gram of a substance by 1 Celsius degree is defined as the **specific heat** of that substance. Joseph Black (1728-1799), a Scottish chemist, first proposed the concept of specific heat in 1760. The specific heat of all substances is measured relative to that of liquid water, which is defined as 1 calorie per gram per Celsius degree (at 15 °C). The specific heat of ice is about 0.5 calorie per gram per Celsius degree (near 0 °C). Specific heats of other familiar substances are listed in Table 4.3 in order of magnitude. The variation in specific heat among different substances

TABLE 4.3 Specific Heat of Some Familiar Substances[a]	
Water	1.000
Wet mud	0.600
Ice (at 0 °C)	0.478
Wood	0.420
Aluminum	0.214
Brick	0.200
Granite	0.192
Sand	0.188
Dry air[b]	0.171
Copper	0.093
Silver	0.056
Gold	0.031

[a]Calories per gram per Celsius degree.
[b]At constant volume.

implies that different materials have different capacities for storing internal energy.

In response to the same input of heat energy, a substance with a low specific heat undergoes a greater rise in temperature than an equivalent mass of a substance with a high specific heat. Water has the greatest specific heat of any naturally occurring substance. From Table 4.3, water's specific heat is about five times that of dry sand. One calorie of heat will raise the temperature of 1 gram of water 1 Celsius degree, whereas 1 calorie of heat will raise the temperature of 1 gram of dry sand by 5 Celsius degrees. This contrast in specific heat helps explain why, at the beach in summer, the sand feels considerably hotter to bare feet than the water (Figure 4.13). The specific heat contrast plus evaporative cooling explains why wet sand feels cooler than dry sand.

THERMAL INERTIA

Water's exceptional capacity to store heat has important implications for weather and climate, especially at middle and high latitudes. A large body of water (such as the ocean or Great Lakes) can significantly influence the climate of downwind localities. The most persistent influence is on air temperature. Compared to an adjacent landmass, a body of water does not warm as much during the day (or in summer) and does not cool as much at night (or in winter). In other words, a large body of water exhibits a greater resistance to temperature change, called **thermal inertia**, than does a landmass. Whereas the greater specific heat of water versus land is the major reason for this contrast in thermal inertia, differences in heat transport also contribute. Sunlight penetrates water to some depth and is absorbed (converted to heat) through a significant volume of water (Chapter 3). But sunlight cannot penetrate the opaque land surface and is therefore absorbed only at the surface. Furthermore, circulation of ocean and lake waters transports heat through great volumes of water, whereas heat is conducted only very slowly into the soil. The input (or output) of equal amounts of heat energy causes a land surface to warm (or cool) more than the equivalent surface area of a body of water.

MARITIME AND CONTINENTAL CLIMATES

Air temperature is regulated to a considerable extent by the temperature of the surface over which the air resides or travels via heat exchanges. With sufficient residence time over a large body of water, an air mass develops the temperature characteristics of the surface water. Hence, places immediately downwind of the ocean experience much less contrast between average winter

FIGURE 4.13
A sign near an entrance to the Great Sand Dunes National Monument in southern Colorado. The contrast in specific heat is one reason why the sand is hotter than the water.

and summer temperatures; such places have a **maritime climate**. Localities at the same latitude (especially mid and high latitudes), but well inland, experience a much greater contrast between winter and summer temperatures; such places have a **continental climate**. That is, in the same latitude belt, summers are cooler and winters are milder in maritime climates than in continental climates.

Consider an example of the contrast in temperature regime between continental and maritime climates (Figure 4.14). The latitude of San Francisco, CA (37.8 degrees N) is almost the same as that of St. Louis, MO (38.8 degrees N) so that the seasonal variation in the amount of incoming solar radiation (due to astronomical

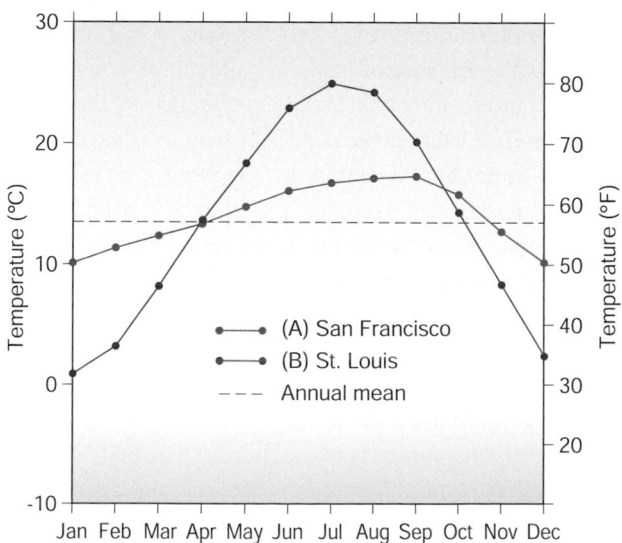

FIGURE 4.14
Variation in monthly mean temperatures for (A) maritime San Francisco, CA and (B) continental St. Louis, MO. The annual mean temperatures for the two cities are nearly identical.

factors) is similar at both places. (San Francisco has an average incoming solar radiation flux of about 196 W/m² while St. Louis receives about 176 W/m².) St. Louis is situated far from the moderating influence of the ocean and its climate is continental. During a recent 30-year period, St. Louis' average summer (June, July, and August) temperature was 25.7 °C (78.2 °F) and its average winter (December, January, and February) temperature was 1.3 °C (34.3 °F), giving an average summer-to-winter seasonal temperature contrast of 24.4 Celsius degrees (45 Fahrenheit degrees). San Francisco, on the other hand, is located on the West Coast, immediately downwind of the Pacific Ocean; its climate is maritime. During the same time period, the average summer temperature at San Francisco was 17.4 °C (63.4 °F) and the average winter temperature was 10.7 °C (51.2 °F), giving an average seasonal temperature contrast of only 6.7 Celsius degrees (12.2 Fahrenheit degrees).

Heat Imbalance: Atmosphere versus Earth's Surface

Weather is not a capricious act of nature and ultimately is a response to unequal rates of radiational heating and radiational cooling within the Earth-atmosphere system. Sensors onboard Earth-orbiting satellites monitor imbalances in rates of radiational heating (due to absorption of solar radiation) and cooling (due to emission of infrared radiation to space). These imbalances are responsible for temperature gradients within the Earth-atmosphere system in response to which the atmosphere circulates and redistributes heat.

Figure 4.15 shows how solar radiation intercepted by planet Earth interacts with the atmosphere and Earth's surface. Numbers represent global annual averages. For every 100 units of solar radiation that enter the upper atmosphere, the Earth-atmosphere sys-

tem scatters or reflects 29.9 units to space, the atmosphere absorbs 22.9 units, and Earth's surface (principally the ocean) absorbs 47.2 units. In response to the surface temperature, Earth's surface emits 116.1 units of infrared radiation. Atmospheric gases and clouds absorb 104.4 units of infrared radiation and emit 97.7 units to Earth's surface (the *greenhouse effect*). A total of 70.1 units of IR radiation are emitted out the top of the atmosphere to space, equal to the amount of solar radiation absorbed by the Earth-atmosphere system (*global radiative equilibrium*).

The global average annual distribution of incoming solar radiation and outgoing infrared radiation implies net warming of Earth's surface and net cooling of the atmosphere (Table 4.4). At Earth's surface, absorption of solar radiation is greater than emission of infrared radiation (+28.8). In the atmosphere, on the other hand, emission of infrared radiation to space is greater than absorption of solar radiation (-28.8). That is, on a global average annual basis, Earth's surface undergoes net radiational heating while the atmosphere undergoes net radiational cooling.

The atmosphere is not actually cooling relative to Earth's surface because radiation is not the only heat transfer mechanism at work. In response to the radiationally induced temperature difference between Earth's surface

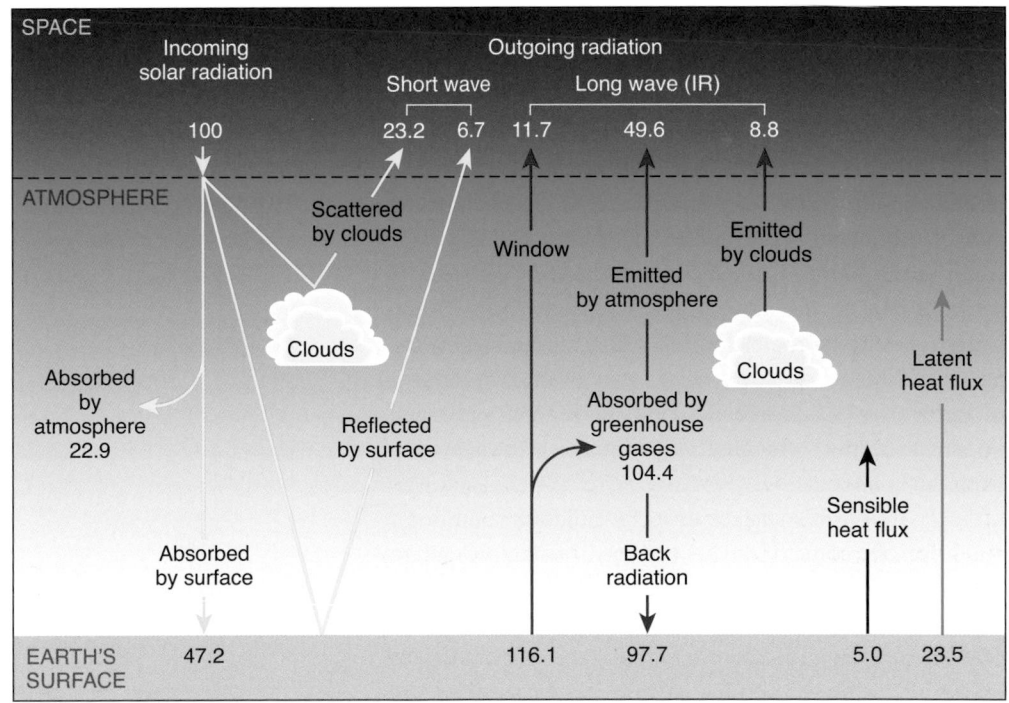

FIGURE 4.15
Globally and annually averaged disposition of 100 units of solar radiation entering the atmosphere. Solar radiation fluxes are depicted at the left, infrared radiation fluxes in the middle, and latent and sensible heat fluxes at the right. [Modified after K.E. Trenberth et al., 2009, "Earth's Global Energy Budget," Bulletin of the American Meteorological Society vol. 90, no. 3, pp. 311-323.]

and atmosphere, heat is transferred from Earth's surface to the atmosphere. A combination of latent heating (phase changes of water) and sensible heating (conduction and convection) is responsible for this heat transfer. As shown in Figure 4.15, on a global annual average basis, 28.5 units of heat energy are transferred from Earth's surface to the atmosphere: 23.5 units (about 82% of the total) by latent heating and 5 units (about 18%) by sensible heating.

LATENT HEATING

As noted earlier in this chapter, latent heating refers to the transfer of heat energy from one place to another as a consequence of phase changes of water. When water changes phase, heat energy is either absorbed from the environment (i.e., melting, evaporation, sublimation) or released to the environment (i.e., freezing, condensation, deposition). As part of the global water cycle, latent heat that vaporizes water at the Earth's surface is transferred in the form of sensible heat to the atmosphere as clouds form. Ocean water covers a large portion of Earth's surface and is the principal source of water vapor that eventually returns to Earth's surface as precipitation. In general, only well inland does most precipitation originate as evaporation from the continents.

As Earth's surface absorbs radiation (both solar and infrared), some of the heat energy is used to vaporize water from the ocean, glaciers, lakes, rivers, soil, and vegetation (*transpiration*) without changing temperatures. The latent heat required for vaporization (evaporation or sublimation) is supplied at the Earth's surface, and heat is subsequently released to the atmosphere during cloud development. Within the troposphere, clouds form as some of the water vapor condenses into liquid water droplets or deposits as ice crystals. During cloud formation, water changes phase and latent heat is released

to the atmosphere. Through latent heating, then, heat is transferred from Earth's surface to the troposphere. In fact, latent heat transfer is more important than either radiational cooling or sensible heat transfer in cooling Earth's surface (Figure 4.16).

TABLE 4.4 **Global Radiation Balance**	
Solar radiation intercepted by Earth	100 units
Solar radiation budget	
Scattered and reflected to space (23.2 + 6.7)	29.9
Absorbed by the atmosphere	22.9
<u>Absorbed at the Earth's surface</u>	<u>47.2</u>
Total	100 units
Radiation budget at the Earth's surface	
Infrared cooling (97.7 − 116.1)	−18.4
<u>Solar heating</u>	<u>+47.2</u>
Net heating	+28.8 units
Radiation budget of the atmosphere	
Infrared cooling (− 49.6 − 8.8 + 104.4 − 97.7)	−51.7
<u>Solar heating</u>	<u>+22.9</u>
Net cooling	−28.8 units
Non-radiative heat transfer: Earth's surface to atmosphere	
Sensible heating (conduction plus convection)	5.0
<u>Latent heating (phase changes of water)</u>	<u>23.5</u>
Net transfer	28.5 units

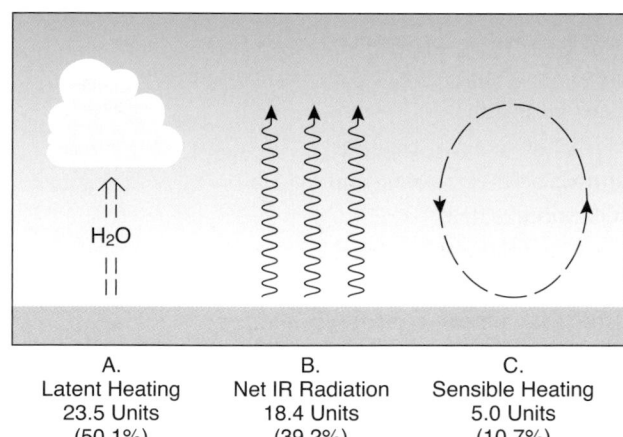

A.	B.	C.
Latent Heating	Net IR Radiation	Sensible Heating
23.5 Units	18.4 Units	5.0 Units
(50.1%)	(39.2%)	(10.7%)

FIGURE 4.16
Earth's surface is cooled via (A) vaporization of water, (B) net emission of infrared radiation, and (C) conduction plus convection. Numbers are global annual averages.

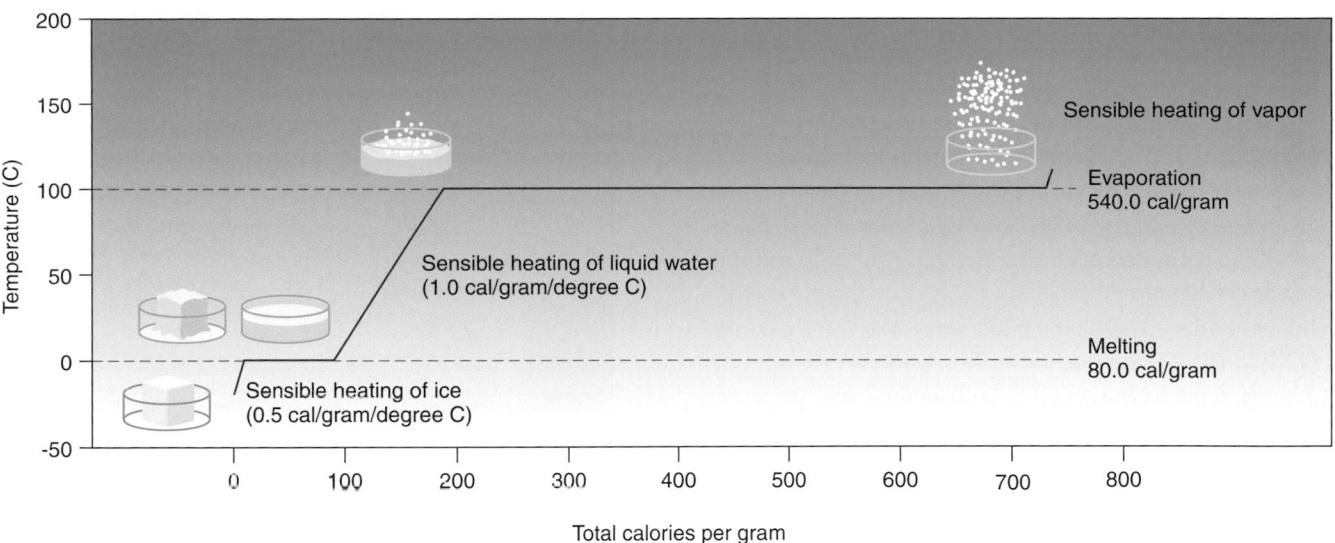

FIGURE 4.17
Heating a 1-gram ice cube causes a rise in temperature plus phase changes.

Unusually large quantities of heat are required to change the phase of water as compared to phase changes of other naturally occurring substances. Consider, for example, the quantity of heat involved in changing a one-gram ice cube at sea-level as it is heated from an initial temperature of −20 °C (−4 °F) to water vapor at a final temperature of 100 °C (212 °F) (Figure 4.17). The specific heat of ice is about 0.5 cal per gram per Celsius degree, which means that 0.5 cal of heat must be supplied for every 1 Celsius degree rise in temperature. Hence, warming our ice cube from −20 °C to 0 °C (−4 °F to 32 °F) requires an input of 10 cal of heat. Once the melting (or freezing) point is reached, an additional 80 cal of heat, called **latent heat of fusion**, must be supplied per gram to break the forces that bind water molecules in the ice phase. The temperature of the water and ice mixture remains at 0 °C until all the ice melts.

Once the ice cube melts completely, additional heating of the water causes the temperature of the liquid water to rise. The specific heat of liquid water is 1 cal per gram per Celsius degree so that 1 cal is needed for every 1 Celsius degree rise in water temperature. While evaporation can occur at any temperature, suppose that no evaporation takes place until the temperature of the system reaches the normal boiling point of 100 °C (212 °F). This 100 Celsius degree rise in temperature requires the addition of 100 calories. The **latent heat of vaporization** varies from about 600 cal per gram at 0 °C (32 °F) to 540 cal per gram at 100 °C (212 °F). When the water boils at sea-level, the temperature of the water will remain at 100 °C (212 °F) until all the liquid is vaporized. Any additional heating will cause the temperature of the vapor to increase. For a 1-gram ice cube at 0 °C to vaporize directly without melting first (*sublimation*), the latent heat of fusion plus the latent heat of vaporization must be supplied to the ice cube. This amounts to 680 cal per gram at 0 °C.

If the process just described is reversed, that is, if the water vapor is cooled until it becomes liquid (*condensation*) and then ice (*freezing*), the water temperature drops and phase changes take place as equivalent amounts of latent heat are released to the environment. When water vapor becomes liquid, the latent heat of vaporization is released to the environment, and when water freezes, the latent heat of fusion is released. If water vapor were to change to ice without first becoming liquid (*deposition*), at 0 °C the combined latent heats of vaporization plus fusion are released to the environment.

SENSIBLE HEATING

Heat transfer via conduction and convection can be monitored (*sensed*) by temperature change; hence, sensible heating encompasses both of these processes. Heat is conducted from the relatively warm surface of the Earth to the cooler overlying air. Heating reduces the density of that air, which is forced to rise by cooler denser air replacing it at the surface (Figure 4.10). In this way, convection transports heat from Earth's surface into the troposphere. Because air is a relatively poor conductor of heat, convection is much more important than conduction as a heat transfer mechanism within the troposphere.

Usually sensible heating combines with latent heating to channel heat from Earth's surface into the troposphere. This happens during thunderstorm development. Updrafts (ascending branches) of vapor-laden air in convection currents often produce **cumulus clouds**, which resemble puffs of cotton floating in the sky (Figure 4.18A). These clouds are sometimes referred to as *fair-weather cumulus* because they seldom produce rain or snow. On the other hand, if atmospheric conditions are favorable (described in Chapter 11), convection currents can surge to great altitudes, and cumulus clouds merge and billow upward to form towering **cumulonimbus clouds**, also known as thunderstorm clouds (Figure 4.18B). In retrospect, two important heat transfer processes (a combination of latent heating and sensible heating) took place last summer when that thunderstorm washed out your ball game or sent you scurrying for shelter at the beach.

At some times and places, heat transfer is directed from the troposphere to Earth's surface, the reverse of the global average annual situation. This reversal in direction of heat transfer occurs, for example, when mild winds blow over cold, snow-covered ground or when warm air moves over a relatively cool water surface. Heat transport from the atmosphere to Earth's surface is the usual situation at night (especially when the sky is clear) when radiational cooling causes Earth's land surface to become colder than the overlying air.

As noted in Chapter 2, nearly all weather is confined to the troposphere, implying that heat transfer by sensible and latent heating operates primarily within the lower atmosphere. Radiational processes dominate heat and temperature distribution above the troposphere.

BOWEN RATIO

The **Bowen ratio,** named after prominent California Institute of Technology astrophysicist Ira Sprague Bowen (1898-1973) describes how the heat energy received at Earth's surface (by absorption of solar and infrared radiation) is partitioned between sensible heating and latent heating. That is,

Bowen ratio = [(sensible heating)/(latent heating)]

At the global scale,

Bowen ratio = [(5 units)/(23.5 units)] = 0.2

A

B

FIGURE 4.18
Latent heating and sensible heating are combined in the formation of (A) cumulus clouds and (B) cumulonimbus (thunderstorm) clouds.

As shown in Table 4.5, the average Bowen ratio varies from one place to another depending on the amount of surface moisture. The wetter the surface, the less important is sensible heating and the more important is latent heating. The Bowen ratio ranges from about 0.1 (one-tenth as much sensible as latent heating) for the ocean to about 5.0 (five times as much sensible as latent

TABLE 4.5
Bowen Ratio[a] of Various Geographical Areas

All Oceans	0.11
Atlantic Ocean	0.11
Pacific Ocean	0.10
Indian Ocean	0.09
All Land	0.96
North America	0.74
South America	0.56
Europe	0.62
Asia	1.14
Africa	1.61
Australia	2.18
Globe	0.20

[a]Ratio of sensible heating to latent heating.

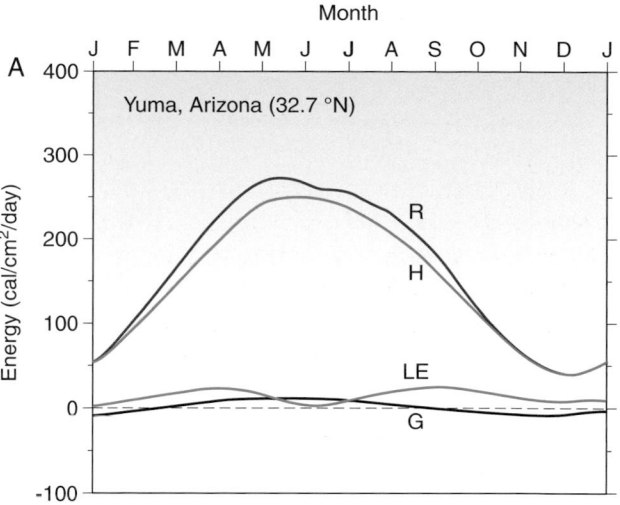

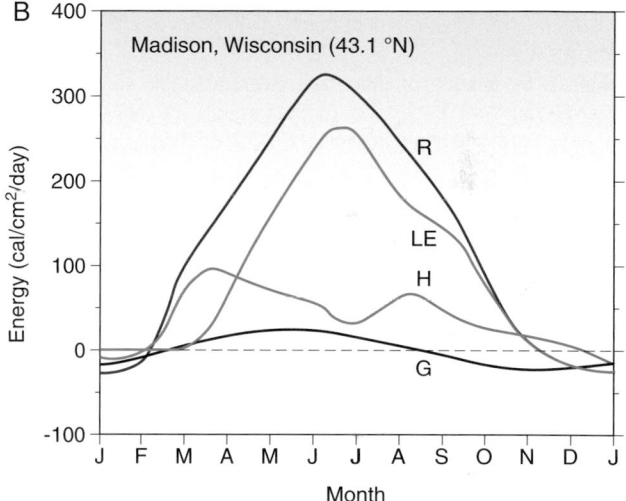

FIGURE 4.19
Surface energy budget through the course of a year at (A) Yuma, AZ, and (B) Madison, WI where R is net radiation absorbed, H is sensible heating (conduction plus convection), LE is latent heating (phase changes of water), and G is storage. [Modified after Sellers, W.D., 1965. *Physical Climatology*. Chicago: The University of Chicago Press, p. 106.]

heating) in deserts. Ocean waters cover much of Earth's surface so it is not surprising that the global Bowen ratio is relatively low (0.2).

The partitioning of net radiation absorbed at Earth's surface into sensible heating and latent heating varies through the course of a year. Consider two examples. Figure 4.19A shows the average annual variation of the surface energy budget for Yuma, AZ. R is the net radiation absorbed, H is the energy used for conduction and convection, LE is the energy used for latent heating, and G is the energy stored in the ground. In this desert locality, note that sensible heating exceeds latent heating throughout the year with the difference between the two reaching a maximum in June and the Bowen ratio is maximum in summer. Figure 4.19B shows the average annual variation of the surface energy budget for Madison, WI. In this much more humid locality, latent heating is greater than sensible heating from April to November and the Bowen ratio reaches a minimum in June and July.

Heat Imbalance: Tropics versus Middle and High Latitudes

On a global scale, imbalances in radiational heating and radiational cooling occur not only between Earth's surface and atmosphere, but also between the tropics and higher latitudes. Because the planet is nearly a sphere, parallel beams of incoming solar radiation strike the tropics more directly than higher latitudes. (That is, solar altitudes are higher in the tropics, but lower at higher latitudes.) At higher latitudes, solar radiation spreads over a greater area and is less intense per unit horizontal surface area than in the tropics.

Emission of infrared radiation by the Earth-atmosphere system also varies with latitude but less than solar radiation. Because air temperatures are generally lower at higher latitudes, IR emission also declines with increasing latitude. (Recall from Chapter 3 that radiation emission is temperature dependent.) Consequently, over the period of a year at higher latitudes, the rate of infrared cooling to space exceeds the rate of warming caused

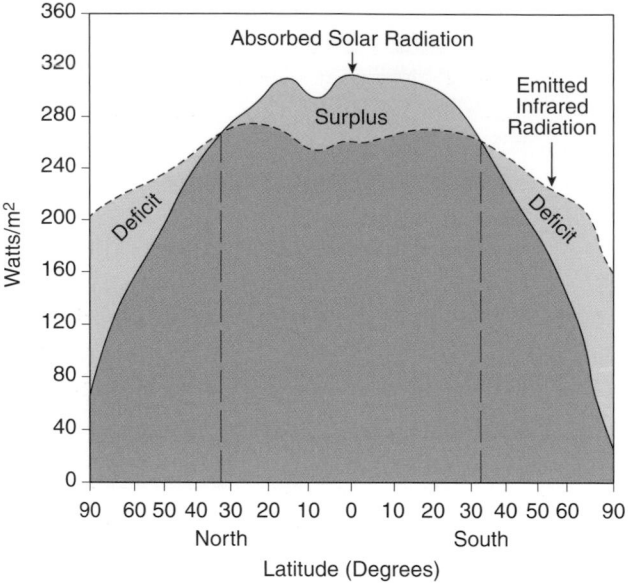

FIGURE 4.20
Variation by latitude of absorbed solar radiation and outgoing infrared radiation based on measurements by sensors flown onboard Earth-orbiting satellites. [NOAA/NESDIS]

by absorption of solar radiation. At lower latitudes the reverse is true, that is, over the course of a year, the rate of solar radiational heating is greater than the rate of infrared radiational cooling (Figure 4.20). Averaged over the globe, incoming energy (absorbed solar radiation) must equal outgoing energy (IR emitted to space) as the planetary average temperature remains nearly constant. That is, the areas under the two curves in Figure 4.20 are equal. This global radiative equilibrium illustrates the *law of energy conservation*. Long-term global climate change, however, alters IR emitted within the Earth-atmosphere system and may shift the planet to a new equilibrium state.

Measurements by sensors onboard Earth-orbiting satellites indicate that the division between regions of net radiational cooling and regions of net radiational warming is close to the 35-degree latitude circle in both hemispheres. By implication, latitudes poleward of about 35 degrees should experience net cooling over the course of a year, while tropical latitudes are sites of net warming. In fact, lower latitudes do not become progressively warmer nor do higher latitudes become cooler because heat is transported poleward from the tropics into middle and high latitudes. This **poleward heat transport** is brought about by (1) air mass exchange, (2) storm systems, and (3) ocean currents. According to research conducted at the National Center for Atmospheric Research (NCAR) at Boulder, CO, atmospheric processes account for 78% of total poleward

heat transport in the Northern Hemisphere and 92% in the Southern Hemisphere.

HEAT TRANSPORT BY AIR MASS EXCHANGE

North-south exchange of air masses transports mainly sensible heat from the tropics into middle and high latitudes. An **air mass** is a huge volume of air covering thousands of square kilometers that is relatively uniform horizontally in temperature and humidity (water vapor concentration). The properties of an air mass largely depend on the characteristics of the surface over which the air mass forms (its *source region*) and travels. Air masses that form at high latitudes over cold, often snow- or ice-covered surfaces are relatively cold. Air masses that form at low latitudes are relatively warm. Air masses that develop over the ocean are humid and those that form over land are relatively dry. Hence, there are four basic types of air masses: cold and humid, cold and dry, warm and humid, and warm and dry (Chapter 10).

Warm air masses that form at lower latitudes flow toward the poles and are replaced by cold air masses that flow toward the equator from source regions at high latitudes. Air masses modify (become cooler or warmer, drier or more humid) to some extent as they move away from their source region, gaining or losing heat energy and/ or moisture in the process. In this north-south exchange of air masses, a net transport of mostly sensible heat energy occurs directed from lower to higher latitudes.

HEAT TRANSPORT BY STORMS

Acquisition and subsequent release of latent heat in storm systems (*cyclones* or *lows*) plays an important role in the poleward transport of heat. At low latitudes, water that evaporates from the warm ocean surface is drawn into the circulation of a developing storm system. As the storm travels into higher latitudes, some of that water vapor condenses into clouds, thereby releasing latent heat to the troposphere. Latent heat of vaporization from low latitudes is thereby conveyed to middle and high latitudes. Because they transport much more water vapor and latent heat, tropical storms and hurricanes are greater contributors to poleward heat transport than ordinary middle latitude (*extratropical*) storms. This mechanism of poleward heat transport is readily apparent in water vapor satellite imagery (Chapters 1 and 6).

HEAT TRANSPORT BY OCEAN CIRCULATION

The ocean contributes to poleward heat transport via wind-driven surface currents and the deeper thermo-haline-driven circulation. Surface water that is warmer than the overlying air is a *heat source* for the atmosphere; that is, heat is transferred from sea to air via conduction, convection, and latent heating. Surface water that is cooler than the overlying air is a *heat sink* for the atmosphere; that is, heat is conducted from air to sea. Warm surface ocean currents, such as the Gulf Stream, flow from the tropics into middle latitudes, supplying heat to the cooler middle latitude troposphere (Figure 4.21). At the same time, cold surface currents, such as the California Current, flow from high to low latitudes, absorbing heat from the relatively warm troposphere and greater solar radiation in the tropics.

The ocean's **thermohaline circulation** is the density-driven movement of water masses, traversing the lengths of the ocean basins. The density of seawater increases with decreasing temperature and increasing salinity. More dense water tends to sink while less dense water rises. The thermohaline circulation transports heat energy, salt, and dissolved gases (e.g., carbon dioxide, oxygen) over great horizontal distances and to great depths in the ocean, and plays an important role in Earth's climate system. In the North Atlantic, for example, a warm surface ocean current flows north and eastward from the Florida Strait. At high latitudes, the surface waters cool, sink, and flow southward as cold bottom waters. This heat transporting circulation is a key component of the ocean's *meridional overturning circulation (MOC)*.

FIGURE 4.21
A composite infrared satellite image showing color-coded sea-surface temperatures. The Gulf Stream is clearly discernible as a ribbon of relatively warm water flowing along the East Coast from Florida north to off the North Carolina coast. [NOAA/NESDIS]

Why Weather?

As we have seen, imbalances in rates of radiational heating and radiational cooling give rise to temperature gradients between (1) Earth's surface and troposphere and (2) low and high latitudes. In response, heat is transported within the Earth-atmosphere system via conduction, convection, cloud development, air mass exchange, and storms. That is, the atmosphere circulates, bringing about changes in the state of the atmosphere (weather). A cause-and-effect chain thus operates in the Earth-atmosphere system, starting with the Sun as the prime energy source and resulting in weather.

We have also seen that within the Earth-atmosphere system, some solar radiation is absorbed, that is, converted to heat, and eventually all of this heat is emitted to space as infrared radiation. Some solar energy is also converted to kinetic energy, the energy of motion, in the circulation of the atmosphere. Kinetic energy is manifested in winds, convection currents, and the north-south exchange of air masses. Circulation (weather) systems do not last indefinitely, however. The kinetic energy of atmospheric circulation ultimately is dissipated as frictional heat as winds blow against Earth's surface. This heat, in turn, is emitted to space as infrared radiation. In summary, the Sun drives the atmosphere: imbalances in solar heating spur atmospheric circulation, which redistributes heat. Hence, solar energy is the ultimate source of kinetic energy in weather systems.

The rate of heat redistribution within the Earth-atmosphere system varies seasonally so that atmospheric circulation and weather also change through the year. When steep temperature gradients prevail across North America, the weather tends to be more dynamic. Storm systems are large and intense, winds are stronger, and the weather is changeable. Such weather is typical of winter, when it is not unusual for daily temperatures in the southern United States to be more than 30 Celsius degrees (54 Fahrenheit degrees) higher than temperatures across southern Canada. When air temperature varies little across the continent, as in summer, the weather tends to be more tranquil, and large-scale weather systems are generally weak and ill-defined. Nevertheless, summer weather is sometimes very active. Intense heating of the ground by the summer Sun coupled with lifting processes often triggers deep convection and development of thunderstorms. Some of these weather systems spawn destructive hail, strong and gusty winds, and heavy rains. However, these systems are usually shorter lived and more localized than winter storms.

Variation of Air Temperature

Air temperature is variable, fluctuating from hour to hour, from one day to the next, with the seasons, and from one place to another. Our discussion of heat transfer processes, Earth's radiation budget, and poleward heat transfer provides some insight as to why air temperature is so variable. The radiation budget plus movements of air masses regulate air temperature locally. Although these two controls actually work in concert, for purpose of study, we initially consider them separately.

LOCAL RADIATION BUDGET

Many factors govern the local radiation budget and air temperature, including the following: (1) latitude along with time of day and day of the year, which determine the solar altitude and the intensity and duration of solar radiation striking Earth's surface; (2) cloud cover, because cloudiness affects the flux of both incoming solar and outgoing terrestrial radiation; and (3) surface characteristics, which determine the albedo and the percentage of absorbed radiation (heat) used for sensible heating and latent heating. Hence, air temperature is generally higher in July than in January (in the Northern Hemisphere), during the day than at night, under clear rather than overcast daytime skies, when the ground is bare instead of snow-covered, and when the ground is dry rather than wet.

The annual temperature cycle (also called the march of mean monthly temperature) reflects the systematic variation in incoming solar radiation over the course of a year (Figure 4.22). In the latitude belt between the Tropics of Cancer and Capricorn, incoming solar radiation varies little through the course of a year so that average monthly air temperatures exhibit minimal seasonal contrast. In fact, in the tropics, the average temperature difference between summer and winter often is less than the average day-to-night temperature contrast. At middle latitudes, solar radiation features a pronounced annual maximum and minimum. At high latitudes, poleward of the Arctic and Antarctic circles, the seasonal difference in solar radiation is extreme, varying from near or at zero in fall and winter to a maximum in spring and summer. The marked seasonal periodicity of incoming solar radiation outside of the tropics accounts for the distinct winter-to-summer temperature contrasts typical of middle and high latitudes.

At middle and high latitudes, the march of mean monthly temperature lags behind the monthly variation in solar radiation so that the warmest and coldest months of the year typically do not coincide with the times of maximum and minimum solar radiation, respectively. The surface air temperatures take time to adjust to seasonal changes in energy flows. Typically, the warmest portion of the year is about a month after the summer solstice, and the coldest part of the year usually occurs about a month after the winter solstice. In the United States, the temperature cycle lags the solar cycle by an average of 27 days. However, in coastal localities with a strong maritime influence (e.g., Florida, the shoreline of New England, and coastal California), the average lag time is up to 36 days. In addition, as we saw earlier in this chapter, the maritime influence reduces the amplitude of the annual march of mean monthly temperature; that is, the winter-to-summer temperature contrast is less in maritime climates.

Over the course of a 24-hour day, surface air temperature responds to regular variations in the flux of radiation. With clear skies and light winds or calm air, the day's lowest (minimum) temperature typically occurs shortly after sunrise, after solar radiation equals outgoing Earth radiation. The day's highest (maximum) temperature is usually recorded in early or mid-afternoon, again when outgoing Earth radiation equals incoming radiation, even though solar radiation peaks around the local solar noon. Air temperature depends on the relative magnitudes of the incoming solar radiation and net outgoing infrared radiation (Figure 4.23). Beginning shortly after sunrise, incoming solar radiation exceeds outgoing IR and the air temperature rises. By early to mid-afternoon, downwelling of IR radiation from the atmosphere (the greenhouse effect) coupled with incoming solar radiation equals upwelling Earth radiation, causing the air temperature to reach its daily maximum. For the remainder of the day, outgoing IR exceeds incoming solar radiation (declining

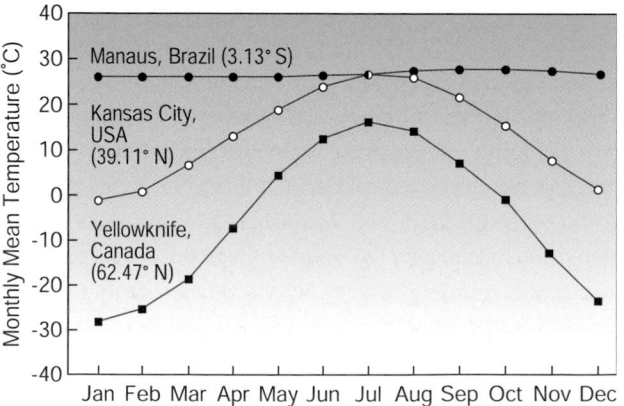

FIGURE 4.22
March of monthly mean temperature at Manaus, Brazil, Kansas City, MO, and Yellowknife, Canada.

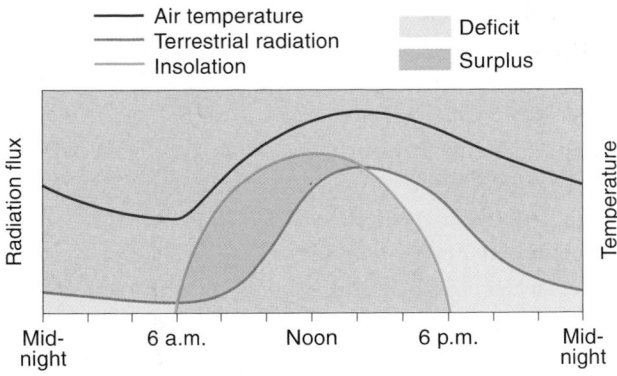

FIGURE 4.23
The variation in air temperature through the course of a 24-hour day depends upon the relative fluxes of incoming solar radiation (insolation) and outgoing infrared radiation

solar altitude) and the air temperature falls. Overnight, the only net radiative flux is outgoing IR radiation, continually cooling the surface and air above.

The risk of sunburn is usually highest around local solar noon (the time of peak solar altitude) and not necessarily during the warmest hours of the day. Incoming solar ultraviolet radiation, the cause of sunburn, is most intense at local noon, but the air temperature typically does not reach its daily maximum until several hours later.

As a further illustration of local radiational controls of air temperature, consider the influence of ground characteristics. All other factors being equal, in response to the same intensity of solar radiation striking Earth's surface, air over a dry surface (e.g., bare soil) warms more than air over a moist or vegetated surface. When the surface is dry, absorbed radiation is used primarily for sensible heating of the air (mainly by conduction and convection of heat from the surface into the overlying air). Hence, the air temperature is higher (and so is the Bowen ratio). On the other hand, when the surface is moist, much of the absorbed radiation is used to evaporate water, so there is less sensible heating and the air temperature is lower.

Dry soil helps explain why unusually high air temperatures often accompany *drought*, a lengthy period of moisture deficit. Soils dry out, crops wither and die, and lakes and other reservoirs shrink. Because less surface moisture is available for vaporization, more of the available heat is channeled into raising the surface temperature, which in tern raises air temperature through conduction and convection. Consider, for example, the severe drought that gripped a ten-state area of the southeastern United States between December 1985 and July 1986. In most places, rainfall was less than 70% of the long-term average, and in the hardest hit areas, portions of the Carolinas, it was less than 40%. By July, many weather stations in the drought-stricken region were setting new high temperature records. Columbia, SC, Savannah, GA, and Raleigh-Durham, NC reported the warmest July on record. Also contributing to record heat was more intense solar radiation reaching the ground, a consequence of less than the usual daytime cloud cover. The same association between exceptionally dry surface conditions and unusually high air temperatures was observed during the severe drought that afflicted Texas and other portions of the Southwest in 2011. At many long-term weather stations, the summer of 2011 was one of the driest and hottest on record.

Snow has a relatively high albedo and substantially reduces the amount of solar radiation that is absorbed at the surface and converted to heat. Furthermore, snow-covered ground reduces sensible heating of the overlying air because some of the available heat is used to vaporize or melt snow. Consequently, a snow cover lowers the day's maximum air temperature. Because snow is also an excellent emitter of infrared radiation, nocturnal radiational cooling is extreme where the ground is snow-covered, especially when skies are clear (minimum greenhouse effect). Cooling near the Earth's surface is further enhanced if winds are very light or the air is calm. Light winds or calm conditions reduce vertical mixing of air, keeping warmer air aloft from mixing downward. On such nights, the air temperature near the surface may be 10 Celsius degrees (18 Fahrenheit degrees) or more lower than if the ground were bare of snow. By reducing both the maximum and minimum daily air temperatures, a snow cover significantly lowers the 24-hr average mean temperature. For these reasons, the extent and duration of a snow cover has an important influence on the mean winter temperature.

COLD AND WARM AIR ADVECTION

Air mass advection refers to the movement of an air mass from one locality to another. With advection, one air mass replaces another air mass having different temperature (and/or humidity) characteristics. **Cold air advection** occurs when the wind transports colder air into a previously warmer area. On a weather map, cold air advection is indicated by winds blowing across regional isotherms from a colder area to a warmer area (arrow A in Figure 4.24). Cold air advection occurs behind a cold front (Chapter 10). **Warm air advection** takes place when the wind blows (across isotherms) from a warmer area to a colder area (arrow B in Figure 4.24). Warm air

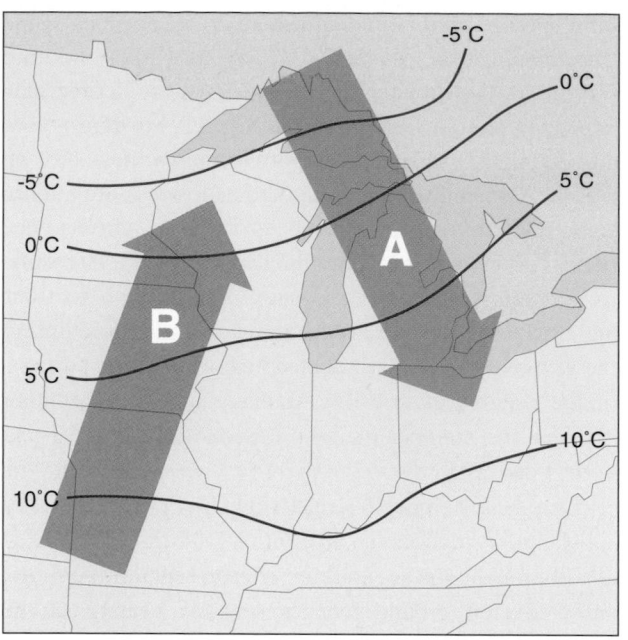

FIGURE 4.24
Cold air advection occurs when (A) horizontal winds blow across regional isotherms from colder areas toward warmer areas, and warm air advection occurs when (B) horizontal winds blow across regional isotherms from warmer areas toward colder areas. Solid lines are isotherms in °C.

advection occurs behind a warm front and ahead of a cold front (Chapter 10). An **isotherm** is a line drawn on a map through localities having the same air temperature. Recall from earlier in this chapter that air mass exchange is a major contributor to poleward heat transport.

The significance of air mass advection for variations in local temperature depends on (1) the initial temperature of the new air mass, and (2) the degree of modification the air mass undergoes as it travels over Earth's surface. For example, a surge of bitterly cold Arctic air loses much of its punch when it travels over ground that is not snow covered, because the Arctic air is warmed from below by sensible heating (conduction and convection). In contrast, modification of an Arctic air mass by sensible heating is minimized when the air mass travels over a cold, snow-covered surface.

So far, we have been describing how horizontal movement of air (advection) might influence air temperature at some locality. However, as we saw in our discussion of convection currents, air also moves vertically. As air ascends and descends, its temperature changes; air cools as it rises but warms as it descends. The reasons for these temperature changes are given in Chapter 5.

Although we have considered the local radiational budget and air mass advection separately, the two actually combine in regulating air temperature. Sometimes air mass advection acts with or against, or even overwhelms, local radiational influences on air temperature. As noted earlier, in response to the local radiation budget, the air temperature usually climbs from a minimum near sunrise to a maximum in early or mid-afternoon. This typical pattern can change, however, if an influx of cold or warm air occurs at the same time. Consider cold air advection. Depending on how cold the incoming air is, air temperatures may climb more slowly than usual, remain steady, or even fall during daylight hours. If cold air advection were extreme, air temperatures may drop precipitously throughout the day, in spite of bright, sunny skies. In another example, as a consequence of strong warm air advection air temperatures may climb through the evening hours, so the day's high temperature would occur at night.

URBAN HEAT ISLAND EFFECT

Anthropogenic activity, including land use, changes in land cover, and energy use, can alter the local radiation budget and thereby the local climate. An example is *urbanization* and the concurrent development and expansion of an **urban heat island**, identifiable by a pattern of closed isotherms encircling a metropolitan area that is significantly warmer than the surrounding rural area. In essence, a city is an island of warmth surrounded by cooler air. Chemist Luke Howard, better known for his classification of clouds, first noted the urban heat island effect in London in the 1810s, although he did not name the phenomenon.

Using land surface temperatures and vegetation data obtained from the Moderate Resolution Imaging Spectroradiometer (MODIS) flown onboard NASA's Terra satellite, Xiaoyang Zhang and colleagues at Boston University reported in 2004 that in 70 eastern North American cities, each with an area greater than 10 km^2 (4 mi^2), springtime land surface temperatures were an average 2.3 Celsius degrees (4.1 Fahrenheit degrees) higher than surrounding rural areas. In late autumn to winter, urban temperatures were on average 1.5 Celsius degrees (2.7 Fahrenheit degrees) higher than in the surrounding areas. In cities, nonagricultural vegetation began to bud 7 days earlier in spring and retained foliage 8 days longer in autumn compared to non-urban areas.

In 2010, Ping Zhang and Marc Imhoff at NASA's Goddard Space Flight Center reported results from a three-year study of satellite-derived land surface temperatures from 42 cities in the northeast United States.

They found that the urban heat island effect caused summer land surface temperatures to be on average 7 to 9 Celsius degrees (13 to 16 Fahrenheit degrees) warmer in cities than in surrounding rural areas. The heat island effect was most pronounced in densely-developed cities with compact urban cores, such as Providence, RI, Washington, DC, Philadelphia, PA, Baltimore, MD, Boston, MA, and Pittsburgh, PA. In contrast, sprawling cities such as Buffalo, NY, experienced a weaker urban heat island effect. This is illustrated through a comparison of Providence and Buffalo (Figure 4.25). Both cities are about the same size, but Providence is considerably more compact and is surrounded by dense forested areas compared to the higher percentage of farmland surrounding Buffalo. (NASA researchers have shown that cities ringed by forests, which cool more than agricultural areas, tend to have a greater temperature gradient between the city and rural fringe and hence a more pronounced heat island effect.) Providence had surface temperatures that were 12.2 Celsius degrees (21.9 Fahrenheit degrees) higher than surrounding rural areas,

compared to a heat island effect of 7.2 Celsius degrees (12.9 Fahrenheit degrees) at Buffalo. The method for comparing cities involves impervious surface area data produced by the USGS-operated Landsat satellite in conjunction with land surface temperature data derived from the MODIS onboard NASA's Aqua and Terra satellites.

The difference in urban/rural temperatures is usually greater at night than during the day and most apparent with calm air or light winds. Under these conditions, the nighttime temperature contrast between a city and its surroundings can be 10 Celsius degrees (18 Fahrenheit degrees), or even higher. When winds are strong and city air mixes with country air, however, the temperature contrast is greatly diminished.

One reason for the urban heat island effect is the relative lack of moisture in cities compared to nearby rural areas. City surfaces (e.g., facades, roofs, pavements, and sidewalks) are made of mostly impervious materials so, to prevent urban flooding, sewer systems intentionally carry away runoff from rain and snowmelt. On the other hand,

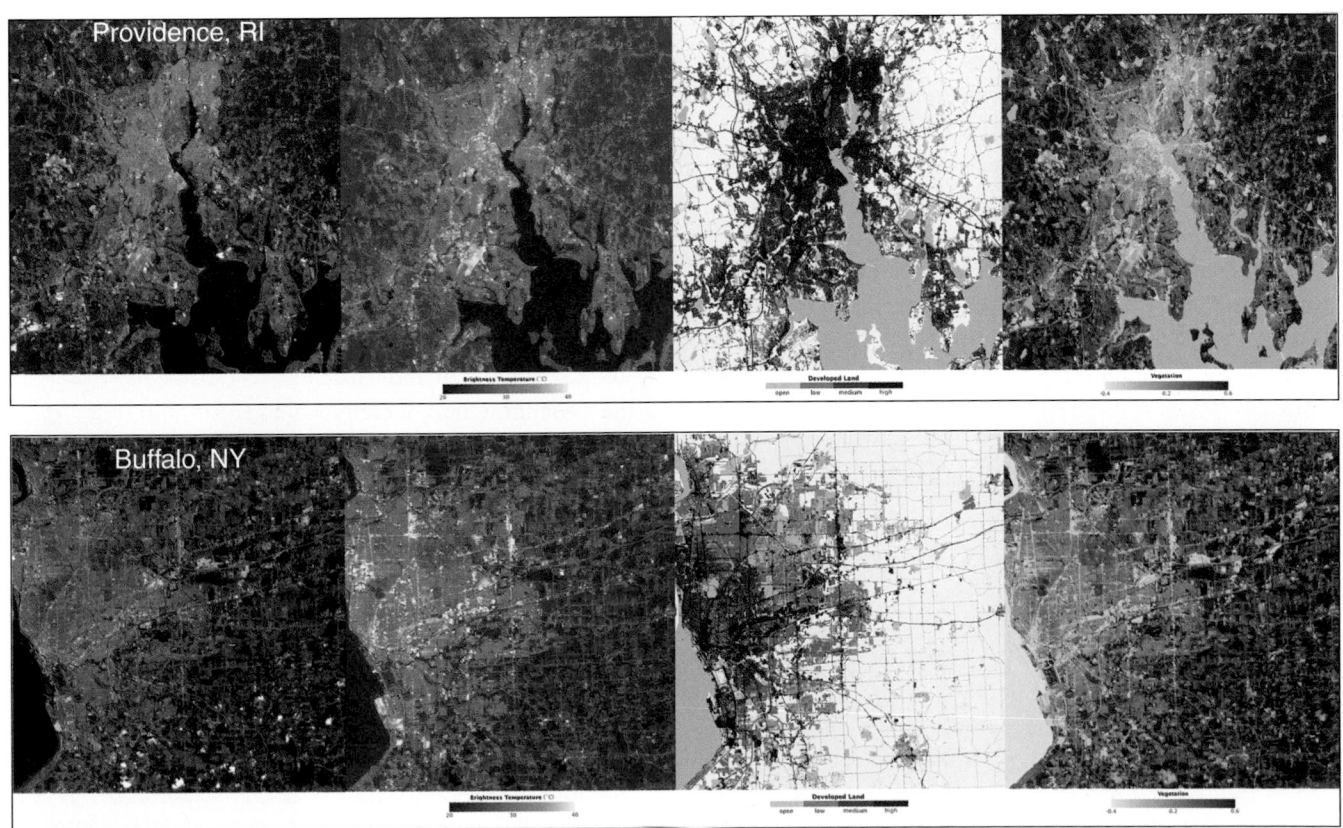

FIGURE 4.25
Satellite-produced maps of Providence. RI (top), and Buffalo, NY (bottom) showing (from left to right) visible light, surface heat, developed land, and vegetation cover. These images highlight the role that differences in development patterns and vegetation cover can have on the magnitude of a city's urban heat island. Though the two cities have the same approximate size, Providence has a significantly stronger heat island. [NASA/Earth Observatory]

the countryside typically has considerable standing water (e.g., lakes, rivers, moist soils) and much more vegetative cover for transpiration (emission of water vapor to the air by plants). The greater moisture outside of cities means that more heat energy is used to evaporate water (latent heating), while in the city, which has less moisture, heat energy goes directly to increasing surface temperatures (sensible heating). The average Bowen ratio is about four times greater in a city than in the countryside.

Other sources of heat (e.g., motor vehicles, space heaters, air conditioners) contribute to an urban heat island. On a cold winter day in New York City, heat from urban sources may approach 100 W/m², equivalent to about 7% of the solar constant. Also, a city's canyon-like terrain of narrow streets and tall buildings produces multiple reflections of sunlight (Figure 4.26), increasing the amount of solar radiation absorbed while the surfaces of cities generally have a lower albedo than the vegetative cover of rural areas. Urban building materials (e.g., concrete, asphalt, and brick) conduct heat more readily than the soil and vegetation of rural areas so the release of heat to the urban atmosphere from streets and the interior of buildings partially offsets radiational cooling, especially at night. In the downtown core of a city, the vertical facades of skyscrapers radiate heat to the surfaces of neighboring buildings rather than the cooler sky, further contributing to warming.

The urban heat island effect may make a modest contribution to global climate change. In 2011, Mark Jacobson and John Ten Hoeve of the Stanford University Atmosphere/Energy Program reported that the urban heat island explains 2 to 4% of the gross global warming observed since the Industrial Revolution. Is it possible that the urban heat island effect could be more significant on a regional level? Baode Chen at the Shanghai Typhoon Institute and colleagues found that urbanization could explain about 24% of the warming observed in eastern China between 1981 and 2007. They examined monthly mean temperature records from 463 weather stations located in a range of urban and rural environments. The greatest increase in the urban heat island effect coincided with a period of rapid urbanization in China.

With the world's future human population growth expected to continue to be greater in cities than rural areas, more people will be exposed to urban-enhanced heating. This has prompted research into strategies that might amend the urban heat island effect. One effective strategy is to construct a *green roof*, coating all or a portion of the upper surfaces of a building with a layer of soil and vegetation. A green roof is more reflective, reduces sensible heating, increases moisture, and can offer better insulation.

A simpler approach is to replace a low albedo roof (e.g., asphalt, tar) with one that has a high albedo (e.g., a roof painted white). Keith Oleson, a scientist at the *National Center for Atmospheric Research (NCAR)* in Boulder, CO, developed a computer model that calculates the flux of solar radiation that would be absorbed or reflected by various urban surfaces in a variety of different cities worldwide. He found that painting every roof white would reduce the urban heat island effect by a third, cooling the world's cities by an average of 0.4 Celsius degrees (0.7 Fahrenheit degrees). In some areas, such as the American Southwest, reducing the solar gain in buildings by making roofs and facades more reflective (Figure 4.27) would significantly reduce the demand for air conditioning, thereby decreasing energy demand and emissions of heat-trapping greenhouse gases. Reducing the urban heat island effect in this way would be most pronounced during the day and in summer. Other scientists have expressed skepticism about the benefit of white roofs. One recent study showed that painting roofs white may actually increase warming by altering atmospheric stability and cloud formation.

FIGURE 4.26
Buildings in New York City. A city's canyon-like terrain of tall buildings and narrow streets produces multiple reflections of sunlight. [Photo by K. Nugnes]

FIGURE 4.27
A tall building in Tucson, AZ. Being highly reflective reduces solar gain. [Courtesy of The Vizant Group]

Conclusions

Heat and temperature are distinct yet closely related quantities. Unequal rates of radiational heating and radiational cooling give rise to temperature gradients within the Earth-atmosphere system. In response to these temperature gradients, heat is transferred from warmer to colder localities via radiation, conduction, and convection. The temperature response of a substance to an addition or loss of heat depends primarily on its specific heat. The contrast in specific heat is one of the primary reasons why the temperature of the surface of a body of water is much less variable than that of land surfaces. This difference in thermal inertia influences the temperature of the overlying air mass so that regions downwind of the ocean (or large lakes) exhibit less temperature contrast between summer and winter than locations far from large water bodies.

Imbalances in radiational heating and cooling within the Earth-atmosphere system are ultimately responsible for the circulation of the atmosphere and ocean. These imbalances occur both vertically (between Earth's surface and the atmosphere) and horizontally (between tropical and higher latitudes). Through atmospheric circulation, heat is redistributed within the Earth-atmosphere system. This enables us to understand how air temperature is regulated by a combination of local radiational controls and air mass advection.

Another major consequence of atmospheric circulation is the formation of clouds that can produce rain, snow, and other forms of precipitation. Before examining cloud and precipitation forming processes in detail, we first consider the behavior of gases, especially as it involves air pressure, another important variable of the atmosphere.

Basic Understandings

- Temperature is directly proportional to the average kinetic energy of the atoms or molecules composing a substance. Heat is the name given to energy transferred from a warmer object to a colder object.
- As a sample of air gains or loses heat, that heat may be used for some combination of changes in temperature, phase of water, or volume of the air sample.
- The Fahrenheit and Celsius temperature scales are based upon the reproducible phase changes in water. While the Fahrenheit temperature scale is still commonly used in the United States, the Celsius temperature scale is more convenient in many scientific and other applications in that a 100-degree interval separates the freezing and boiling points of pure water at sea level and its degree unit is equivalent to a Kelvin unit. The Kelvin scale is based on absolute zero and is a more direct measure of average kinetic molecular activity than either the Fahrenheit or Celsius temperature scales. An object at absolute zero (0 kelvin) emits no electromagnetic radiation. Heat energy is quantified as calories, joules, or British thermal units.

- Common types of thermometers are liquid-in-glass, electronic, and bimetallic strips. Thermometers should be mounted where they are well ventilated but sheltered from precipitation, direct sunshine, and the night sky.

- Heat always flows in response to temperature gradients from locations of higher temperature to locations of lower temperature. This is a consequence of the second law of thermodynamics. Heat transfer occurs via radiation, sensible heating (conduction and convection), as well as by latent heat (accompanying phase changes of water).

- All objects absorb and emit electromagnetic radiation, an energy transport mechanism that does not require a physical medium. If absorption exceeds emission (radiational heating), the temperature of an object rises, and if emission exceeds absorption (radiational cooling), the temperature of an object falls. At radiative equilibrium, absorption balances emission and the temperature of the object is constant.

- As a rule, solids (especially metals) are better conductors of heat than are liquids, and liquids are better conductors than gases. Motionless air is a very poor conductor of heat. Convection is the transport of heat within a fluid via motion of the fluid itself and is much more important than conduction in transporting heat within the troposphere.

- When water changes phase, heat is either absorbed from or released to the environment. Latent heat is absorbed during melting, evaporation, and sublimation. Latent heat is released during freezing, condensation, and deposition.

- The temperature response to an input or output of heat differs from one substance to another depending primarily on the specific heat of each substance. Water bodies, such as the ocean or large lakes, exhibit less temperature variability from day to night, and from summer to winter, than do landmasses primarily because the specific heat of water is greater than that of land. Also, solar radiation penetrates water but soil is opaque to solar radiation, and water circulates. The winter-to-summer temperature contrast is greater in continental climates than in maritime climates. In addition, compared to continental locales, maritime locales experience a greater lag in the temperature response to seasonal variations in incoming solar radiation.

- On a global average annual basis, radiational cooling is greater than radiational heating of the atmosphere. On the other hand, radiational heating is greater than radiational cooling of Earth's surface. In response, heat is transported from the warmer Earth's surface to the cooler troposphere via latent heating (vaporization of water followed by cloud development) and sensible heating (conduction plus convection).

- The ratio of sensible heating to latent heating, the Bowen ratio, depends on the amount of moisture at the Earth's surface. The Bowen ratio is relatively high for dry surfaces and relatively low for wet surfaces. The global Bowen ratio is relatively low at 0.20 primarily because much of the planet is covered by water that is subject to evaporation.

- Poleward of about 35 degrees latitude, over the course of a year, the rate of cooling due to infrared emission to space is greater than the rate of warming due to absorption of incoming solar radiation. In tropical latitudes, on the other hand, the rate of warming due to absorption of solar radiation is greater than the rate of cooling due to emission of infrared radiation. Poleward heat transport within the Earth system is the consequence.

- Poleward heat transport is brought about by north-south exchange of air masses, release of latent heat in storm systems, and ocean circulation (wind-driven surface currents and the thermohaline circulation).

- Energy from the Sun drives the circulation of the atmosphere. Because its receipt is not uniform, heat imbalances occur within the Earth-atmosphere system. Some solar energy is converted to the kinetic energy of atmospheric circulation, which is ultimately dissipated as frictional heat that is radiated to space.

- The radiation budget plus air mass advection govern variations in local air temperature. The latitude of a locale has a major influence on the radiation budget as it is related to the amount of incident solar radiation. The local radiation budget varies with time of day, time of the year, cloud cover, and Earth's surface properties. Air mass advection occurs wherever warm air replaces cold air or cold air replaces warm air. Cold or warm air advection reinforces, compensates for, or even overwhelms the influence of the local radiation budget on air temperature.

- At middle and high latitudes, the march of mean monthly temperature lags behind the monthly variation in solar radiation so that the warmest and coldest months of the year do not coincide with the times of maximum and minimum incoming solar radiation.

- With clear skies, light winds or calm air, the 24-hr daily minimum air temperature typically occurs near sunrise and the maximum air temperature usually is recorded in the early to mid afternoon.

- Record high air temperatures often accompany a prolonged and intense drought.

- By reducing both the maximum and minimum air temperatures, a snow cover significantly lowers the 24-hr mean temperature. For this reason, the average seasonal duration of snow-covered ground influences the climate.

- Mean annual air temperature is a degree or two higher in a city than in the surrounding countryside. Contributing to the formation of an urban heat island are several factors. Compared to rural areas, cities have (1) less standing water and moist surfaces (higher Bowen ratio), (2) a greater concentration of heat sources (e.g., motor vehicles), (3) a lower surface albedo, (4) multiple reflections of sunlight within the cityscape, and (5) component materials that more readily conduct heat (i.e., brick, asphalt, concrete) and store heat for release at a later time.

Enduring Ideas

- Temperature is directly proportional to the average kinetic energy of atoms and molecules composing a substance and is measured with a thermometer. Heat is energy that flows in response to a temperature gradient from where it is warmer to where it is colder. Within the Earth-atmosphere system, heat is transferred via radiation, conduction, and convection.
- When water changes phase, heat energy is either absorbed from the environment (i.e., melting, evaporation, sublimation) or released to the environment (i.e., freezing, condensation, deposition).
- The exceptionally high specific heat of water is the principal reason maritime climates have a smaller diurnal and annual temperature range than continental climates.
- The Sun is the ultimate source of kinetic energy in weather systems. Imbalances in solar heating give rise to temperature gradients between Earth's surface and troposphere, and between low and high latitudes, spurring atmospheric circulation (the motion of air).
- The air temperature is governed by a combination of the local radiation budget (involving both incoming solar radiation and outgoing infrared radiation) and air mass advection.

Key Terms

temperature	radiational heating	latent heat of vaporization
internal energy	radiational cooling	cumulus clouds
heat	conduction (of heat)	cumulonimbus clouds
absolute zero	heat conductivity	Bowen ratio
calorie (cal)	convection	poleward heat transport
British thermal unit (Btu)	sensible heating	air mass
thermometer	latent heat	thermohaline circulation
thermograph	latent heating	air mass advection
response time	specific heat	cold air advection
temperature gradient	thermal inertia	warm air advection
second law of thermodynamics	maritime climate	isotherm
radiation	continental climate	urban heat island
absorption	latent heat of fusion	

Review

1. Describe the relationship between the kinetic energy of the atoms or molecules composing a substance and the temperature of that substance.
2. The Kelvin scale has no negative values. Explain why.
3. What is meant by the *response time* of a thermometer? What type of thermometer has the fastest response time?
4. Explain how heat transfer follows the second law of thermodynamics.
5. Distinguish between sensible heating and latent heating of the atmosphere. Which is more important on a global annual average basis?
6. During which phase changes of water is latent heat released to the environment?
7. On a global annual average basis, what is the most important process in cooling Earth's surface?
8. Explain how thunderstorms transfer heat from Earth's surface to the middle and upper troposphere.
9. Describe the mechanisms involved in poleward heat transport within the Earth-atmosphere system.
10. Provide some examples of how Earth's surface properties influence air temperature in the lower troposphere.

Critical Thinking

1. A traffic sign along an Ohio highway warns motorists that a bridge surface freezes before the road surface. Why does the bridge surface freeze first?
2. In northern climates in winter, how does the depth of the snow cover influence the depth to which the ground freezes?
3. Compare the advantages and disadvantages of the Celsius versus Fahrenheit temperature scales for weather observation and forecasting.
4. At what air temperature is the reading on the Celsius scale the same as that on the Fahrenheit scale?
5. Convert the following Celsius temperatures to Fahrenheit temperatures: 0°C, 100°C, 5°C, 20°C, -10°C.
6. Calculate the temperature change in Celsius degrees when (a) 10 cal of heat is added to 5 g of water; (b) 100 cal of heat is added to dry sand having a mass of 100 g; (c) 15 cal of heat is added to 5 g of ice at -10°C.
7. Air, especially still air, is a poor conductor of heat. Present some examples of how this property of air is used to conserve energy for space heating and cooling.
8. Compute the number of heating degree-days on a day when the minimum air temperature was 13°F and the maximum air temperature was 26°F.
9. How would you expect the Bowen ratio of agricultural land to change as a drought intensifies? Explain your response.
10. How and why would the temperature of an Arctic air mass change as the air mass moves from snow-covered ground to bare ground?

Managing Weather Risk

Businesses commonly purchase insurance to protect themselves against losses caused by high risk, low probability weather extremes (e.g., hurricanes, floods). But how can businesses protect themselves against decreased revenues arising from low risk, high probability weather events (e.g., mild winter, exceptionally cool summer, decreased snowfall)? These businesses can minimize their losses by purchasing a weather derivative, a risk-management contract designed to protect a weather-sensitive business against weather-related losses.

According to the Weather Risk Management Association (WRMA), there are a series of steps fundamental to a weather-sensitive business establishing a risk management solution. First, the business identifies the critical weather variable(s) and then quantifies the impacts of the weather on its revenues, margins, profits and/or costs. The business finds a reliable, neutral source of historical and current weather data (e.g., as archived by NOAA's National Weather Service) and determines the time period

FIGURE 1
The Chicago Mercantile Exchange, where most weather derivative contracts are traded. [Photo by liz_noise//Creative Commons Attribution 2.0 Generic license]

of the weather variable's influence. The business then quantifies the relationship between changes in the weather variable and impact on the weather-sensitive financial parameter and translates the sensitivity in financial parameter changes into terms of the weather variable.

There are three main types of weather risk management programs: aggregate measures of weather, adverse days, and adverse events. Aggregate measures of weather include variables such as average temperatures, cumulative heating degree-days, and total snowfall. Adverse days refer to the number of days when a particular weather variable is above or below a critical value, and adverse events describe a program sensitive to the occurrence of a specific weather condition during the contract period.

The business transfers its risk into a weather derivative contract with a risk taker, often a hedge fund, insurance company, or energy trader. The business pays a premium to the risk taker and both parties establish a payment plan to the business should adverse weather conditions occur during the contract period. Most weather derivative contracts are traded on the Chicago Mercantile Exchange (Figure 1).

Consider the following example of an aggregate measures of weather contract. An electrical utility has decreased revenues during a mild winter due to a reduction in the number of kilowatt hours (KWHs) consumed by its customers as well as the decreased price per KWH in the energy trading market. In order to hedge against future mild winters, the utility company purchases a weather derivative. The derivative buyer (DB) and derivative seller (DS) agree that the contract will be based on heating degree-days. (In this example, DB is the electrical utility company and DS is the risk taker.) As detailed in this chapter's second *For Further Exploration*, heating degree-days are a measure of household energy consumption for space heating. Heating degree-days are computed for days when the mean daily outdoor temperature is less than 65 °F and cumulative totals of degree-days are maintained throughout the heating season.

Prior to the contract period, DB pays DS a premium to assume DB's risk. Meteorologists working for DS prepare a probabilistic forecast of seasonal heating degree-days for DB's location based on climatological data or long-range weather outlooks issued by the National Weather Service (Figure 2). DB and DS agree on a threshold value for the number of heating degree-days accumulated during the period of the contract (i.e., the upcoming winter heating season) and a formula for payment. If the actual number of heating degree-days is less than the threshold value, DS pays DB an amount based on the difference between the threshold and actual values and the effect of that difference on consumer demand for electricity. Thereby, the electrical utility company offsets its diminished revenue. If the actual number of heating degree-days is greater than or equal to the threshold value, no payment is made to DB, and the premium becomes profit for DS. In this case, even though the electrical utility pays a premium, they are protected against large revenue losses.

The weather risk market started in the fall of 1997 when the former energy trader Enron Corporation agreed to pay the utility Koch Industries Inc. $10,000 for each degree in temperature below the winter average. The market experienced considerable growth during the winters of 1997-98 and 1998-98, which were influenced by a strong El Niño and La Niña, respectively. According to the WRMA, the market experienced a lull during the energy crisis in 2001-02, and then increased 100-fold from 2004-07 until declining during the financial downturn of 2008-09. The market for customized weather derivatives grew by nearly 30% from 2010 to 2011 with the overall market increasing by 20%. The market has expanded in recent years and now encompasses a variety of sectors in addition to energy, including agriculture, construction, transportation, and entertainment.

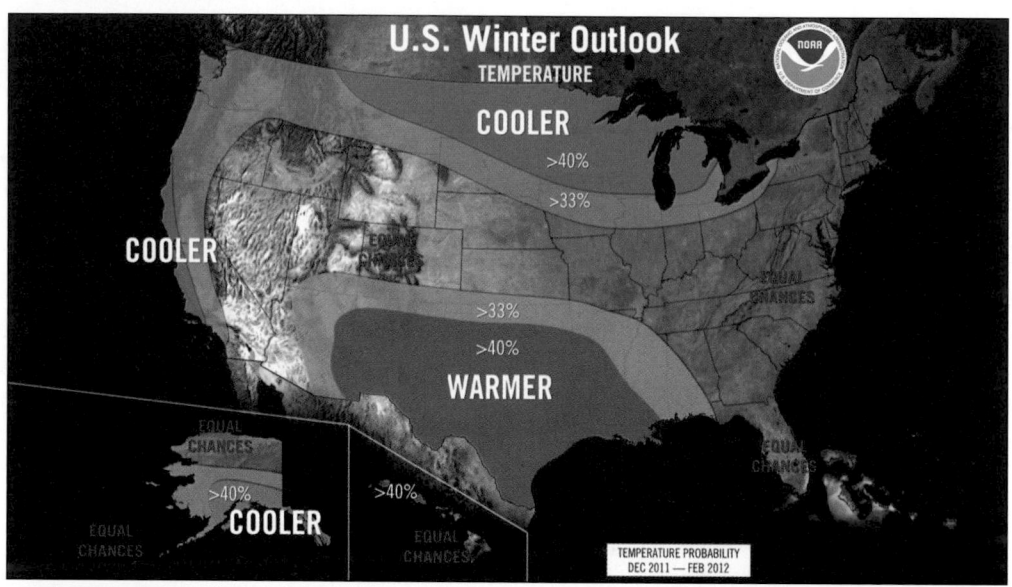

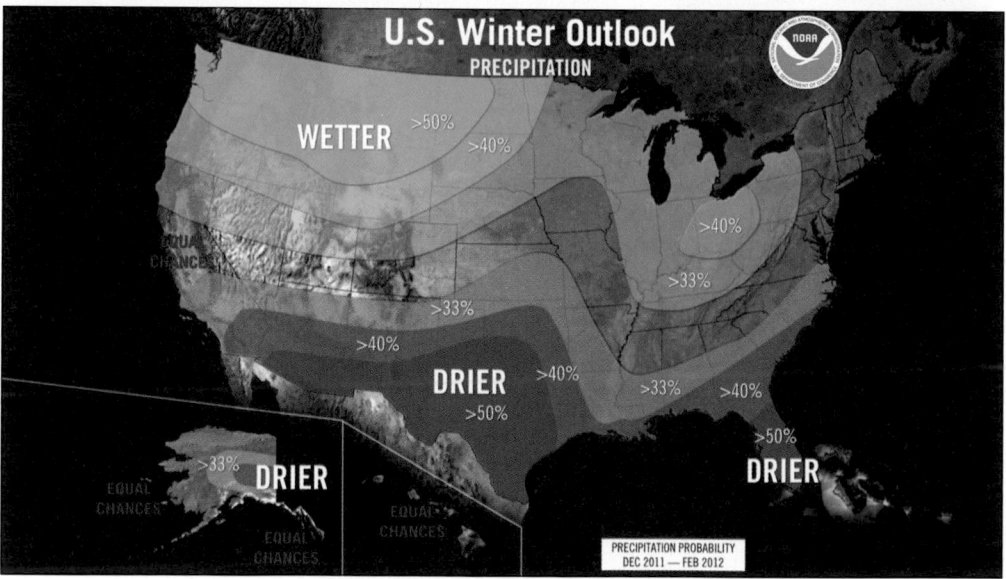

FIGURE 2
Temperature (top) and precipitation (bottom) outlooks generated by NOAA's National Weather Service. [NOAA]

Heating and Cooling Degree-Days

In some regions of the country, television and newspaper weather summaries routinely report heating or cooling degree-day totals in addition to daily maximum and minimum temperatures. Heating and cooling degree-days are indicators of household energy consumption for space heating and cooling, respectively.

In the United States, *heating degree-days* are based on the Fahrenheit temperature scale and are computed only for days when the average outdoor air temperature is lower than 65 °F (18 °C). Heating engineers who formulated this index early in the 20th century found that when the average outdoor temperature drops below 65 °F, space heating is required in most buildings to maintain an average indoor air temperature of 70 °F (21 °C). The average daily temperature is the simple arithmetic average of the 24-hr maximum and minimum air temperatures. Each degree of average temperature below 65 °F is counted as *one heating degree-day*. Subtracting the average daily temperature from 65 °F yields the number of heating degree-days for that day. Suppose, for example, that this morning's low temperature was 36 °F, and this afternoon's high temperature was 52 °F. Today's average temperature would then be 44 °F, for a total of 21 heating degree-days (65 − 44 = 21). It is usual to keep a running total of heating degree-days, that is, to add degree-days for successive days through the heating season (compiled from 1 July of one year through 30 June of the next).

Fuel distributors and power companies closely monitor heating degree-days. Fuel oil dealers base fuel use rates on cumulative degree-days and schedule deliveries accordingly. Natural gas and electrical utilities anticipate power demands on the basis of degree-day totals, and implement priority use policies on the same basis when capacity fails to keep pace with demand.

The map in Figure 1 is a plot of the average annual heating degree-day totals over the coterminous United States. Outside of mountainous areas, regions of equal heating degree-day totals tend to parallel latitude lines with degree-day totals increasing poleward. As an example, the average annual space-heating requirement in Chicago (6100 heating degree-days) is about four times that of New Orleans (1500 heating degree-days). If per unit fuel costs were the same in both cities, then, in an average winter, Chicago homeowners can expect to pay four times as much for space heating as homeowners in New Orleans. This assumes that buildings in the two cities are comparable in structure and insulation.

Cooling degree-days are computed only for days when the average outdoor air temperature is higher than 65 °F (although higher base temperatures are sometimes used). Supplemental air conditioning may be needed on such days.

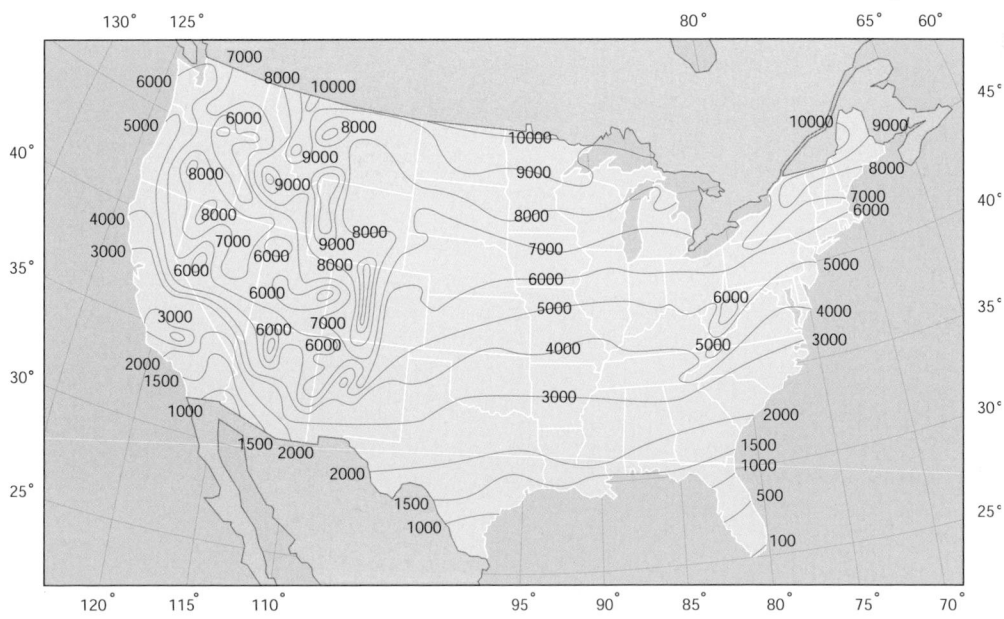

FIGURE 1
Average annual heating degree-day totals over the lower 48 states using a base of 65 °F.

Again, a cumulative total is maintained through the cooling season (1 January through 31 December). Across the U.S., average annual cooling degree-day totals range from less than 500 in the northern tier states and along much of the Pacific coast to more than 4000 in South Texas, South Florida, and the Desert Southwest.

Indices of heating and cooling requirements are based only on outdoor air temperatures and do not take into account other weather elements, such as air circulation and humidity, which also influence human comfort and demands for space heating and cooling. Heating and cooling degree-days are therefore only approximations of residential fuel demands for heating and cooling.

Wind Chill

At low air temperature, the wind increases human discomfort outdoors and heightens the danger of *frostbite*, the freezing of body tissue, or *hypothermia,* a potentially lethal condition brought on by a drop in the temperature of the body's vital organs (e.g., heart, lungs). Air in motion is more effective than still air in transporting heat away from the body. To account for the effect of wind on human comfort and wellbeing, weather reports during winter in northern and mountainous regions include the *wind chill equivalent temperature (WET)* along with the actual air temperature.

A very thin layer of motionless air (thickness measured in millimeters) next to the skin helps insulate the body from heat loss to the environment. Within this *boundary layer*, heat loss is through the very slow process of conduction. (Recall that calm air is a poor conductor of heat.) The boundary layer is thickest and offers maximum insulation when the wind speed is under 0.5 mph (0.2 m per sec). Because water vapor molecules diffuse slowly through the skin's boundary layer, the rate of evaporative cooling is also relatively slow. Hence, even at very low air temperatures, people who are appropriately clothed are comfortable as long as winds are light. As winds strengthen, however, the boundary layer becomes thinner and the rate of heat and water vapor transfer from the skin surface increases. Thinning of the boundary layer (and accelerated heat loss) is most pronounced at low to moderate wind speeds. As wind speed increases above 35 mph (55 km per hr), incremental heat loss becomes relatively small as the boundary layer approaches its minimum thickness and is essentially eroded away.

The WET, also called the *wind chill index*, is a measure of the rate of body heat loss due to a combination of wind and low air temperature. The original index was based upon research conducted in Antarctica by P.A. Siple and C.F. Passel during the winter of 1941. Their original objective was to measure the time required for water to freeze under various weather conditions. Only later did the idea arise to develop a wind chill index based upon this research. Their experiments using a small water-filled non-insulated plastic cylinder provide, at best, a rough approximation of the thermal response of a human body dressed appropriately for cold and windy weather. Their index also fails to account for the appreciable warming that occurs when the body absorbs solar radiation—Siple and Passel made their measurements in darkness.

In the winter of 2001-02, the National Weather Service (NWS) and the Meteorological Services of Canada (MSC) implemented a major revision of the wind chill index (Figure 1). The new index incorporates recent advances

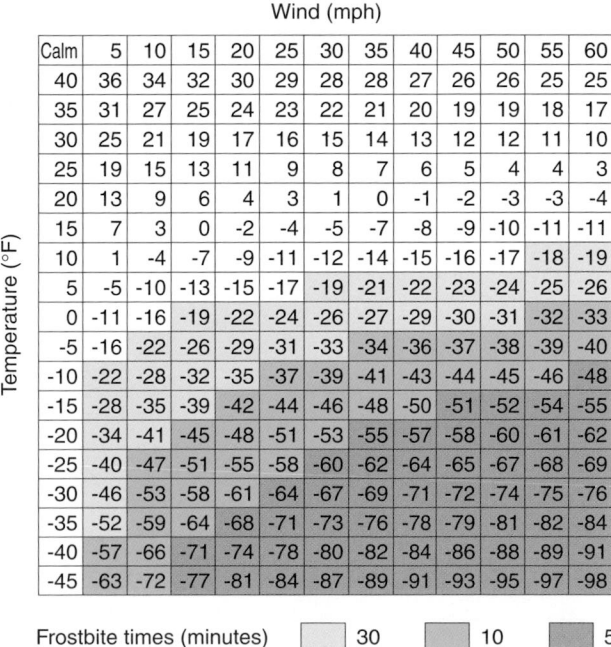

FIGURE 1
Wind chill index introduced by NOAA's National Weather Service in the winter of 2001-02.

in science, technology, and computer modeling to provide a more accurate, understandable, and useful formula for representing the combined hazard of wind and low air temperatures. The new wind chill index uses wind speed adjusted for the average height above ground (1.5 m or 5 ft) of a human's face instead of standard anemometer height (10 m or 33 ft), applies modern heat transfer theory, and assumes the worst case scenario for solar radiation (clear night sky). The formula for computing the WET in degrees Fahrenheit is:

$$\text{WET (}^\circ\text{F)} = 35.74 + 0.6215T - 35.75 \, (V^{0.16}) + 0.4275T \, (V^{0.16})$$

where T is the temperature in °F and V is the wind speed in miles per hour. This formula is the basis for the wind chill Table. (For computing wind chill, calm is defined as 3 mph or less.)

Suppose, for example, the air temperature is 35 °F and the air is calm; the WET (35 °F) is the same as the actual air temperature. If the wind strengthens to 25 mph with no change in actual air temperature, the WET drops to 23 °F. Contrary to popular opinion, this does not mean that the temperature of exposed skin actually drops to 23 °F. Skin temperature can drop no lower than the actual air temperature, which in this example is 35 °F. When skin temperature equals the air temperature, the temperature gradient between the skin surface and adjacent air is zero and no heat is exchanged between the skin and the air. Even though the WET may fall below the freezing point, skin will not freeze if the actual air temperature is above the freezing point.

On a cold and windy day, an exposed body part loses heat to the environment at the same rate that it would if the air were calm and the actual air temperature equaled the wind chill equivalent temperature. In fact, many combinations of air temperature and wind speed produce about the same rate of heat loss. For instance, heat loss is essentially the same for air temperature/wind speed combinations of 15 °F/5 mph and 25 °F/35 mph. In both cases, the wind chill equivalent temperature is 7 °F.

During cold and blustery weather (Figure 2), if the human body cannot supply heat to the skin at a rate sufficient to compensate for heat loss, skin temperature declines. Initially a person has a sense of discomfort. If the rate of heat loss

FIGURE 2
Person bending into the wind during a Midwest blizzard in March 1966. Cold and blustery weather can cause discomfort and heat loss. [NOAA's National Weather Service (NWS) Collection]

from exposed skin causes skin temperature to fall to subfreezing levels, then frostbite may ensue. Particularly susceptible to frostbite are body parts that are usually exposed and have a relatively high surface-to-volume ratio, such as the ears, nose, and fingers. Frostbite may occur in 30 minutes or less at wind chill values of -18 °F (-28 °C) or lower (Figure 1). In extreme circumstances, the rate of heat loss from the body may be great enough to cause life-threatening hypothermia.

Some of the Earth's most extreme wind chill values are measured on Antarctica in winter. Scientists from around the world spend months at established international scientific research stations. In Figure 3A, Dr. Alan Robock, Distinguished Professor of Meteorology at Rutgers University, wears Antarctic protective gear to prevent frostbite. Figure 3B shows a weather station display inside an Antarctic research station showing a wind speed of 21 kts (24 mph) measured by an outdoor sensor, and a calculated wind chill value of -59 °C (-74 °F). Using Figure 1, we can infer from these values an outdoor temperature of about -38 °C (-37 °F).

A

B

FIGURE 3
(A) Dr. Alan Robock, wears Antarctic protective gear to prevent frostbite in extremely low Antarctic temperatures. (B) Weather station (display screen indoors) measuring an outside wind chill of -59 °C (-74 °F) in Antarctica on 23 August 2004. [Photos copyright Alan Robock]

An F-16 Aggressor soars above snow capped mountains in the Joint Pacific Alaska Range Complex. To protect airmen from the insufficient oxygen available at such high altitudes (and such low barometric pressure), a pressure suit is necessary. [U.S. Air Force photo/Staff Sgt. Christopher Boitz]

AIR PRESSURE

Chapter Highlights

Case-in-Point
   Air Pressures on Mount Everest
Defining Air Pressure
Air Pressure Measurement
Air Pressure Units
Variation in Air Pressure with Altitude
Horizontal Variations in Air Pressure
   Influence of Temperature and Humidity
   Influence of Diverging and Converging
      Winds
Highs and Lows
The Gas Law
Expansional Cooling and Compressional
   Warming
   Conservation of Energy
   Adiabatic Process
Conclusions
Basic Understandings/Enduring Ideas
Key Terms/Review/Critical Thinking
For Further Exploration
   Human Responses to Changes in Air
      Pressure
   Comparing Air and Water Pressure
   Determining Altitude from Air Pressure

Learning Objectives

Define air pressure.
Identify the advantages of an aneroid barometer over a mercury barometer.
Explain the significance of air pressure tendency for local weather forecasting.
Describe how air pressure and air density change with altitude.
Explain how and why meteorologists adjust air pressure readings to sea level.
Describe how air temperature and water vapor concentration influence the density of air and air pressure at Earth's surface.
Demonstrate how divergence and convergence of horizontal winds can cause changes in air pressure.
Discuss how surface air pressure varies with different types of weather systems.
Show how the gas law applies to the atmosphere.
Explain why ascending air cools whereas descending air warms.
Define an adiabatic process.
Distinguish between the dry adiabatic lapse rate and the moist adiabatic lapse rate.

What is the significance of horizontal and vertical variations in air pressure?

Case-in-Point

Air Pressures on Mount Everest

At 8850 m (29,035 ft) above sea level, the summit of Mount Everest is the highest mountain peak in the world (Figure 5.1). This massive glacier-encrusted mountain is located at the edge of the Tibetan Plateau on the border between Tibet and Nepal in the central Himalayas. In 1856, the mountain was named for Sir George Everest (1790-1866), surveyor general of India. For most Nepali people, however, the mountain is Sagarmatha ("forehead in the sky") whereas the Sherpa people of northern Nepal call it Chomolungma ("Goddess Mother of the World"). Local peoples revered the mountain as sacred and did not attempt to scale it prior to the 20th century. For many years, travel restrictions on foreign visitors, brutal weather, the

FIGURE 5.1
The peak of Mt. Everest. [Photo by Geoff Childs, Washington University in St. Louis]

threat of massive snow and ice avalanches on the mountain, plus the thin air at higher elevations were insurmountable barriers for climbers intent on conquering the peak. The summit of Mount Everest was not reached until 29 May 1953 when scaled by Sir Edmund Hillary (1919-2008) of New Zealand and Sherpa Tenzing Norgay (1914-1986) of Nepal. From then until the close of the 2011 climbing season, there have been 5532 total ascents to the summit. Unfortunately, some 223 lives have been lost during that same time period.

The latitude of Mount Everest (28 degrees N) is about the same as Tampa, FL, but the rapid decline of air temperature with elevation means that the upper reaches of the mountain are perpetually at subfreezing temperatures. Estimated January mean temperature at the summit of Mount Everest is −36 °C (−33 °F). In July, the warmest month, the mean temperature is about −19 °C (−2 °F). From June through September, moist winds blowing inland from the Indian Ocean (the wet monsoon) shroud the mountain in thick clouds and heavy snows. From November through February, the upper tropospheric jet stream (a narrow corridor of very strong winds) dips down from the north, displacing the monsoon flow, and bringing winds to Mount Everest that frequently blow in excess of hurricane force. High winds combined with low temperatures make frostbite and hypothermia continual hazards for climbers (Chapter 4). Because of the threat of high winds through April and the onset of heavy snows by June, most ascents of Mount Everest take place in May.

Climbers must breathe air that is progressively thinner with increasing elevation; that is, the number of air molecules per unit volume decreases. Although the ratio of nitrogen (N_2) to oxygen (O_2) remains constant at about 4 to 1, the decline in the amount of oxygen (coupled with the need for more oxygen due to stress caused by extreme weather) can quickly lead to life-threatening conditions. The decline in air density with altitude is accompanied by a decrease in air pressure (the weight per unit area of the overlying column of air). At the summit of Mount Everest, the partial pressure exerted by oxygen is only a third of its sea level value. Without a supplemental oxygen supply, people cannot survive for very long at the summit. Climbers attempt to cope with these extreme conditions by ascending gradually to higher altitudes—taking several weeks for their bodies to adjust to the thin air. But even with *acclimatization* (discussed in this chapter's first *For Further Exploration*), the summit of Mount Everest is so high and the air so thin that until recently experts considered survival near or at the summit to be impossible without a supplemental oxygen supply. The experts were proved wrong in 1978 when the Italian mountaineer Reinhold Messner and Austrian mountaineer Peter Habeler reached the summit without a supplemental oxygen supply and survived. In spite of this remarkable feat, the extreme environment at the summit of Mount Everest remains very close to the physiological limits of human life.

In describing the state of the atmosphere, television and radio weathercasts usually cite the latest air pressure reading along with air temperature, relative humidity, and other weather elements. Although we are physically aware of changes in temperature and humidity, we do not sense changes in air pressure as readily. If we follow air pressure reports over a period of time, however, we quickly learn that important changes in weather can be associated with relatively small variations in air pressure.

In this chapter, we examine the properties of air pressure, how air pressure is measured, and the reasons for spatial and temporal variations in air pressure. We also consider how air responds to changes in pressure when moving vertically (up and down) within the atmosphere. In later chapters, we describe how variations in air pressure contribute to the circulation of the atmosphere and weather.

Defining Air Pressure

Because the atmosphere consists of molecules that are moving rapidly in all directions and have mass, air exerts a force on the surface of all objects that it contacts. Air pressure represents a measure of that force per unit surface area. A **force** is defined as a push or pull on an object and is computed as mass multiplied by acceleration. Molecules composing air are always in rapid, random motion, and each molecule exerts a force as it collides with other molecules, including those on the surface of a solid (e.g., the ground) or liquid (e.g., the ocean). In one millionth of one second, billions upon billions of gas molecules bombard every square centimeter of Earth's surface. The total air pressure is the cumulative force of a multitude of molecules colliding with a unit surface area of any object in contact with air.

Each gaseous component of air contributes to the total air pressure as described by Dalton's law of partial pressures, named for its discoverer the British scientist John Dalton (1766-1844). According to **Dalton's law**, the total pressure exerted by a mixture of gases equals the sum of the pressures produced by each constituent gas; that is, each gas species in the mixture acts independently of all the other molecules. Stated another way, each gas exerts a pressure as though it were the only gas present.

The pressure produced by the gas molecules composing air depends on (1) the mass of the molecules, and (2) the kinetic molecular activity. In the larger sense, we can think of **air pressure** at a given location on the Earth's surface as the weight per unit area of the column of air above that location. The pressure at any point within the atmosphere is equal to the weight per unit area of the atmosphere above that point. *Weight* is the force exerted by gravity on a mass, that is,

weight = (mass) × (acceleration of gravity)

The average air pressure at sea level is equivalent to the weight exerted by Earth's gravity on a horizontal surface area of one square cm by a mass of approximately 1.0 kg (or equivalently to 14.7 lb per square in.). A column of water about 10 m (33 ft) deep produces this same pressure at its base. (For a comparison of air pressure with water pressure, see this chapter's second *For Further Exploration*.) Hence, the total weight of the atmosphere on the roof of a typical three-bedroom ranch-style house at sea level is about 2.1 million kg (4.6 million lb), equivalent to the combined weight of about 1500 full-size autos. Why doesn't the roof collapse under all that weight? The reason is that air pressure at any point is the same in all directions (up, down, and sideways); that is, air is pushing up from under the roof with the same pressure. Air pressure within the house exactly counterbalances air pressure outside the house so that the net air pressure acting on the roof is zero. This pressure balance (or equilibrium) is a prevailing condition in the atmosphere because at any point in a gas, or a mixture of gases, pressure is exerted equally in all directions.

Air Pressure Measurement

The **barometer** is the instrument used to measure air pressure and monitor its changes. The basic types are the mercury barometer and the aneroid barometer.

The standard, though cumbersome, of the two is the **mercury barometer**, invented in 1643 by Evangelista Torricelli (1608-1647), an Italian mathematician and student of Galileo. The instrument consists of a glass tube a little less than 1.0 m (39 in.) in length, sealed at one end, open at the other end, and initially filled with mercury, which is about 13 times denser than liquid water. The open end of the tube is inverted into a small open container of mercury, as shown in Figure 5.2. Mercury settles down the tube (and into the container) until the pressure of the mercury column exactly balances the pressure of the atmosphere acting on the surface of the mercury in the open container.

The average air pressure at sea level will support the mercury column in the tube to a height of 760 mm (29.92 in.). The height of the mercury column changes as air pressure changes. Falling air pressure allows the

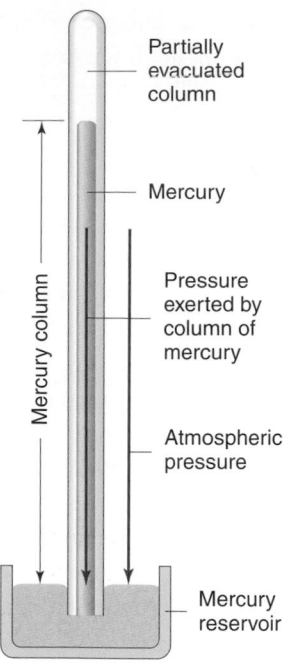

FIGURE 5.2
Schematic drawing of a mercury barometer indicating that the weight of the column of mercury is balanced by the weight of the air column.

mercury column to drop, whereas increasing air pressure forces the mercury column to rise, with the height of the mercury column directly proportional to air pressure. This is the origin of the common practice of expressing air pressure in units of length of the mercury column. Air pressure readings by a mercury barometer require adjustments for (1) the expansion and contraction of mercury that accompany changes in temperature, and (2) the slight variation of gravity with latitude and altitude that affects the weight of the mercury column. By convention, readings are adjusted to standard conditions of 0 °C (32 °F) and 45 degrees latitude at sea level.

An **aneroid** (nonliquid) **barometer** is more portable but less precise than a mercury barometer. It often consists of a flexible sealed chamber from which much of the air has been evacuated (Figure 5.3A). An internal spring prevents the chamber from collapsing. As air pressure changes, the chamber flexes, compressing when air pressure rises and expanding when air pressure drops. A series of gears and levers transmits and magnifies these movements to a pointer on a dial, which is calibrated to read in equivalent millimeters (or inches) of mercury, or to read directly in units of air pressure (Figure 5.3B). The latest aneroid barometers provide direct digital readouts (and even have been incorporated into wristwatches). Increasingly, new aneroid barometers are electronic, that is, they depend on the effect of air pressure on electrical properties of a crystalline substance. The barometric pressure sensors in the automated surface weather stations operated by NOAA's National Weather Service are digital pressure transducers.

Air pressure readings are usually taken indoors in order to maintain a relatively constant temperature that is close to standard conditions. Indoor air pressure of most buildings quickly adjusts to changes in outdoor air pressure. The instrument should be anchored to a sturdy

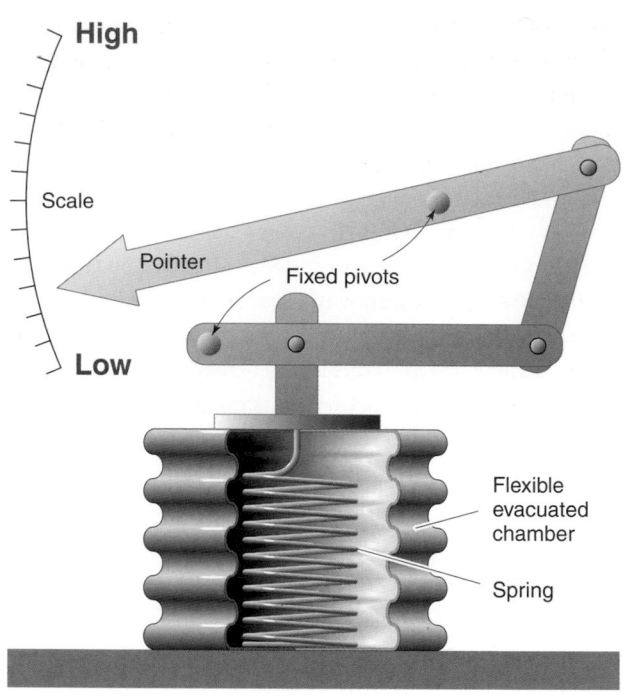

A

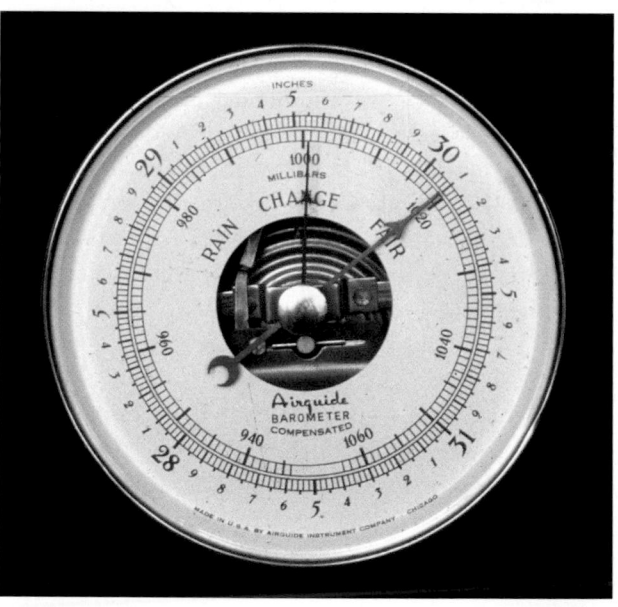

B

FIGURE 5.3
An aneroid barometer, a portable instrument used to monitor air pressure. (A) schematic drawing of internal view and (B) external view. [Adapted from Snow, et al., "Basic Meteorological Observations for Schools: Atmospheric Pressure." *Bulletin of the American Meteorological Society* 73(1992):785]

wall not exposed to direct sunlight. Mercury barometers must be vertical and most aneroid barometers are designed to be read in a vertical orientation. In reading a dial-type aneroid barometer, it is a good idea to first gently tap the barometer because friction in the mechanism may cause the pointer to stick.

Aneroid barometers intended for home use typically have dials with legends, such as *fair*, *changeable*, and *stormy* corresponding to certain ranges of air pressure. These designations should not be taken literally because a given air pressure reading does not always correspond to a specific type of weather. Much more useful than these legends for local weather forecasting is **air pressure tendency**, that is, the change in air pressure over a specific time interval, such as several hours. Rising air pressure usually means continued fair or clearing weather, whereas falling air pressure generally signals the approach of stormy weather. For determining pressure tendency, some aneroid barometers are equipped with a second pointer that serves as a reference marker. By turning the knob on the barometer face, the user sets the second pointer to correspond to the current air pressure reading. At a later time (perhaps in 3 hrs or less during changeable weather), the user can observe the new pressure reading and compare it with the earlier set reading to determine the air pressure tendency. An aneroid barometer may be linked to a pen that records on a clock-driven drum chart. This instrument, called a **barograph**, provides a continuous trace of air pressure variations with time, making it easier to determine pressure tendency (Figure 5.4).

FIGURE 5.4
A barograph provides a continuous trace of air pressure variations with time.

TABLE 5.1 Conversion Factors for Units of Air Pressure
1 bar = 1000 millibars (mb)
1 mb = 0.02953 in. of mercury
1 inch of mercury = 33.86 mb
1 kilopascal (kPa) = 1000 pascals (Pa)
1 hectopascal (hPa) = 100 Pa
1 mb = 1 hPa
1 inch of mercury = 33.86 hPa
1 inch of mercury = 25.4 mm of mercury

Air Pressure Units

On television and radio weathercasts, air pressure readings are usually reported in units of length (millimeters or inches of mercury). A scientifically more appropriate practice is to express air pressure in the fundamental units of pressure. The *pascal (Pa)* is the standard metric unit of pressure with the average air pressure at sea level of 101,325 Pa, 1013.25 hectopascals (hPa), or 101.325 kilopascals (kPa). U.S. meteorologists traditionally designate air pressure in *millibar (mb)* units, where 1 mb equals 1 hPa or 100 Pa. In turn, 1 mb is the equivalent of 0.02953 in. of mercury (Table 5.1). Average sea level air pressure is 1013.25 mb.

The usual worldwide range in sea-level air pressure is roughly 970 to 1040 mb (28.64 to 30.71 in. of mercury). The lowest sea-level air pressure ever recorded was 870 mb (25.69 in. of mercury), measured on 12 October 1979 in the eye of Typhoon Tip over the Pacific Ocean northwest of Guam. Air pressure no doubt has been lower than 870 mb in some tornadoes, but such low readings have never been confirmed directly because tornadic winds are so strong that they destroy barometers. The highest sea-level air pressure ever recorded was 1083.8 mb (32.01 in. of mercury) at Agata, Siberia on 31 December 1968 and was produced by an extremely cold, dense air mass.

Variation in Air Pressure with Altitude

We know from inflating an automobile or bicycle tire that air is compressible; that is, its volume and density are variable. The pull of gravity compresses the atmosphere so that the maximum **air density** (mass of molecules per unit volume) is at Earth's surface. In other words, the atmosphere's gas molecules are most closely spaced at the Earth's surface, and the spacing between molecules increases with increasing altitude. The number of gas molecules per unit volume, the **number density** of air, thus decreases with altitude. Thinning of air is so rapid within the lower atmosphere that at an altitude of 6.5 km (4 mi), air density is about 50% of its average sea-level value and at 18 km (11 mi), it is about 10%.

Thinning of air with altitude is accompanied by a decline in air pressure (Figure 5.5). The French mathematician Blaise Pascal (1623-1662) was the first to propose that air pressure decreases with increasing altitude. Poor health prevented Pascal from verifying his hypothesis by actual field measurements, so he asked his brother-in-law, Florin Perier, for assistance. Perier agreed, and on 19 September 1648, Perier along with some clerics and laymen from his hometown set off with barometer in hand and climbed to the summit of Puy-de-Dôme (elevation, 1464 m or 4800 ft) in the Auvergene region of southern France. At a pre-arranged time, Pascal at the base of the mountain and Perier at the summit took barometer readings. Those readings confirmed Pascal's hypothesis of declining air pressure with increasing altitude.

Because air is compressible, the rate of air pressure drop with altitude is not uniform; it is greatest

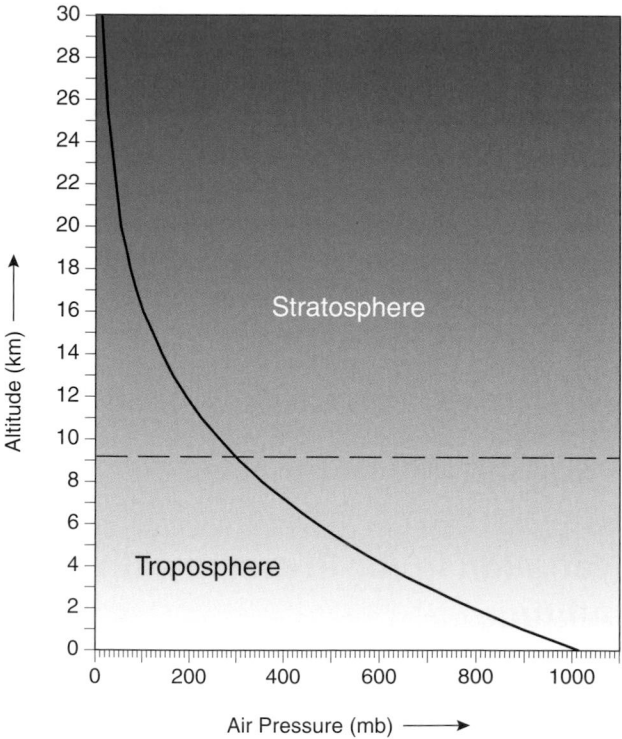

FIGURE 5.5
Average variation in air pressure in millibars (mb) with altitude (km). The average air pressure at sea level is 1013.25 mb.

at Earth's surface and then becomes more gradual aloft. For example, air pressure decreases about 25% in the first 2500 m (8200 ft) of altitude, but a further ascent of 3000 m (9800 ft) is required for an additional 25% drop in air pressure. In the lower troposphere, air pressure usually drops by roughly one millibar for every 10 m of ascent. However, as we will see shortly, the actual rate depends on the air temperature and humidity.

Vertical profiles of average air pressure (Figure 5.5) and temperature (Figure 2.17) are based on the standard atmosphere, a model of the real atmosphere which was initially used to assess aircraft performance. The **standard atmosphere** is the state of the atmosphere averaged for all latitudes and seasons. It features a fixed sea-level air temperature (15 °C or 59 °F) and pressure (1013.25 mb) together with fixed vertical profiles of air temperature, air pressure, and air density (Table 5.2). However, the actual altitude where the air pressure is a specific value (e.g., 500 mb) varies with time and location as temperature and pressure depart from standard conditions. In fact, meteorologists routinely draw upper-air weather maps on which are plotted contours of elevation of an *isobaric surface*, that is, an imaginary surface where the air pressure is the same everywhere. For example, contour maps are constructed for the 300-mb, 500-mb, and 850-mb levels. Observational data for these maps are obtained primarily from radiosondes (Chapter 2).

Although air pressure and air density drop with increasing altitude, it is impossible to specify an altitude where Earth's atmosphere definitely ends. That is, no one altitude can be identified as the top of the atmosphere and the beginning of interplanetary space. At best, we can describe the vertical extent of the atmosphere in terms of the relative distribution of its mass with altitude. Half of the atmosphere's mass is between Earth's surface and an average altitude of about 5500 m (18,000 ft). About 99% of the atmosphere's mass is below 32 km (20 mi). Above 80 km (50 mi), that is, above the *homosphere*, the relative proportions of atmospheric gases change markedly, and by about 1000 km (620 mi), the atmosphere merges with the highly rarefied interplanetary gases, mainly hydrogen and helium.

If the density of Earth's atmosphere were uniform throughout (a *homogeneous atmosphere*) like a liquid, the atmosphere would have a well-defined top. Assuming a surface temperature of 0 °C (32 °F), a homogeneous atmosphere would have a thickness of only about 8000 m (26,000 ft). In such an atmosphere, the temperature would drop with altitude at a rate of 34 Celsius degrees per 1000 m, the *autoconvective lapse rate*. In the real atmosphere, an autoconvective lapse rate is possible in a thin layer of uniformly mixed air overlying the intensely hot surface of a desert, most likely in the afternoon.

From a somewhat different perspective, at an altitude of only 32 km (20 mi), air pressure is less than 1% of its average sea-level value. The rapid vertical pressure drop in the lower reaches of the atmosphere means that appreciable variations in surface air pressure accompany relatively minor changes in land elevation. For example, the average air pressure at Denver, the 'mile-high city', is about 83% of the average air pressure at Boston, located just above sea level. As noted in this chapter's Case-in-Point, the expansion and thinning of air that accompany the decline in air pressure with altitude can cause discomfort

	TABLE 5.2	The Standard Atmosphere			
Altitude (km)	Temperature (°C)	Pressure (mb)	P/P_o*	Density (kg/m³)	D/D_o*
30.00	−46.6	11.97	0.01	0.02	0.02
25.00	−51.6	25.49	0.03	0.04	0.03
20.00	−56.5	55.29	0.05	0.09	0.07
19.00	−56.5	64.67	0.06	0.10	0.08
18.00	−56.5	75.65	0.07	0.12	0.09
17.00	−56.5	88.49	0.09	0.14	0.12
16.00	−56.5	103.52	0.10	0.17	0.14
15.00	−56.5	121.11	0.12	0.20	0.16
14.00	−56.5	141.70	0.14	0.23	0.19
13.00	−56.5	165.79	0.16	0.27	0.22
12.00	−56.5	193.99	0.19	0.31	0.25
11.00	−56.4	226.99	0.22	0.37	0.30
10.00	−49.9	264.99	0.26	0.41	0.34
9.50	−46.7	285.84	0.28	0.44	0.36
9.00	−43.4	308.00	0.30	0.47	0.38
8.50	−40.2	331.54	0.33	0.50	0.40
8.00	−36.9	356.51	0.35	0.53	0.43
7.50	−33.7	382.99	0.38	0.56	0.45
7.00	−30.5	411.05	0.41	0.59	0.48
6.50	−27.2	440.75	0.43	0.62	0.50
6.00	−23.9	472.17	0.47	0.66	0.54
5.50	−20.7	505.39	0.50	0.70	0.57
5.00	−17.5	540.48	0.53	0.74	0.60
4.50	−14.2	577.52	0.57	0.78	0.63
4.00	−11.0	616.60	0.61	0.82	0.67
3.50	−7.7	657.80	0.65	0.86	0.70
3.00	−4.5	701.21	0.69	0.91	0.74
2.50	−1.2	746.91	0.74	0.96	0.78
2.00	2.0	795.01	0.78	1.01	0.82
1.50	5.3	845.59	0.83	1.06	0.86
1.00	8.5	898.76	0.89	1.11	0.91
0.50	11.8	954.61	0.94	1.17	0.95
0.00	15.0	1013.25	1.00	1.23	1.00

* P/P_o = ratio of air pressure to its sea-level value; D/D_o = ratio of air density to its sea-level value.

and even serious illness or death for visitors to high altitudes. For more on this topic, refer to this chapter's first *For Further Exploration*.

Very low air density at high altitudes also has interesting implications for air temperature and heat transfer. In the thermosphere, the highest thermal subdivision of the atmosphere (Figure 2.17), individual atoms and molecules move about with average kinetic energy indicative of very high temperatures. However, there are so few atoms and molecules per unit volume that very little heat is transferred. In spite of temperatures that approach 1200 °C (2200 °F) in the thermosphere, heat is not readily conducted to cooler bodies. For example, satellites orbiting at these altitudes do not acquire such high temperatures.

Because air pressure drops with increasing altitude at rates that can be readily determined, an aneroid barometer can be calibrated to monitor altitude. Such an instrument is a type of **altimeter**. For more on this, see this chapter's third *For Further Exploration*.

Horizontal Variations in Air Pressure

Surface air pressure differs from one place to another, and these variations are not always due to differences in the elevation of the land. In fact, meteorologists are more interested in air pressure variations that arise from influences other than land elevation. Hence, weather observers determine an equivalent sea-level air pressure value; that is, for stations located above sea level, they adjust local air pressure readings to approximately what the air pressure would be if the station were actually located at sea level. The simplest adjustment assumes an imaginary column of air having the properties of the standard atmosphere extending from the station down to sea level. When this *reduction to sea level* is carried out everywhere, air pressure is observed to vary from one place to another (Figure 5.6) and fluctuate from day to day, and even from hour to hour (Figure 5.7).

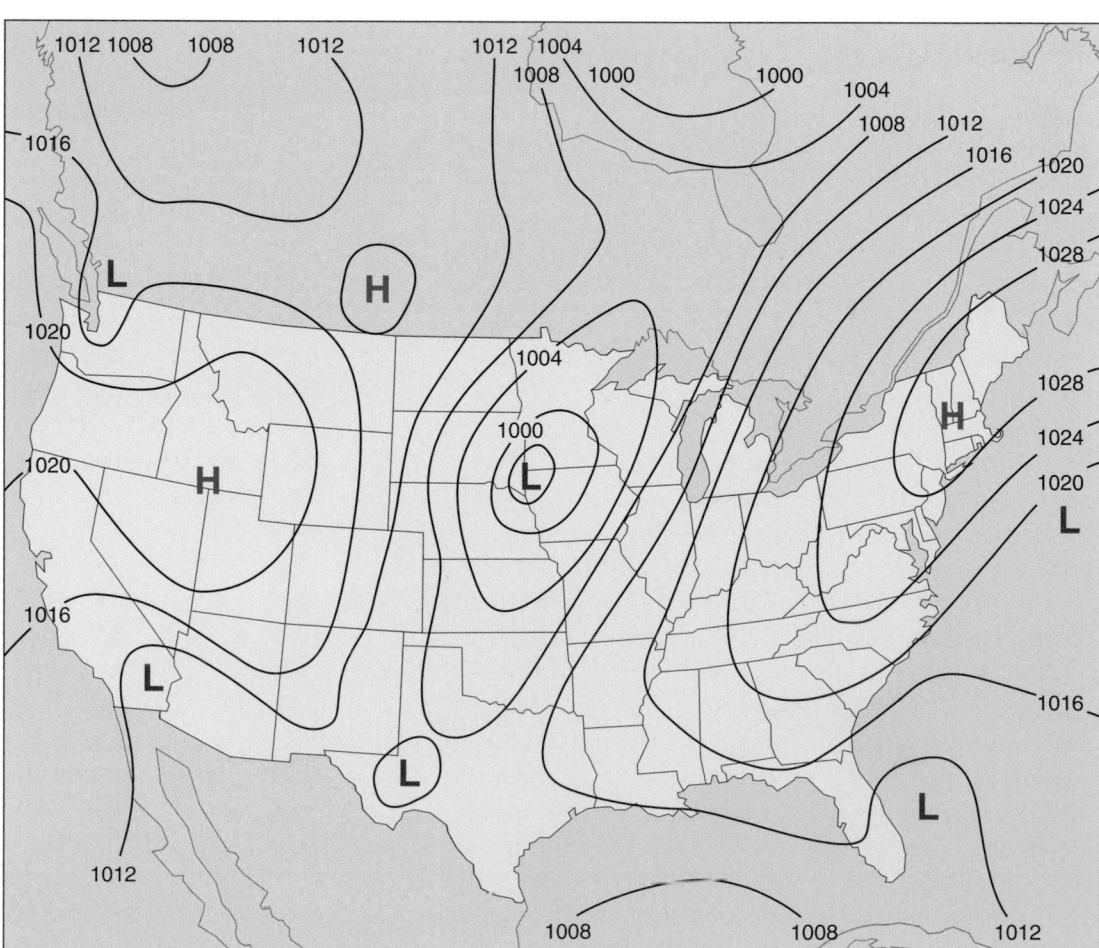

FIGURE 5.6
A surface weather map shows spatial variations in air pressure (reduced to sea level). Dark lines are isobars (in mb) passing through localities having the same air pressure. *L* is plotted where air pressure is relatively low and *H* is plotted where air pressure is relatively high.

Time

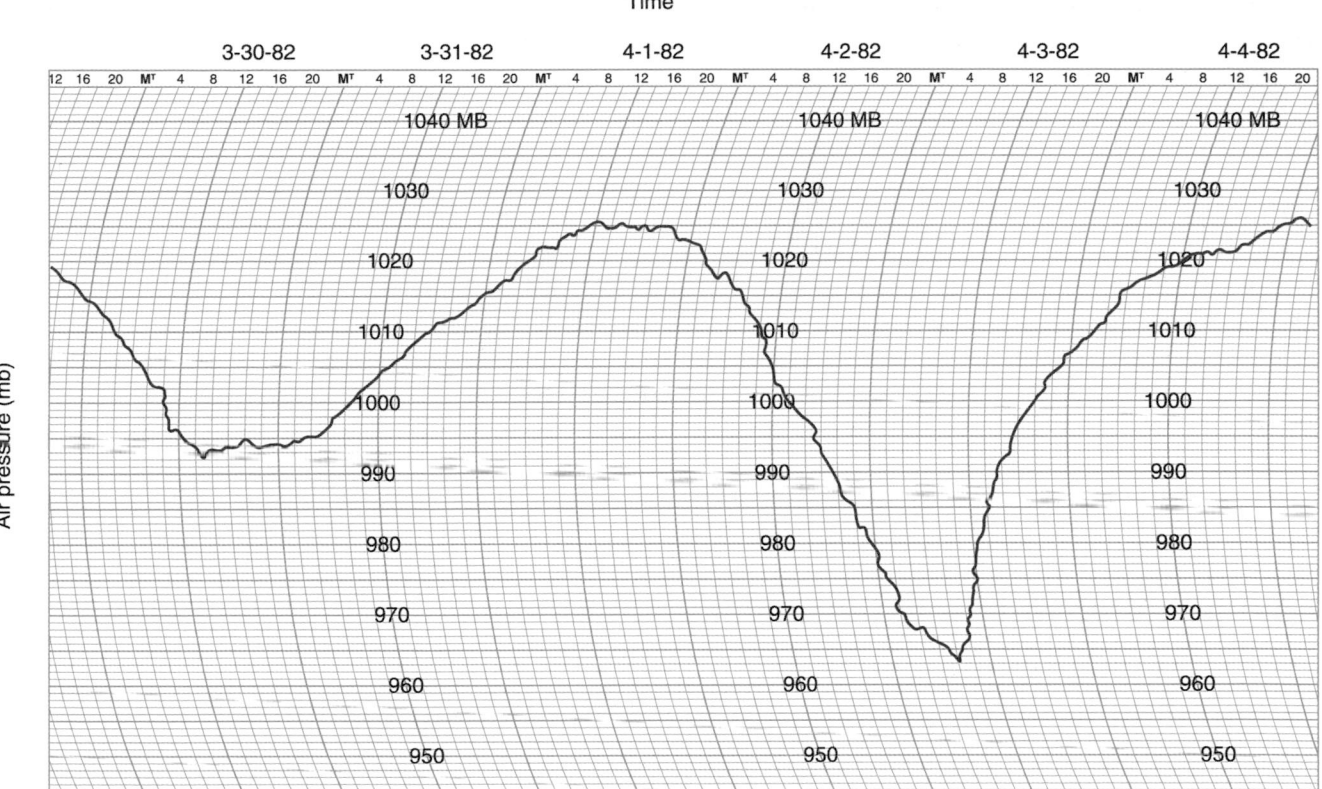

M = midnight on scale at top of record; time lines are in 2 hour increments

FIGURE 5.7
A trace from a barograph showing the variation in air pressure in millibars reduced to sea level at Green Bay, Wisconsin, from 30 March through 4 April 1982. During the period from 2 to 3 April, Green Bay was under the influence of a very intense low pressure system. Note that significant changes in air pressure occur from day to day and even from one hour to the next.

Air pressure readings, reduced to sea level, range in value from one place to another, but the rate of this variation horizontally is much less than the rate at which air pressure drops with altitude. Except in tornadoes or intense hurricanes, the horizontal pressure gradient at sea level rarely exceeds 3 mb per 200 km (125 mi), which is equivalent to the vertical change in air pressure experienced in the lowest 30 m (100 ft) of the troposphere. Nonetheless, these relatively small horizontal changes in surface air pressure can be accompanied by significant changes in weather.

In middle latitudes, a never ending procession of different air masses brings about changes in air pressure and weather. Recall that an *air mass* is a huge volume of air that is relatively uniform horizontally in temperature and water vapor concentration. As air masses move across the face of the Earth, surface air pressures fall or rise, and the weather changes. As a general rule, the weather becomes stormy when air pressure falls but clears or remains fair when air pressure rises.

Why do some air masses exert greater pressure than other air masses? One reason is the contrast in air density that arises from differences in air temperature or water vapor concentration. As a rule, temperature has a much more important influence on air density and pressure than does water vapor concentration. Also, the pressure of an air column may change due to net convergence of winds, which increases the amount of air and the pressure exerted by the column. In contrast, net divergence of winds decreases the amount of air and the pressure exerted by the column.

INFLUENCE OF TEMPERATURE AND HUMIDITY

Recall from Chapter 4 that temperature is a measure of the average kinetic energy of molecules. Rising air temperature corresponds to an increase in average kinetic molecular activity. If air is heated within a closed container, such as a rigid metal can, we would expect the air pressure acting on the internal walls of the container to rise as the increasingly energetic molecules bombard the walls with greater force. Here we are considering pressure as the force exerted by the molecules colliding with a unit surface area of the container's walls. The air density inside the container does not change because no air is added to or removed from the container and the air volume does not change. By contrast, except for Earth's surface, the atmosphere is not confined, so air is free to expand and contract. That is, in the atmosphere, air density is variable.

Within the atmosphere, when air is heated (e.g., by absorption of radiation, conduction, convection, or condensation or depostition of water vapor), air density usually decreases because the greater activity of the heated molecules increases the spacing between neighboring molecules. As a column of air is heated, the number of molecules per unit volume decreases; that is, air density decreases as with the air in hot air balloons (Figure 5.8). A lowered total mass in the column would exert less pressure on Earth's surface.

The greater density of cold air versus warm air at the same pressure affects the rate at which air pressure drops with increasing altitude. Air pressure drops more rapidly with altitude within a cold (denser) column of air than in a warm (less dense) column of air. Beginning with equal pressures at Earth's surface, equivalent pressure surfaces (e.g., the 500-mb level) occur at a lower altitude in a cold column of air than in a warm column of air. Within the troposphere, isobaric surfaces (surfaces of constant pressure) slope downward from the relatively warm tropics toward relatively cold high latitudes. As noted in this chapter's third *For Further Exploration*, the sloping of isobaric surfaces has important implications for altimetry.

Increasing humidity affects air density in the same way as rising air temperature; that is, the greater the concentration of water vapor, the less dense is the air. This statement is contrary to the popular perception that humid air feels "heavier" than dry air. Although hot, muggy air may weigh heavily on a person's disposition, humid air is, in fact, less dense than dry air at the same temperature and pressure. Water vapor reduces the density of air because the molecular weight of water is less than the average molecular weight of dry air.

Equal volumes of all gases (or mixtures of gases) measured under the same conditions of temperature and pressure contain the same number of molecules. This is

FIGURE 5.8
Hot air balloons ascend within the atmosphere because heated air within the balloons is less dense than the cooler air surrounding the balloons. [Courtesy of NASA/Goddard Space Flight Center]

Avogadro's law, first stated in 1811 by the Italian chemist Amedeo Avogadro (1776-1856). Avogadro's law implies that when water molecules enter the atmosphere as a vapor while the temperature and pressure of the air remains constant, they displace other gas molecules, principally nitrogen and oxygen. The molecular weight of water (H_2O) is 18 atomic mass units whereas the mass-weighted mean molecular weight of dry air (mostly N_2 and O_2) is about 29 atomic mass units. With all other factors equal, as the water vapor concentration in air increases (say from the evaporation of water from Earth's surface), the net effect is for air density to decrease. For equal volumes at the same temperature, then, a column of humid air exerts less pressure than a column of relatively dry air.

Cold, dry air masses are denser and usually produce higher surface pressures than warm, humid air masses. Warm, dry air masses, in turn, often exert higher surface pressures than equally warm, but more humid, air masses. Hence, a change in surface air pressure usually accompanies the replacement of one air mass by another of different temperatures, that is, by *cold air advection* or *warm air advection*. *Air mass modification* (changes in air mass temperature and/or water vapor concentration) also can alter the surface air pressure. These modifications may occur when an air mass travels over a different surface type (e.g., from cold snow cover to warmer bare ground) or, if the air mass is stationary, when the air is locally heated or cooled. In Chapter 4, we examined the regulation of air temperature by the local radiation budget plus air mass advection. The above discussion implies that local conditions and air mass advection can also influence surface air pressure.

INFLUENCE OF DIVERGING AND CONVERGING WINDS

In addition to variations in air temperature and (to a lesser extent) humidity, convergence or divergence of winds may also bring about changes in surface air pressure. Convergence or divergence is produced by a circulation pattern in which (1) horizontal winds blow toward or away from some location, or (2) wind speed changes in a downstream direction. We consider the first mechanism here and the second in Chapter 8.

Suppose that at Earth's surface, horizontal winds blow away from a column of air in all directions, as in the lower part of Figure 5.9. This is an example of **diverging winds**. At the same time, horizontal winds aloft blow toward the air column; this is an example of **converging winds**. Within the air column, air descends from above and takes the place of air diverging at the surface. If more air diverges at the surface than converges aloft, the amount of air in the column decreases and the air density and surface air pressure decrease. On the other hand, if more air converges aloft than diverges at the surface, the amount of air in the column increases and the density of the air column and surface air pressure increase. In later chapters, we discuss in greater detail how divergence and convergence of air cause air pressure changes within weather systems.

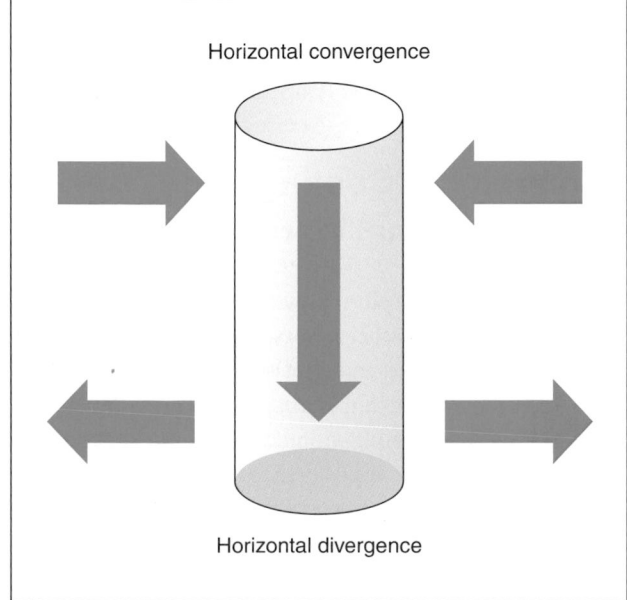

FIGURE 5.9
In this vertical cross section through the troposphere, horizontal winds diverging at the surface and converging aloft induce descending air motion.

Highs and Lows

We now have more insight into the meaning of those H (or High) and L (or Low) symbols plotted on surface weather maps (Figure 5.6). To locate Highs and Lows, a map is drawn to reveal the surface air pressure pattern. First, simultaneous air pressure readings at all weather stations are adjusted to sea level and plotted on a surface weather map. Then isobars are drawn. An **isobar** is a line passing through locations having the same air pressure. Drawing isobars usually requires interpolation between reporting weather stations. By U.S. convention, isobars are drawn at 4-mb intervals (e.g., 996 mb, 1000 mb, 1004 mb) whereas in Europe the convention is to use 5-mb intervals.

A High or H symbol is used to designate places where sea-level air pressure is relatively high compared to the air pressure in surrounding areas. A large-scale High is also known as an *anticyclone*. A Low or L symbol signifies regions where sea-level air pressure is relatively low compared to surrounding areas. A large-scale Low is also known as a *cyclone*. For reasons presented in Chapter 8, usually a High is a fair weather system, and a Low is a stormy weather system. Viewed from above in the Northern Hemisphere, surface winds blow clockwise and diverge outward from the center of a High. Viewed from above in the Northern Hemisphere, surface winds blow counterclockwise about the center of a Low and converge inward. Winds are strong where isobars are closely spaced because of greater horizontal pressure differences over distance, but are weak where isobars are widely spaced as the horizontal pressure differences over distance are less. Isobars are typically more closely spaced near the center of a Low than near the center of a High.

The Gas Law

To this point, we have described the state of the atmosphere in terms of variations in temperature, pressure, and density. These important properties of air are known collectively as **variables of state**; their magnitudes change from one place to another across Earth's surface, with altitude above Earth's surface, and with time.

At typical temperatures and pressures encountered in dry clean air at sea level, the number density is approximately 25 billion molecules per cubic centimeter and the mass density is approximately 1.23 kg per cubic meter (0.1 lb per cubic ft). As we saw in Chapter 4, temperature is directly proportional to the average kinetic energy of the individual atoms or molecules composing a substance. The kinetic energy of a molecule is a function of both its mass and speed. At sea level, with normal temperatures and pressures, an air molecule moves at an average speed of approximately 460 m per sec (about 1000 mph).

The three variables of state are interrelated through the **gas law**. Although the gas law was derived for a single ideal gas, the law provides a reasonably accurate description of the behavior of air, which is a mixture of many different gases. By definition, an **ideal gas** follows the *kinetic molecular theory* precisely. That is, an ideal gas is made up of a very large number of molecules that are in rapid and random motion. Molecules in motion experience perfectly elastic collisions, so they lose no momentum. Furthermore, molecules are so small that the attractive forces between them are negligible.

An ideal gas in a confined volume follows Charles' law and Boyle's law exactly, whereas a real gas behaves only approximately as these laws predict. Named for the French scientist Jacques Charles (1746-1823), **Charles' law** holds that with constant pressure, the absolute temperature (in kelvins) of an ideal gas is inversely proportional to the density of the gas. As a sample of gas is heated, the gas expands and its density decreases. According to **Boyle's law**, named for the Irish physicist and chemist Robert Boyle (1627-91), when the temperature is held constant, the pressure and density of an ideal gas are directly proportional. As the pressure on a sample of an ideal gas increases, its volume decreases and its density increases. Charles' law and Boyle's law are combined as the *ideal gas law.*

Simply put, the ideal gas law states that the pressure exerted by a gas is directly proportional to the product of its density and temperature (expressed in kelvins). As a word equation, the ideal gas law becomes

pressure = (gas constant) × (density) × (temperature)

The *gas constant* is an experimentally derived number that converts a proportional relationship to an equation. The constant of proportionality varies depending on the specific gas. However, by assigning the constant a certain value, this law describes approximately the behavior of dry air (air minus water vapor). With an additional adjustment, the gas law represents the behavior of the variables of state for humid air as long as no condensation or deposition takes place.

From the gas law equation, the following holds:

1. The density of air within a rigid closed container remains constant because the volume of the container and the mass of its contents are fixed. An increase in pressure exerted by the gas molecules on the interior walls of the container accompanies a rise in the temperature of the gas as the molecular speed increases. Hence, according to the gas law, the pressure and temperature are directly proportional at fixed density.

2. Consider an **air parcel**, an arbitrary amount of air useful for visualizing atmospheric processes, containing a fixed number of molecules. Its volume can change, but its mass remains the same, meaning that the density changes. If the temperature is held constant, compressing the air parcel causes its density to increase because its volume decreases (while its mass remains constant). Hence, at a fixed temperature, pressure is directly proportional to density.

3. For the same air parcel at constant pressure, a rise in temperature is accompanied by a decrease in density. The density decreases because the more energetic molecules cause the parcel to expand, that is, its volume increases while its mass remains constant. Hence, at a fixed pressure, temperature is inversely proportional to density.

In the atmosphere, however, the situation is more complicated because all three variables of state can change simultaneously. Hence, as the air temperature rises (perhaps in response to radiational heating), air expands, air density decreases, and air pressure at the Earth's surface falls. On the other hand, in winter, it is usual for air temperature to drop (not rise) as the surface air pressure rises. The gas law is satisfied in the atmosphere because air density increases as the temperature drops.

Expansional Cooling and Compressional Warming

In this section, we explore how the variables of state respond when an air parcel moves vertically (up or down) within the atmosphere and experiences a change in pressure. As noted earlier in this chapter, air pressure always decreases with increasing altitude within the atmosphere. One of the important consequences of the decline of air pressure with altitude is the expansion or compression of air parcels as they move up or down. As an air parcel ascends and expands, its temperature drops; as an air parcel descends and is compressed, its temperature rises. As will be discussed in Chapter 6, expansional cooling and compressional warming are major factors in the development and dissipation of clouds.

Whenever a gas (or mixture of gases) expands, the temperature of the gas drops; this process is known as **expansional cooling**. Expansional cooling explains why air released through the open valve of a bicycle tire is cool to the touch. Air pressure within the tire (perhaps 60 lb per square in.) is as much as four times greater than the atmospheric pressure outside the tire (around 15 lb per square in.). Air streaming through an open tire valve into the atmosphere does work as it pushes aside air that formerly occupied the volume into which it expands. Work requires energy, and the energy used in the work of expansion is drawn from the internal energy of the air so the temperature of the expanding air drops. Air that ascends within the atmosphere also undergoes expansional cooling in response to falling air pressure. As an air parcel ascends in the atmosphere, it expands in much the same way as a helium-filled balloon as it drifts skyward. Assuming that the parcel does not exchange heat with its environment, changes in volume translate into changes in the parcel's temperature. A rising air parcel expands due to the lowering surrounding pressure and its temperature drops.

Conversely, as the pressure on an air parcel increases, the air parcel is compressed and its temperature rises, a process known as **compressional warming**. A familiar example of compressional warming is the heating of the cylinder wall of a tire pump as air is pumped (compressed) into a tire. The work of compressing air is converted into heat. As an air parcel descends within the atmosphere, it is compressed because of increased surrounding pressure and its temperature rises.

In a more general sense, imagine that upward or downward currents of air consist of continuous streams of individual air parcels. The temperature behavior of an air current is the same as that of its component air parcels. In summary, ascending currents of air (such as the updraft in a convective circulation) cool whereas descending currents of air (such as the downdraft in a convective circulation) warm.

CONSERVATION OF ENERGY

The *law of energy conservation* (also known as the *first law of thermodynamics*) applies to expansional cooling and compressional warming of air parcels moving vertically in the atmosphere. According to this law, we can account for all the original energy in any physical or biological system regardless of the energy transformations that take place. Heat energy gained by an air parcel is either added to the parcel's store of internal energy or is used to do work on the parcel. Conversely, heat energy that is released by an air parcel is either subtracted from

the internal energy of the parcel or is the consequence of the parcel doing work on its surroundings.

We explore this concept further by examining separately (1) internal energy and (2) work done on or by a dry air parcel. As noted in Chapter 4, *internal energy* encompasses all the energy in a substance, that is, the kinetic energy of atoms and molecules plus the potential energy arising from forces between atoms and molecules. A change in the internal energy of an air parcel is directly proportional to a change in temperature of the parcel (with the specific heat being the constant of proportionality). A transfer of heat to the parcel raises its temperature whereas a transfer of heat from the parcel lowers its temperature.

Work done on or by a dry air parcel consists of compression or expansion of the air parcel. Consider, for example, a sealed cylindrical container of air equipped at one end with a piston that can compress the air sample or allow the air sample to expand. If we move the piston to compress the air, we use energy to do work on the air sample (Figure 5.10A). If we then release the piston, the air sample expands and works against (pushes) the piston (Figure 5.10B). When the air is compressed, work is

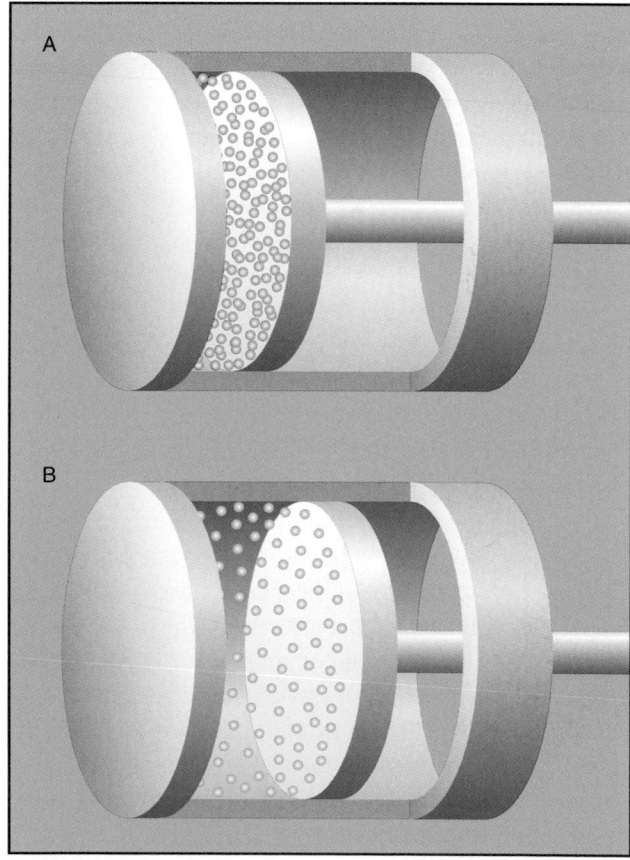

FIGURE 5.10
A sealed cylindrical container is equipped at one end with a piston that (A) compreses the air, and (B) allows the air to expand.

done on it but when the air expands, the air does work on its surroundings. Energy is either supplied (during compression) or released (during expansion). Hence, according to the law of energy conservation, the heat flow into and out of an air parcel is accounted for by changes in internal energy (temperature), the work of expansion or compression of air, or both.

ADIABATIC PROCESS

With our application of the law of energy conservation to air parcels moving vertically within the atmosphere, we can now turn to the adiabatic assumption. During an **adiabatic process**, no heat is exchanged between an air parcel and its surroundings. Because air often moves vertically relatively rapidly in large quantities, it can be assumed to behave adiabaticaly. That is, while ascending or descending within the atmosphere, the air parcel is neither heated nor cooled by radiation, conduction, phase changes of water, or mixing with the environment. The temperature of an ascending or descending unsaturated air parcel changes in response to expansion or compression only.

Adiabatic cooling of ascending unsaturated air amounts to 9.8 Celsius degrees per 1000 m (5.5 Fahrenheit degrees per 1000 ft) of ascent. This is the **dry adiabatic lapse rate** (Figure 5.11). The "dry" designation can be misleading because ascending and descending air parcels contain some water vapor. The dry adiabatic lapse rate applies to any parcel that is not saturated with water vapor (because no condensation or evaporation can take place during a dry adiabatic process). On the other hand, the atmosphere compresses descending air, and the air temperature rises 9.8 Celsius degrees for every 1000 m (5.5 Fahrenheit degrees per 1000 ft) of descent. Hence, the dry adiabatic lapse rate describes the temperature change of unsaturated air as it moves vertically (up or down) within the atmosphere.

Should rising air parcels cool to the point that the relative humidity reaches 100% (saturation) and condensation or deposition occurs, the ascending air no longer cools at the dry adiabatic rate (Figure 5.12). Latent heat that is released to the environment during ongoing condensation or deposition *partially* counters expansional cooling. Consequently, an ascending saturated (cloudy) air parcel cools at a lower rate than an ascending unsaturated (clear) air parcel. Rising saturated air parcels cool at the **moist adiabatic lapse rate**.

Although the magnitude of the dry adiabatic lapse rate is constant, the moist adiabatic lapse rate varies with temperature. Warm saturated air has much more

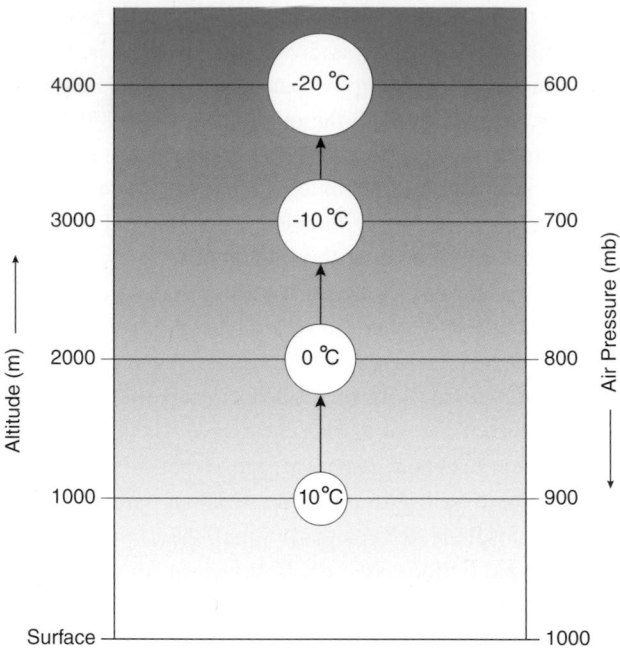

FIGURE 5.11
The dry adiabatic lapse rate describes the expansional cooling of ascending unsaturated (clear) air parcels. Here, the magnitude of the dry adiabatic lapse rate is rounded off to 10 Celsius degrees per 1000 m.

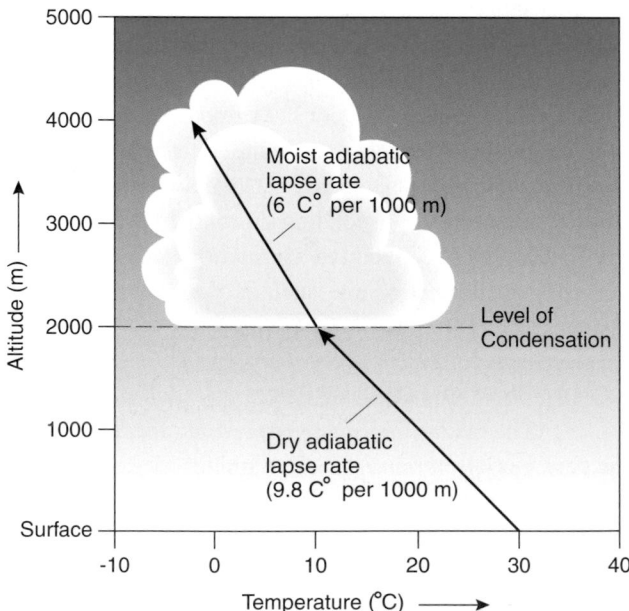

FIGURE 5.12
Ascending unsaturated (clear) air parcels cool at the dry adiabatic lapse rate whereas ascending saturated (cloudy) air parcels cool at the lesser moist adiabatic lapse rate.

water vapor than cool saturated air. Also, the amount of water vapor in saturated air increases exponentially with increasing temperature. That is, the difference in water vapor accompanying a 10 Celsius degree increase at higher temperatures is much greater than the difference associated with a 10 Celsius degree temperature increase at lower temperatures. Thusly, for the same temperature change, the greater condensation or deposition in warm saturated air releases more latent heat than condensation or deposition in cool saturated air. The greater the quantity of latent heat released, the more the expansional cooling is offset, and the smaller is the magnitude of the moist adiabatic lapse rate. The moist adiabatic lapse rate ranges from about 4 Celsius degrees per 1000 m (2.2 Fahrenheit degrees per 1000 ft) for very warm saturated air to very near 9.8 Celsius degrees per 1000 m (5.5 Fahrenheit degrees per 1000 ft) for very cold saturated air. For convenience, we use an average value of 6 Celsius degrees per 1000 m (3.3 Fahrenheit degrees per 1000 ft) for the moist adiabatic lapse rate.

Conclusions

We now have a working definition of air pressure, and we have examined the causes of spatial and temporal variations in air pressure within the Earth-atmosphere system. Air pressure drops rapidly with altitude in the lower troposphere and then more gradually aloft. At great altitudes, the atmosphere gradually merges with the gases of interplanetary space so that the atmosphere has no clearly defined top. Adjusting a station's barometer readings to sea level eliminates the influence of weather station elevation on air pressure. Then, surface air pressure depends on air density, which in turn, is governed by air temperature and, to a lesser extent, by the concentration of water vapor in air. Diverging or converging winds may also affect air density and surface air pressure.

The range of air temperature and air pressure in the Earth-atmosphere system allows water to exist in all three phases. In Chapter 4, we saw how phase changes of water help transfer heat energy within the Earth-atmosphere system. Changes in air temperature also trigger the phase changes of water that cause clouds to form or to dissipate. In the next chapter, we take a closer look at water within the atmosphere with emphasis on cloud-forming processes.

Basic Understandings

- The pressure (force per unit area) exerted by the atmosphere depends on the pull of gravity along with the mass and kinetic energy of the gas molecules that compose air. We can think of air pressure as the weight of a column of air acting on a unit area of Earth's surface. The air pressure at any specified altitude is equal to the weight per unit area of the air column above that altitude.

- At any specified point within the atmosphere, air pressure of the same magnitude is exerted in all directions.

- A barometer is the standard weather instrument for monitoring changes in air pressure. A mercury barometer is more accurate but not as portable as an aneroid barometer. The aneroid barometer can be calibrated to measure altitude.

- Air pressure and air density decrease rapidly with increasing altitude in the lower atmosphere and then more gradually aloft. The atmosphere has no clearly defined upper boundary; rather, Earth's atmosphere gradually merges with the highly rarefied hydrogen/helium atmosphere of interplanetary space. About 50% of the mass of the atmosphere occurs below an altitude of 5.5 km (3.5 mi), and 99% below an altitude of 32 km (20 mi).

- Barometer readings are adjusted to sea level to permit a comparison of all simultaneous air pressure readings at the same elevation, that is, to remove the influence of station elevation on air pressure.

- Within the atmosphere, at constant pressure, air density is inversely proportional to air temperature. Hence, all other factors being equal, cold air masses are denser and exert higher pressure at Earth's surface than do warm air masses.

- Air pressure drops more rapidly with altitude in a column of cold air than in a column of warm air. Hence, an isobaric surface (e.g., the 500-mb level) usually slopes downward toward colder air.

- Within the atmosphere, air density is also inversely proportional to water vapor concentration. Hence, at equivalent temperatures, dry air masses are denser than humid air masses. As a rule, temperature has a greater influence on air density and air pressure than does humidity.

- Air pressure may fluctuate in response to divergence or convergence of moving air, which is produced by changes in wind speed or direction.

- Important changes in weather often accompany relatively small changes in air pressure at Earth's surface. As a rule, high or rising pressure signals clearing or continued fair weather, whereas low or falling pressure means stormy weather.

- Variables of state of the atmosphere (temperature, pressure, and density) are interrelated through the gas law. Although the gas law was derived for a single ideal gas, the law provides a reasonably accurate description of the behavior of air, which is a mixture of many different gases. By definition, an ideal gas follows the *kinetic molecular theory* precisely.

- Charles' law holds that with constant pressure, the absolute temperature of an ideal gas is inversely proportional to the density of the gas. As a sample of gas is heated, the gas expands and its density decreases. According to Boyle's law, when the temperature is held constant, the pressure and density of an ideal gas are directly proportional. As the pressure on an ideal gas increases, its volume decreases and its density increases. Charles' law and Boyle's law are combined as the ideal gas law.

- Air pressure always declines with increasing altitude. Hence, as air ascends in the atmosphere, it expands and cools. Conversely, as air descends in the atmosphere, it is compressed and warms.

- The law of energy conservation applies to expansional cooling and compressional warming of air parcels or currents moving vertically in the atmosphere. Heat energy gained by an air parcel is either added to the parcel's store of internal energy or is used to do work on the parcel. Conversely, heat energy that is released by an air parcel is either subtracted from the internal energy of the parcel or is the consequence of the parcel's work on its surroundings.

- During an adiabatic process, no heat is exchanged between an air parcel and its environment. Adiabatic cooling of ascending unsaturated air amounts to 9.8 Celsius degrees per 1000 m (5.5 Fahrenheit degrees per 1000 ft) of ascent. This is the dry adiabatic lapse rate. Because of the release of latent heat accompanying phase changes of water vapor, an ascending saturated (cloudy) air parcel cools more slowly than an ascending unsaturated (clear) air parcel. Rising saturated air parcels cool at the moist adiabatic lapse rate averaging 6 Celsius degrees per 1000 m (3.3 Fahrenheit degrees per 1000 ft).

Enduring Ideas

- Air pressure can be thought of as the weight per unit area of the atmosphere above. Pressure always decreases with altitude more rapidly in a column of cold air than warm air. Therefore, an isobaric surface within the atmosphere usually slopes downward toward colder air.

- Air pressure readings are obtained by a barometer and are reduced to sea level in order to eliminate the influence of land elevation. Isobars, lines passing through locations having the same air pressure, plotted on a surface weather map reveal air pressure patterns, including the locations of centers of low and high pressure.

- Pressure exerted by an air mass depends upon its temperature and water vapor concentration. Temperature has the stronger influence on air pressure, and cold air masses are denser and generally exert higher pressure at Earth's surface than do warm air masses. At the same temperature, dry air masses are denser and usually exert more surface pressure than do humid air masses. Pressure at a location is influenced by air mass advection, air mass modification, and net divergence or convergence of winds in the air column above that location.

- Following the law of energy conservation, the temperature of an ascending or descending unsaturated air parcel changes in response to expansion (cooling) or compression (warming). Rising saturated air parcels cool at the moist adiabatic lapse rate, where latent heat released to the environment through condensation or deposition partially counters expansional cooling.

Key Terms

force	standard atmosphere	Charles' law
Dalton's law	altimeter	Boyle's law
air pressure	Avogadro's law	expansional cooling
barometer	diverging winds	compressional warming
mercury barometer	converging winds	law of energy conservation
aneroid barometer	isobar	internal energy
air pressure tendency	variables of state	adiabatic process
air density	gas law	dry adiabatic lapse rate
number density	ideal gas	moist adiabatic lapse rate

Review

1. How does Dalton's law apply to the atmosphere?
2. Provide a definition of *air pressure* that applies to Earth's surface and any altitude within the atmosphere.
3. Compare the advantages and disadvantages of a mercury barometer versus an aneroid barometer.
4. Explain how air pressure tendency can be a useful indicator of future weather.
5. Air is a compressible mixture of gases. How does this property of air affect the rate at which air pressure decreases with increasing altitude?
6. How does a change in temperature affect air density? How does a change in humidity affect air density?
7. Why does a cold, dry air mass exert a greater surface air pressure than an equally cold but more humid air mass?
8. Why are surface air pressure readings adjusted to what they would be if the weather station were actually located at sea level?
9. Distinguish between Charles' law and Boyle's law.
10. Why is the dry adiabatic lapse rate greater than the moist adiabatic lapse rate?

Critical Thinking

1. Is there some altitude which clearly marks the top of the atmosphere? Explain your response.
2. On televised weathercasts, air pressure often is reported in units of length (inches) rather than units of pressure (millibars). Explain why.
3. On a particularly warm and humid evening, a sportscaster comments that baseballs hit to the outfield "will not carry far in this heavy air." How valid is the sportscaster's observation?
4. If water is used in place of mercury in a glass-tube barometer, what is the required height of the tube? The density of mercury (Hg) is 13.6 g/cm^3 and the density of water is 1.0 g/cm^3.
5. At Minneapolis, MN, the air pressure typically drops more rapidly with altitude on a cold day in January than on a warm day in July. Explain the difference.
6. The tropopause is higher over the tropics than over middle latitudes. Explain why.
7. A jet aircraft is cruising at 600-mb level, that is, at the altitude where the air pressure is 600 mb. What fraction of the atmosphere's mass is below the aircraft?
8. An unsaturated air parcel has an initial temperature of 15°C. If the parcel is lifted 1500 m and remains unsaturated, what is its new temperature?
9. What is an adiabatic process? Do adiabatic processes occur in the atmosphere? If so, provide an example.
10. Why do temperature changes accompany the vertical (up and down) motions of air parcels?

Human Responses to Changes in Air Pressure

Visitors to mountain elevations above about 2500 m (8000 ft) may develop symptoms of *acute mountain sickness*: headache, shortness of breath, fatigue, insomnia, and nausea. Such illness is not uncommon—perhaps one of every four people who ascend to high elevations experiences one or more of these distressing symptoms. Some people who spend more than 36 to 72 hrs at elevations greater than about 2800 m (9000 ft) develop *high-altitude pulmonary edema*. Symptoms of this potentially life-threatening condition are somewhat similar to those of pneumonia, that is, severe cough, shortness of breath, lethargy, mild fever, and a buildup of fluid (edema) in the lungs. Onset of high-altitude pulmonary edema requires prompt medical attention.

Breathing air that does not supply adequate oxygen (O_2) can bring on altitude sickness. Thinning of air that accompanies the fall in air pressure with altitude also means a rapid decrease in the *number density* of the atmosphere's principal gases; that is, a decline in the number of nitrogen (N_2) and oxygen (O_2) molecules per unit volume of air. At altitudes of 2450 m to 2750 m (8000 ft to 10,000 ft), oxygen's number density is only about 74% to 71%, respectively, of its value at sea level. Less oxygen per unit volume of air may lead to *hypoxia,* a deficiency of oxygen in the body. Symptoms of mild hypoxia are commonly known as mountain sickness, whereas symptoms of more severe (acute) hypoxia constitute high altitude pulmonary edema.

As a person ascends to high altitudes (Figure 1), the body attempts to compensate for the decline in O_2 by altering breathing. The breathing rate accelerates and breathing becomes deeper to increase the amount of oxygen entering the lungs per unit time. Accelerated and deeper breathing provides no relief from hypoxia unless adequate oxygen is delivered to the lungs and quickly transported to body tissues. Hence, the heart rate and stroke volume (the quantity of blood pumped out of the heart during each contraction) also increase. Although the quantity of oxygen per unit volume of blood remains essentially constant, the heart propels more blood (and hence, more oxygen) per unit time to body tissue.

FIGURE 1
Air Force pararescue reservists climb Mount Hood, OR while searching for two lost climbers during a search and rescue scenario. [Courtesy U.S. Air Force Reserve Command]

The easiest way to alleviate mountain sickness is to descend to and remain at a lower altitude, especially at night. The altitude at which a person sleeps is more important than the altitude reached during the day. If a person remains at a high altitude, emergency adjustments by the heart and the muscles that control breathing reduce the symptoms of hypoxia while other longer-term changes take place in the body. Acute mountain sickness should be treated more agressively because of its life-threatening potential. Symptoms of mountain sickness gradually subside after several days as additional hemoglobin (molecules that pick up oxygen from air in the lungs) and red blood cells (cells that transport hemoglobin molecules in the bloodstream) are manufactured in the bone marrow. Although increased cardiac output generally tapers off within a week or so, accelerated breathing persists as long as a person remains at high altitude. The body thus adjusts to low oxygen levels that otherwise could cause serious health problems; this is an example of *acclimatization*.

A person can hasten acclimatization to high altitudes. Although a person may be eager to put in a full day of skiing or backpacking upon arriving in the mountains, he or she is well advised to minimize physical activity during the first day or two. Here the old adage *the impatient becomes the patient* applies. Overexertion often results in more severe and persistent symptoms. Because alcohol aggravates mountain sickness, its use should be avoided, particularly during the first 48 hours at high altitudes.

Some people are better able to acclimatize than others because of differences in their genetic makeup and general health. Nonetheless, there is a limit to acclimatization. Long-term residence at high altitudes cannot fully restore the body's capacity to perform work at the pace that is possible at sea level. At altitudes above about 5200 m (17,000 ft), without supplementary oxygen, nothing can be done to prevent a continual decline in all bodily functions. Hence, as described in this chapter's Case-in-Point, the summit of Mount Everest, which at 8848 m (29,029 ft) above sea level is the world's highest peak, is very close to (if not beyond) the limits of human survival.

Symptoms of hypoxia become more severe as the air thins; hence, regulations regarding supplemental oxygen supply begin for civil aircraft of U.S. registry operating at altitudes above 3810 m (12,500 ft). In actual practice, commercial aircraft cabins are pressurized beginning at takeoff and remain pressurized throughout the flight. A cabin is typically pressurized to about 75% of average sea-level air pressure. Although the aircraft cabin is pressurized, people commonly feel the effects of changing air pressure in a rapidly ascending or descending aircraft by a popping sensation in their ears. Rapid ascent or descent in an elevator or in a car on mountain roads often produces the same sensation. Ear popping is symptomatic of a natural process that helps protect the eardrum from damage.

The eardrum separates the outer ear from the middle ear chamber (Figure 2). As an aircraft takes off and cabin pressure drops, air pressure in the outer ear declines. As air pressure in the outer ear changes, the eardrum is distorted unless a compensating pressure change takes place in the middle ear. If the pressure does not equalize between the outer and middle ear, the eardrum bulges outward. On the other hand, when an aircraft descends and cabin pressure increases,

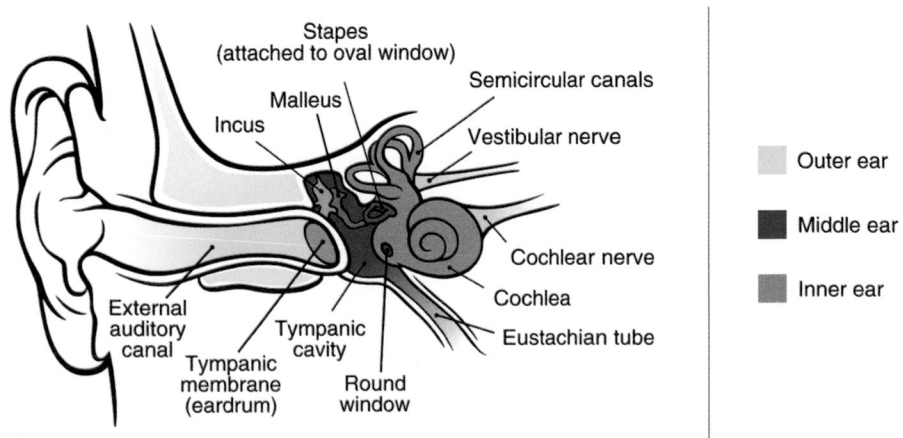

FIGURE 2

Anatomy of the ear. [Adapted from drawing by Chittka L, Brockmann, Perception Space—The Final Frontier, A PLoS Biology Vol. 3, No. 4, e137 doi:10.1371/journal.pbio.0030137]

air pressure in the outer ear increases. Without a compensating pressure change in the middle ear, the eardrum bulges inward. In both cases, deformation of the eardrum not only causes physical discomfort, but a bulging eardrum does not vibrate efficiently so that sounds are muffled. If the air pressure difference between the middle ear and outer ear continues to increase, the eardrum could rupture, perhaps causing permanent hearing loss.

Fortunately, the body has a natural mechanism that regulates air pressure in the middle ear. The *Eustachian tube* connects the middle ear to the upper throat region which, in turn, leads to the outside via the oral and nasal cavities. Normally, the Eustachian tube is closed where it enters the throat, but it opens if a sufficient air pressure difference develops between the middle ear and the throat. When the Eustachian tube opens, air pressure in the middle ear quickly comes into equilibrium with the external air pressure and the eardrum pops back to its normal shape. Vibrations of the eardrum associated with its rapid change in shape are what we hear as *ear popping*, the body's way of preventing a permanent hearing loss when a person experiences a rapid change in air pressure. To reduce discomfort, yawning or swallowing hastens the opening of the Eustachian tube. For this reason, travelers in aircraft or on mountain roads are advised to chew gum because the swallowing that accompanies gum chewing helps to open the Eustachian tube.

When the body's Eustachian tubes are not properly functioning due to sinus blockage, flying can be quite painful. A blocked Eustachian tube prevents the equalization in pressure between the middle ear chamber and the external surroundings. This can lead to severe pain, dizziness, and possible rupture of the eardrum. Persons flying with sinus blockage due to colds, allergies, or infections are well advised to consider taking an over-the-counter decongestant pill prior to flying, using a nasal decongestant spray, and chewing gum and swallowing frequently during ascent and descent. Bottles or pacifiers may help infants and young children. Other methods for alleviating discomfort are to plug your nose and swallow (when ascending in aircraft) or to plug your nose, close your mouth, and gently attempt to exhale (on descent). Both of these methods attempt to bring the pressures into equilibrium.

In summary, people can be sensitive to the effects of changing air pressure caused by visits to mountain elevations or experiencing rapid ascent and descent in aircraft. Human health problems, such as clogged sinus passages, can exacerbate these sensitivities. Breathing air at high elevations (lower pressures) that does not supply adequate oxygen can bring on altitude sickness, and visitors to these higher elevations are wise to properly acclimatize. Ear popping is the body's natural response to changes in air pressure caused by quick ascent or descent, for example within an aircraft, a vehicle on mountain roads, or an elevator.

Comparing Air and Water Pressure

Ocean water exerts pressure (i.e., force per unit area) as does the atmosphere. Liquids and gases are fluids and, as such, exhibit similar physical behavior. As detailed earlier in this chapter, atmospheric pressure can be thought of as the weight of a column of air acting over a unit area at the base of the column. By convention, standard atmospheric pressure is the average air pressure at sea level at 45 degrees latitude and an air temperature of 15 °C (59 °F). This pressure is equivalent to the weight exerted by a mass of 1.03 kg on a square cm (14.7 lb per square in.) or 1013.25 millibars (mb). Water is much denser than air so that a column of equivalent height produces much greater pressure.

A column of fresh water with a height of 10.33 m (33.9 ft) exerts a pressure at its base that approximately equals one standard atmosphere. A barometer made with water would have a liquid column 33.9 ft high! Hence, the pressure at the deepest known place in the ocean, the Challenger Deep in the Mariana Trench in the Western Pacific where the ocean depth is about 11,000 m (36,100 ft), is more than 1000 times greater than standard atmospheric pressure. When determining pressure, water can be approximated as an incompressible fluid; that is, water density does not vary significantly with increasing pressure. To a good first approximation, the pressure at any point in a water column is directly related to depth and the relationship is linear; that is, doubling the depth doubles the pressure. At a depth of 10.33 m, the water pressure is 1.03 kg per square cm. At ten times this depth, 103.3 m (339 ft), the water pressure is ten times as great, that is, 10.3 kg per square cm.

Pressure has long been used as a measure of depth in the ocean because pressure is much easier to measure in situ than depth, and to a good approximation (error of less than 2%), pressure is equivalent to depth. Ocean scientists, like their counterparts in atmospheric science, commonly use the bar and its derivatives as a standard unit of pressure. The water pressure expressed in decibars (0.1 bar) is numerically equivalent to the water depth expressed in meters. The interchangeability of the two measures (i.e., water pressure in decibars and water depth in meters) greatly simplifies data analysis. For example, the water pressure at 100 m (330 ft) depth is about 100 decibars.

Excessively high pressures encountered in ocean water test the limits of human survival. As of this writing, the deepest scuba dive record was set in 2005 by Nuno Gomez of South Africa, who descended to 318.25 m (1044.13 ft) in the Red Sea. At this depth, the pressure was about 318 decibars, or 31.8 times greater than the average air pressure at sea level. Scuba divers wear a special breathing apparatus to equalize the pressure in all air spaces with the surrounding water pressure and provide the proper mixture of gases. Gomez descended to the record depth in about 20 minutes, and ascended to the surface in a series of stages lasting 12 hrs. Divers must ascend slowly (Figure 1) in order to gradually release water pressure on the body, and avoid potentially life-threatening decompression sickness, also known as "the bends", where gas bubbles (mainly nitrogen) form in the body.

FIGURE 1
Scuba divers during a 20-ft decompression stop after a deep dive. [NOAA Flower Garden Banks National Marine Sanctuary]

Determining Altitude from Air Pressure

Altimetry is the determination of altitude above mean sea level. A *pressure altimeter* is an aneroid barometer that is graduated in increments of altitude. The graduation (that is, the calibration of altitude against air pressure) is prescribed by the *standard atmosphere*, described elsewhere in this chapter. At any time and place, however, the real atmosphere usually differs from the standard atmosphere so that an altimeter typically does not give the true altitude. The *indicated altitude* (the altimeter reading) is the same as the *true altitude* only when air pressure and temperature match the standard atmosphere. Unless adjustments are made, the discrepancy between indicated and true altitudes can pose serious problems, especially for private aircraft during the crucial takeoff and landing phases of flight or when flying over mountainous terrain.

Changes in surface air pressure en route are one cause of differences between indicated and true altitudes. For example, as an aircraft travels toward a destination reporting a lower surface air pressure than its departure point, the altimeter will read higher than the true altitude. Hence, aircraft altimeters are equipped with a movable scale that enables the pilot to adjust altimeter readings. The Federal Aviation Administration (FAA) requires all aircraft flying below an altitude of 5500 m (18,000 ft) to calibrate altimeters to surface air pressure radioed from flight service stations en route. (Above 5500 m, aircraft fly along an isobaric surface with the altimeter zeroed at the standard sea-level pressure of 1013.25 mb.)

In-flight adjustments of altimeters to surface conditions, however, do not correct for pressure variations that arise en route principally from temperature variations within the air column beneath the aircraft. Cold air is denser than warm air so that air pressure drops more rapidly with altitude in cold air than in warm air. Within a column of cold air, a given air pressure occurs at a lower altitude than does the same air pressure in a column of warm air (Figure 1). This means, for example, that as an aircraft flies into a column of air that is warmer than specified by the standard atmosphere, the altitude indicated by the altimeter will be lower than the true altitude. On the other hand, the altimeter onboard an aircraft flying into air colder than specified by the standard atmosphere will read too high.

The danger, of course, is that an erroneous altimeter reading may impede a pilot's ability to clear an obstacle such as a mountain peak. In practice, this hazard can be greatly reduced by an onboard computer that measures air temperature at flight level and makes appropriate adjustments to the altimeter reading. Note that this correction should be based on the mean temperature of the air column so that the error is reduced but not eliminated.

In summary, differences between altimeter readings and true altitude arise from en route changes in surface air pressure and/or average temperature in the air column beneath the aircraft. Even with adjustments in altimeter readings, pilots are well advised to follow the adage *"cold or low, look out below."* Hence, pilots should always select a flight altitude that will allow for a margin of safety, especially when flying over mountainous terrain or during conditions of restricted visibility.

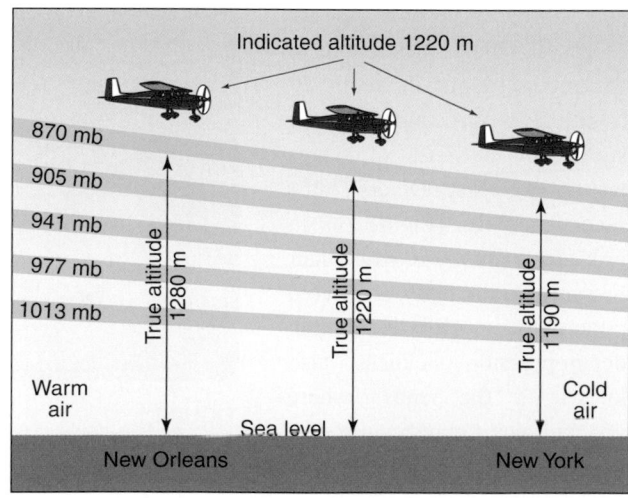

FIGURE 1

An aircraft *pressure altimeter* is initially calibrated based on the relationship between air pressure and altitude as specified by the standard atmosphere. But as the aircraft flies into colder or warmer air, the indicated altitude may differ from the true altitude because air pressure drops more rapidly with altitude in cold air than in warm air.

An alternative to a pressure altimeter is a *radio* (or *electronic*) *altimeter*, which is flown onboard most commercial aircraft. This instrument emits pulses of radio waves (or a continuous signal) from the plane to the ground, where the signal is reflected back to the aircraft. Altitude is calibrated in terms of the time elapsed between emission and reception of the signal. The longer the signal takes, the greater is the indicated altitude of the aircraft. Recent advances in GPS (Global Positioning System) technology, which provides accurate elevation information, have caused GPS to be adopted as an alternative method of altitude determination.

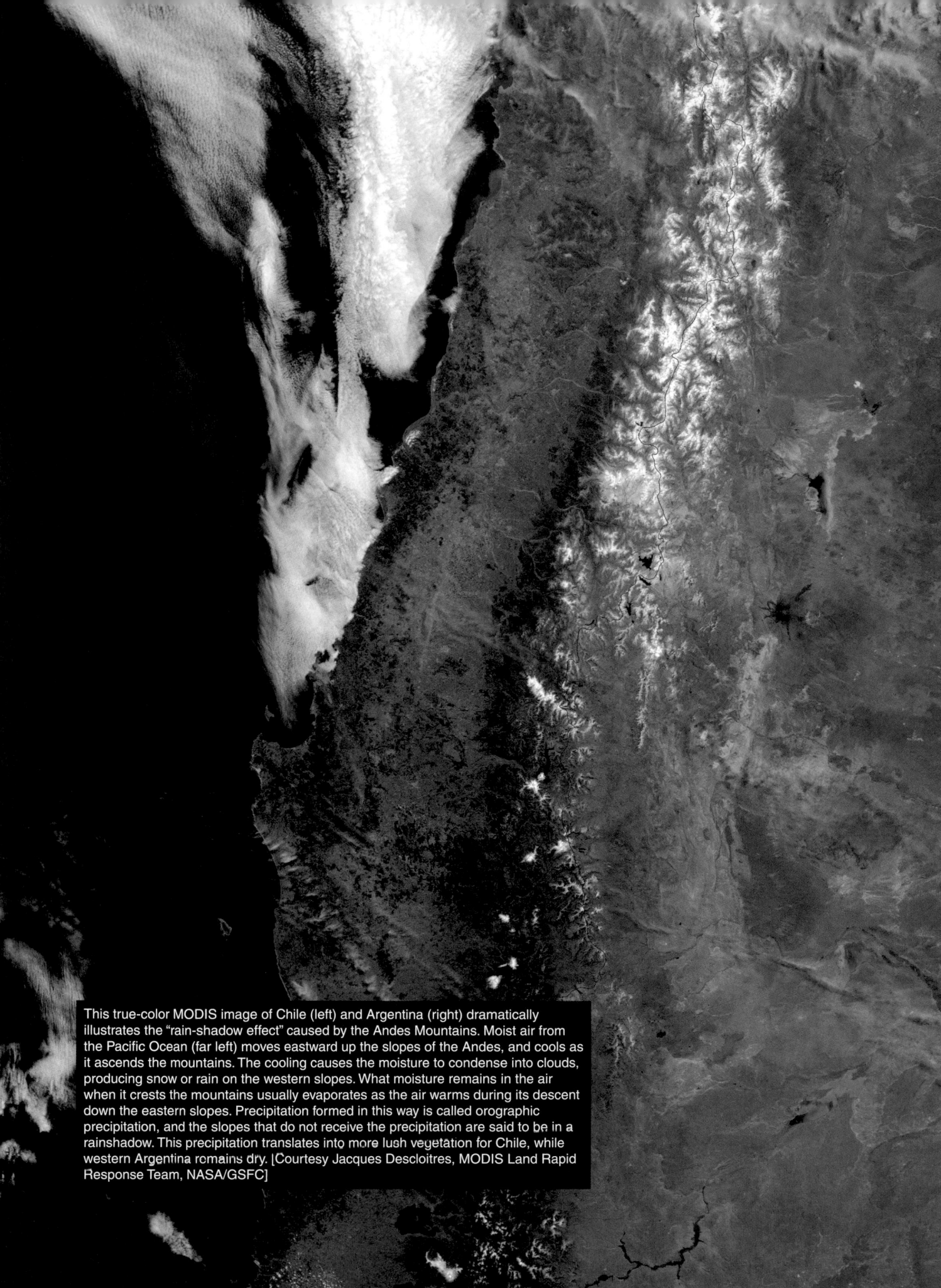

This true-color MODIS image of Chile (left) and Argentina (right) dramatically illustrates the "rain-shadow effect" caused by the Andes Mountains. Moist air from the Pacific Ocean (far left) moves eastward up the slopes of the Andes, and cools as it ascends the mountains. The cooling causes the moisture to condense into clouds, producing snow or rain on the western slopes. What moisture remains in the air when it crests the mountains usually evaporates as the air warms during its descent down the eastern slopes. Precipitation formed in this way is called orographic precipitation, and the slopes that do not receive the precipitation are said to be in a rainshadow. This precipitation translates into more lush vegetation for Chile, while western Argentina remains dry. [Courtesy Jacques Descloitres, MODIS Land Rapid Response Team, NASA/GSFC]

HUMIDITY, SATURATION, & STABILITY

Chapter Highlights

Case-in-Point
  Atmospheric Rivers
Global Water Cycle
  Transfer Processes
  Global Water Budget
How Humid Is It?
  Vapor Pressure
  Mixing Ratio, Specific Humidity, and
    Absolute Humidity
  Saturated Air
  Relative Humidity
  Dewpoint
  Precipitable Water
Monitoring Water Vapor
  Humidity Instruments
  Water Vapor Satellite Imagery
How Air Becomes Saturated
Atmospheric Stability
  Soundings
  Stüve Diagram
Lifting Processes
Conclusions
Basic Understandings/Enduring Ideas
Key Ideas/Review/Critical Thinking
For Further Exploration
  Heat, Humidity and Human Comfort
  Measuring Evaporation
  Atmospheric Stability and Air Quality

Learning Objectives

Identify the principal reservoirs in the global
  water cycle.
Distinguish among the various phase
  changes of water.
Describe the role of latent heat in phase
  changes of water.
List some of the implications of the global
  water budget.
Describe ways in which the water vapor
  component of air is quantified.
Compute the relative humidity from either
  the mixing ratio or the vapor pressure.
Explain why and how relative humidity is
  temperature dependent.
Explain why the average precipitable water
  value generally decreases with increasing
  latitude.
Describe how a sling psychrometer is used to
  determine relative humidity.
Explain how atmospheric stability is
  determined.
Plot a sounding on a Stüve thermodynamic
  diagram.
Describe how atmospheric stability affects
  vertical motion of air and cloud formation.
Identify the various lifting processes that
  operate within the atmosphere that bring
  about the formation of clouds.
Describe how a mountain range can influence
  the pattern of clouds and precipitation.

How does the cycling of water in the Earth-atmosphere system
help maintain a habitable planet?

Case-in-Point

Atmospheric Rivers

Atmospheric river (AR) is a recently coined term that describes a relatively narrow band of concentrated water vapor transport in the lower atmosphere, responsible for most of the horizontal flow of water vapor outside of the tropics (Figure 6.1). Especially in the Pacific coast states, ARs play an important role in the fresh water supply, extreme precipitation events, and major flooding. Most landfalling ARs are weak and deliver beneficial rains and snows. For example, ARs are essential contributors to the water supply of the West Coast states where an estimated 30% to 50% of the average annual precipitation comes from a half dozen or so landfalling ARs. This precipitation builds the mountain snowpack in winter that then supplies fresh water to semi-arid regions via snowmelt during the dry season or drought.

The term originally was coined by research scientists Reginald Newell and Yong Zhu of the Massachusetts Institute of Technology in the early 1990s. They found that ARs are responsible for more than 90% of the global meridional (north/south) transport of water vapor. At any given time this flow is concentrated in 3 to 5 narrow bands. ARs contribute in a major way to the ocean-to-land flux of water that helps drive the runoff from land-to-ocean. Hence, atmospheric rivers are key components of the global water cycle.

Typically an AR is centered in the lower troposphere about 3000 m (10,000 ft) above Earth's surface, has a width of less than 500 km (310 mi), and is at least 2000 km (1240 mi) long—sometimes winding over an entire ocean basin. At the core of an AR, the wind speed is greater than 45 km per hr (30 mph), and a strong AR transports the equivalent of about 7.5 to 15 times the average daily discharge of water at the mouth of the Mississippi River. Depending on its strength, a landfalling AR can have impacts that are positive (e.g., beneficial rains, snows contributing to the fresh water supply) or negative (e.g., torrential rains, flooding, mudslides, property damage, loss of life).

ARs that affect the weather of the U.S. Pacific coast states form, move, and develop with winter storms that take shape over the North Pacific Ocean and are visible from space as huge comma shaped swirls of clouds. A cold front aligns along the tail of the comma, arcing southwestward toward Hawaii and beyond. As the storm tracks eastward, warm humid air flows

FIGURE 6.1
An enhanced GOES-11 water vapor satellite image from 20 December 2010 showing an atmospheric river crossing the North Pacific and making landfall along the California coast. [Naval Research Laboratory]

poleward ahead of the cold front and water vapor is concentrated into narrow ribbons in the warm sector of the storm. This warm, humid flow constitutes an atmospheric river. Occasionally an AR dips southward and entrains moisture and heat directly from the Pacific subtropics and tropics. Such an AR is more likely to produce torrential rains and heavy snows as it extends into mid-latitudes and ultimately makes landfall. The *Pineapple Express* is an example of this type of AR that originates over the waters near Hawaii, flows toward the northeast, and makes landfall along the California coast.

Atmospheric rivers make landfall in many mid-latitude western coastal regions including the West Coast of North America, Western Europe, and the West Coast of Africa. ARs are also known to track northward out of the Gulf of Mexico as well as along the U.S. Eastern Seaboard. For example, major flooding linked to landfalling ARs occurred in Nashville, TN, in May 2010 and the Carolinas in October 2010.

The most extreme precipitation and greatest flood potential develop where landfalling ARs encounter coastal and inland mountain ranges and undergo *orographic lifting* of very humid air. Rain and snow from ARs may trigger mudslides and cause loss of life and considerable property damage. From 17 to 22 December 2010, an atmospheric river produced up to 66 cm (26 in.) of precipitation in California and up to 520 cm (17 ft) of snow in the Sierra Nevada Mountains. From 7 to 11 January 2005, as much as 60 cm (24 in.) of rain fell in the San Gabriel Mountains north of Los Angeles, the result of a strong onshore AR coupled with orographic lifting of the very humid air. Flooding caused a massive mudslide in La Conchita, CA, taking 10 lives (Figure 6.2). Property damage amounted to hundreds of millions of dollars. Huge snowfalls of 1.5 to 2.4 m (5 to 8 ft) in depth occurred in the Lake Tahoe area of the Sierra Nevada Mountains.

Major flooding on most rivers from California to Washington State is linked to landfalling atmospheric rivers. All seven floods since October 1997 on the Russian River, the southward flowing river that drains Sonoma and Mendocino Counties in Northern California, are directly attributable to heavy rainfall associated with ARs. Furthermore, since the 1970s the 10 largest winter floods in the United Kingdom have been linked to atmospheric rivers.

FIGURE 6.2
Heavy rains associated with a landfalling atmospheric river caused a massive mudslide in La Conchita, CA, in January 2005. The mudslide pushed houses in its path rather than flowing around or through them. The left part of the house became detached from the right part. [USGS, from the report *Landslide Hazards at La Conchita, California* by Randall W. Jibson]

Water occurs in all three phases in the atmosphere: as water vapor (an invisible gas), and as aggregates of tiny ice crystals and water droplets visible as clouds. The total amount of water in the atmosphere is relatively small and most of that is confined to the lower portion of the troposphere. If all the water vapor were removed from the atmosphere as rain and distributed uniformly over the globe, this water would cover Earth's surface to a depth of only about 2.5 cm (1.0 in.). Water continually enters the atmosphere as vapor from reservoirs of water at Earth's surface (e.g., ocean, lakes, soil, vegetation) while water continually leaves the atmosphere, returning to Earth's surface as rain, snow, and other forms of precipitation, or by condensation. On average, the residence time of a water molecule is about 10 days in the atmosphere and on the order of hours in a cloud. The exchange of water between Earth's surface and atmosphere is an essential component of the global water cycle.

In this chapter, we consider how the global water cycle functions, particularly as it relates to the transfer of water between Earth's surface and atmosphere. We learn how to quantify the water vapor component of air, how air becomes saturated through uplift and expansional cooling, and how atmospheric stability influences the ascent of air. All this is important because as air nears saturation, clouds are more and more likely to form, and clouds are prerequisites to precipitation.

Global Water Cycle

The total amount of water on Earth is neither increasing nor decreasing, although natural processes continually generate and break down water at essentially equal rates. Water vapor accounts for perhaps half of all gases emitted during a volcanic eruption; some of this water was originally sequestered in magma and solid rock, and the rest is recycled surface and groundwater. Volcanic activity is more or less continuous on Earth and adds to the supply of water. Also, a minute amount of water is added to Earth by comets, asteroids, meteorites and other extraterrestrial debris that continually bombard the upper atmosphere (Chapter 2). At the same time, intense solar radiation entering the upper atmosphere, converts (*photo-dissociates*) a small amount of water vapor into its constituent hydrogen and oxygen atoms, which may escape to space. Also, water chemically reacts with other substances and thereby is locked up in various compounds. Some of the water that is removed may be released eventually through other chemical reactions involving the new compounds. Annually, additions of water from volcanic eruptions and extraterrestrial sources roughly equal losses of

TABLE 6.1 Water Stored in Reservoirs of the Global Water Cycle	
Reservoir	*Percent of total water*
Ocean	97.20
Ice sheets and glaciers	2.15
Groundwater	0.62
Lakes (freshwater)	0.009
Inland seas, saline lakes	0.008
Soil moisture	0.005
Atmosphere	0.001
Rivers and streams	0.0001
Living Organisms	0.00004

water through photodissociation of water vapor and chemical reactions. This balance of give and take has prevailed on Earth for perhaps hundreds of millions of years.

The essentially fixed quantity of water on Earth is distributed among various reservoirs (Table 6.1). The ocean is the largest of these reservoirs by far, accounting for 97.2% of all water on the planet; most of the rest is tied up as ice sheets up to 3 km (1.8 mi) thick that cover most of Antarctica and Greenland. Relatively small amounts of water occur in inland seas and lakes, and occupy the tiny pore spaces and fractures within soil, sediment, and bedrock (soil moisture and groundwater). Even smaller amounts of water occur in the atmosphere as clouds, precipitation particles, and invisible water vapor. Rivers and streams and living organisms (plants and animals) comprise a minute amount of Earth's water.

The ceaseless movement of water through and among the various reservoirs on a planetary scale is known as the **global water cycle** (Figure 6.3). In brief, water vaporizes from ocean and land surfaces to the atmosphere, where winds can transport water vapor to other locations, sometimes thousands of kilometers away (Figure 6.4). (Recall our discussion of atmospheric rivers in this chapter's Case-in-Point.) Clouds form, and then rain, snow and other forms of precipitation may fall from clouds to Earth's surface, recharging the ocean and terrestrial (land-based) reservoirs of water. From terrestrial reservoirs, water flows into the ocean basins. The continuity of the global water cycle is captured in a verse from Ecclesiastes 1:7: "Every river flows into the sea, but the sea is not yet full. The waters return to where the rivers began and start all over again."

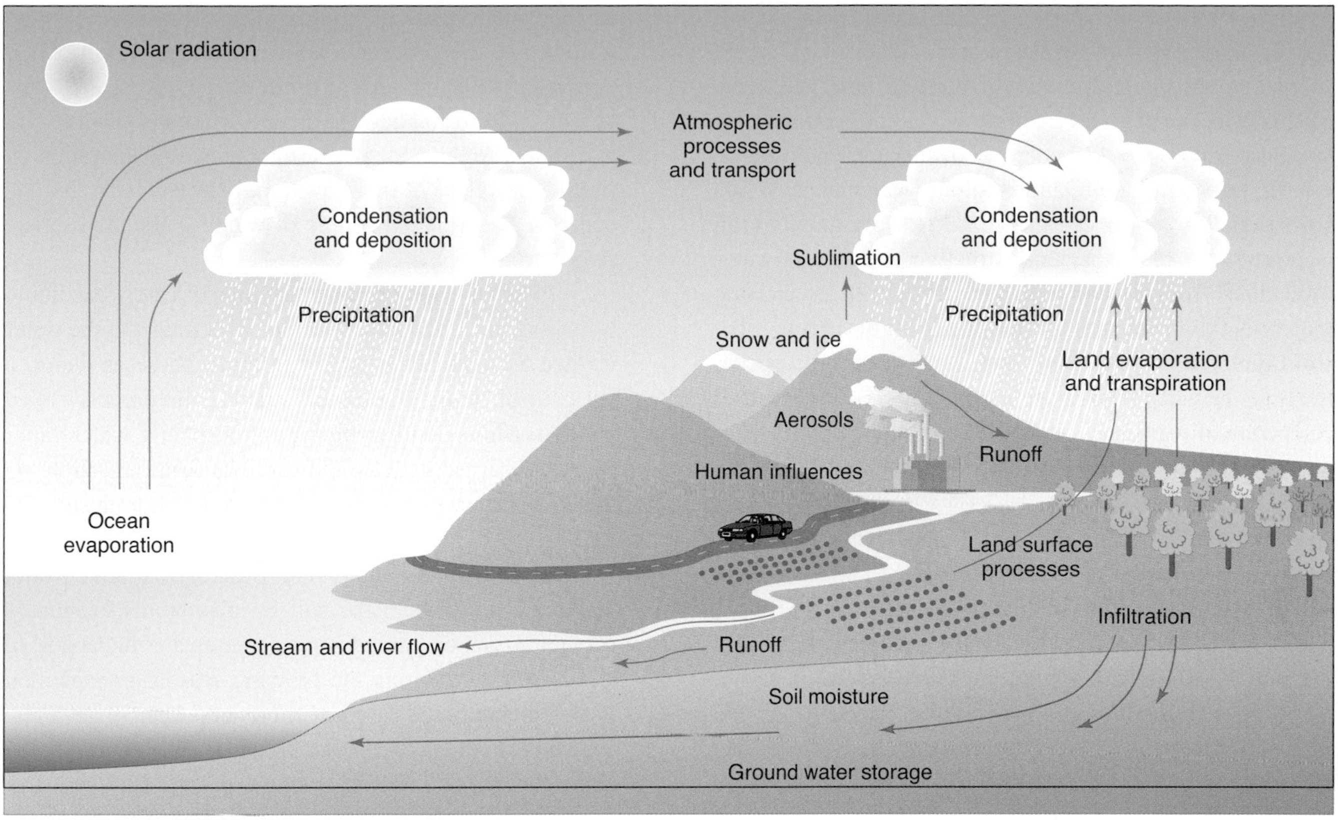

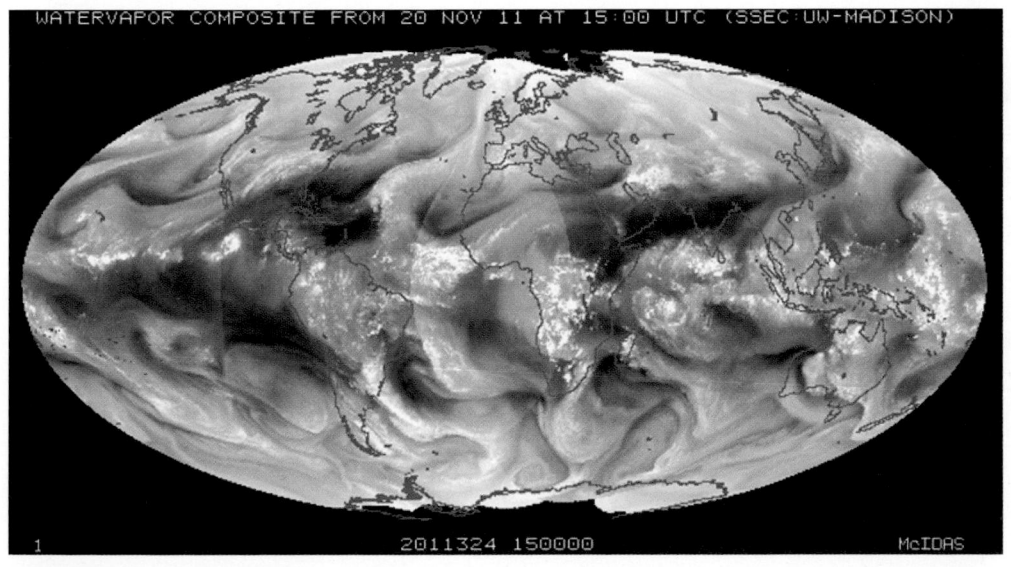

FIGURE 6.3
The global water cycle is a continuous flow of water and energy among oceanic, terrestrial (land-based), and atmospheric reservoirs.

FIGURE 6.4
Global-scale water vapor image showing evidence of long range transport. [Courtesy of University of Wisconsin-Madison Space Science and Engineering Center (SSEC)]

The Sun drives the global water cycle. As we saw in Chapter 4, some of the radiation that strikes Earth's surface is absorbed, that is, converted to heat, and some of this heat is used to evaporate water (or sublimate ice and snow). If water did not vaporize, there would be no clouds, no precipitation, and no global water cycle. While solar radiation powers the global water cycle, gravity is an important player in that this attractive force keeps water molecules from escaping to space, as well as causing water to fall from the sky as precipitation and to flow from the continents to the ocean. This section focuses on the links between the atmospheric, oceanic, and terrestrial reservoirs of water.

TRANSFER PROCESSES

As part of the global water cycle, water is transferred between Earth's surface and the atmosphere via phase changes (evaporation, condensation, transpiration, sublimation, and deposition) and by precipitation. Within the ranges of air temperature and air pressure on Earth, all three phases of water can coexist naturally. At the interface between liquid water and air (e.g., lake or sea surface), water molecules continually change phase: some crossing the interface from water to air and others from air to water. If more water molecules enter the atmosphere as vapor than return as liquid, a net loss occurs in liquid water mass; this process is known as **evaporation**. Water evaporates from the surface of the ocean, lakes, and rivers as well as from soil and the damp surfaces of plant leaves and stems. About 85% of the total annual evaporation on Earth takes place at the seawater/air interface so that the ocean is the principal source of water in the atmosphere (Figure 6.5).

On the other hand, at the interface between liquid water and air, if more water molecules return to the water surface as liquid than enter the atmosphere as vapor, a net gain of liquid water mass results, the process called **condensation**. Condensation of atmospheric water vapor can be observed when small droplets form and grow on the cold surface of a glass of iced tea on a humid summer day.

Transpiration is the process whereby water that is taken up from the soil by plant roots eventually escapes as vapor through tiny pores on the underside of leaves. On land during the growing season, transpiration is considerable and is often more important than direct evaporation of water in supplying water vapor to the atmosphere. For example, a single hectare (2.5 acres) of corn typically transpires about 34,000 liters (L) (8800 gal) of water per day. Annually, a mature oak tree may transpire

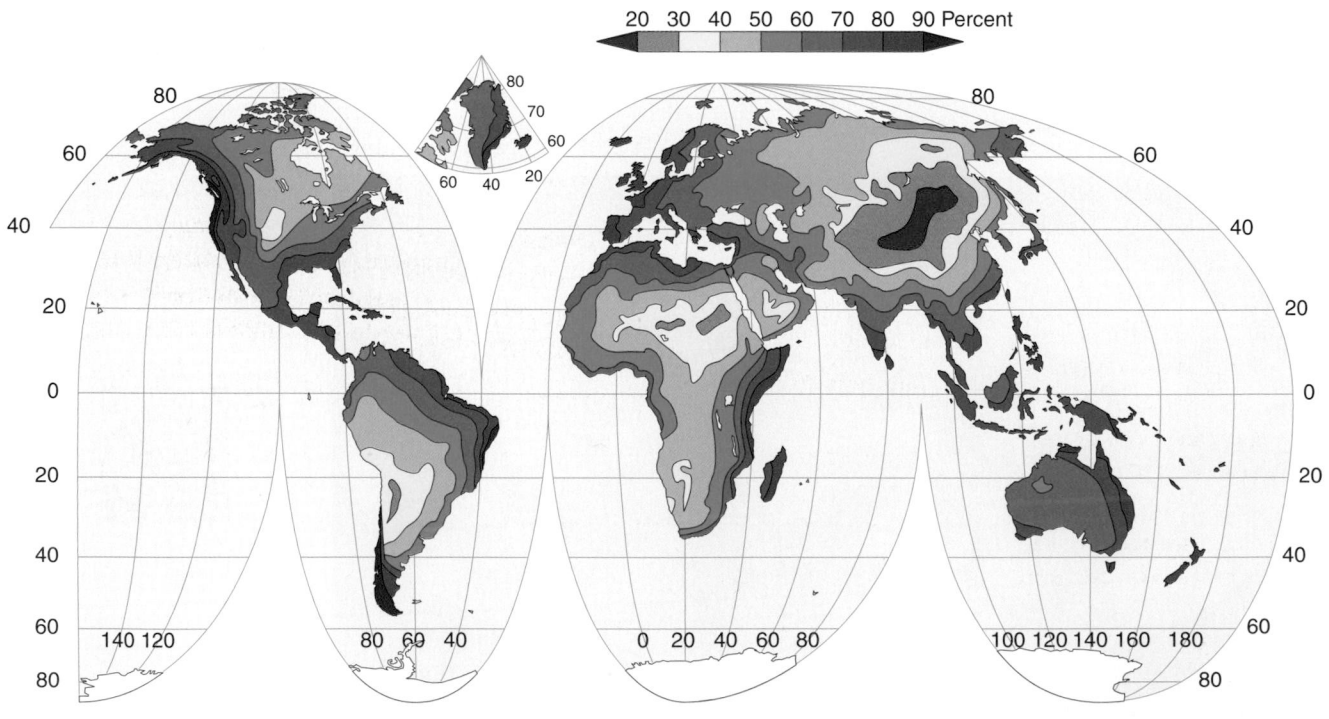

FIGURE 6.5
The percentage of annual precipitation over land that originally vaporized from the ocean, averaged over 15 years. In many land areas, the principal source of water for precipitation is evaporation from the ocean. [Adapted from World Climate Research Programme, Global Energy and Water Cycle Experiment]

more than 150,000 L (40,000 gal) of water. Measurements of direct evaporation from Earth's surface plus transpiration are often combined as **evapotranspiration**.

At the interface between ice and air (e.g., the surface of snow cover), water molecules are also continually changing phase: from ice directly to vapor and from vapor directly to ice. If more water molecules enter the atmosphere as vapor than transition to ice, a net loss of ice mass occurs. **Sublimation** is the process whereby ice or snow becomes vapor without first becoming a liquid. Sublimation explains the gradual disappearance of snow and ice on sidewalks and roads even while the air temperature remains well below freezing. On the other hand, if more atmospheric water molecules transition to ice than move from ice to vapor, a net gain of ice mass results. **Deposition** is the process whereby water vapor becomes ice without first becoming a liquid. During a winter night, formation of frost on automobile windows is an example of deposition.

Condensation or deposition within the atmosphere produces clouds. **Precipitation** is water in liquid, frozen or freezing form (i.e., rain, drizzle, snow, ice pellets, hail, and freezing rain) that falls from clouds under the influence of gravity and reaches Earth's surface, where most of it eventually vaporizes back into the atmosphere.

Evaporation (or sublimation) purifies water entering the vapor phase and condensation of this vapor produces purified liquid water. As water vaporizes from Earth's surface, all suspended and dissolved substances such as sea salts and other contaminants are left behind. Through this natural cleansing mechanism, salty ocean water is the source of much of what eventually falls as freshwater precipitation that replenishes reservoirs on Earth's surface. Purification of water through phase changes is known as **distillation**.

GLOBAL WATER BUDGET

Return of water from the atmosphere to the land and ocean via condensation, deposition, and precipitation completes an essential subcycle of the global water cycle. To learn more about this subcycle, compare the transfer of water between the continents and the atmosphere with that between the ocean and the atmosphere. The balance sheet for the inputs and outputs of water to and from the various global reservoirs is called the **global water budget** (Table 6.2).

Over the course of a year, the volume of precipitation (rain plus melted snow) that falls on land exceeds the total volume of water that vaporizes (via evaporation, transpiration, and sublimation) from land by about one-third. This imbalance occurs because landmasses favor certain precipitation-forming mechanisms. Over the same period, the volume of precipitation falling on the ocean is less than the volume of water that evaporates from the ocean. Evaporation is more important because the ocean surface is a nearly limitless source of water vapor. The global water budget thus indicates an annual net gain of water mass on the continents and an annual net loss of water mass from the ocean. The annual excess of water on the continents essentially equals the deficit from the ocean. To restore this balance and complete the global water cycle, excess water on land drips, seeps and flows by gravity back to the sea. The net flow of water from land to sea implies a net flow of water within the atmosphere directed from sea to land. As described in this chapter's Case-in-Point, *atmospheric rivers* play an important role in this transport of water from sea to land as part of the global water cycle.

Precipitation strikes the ground directly or it may be intercepted by vegetation and then evaporates or drips to the ground (Figure 6.6). Also, some trees collect moisture from drifting fog or low clouds and that water drips to the ground (known as *fog drip*). Once water reaches Earth's land surface, it follows various pathways. Some water vaporizes directly back into the atmosphere while some is temporarily stored in lakes, snow and ice fields, or glaciers. The remainder either flows on the surface as rivers or streams (*runoff component*) or seeps into the ground as soil moisture or groundwater (*infiltration component*). About one-third of the precipitation that falls on land runs off to

	TABLE 6.2	
	Global Water Budget	
Source	*Cubic meters per year*	*Gallons per year*
Precipitation on the ocean	$+3.24 \times 10^{14}$	$+85.5 \times 10^{15}$
Evaporation from the ocean	-3.60×10^{14}	-95.2×10^{15}
Net loss from the ocean	-0.36×10^{14}	-9.7×10^{15}
Precipitation on land	$+0.98 \times 10^{14}$	$+26.1 \times 10^{15}$
Evapotranspiration from land	-0.62×10^{14}	-16.4×10^{15}
Net gain on land	$+0.36 \times 10^{14}$	$+9.7 \times 10^{15}$

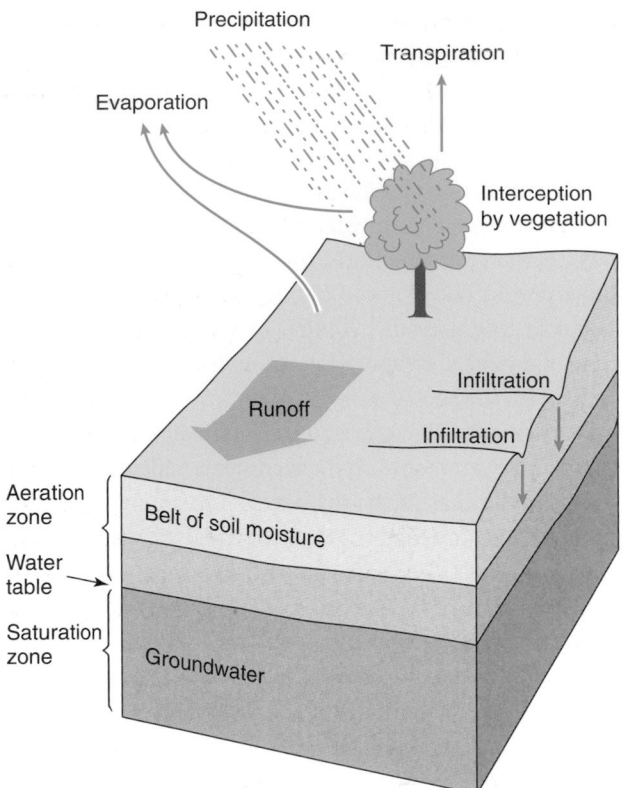

FIGURE 6.6
Various pathways taken by precipitation falling on land.

the ocean. The ratio of the portion of water that infiltrates the ground to the portion that runs off depends on rainfall intensity, vegetation, topography, and physical properties of the intercepting land surface. For example, rain falling on mountainous terrain usually runs off quickly. Likewise, rain falling on frozen ground or city streets mostly runs off whereas rain falling on unfrozen sandy soil readily soaks into the ground.

Rivers and streams plus their tributaries drain a fixed geographical area known as a *drainage basin* (or *watershed*). The quantity and quality of water flowing in a river depends on the climate, vegetation, topography, geology, and land use in its drainage basin. For example, where the climate features distinct rainy and dry seasons, stream flow can vary considerably through the year. A drainage basin may also include lakes, wetlands, glaciers, and other temporary impoundments of surface water.

How Humid Is It?

It's not the heat, it's the humidity is a popular statement that attributes the discomfort a person feels on a hot, muggy day to the water vapor component of air. **Humidity** is a general term referring to any one of many different ways of describing the amount of water vapor in the air. Humidity is an important determinant of our physical comfort, as discussed in this chapter's first *For Further Exploration*. Experience tells us that humidity varies with the season, from one day to the next, within a single day, and from one place to another. In most places summer days feel more humid than a typical winter day. In cold regions, dry winter air also causes some discomfort. This section covers the various quantitative measures of humidity.

VAPOR PRESSURE

When water enters the atmosphere as vapor, water molecules disperse and mix with the other gases composing air (mostly nitrogen and oxygen) thereby contributing to the total pressure exerted by the atmosphere. The amount of pressure produced by water vapor molecules is a measure of the humidity: the more water vapor, the greater the pressure exerted by water molecules. Water vapor's contribution to the total air pressure, as described by *Dalton's law* of partial pressures (Chapter 5), is known as **vapor pressure**.

As the water vapor content of the atmosphere increases, the vapor pressure also increases. Water vapor is a highly variable component of air but composes at most no more than about 4% of the atmosphere's mass in the lowest kilometer—even in the sultry air over the tropical ocean and rainforests. The total pressure exerted by all atmospheric gases at sea level averages approximately 1000 mb. This means that the vapor pressure is very unlikely anywhere to exceed 40 mb (4% × 1000 mb = 40 mb) at sea level, and in most places the vapor pressure is considerably less than 40 mb.

MIXING RATIO, SPECIFIC HUMIDITY, AND ABSOLUTE HUMIDITY

Other ways of quantifying humidity are the mixing ratio, specific humidity, and absolute humidity. **Mixing ratio** is defined as the mass of water vapor per mass of the remaining dry air, usually expressed as so many grams of water vapor per kilogram of dry air. Typically the mixing ratio at sea level is less than 40 grams per kilogram. **Specific humidity** is defined as the ratio of the mass (in grams) of water vapor to the mass (in kilograms) of the air including the water vapor (that is, the combined mass of dry air plus water vapor totals one kilogram). Values of specific humidity and mixing ratio are so close that they usually can be considered equivalent.

Absolute humidity is defined as the mass of water vapor per unit volume of humid air, typically expressed as grams of water vapor per cubic meter of air. In other words, absolute humidity is the density of the water vapor component of air. Although conceptually simple, absolute humidity has limited application in meteorology because the volume of a parcel of air may change causing its absolute humidity to vary even though no water vapor is gained or lost by the parcel. For example, an air parcel's volume increases when it is heated or lifted in the atmosphere. Even though no water vapor enters or leaves the air parcel, the amount of water vapor per unit volume decreases, that is, the absolute humidity decreases.

SATURATED AIR

Regardless of how humidity is quantified, the amount of water vapor in air at a specified temperature has an upper limit. At its maximum humidity, air is described as *saturated* with water vapor. Earlier in this chapter, we described a two-way exchange of water molecules at the interface between water and air (or between ice and air). Water molecules are in a continual state of flux between the liquid (or ice) and vapor phases. During evaporation, more water molecules become vapor than return to the liquid phase, and during condensation, more water molecules return to the liquid phase than enter the vapor phase. Eventually, a dynamic equilibrium may develop such that the flux of water molecules is the same in both directions; that is, liquid water becomes vapor at the same rate that water vapor becomes liquid. At equilibrium, above a plane surface of liquid water or ice, the air is saturated with water vapor. The vapor pressure at equilibrium is called the **saturation vapor pressure**, the mixing ratio is the **saturation mixing ratio**, the specific humidity is the **saturation specific humidity**, and the absolute humidity is the **saturation absolute humidity**.

Altering the temperature disturbs this dynamic equilibrium at least temporarily. Heating water to a higher temperature causes the average kinetic energy of individual water molecules to increase so that they more readily escape the water surface as vapor. Initially, evaporation prevails. With a sufficient supply of water, and as long as water vapor is not continually carried away by the wind, eventually a new dynamic equilibrium is established. That is, the flux of water molecules becoming liquid again balances the flux of water molecules becoming vapor. This new equilibrium is achieved with a greater concentration of water vapor in the air at the higher temperature. Hence, raising the air temperature increases the saturation vapor pressure, saturation mixing ratio, saturation specific humidity, and saturation absolute humidity.

Conversely, with a drop in water temperature, the average kinetic energy of individual water molecules decreases and molecules less readily escape the water surface as vapor. Initially, at the lower temperature condensation prevails but eventually a new equilibrium is established; that is, the flux of water molecules becoming vapor again balances the flux of water molecules becoming liquid. This new equilibrium is achieved with less water vapor in the air at the lower temperature.

Ultimately, the water vapor component of air depends on the rate of vaporization of water, which is regulated chiefly by temperature, and the presence of a water supply. The dependence of the saturation vapor pressure on temperature is shown in Figure 6.7 and Table 6.3. The variation of the saturation mixing ratio with temperature is shown in Figure 6.8 and Table 6.4. For both measures of humidity, the relationship with temperature is non-linear. Because the mixing ratio depends on the mass of the air, the saturation mixing ratio also varies with pressure. Extensive laboratory work by 19th century steam engineers and scientists is the source of these data.

Note that for temperatures below freezing, two different values are given: one over supercooled water and the other over ice. Water that is *supercooled* remains liquid at subfreezing temperatures. As a general rule, the saturation vapor pressure and saturation mixing ratio approximately double in value for every 10 Celsius degree (20 Fahrenheit degree) rise in temperature. At 100 °C (212 °F) at sea level, the saturation vapor pressure is 1013.25 mb, the same as the standard sea level air pressure. Water boils when the saturation vapor pressure and ambient (surrounding) air pressure are equal.

The relationship between temperature and saturation vapor pressure (or saturation mixing ratio) is popularly interpreted to mean that warm air can "hold" more water vapor than can cold air. That is, air is likened to a sponge that can soak up only so much water depending on the temperature. While this may appear intuitively appealing, this analogy is misleading. Air does not literally hold water vapor like a sponge; rather, water vapor exists independently of the other gases that form the mixture known as air. Recall from Dalton's law that each gas in a mixture of gases exerts a pressure as though it were the only gas present. Water vapor is just one of the many gases that compose air. If water vapor were added to air at constant pressure and temperature, it follows that water vapor displaces some of the other gaseous components of air. Because temperature largely governs the rate of vaporization of water, the saturation vapor pressure (or saturation mixing ratio) is actually a measure of water's vaporization rate. At any specified temperature, the saturation vapor pressure (or saturation

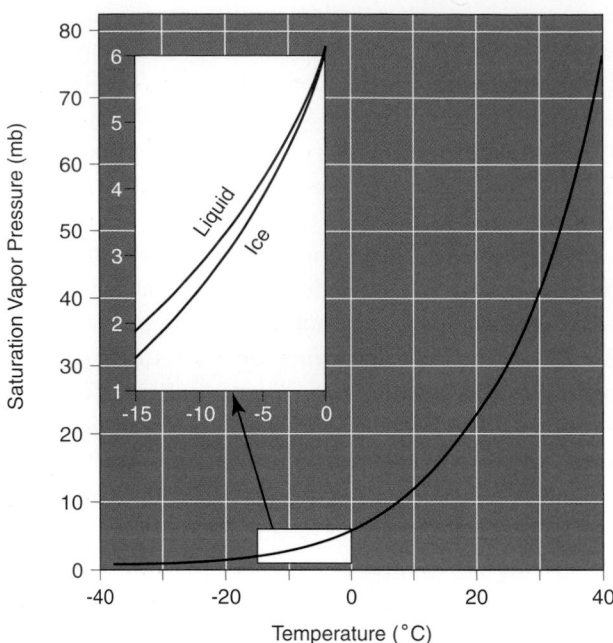

FIGURE 6.7
Variation in saturation vapor pressure with changing air temperature. Note that at subfreezing temperatures, the saturation vapor pressure is greater over supercooled water than over ice.

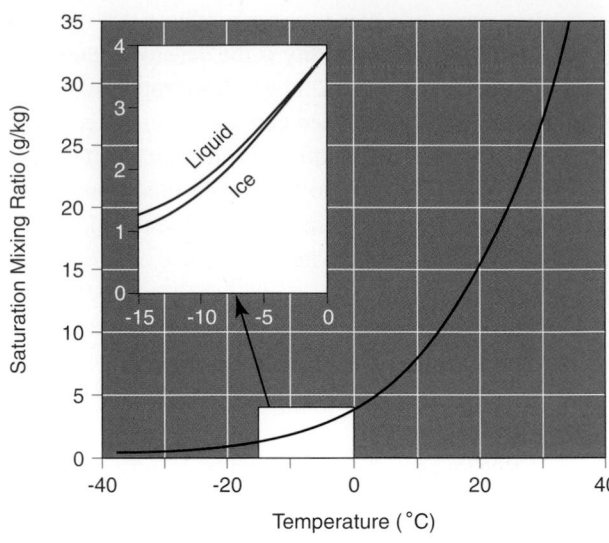

FIGURE 6.8
Variation in saturation mixing ratio with changing air temperature (at a pressure of 1000 mb). Note that at subfreezing temperatures, the saturation mixing ratio is greater over supercooled water than over ice.

TABLE 6.3 **Variation of Saturation Vapor Pressure with Temperature**		
Temperature	*Saturation Vapor Pressure (mb)*	
°C (°F)	*Over water*	*Over ice*
50 (122)	123.40	
45 (113)	95.86	
40 (104)	73.78	
35 (95)	56.24	
30 (86)	42.43	
25 (77)	31.67	
20 (68)	23.37	
15 (59)	17.04	
10 (50)	12.27	
5 (41)	8.72	
0 (32)	6.11	6.11
−5 (23)	4.21	4.02
−10 (14)	2.86	2.60
−15 (5)	1.91	1.65
−20 (−4)	1.25	1.03
−25 (−13)	0.80	0.63
−30 (−22)	0.51	0.38
−35 (−31)	0.31	0.22
−40 (−40)	0.19	0.13

TABLE 6.4 **Variation of Saturation Mixing Ratio with Temperature (at 1000 mb pressure)**		
Temperature	*Saturation mixing ratio (g/kg)*	
°C (°F)	*Over water*	*Over ice*
50 (122)	88.12	
45 (113)	66.33	
40 (104)	49.81	
35 (95)	37.25	
30 (86)	27.69	
25 (77)	20.44	
20 (68)	14.95	
15 (59)	10.83	
10 (50)	7.76	
5 (41)	5.50	
0 (32)	3.84	3.84
−5 (23)	2.64	2.52
−10 (14)	1.79	1.63
−15 (5)	1.20	1.03
−20 (−4)	0.78	0.65
−25 (−13)	0.50	0.40
−30 (−22)	0.32	0.24
−35 (−31)	0.20	0.14
−40 (−40)	0.12	0.08

mixing ratio) would have the same value regardless of the presence or absence of other gases in the atmosphere.

RELATIVE HUMIDITY

Relative humidity is the water vapor measure frequently reported by television and radio weathercasters and is probably the one most familiar to most of us. **Relative humidity** compares the actual amount of water vapor in the air with the amount of water vapor that would be present if that same air were saturated. Relative humidity (RH) is expressed as a percentage and can be computed from either the vapor pressure or mixing ratio. That is,

$$RH = \frac{\text{(vapor pressure)}}{\text{(saturation vapor pressure)}} \times 100\%$$

or

$$RH = \frac{\text{(mixing ratio)}}{\text{(saturation mixing ratio)}} \times 100\%$$

When the actual concentration of water vapor in air equals the water vapor concentration at saturation, the relative humidity is 100%; that is, the air is saturated with respect to water vapor.

Consider an example of how relative humidity is computed. Suppose that the air temperature is 10 °C (50 °F) and the vapor pressure is 6.1 mb. From Table 6.3, we determine that the saturation vapor pressure of air at 10 °C is 12.27 mb. Using the formula above, we compute a relative humidity of about 50%, that is,

$$RH = [(6.1 \text{ mb})/(12.27 \text{ mb})] \times 100\% = 49.7\%$$

At constant temperature and pressure, the relative humidity varies directly with the vapor pressure (or mixing ratio); that is, the relative humidity increases as water vapor is added to air as long as the air temperature and pressure remain constant. Because the saturation vapor pressure varies directly with temperature, the relative humidity varies inversely with temperature. If no water vapor were added to or removed from unsaturated air, the relative humidity increases as the temperature drops, and decreases as the temperature rises. Consider a common example. With a clear sky and calm air or light winds, the air temperature usually rises from a minimum near sunrise to a maximum during early to mid-afternoon and then falls through the evening hours and overnight (Chapter 4). If the amount of water vapor in air remains essentially constant throughout the day, then the relative humidity will vary inversely with air temperature.

As shown in Figure 6.9, the relative humidity is highest when the air temperature is lowest and the relative humidity is lowest when the temperature is highest. After sunrise, as the air warms, the relative humidity drops because the saturation vapor pressure increases as the air temperature rises.

In the previous example, the relative humidity responded only to variations in local air temperature; the air was calm or winds were light so there was no significant cold or warm air advection. The situation is made more complicated by air mass advection, which can influence both local air temperature and vapor pressure. For example, when a warm and humid air mass replaces a cool and dry air mass, both rising temperatures and increasing vapor pressure affect the local relative humidity.

DEWPOINT

Dewpoint, cited on some television and radio weathercasts, is another useful measure of humidity. **Dewpoint** is the temperature to which air must be cooled at constant pressure to achieve saturation of air relative

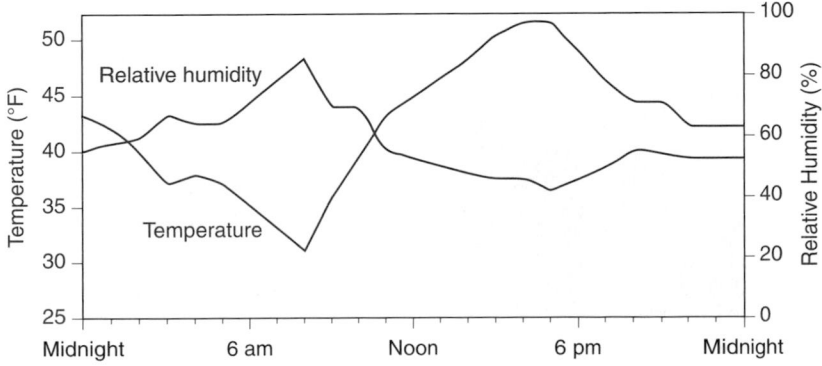

FIGURE 6.9
Temperature and relative humidity on 9 February 2009 at Washington National Airport, VA. The vapor pressure varies only slightly through the course of the day so that the relative humidity varies inversely with the temperature. The relative humidity increases as the temperature drops and decreases as the temperature rises. [Data from the University of Wyoming]

to liquid water. While the relative humidity gives the percentage of how humid air is compared to if that same air were saturated, dewpoint directly measures the water vapor content of air. The higher the dewpoint, the greater is the concentration of water vapor in air. Cooling unsaturated air at constant pressure increases its relative humidity. When the relative humidity reaches 100%, the air is saturated and the air temperature is the same as the dewpoint. Warming the air without adding water vapor via evaporation lowers the relative humidity and increases the difference between the actual air temperature and the dewpoint.

Dew consists of tiny droplets of water formed when water vapor condenses on a cold surface such as flower petals on a clear, calm night (Figure 6.10). As discussed in Chapter 7, dew forms as a consequence of radiational cooling. (Dew that freezes after forming is known as *white dew* because it appears as white beads of ice.) Dew is not a form of precipitation because it does not fall from clouds to Earth's surface. For water vapor to condense as dew on the surface of an object, the temperature of that surface must cool below the dewpoint. When cooling at constant pressure produces saturation at an air temperature below freezing, water vapor deposits as **frost**, ice crystals that form on exposed surfaces such as plants (Figure 6.11). The air temperature at which further cooling will cause frost to form is known as the **frost point**.

From the above discussion, dewpoint is an ideal measure of atmospheric humidity. Average daily, monthly, or annual dewpoints can be computed from hourly dewpoint readings made at automated weather stations. As shown in Figure 6.12, on average, in winter a steep dewpoint gradient prevails from the Pacific and Gulf of Mexico coasts to central Canada. The dewpoint

FIGURE 6.11
Frost-covered leaves. [Photo by Michelle Bambary]

(actually frost point) is exceptionally low in polar and Arctic air masses that invade broad regions of the United States in winter. In summer, dewpoints are highest at localities bordering the Gulf of Mexico and along the southeast coast north to Delaware. In that area, the July mean dewpoint typically ranges between 21 °C and 24 °C (70 °F and 75 °F). During the oppressive heat waves that sometimes sweep over the continent east of the Rocky Mountains, the dewpoint may top the low 20s Celsius (low 70s Fahrenheit) even in the northern states. For example, on 30 July 1999, during a particularly humid heat wave, the dewpoint at Milwaukee, WI tied a record high of 27.8 °C (82 °F). Summer dewpoints are lowest in the Rocky Mountain States and the American Southwest. From northwest New Mexico northward into western Montana, the July mean dewpoint is generally in the single digits Celsius (40s Fahrenheit).

PRECIPITABLE WATER

Precipitable water is another way of describing the amount of water vapor in the atmosphere. But unlike the other humidity measures that represent the amount of water vapor at a specific place, **precipitable water** is the depth of water that would be produced if all the water vapor in a vertical column of air were condensed into liquid water. The air column is usually taken to extend from Earth's surface to the top of the troposphere, the portion of the atmosphere where most water vapor occurs. A reasonably good measure of precipitable water is obtained from *in situ* measurements by radiosondes or remotely by Earth-orbiting satellites operating in the sounding mode.

As noted earlier in this chapter, if all the water vapor in the global atmosphere condensed, a layer of

FIGURE 6.10
Morning dew on a flower. [Photo by Jon Nese, The Pennsylvania State University]

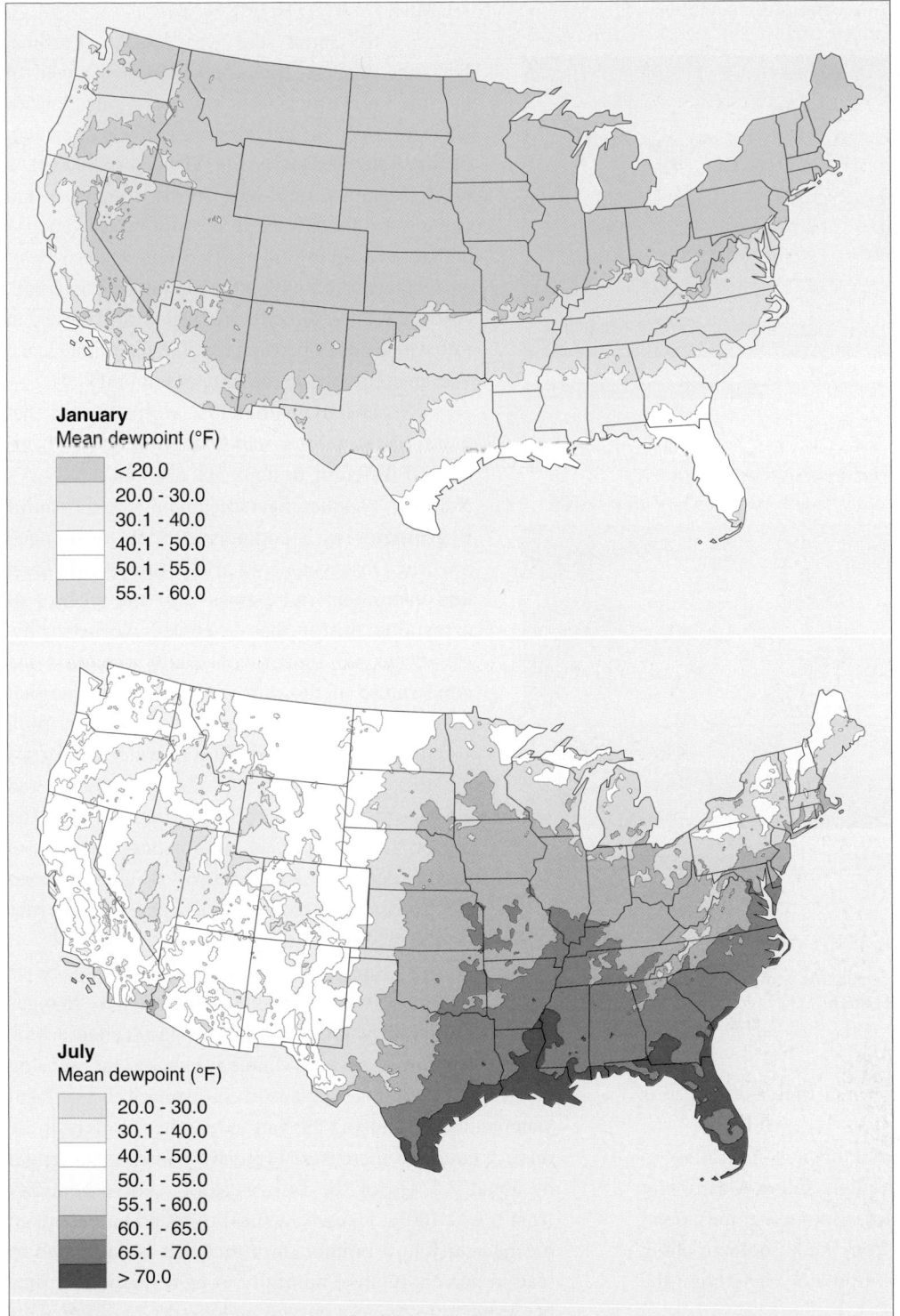

FIGURE 6.12
Average surface dewpoint in the coterminous United States for January (top map) and July (bottom map).

water would cover the entire Earth's surface to a depth of 2.5 cm (1.0 in.). The global pattern of monthly average precipitable water (in cm) is plotted in Figure 6.13 for January (top) and in Figure 6.13 for July (bottom). Beyond the tropics, the average precipitable water decreases with latitude in response to the decline in mean air temperature toward the poles; that is, the rate of evaporation and amount of precipitable water are lower in colder regions. Precipitable water declines from more than 4.0 cm (1.6 in.) in the humid tropics to less than 0.5 cm (0.2 in.) at the polar latitudes. Seasonal variations in precipitable water also occur in middle and high latitudes, with highest values

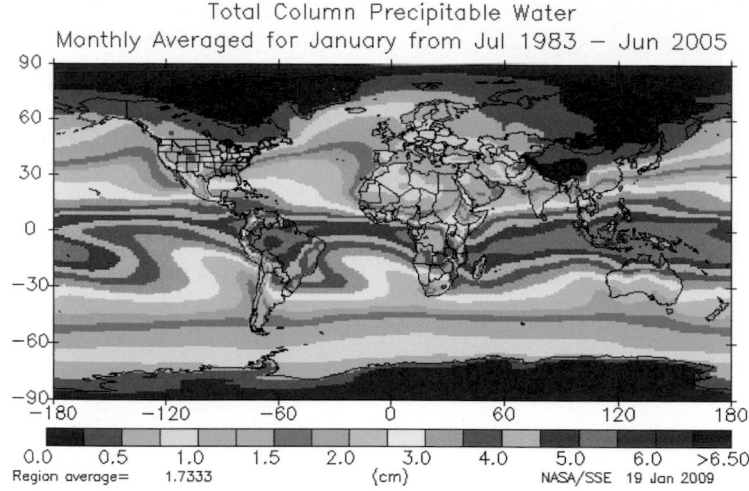

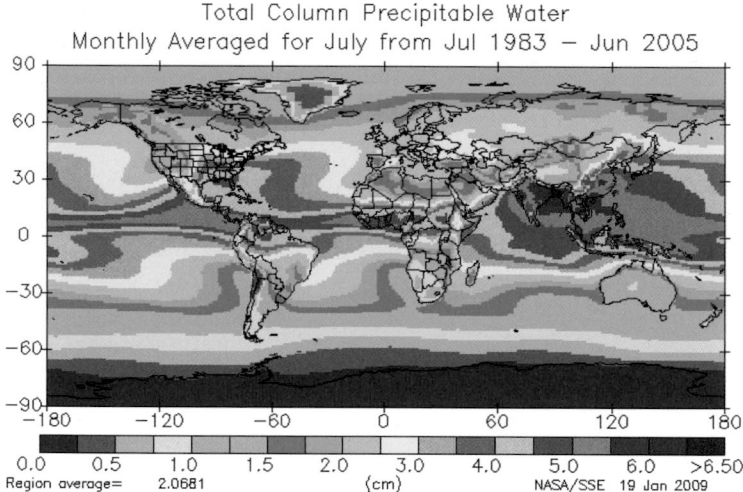

FIGURE 6.13
Global map of long-term average precipitable water in cm for January (top map) and July (bottom map). [Courtesy of NASA]

generally occurring in summer when the troposphere is warmest.

Precipitable water is not always indicative of the amount of precipitation that might fall at a particular location. Numerous other factors, including horizontal advection of water vapor to or from the air column along with recycling of water locally, ultimately determine the amount of precipitation.

Monitoring Water Vapor

A variety of techniques are used to monitor atmospheric water vapor both directly and remotely (by satellite). For information on measuring evapotranspiration, see this chapter's second *For Further Exploration*.

HUMIDITY INSTRUMENTS

In about the year 1450, Cardinal Nicholas of Cusa (*ca.* 1400-1464) conceived of the first known design in western civilization for an instrument to gauge the water vapor content of air. Later, Leonardo da Vinci (1452-1519), a creative genius, improved on the original design using a simple balance to measure water vapor in air. He set a small wad of dry cotton on one side of a scale, balanced with a weight on the other side. As the cotton absorbed water vapor from the air, an imbalance developed and the amount of imbalance was an approximate measure of humidity.

The **hygrometer** is an instrument that measures the water vapor concentration of air; several different designs are available. NOAA's National Weather Service employs a **dewpoint hygrometer** as a component of its Automated Surface Observing System (Figure 6.14). With this instrument, air passes over the surface of a metallic mirror that is cooled electronically. An electronic sensor continually monitors the temperature of the mirror at the same time that an infrared beam is pointed at the mirror. With sufficient cooling, a thin film of water condenses on the mirror changing its reflectivity and altering the reflection of the infrared beam. The mirror temperature is automatically recorded as the dewpoint. The mirror is then warmed electronically to evaporate the dew and prepare for the next measurement.

Some hygrometers take advantage of the sensitivity of organic materials to changes in humidity. One common design, called a **hair hygrometer**, uses human hair as the sensing element. Cells in the hair adsorb (collect on the surface) water and swell causing the hair to lengthen slightly as the relative humidity increases. Typically, hair changes length by about 2.5% over the full range of relative humidity from 0% to 100%. Usually, a sheaf of blond hair is linked mechanically to a pointer on a dial that is calibrated to read in percent relative humidity. A hair hygrometer may be designed to move a pen on a clock-driven drum. This instrument, called a **hygrograph**, provides a continuous trace of fluctuations in relative humidity with time. Unfortunately, hair hygrometers do not measure extremes in relative humidity accurately nor do they respond quickly to rapid changes in humidity.

An **electronic hygrometer** is based on changes in the electrical resistance of certain chemicals as they

FIGURE 6.14
The temperature/dewpoint sensor (hygrothermometer) is a component of NOAA's National Weather Service's Automated Surface Observing System (ASOS). [Courtesy of NOAA]

adsorb water vapor from the air. The adsorbing element or hygroscopic chemical (e.g., lithium chloride) may be a thin carbon coating on a glass or plastic strip. The more humid the air, the more water adsorbed, and the lower is the resistance to an electric current passing through the sensing element. Variations in electrical resistance are calibrated in terms of percent relative humidity or dewpoint. An electronic hygrometer is flown onboard radiosondes (Chapter 2).

For more than a century, most meteorological observations of water vapor utilized a psychrometer, an instrument that provides an indirect measure of relative humidity. A **psychrometer** consists of two identical liquid-in-glass thermometers mounted side by side with the bulb of one thermometer wrapped in a muslin wick. To take a reading, the wick-covered bulb is first soaked in water (preferably distilled) and then the instrument is ventilated. A *sling psychrometer* has an attached handle that allows it to be whirled about with ease (Figure 6.15), whereas a small fan is used to ventilate an *aspirated psychrometer*. The *dry-bulb thermometer* of the instrument measures the ambient air temperature. Water vaporizes from the muslin wick into the air streaming past the *wet-bulb thermometer* and evaporative cooling lowers the reading on this thermometer to the **wet-bulb temperature**. As ventilation continues by whirling or aspiration, the reading on the wet-bulb thermometer eventually steadies. The drier the air, the greater the evaporation, and the lower is the wet-bulb temperature compared to the dry-bulb temperature. The difference between the dry-bulb temperature and the wet-bulb temperature, known as the **wet-bulb depression**, is calibrated in terms of percent relative humidity on a psychrometric table (Table 6.5). This table was developed

more than a century ago using a combination of theoretical considerations and extensive field experiments.

Consider an example of how a psychrometer is used to determine the relative humidity. Suppose that the air temperature (dry-bulb thermometer reading) is 20 °C (68 °F) and the wet-bulb temperature is 15 °C (59 °F) resulting in a wet-bulb depression (dry-bulb reading minus wet-bulb reading) of 5 Celsius degrees (9 Fahrenheit degrees). From Table 6.5, the relative humidity is 58%. When the wet-bulb depression is zero, the air is saturated; that is, the relative humidity is 100%.

A major drawback of a psychrometer is the difficulty of taking measurements at low and subfreezing temperatures. At low air temperatures, small differences in the wet-bulb depression correspond to greater increments in relative humidity than is the case at high air temperatures. This situation calls for greater precision in reading the instrument at low air temperatures. At freezing, ice forms on the wick inhibiting evaporative cooling. For this reason, an electronic or hair hygrometer is the better choice at air temperatures below 0 °C (32 °F). Furthermore, in very dry air the wick may dry out prior to reaching the full wet-bulb depression value.

The wet-bulb temperature is not the same as the dewpoint. The wet-bulb temperature is determined by inducing evaporative cooling. The process of adding water vapor to the air surrounding the wet-bulb thermometer raises the temperature at which dew will form so that, except at saturation, the wet-bulb temperature is higher than the dewpoint. At saturation, the dewpoint, wet-bulb, and ambient air temperatures are the same. The dewpoint also can be obtained from measurements of the dry-bulb

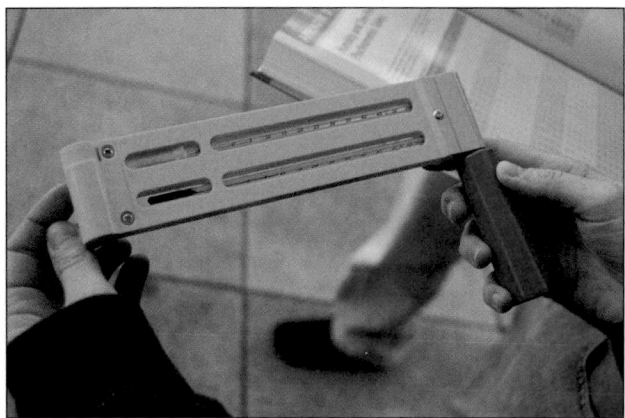

FIGURE 6.15
When whirled about, a sling psychrometer measures the actual air temperature (dry-bulb reading) and the wet-bulb temperature. The relative humidity and dewpoint can then be determined from psychrometric tables (Tables 6.5 and 6.6). [Photo courtesy of Millersville University Meteorology Program]

TABLE 6.5
Psychrometric Table: Relative Humidity (Percent)

Dry-bulb Temp (°C)	0.5	1.0	1.5	2.0	2.5	3.0	3.5	4.0	4.5	5.0	7.5	10.0	12.5	15.0	17.5
−10.0	85	69	54	39	24	10	—	—	—	—	—	—	—	—	—
−7.5	87	73	60	48	35	22	10	—	—	—	—	—	—	—	—
−5.0	88	77	66	54	43	32	21	11	1	—	—	—	—	—	—
−2.5	90	80	70	60	50	41	37	22	12	3	—	—	—	—	—
0.0	91	82	73	65	56	47	39	31	23	15	—	—	—	—	—
2.5	92	84	76	68	61	53	46	38	31	24	—	—	—	—	—
5.0	93	86	78	71	65	58	51	45	38	32	1	—	—	—	—
7.5	93	87	80	74	68	62	56	50	44	38	11	—	—	—	—
10.0	94	88	82	76	71	65	60	54	49	44	19	—	—	—	—
12.5	94	89	84	78	73	68	63	58	53	48	25	4	—	—	—
15.0	95	90	85	80	75	70	66	61	57	52	31	12	—	—	—
17.5	95	90	86	81	77	72	68	64	60	55	36	18	2	—	—
20.0	95	91	87	82	78	74	70	66	62	58	40	24	8	—	—
22.5	96	92	87	83	80	76	72	68	64	61	44	28	14	1	—
25.0	96	92	88	84	81	77	73	70	66	63	47	32	19	7	—
27.5	96	92	89	85	82	78	75	71	68	65	50	36	23	12	1
30.0	96	93	89	86	82	79	76	73	70	67	52	39	27	16	6
32.5	97	93	90	86	83	80	77	74	71	68	54	42	30	20	11
35.0	97	93	90	87	84	81	78	75	72	69	56	44	33	23	14
37.5	97	94	91	87	85	82	79	76	73	70	58	46	36	26	18
40.0	97	94	91	88	85	82	79	77	74	72	59	48	38	29	21

TABLE 6.6
Psychrometric Table: Dewpoint Temperature (°C)

Dry-bulb Temp (°C)	0.5	1.0	1.5	2.0	2.5	3.0	3.5	4.0	4.5	5.0	7.5	10.0	12.5	15.0	17.5
−10.0	−12.1	−14.5	−17.5	−21.3	−26.6	−36.3	—	—	—	—	—	—	—	—	—
−7.5	−9.3	−11.4	−13.8	−16.7	−20.4	−25.5	−34.4	—	—	—	—	—	—	—	—
−5.0	−6.6	−8.4	−10.4	−12.8	−15.6	−19.0	−23.7	−31.3	−78.6	—	—	—	—	—	—
−2.5	−3.9	−5.5	−7.3	−9.2	−11.4	−14.1	−17.3	−21.5	−27.7	−41.3	—	—	—	—	—
0.0	−1.3	−2.7	−4.2	−5.9	−7.7	−9.8	−12.3	−15.2	−18.9	−23.9	—	—	—	—	—
2.5	1.3	0.1	−1.3	−2.7	−4.3	−6.1	−8.0	−10.3	−12.9	−16.1	—	—	—	—	—
5.0	3.9	2.8	1.6	0.3	−1.1	−2.6	−4.2	−6.1	−8.1	−10.4	−47.7	—	—	—	—
7.5	6.5	5.5	4.4	3.2	2.0	0.7	−0.8	−2.3	−4.0	−5.8	−21.6	—	—	—	—
10.0	9.1	8.1	7.1	6.0	4.9	3.8	2.5	1.2	−0.2	−1.8	−12.8	—	—	—	—
12.5	11.6	10.7	9.8	8.8	7.8	6.7	5.6	4.5	3.2	1.9	−6.8	−28.2	—	—	—
15.0	14.2	13.3	12.5	11.6	10.6	9.6	8.6	7.6	6.5	5.3	−1.9	−14.5	—	—	—
17.5	16.7	15.9	15.1	14.3	13.4	12.5	11.5	10.6	9.6	8.5	2.3	−7.0	−35.1	—	—
20.0	19.3	18.5	17.7	16.9	16.1	15.3	14.4	13.5	12.6	11.6	6.1	−1.4	−14.9	—	—
22.5	21.8	21.1	20.3	19.6	18.8	18.0	17.2	16.3	15.5	14.6	9.6	3.2	−6.3	−37.5	—
25.0	24.3	23.6	22.9	22.2	21.4	20.7	19.9	19.1	18.3	17.5	12.9	7.3	−0.2	−13.7	—
27.5	26.8	26.2	25.5	24.8	24.1	23.3	22.6	21.9	21.1	20.3	16.1	11.1	4.7	−4.7	−31.7
30.0	29.4	28.7	28.0	27.4	26.7	26.0	25.3	24.6	23.8	23.1	19.1	14.5	9.0	1.6	−11.1
32.5	31.9	31.2	30.6	29.9	29.3	28.6	27.9	27.2	26.5	25.8	22.1	17.8	12.8	6.6	−2.4
35.0	34.4	33.8	33.1	32.5	31.9	31.2	30.6	29.9	29.2	28.5	24.9	21.0	16.4	11.0	3.9
37.5	36.9	36.3	35.7	35.1	34.4	33.8	33.2	32.5	31.9	31.2	27.7	24.0	19.8	14.9	8.9
40.0	39.4	38.8	38.2	37.6	37.0	36.4	35.8	35.1	34.5	33.9	30.5	26.9	23.0	18.5	13.3

temperature and the wet-bulb depression (Table 6.6). For example, if the dry-bulb temperature is 20 °C (68 °F) and the wet-bulb depression is 5 Celsius degrees (9 Fahrenheit degrees), then the dewpoint is 11.6 °C (53 °F). In this case, the wet-bulb temperature is 15 °C (59 °F), some 3.4 Celsius degrees (6 Fahrenheit degrees) higher than the dewpoint.

WATER VAPOR SATELLITE IMAGERY

Water vapor satellite imagery is a valuable remote sensing product for tracking the broad scale distribution and movement of water vapor within the atmosphere (Figure 6.16). Water vapor is invisible and does not appear on visible satellite imagery. But water vapor efficiently absorbs and emits in certain wavelength ranges of infrared radiation that are readily detected by a special infrared sensor onboard a weather satellite. The sensor that produces conventional infrared (thermal) satellite images is sensitive to infrared radiation emitted by the Earth-atmosphere system at wavelengths within atmospheric windows, that is, within wavelength bands from 10.5 to 12.6 micrometers. Recall from Chapter 3 that *atmospheric windows* refer to wavelength bands of upwelling infrared radiation not absorbed by the clear atmosphere. On the other hand, the satellite water vapor sensor detects infrared radiation at a wavelength of 6.7 micrometers, which is strongly absorbed and emitted by water vapor in the atmosphere.

Water vapor imagery displays only water vapor and clouds occurring at altitudes above about 3000 m (10,000 ft). Although this type of satellite imagery does not show water vapor (or clouds) at lower altitudes, it does detect otherwise transparent water vapor in the middle troposphere and can show moisture gathering in broad scale circulation patterns prior to the development of extensive cloudiness in an incipient storm system. Water vapor imagery typically utilizes a gray scale. At one extreme, black indicates little or no water vapor, whereas at the other extreme, milky white signals a relatively high concentration of water vapor. Clouds appear as bright white blotches.

Sequential water vapor images viewed in rapid succession to detect motion show water vapor being transported horizontally as huge swirling plumes, often originating over tropical seas. A typical **water vapor plume** (Figure 6.17) can be several hundred kilometers wide and thousands of kilometers in length. Plumes supply moisture to hurricanes, thunderstorm clusters, and winter storms. (An *atmospheric river* is a low-level component of a water vapor plume.) Water vapor imagery also reveals something about the vertical (up and down) motion of air. Most water vapor originates at the surface of the Earth so that air is moving upward in humid regions (appearing milky white) and downward in dry regions (appearing black).

Since the late 1990s, scientists have also measured atmospheric water vapor remotely using passive microwave sensors onboard polar-orbiting satellites. Because the amount of microwave radiation emitted by Earth's surface varies with the amount of water vapor between Earth's surface and the satellite, a microwave sensor (e.g., the *Special Sensor Microwave Imager, SSM/I*) can measure the water vapor present throughout the air column. This is known as *vertically integrated water vapor*. Measurements are expressed in cm as if all the water vapor in the atmospheric column were condensed into a layer of liquid water (i.e., precipitable water).

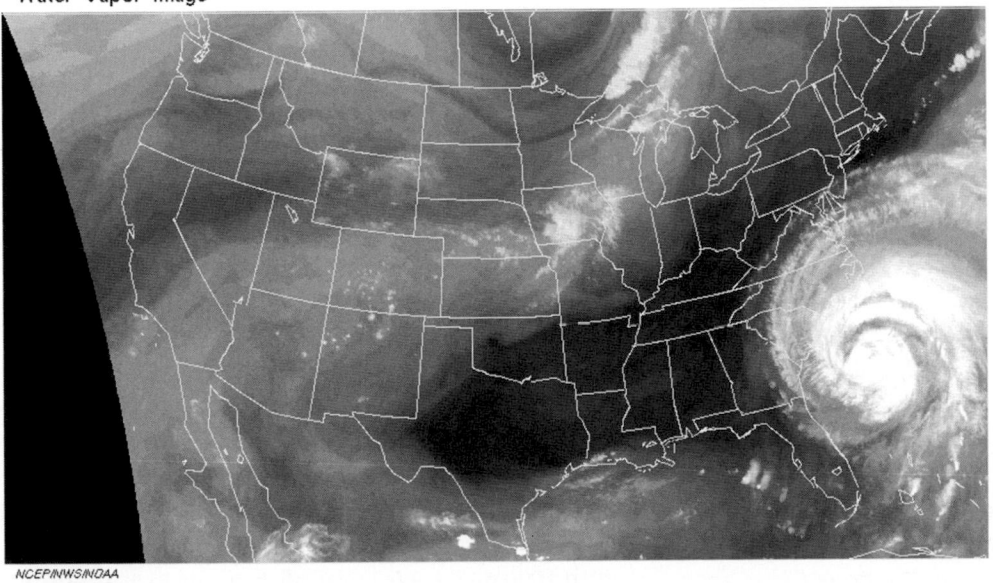

Water Vapor Image   1815Z 26 AUG 2011

NCEP/NWS/NOAA

FIGURE 6.16
Water vapor imagery from satellites portrays atmospheric moisture at altitudes above 3000 m (10,000 ft) as in this August 2011 image of Hurricane Irene. Milky white indicates relatively high concentrations of water vapor; black signals little or no water vapor; and bright white blotches are clouds. For the most current visible satellite image, go to the course website and in the "Satellite" section - select "Water Vapor - Latest.".

Water Vapor Image   1115Z 07 MAR 2009

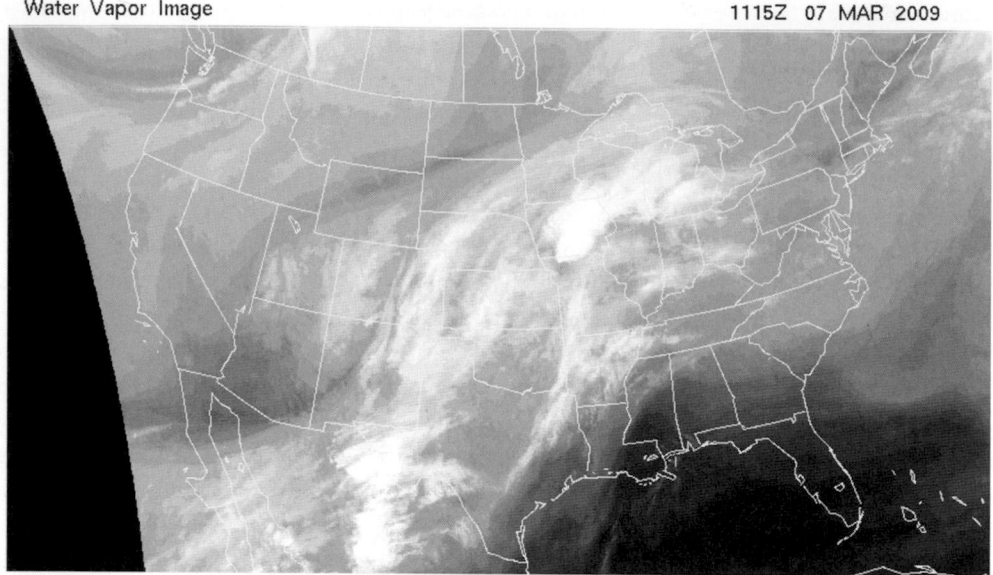

NCEP/NWS/NOAA

FIGURE 6.17
A water vapor image portraying long-distance transport of water vapor in a plume off the Pacific and through Mexico, feeding thunderstorms along a cold front, in the central U.S.

The SSM/I works well over the ocean, providing frequent global updates. Microwave imagery has been successfully applied to learn about the development, physical characteristics, and positive and negative impacts of atmospheric rivers (ARs), which are corridors of enhanced water vapor transportation in the lower atmosphere, as described in this chapter's *Case-in-Point*. So far, research on ARs has focused mainly on the eastern Pacific Ocean and western North America where landfalling ARs can bring beneficial precipitation or torrential rains, flooding, mudslides, property damage, and loss of life.

How Air Becomes Saturated

The relative humidity is variable and as it approaches 100%, condensation or deposition of water vapor becomes more and more likely. Condensation or deposition within the atmosphere produces clouds so the probability of cloud development increases as the relative humidity nears saturation. A **cloud** is a visible aggregate of tiny water droplets and/or ice crystals suspended in the atmosphere. (Under special circumstances, the relative humidity can rise slightly higher than 100% without water vapor changing phase; in that case air is *supersaturated*.)

What causes the relative humidity to increase? This is an important question because clouds are required for precipitation. Without precipitation, the global water cycle would not exist (and neither would we). The relative humidity of unsaturated air increases: (1) when air is cooled—the saturation vapor pressure decreases while the actual vapor pressure remains constant (or equivalently, the air temperature falls to the dewpoint); (2) when water vapor is added to the air at constant temperature; the vapor pressure increases while the saturation vapor pressure remains constant. In this chapter, we focus on the first process in which expansional cooling of air results in saturation.

Expansional cooling is the principal means whereby clouds form in the atmosphere. Recall from Chapter 5 that ascending air expands and cools whereas descending air is compressed and warms. As ascending currents of unsaturated (clear) air cool, the relative humidity increases and approaches saturation. At or near saturation, clouds usually form. On the other hand, descending currents of air warm, the relative humidity decreases, and existing clouds vaporize. Within the atmosphere, expansional cooling and compressional warming of unsaturated air are essentially *adiabatic processes*. This enables us to accurately predict the change in temperature of unsaturated air parcels as they ascend or descend within the atmosphere. Recall from Chapter 5 that the dry adiabatic lapse rate is 9.8 Celsius degrees per 1000 m (5.5 Fahrenheit degrees per 1000 ft) whereas the average moist adiabatic lapse rate is about 6 Celsius degrees per 1000 m (3.3 Fahrenheit degrees per 1000 ft).

Regardless of temperature, the relative humidity of ascending saturated (cloudy) air remains constant at 100%. As saturated air expands and cools, however, its saturation mixing ratio declines and so does its actual mixing ratio. That is, some water vapor is converted to water droplets (condensation) or ice crystals (deposition). Consider an example. Saturated air at 10 °C (50 °F) has a mixing ratio of about 7.8 g/kg (Table 6.4). If that air is lifted 1000 m, it cools moist adiabatically to 4 °C (39 °F). At that temperature, the saturation mixing ratio is about 5.1 g/kg. Because the relative humidity remains at 100%, the mixing ratio must drop from 7.8 g/kg to about 5.1 g/kg. This means that 2.7 g/kg of water vapor condensed to cloud droplets. Some of these cloud droplets may grow into raindrops and fall out of the cloud (Chapter 7).

As long as an air parcel continues to ascend, it expands and its temperature drops. Buoyant forces arising from density differences between a rising air parcel and the surrounding (ambient) air may either enhance or suppress vertical motion of air. The net effect depends on the stability of the atmosphere.

Atmospheric Stability

At constant pressure, the higher the temperature of an air parcel, the lower is its density. Parcels that are warmer (less dense) than the ambient air are buoyed upward by the denser surrounding air, whereas parcels that are cooler (denser) than the ambient air sink. An air parcel continues to rise or sink until it reaches ambient air of the same density (i.e., the same temperature).

We determine **atmospheric stability** by comparing the temperature change of an ascending or descending air parcel with the temperature profile, or *sounding*, of the

ambient air through which the parcel ascends or descends. As we have seen, the cooling rate of a rising air parcel depends on whether the parcel is saturated (moist adiabatic lapse rate) or unsaturated (dry adiabatic lapse rate), whereas the warming rate of any descending air parcel is essentially 9.8 Celsius degrees per 1000 m. Recall from Chapter 2 that balloon-borne radiosondes routinely monitor the vertical temperature profile (sounding) of the ambient air up to an altitude of about 30,000 m (100,000 ft) where the balloon typically bursts.

Within a **stable air layer**, an ascending air parcel becomes cooler (denser) than the ambient air, and a descending air parcel becomes warmer (less dense) than the ambient air (Figure 6.18). Any upward or downward displacement of an air parcel in stable air gives rise to forces that will tend to return the parcel to its original altitude. Within an **unstable air layer**, on the other hand, an ascending air parcel remains warmer (less dense) than the ambient air and continues to ascend, whereas a descending air parcel remains cooler (denser) than the ambient air and continues to descend (Figure 6.19).

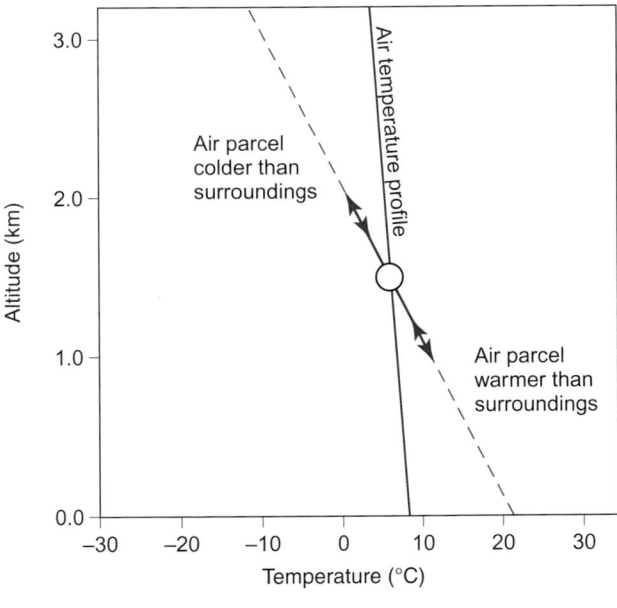

FIGURE 6.18
Within this air layer, an unsaturated air parcel (represented by a circle) initially has the same temperature as the surrounding air. The parcel temperature can be read from the sounding (solid purple line) acquired by a radiosonde. If the parcel is displaced upward, it expands and cools at the dry adiabatic rate (represented by the dashed red line). The parcel is now colder and denser than the surrounding air and descends back to its original altitude. If the parcel is displaced downward within the air layer, it will undergo compressional warming at the dry adiabatic rate. The parcel is now warmer and less dense than its surroundings and ascends to its original altitude. Such an air layer which inhibits vertical motion of air parcels is described as stable.

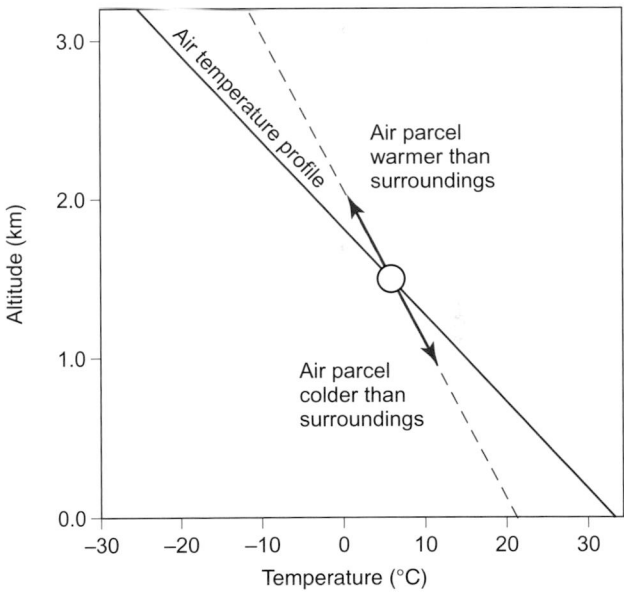

FIGURE 6.19
Within this air layer, an unsaturated air parcel (represented by a circle) initially has the same temperature as the surrounding air. The parcel temperature can be read from the sounding (solid purple line) acquired by a radiosonde. If the parcel is displaced upward, it expands and cools at the dry adiabatic rate (represented by the dashed red line). Now the parcel is warmer and less dense than the surrounding air and continues its ascent away from its original altitude. If the parcel is displaced downward, it will undergo compressional warming at the dry adiabatic rate but will be colder and denser than its surroundings and continues its descent. Such an air layer which enhances vertical motion of air parcels is described as unstable.

SOUNDINGS

Although lapse rates for both unsaturated and saturated air parcels moving vertically within the atmosphere are essentially constant, the actual temperature profile (sounding) of the ambient air changes significantly, sometimes even from one hour to the next. Changes in the sounding often are accompanied by changes in atmospheric stability. Soundings (and stability) can change as a consequence of (1) local radiational heating or cooling, (2) air mass advection, or (3) large-scale ascent or descent of air.

On a clear and calm night, radiational cooling chills the ground, which in turn, cools and stabilizes the air immediately above the surface. On the other hand, during the day, intense solar radiation heats the ground, which in turn, heats and destabilizes the overlying air. An air mass is stabilized as it travels over a colder surface (e.g. snow-covered ground), but is destabilized as it flows over a warmer surface. Through radiation or advection, atmospheric stability changes because the temperature profile (i.e., the sounding) changes. Air is stabilized when cooling from below reduces the temperature lapse rate (the rate at which air temperature varies with increasing altitude), and air is destabilized when heating from below increases the temperature lapse rate.

Generally, an air layer becomes more stable when it descends (subsides) but less stable when it ascends. When an air layer subsides, the upper portion undergoes more compressional warming than does the lower portion so that the lapse rate within the air layer decreases and stability increases. On the other hand, the lower portion of an ascending air layer is usually more humid than the upper portion and so achieves saturation sooner. Release of latent heat in the lower portion of the air layer increases the lapse rate so that the stability decreases. Further complicating matters, stability can change with altitude. For example, a stable air layer may be sandwiched between unstable air layers. Different types of air mass advection occurring at different altitudes within the troposphere are often responsible for such variations in stability.

Figure 6.20 summarizes atmospheric stability conditions for a variety of soundings, along with the dry adiabatic and average moist adiabatic lapse rates. If the sounding indicates that the temperature of the ambient air is dropping more rapidly with altitude than the dry adiabatic lapse rate (that is, more than 9.8 Celsius degrees per 1000 m), then the ambient air is unstable for both saturated and unsaturated air parcels. This situation is known as **absolute instability**. If the sounding lies between the dry adiabatic and moist adiabatic lapse rates,

conditional stability prevails; that is, the air layer is stable for unsaturated air parcels and unstable for saturated air parcels. With this relatively common situation, unsaturated air must be forced upward in order to reach saturation. But once saturation is achieved, the now cloudy air cools at the lower moist adiabatic lapse rate.

An air layer is stable for both saturated and unsaturated air parcels when the sounding indicates any of the following conditions: (1) the temperature of the ambient air drops more slowly with altitude than the moist adiabatic lapse rate; (2) the temperature does not change with altitude (*isothermal*); (3) the temperature increases with altitude (*temperature inversion*). Any one of these

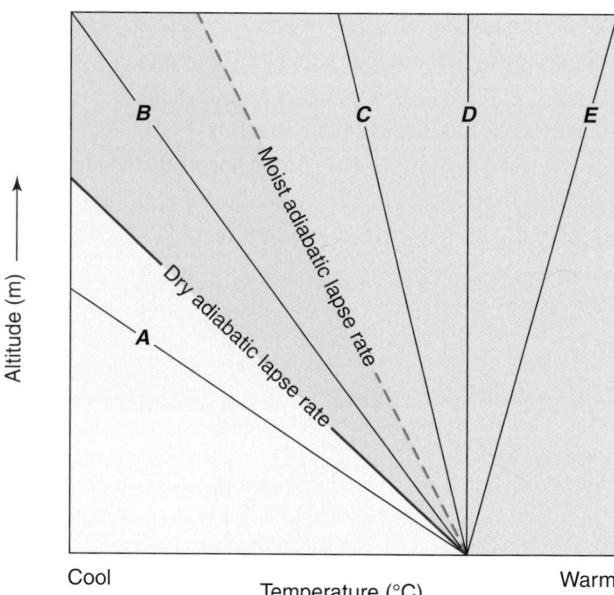

Sounding	Clear air	Cloudy air	Category	
A	Unstable	Unstable	Absolute instability	
B	Stable	Unstable	Conditional stability	
C	Stable	Stable	Lapse	Absolute stability
D	Stable	Stable	Isothermal	
E	Stable	Stable	Inversion	

FIGURE 6.20

Air stability is determined by comparing the temperature (density) of an ascending or descending air parcel with the temperature (density) of the surrounding air (sounding). In the vertical atmosphere view, an air parcel at the Earth's surface initially has the same temperature as its surroundings. If the parcel is lifted, its temperature drops either dry adiabatically if it is unsaturated (clear air), or moist adiabatically if it is saturated (cloudy air). The temperature of the surrounding air is determined by radiosonde measurements. Depending on the specific sounding (A through E), the atmosphere is either stable or unstable. Stability conditions for each of the five sample soundings are summarized in the table.

three types of temperature profiles indicates **absolute stability**.

What happens when a sounding coincides with either the dry or moist adiabatic lapse rate? A sounding that equals the dry adiabatic lapse rate is neutral for unsaturated air parcels and unstable for saturated air parcels. A sounding that is the same as the moist adiabatic lapse rate is neutral for ascending saturated air parcels but stable for either ascending or descending unsaturated air parcels. Within a **neutral air layer**, a rising or descending air parcel has the same temperature (and density) as its surroundings. Hence, a neutral air layer neither impedes nor spurs upward or downward motion of air parcels.

In summary, atmospheric stability influences weather by affecting the vertical motion of air. Stable air suppresses vertical motion whereas unstable air enhances vertical motion, convection, expansional cooling, and cloud development. Because stability also affects the rate at which polluted air mixes with clean air, stability is an important factor in assessing air pollution potential. We examine this issue in this chapter's third *For Further Exploration*.

STÜVE DIAGRAM

Soundings and adiabatic processes in the atmosphere can be displayed graphically on a thermodynamic diagram. Although several types are available, in this book we use the **Stüve thermodynamic diagram**.

The Stüve diagram is designed to permit evaluation of various atmospheric processes using graphical techniques (Figure 6.21). Coordinates on this diagram consist of temperature in degrees Celsius and Fahrenheit (horizontal axis), increasing from left to right, and pressure in millibars (right vertical axis), decreasing upward. Also marked on some diagrams are altitudes in km and ft (left vertical axis) that correspond to pressure levels, based on the standard atmosphere (Chapter 5). The dry adiabatic lapse rate is represented by a family of straight red lines slanting from lower right toward upper left. Often these lines are called *dry adiabats*. An ascending unsaturated air parcel follows a dry adiabat (either along one of the drawn adiabats or one that parallels the printed lines) from its initial temperature and pressure to its final temperature and pressure. Consider an example.

Suppose that the temperature of an unsaturated air parcel is 20 °C (68 °F) and the air pressure is 1000 mb. These initial conditions plot as point A on the Stüve diagram in Figure 6.22. Now simulate expansional cooling by moving the air parcel upward along a dry adiabat to the 850-mb level. The temperature of the parcel at 850 mb is about 8 °C (46 °F) (point B in Figure 6.22). Now continue along the dry adiabat to a pressure of 700 mb where the air parcel temperature is about −8 °C (18 °F) (point C in Figure 6.22). Finally, move the air parcel back along the same dry adiabat to 1000 mb to simulate descent of the parcel and compressional warming. Because the

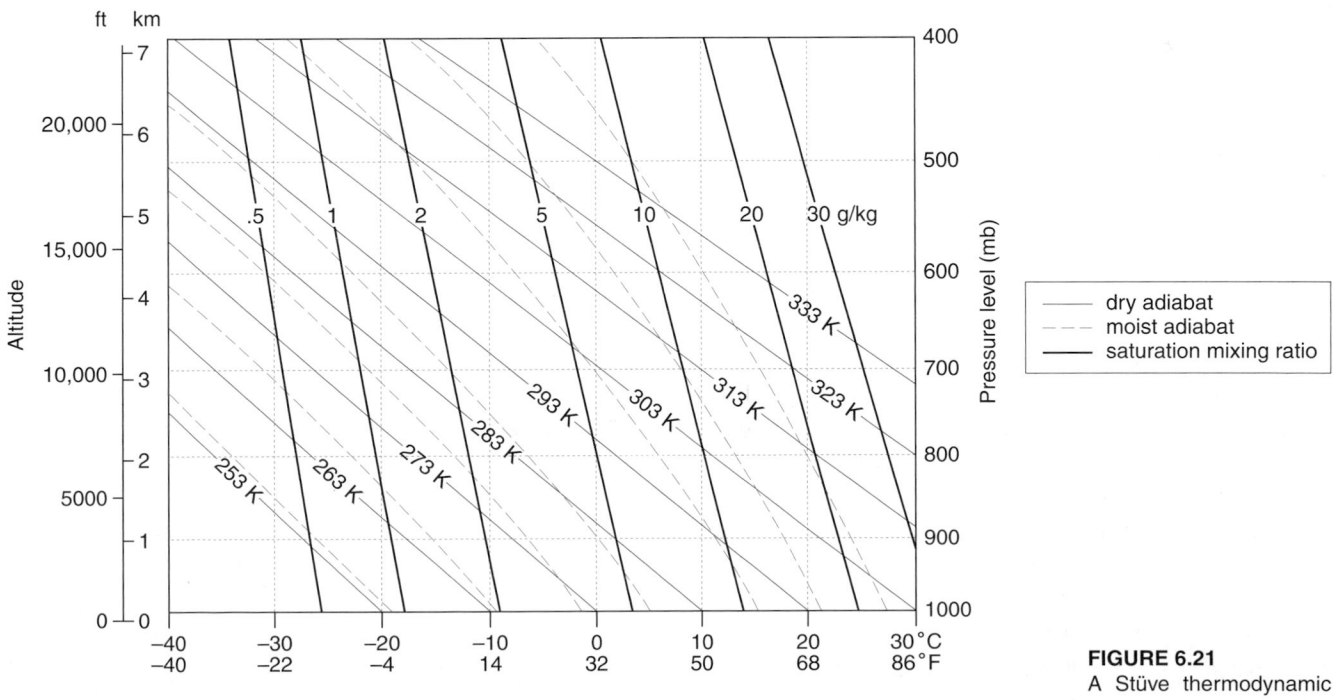

FIGURE 6.21
A Stüve thermodynamic diagram.

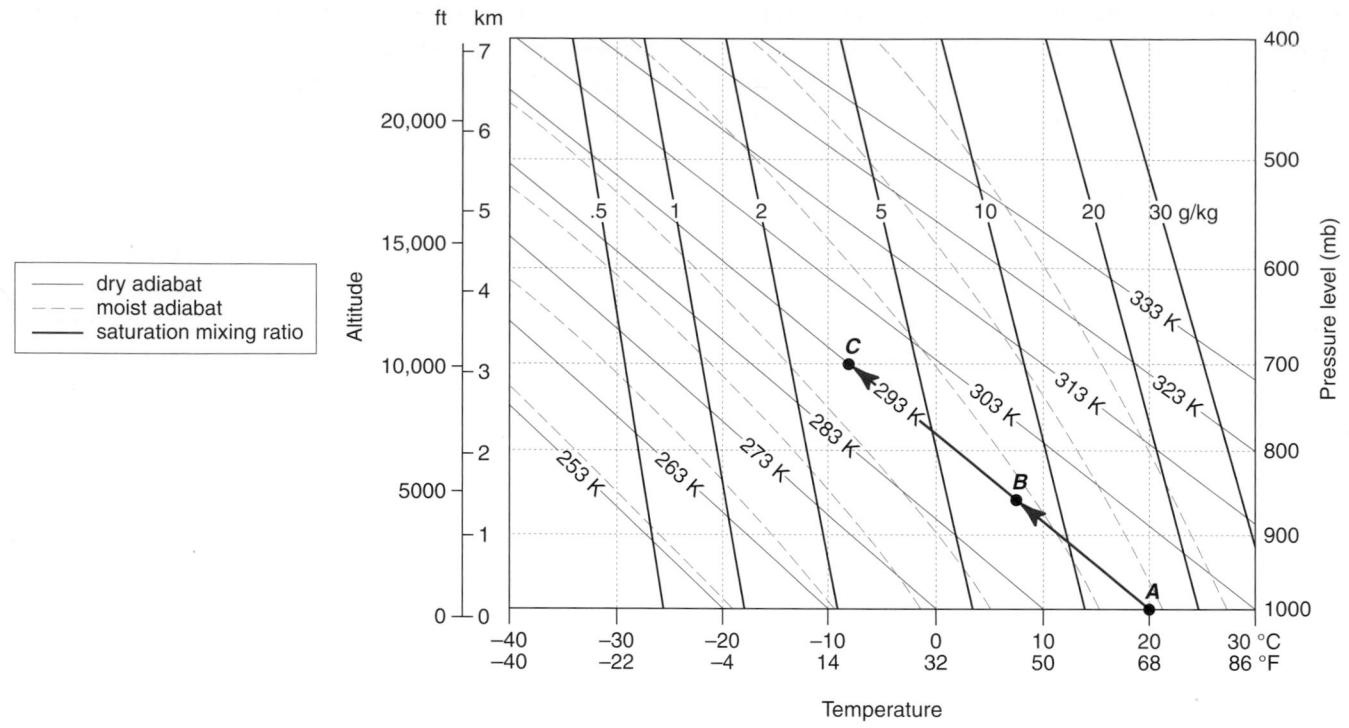

FIGURE 6.22
On a Stüve thermodynamic diagram, an unsaturated air parcel at point A is subject to a dry adiabatic expansion to point B (850 mb) and then to point C (700 mb).

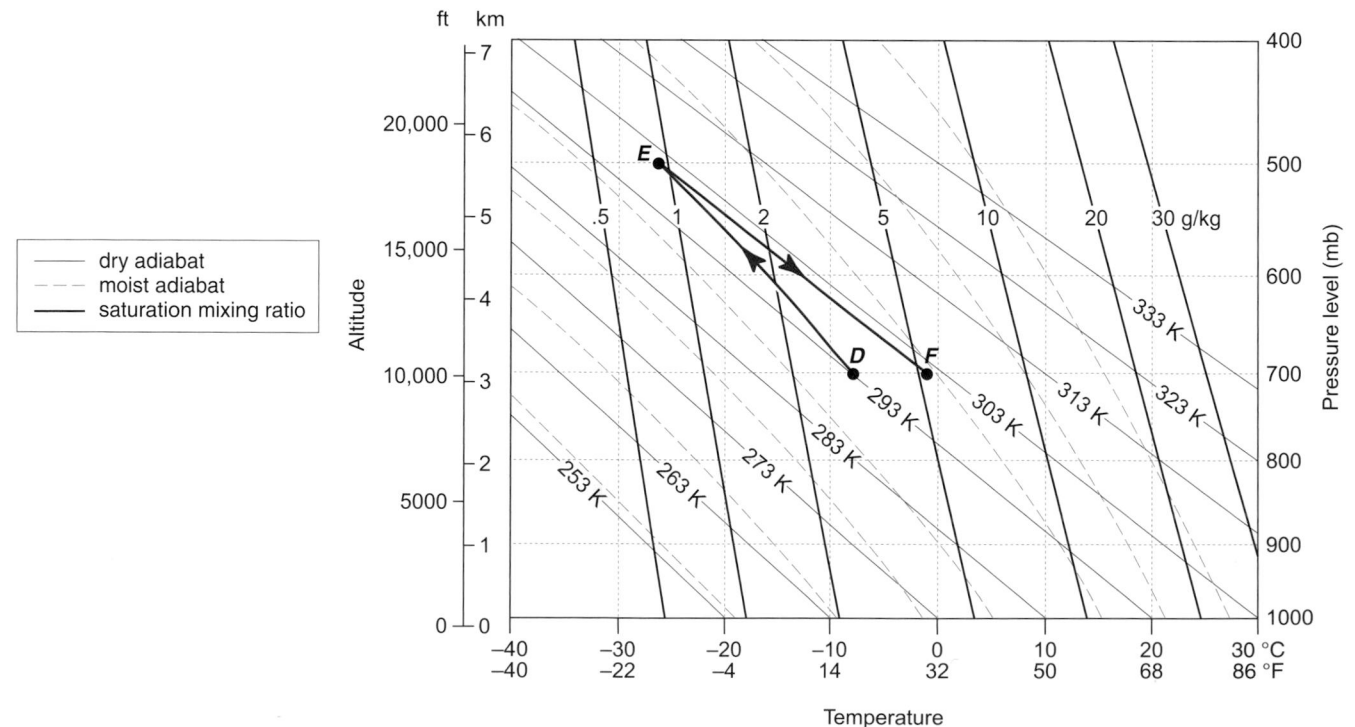

FIGURE 6.23
On a Stüve thermodynamic diagram, a saturated air parcel at point D (700 mb) is subject to a moist adiabatic expansion to point E (500 mb). The parcel is then subject to a dry adiabatic compression to point F (700 mb).

air parcel warms during descent at the same rate as it cooled during ascent, at the 1000-mb level, the parcel's temperature is at its original value of 20 °C (68 °F). In this case, the parcel remained unsaturated; water vapor did not change phase so that latent heat was not a factor in altering the parcel's temperature.

Moist adiabatic lapse rates appear on a Stüve diagram as a set of gently curving dashed lines, also known as *moist adiabats*. Moist adiabats become nearly indistinguishable from dry adiabats at very low pressure and low temperature—where very little water vapor exists and latent heat release is minimal. An ascending saturated air parcel follows a moist adiabat (either along one of the drawn moist adiabats or one that parallels the printed lines) from its initial temperature and pressure to its final temperature and pressure. Consider an example.

Suppose that a parcel of saturated air at the base of a cloud has a temperature of −8 °C (18 °F) at a pressure of 700 mb (point D in Figure 6.23). As the parcel ascends along a moist adiabat to 500 mb, its temperature drops to −27 °C (−17 °F) (point E in Figure 6.23). Assume that the water that condensed during the ascent immediately fell out of the parcel (as precipitation). If the parcel then descends back to the 700-mb level, it follows a dry adiabat and is compressionally warmed to −1 °C (30 °F) (point F in Figure 6.23). The parcel is now some 7 Celsius degrees (12 Fahrenheit degrees) warmer than it was when it started its ascent because of the release of latent heat due to phase changes of water vapor during its ascent.

Plotting a sounding on a Stüve diagram enables meteorologists to determine the stability of various layers within the atmosphere and assess the potential for deep convection and cloud development. Meteorologists also use plotted values of saturation mixing ratio (slanted black lines labeled in gm per kg) to determine relative humidity and saturation levels of rising air parcels.

Lifting Processes

Ascending unsaturated air cools and its relative humidity increases. With sufficient ascent and expansional cooling, the relative humidity nears 100% and condensation or deposition begins, thus forming clouds. What causes air to rise? Air rises (1) as the ascending branch of a convection current, (2) along the surface of a front, (3) up the slopes of a hill or mountain, or (4) where surface winds converge.

As we saw in Chapter 4, sensible heating (conduction and convection) is a means of heat transfer between Earth's surface and atmosphere. The Sun heats Earth's surface, which then heats the overlying air by

conduction. The heated air expands and its density decreases. Cool, dense air sinks toward Earth's surface and displaces the less dense heated air. The heated air rises, expands, and cools. Cumulus clouds may form where convection currents ascend and the sky is generally cloud-free where convection currents descend (Figure 6.24). The higher the altitude reached by ascending convection currents, the greater the amount of expansional cooling, and the more likely it is that clouds (and precipitation) will form. Convection currents that soar to great altitudes

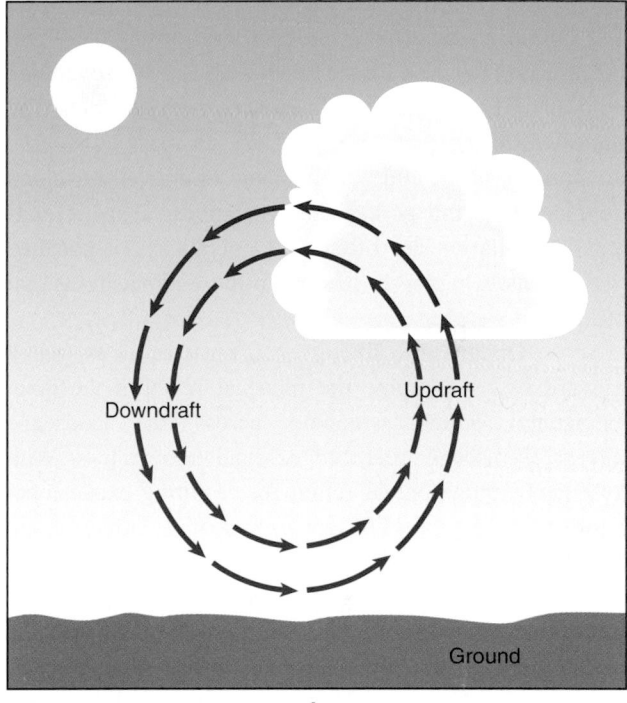

A

B

FIGURE 6.24
A convection current consists of (A) an updraft and a downdraft; (B) cumulus clouds form where air ascends while the surrounding sky is cloud-free where the air descends.

within the troposphere typically spawn thunderstorms (*cumulonimbus clouds*).

Clouds and precipitation are often triggered by **frontal uplift**, which occurs where contrasting air masses meet. Recall that a *front* is a narrow zone of transition between two air masses that differ in temperature and/or humidity (Chapter 1). A warm and humid air mass is less dense than a cold and dry air mass and hence, as a cold air mass retreats, the warm air advances by riding up and over the cold air (Figure 1.8A). The leading edge of the advancing warm air at Earth's surface is known as a *warm front*. In contrast, cold and dry air displaces warm and humid air by sliding under it and forcing the warm air upward (Figure 1.8B). The leading edge of advancing cold air at Earth's surface is known as a *cold front*. The net effect of the replacement of one air mass by another air mass is uplift and expansional cooling of air, cloud development, and perhaps rain or snow. Hence, clouds and precipitation are often (but not always) associated with fronts. Chapter 10 has much more about fronts and associated weather.

Orographic lifting occurs where air is forced upward by topography, the physical relief of the land. Horizontal winds sweeping across the landscape alternately ascend hills and descend into valleys. With sufficient topographical relief, the resulting expansional cooling and compressional warming of air affects cloud and precipitation development.

A mountain range that intercepts the prevailing winds forms a natural barrier that is responsible for a cloudier, wetter climate on one side of the range than on the other side (Figure 6.25). Air that is forced to ascend the *windward slopes* (facing the oncoming wind) expands and cools, which increases its relative humidity. With sufficient cooling, saturation is achieved and clouds and precipitation develop. The altitude at which the rising air becomes saturated and clouds form is known as the **lifting condensation level (LCL)** and usually corresponds to the base of the clouds. Meanwhile, air that descends the *leeward slopes* (downwind side) is compressed and warms, raising its saturation vapor pressure while the remaining cloud particles evaporate or sublimate. The existing clouds vaporize and the relative humidity of descending air decreases so that precipitation is less likely. In this way, mountain ranges induce contrasting climates: moist climates on the windward slopes and dry climates on the leeward slopes. Dry conditions often extend many hundreds of kilometers downwind of a prominent mountain range; this region is known as a **rain shadow**. The Rocky Mountain rain shadow, for example, extends eastward

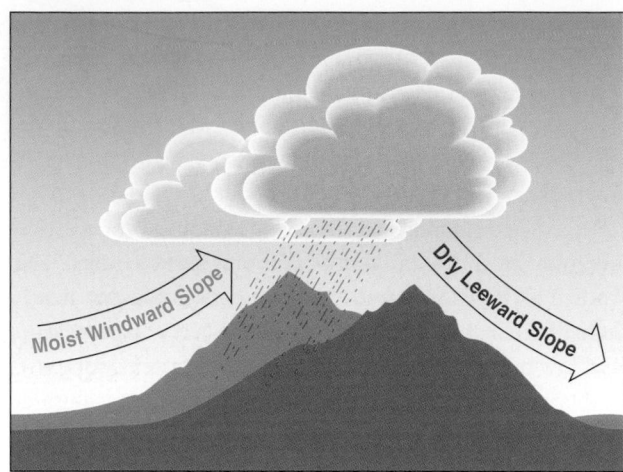

FIGURE 6.25
Winds ascend along the windward slopes of a mountain range and descend along the leeward slopes. The climate is wetter along the windward slopes than the leeward slopes.

from the Continental Divide to about the 100th meridian (100 degrees W longitude), traditionally considered to be the boundary between dry land or irrigated agriculture to the west and rain-fed agriculture to the east.

An orographically induced contrast in precipitation is apparent in the Pacific Northwest, where the north-south trending Coastal and Cascade Mountain Ranges intercept the prevailing west-to-east flow of humid air from off the Pacific Ocean (Figure 6.26). Exceptionally rainy conditions prevail in western Washington and Oregon, whereas semiarid conditions characterize much of the eastern portion of those states. For example, the average annual precipitation (rain plus melted snow) at Astoria, OR, on the Pacific coast is 172 cm (67.7 in.) but only 20.8 cm (8.2 in.) at Yakima, WA, in the rain shadow of the Cascade Mountains. This climate contrast affects the indigenous plant and animal communities, domestic water supply, demand for irrigation water, types of crops that can be grown, and requirements for human shelter.

The influence of topography and prevailing wind direction on precipitation patterns is also impressive on the mountainous islands of Hawaii, where a few volcanic peaks top 4000 m (13,000 ft). Over the ocean waters surrounding the islands, estimated mean annual rainfall is uniformly between about 560 mm (22 in.) and 700 mm (28 in.). On the islands, subjected to relatively steady winds from the northeast, however, mean annual precipitation is much more variable, ranging from only 190 mm (7.5 in.) leeward of the Kohala Mountains on the Big Island of Hawaii to 11,990 mm (39.3 ft) on the windward slopes of Mount Waialeale on the island of Kauai.

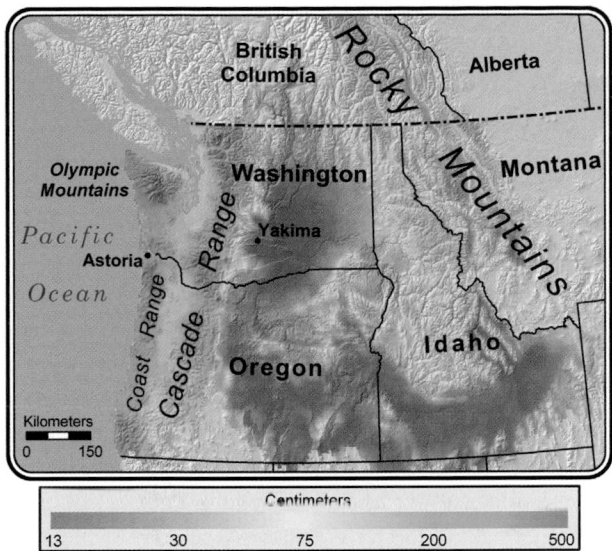

FIGURE 6.26
Mean annual precipitation (from 1971-2000) in areas of the Pacific Northwest. [Courtesy of Robert A. Norheim, Climate Impacts Group, University of Washington]

Another mechanism responsible for uplift and cloud formation is convergence of surface winds. Associated upward air motion means expansional cooling, increasing relative humidity, and eventually cloud and perhaps precipitation formation. In later chapters, we describe weather systems in which convergence of surface winds plays an important role in triggering stormy weather. For example, converging surface winds are largely responsible for the cloudiness and precipitation associated with a low-pressure system (cyclone). Also, converging sea breezes is a major factor in the relatively high frequency of thunderstorms in central Florida (Chapter 11).

Fronts, topography, or converging winds can force stable unsaturated air upward to the point that clouds develop. If the air is conditionally stable, clouds will surge upward and precipitation may be heavy. On the other hand, if air is absolutely stable, clouds will show little vertical development and precipitation is likely to be light if it occurs at all.

Conclusions

The atmosphere is one of the several reservoirs of water in the global water cycle. Water vapor enters the atmosphere from Earth's surface via evaporation of liquid water, transpiration by plants, and sublimation of ice and snow. Water moves from the atmosphere to Earth's surface via precipitation, condensation (dew), and deposition (frost). The amount of water vapor in air (humidity) can be expressed as vapor pressure, mixing ratio, specific humidity, absolute humidity, relative humidity, dewpoint, or precipitable water. The saturation vapor pressure and saturation mixing ratio increase with rising air temperature so that the water vapor component is greater in saturated warm air than in saturated cold air. On the other hand, the relative humidity of an unsaturated air parcel varies inversely with temperature.

Expansional cooling is the principal means whereby air becomes saturated with respect to water, and clouds form. As ascending air expands and cools, its relative humidity nears 100%, and water vapor condenses (or deposits) as aggregates of tiny water droplets (or ice crystals) that are visible as clouds. Conversely, descending air is compressed and warms, existing clouds vaporize, and clear-air relative humidity drops.

Ascending air parcels cool at the dry adiabatic lapse rate if unsaturated and at the moist adiabatic lapse rate if saturated. Atmospheric stability influences vertical motion in the atmosphere such that stable ambient air suppresses convection and the associated ascent of air parcels and cloud development, whereas unstable ambient air enhances ascent of air parcels and favors cloud development. Atmospheric stability can be determined by comparing the cooling rate of ascending air parcels with the temperature profile (sounding) of the ambient air through which air parcels ascend. Air ascends in convection currents, along fronts, up the slopes of mountain ranges, and in response to converging surface winds.

We continue our study of water in the atmosphere in the next chapter with a focus on cloud and precipitation forming processes.

Basic Understandings

- The global water cycle is the ceaseless circulation of a fixed quantity of water among Earth's oceanic, atmospheric, and terrestrial reservoirs. Powered by solar radiation and influenced by gravity, water moves between Earth's surface and the atmosphere via evaporation, condensation, transpiration, sublimation, deposition, and precipitation. Latent heat energy accompanies these processes.
- The global water budget indicates an annual surplus of precipitation over evapotranspiration on the land and more evaporation than precipitation over the ocean, implying a net flow of water from land to sea. Precipitation

falling on land vaporizes, infiltrates the ground, is taken up by plant roots and transpired to the atmosphere, is temporarily stored in lakes, ponds, ice fields, and glaciers, and/or runs off into rivers or streams.

- Vapor pressure, mixing ratio, specific humidity, and absolute humidity are direct measures of the water vapor component of air. At a specific temperature, relative humidity compares the actual amount of water vapor in air to the amount of water vapor at saturation.

- Saturation vapor pressure and saturation mixing ratio signal an equilibrium in the flux of water molecules entering and leaving the vapor phase. Both parameters increase with rising temperature because temperature largely regulates the rate of vaporization of water.

- When the water vapor concentration of air equals the water vapor concentration at saturation, the relative humidity is 100%. For unsaturated air on a calm day, the relative humidity varies inversely with temperature; that is, the relative humidity is highest when the air temperature is lowest (near sunrise), and the relative humidity is lowest when the air temperature is highest (early to mid-afternoon).

- Dewpoint is the temperature to which air is cooled at constant pressure to achieve saturation. Dew consists of tiny droplets of water that condense onto cold surfaces from air that has been chilled to its dewpoint and below. When cooling at constant pressure produces saturation at air temperatures below freezing, water vapor deposits on cold surfaces as frost, ice crystals that form on the exposed surfaces of objects. The air temperature at which frost begins forming is called the frost point. Tropical air masses typically have higher dewpoints than polar air masses.

- Precipitable water is the depth of water that would be produced if all the water vapor in a vertical column of air, extending from Earth's surface to the tropopause, were condensed into liquid water. Generally, the amount of precipitable water decreases with increasing latitude in response to lowering average temperatures and reduced evaporation.

- Hygrometers are instruments that directly measure the water vapor concentration of air. They include dewpoint hygrometers, hair

hygrometers, and electronic hygrometers. A less direct measure of humidity is provided by a psychrometer consisting of dry-bulb and wet-bulb thermometers mounted side-by-side. The greater the wet-bulb temperature depression, the lower is the relative humidity. When the wet-bulb and dry-bulb temperatures are equal, the relative humidity is 100%, indicating saturated air.

- Water vapor imagery obtained by sensors onboard satellites portrays water vapor and clouds at altitudes above about 3000 m (10,000 ft) and is derived from infrared radiation emitted by atmospheric water at a wavelength of 6.7 micrometers. Water vapor plumes revealed in water vapor images are particularly valuable for tracking the long-distance transport of moisture plumes within the atmosphere.

- Since the 1990s, scientists have used microwave satellite imagery to locate and study the properties of atmospheric rivers. When these corridors of enhanced water vapor flow make landfall they can produce torrential rains that are responsible for destructive flooding in the U.S. Pacific coast states.

- The relative humidity of unsaturated air increases when the air is cooled (lowering the saturation vapor pressure) or when water vapor is added to the air (increasing the vapor pressure) at constant temperature. As the relative humidity nears 100%, clouds are increasingly likely to form.

- Expansional cooling is the principal means whereby clouds form in the atmosphere. As ascending currents of unsaturated (clear) air cool, the relative humidity increases and approaches saturation. At or near saturation, clouds form.

- Atmospheric stability is determined by comparing the temperature (or density) of an air parcel moving vertically (up or down) within the atmosphere with the temperature (or density) of the surrounding (ambient) air. Radiosondes and satellite sensors monitor the variation of ambient air temperature with altitude (a sounding).

- Stable air inhibits vertical motion, convection, and cloud formation. Unstable air enhances vertical motion, convection, and cloud formation. Local radiational heating or radiational cooling, air mass advection, or large-scale ascent or subsidence can change the stability of an air mass.

- Expansional cooling and cloud development occur through uplift of air in convection currents, along fronts or mountain slopes, or where surface winds converge. Clouds form in the updraft of convection currents but dissipate or fail to develop in the downdraft. The ascent of warmer, less dense air along a front may give rise to clouds and precipitation.

- Winds ascend along the windward slopes of a mountain range and descend along the leeward slopes. For this reason, the windward slopes tend to be cloudier and wetter than the leeward slopes. A rain shadow with semiarid to arid conditions may extend hundreds of kilometers downwind of a prominent mountain range.

Enduring Ideas

- The global water cycle is powered by solar radiation, which drives large-scale evaporation of ocean water, and also by gravity. As part of the global water cycle, water is transferred between Earth's surface and atmosphere via phase changes and precipitation. The net flow of water on Earth's surface is from land to sea whereas it is from sea to land in the atmosphere.
- There are many ways of quantifying water vapor in air, the most familiar being relative humidity and the dewpoint. The amount of water vapor in air at a certain temperature has an upper limit known as saturation. Ascending currents of unsaturated air undergo expansional cooling often leading to saturation and cloud formation. Water vapor satellite imagery portrays the broad scale movement of water vapor in the atmosphere and specialized instruments provide direct measurements.
- Local radiational heating or cooling, air mass advection, or large-scale ascent or descent of air can change atmospheric stability. A thermodynamic diagram (e.g., Stüve) permits evaluation of stability and various atmospheric processes such as the expansional cooling of ascending air. Air rises in convection currents, along or just ahead of fronts, up the windward slopes of hills and mountains, and where surface winds converge.

Key Terms

atmospheric river (AR)	saturation vapor pressure	wet-bulb temperature
global water cycle	saturation mixing ratio	wet-bulb depression
evaporation	saturation specific humidity	water vapor imagery
condensation	saturation absolute humidity	cloud
transpiration	relative humidity	atmospheric stability
evapotranspiration	dewpoint	stable air layer
sublimation	dew	unstable air layer
deposition	frost	absolute instability
global water budget	frost point	conditional stability
drainage basin	precipitable water	absolute stability
watershed	hygrometer	neutral air layer
humidity	dewpoint hygrometer	Stüve thermodynamic diagram
vapor pressure	hair hygrometer	frontal uplift
mixing ratio	hygrograph	orographic lifting
specific humidity	electronic hygrometer	lifting condensation level (LCL)
absolute humidity	psychrometer	rain shadow

Review

1. Identify and describe the various phase change processes involving water.
2. Explain how the Sun drives the global water cycle.
3. How does distillation convert seawater to fresh water?
4. Define vapor pressure. How does Dalton's law of partial pressures apply to vapor pressure?
5. If the mixing ratio equals the saturation mixing ratio, what is the relative humidity?
6. On a clear calm day, why does the relative humidity usually decrease from a maximum shortly after sunrise to a minimum in the early or mid afternoon?
7. Why does the amount of precipitable water vary with the mean temperature of the troposphere?
8. Describe the principal mechanism whereby clouds form in the atmosphere.
9. Why are clouds and precipitation more likely on the windward slopes of a mountain range than the leeward slopes?
10. What is the significance of a rain shadow for agriculture?

Critical Thinking

1. Provide some examples of how the physical characteristics of Earth's surface affect the ratio of how much rainwater (or snow melt) runs off versus how much infiltrates the ground.
2. How and why does the moisture content of the top soil affect the maximum air temperature in the lower troposphere?
3. In late autumn or early winter, cold air flows on northwest winds from snow-covered ground across the ice-free waters of Lake Superior. Describe the changes in temperature, vapor pressure, and stability of the cold air as it flows over the warmer surface waters of the lake.
4. Are convective clouds (e.g., cumulus) more likely to form over snow-covered ground or bare ground? Explain your reasoning.
5. Determine the relative humidity if the . . .
    a. vapor pressure is 6 mb and the saturation vapor pressure is 24 mb.
    b. vapor pressure is 12 mb and the saturation vapor pressure is 12 mb.
    c. mixing ratio is 10 g/kg and the temperature is 25°C.
6. Under what atmospheric condition are the actual air temperature, dewpoint, and wet-bulb temperature the same?
7. A saturated air parcel ascends in the atmosphere. What is the parcel's initial relative humidity? What happens to the value of the relative humidity of the saturated parcel as it continues to ascend? What does this imply about the vapor pressure of the ascending saturated parcel?
8. Determine whether the following soundings obtained from radiosonde launches are stable, unstable, or neutral for both saturated and unsaturated air parcels. . .
    a. -7 C°/1000 m
    b. +8C°/1000 m
    c. -12C°/1000 m
    d. -4C°/1000 m
9. How does stability affect the vertical development of cumuliform clouds?
10. Explain how converging surface winds can cause clouds to form.

Heat, Humidity and Human Comfort

The relationship between the vapor pressure of ambient air and evaporative cooling helps explain why people usually are more uncomfortable when the weather is both hot and humid. Cooling of the skin surface when perspiration evaporates (evaporative cooling) is a natural process that enhances heat loss from the body. With increasing humidity (vapor pressure), however, the evaporation rate decreases and evaporative cooling is less thereby hampering the body's ability to maintain a nearly constant core temperature of approximately 37 °C (98.6 °F). (The body's *core* refers to vital organs such as the heart and lungs.) In contrast, drier air promotes evaporative cooling. At air temperatures above 25 °C (77 °F), most people feel more comfortable when the air is dry rather than humid.

A combination of high temperature and high humidity adversely affects everyone. Subjected to hot and humid weather, people generally are more irritable and less able to perform physical and mental tasks, the efficiency of factory workers declines, and students do not concentrate as well. Extreme heat and humidity are a potentially lethal combination when these weather conditions overtax the body's capacity to maintain its core temperature. Heat and humidity can also exacerbate other health problems and lead to premature death.

In mid-July 1995, the high air temperature combined with unusually high humidity to push the *heat index* to record high levels over the Midwest and in cities along the East Coast. More than 1000 people lost their lives, which is more than the average annual number of fatalities in the U.S. from floods, hurricanes, and tornadoes combined. Chicago was particularly hard hit during the week of 14 to 20 July 1995, when 739 more Chicagoans died than expected in normal conditions. (Public safety officials estimate the number of heat and humidity deaths based on the *excess death rate*, that is, the number of reported deaths minus the typical number of deaths expected over a specified period.) According to the U.S. Centers for Disease Control and Prevention, individuals most vulnerable to the hazard of heat and humidity are the elderly who live alone, do not leave home daily, lack access to transportation, are sick or bedridden, lack air conditioning, and do not have nearby social contacts. While the July 1995 heat wave was an extreme, the National Weather Service estimates that episodes of extreme heat, usually accompanied by high humidity, take the lives of about 115 people annually in the U.S. (based on records from 2001 to 2010). Consequently, extreme heat is the nation's most deadly weather hazard.

In the summer of 2003, rather than a combination of high temperatures and humidity, extreme heat alone was responsible for one of Europe's deadliest heat waves in more than a century. A 2006 report by the Earth Policy Institute placed the heat-related death toll at more than 52,000, while the estimates of mortality in Italy and France, using the *excess death rate* method, exceeded 18,000 and 15,000, respectively.

Persistent episodes of hot, dry weather began in late spring 2003 and, in some parts of Europe, lasted until early October. Unusually warm weather prevailed across the British Isles to the Iberian Peninsula and eastward into Germany and Italy. From Spain to Hungary, the average June temperature was 3 to 5 Celsius degrees (5.4 to 9 Fahrenheit degrees) above the long-term average. The heat wave during the first half of August was particularly severe with some localities establishing new all-time maximum temperature records. During July and August, from southern Spain to central France, the daily maximum temperature topped 34 °C (93 °F) on 30 to 50 days, considerably more than normal. In Paris, the maximum daily temperature was greater than 38 °C (100 °F) for six days. From 1 June to 31 August, the maximum daily temperature was above average on all but eight days. The all-time August maximum temperature, set in 1911, was broken by almost 2 Celsius degrees (3.6 Fahrenheit degrees) six times during the month. Not only are most Europeans unaccustomed to coping with such extreme heat, the accompanying dry weather contributed to wildfires and significantly cut crop yields.

Hot weather, regardless of humidity, poses a serious hazard for young children left unattended in motor vehicles. On average, between 29 and 37 children die each year in the U.S. from vehicle-related hyperthermia (based on 1995-2009 news reports). Just over half the cases involve a child forgotten in a vehicle, while about a quarter of the children were intentionally left (e.g., while a parent ran an errand unaware how quickly the car would heat). A car in direct sunlight with no ventilation (windows closed), can have maximum temperatures in the passenger cabin of 70 °C (158 °F) or higher.

Information about how quickly the temperature can rise inside a vehicle demonstrates to the public the risk of vehicle-related hyperthermia. Andrew Grundstein and colleagues from the University of Georgia in Athens, GA, investigated the temperature rise in a vehicle exposed to meteorological and other environmental conditions, as well as using a human heat balance model for the potential impact of heat stress on a child left unattended. The experiment focused on 14 clear days in Athens, GA, from April through August 2007, with a car parked on an open asphalt-covered lot in direct exposure to sunshine and no ventilation (windows closed). These conditions favored maximum heating within the passenger cabin. At mid-day, the flux of incoming solar radiation increases as infrared radiation is emitted. Eventually, an equilibrium is achieved at very high temperatures within the vehicle. While the initial ambient air temperature at 1100 EDT ranged from 15 °C to 34 °C (59 °F to 93 °F), the cabin temperature reached a maximum of 43 °C to 62 °C (109 °F to 144 °F) by 1300 EDT. (For more on this study, refer to A. Grundstein *et al.*, 2010, *Bulletin of the American Meteorological Society*, September, pp. 1183-1191.) The hazard depends on both the initial ambient air temperature and when the vehicle is parked outside, but, regardless, a child should never be left alone in the passenger cabin. Pets face the same hazards if left in parked vehicles unattended.

Scientists have experimented with a variety of indexes in an effort to gauge the combined influence of temperature and humidity on human wellbeing as well as advise people of the potential hazards of heat stress. Since the summer of 1984, the National Weather Service has regularly reported the *heat index*, or *apparent temperature index*, which was developed by R.G. Steadman in 1979. The heat index accounts for the increasing difficulty for a body to dissipate heat in the environment as the relative humidity increases. As illustrated in Figure 1A, at an air temperature of 90 °F (32 °C) and a relative humidity of 60%, the body loses heat to the environment at the same rate as if the air temperature were 105 °F (41 °C) with a relative humidity of 10%. In both cases, the *apparent temperature*

A

Relative humidity (%)

Air temperature (°F)	0	5	10	15	20	25	30	35	40	45	50	55	60	65	70	75	80	85	90	95	100
120	107	111	116	123	130	139	148														
115	103	107	111	115	120	127	135	143	151												
110	99	102	105	108	112	117	123	130	137	143	150										
105	95	97	100	102	105	109	113	118	123	129	135	142	149								
100	91	93	95	97	99	101	104	107	110	115	120	126	132	138	144						
95	87	88	90	91	93	94	96	98	101	104	107	110	114	119	124	130	136				
90	83	84	85	86	87	88	90	91	93	95	96	98	100	102	106	109	113	117	122		
85	76	79	80	81	82	83	84	85	86	87	88	89	90	91	93	95	97	99	102	105	108
80	73	74	75	76	77	77	78	79	79	80	81	81	82	83	85	86	86	87	88	89	91
75	69	69	70	71	72	72	73	73	74	74	75	75	76	76	77	77	78	78	79	79	80
70	64	64	65	65	66	66	67	67	68	68	69	69	70	70	70	70	71	71	71	71	72

B

Category	Apparent temperature*	Heat syndrome
I	130 °F (54 °C) or higher	Extreme danger; heat stroke imminent.
II	105 to 130 °F (41 to 54 °C)	Danger; heat cramps or heat exhaustion likely. Heat stroke possible with prolonged exposure and physical activity.
III	90 to 105 °F (32 to 41 °C)	Heat cramps or heat exhaustion possible with prolonged exposure and physical activity.
IV	80 to 90 °F (27 to 32 °C)	Fatigue possible with prolonged exposure and physical activity.

Source: NOAA National Weather Service
*Apparent temperature combines the effects of heat and humidity on human comfort.

FIGURE 1
(A) Apparent temperature index and (B) hazards posed by heat stress for range of apparent temperatures. [Courtesy of NOAA.]

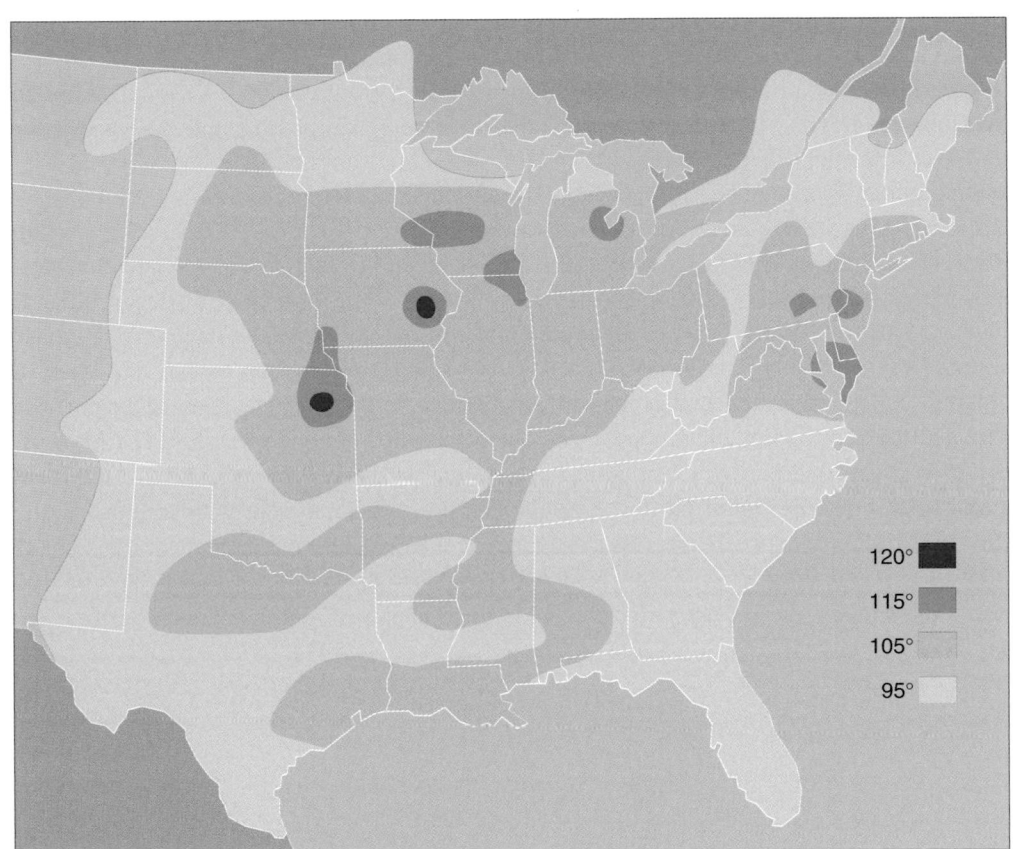

FIGURE 2
Maximum apparent temperature (in °F) recorded during the heat wave of 10-16 July 1995. [NOAA, NWS, Climate Prediction Center]

is 100 °F (38 °C). Examination of Figure 1A reveals that various combinations of air temperature and relative humidity produce the same apparent temperature, implying that all these temperature/relative humidity combinations have the same impact on human comfort and well-being.

Based on the severity of potential health impacts, the heat index is divided into four categories, which, with the associated symptoms, are listed in Figure 1B. Apparent temperatures between 80 °F and 90 °F (27 °C and 32 °C) in category IV pose the least hazard but physical activity is still more fatiguing than usual and caution is advised. At the opposite end of the scale, apparent temperatures in category I are greater than 130 °F (53 °C) and pose the greatest health risk, the danger is extreme and heatstroke (*hyperthermia*) may be imminent. During the heat wave of 10 to 16 July 1995, the highest heat index values were in the range of 115 °F to 120 °F (46 °C to 49 °C) (Figure 2).

The significance of a specific apparent temperature for personal wellbeing depends upon the assumptions that form the basis of Steadman's index as well as age, general health, and body characteristics of an individual. Steadman's heat index assumes a person is resting in the shade with light winds, yet exercise and exposure to direct sunlight add considerably to the heat load the body must dissipate to maintain a normal core temperature. For example, the apparent temperature may be as much as 8 Celsius degrees (14 Fahrenheit degrees) higher in the Sun than the shade. Furthermore, if the ambient air temperature is higher than skin temperature, the wind will convey more heat to the body than away from the body. Any combination of physical activity, exposure to direct sunlight, or hot winds elevates the danger suggested by the apparent temperature.

Where do people experience the most uncomfortable (and potentially life-threatening) summer weather in the U.S.? In view of the adverse impact of high humidity on evaporative cooling, many of us would probably point to locales bordering the Gulf of Mexico where high temperature and high humidity are the norm. For example, during a summer heat wave, a typical temperature/relative humidity combination in New Orleans is 90 °F (32 °C)/60%. According to Figure 1, these conditions translate into an apparent temperature of 100 °F (38 °C), a category III hazard. During such weather conditions, most people would experience some discomfort and everyone is well advised to closely monitor his or her level of physical activity.

Although the relative humidity is considerably lower, much higher ambient air temperatures in the deserts of Arizona and California translate into greater heat stress than along the more humid Gulf Coast. For example, residents of Phoenix, AZ, experience summer afternoon temperatures that approach 110 °F (43 °C) with a relative humidity of 20%. This combination of heat and humidity gives an apparent temperature of 112 °F (44 °C), a category II hazard. In general, residents of the eastern half of North America associate greater discomfort with the muggy days of summer whereas residents of the American Southwest ascribe their days of greatest discomfort to the very hot, dry desert summer.

Low humidity also influences human comfort. As the relative humidity declines, the vapor pressure gradient between the skin and the surrounding air increases, and the evaporative cooling rate rises. The lower the relative humidity, the greater the vapor pressure gradient and, consequently, the higher the rate of heat loss from the body via evaporative cooling. As shown in Figure 1A, for a specific ambient air temperature with successively lower values of relative humidity, the difference between apparent temperature and ambient air temperature decreases until the apparent temperature is lower than the ambient temperature. At and below this threshold, the higher rate of evaporation causes excessive heat loss. That is, enhanced evaporative cooling of the skin at low relative humidity creates the sensation that the air is cooler than it actually is.

Measuring Evaporation

The simplest and least expensive instrument for measuring evaporation is a *pan evaporimeter*, a large cylindrical metal pan filled with fresh water and equipped with a water height gauge, rain gauge, and floating thermometer (for monitoring temperature at the air/water interface). Use of these devices for measuring evaporation dates back to the 19th century but the instrument was not standardized until 1951. The standard NOAA National Weather Service evaporation pan (often found at agricultural experiment stations) measures 121.9 cm (48 in.) in diameter and is 25.4 cm (10 in.) deep (Figure 1). Changes in water level are recorded as long as the temperature is high enough for the water surface to remain ice-free. The drop in water level from one day to the next (minus any rainfall) is recorded as evaporation. Pan sites are typically fenced to prevent wildlife, dogs, and livestock from drinking,

A more sophisticated instrument for estimating direct evaporation plus transpiration (evapotranspiration) is the lysimeter. A *lysimeter* consists of an enclosed block of soil with a natural vegetation cover (usually grass) that is seated on a scale. Continuous measurements are made of rainfall and water percolating through the soil. Day-to-day changes in the weight of the soil block not accounted for by rainfall, runoff, or percolation are attributed to evapotranspiration.

A lysimeter can also be used to measure potential evapotranspiration, a useful parameter in arid and semiarid localities. *Potential evapotranspiration* refers to the maximum possible water loss (to vapor) given the available heat energy (required for latent heating). Potential evapotranspiration is measured in the same way as evapotranspiration except that the soil/vegetation surface of the lysimeter is continually irrigated. Under these conditions, evapotranspiration proceeds at the maximum possible rate. Lysimeters are even sunk into a snow pack to monitor sublimation and evaporation (of meltwater).

Numerical models are used to compute evapotranspiration based on direct measurements or estimates of selected environmental factors that influence the rate of vaporization of water. Evapotranspiration increases with rising temperature, increasing wind speed, and as the difference in vapor pressure between a water surface and the overlying air increases. Transpiration also varies with dewpoint, intensity of incident solar radiation, length of daylight, type of vegetation, stage in a plant's life cycle, and soil moisture.

FIGURE 1
Standard National Weather Service evaporation pan. [NOAA, NWS Training Center]

Atmospheric Stability and Air Quality

On the morning of 26 October 1948, a fog blanket reeking of pungent sulfur dioxide (SO_2) fumes spread over the town of Donora nestled in southwestern Pennsylvania's Monongahela River Valley. Before the fog lifted some five days later, almost half of the area's 14,000 inhabitants had fallen ill and 20 had died. That killer fog resulted from a combination of mountainous terrain and weather conditions that trapped and concentrated deadly effluents from the community's steel mill, zinc smelter, and sulfuric acid plant.

Air pollutants are especially dangerous when atmospheric conditions inhibit their dilution. Once pollutants enter the atmosphere through smokestacks or exhaust pipes, their concentrations usually begin to decline as they mix with cleaner air. The more thorough the mixing, the more rapid is the rate of dilution. When conditions in the atmosphere favor rapid dilution, the impact of air pollution is usually minor. On other occasions, called *air pollution episodes*, conditions in the atmosphere minimize dilution, and the impact can be severe, especially on human health. The two weather factors that most influence the rate of dilution are wind speed and atmospheric stability.

Intuitively, we know that air is likely to mix more thoroughly on a windy day than on a calm one. When it is windy, turbulent eddies accelerate dilution by mixing polluted air and cleaner air. But when the air is calm, dilution occurs by the much slower process of *molecular diffusion*, that is, dispersal at the molecular level. As a general rule, doubling the wind speed cuts the concentration of air pollutants in half (Figure 1). Some weather systems favor light winds and thus inhibit dispersal of contaminants. Over a broad region about the center of an *anticyclone* (a high pressure system), for example, winds are light or the air is calm, and pollutants do not disperse readily. On the other hand, within a *cyclone* (a low pressure system), stronger winds mean more rapid dilution of polluted air. In addition, rain or snow often associated with a cyclone cleans the air by washing pollutants to Earth's surface.

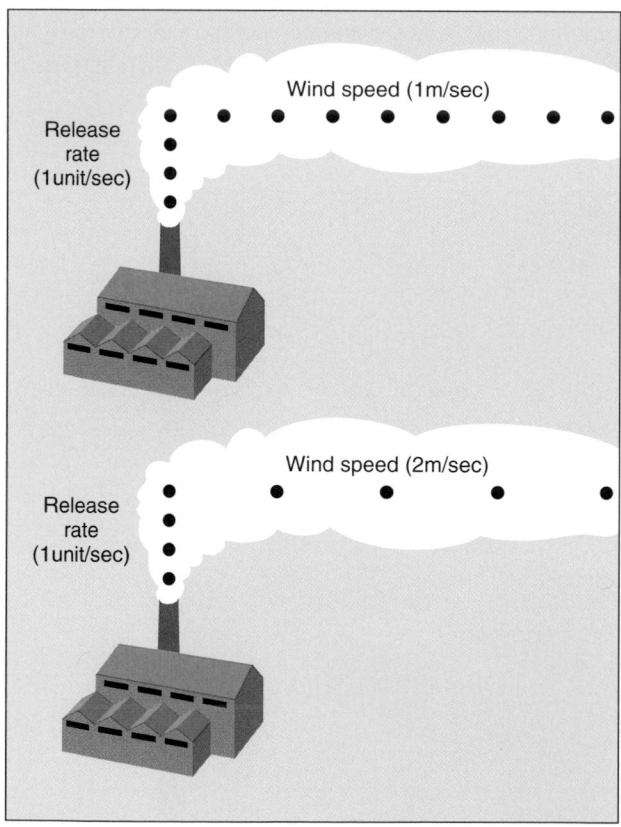

FIGURE 1
Doubling the wind speed from 1 m per sec to 2 m per sec increases the spacing between puffs of smoke by a factor of two, thereby reducing pollutant concentrations by one half.

Atmospheric stability is a second factor that influences the rate at which polluted air mixes with clean air. Stability affects vertical motion within the atmosphere so that convection and turbulence are enhanced when the air is unstable but inhibited when the air is stable. A parcel of polluted air emitted into an unstable air environment undergoes more mixing than does the same parcel of polluted air emitted into a stable air environment. Stable air inhibits the vertical transport of air pollutants, and a layer of stable air aloft may act as a lid over the lower troposphere, trapping air pollutants near Earth's surface. Steady emission of contaminants into stable air results in the accumulation of pollutants.

Mixing depth is the vertical distance between Earth's surface and the altitude to which convection currents (i.e., mixing) reach. When mixing depths are great (many kilometers, for example), pollutants readily mix with the abundant volume of clean air and dilution is enhanced. When mixing depths are shallow (less than 1000 m, for example), air pollutants are restricted to a smaller volume of air, and concentrations may build to unhealthy levels. When ambient air is stable, convection and turbulence are suppressed and mixing depths are shallow. When ambient air is unstable, convection and turbulence are enhanced and mixing depths increase. Because solar heating triggers convection, mixing depths tend to be greater in the afternoon than in the morning, during the day than at night, and in summer than in winter.

We can sometimes estimate the relative stability of air layers by observing the behavior of a plume of smoke rising from a smokestack. If the smoke enters unstable air the plume undulates. In general, this type of plume behavior indicates that polluted air is mixing readily with the surrounding cleaner air and being diluted. The net effect is improved air quality (except where the plume temporarily loops to the ground). On the other hand, a plume of smoke that flattens and spreads slowly downwind indicates very stable conditions and minimal dilution.

An air pollution episode is most likely when a persistent *temperature inversion* develops. Within an air layer characterized by a temperature inversion, air temperature increases with altitude; that is, warmer, lighter air overlies cooler, denser air. (A temperature inversion is the *inverse* of the usual temperature profile of the troposphere.) This extremely stable stratification strongly inhibits mixing and dilution of pollutants. A temperature inversion forms by (1) subsidence (sinking) of air, (2) extreme radiational cooling, or (3) some cases of air mass advection. The inversion may occur aloft or near Earth's surface.

A *subsidence temperature inversion* forms a lid over a broad area, often encompassing several states at once (Figure 2). It develops during a period of fair weather when a warm anticyclone stalls. An anticyclone produces subsiding air that is warmed by compression. Subsiding warm air is prevented from reaching Earth's surface by the *mixing layer*, the portion of the troposphere in which convection thoroughly mixes the air. The air temperature within the mixing layer drops with altitude, but air just above the mixing layer, having been warmed by sinking and compression, is warmer than air at the top of the mixing layer. A temperature inversion marks a transition between the mixing layer below and the compressionally warmed air above. During these atmospheric conditions, air pollutants are distributed throughout the mixing layer, but no higher than the temperature inversion. Pollutants are confined to a relatively small volume of air, and continual emission will elevate concentrations.

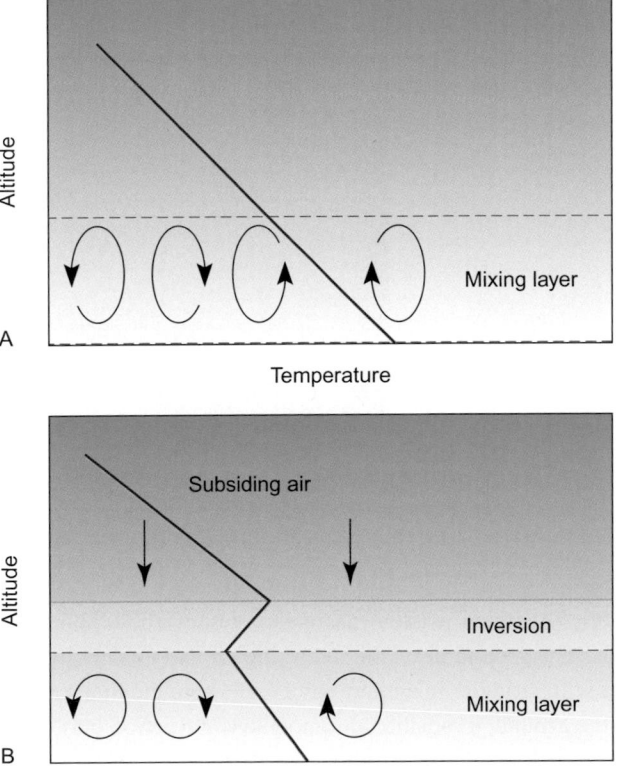

FIGURE 2
A temperature inversion can develop aloft through subsidence of air. Compare a sounding prior to subsidence (A) with a sounding during subsidence (B). The temperature inversion acts as lid over the lower atmosphere, trapping air pollutants.

A *radiational temperature inversion* is usually more localized than a subsidence temperature inversion (Figure 3). With clear night skies, radiational cooling chills both the ground and the air above the ground. However, the ground is a better emitter of infrared radiation than is the overlying air, so the ground surface cools faster than the air above it. Heat is conducted from the warmer overlying air to the cooler ground. Chilling of the air from below means that the coldest air is next to the ground and the air temperature increases with altitude; that is, a low-level temperature inversion forms. This process requires light winds or calm conditions because stronger winds would mix the air and prevent the stratification of warm air over cold air. A fresh snow cover often strengthens the inversion and hence, the stability.

A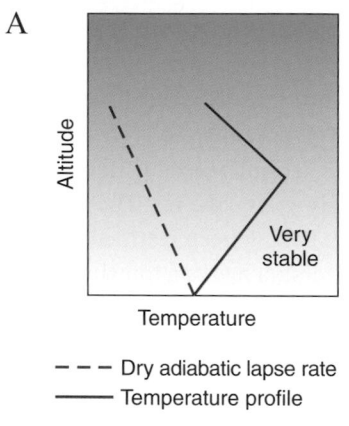

B

FIGURE 3
(A) A low-level temperature inversion caused by nocturnal radiational cooling. (B) Steam emitted from a power plant on a clear, cold January morning was initially warmer than the ambient air and rose. When the temperature of the steam became equal to the temperature of the ambient air (within the low-level temperature inversion), it spread out horizontally and diffused. [Photo by K.A. Nugnes]

A radiational temperature inversion usually disappears within a few hours after sunrise as the Sun heats the ground and the wind strengthens. The ground heats the overlying air and eventually reestablishes a normal sounding in which air temperature decreases with altitude. In winter, however, where the Sun's rays are relatively weak and a highly reflective layer of snow covers the ground, a radiational temperature inversion may persist for days or even weeks.

Advection of air masses can also give rise to temperature inversions and affect air quality. This sometimes happens at the base of the Rocky Mountains (Figure 4). A west wind is compressionally warmed as it descends the leeward slopes of the mountain range but along the foot of the mountains, surface winds from the north advect a shallow layer of cold air southward. Hence, a temperature inversion aloft separates the relatively warm air above from the surface layer of cold air. Although temperature inversions also characterize warm and cold fronts, these inversions have little adverse effect on air quality because fronts are in motion and the accompanying precipitation washes pollutants from the air.

In Donora, a persistent temperature inversion developed, and the smoke emitted from the community's steel mill, zinc smelter, and sulfuric acid plant underwent minimal dilution and was essentially trapped near the surface. This tragic event, along with others, spurred development of the U.S. Clean Air Act for reducing pollution, as well as greater awareness about the relationship between atmospheric conditions and air quality.

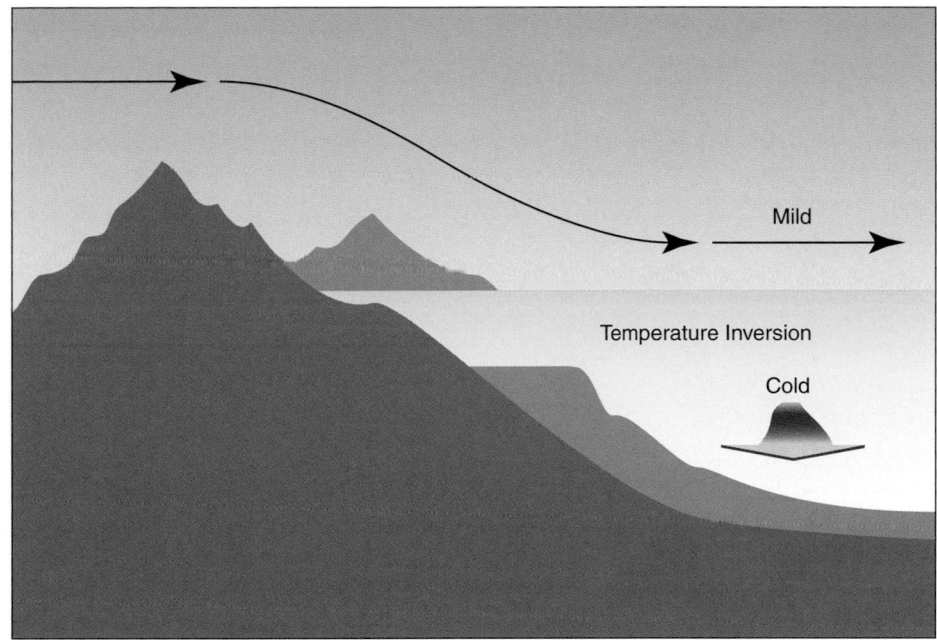

FIGURE 4
A temperature inversion develops aloft adjacent to the leeward slopes of the Rocky Mountains. A west-to-east air flow descends the leeward slopes, is warmed by compression, and overlies a layer of colder air advected southward by north winds.

Cattle in West-Central Brazil graze under threatening skies with the National Center for Atmospheric Research's S-Pol radar (collecting data for the Tropical Rainfall Measuring Mission in 1998) in the foreground. [Courtesy University Corporation for Atmospheric Research. Photo by Scott Ellis.]

CLOUDS, PRECIPITATION, & WEATHER RADAR

Chapter Highlights

Case-in-Point
    Aircraft Contrails and Cloud Cover
Cloud Formation
    The Curvature Effect
    Role of Nuclei
    Supercooled Water
Cloud Classification
    High Clouds
    Middle Clouds
    Low Clouds
    Clouds Having Vertical Development
    Unusual Clouds
Fog
Precipitation Processes
    Terminal Velocity
    Warm-Cloud Precipitation
    Cold-Cloud Precipitation
    Holes and Canals in Clouds
Forms of Precipitation
Acid Deposition
Weather Radar: Locating Precipitation
    Reflectivity Mode
    Velocity (Doppler) Mode
    Dual-Polarization Weather Radar
    Phased Array Weather Radar
Measuring Precipitation
    Rain and Snow Gauges
    Remote Sensing of Precipitation
Conclusions
Basic Understandings/Enduring Ideas
Key Terms/Review/Critical Thinking
For Further Exploration
    Clouds by Mixing
    Rainmaking
    *When Is It Too Cold or Too Warm to
        Snow?*

Learning Objectives

Explain the role of nuclei in cloud development.
List the sources and types of cloud nuclei.
Explain the significance of hygroscopic nuclei.
Describe how clouds are classified.
Distinguish among the principal cloud types.
Identify the various fog-forming processes.
List conditions required for extreme nocturnal radiational cooling.
Explain the significance of terminal velocity in the formation of precipitation.
Distinguish between warm clouds and cold clouds.
Describe the collision-coalescence process.
Describe the Bergeron-Findeisen process.
Distinguish among the various solid and liquid forms of precipitation.
Explain how weather radar detects precipitation.
Distinguish between the reflectivity and velocity modes for weather radar.
Describe how rain and snow are measured directly and remotely.

How do clouds and precipitation form?

Aircraft Contrails and Cloud Cover

A familiar sight in the daytime clear or partly cloudy sky is **contrails**, bright white streamers of ice crystals that can form in the exhaust of jet aircraft flying in low temperature environments, mainly at high altitudes (Figure 7.1). Contrails modify cloud cover along heavily traveled air corridors between major urban areas, with possible implications for weather and climate (Figure 7.2). According to some studies, contrails already cover 5% of the sky in some major flight corridors of the eastern United States. A contrail, short for *condensation trail*, develops when the hot humid air exhausted from a jet engine mixes with the cold drier air of high altitudes. (A similar process is the reason you can *see your breath* when exhaling outdoors on a cold day.) Depending on atmospheric conditions, contrails may dissipate (sublimate) within minutes or hours, or they may spread laterally to form persistent wispy cirrus clouds. For more details, see this chapter's first *For Further Exploration, Clouds by Mixing*.

Atmospheric scientists hypothesized that increased cloud cover caused by contrails may affect the local radiation budget. During daylight hours, contrails and their cirrus byproducts reflect incoming solar radiation, thereby cooling Earth's surface. At night, contrails and associated cirrus clouds absorb and emit infrared radiation radiating from below, thereby enhancing the greenhouse effect and elevating Earth's surface temperature. Cooler days and warmer nights reduce the *diurnal temperature range (DTR)*, the difference between the day's maximum and minimum temperatures. Furthermore, contrails may stimulate precipitation locally by supplying ice crystal nuclei for lower clouds.

Scientists had an opportunity to test their hypothesis regarding the impact of contrails on the radiation budget immediately after the terrorist attacks of 11 September 2001 when the Federal Aviation Administration (FAA) ordered a three-day shutdown of all commercial air traffic in the United States. Almost immediately, contrails began to dissipate. David J. Travis, a climate scientist at the University of Wisconsin-Whitewater, and colleagues at Pennsylvania State University analyzed the effects of a contrail-free sky on surface temperatures at 4000 U.S. weather stations. They found that the DTR during the three-day period was about 1.1 Celsius degree (2 Fahrenheit

FIGURE 7.1
Contrails form when the hot, humid exhaust from jet aircraft engines mixes with the cold, drier ambient air. [Photo by Judy Peterson, from *The Stories Clouds Tell* (AMS) by Margaret (Peggy) LeMone]

degrees) higher than the long-term (1971-2001) average. Furthermore, the DTRs for the three-day periods preceding and following the shut-down were below the long-term average. These findings prompted Travis and his colleagues to attribute the increase in DTR during the grounding of aircraft to the absence of contrails. With global aircraft traffic expected to increase by 2% to 5% per year through 2050, some scientists predict that the contribution of contrails to climate change will become more significant.

In 2008, however, researchers at Texas A&M University and NASA's Langley Research Center took another look at whether the absence of contrails was responsible for the increase in DTR during 11-14 September 2001. They examined cloud cover, humidity, winds, and surface temperatures over the coterminous U.S. during the period 1971-2001. They concluded that the increase in DTR during the grounding of aircraft was likely the result of anomalies in low cloud cover rather than the absence of contrails. Low clouds are thicker and warmer than high clouds, or contrails, and reflect more sunlight to space at the same time emitting more infrared radiation to Earth's surface. Furthermore, the researchers pointed out that the DTR during 11-14 September was within the range of natural climate variability. There is still much research to be done regarding the relationship between contrails and DTR.

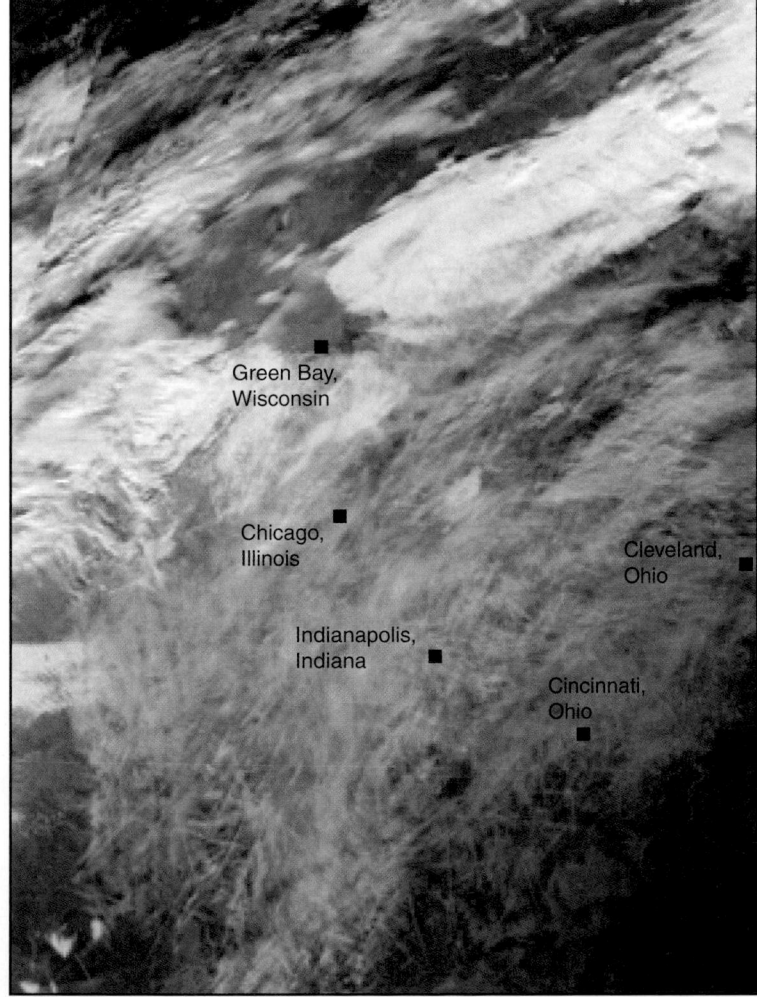

FIGURE 7.2
MODIS image of the midwestern U.S. on 25 November 2006, obtained by NASA's Terra satellite. Note the high concentration of discernible contrails in the southern half of the image. [Courtesy of NASA Earth Observatory]

Clouds whisk across the sky in ever-changing patterns of white and gray; fog lends a feeling of eerie calmness to a dreary day. Clouds and fog are products of condensation or deposition of water vapor within the atmosphere. Most clouds are the consequence of saturation brought about by uplift and expansional cooling of air. Fog is a cloud that forms near or in contact with Earth's surface; most fogs develop when the lowest layer of air is chilled to saturation via radiational or advective cooling.

Cloud formation is an essential part of the global water cycle because without clouds there would be no rain or snow. And yet most clouds—even those associated with large storm systems—do not produce precipitation. The principal topics of this chapter are the development and classification of clouds and fog, plus the formation and types of precipitation. We also examine weather radar, a valuable tool for locating and tracking areas of precipitation and for monitoring air motion within weather systems. This chapter closes with a summary of the various direct and remote methods for measuring precipitation.

Cloud Formation

Water vapor is an invisible component of air, but its condensation and deposition products (water droplets and ice crystals) are visible. A **cloud** is the visible product of condensation or deposition of water vapor within the atmosphere; it consists of a large and visible aggregate of minute water droplets and/or ice crystals. This section covers the process of cloud development.

Clouds are increasingly likely to form as air nears saturation (Chapter 6). But when scientists try to simulate this process in the laboratory using a clean-air chamber, a cloud does not form even if the relative humidity is elevated well above 100%, that is, to supersaturation values. A clean-air chamber is an enclosed container of air from which all solid and liquid particles (aerosols) have been filtered out. Introducing water vapor into the chamber raises the relative humidity but even if the relative humidity is elevated to as high as 200%, no cloud forms. How is that possible?

We expect that once air becomes saturated (relative to a flat-water surface), condensation would initially produce extremely small droplets and those droplets would eventually grow into cloud droplets through additional condensation. However, droplets will not form in the first place in a clear-air environment unless the relative humidity rises to extraordinarily high levels of supersaturation.

THE CURVATURE EFFECT

Tiny droplets cannot form in a cloud chamber without a supersaturated environment because of the **curvature effect**. The curvature of a water surface affects the ability of water molecules to escape (vaporize) from the water surface. The smaller the droplet, the greater is the concentration of surrounding water vapor that is necessary for the droplet to grow. The curvature of the surface of a spherical water droplet increases as the radius of the droplet decreases; that is, the surface of a small droplet has a greater curvature than the surface of a large droplet. (At the limit, a flat-water surface has no curvature.) With increasing curvature, molecules that form on the surface of a droplet have fewer neighboring molecules and are more weakly bonded. Molecules composing a flat-water surface have the greatest number of neighboring molecules and are most strongly bonded. Bond strength affects the flux of water molecules from the liquid to vapor phase. For this reason, water molecules more readily escape a water droplet than a flat-water surface and water molecules more readily escape a small water droplet than a large water droplet. Consequently, at the same temperature, the saturation vapor pressure surrounding a small water droplet is greater than that surrounding a large droplet.

Values of saturation vapor pressure listed in Table 6.3 are defined for air over a flat surface of fresh water and are the basis for computing relative humidity (Chapter 6). The saturation vapor pressure increases with increasing curvature (and decreasing droplet size). For droplets having a radius of 0.001 micrometer, the saturation vapor pressure at a given temperature is about 3.4 times that specified in Table 6.3, which computes to a relative humidity of 340%. On the other hand, for droplets having a radius greater than 1.0 micrometer, the difference between the saturation vapor pressure surrounding the droplet and the values listed in Table 6.3 is minimal and computes to a relative humidity only slightly higher than 100%.

The bottom line: In clean air, water vapor will not condense into tiny water droplets unless the air is supersaturated to very high levels. In the real atmosphere, the relative humidity never even approaches such great supersaturation values, so tiny droplets would not form or grow. What then makes cloud formation possible?

ROLE OF NUCLEI

The clean-air chamber experiment described above failed to produce a cloud because the air was not sufficiently supersaturated. In the real atmosphere, supersaturated air is not needed for clouds to develop; in fact, cloud droplets

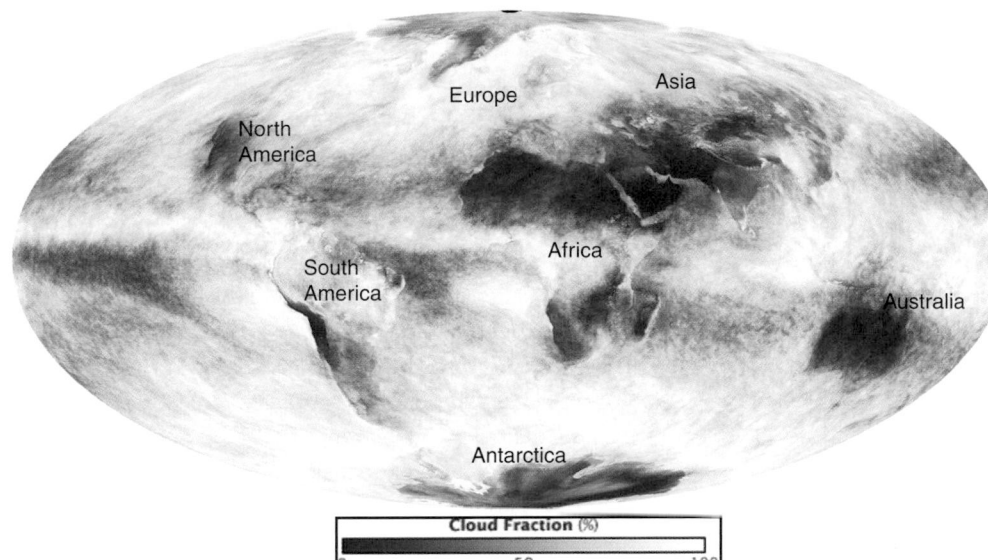

FIGURE 7.3
Average percentage of areal cloud coverage for 1-31 October 2009. Color scale ranges from blue (no clouds) to white (totally cloudy) representing the portion of each pixel that was covered by clouds. Measurements obtained by the Moderate Resolution Imaging Spectrometer (MODIS) on NASA's Terra satellite. [NASA Earth Observatory image by Kevin Ward, based on data provided by the NASA Earth Observation (NEO) Project]

readily form as the relative humidity nears 100%. And clouds are abundant; at any given time shrouding about 60% of the planet (Figure 7.3). How do cloud droplets (or ice crystals) form? Suspended in the atmosphere are solid and liquid particles known as **nuclei** that have radii greater than 1.0 micrometer and favor condensation (or deposition) of water vapor. Once condensation (or deposition) starts, they become comparably sized cloud droplets (or ice crystals). Additional growth of these droplets (or ice crystals) by condensation (or deposition) is more likely because they can grow in an environment with a relative humidity that exceeds 100% by only 1% or so. Most cloud particles have radii of 2 to 10 micrometers.

Nuclei, essential for cloud formation, are abundant in the atmosphere and are continually cycled into the atmosphere from Earth's surface. Sources of nuclei include volcanic eruptions, wind erosion of soil, forest fires, and ocean spray. When sea waves break, drops of salt water enter the atmosphere and the water evaporates leaving behind tiny sea-salt crystals that function as nuclei (Figure 7.4). Emissions from domestic and industrial chimneys and exhaust pipes also contribute nuclei to the atmosphere.

Depending on the product (liquid water droplets or ice crystals), a distinction is made between cloud condensation nuclei and ice-forming nuclei. **Cloud condensation nuclei (CCN)** promote condensation of water vapor at temperatures both above and below the freezing point of water. Within the atmosphere, water vapor can condense into cloud droplets that remain liquid even at temperatures well below 0 °C (32 °F). Droplets at such temperatures are described as *supercooled* (discussed

later in this chapter). **Ice-forming nuclei (IN)** are much less common than CCN and most promote formation of ice crystals only at temperatures well below freezing. The most efficient ice-forming nuclei are substances having a crystal structure similar to that of ice and most are almost insoluble in water. Clayey soil particles are excellent ice-forming nuclei. The two types of ice-forming nuclei are freezing nuclei and deposition nuclei. *Freezing nuclei* are particles on which water vapor condenses and subsequently freezes; most are active at temperature below −9 °C (16 °F). However, biological ice nucleators are known to be very active at higher temperatures. Some bacterial plant pathogens catalyze ice formation at temperatures near −2 °C (28.4 °F). *Deposition nuclei* are particles on which water vapor deposits directly as ice (without first becoming a liquid) and are not fully active until the temperature drops below about −20 °C (−4 °F).

FIGURE 7.4
Breaking sea waves are a source of salt crystals that serve as hygroscopic nuclei for cloud development.

Hygroscopic nuclei are a special category of cloud condensation nuclei that possess a chemical attraction for water molecules. Condensation begins on hygroscopic nuclei at a relative humidity well under 100%. Magnesium chloride ($MgCl_2$), a salt in sea-spray, can promote condensation at a relative humidity as low as 70%! Hence, clouds form more readily where hygroscopic nuclei are abundant.

Urban-industrial areas have many sources of hygroscopic nuclei and this helps explain why localities downwind of large cities tend to be somewhat cloudier and rainier than upwind localities. The *Metropolitan Meteorological Experiment (METROMEX)* conducted in the 1970s demonstrated that during summer months the average rainfall was 5% to 25% greater within St. Louis, MO and areas about 50 to 75 km (31 to 47 mi) downwind of the St. Louis, than it was in areas upwind of the city. In more recent years, similar urban effects on precipitation were detected in Tokyo, Phoenix, and other cities. Besides being a source of hygroscopic nuclei, cities also spur cloud and precipitation development by contributing water vapor (raising the relative humidity) and heat (adding to the buoyancy of air). Furthermore, the relative roughness of a city surface enhances convergence of surface winds, updrafts, expansional cooling, and cloud development (Chapter 8).

Interestingly, not all ice crystals form on the traditional nuclei described above. While sampling ice crystals in cold clouds (i.e., clouds at subfreezing temperatures) from aircraft, scientists discovered that the concentration of ice crystals is highly variable within the cloud and ice crystals are 10 to 1000 times more numerous than nuclei. This discrepancy is likely explained by the presence of fragments of ice crystals that break off during midair collisions. These ice fragments then function as additional nuclei.

SUPERCOOLED WATER

About 1783, the Swiss naturalist Horace de Saussure (1740-1799) demonstrated **supercooling** of water; that is, he was able to cool fresh water below its normal freezing point of 0 °C (32 °F) and the water remained liquid. Sixty-seven years later while on a balloon ascent the French chemist Jean Barrel confirmed the existence of supercooled cloud droplets in the free atmosphere. Cloud condensation nuclei promote condensation of water vapor at temperatures both above and below 0 °C. How can cloud droplets remain liquid even at temperatures well below freezing?

Freezing of fresh water in the absence of already existing ice begins with the linking of water molecules to form a tiny ice structure known as an *ice embryo*. Once an ice embryo grows larger than a certain critical size, its growth continues and freezing begins. Within the very small droplets composing clouds, kinetic molecular activity is sufficient to inhibit growth of ice embryos to the critical size so that water droplets remain liquid even at temperatures well below freezing. As the droplet temperature falls, however, the average kinetic molecular activity diminishes, the number of ice embryos increases, and the probability of an ice embryo reaching the critical size increases. Formation of ice embryos of critical size due to the chance aggregation of water molecules is known as *homogeneous* (or *spontaneous*) *nucleation*. Laboratory experiments demonstrate that if no foreign particles (aerosols) are present that would act as ice-forming nuclei, homogeneous nucleation of a cloud droplet can only occur at temperatures lower than −35 °C (−31 °F). A cloud droplet can cool to as low as −39 °C (−38.2 °F) without freezing, but any additional cooling always leads to homogeneous nucleation. This threshold temperature is called the *Schaefer point*, named for Vincent J. Schaefer (1906-1993), a pioneer in cloud physics research at the General Electric Research Laboratory in Schenectady, NY.

If a supercooled water droplet contains (or comes in contact with) foreign particles that are ice-forming nuclei, the cloud droplet will freeze at a temperature below the freezing point but well above the Schaefer point. The foreign particle forms an ice embryo. The growth of the embryo causes the droplet to freeze. This process is referred to as *heterogeneous nucleation*.

Cloud Classification

Clouds come in an infinite variety of shapes and sizes, something that is obvious from even a cursory glance at the sky. Characteristics that are common to different clouds form the basis for their classification. Cloud classification is useful because different types of clouds tell us something about atmospheric processes responsible for their formation and may provide clues as to future weather. British naturalist Luke Howard (1772-1864) was among the first to devise a classification of cloud types. Formulated in 1802-03, the essentials of Howard's scheme are still used today.

Clouds are classified on the basis of (1) general appearance (texture) as observed from the ground, (2) altitude of cloud base, (3) temperature, or (4) composition (Table 7.1). Based on a cloud's general appearance, the simplest distinction is among cirriform, stratiform, and cumuliform clouds. A **cirriform cloud** is wispy or fibrous,

---

TABLE 7.1
Summary of Cloud Classification Schemes

General Appearance

    Cirriform
    Stratiform
    Cumuliform

Altitude

    High
    Middle
    Low
    Significant Vertical Development

Temperature

    Warm
    Cold

Composition

    Ice crystals/supercooled water droplets
    Water droplets

TABLE 7.2
Cloud Classification

Cloud group	Altitude of base (m)[a]
High clouds	
Cirrus (Ci)	5000-13,000
Cirrostratus (Cs)	5000-13,000
Cirrocumulus (Cc)	5000-13,000
Middle clouds	
Altostratus (As)	2000-5000
Altocumulus (Ac)	2000-5000
Low clouds	
Stratus (St)	surface-2000
Stratocumulus (Sc)	surface-2000
Nimbostratus (Ns)	surface-2000
Clouds with vertical development	
Cumulus (Cu)	to 3000
Cumulonimbus (Cb)	to 3000

[a]Average for middle latitudes.

---

a **stratiform cloud** is layered, and a **cumuliform cloud** is heaped or puffy. On the basis of altitude of cloud base, a distinction is made among high, middle, and low clouds, and clouds having significant vertical development. The altitude of a cloud is important in that it affects its temperature and ultimately its composition. High, middle, and low clouds generally occur in layers and are produced by gentle uplift of air (typically less than 5 cm per sec or 0.1 mph) over a broad geographical area. Layered clouds often form along and ahead of a warm front. Clouds having significant vertical development usually cover smaller areas but are associated with much more vigorous uplift (sometimes in excess of 30 m per sec or 70 mph). These vertically developed clouds may form along a cold front.

Clouds are described as warm or cold. The temperature in a **warm cloud** is greater than 0 °C (32 °F) whereas the temperature in a **cold cloud** is at or less than 0 °C (32 °F). Because the average air temperature decreases from sea level to the tropopause and ascending cloudy air expands and cools, high clouds are colder than low clouds. Vertically developed clouds that surge to great altitudes, such as thunderstorm clouds (cumulonimbus), may be warm near their base but cold aloft. To a large

extent, the ambient air temperature at cloud level dictates the composition and appearance of clouds. Cold clouds are made up of ice crystals and/or supercooled water droplets; ice crystal clouds have wispy edges. Warm clouds are made up of water droplets and have sharply defined boundaries.

Scientists have identified ten fundamental cloud types, organized by altitude of cloud base in middle latitudes (Table 7.2). The actual altitude of cloud base varies seasonally and with latitude. For example, the base of high clouds may be as low as 3000 m (10,000 ft) in polar latitudes and as high as 18,000 m (60,000 ft) in the tropics. Furthermore, certain qualifying terms are applied to clouds having special characteristics; some of these are listed in Table 7.3.

HIGH CLOUDS

High clouds have bases at altitudes above 5000 m (16,000 ft) where average temperatures are typically below −25 °C (−13 °F). At such low temperatures, clouds are composed of ice crystals almost exclusively, explaining their fibrous appearance in the presence of sunlight. Their names include the prefix *cirro* from the Latin meaning "a curl of hair."

TABLE 7.3
Special Characteristics of Cloud Types

Characteristic	Meaning	Applied to
Castellanus	Tower-like vertical development	Cirrocumulus, Altocumulus
Congestus	Crowded in heaps	Cumulus
Fractus	Broken, ragged, torn	Stratus, Cumulus
Humilis	Little vertical development, flat base	Cumulus
Incus	Anvil-shape	Cumulonimbus
Lenticularis	Lens-shaped	Cirrostratus, Altocumulus, Stratocumulus
Mammatus	Hanging protuberances	Cumulonimbus
Pileus	Cap-shaped cloud above	Cumuliform
Uncinus	Hook-shaped	Cirrus

Cirrus (Ci) clouds are nearly transparent and occur as delicate silky strands, sometimes called mares' tails (Figure 7.5A). Strands are actually streaks of falling ice crystals being blown laterally by strong winds. Like cirrus clouds, **cirrostratus (Cs) clouds** are also nearly transparent, so the Sun or Moon readily shines through them. They form a thin, white veil or sheet that partially or totally covers the sky (Figure 7.5B). **Cirrocumulus (Cc) clouds** consist of small, white, rounded patches arranged in a wavelike or mackerel (resembling fish scales) pattern (Figure 7.5C). Rarely do these clouds cover the entire sky. As a rule, no high cloud is thick enough to prevent objects on the ground from casting shadows during daylight hours.

High clouds, especially cirrostratus, are often the first clouds to appear in advance of an approaching middle latitude storm system. Cirrus clouds sometimes develop in association with the vertical motions around the upper tropospheric jet stream (Chapter 9). At times, cirriform clouds can also result from portions of clouds, particularly the anvils of cumulonimbus clouds, being blown off from the tops of thunderstorm cells, by strong upper tropospheric winds.

MIDDLE CLOUDS

The base of middle clouds typically occurs at altitudes ranging from 2000 to 5000 m (6600 to 16,000 ft). Temperatures in these clouds are generally between 0 °C and −25 °C (32 °F and −13 °F); they are composed of supercooled water droplets or a mixture of supercooled water droplets and ice crystals. Their names include the prefix *alto*. **Altostratus (As) clouds** occur as uniformly gray or bluish white layers that totally or partially cover the sky (Figure 7.6A). Usually they are sufficiently thick that the Sun is only dimly visible, as if viewed through frosted glass. (As a rule, clouds must have a thickness of at least 50 m or about 150 ft to completely block out

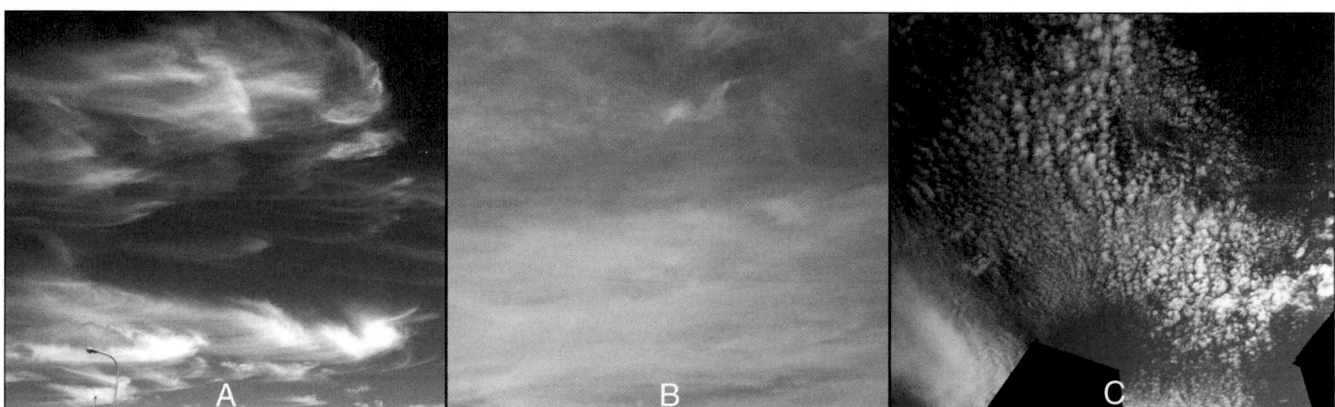

FIGURE 7.5
High stratiform clouds are composed of mostly ice crystals: (A) cirrus, (B) cirrostratus, and (C) cirrocumulus.

FIGURE 7.6
Middle stratiform clouds: (A) altostratus, (B) altocumulus [photo by Lin Chambers, NASA], are composed of ice crystals, water droplets, or a combination of the two.

FIGURE 7.7
Low stratiform clouds: (A) stratocumulus, (B) stratus. [Photos by K.A. Nugnes]

the Sun.) **Altocumulus (Ac) clouds** consist of roll-like patches or puffs that often form waves or parallel bands (Figure 7.6B). They are distinguished from cirrocumulus by the larger size of the cloud patches and by sharper edges indicating the presence of water droplets rather than ice crystals. Altocumulus may occur as several distinct layers simultaneously and rarely produce precipitation that reaches the ground. If they appear early on a warm, humid summer day, there is a chance of an afternoon thunderstorm.

LOW CLOUDS

Low clouds have bases ranging from Earth's surface (fog) up to an altitude of perhaps 2000 m (6600 ft). Low clouds that form at temperatures greater than −5 °C (23 °F) are composed of mostly water droplets. **Stratocumulus (Sc) clouds** consist of large, irregular puffs or rolls separated by areas of clear sky (Figure 7.7A). Seldom does significant precipitation fall from stratocumulus clouds. **Stratus (St) clouds** appear as a uniform gray layer stretching from horizon to horizon

(Figure 7.7B). Fog occurs where stratus meets the ground, and when fog lifts, the stratus cloud deck may break up into stratocumulus.

Usually only drizzle falls from stratus clouds, but significant amounts of rain or snow may fall from the much thicker **nimbostratus (Ns) clouds**. Viewed from the ground, nimbostratus resemble stratus except that they are darker gray and have a less uniform, more ragged base. Precipitation from nimbostratus tends to be light to moderate and continuous for 12 hrs or longer. By contrast, relatively brief but heavy showery-type precipitation is often associated with a cumulonimbus (thunderstorm) cloud. Gentle rainfall from nimbostratus more readily infiltrates the soil, whereas the intense downpour of a thunderstorm quickly saturates the upper soil and then mostly runs off, perhaps causing flash flooding (Chapter 11).

Stratus and nimbostratus clouds often are found close to an approaching warm front, where warm and humid air is lifted over a dome of cooler air ahead of the front. Stratocumulus clouds may represent either eroding stratus clouds or are caused by unstable atmospheric conditions in the lower troposphere as cold air moves over relatively warm ground in the wake of a departing low pressure system.

CLOUDS HAVING VERTICAL DEVELOPMENT

Clouds having significant vertical development form in the updrafts of convection currents that can surge to great altitudes in the troposphere. The altitude at which condensation begins to occur through convection is known as the **convective condensation level (CCL)** and coincides with the altitude of cumuliform cloud base, typically between 1000 and 2000 m (3600 and 6600 ft). **Cumulus (Cu) clouds** are relatively small puffy clouds that resemble balls of cotton floating in the sky on a fair weather day (Figure 7.8A). Solar heating drives convection so that cumulus cloud development often follows the daily variation of incoming solar radiation. On a fair day, cumulus clouds begin forming by middle to late morning after the Sun has warmed the ground and initiated convection. Cumulus sky cover is most extensive by mid afternoon, usually the warmest time of day. If cumulus clouds show some vertical growth, these normally fair-weather clouds may produce a brief, light shower of rain or snow. As sunset approaches, convection weakens, and cumulus clouds begin dissipating (that is, they vaporize).

Where convection is suppressed, so too is the development of cumulus clouds. Relatively cold surfaces chill and stabilize the overlying air and inhibit convection so that cumulus clouds do not readily form over snow-covered surfaces. Cold water surfaces also suppress convection. This effect is sometimes observed along the shores of the Great Lakes during late spring and early summer afternoons when, on average, lake surface temperatures are lower than temperatures of the adjacent land surface. Fair-weather cumulus clouds develop over the warm land but not over the relatively cold lake surface (Figure 7.9). In fact, rows of cumulus clouds may form a distinct boundary along the shoreline.

Once cumulus clouds form, the stability profile of the troposphere determines the extent of vertical cloud development. If ambient air aloft were stable, vertical motion would be inhibited and cumulus clouds would exhibit little vertical growth. Under these conditions, the weather is likely to remain fair. On the other hand, if ambient air aloft were unstable for saturated (cloudy) air, then vertical motion would be enhanced, and the tops

FIGURE 7.8
Cumuliform clouds: (A) cumulus [Photo by Stephen Corfidi, NOAA/NWS/SPC], (B) cumulus congestus [Photo by K.A. Nugnes], (C) cumulonimbus. [Photo by Robert Henson, *The Stories Clouds Tell* (AMS)]

of cumulus clouds would surge upward. If the ambient air were unstable to great altitudes, the entire cloud mass would take on a cauliflower appearance as it builds into a towering **cumulus congestus cloud** (Figure 7.8B) or perhaps a **cumulonimbus (Cb) cloud** (Figure 7.8C). Cumulonimbus clouds have tops that sometimes reach altitudes of 20,000 m (60,000 ft) or higher. The upper portion of a cumulonimbus cloud, with its familiar anvil shape, is typically composed of mostly ice crystals, the middle portion may be supercooled water droplets or a mixture of supercooled water droplets and ice crystals, while the lower portion of the cloud consists of water droplets.

Although this cloud classification category is usually reserved for cumuliform clouds, altostratus and nimbostratus also may be considered clouds having vertical development. Normally classified as middle and low clouds, respectively, based on cloud-base altitude, altostratus and nimbostratus often are so thick that they extend through more than one level. Furthermore, cumuliform and stratiform clouds often appear in the sky at the same time. For example, a layer of high cirrus may overlie low-level cumulus clouds. The cirrus may have developed in advance of an approaching surface warm front and is too thin to significantly weaken the flux of incoming solar radiation. The Sun heats the ground, triggering convection and development of cumulus clouds.

When more than one cloud layer appears in the sky at the same time, it can be instructive to visually track the direction of movement of each cloud layer. Often clouds at different altitudes move in different directions

FIGURE 7.10
Measurements obtained by the Moderate Resolution Imaging Spectrometer (MODIS) on NASA's Terra satellite show cloud streets over Lake Superior. [NASA]

at different speeds. Because most clouds move with the wind, this observation indicates that, at least within the troposphere (where most clouds are found), the horizontal wind changes direction and speed with altitude. Any change in wind direction or wind speed with distance is known as **wind shear**. Strong vertical shear in horizontal wind speed or direction may cause clouds (or cloud elements) to line up as rows that often extend for distances of hundreds of kilometers. Such **cloud streets** (Figure 7.10) are observed in cumulus, stratocumulus, and cirrocumulus clouds. They develop where the air is humid and the axes of convection cells align with the average wind direction.

UNUSUAL CLOUDS

Chances are that all of us have seen most of the clouds described above because the circumstances leading to their formation are quite common. Other, more unusual clouds are formed by special atmospheric conditions that in some cases arise from orographic effects.

Many people have reported certain disk-shaped clouds as UFOs (*Unidentified Flying Objects*). Observing such clouds over a period of time reveals an intriguing characteristic besides their shape: They are almost motionless and differ from many other types of clouds that are carried along by winds aloft. These disk or lens-shaped middle level clouds are **altocumulus lenticularis clouds** (Figure 7.11) generated when a mountain range disturbs large-scale horizontal winds. As strong horizontal winds encounter a mountain range, the wind is deflected up the windward slopes and down the leeward slopes. This occurs, for example, as the prevailing westerlies, blowing from west to east, cross the Sierra Nevada Mountains of California or the Front Range of the Rocky Mountains. If the air ascending the windward slopes is stable, then the

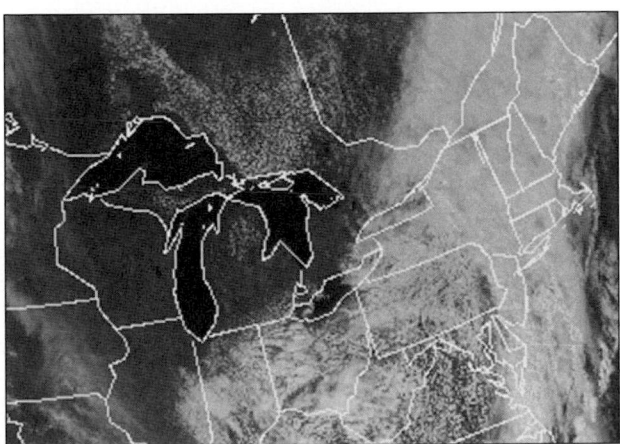

FIGURE 7.9
In this visible satellite image from 4 May 2011, clusters of cumulus clouds develop over the relatively warm land surfaces of northern Michigan and southern Canada, but not over the adjacent cooler surfaces of the western Great Lakes.

FIGURE 7.11
Lenticular clouds over Mt. Rainier, WA, on 2 February 2011. [Photo by Ken Smith]

wind is deflected into a wave-like pattern that can extend hundreds of kilometers downwind of the mountain crest (Figure 7.12).

If the air is sufficiently humid, lenticular clouds form at the crests of the waves and are absent in the troughs. Where air ascends within a wave, expansional cooling leads to condensation (or deposition) and clouds form. Where air descends within a wave, compressional warming leads to evaporation (or sublimation) and clouds dissipate. Clouds formed in this way directly over the mountain are known as **mountain-wave clouds**. (If these clouds extend just downwind of the mountain peak, they are referred to as *banner clouds*, as shown in Figure 7.13.) Lenticular clouds formed in the wave crests downwind of the mountain are known as **lee-wave clouds**.

Mountain-wave and lee-wave clouds are stationary because the wave itself is stationary, even though winds passing through the waves are strong. A stationary wave within the atmosphere is known as a *standing wave*. A waterfall on a river is a good analogy. In the plunge pool at the base of the waterfall, the water is very turbulent, but if we follow the river downstream, we quickly leave the turbulent region behind. The river water flows ceaselessly, but the turbulent segment remains stationary at the foot of

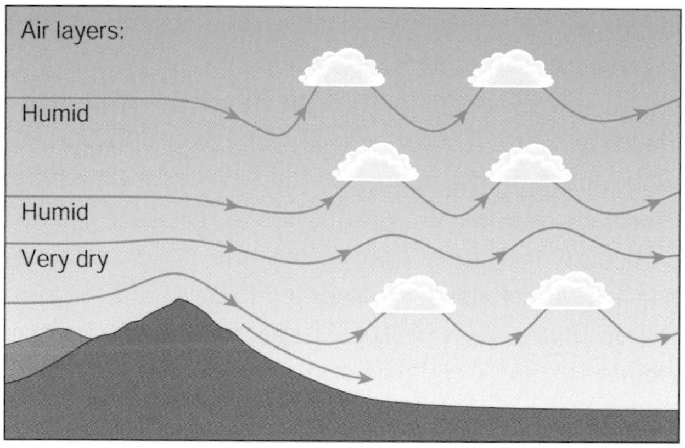

FIGURE 7.12
A mountain range deflects the horizontal wind into a wave-like pattern that extends downwind of the summit. If the air is sufficiently humid, clouds develop in the wave crests where ascending air expands and cools but clouds are absent in wave troughs where descending air is compressed and warms.

FIGURE 7.13
Banner cloud over Grand Teton in northwestern Wyoming. [Photo by John Spencer, Southwest Research Institute]

the waterfall, just as a standing wave in the atmosphere remains stationary to the lee of a mountain range.

In some cases, a rotor-like circulation develops under the crest of a standing wave. Air rotates about a horizontal axis that roughly parallels the mountain range and sometimes a cloud (known as a *rotor cloud*) forms in the ascending branch. Turbulence associated with the rotor can be a major hazard to aviation.

In a stable atmosphere, waves and wave-type clouds also can develop along the interface between two layers of air that are moving horizontally at different speeds. In Figure 7.14A, a layer of relatively warm air overlies a layer of colder, denser air. The warm air is moving horizontally more rapidly than the cold air. This vertical shear in the horizontal wind gives rise to *Kelvin-Helmholz waves* along the boundary between the two air layers. (Waves formed in this way are analogous to wind-driven water waves that form along the boundary between the ocean and atmosphere.) Clouds that develop at the crests of Kelvin-Helmholz waves consist of a train of regularly spaced curls and are known as *billow clouds* (Figure 7.14B).

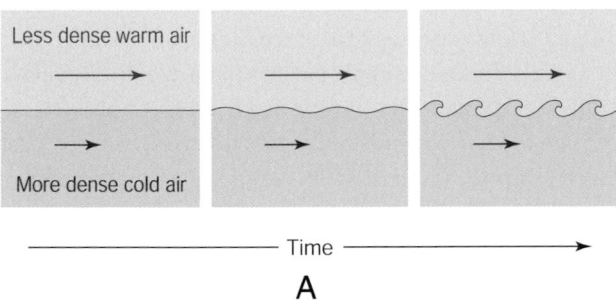

FIGURE 7.14
(A) Formation of Kelvin-Helmholtz waves along the interface between two layers of air that contrast in density. (B) Billow clouds may form in the wave crests. [Photo by Mila Zinkova/License: Creative Commons Attribution ShareAlike 2.5]

Waves traveling horizontally within the atmosphere are also responsible for patterns exhibited by stratocumulus and cirrocumulus (as well as by varieties of altocumulus in addition to lenticularis). These waves are not linked to mountain ranges and propagate at different altitudes within the troposphere. Such waves produce bands of clouds aligned either parallel or perpendicular to the wind direction.

Almost all water vapor is confined to the troposphere, so it is not surprising that most clouds occur below the tropopause. Mountain-wave clouds sometimes develop in the stratosphere above high mountain crests and occasionally the top of an intense cumulonimbus cloud surges through the tropopause and into the lower stratosphere. Other examples of clouds in the upper atmosphere are noctilucent clouds and nacreous clouds.

Somewhat mysterious are cirrostratus-like **noctilucent clouds** that form in the upper mesosphere above an altitude of about 80 km (50 mi). They are seldom observed, and then only at high latitudes (mostly above 50 degrees in both hemispheres) in summer. They are too thin to be visible during the day and only show up at night. From well below the horizon, the Sun illuminates noctilucent (*night-shining*) clouds, imparting to them a luminous blue-white or silvery white appearance that stands out against the black night sky (Figure 7.15). They are likely composed of tiny water-ice crystals formed on meteoric dust particles.

How is it possible for ice clouds to form in the upper atmosphere where there is so little water vapor (about 10 parts per million by volume)? The key is the exceptionally low air temperature in the upper mesosphere. As noted in Chapter 2, atmospheric temperatures are lowest at the mesopause. January temperatures generally range from −53 °C (−64 °F) to −43 °C (−46 °F) and July temperatures may drop as low as −173 °C (−279 °F). At such low temperatures, ice clouds can form at extremely low vapor pressures.

Nonetheless, this begs the question of where the water vapor comes from. Some is ejected into the upper atmosphere during violent volcanic eruptions. In fact, noctilucent clouds were first observed (in Germany and Russia) about two years after the violent 1883 eruption of the Indonesian volcano Krakatoa that blasted gases and ash to altitudes as great as 80 km (50 mi). The most important source of water vapor in the mesosphere may be the flux of water vapor from the lower atmosphere to the mesosphere associated with the meridional (north/south) summer circulation. Yet another source of water

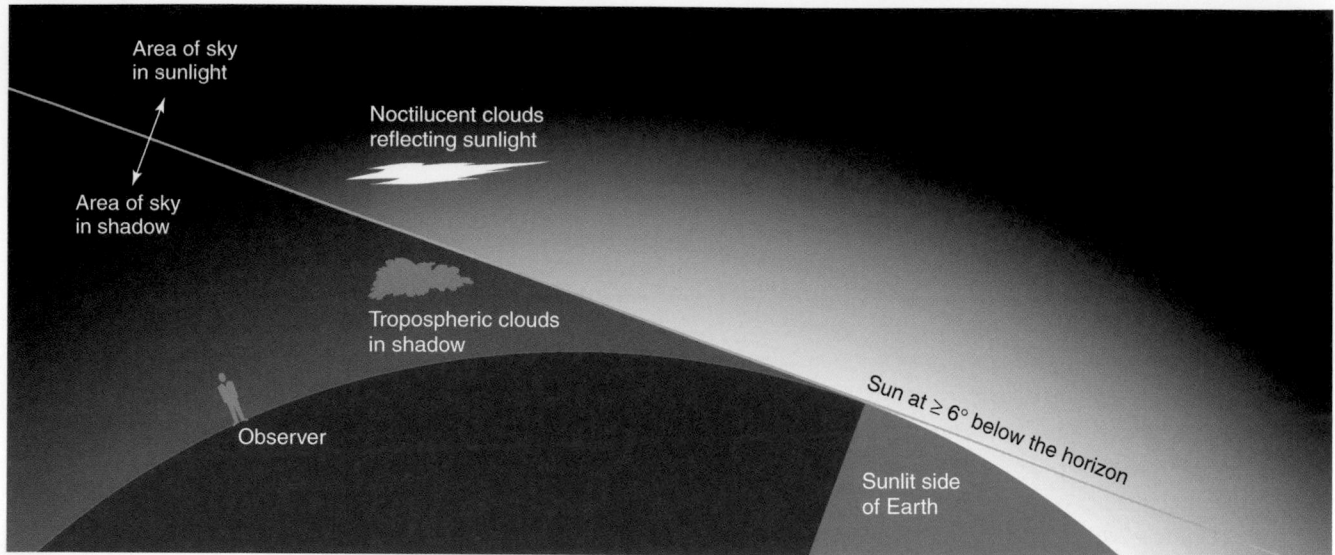

FIGURE 7.15
Noctilucent clouds are visible after sunset or prior to sunrise. These unusual clouds occur at such great altitudes (in the upper mesosphere) that they are illuminated by the rays of the Sun that is below the observer's horizon.

vapor may be chemical reactions involving methane (CH_4). Atmospheric circulation transports methane from the troposphere into the stratosphere where it is broken down by solar ultraviolet radiation, releasing hydrogen that is oxidized to water vapor through complex chemical reactions. Some of that water vapor is transported to the upper mesosphere and deposited as ice crystals on meteoric dust particles to form noctilucent clouds.

Curiously, the frequency of sightings of noctilucent clouds has been increasing since they were first reported in 1885. They are also being sighted at lower latitudes (as far south as Colorado). During the same period, the concentration of atmospheric methane has also been increasing (Chapter 3). G.E. Thomas of the University of Colorado's Laboratory for Atmospheric and Space Physics, suspects that the present upward trend in atmospheric methane and sightings of noctilucent clouds may be related.

Nacreous clouds, also known as *polar stratospheric clouds (PSC)*, occur in the stratosphere at altitudes of about 20 to 30 km (12 to 19 mi) where temperatures favor water in either the solid or supercooled state. These cirrus or altocumulus lenticularis clouds form on sulfuric acid nuclei possibly of volcanic origin. Because of their soft, pearly luster, these rarely seen clouds are also known as *mother-of-pearl clouds*. They are best viewed at high latitudes (e.g., Antarctica, Alaska, Scotland, Scandinavia) in winter when illuminated by the setting Sun. As noted in Chapter 3, these clouds probably play a major role in the formation of the Antarctic ozone hole.

Fog

Fog is a visibility-restricting suspension of tiny water droplets or ice crystals (called *ice fog*) in an air layer next to Earth's surface. Simply put, fog is a cloud in contact with the ground. By international convention, fog is defined as restricting visibility to 1000 m (3250 ft) or less; otherwise the suspension is called **mist**. (The popular definition of mist is *light drizzle*.) The presence of *dense fog* reduces visibility to 100 m (330 ft) or less. Fog may develop when air becomes saturated through radiational cooling, advective cooling, addition of water vapor, or expansional cooling.

In restricting visibility, fog adversely affects travel by motor vehicles, ship, and aircraft. Fog tends to be a major problem in some coastal areas. The world's foggiest land areas are Point Reyes, CA, and Argentia, Newfoundland. In both locations, fog occurs a little more than 200 days per year. Another exceptionally foggy place in the U.S. is Cape Disappointment, WA, at the mouth of the Columbia River, where dense fog occurs on average about 2556 hrs annually (30% of the time). Moose Peak Lighthouse on Maine's Mistake Island has about 1560 hrs of dense fog per year.

With a clear night sky, light winds, and an air mass that is humid near the ground and relatively dry aloft, radiational cooling may cause the air near the ground to approach saturation. When this condition occurs, a ground-level cloud forms and is called **radiation fog** (Figure 7.16). High humidity at low levels within the air mass is usually

FIGURE 7.16
Radiation fog seen here in the distance develops as a consequence of radiational cooling on a night when there is some air movement and the atmosphere is humid at low levels. [Photo by Matthias Süßen/License: Creative Commons ShareAlike 2.5}

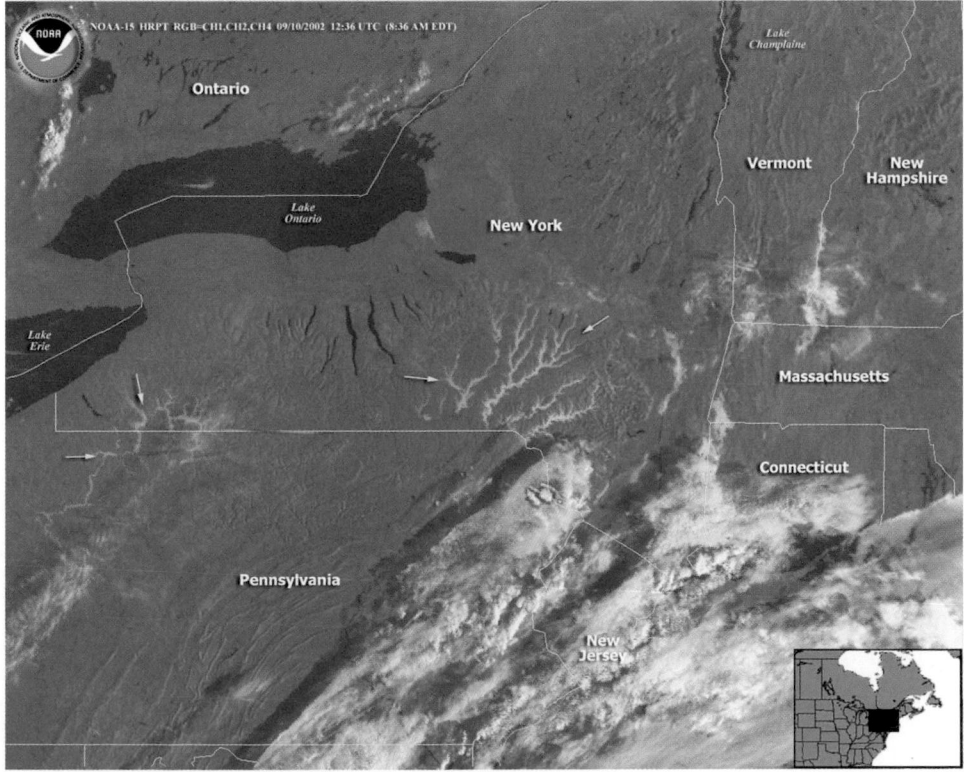

FIGURE 7.17
Visible satellite image showing morning fog in river valleys in parts of Pennsylvania and upstate New York (between the yellow arrows). [NOAA]

due to evaporation of water from a moist surface. Hence, radiation fog is most common over marshy areas or where the soil has been saturated by recent rainfall or snowmelt.

Light winds, rather than calm air, favor development of radiation fog. Light winds produce a slight mixing that more effectively transfers heat from the layer of humid air to the relatively cold ground. Consequently, the entire air layer is chilled below the dewpoint (Chapter 6). If air were calm, however, the lack of mixing would mean heat transfer by conduction alone. Because motionless air is a poor conductor of heat, only a very thin layer of air immediately in contact with the ground would be cooled to saturation. Hence, calm conditions favor dew or frost rather than radiation fog. On the other hand, if winds become too strong, the humid air at low levels mixes with the drier air aloft, the relative humidity drops, and radiation fog disperses or fails to develop. Air that is chilled by radiational cooling is relatively dense and, in hilly terrain, drains downslope and settles in low-lying areas such as river valleys and wetlands. Because of this air movement, known as *cold-air drainage*, hilltops may be clear of fog while fog is thick and persistent in deep valleys (Figure 7.17).

Once a fog bank forms at night, radiational cooling becomes a little more complicated. The top surface of the fog bank very efficiently emits infrared (IR) radiation skyward. Fog droplets absorb some of the IR emitted by the ground and the droplets, in turn, emit IR to the ground so that the ground cools more slowly. The maximum cooling rate shifts to the top of the fog bank, the fog layer thickens, and visibility continues to drop.

Often radiation fog lasts for only a few hours after sunrise. Fog gradually thins and disperses as saturated air at low altitudes mixes with drier air above the fog bank; the relative humidity drops and fog droplets vaporize. Mixing may be caused by convection triggered by solar heating of the ground or by strengthening of regional winds. In winter, however, the top surface of a fog layer readily reflects the weak rays of the Sun and radiation fog may linger. For example, in the valleys of the Great Basin and in California's San Joaquin Valley, winter radiation fogs may persist for many days to weeks at a time.

Fog sometimes accompanies air mass advection. The temperature and the water vapor concentration of an air mass depend on the nature of the surface over which the air mass forms and travels. As an air mass moves from one place to another, termed *air mass advection*, those characteristics change, partly because of the modifying influence of the surfaces over which the air mass travels. When the advecting air passes over a relatively cold surface, the air mass may be chilled to saturation in its lowest layers. This type of cooling is known as *advective cooling* and occurs, for example, in early spring in the northern states when mild, humid air flows over relatively cold, snow-covered ground. Snow on the ground may chill the overlying air to the dewpoint and fog develops. Fog formed by advective cooling is known as **advection fog**.

Advection fog also develops when warm and humid air streams over the relatively cold surface of a lake or the ocean. Persistent, dense fogs develop in this way over the Great Lakes in summer (Figure 7.18). Thick sea fog forms when mild maritime air from over the warm Gulf Stream flows northward and encounters the frigid waters of the Grand Banks off Newfoundland.

Nocturnal radiational cooling sometimes combines with air mass advection to produce fog. An initially dry air mass becomes more humid in its lower levels following a long trajectory over the open waters of a lake or the ocean. Over land areas, downwind from the water body, if the night sky is clear, the modified air mass is subjected to extreme radiational cooling and fog develops. San Francisco's famous fog forms in this way. Onshore winds transport cool, humid air into the city where nocturnal radiational cooling chills the air to saturation.

Steam fog (sometimes called *Arctic sea smoke*) is fog that develops in late fall or winter when extremely

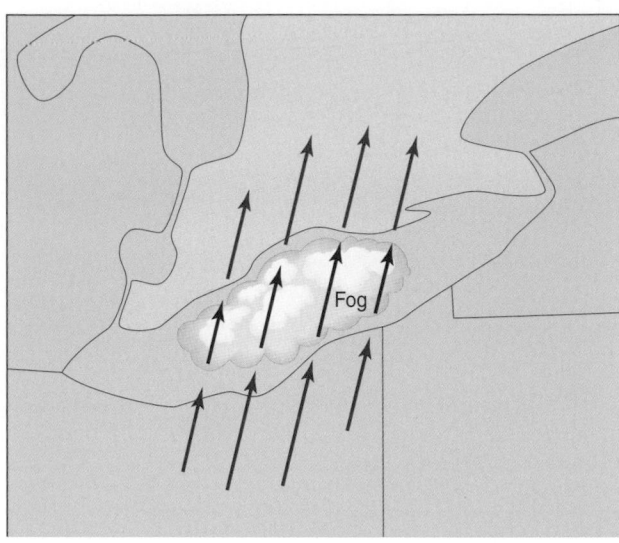

FIGURE 7.18
In summer, fog forms over the Great Lakes when relatively warm and humid air passes over the cold water surface and is chilled to its dewpoint.

cold and dry air flows over a large unfrozen body of water (Figure 7.19). Evaporation and sensible heating cause the lower portion of the air mass to become more humid and warmer than the air above. Heating from below destabilizes the air and the consequent mixing of mild, humid air with cold, dry air brings the air to saturation and fog forms. Because the air is destabilized, fog appears as rising streamers resembling smoke or steam. Steam fog also develops on a cold day over a heated outdoor swimming pool or hot tub and sometimes over a wet highway or field when the Sun comes out after a rain shower.

Steam fog over Lake Superior
Air temperature: 6 below zero

FIGURE 7.19
Steam fog at Presque Isle Park in Marquette, MI, on the shore of Lake Superior. [Photo by Donald Rolfson]

Another perspective on the origin of steam fog considers the distribution of vapor pressure over the water surface. Within about 1 to 2 m (3 to 7 ft) of the water surface, the vapor pressure is relatively high because of evaporation of water into the heated air. Above this shallow humid air layer, the air is colder and the vapor pressure is relatively low. A vertical vapor pressure gradient develops over the water body and, in response, water molecules stream (diffuse) upward. (As we will see in Chapter 8, gases move from high toward low pressure in response to a pressure gradient.) As the humid air streams upward, it expands and cools and the water vapor condenses into tiny droplets that we see as steam fog. Rising streamers encounter drier air and evaporate perhaps 5 to 10 m (15 to 35 ft) above the water surface.

Fog may develop on hillsides or mountain slopes as a consequence of the upslope movement of humid air. Ascending humid air undergoes expansional cooling and eventually reaches saturation. Any further ascent of the saturated air produces **upslope fog** (Figure 7.20). In the coastal mountain ranges of California, upslope fog sometimes overtops the range crest and spreads as stratus clouds over the leeward valleys. A cloud layer formed in this way is known as *high fog*.

Precipitation Processes

Clouds are no guarantee that it will rain or snow. Nimbostratus and cumulonimbus clouds produce the bulk of precipitation, but most clouds do not yield any significant rain or snow. This is because a special combination of circumstances is required for precipitation to develop. **Precipitation** is water in solid or liquid form that falls from clouds to Earth's surface under the influence of gravity. Key to understanding the mechanisms of precipitation formation is the concept of terminal velocity, the speed of a falling particle. We begin this section by examining terminal velocity and then cover the two mechanisms of precipitation formation: the collision-coalescence process and the Bergeron-Findeisen process.

TERMINAL VELOCITY
Terminal velocity is the constant speed attained by a particle falling through a motionless fluid such as water or air (Figure 7.21). The speed of a falling cloud droplet or ice crystal (or any other particle) in calm air is regulated by (1) **gravity**, the force that accelerates the particle directly downward toward Earth's surface, and (2) the resistance offered by the air through which the particle is falling. A downward accelerating particle meets increasing air

FIGURE 7.20
Upslope fog formed when humid air is forced to ascend the windward slopes of a hill or mountain. Expansional cooling brings the air to saturation. [Photo by Hyrum Wright, Brigham Young University]

resistance while gravity remains essentially constant. The magnitude of the upward resisting force eventually equals gravity; that is, the two opposing forces come into balance. When forces are balanced, the downward moving particle attains a constant speed. According to **Newton's first law of motion**, an object in constant straight-line motion or at rest remains that way unless acted upon by an unbalanced force. In this case, air resistance (directed upward) balances gravity (acting downward) and the particle continues moving downward at constant speed. That speed is the particle's terminal velocity.

Terminal velocity helps explain why the tiny particles that compose clouds (water droplets and ice crystals) will not fall to Earth's surface as precipitation unless they undergo considerable growth. Generally,

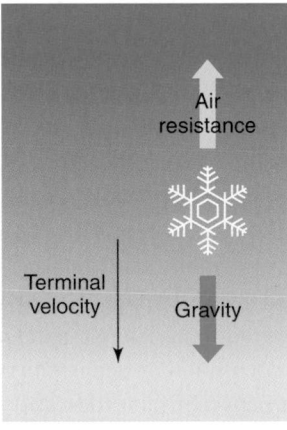

FIGURE 7.21
Terminal velocity is the constant downward-directed speed of a particle within a motionless fluid due to a balance between gravity (acting downward) and fluid (air) resistance (acting upward).

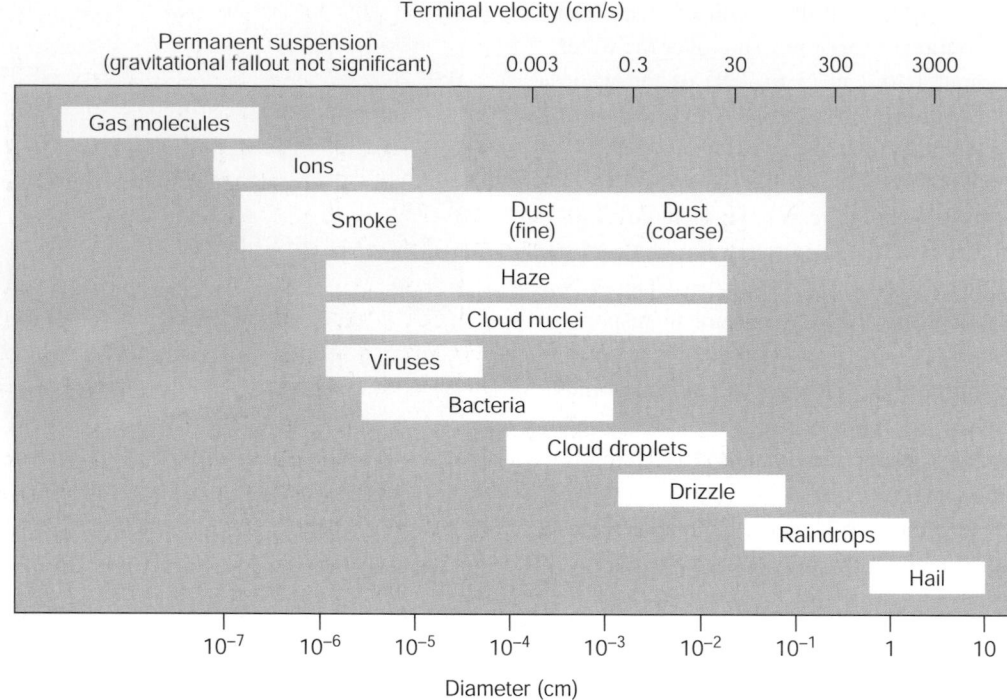

FIGURE 7.22
The terminal velocity of a particle falling through air increases with the size of the particle.

terminal velocity increases with the size of the particle (Figure 7.22). For a particle to descend from a cloud to Earth's surface as precipitation, the particle's terminal velocity must be greater than the rising motion of air within the cloud (updrafts). Cloud water droplets and ice crystals are so small (typically 10 to 20 micrometers in diameter) that their terminal velocities are very low (usually only 0.3 to 1.2 cm per sec). Hence, even weak updrafts will keep them suspended in clouds. If droplets or ice crystals fall through the base of a cloud, their terminal velocity is so slow that it would take 24 hrs or longer for the cloud particles to reach Earth's surface. Long before that happens, the particles would vaporize in the unsaturated air under the cloud. (The relative humidity of the air below cloud base is less than 100%.)

For clouds to precipitate, cloud particles must grow large enough that their terminal velocities overwhelm the updrafts. This is no minor task! It takes the water content of about 1 million cloud droplets (having diameters of 10 to 20 micrometers) to form a single raindrop (about 2 mm in diameter). Condensation or deposition alone cannot account for the formation of raindrop-sized particles. Cloud droplets grow into raindrops through collision-coalescence of cloud droplets in warm clouds whereas a combination of the Bergeron-Findeisen process and collision-coalescence causes ice crystals in cold clouds to grow into snowflakes.

WARM-CLOUD PRECIPITATION

In the 1940s, radar analysis of thunderstorms revealed that plentiful rain could fall from clouds that were too warm to contain ice. In a warm cloud (at temperatures above 0 °C), droplets may grow by colliding and coalescing (merging) with one another in the **collision-coalescence process**. This process takes place in a cloud composed of a mixture of droplets of different sizes, ideally with some droplets having diameters of at least 20 micrometers. Cloud droplets of unequal diameters have different terminal velocities, greatly increasing the likelihood that faster falling (larger) droplets will overtake, and then collide and coalesce with slower falling (smaller) droplets in their paths. Laboratory simulations demonstrate that colliding droplets will not coalesce unless they are significantly different sizes. Larger droplets form initially on larger nuclei (e.g., sea salt particles), through random collisions between smaller droplets, or via mixing between cloud droplets and the surrounding drier air.

Through repeated collisions and coalescence, droplets grow larger (Figure 7.23). On the other hand, cloud droplets of uniform diameter have essentially the same terminal velocities so that collisions are infrequent. Through collision-coalescence, a warm cloud droplet may grow so large that it becomes unstable and splits into smaller droplets that go on to coalesce with other droplets

FIGURE 7.23
Within a warm cloud, a relatively large droplet falls through a cloud of much smaller droplets. The larger droplet falls faster and collides with the smaller droplets in its path and grows by coalescence. This is the collision-coalescence process of precipitation formation.

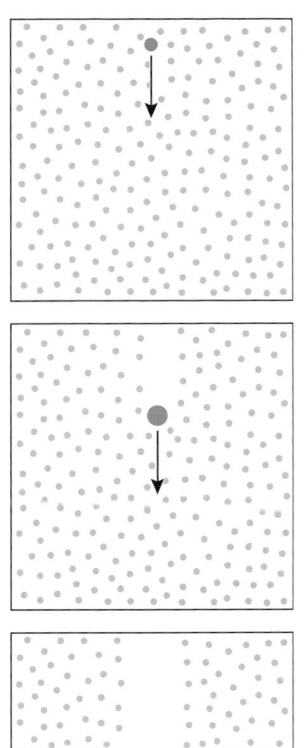

· Cloud droplets
● Rain drop

in a kind of chain reaction. Eventually, droplets become sufficiently large that they survive a fall from clouds to Earth's surface as raindrops.

A key factor in precipitation formation through collision-coalescence is collision efficiency. *Collision efficiency* refers to the fraction of all droplets in the path of a falling larger droplet that comes in contact with the larger droplet. Collision efficiency varies with the size of the larger droplet and is minimal (less than 10%) for droplets less than 40 micrometers in diameter but approaches 100% for droplets greater than 80 micrometers in diameter.

COLD-CLOUD PRECIPITATION

Although precipitation formation through the collision-coalescence process in warm clouds is important, especially in the tropics, most precipitation falling in middle and high latitudes originates in cold clouds (clouds or portions of clouds at temperatures below 0 °C) that contain some ice crystals. At middle and high latitudes, clouds routinely reach altitudes where the temperature is below freezing regardless of season. Precipitation is most

likely to fall from a cloud initially composed of a mixture of supercooled water droplets and ice crystals. In 1911, the German meteorologist Alfred Wegener (1880-1930), one of the original proponents of *continental drift*, first proposed that ice crystals grow at the expense of the surrounding supercooled water droplets in response to differences in saturation vapor pressure between water and ice surfaces. Around 1930, the Swedish meteorologist Tor Bergeron (1891-1977) contributed to an understanding of this process and the German physicist Walter Findeisen (1909-1945) subsequently elaborated on the concept. For this reason, the mechanism of precipitation formation in cold clouds is called the **Bergeron-Findeisen process**, also known as the *ice-crystal process*.

As noted above, cloud condensation nuclei are much more abundant than ice-forming nuclei and are more efficient in forming a cloud particle so that a mixed cloud (at least initially) consists of far more supercooled water droplets than ice crystals. Supercooled water droplets quickly vaporize as ice crystals grow. At subfreezing temperatures, water molecules more readily vaporize from a liquid (supercooled) water surface than from an ice surface because water molecules are more strongly bonded in the solid phase than the liquid phase. At the same subfreezing temperature, the saturation vapor pressure is greater over supercooled water than over ice. For temperatures below 0 °C (32 °F), two values of the saturation vapor pressure are given in Table 6.3, one over supercooled water, the other over ice.

Within a cloud composed of a mixture of ice crystals and supercooled water droplets, a vapor pressure that is saturated for water droplets is actually supersaturated for ice crystals at the same temperature. Suppose, for example, that a cloud of supercooled water droplets at −15 °C (5 °F) is saturated relative to water (i.e., relative humidity = 100%). According to Table 6.3, the saturation vapor pressure is 1.91 mb. Imagine the sudden appearance of an ice crystal in the same region of the cloud. Table 6.3 indicates that the saturation vapor pressure over ice at −15 °C (5 °F) is 1.65 mb so that air surrounding the ice crystals is supersaturated (relative humidity = 116%). Water vapor deposits on the ice crystals and the crystals grow. Deposition lowers the vapor pressure surrounding the water droplets to unsaturated conditions relative to the droplets, causing the droplets to evaporate. Also, a lower vapor pressure in the air surrounding the ice crystals produces a pressure gradient such that water vapor molecules diffuse toward the ice. Hence, through the Bergeron-Findeisen process, ice crystals grow at the expense of supercooled water droplets.

FIGURE 7.24
Within a cold cloud, ice crystals grow at the expense of supercooled water droplets. As they grow larger, ice crystals fall faster and collide with droplets and other ice crystals in their paths. Eventually, they may grow large enough to fall out of the cloud as snowflakes. This is the Bergeron-Findeisen process of precipitation formation.

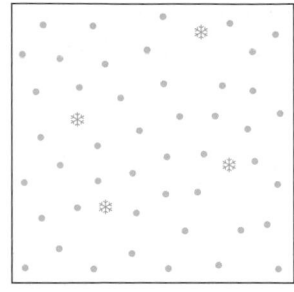

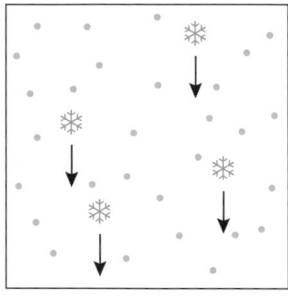

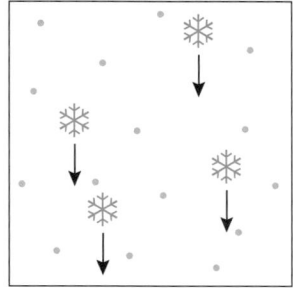

· Supercooled cloud droplets
※ Ice crystals

FIGURE 7.25
Much of the rain falling from the base of a distant thunderstorm evaporates in the relatively dry air beneath the cloud. The shaft of falling raindrops is known as virga.

Growth of ice crystals is accompanied by an increase in their terminal velocities. Larger, faster falling ice crystals overtake, collide and agglomerate (mass together) with smaller, slower falling ice crystals and supercooled water droplets (Figure 7.24). Collisions with droplets cause riming of ice crystals. Colliding supercooled droplets adhere to each other forming two or more ice crystals that grow still larger and fall out of the cloud base as snowflakes. If air temperatures are below freezing at least much of the way to the ground, ice crystals reach Earth's surface as snowflakes. If air temperatures are above freezing in the lower reaches of the cloud or between the cloud and Earth's surface, snowflakes melt into raindrops.

Regardless of the mechanism of formation, once a falling raindrop or a snowflake leaves the base of a cloud, it enters unsaturated air and begins to vaporize. The longer the distance to Earth's surface, (i.e., the higher the cloud base) and the lower the relative humidity of air beneath the clouds, the greater is the quantity of rain or snow that vaporizes. This is one of the reasons why highlands, being closer to the base of clouds, usually receive more precipitation than lowlands.

Viewed from a distance, a shaft of raindrops, ice particles, or snowflakes falling from a cloud may appear as dark curtains or streaks against a bright background (Figure 7.25). **Virga** is the name given to such streaks when much or all the water or ice particles vaporize before reaching Earth's surface. This is a common sight in the Great Basin of the American West where much potential precipitation vaporizes while falling through the very dry air beneath cloud base.

For a discussion of efforts to increase precipitation by stimulating natural precipitation processes, refer to this chapter's second *For Further Exploration, Rainmaking.*

HOLES AND CANALS IN CLOUDS

Distinctive holes or canals that appear in an otherwise uniform midlevel cloud layer (Figure 7.26) are referred to as *hole-punch clouds* (circular voids) or *canal clouds* (linear voids). Holes in clouds are not unusual, numerous published reports and photographs of them have appeared since the early 1940s. Cloud physicist P.V. Hobbs (1936-2005) of the University of Washington attributed at least some holes to aircraft flying through thin cloud layers, most likely altocumulus clouds, composed of supercooled water droplets.

Holes are generated from inadvertent seeding of clouds by ice particles produced when aircraft fly through the cloud layer. Spontaneous freezing of cloud droplets occurs through sudden adiabatic expansion and cooling of air as it flows behind the aircraft propeller tips or over the upper surfaces of jet aircraft wings. Injection of ice particles into a cloud of supercooled water droplets triggers the Bergeron-Findeisen precipitation process whereby ice crystals, which begin forming at cloud temperatures of

FIGURE 7.26
(A) Hole punch cloud and (B) canal clouds generated through inadvertent seeding of clouds by ice particles. Aircraft ascending or descending through a cloud layer can create holes; aircraft moving horizontally through a cloud layer can create canals. [Top photo by Lin Chambers/NASA. Bottom photo by Dollsworth]

−10 °C to −20 °C (14 °F to −4 °F), grow at the expense of the supercooled water droplets (as described in the previous section). Snow develops and falls from the cloud as streamers, creating a void. The release of latent heat accompanying the growth of ice crystals adds to buoyancy and invigorates updrafts, further adding to ice crystal growth that expands holes and canals. Aircraft ascending or descending through a cloud generates nearly circular holes whereas aircraft flying horizontally through a cloud layer creates canals. Formation of holes and canals may enhance precipitation near major airports.

In the 1 July 2011 issue of *Science*, Andrew J. Heymsfield of the National Center for Atmospheric Research (NCAR) in Boulder, CO, and colleagues reported on aircraft-induced holes and canals in clouds over Texas and adjacent states on 29 January 2007. At an altitude near 7700 m (25,300 ft), cloud thickness of 150 m (490 ft), and cloud top temperature of −30 °C (−22 °F), cloud particles were mostly liquid. Some of the 92 voids observed on GOES satellite imagery had lengths exceeding 100 km (62 mi) and persisted for four hours or longer. Researchers identified 75% of the hole-producing aircraft, ranging from large passenger jets and turboprops to propeller/piston aircraft.

Forms of Precipitation

Within the atmosphere, some water vapor condenses or deposits forming clouds, and through the collision-coalescence and Bergeron-Findeisen processes, some clouds precipitate. Depending upon environmental conditions, precipitation occurs in different forms. Included are liquid precipitation (rain, drizzle), freezing precipitation (freezing rain, freezing drizzle), and frozen precipitation (snow, snow pellets, snow grains, ice pellets, hail).

Rain consists of liquid water drops that fall mostly from nimbostratus and cumulonimbus clouds. Air resistance causes raindrops to fall as slightly flattened spheres (not teardrops as popularly portrayed) with diameters generally in the range of 0.5 to 6 mm (0.02 to 0.2 in.). At greater diameters, drops are unstable and break up into smaller drops. (Imagine what life would be like if raindrops grew to the size of basketballs!) Water's surface tension keeps small drops together, but once a critical size is reached, surface tension is incapable of maintaining the drop. (*Surface tension* is the attraction between molecules at or near the surface of a liquid.) Research meteorologists sampling warm-cloud raindrops over Hilo, HI discovered an exception to this rule. They reported raindrops as large as 8 mm (0.3 in.) in diameter (about the size of a pea).

In this case, strong steady updrafts in convective clouds prolonged the period of collision-coalescence, thereby enabling warm drops to grow to extraordinary sizes. At middle and high latitudes, raindrops usually begin as snowflakes (or sometimes would-be hailstones) that melt and coalesce as they descend through air at temperatures above 0 °C (32 °F).

Drizzle consists of liquid water drops having diameters between 0.2 and 0.5 mm (0.01 and 0.02 in.) that drift very slowly toward Earth's surface. Drizzle originates mostly in stratus clouds that are so low and thin that droplets undergo only limited growth by collision-coalescence. Drizzle often occurs with fog and contributes to poor visibility.

Snow is an agglomeration of ice crystals, in the form of flakes, that fall from clouds through the atmosphere to Earth's surface. Snowflakes can be single crystals but commonly are conglomerations and crystals can be rimmed. The shape and size of snowflake crystals vary widely depending upon the changing temperature and humidity in the portion of the cloud where they form and grow, at the expense of supercooled water droplets. (Snowflakes are not frozen raindrops.) The most common shape is a fern-like branching star called a dendrite but ice crystals also occur as needles, plates, and columns (Figure 7.27). The smallest snow crystals are mostly hexagonal prisms, but as crystals grow, branches extend outward from the corners producing more complex shapes.

All snowflake crystals exhibit hexagonal (six-fold) symmetry deriving from the hexagonal lattice formed by the water molecules in an ice crystal (Figure 7.28). The hexagonal symmetry of snowflake crystals captured the interest of Chinese scholars, keen on understanding nature and mathematics, as far back as 135 BCE. In 1611, the German mathematician Johannes Kepler (1571-1630) authored a booklet in which he unsuccessfully attempted to explain why snowflakes always have six-fold symmetry. It was not until three centuries later and the development of X-ray crystallography that ice-crystal symmetry was explained in terms of the ice lattice.

No two snowflakes follow the exact same path through a cloud, where their growth is very sensitive to small changes in temperature and water vapor, creating a huge variety of snowflake shapes (especially among large complex snowflakes). However, this does not preclude the possibility of two identical ice crystals occurring in a cold cloud, such as experienced during a research flight over Wausau, WI, in 1988. Nancy Knight (1922-2011) of the National Center for Atmospheric Research (NCAR) in Boulder, CO, serendipitously found two identical ice

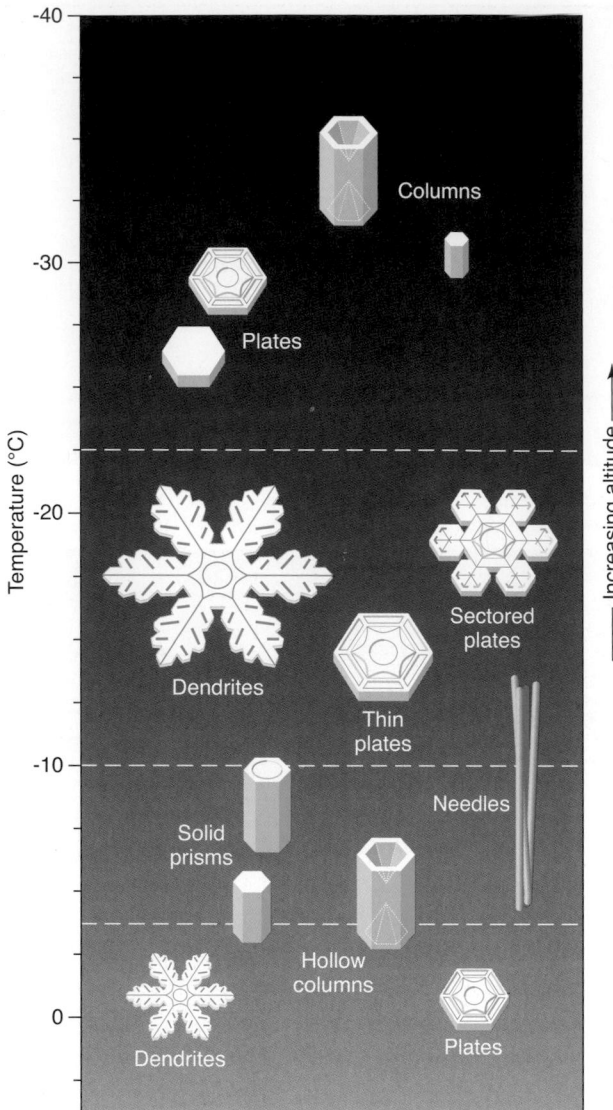

FIGURE 7.27
Snowflakes take on a variety of forms depending to a certain extent on cloud temperature.

crystals while collecting samples from an aircraft at an altitude of 6000 m (20,000 ft). Both crystals were hollow hexagonal prisms.

Snowflake size depends in part on the availability of water vapor during crystal growth. At very low air temperatures, water vapor concentration is relatively low so snowflakes are small. Collision efficiency, as flakes fall through the atmosphere, also affects snowflake size. At air temperatures near or slightly above freezing, snowflakes more readily stick together (agglomerate) after colliding. Their aggregate diameters have been known to reach 5 to 10 cm (2 to 4 in.) or greater. Appearance of such large flakes may indicate that snow is about to turn to rain.

FIGURE 7.28
A physical model of the crystal lattice of ice. Each water molecule is bound tightly to its neighbors but intermolecular bonds are elastic so that molecules vibrate about fixed locations in the lattice. For this reason, an ice cube retains its shape. The ordered arrangement of water molecules in the crystal lattice is responsible for the hexagonal structure of ice crystals. The internal framework of ice is an open network of water molecules, so that the molecules in ice crystals are not as closely packed as a similar number of molecules in liquid water. This property causes ice to be less dense than liquid water.

Snow pellets and snow grains are related to snowflakes. **Snow pellets** (formerly called *soft hail* or *graupel*) are soft conical or spherical particles of ice having diameters of 2 to 5 mm (0.08 to 0.2 in.). They form when supercooled cloud droplets collide and freeze on an ice crystal. **Snow grains** (also known as *granular snow*) are flat or elongated opaque white particles of ice, usually less than 1.0 mm (0.04 in.) in diameter, that originate in much the same way as drizzle except they freeze prior to reaching the ground.

Is it ever too cold to snow? When is it too warm to snow? For answers to these questions, see this chapter's third *For Further Exploration*.

Ice pellets, commonly called **sleet**, are spherical or irregularly shaped transparent or translucent particles of ice that are 5 mm (0.2 in.) or less in diameter. Ice pellets form when snowflakes partially or completely melt as they fall through above-freezing air beneath cloud base (Figure 7.29). These raindrops or partially melted snowflakes then fall into a relatively thick layer of subfreezing air where they refreeze into ice particles prior to striking the ground.

Freezing rain (or *freezing drizzle*) consists of rain (or drizzle) drops that become supercooled, and at least partially freeze on contact with cold surfaces (at subfreezing temperatures) to form a coating of ice (*glaze*) on roads, tree branches, and other exposed surfaces. Even a thin coating

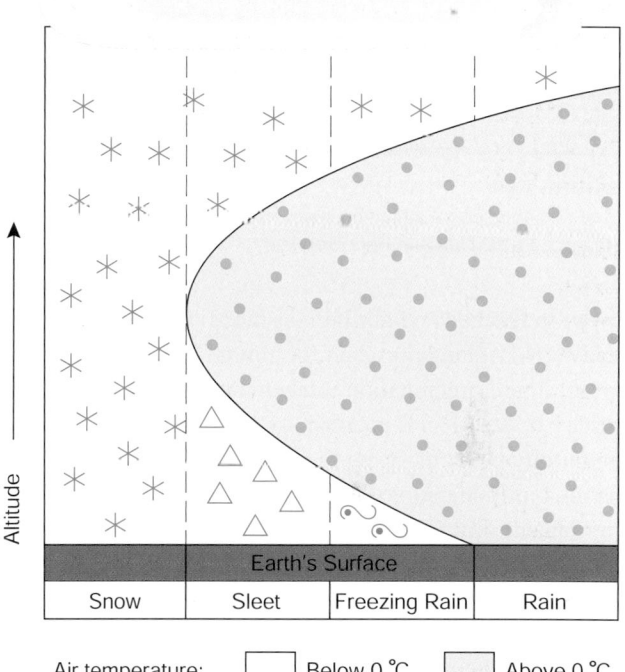

FIGURE 7.29
Schematic cross-section of the lower atmosphere showing temperature conditions for formation of frozen, freezing, and liquid forms of precipitation.

of ice can create hazardous driving and walking conditions. Sometimes glaze grows so thick that its weight brings down tree limbs (Figure 7.30) and snaps power lines. Freezing rain develops in much the same way as sleet except that the layer of subfreezing air at Earth's surface is shallower (Figure 7.31). If surface temperatures are close to freezing, the supercooled drops partially freeze, producing a wet and very slippery surface. We can readily distinguish ice pellets from freezing rain: ice pellets bounce when striking the ground whereas freezing raindrops do not bounce. The ground must be at or below freezing for freezing rain to occur, while snow and ice pellets may accumulate on surfaces at air temperatures above 0 °C.

Freezing rain usually occurs during a relatively brief interval (several hours) between episodes of snow and ordinary rain. Exceptionally long periods of freezing rain can occur, however. In January 1998, a freezing rain

FIGURE 7.30
Freezing rain accumulation breaking tree branches in Shenandoah National Park. [National Park Service]

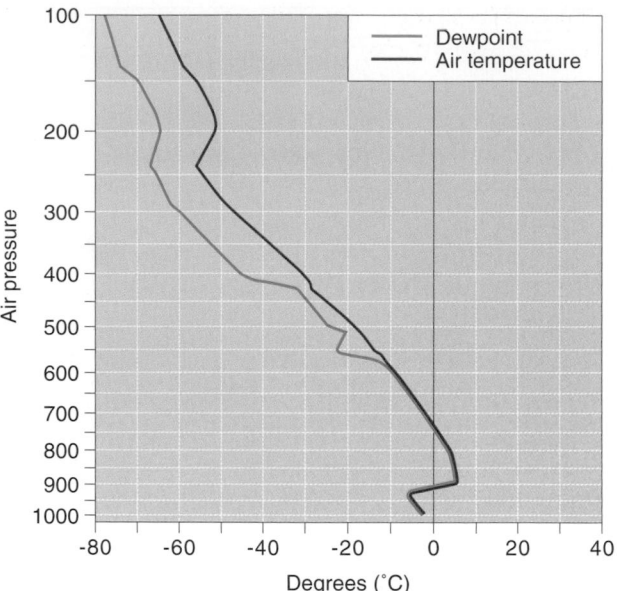

FIGURE 7.31
Radiosonde soundings of temperature and dewpoint indicating conditions favorable for freezing rain. Frozen precipitation particles melt in the layer from about 750 to 900 mb, becomes supercooled droplets below 900 mb, and freeze on ground surfaces.

event in Quebec and northern sections of New York State and New England persisted for almost five days. In many areas, liquid precipitation totaled about 10 cm (4 in.) and turned to ice. Thick coatings of ice brought down trees and utility poles blocking roads and cutting off power for about 4 million people (1 million in the U.S. and 3 million in Canada). For Canada, this was the nation's most costly natural disaster with insurance companies paying out about $1 billion.

Freezing rain events can be highly localized and persistent where the terrain is hilly or mountainous such as the Appalachians of Pennsylvania and West Virginia. Memorable freezing rain events have also occurred in the Columbia Gorge near Portland, OR. Cold, dense air drains into valley bottoms. Rain that falls from warm air aloft into this shallow layer of subfreezing air becomes supercooled and forms a glaze on cold surfaces. This situation often persists until the wind shifts and flushes the subfreezing air out of the valley bottom.

Glaze can develop on a road surface even without supercooling of raindrops. A prolonged period of extremely cold winter weather will chill a road surface to temperatures well below freezing. A sudden shift in weather pattern causes the air temperature to rise above the freezing point while the road surface temperature remains below freezing. Rain (not supercooled) freezes on the road surface forming a glaze that is just as hazardous as if it were produced by freezing rain or drizzle.

Hail consists of balls or jagged lumps of ice, often characterized by concentric internal layering resembling the structure of an onion (Figure 7.32). Layering is related to the various moisture collection environments that the hail encounters as it travels through the cloud.

Hail forms within intense thunderstorms (cumulonimbus clouds) characterized by vigorous updrafts, an abundant supply of supercooled water droplets, and great vertical cloud development. Updrafts transport ice pellets into the middle and upper portions of the cumulonimbus cloud. Along the way, ice pellets grow by collecting supercooled water droplets, and eventually become too heavy to be

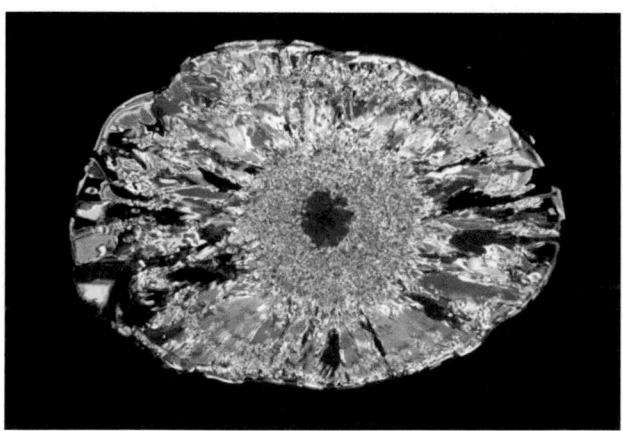

FIGURE 7.32
A hailstone consists of internal concentric layers of clear and opaque ice as shown in this cross section photographed in polarized light. [National Center for Atmospheric Research/ University Corporation for Atmospheric Research/National Science Foundation]

supported by updrafts. Ice pellets then descend through the cloud, exit the cloud base, and enter air that is typically above the freezing point. Pellets begin melting, but if large enough initially, some ice will survive the journey to Earth's surface as *hailstones*.

Most hailstones are harmless granules of ice less than 1.0 cm in diameter, but violent thunderstorms may spawn destructive hailstones the size of golf balls or larger. In most of the U.S., large hail having a diameter of 0.75 in. (2 cm) or greater is one of the criteria whereby NOAA's National Weather Service designates a thunderstorm as severe. (In the nation's Central Region, a hailstone diameter of 1.0 in. (2.5 cm.) is the threshold for a severe thunderstorm.) Hail is usually a spring or summer phenomenon that is particularly devastating for crops. There is more about hail and severe thunderstorms in Chapter 11.

Acid Deposition

As described in Chapter 6, the atmospheric sub-cycle of the global water cycle purifies water. That is, when water evaporates (or ice sublimates), dissolved and suspended materials are left behind. Cloud particles and precipitation are slightly acidic because they naturally dissolve some atmospheric carbon dioxide, producing weak carbonic acid (H_2CO_3). However, where the atmosphere contains elevated levels of sulfur dioxide (SO_2) or nitrogen oxides (NO_x), common combustion products of fossil fuels, cloud and fog particles and precipitation may become significantly more acidic than normal with adverse effects on ecosystems (especially freshwater types).

Acid deposition, the focus of this section, refers to the delivery of acidic particles, usually sulfuric acid (H_2SO_4) and nitric acid (HNO_3), to Earth's surface. With acid deposition, a distinction is made between dry and wet deposition. With *dry deposition*, particulates and gases gravitationally settle to the ground in dry form and then react with water (especially during dew and frost formation) increasing its acidity. With *wet deposition*, airborne particles participate in a complex set of chemical reactions involving sunlight, water vapor, and other gases, converting to tiny droplets of dilute sulfuric and nitric acid. These acidic particles affect atmospheric chemistry by slowly settling to Earth's surface, adhering to cloud particles or fog droplets (*acid fog*), or acting as cloud condensation nuclei (CCN). Should precipitation develop, acids dissolve in precipitation or are swept up by it increasing the acidity possibly to 200 times more than normal. Furthermore, in the absence of precipitation,

sulfuric acid droplets convert to acidic aerosols (sulfate particles) that reduce visibility and cause human health problems when inhaled.

An **acid** is a hydrogen-containing compound that releases positively charged hydrogen ions (H^+) when dissolved in water. Strong acids more readily release hydrogen ions than weak acids. An **alkaline substance** (or *base*) releases negatively charged hydroxyl ions (OH^-) when dissolved in water and may also be weak or strong. Pure water has properties of both acids and bases as water molecules (H_2O) continually break up (into hydrogen and hydroxyl ions) and re-form. That is,

$$H_2O \leftrightharpoons H^{+1} + OH^{-1}$$

The acidity of a substance is expressed as pH, a measure of its hydrogen ion concentration. On the **pH scale**, the pH increases from 0 to 14 as the hydrogen ion concentration decreases (Figure 7.33). A pH of 7 is neutral with an equal number of H^{+1} and OH^{-1} ion while a pH above 7 is increasingly alkaline and a pH below 7 is increasingly acidic. The pH scale is logarithmic, each unit increment corresponding to a tenfold change in acidity or alkalinity. Hence, for example, a two-unit drop in pH (e.g., from 5.0 to 3.0) represents a hundredfold (10×10) increase in acidity.

The normal pH of rainwater saturated with CO_2 is 5.6 and any rain (or *snow*) having a pH below this value is designated **acid rain** (or *acid snow*). Back in the 1960s, ecologist Gene E. Likens of the *Cary Institute of Ecosystem Studies*, Millbrook, NY, was among the first to sound the alarm regarding acid rain in the United States. Likens and colleagues reported an increase in the acidity of rainfall over the eastern U.S. between 1955 and 1973. Their findings were later confirmed and updated by measurements made by the National Atmospheric Deposition Program in the U.S. and by the Canadian Network for Sampling Precipitation. Rain and snow were most acidic in the northeastern United States and adjacent portions of Canada, downwind of major stationary sources of acid deposition precursors (e.g., power plant emissions in the Ohio River Valley).

Acid rain (and *acid fog*) has been implicated, along with other air pollutants (especially ozone), in the decline and dieback of coniferous forests in West Germany and in the Appalachian Mountains from North Carolina to New England. Acid rain leaches minerals (e.g., calcium and potassium) from the soil, decreasing the nutrients necessary for plant growth. Furthermore, acid rain mobilizes aluminum (Al) in the soil, and trees take up

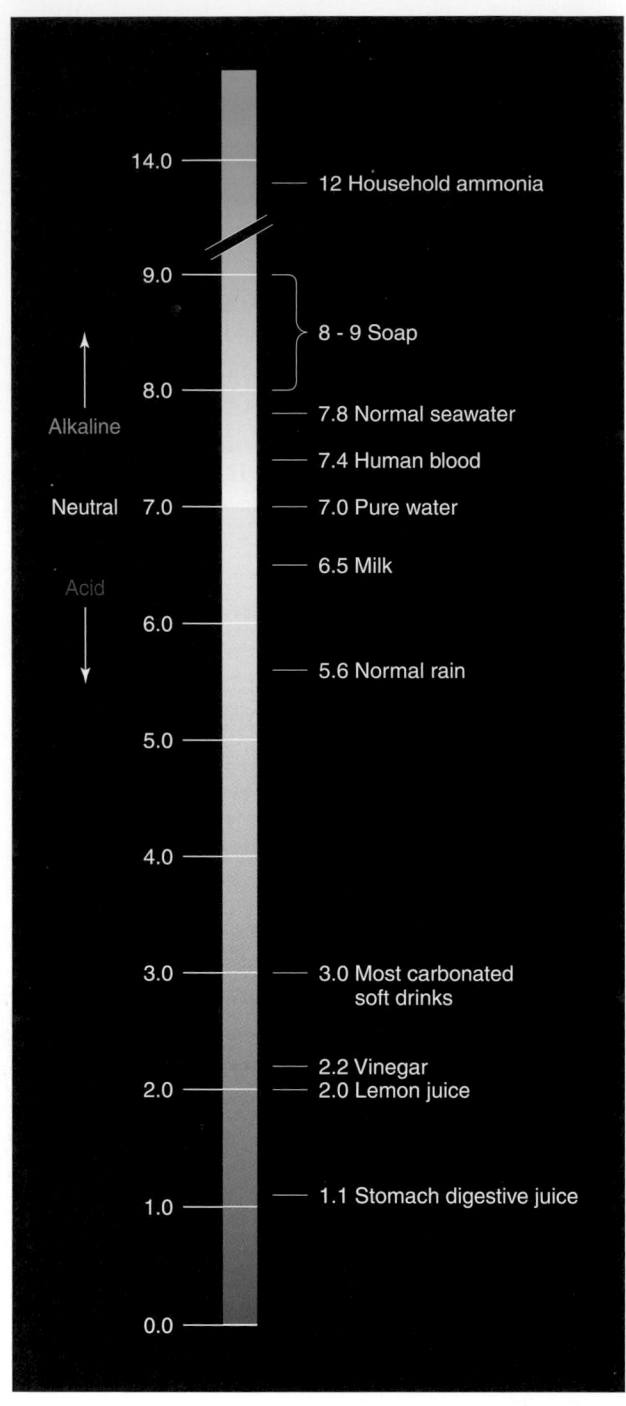

FIGURE 7.33

Acidity and alkalinity are expressed in terms of pH, a measure of the hydrogen ion concentration in water. The pH increases from 0 to 14 as the hydrogen ion concentration decreases. On the pH scale, pure water has a pH of 7, which is considered neutral; a pH above 7 is increasingly alkaline and a pH below 7 is increasingly acidic. Normal rain water (in equilibrium with atmospheric carbon dioxide) has a pH of 5.6.

this harmful chemical substitute in the place of calcium. The acceleration in the weathering of building materials, especially limestone, marble, and concrete, is caused by the increased acidity in rain, as is the faster corrosion of metals. Where soils are thin and the bedrock is non-carbonate (neither limestone nor dolostone), acid rains and snowmelt are not neutralized (buffered) and adversely impact waterways and aquatic ecosystems. Acid drainage lowers the pH of lakes and streams to levels that disrupt the reproductive cycles of fish. Increased acidity caused the decline or elimination of fish populations in lakes and streams in Norway, Sweden, eastern Canada, and the northeastern United States. As waters became more acidic, biodiversity declined.

In response to the acid deposition problem, the U.S. Congress created the *Acid Rain Program (ARP)* under Title IV of the *1990 Clean Air Act Amendments* in 1990. The goal of this legislation was to significantly lower emissions of sulfur dioxide (SO_2) and nitrogen oxides (NO_x), the principal precursors of acid deposition, emitted as byproducts of mostly coal-fired power plants. According to the U.S. Environmental Protection Agency, *Acid Rain and Related Programs, 2007 Progress Report,* the electric power industry accounted for about 69% of annual SO_2 emissions and 20% of total annual NO_x released to the atmosphere.

The Acid Rain Program imposed a declining annual cap on emissions of SO_2 and NO_x from electric generating units in the contiguous U.S. Since its inception, the ARP has significantly reduced emissions of acid deposition precursors. By 2009, SO_2 emissions had dropped 64% compared to 1990 levels, and at 5.7 million tons, was well below the final annual emissions cap of 8.95 set for compliance in 2010. From 1995 to 2009, annual NO_x emissions declined by about 4.1 million tons, a net reduction of 67%. The 2 million tons of NO_x emitted from sources targeted by ARP in 2009 was 6.1 million tons less than the level projected for 2000 without ARP, or more than triple the required ARP NO_x emissions reduction objective.

Even as the electric power industry worked to achieve full compliance with the provisions of the ARP, signs of recovery appeared in many acidified lakes, ponds and streams in New England, the Adirondack Mountains of New York State, and the northern Appalachian Plateau. As waters become less acidic, life is returning to waterways once considered dead. Aquatic organisms such as algae, plankton and even fish are returning to many Adirondack lakes that had been clear and lifeless.

Weather Radar: Locating Precipitation

As noted in Chapter 2, **radar** (acronym for *ra*dio *d*etection *a*nd *r*anging) is a valuable remote sensing tool for determining the location, movement, and intensity of areas of precipitation. Radar emits microwave signals and receives reflected signals from targets as its beam continually scans a large volume of the lower atmosphere. Today's weather radar can detect a tornado's circulation as it develops within its parent thunderstorm cloud, the spiral bands of a hurricane before they sweep onshore, and microbursts that can be hazardous to aircraft. Weather radar also monitors rainfall rates and provides cumulative rainfall totals that may forewarn of flash flooding.

As of this writing, a network of 160 *WSR-88D* weather radar units monitors conditions over much of the nation, with 122 operated by the National Weather Service, and 38 by the Department of Defense and the Federal Aviation Administration. *WSR-88D* radar operates in either the reflectivity or velocity mode. In the *reflectivity mode*, the radar signal detects areas of precipitation whereas in the *velocity mode*, the radar determines air motions directly toward or away from the radar associated with the circulation within a weather system.

REFLECTIVITY MODE

The *WSR-88D* weather radar emits short pulses of microwave energy having wavelengths of 10.0 to 11.1 cm (Figure 7.34). At these wavelengths, radar signals are reflected (or more precisely, scattered) by rain, snow, or hail but not significantly by the tiny water droplets or ice crystals that compose clouds. That is, weather radar is designed to detect precipitation-size particles but not the parent clouds. Falling precipitation particles reflect some of the radar signal back to a receiving unit where it is electronically processed and displayed on a computer screen as a **radar echo**.

Even when no precipitation is falling, the *WSR-88D* weather radar can detect boundaries within the atmosphere. Operating in the very sensitive *clear-air mode*, radar signals detect dust particles or swarms of insects that tend to collect along boundaries within air masses (e.g., an outflow boundary from a distant thunderstorm). Birds feeding on the insects also produce radar echoes and thereby make the radar display of boundaries more pronounced. These boundaries may be important because they are potential sites for future development of thunderstorms (Chapter 11).

The weather radar sending and receiving unit is a dish-type antenna housed in a radome (Figure 7.35). A *radome* is a spherical structure composed of fiberglass that protects the radar antenna from wind and weather but is transparent to radar signals. Pulsed radar signals are sent out and received hundreds of times each second as the antenna sweeps out a spiraling volume scan of the sky every 5 minutes. The radar combines several single up and down elevation scans (usually ranging from 0.5° to 19.5° above the horizon) with a 360-degree sweep in the horizontal direction to complete one volume scan. The product is a map of radar echoes representing the precipitation distribution (if any) in the region surrounding the radar unit. The speed of the radar pulse is known so that the elapsed time between emission and reception of a radar signal can be calibrated to give the distance to the precipitation.

The strength (intensity) of a radar echo depends on the reflectivity of the targeted precipitation particles with hailstones and large raindrops exhibiting the greatest reflectivity. The concentration of raindrops along the

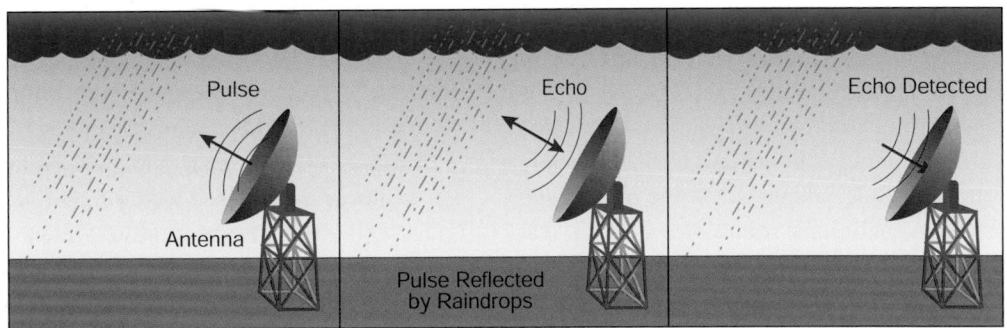

FIGURE 7.34
Operating in the reflectivity mode, weather radar continually sends out pulses of microwave energy that are scattered by raindrops, hailstones, and snowflakes back to a receiving unit that processes and displays radar echoes on a viewing screen.

FIGURE 7.35
This radome for a *WSR-88D* radar unit houses a rotating radar dish antenna. [NOAA/NWS Miami Weather Forecast Office]

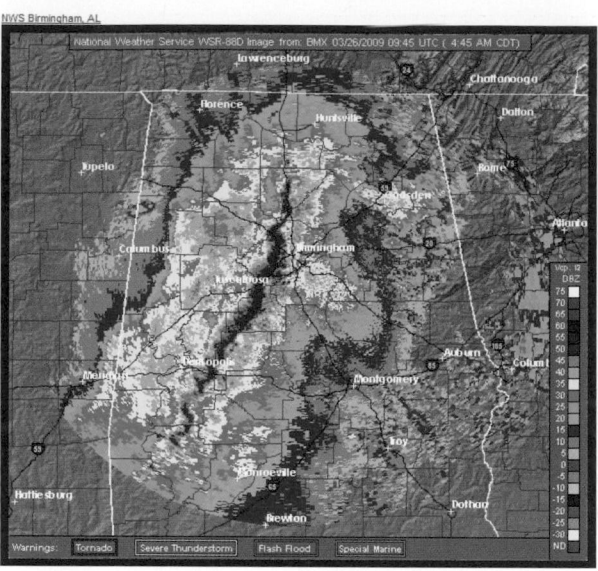

FIGURE 7.36
A sample radar reflectivity product (at 0945 Z 26 March 2009 from Birmingham, AL) in which echo intensity is graduated by color.

path of the radar beam also affects echo intensity but to a lesser extent. Echo intensity is an index of rainfall rate and can be used to identify severe thunderstorm cells, which often contain large hail. Radar systems electronically display echo intensity via a color scale. By one convention (often seen on televised weathercasts), red indicates very heavy rain (or hail) and, at the other end of the scale, light green (or sometimes blue) signifies very light rain (Figure 7.36). Other color schemes are sometimes used to portray snow (e.g., white) or a mixture of rain and snow (e.g., pink). As described in more detail later in this chapter, weather radar can also produce maps of rainfall totals over a specified time period.

Not all radar echoes are caused by meteorological phenomena. Nearby objects fixed on the ground, such as buildings or smokestacks, also reflect radar signals; such radar echoes constitute **ground clutter**. A ground clutter pattern produced in this way is unique to a particular radar site, appears all the time, and hence is readily distinguished from precipitation echoes. Typically, this constant ground clutter echo pattern is electronically subtracted from reflectivity displays. On the other hand, special atmospheric conditions sometimes develop that give rise to ground clutter that may be mistaken for precipitation. For example, a strong temperature inversion may cause the outgoing radar signal to bend downward so that it strikes the ground. Such ground clutter differs from precipitation echoes in that it is stationary, stronger, and grainier in appearance. Often, these *anomalous*

propagation patterns appear just prior to sunrise when a strong low-level temperature inversion is present.

Sometimes a radar screen shows a broad area of echoes, but weather stations located in the same area report no precipitation. In this case, the radar is detecting echoes from rain and/or snow that falls from clouds but vaporizes completely in the relatively dry air beneath cloud base. (Recall our earlier discussion of *virga*.)

VELOCITY (DOPPLER) MODE

WSR-88D is the first weather radar to operate in both velocity and reflectivity modes. In the velocity mode, the radar system utilizes the Doppler effect to determine horizontal air motions within a weather system based upon the detected motions of precipitation particles assumed to be tracers of the wind. The Doppler effect is named for Johann Christian Doppler (1803-1853), the Austrian physicist who was the first to explain the phenomenon in 1842. For this reason, radar with velocity-detection capability is often referred to as a **Doppler radar**. The Doppler effect is also employed in devices that measure the speed of a pitched ball or motor vehicle traffic.

The **Doppler effect** refers to a shift in the frequency of sound waves or electromagnetic waves emanating from a moving source. For example, the pitch (frequency) of a train whistle sounds higher as a train approaches and drops off as the train pulls away. As shown schematically in Figure 7.37A, the crests of sound waves radiate outward from a stationary source as evenly spaced

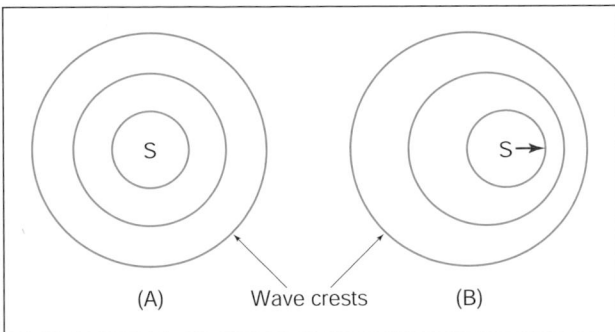

FIGURE 7.37
The Doppler effect refers to the shift in the frequency of sound or electromagnetic waves that accompanies the motion of the wave source(s) or wave receiver. (A) A sound wave source is stationary and wave frequency is uniform everywhere. (B) The wave source (e.g., train whistle) is in motion so that wave frequency is greater ahead of the source than behind the source.

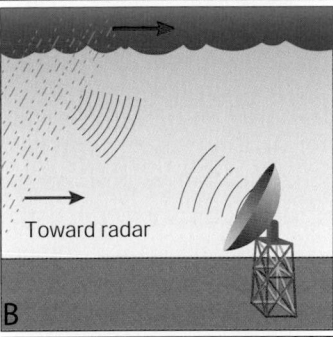

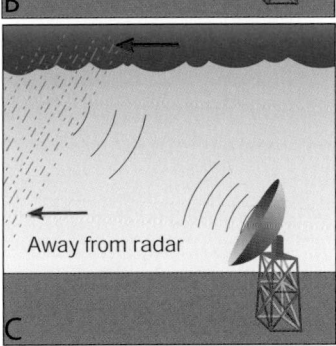

FIGURE 7.38
Doppler radar operating in three different weather situations. (A) With a stationary rain shower, the frequency of the returned signal is the same as that of the emitted echo. (B) If the rain shower approaches the radar, the frequency of the return echo is greater than that of the emitted signal. (C) If the rain shower moves away from the radar, the frequency of the return echo is less than that of the emitted signal. [Adapted from SAM, NOAA/Environmental Research Laboratories/Forecast Systems Laboratory]

concentric circles; that is, the wave frequency is uniform. If the source is moving, however, as in Figure 7.37B, wave crests become more closely spaced in the direction in which the source is moving so that the frequency is higher ahead of the source but lower behind the source.

Doppler radar monitors the motion of precipitation particles moving directly toward or away from the radar antenna. The frequency of the radar signal shifts slightly between emission and the return signal (*echo*) and this frequency shift is calibrated in terms of the motion of the particle (Figure 7.38). No frequency shift occurs for components of particle motion that are perpendicular to the radar beam; that is, Doppler radar cannot detect air motion at right angles to the radar beam. When air is moving in directions other than directly toward or away from the path of the radar signal, only the component of motion toward or away from the radar is detectable. Doppler radar displays are color-coded so that greens and blues (cold colors) indicate motion directly toward the radar and reds and yellows (warm colors) indicate motion directly away from the radar (Figure 7.39).

Doppler radar enables meteorologists to better provide the public with advance warning of hazardous weather including severe thunderstorms, tornadoes, and potentially flooding rains. The geographical range of radar is limited by the curvature of the Earth; that is, radar signals travel along downward-bending paths somewhat less curved than the underlying Earth's surface. Radar sees well beyond the visual horizon but its beam gains altitude with increasing distance from the radar antenna so that when operating in the reflectivity mode, the radar's maximum effective range is about 460 km (285 mi). For other reasons when operating in the velocity (Doppler)

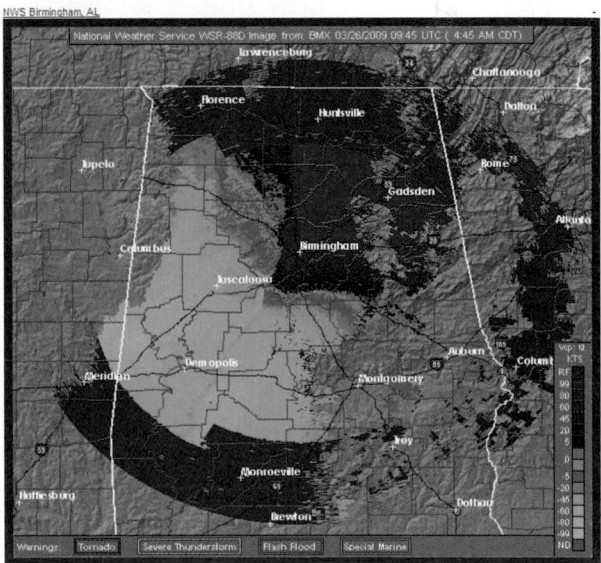

FIGURE 7.39
A sample Doppler weather radar image (at the same time as Figure 7.36). Greens and blues (cold colors) indicate motion directly toward the radar whereas reds and yellows (warm colors) indicate motion directly away from the radar.

mode, the radar's maximum range decreases to about 230 km (143 mi).

DUAL POLARIZATION WEATHER RADAR

The entire fleet of *WSR-88D* radars is undergoing a major upgrade to dual-polarization technology that is expected to be complete in 2013. This technology provides weather forecasters with more precise information on precipitation type, accumulation, and threat identification, enabling them to more accurately diagnose severe weather. Current *WSR-88D* radars send out a horizontal electromagnetic wave in one direction, giving a horizontal measurement of an object such as a raindrop. Dual polarization refers to electromagnetic waves transmitted in two dimensions, both horizontal and vertical. As the radar receives both reflected waves, a computer program separates the waves into horizontal and vertical components, thus providing a 2-D picture of the size and shape of an object, as well as the distribution of many objects (e.g., raindrop size distribution). Dual-polarization radar has many advantages over traditional radar. For example,

it allows forecasters to clearly identify rain, hail, snow, or ice pellets within a radar display, and filter out non-meteorological targets (e.g., birds, bugs, ground targets). It can detect tornado debris, giving meteorologists a high degree of confidence that a radar-detected tornado is actually on the ground. For example, on 29 February 2011, dual-polarization radar at Springfield, MO, provided a 25 min. lead time on the Branson, MO, tornado, double the national average. The radar also helps greatly improve rainfall estimation. The Newport/Morehead City, NC, dual-polarization radar detected never before seen details in the landfall of Hurricane Irene on 27 August 2011. Preliminary reports show that the dual-polarization products estimated Irene's total precipitation better than radar reflectivity alone. Figure 7.40 displays dual-polarization weather products from Irene's landfall. The reflectivity display (A) and velocity display (B) are the same as traditional radar. A new differential reflectivity product (C) displays more intense convection northeast of the hurricane's eye. The specific differential phase product (D) pinpoints regions of heaviest rain within the hurricane.

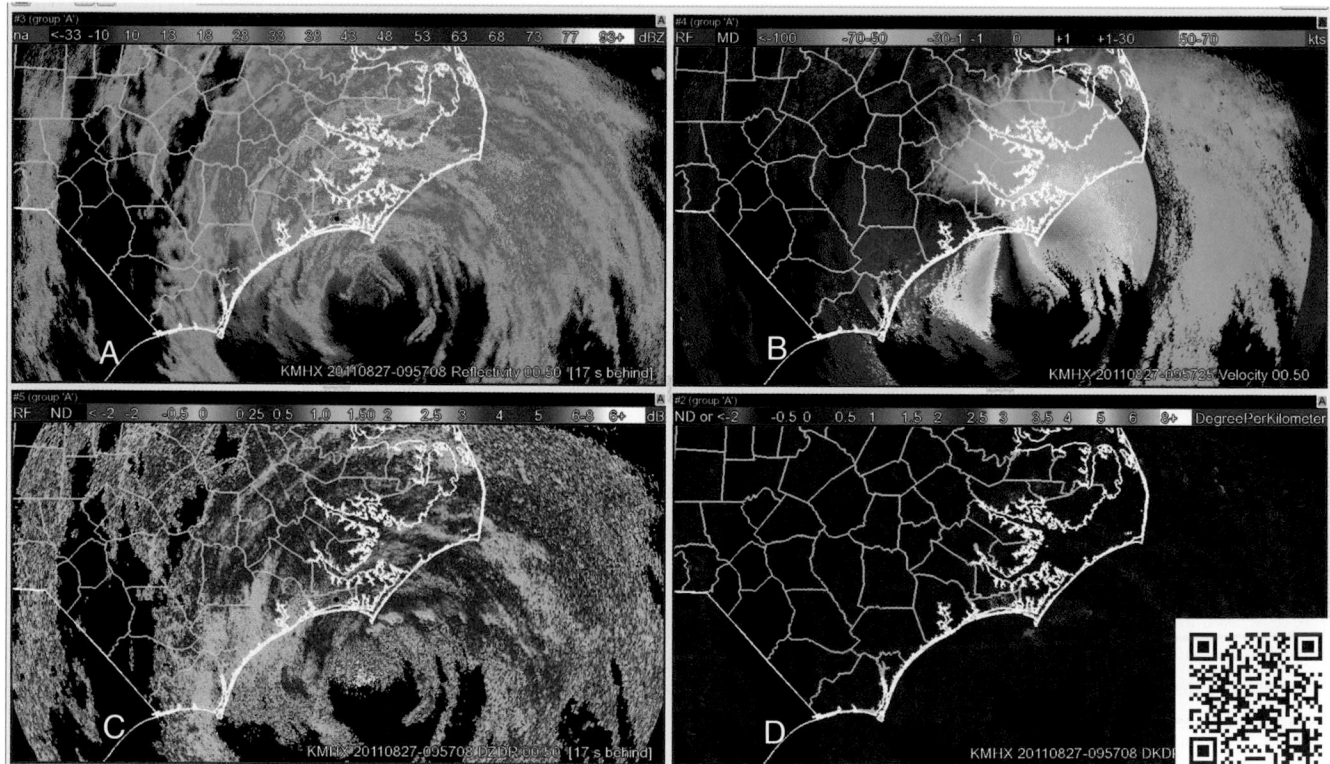

FIGURE 7.40

A dual-polarization weather radar product display from Hurricane Irene's landfall on the Outer Banks of North Carolina on 27 August 2011. (A) Reflectivity and (B) velocity remain an important part of the product package. In the velocity display, the blues and yellows represent greatest motion towards and away from the radar, respectively. (C) The dual-polarization Differential Reflectivity product shows the most intense convective cells (where rain and winds are maximized) located northeast of the eye as green to yellow/red shadings. (D) The dual-polarization Specific Differential Phase product pinpoints the area of heaviest rains as pink areas primarily to the north and northeast of the eye and in Irene's outer bands. [NOAA/NWS]

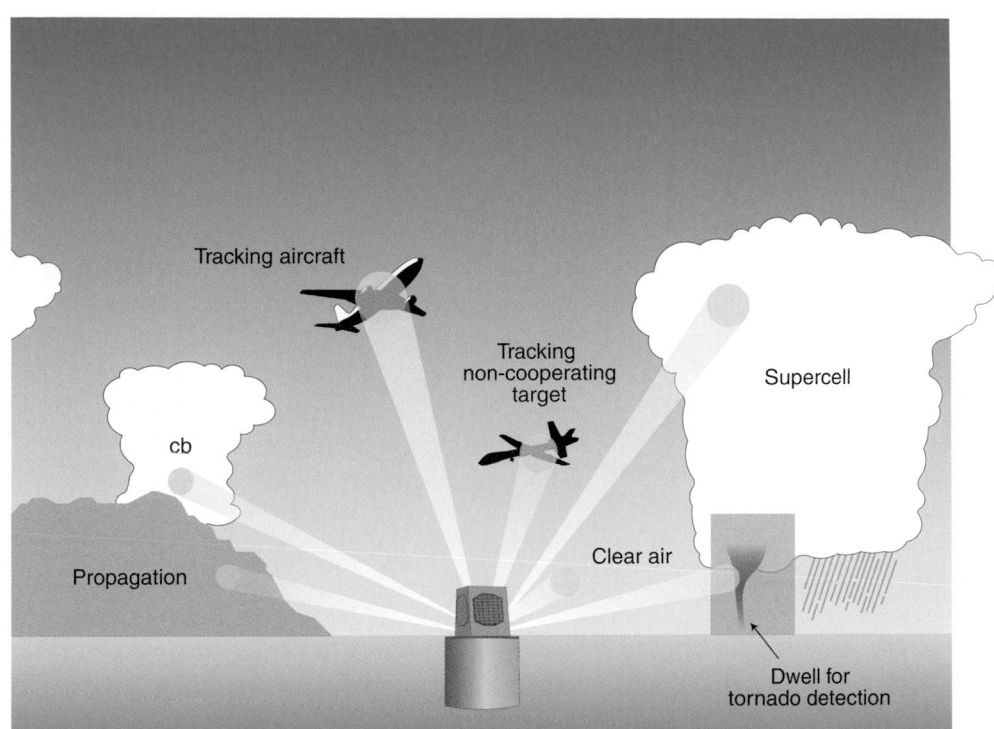

FIGURE 7.41
Diagram of phased-array weather radar, showing from left: full-volume continuous scan through a developing cumulonimbus cloud, (cb) full-volume continuous scan through the planetary boundary layer (clear air) for mapping winds, detection and tracking of aircraft including non-cooperative targets, full-volume continuous scan through a supercell storm, and long-dwell scan through a region of a potential tornado. [Adapted from NOAA/National Severe Storms Laboratory]

PHASED ARRAY WEATHER RADAR

The long-term future of weather radar lies in *phased array weather radar* (Figure 7.41), which uses multiple beams sent out simultaneously. The entire atmosphere is scanned much faster (about 30 seconds instead of the current 5 -7 minutes) and the radar has the ability to quickly focus on a severe weather feature, such as a possible tornado. In addition, phased array weather radar provides a more detailed examination of storm evolution and will help scientists refine conceptual and numerical weather models. Researchers predict that phased array weather radar could double lead times for severe weather warnings and improve icing forecasts for aviation.

Measuring Precipitation

Rainfall and snowfall are routinely measured in terms of depth of accumulation over a specified interval of time, usually hourly and every 6 hrs and 24 hrs. Measurements are made directly by collecting a volume of rain or snow in a rain or snow gauge or estimated remotely by weather radar or satellite sensors.

RAIN AND SNOW GAUGES

Today precipitation is collected and measured essentially in the same way as it was more than 2000 years ago: a container open to the sky. The first reference to a rain gauge appears in the Indian manuscript *Arthasastra*, authored by Kautila (also known as Vishnugupta Chanakya) sometime in the 4th century BCE. The gauge consisted of a 45-cm (18-in.) diameter bowl. Kautila, India's Chancellor of Exchequor, formulated a plan for taxing lands based on local rainfall. Rain gauges were reported in Palestine in the 1st century CE. The first snow gauges appeared in China in 1287 and by the middle of the 15th century rain gauges were introduced in Korea. The first reported use of a rain gauge in Europe was in 1639 by the Italian Benedetto Castelli (1577-1644).

A standard non-recording **rain gauge** design consists of a cylinder equipped with a cone-shaped funnel at the top (Figure 7.42). The funnel directs rainwater into a narrower cylinder seated inside the larger outer cylinder. The funnel and narrow cylinder configuration magnifies the scale of measurement so that the instrument can resolve rainfall into increments as small as 0.01 in. (0.25 mm). A simple graduated stick is used to measure the depth of water that accumulates in the inner cylinder. (The stick is graduated in inches in the U.S. but in millimeters in Canada and most other nations.) Rainfall of less than 0.005 in. (0.1 mm) is recorded as a *trace*. Rainfall is usually measured at some fixed time once every 24 hrs and the gauge is then emptied. This type of rain gauge must be read manually and many of the NWS Cooperative Observing Stations across the U.S. still rely upon these gauges.

FIGURE 7.42
A standard National Weather Service non-recording rain gauge.

FIGURE 7.43
Weighing-bucket rain gauge recently installed as a component of the National Weather Service Automated Surface Observing System. The gauge is surrounded by a wind shield to provide greater accuracy in windy conditions, especially when the gauge is collecting snow.

Continuous monitoring of rainfall is often useful, especially in flood-prone areas. Accordingly, some rain gauges are designed to provide a cumulative record of rainfall through time, thereby yielding rainfall rates, an indicator of precipitation intensity. Two common designs are the weighing-bucket and tipping-bucket rain gauges. These recording rain gauges can be used in remote locations. A **weighing-bucket rain gauge** calibrates the weight of accumulating rainwater in terms of water depth. This instrument has a device that marks a chart on a clock-driven drum or sends an electronic signal to a computer for processing. At subfreezing temperatures, antifreeze in the collection bucket or a heater melts snow or ice pellets producing a cumulative record of melt water.

A **tipping-bucket rain gauge** consists of a free-swinging container partitioned into two compartments, each of which can collect the equivalent of 0.01 in. of rainfall. Each compartment alternately fills with water, tips and spills its contents, and trips an electric switch that either marks a chart on a clock-driven drum or sends an electrical pulse to a computer for recording. Compared to a weighing-bucket rain gauge, a tipping-

bucket rain gauge does not perform as well at subfreezing air temperatures or at very high rainfall rates. A heated tipping-bucket gauge was a standard component of the NWS Automated Surface Observing System (ASOS) before it was recently replaced by a weighing-bucket precipitation gauge (Figure 7.43). ASOS sites owned by the FAA still use the tipping-bucket rain gauge.

For snow, meteorologists and hydrologists are interested in (1) the depth of snow that falls during the period between observations, (2) the melt water equivalent of that snowfall, and (3) the depth of snow on the ground at observation time. New snowfall accumulates on a simple wooden board placed on top of the old snow cover. Using a ruler graduated in tenths of an inch (or a meter stick graduated in centimeters), snow depth is measured to the board. The board is then swept clean and moved to a new location. Snowfall measurements, reported to the nearest 0.1 in., are made at regularly scheduled times, typically once daily.

The melt-water equivalent of new snowfall can be determined by measurements from a weighing-bucket gauge or by melting the snow collected in a non-recording gauge (with the funnel and inner cylinder removed). The average density of fresh snow is about 8% of water (0.08 gram per cubic cm) so as a very general rule, 33 cm (13 in.) of fresh snow melts down to 2.5 cm (1.0 in.) of water. However, the actual snow/melt-water ratio varies considerably depending on the crystalline form of the snowflakes, and the temperature of the air through which the snow falls. Snow consisting of columnar crystals can be relatively dense whereas snow made up of dendritic crystals (typical of lake-effect snow) has relatively low density, with the ratio of snowfall depth to melt-water depth as great as 50 to 1. Wet snow falling at surface air temperatures at or above 0 °C (32 °F) has much greater water content than dry snow falling at very low surface air temperatures. The ratio of snow depth to melt-water depth may vary from 3 to 1 for very wet snow to 30 to 1 or higher for dry fluffy snow. As a snow cover ages, snow settles, air-filled space decreases, snow converts to tiny granules of ice, and snow density increases.

Snow depth represents an accumulation of any fresh snow plus the remnants of past snowfalls. The depth of snow on the ground is usually determined using a graduated rule inserted vertically into the snow at several representative locations. These measurements are averaged, rounded to the nearest inch, and reported daily. Care must be taken to avoid areas where the wind whipped the snow into drifts.

Rainfall and snowfall can vary considerably from one place to another, especially when produced in convective showers and thunderstorms. Precipitation gauges are sited so that measurements are as accurate and representative as possible. Instruments must be sheltered from strong winds, which cut the collection efficiency of rain gauges. Winds tend to accelerate over the open top of a gauge, reducing the amount of rain that enters the instrument. One study found that winds blowing at 16 km per hr (10 mph) reduced the amount of rainwater collected by 10% compared to the amount that would have accumulated in the gauge if the air were calm. Many rain gauges are equipped with wind shields to eliminate or at least reduce the influence of wind eddies on the catch. Such a shield may consist of attached free-hanging slats, placed around the orifice of the gauge (Figure 7.43). Precipitation gauges should not be placed in windswept locations, but must be sited well away from buildings and tall vegetation that might shield the instrument from precipitation. As a general rule, a gauge should be no closer to an obstacle than four times the height of the obstacle.

Traditionally, atmospheric scientists derive spatial patterns of atmospheric variables from measurements made by instruments at discrete points, that is, at weather stations. This approach works reasonably well for variables such as air temperature and pressure that vary continuously from one place to another. Hence, the temperature is likely to be near 65 °F at a location half way between two weather stations, one reporting 60 °F and the other reporting 70 °F. Basing spatial patterns on measurements at discrete points does not work as well for discontinuous variables such as precipitation. Often precipitation exhibits considerable spatial variability. Hence, it does not always follow that moderate amounts of rain fell at a location halfway between two weather stations, one reporting heavy rainfall and the other reporting light rainfall. Furthermore, regular precipitation measurements are missing from vast stretches of the ocean where significant amounts of precipitation fall. To help remedy this situation, scientists developed radar technologies and satellite-borne instruments that provide more continuous coverage of precipitation over broad areas.

REMOTE SENSING OF PRECIPITATION

The *WSR-88D* weather radar determines not only the location and intensity of rainfall but also provides estimates of the total amount of rainfall. Whereas networks of precipitation gauges measure rainfall at discrete locations often more than 100 km (60 mi) apart, radar measures rainfall within continuous volume scans of the atmosphere centered on the radar site. A special computer algorithm uses radar monitoring of rainfall rate to generate color-coded maps of rainfall totals over a specific period of time (Figure 7.44).

To estimate rainfall, the *WSR-88D* radar measures the reflectivity obtained from a volume scan that combines the lowest four tilt angles (0.5°, 1.5°, 2.5°, and 3.5° above the horizon) as frequently as every six minutes. Reflectivity is proportional to the *surface area* of raindrops whereas rainfall rate is proportional to the *volume* of the raindrops. Reflectivity data are converted to rainfall rates by estimating the raindrop size distribution and results are then checked against selected precipitation gauge measurements on the ground. If properly calibrated, radar can provide a better spatial representation of hourly precipitation amounts than networks of precipitation gauges. This guidance information is particularly valuable for flash flood forecasting (Chapter 11).

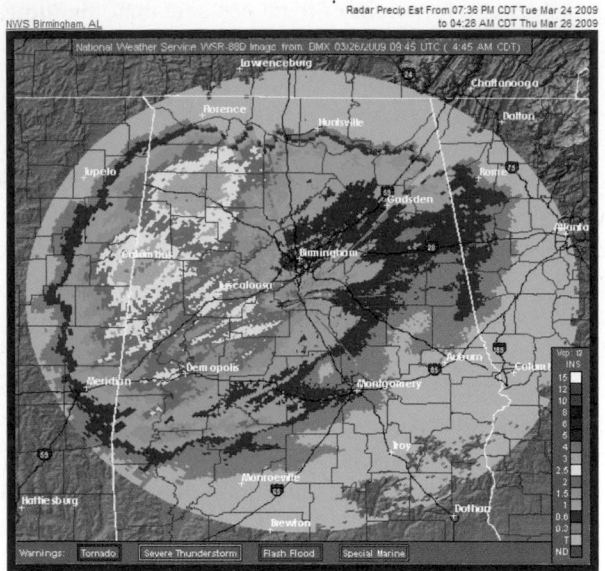

FIGURE 7.44
Cumulative rainfall (from 0036 Z 24 March 2009 to 0945 Z 26 March 2009) as determined by computer analysis of radar echoes.

Satellite sensors are also valuable remote sensing tools for precipitation. The *Tropical Rainfall Measuring Mission (TRMM)* utilizes satellite technology primarily to measure rainfall over the area between latitudes 40 degrees N and 40 degrees S. Knowledge of tropical rainfall and the associated flow of heat energy promises to improve our understanding of the global climate system and make possible the development of more realistic global climate models. About two-thirds of all precipitation falls in the tropics and much of the planetary-scale heat transport originates in the tropics. This satellite mission is also helping scientists detect the onset and evolution of El Niño and La Niña (Chapter 9). Launched in November 1997, TRMM is a joint venture of the National Aeronautics and Space Administration (NASA) and Japan's National Space Development Agency.

The TRMM satellite, in orbit about 350 km (215 mi) above Earth's surface, utilizes three sensors to measure rainfall remotely: radar, a microwave imager, and a visible/IR scanner. Satellite-based precipitation radar (PR) is similar to ground-based radar except that the signal has a much shorter wavelength. PR obtains vertical profiles of rain and snow from Earth's surface to an altitude of about 20,000 m (65,500 ft) and has a horizontal resolution on Earth's surface of 4 km (2.5 mi). The TRMM Microwave Imager (TMI) measures the minute flux of microwave energy emitted by the Earth-atmosphere system. The instrument can "see through"

clouds and measure sea-surface temperatures. Variations in the amount of energy received at different wavelengths are interpreted to measure water vapor, cloud water, and rainfall intensity. Similar sensors have been flown onboard polar-orbiting Defense Meteorological Satellites since 1987. The TRMM Visible and Infrared Scanner (VIRS) monitors five spectral channels from visible to infrared (between 0.63 and 12 micrometers). The height of convective clouds (determined from IR-derived cloud-top temperature) is the basis for rainfall estimates; that is, the higher the cloud, the greater the rainfall. The same type of precipitation-measuring scanner is used routinely on Polar Orbiting Environmental Satellites (POES) and Geostationary Operational Satellites (GOES) for weather analysis and forecasting. (Recall also our discussion of atmospheric rivers in the Case-in-Point of Chapter 6.)

Scientists also measure snow cover remotely, information that is valuable for water resource managers and flood forecasters. In mountainous areas where snowfall is considerable, the melt water equivalent of snow can be determined remotely using snow pillows. A *snow pillow* is a device that is filled with antifreeze solution and fitted with a manometer (an instrument that measures pressure changes) that is calibrated to give the water equivalent from the weight of the overlying snow cover. Data are radioed to a satellite and from there to a data center for downloading and analysis. The *Airborne Gamma Radiation Snow Survey Program* also provides remote sensing of the water content of snow packs. At designated times, low-flying aircraft measure terrestrial gamma ray emission along selected flight lines. Gamma radiation emitted by Earth materials attenuates as the snow pack thickens. Comparison of gamma emission over the same flight line with and without snow cover is calibrated in terms of the meltwater equivalent of snow (or soil moisture).

Conclusions

As air nears saturation, water vapor begins condensing or depositing on airborne nuclei. With continued cooling, usually due to expansion of rising air, clouds form. Clouds are distinguished on the basis of appearance (cirriform, stratiform, cumuliform), altitude of their base (high, middle, low, having significant vertical development), temperature (cold, warm), and composition (water droplets, ice crystals). Fog is a cloud in contact with Earth's surface and forms when air becomes saturated at low levels through radiational cooling, chilling of warm humid air as it moves over a relatively cold surface, cold air advection over a relatively warm, wet surface, or expansional cooling that accompanies upslope motion of air in mountainous terrain.

Cloud droplets and ice crystals are much too small to fall to Earth's surface as precipitation. Through the collision-coalescence process in warm clouds, relatively large droplets collide with smaller droplets and coalesce, creating droplets large enough to survive the fall to Earth's surface as precipitation. Through the Bergeron-Findeisen process in cold clouds, ice crystals grow by deposition and become large enough to fall. Precipitation occurs as rain, drizzle, snow, ice pellets (sleet), freezing rain (or freezing drizzle), and hail. Precipitation type depends upon cloud conditions and the temperature of the air column through which the precipitation falls to Earth's surface.

A variety of direct and remote sensing techniques are employed by scientists to monitor and measure precipitation. Weather radar sends out pulses of microwave energy that locate areas of precipitation and determine the intensity of rainfall. Operating in the Doppler mode, weather radar can monitor air motions within weather systems, making it possible to warn the public in advance of the potential for severe weather. Instruments onboard Earth-orbiting satellites are also used to estimate rainfall over tropical and subtropical latitudes.

Atmospheric circulation plays a key role in bringing air to saturation and triggering cloud development. The next five chapters cover the many atmospheric circulation systems and their associated weather. Chapter 8 begins with a discussion of the forces that drive and shape atmospheric circulation.

Basic Understandings

- Clouds, the visible product of condensation or deposition within the atmosphere, are composed of large numbers of tiny water droplets, ice crystals, or a combination of the two.
- As the relative humidity approaches 100%, condensation and deposition occur on nuclei, tiny solid and liquid particles suspended in the atmosphere. Cloud condensation nuclei are much more abundant than ice-forming nuclei (i.e., freezing nuclei and deposition nuclei). Most ice-forming nuclei are active at temperatures well below the freezing point.
- Many condensation nuclei are hygroscopic; that is, they have a chemical affinity for water and induce condensation at a relative humidity less than 100%.

- Clouds are classified by general appearance (cirriform, stratiform, cumuliform), altitude of base (high, middle, low, significant vertical development), temperature (cold, warm), and composition (water droplets, ice crystals).
- High, middle, and low clouds are produced by relatively gentle uplift of air over a broad geographical area. These clouds are layered, that is, stratiform. Clouds exhibiting significant vertical development are the consequence of more vigorous uplift in more restricted geographical areas and are heaped or puffy in appearance, that is, cumuliform.
- High clouds are composed of mainly ice crystals, whereas middle and low clouds are mostly water droplets or a mixture of ice crystals and supercooled water droplets.
- Nimbostratus and cumulonimbus are the principal precipitation-producing clouds. Precipitation from nimbostratus clouds is typically lighter and lasts longer than showery precipitation from cumulonimbus clouds.
- Atmospheric stability determines whether cumulus clouds build vertically into cumulus congestus or cumulonimbus clouds. The convective condensation level corresponds to the base of cumuliform clouds.
- Lee-wave clouds develop downwind of a prominent mountain range and are nearly stationary. Noctilucent and nacreous clouds are among the very few clouds that occur above the troposphere. Noctilucent clouds develop in the upper mesosphere and may be composed of ice deposited on meteoric dust particles. Nacreous clouds form in the upper stratosphere where temperatures favor water in either the solid or supercooled state. They are cirrus or altocumulus lenticularis that form on sulfuric acid nuclei possibly of volcanic origin.
- Fog is a visibility-restricting suspension of tiny water droplets or ice crystals (called ice fog) in an air layer next to Earth's surface. Simply put, fog is a cloud in contact with the ground. By international convention, fog is defined as restricting visibility to 1000 m (3250 ft) or less. Based on mode of origin, fog is classified as radiation fog, advection fog, steam fog, or upslope fog.
- Because of their relatively low terminal velocities, cloud droplets and ice crystals remain suspended in the atmosphere

indefinitely unless they vaporize or undergo significant growth.

- The collision-coalescence process and Bergeron-Findeisen process describe how cloud particles grow sufficiently large to counter updrafts and, under the influence of gravity, fall to Earth's surface as precipitation.

- Precipitation in warm clouds (temperatures above freezing) occurs by the collision-coalescence process, which requires the presence of relatively large cloud droplets that grow through collision and coalescence with smaller cloud droplets.

- The Bergeron-Findeisen process requires the coexistence of ice crystals, supercooled water droplets, and water vapor in cold clouds. At the same subfreezing temperature, the saturation vapor pressure surrounding a supercooled water droplet is higher than the saturation vapor pressure surrounding an ice crystal. Hence, air that is saturated for supercooled droplets is supersaturated for ice crystals. Ice crystals grow by deposition while water droplets vaporize.

- The bulk of precipitation that falls at middle and high latitudes originates in cold clouds composed of a mixture of ice crystals and supercooled water droplets. Once the precipitation process begins, collision and coalescence promote growth of cloud particles into precipitation particles.

- Principal forms of precipitation are rain, drizzle, snow, ice pellets (sleet), freezing rain, and hail. The form of precipitation depends on the source cloud and the temperature profile (sounding) of the air beneath the cloud.

- Most rain and snow fall from relatively thick clouds (nimbostratus or cumulonimbus) whereas drizzle falls from relatively thin clouds (stratus). Ice pellets are raindrops that freeze prior to reaching the ground, whereas freezing rain consists of supercooled drops that freeze on contact with surfaces at subfreezing temperatures. Hail is produced by intense thunderstorms having vigorous updrafts, an abundant supply of supercooled water droplets, and great vertical development.

- Cloud particles and precipitation are slightly acidic because they naturally dissolve some atmospheric carbon dioxide, producing weak carbonic acid. Acid deposition takes the form of dry deposition or wet deposition. Sulfuric acid and nitric acid are converted from sulfer dioxide and oxides of nitrogen, emitted as a byproduct of fossil fuel combustion, and dissolve in precipitation or are swept up by it increasing the acidity (lower pH).

- Weather radar is used to determine the location, intensity, and movement of areas of precipitation. Operating in the reflectivity mode, radar echo strength increases with precipitation intensity. Operating in the Doppler mode, weather radar can be used to determine the detailed motion of air within a weather system.

- Precipitation is measured directly by collecting a volume of rain or snow in a rain or snow gauge or estimated remotely using weather radar or sensors onboard Earth-orbiting satellites.

Enduring Ideas

- Clouds are the visible product of condensation or deposition of water vapor on nuclei suspended in the atmosphere. Clouds are classified on the basis of general appearance, altitude of cloud base, temperature, or composition. Fog is a cloud in contact with the ground and may develop via radiational cooling, advective cooling, addition of water vapor, or expansional cooling.
- Most clouds do not produce precipitation. Terminal velocity must exceed the cloud updraft for cloud particles to fall toward Earth's surface. Cloud particles undergo the necessary growth to fall as precipitation via the collision-coalescence process operating in warm clouds (temperatures greater than 0 °C or 32 °F) and the Bergeron-Findeisen process typically combined with collision-coalescence in cold clouds (temperatures less than 0 °C or 32 °F). Most precipitation in middle and high latitudes falls from cold clouds, and can be in liquid, freezing, or frozen form. Precipitation is measured directly by gauges and remotely by weather radar and sensors on Earth-orbiting satellites.
- The criterion for precipitation to be acidic is a pH lower than 5.6 (the normal pH for rainwater saturated with CO_2). Acid rain threatens aquatic ecosystems and forests and corrodes some building materials. A successful reduction in the electric power industry's emissions of the principal acid rain precursors (sulfur dioxide and oxides of nitrogen) has led to some recovery in acidified lakes and streams.

Key Terms

contrails	cumulus (Cu) clouds	rain
cloud	cumulus congestus cloud	drizzle
curvature effect	cumulonimbus (Cb) cloud	snow
nuclei	wind shear	snow pellets
cloud condensation nuclei	cloud streets	snow grains
(CCN)	altocumulus lenticularis clouds	ice pellets
ice-forming nuclei	mountain-wave clouds	sleet
hygroscopic nuclei	Kelvin-Helmholz waves	freezing rain
supercooling (of water)	noctilucent clouds	hail
cirriform cloud	nacreous clouds	acid deposition
stratiform cloud	fog	acid
cumuliform cloud	mist	alkaline substance
warm cloud	radiation fog	pH scale
cold cloud	cold air drainage	acid rain
cirrus (Ci) clouds	advection fog	radar
cirrostratus (Cs) clouds	steam fog	radar echo
cirrocumulus (Cc) clouds	upslope fog	ground clutter
altostratus (As) clouds	precipitation	Doppler radar
altocumulus (Ac) clouds	terminal velocity	Doppler effect
stratocumulus (Sc) clouds	gravity	rain gauge
stratus (St) clouds	Newton's first law of motion	weighing-bucket rain gauge
nimbostratus (Ns) clouds	collision-coalescence process	tipping-bucket rain gauge
convective condensation level	Bergeron-Findeisen process	
(CCL)	virga	

Review

1. What conditions are required for a cloud to form in the atmosphere?
2. What role is played by cloud condensation nuclei (CCN) in the formation of a cloud? Also, identify some of the natural sources of CCN.
3. What is the significance of hygroscopic nuclei in cloud formation?
4. Describe the general relationships among the altitude, temperature, and composition of stratiform clouds.
5. Explain how lee-wave clouds form downwind from a mountain range.
6. Identify the unique features of a noctilucent cloud.
7. Distinguish between atmospheric conditions responsible for development of radiation fog versus conditions that favor dew formation.
8. Describe the process involved in the formation of warm-cloud precipitation.
9. What is the role of supercooled water droplets in the formation of cold-cloud precipitation?
10. Distinguish between freezing rain and sleet (ice pellets).

Critical Thinking

1. How is it possible for cloud droplets to become supercooled?
2. Explain why rain is normally slightly acidic in a non-polluted atmosphere. Define *acid rain*.
3. How does weather radar operating in the reflectivity mode locate and plot the motion of areas of precipitation?
4. What is the value of weather radar operating in the Doppler mode in forecasting severe thunderstorms, including those that may spawn tornadoes?
5. What is the significance of the Schaefer point?
6. Identify the various atmospheric conditions that influence the height of the base of a cumuliform cloud. What controls the altitude to which a cumulonimbus cloud builds in the atmosphere?
7. Explain why cumuliform clouds are relatively rare over cold-water surfaces and snow-covered surfaces.
8. In hilly terrain, radiation fog is most common in low areas such as marshes and river valleys. Please explain why.
9. At the same sub-freezing temperature, the saturation vapor pressure surrounding a supercooled water droplet is greater than the saturation vapor pressure surrounding an ice crystal. Why the difference and why is this important?
10. What are the advantages of the new dual-polarization weather radar over conventional Doppler weather radar?

Clouds by Mixing

Outdoors on a cold winter day, you sometimes "see" your breath when you exhale. You are actually seeing a small cloud formed by the mixing of your warm, humid breath with the colder, drier ambient air. Mixing of air masses (or air layers within the same air mass) that differ in temperature and vapor pressure is another mechanism whereby clouds can form in the atmosphere. Surprisingly, a cloud may form even though the two air masses are initially unsaturated.

As emphasized in Chapter 6, the saturation vapor pressure increases as temperature increases, and the rate of increase accelerates with rising temperature. This relationship is depicted schematically in Figure 1 where temperature is plotted on the horizontal axis and vapor pressure is plotted on the vertical axis. The average temperature and average vapor pressure of a specific air mass plot as a single point on the diagram. An air mass that plots below the saturation vapor pressure curve is unsaturated, on the curve is saturated, and above the curve is supersaturated.

Suppose that two different unsaturated air masses plot as points *A* and *B* on the diagram. It is reasonable to assume that mixing the two air masses would produce a new air mass having properties (average temperature and vapor pressure) that would plot somewhere along a straight line connecting the points *A* and *B*. The precise location of that point (the mixture) along the line depends on the relative volumes of the two air masses. In any case, the mixture of the two air masses is unsaturated so that no cloud forms. Consider, however, two other unsaturated air masses, *C* and *D*, one cold and dry, the other warm and humid. In this case the straight line linking *C* and *D* intersects the saturation vapor pressure curve. The new air mass resulting from the mixing of equal volumes of air masses *C* and *D* is supersaturated, so a cloud would form until the vapor pressure is reduced to the saturation value.

This description of cloud formation by mixing provides insight as to the cause of jet aircraft contrails (Figure 7.1). The exhaust from jet engines mixes with the ambient air and may form a *contrail* (i.e., condensation trail) behind the aircraft. Heat and water are among the combustion products of jet engines so that the exhaust is hot and humid. Turbulence in the wake of a jet engine promotes the mixing of the exhaust with ambient air. If the ambient air has the appropriate combination of vapor pressure and temperature, the mixture will be saturated and a contrail forms. Such conditions are most likely in the upper troposphere where commercial jetliners travel.

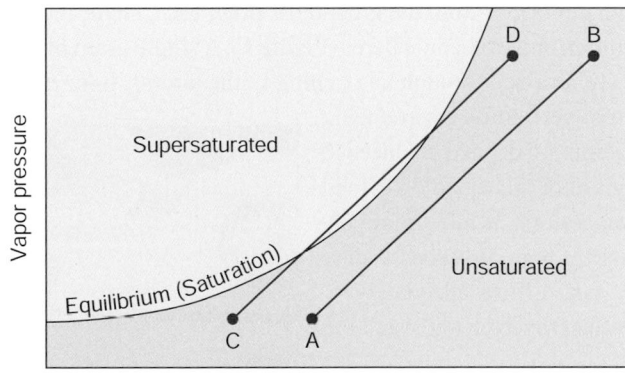

FIGURE 1
Variation of saturation vapor pressure with temperature. The properties of an air mass that plots as a point above the curve is supersaturated, on the curve is saturated, and below the curve is unsaturated.

Rainmaking

Since the mid-1940s, considerable research has gone into developing methods of enhancing precipitation by cloud seeding. *Cloud seeding* is an attempt to stimulate natural precipitation processes by injecting nucleating agents into clouds. Cloud seeding also has been used in attempts to reduce the size of hailstones, dissipate fog, and lessen the effects of lightning. Most cloud seeding is directed at cold clouds, although recent years have seen growing interest in warm cloud seeding.

Discovery of cloud seeding was serendipitous. In the 1940s, the Nobel Prize winning chemist Irving Langmuir (1881-1957) and his assistant Vincent J. Schaefer were conducting experiments at the General Electric Research Laboratory in Schenectady, NY, on ways to prevent the build-up of ice on the surface of airplane wings. On a summer day in 1946, Schaefer was investigating how to cause supercooled cloud droplets to freeze prior to fusing to the wing. Working with supercooled water contained in a cold chamber (a chest-type freezer open at the top), the hot weather forced him to chill the chamber, which he did with dry ice (solid CO_2). To Schaefer's amazement, the supercooled water in the chamber converted to a cloud of ice crystals. He realized that this might be a way to stimulate the Bergeron-Findeisen precipitation process (described earlier in this chapter). On 13 November 1946, General Electric scientists confirmed this new technique of cloud seeding when they dumped dry ice particles from an aircraft into a stratus cloud deck composed of supercooled droplets. Almost immediately, the seeded region of the cloud deck converted to ice crystals.

The day after the test of dry ice by Langmuir and Schaefer, Bernard Vonnegut (1914-1997) discovered that microscopic silver iodide (AgI), a substance with crystal properties similar to water ice, could also be used as a nucleating agent. He conducted the first field test using silver iodide in 1948. Silver iodide crystals are freezing nuclei that are active at $-5\ °C$ (23 °F) and below.

Typically, the seeding (nucleating) agent is either silver iodide or else dry ice, which has a temperature of about $-80\ °C$ ($-112\ °F$). Occasionally, extremely cold liquid propane is used. Dry ice pellets are so cold that within a cloud they cause surrounding supercooled water droplets to freeze, which then grow into ice crystals through the Bergeron-Findeisen process followed by collision and agglomeration. Furthermore, latent heat released when supercooled water droplets freeze makes the cloud slightly warmer and more buoyant, stimulating additional cloud growth that can lead to more precipitation. The objective of seeding cold clouds is to stimulate the Bergeron-Findeisen process in clouds that are deficient in ice crystals.

Clouds can be seeded via aircraft or from the ground. In either case, silver iodide is released into updrafts at the cloud base from liquid fuel generators or pyrotechnic flares (Figure 1). A single gram of silver iodide can produce as many as 10^{15} particles. Dry ice pellets are dropped through an opening in the aircraft floor directly into a cloud's updraft. Dry ice pellets are less efficient than silver iodide at producing ice crystals; as much as 2000 grams of dry ice are needed to match the output of just one gram of silver iodide. Cloud seeding from ground-based generators, while relatively inexpensive, can be less satisfactory than airborne seeding because the seeding agent may not diffuse adequately or reach sufficient altitudes to be effective. Nonetheless, both methods are used.

The traditional method of warm cloud seeding is to inject hygroscopic substances (e.g., sea salt crystals) into clouds to stimulate formation of relatively large cloud droplets with greater terminal velocities that can grow into raindrops by collision and coalescence. In the early 1990s, this technique was given a boost with development of a new cloud-seeding flare (based on a U.S. Navy fog-producing flare) that, when ignited, releases large quantities of microscopic salt particles (mostly potassium chloride, KCl). A series of flares mounted on an aircraft wing releases salt

FIGURE 1
A Cessna 210 airplane, rebuilt for cloud seeding, with a detailed view of a silver iodide generator. [Photo by Christian Jansky/ License: Creative Commons Attribution ShareAlike 2.5]

particles as the plane flies just below the cloud base, so updrafts will carry salt particles into the (cumuliform) cloud. Initial results from experimental use of this technique in South Africa and Mexico suggest that flares are more effective than the old methods of liquid sprays. In August 2000, Roelof Bruintjes and his colleagues at the National Center for Atmospheric Research (NCAR) reported statistically significant rainfall enhancement using this technique during a 3-year cloud seeding experiment carried out over drought-prone northern Mexico.

The Sierra Cooperative Pilot Project (SCPP) is an example of a long-term, ongoing study of cloud seeding. Scientists seed *orographic clouds* over the windward western slopes of California's Sierra Nevada Mountains (Figure 2) to increase snowfall and thicken mountain snow pack. The consequent increase in spring runoff is intended to help California meet its growing domestic and agricultural water demands. The American River Basin, just west of Lake Tahoe, is the principal SCPP study site and January through March the primary seeding season. Clouds targeted for seeding are rich in supercooled water droplets but deficient in ice crystals, which SCPP scientists identify with an aircraft outfitted with sophisticated instruments to measure the size and concentration of cloud and precipitation particles. Another aircraft seeds selected cold clouds with silver iodide or dry ice pellets. Meanwhile, on the ground, an array of precipitation gauges, weather radar, and other weather instruments monitor the effectiveness of seeding.

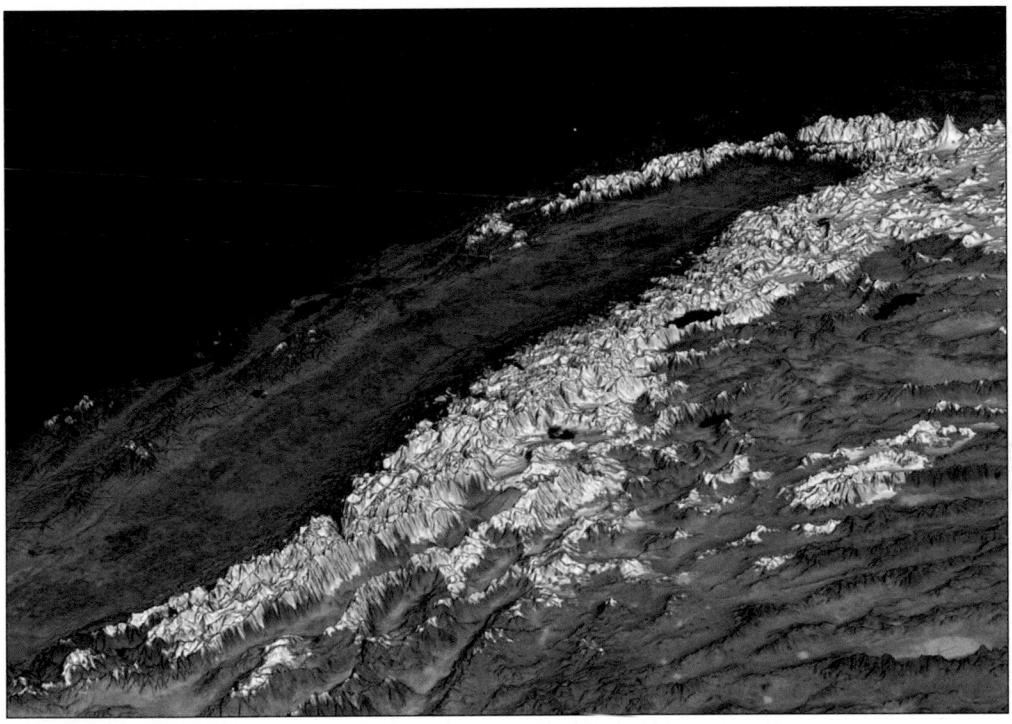

FIGURE 2
Snow cover on California's Sierra Nevada mountains. [NASA]

Does cloud seeding work? In the 1970s and early 1980s, NOAA scientists conducted a statistically rigorous experiment designed to test the effectiveness of cloud seeding. The experiment involved seeding cumulus clouds over southern Florida. The testing period was evenly divided between days when clouds were seeded with silver iodide crystals and days when clouds were seeded with inert (chemically inactive) sand grains. Only after the experiment was completed and the results analyzed were participating scientists informed of the specific days when silver iodide was used. This procedure was intended to ensure both an unbiased selection of clouds to be seeded and an unbiased interpretation of results.

The Florida seeding experiment was divided into two phases. Results from an initial *exploratory* phase were to be either verified or rejected by a later *confirmatory* phase when seeding was repeated. Results of the first phase were very encouraging, showing a 25% increase in rainfall on days when silver iodide was the seeding agent compared to days when sand was the bogus seeding agent. This finding was statistically significant at the 90% level, meaning that there is only a 10% probability that the increased rainfall was a chance occurrence. However, the success of the initial phase was not replicated during the second phase when seeding brought no statistically significant increase in rainfall.

Although cloud seeding is successful in some instances, the actual amount of additional precipitation produced by cloud seeding and the advisability of large-scale seeding efforts are controversial. Some cloud seeders claim to increase precipitation by 15% to 20%, but cannot answer if the rain or snow that follows cloud seeding would have fallen anyway. Separating the effects of cloud seeding from the natural variability in cloud processes remains a major challenge.

Even if successful, cloud seeding may merely redistribute a fixed supply of precipitation, so that more precipitation in one area might mean a compensating reduction in another area. For example, rainmaking might benefit agricultural interests on the High Plains of eastern Colorado but also deprive wheat farmers of rain in the downwind states of Kansas and Nebraska. Cloud seeders may also miss the target area. These and other unintended consequences of cloud seeding have not been convincingly demonstrated anymore than seeding itself. If such unintended effects do occur, legal battles would cross political boundaries and involve residents of adjacent counties, states, provinces, and countries.

The potential for cloud seeding success is greatest in areas already receiving considerable precipitation, as in the SCPP California effort. Unfortunately, the potential for successful cloud seeding in arid or drought-prone areas is less than promising because the necessary atmospheric conditions usually are not present due to insufficient atmospheric moisture or lack of suitable lifting mechanisms to produce clouds. In an *Information Statement* on cloud seeding issued in November 2010, the American Meteorological Society cautioned against relying exclusively on cloud seeding for relief during a drought. Cloud seeding has the potential of being more effective in ameliorating the impact of a drought if it is a component of an opportunistic long-term water management plan that includes soil moisture conservation, improving crop land, and increasing the supply of water in storage.

When Is It Too Cold or Too Warm to Snow?

During an episode of particularly frigid winter weather, some will claim that it's too cold to snow and, in fact, the coldest weather is often accompanied by fair skies associated with an Arctic high pressure system. Climate records of the northern United States and Canada indicate that snowfall totals decline with falling temperature. For some north-central U.S cities (e.g., Billings, MT, Huron, SD, and Minneapolis-St. Paul, MN), March on average is both the snowiest and the mildest of the months December through March. In Canada, outside of the mountains, average annual snowfall declines from more than 400 cm (160 in.) in southern, relatively mild parts of maritime Newfoundland to less than 100 cm (40 in.) along the frigid shores of the Arctic Ocean. Though total snowfall decreases with falling temperature, snow is possible even at extremely low air temperatures. In the bitter cold, however, snowflakes are small and accumulations meager.

The relatively small amount of water vapor in very cold air means that comparatively little water is available for precipitation (i.e., low amounts of *precipitable water*). Recall from Chapter 6 that the saturation vapor pressure drops rapidly as air temperature falls. For example, the water vapor concentration in saturated air at −30 °C (−22 °F) is only about 12% of the water vapor concentration in saturated air at −5 °C (23 °F). Hence, the amount of water potentially available for precipitation decreases with dropping air temperature.

When the temperature of the lower atmosphere is within a few degrees of the freezing point, the potential amount of water that can precipitate as snow is at a maximum and the heaviest snowfalls typically occur. For the same reason, moderate or heavy snowfall is very unlikely when the temperature of the lower atmosphere falls below −20 °C (−4 °F). But even in the coldest regions of the globe where precipitable water amounts are lowest, some snow falls. For example, an average annual 5 cm (2 in.) of snow falls on the high interior plateau of Greenland where average annual temperatures are below −30 °C (−22 °F).

When is it too warm to snow? Surprisingly, snow can fall to the ground even when the near-surface air temperature is as high as 10 °C (50 °F). For this to happen, the *wet-bulb temperature* must remain below 0 °C (32 °F) so that the relative humidity is very low. (Recall from Chapter 6 that the wet-bulb temperature is the lowest ambient air temperature that can be achieved via evaporative cooling.) For example, if the air temperature is 5 °C (41 °F), the relative humidity must be under 32% for the wet-bulb temperature to be subfreezing (see Table 6.5).

Imagine that rain drops and snowflakes fall from clouds into a layer of air below the cloud base that is relatively dry with above freezing temperatures but a wet-bulb temperature below freezing. The rain drops would rapidly vaporize and the snowflakes partially melt. The energy to melt the snow (latent heat of fusion) and evaporate the rain (latent heat of evaporation) comes from the sensible heat of the air, cooling it. Air temperature can fall to the wet-bulb temperature, that is below freezing supporting snowfall. While the wet-bulb temperature remains nearly constant, the addition of water vapor raises the dewpoint until the temperature reaches 0 °C (32 °F) and the air layer is saturated. Hence, under the conditions described, what started out as rain (or a mixture of rain and snow) turns to snow. This is most likely to happen if the precipitation is moderate to heavy, as the more snowflakes that are melting and vaporizing, the ambient air is cooled.

A windstorm with sustained winds in excess of 50 mph (80 kph), battered the coasts of Washington and Oregon for over 48 hours during December 1-3, 2007. Flooding from 48-ft (15 m) waves led to the closure of all East-West roads through the Coast Range into the Willamette Valley, and cut power to the area for at least 4 days. Photo shows beach sand blown inland partially submerging some homes. [Courtesy NOAA]

WIND & WEATHER

Chapter Highlights

Case-in-Point
    Sinking of the Edmund Fitzgerald
Forces Governing the Wind
    Pressure Gradient Force
    Centripetal Force
    Coriolis Effect
    Friction
    Gravity
    Summary
Wind: Joining Forces
    Hydrostatic Equilibrium
    Geostrophic Wind
    Gradient Wind
    Surface Winds in Highs and Lows
Continuity of Wind
Monitoring Wind Speed and Direction
Scales of Atmospheric Circulation
Conclusions
Basic Understandings/Enduring Ideas
Key Terms/Review/Critical Thinking
For Further Exploration
    Lake-Effect Snow
    Wind Profilers
    Wind Power

Learning Objectives

Distinguish among the directional
    components of the wind.
Identify the forces that initiate and shape the
    circulation of air.
Explain how horizontal air pressure gradients
    develop.
Describe how the pressure gradient force
    affects the motion of air.
Explain the source of the centripetal force.
Present the basic reason for the Coriolis
    Effect and describe how the magnitude of
    the Coriolis Effect varies with latitude.
Explain why the Coriolis Effect acts in
    opposite directions in the Northern and
    Southern Hemispheres.
Demonstrate why gravity influences vertical
    motion and not horizontal motion.
Present Newton's first and second laws of
    motion.
Describe the balance of forces in hydrostatic
    equilibrium.
Summarize the interaction of forces in
    geostrophic and gradient winds.
Explain how friction influences the magnitude
    and direction of surface winds.
Compare and contrast the circulation in
    cyclones and anticyclones in the Northern
    and Southern Hemispheres.
Explain why stormy weather is associated
    with cyclones and fair weather with
    anticyclones.
Describe how wind speed and direction are
    monitored.

What forces control the speed and direction of the wind?

Case-in-Point

Sinking of the *Edmund Fitzgerald*

A ballad made popular by Gordon Lightfoot memorialized the November 1975 sinking of the American ore carrier *Edmund Fitzgerald* (Figure 1) with the loss of all 29 crewmembers in the storm-tossed waters of Lake Superior. At the time, the 222-m (729-ft) long ship was the largest ore carrier on the Great Lakes. Early on the afternoon of 9 November, the ship, fully loaded with 26,116 tons of iron ore (taconite) pellets, left the Duluth-Superior harbor at the far western end of Lake Superior on a northeast course at about 20 knots (23 mph). The ship's destination was a steel plant on Zug Island in the Detroit River.

At 6 a.m. (CST) on 9 November, a cyclone (low-pressure system) began organizing over central Kansas. From there, the intensifying storm tracked toward the northeast and at 6 a.m. on 10 November its center passed near La Crosse, WI, and at noon was centered just west of Marquette, MI. By this time the storm's central pressure had dropped to 982 mb and gale-force northeast winds were sweeping the eastern end of Lake Superior. Winds gusted to 115 km per hr (71 mph) at Sault Ste. Marie, MI.

At 1 a.m. (CST) on 10 November, the *Edmund Fitzgerald* reported northeast winds at 97 km per hr (60 mph) with waves to 3 m (10 ft). At 7 a.m., the ship was about 73 km (45 mi) north of Copper Harbor, MI and reporting northeast winds at 65 km per hr (40 mph). With the storm bearing down on the *Edmund Fitzgerald*, the ship's captain, Ernest McSorley, changed course to the east and then southeast hugging the north shore of Lake Superior. McSorley thought that this course would take the ship through waters that were sheltered from the strong northeast winds. That afternoon, the storm passed over the *Edmund Fitzgerald*. In the evening, as the storm center neared Moosonee, ON, near the southern shore of James Bay, winds on Lake Superior shifted from northeast to north and then northwest and west. The longer fetch of the northwest and west winds over the lake caused the waves to grow. (*Fetch* is the distance the wind blows over a continuous water surface.) Another ship within several kilometers of the *Edmund Fitzgerald* estimated wind speeds of 94 km per hr (58 mph) gusting to 137 km per hr (85 mph) with waves of 3.5 to 5 m (12 to 16 ft).

At approximately 6:15 p.m. (CST), the *Edmund Fitzgerald* sank in 160 m (530 ft) of water about 27 km (17 mi) north-northwest of Whitefish Point, MI. More than 1000 ships have sunk in the Great Lakes and most of these wrecks were weather-related. The *Edmund Fitzgerald* was the largest ever to go down.

The precise cause of the sinking of the *Edmund Fitzgerald* is the subject of considerable debate. While weather certainly played an important role, research in the years since the tragedy suggests that structural deficiencies and poor condition of the ship may have also played a key role, as well as a possible grounding or near-grounding of the vessel near Caribou Island.

FIGURE 8.1

The *Edmund Fitzgerald*. [Watercolor painting with pen and ink by Leo Kuschel of Taylor, MI]

Some weather systems are responsible for clear skies, light winds, and frosty mornings, whereas others bring ominous clouds, precipitation, and biting winds. Some weather systems trigger brief showers, whereas others are accompanied by persistent fog and drizzle. Certain weather systems dominate the weather over thousands of square kilometers for weeks at a time. Different weather systems bring different types of weather depending on the air circulation (wind and vertical motions) that characterizes each system.

Wind, the principal focus of this chapter, is the local motion of air measured relative to Earth's surface. The atmosphere is an integral and coupled component of the rotating planet. Once every 24 hrs, all points on Earth's surface and in the atmosphere (except right at the poles) complete a circular path in space. Because the distance between Earth's surface and the axis of rotation decreases with increasing latitude, the circumference of that path also decreases with increasing latitude. Hence, the speed at which fixed points on Earth's surface progress eastward decreases with increasing latitude; the speed drops from 1670 km per hr (1035 mph) at the equator to 834 km per hr (517 mph) at 60 degrees N and S, and to zero at the poles. We are unaware of this rapid motion because Earth is our reference when measuring our motions. In meteorology, we are interested in air motion measured relative to Earth's surface (i.e., the wind).

In Chapter 4, we saw how unequal rates of radiational heating and cooling produce temperature gradients within the Earth-atmosphere system. In response to temperature gradients between Earth's surface and the troposphere as well as between the tropics and higher latitudes, the atmosphere (and ocean) circulates and heat is redistributed. This chapter covers the various forces that either initiate or modify atmospheric circulation, that is, the wind. First, each force is examined separately as if it acted independently of all the other forces. Forces are then combined to demonstrate how together they drive atmospheric circulation.

Forces Governing the Wind

Several forces influence wind speed and direction. A **force** is defined as a push or pull that can cause an object at rest to move or that alters the movement of an object already in motion. Like the wind, a force has both direction and magnitude (a *vector* quantity).

In examining the various forces that govern the wind, it is useful to apply each force to a parcel that is a unit mass (e.g., single kilogram) of air. Imagine the wind to be a continuous stream of air consisting of discrete air parcels. Assume that any force acting on an air parcel has the same influence on a stream of air parcels, in other words, on the wind. Hence, in describing each force that affects air motion, we can examine the force per unit mass of air.

In meteorology, *force* and *acceleration* are terms often used interchangeably because it is common to treat forces as acting on a unit mass of air. This practice follows from **Newton's second law of motion**, where

$$\text{force} = (\text{mass}) \times (\text{acceleration}).$$

If the air parcel is a unit mass, a force per unit mass is numerically equivalent to acceleration. Although the terms are numerically equivalent, they are not the same. Acceleration (a change in velocity) is actually a response to a force. A force acts on an air parcel to accelerate that parcel. Furthermore, acceleration involves a change in speed or direction or both. We have more on this later.

Forces acting on air parcels either initiate or modify motion and are the consequence of (1) an air pressure gradient, (2) the centripetal force, (3) the Coriolis Effect, (4) friction, and (5) gravity. Actually, the centripetal force is not an independent force but occurs as a consequence of other forces.

PRESSURE GRADIENT FORCE

A *gradient* is simply a change in some property over distance. An **air pressure gradient** exists whenever air pressure varies from one place to another. As noted in Chapter 5, spatial variations in air pressure can arise from contrasts in air temperature (principally), differences in water vapor concentration, or both. In addition, winds diverging or converging over a broad region can bring about air pressure changes by redistributing the amount of air in a vertical column of the atmosphere and thereby inducing an air pressure gradient.

Air pressure gradients occur both horizontally and vertically within the atmosphere. A horizontal pressure gradient refers to air pressure changes along a surface of constant altitude (e.g., mean sea level). A vertical pressure gradient refers to the air pressure change directly above Earth's surface; it is a permanent feature of the atmosphere. Air pressure is determined by the weight of the overlying air, so air pressure is greatest at the Earth's surface and always decreases with altitude. Horizontal air pressure gradients can be determined on weather maps from patterns of **isobars**, lines joining points having the same air pressure. But first, to eliminate the influence of station elevation on air pressure, the barometer reading at each weather station

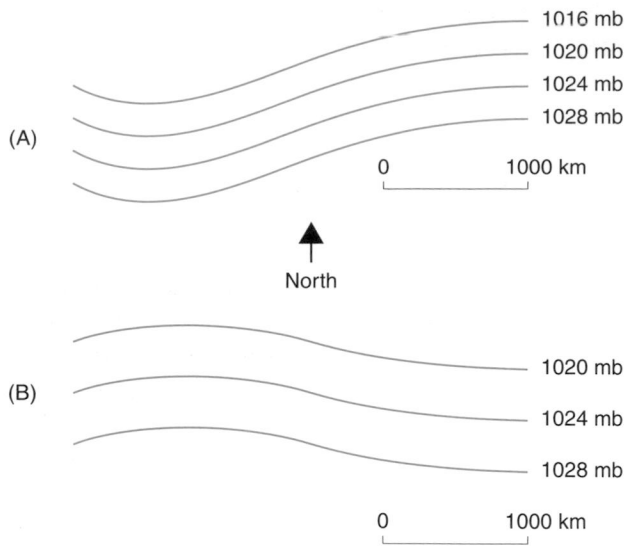

(A)

1016 mb
1020 mb
1024 mb
1028 mb

0   1000 km

↑
North

(B)

1020 mb

1024 mb

1028 mb

0   1000 km

FIGURE 8.2
In these map views the horizontal air pressure gradient is (A) relatively steep where isobars are closely spaced and (B) weaker where isobars are farther apart. Note that isobars are lines joining places of equal air pressure. Here the contour interval (difference between successive isobars) is 4 mb.

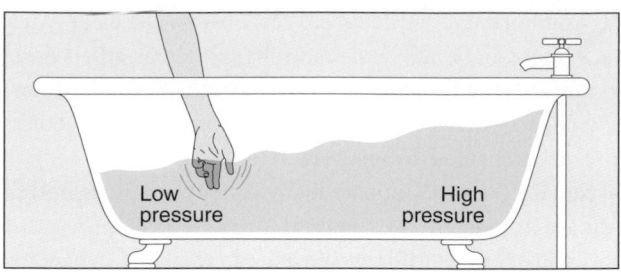

Low pressure   High pressure

FIGURE 8.3
Sloshing water back and forth from one end of a tub to the other end creates water pressure gradients along the bottom of the tub. The water pressure gradient along the tub bottom is analogous to a horizontal air pressure gradient in the atmosphere. In response to a pressure gradient, water (or air) flows from an area of higher pressure toward an area of lower pressure.

is adjusted to sea level (Chapter 5). By U.S. convention, isobars are drawn at 4-mb (4-hPa) intervals and interpolation between weather stations is always necessary.

An isobaric analysis is used to locate centers of high and low pressure on a weather map and to determine the magnitudes of the horizontal air pressure gradients between and within weather systems. Where isobars are closely spaced (Figure 8.2A), air pressure changes rapidly with distance between isobars, and the pressure gradient is described as steep or strong. More widely spaced isobars (Figure 8.2B) indicate that air pressure changes more gradually with distance between isobars, and the pressure gradient is weaker. Note that air pressure gradients are always measured in the direction of greatest change, that is, *perpendicular* to isobars.

How do air pressure gradients influence the motion of air? Consider an analogous situation. Suppose a bathtub is partially filled with water, as shown in Figure 8.3. Sloshing the water back and forth from one end of the tub to the other creates rapidly changing water pressure gradients along the tub bottom. Most of the time, the water surface is not horizontal. The water pressure acting on the bottom is high where the water is relatively deep and low where the water is relatively shallow. If we stop agitating the water, the sloshing back and forth dampens, and the undulating water surface returns to horizontal. The water pressure along the bottom is then the same from one end of the tub to the other.

Hence, in response to a water pressure gradient, water flows from one end of the tub (where the bottom water pressure is greater) to the other end (where the bottom water pressure is less), thereby acting to eliminate the pressure gradient.

Similarly, when a horizontal air pressure gradient develops, air flows in a direction to eliminate the pressure gradient. The force that causes air parcels to move as the consequence of an air pressure gradient, known as the **pressure gradient force**, always acts directly perpendicular to isobars and toward lowest pressure. The magnitude of the pressure gradient force is inversely related to the spacing of isobars. The wind is relatively strong where the pressure gradient is steep (closely spaced isobars), light where the pressure gradient is weak (widely spaced isobars), and the air is calm where there is no pressure gradient.

CENTRIPETAL FORCE

Isobars plotted on a surface weather map are almost always curved, indicating that the pressure gradient force changes direction from one place to another. Consequently, the horizontal wind blows in curved paths. Curved motion indicates the influence of the centripetal force.

A simple demonstration illustrates the centripetal force. A rock is tied to a string and then whirled about so that the tethered rock travels a circular orbit of constant radius (Figure 8.4). The string exerts an inward force on the rock, confining it to a curved (circular) path. At any instant, the force is directed perpendicular to the direction of motion of the rock and toward the center of the circular orbit. For this reason, the force is known as the **centripetal** (*center-seeking*) **force**. If we cut the string, the centripetal force no longer operates; that is, a force no longer confines the rock to a curved path. As seen from

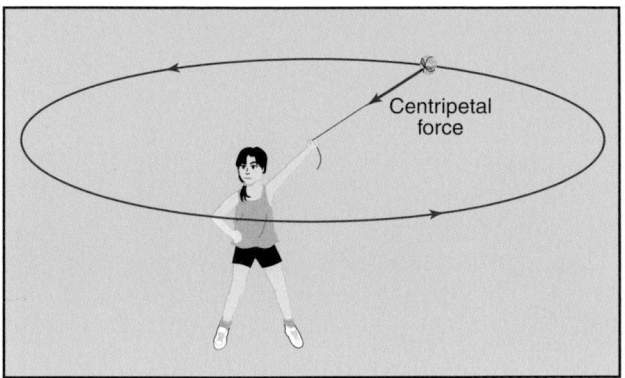

FIGURE 8.4
A rock tied to a string follows a circular path as it is whirled about. Centripetal (*center-seeking*) is the name of the force that confines an object to a curved path. If the string is cut, the centripetal force is eliminated, and the rock moves in a straight line (tangent to the circular path) as seen from above.

above, the rock flies off in a straight line as described by *Newton's first law of motion*; that is, an object in straight-line, un-accelerated motion remains that way unless acted upon by an unbalanced force.

An unbalanced force causes acceleration. We usually think of acceleration as a change in speed, as when an automobile speeds up. But velocity is a vector quantity; that is, it has both magnitude and direction. Acceleration is also a vector quantity and consists of a change in either speed or direction, or both. In our rock-on-a-string example, centripetal force is responsible only for a continual change in the direction of the rock (curved rather than straight-line path); the rock neither speeds up nor slows down. The acceleration imparted to a unit mass by the centripetal force is directed toward the center of curvature, with a magnitude directly proportional to the square of the speed and inversely proportional to the radius of curvature of the path. That is, increasing the rotation rate, which increases the forward speed of the rock, requires a larger inward centripetal force, as would a shorter string (or radius of curvature).

The centripetal force is not an independent force; rather, it arises from the action of other forces and may be the consequence of imbalances in other forces. In our rock-on-a-string example, the tension of the string is responsible for the centripetal force. Consider another example. Suppose you are a passenger in an automobile that rounds a curve at high speed. You are aware of a force that pushes you outward from the turning automoblie. Actually, you experience the tendency for your body to continue moving in a straight path while the vehicle follows a curved path. In this case, the frictional resistance

of the tires against the pavement provides the centripetal force, keeping the vehicle, with you in it, on the road.

A centripetal force operates whenever air parcels follow a curved path. As we will see later in this chapter, the centripetal force arises from an imbalance in other forces operating in the atmosphere.

CORIOLIS EFFECT

Imagine that you are located in space a few thousand kilometers away at some fixed point, looking towards Planet Earth. Over many hours, you follow the track of an upper-level low pressure system, clearly identifiable by a mass of clouds slowly swirling about a central point. From your perspective, the upper-level low (storm) center appears to be moving in a straight line at constant speed. (Note that a geostationary satellite does not provide this perspective because the satellite is in a coordinate system fixed with respect to the Earth below.) At the same time, an observer on Earth is also tracking the storm using radar and surface weather observations. From that observer's perspective, the storm center appears to follow a curved path. Surprisingly, both descriptions of the storm's track are correct!

The two descriptions are correct because the two observers used different frames of reference in following the storm's movements. The Earthbound observer's frame of reference is the solid Earth to which the familiar north-south, east-west, and up-down coordinate system is attached. To the Earthbound observer, it is not obvious that this coordinate system is rotating because it and the observer rotate along with the turning Earth. Viewed from space, however, it is evident that the Earthbound coordinate system actually shifts as Earth rotates (Figure 8.5). In fact, Earth and the coordinate system rotate under the storm (or any other object moving freely over Earth's surface). Meanwhile, from your location in space, you followed the storm's movement with respect to a non-rotating coordinate system, fixed in space with respect to the background stars. In summary, the difference in observed storm tracks (curved versus straight) arises from the differences in coordinate systems, rotating versus non-rotating.

Recall our earlier discussion of *Newton's first law of motion* and the centripetal force. We saw that curved motion implies that an unbalanced force is operating, whereas steady, straight-line motion implies a balance of forces. Applying this law to our storm track example, we conclude that an unbalanced force operates when we use the Earthbound rotating coordinate system, whereas forces are balanced when we use the non-rotating coordinate system

FIGURE 8.5
Viewed from a fixed point in space, the familiar north-south, east-west frame of reference rotates eastward in space as Earth rotates on its axis. Rotation of this coordinate system gives rise to the Coriolis Effect.

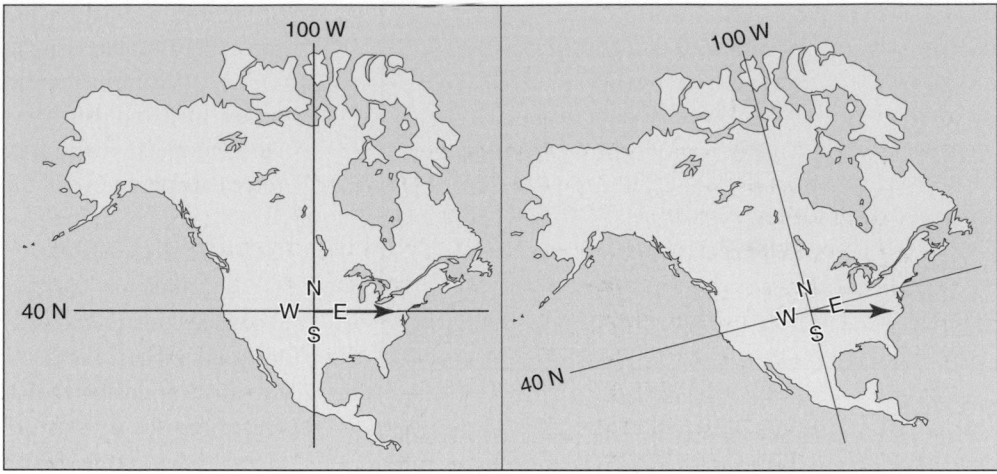

fixed in space. Hence, changing our frame of reference (coordinate system) from non-rotating to rotating introduces an apparent force responsible for curved motion.

Deflection of moving objects viewed from within a rotating frame of reference was first explained by Italian scientists Giovanni Riccioli (1598-1671) and Francesco Grimaldi (1618-1663). Almost two centuries later, in 1835, the French mathematician Gaspard-Gustave de Coriolis (1792-1843) first described the phenomenon mathematically. For this reason, the deflection experienced with the Earthbound rotating coordinate system is referred to as the **Coriolis Effect** and the apparent force invented to describe its magnitude and direction is called the Coriolis Force. Wind direction and speed are measured with respect to the north-south, east-west, and up-down frame of reference fixed to the rotating planet. Therefore, we must consider the Coriolis Effect in explaining air circulation. The Coriolis Effect deflects the horizontal wind to the right of its initial direction in the Northern Hemisphere and to the left of its initial direction in the Southern Hemisphere (Figure 8.6).

Why is the Coriolis Effect reversed between the hemispheres so that large-scale winds in the Southern Hemisphere swerve to the left rather than to the right? This reversal is related to the difference in an observer's sense of Earth's rotation in the two hemispheres. To an observer looking down from high above the North Pole, the planet rotates counterclockwise, whereas to an observer high above the South Pole, the planet rotates clockwise. For the observer measuring motion relative to a coordinate system anchored to the rotating Earth, this reversal in the apparent direction of rotation between the two hemispheres translates into a reversal in the direction of the Coriolis Effect.

The Coriolis Effect influences the wind blowing in any direction, and the amount of deflection varies significantly with latitude (as the sine of the latitude). Earth's rotation on its axis imparts a rotation to our Earth-bound frame of reference that increases in magnitude from zero at the equator to a maximum at the poles. This variation with latitude can be understood by visualizing the daily rotation of towers about their vertical axis when located at different latitudes. In a 24-hr day, Earth completes one rotation, as would a tower located at the North or South Pole. In the same period, a tower at the equator would not rotate at all around its vertical axis

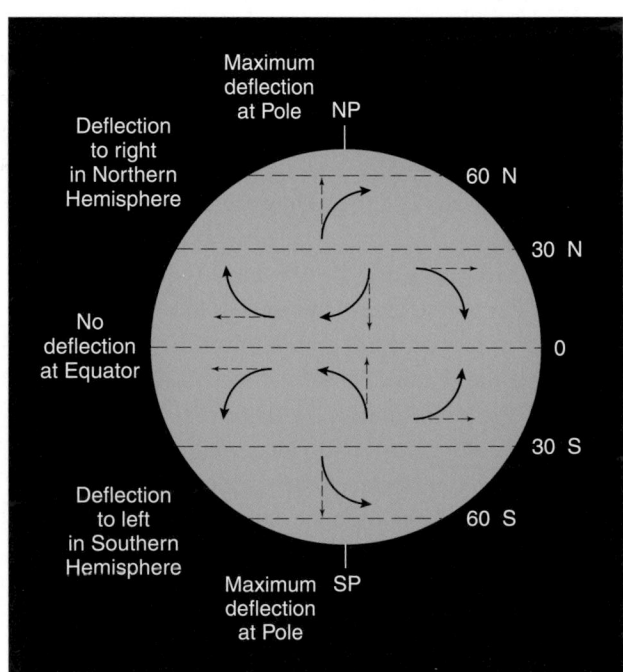

FIGURE 8.6
Large-scale winds are deflected to the right of their initial direction in the Northern Hemisphere and to the left of their initial direction in the Southern Hemisphere. This deflection, known as the Coriolis Effect, is maximum at the poles and zero at the equator.

because of its orientation perpendicular to Earth's axis of rotation. At any latitude in between, some rotation of a tower occurs but not as much as at the poles. The Coriolis Effect is thus latitude-dependent: the Coriolis Effect is zero at the equator and increases poleward.

The magnitude of the Coriolis Effect also varies with wind speed and spatial scale of atmospheric circulation. The Coriolis Effect increases as the wind strengthens because, in the same time interval, faster moving air parcels cover greater distances than slower moving air parcels. The longer the trajectory, the greater is the rotation of the underlying Earth. The Coriolis Effect significantly influences the wind only in large-scale weather systems, that is, systems larger than ordinary thunderstorms. Large-scale weather systems also have longer life expectancies than small-scale systems so that air parcels travel greater distances over longer periods of time, which allows the impact of Earth's rotation to manifest itself.

A rotational motion usually accompanies the draining of water from a sink or a bathtub. It is a popular misconception that the direction of this rotation (clockwise or counterclockwise) is consistently in one direction in the Northern Hemisphere and in the opposite direction in the Southern Hemisphere, presumably because of the Coriolis Effect. At the very small scale represented by the water in a sink or bathtub, the magnitude of the Coriolis Effect is simply too small to significantly influence the direction of rotation. Drainage direction is more likely a consequence of some pre-existing motion of the water or the shape of the sink or bathtub and may be either clockwise or counterclockwise regardless of hemisphere.

FRICTION

Friction is the resistance that an object or medium encounters as it moves in contact with another object or medium. We are all familiar with the friction associated with solid objects, as when we attempt to slide a heavy appliance across the floor. But friction also affects fluids, both liquids and gases. The friction of fluid flow, known as **viscosity**, is of two types: molecular viscosity and eddy viscosity. One source of fluid friction is the random motion of molecules composing a liquid or gas; this type of fluid friction is called **molecular viscosity**. Considerably more important, however, is fluid friction that arises from much larger irregular motions, called *eddies*, which develop within fluids; this type of fluid friction is known as **eddy viscosity**.

A swiftly flowing stream illustrates the effects of eddy viscosity. Rocks in the streambed obstruct the flow of water causing the current to break into eddies immediately downstream of the rocks (Figure 8.7). Eddies, visible as swirls of water, tap some of the stream's kinetic energy so that the stream slows. In an analogous manner, obstacles on Earth's surface such as trees and houses break the wind into eddies of various sizes to the lee of each obstacle. Consequently, the near surface wind slows.

FIGURE 8.7
Rocks shown in the riverbead of the Jarbridge River in Idaho are obstacles that break the current into turbulent eddies downstream from the rocks, slowing the river flow. A similar frictional interaction takes place as wind encounters obstacles on Earth's surface. [Bureau of Land Management File Photo]

FIGURE 8.8
A snow fence slows the wind, reducing its ability to transport snow in suspension. Hence, snow accumulates immediately downwind from the snow fence. [Courtesy of Franklin County, Ohio Engineer's Office]

A snow fence provides a practical illustration of the frictional slowing of the wind (Figure 8.8). Snow fences are designed to trap wind-blown snow, in some instances to prevent snow from drifting onto a nearby highway and in others to keep the soil snow-covered. (Snow that accumulates downwind of a snow fence ensures a supply of soil moisture and insulates the soil from freezing to great depths.) A snow fence taps some of the wind's kinetic energy by breaking the wind into small eddies. The wind speed diminishes, losing some of its snow-transporting ability, and snow accumulates on the downwind side of the fence.

The rougher the surface of the Earth, the greater is the eddy viscosity of the wind. A forest thus offers more frictional resistance to the wind than does the smoother surface of a freshly mowed lawn. Eddy viscosity diminishes rapidly with altitude above Earth's surface, away from obstacles on the ground mainly responsible for frictional resistance. Hence, horizontal wind speed increases with altitude. This explains the advantage of positioning a wind turbine at as high an elevation above surrounding land as possible. Above an average altitude of about 1000 m (3300 ft), friction is such a minor force that has little impact on the smooth flow of air. The atmospheric zone to which frictional resistance (eddy viscosity) is essentially confined is called the **atmospheric boundary layer**.

Turbulence is fluid flow characterized by eddy motion. Various obstacles on Earth's surface exert a drag on the wind and are sources of eddies. Irregular fluid flow that originates in this way is known as *mechanical turbulence*. In addition, eddies develop in air as a consequence of solar heating of the ground; irregular fluid flow that originates in this way is known as *thermal turbulence*. Convection is an example of thermal turbulence within the atmosphere. In actual practice, it is virtually impossible to distinguish between the two sources of turbulent eddies (mechanical or thermal). Regardless of source, we experience turbulent eddies as gusts of wind. The gustiness of the wind often varies with the time of day; that is, gusts tend to be strongest during the warmest hours of the day. Pilots of light aircraft are well aware of the changes in turbulence between day and night.

GRAVITY

Air parcels, like all other objects with mass, are subject to the force of **gravity**, and are pulled towards Earth. Gravity is the net result of two forces working together: gravitation and centripetal force. *Gravitation* is the force of attraction between Earth and some object; its magnitude is directly proportional to the product of the masses of Earth and the object, but inversely proportional to the square of the distance between their centers of mass. The much weaker centripetal force is imparted to all objects because of their rotation with Earth on its axis. Combined as gravity, the two forces accelerate a unit mass of any object directly downward toward Earth's surface at the rate of 9.8 m per sec each second.

Gravity always acts directly downward. For this reason, gravity, unlike the Coriolis Effect and friction, does not modify the horizontal wind. Gravity influences air that is ascending or descending, such as the updrafts and downdrafts in convection currents, and gravity is responsible for the downhill drainage of cold, dense air.

SUMMARY

We have now examined individually the various forces that influence horizontal and vertical air motion, and we can draw the following conclusions:

1. The horizontal pressure gradient force, which is responsible for initiating essentially all air motion, accelerates air parcels perpendicular to isobars away from regions of high air pressure and toward regions of low air pressure. The magnitude of the force is directly proportional to the pressure gradient; that is, the closer the spacing of isobars, the greater is the magnitude of the pressure gradient force.

2. A centripetal force is an imbalance of actual forces and exists whenever the wind describes a curved path. It is responsible for a change in wind direction, but not wind speed. This force, which is

always pointed inward toward the axis of rotation, has a magnitude that is proportional to the parcel's speed, but inversely proportional to the radius of curvature.

3. The Coriolis Effect arises from the rotation of Earth on its axis and deflects large-scale winds to the right of their initial direction in the Northern Hemisphere, but to the left of their initial direction in the Southern Hemisphere. The Coriolis Effect increases with latitude from zero at the equator to a maximum at the poles and is directly proportional to wind speed.

4. Friction always opposes motion, acting opposite to the wind direction and increasing with increasing surface roughness. Friction slows horizontal winds blowing within about 1000 m (3300 ft) of Earth's surface.

5. Gravity pulls all air downward toward the Earth's surface and because of its vertical direction does not modify the horizontal wind.

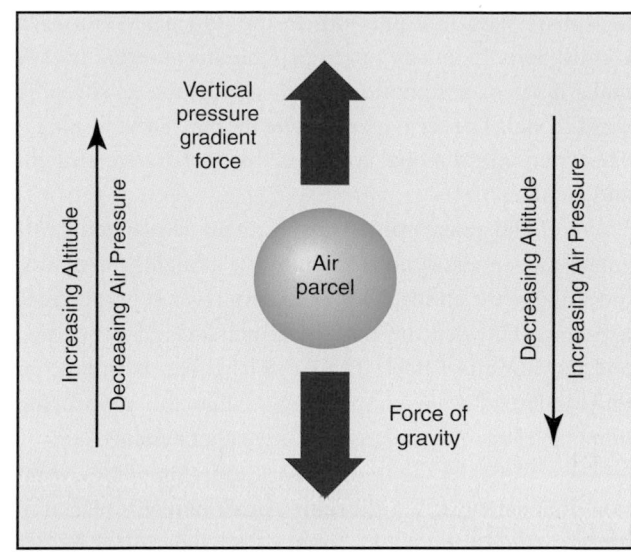

FIGURE 8.9
In the atmosphere, hydrostatic equilibrium is the balance between the upward-directed pressure gradient force and the downward-directed force of gravity.

Wind: Joining Forces

To this point in our discussion, we have examined forces operating in the atmosphere as if each force acted independently of all the others. In reality, these forces interact with one another in governing both wind speed and direction. In some cases, two or more forces achieve a balance or equilibrium. From *Newton's first law of motion*, when the forces acting on an air parcel are balanced, the parcel either remains stationary or continues to move along a straight path at constant speed. When forces are balanced, the net acceleration is zero.

In this section, we build an increasingly sophisticated model of atmospheric motion by examining how forces interact in the atmosphere to control the vertical and horizontal motion of air, that is, the wind. These interactions result in (1) hydrostatic equilibrium, (2) the geostrophic wind, (3) the gradient wind, and (4) surface winds (horizontal winds within the atmospheric boundary layer).

HYDROSTATIC EQUILIBRIUM

In developing a realistic model of atmospheric motion, we will assume that the atmosphere is in a state of hydrostatic equilibrium. This assumption allows us to focus on horizontal motions of air.

Air pressure always decreases with increasing altitude so that a vertical air pressure gradient is a permanent feature of the atmosphere. As shown schematically in Figure 8.9, the force due to this pressure gradient is directed upward from higher air pressures at Earth's surface toward lower air pressures aloft. If this force acted alone, the vertical pressure gradient force would accelerate air away from Earth, and we would be left gasping for breath. However, except for relatively brief periods in some small-scale violent weather systems (e.g., severe thunderstorms), the atmosphere's vertical pressure gradient force is almost balanced by the equal but oppositely directed force of gravity. An actual balance of these two forces is known as **hydrostatic equilibrium**.

As noted earlier, when forces are in balance, the acceleration is zero; that is, there is no change in speed or direction. Hydrostatic equilibrium, however, does not preclude vertical (up or down) motions of air. With balanced forces, ascending air parcels continue moving upward at constant velocity while descending air parcels continue moving downward at constant velocity. Slight deviations from hydrostatic equilibrium cause air parcels to change speed (accelerate) vertically.

GEOSTROPHIC WIND

The existence of air pressure differences in a horizontal direction would result in a wind blowing from high pressure toward low pressure. Air moves in this direction on a local scale, such as during a sea or lake breeze when cooler, denser air situated over a body of water moves inland. However, inspection of upper-air weather maps shows that horizontal winds above the atmospheric boundary layer tend to blow parallel to isobars with low pressure to the left in the Northern Hemisphere. In an attempt to explain this observation, we make a set of assumptions contained in the geostrophic wind model. The term *geostrophic* means "Earth turning." This frictionless model assumes that isobars are straight and parallel.

The **geostrophic wind** is an un-accelerated, horizontal movement of air that follows a straight path at altitudes above the atmospheric boundary layer. It results from a balance between the horizontal pressure gradient force and the Coriolis Effect. The Coriolis Effect is significant only in broad-scale circulations so that the geostrophic wind develops only in large-scale weather systems.

Consider the traditional description of the evolution of geostrophic equilibrium: An air parcel is placed in a preexisting horizontal pressure field where isobars are straight and parallel (Figure 8.10). In response to the horizontal pressure gradient force (P_H), the air parcel initially accelerates directly across isobars from high pressure toward low pressure. As the air parcel accelerates, however, the Coriolis Effect (C) comes into play, strengthens, and causes the air parcel to turn gradually to the right of its initial flow direction (in the Northern Hemisphere). The Coriolis Effect changes direction as the parcel turns, but always remains at right angles to the parcel's direction of motion. The parcel continues turning until the two forces are acting in opposite directions and balance one another, known as *geostrophic equilibrium*. The geostrophic wind

blows at a constant speed in a straight path parallel to isobars with the lowest air pressure to the left of the direction of air motion. In the Southern Hemisphere, where the sense of Earth's rotation is opposite to that of the Northern Hemisphere, the Coriolis Effect causes the air parcel to turn to the left until the flow is parallel to isobars and the lowest air pressure is to the right.

Numerical simulations of the interactions of the pressure gradient force and the Coriolis Effect conducted by J.A. Knox and his colleagues at NASA's Goddard Institute for Space Studies show that air parcels actually undergo an oscillatory motion along the curved path (dashed blue line) shown in Figure 8.10. The oscillatory motion gradually dampens as air parcels approach geostrophic equilibrium. This so-called *inertial oscillation* is an interaction of the pressure field with the motion of the air parcels and may be an important feature of some weather systems.

GRADIENT WIND

A more common model of atmospheric flow is the **gradient wind** that shares many of the same characteristics as the geostrophic wind. It is also large-scale, horizontal, and frictionless, and blows parallel to isobars. However, the important distinction between the two models is that the geostrophic wind blows in a straight path, whereas the path of the gradient wind is curved. Forces are not balanced in the gradient wind because a centripetal force constrains air parcels to a curved trajectory. Recall from our earlier discussion that the centripetal force changes only the direction and not the speed of an air parcel. The horizontal pressure gradient force, the Coriolis Effect, and the centripetal force interact in the gradient wind.

A gradient wind develops at altitudes above the atmospheric boundary layer around a dome of high air pressure, called an **anticyclone** (or *High*), or around a center of low air pressure, called a **cyclone** (or *Low*). In an idealized anticyclone, isobars form a series of concentric circles about the location of highest air pressure, as shown in Figure 8.11. Under gradient wind conditions, the horizontal pressure gradient force (P_H) is directed radially outward, away from the center of the high. The Coriolis Effect (C) is directed inward. The Coriolis Effect is slightly greater than the pressure gradient force, with the difference between the two forces giving rise to the inward-directed centripetal force (C_E). (This is what was meant earlier when we indicated that a centripetal force results from an imbalance of other forces.) Viewed from above in the Northern Hemisphere, the gradient wind in an anticyclone blows clockwise and parallel to isobars above the atmospheric boundary layer.

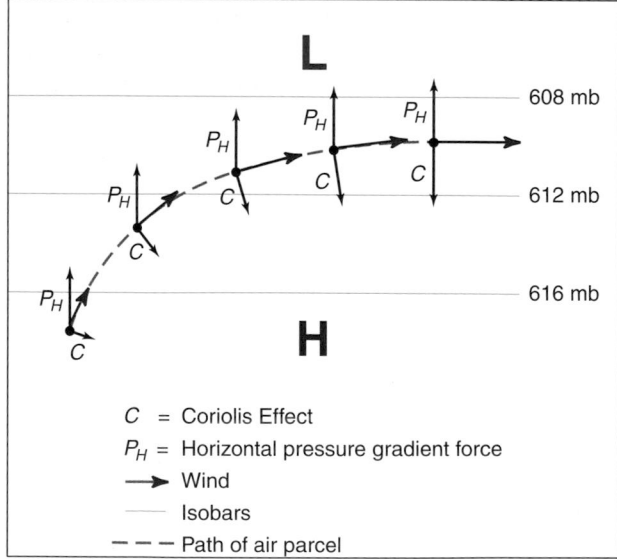

FIGURE 8.10
This diagram approximates the evolution of geostrophic equilibrium, a balance between the horizontal pressure gradient force and the Coriolis Effect. The geostrophic wind blows parallel to straight isobars at altitudes above the atmospheric boundary layer.

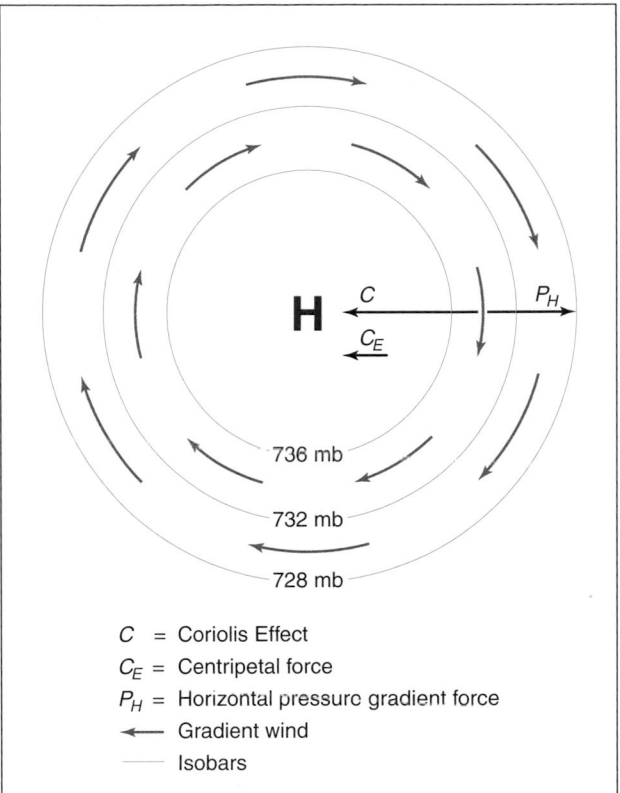

FIGURE 8.11
Viewed from above in the Northern Hemisphere, the gradient wind blows clockwise and parallel to isobars in an anticyclone. In this idealized case, isobars form a pattern of concentric circles.

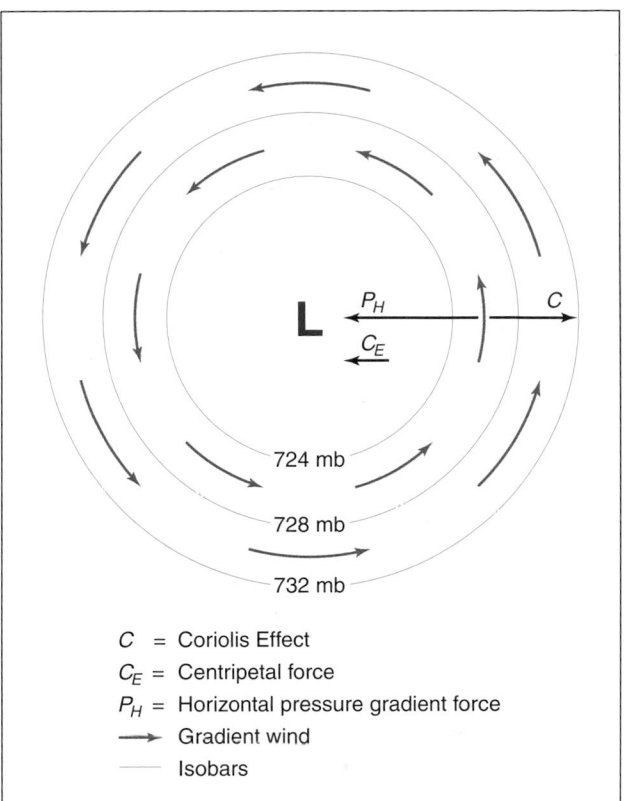

FIGURE 8.12
Viewed from above in the Northern Hemisphere, the gradient wind blows counterclockwise and parallel to isobars in a cyclone. In this idealized case, isobars form a pattern of concentric circles.

In an idealized cyclone, isobars form a series of concentric circles about the location of lowest air pressure. As indicated in Figure 8.12, the horizontal pressure gradient force (P_H) is directed inward toward the cyclone center, and the Coriolis Effect (C) is directed radially outward from the center of the low. The pressure gradient force is slightly greater than the Coriolis Effect, with the difference being equal to the inward-directed centripetal force (C_E). Viewed from above the Northern Hemisphere, the gradient wind in a cyclone blows counterclockwise and parallel to isobars above the atmospheric boundary layer.

The geostrophic and gradient wind models only approximate the actual behavior of horizontal winds above the atmospheric boundary layer. Nonetheless, these approximations are quite useful, and atmospheric scientists routinely rely on such approximations in analyzing weather maps.

SURFACE WINDS IN HIGHS AND LOWS

Geostrophic and gradient winds are frictionless; that is, they occur at altitudes where frictional resistance is insignificant. How does friction (surface roughness) affect horizontal winds within the atmospheric boundary layer? Intuitively, we know that friction should slow the wind, but in addition, friction interacts with the other forces and alters the wind direction.

As shown in Figure 8.13, for large-scale air motion along a straight path, the frictional force (F) combines with the Coriolis Effect (C) to balance the horizontal pressure gradient force (P_H). Friction always acts directly opposite (180 degrees to) the wind direction whereas the Coriolis Effect is always at a right angle (90 degrees) to the wind direction. Friction slows the wind and thereby weakens the Coriolis Effect so that the Coriolis no longer balances the horizontal pressure gradient force. The horizontal pressure gradient force (P_H) is balanced by the resultant (R) of the Coriolis (C) plus friction (F). Friction (due to the roughness of Earth's surface) slows the horizontal wind and shifts the wind direction across isobars and toward low pressure.

The angle between near-surface wind direction and isobars depends on the roughness of Earth's surface. That angle varies from 10 degrees or less over relatively

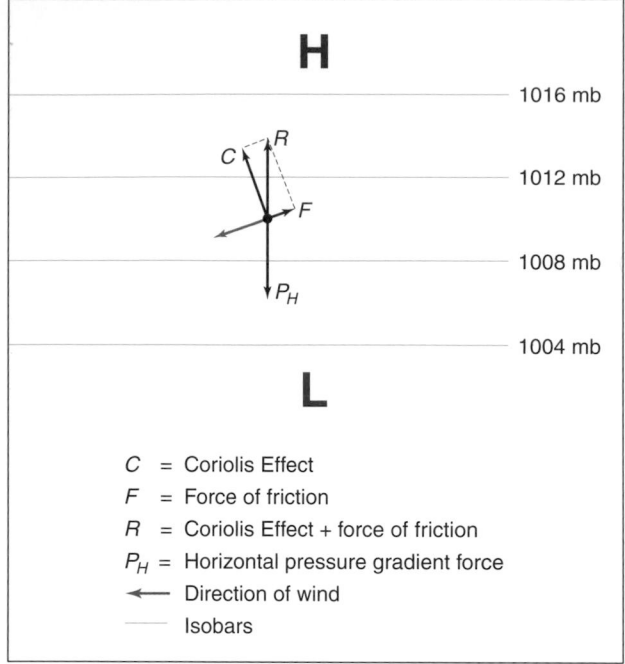

FIGURE 8.13
Within the atmospheric boundary layer, friction slows the wind and shifts the wind across isobars toward low pressure.

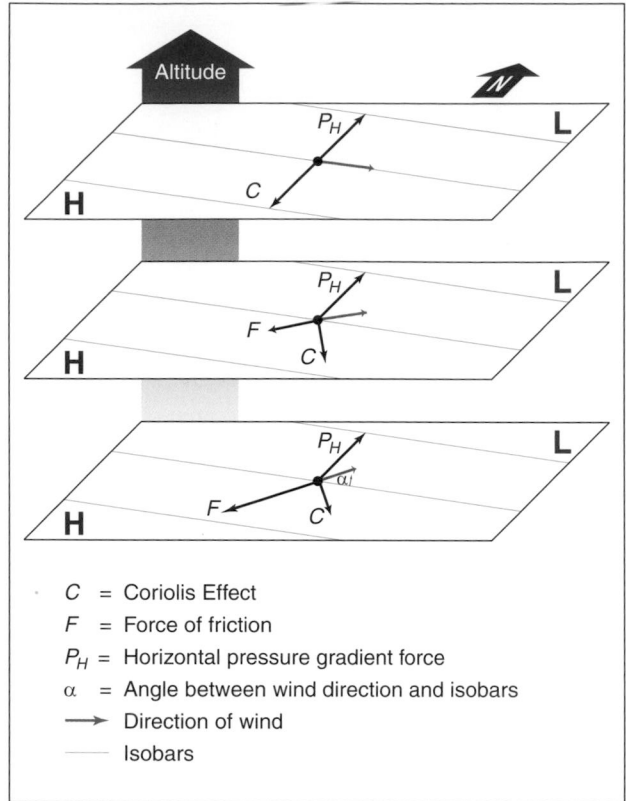

FIGURE 8.14
For the same horizontal air pressure gradient, the angle between the wind direction and isobars decreases with altitude within the atmospheric boundary layer.

smooth surfaces, such as over the ocean, where friction is minimal, to almost 45 degrees over rough terrain, where friction is greater, such as over a forest. As noted earlier, friction's influence on the horizontal wind diminishes with altitude and where it becomes negligibly small marks the top of the atmospheric boundary layer. Thus horizontal winds strengthen with altitude through the atmospheric boundary layer. Furthermore, the angle between wind direction and isobars is greatest near Earth's surface, decreases with altitude, and is essentially zero at the top of the atmospheric boundary layer (Figure 8.14). Above the atmospheric boundary layer, the horizontal wind is either geostrophic (where isobars are straight) or gradient (where isobars are curved).

How does surface roughness affect horizontal surface winds blowing in an anticyclone and cyclone? As with surface winds in a pressure field in which isobars are straight and parallel, friction slows anticyclonic and cyclonic winds and combines with the Coriolis Effect to shift winds so that they blow across isobars and toward low pressure. Viewed from above, surface winds in a Northern Hemisphere anticyclone blow clockwise and spiral outward, as shown in Figure 8.15, and surface winds in a Northern Hemisphere cyclone blow counterclockwise and spiral inward, as shown in Figure 8.16.

The pattern of surface winds associated with a cyclone is the basis for a simple rule of thumb for locating the center of a cyclone. In a Northern Hemisphere location, if you stand with your back to the wind and then turn approximately 45 degrees to your right, the cyclone center will be located to your left. This rule is a modification of an observation first stated in 1857 by Dutch meteorologist Christopher H. D. Buys-Ballot (1817-1890). It must be applied with caution, however, because isobars are not always circular in cyclones and large-scale surface winds may be modified by local atmospheric circulation systems such as a sea breeze or a valley breeze (Chapter 10). In addition, winds may be channeled through mountain passes or in urban canyons in ways that do not conform to this model.

In the Southern Hemisphere, anticyclonic and cyclonic winds blow opposite their Northern Hemisphere counterparts. The difference is due to the Coriolis Effect acting to the left of the direction of motion in the Southern Hemisphere, opposite to its deflection in the Northern Hemisphere. Viewed from above in the Southern

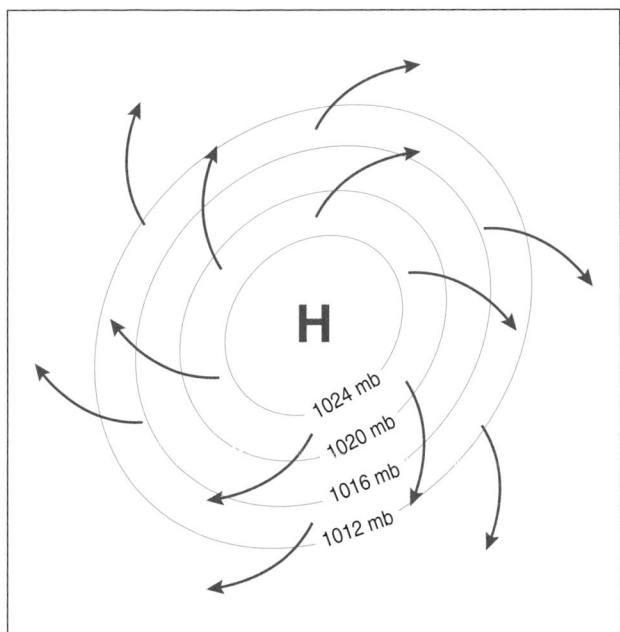

FIGURE 8.15
Viewed from above in the Northern Hemisphere, surface winds blow clockwise and outward in an anticyclone.

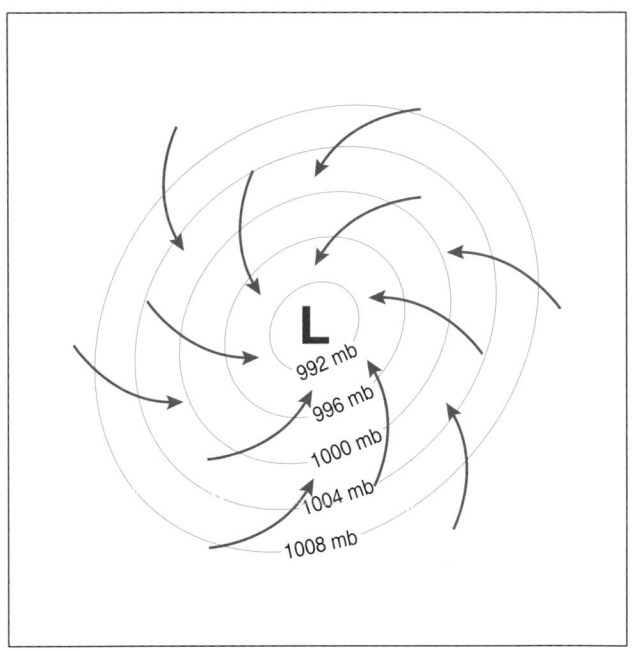

FIGURE 8.16
Viewed from above in the Northern Hemisphere, surface winds blow counterclockwise and inward in a cyclone.

Hemisphere, surface winds in a cyclone blow clockwise and spiral inward, whereas surface winds in an anticyclone blow counterclockwise and spiral outward. Above the atmospheric boundary layer, Southern Hemisphere cyclonic winds blow clockwise and parallel to isobars, whereas Southern Hemisphere anticyclonic winds blow counterclockwise and parallel to isobars.

A glance at almost any national weather map reveals that isobars seldom describe lengthy straight segments or circular patterns (Figure 8.17). Isobars often

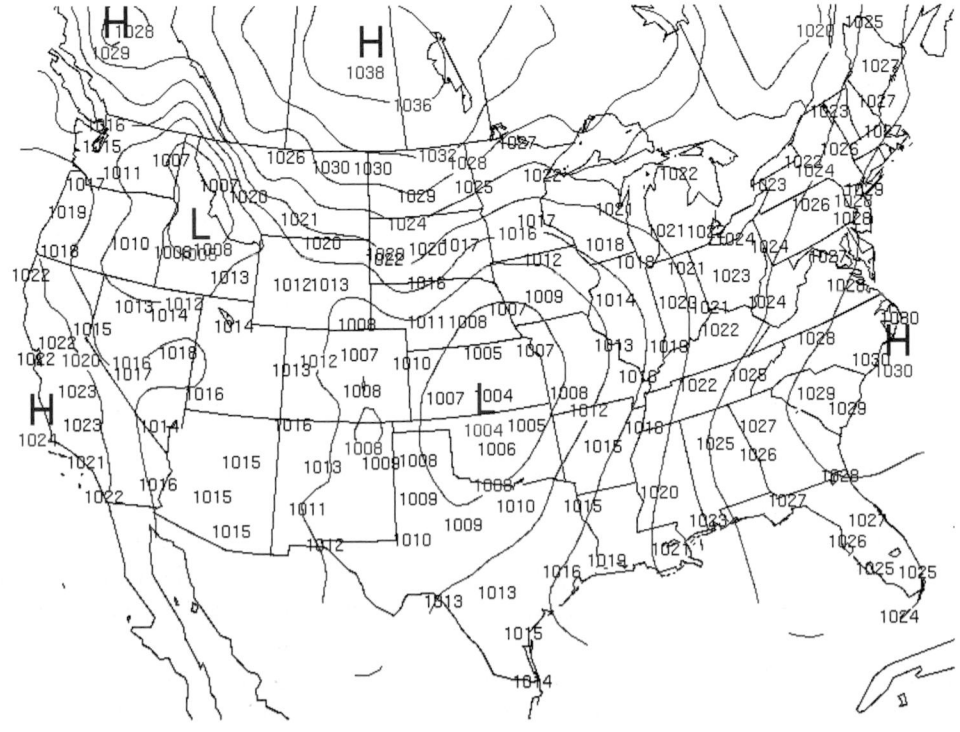

FIGURE 8.17
On a typical surface weather map, isobars exhibit anticyclonic curvature (*ridges*) and cyclonic curvature (*troughs*).

form more complicated patterns of *ridges* (anticyclonic curvature) and *troughs* (cyclonic curvature). In ridges and troughs, winds tend to parallel isobars above the atmospheric boundary layer and cross isobars toward low pressure near Earth's surface. An additional consideration in analyzing isobaric patterns for wind is the spacing of isobars. As noted earlier, the greater the air pressure gradient, the faster is the wind. Where isobars are closely spaced, the geostrophic and gradient winds are relatively strong. Where isobars are widely spaced, these winds are weak. The same rule applies to surface winds.

Continuity of Wind

Like all fluids, air is continuous and this *continuity* implies a link between the horizontal and vertical components of the wind. For example, surface winds are forced to follow Earth's undulating topography, ascending hills and descending into valleys. In addition, uplift occurs along frontal boundaries as one air mass advances and either overrides or pushes under another retreating air mass (Chapters 1 and 6). Having examined the horizontal circulation in anticyclones and cyclones, we can identify other important connections between the horizontal and vertical components of the wind.

As noted earlier in this chapter, surface winds in a Northern Hemisphere anticyclone spiral clockwise and outward from its high pressure center. Consequently, horizontal surface winds diverge away from the center of the high. A vacuum does not develop at the center, however, because air descends to Earth's surface and replaces the air that is diverging. Aloft, horizontal winds converge above the center of the surface high

to replace the descending air (Figure 8.18). Recall from Chapter 6 that adiabatic compression raises the temperature and saturation vapor pressure of descending air, causing existing clouds to vaporize, and the relative humidity to lower. Skies therefore tend to be clear within anticyclones, and anticyclones are appropriately described as fair weather systems. Furthermore, within an anticyclone, the horizontal air pressure gradient is typically very weak over a broad area around the center of the system. Light winds or calm air coupled with clear skies and low humidity favor intense nocturnal radiational cooling. Air adjacent to the ground may be chilled to saturation so that nighttime dew, frost, or radiation fog may develop (Chapter 7). As discussed in Chapter 10, air masses develop under large, slow moving high pressure systems because the air is modified by the underlying surface.

Surface winds in a Northern Hemisphere cyclone spiral counterclockwise and inward. Surface winds therefore converge toward the center of a low. Air does not simply pile up at the center; rather, air ascends in response to converging surface winds and diverging winds aloft (Figure 8.19). Recall from Chapter 6 that adiabatic expansion of ascending air lowers the temperature and saturation vapor pressure, thereby increasing the relative humidity of unsaturated air. Clouds and precipitation may eventually develop, so that cyclones are typically stormy weather systems. Because air flows into a low pressure system from all directions, at middle and high latitudes, these weather systems tend to bring together different air masses forming fronts.

Continuity of the wind also means that vertical motion can be induced by downwind changes in surface

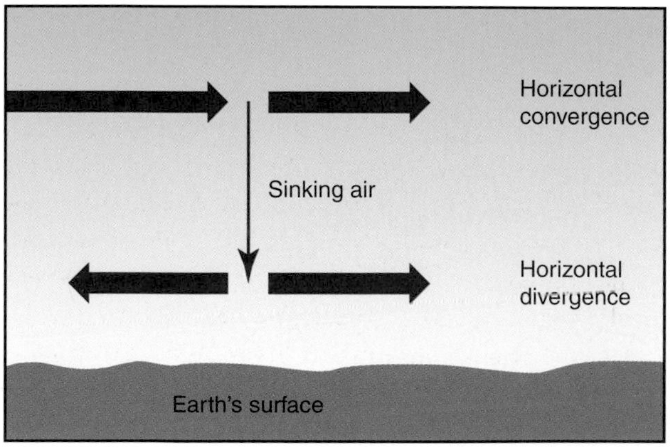

FIGURE 8.18
In this idealized vertical cross-section of an anticyclone, horizontal winds converge aloft, air descends, and surface winds diverge.

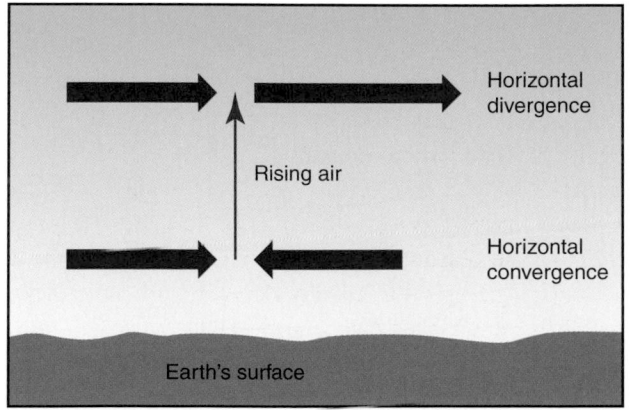

FIGURE 8.19
In this idealized vertical cross-section of a cyclone, surface winds converge, air ascends, and winds aloft diverge.

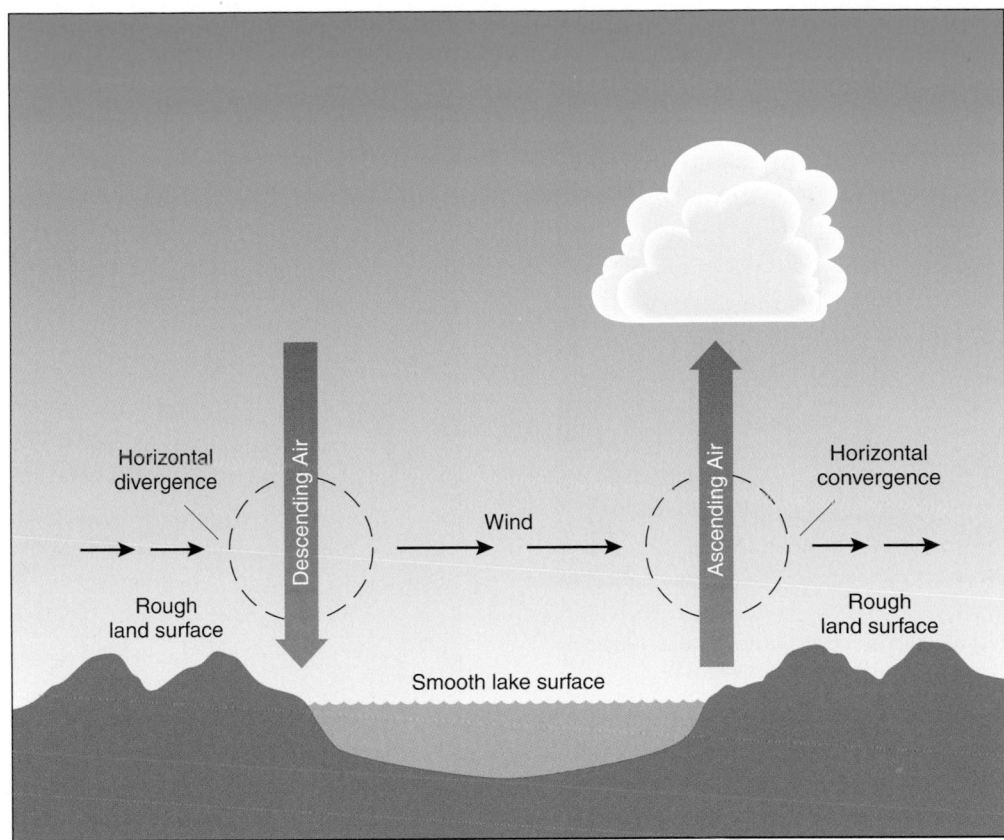

FIGURE 8.20
Surface winds accelerate and undergo horizontal divergence when blowing from a rough to a smooth surface (e.g., from land to water). Surface winds slow and undergo horizontal convergence when blowing from a smooth to a rough surface (e.g., from water to land). Divergence of surface winds causes air to descend whereas convergence of surface winds causes air to ascend.

roughness. The rougher the Earth's surface, the greater the resistance it offers to horizontal winds. When the horizontal wind blows from a rough surface to a relatively smooth surface, as when it blows from land to sea, the wind accelerates. As shown in Figure 8.20, this acceleration causes the wind to diverge (stretch), thereby inducing downward motion of air (an example of *speed divergence*). In contrast, when the horizontal wind blows from a smooth to a rough surface, the wind slows and converges (piles up), thereby inducing upward air motion (an example of *speed convergence*). This is one reason why, along a coastline, cumuliform clouds (e.g., cumulus) tend to develop with an onshore wind (directed from sea to land) and tend to dissipate with an offshore wind (directed from land to sea). Frictionally-induced convergence of surface winds also plays an important role in the development of lake-effect snow, as discussed in this chapter's first *For Further Exploration*.

Monitoring Wind Speed and Direction

A distinction is usually made between *horizontal* (east-west and north-south) and *vertical* (up-down) components of the wind. Except in small, intense weather systems such as thunderstorms, the magnitude of vertical air motion is typically only 1% to 10% of the horizontal wind speed. Nonetheless, as demonstrated in Chapter 6, the vertical component of the wind plays the key role in cloud formation with ascending air that promotes expansional cooling and in creating broad expanses of fair weather by the compressional heating of sinking air. Furthermore, as described above, vertical and horizontal components of the wind are linked so that a change in one may be accompanied by a change in the other.

The most common wind-monitoring instruments are designed to measure only the horizontal component of

FIGURE 8.21
A traditional iron cast wind vane showing a south wind. [Photo by Unisouth/License: Creative Commons ShareAlike 3.0]

FIGURE 8.22
The Space Shuttle Endeavour is framed by a wind sock at launch pad 39A at NASA's Kennedy Space Center in Cape Canaveral, Florida on 11 July 2009. [NASA/Bill Ingalls]

the wind. For some specialized research purposes, very sensitive instruments are available that measure vertical wind speeds or a combination of vertical and horizontal wind components of air in motion. An ordinary **wind vane** consists of a free-swinging horizontal shaft with a vertical plate at one end and a counterweight (arrowhead) at the other end (Figure 8.21). The counterweight always points directly into the wind. Another design is the airport **windsock**, a cone-shaped cloth sleeve that is open at both ends (Figure 8.22). The larger end of the sock is held open by a metal ring that is attached to a pole and is free to rotate. Air enters the larger opening and stretches the sleeve downwind.

Wind direction is always designated as the direction *from which* the wind blows. A wind blowing from the east toward the west is described as an east wind whereas a wind blowing from the northwest to the southeast is a northwest wind. A wind vane may be linked electronically or mechanically to a dial that is calibrated to read in points of the compass or in degrees. Measured clockwise from true north, an east wind is specified as 90 degrees, a south wind as 180 degrees, a west wind as 270 degrees, and a north wind as 360 degrees. Wind direction is reported as 0 degrees only during calm conditions (when the wind speed is zero).

Wind speed can be estimated by observing the wind's effect on lake or ocean surfaces or on land-based flexible objects such as trees. Such observations are the basis of the **Beaufort scale**, which is a graduated

sequence of wind strength ranging from 0 for calm air to 12 for hurricane-strength winds (Table 8.1). The scale bears the name of Sir Francis Beaufort (1774-1857), who developed it in the early 1800s while a ship's commander in the British Navy. Beaufort's goal was to standardize terms used by sailors to describe the state of the sea under various wind conditions. In 1838, after some revision, the British Navy adopted the Beaufort scale, and in 1853, it was sanctioned for international use by seafarers. Later, when the scale was extended from sea to land, it was necessary to develop wind speed equivalents for each Beaufort number. A uniform set of equivalents was adopted in 1926 and revised slightly in 1946. The Beaufort scale is still in use today. Although crude by today's standards, some modern-day mariners prefer Beaufort numbers to onboard instrument measurements of wind speed at sea.

A **cup anemometer** consists of 3 or 4 open hemispheric or cone shaped cups mounted to spin

TABLE 8.1
Beaufort Scale of Wind Force

Beaufort Number	General description	Land and sea observations for estimating wind speed	Wind speed 10 m above ground (km per hr)
0	Calm	Smoke rises vertically; sea like mirror.	<1
1	Light air	Smoke but not wind vane shows direction of wind; slight ripples at sea.	1-5
2	Light breeze	Wind felt on face, leaves rustle, wind vane moves; small, short wavelets.	6-11
3	Gentle breeze	Leaves and small twigs move constantly, small flags extended; large wavelets, scattered whitecaps.	12-19
4	Moderate breeze	Dust and loose paper raised, small branches moved; small waves, frequent whitecaps.	20-28
5	Fresh breeze	Small leafy trees swayed; moderate waves.	29-38
6	Strong breeze	Large branches in motion, whistling heard in utility wires; large waves, some spray.	39-49
7	Near gale	Whole trees in motion; white foam from breaking waves.	50-61
8	Gale	Twigs break off trees; moderately high waves of great length.	62-74
9	Strong gale	Slight structural damage; crests of waves begin to roll over, spray may impede visibility.	75-88
10	Storm	Trees uprooted, considerable structural damage; sea white with foam, heavy tumbling of sea.	89-102
11	Violent storm	Very rare, widespread damage; unusually high waves.	103-118
12	Hurricane	Very rare, disastrous; much foam and spray greatly reduce visibility.	119 and over

FIGURE 8.23
A cup anemometer measures wind speed on Craft Island, WA. The greater the wind speed, the faster the cups spin. [Photo by Walter Siegmuind/License: Creative Commons ShareAlike 3.0]

horizontally on a vertical shaft (Figure 8.23). At least one open cup faces the wind at any point in time. The rotation rate of the cups is calibrated to read in m per sec, km per hr, or knots. (One *knot* = 1 nautical mph = 0.515 m per sec = 1.15 statute mph.) Several other types of anemometers are available, including the very sensitive **hot-wire anemometer**. With this instrument, the wind blows past a heated wire or wires and the heat lost to moving air is calibrated in terms of wind speed. Some wind sensors, called **aerovanes** (Figure 8.24), are designed to measure both wind speed and direction. This instrument has a three or four blade propeller that spins at a rate proportional to the wind speed and a fin on its back that causes it to turn into the wind, thereby indicating the wind direction. Electronic sensors in the

FIGURE 8.24
Project Vortex mesonet vehicle aerovane for wind speed and direction. [NOAA Photo Library; OAR/ERL/National Severe Storms Laboratory]

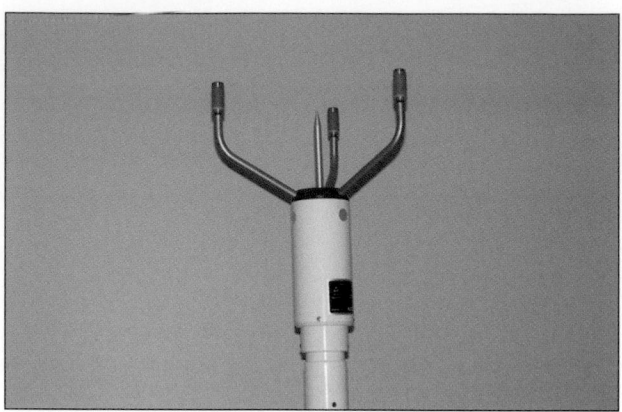

FIGURE 8.25
A sonic anemometer is based on the effect of wind on the propagation of sound waves. This instrument recently replaced cup anemometers at NWS Automated Surface Observing Systems (ASOS).

instrument are connected to a recording computer or digital display panel.

Another wind measuring instrument is based on the effect of wind on the propagation of sound waves (Chapter 14). A **sonic anemometer** consists of three arms that send and receive ultrasonic pulses (Figure 8.25). The travel times of sound waves with and against the wind are translated into wind speed and direction. Sonic anemometers have recently replaced cup anemometers as the standard wind sensor in the National Weather Service's Automated Surface Observing System (ASOS).

Recording a continuous trace of wind speed and direction can be informative. Wind vanes and anemometers can be linked to pens that record on a paper chart that is attached to a clock-driven drum. For example, the trace in Figure 8.26 indicates considerable variation in both wind direction and wind speed with time. Gusts and lulls in the wind speed, along with changes in wind direction, indicate turbulent air. More commonly today, a computer records the output from wind vanes and anemometers and readings are displayed digitally.

Ideally, a wind vane and anemometer system should be mounted on a tower so that the instruments monitor horizontal winds 10 m (33 ft) above the ground, the standard height for National Weather Service anemometers. Rooftop locations should be avoided because the wind tends to accelerate as it flows over buildings. In addition, the system should be sited well away from (1) structures that might shelter the instruments and (2) any obstacles that might channel (and thus accelerate) the wind or alter its direction.

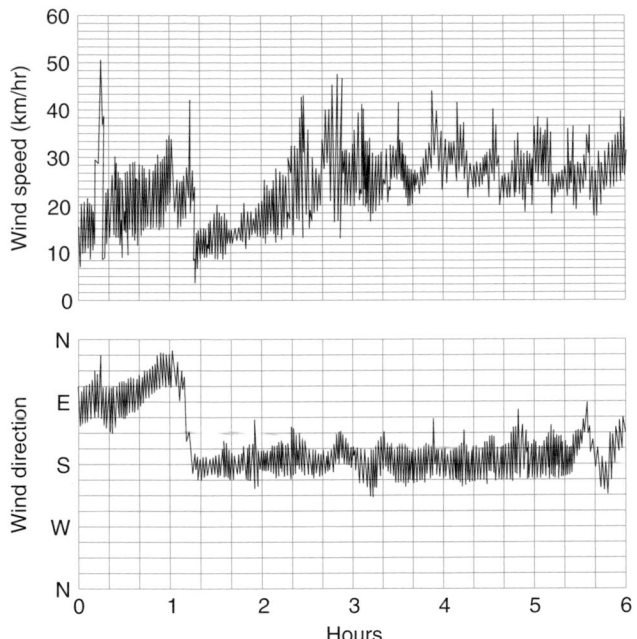

FIGURE 8.26

Continuous trace of time variations in wind speed and direction over a six-hour period.

Wind vanes and anemometers monitor winds near Earth's surface but primarily over land. In the past, information on winds at sea came from infrequent and sometimes unreliable measurements by ships and instrumented buoys. Today, scatterometers flown on satellites (or aircraft) monitor near-surface ocean wind speeds and directions accurately and continually. A *scatterometer* is a radar system that emits pulses of microwave energy to the sea surface where waves backscatter some of the energy to the instrument (an echo). The stronger the wind, the higher the waves, and the greater is the backscattering. Sea-surface wind vectors are determined from the strength and pattern of the echoes.

How are winds measured aloft? In Chapter 2, we saw that winds in the troposphere are measured by tracking the movements of a balloon-borne radiosonde (a *rawinsonde observation*). In the 1970s, sensors onboard geostationary satellites began to indirectly measure winds aloft over the ocean and other regions where weather observations are sparse. A sequence of satellite images of the same Earth view formatted as an animated loop is used to determine the speed of individual recognizable cloud elements that are assumed to be carried by the wind at a given level in the atmosphere. Since the early 1990s, the Doppler Effect has been used in *wind profilers* to monitor winds up to an altitude of about 16,000 m (52,500 ft). For more on these wind profilers, refer to this chapter's second *For Further Exploration*. This chapter's third *For Further Exploration, Wind Power*, describes how modern technology is tapping the energy of air in motion.

Scales of Atmospheric Circulation

Although the atmosphere is a continuous fluid, for convenience of study we subdivide atmospheric circulation into discrete weather systems that operate at different spatial and temporal scales (Table 8.2). The large-scale wind belts encircling the planet (polar easterlies, midlatitude westerlies, and trade winds) are global or **planetary-scale systems**. **Synoptic-scale systems** are continental or oceanic in scale; extratropical cyclones, hurricanes, and air masses are examples. **Mesoscale systems** include thunderstorms and sea and lake breezes, circulation systems that are so small that they may influence the weather in only a portion of a large city or county. A weather system covering a very small area such as several city blocks or a small town represents the smallest spatial subdivision of atmospheric motion, **microscale systems**. A weak tornado is an example of a microscale system.

Circulation systems not only differ in spatial scale, they also contrast in life expectancy. Essentially the same pattern of planetary-scale circulation may persist for weeks or even months (Chapter 9). Synoptic-scale systems typically last for several days to a week or so as they travel

TABLE 8.2 Scales of Atmospheric Circulation			
Circulation	*Space scale*	*Time scale*	*Example*
Planetary scale	3000 to 40,000 km	weeks to month	trade winds
Synoptic scale	1000 to 3000 km	days to a week	Arctic high
Mesoscale	10 to 1000 km	hours to a day	severe thunderstorm
Microscale	1 m to 10 km	seconds to an hour	tornado

over distances of thousands of kilometers. Mesoscale systems usually complete their life cycles in a matter of hours to perhaps a day, whereas microscale systems might persist for minutes or less. Other differences exist among the various scales of atmospheric circulation systems. At the micro- and meso-scale, vertical wind speeds may be comparable in magnitude to horizontal wind speeds. At the synoptic and planetary scales, however, horizontal winds are considerably stronger than vertical flow. Furthermore, at the micro- and meso-scale, the Coriolis Effect is usually negligibly small. By contrast, the Coriolis Effect is very important in synoptic- and planetary-scale circulation systems.

Each smaller scale weather system is part of, and dependent on, larger scale atmospheric circulation; that is, the various scales of atmospheric motion form a kind of hierarchy. For example, extreme nocturnal radiational cooling requires a synoptic weather pattern that favors clear skies and light winds or calm air. At the microscale, such weather conditions may lead to formation of frost or radiation fog in a river valley. But regardless of size, all scales of atmospheric circulation are ultimately governed by the boundary conditions of the system. In Chapters 9 through 12, we apply basic understandings of the atmosphere in examining weather systems operating at all scales.

Conclusions

Unequal rates of radiational heating and radiational cooling within the Earth-atmosphere system are responsible for gradients in temperature. In response to those temperature gradients, the atmosphere circulates and thereby, heat energy is converted to kinetic energy. In this chapter, we have examined the various forces that initiate and shape atmospheric circulation (the wind). Note that the pressure gradient force and gravity would exist even if air were not in motion, with the other forces (centripetal, Coriolis, and friction) coming into play only after air is in motion.

From everyday experience, we are aware of the end product of the forces described in this chapter, namely the wind, but we are not readily aware of the individual forces. In learning about atmospheric forces and how they interact, we see that each force is bound by certain constraints. For example, friction is important only in the lower troposphere, and the Coriolis Effect always deflects large-scale winds to the right in the Northern Hemisphere. We built a realistic model of atmospheric motion that demonstrates why and how winds circulate around high and low pressure systems.

In the following chapters, our awareness of these and other constraints will aid in understanding the characteristics of the various weather systems; for example, why hurricanes do not form at the equator and why winds in a tornado may blow in either a clockwise or counterclockwise direction. We are now ready to begin our discussion of the characteristics of the various circulation systems, beginning with those operating at the planetary scale.

Basic Understandings

- Wind is the movement of air measured relative to Earth's surface. Wind has both direction and magnitude and is usually divided into horizontal and vertical components, although atmospheric scientists often use the term "wind" when referring to the horizontal movement of air.

- The horizontal wind is governed by interactions of the pressure gradient force, centripetal force, Coriolis Effect, friction, and gravity.

- The pressure gradient force initiates air motion and arises in part from spatial variations in air temperature and, to a lesser extent, water vapor concentration. In response to pressure gradients, air accelerates away from areas of relatively high air pressure, perpendicular to isobars, and toward areas of relatively low air pressure.

- The centripetal force arises from an imbalance in other forces and operates whenever the wind follows a curved path. The centripetal force is responsible for changes in wind direction but not changes in wind speed.

- The Coriolis Effect arises from Earth's rotation on its axis. Wind is deflected to the right of its initial direction in the Northern Hemisphere, but to the left in the Southern Hemisphere. The force invented to describe and quantify this deflection is zero at the equator and increases with latitude to a maximum at the poles. The Coriolis Effect also increases as wind speed increases and is most important in large-scale (planetary- and synoptic-scale) circulation systems.

- Friction affects horizontal winds blowing within about 1000 m (3300 ft) of Earth's surface (the atmospheric boundary layer). Obstacles on Earth's surface slow the wind by breaking it into turbulent eddies.

- Gravity always pulls objects directly downward and is important in the vertical motion of air (e.g., cold air drainage).

- Hydrostatic equilibrium is the balance between the upward-directed pressure gradient force resulting from the decrease of air pressure with altitude and the downward-directed force of gravity. Slight deviations from hydrostatic equilibrium cause air to accelerate upward or downward.

- The geostrophic wind is a steady, horizontal wind that blows in a straight path parallel to isobars at altitudes above the atmospheric boundary layer. The geostrophic wind results from a balance between the horizontal pressure gradient force and the Coriolis Effect.

- The gradient wind is a horizontal wind that parallels curved isobars at altitudes above the atmospheric boundary layer. The centripetal force operates in the gradient wind as the result of an imbalance between the horizontal pressure gradient force and the Coriolis Effect. Viewed from above in the Northern Hemisphere, the gradient wind blows clockwise in anticyclones (*Highs*) and counterclockwise in cyclones (*Lows*).

- In large-scale (synoptic- and planetary-scale) circulation systems, friction slows the near-surface wind and interacts with the Coriolis Effect to shift the wind direction across isobars and toward lower pressure.

- Within the atmospheric boundary layer, viewed from above, horizontal winds blow clockwise and spiral outward in Northern Hemisphere anticyclones but counterclockwise and inward in Northern Hemisphere cyclones.

- In an anticyclone, horizontal divergence of surface winds causes air above to descend and warm by compression. Hence, an anticyclone is generally a fair-weather system. In a cyclone, horizontal convergence of surface winds causes ascending air, which expands and cools. Hence, a cyclone is frequently a stormy weather system.

- Along a coastline, surface winds blowing offshore speed up and undergo horizontal divergence (inducing descending air), whereas onshore winds are slowed by friction and undergo horizontal convergence (inducing ascending air).

- A wind vane and anemometer are the usual instruments for monitoring surface wind direction and speed. Winds aloft can be determined remotely by tracking rawinsondes, precipitation particles (by Doppler radar), and cloud elements (from sensors on Earth-orbiting satellites).

- Atmospheric circulation is typically divided into four spatial/temporal scales: planetary, synoptic, mesoscale, and microscale.

Enduring Ideas

- Forces have both magnitude and direction. The forces that either initiate or modify the wind include the pressure gradient force, centripetal force, Coriolis Effect, friction, and gravity. Horizontal winds strengthen with altitude through the atmospheric boundary layer and the angle between wind direction and isobars is greatest near Earth's surface. Above the boundary layer, the wind is geostrophic where isobars are straight and gradient where isobars are curved.

- Northern Hemisphere surface winds blow clockwise and outward about a high and counterclockwise and inward about a low. Air is continuous, thus it descends in anticyclones, which are typically fair weather systems, and ascends in cyclones, which are typically stormy weather systems. Highs and lows are the major synoptic-scale weather makers of middle and high latitudes.

- Wind direction is always designated as the direction from which the wind blows. Specialized instruments monitor wind speed and direction.

Key Terms

wind	eddy viscosity	windsock
force	atmospheric boundary layer	Beaufort scale
Newton's second law of motion	turbulence	cup anemometer
air pressure gradient	gravity	hot-wire anemometer
pressure gradient force	hydrostatic equilibrium	aerovanes
centripetal force	geostrophic wind	sonic anemometer
Coriolis Effect	gradient wind	planetary-scale system
friction	anticyclone	synoptic-scale systems
viscosity	cyclone	mesoscale systems
molecular viscosity	wind vane	microscale systems

Review

1. Provide a definition of wind.
2. What causes horizontal air pressure gradients? How do air parcels respond to a horizontal air pressure gradient?
3. What is the relationship between the horizontal wind speed and the spacing of isobars on a surface weather map?
4. Why does the Coriolis Effect reverse direction between the Northern and Southern Hemispheres?
5. Describe how the Coriolis Effect varies with wind speed and latitude.
6. How does the roughness of Earth's surface affect horizontal wind speed and direction within the atmospheric boundary layer?
7. Provide an example of how gravity influences air motion.
8. What forces are balanced in the geostrophic wind?
9. Why does the circulation within an anticyclone favor generally fair weather?
10. Why does the circulation within a cyclone usually bring cloudy, stormy weather?

Critical Thinking

1. Why is radiation fog more likely near the center of an anticyclone than near the center of a cyclone?
2. Distinguish between the geostrophic wind and the gradient wind.
3. The pattern of horizontal winds blowing about the center of a high-pressure system implies the existence of a centripetal force. Explain why.
4. Why are horizontal winds associated with a sloping pressure surface (e.g., 700-mb surface)?
5. Describe the relationship between a high pressure system and an air mass.
6. Along a coastline, cumuliform clouds are more likely with an onshore wind (directed from water to land) than an offshore wind (directed from land to water). Explain why.
7. Upper-air support for a developing cyclone requires horizontal divergence. Explain why.
8. Suppose that a cyclone is centered over St. Louis, MO. Describe the type of air mass advection to the southeast and to the northwest of the storm center.
9. In view of Newton's first law of motion, is the gradient wind a consequence of balanced forces? Explain your answer.
10. What is hydrostatic equilibrium? Is vertical motion of air possible with hydrostatic equilibrium? Explain your answer.

Lake-Effect Snow

A highly localized fall of snow immediately downwind of an open lake at middle latitudes of the Northern Hemisphere is known as *lake-effect snow*. Typically, such snows extend inland only a few tens of kilometers. In fact, the snowfall is often confined to such a small area that weather stations might not detect the event except by radar or local observer reports. Residents of the affected area, however, may be swamped by snow. On 21 January 1994, Adams, NY at the eastern end of Lake Ontario, received 152 cm (60 in.) of road-clogging snow in only 18 hrs. Such an extreme lake-effect snowfall is sometimes called a *snowburst*.

Lake-effect snow is most common in late autumn and early winter when lake surface temperatures are still relatively mild. As an early season outbreak of cold (Arctic) air streams over the lake, water readily evaporates and raises the vapor pressure of the lowest portion of the advecting air mass. (As described in Chapter 7, this same process may also produce *steam fog*.) In addition, the warmer lake water heats the advecting cold air from below, reducing its stability and enhancing convection and cloud development. Often this is all that is needed to trigger snowfall over the lake.

As the modified (milder, more humid, and less stable) air flows toward the lake's lee (downwind) shore, the contrast in surface roughness between the lake and land becomes important. The rougher land surface slows onshore winds, and the consequent horizontal speed convergence induces ascent of air, further development of clouds, and lake-effect snow. The topography of the downwind shore also affects the amount of precipitation—hilly terrain forces greater uplift and enhances snowfall.

The frequency and intensity of lake-effect snows hinge on the degree of air mass modification, which in turn, depends on (1) the temperature contrast between the relatively warm lake surface and overlying cold air, and (2) the distance the wind travels over open water (*fetch*). Field studies conducted in the Great Lakes region indicate that lake-effect snow is most likely when the air temperature difference between the lake-surface and the 850-mb level (at an altitude of about 1500 m or 5000 ft) is greater than 13 Celsius degrees (23 Fahrenheit degrees) and the temperature contrast between the lake-surface and the adjacent land surface exceeds 10 Celsius degrees (18 Fahrenheit degrees). Ideally, the fetch of the wind must be at least 160 km (100 mi) and the horizontal wind direction must exhibit very little shear with altitude (varying by less than 30 degrees between the lake-surface and an altitude of about 3000 m or 10,000 ft).

To the lee of the Great Lakes, the bulk of lake-effect snow falls between mid-November and mid-January, the period of greatest temperature contrast between lake-surface and the overlying air. Cold air usually sweeps into the Great Lakes region on north to northwest winds. Considering the maximum possible fetch, the greatest potential for substantial lake-effect snows is along the downwind southern and eastern shores of the Great Lakes. In these snowbelts, lake-effect snow accounts for a substantial portion of total seasonal snowfall (Figure 1). Snowbelts are located in western New York, northwestern Pennsylvania, northeastern Ohio, southwestern and central Ontario, northwestern and north-central Indiana, western Lower Michigan and the Upper Peninsula of Michigan. Because of the lake-effect, the greatest average seasonal snowfall east of the Rockies occurs on the Tug Hill Plateau of New York (east of

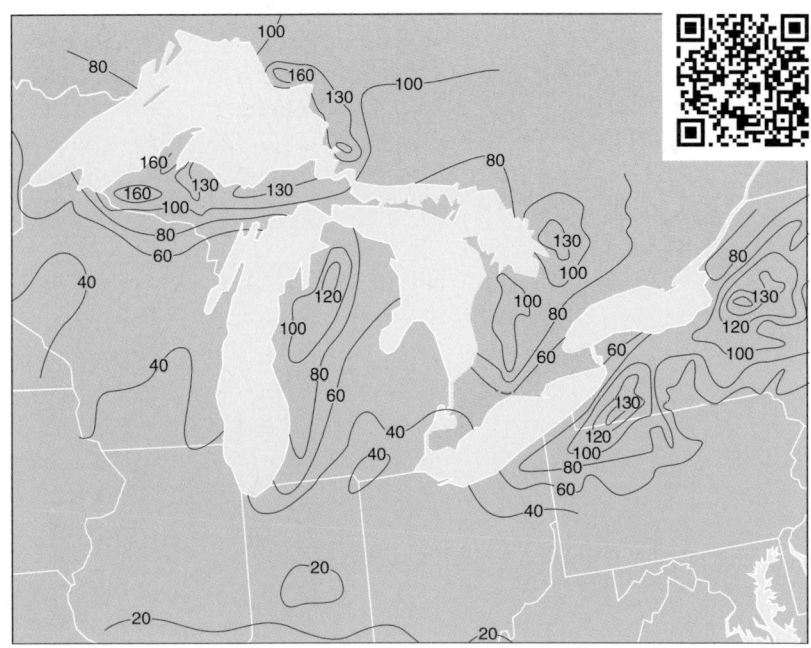

FIGURE 1

Average annual snowfall in the Great Lakes region (in inches). Note that the greatest snowfall totals occur downwind of the lakes and are largely due to lake-effect snows. [NOAA data]

Lake Ontario) and Upper Michigan (downwind of Lake Superior). In these locales, average seasonal snowfall typically exceeds 500 cm (200 in.). Delaware in Upper Michigan's Keweenaw County set a single-season record of 992 cm (390.4 in.) in 1978-79.

Occasionally, lake-effect snows develop on the western shores of the lakes which are normally upwind. For example, an early winter extratropical cyclone tracking through the lower Great Lakes region may produce strong northeast winds which blow onshore on the western shores. In this case, sometimes distinguished as *lake-enhanced snow*, the lake's snow-generating mechanism adds to the snowfall produced by the cyclone. This can mean paralyzing accumulations of snow for the Milwaukee-Chicago metropolitan areas, for example.

Strong cold winds blowing over the Great Lakes from the west and northwest produce lake-effect snow that falls from cloud bands that are oriented parallel to the wind direction. These *wind-parallel snow bands* are shown in the visible satellite image in Figure 2. Lake-effect snow can also develop with relatively weak regional winds. With a sufficient temperature contrast between water and land surfaces (more than 10 Celsius degrees or 18 Fahrenheit degrees), a horizontal air pressure gradient develops between the land and lake with higher pressure over the colder land surface and lower pressure over the warmer lake surface. In response to this pressure gradient, a cold wind develops that is directed offshore (called a *land breeze*). The leading edge of the land breeze is like a miniature cold front that forces upward the warmer air over the lake; clouds form and snow develops in bands that are oriented parallel to the shoreline. *Shore-parallel snow bands* are shown in the visible satellite image in Figure 3.

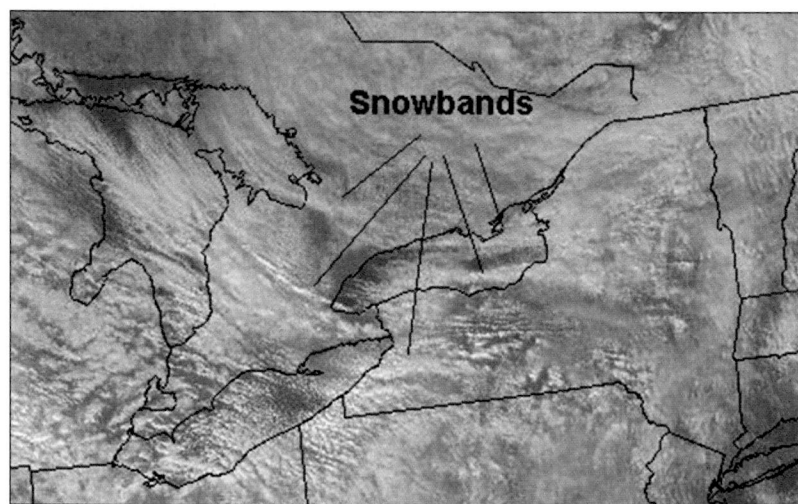

FIGURE 2
A visible satellite image showing widespread wind-parallel snow bands.

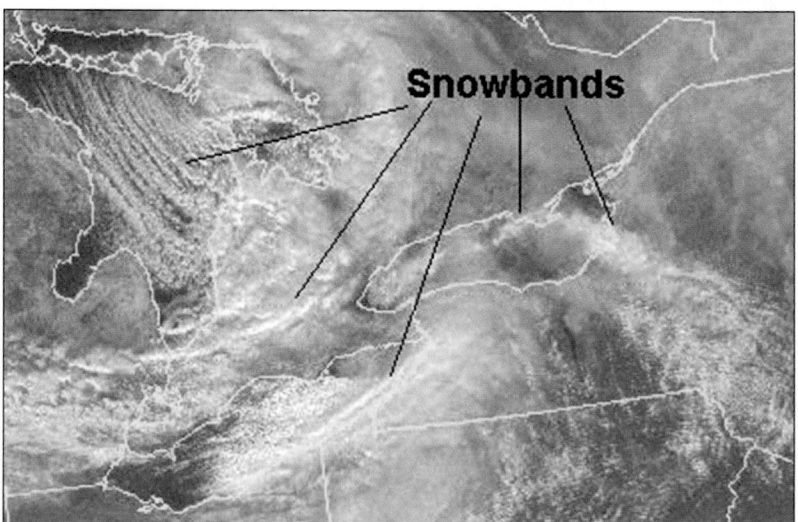

FIGURE 3
A visible satellite image showing shore-parallel snow bands along Lakes Erie and Ontario.

FIGURE 4
On December 23, 2001, less than two months before the start of the 2002 Winter Olympics, snow blankets Salt Lake City and the surrounding area. The Great Salt Lake (center) often contributes to the region's snowfall through the lake-effect.

Where shore-parallel snow bands drift onshore, they usually produce greater snowfalls than wind-parallel snow bands.

In addition to the Great Lakes region, lake-effect snows also affect the valleys southeast of Utah's Great Salt Lake. Each year perhaps six significant lake-effect snowfalls occur in the Tooele and Salt Lake valleys, bowl-shaped depressions that slope up and away from Great Salt Lake (Figure 4). Both valleys parallel the northwest-southeast-trending long axis of the lake. Which of the two valleys receives the heavier snowfall depends on wind direction. Cold air on northwest winds is moderated by contact with the relatively warm lake surface, and downwind, the mountain valley topography forces horizontal convergence and uplift. Clouds billow upward and locally release bursts of heavy snow.

Great Salt Lake snows are generally less intense than their Great Lakes counterparts. Major reasons are the smaller area of the lake and the shorter fetch of winds blowing over the lake, so that air masses do not modify as much. The waters of Great Salt Lake cover an area that is only about 13% of that of Lake Ontario, the smallest of the five Great Lakes. Maximum fetch on Great Salt Lake is only about 120 km (75 mi), whereas maximum fetches on the Great Lakes are many hundreds of kilometers.

Similar localized snowfalls sometimes accompany onshore winds along Atlantic coastal localities. Cape Cod, MA, for example, may experience this "ocean effect" snow with northerly winds. On rare occasions, north winds following the passage of a coastal low pressure system have even produced "bay-effect" snows over eastern Virginia off the Chesapeake Bay.

Wind Profilers

A special application of Doppler radar is providing meteorologists with much more detailed surveillance of winds aloft than is possible by tracking radiosondes (a *rawinsonde observation*). A *wind profiler* consists of a wire-mesh antenna about half the size of a tennis court that sends radar signals (having wavelengths of 33 cm to 6 m) upward into the atmosphere (Figure 1). The signal detects changes in atmospheric density caused by turbulent mixing of volumes of air that differ slightly in temperature and humidity. Fluctuations in the index of refraction are used as a tracer of both the horizontal and vertical components of the wind.

In May 1992, the National Oceanic and Atmospheric Administration (NOAA) announced that the nation's first wind profiler network was up and operating. The inaugural network consisted of 31 wind profilers in the central U.S. and 1 in Alaska. As of this writing, the wind profiler network is composed of 35 stations in 18 states including Alaska. Each wind profiler monitors wind speed and direction at 72 different levels up to an altitude of 16,250 m (53,314 ft). Data are automatically collected every 6 minutes and transmitted to a quality-control hub in Boulder, CO. Wind data are then averaged over 1-hr periods and made available to National Weather Service Forecast Offices across the nation.

The principal advantage of a wind profiler over a radiosonde is the much greater frequency of wind observations. As noted in Chapter 2, radiosondes are launched once every 12 hrs. Winds aloft can change significantly in much shorter periods. Such changes could, for example, alter the track of a winter cyclone so that a city in the storm's path receives snow instead of rain. More detailed monitoring of winds aloft also permits commercial airlines to better plan flight routes that avoid strong headwinds and take advantage of tailwinds. In recent years, temperature profiling and moisture-sensing capabilities have been added to some wind profilers.

FIGURE 1
A wind profiler located in northwestern Missouri.

Wind Power

Harnessing the kinetic energy of the wind is a technology that dates to the 12th century in portions of the Middle East where water power was not available. In North America, the energy crisis of the 1970s spurred renewed interest in this energy source. More recent efforts to reduce our nation's dependency on fossil fuels in view of concerns over possible global climate change have rekindled interest in *wind power*. Today, scientists and engineers employ modern aerodynamic principles and space-age materials in designing and constructing modern wind-driven turbines that convert some of the wind's kinetic energy into electricity.

In Chapters 3 and 4, we saw how the Sun drives the atmosphere. Only about 2% of the solar energy that reaches Earth is ultimately converted to the kinetic energy of wind, but that is still a tremendous amount of energy. Theoretically, windmill blades can convert a maximum of about 60% of the wind's energy into mechanical energy. In actual practice, however, scientists reported in 2007 that wind generators that began operating before 1998 extract on average about 22% of the wind's energy. Those put into operation after 1998 are more efficient, reaching 36% capacity in 2004 and 2005. Furthermore, a wind turbine delivers most of its electrical output when wind speeds are 40 km per hr (25 mph) or higher. The amount of power that a wind turbine can extract from air in motion varies directly with (1) the cube of the instantaneous wind speed at hub height (V^3), (2) the area swept out by the windmill blade, and (3) air density.

Wind speed is by far the most important consideration in evaluating a location's wind energy potential. Even small changes in wind speed translate into great variations in energy harvested. For example, doubling the wind speed (a common occurrence) multiplies the available wind power by a factor of eight ($2 \times 2 \times 2$). Wind speed usually increases with altitude above Earth's surface, especially within the lowest meters of the atmosphere, so that a wind turbine is mounted on the top of a tower to take advantage of the stronger winds at that elevation. As a general rule, a minimum of several years of detailed wind monitoring is needed for a preliminary evaluation of the wind power potential of any site. Also, the long-term climate record should be checked for the frequency of potentially destructive winds. Wind data from nearby weather stations can be very useful as long as care is taken in extrapolating wind data between localities and from one elevation to another. (Recall that standard National Weather Service anemometers are located 10 m or 33 ft above the ground.)

At ordinary speeds, wind is a relatively diffuse energy source, comparable in magnitude to incoming solar radiation. Hence, a wind turbine's power generation potential also depends on the area swept out by the windmill blades; larger windmill blades can harvest more energy. Until recent decades, design and strength-of-materials problems limited the size of wind turbines. Today, availability of stronger and lighter materials for blades is making possible larger wind turbines that can generate as much as 0.5 megawatt of electricity. One wind turbine design uses three 23-m (75-ft) blades (Figure 1).

Tapping the wind's energy for electric power generation faces several limiting factors. Wind speed and direction vary with exposure of the site, roughness of the terrain, and season of the year. The most formidable

FIGURE 1
This windmill farm in northeastern Wisconsin is designed to meet the electrical energy needs of about 3600 households.

obstacle to the development of wind power potential is the inherent variability of the wind so that the electrical output of a wind turbine also varies. Hence, a wind power system must include a means of storing the energy generated during gusty periods for use when the wind is light or the air is calm. Banks of batteries may serve this purpose. Wind power has its greatest immediate potential in regions where average winds are relatively strong and consistent in direction. Figure 2 is the U.S. Department of Energy (DOE) National Renewable Energy Laboratory (NREL) and AWS Truepower map of predicted mean annual wind speeds at 80 m, the minimum height in which utility-scale, land-based turbines are generally installed. Areas with wind resources suitable for wind developement typically have average 80-m wind speeds of 6.5 m/s or greater. Favorable regions include the western High Plains, the Pacific Northwest coast, portions of coastal California, the Great Lakes region, the south coast of Texas, and exposed summits and passes in the Rockies and Appalachians. About 60% of the nation's installed wind power occurs along a corridor from Texas to the Great Lakes (Figure 3). Shallow offshore locations are also being considered as potential sites for wind turbines.

Economy of scale favors centralized arrays of many wind turbines, called *windmill farms*, over individual household wind turbines. Windmill farms consist of a dozen or more super wind turbines, each capable of producing as

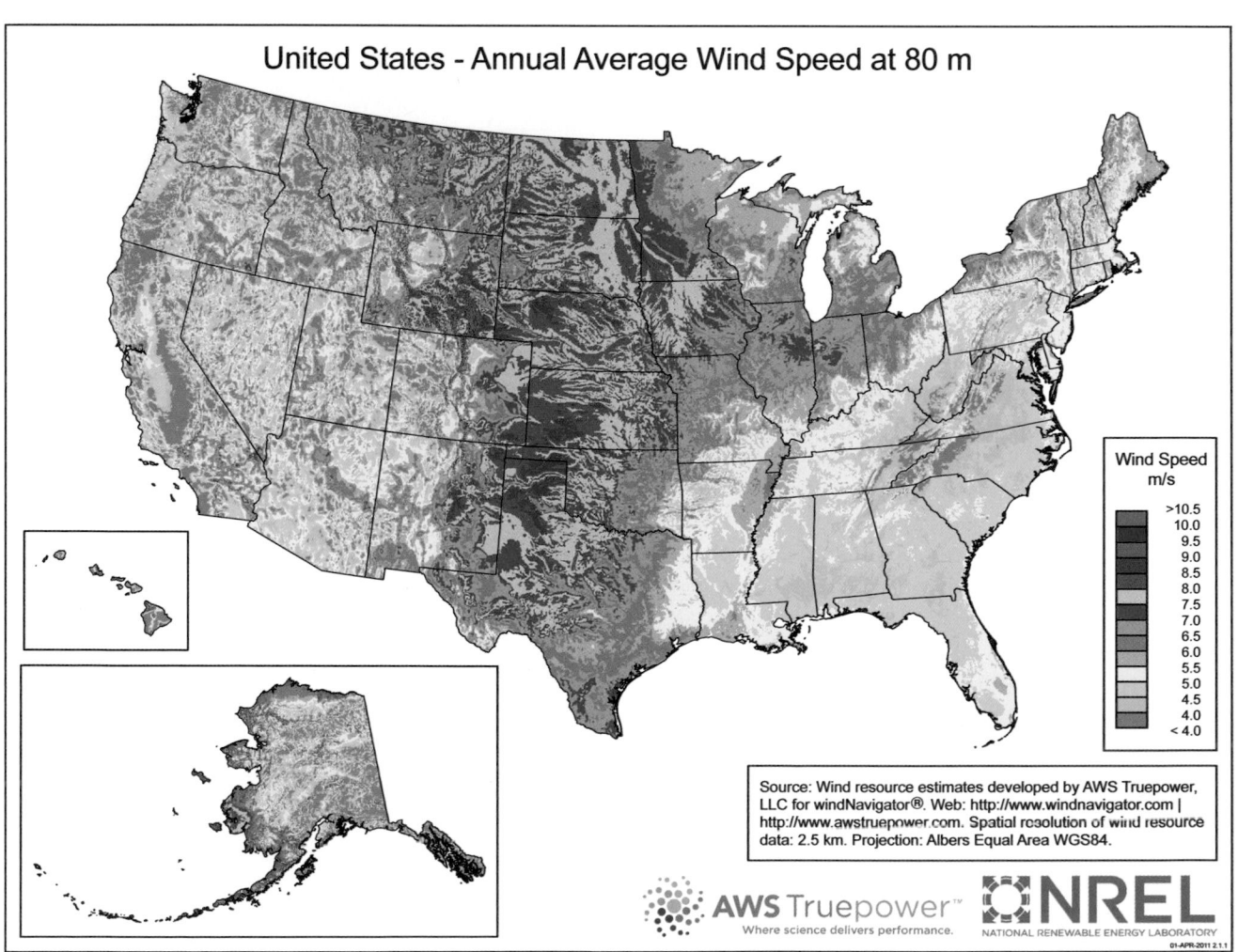

FIGURE 2
Wind resource map showing predicted mean annual wind speeds at 80 m height. [DOE/NREL and AWS Truepower]

much as 0.5 megawatt of electricity. Windmill farms currently operating in California supplement conventional sources by feeding electricity directly to power grids. Since 1981, more than 16,000 turbines have been installed in windy mountain passes in California. According to the California Energy Commission, in 2010, windmill farms in California generated 6172 gigawatt-hours of electricity, about 3% of the state's gross system power.

As of this writing, wind energy supplied more than 2% of the nation's electrical needs. As of 30 September 2011, more than 43,635 megawatts of wind power was installed across the United States. Nationwide, wind power is growing at an average rate of about 25% per year. DOE is leading a joint effort to increase wind's contribution to 20% by 2030. The DOE *20% Wind Energy by 2030: Increasing Wind Energy's Contribution to U.S. Electricity Supply* report explains that reaching this goal will require improved turbine technology, significantly expanded transmission infrastructure, enhanced manufacturing capability, and large markets to purchase and use wind energy. Meeting the 20% goal would prevent the production of 7.6 cumulative gigatons of carbon dioxide in the electric sector during the period 2007 to 2030, as well as reducing cumulative water consumption by 8% (4 trillion gallons). The initial investment to reach the 20% goal would be higher than for other energy sources, but wind power offers lower ongoing energy costs for the future.

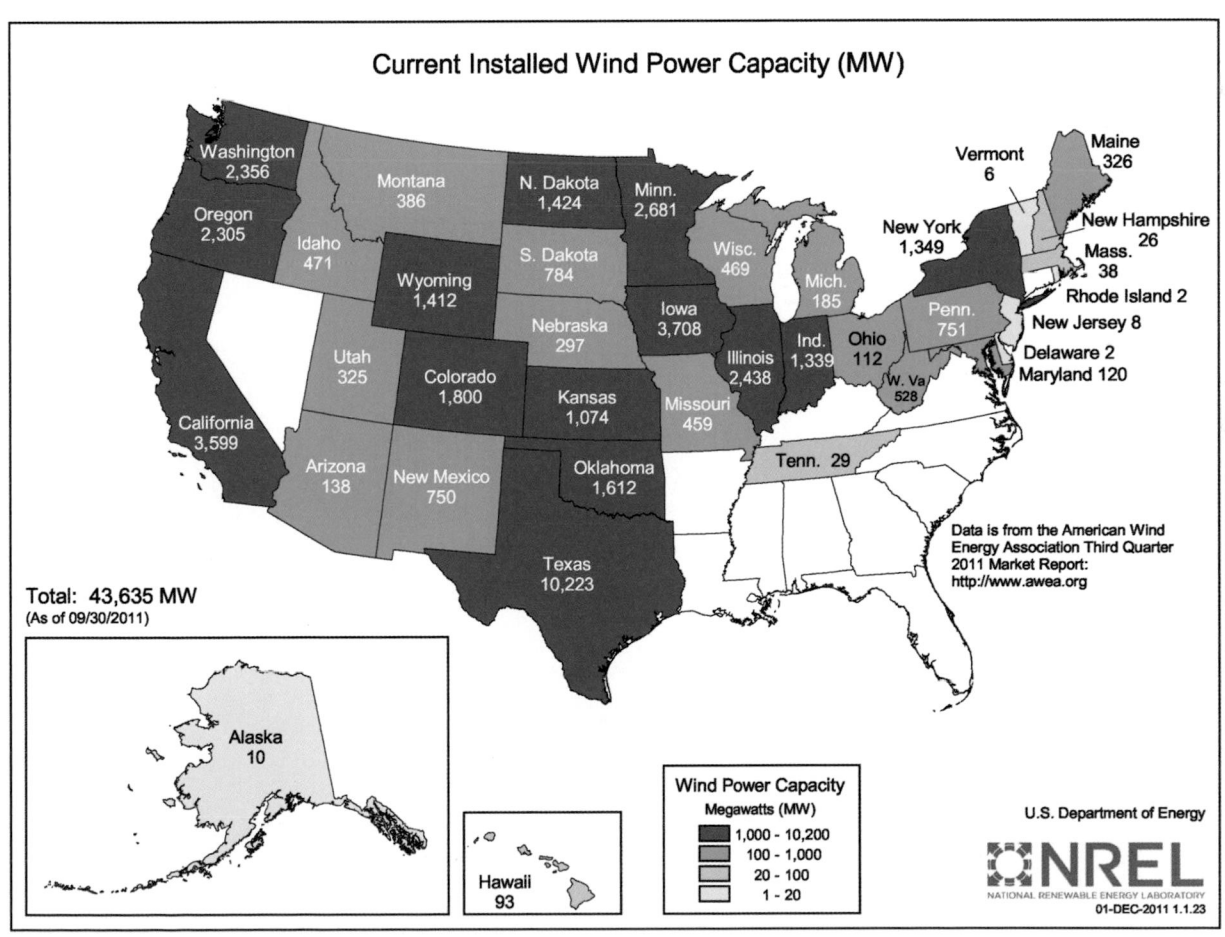

FIGURE 3
Installed wind power capacity map as of 2011. [DOE/NREL]

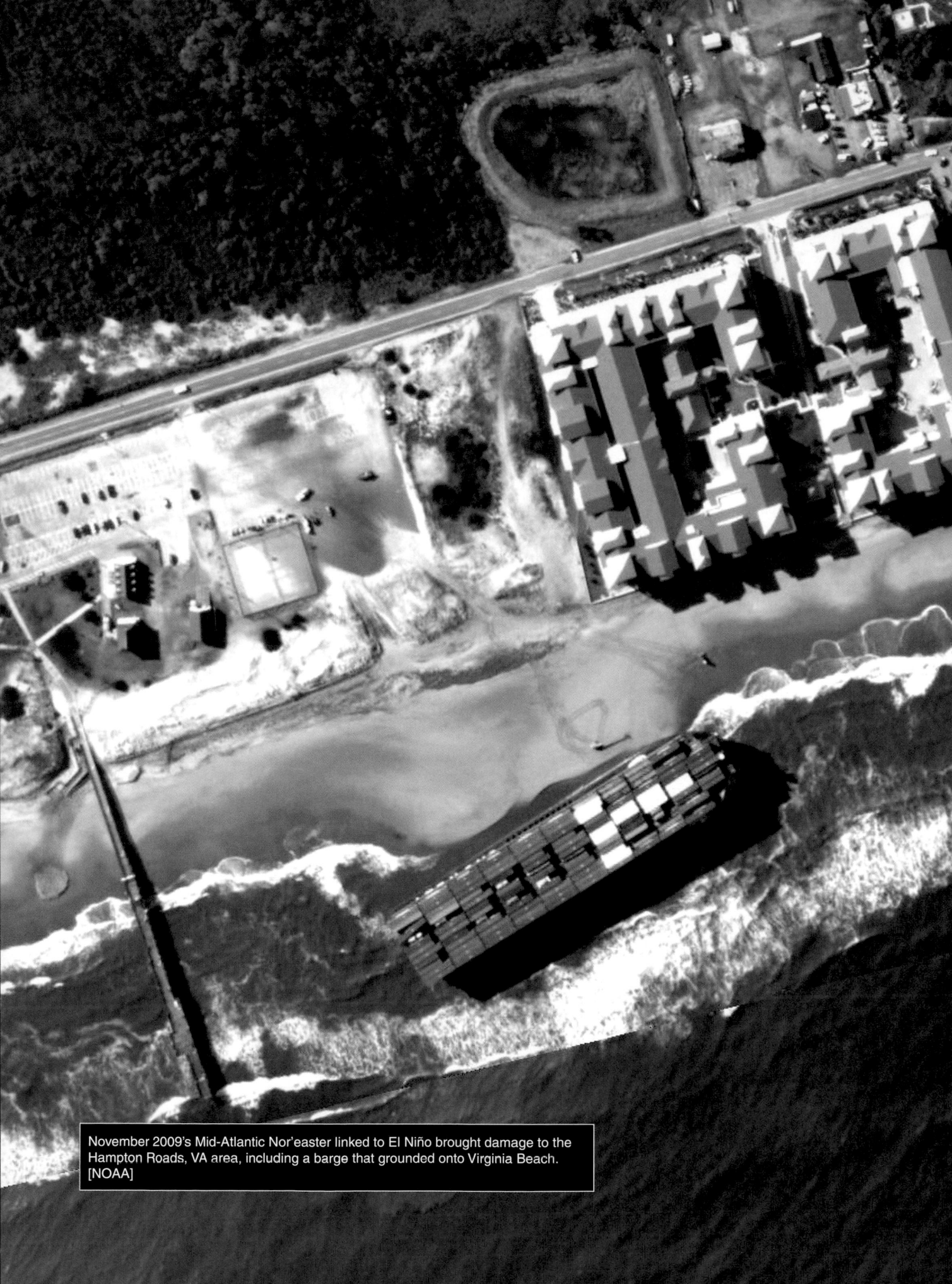

November 2009's Mid-Atlantic Nor'easter linked to El Niño brought damage to the Hampton Roads, VA area, including a barge that grounded onto Virginia Beach. [NOAA]

ATMOSPHERE'S PLANETARY CIRCULATION

Chapter Highlights

Case-in-Point
   Drought in the Sahel of West Africa
Idealized Circulation Pattern
Features of the Planetary-Scale Circulation
   Pressure Systems and Wind Belts
   Winds Aloft
   Trade Wind Inversion
   Seasonal Shifts
   Ocean Surface Currents
Monsoon Circulation
Long Waves in the Westerlies
   Zonal and Meridional Flow Patterns
   Blocking Systems and Weather Extremes
Jet Streams
Cyclone Development
El Niño and La Niña
   Ekman Transport
   *Neutral Conditions in the Tropical
     Pacific*
   Warm Phase
   Cold Phase
   Predicting and Monitoring ENSO
   Frequency of El Niño and La Niña
North Atlantic Oscillation
Arctic Oscillation
Pacific Decadal Oscillation
Conclusions
Basic Understandings/Enduring Ideas
Key Terms/Review/Critical Thinking
For Further Exploration
   Defining Drought
   *Historic El Niño Episodes of 1982-83 and
     1997-98*

Learning Objectives

Identify the principal components of the atmosphere's planetary-scale circulation.

Describe how semi-permanent subtropical anticyclones are linked to trade winds.

Describe the linkage between semi-permanent subtropical anticyclones and westerlies.

Compare and contrast planetary winds at Earth's surface with those in the mid and high troposphere.

Describe the seasonal changes in the planetary-scale circulation.

Demonstrate how the prevailing planetary-scale circulation influences climate.

Explain how prevailing winds set up ocean surface currents.

Identify the various factors that contribute to a monsoon circulation.

Contrast weather patterns associated with zonal and meridional flow in westerlies.

Explain the relationship between blocking circulation patterns and weather extremes.

Describe the linkage between the polar front and the middle latitude jet stream.

Explain the role of the jet stream and upper-air troughs in the development of extratropical cyclones.

Identify the changes that take place in the tropical Pacific ocean and atmosphere during El Niño and La Niña.

Describe the relationship between El Niño and the Southern Oscillation.

Define teleconnection and provide examples.

What are the principal features of planetary-scale atmospheric circulation and how do they affect weather and climate?

Drought in the Sahel of West Africa

Perhaps nowhere in the world has prolonged drought caused more human misery than sub-Saharan Africa, much of which is known as the Sahel. The *Sahel*, an Arabic word meaning "shore" or "border," is the transition zone in West Africa between the northern Sahara Desert and the southern rainforest of the Guinea coast. As shown on the map in Figure 9.1, the Sahel of West Africa includes all or part of (from west to east) Senegal, Mauritania,

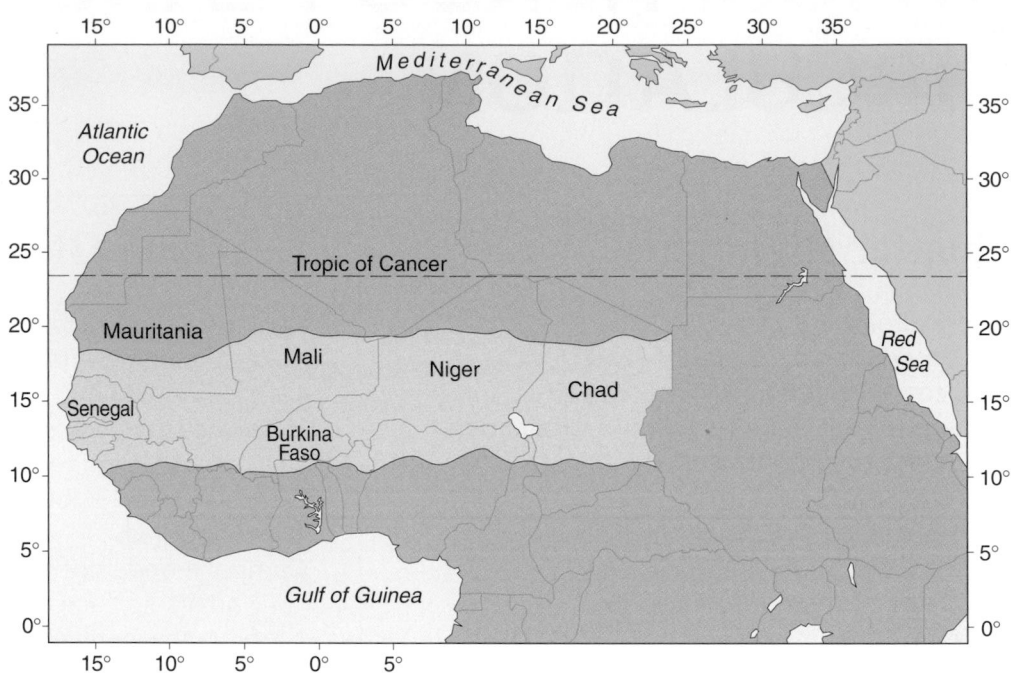

FIGURE 9.1
The Sahel of West Africa.

Mali, Burkina Faso, Niger, and Chad. These are among the poorest nations on Earth. Low average annual rainfall, considerable year-to-year variability in rainfall, and prolonged droughts bring considerable hardship to the people of the Sahel. Because so many depend on subsistence agriculture for their livelihood, they are particularly vulnerable to the negative impacts of drought. Droughts have forced them off lands that, even in the best of times, are marginal for survival of crops and livestock. Forced to migrate into cities in search of food and work, many people end up in refugee camps. When conditions worsen, the few remaining crops wither, livestock become emaciated and the people starve.

The West African monsoon circulation governs the semi-arid climate of the Sahel, with a rainy "high Sun" season (Northern Hemisphere summer) and a dry "low Sun" season (Northern Hemisphere winter). From north to south, the average length of the rainy season and mean annual rainfall increases across the Sahel. At the northern edge of the Sahel (about 18 degrees N), the rainy season usually begins in June and may last only a month or two; mean annual rainfall typically is less than 100 mm (about 4 in.). At the same time, in the extreme southern Sahel (about 10 degrees N), the rainy season begins as early as April and persists for up to 5 or 6 months. Mean annual rainfall exceeds 500 mm (about 20 in.).

Shifts between the dry season and rainy season in the Sahel are linked to changes in the prevailing winds associated with the north/south migration of the *intertropical convergence zone (ITCZ)*, a planetary-scale band of convective clouds and showers that marks the convergence of the trade winds. As the ITCZ follows the Sun, its northward surge in spring triggers rainfall whereas its southward shift in autumn brings the rainy season to an end. During the rainy season, surface winds blow from the southwest, transporting humid air inland from the Atlantic Ocean. During the dry season, the Sahel is under the influence of the dry eastern flank of the *Bermuda-Azores subtropical anticyclone* and surface winds blow from the dry north and northwest.

In a summer when the ITCZ fails to move as far north as usual, arrives late, or shifts southward, early rainfall is below average and a succession of such summers brings drought. In the

20th century, the people of the Sahel endured three major, persistent droughts: 1910-1914, 1940-1944, and the late 1960s through the late 1990s. Rainfall during the 1961-1990 climatological averaging period was 20% to 40% lower than it was during the prior three decades, constituting the greatest 30-year anomaly in precipitation recorded anywhere in the world. Drought and famine claimed over 600,000 lives in 1972-1975 and about the same number again in 1984-1985. The region's long-term climate record reconstructed from historic accounts of lake-level fluctuations and landscape changes indicates that intervals of severe drought, lasting for decades to centuries, have been common in the Sahel.

Drought in the Sahel is the product of interactions involving the atmosphere, ocean, land-use, and human activity. In recent decades, scientists have attributed severe drought in the Sahel to changes in atmospheric circulation (i.e., failure of the West African monsoon) arising from anomalies in sea-surface temperatures (SST) in the eastern tropical Atlantic, south of the equator, and southwest of West Africa (e.g., the Gulf of Guinea) as well as in the Indian Ocean. Human activity (i.e., overgrazing, conversion of woodland to agriculture) may also contribute to persistent drought in the Sahel by reinforcing dry conditions.

Overgrazing and deforestation alter the regional radiation budget by denuding the soil's vegetative cover and degrading the quality and moisture holding capacity of soils. Soil albedo increases, moisture supply to the atmosphere decreases, the surface cools, convection weakens, and rainfall decreases. Satellite surveillance confirms that successive phases of vegetation growth (during the rainy season) and vegetation senescence (during the following dry season) alter the land surface albedo. Bare soil tends to be brighter than vegetation so that an increase in vegetative cover is accompanied by a decrease in surface albedo. Parts of the Sahel probably are undergoing *desertification*, a positive feedback mechanism in which arable land is converted to desert due to a combination of climate change and poor land management practices. At present, the extent to which desertification contributes to climate variability (including drought) in the Sahel is not known.

What does the future hold for the Sahel? Today's climate models disagree on future trends in rainfall in the Sahel. For example, Kerry Cook and Edward Vizy of Cornell University selected three models from the 2007 *IPCC Fourth Assessment Report* that best reproduced the climate of the 20th century. One model predicted relatively wet conditions in the Sahel for the entire 21st century, another predicted severe drying late in the century, and the third called for modest drying throughout the period.

To improve prediction of the West African monsoon and better understand feedback mechanisms involving seasonal changes in rainfall, vegetation, and surface albedo in the Sahel, Phase I of the *African Monsoon Multidisciplinary Analysis (AMMA)* field experiment was conducted from 2002-2010. Phase II of the AMMA extends from 2010-2020. The AMMA is a consortium of more than 140 European, American, and African agencies and institutions that monitors almost all aspects of the West African monsoon including rainfall, cloud properties, dust transport, and SST in the Gulf of Guinea. AMMA Phase I made major progress in understanding the continental water cycle of Africa, observing Gulf of Guinea interactions with the monsoon system, understanding the role of land-atmosphere interaction in weather and climate prediction, discerning West African gas and aerosol emissions, and achieving measurement programs for health, agricultural, and water systems. AMMA Phase I made huge improvements to the upper-air (radiosonde) network and coordination of West African observing stations. A greater density of surface and upper-air stations promises to improve the performance of climate models by providing more data for initializing them and verifying their predictions. The scientific priorities for AMMA Phase II involve researching society, environment, and climate interactions, including improvements to weather and climate forecasting, and enhancing knowledge of the African monsoon system.

Ultimately, the planetary-scale atmospheric circulation is responsible for the development and displacement of smaller scale weather systems. Experience indicates that many of the weather systems, such as cyclones and anticyclones, which affect middle latitudes, travel from west to east whereas in the tropics, weather systems such as hurricanes and tropical storms usually track from east to west. Appropriately, our description of atmospheric circulation systems begins with the global scale. This chapter covers the pressure systems and wind belts that operate at the planetary scale, with special emphasis on circulation patterns of the middle latitude west winds (the westerlies). We also cover anomalous variations in the planetary circulation regime that result in El Niño and La Niña, where atmosphere/ocean interactions in the tropical Pacific have ramifications for weather and climate in many other regions of the world.

Idealized Circulation Pattern

The circulation of the atmosphere operating at the planetary-scale is shaped by boundary conditions that involve the forces discussed in Chapter 8. These boundary conditions include the flux of incoming solar radiation, outgoing infrared radiation, Earth's rotation on its axis (the Coriolis Effect), and the physical properties of Earth's surface (e.g., land versus ocean).

To understand how the planetary-scale circulation is shaped, we begin with an idealized model of planet Earth. Imagine Earth as a non-rotating sphere with a uniform solid surface while keeping in mind where the equator and poles would be located if it were rotating. Assume that the Sun heats the equatorial regions more intensely than the poles as it does on the real Earth. A temperature gradient develops between the equator and poles. In response, two huge convection cells form, one in each hemisphere (Figure 9.2A). Cold, dense air sinks at the poles

and flows at the surface toward the equator, where it forces warm, less dense air to rise. Aloft, the equatorial air flows toward the poles, completing the convective circulation.

If our idealized planet begins to rotate as the real Earth rotates, the Coriolis Effect comes into play (Figure 9.2B). In the Northern Hemisphere, surface winds observed relative to Earth's surface shift to the right and blow toward the southwest, and in the Southern Hemisphere, surface winds shift to the left and blow toward the northwest. Hence, on our hypothetical planet, surface winds would initially blow counter to the planet's direction of rotation, which is from west to east. As the planet

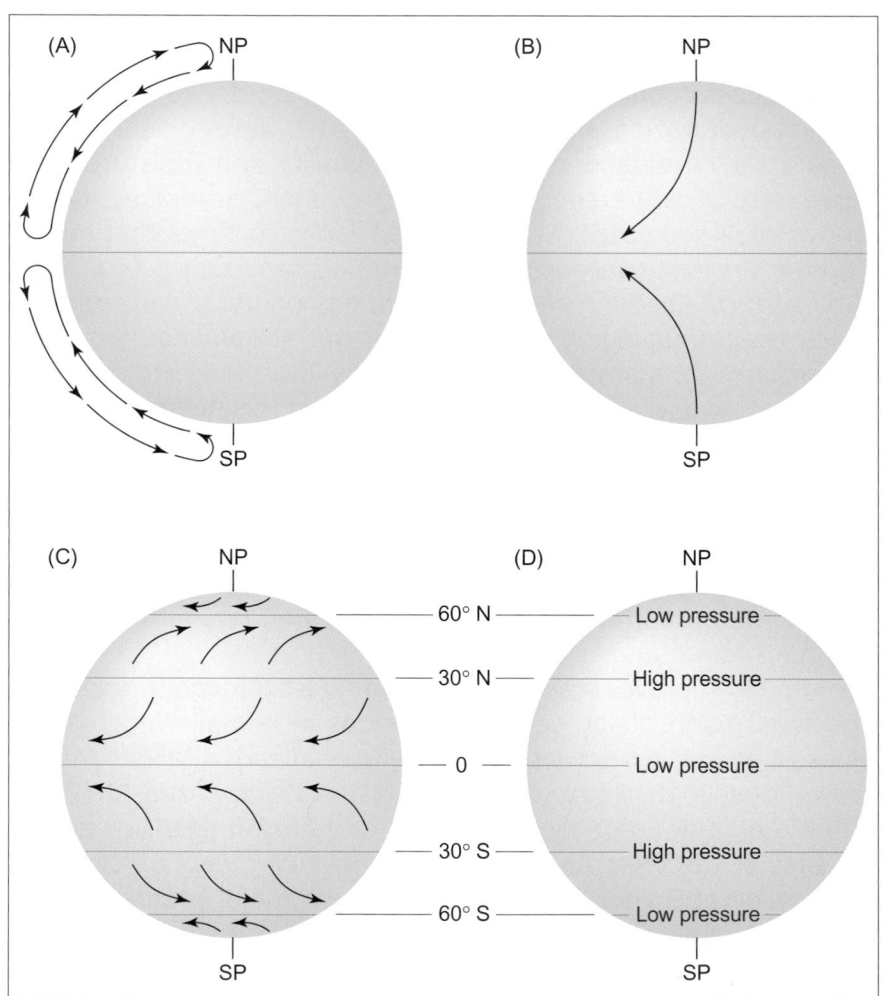

FIGURE 9.2

Planetary-scale air circulation on a highly idealized spherical model of Earth featuring a uniform solid surface. (A) If the sphere is initially non-rotating, huge convection currents develop on the sunlit portion of the planet's two hemispheres so that air circulates between the hot equator and cold poles. (B) With a rotating Earth, surface winds blow from the northeast in the Northern Hemisphere and from the southeast in the Southern Hemisphere owing to the Coriolis Effect. (C) In reality, surface winds divide into three belts in each hemisphere of the rotating planet. (D) Zones of converging and diverging surface winds give rise to east-west belts of low pressure and high pressure.

speeds up to Earth's current rotational rate, the simple hemispherical wind pattern would split into three belts in each hemisphere, so that some winds blow with and some winds blow against the planet's rotational direction (Figure 9.2C). In the Northern Hemisphere, prevailing surface winds are from the northeast between the equator and 30 degrees latitude, from the southwest between 30 and 60 degrees, and from the northeast between 60 degrees and the North Pole. In the Southern Hemisphere, prevailing surface winds blow from the southeast between the equator and 30 degrees, from the northwest between 30 and 60 degrees, and from the southeast between 60 degrees and the South Pole.

Surface winds converge along the equator and along the 60-degree latitude circles. Convergence leads to ascending air, expansional cooling, cloud development, and precipitation. These surface convergence zones are belts of relatively low surface air pressure (Figure 9.2D) because diverging air in the upper troposphere has the net effect of removing air from a column. On the other hand, surface winds diverge at the poles and along the 30-degree latitude circles. In these regions, air descends, is compressed and warms, and the weather is generally fair. These divergence zones are belts of relatively high surface air pressure as corresponding convergence aloft has the net effect of adding air to a column.

If the continents and ocean basins are added to our idealized Earth, the temperature characteristics of Earth's surface become more complicated, and so do the planetary-scale air pressure pattern and winds. Some of the pressure belts break into separate cells, and important contrasts in air pressure develop over land versus sea. Now our idealized model of planetary-scale atmospheric circulation more closely approximates the actual time-averaged circulation pattern. In the next section, we describe the principal features of the planetary-scale atmospheric circulation.

Features of the Planetary-Scale Circulation

Maps of global average air pressure at sea level for January and July reveal several areas of relatively high and low air pressure (Figure 9.3). These are

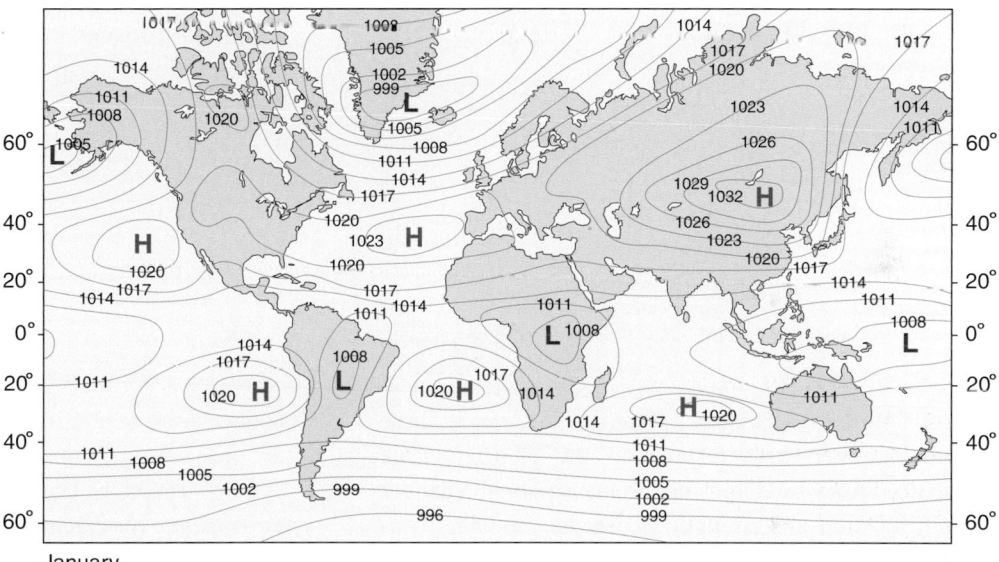

January

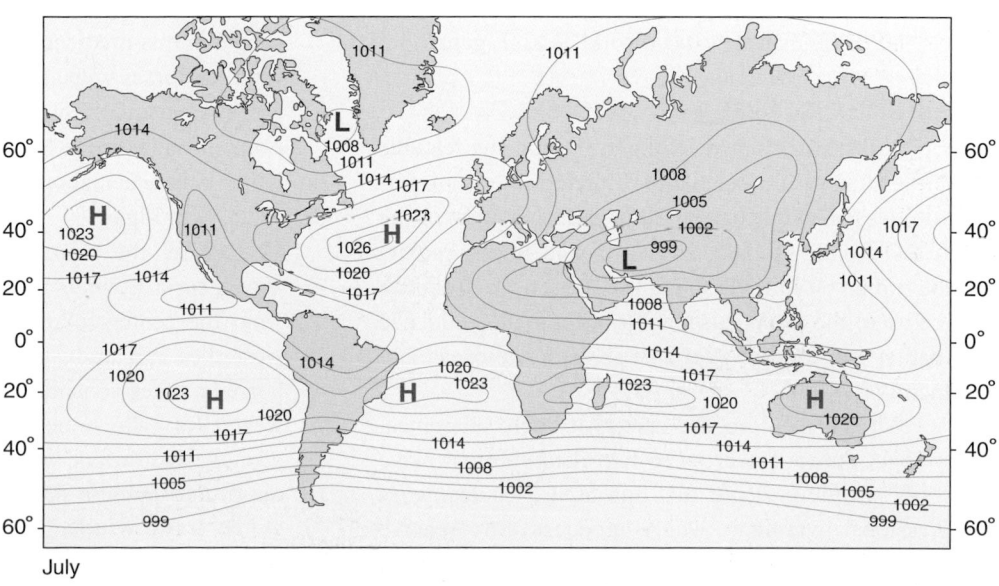

July

FIGURE 9.3
Mean sea-level air pressure during January and July. Contour lines are isobars in millibars (mb).

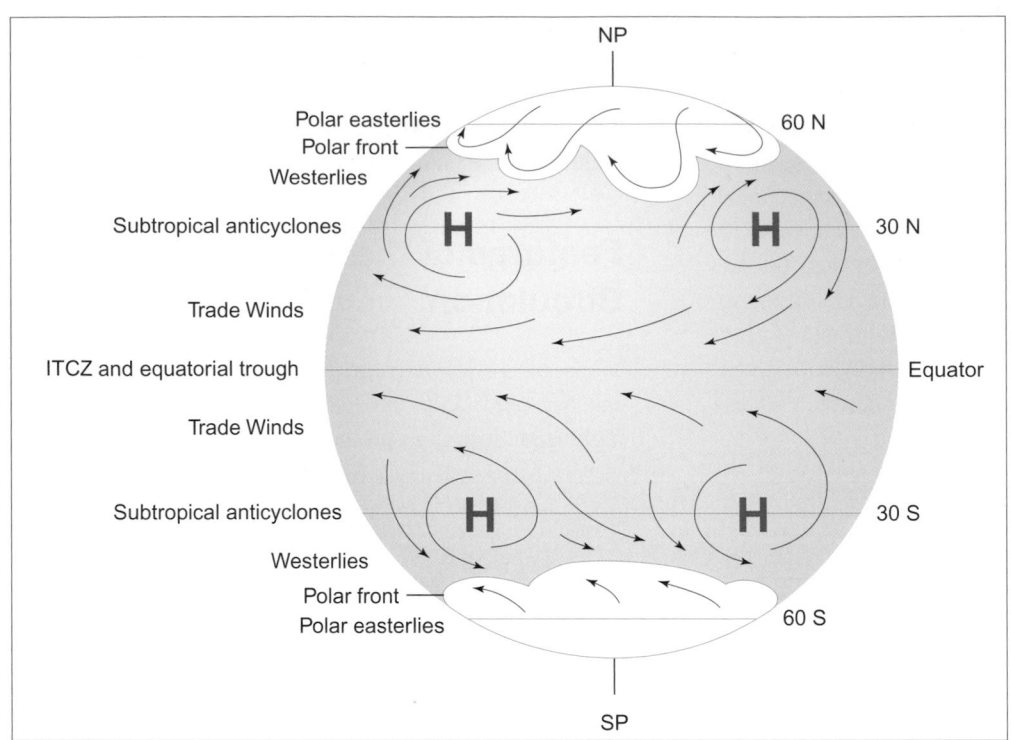

FIGURE 9.4
Schematic representation of the planetary-scale surface circulation of the atmosphere.

semi-permanent pressure systems. Although these systems are persistent features of the planetary-scale circulation, they undergo important seasonal changes in both location and strength, hence the modifier *semi-permanent*. Pressure systems include subtropical anticyclones, the intertropical convergence zone (ITCZ), subpolar lows, and polar highs. These pressure systems are in turn, linked by planetary-scale wind belts (Figure 9.4).

PRESSURE SYSTEMS AND WIND BELTS

Subtropical anticyclones are imposing features of the planetary-scale circulation centered over subtropical latitudes (on average, near 30 degrees N and S) of the North and South Atlantic, the North and South Pacific, and the Indian Ocean. These highs extend from the ocean surface up to the tropopause, and exert a major influence on weather and climate over vast areas of the ocean and continents.

Stretching from the center of each subtropical anticyclone, outward over its eastern flank, are extensive areas of subsiding stable air. Subsiding air undergoes compressional warming, which produces low relative humidity and generally fair skies. The world's major subtropical deserts, including the Sahara of North Africa and the Sonora of Mexico and southwest United States,

are located under the eastern flanks of subtropical anticyclones. On the far western portions of the subtropical highs, however, subsidence is less, the air is not as stable, and episodes of cloudy, stormy weather are more frequent. The contrast in climate between the eastern and western flanks of a subtropical high is apparent across southern North America. The weather of the American Southwest (on the eastern side of the North Pacific high, also called the *Hawaiian high*) is considerably drier than the weather of the American Southeast (on the western side of the North Atlantic high, also called the *Bermuda-Azores high*).

As is typical of anticyclones, a subtropical high features a weak horizontal air pressure gradient over a broad area surrounding the system's center. Hence, surface winds are very light or the air is calm over extensive areas of the subtropical ocean. This situation played havoc with ancient sailing ships, which were becalmed for days or even weeks at a time. Ships setting sail from Spain to the New World were often caught in this predicament, forcing crews to jettison their cargo of horses when supplies of water and food ran low. For this reason, early mariners referred to this region of calm air as the **horse latitudes**, a name now applied to all latitudes between about 30 and 35 degrees N and S under subtropical highs.

In the Northern Hemisphere, viewed from above, surface winds blow clockwise and outward, away from the centers of the subtropical highs, forming the westerlies and trade winds. Surface winds north of the horse latitudes constitute the highly variable **midlatitude westerlies** (which on average actually blow from the southwest). Surface winds blowing from the northeast out of the southern flanks of the anticyclones are known as the **trade winds**. The trades are the most persistent winds on the planet, in some regions blowing from the same direction more than 80% of the time. Analogous winds develop in the Southern Hemisphere, but recall that the Coriolis deflection is to the left in the Southern Hemisphere. A counterclockwise and outward surface airflow thus causes southeast trade winds on the northern flanks of the Southern Hemisphere subtropical highs and a belt of northwesterly winds on the southern flanks.

Mariners were aware of the westerlies and trades of the North Atlantic at least as early as the 15th century. On his venture to find a route to India, Christopher Columbus (1451-1506) took advantage of what we now know is the circulation about the Bermuda-Azores subtropical high. During his westward voyage, Columbus first sailed southward from Spain along the northwest African coast into the northeast trade winds that eventually took his ships westward to the Caribbean. On his return trip to Spain, he sailed north into the westerlies.

Trade winds of the two hemispheres converge into a broad east-west equatorial belt of light and variable winds, called the **doldrums**. In that belt, ascending air induces cloudiness and rainfall. The most active weather develops along the **intertropical convergence zone (ITCZ)**, a discontinuous low-pressure belt with thunderstorms paralleling the equator. The average location of the ITCZ corresponds approximately to the latitude where Earth's mean annual surface temperature is highest, the **heat equator**. Primarily because more land is in the Northern Hemisphere than in the Southern Hemisphere, the world-wide average location of the heat equator is near 10 degrees N latitude.

On the poleward side of the subtropical highs, surface westerlies flow into regions of low pressure, found primarily over subpolar ocean basins. In the Northern Hemisphere, there are two **subpolar lows**: the *Aleutian low* over the North Pacific Ocean and the *Icelandic low* over the North Atlantic Ocean. These pressure cells mark the convergence of the midlatitude southwesterlies with the polar northeasterly winds. By contrast, in the Southern Hemisphere, the midlatitude northwesterlies and the polar southeasterlies converge along a nearly continuous belt of low pressure surrounding the Antarctic continent (the *Antarctic circumpolar vortex*).

Surface westerlies meet and override the polar easterlies along the **polar front**. Recall that a *front* is a narrow zone of transition between air masses that differ in temperature, humidity, or both. In this case, dense, cold air masses flowing toward the equator meet milder, less dense air masses moving toward the pole. The polar front is not continuous around the globe; rather, it is well-defined in some areas and not in others, depending on the temperature contrast across the front. Where that temperature gradient across the boundary between air masses is great, the front

is well defined and is a potential site for development of extratropical cyclones (Chapter 10). On the other hand, where the air temperature contrast is minimal, the polar front is poorly defined and inactive or non-existent. The polar front is usually apparent on a surface weather map of North America, dividing colder air to the north from warmer air to the south. At any time, segments of the polar front may be stationary or moving northward as a warm front or southward as a cold front.

At high latitudes, air subsides and diverges at the surface away from the centers of shallow, cold anticyclones. In the Northern Hemisphere, **polar highs** are well developed only in winter over the continental interiors. In the Southern Hemisphere, cold highs persist over the glacier-bound Antarctic continent year-round.

WINDS ALOFT

What is the pattern of the planetary-scale winds aloft, that is, in the middle and upper troposphere? As noted earlier, air subsides in subtropical anticyclones, sweeps toward the equator as the surface trade winds, and then ascends in the ITCZ. Aloft, in the middle and upper troposphere, winds blow poleward, away from the doldrums and into the subtropical highs. The Coriolis Effect shifts these upper-level winds toward the right in the Northern Hemisphere (southwest winds), and toward the left in the Southern Hemisphere (northwest winds). In the tropics, therefore, winds aloft blow in a direction opposite that of the underlying surface trade winds.

The north-south vertical profile of this low-latitude circulation resembles a huge convection current (Figure 9.5). This circulation is known as the **Hadley cell**, named for the English meteorologist

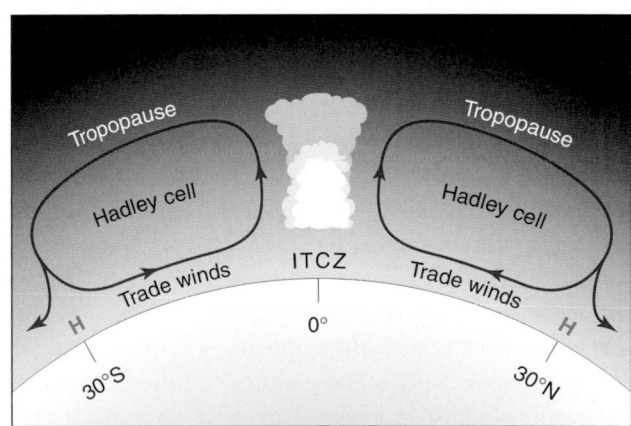

FIGURE 9.5
Idealized north-south vertical cross-section of the Hadley cell circulation in tropical latitudes of the Northern and Southern Hemispheres. Air rises in the intertropical convergence zone (ITCZ) and sinks in the subtropical anticyclones.

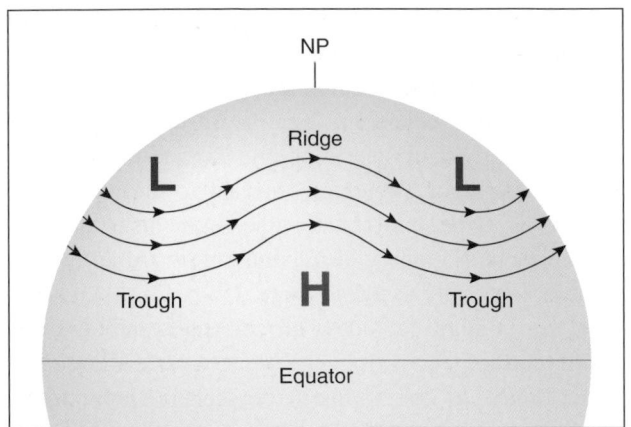

FIGURE 9.6
In the middle and upper troposphere, the Northern Hemisphere westerlies blow from west to east in a wave-like pattern of ridges (clockwise turns) and troughs (counterclockwise turns).

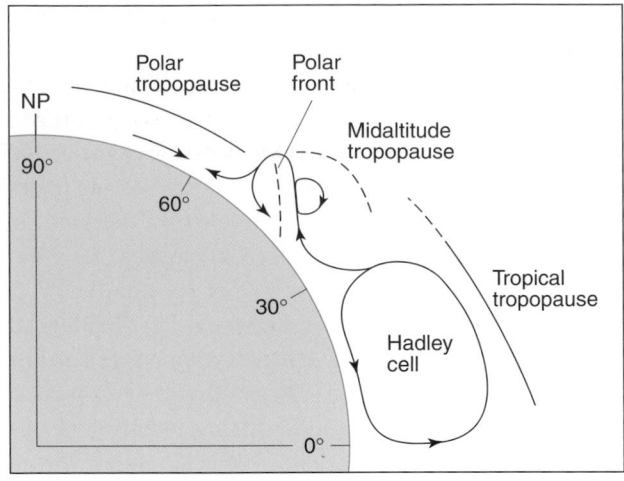

FIGURE 9.7
Vertical cross-section of the north-south (meridional) component of the atmospheric circulation of the Northern Hemisphere. Note that the tropopause occurs in three segments, occurring at highest altitudes in the tropics and lowest altitudes in the polar regions. The vertical scale is greatly exaggerated.

George Hadley (1685-1768), who first proposed its existence in 1735. Hadley cells are situated on either side of the ITCZ and extend poleward to the subtropical highs. It was once proposed that separate cells, similar to Hadley cells, occurred in both middle and polar latitudes, but upper-air monitoring does not provide definitive evidence of such well-defined cells.

Aloft in middle latitudes, winds blow from west to east in a wavelike pattern of ridges (clockwise turns) and troughs (counterclockwise turns), as shown in Figure 9.6. These winds are responsible for the development and movement of the synoptic-scale weather systems (highs, lows, air masses, and fronts) discussed in detail in Chapter 10. Also, their north/south components contribute to poleward heat transport (Chapter 4). Because the upper-air westerlies are so important for the weather of middle latitudes, we examine them more extensively in a separate section of this chapter.

Figure 9.7 is a vertical cross-sectional profile of prevailing winds in the troposphere from the North Pole to the equator. In this perspective, we are viewing only the north-south and vertical (up-down) components of the wind and are neglecting the west-east component. The altitude of the tropopause is directly related to the mean air temperature of the troposphere; that is, the lower the mean temperature, the lower is the altitude of the tropopause. Note that the altitude of the tropopause does not steadily increase from pole to equator but occurs in discrete steps. The tropical tropopause is at a higher altitude than the midlatitude tropopause, which, in turn, is at a higher altitude than the polar tropopause.

TRADE WIND INVERSION

The circulation on the eastern flank of subtropical anticyclones gives rise to the **trade wind inversion**, a persistent and climatically significant feature of the planetary-scale circulation over the eastern portions of tropical ocean basins. Key to formation of the trade wind inversion is the descending branch of the Hadley cell. Air subsiding in the subtropical highs as part of the Hadley circulation is warmed by compression and its relative humidity decreases. Descending air encounters the *marine air layer*, a shallow layer of air that overlies the ocean surface. Where sea-surface temperatures (SST) are relatively low, the marine air layer is cool, humid, and stable. Where sea-surface temperatures are relatively high, the marine air layer is warm, more humid, less stable, and well mixed by convection.

Air subsiding from above is warmer (and much drier) than the upper portion of the marine air layer. The trade wind inversion forms at the altitude where air subsiding from above meets the top of the marine air layer. As noted in Chapter 6, the air temperature increases with increasing altitude within a temperature inversion. Air characterized by a temperature inversion is extremely stable and strongly inhibits vertical motion of air. The trade wind inversion is an elevated temperature inversion that essentially acts as a lid over the lower atmosphere.

As noted above, most subsidence occurs on the eastern flank of a subtropical high. A trade wind inversion develops to the east and southeast of the center of a subtropical high, in the region dominated by the trade winds. The inversion is highest near the center of the high and gradually slopes downward toward the western coasts of the continents. For example, the trade wind inversion slopes downward from an average altitude of about 2000 m (6560 ft) over Hawaii to about 800 m (2400 ft) over coastal southern California.

Because of its extreme stability, the trade wind inversion limits the vertical development of convective clouds and rainfall. As we will see in Chapter 12, this inhibits formation of tropical storms and hurricanes. The inversion also limits orographic precipitation (Chapter 6). For example, some high volcanic peaks on the Hawaiian Islands poke through the trade wind inversion into the dry air above. Because of orographic lifting, rainfall generally increases with elevation along windward slopes of the volcanic mountains but only up to the trade wind inversion. Above the inversion, conditions are dry.

SEASONAL SHIFTS

Between winter and summer, important changes take place in the planetary-scale circulation with implications for climates in various parts of the world. Pressure systems, the polar front, the planetary wind belts, and the ITCZ follow the Sun, shifting toward the poles in a hemisphere's spring and toward the equator in its autumn. Because the seasons are six months apart in the two hemispheres, the planetary-scale systems of both hemispheres move north and south in tandem. In addition, the strength of pressure cells varies seasonally. Subtropical anticyclones, such as the Bermuda-Azores high, exert higher surface pressures in summer than in winter when they are displaced poleward. The Icelandic low deepens in winter and greatly weakens in summer, and, though

well developed in winter, the Aleutian low disappears in summer.

Seasonal reversals in surface air pressure also occur over the continents. These pressure changes stem from the contrast in solar heating of land versus ocean. For reasons presented in Chapter 4, the ocean surface exhibits smaller temperature variations over the course of a year than does Earth's land surface. Hence, continents at middle and high latitudes are dominated by relatively high pressure in winter and relatively low pressure in summer. In winter, in response to extreme radiational cooling, cold anticyclones develop over northwestern North America and over the interior of Asia, the most prominent of which is the massive Siberian high. In summer, in response to intense solar heating, a belt of low pressure forms across North Africa and from the Arabian Peninsula eastward into Southeast Asia. Warm low-pressure cells (called *thermal lows*) also develop in summer over arid and semi-arid regions of Mexico and the Southwestern United States.

Seasonal changes in the planetary-scale wind belts, pressure systems, and the ITCZ leave their mark on the world's climates. Northward migration of the ITCZ triggers summer monsoon rains in Central America, North Africa, India, and Southeast Asia (discussed in the following section). As shown in Figure 9.8, with the exception of the Indian Ocean basin, north-south movements of the ITCZ are generally greater over continents than over the ocean. Anchoring of the ITCZ over the ocean is a consequence of the ocean's greater thermal inertia.

The influence of the subtropical anticyclones on precipitation regimes is illustrated by the contrast in the march of mean monthly precipitation between Charleston, SC, versus San Diego, CA (Figure 9.9).

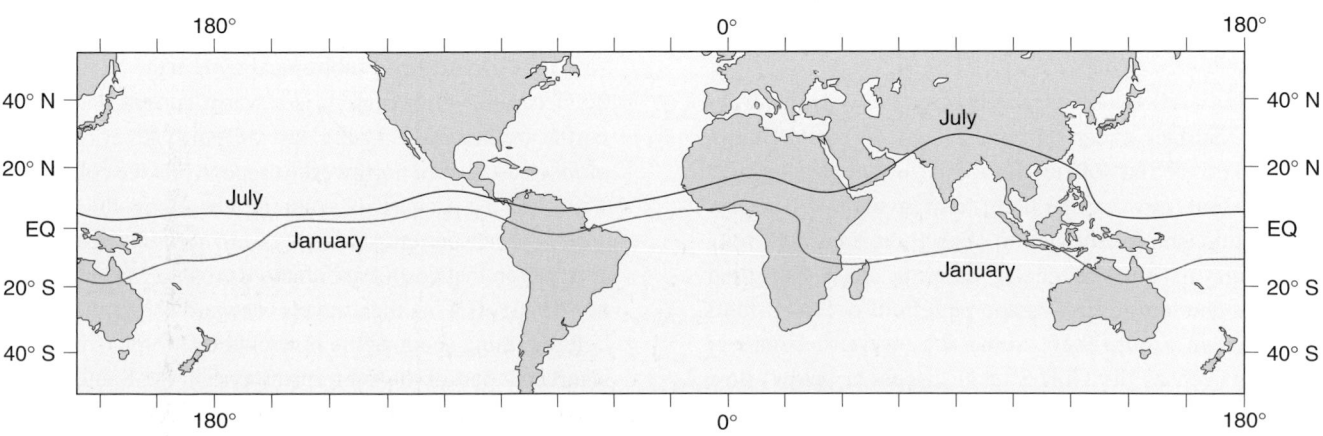

FIGURE 9.8
The intertropical convergence zone (ITCZ) follows the Sun, reaching its most northerly latitudes in July and its most southerly latitudes in January.

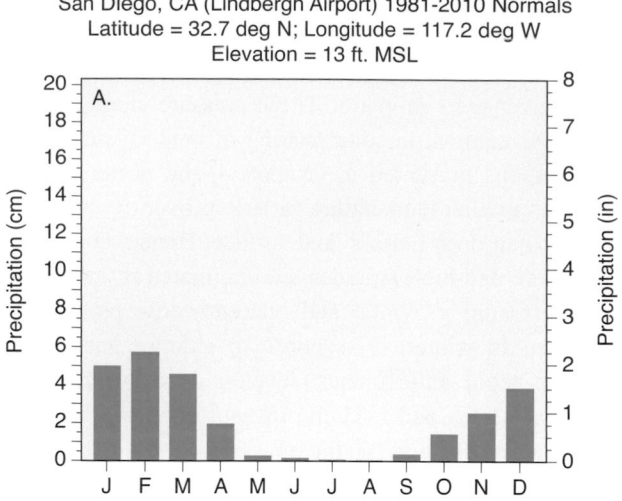

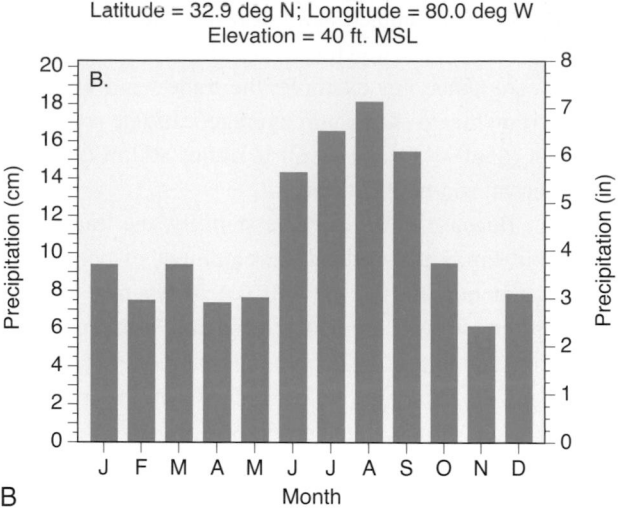

FIGURE 9.9

Mean monthly precipitation (rain plus melted snow) in centimeters and inches for (A) San Diego, CA, and (B) Charleston, SC for the period 1981-2010. San Diego has a distinct dry season (summer) when its weather is dominated by subsiding air on the eastern flank of the Hawaiian subtropical high.

Although the two cities have about the same latitude, San Diego has a distinct rainy season (winter) and dry season (summer), whereas Charleston receives abundant precipitation throughout the year (with a peak in June-September). In summer, San Diego is under the dry eastern flank of the Hawaiian high, while Charleston is on the receiving end of the humid flow on the western flank of the Bermuda-Azores high. Consequently, San Diego is dry while Charleston has frequent episodes of convective showers, especially during late spring and summer. After the subtropical highs shift equatorward in autumn, the weather of both cities is influenced by widespread precipitation associated with west-to-east moving extratropical cyclones so that winters are relatively wet in both places.

OCEAN SURFACE CURRENTS

Surface ocean currents are wind-driven. Hence, the horizontal movement of ocean surface waters to a large extent mirrors the long-term average planetary-scale atmospheric circulation. Early on, mariners took advantage of surface ocean currents to hasten their voyage. The long-term average pattern of ocean currents is plotted in Figure 9.10. Some (i.e., western boundary currents such as the Gulf Stream discussed below) flow faster than others. Winds associated with passing storm systems disturb the ocean surface and can cause the local flow of ocean currents to deviate temporarily from the long-term average pattern.

As shown in Figure 9.10, large-scale circulation regimes, called gyres, appear in each of the major ocean basins and are associated with the prevailing winds that help set up the currents. The trades winds and the westerlies associated with the semi-permanent subtropical highs drive the **subtropical gyres**. These gyres are centered near 30 degrees latitude in the North and South Atlantic, North and South Pacific, and Indian Oceans. Subtropical gyres are similar in the Northern and Southern Hemisphere except that they rotate in opposite directions because the Coriolis Effect acts in opposite directions in the two hemispheres. Viewed from above, subtropical gyres rotate in a clockwise direction in the Northern Hemisphere and in a counterclockwise direction in the Southern Hemisphere.

As part of the subtropical gyre in the North Atlantic Ocean, the Gulf Stream is a warm current that moves northward along the East Coast of North America before turning east toward northwest Europe, while the cold southward flowing Canary Current moves along the African coast. Surface currents flow westward across the equatorial Atlantic. In the North Pacific, the Kuroshio Current moves northward off Asia, then travels eastward across the Pacific before turning south as the now cold California Current. A significant part of the subtropical gyre in the South Pacific Ocean is the cold northward flowing Peru Current, which then helps form a westward flowing equatorial current.

Sub-polar gyres, smaller than their subtropical counterparts, occur at high latitudes of the Northern

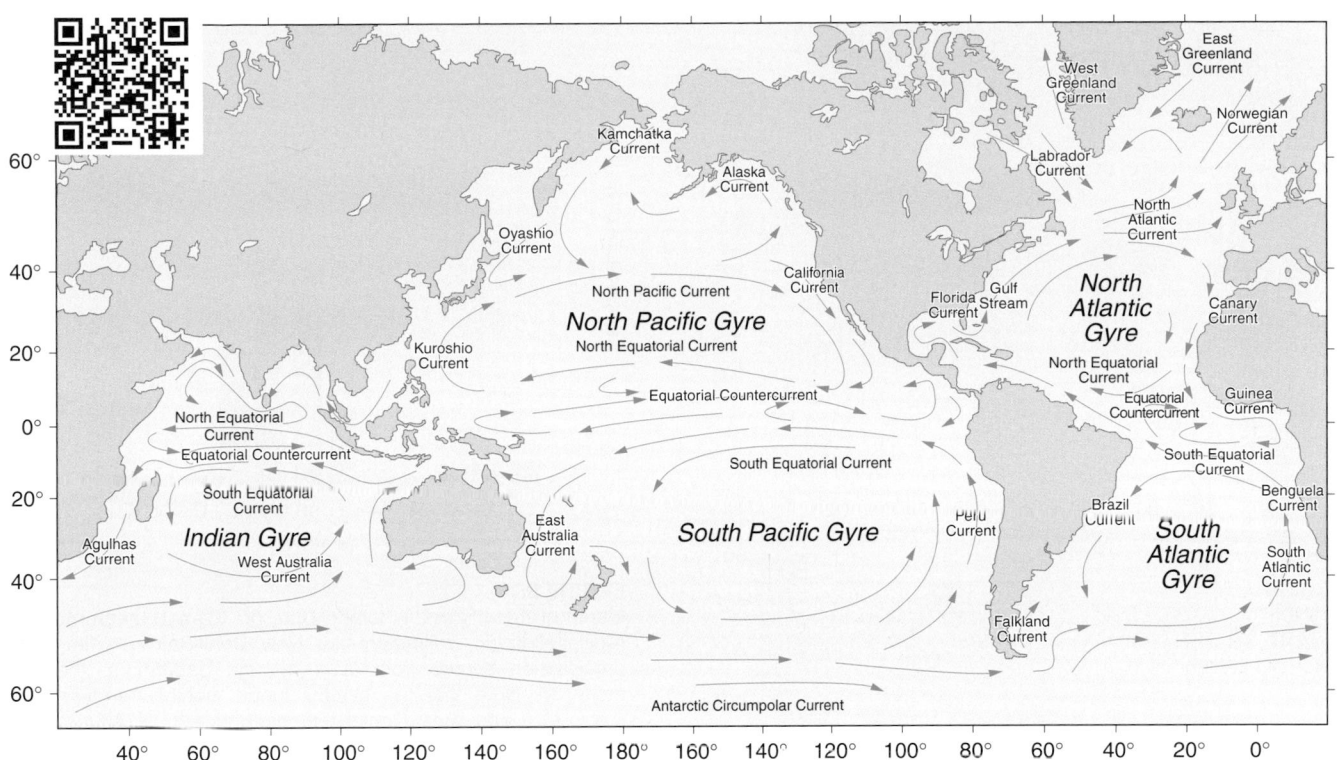

FIGURE 9.10
Long-term average pattern of wind-driven ocean-surface currents. Gyres in the ocean basin are driven by the planetary-scale atmospheric circulation.

Hemisphere; they are the gyre south of Alaska in the far North Pacific Ocean and gyres south of Greenland in the far North Atlantic. The counterclockwise surface winds in the Aleutian and Icelandic sub-polar low pressure systems drive the sub-polar gyres. Hence, viewed from above, the rotation in the these sub-polar gyres is opposite that of the Northern Hemisphere subtropical gyres.

Monsoon Circulation

Monsoon is derived from the Arabic word *mausim* for season. A **monsoon circulation** characterizes regions where seasonal reversals in prevailing winds cause wet summers and relatively dry winters. Monsoon was first applied to winds over the Arabian Sea, which blow from the northeast for about six months and then from the southwest for another six months. In this section, we describe the vigorous monsoon circulation over portions of Africa and Asia, where more than 2 billion people depend on monsoon rains for their drinking water and agriculture. Over much of India, monsoon rains (falling mostly between June and September) account for 80% or more of total annual precipitation. We then discuss the weaker monsoon that affects the American Southwest.

What causes the Asian and African monsoons? As first proposed in 1686 by the English astronomer Edmund Halley (1656-1742), monsoons depend on seasonal contrasts in the heating of land and ocean surfaces. The ocean has a greater *thermal inertia* than does the land (Chapter 4). In spring, relatively cool air over the ocean and relatively warm air over the land give rise to a horizontal air pressure gradient directed from sea to land, generating an onshore flow of humid air during the warm summer season (Figure 9.11A). Over the land, intense solar heating triggers convection. Hot, humid air rises, and consequent expansional cooling leads to condensation, cloud development, and rain. Release of latent heat intensifies the buoyant uplift, triggering even more clouds and rainfall. Aloft, the air spreads seaward and subsides over the relatively cool ocean surface, thus completing the monsoon circulation. The trajectory of monsoon winds is sufficiently long and persistent to be influenced by the Coriolis Effect. Surface monsoon winds are deflected to the right in the Northern Hemisphere and to the left in the Southern Hemisphere.

By early autumn, radiational cooling chills the land more than the adjacent ocean surface, resulting in higher surface pressures over land and setting up a

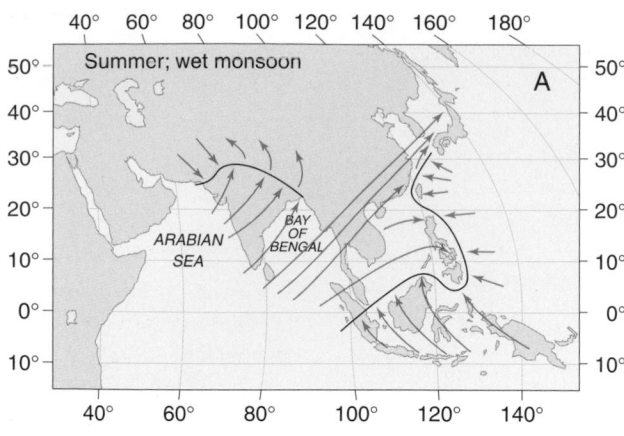

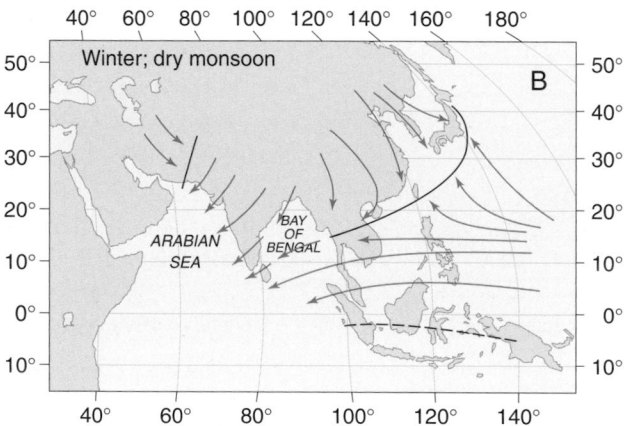

FIGURE 9.11
Streamlines showing surface wind patterns over Asia during (A) wet monsoon (summer) and (B) dry monsoon (winter).

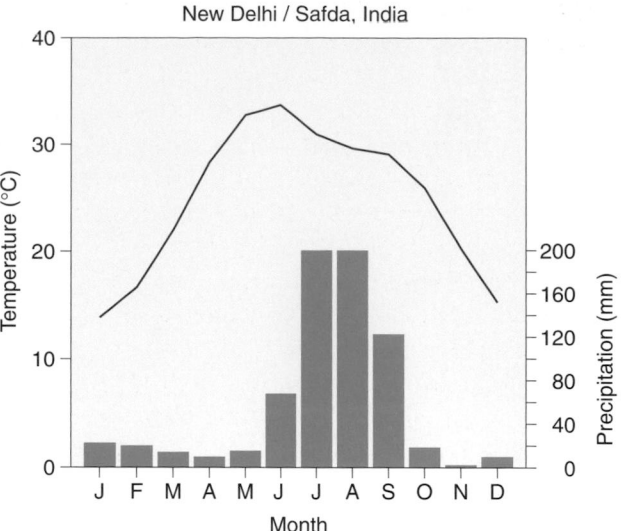

FIGURE 9.12
March of mean monthly temperature (in °C) and mean monthly precipitation (in millimeters) for New Dehli/Safda, India. Most precipitation occurs from June through September, the wet monsoon. Note that the highest mean monthly temperatures are in May and June. Cooler temperatures in July, August, and September are due to considerable cloudiness during that time of year. [GHCN data supplied by E. J. Hopkins]

horizontal air pressure gradient directed from land to sea. Air subsides over the land, and dry surface winds sweep seaward (Figure 9.11B). Air rises over the relatively warm ocean surface and aloft drifts landward, completing the winter monsoon circulation. Over land, therefore, the summer monsoon is wet whereas the winter monsoon is dry (Figure 9.12).

Topography complicates the monsoon circulation and the geographical distribution of rainfall. For example, the immense uplifted Tibetan plateau strongly influences the Asian monsoon because its elevation tops 4000 m (13,000 ft) over a broad area. In winter, the westerly jet stream (a corridor of very strong winds in the westerlies) splits into two branches, one to the south and the other to the north of the plateau. The southerly branch steers cyclones that originate over the Mediterranean Sea across northern India, bringing significant precipitation to that region. Meanwhile, the rest of India experiences the dry monsoon flow. In spring, the southern branch of the jet

stream weakens and by late May shifts northward over the plateau. It is not until this happens that the moist monsoon flow begins over India.

Monsoon rainfall is neither uniform nor continuous. On the contrary, the rainy season typically consists of a sequence of active and dormant phases. During a *monsoon active phase*, the weather is mostly cloudy with frequent deluges of rain, but during a *monsoon dormant phase*, the weather is sunny and hot. The monsoon shifts from active to dormant phases as bands of heavy rainfall surge inland. Heavy rains first strike coastal areas and soak the ground. As the soil becomes saturated with water, more solar radiation is used for evaporation and less is available for sensible heating. Coastal areas cool, uplift weakens, and skies partially clear (monsoon dormant phase). Meanwhile, the area of maximum heating, most vigorous uplift, and heavy rains shifts inland. Back in the coastal areas, however, the hot Sun eventually dries the soil, sensible heating intensifies, uplift strengthens, and rains resume (monsoon active phase). This sequence of active and dormant phases is repeated about every 15 to 20 days during the wet monsoon.

Solar radiation, land and water distribution, and topography impose some regularity on the monsoon circulation and monsoon climates; that is, summers are wet and winters are dry. The planetary-scale circulation (especially shifts of the ITCZ) and the strength and distribution of convective activity, however, vary from year to year. These variations mean that the intensity and duration of monsoon rains change from one year to the next. Consequently, monsoon failure and drought is always possible in monsoon climates. We considered a particularly tragic example of this in the chapter's Case-in-Point, *Drought in the Sahel of West Africa.*

The *North American Monsoon System (NAMS)*, also called the **Southwest Monsoon**, is a prominent feature of the climate of the American Southwest including Arizona, New Mexico, and parts of southern Colorado. The monsoon is responsible for a dramatic increase in rainfall to this region mainly during July and August. Many locales in New Mexico, for example, receive 40% to 50% of their total annual precipitation during these two months.

The Southwest Monsoon originates over northern Mexico during May and June. In fact, the Southwest Monsoon is actually the northern extension of the Mexican monsoon which is responsible for up to 70% of annual precipitation in portions of Mexico. Intense solar radiational heating of the ground raises the temperature of the overlying air, reducing the air pressure at low levels of the troposphere over a broad area (a *thermal low*). Meanwhile a warm high pressure system (the *monsoon high*) develops aloft in the middle and upper troposphere.

A horizontal pressure gradient develops at low levels, which is directed from high pressure over the waters of the Gulf of California and the Gulf of Mexico toward low pressure over interior Mexico. In response to this pressure gradient, warm humid air moves inland giving rise to considerable cloudiness and rainfall. Southerly winds on the western flank of the monsoon high advect moisture northward between the Sierra Madre Mountains to the east and smaller mountains in Baja California to the west (Figure 9.13).

The monsoon high shifts northward and by July is centered in New Mexico. Winds across the American Southwest shift from the dry westerlies to a humid flow from the south and southeast with some of the flow originating over the Gulf of California and some from the Gulf of Mexico. With this wind shift the Southwest Monsoon is underway. As a rough guide, the monsoon is considered to begin in Phoenix, AZ, when the dewpoint climbs to 13 °C (55 °F) or higher for three consecutive days. The average date for the start of the monsoon at Phoenix is 7 July.

During the monsoon, rainfall across the American Southwest is not continuous but consists of isolated showers and thunderstorms, with the heaviest rains in the mountains. Subsiding dry air associated with the monsoon high is responsible for frequent dry episodes. Nonetheless, the Southwest Monsoon, which is sometimes augmented by the remnants of tropical cyclones tracking inland from the eastern Pacific Ocean, is responsible for 70% to 80% of total annual precipitation across Arizona and the Sonora Desert, as exemplified by mean monthly precipitation and average daily dewpoint in Tucson, AZ (Figure 9.14). The Southwest Monsoon usually ends in September with the return of the dry westerlies.

The Southwest Monsoon is the source of water for dry land farming, ranching, and wildfire control but the timing and strength of the monsoon vary each year. A weak monsoon results in water shortages and greater susceptibility to forest and brush fires. Also, heavy rains cause flash flooding of waterways that are usually dry most of the year. In New Mexico, flash floods are most likely in July and August.

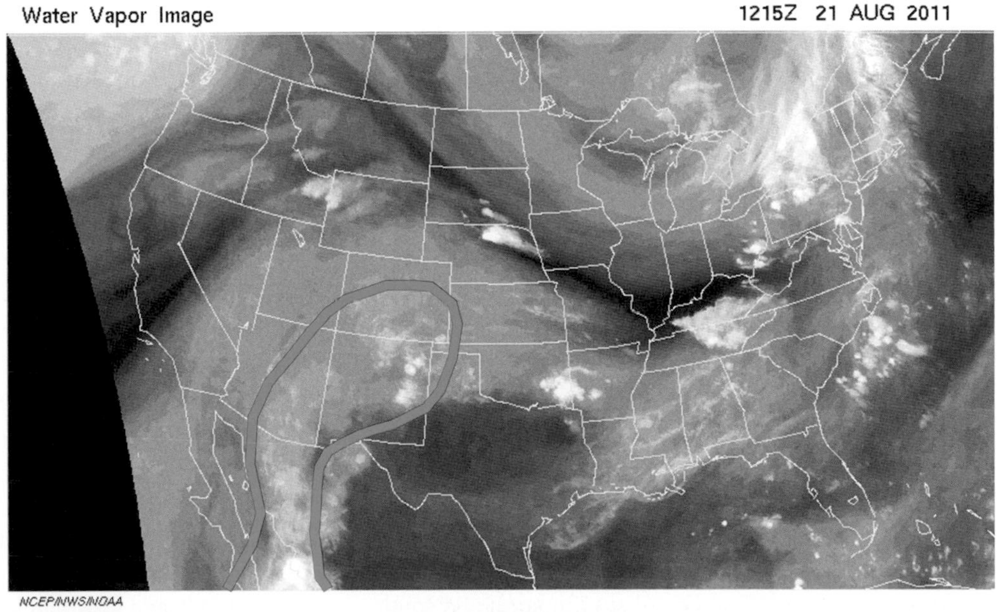

Water Vapor Image · 1215Z 21 AUG 2011

NCEP/NWS/NOAA

FIGURE 9.13
Water vapor image portraying the Southwest Monsoon, as the northern extension of the Mexican monsoon. Areas of high water vapor content are shown over Arizona, New Mexico, and Colorado from a plume originating over the Gulf of California. [NCEP/NWS/NOAA]

Tucson, AZ (International Airport) 1981-2010 Normals

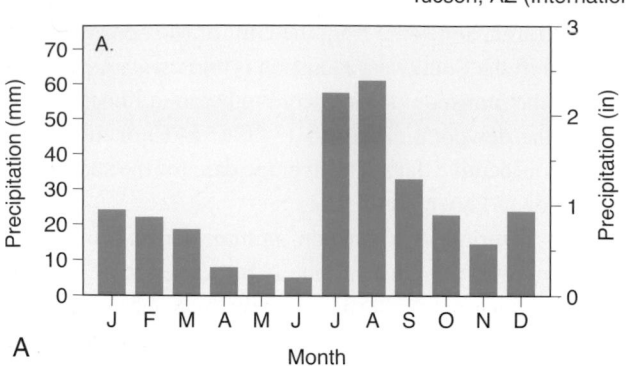

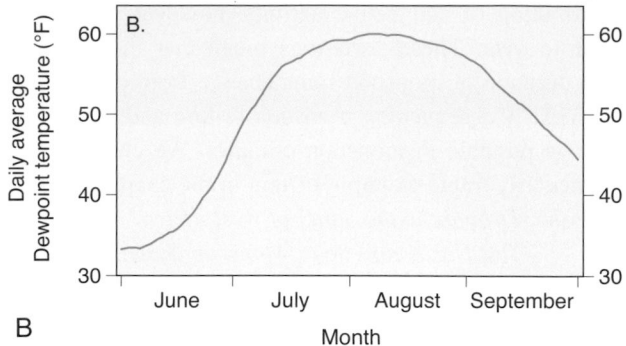

FIGURE 9.14
(A) Mean monthly precipitation (based on 1981-2010) at Tucson International Airport, AZ, showing the abrupt increase in rainfall from June to July signaling the onset of the Southwest Monsoon. (B) Average daily dewpoint at Tucson, showing a dramatic increase between June and July. [NOAA data]

Long Waves in the Westerlies

The middle latitude westerlies of the Northern Hemisphere merit special attention here because they govern the weather over much of North America. As noted earlier, in the middle and upper troposphere, the westerlies blow from west to east about the hemisphere in wavelike patterns of ridges and troughs. Winds exhibit a clockwise (anticyclonic) curvature in ridges, and a counterclockwise (cyclonic) curvature in troughs. Between two and five waves typically encircle the hemisphere at any one time. These long waves are called **Rossby waves**, after Carl-Gustaf Rossby (1898-1957), the Swedish-American meteorologist who discovered them in the late 1930s. Rossby waves characterize the westerlies above the 500-mb level, that is, above the altitude where the atmospheric pressure is 500 mb. At lower levels, waves are distorted by friction, topographic irregularities of Earth's surface, and warm and cold water surfaces.

Atmospheric scientists describe the upper-air westerlies in terms of (1) wavelength (distance between successive troughs or, equivalently, successive ridges), (2) wave amplitude (north-south extent), and (3) number of waves encircling the hemisphere. All three wave characteristics change with time and, as a direct consequence, the weather also changes. The westerlies are more vigorous in winter than in summer. In winter, they exhibit fewer waves of longer length and greater amplitude. These seasonal changes stem from variations in the north-south air pressure gradient, which is steeper in winter because of the greater temperature contrast between north and south at that time of year. In summer, north-south temperature differences are less, air pressure gradients are weaker and the westerlies are displaced poleward.

ZONAL AND MERIDIONAL FLOW PATTERNS

The weaving westerlies have two components of motion: a north-south wind superimposed on a west-to-east wind. The north-south airflow is the westerlies' meridional component and the west-to-east airflow is its zonal component. The meridional component of Rossby waves brings about a north-south exchange of air masses. In the Northern Hemisphere, winds blowing from the southwest convey warm air masses toward the northeast, and winds blowing from the northwest transport cold air masses southeastward. Cold air is exchanged for warm air and heat is transported poleward (Chapter 4). However, as Rossby waves change length, amplitude, and number, concurrent changes take place in the advection of air masses. Consider some examples.

Occasionally, the westerlies blow almost directly from west to east, nearly parallel to latitude lines, with only a weak meridional component (Figure 9.15A). This is known as a **zonal flow pattern** in which north-south exchange of air masses is minimal. Cold air masses stay to the north while warm air masses remain in the south. At the same time, air from over the Pacific Ocean floods the coterminous United States and southern Canada. **Pacific air** dries out to some extent as it passes over the western mountain ranges, and is compressed and warmed adiabatically as it descends onto the Great Plains, spreading relatively mild air and generally fair weather east of the Rocky Mountains.

At other times, the westerlies exhibit considerable amplitude and flow in a pattern of deep troughs and sharp ridges (Figure 9.15B). In this **meridional flow pattern**, masses of cold air surge southward and warm air streams northward. Greater west-to-east temperature contrasts de-

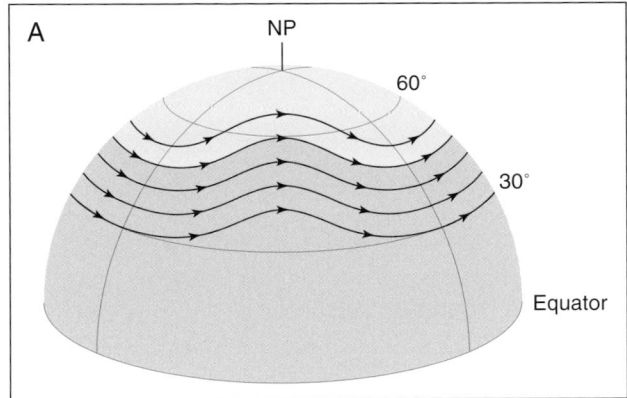

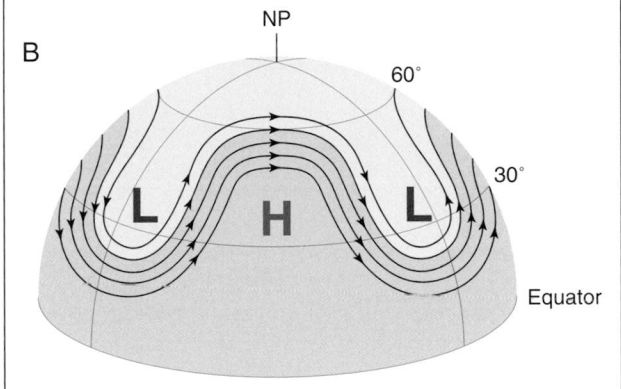

FIGURE 9.15
Midlatitude westerlies exhibit (A) a zonal flow pattern when winds blow almost directly from west to east, with only a weak meridional (north-south) component, (B) a meridional flow pattern when west to east winds have a strong meridional (north-south) component.

velop over the United States and southern Canada. Where contrasting air masses collide, warm air overrides cold air, and the stage is set for the development of extratropical cyclones that are then steered by the westerlies.

These two illustrations of Rossby wave configurations are opposite extremes of a wide range of possible westerly wind patterns, each featuring different components of meridional and zonal flow. More complicated is a **split flow pattern**, in which westerlies to the north have a wave configuration different from that of westerlies to the south. For example, winds may be zonal across central Canada while winds are meridional over much of the coterminous United States.

The westerly wind pattern typically shifts back and forth between dominantly zonal and dominantly meridional flow. For example, zonal flow might persist for a week and then give way to a more meridional flow that lasts for a few weeks, before returning to zonal flow. The transition from one wave pattern to another is usually abrupt, sometimes taking place within a single day. Such abruptness poses a challenge to weather forecasters because a sudden shift in the upper-air winds may divert a cyclone toward or away from a locality or cause an unanticipated influx of colder air that could, for example, change rain to snow.

Unfortunately for the long-range weather forecaster, shifts in westerly wave patterns appear to have no regularity; that is, no predictable zonal/meridional cycle is readily apparent. The only observation useful to forecasters is that meridional flow patterns tend to persist for longer periods than zonal patterns. During the winter of 1976-77, for example, a strong meridional wave pattern persisted over North America from late October through mid-February. Northwesterly winds brought surge after surge of bitterly cold Arctic air into the midsection of the United States. The result was one of the coldest winters of the 20th century for that region. During the same period, persistent winds from the southwest brought unseasonably mild air to far-western North America, including Alaska.

BLOCKING SYSTEMS AND WEATHER EXTREMES

For the continent as a whole, North American weather is more dramatic when the westerlies are strongly meridional. Sometimes meanders of the westerlies become so great that huge whirling masses of air actually separate from the main westerly air flow. This situation, shown schematically in Figure 9.16, is analogous to whirlpools that form in rapidly flowing rivers. In the atmosphere, cutoff masses of air whirl in either a cyclonic or an anticyclonic direction (viewed from above). A *cutoff Low* or a *cutoff High* that prevents the usual west-to-east movement of weather systems is referred to as a **blocking system**. Because a blocking circulation pattern tends to persist for extended periods (lasting several weeks or longer), extremes of weather such as drought or flooding rains or excessive heat or cold can result. Consider some examples.

In the spring and early summer of 1986, an unusually severe and prolonged drought affected a ten-state region in the southeastern United States. (For a discussion of what constitutes drought, refer to this chapter's first *For Further Exploration*.) In the hardest hit areas, precipitation totals for the eight months prior to August 1986 were less than 40% of the long-term average. The primary reason for the May-to-early-July episode of drought was a persistent westerly long-wave pattern that featured a trough anchored off the East Coast near longitude 65 degrees W and, upstream (to the west), a ridge over the Southeast. In effect, the center of the Bermuda-Azores subtropical anticyclone extended its influence well west of its usual domain. This persistent circulation

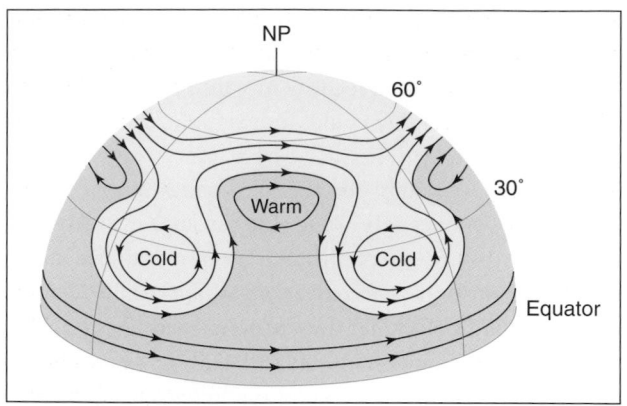

FIGURE 9.16
Blocking pattern in the middle latitude westerlies in which huge pools of rotating air are cut off from the main west to east air flow. The pool of relatively cool air rotating in a counterclockwise direction is a cutoff low, and the pool of relatively warm air rotating in a clockwise direction is a cutoff high. The latter system is sometimes referred to as an *omega block* because of its resemblance to the Greek letter.

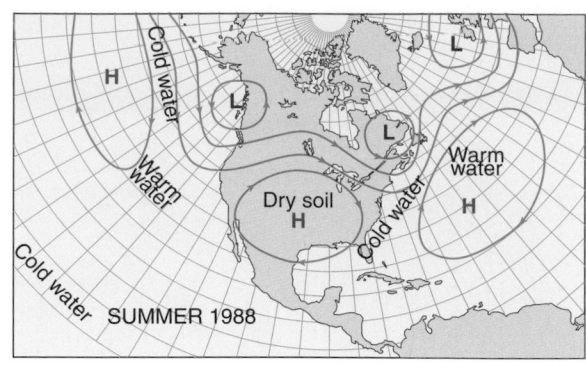

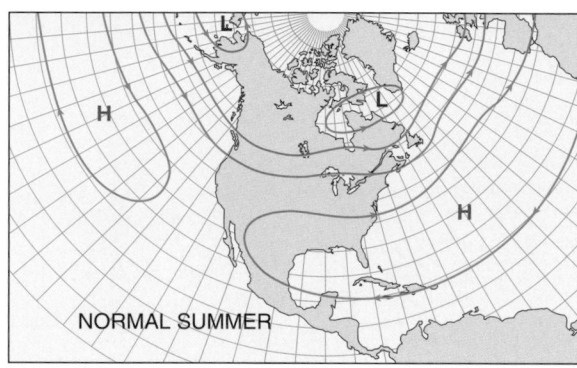

FIGURE 9.17
Prevailing circulation pattern in the mid to upper troposphere (A) during the summer of 1988 as compared to (B) the long-term average circulation pattern. The blocking warm anticyclone over the central U.S. contributed to severe drought.

pattern brought day after day of subsiding stable air that suppressed the convective activity that normally brings the Southeast abundant spring and summer rainfall.

The drought that affected the Midwest and Great Plains during the spring and summer of 1988 is another example of the linkage between a weather extreme and a blocking weather pattern. In Figure 9.17, the major upper-air circulation features that dominated the summer of 1988 are contrasted with long-term average conditions. From early May through mid-August, the prevailing westerlies were more meridional than usual and featured a huge stationary, warm high-pressure system over the nation's midsection and troughs over the West Coast and East Coast. The belt of strongest westerlies was displaced north of its usual location so that moisture-bearing weather systems were diverted into central Canada, well north of their usual paths. In the Corn Belt, the May through June period was the driest since 1895. By late July, drought was categorized as either severe or extreme over 43% of the land area of the coterminous United States. By the end of the growing season, the impact on the nation's grain harvest was severe: corn production was down 33%, soybeans 20%, and spring wheat more than 50%. The National Climatic Data Center (NCDC) estimated that the 1988 drought was one of the most costly weather-related disasters to hit the nation in decades, resulting in total damage of $71.2 billion (in 2007 dollars).

A blocking circulation pattern during the summer of 1993 was responsible for record flooding in the Midwest and drought over the Southeast (Figure 9.18). A cold upper-air trough stalled over the Pacific Northwest

and northern Rocky Mountains, bringing unseasonably cool weather to those regions. Meanwhile, the Bermuda-Azores high shifted west of its usual location over the subtropical Atlantic, causing the worst drought since 1986 over the Carolinas and Virginia. Between the northwestern trough and the subtropical high to the southeast, the principal storm track and an unseasonably strong jet stream weaved over the Midwest. This circulation pattern persisted through June, July, and part of August, bringing a nearly continual procession of clusters of thunderstorms to that region. Across Iowa, Illinois, and Wisconsin, the June-July period was the wettest on record. Meanwhile over the Southeast, subsiding, stable air inhibited thunderstorm development so that hot and dry conditions prevailed through most of the summer.

Heavy rains falling on the drainage basins of the Missouri and Upper Mississippi River valleys saturated soils and triggered excessive runoff, creating all-time record river crests, and causing flooding that impacted all or part of nine states. Setting the stage for flooding in the upper Mississippi River valley was the wet autumn of 1992,

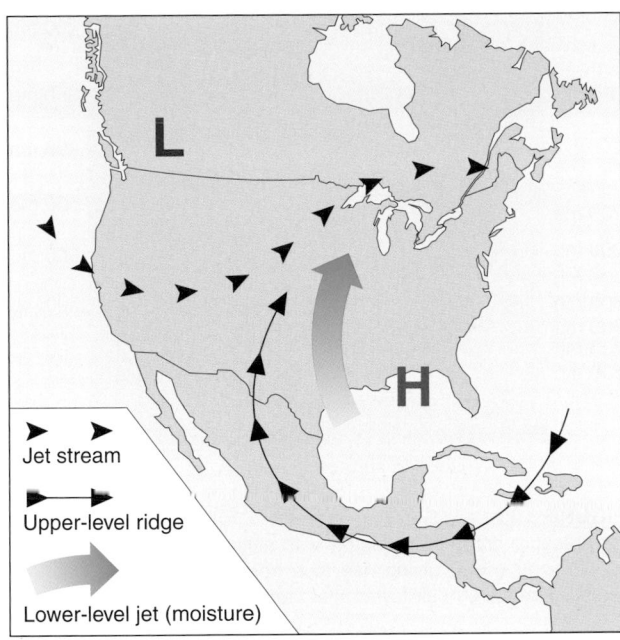

FIGURE 9.18
Principal features of the prevailing atmospheric circulation pattern during the summer of 1993.

heavy winter snowfall, and abundant spring snowmelt. The worst flooding occurred between Minneapolis, MN, and Cairo, IL, on the Mississippi River, and between Omaha, NE, and St. Louis, MO, on the Missouri River. Flooding was not only unprecedented in magnitude but also in persistence with some places reporting more than one record river crest. At St. Louis, the Mississippi River remained above the previous record crest for three weeks. Societal and economic impacts of the 1993 flood were devastating. In many cities, such as Des Moines, IA, and St. Joseph, MO, the freshwater supply was cut off. Barge traffic was halted on the Upper Mississippi River and a portion of the Missouri River from late June to early August. At least 50,000 homes were damaged or destroyed; and more than 4 million hectares (10 million acres) of cropland were inundated. All told, property damage totaled $30.2 billion (in 2007 dollars), about one-third in crop losses. The death toll from flooding was 48.

In the summer of 1993, some parts of the Midwest received more than twice the long-term average seasonal rainfall while some localities in the Southeast received less than half of their long-term average seasonal rainfall (Figure 9.19). As a whole, the Southeast experienced its second driest July on record, and from Alabama and Georgia north to Tennessee and Virginia, July 1993 was the hottest on record. The combination of drought and heat stress caused severe crop damage, especially in South Carolina where over 95% of the corn crop and 70% of the soybean crop were lost. Total crop losses were estimated at $1.4 billion (in 2007 dollars). Deaths due to heat stress along the Eastern Seaboard topped 100.

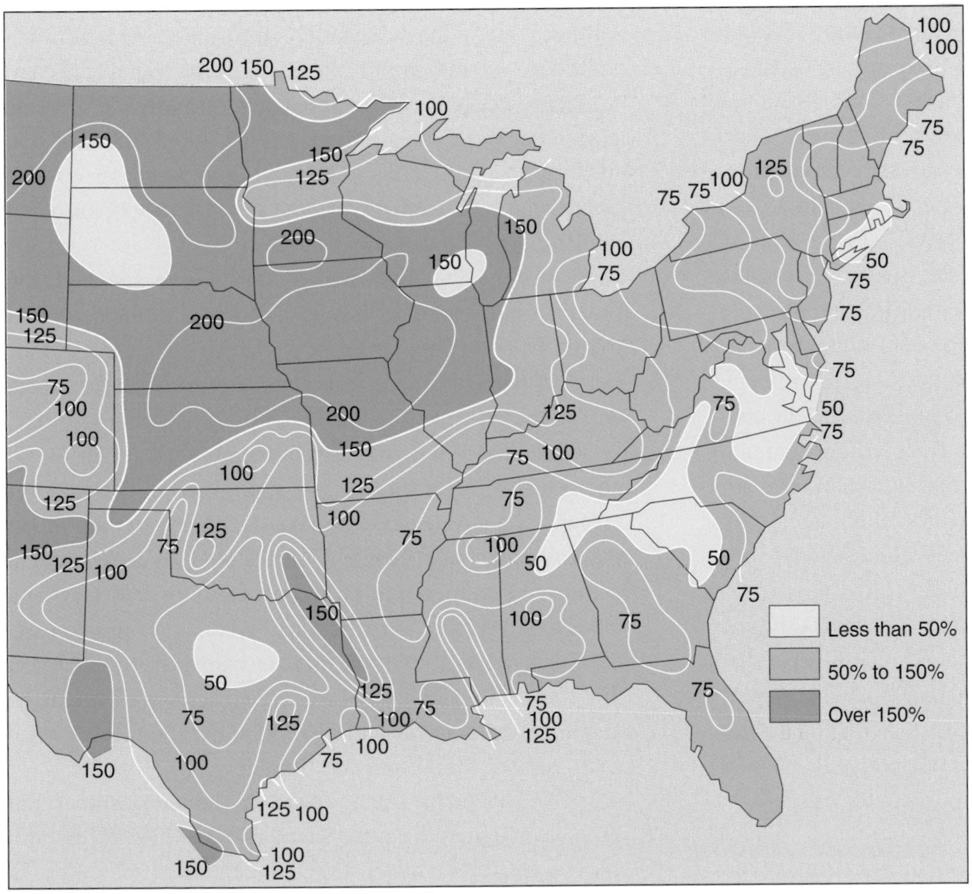

As illustrated by weather events of the summers of 1986, 1988, and 1993, a persistent meridional flow pattern in the westerlies can block the usual movement of weather systems and cause weather extremes such as drought or flooding rains.

FIGURE 9.19
Total rainfall for the period June-August 1993, expressed as a percentage of the long-term average. [From NOAA, Climate Analysis Center]

In fact, any westerly wave pattern, meridional or zonal, can cause extremes in weather if the pattern persists for a sufficient length of time. A persistent westerly wave pattern means the same type of air mass advection, the same storm tracks, and basically the same weather type.

Jet Streams

Within the atmosphere are relatively narrow corridors of very strong winds, known as *jet streams*. In middle latitudes, the most prominent jet stream is located above the polar front in the upper troposphere between the mid-latitude tropopause and the polar tropopause. Because of the close association with the polar front, it is known as the **polar front jet stream**. This jet follows the meandering path of the planetary westerly waves and attains wind speeds that frequently top 160 km per hr (100 mph). This jet stream is a helpful tail wind for eastbound high-altitude aircraft, but a hindering head wind for westbound flights.

Why is a jet stream associated with the polar front? The polar front is a narrow zone of transition between relatively cold and warm air masses. The link between a jet stream near the tropopause and a horizontal air temperature gradient at the Earth's surface hinges on the influence of air temperature on air density. Cold air is denser than warm air, so air pressure drops more rapidly with altitude in a column of cold air than it does in a column of warm air (Figure 9.20). Hence, even if the air pressure at the Earth's surface is nearly the same everywhere, with increasing altitude a horizontal pressure gradient develops between adjacent cold and warm air masses. At any specified altitude within the troposphere, the pressure is higher in the warm air mass than in the cold air mass. In response to this horizontal air pressure gradient, air accelerates away from high pressure (the warm air column) and toward low pressure (the cold air column). Simultaneously, the Coriolis Effect comes into play, increases as air speed increases, and eventually balances the horizontal pressure gradient force. Consequently, the wind blows parallel to isotherms (that is, generally parallel to the underlying surface front) with the cold air to the left of the direction of motion in the Northern Hemisphere. The pattern is reversed in the Southern Hemisphere due to the Coriolis Effect, with cold air on the right.

With the horizontal air pressure gradient strengthening with increasing altitude, the horizontal wind speed increases with altitude. Wind speed is highest near the tropopause because at altitudes above the tropopause (in the stratosphere) the horizontal temperature gradient

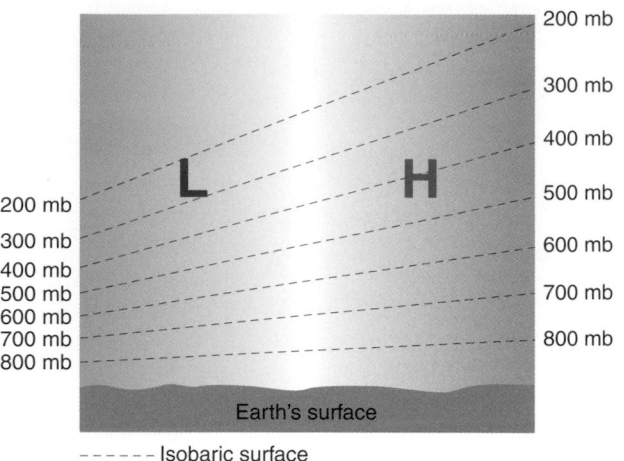

FIGURE 9.20
Air pressure drops more rapidly with altitude in cold air (left) than in warm air (right) giving rise to a horizontal pressure gradient directed from warm air toward cold air above Earth's surface.

reverses direction. Absorption of sunlight by ozone (O_3) heats the polar stratopause more continuously than in the tropics. In the troposphere, the coldest air is at higher latitudes, but in the stratosphere, the coldest air is at lower latitudes. In the stratosphere, the horizontal pressure gradient weakens with increasing altitude so that the horizontal wind speed also weakens with altitude above the tropopause. Hence, the maximum wind speed (the polar front jet stream) is located near the tropopause and above the polar front.

Like the front with which it is associated, the polar front jet stream is not uniformly defined around the globe. The polar front is well defined where surface horizontal temperature gradients are particularly steep, and so the jet stream winds are stronger. Such a segment, in which the wind may strengthen by as much as an additional 100 km per hr (62 mph), is known as a **jet streak**. The strongest jet streaks develop in winter along the east coasts of North America and Asia where the contrast in temperature between snow-covered land and ice-free sea surface is particularly great. Over those areas, jet streak wind speeds on rare occasions have exceeded 350 km per hr (217 mph). A typical jet streak might be 160 km (100 mi) wide and 2 to 3 km (1 to 2 mi) thick, with a length of several hundred kilometers.

The polar front jet stream undergoes seasonal shifts, strengthening in winter (when the north-south air temperature contrast is greatest) and weakening in summer (when temperature contrasts are less). As shown in Figure 9.21, the average summer location of the polar front jet stream is across southern Canada, and the average

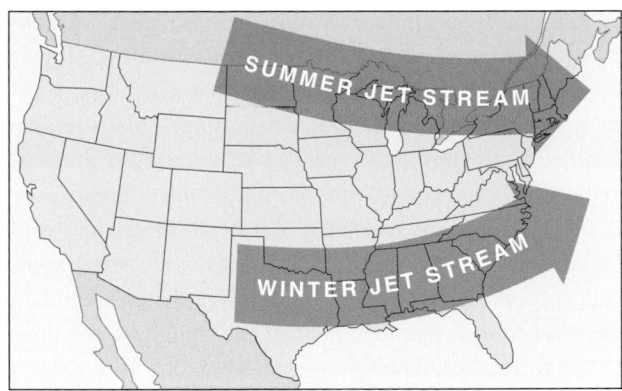

FIGURE 9.21
Approximate average location of the polar front jet stream in winter (December to March) and summer (June to September).

winter position is across the southern United States. These locations represent long-term averages; the jet stream actually weaves over a considerable range of latitude from week to week, and even from one day to the next. As a general rule, when the polar front jet stream is south of your location, the weather tends to be relatively cold, and when the polar front jet stream is north of your location, the weather tends to be relatively warm.

The polar front jet stream is not the only jet stream. The **subtropical jet stream** occurs near the break in the tropopause between tropical and middle latitudes, on the poleward side of the Hadley cell. It is strongest in winter and less variable with latitude than its northerly counterpart. Other jet streams include the *tropical easterly jet*, a feature of the summer circulation located at about 15 degrees N over North Africa, and south of India and Southeast Asia. In addition, in summer a *low-level jet*, perhaps several hundred meters above Earth's surface, surges up the Mississippi River Valley and contributes to the development of nocturnal thunderstorms (Chapter 11).

Cyclone Development

Certain characteristics of the westerlies, including troughs and jet streaks, support the development of extratropical cyclones. A **short wave** is a ripple

superimposed on Rossby long waves. Although Rossby waves (and associated troughs and ridges) usually drift very slowly eastward, short waves usually propagate rapidly along the Rossby waves. Whereas five or fewer long waves encircle the hemisphere, there may be a dozen or more short waves.

Both short waves and long waves in the westerlies contribute to cyclone development by inducing divergence of horizontal winds aloft, that is, in the upper troposphere. As described in Chapter 8, patterns of divergence and convergence of winds aloft are linked to vertical air motion and divergence and convergence of surface winds. Diverging horizontal winds characterize a cyclone aloft while horizontal winds converge at the surface. For the same isobar spacing (air pressure gradient), gradient and geostrophic wind speeds are not equal; that is, anticyclonic gradient winds are stronger than geostrophic winds, and cyclonic gradient winds are weaker than geostrophic winds. Hence, as shown in Figure 9.22, westerly winds tend to weaken (A) downstream of a ridge and speed up (B) downstream of a trough. This produces horizontal divergence of mid- to upper-tropospheric winds to the east of a trough (and west of a ridge). Short and long waves thus favor cyclone development by inducing horizontal divergence aloft. Figure 9.23 portrays the relationship between an upper-air wave and surface cyclone and anticyclone.

Jet streaks also play an important role in the generation and maintenance of extratropical cyclones. Air flowing through a jet streak changes speed and direction; these changes induce a pattern of horizontal divergence and horizontal convergence. Consider Figure 9.24, a schematic representation of a straight jet streak viewed from above. Blue lines are *isotachs*, lines of

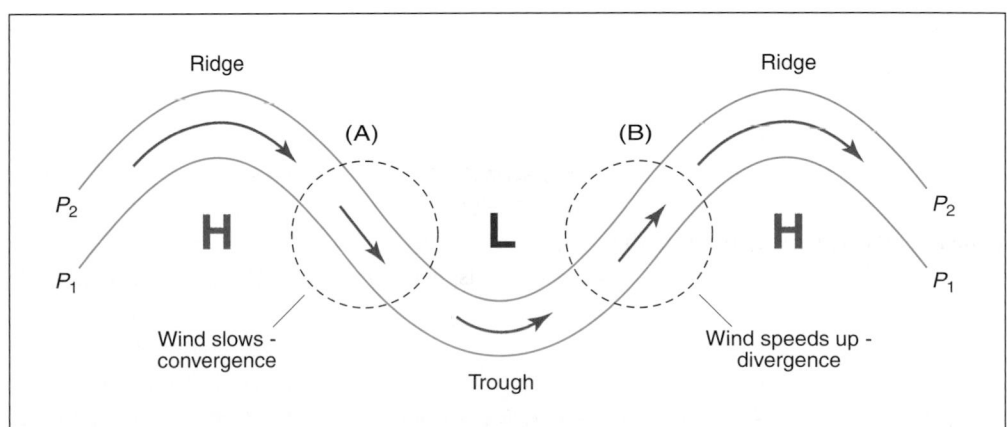

FIGURE 9.22
In flowing into a trough, Westerlies slow, inducing speed convergence aloft. In flowing into a ridge, the Westerlies accelerate, inducing speed divergence aloft. Solid lines are isobars; P1 is greater than P2.

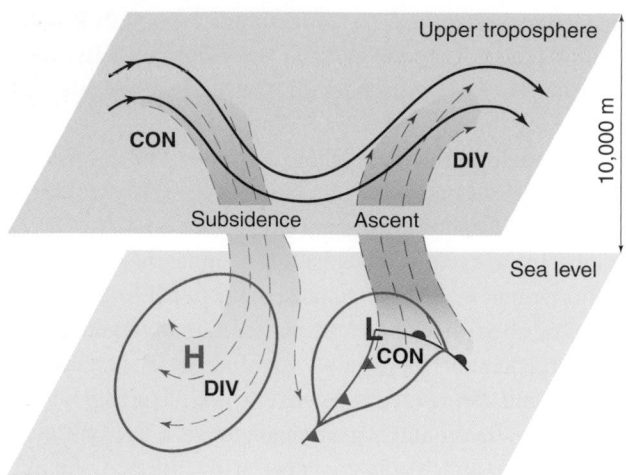

CON = Convergence
DIV = Divergence

FIGURE 9.23
Schematic representation of the relationship between waves in the westerlies and a surface high and low.

equal wind speed (in km per hr). Air accelerates as it enters the jet streak and decelerates as it exits the jet streak. Associated with these changes in wind speed is diverging and converging air. Viewed from above, a jet streak can be divided into four quadrants: left-rear, right-rear, left-front, and right-front. Horizontal divergence occurs in both the left-front and right-rear quadrants, and horizontal convergence takes place in both the right-front and the left-rear quadrants.

A jet streak provides *upper-air support* for a cyclone by contributing horizontal divergence aloft. For

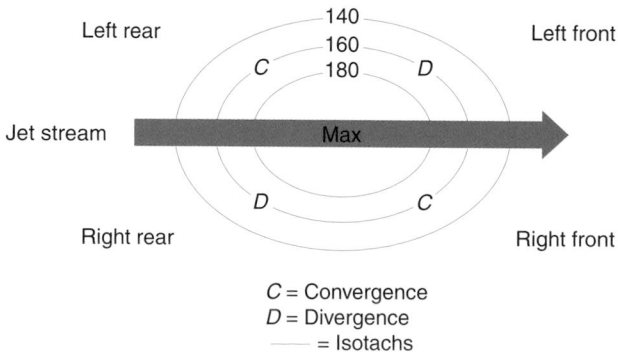

C = Convergence
D = Divergence
——— = Isotachs

FIGURE 9.24
Map view of a jet streak, a segment of accelerated winds within the polar front jet stream with associated regions of (D) horizontal divergence and (C) horizontal convergence aloft. In this view from above, a jet streak is outlined by *isotachs*, lines of equal wind speed (in km per hr). In this case of a straight jet streak, the strongest horizontal divergence is in the left-front quadrant, supplying upper-air support for the development of an extratropical cyclone.

an essentially straight jet streak, the strongest horizontal divergence occurs in the left-front quadrant, so it is under this sector of the jet streak that a cyclone typically has the best chance of developing. This rule also applies to a cyclonically curved jet streak; for an anticyclonically curved jet streak, the strongest divergence is in the right-rear quadrant. Although the polar front jet stream weaves with the westerlies, a jet streak typically progresses from west to east at a faster pace than the west to east displacement of the troughs and ridges in the westerlies. The strongest divergence aloft occurs when a jet streak is situated on the east side of a trough. Hence, cyclone development is much more likely when a jet streak is positioned to generate divergence on the east side of a trough.

El Niño and La Niña

To this point in the chapter we have focused on average atmospheric circulation patterns. Like any fluid flow, the atmosphere can become disturbed and exhibit variability. Planetary-scale oscillations in the atmosphere occur, namely the El Niño Southern Oscillation, which includes El Niño and La Niña, the North Atlantic Oscillation, Arctic Oscillation, and Pacific Decadal Oscillation. These have regional and sometimes worldwide impacts on weather and climate.

Some middle latitude weather extremes such as drought or unusually heavy rainfall are linked to episodic and sometimes significant changes in atmospheric and oceanic circulation in the tropics. One of the best known of these circulation anomalies occurs in the tropical Pacific and is known as **El Niño**. During El Niño, trade winds weaken, sea-surface temperatures (SST) rise well above long-term averages over the central and eastern tropical Pacific, and areas of heavy rainfall shift from the western into the central tropical Pacific. Sometimes (but not always) following El Niño is **La Niña**, an episode of exceptionally strong trade winds across the tropical Pacific with lower than usual SST in the central and eastern tropical Pacific. Based on changes in SST in the central and eastern tropical Pacific, some scientists refer to El Niño as the *warm phase* and La Niña as the *cold phase* of this tropical atmosphere/ocean interaction.

Broad-scale changes in SST patterns over the tropical Pacific that accompany El Niño and La Niña influence the prevailing circulation of the atmosphere in middle latitudes, especially in winter. Weather extremes that most often accompany El Niño are essentially opposite those that usually occur during La Niña.

EL NIÑO AND LA NIÑA

Initially, El Niño was named by fishermen for the unusually warm south-flowing ocean current, and accompanying poor fishing conditions, off the coast of Peru and Ecuador. El Niño (the boy) arrives around Christmas and refers to the Christ child. Typically, these warm water episodes last a month or two before SST and the fisheries return to normal. However, every 3 to 7 years, El Niño would persist for a year or perhaps longer than 18 months. Over vast stretches of the tropical Pacific, there were significant changes in SST creating major shifts in planetary-scale atmospheric and oceanic circulations, and collapse of important South American fisheries. Today, the term El Niño is reserved for the long-lasting ocean/atmosphere anomalies.

The first steps in understanding El Niño came not from South America or the Pacific, but the Indian monsoon failure in 1899-1900, when more than a million lives were lost. In response, Britain appointed Englishman Sir Gilbert Walker (1868-1958) to Director General of Observatories in India and charged him with developing a method to predict the Indian monsoon. For twenty years, he extensively studied the relationship between monsoon rains and weather conditions around the world and in 1924, discovered the **Southern Oscillation**, a seesaw variation in air pressure across the tropical Indian and Pacific Oceans. When air pressure is high in the eastern tropical Pacific, it is low west of the International Dateline over the western tropical Pacific and the Indian Ocean, and monsoon rains are plentiful over India. With the opposite pressure pattern (high pressure west of the International Dateline and low pressure east of the dateline), Indian monsoon rains are lighter than usual.

Today, the **Southern Oscillation Index (SOI)** is based on the difference in air pressure between Darwin on the north coast of Australia at 12 degrees S, 130 degrees E, and Tahiti, an island in the central south Pacific at 18 degrees S, 149 degrees W. When air pressure is anomalously low at Darwin, it is anomalously high at Tahiti, and when high at Darwin, it is low at Tahiti (Figure 9.25). (Some questions surround the quality of air pressure readings at Tahiti prior to 1935.)

Another four decades passed before the Southern Oscillation was linked to El Niño. In 1966, the Norwegian-American meteorologist Jacob Bjerknes (1897-1975) demonstrated how El Niño and the Southern Oscillation interact by analyzing oceanic/atmospheric observations gathered from the tropical Pacific during the *International Geophysical Year*, 1957-58, which fortuitously coincided with a strong El Niño. As the air

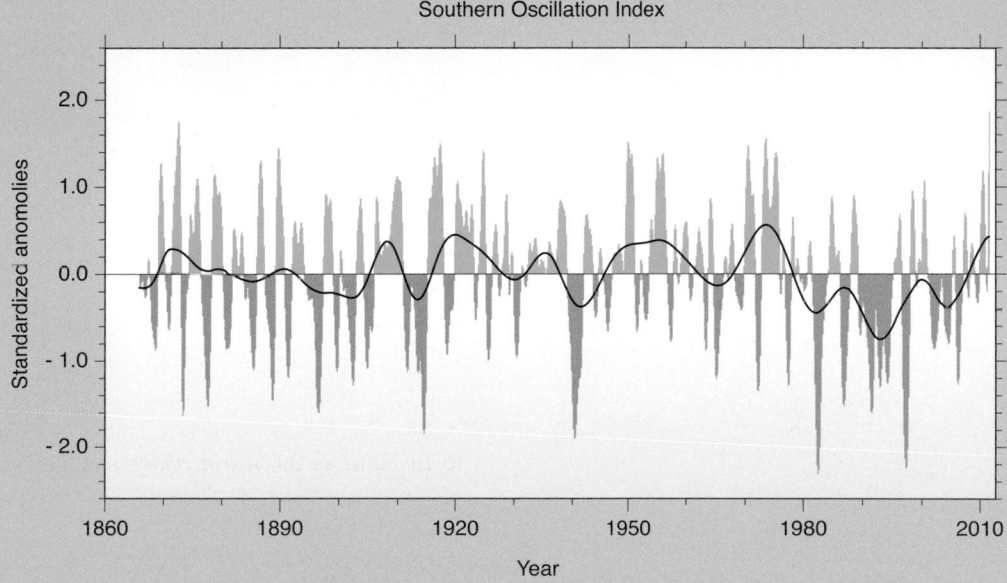

FIGURE 9.25
Variation in the Southern Oscillation Index based on monthly mean sea level pressure anomalies at Darwin, Australia. Strongly positive values of the index indicate La Niña conditions whereas strongly negative values of the index indicate El Niño conditions. [NOAA, Earth System Research Laboratory, Physical Science Division]

pressure gradient changes across the tropical Pacific, with air pressure rising to the west and falling to the east, the SST pattern also shifts. At the same time, the weakening pressure gradient across the tropical Pacific heralds the slackening of the trade winds, and an El Niño episode is underway. This relationship between El Niño and the Southern Oscillation is known by the acronym **ENSO**.

ENSO is a *coupled* phenomenon in that its variability in Earth's climate system cannot be explained as exclusively an oceanic or atmospheric event. *Coupled* refers to more than merely something that occurs in both the ocean and atmosphere; in this case, the phenomenon depends on feedbacks between the ocean and atmosphere. Changes in ocean conditions (primarily SST) can drive changes in atmospheric circulation that affect cloud and precipitation patterns. What is unique about ENSO is the strong coupling: changes in the ocean drive changes in the atmosphere which then feedback and further alter the ocean, and so on.

Not until the El Niño of 1982-83, one of the two most intense of the 20th century, was the potential worldwide impact of ENSO realized. That episode spurred development of numerical models to simulate ENSO as well as deployment of a network of instrumented buoys and satellites to provide advance warning of a developing El Niño. Also, the last three decades have seen increasing interest in La Niña (the girl), the name coined in the mid-1980s for an atmosphere/ocean interaction essentially opposite El Niño, although typically less intense. The cold La Niña can be thought of as the opposite extreme of the *ENSO cycle*.

El Niño or La Niña are now routinely incorporated into long-range seasonal weather outlooks worldwide because of their importance in year-to-year climate variability (Chapter 13). Such outlooks identify areas of expected anomalies in temperature and precipitation, and guide development of regional water management and agricultural strategies. Adoption of these strategies helps lessen the impact of attendant weather extremes on water supply and food production.

EKMAN TRANSPORT

El Niño and La Niña represent significant departures from the long-term average or *neutral* atmosphere/ocean conditions in the tropical Pacific. Understanding El Niño and La Niña first requires a look at neutral conditions, beginning with a summary of Ekman transport and its influence on sea-surface temperature.

In some coastal areas of the ocean, the combination of persistent winds, Earth's rotation (Coriolis Effect), and restrictions on lateral movements of water caused by shorelines and shallow bottoms induces upward (upwelling) and downward (downwelling) movements of water that affect surface water temperatures. In the tropical Pacific, changes in upwelling are responsible for shifts in sea-surface temperatures that characterize neutral, El Niño, and La Niña episodes.

If Earth did not rotate, frictional coupling between moving air and the ocean surface would push a thin layer of water in the same direction as the wind. This surface layer in turn would drag the layer beneath it, putting it into motion. This interaction would propagate downward through successive ocean layers, like pushing on cards in a deck, each moving forward at a slower speed than the layer above. However, because Earth rotates, the shallow layer of surface water set in motion by the wind is deflected to the right of the wind direction in the Northern Hemisphere and to the left of the wind direction in the Southern Hemisphere (the *Coriolis Effect*). Except at the equator, where the Coriolis Effect is zero, each layer of water put into motion by the layer above shifts direction because of Earth's rotation.

Using arrows, or vectors, to represent the direction and speed of layers of water at successive depths in the ocean, a simplified model of the three-dimensional current pattern caused by a steady horizontal wind emerges (Figure 9.26A). This model is known as the **Ekman spiral**, named for the Swedish physicist V. Walfrid Ekman (1874-1954) who first described it mathematically in 1905. Ekman based his model on observations made by the Norwegian explorer Fridtjof Nansen (1861-1930). Nansen was interested in learning about the ocean currents of the polar seas. In 1893, he allowed his 39-m (128-ft) wooden ship, the *Fram*, to freeze into the Arctic pack ice about 1100 km (685 mi) south of the North Pole. His goal was to drift with the ice and cross the North Pole thereby determining how ocean currents affect the movement of pack ice. The *Fram* remained locked in pack ice for 35 months but came no closer than about 394 km (245 mi) to the North Pole. As the *Fram* slowly drifted with the ice, Nansen noticed that the direction of ice and ship

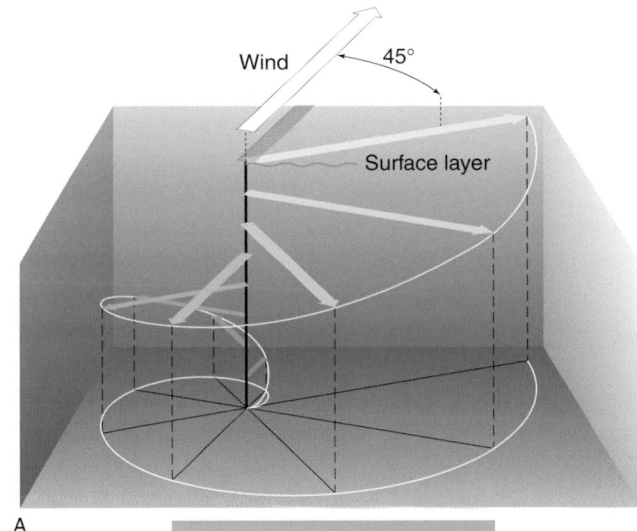

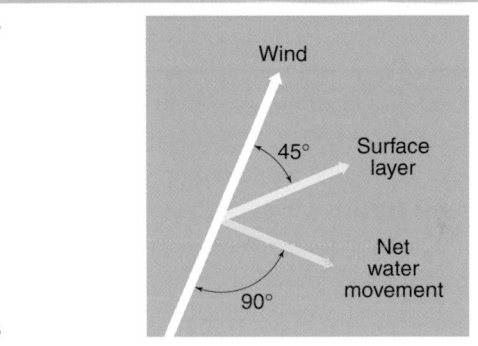

FIGURE 9.26
The Ekman spiral describes how the horizontal wind sets surface waters in motion. (A) As represented by horizontal vectors, the speed and direction of water motion change with increasing depth. (B) Viewed from above in the Northern Hemisphere, the surface layer of water moves at 45 degrees to the right of the wind direction. The net transport of water through the entire wind-driven column (i.e., the Ekman transport) is 90 degrees to the right of the wind. Because of the reversal of the Coriolis deflection in the Southern Hemisphere, Ekman transport is 90 degrees to the left of the wind south of the equator.

movement was consistently 20 to 40 degrees to the right of the prevailing wind direction.

The Ekman spiral indicates that the direction and speed of water motion change with increasing depth. In an ideal case, a steady wind blowing across an ocean of unlimited depth and extent causes surface waters to move at an angle of 45 degrees to the right of the wind in the Northern Hemisphere (45 degrees to the left of the wind in the Southern Hemisphere). Each successively lower layer moves more toward the right and at a slower speed. At a depth of about 100 to 150 m (330 to 500 ft), the Ekman spiral has gone through less than half a turn. Yet water moves so slowly (about 4% of the surface current) that this depth is considered to be the lower limit of the wind's influence on the movement of ocean waters.

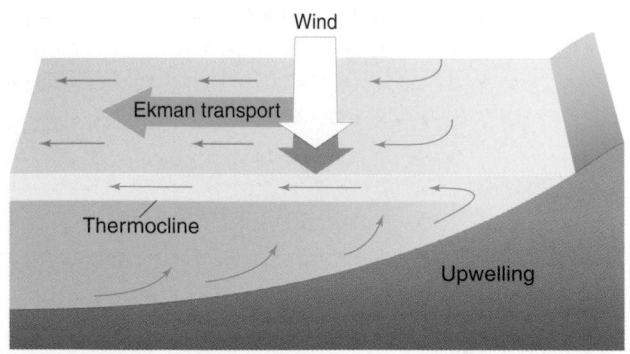

FIGURE 9.27
Where Ekman transport moves surface waters away from the coast, surface waters are replaced by water that wells up from below in the process known as upwelling.

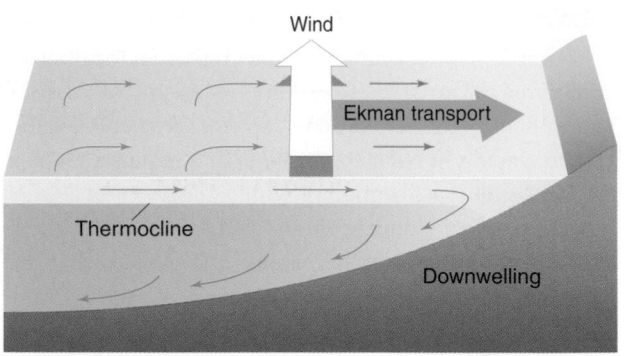

FIGURE 9.28
Where Ekman transport moves surface waters toward the coast, the water piles up and sinks in the process known as downwelling.

In the Northern Hemisphere, the Ekman spiral predicts net water movement through a depth of about 100 to 150 m (330 to 500 ft) at 90 degrees to the right of the wind direction (Figure 9.26B). That is, if one adds up all the arrows in Figure 9.26A, the resulting flow is at 90 degrees to the right of the surface wind direction. In the Southern Hemisphere, the net water movement is 90 degrees to the left of the surface wind direction. This net transport of water due to coupling between the surface wind and water is known as **Ekman transport**, and is an important type of flow in the surface layer of the ocean (the *mixed layer*).

The real ocean departs from the idealized conditions represented by the Ekman spiral; that is, wind-induced water movements often differ appreciably from theoretical predictions. To arrive at 45 degrees as the angle between the directions of the surface waters and surface winds, Ekman made many simplifying assumptions. In the real ocean, however, the angle is closer to 15 to 20 degrees at the most, regardless of the water depth. In shallow water, for example, the water depth is insufficient for a full spiral to develop so that the angle between the horizontal wind direction and surface-water movement can be as little as 15 degrees. As the water deepens, the angle increases and approaches 45 degrees.

Where Ekman transport moves surface waters away from the coast, those waters are replaced by water that wells up from below in a process known as **upwelling** (Figure 9.27). Where Ekman transport moves surface waters toward the coast, the surface water piles up and sinks in the process known as **downwelling** (Figure 9.28). Upwelling is also responsible for the tongue of relatively cool surface waters along the equator in the eastern tropical Pacific. Near the equator, the northeast trade winds of the Northern Hemisphere converge with the southeast trade winds of the Southern Hemisphere. The associated Ek-

man transport (although weak because of minimal Coriolis Effect) causes surface waters to diverge away from the equator and colder water wells up from below replacing the departing surface waters. *Equatorial upwelling* (Figure 9.29) produces a strip of relatively low sea-surface temperatures along the equator from the coast of South America westward to near the International Dateline.

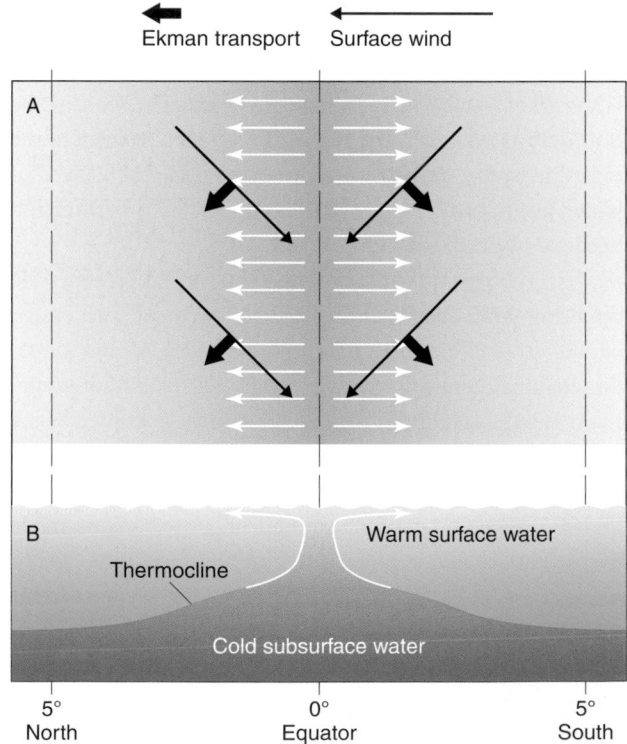

FIGURE 9.29
Equatorial upwelling. (A) In this plan view of the ocean from 5 degrees S to 5 degrees N, the trade winds of the two hemispheres are shown to converge near the equator. The consequent Ekman transport away from the equator gives rise to upwelling as shown in (B) a vertical cross section from 5 degrees S to 5 degrees N.

NEUTRAL CONDITIONS IN THE TROPICAL PACIFIC

The air pressure gradient directed from high air pressure over the central and eastern tropical Pacific and low air pressure over the western tropical Pacific helps strengthen the trade winds, which blow from the northeast in the Northern Hemisphere and from the southeast in the Southern Hemisphere. The greater the air pressure gradient from east to west, the stronger the general east to west flow.

South of the equator, prevailing winds blow from the south or southwest along the west coast of South America. Ekman transport drives warm surface waters to the left (westward), away from the coast (Figure 9.30). As the warm, nutrient-poor surface waters are pushed off-shore, the cold, nutrient-rich waters well up from below the *thermocline* (the transition zone between cold water below and warm water above) which is only 50 to 100 m (165 to 325 ft) deep along the coast. Although this zone of *coastal upwelling* is narrow, typically less than 15 km or 10 mi wide, the abundance of nutrients carried into the photic zone spurs an explosive growth of phytoplankton and supports a diverse ecosystem and highly productive fishery.

Meanwhile, relatively warm surface waters are driven by these same southwest trade winds westward toward Indonesia and northern Australia along the north and south equatorial currents. This wedge of warm water increases the depth of the thermocline to 150 m (490 ft) and raises sea level in the western tropical Pacific. That warm water, piled higher by trans-Pacific trade winds, also expands when heated so that sea level is 60 cm (2 ft) higher in the west than in the east with an 8 Celsius degrees (14.4 Fahrenheit degrees) contrast in SST.

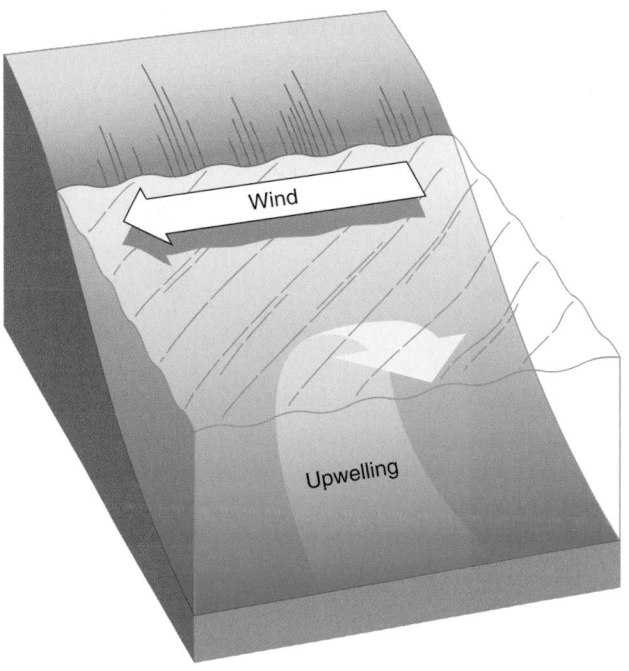

FIGURE 9.30
Off the Southern Hemisphere portion of the northwest coast of South America, Ekman transport of surface waters is to the left of the wind direction and away from the coastline. Consequently, cold, nutrient-rich bottom waters upwell to the surface.

Waters with high sea-surface temperatures heat the overlying air in the western tropical Pacific, further lowering surface air pressure and strengthening convection. Water vapor condenses into towering cumulonimbus (thunderstorm) clouds that produce heavy rainfall (Figure 9.31). Aloft this air flows back eastward and sinks over the cooler waters of the eastern tropical Pacific.

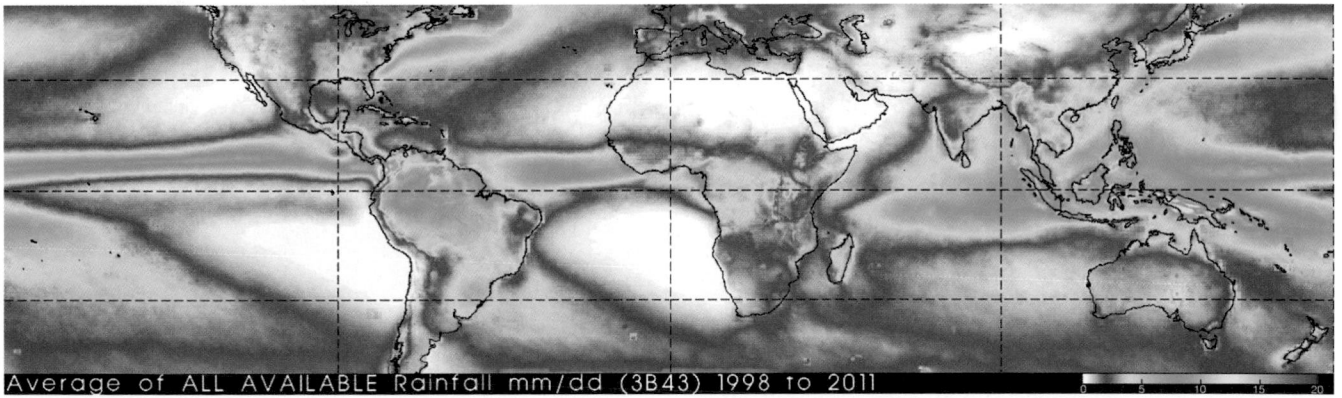

Average of ALL AVAILABLE Rainfall mm/dd (3B43) 1998 to 2011

FIGURE 9.31
Benchmark average rainfall in millimeters per day (mm/d) for the 13-year period 1998 through 2011. The heaviest rainfall is in the western tropical Pacific where sea-surface temperatures are highest. These data were obtained from the TRMM Microwave Imager and IR sensors onboard geosynchronous satellites supplemented by conventional rain gauge measurements. [Source: NASA, Tropical Rainfall Measuring Mission (TRMM)]

Normal Conditions

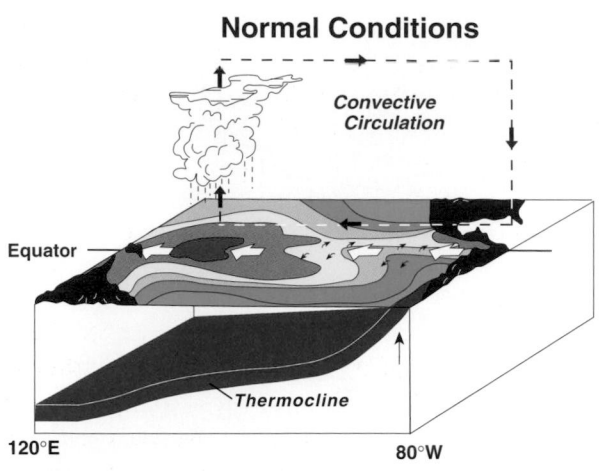

FIGURE 9.32
Schematic block diagram showing ocean/atmosphere conditions in the tropical Pacific during normal (neutral) episodes. Red indicates areas of highest sea-surface temperatures. White arrows are surface ocean currents. [NOAA, Pacific Marine Environmental Laboratory (PMEL), Tropical Atmosphere Ocean Project]

El Niño Conditions

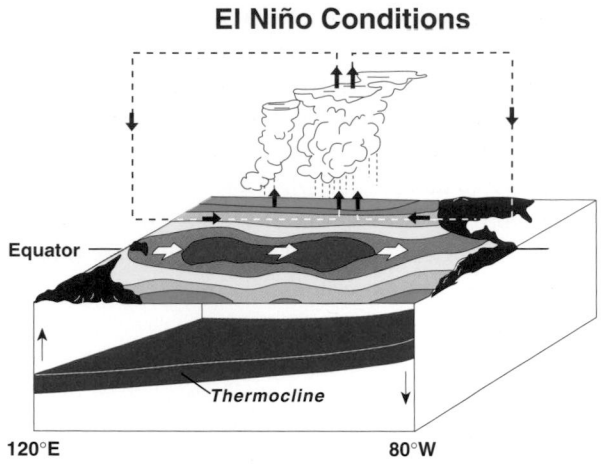

FIGURE 9.33
Schematic block diagram showing ocean/atmosphere conditions in the tropical Pacific during El Niño conditions. Red indicates areas of highest sea-surface temperatures. White arrows are surface ocean currents. [NOAA, Pacific Marine Environmental Laboratory (PMEL), Tropical Atmosphere Ocean Project]

There, low SST in the central and eastern tropical Pacific chill the overlying air, further raising surface air pressure and suppressing convection. Sinking air is compressed and warmed so that clouds vaporize or fail to develop.

High SST in the western tropical Pacific lower the surface air pressure, whereas low SST in the eastern tropical Pacific raise the surface air pressure. Hence, during neutral conditions, the east-west SST gradient reinforces the trade winds by strengthening the east-west pressure gradient. In flowing over the ocean surface, the trade winds become warmer and more humid. In the western tropical Pacific warm humid air rises, expands, and cools (Figure 9.32). This completes the large convective-type circulation known as the **Walker Circulation**, named for Sir Gilbert Walker.

WARM PHASE

As part of the Southern Oscillation, air pressure falls over the eastern tropical Pacific and rises over the western tropical Pacific, which can trigger an El Niño episode. As the air pressure gradient weakens across the tropical Pacific, trade winds slacken and, during an intense El Niño, trade winds west of the International Dateline may reverse direction and blow toward the east.

In response to these shifts in atmospheric circulation, surface ocean currents, SST, sea level, thermocline depth, and upwelling all shift (Figure 9.33). The relaxed trade winds cause the ocean surface currents to weaken and, occasionally, reverse direction as well, flowing eastward. Hence, the thick, warm surface water normally in the west drifts slowly eastward, taking several

months to reach the Americas where it is deflected by the continents to as far north as Canada and as far south as central Chile. In the western tropical Pacific, higher air pressure causes SST to drop, which along with the lack of trade winds, causes sea-level to fall and the thermocline to rise. The lower air pressure in the eastern tropical Pacific causes SST to rise, which increases sea-level and, without the trade winds to produce upwelling, the thermocline deepens (Figure 9.34).

The impacts of these environmental changes can be severe on marine ecosystems. Reduced upwelling

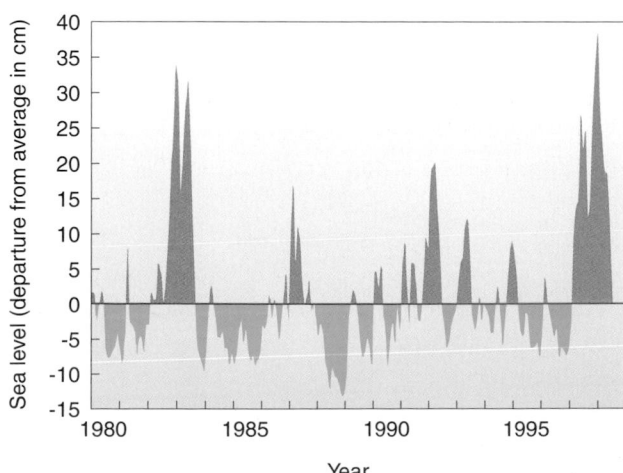

FIGURE 9.34
Sea level record at Galapagos in the eastern tropical Pacific based on tide gauge records and expressed in cm as departure from the long-term average. Relatively high sea levels correspond to El Niño episodes. [NOAA, PMEL]

means the cold, nutrient-rich waters along the coast of Ecuador and Peru remain below, causing phytoplankton to decline and fish populations to plummet. Peru's fishing industry was booming in the 1950s and 1960s, and by 1970 was one of the largest in the world, accounting for 20% of the total global catch of anchovies and a third of the nation's foreign income. However, over-fishing and the 1972-73 El Niño decimated the Peruvian fishery and it has yet to recover.

Warmer surface waters can also stress coral reefs living in shallow tropical waters. Responding to elevated sea-surface temperatures, coral expels zooxanthallae, the symbiotic microscopic algae that supply coral with oxygen and some organic compounds produced through photosynthesis. Without zooxanthallae, coral polyps have little pigmentation and appear nearly transparent on the coral's white skeleton, a condition known as **coral bleaching**. Excessive bleaching can kill coral polyps, destroying habitats for a great variety of marine organisms. In addition, bleaching makes coral more vulnerable to infectious diseases. Coral bleaching was widespread and severe during the 1997-98 El Niño/La Niña episodes and in 2005 and 2010. Global warming is likely to bring more frequent and severe spikes in SST and more extensive coral bleaching.

Anomalous weather patterns develop in the tropics and subtropics in response to low SST in the western tropical Pacific, high SST in the central and eastern tropical Pacific, and changes in the trade winds of El Niño. Normally, abundant rainfall comes from the winds blowing onshore over Indonesia, but during El Niño, prevailing winds are directed offshore and Indonesia is dry. El Niño has also brought droughts to India, eastern Australia, northeastern Brazil, and southern Africa. Meanwhile, warmer surface waters off the west coast of South America spur convection and heavy rainfall along the normally dry coastal plain, causing flash flooding. Wetter conditions tend to occur in southern Brazil, Uruguay, and equatorial East Africa.

Martin Hoerling, a NOAA meteorologist, and colleagues found that central India endured 10 severe droughts during the monsoon rainy season in El Niño years for the period 1871 to 2002. When the greatest SST anomalies during El Niño occurred in either central or eastern equatorial Pacific, drought would dominate central India. All 10 droughts occurred during El Niño episodes. However, not all El Niño events during the 132-year period were accompanied by drought in central India. In fact, in 13 cases rainy season precipitation was at or slightly greater than the long-term average. In examining the various cases of drought, Hoerling and colleagues found that drought occurred in central India when the highest SST were found in the central equatorial Pacific. Numerical models predicted that these exceptionally warm waters would produce warm, humid air that would rise high in the tropical troposphere while losing much of its moisture, then moved over central India, and subsided. The subsiding air inhibited development of clouds and precipitation and soon a drought would be underway.

Typically, El Niño also brings dry weather to the Hawaiian Islands. The North Pacific subtropical anticyclone shifts so that the descending air is closer to the Islands, creating a persistent dry weather pattern. Almost all of Hawaii's major droughts during the 20th century coincided with an El Niño event.

The intensity, frequency, and spatial distribution of tropical cyclones (e.g., tropical storms, hurricanes) are also influenced by El Niño. The extensive area of warmer water over the eastern tropical Pacific allows hurricanes to travel farther north and west than usual, altering their intensity and location in the Pacific and Indian Oceans. In the North Atlantic Ocean, stronger than usual winds aloft inhibit the development of tropical cyclones, and those few that do develop usually are weaker and short lived (Chapter 12).

El Niño has a ripple effect on the weather of middle latitudes, especially in winter. A link between changes in atmospheric circulation patterns in widely separated regions of the world, often thousands of kilometers apart, is known as a **teleconnection**. Fueled by latent heat released during deep convection and buildup of thunderstorms in the tropical troposphere, teleconnections help shape the planetary-scale circulation. During El Niño when high SST in the central and eastern tropical Pacific heats and destabilizes the troposphere, wind and weather patterns shift worldwide. A teleconnection governs the strength and direction of atmospheric circulation, the course of jet streams, storm tracks, and moisture transport by winds at higher latitudes. Like large boulders that generate eddies in a swiftly flowing stream, towering thunderstorms build high into the tropical troposphere and deflect the upper air winds. Moving the boulders will displace the train of eddies. In the same way, a shift in the location of the principal area of convection eastward over the tropical Pacific redirects the atmospheric circulation.

During typical El Niño winters, prevailing storm tracks bring abundant rainfall and cooler than usual conditions to the Gulf Coast states, from Texas to Florida. Over the northern U.S. and Canada, prevailing winds blow from the west, moving cold air masses eastward across the Arctic and northern Canada. Persistence of this circulation pattern prevents cold air masses from invading south, so that mild weather prevails over much of Canada, Alaska, and parts of the northern U.S. West-to-east flow in the westerlies also diminishes the usual spring contrast between warm, humid air masses moving northeastward from the Gulf of Mexico and cold, dry air masses sweeping southeastward from Canada. Hence, severe thunderstorms and tornadoes may be less frequent than usual in the Ohio and Tennessee River Valleys.

Although some weather extremes almost always accompany El Niño, no two events are the same because El Niño is only one of many factors that influence inter-annual variability of climate. In southern California, for example, heavy winter rains (snows at higher elevations) have occurred during some but not all El Niño events. Record heavy rainfall in Southern California during January 1995 was linked to a southerly shift of the jet stream and storm track over the eastern Pacific. In that case, a change in atmospheric circulation associated with El Niño was the culprit. Whereas the 1982-83 El Niño brought severe drought to eastern Australia, dry conditions in Australia during the 1997-98 El Niño were far less severe. For more on these two historic El Niño episodes, see this chapter's second *For Further Exploration*.

COLD PHASE

A period of unusually strong trade winds and exceptionally vigorous upwelling in the eastern tropical Pacific, La Niña is an exaggeration of neutral conditions and the opposite of El Niño, which it can, but does not always, follow (Figure 9.35). During La Niña, the air pressure is greater over colder surface waters of the central and eastern tropical Pacific and lower over warmer surface waters of the western tropical Pacific. SST anomalies over the eastern tropical Pacific typically have a greater magnitude during El Niño than during La Niña. SST usually rise 5 to 6 Celsius degrees (9 to 11 Fahrenheit degrees) above the long-term average during an intense El Niño but drop only 2 to 3 Celsius degrees (3.6 to 5.4 Fahrenheit degrees) below the long-term average during a strong La Niña.

Accompanying La Niña are worldwide weather extremes that are often opposite those observed during El Niño. As with El Niño, the most consistent middle latitude teleconnections appear in winter. In the tropical Pacific, lower than usual SST in the east inhibit rainfall and higher than usual SST in the west enhance rainfall in Indonesia, Malaysia, and northern Australia during the Northern Hemisphere winter and the Philippines during the North-

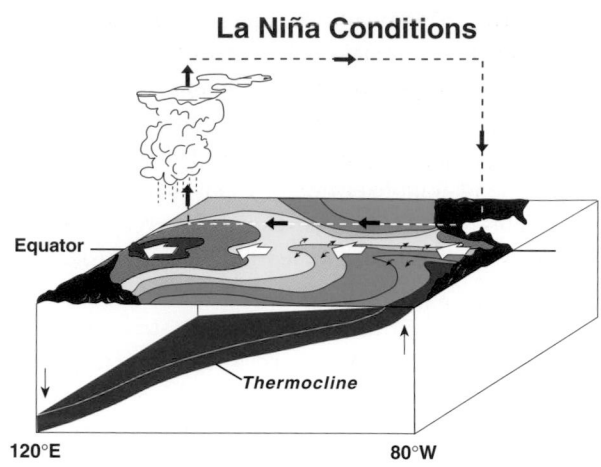

FIGURE 9.35
Schematic block diagram showing ocean/atmosphere conditions in the tropical Pacific during La Niña conditions. Red indicates areas of highest sea-surface temperatures. White arrows are surface ocean currents. [NOAA, Pacific Marine Environmental Laboratory (PMEL), Tropical Atmosphere Ocean Project]

ern Hemisphere summer. Elsewhere around the globe, the Indian monsoon rainfall (in summer) tends to be heavier than average (especially in northwest India) and wet conditions prevail over southeastern Africa and northern Brazil (during the Northern Hemisphere winter). Southern Brazil to central Argentina experiences a dry winter. In addition, weak winds aloft during La Niña favor tropical cyclone formation in the Atlantic Basin.

Across middle latitudes of the Northern Hemisphere, westerlies tend to be more meridional during La Niña. These winds steer cold air masses toward the southeast and warm air masses toward the northeast. Occasionally, a meridional flow pattern becomes so extreme that a blocking pattern (discussed earlier in this chapter) develops. For example, a blocking pattern in the westerlies was responsible for the severe summer drought that afflicted the central U.S. during the La Niña year of 1988.

In the spring, a more meridional flow pattern in the westerlies increases the likelihood of severe thunderstorms and tornadoes across the central U.S. by bringing together air masses with great contrasts in temperature and humidity, a key ingredient for development of severe storms. Also in the U.S., La Niña tends to be accompanied by below average winter precipitation and mild temperatures in a band from the Southwest, through the central and southern Rockies, eastward to the Gulf Coast. In the Pacific Northwest, winter tends to be cool and wet. Lower than usual winter temperatures also occur east of the northern Rockies and north-central states.

The La Niña of 2011, the sixth strongest dating back to 1949, was responsible for the classic signature of drought in the Southwest (Figure 9.36A) and floodwaters in the Midwest and South (Figure 9.36B). Texas and parts of several surrounding states (New Mexico, Oklahoma, and Louisiana) experienced drought that at its worst was characterized as "exceptional" over 86% of Texas. March through September, 2011, was the driest 7-month span on record for the Lone Star State. John Nielsen-Gammon, Texas State Climatologist, reported that the state's average rainfall from October through April was 14.78 cm (5.82 in.), breaking the previous record of 14.86 cm (5.85 in.) set in 1918. Exacerbating the situation, the summer was one of the warmest on record and wildfires burned over thousands of square kilometers of land (Figure 9.36C). The agricultural impact was severe with farm fields, cattle ponds, and bayous drying up. For more imagery related to the 2011 Southwest drought, see this chapter's first *For Further Exploration*.

FIGURE 9.36
(A) Browned out corn field near Los Fresnos, TX, June 21, 2011. [NOAA] (B) The Missouri river flooding in Bellevue, NE, June 22, 2011. (C) The Cass County-Bear Creek Fire which burned over 50,000 acres in Texas. [Dave Hall/Office of Emergency Management, Texarkana, TX]

PREDICTING AND MONITORING ENSO

Scientists have developed numerical models that simulate the onset, evolution, and decay of El Niño and La Niña. These models approximate oceanic processes that alter sea-surface temperatures, and the atmospheric response, including convection, clouds, and winds. Forecasters rely on two basic types of numerical models to predict El Niño or La Niña: empirical (or statistical) models and dynamical models. An *empirical model* compares current and evolving oceanic and atmospheric conditions with observational data from previous episodes over the prior 40 years. The similarity between past and present conditions is the basis for an empirical model prediction. A *dynamical model* uses mathematical equations to simulate interactions or couplings involving the atmosphere, ocean, and land.

Reliable observational data from the tropical Pacific Ocean and atmosphere are essential for detecting a developing El Niño or La Niña, and initializing numerical models. The observational data represent the initial conditions, used as a starting point for predicting future states of the ocean and atmosphere, and for verifying the model predictions and results. This is especially important for dynamical models, which depend on input of reliable data for their complex coupled equations. Then, as El Niño or La Niña unfold, data are continuously assimilated into the models to correct or "nudge" the model.

To improve understanding, detection, and prediction of ENSO related variability by producing the observational data needed for models, monitoring systems were deployed in the tropical Pacific as part of the 10-year (1985-94) international *Tropical Ocean Global Atmosphere (TOGA)* study. By December 1994, the ENSO Observing System was fully operational. The ENSO Observing System consists of an array of moored and drifting instrumented buoys, island and coastal tide gauges, ship-based measurements, and satellites (Figure 9.37).

TAO *(Tropical Atmosphere/Ocean)*, one component of the ENSO Observing System, is an array of moored buoys, small instrumented platforms, in the tropical Pacific Ocean (Figure 9.38). Buoys are strategically placed within 8 degrees N, 8 degrees S, 95 degrees W, and 137 degrees E. As shown in Figure 9.39, approximately 70 deep-sea moorings, renamed TAO/TRITON in 2000, measure atmospheric and oceanic variables, including air temperature, wind, relative humidity, and sea-surface and subsurface temperatures at 10 depths in the upper 500 m (1640 ft) of the ocean. Several newer moorings also have salinity sensors, along with additional meteorological sensors. Five moorings along the equator use Subsurface Acoustic Doppler Current Profilers to measure ocean current velocity. Observational data are transmitted to NOAA's Pacific Marine Environmental Laboratory (PMEL) in Seattle, WA, via a NOAA polar-orbiting satellite, and are available in near real-time on the Internet.

Remote sensing by satellite, such as NOAA and NASA satellites that monitor cloud cover and map SST, plays an important role in providing early warning of an evolving El Niño or La Niña. The *TOPEX/Poseidon* satellite, a joint mission between NASA and the *Centre National d'Etudes Spatiales (CNES)* in France, provided images of ocean surface topography (sea level) at 10-day

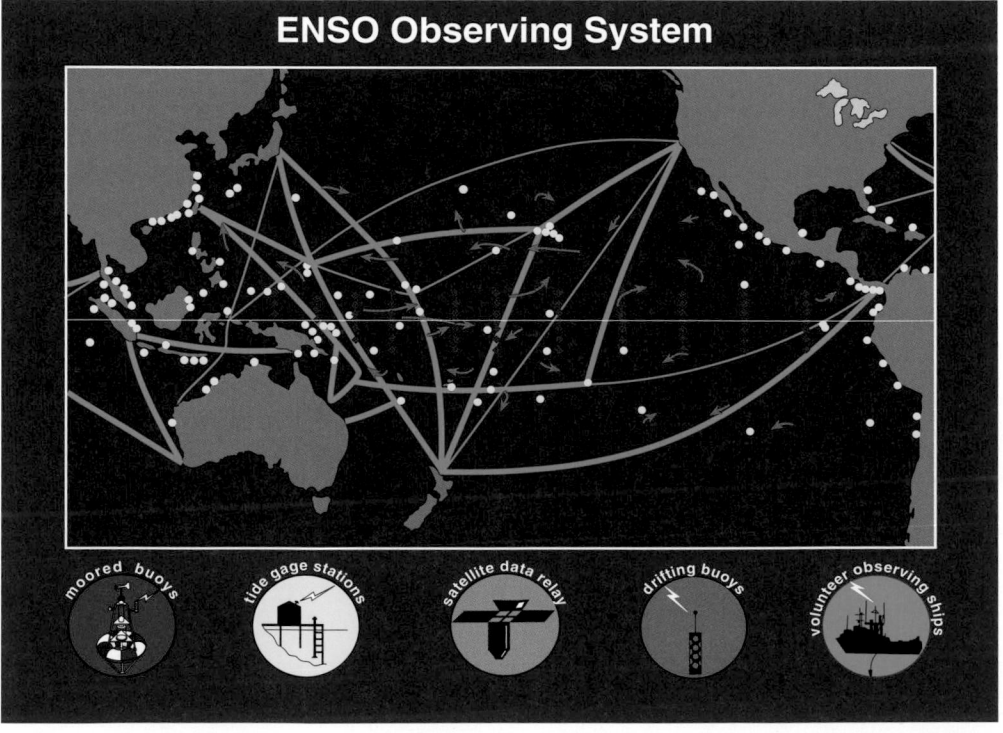

FIGURE 9.37
Components of the ENSO Observing System provide advance warning and monitor the development and decay of El Niño and La Niña events. [NOAA PMEL]

FIGURE 9.38
An instrumented TAO moored buoy undergoing testing at NOAA's Pacific Marine Environmental Laboratory in Seattle, WA. An array of similar moored buoys gathers oceanic and atmospheric data from the tropical Pacific as part of the ENSO Observing System.

intervals from 1992 until January 2006 after completing 62,000 orbits. In December 2001, NASA and CNES launched *Jason 1*, successor to TOPEX/Poseidon, and in June 2008, *Jason 2* was launched into the same orbit. This latest Ocean Surface Topography Mission (OSTM) is an international collaborative effort involving NASA, NOAA, CNES, and the *European Organisation for the Exploitation of Meteorological Satellites (EUMETSAT)*. Designed to operate for at least three years, *Jason 2* extended the continuous climate record of precise sea surface height measurements into the decade of the 2010s using

the next generation of more accurate instruments. Radar altimeters onboard these satellites bounce microwaves off the ocean surface to obtain precise distance measurements between satellite and the sea surface, producing images of sea surface height. Elevated topography (hills) can indicate warmer water whereas areas of low topography (valleys) can indicate colder water (Figure 9.40). Such images can be used to calculate ocean surface currents based on the geostrophic assumption and to identify and track El Niño and La Niña.

The joint U.S.-Japanese *Tropical Rainfall Measuring Mission (TRMM)*, launched in November 1997, also detects the onset and evolution of El Niño and La Niña. The TRMM satellite, as described in Chapter 7, uses active radar and passive microwave energy sensors to monitor clouds, precipitation, and radiation between 40 degrees N and 40 degrees S. Uniquely, the TRMM Microwave Imager (TMI) can "see through" clouds to measure SST. The TMI uses microwaves emitted by the sea surface to characterize the IR emission and determine the radiation temperature. However, while microwaves pass through clouds with little attenuation, they are strongly scattered and absorbed by rainfall so TMI can measure SST only during fair weather.

Increasingly accurate predictions of El Niño and La Niña are enabling better handling of the impacts of interannual climate variability. Greater forecasting skill allows for informed strategic planning in agriculture, fisheries, and water resource management. Consider an example: Peru's economy, like that of most developing nations, is sensitive to climate variability. In Peru, El Niño is bad for fishing and often accompanied by destructive flooding while La Niña benefits fishing but may also bring drought and crop failure. Advance warning of an impending

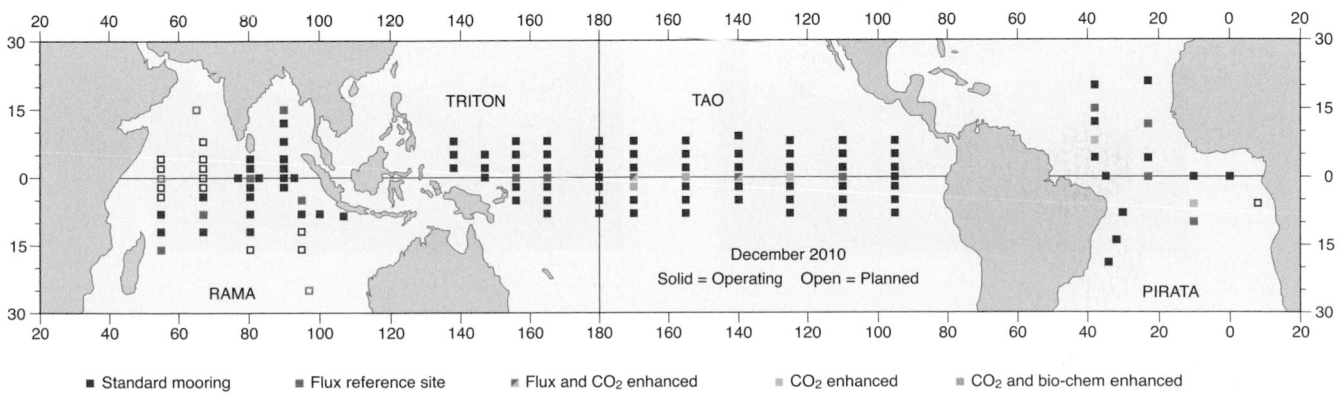

FIGURE 9.39
Global tropical deep-sea mooring locations, including TAO/TRITON, as of December 2010.

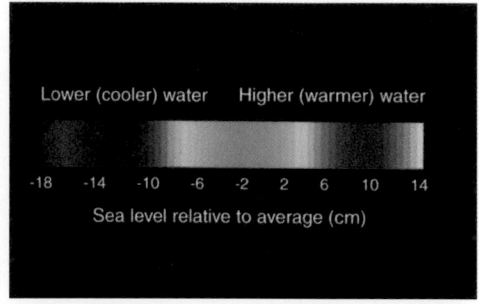

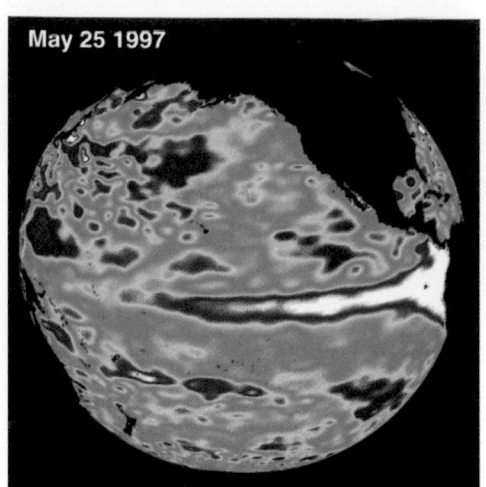

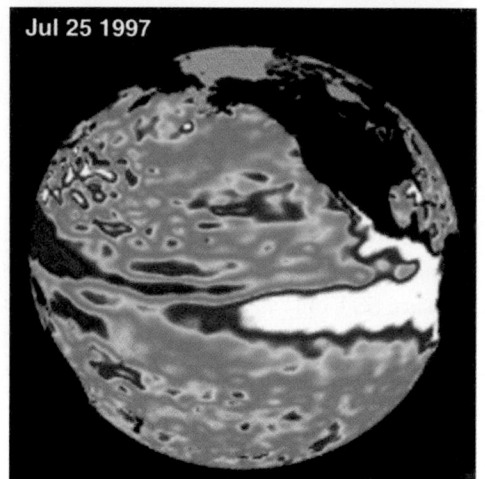

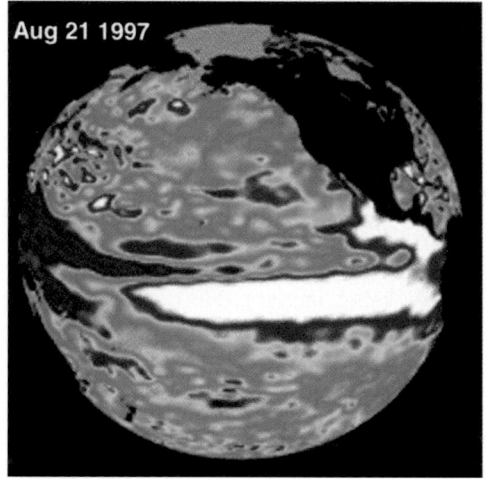

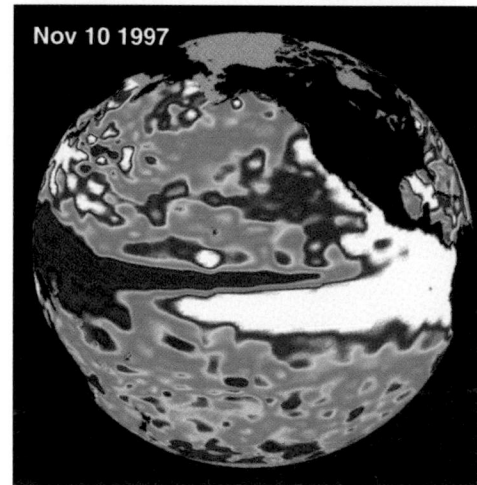

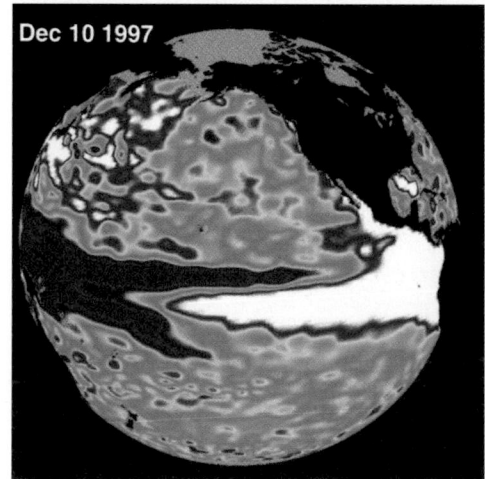

FIGURE 9.40

Evolution of the 1997-98 El Niño as derived from changes in ocean surface height (compared to the long-term average) as measured by altimeter sensors onboard the TOPEX/Poseidon satellite. On the color scale, whites and reds indicate elevated areas (warmer than normal water). In the white areas, the sea surface is 14 to 32 cm (6 to 13 in.) above normal; in the red areas it is about 10 cm (4 in.) above normal. Green indicates normal sea level whereas purple corresponds to areas that are at least 18 cm (7 in.) below normal sea level (colder than normal water). [Courtesy of NASA Goddard Space Flight Center]

El Niño or La Niña prior to the start of the growing season allows agricultural interests and government officials to consult on what crops to plant. If the forecast calls for El Niño, rice is favored over cotton because rice thrives during a wet growing season whereas cotton is more drought-tolerant and therefore is more suitable when La Niña is forecasted. Also, in anticipation of the heavy rainfall likely to accompany a full-blown El Niño, water resource managers can direct the gradual draw down of reservoirs to reduce flooding.

FREQUENCY OF EL NIÑO AND LA NIÑA

In September 2003, NOAA scientists provided an index for operational definitions of El Niño and La Niña, based on sea-level air pressure, zonal (east-west), and meridional (north-south) components of the surface wind, surface air temperature, sky cloud cover, and sea-surface temperature measured in the tropical Pacific (Figure 9.41). Sea-surface temperatures are mapped in an area bound by 120 degrees W and 170 degrees W, and 5 degrees N and 5 degrees S, which includes the equatorial cold tongue. By this index, El Niño is characterized by a *positive* SST departure from normal, and La Niña by a *negative* SST departure, greater than or equal to 0.5 Celsius degrees averaged over three consecutive months, based on the 1971-2000 average.

In February 2009, NOAA's Climate Prediction Center launched its ENSO Alert System. An El Niño or La Niña *watch* is issued when conditions in the equatorial Pacific are favorable for their development within three months. An El Niño or La Niña *advisory* is issued when El Niño or La Niña conditions have already developed and are expected to continue.

La Niña does not always follow El Niño but is most likely to develop after a particularly intense El Niño. An intense El Niño conveys great amounts of heat from the eastern tropical Pacific into higher latitudes, setting the stage for La Niña. Additionally, the unusually cold water located just below the warm surface water is poised to well up to the surface as soon as the trade winds strengthen again.

While El Niño episodes outnumber La Niña episodes, La Niña tends to last longer, and the balance of the time neutral or near-neutral conditions prevail. During the second half of the 20th century, El Niño conditions prevailed 31% of the time and La Niña occurred 23% of the time. The most recent El Niño was a major episode in 2009-10, followed by La Niña in July 2010 that persisted through April 2011 and redeveloped in September 2011. At the time of this writing, a transition to neutral conditions was expected (Table 9.1).

As noted earlier, associated with El Niño and La Niña episodes are *teleconnections*, linkages between weather changes occurring in widely separated regions of the globe, sometimes thousands of kilometers apart. Other teleconnections involve the North Atlantic Oscillation (NAO), the Arctic Oscillation (AO), and the Pacific Decadal Oscillation (PDO). In the following sections, we describe these teleconnections and their contributions to climate variability.

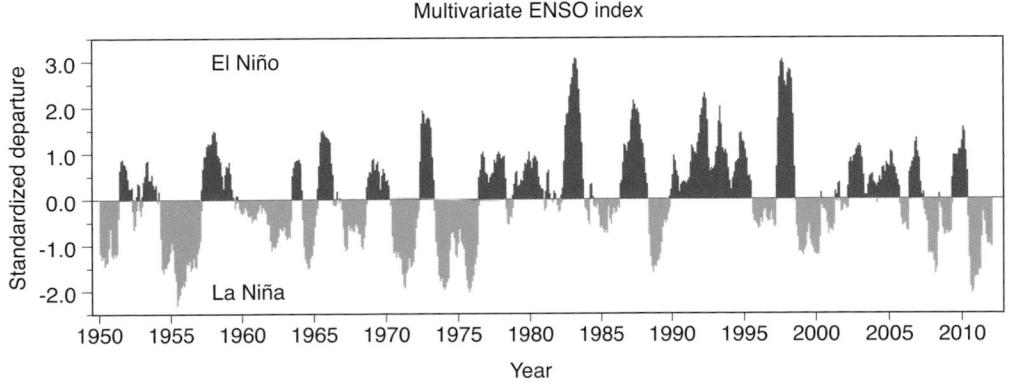

FIGURE 9.41

Variations in the Multivariate ENSO Index showing the sequence of El Niño and La Niña events since 1950. The Index is based on six variables measured in the tropical Pacific: sea-level air pressure, zonal (east-west) component of surface wind, meridional (north-south) component of surface wind, surface air temperature, sky cloud cover, and sea-surface temperature. Sea-surface temperature anomalies (departures from 1971-2000 averages) are measured for the area in the tropical Pacific Ocean between 5 degrees N and 5 degrees S latitude and from 120 degrees W to 170 degrees W longitude. Warm anomalies (greater than about 0.5 Celsius degree) generally indicate El Niño whereas cold anomalies (less than about −0.5 Celsius degree) generally indicate La Niña. [NOAA/ESRL/Physical Science Division-University of Colorado at Boulder/CIRES/CDC]

TABLE 9.1
El Niño and La Niña Events since 1950

__El Niño__

2009-10, 2006-07, 2004-05, 2002-03, 1997-98, 1994-95, 1991-92, 1986-88, 1982-83, 1977-78, 1976-77, 1972-73, 1969-70, 1968-69, 1965-66, 1963-64, 1957-58, 1951

__La Niña__

2011-12, 2010-11, 2008, 2007-08, 2000-01, 1998-2000, 1995-96, 1988-89, 1984-85, 1973-76, 1970-72, 1964-65, 1954-57, 1949-51

North Atlantic Oscillation

Over the North Atlantic Ocean the time-averaged, planetary-scale atmospheric circulation features a subpolar low pressure system located between Iceland and Greenland, the *Icelandic low*, and a massive subtropical anticyclone centered near 30 degrees N that stretches from Bermuda to near the Azores, the *Azores high*. The **North Atlantic Oscillation (NAO)**, also discovered by Sir Gilbert Walker in the 1920s, is a seesaw variation in sea-level air pressure between the Azores and Iceland. When air pressure is higher than the long-term average over the Azores, it is lower than the long-term average over Iceland and vice versa. The air pressure gradient between the Azores high and the Icelandic low governs the strength and direction of the westerly winds, the middle latitude jet stream, and storm tracks across the North Atlantic.

The North Atlantic Oscillation accounts for much of the variability in the weather of the North Atlantic region, influencing precipitation and air temperature patterns primarily in winter (December, January, and February) over eastern North America, much of Europe, North Africa, and on some

occasions as far north as Siberia. The *NAO Index* is directly proportional to the strength of the North Atlantic air pressure gradient. Its magnitude is based on the seasonal average air pressure difference between Gibraltar and Reykjavik, Iceland (Figure 9.42).

When the NAO Index is high (positive), stronger than usual winter winds blow across the North Atlantic, advecting cold air masses over eastern Canada and the U.S. so that winters tend to be colder than usual in that region. But cold air masses modify considerably as they move over the mild ocean surface, warming and becoming more humid so that downstream, over western and central Europe, winters are milder and wetter whereas summers are relatively cool. Meanwhile, winters tend to be dry in the Mediterranean region.

On the other hand, when the NAO Index is low (negative), steering winds over the North Atlantic shift southward toward the Mediterranean and provide more rainfall to southern Europe and North Africa. Over northern Europe, winters are colder than usual and summers are characterized by heat waves and reduced rainfall. Meanwhile, wet and mild conditions prevail from the Mediterranean eastward into the Middle East. The eastern U.S. and Canada tend to experience mild winters while in the southeast U.S. winters are colder than usual.

The NAO Index exhibits no particular periodicity and can vary significantly from one year to the next and from decade to decade; it is much less regular than the ENSO cycle. Prolonged periods (e.g., several successive months) of positive or negative NAO Index are common.

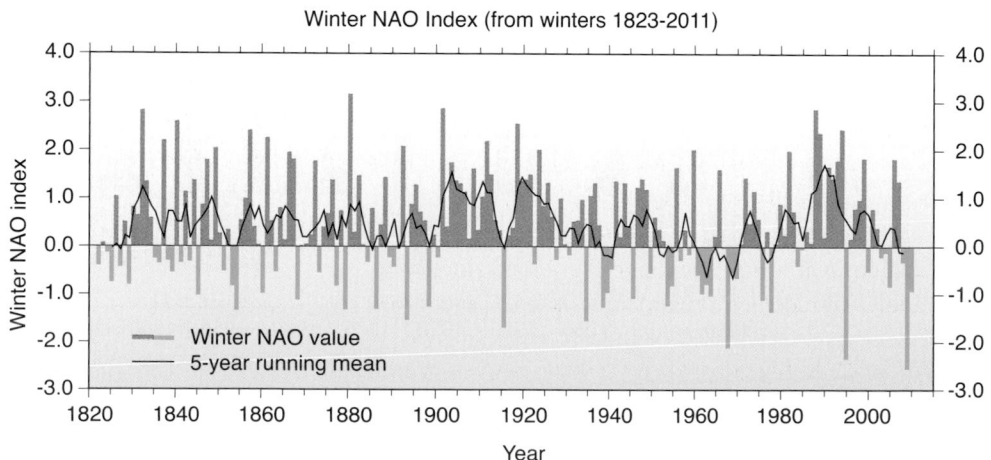

FIGURE 9.42
Record of the North Atlantic Oscillation (NAO) during winter (December to March) through 2010-11, based on the difference between the normalized sea-level air pressure at Gibraltar and the normalized sea-level air pressure over southwest Iceland. Solid black line is a running mean. [From Tim Osborn, Climate Research Unit, University of East Anglia, Norwich, UK]

The NAO Index was generally low from the mid-1950s through 1978-79. This was followed by an abrupt transition to a positive NAO Index during the winter of 1979-80 that persisted through 1994-95. A return to strongly negative NAO occurred from November 1995 to February 1996 when the NAO Index trended generally downward into the winter of 2010-11.

Changes in winter moisture supply associated with NAO have had varied impacts in Europe and North Africa. During recent decades of relatively high NAO Index, wet winters elevated the hydroelectric power potential in the Scandinavian nations, lengthened the growing season over northern Eurasia, but also diminished the snow cover for winter recreation. Meanwhile, a moisture deficit has been the problem in the Iberian Peninsula, the watershed of the Tigris and Euphrates Rivers, and the Sahel of West Africa. We have more to say about the relationship between the NAO and the extreme weather in the winter of 2011-12 in Chapter 15.

Arctic Oscillation

First identified in 1951 by Edward Lorenz (1917-2008) and related to the North Atlantic Oscillation, the **Arctic Oscillation (AO)** is a seesaw variation in sea-level air pressure between the Arctic and the portion of the middle latitudes centered from about 37 to 45 degrees N. Changes in the surface air pressure gradient alter the strength of the band of winds aloft (the *polar vortex*) blowing in a mostly zonal pattern counterclockwise (viewed from above) around the Arctic. Strengthening and weakening of these polar winds impact winter weather in middle latitudes and contribute to climate variability and changes in ocean circulation. The *AO Index* is a measure of the surface air pressure pattern and the degree to which Arctic air plunges into middle latitudes.

The AO Index shifts between positive and negative phases (Figure 9.43). When the AO Index is in its *positive phase*, the pressure pattern is lower in the polar region and winds encircling the Arctic (including the middle latitude jet stream) are stronger and blow consistently from west to east. These stronger winds act as a dam to impede the southeastward flow of Arctic air. Higher pressure at middle latitudes shift ocean storms farther north and these changes in the circulation pattern bring wetter (snowier) winter weather to Alaska, Scotland and Scandinavia while conditions are drier in the western U.S. and the Mediterranean. The middle latitude westerlies also strengthen and blow more directly from west to east, flooding much of the U.S. with relatively mild air

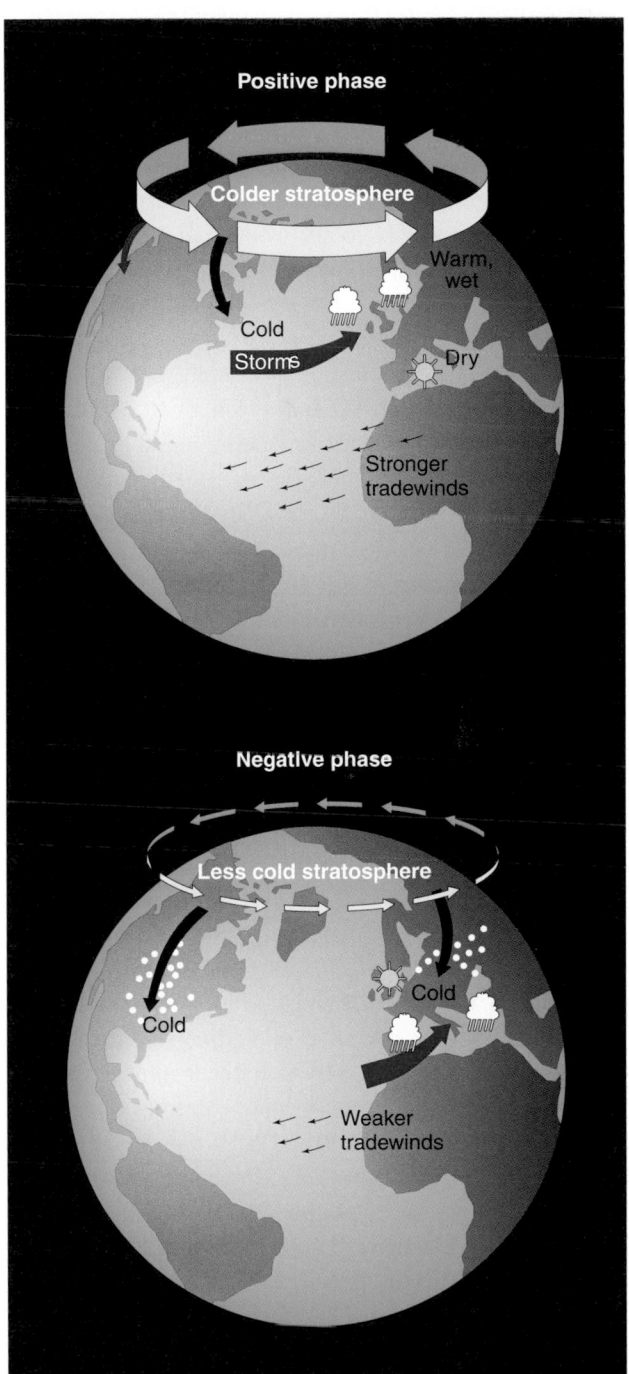

FIGURE 9.43
Atmospheric circulation changes between the positive phase (top) and negative phase (bottom) of the Arctic Oscillation. [From J. Wallace, University of Washington]

that originates over the Pacific Ocean. Average winter temperatures are higher than usual, especially east of the Rocky Mountains, and major snowstorms are less likely. Meanwhile, Greenland and Newfoundland are colder than normal.

During the *negative phase* of the AO Index, air pressure is relatively high over the polar region, the zonal component of the wind is weaker, and the polar vortex circulation is not as vigorous as usual. This allows greater movement of bitterly cold Arctic air masses out of their far northern source regions and southeastward into middle latitudes. The result is colder than usual winter weather for most of the U.S., Northern Europe, Russia, China, and Japan. Heavy lake-effect snows downwind from the Great Lakes are more frequent and nor'easters are more likely along the U.S. Eastern Seaboard (Chapter 10).

Although the AO Index can fluctuate between positive and negative phases on daily, monthly, seasonal and annual time scales, there are extended periods when either the positive or negative phase dominates winter. In the 1960s, the negative phase of the Arctic Oscillation prevailed. Beginning in the 1970s, the AO tended to be more positive, a trend that is consistent with observed climate fluctuations in middle latitudes (e.g., less frequent episodes of extreme cold and major snowstorms). Furthermore, during this recent episode of dominantly positive AO phase, winds have been delivering warmer than usual air and ocean water into the Arctic. As we will see in Chapter 15, this may help explain the recent shrinkage of Arctic ice cover.

In February 2010, the AO Index declined to its most negative monthly mean value since reliable records of the Index began in 1950. As discussed in the Case-in-Point of Chapter 1, during that month three historic snowstorms struck the U.S. mid-Atlantic region from Washington, DC, north to Philadelphia, PA. In another example of the association of weather extremes with negative AO Index, between 1950 and 2010, 9 of the 10 coldest Januaries on record in New York City coincided with very negative values of the AO Index. While the climate record may indicate a tendency toward more weather extremes during winters when the AO Index is exceptionally low, this correlation does not always hold.

Pacific Decadal Oscillation

The **Pacific Decadal Oscillation (PDO)** is a long-lived variation in climate over the North Pacific and North America. Sea-surface temperatures fluctuate between the north central Pacific (north of 20 degrees N) and the west coast of North America. In 1996, the PDO was discovered and named by fisheries scientist Steven R. Hare who was investigating the relationship between Alaskan salmon production and the Pacific climate.

SST patterns in the PDO are analogous to ENSO warm and cold phases. During a PDO *warm phase*, sea-surface temperatures are lower than usual over the broad central interior of the North Pacific Ocean and above average in a narrow strip along the coasts of Alaska, the Pacific Northwest, and western Canada. In an interesting parallel to what happens off the coast of Ecuador and Chile during El Niño, the layer of relatively warm surface waters off the Pacific Northwest Coast significantly reduces upwelling of nutrient-rich bottom water. Populations of phytoplankton and zooplankton plummet and juvenile salmon migrating to coastal areas from streams and rivers starve. On the other hand, during a PDO *cold phase*, SST are higher in the North Pacific interior and lower along the coast associated with the return of nutrients and salmon.

Key to the climatic impact of PDO is the strength of the subpolar *Aleutian low*, which persists through the winter off the Alaskan coast. During a PDO *warm phase*, the Aleutian low is well developed and its strong counterclockwise winds steer mild and relatively dry air masses into the Pacific Northwest. Winters tend to be mild and dry, and water supplies suffer from reduced mountain snow pack. But during a PDO *cold phase*, the Aleutian cyclone is weaker so that cold, moist air masses more frequently invade the Pacific Northwest. Winters are colder and wetter, and the mountain snowpack is thicker. PDO phases tend to shift every 20 to 30 years (Figure 9.44). Cold phases prevailed from 1890-1924 and again from 1947-1976 whereas warm phases prevailed from 1925-1946 and 1977 through the mid-1990s.

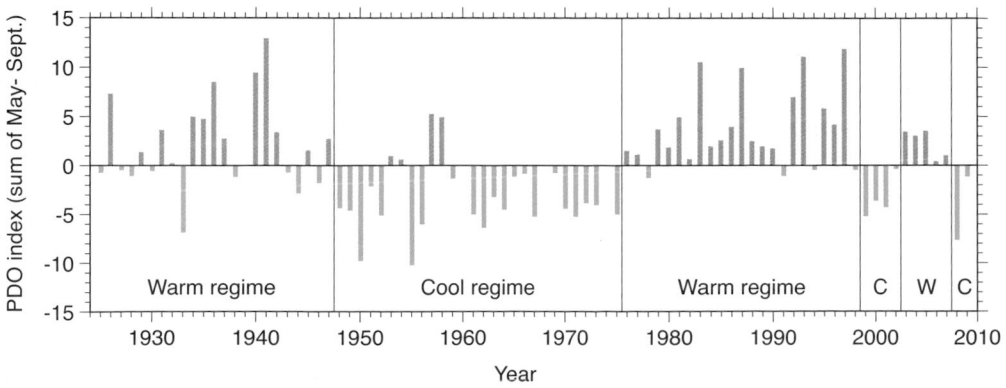

FIGURE 9.44
Variations in the Pacific Decadal Oscillation between warm and cool phases or regimes.

Conclusions

We have examined the time-averaged characteristics of the planetary-scale atmospheric circulation. The main features of that circulation are the intertropical convergence zone (ITCZ), the trade winds, semi-permanent subtropical anticyclones, the westerlies, polar front, subpolar lows, polar easterlies, and polar highs. Aloft (in the middle and upper troposphere), winds blow counter to the surface trades in tropical latitudes and in a west-to-east wave pattern of troughs and ridges at middle and high latitudes. Through the course of a year, these components of the planetary-scale circulation shift in tandem north and south as they follow the Sun.

A monsoon circulation is characterized by wet summers associated with onshore surface winds and dry winters associated with offshore surface winds. The contrast in thermal inertia between land and sea plus orographic lifting play important roles in monsoon climates. This chapter focused on the linkages between the meandering westerly winds aloft and the weather of middle latitudes. Specifically, the westerlies generate horizontal divergence aloft necessary for the development of cyclones, to steer storm systems, and to control air mass advection and poleward heat transport. In the mid and upper troposphere, westerly waves as seen from above vary in length, number and amplitude, shifting between zonal and meridional flow patterns. When the flow is very meridional, cyclones and anticyclones can become cutoff from the main west to east flow pattern, often giving rise to weather extremes.

In describing El Niño and La Niña, we emphasized the potential linkages (teleconnections) between the sea surface conditions in the tropical Pacific and weather extremes such as drought at higher latitudes. Other teleconnections involve the North Atlantic Oscillation (NAO), the Arctic Oscillation (AO), and the Pacific Decadal Oscillation. Our examination of atmospheric circulation continues in the next chapter with the focus shifting to the synoptic-scale weather systems of middle latitudes, i.e., air masses, fronts, extratropical cyclones, and anticyclones.

Basic Understandings

- The principal features of the atmosphere's planetary-scale circulation are the intertropical convergence zone (ITCZ), trade winds, subtropical anticyclones, westerlies of middle latitudes, subpolar lows, polar front, polar easterlies, and polar highs. The ITCZ, wind belts, subtropical highs, and polar front follow the Sun's annual north and south shift relative to the equator.

- Trade winds blow out of the equatorward flanks of the subtropical anticyclones, while the westerlies blow out of their poleward flanks. The east side of a subtropical anticyclone features subsiding air and dry climates, whereas the west side is more humid and receives more rainfall.

- Subsiding air on the eastern flank of semi-permanent subtropical anticyclones produces the trade wind inversion, a persistent and climatically important feature of the planetary-scale circulation that limits the depth of convection over the eastern portions of tropical ocean basins.

- Aloft, in the tropical troposphere, winds flow counter to surface winds, completing the Hadley cell circulation. At higher latitudes, upper-air winds meander from west to east as Rossby long waves, each wave consisting of a ridge (clockwise turn) and a trough (counterclockwise turn).

- Contrasts in Earth's surface temperatures in winter favor relatively high air pressure over cold continents and low air pressure over the warmer ocean. In summer, this pattern reverses, with relatively low pressure prevailing over continents and high pressure over the relatively cooler ocean.

- Surface ocean currents are wind-driven so that the horizontal flow of ocean surface currents to a large extent mirrors the long-term average planetary-scale atmospheric circulation. The trade winds and westerlies drive the ocean's subtropical gyres.

- In subtropical latitudes, a monsoon circulation is responsible for wet summers and dry winters. Differences in solar heating of land versus sea, topography, and seasonal shifts in the planetary-scale circulation (e.g., the ITCZ) play key roles in the monsoon circulation.

- The Southwest Monsoon affects the American Southwest including Arizona, New Mexico, and parts of southern Colorado, bringing a dramatic increase in rainfall to this region mainly during July and August. The Southwest Monsoon originates over northern Mexico during May and

June and is actually the northern extension of the Mexican monsoon which is fed by moisture that originates over the Gulf of California and the Gulf of Mexico.

- The wave pattern exhibited by the upper-air westerlies varies in length, amplitude, and number. At one extreme, westerlies can be strongly zonal, that is, they blow mostly from west to east with little latitudinal amplitude. At the other extreme, westerlies can be strongly meridional, that is, they blow from west to east and exhibit considerable amplitude. Shifts between zonal and meridional flow patterns can be abrupt, altering the north-south air mass exchange, poleward heat transport, and storm tracks.

- Cutoff *Lows* and cutoff *Highs* block the usual west to east progression of weather systems and may lead to weather extremes such as drought, excessive rainfall, or extended periods of unusually high or low air temperature.

- The polar front jet stream is a narrow corridor of high-speed winds within the westerlies; it is located near the tropopause and above the polar front. A jet streak is a region of exceptionally strong jet stream winds situated over a well-defined segment of the polar front.

- Cyclone development is most likely under the left front quadrant of a jet streak and to the east of an upper-level trough. Upper-air support (horizontal divergence) is strongest in that quadrant.

- El Niño refers to anomalous atmospheric and oceanic circulation regimes associated with a lengthy period of higher than normal sea-surface temperatures over a vast area of the central and eastern tropical Pacific. El Niño is accompanied by weather extremes in tropical latitudes and in other areas of the globe. La Niña is the term coined for atmosphere/ocean conditions essentially opposite El Niño.

- The Southern Oscillation is a seesaw variation in air pressure across the tropical Indian and Pacific Oceans. When air pressure is low over the Indian Ocean and the western tropical Pacific, it is high east of the International Dateline in the eastern tropical Pacific. An El Niño episode begins when the air pressure gradient across the tropical Pacific begins to weaken, heralding the slackening of the trade winds.

- Ekman transport plays an important role in governing sea-surface temperatures (SST) by affecting upwelling and downwelling of ocean water. Viewed from above, the wind-driven horizontal motion of successively lower layers of ocean water forms a spiral; the current gradually turns and weakens with increasing depth. The Ekman spiral transports near-surface waters at an angle of about 90 degrees to the right of the wind direction in the Northern Hemisphere and to the left in the Southern Hemisphere.

- During neutral (long-term average) conditions, relatively cool surface waters in the central and eastern tropical Pacific chill the overlying air and suppress convection so that rainfall is light in that region as well as along the adjacent western coastal plain of South America. Meanwhile, over the western tropical Pacific, relatively warm surface waters heat the overlying air, destabilizing the troposphere, strengthening convection, and giving rise to heavy rainfall.

- In the eastern tropical Pacific as El Niño evolves, sea-surface temperatures (SST) rise, sea level climbs, and the depth to the thermocline increases. Meanwhile, SST drop, sea level falls, and depth to the thermocline decreases in the western tropical Pacific. Conditions are drier than usual in the western tropical Pacific and wetter than normal in the central tropical Pacific.

- La Niña is a period of unusually strong trade winds, exceptionally vigorous upwelling, and lower than usual sea-surface temperatures in the eastern tropical Pacific. Weather extremes associated with La Niña are essentially opposite those that tend to occur during El Niño.

- Teleconnections are linkages between weather changes occurring in widely separated regions of the globe. Teleconnections are observed with El Niño, La Niña, the North Atlantic Oscillation (NAO), the Arctic Oscillation (AO), and the Pacific Decadal Oscillation (PDO).

- The North Atlantic Oscillation is a seesaw variation in sea-level air pressure between the Azores and Iceland that governs the strength and direction of the westerlies, middle latitude jet stream, and storm tracks across the North Atlantic.

- The Arctic Oscillation (AO) is a seesaw variation in sea-level air pressure between the Arctic and middle latitudes. Associated changes in the polar vortex and middle latitude jet stream influence the degree of southeasterly flow of bitterly cold Arctic air in winter.

- The Pacific Decadal Oscillation (PDO) is a long-lived variation in climate over the North Pacific Ocean and North America involving changes in the strength of the Aleutian subpolar low, which persists through winter off the Alaskan coast.

Enduring Ideas

- Planetary-scale pressure systems and wind belts govern the development and movement of smaller-scale weather systems. In subtropical latitudes, a monsoon circulation causes wet summers and dry winters. In middle latitudes, waves in the westerlies set the stage for the perpetual progression of high and low pressure systems. In North America, the weather is more dramatic (and sometimes extreme) when the westerlies are meridional as compared to zonal flow.
- The strength of the polar front jet stream, located near the tropopause, is directly proportional to the magnitude of the underlying horizontal air temperature gradient. Hence, the jet stream is generally strongest in winter when the north-south temperature gradient is greatest. Jet streaks provide upper-air support for a cyclone by contributing horizontal divergence aloft.
- Some middle latitude weather extremes are linked to circulation anomalies in the tropical Pacific Ocean, most prominently El Niño and La Niña. El Niño is a period of weaker trade winds and well above-average SST over the central and eastern tropical Pacific. La Niña features exceptionally strong trade winds and lower than usual SST. The ENSO Alert System issues watches and advisories for development and continuation of El Niño and La Niña conditions.
- Teleconnections are linkages between weather conditions occurring in widely separated regions of the world, such as those observed with El Niño and La Niña. Other teleconnections involve the North Atlantic Oscillation (NAO), Arctic Oscillation (AO), and the Pacific Decadal Oscillation (PDO).

Key Terms

semi-permanent pressure systems	sub-polar gyres	Southern Oscillation
subtropical anticyclones	monsoon circulation	Southern Oscillation Index
horse latitudes	Southwest Monsoon	ENSO
midlatitude westerlies	Rossby waves	Ekman spiral
trade winds	short wave	Ekman transport
doldrums	zonal flow pattern	upwelling
intertropical convergence zone (ITCZ)	Pacific air	downwelling
	meridional flow pattern	Walker Circulation
heat equator	split flow pattern	coral bleaching
subpolar lows	blocking system	teleconnection
polar front	polar front jet stream	ENSO Observing System
polar highs	jet streak	ENSO Alert System
Hadley cell	subtropical jet stream	North Atlantic Oscillation (NAO)
trade wind inversion	El Niño	Arctic Oscillation (AO)
subtropical gyres	La Niña	Pacific Decadal Oscillation (PDO)

Review

1. What is the relationship between semi-permanent subtropical anticyclones and the location of the world's subtropical deserts?
2. How are the trade winds and westerlies linked to the semi-permanent subtropical anticyclones?
3. Describe the weather along the intertropical convergence zone (ITCZ).
4. How do the Aleutian and Icelandic lows change between winter and summer?
5. Why does the climate of Southern California feature a wet winter and dry summer?
6. How are the subtropical gyres of the ocean basins associated with the prevailing planetary-scale wind belts?
7. Describe the seasonal shifts in the latitude of the polar front and the semi-permanent subtropical highs.
8. What factors contribute to the development of the Southwest Monsoon?
9. Explain the association between weather extremes and a blocking pattern in the mid-latitude westerlies.
10. Why does a jet stream occur over the polar front?

Critical Thinking

1. How does a jet streak provide upper-air support for the development of an extratropical cyclone?
2. Explain why a ridge in the 500-mb flow pattern is associated with an underlying mass of relatively warm air.
3. Seasonal shifts in the location of the intertropical convergence zone (ITCZ) are greater over the continents than over the ocean. Explain why.
4. Describe the steps in the development of the Southwest Monsoon.
5. How is a cut-off high pressure system in the atmosphere analogous to a whirlpool that develops in a swiftly flowing stream or river?
6. Explain why in winter a meridional flow pattern in the westerlies over the coterminous U.S. is more likely to favor development of extratropical cyclones than is a zonal pattern in the westerlies.
7. Why is ENSO referred to as a *coupled* phenomenon?
8. Explain how Ekman transport produces coastal and equatorial upwelling?
9. By examining the surface temperature pattern across North America, how can you determine the likely locations of the polar front and jet stream?
10. Scientists are developing techniques to successfully predict El Niño and La Niña episodes months or more in advance. What are the advantages of such predictions?

Defining Drought

Drought is an extended interval of abnormally dry weather sufficiently prolonged for the lack of water to cause a serious imbalance (i.e., crop damage, water supply shortage, etc.) in the affected area (Figure 1). Drought also affects the supply of water for domestic and industrial use and hydroelectric power generation. Soils dry out; crops wither and die; lakes and other reservoirs shrink; and river and stream flow slackens thereby impeding navigation. Unusually high air temperatures often accompany summer drought adding to the stress on crops.

A drought usually begins gradually and without warning. In fact, it is difficult if not impossible to tell whether a spell of dry weather actually signals the onset of drought. Similarly, the end of a drought is always uncertain because one rain event, even if substantial, does not necessarily break a drought. Furthermore, whether a dry spell is a drought depends on its impact so that a distinction is made among hydrologic drought, agricultural drought, and meteorological drought.

For water resource interests, a *hydrologic drought* is a period of moisture deficit that reduces stream or river discharge and groundwater supply to levels that adversely affect water-based activities such as irrigation, barge traffic, or hydroelectric power generation. Hydrologic drought develops when the water supply is inadequate during one or successive water years. A *water year* is defined as extending from 1 October of one year through 30 September of the next year. *Agricultural drought* depends on the shorter-term supply of rainfall and soil moisture for crops during the growing season. Complicating the criterion for agricultural drought is the fact that different crops have different water requirements, and the water needs of a specific crop species change as the crop progresses through its life cycle.

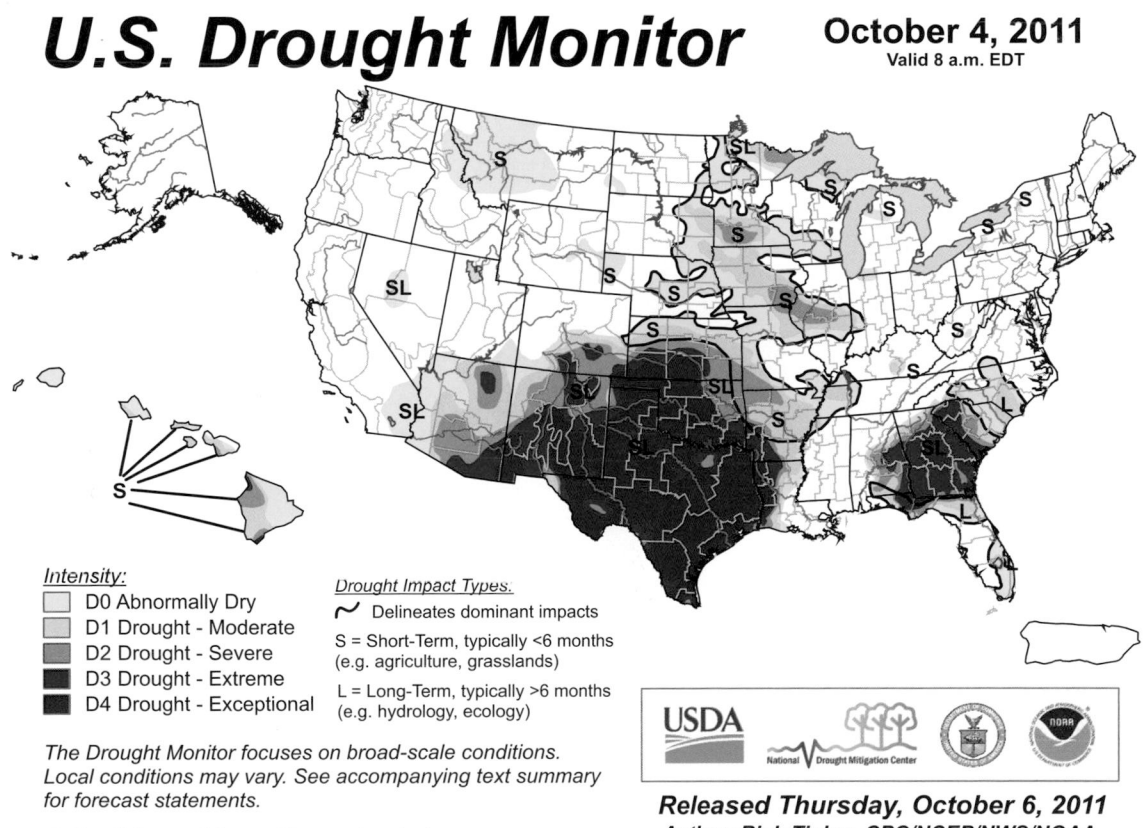

FIGURE 1
A sample U.S. Drought Monitor indicating the intensity of drought and the impacts. This is from the height of the 2011 drought in Texas and portions of surrounding states. [Courtesy of National Drought Mitigation Center]

Inadequate moisture at a critical stage of crop growth and maturation, especially over successive growing seasons, may constitute agricultural drought. Hydrologic and agricultural droughts do not always coincide.

Varying criteria have been used to define *meteorological drought*. Some meteorologists define drought as a period when the seasonal or annual precipitation falls below a certain threshold percentage (e.g., 85%) of the long-term (e.g., 30-year) average. But basing the criterion for drought on precipitation alone ignores the influence of temperature and wind on the evaporation rate. Higher temperatures and/or stronger winds increase the rate of evaporation, exacerbating the severity of drought.

One of the most popular drought indicators is incorporated in the *Palmer Drought Severity Index* (Figure 2). The Palmer Index uses temperature and rainfall data in a formula that gauges unusual dryness or wetness over extended intervals from months to years. NOAA's National Weather Service and the U.S. Department of Agriculture jointly compute the Palmer Index every week for each of the 344 climatic divisions across the U.S. A Palmer Index map portrays those divisions experiencing drought with negative values, while those regions receiving excess precipitation have positive index values. Index values range from greater than +4.00 for extremely wet conditions to under –4.00 for extreme drought; zero indicates long-term average moisture levels.

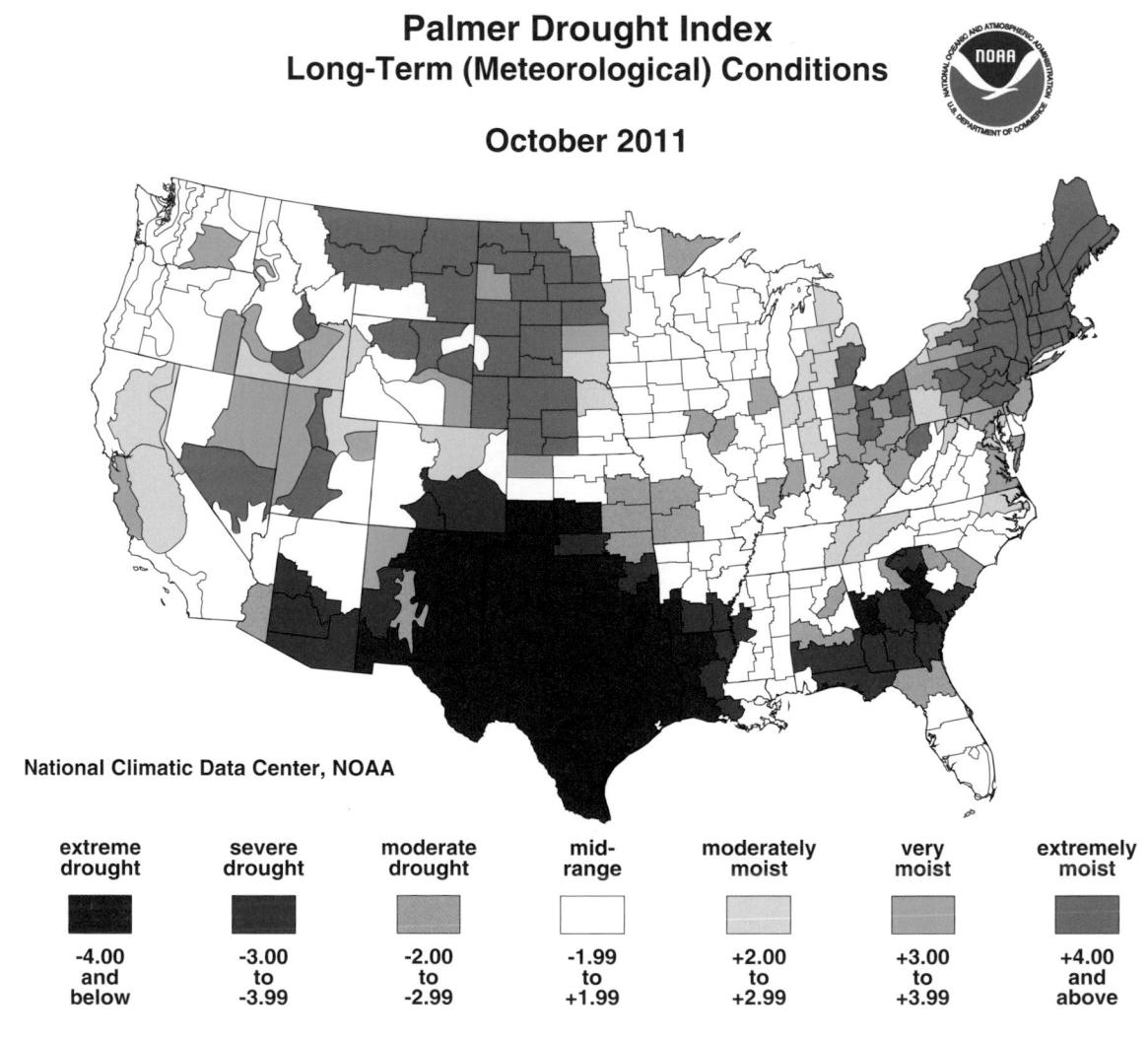

Palmer Drought Index
Long-Term (Meteorological) Conditions

October 2011

National Climatic Data Center, NOAA

extreme drought	severe drought	moderate drought	mid-range	moderately moist	very moist	extremely moist
-4.00 and below	-3.00 to -3.99	-2.00 to -2.99	-1.99 to +1.99	+2.00 to +2.99	+3.00 to +3.99	+4.00 and above

FIGURE 2
A sample Palmer Drought Severity Index Map by Climate Divisions. [NOAA/NESDIS/NCDC]

Historic El Niño Episodes of 1982-83 and 1997-98

In 1982-83, the weather seemed to go wild in many parts of the world, with a total worldwide impact consisting of thousands of deaths and an estimated $13 billion in property damage. Excessive rains from mid-November 1982 until late January 1983 caused the worst flooding of the century in usually arid Ecuador. In a span of only three months, strong winds and torrential rains produced by six tropical cyclones lashed the islands of French Polynesia, a region that averages only one tropical cyclone strike about every 5 years. At the other extreme, drought parched eastern Australia, Indonesia, and southern Africa, with huge drought-related wildfires breaking out in Australia and Borneo. Australia's drought was the nation's worst in 200 years, causing $2 billion in crop losses and the deaths of millions of sheep and cattle. Meanwhile, persistent drought in the Sahel of Africa only grew worse. Over North America, the winter storm track shifted hundreds of kilometers south, bringing episodes of destructive high winds and heavy rains to portions of California. Flooding rains also caused havoc across the southeastern United States. In the northern U.S., ski resorts experienced a snow drought and considerable economic loss as a consequence.

Just prior to these worldwide weather extremes, the ocean circulation off the northwest coast of South America changed drastically, with dire implications for marine productivity. Along the coast of Ecuador and Peru, plankton populations plunged to about 5% of their normal level. The decline in plankton reduced the numbers of anchovy, which feed on plankton, to a record low. Other fish dependent on anchovy, such as jack mackerel, suffered a similar fate, and the commercial fisheries off the coast of Ecuador and Peru collapsed. With the decline in fish populations, marine birds (e.g., frigate birds and terns) and marine mammals (e.g., fur seals and sea lions) also experienced major population declines, food scarcity causing breeding failures as well as migration or starvation.

The impact of these 1982-83 environmental changes was particularly severe on the ecology of the remote island of Kiritimati (Christmas Island), located at 2 degrees N, 157 degrees W, in the central tropical Pacific. An estimated 17 million seabirds, which feed on fish and squid, normally nest on the island, but scientists arriving on Kiritimati in November 1982 discovered that almost all adult birds had abandoned the island, leaving their young to starve. Apparently, sharp declines in food sources forced fish and squid to search for better feeding grounds, and adult birds followed their prey. By July 1983, with the return of more typical ocean circulation came a resurgence in plankton populations, adult birds returned to Kiritimati, nesting began again, and sea bird populations started to recover.

At first, scientists attributed the weather extremes to the violent eruption of the Mexican volcano El Chichón in March-April 1982. But it quickly became apparent that the worldwide weather extremes were linked to large-scale atmosphere/ocean circulation changes in the tropical Pacific and soon a new scientific term was added to the public's vocabulary: El Niño. The 1982-83 El Niño spurred further research on atmosphere-ocean circulation changes, and the deployment of an array of in situ and remote sensing instrument platforms in the tropical Pacific to provide advance warning of the development of El Niño. In the late 1990s, when El Niño returned in its full fury, the global community was better prepared thereby lessening the impact.

With weather extremes and outbreaks of disease from standing water, such as malaria and cholera, claiming an estimated 22,000 lives worldwide and causing $36 billion in economic losses, the 1997-98 El Niño rivaled the 1982-83 El Niño as the most intense of the 20th century or since. This El Niño developed rapidly in early 1997, with trade winds weakening until they reversed direction in the western tropical Pacific, and equatorial upwelling ceased during the Northern Hemisphere summer of 1997. A pool of exceptionally warm surface waters (SST greater than 29 °C or 84 °F) migrated eastward from the western tropical Pacific, and equatorial upwelling ceased that summer in the Northern Hemisphere. By that fall in the eastern tropical Pacific, the SST was least 5 Celsius degrees (9 Fahrenheit degrees) higher than the long-term average, setting record highs each month from June through September. As 1997 drew to a close, the thermocline flattened across the tropical Pacific, rising 20 to 40 m (65 to 130 ft) in the west and falling more than 90 m (295 ft) in the east. Somewhat lower SST in the west and much higher than usual SST in the east, weakened the east-west SST gradient and the already weakened trade winds.

The 1997-98 El Niño came to an abrupt end in mid-May 1998. Trade winds strengthened rapidly; upwelling resumed along the equator and off the northwest coast of South America. The SST over the eastern tropical Pacific plummeted in response to upwelling of very cold water. At one location along the equator (125 degrees W), the SST fell 8 Celsius degrees (14 Fahrenheit degrees) in only four weeks.

The day following the first great snowstorm of February 2010: a tree in Fairfax, Virginia looks like a snow sculpture. [Photo by Neal Kaske. Courtesy NOAA]

WEATHER SYSTEMS OF MIDDLE LATITUDES

Chapter Highlights

Case-in-Point
  Early Observations on East Coast Cyclones
Air Masses
  North American Types and Source Regions
  Air Mass Modification
Frontal Weather
  Stationary Front
  Warm Front
  Cold Front
  Occluded Fronts
  Summary
Extratropical Cyclones
  Life Cycle
  Conveyor Belt Model
  Cyclone Weather
  Principal Cyclone Tracks
  Cold Side/Warm Side
  Winter Storms
  Cold- and Warm-Core Systems
Anticyclones
  Arctic and Polar Highs
  Warm Highs
  Anticyclone Weather
Local and Regional Circulation Systems
  Sea (or Lake) Breeze and Land Breeze
  Mountain Breeze and Valley Breeze
  Chinook Wind
  Santa Ana Wind
  Katabatic Wind
  Desert Winds
  Heat Burst
Conclusions
Basic Understandings/Enduring Ideas
Key Terms/Review/Critical Thinking
For Further Exploration
  The Case of the Missing Storm
  Nor'easters
  *Santa Ana Winds and California Fire
    Weather*

Learning Objectives

Identify the various air masses that regularly form or travel over North America.

Describe how and why air masses modify as they travel from their source region.

Identify and distinguish among the various types of fronts.

Distinguish between warm frontal weather and cold frontal weather.

Explain why fronts are associated with extratropical cyclones but not with anticyclones.

Sketch and describe the principal stages in the life cycle of an extratropical cyclone.

Describe the components of the conveyor-belt model of an extratropical cyclone.

Describe the linkage between a surface cyclone and the westerly flow aloft.

Identify the principal storm tracks across North America.

Distinguish between cold-core lows and warm-core lows.

Distinguish between cold-core highs and warm-core highs.

Describe the types of air mass advection associated with an anticyclone.

Explain the relationship between a cold-core anticyclone and a dome of Arctic air.

Compare and contrast sea and land breezes.

Explain diurnal variations of winds in deserts and mountainous areas.

What systems shape the weather of middle latitudes?

Case-in-Point

Early Observations on East Coast Cyclones

Extratropical cyclones are major weather makers within the middle latitudes. When viewed from above in the Northern Hemisphere, their surface winds blow counterclockwise and spiral inward while winds aloft provide upper-air support (horizontal divergence) and determine their track. Today, these complex rotational and translational motions of cyclones have been verified by satellite sensors and observations from weather stations.

Long before that technology, Daniel Defoe (1660-1731), the English journalist and novelist, was the first to propose that storms generally track from west to east in middle latitudes. He did an extensive study of a great storm that lashed the British Isles on 7-8 December 1703 and received reports that several days earlier a similar storm had ravaged the East Coast of North America. Defoe assumed that the storms were the same and concluded that the system had moved across the North Atlantic.

Among his many notable achievements, Benjamin Franklin (1706-1790) (Figure 10.1) also is credited with being the first American

FIGURE 10.1
Portrait of Benjamin Franklin by artist David Martin. [Credit: Library of Congress, LC-USZC4-3576]

to discover that storms usually move in an easterly or northeasterly direction. On the evening of 21 October 1743, Franklin was in Philadelphia, anticipating the viewing of a lunar eclipse at 9 p.m. However, winds picked up from the northeast and thick clouds rolled in, blocking his view. Later, in the mail, Franklin received a newspaper from Boston, 500 km (310 mi) to the northeast of Philadelphia, which told of the spectacular eclipse on 21 October, although a storm brought cloudiness and windy conditions by midnight. Like most of his contemporary natural scientists, Franklin assumed coastal storms that produced winds blowing from the northeast (a *nor'easter*) approached from a northeast direction so the storm should have come from the Boston area, obscuring that night sky before his. That the storm clouds obscured the eclipse in Philadelphia but arrived in Boston's night sky several hours later suggested that the storm had tracked from the southwest toward the northeast. Intrigued by his discovery, Franklin contacted correspondents along the Eastern Seaboard concerning this and previous storms. He concluded that the wind direction in a storm was not an indication of the storm's direction of movement.

In the early 1920s, scientists at the Geophysical Institute in Bergen, Norway, used a synthesis of prior research and data from telegraph-linked observational networks to develop the *Norwegian cyclone model*. This conceptual model signaled a major step forward in understanding the structure and evolution of mid-latitude cyclones from inception to decay and their associated fronts, clouds, and precipitation patterns. Advances in atmospheric science, the advent of vertical sounding of the atmosphere and remote sensing by satellites led to revisions in the original model, and the model remains very useful in conceptualizing the role of extratropical cyclones in regulating Earth's energy and moisture budget.

The ceaseless succession of synoptic-scale weather systems is largely responsible for the considerable day-to-day variability of weather in middle latitudes. In the Northern Hemisphere, middle latitudes extend from the Tropic of Cancer north to the Arctic Circle. The weather of middle latitudes is particularly dynamic because of the migration of cyclones and anticyclones that are embedded in the prevailing westerlies of the planetary-scale atmospheric circulation (Chapter 9). This latitude belt features the greatest meridional temperature contrast in the troposphere. Recall from Chapter 4 that because of imbalances between radiational heating and cooling, high latitudes experience a net heat energy deficit while the tropics have a net heat energy surplus. As a consequence, the greatest poleward transport of sensible and latent heat by atmospheric and oceanic circulations occurs in middle latitudes.

In this chapter, we examine the circulation features and weather conditions that accompany the major weather systems of middle latitudes. We cover air masses, fronts, cyclones (Lows), and anticyclones (Highs). In addition, this chapter includes selected local and regional circulation systems. All these weather systems are interrelated as their development and movements are ultimately governed by the planetary-scale circulation pattern.

Air Masses

An **air mass**, a huge expanse of air covering thousands of square kilometers, is relatively uniform horizontally in temperature and humidity, with both quantities typically decreasing with altitude. The properties of an air mass depend upon the type of surface over which it develops, its *source region*. An air mass source region features nearly homogeneous surface characteristics over a broad area with little topographic relief, such as a great expanse of snow-covered ground or a vast stretch of ocean waters. To become uniform in temperature and humidity, an air mass must reside in its source region for several days to weeks.

Often, large semi-permanent high pressure systems such as the *Siberian high* dominate air mass source regions (Chapter 9). Subsidence of air and horizontal divergence of surface winds in an anticyclone favor uniform properties of air in the system. Light winds inhibit mixing and together with the long residence time that is typical of high pressure systems, the distinct thermal properties of the surface are transferred to the air mass. Air mass source regions are typically not found where cyclones develop or along prevailing storm tracks.

Air masses are classified as either cold (polar, abbreviated as *P*) or warm (tropical or *T*), and either dry (continental or *c*) or humid (maritime or *m*). Air masses that form over cold snow-covered surfaces of high latitudes are relatively cold, whereas those that develop over the warm surfaces of low latitudes are relatively warm. Air masses that form over land tend to be relatively dry, whereas those that develop over the ocean are relatively humid. Based on combinations of temperature and humidity, the four basic types of air masses are: cold and dry, *continental polar* (*cP*); cold and humid, *maritime polar* (*mP*); warm and dry, *continental tropical* (*cT*); and warm and humid, *maritime tropical* (*mT*). A fifth air mass type, *Arctic* (*A*) air, is dry like continental polar air but colder.

NORTH AMERICAN TYPES AND SOURCE REGIONS

All the air mass types listed above occur over North America, forming over either the landmass or surrounding ocean water in certain characteristic regions. Source regions for these air masses are plotted in Figure 10.2.

Continental tropical air (cT) develops primarily in summer over the subtropical deserts of Mexico and the southwestern United States and is hot and dry. **Maritime tropical air (mT)** is very warm and humid because its source regions are tropical and subtropical seas (e.g., Gulf of Mexico). This air mass retains these properties year-round and is responsible for oppressive summer heat and humidity east of the Rocky Mountains. The source regions for **maritime polar air (mP)** are the cold ocean waters of the North Pacific Ocean and North Atlantic Ocean, especially north of 40 degrees N. Along the West Coast, *mP* air brings heavy winter rains, snows in the mountains, and persistent coastal fogs in summer. Dry **continental polar air (cP)** develops over the northern interior of North America. In winter, *cP* air is typically very cold because the ground in its source region is often snow covered, length of daylight is short, solar radiation is weak, and radiational cooling is extreme. In summer, when the snow-free source region warms in response to extended hours of bright sunshine, *cP* air is quite mild and pleasant.

Arctic air (A) forms over the snow- or ice-covered regions of Siberia, the Arctic Basin, Greenland, and North America, north of about 60 degrees N, in much the same way as continental polar air, but in a region that receives very little solar radiation in winter, although it still radiates strongly to space in the infrared. These exceptionally cold and dry air masses are responsible for

FIGURE 10.2
Source regions of North American air masses. The temperature and humidity of an air mass depend on the properties of its source region.

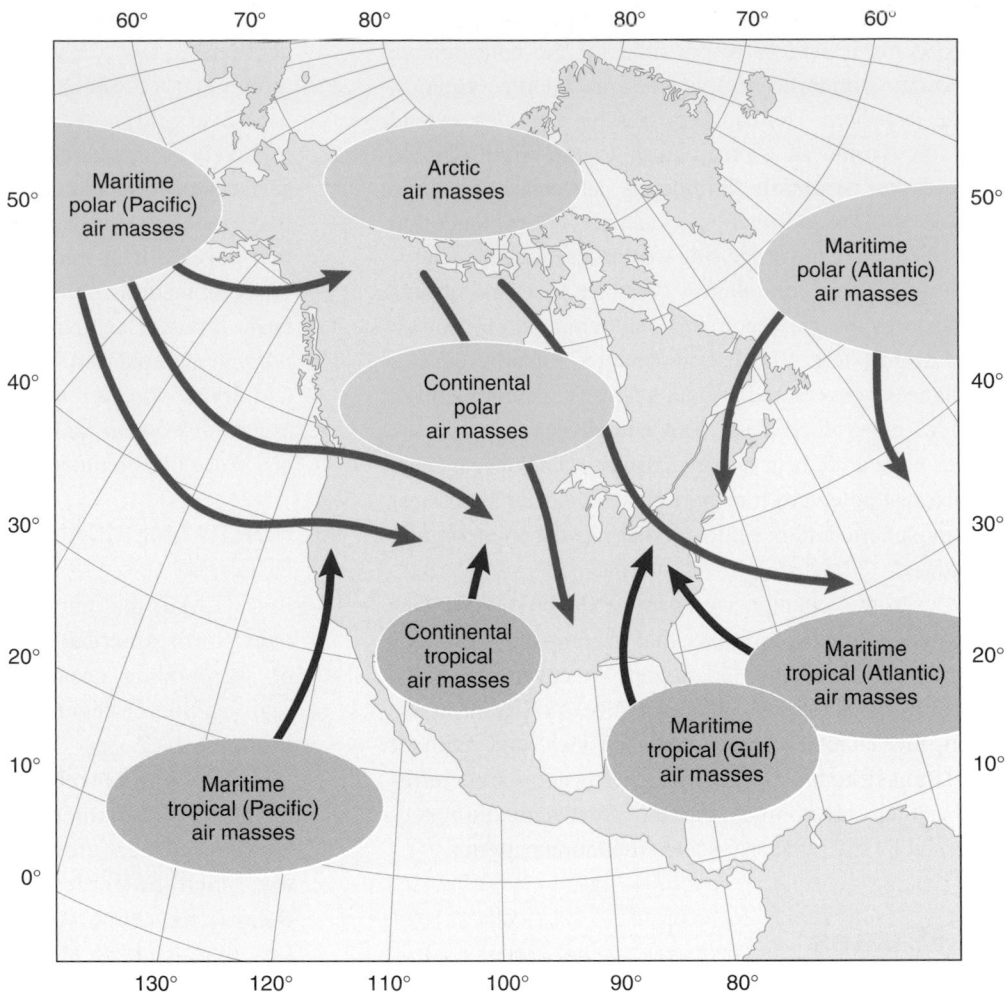

the bone-numbing winter cold waves that sweep across the Great Plains, at times penetrating as far south as the Gulf of Mexico and central Florida. For example, the *Siberian Express* is the name given to an Arctic air mass that forms over Siberia, crosses over the North Pole into Canada, and plunges onto the Great Plains and then south and eastward.

Air masses differ not only in temperature and humidity but also in stability. As noted in Chapter 6, stability is an important property of air because it influences vertical motion and the consequent development of clouds and precipitation. Table 10.1 is a list of the usual stability, temperature, and humidity characteristics of North American air masses within their source regions.

AIR MASS MODIFICATION

As air masses move from place to place, their properties modify, some air masses changing more than others. Changes may occur in temperature, humidity, and/ or stability. **Air mass modification** occurs primarily by

(1) exchange of heat or moisture, or both, with the surface over which the air mass travels; (2) radiational heating or cooling; and (3) adiabatic heating or cooling associated with large-scale vertical motion.

In winter, as a *cP* air mass travels southeastward from Canada into the coterminous United States, its temperature can modify rapidly. Although daily minimum temperatures in the Northern Plains might dip well below −18 °C (0 °F), after the polar air arrives in the southern states minimum temperatures may not drop much below the freezing point. Rapid air mass modification occurs because, outside of its source region, polar air is usually colder than the ground over which it travels. The Sun warms the snow-free ground, and the warmer ground heats the bottom of the air mass, destabilizing it and triggering convection currents that distribute heat vertically within the air mass.

A similar process of heating from below and destabilization occurs when a *cP* air mass crosses the East Coast and travels over the relatively warm waters of the

TABLE 10.1
Stability, Temperature, and Humidity Characteristics of North American Air Masses

Air mass type	Source region stability		Characteristics	
	Winter	Summer	Winter	Summer
A	Stable		Bitter cold, dry	
cP	Stable	Stable	Very cold, dry	Cool, dry
cT	Unstable	Unstable	Warm, dry	Hot, dry
mP (Pacific)	Unstable	Unstable	Mild, humid	Mild, humid
mP (Atlantic)	Unstable	Stable	Cold, humid	Cool, humid
mT (Atlantic)	Unstable	Unstable	Warm, humid	Warm, humid

western Atlantic. In addition, evaporation from the sea surface increases the water vapor concentration of the air mass. Saturation is readily achieved in the cold, humid, and relatively unstable lower levels of the air mass. Extensive areas of low clouds and fog develop. These clouds usually are organized into narrow rows, known as *cloud streets* (Figure 7.10), oriented parallel to the wind direction.

When *cP* air travels over snow-covered ground, however, modification is less because much of the incoming solar radiation is reflected rather than converted to heat (via absorption). The relatively cold snow surface chills the air, increasing stability and weakening convection. Extreme nocturnal radiational cooling plays an important role in inhibiting modification of the *cP* air mass, especially when nights are long.

A tropical air mass does not modify as readily as a polar air mass because, outside of its source region, tropical air is often warmer than the ground over which it travels. The bottom of the air mass cools by contact with the ground; this cooling stabilizes the air mass by increasing the density of the near-surface air and suppresses convection currents. Hence, cooling is restricted to the lowest portion of the air mass. By contrast, if a tropical air mass moves over a warmer surface, the air mass can become even warmer. Thus, a winter cold wave loses much of its punch as it pushes southward from Canada, but a summer heat wave can retain its warmth as it journeys from the Gulf of Mexico northward into southern Canada.

Air masses undergo significant changes in temperature and humidity through orographic lifting

(Chapter 6). When cool and humid *mP* air sweeps inland from off the Pacific Ocean, the air is forced to ascend the windward slopes of coastal mountain ranges such as the Cascades and the Sierras and expands and cools adiabatically. Cooling can bring air to saturation and trigger condensation or deposition (cloud development) and precipitation, primarily on the windward slopes. Heating by the latent heat released during condensation (and deposition) partially offsets adiabatic cooling. Then, as the air mass descends the leeward slopes into the Great Basin, it warms adiabatically and clouds vaporize. Some evaporative cooling is associated with the cloud dissipation, but net heating (and drying) prevails as compressional heating is greater. Because water precipitated out on the windward side of the mountains, descending leeward air loses its saturation at a higher altitude than the altitude at which saturation was achieved by air ascending the windward slopes. Descending unsaturated air, subjected to compressional heating only, warms at the greater dry adiabatic rate.

The same modification processes are repeated as the air mass is forced to flow up and over the Rockies. Eventually, the air mass emerges on the Great Plains considerably milder and drier than the original *mP* air mass. East of the Rockies, such an air mass is described as modified **Pacific air**. When the westerly wave pattern aloft is dominantly zonal (Chapter 9), Pacific air floods the eastern two-thirds of the United States and southern Canada. Much of eastern North America experiences lengthy episodes of mild and generally dry weather because polar air masses stay far to the north, while tropical air masses stay to the south.

Frontal Weather

A **front** is a narrow zone of transition between air masses that differ in density usually because of temperature contrasts; for this reason, we use the nomenclature *cold front* and *warm front*. However, density differences may also arise from contrasts in humidity. Although the transition zone associated with an actual front may be a hundred or more kilometers wide, traditionally a line representing a front is drawn on a weather map where the warm (less dense) edge of the transition zone meets the surface. Air temperatures are nearly constant on the warm side of the front, but fall with distance from the front to a region of nearly uniform temperature in the cold air mass.

A front is also associated with a trough in the sea-level pressure pattern, a corresponding wind shift, and converging winds. Where contrasting air masses meet, the colder (denser) air forces the warmer (less dense) air to ascend. Often the ascending air cools sufficiently (via expansion) that clouds and precipitation develop. Depending on the slope of the front in the vertical plane and the motion of air relative to the front, frontal weather may be confined to a narrow band (tens of kilometers wide), or it may extend over a broad region (hundreds of kilometers wide). In addition, the slope of the front influences the types of clouds and precipitation forming along the front. In this section, we describe the four basic types of synoptic-scale fronts: stationary, warm, cold, and occluded.

STATIONARY FRONT

A **stationary front** exhibits little or no forward movement. For example, a stationary front develops along the Front Range of the Rocky Mountains when a shallow pool of polar air (typically about 1000 m thick) surges south and southwestward out of the Prairie Provinces of Canada. The vertical extent of the air mass is below the level of the mountain passes, so the shallow, cold, dry air mass abuts the mountain range and can push no further westward; therefore, its leading edge is marked by a stationary front paralleling the Continental Divide, the crest of the mountain range. Milder air remains in the Great Basin to the west of the Rockies. On occasion, the cold air mass may be sufficiently deep that cold air pours westward through mountain passes into the Great Basin. Similarly, a stationary front can form when any type of preexisting front aligns with the upper-level wind pattern. Under the appropriate conditions, a stationary front can also form along a boundary in the surface temperature pattern, such as a coastline or the edge of a regional snow cover.

Many features of a stationary front are common to all fronts. As shown in a vertical cross section in Figure 10.3, a front slopes from Earth's surface toward colder air (or more precisely, denser air). A front lies in a trough in the pressure pattern on any horizontal surface intersecting the front; this is especially evident in the sea-level isobars drawn on a surface weather map. Recall that the surface wind blows across the isobars toward lower pressure, due to friction. Winds on the two sides of a front exhibit different directions and often different speeds.

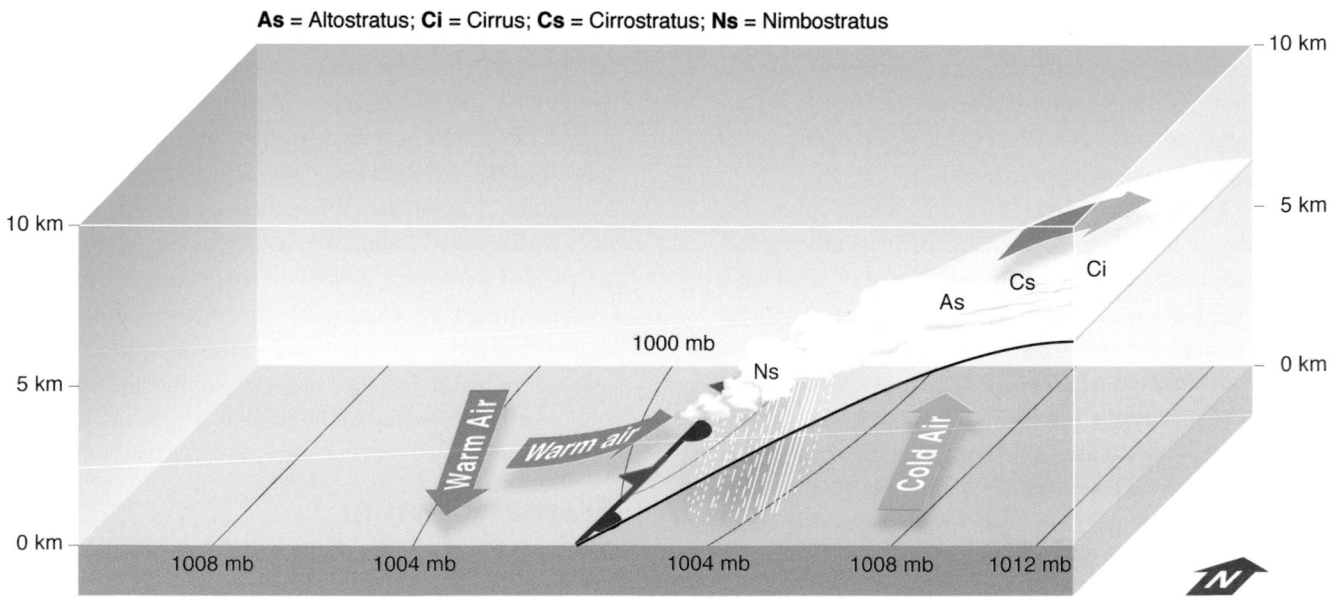

FIGURE 10.3
A stationary front; the vertical scale is greatly exaggerated.

The differences in wind direction and speed across a front are usually associated with convergence, which leads to upward motion, clouds, and perhaps precipitation.

A stationary front does not always have a broad region of associated clouds and precipitation as depicted in Figure 10.3. Frontal weather can vary considerably from case to case, depending on the supply of water vapor and the specifics of air motion relative to the front. In cases that do produce precipitation, the rain or snow falls mostly on the cold side of the stationary front. Warm humid air flows up and over the cooler air mass, more or less along the sloping frontal surface. Ascending air cools by expansion and becomes saturated, which triggers condensation and perhaps precipitation. This situation is often referred to as **overrunning** and can result in an extended period of relatively widespread cloudiness, drizzle, light rain, or light snow.

WARM FRONT

If a stationary front begins moving such that the warm (less dense) air advances while the cold (more dense) air retreats, the front becomes a **warm front**. The overall characteristics of a warm front (Figure 10.4) are very similar to those of the stationary front.

Differences between a warm front and a stationary front are evident in a comparison of Figures 10.4 and 10.3. The slope of the warm frontal surface (ratio of vertical rise to horizontal distance) is less near Earth's surface because surface roughness (friction) has slowed the warm front. Winds on the warm side of the front are quite similar in both instances, but air on the cold side of the warm front is retreating. Thus, the warm air advances relative to Earth's surface, rather than just gliding up and over the cold air as in the case of a stationary front.

As a warm front approaches, clouds develop, gradually lowering and thickening in the following general sequence: cirrus, cirrostratus, altostratus, nimbostratus, and stratus. This sequence of stratiform clouds reflects gentle uplift associated with overrunning in a stable environment. The initial wispy cirrus clouds may appear more than 1000 km (620 mi) in advance of the surface warm front (Figure 10.5). Slowly, clouds spread laterally and form thin sheets of cirrostratus, turning the sky a bright milky white. The tiny ice crystals comprising these high clouds (bases above 7000 m or 23,000 ft) may reflect and refract sunlight to produce halos or sundogs (Chapter 14). Appearance of these optical phenomena may herald the approach of stormy weather a few days in advance.

In time, cirrostratus clouds give way to altostratus, thin clouds with bases at altitudes of 2000 to 7000 m (6500 to 23,000 ft). Soon after altostratus clouds thicken enough to block out the Sun (or Moon), additional lowering of the cloud deck often occurs, accompanied by light rain or snow. Steady precipitation falls from low,

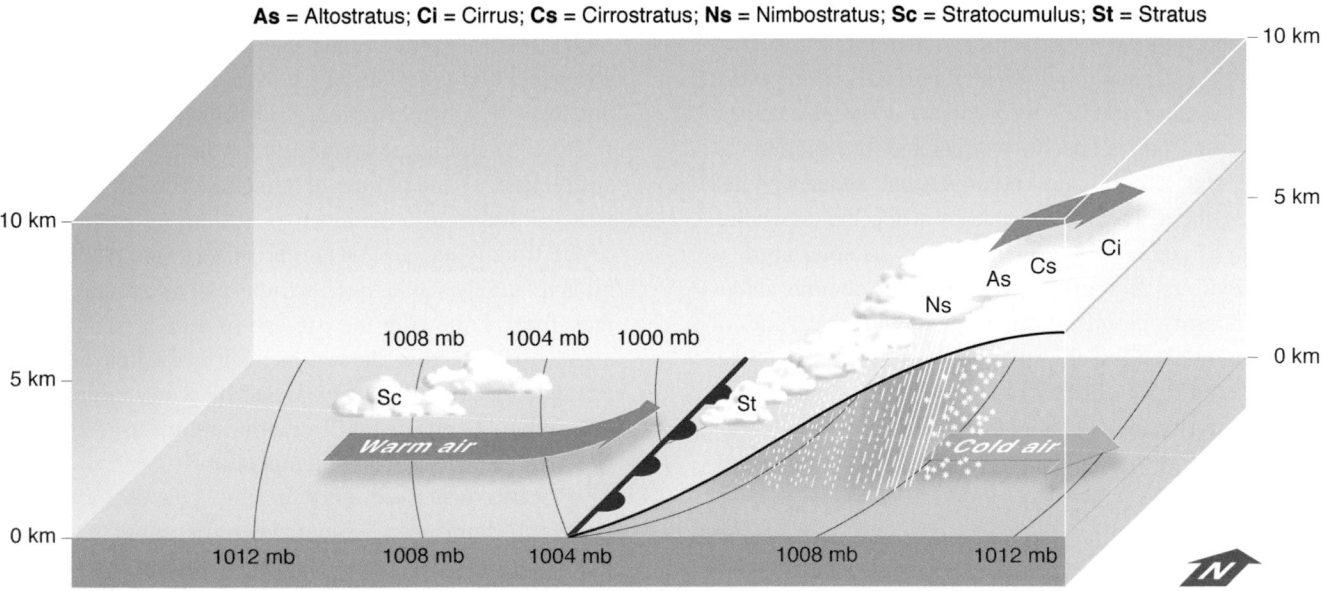

As = Altostratus; **Ci** = Cirrus; **Cs** = Cirrostratus; **Ns** = Nimbostratus; **Sc** = Stratocumulus; **St** = Stratus

FIGURE 10.4
A warm front is the leading edge of a warm air mass that advances by overrunning denser colder and/or drier air; the vertical scale is greatly exaggerated.

FIGURE 10.5
Thin wispy cirrus clouds may appear more than 1000 km (620 mi) in advance of a surface warm front.

gray nimbostratus clouds and persists until the warm front finally passes, a period that may exceed 24 hrs. Copious amounts of rain may fall ahead of the surface warm front, and because precipitation intensity is usually only light to moderate, much of the water infiltrates the soil (as long as the ground is unfrozen and not already saturated with water). If the air is cold enough for the precipitation to fall in the form of snow, accumulations may be substantial.

Just ahead of the surface warm front, steady precipitation associated with nimbostratus clouds usually gives way to drizzle falling from low stratus clouds (bases below 2000 m or 6500 ft) and sometimes fog. **Frontal fog** develops when rain falling through the shallow layer of cool air at the ground evaporates and increases the water vapor concentration to saturation. After the warm front finally passes, frontal fog dissipates and skies clear, at least partially, because the zone of overrunning has also passed. The weather tends to become warmer and more humid.

The cloud and precipitation sequence just described for a warm front applies when the advancing warm air is stable. If the warm air is unstable, uplift is more vigorous and often gives rise to cumulonimbus (thunderstorm) clouds embedded within the zone of overrunning ahead of the surface warm front. Lightning, thunder, and brief periods of heavy rainfall, or perhaps snowfall, may punctuate the otherwise steady fall of light-to-moderate precipitation.

COLD FRONT

An air mass boundary becomes a **cold front** if it begins to move in such a way that colder (more dense) air displaces warmer (less dense) air. Over North America in winter, the temperature contrast is typically greater across a cold front than across stationary or warm fronts.

In summer, however, maximum air temperatures on either side of a cold front are sometimes essentially the same. When this happens, the density contrast between the two air masses arises chiefly from differences in humidity rather than temperature. At the same temperature and pressure, drier air is denser than more humid air (Chapter 6). Following passage of a cold front, both temperature and humidity usually decrease.

Friction associated with the roughness of Earth's surface slows the advancing cold air and is responsible for the nose-shaped vertical profile that characterizes a cold front (Figure 10.6). The slope of a cold frontal surface is steeper (1:50 to 1:100) than the slope of a warm frontal surface (1:150 on average). Because of the steep frontal slope and the typical flow aloft, across the front, from the cold to the warm side, uplift is restricted to a narrow zone at or near the cold front's leading edge. Low-level air motion also differs from that in warm fronts and stationary fronts; the low-level air motion in the cold air is, at least in part, toward the front and forces the warm air aloft.

If the cold front advances at a moderate but steady pace, around 30 km per hr (20 mph), the type of frontal weather will depend on the stability of the warmer air. Brief showers are likely as uplift along the steep cold front is sufficiently vigorous to produce a narrow band of cumuliform clouds and convective showers at or just ahead of the front. If the warm air ahead of the front is unstable, uplift is more vigorous, giving rise to towering cumulus congestus clouds that build into cumulonimbus clouds with cirrus clouds blown downstream by winds at high altitudes. These thunderstorms may be accompanied by strong and gusty surface winds, hail, or other severe weather. If a well-defined cold front progresses at a relatively rapid pace, say 45 km per hr (28 mph), then a **squall line**, a band of intense thunderstorms, may develop either at the front or as much as 300 km (180 mi) ahead of the front (Chapter 11). On the other hand, if the warm air is relatively stable, nimbostratus and altostratus clouds may form. Following the passage of a cold front, colder air sweeping across the warmer surface often becomes unstable, leading to gusty westerly winds along with stratocumulus clouds and intermittent light rain or snow showers (depending on the temperature).

As discussed later in this chapter, a typical Northern Hemisphere cold front trails south or south-westward from the center of an extratropical cyclone and progresses from west to east. In winter, cold fronts often drop southward out of Canada onto the Great Plains, even traveling as far south as the Gulf of Mexico. Southward or southwestward moving cold fronts occur east of the

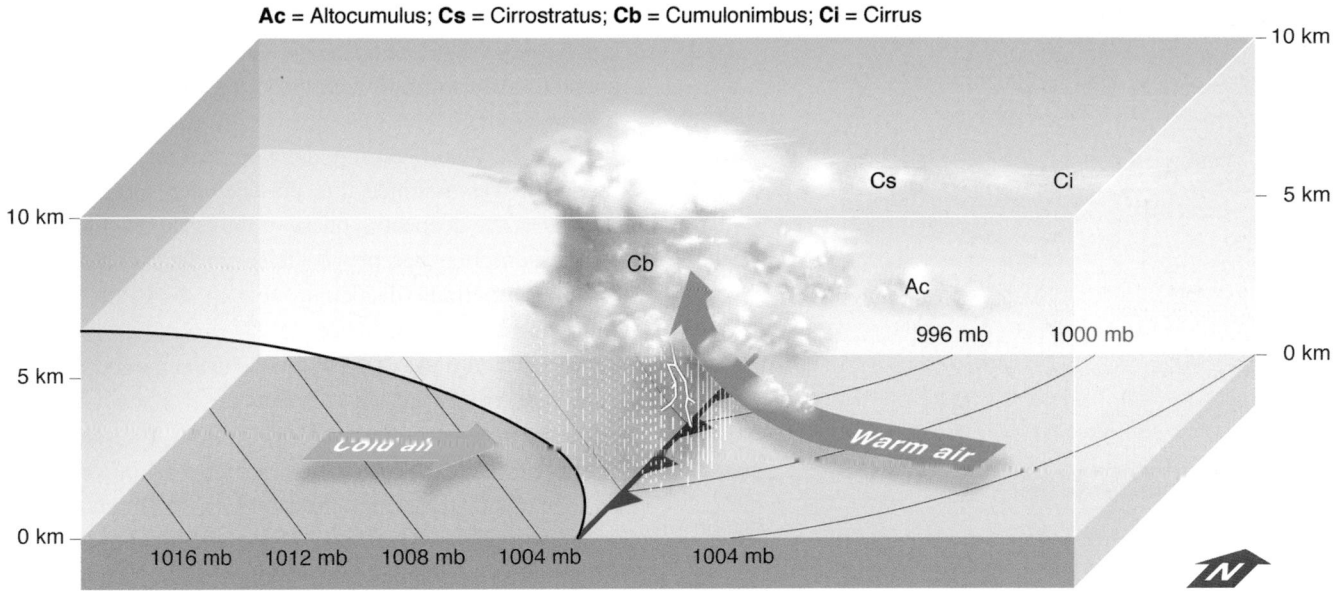

Ac = Altocumulus; **Cs** = Cirrostratus; **Cb** = Cumulonimbus; **Ci** = Cirrus

FIGURE 10.6
A cold front is the leading edge of a cold and/or drier air mass that advances by displacing warmer and/or more humid air; the vertical scale is greatly exaggerated.

Appalachians in New England, but in that region, they are seen most frequently in summer and fall and are referred to as **back-door cold fronts**. These fronts often usher in welcome relief from hot weather, in the form of *cP* air from Canada or *mP* air from the Atlantic Ocean. The Appalachian Mountains block the westward progression of the cooler air, which then moves farther south on the eastern side of the Appalachians along the Piedmont and coastal plains (Figure 10.7).

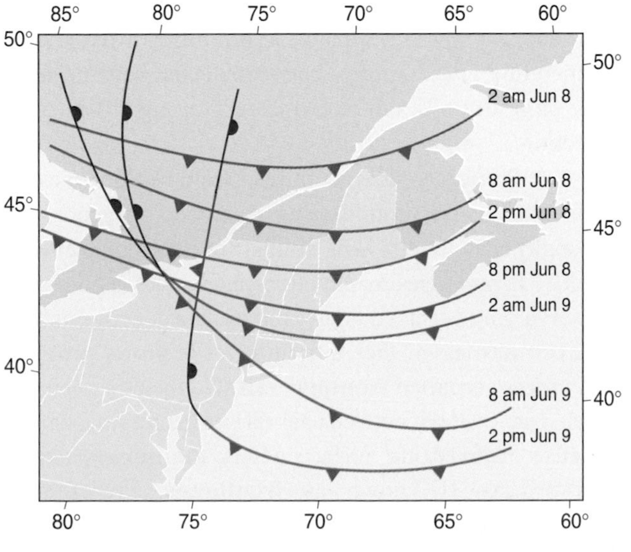

OCCLUDED FRONTS

Late in its life cycle, an extratropical cyclone moves into colder air and a front forms, known as an **occluded front**, or simply an *occlusion*. According to the classic model, an occluded front forms when a faster moving cold front catches up to a slower moving warm front. There are two types of occlusions, distinguished by the temperature contrast between the air behind the cold front and the air ahead of the warm front: cold occlusion and warm occlusion. If air behind the advancing cold front (*cP*) is colder than the cool air (*mP*) ahead of the warm front, the cold air slides under and lifts the air in the warm sector, the cool air, and the warm front (Figure 10.8). This motion produces a *cold occlusion* that has the characteristics of a cold front at the surface, but the temperature contrast between the cold and cool air masses is less than across a typical winter cold front. Weather ahead of the occlusion is similar to that in advance of a warm front, but the actual frontal passage may be marked by more showery conditions, similar to a cold front. This type of occlusion is most common over the eastern half of the North American continent, where the coldest air follows behind the front on northwest winds.

FIGURE 10.7
A back-door cold front moves across New England toward the south and southwest.

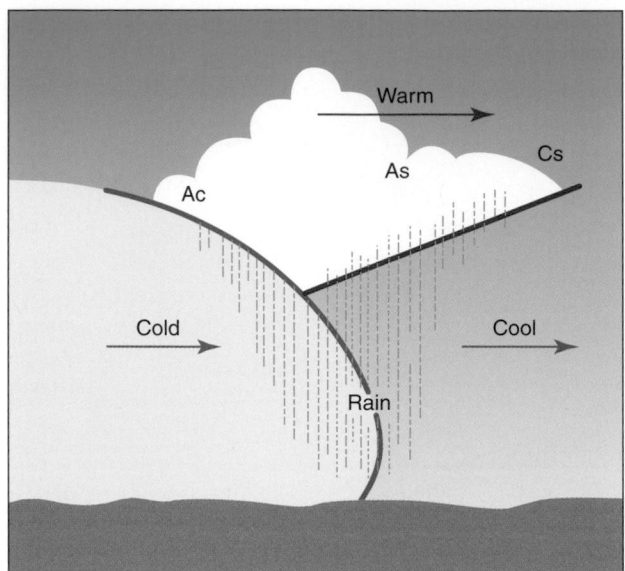

FIGURE 10.8
Schematic vertical cross-section of a cold-type occlusion with the vertical scale greatly exaggerated.

A *warm occlusion* develops when air behind the advancing cold front is not as cold as the air ahead of the warm front (Figure 10.9). This type of occlusion often occurs in the northerly portions of western coasts, such as in Europe or in the Pacific Northwest. In this case, the air behind the cold front is relatively mild (*mP*), having traversed ocean waters, whereas the air ahead of the warm front is relatively cold (*cP*), having traveled over

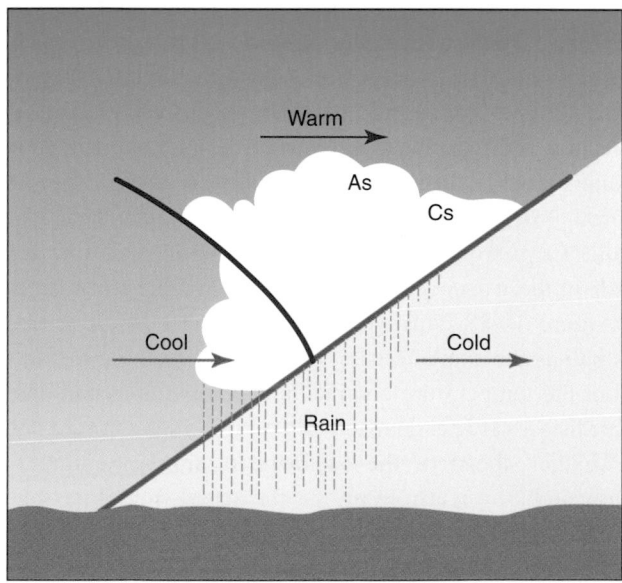

FIGURE 10.9
Schematic vertical cross-section of a warm-type occlusion with the vertical scale greatly exaggerated.

land. With this type of occlusion, the air behind the cold front slides under the warm air but rides over the colder air. Weather ahead of a warm occlusion is similar to that ahead of a warm front, with the surface front behaving as a warm front.

Early on, during development of the cyclone model, it was assumed that the occluded front signaled the end of the deepening phase of the cyclone's life cycle. This assumption has proven false. In fact, extratropical cyclones continue deepening for up to 12 to 24 hrs following onset of occlusion.

Regardless of type, an occlusion can be difficult to locate from surface weather observations because the temperature contrast across the front is often small, precipitation occurs over a broad region masking the front, and the associated trough may not be as pronounced as it is for a cold or warm front. However, satellite imagery shows that an occluded front can be as sharply defined as a cold front in oceanic weather systems.

SUMMARY

At middle and high latitudes, the lower troposphere is comprised of air masses that contrast in temperature and/or humidity. Narrow transition zones, called fronts, border air masses. The motion of a front is related to the lower tropospheric motion on the cold side of the front and fronts are classified on the basis of the movement of the cold air mass. Cold air retreats ahead of a warm front, advances behind a cold front, and moves parallel to a stationary front. Clouds and precipitation develop along fronts only when a significant density contrast exists between air masses, and the supply of water vapor is adequate. With little difference in temperature and humidity between air masses, the front may pass virtually unnoticed except for a shift in wind direction.

Properties that define a front (differences in temperature and/or humidity, wind shift, convergence, and a trough) change with time similar to the way air masses modify. If processes in the atmosphere, such as a region of converging surface winds, increase the density contrast between air masses, then a front forms or grows stronger; this process is called **frontogenesis**. On the other hand, if the density contrast between air masses decreases, perhaps because of diverging surface winds, the front weakens; this process is known as **frontolysis**. Precipitation associated with the front also tends to increase or diminish in intensity as the front strengthens or weakens. Frontal weather occurs in combination with cyclones. We explore this relationship next.

TABLE 10.2
Distinguishing Characteristics of a Mature Extratropical Cyclone

Feature	Typical value
Diameter	1500–3000 km (1000–2000 mi)
Maximum winds	55–70 km per hr (35–45 mph)
Time of occurrence	All seasons; especially winter (DJFM) in the Northern Hemisphere
Frequency	Approximately 3 per day, poleward of 30 degrees N
Lifetime	3–4 days
Energy source	Dynamic forcing (potential to kinetic energy)
Region of origin	Middle latitudes, polar front
Direction of motion	Eastward or northeastward
Lateral speed	30–65 km per hr (20–40 mph)

Extratropical Cyclones

Frontal weather occurs in combination with an **extratropical cyclone**, a major weather maker of middle and high latitudes. (See summarization by Edward J. Hopkins, University of Wisconsin-Madison in Table 10.2.) Viewed from above in the Northern Hemisphere, surface winds blow counterclockwise and inward toward the low-pressure center of a cyclone (Chapter 8). Surface winds converge, air ascends, expands, and cools, resulting in clouds and precipitation. This section explores the life cycle and characteristics of these synoptic-scale weather systems.

LIFE CYCLE

During World War I and continuing into the 1920s, scientists at the Geophysical Institute in Bergen, Norway conducted ground-breaking studies of synoptic-scale weather systems. Led by Vilhelm Bjerknes (1862-1951) and including his son Jacob Bjerknes (1897-1975) and Swedish meteorologist Tor Bergeron (1891-1977), researchers investigated air masses, fronts, and cyclones. They were the first to describe the basic stages in the life cycle of an extratropical cyclone and for this reason this conceptual model is referred to as the **Norwegian cyclone model**. The model was derived primarily from surface weather observations; at the time researchers were unable to monitor winds aloft routinely. Subsequent advances in atmospheric monitoring techniques, especially remote sensing by satellite, have verified the main features of the Norwegian cyclone model. Amazingly, the model remains a close approximation of our current understanding of middle latitude cyclones, even though individual cyclones may not follow the model precisely.

With adequate *upper-air support*, an extratropical cyclone can form and intensify. **Cyclogenesis**, the birth of a cyclone, usually takes place along the polar front directly under an area of strong horizontal divergence in the upper troposphere. As noted in Chapter 9, strong horizontal divergence aloft occurs to the east of an upper-level trough and under the left-front quadrant of a jet streak. If horizontal divergence aloft removes more mass from a column of air than is supplied by converging surface winds, the air pressure at the bottom of the column falls. Consequently, a horizontal air pressure gradient develops in the lower reaches of the atmosphere and a cyclonic circulation begins; that is, a storm is born. Westerlies aloft then steer and support the cyclone as it progresses through its life cycle (Figure 10.10).

Just prior to the formation of a typical cyclone, the polar front is often stationary with surface winds blowing parallel to the front. As the surface air pressure drops, surface winds respond and turn toward the front and converge as segments of the front begin to move (Figure 10.10A). West of the center of lowest sea-level air pressure, the polar front pushes toward the southeast as a *cold front*. East of the low center, the polar front advances northward as a *warm front*. The minimum pressure for the incipient low might be near 1000 mb, with a single closed isobar on a standard surface weather map, with a 4-mb interval between isobars. Satellite imagery shows that the narrow cloud band associated with the stationary front develops a

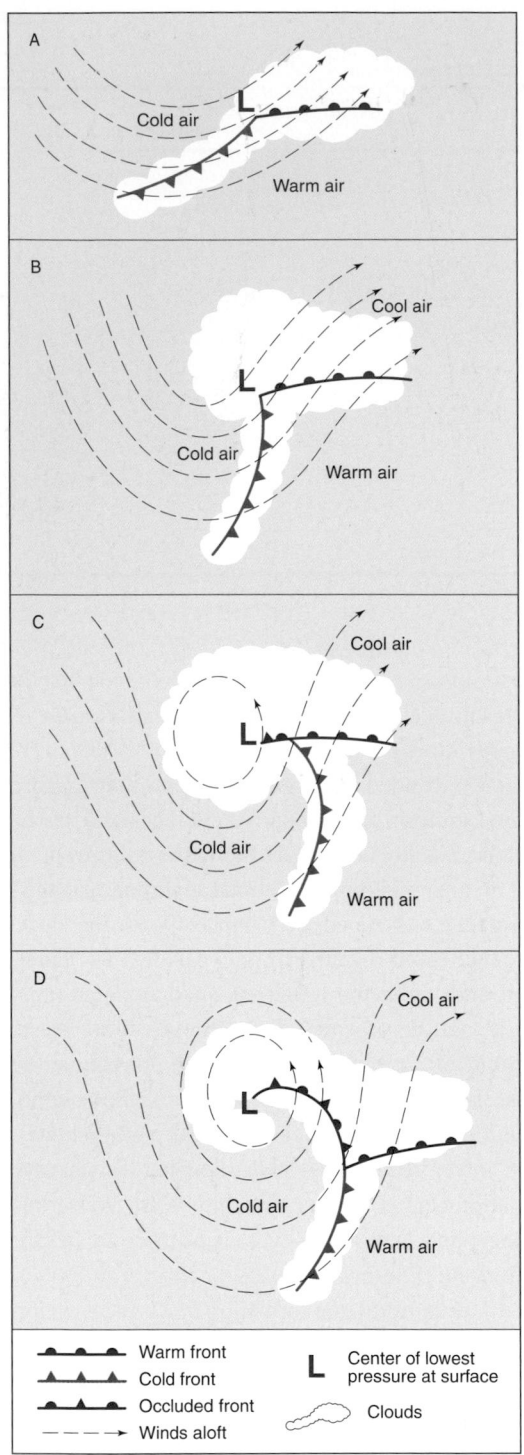

FIGURE 10.10
The life cycle of an extratropical cyclone (low-pressure system): (A) incipient cyclone, (B) wave cyclone, (C) beginning of occlusion, and (D) bent-back occlusion. As a wave cyclone, the system's center is located east of the upper-level trough; at occlusion, the system's center is under the upper-level trough. Air temperatures refer to conditions at Earth's surface.

bulge at the low center and extends along the developing warm front. The upper-level circulation pattern depicted in Figure 10.10A shows a trough to the west of the surface low, a position that is favorable for further development (*deepening*) of the system. With more horizontal divergence aloft than convergence near the surface, the surface pressure at the cyclone center continues to fall.

If the circulation pattern in the upper troposphere does not favor further development of the cyclone, the low center typically ripples along the stationary front without deepening, producing cloudiness and light precipitation along the way. Such cyclones affect only a small area and typically travel along the front at 50 to 70 km per hr (30 to 45 mph). On the other hand, if atmospheric conditions support development of the incipient cyclone, the central pressure continues to drop, the associated horizontal pressure gradient becomes steeper, and counterclockwise winds strengthen. The upper-level trough also frequently deepens while remaining to the west of the low center, as cold air is brought into the system from the northwest.

The *warm sector* of the cyclone (the area between the warm and cold fronts, occupied by warm air at the surface) becomes better defined during the early stage of storm development. At this stage in the cyclone's life cycle, fronts form a pronounced wave pattern (Figure 10.10B), hence the descriptive name *wave cyclone*. The surface cyclone may now have a central pressure approaching 990 mb, identified by perhaps three closed isobars on a weather map, and typically is moving eastward or northeastward at 40 to 55 km per hr (25 to 35 mph). A large-scale comma-shaped cloud pattern, known as a **comma cloud**, typically appears in satellite images at this stage, reflecting the strengthening of the system's circulation (Figure 10.11). The head of the comma extends from the low center to the northwest, with its tail trailing southward or southwestward along the cold front. Extensive stratiform cloudiness caused by overrunning appears north of the warm front.

The cold front generally moves faster than the warm front so that the angle between the two fronts gradually closes; that is, the area occupied by the surface warm sector diminishes. As the cold front closes in on the warm front, an *occluded front* begins forming near the low center, forcing the warm air aloft and causing the warm sector at the surface to occupy a smaller area away from the surface low center. Figure 10.10C represents the beginning of the occlusion stage of the cyclone's life cycle. The upper-level pattern now features a closed circulation almost above the surface cyclone. When the upper-level low center is directly over the surface low

FIGURE 10.11
The comma-shaped pattern of clouds on this infrared satellite image is characteristic of a well-developed extratropical cyclone. Note the *dry slot* (black shading) to the south of the system's center over Nebraska.

center, the system is described as *vertically stacked.* Dry air descending behind the cold front is drawn nearly into the center of the cyclone as a **dry slot** that separates the cloud band along the cold front from the comma head, now more west than northwest of the low center. The central pressure of the cyclone has dropped significantly to perhaps 985 mb in the time that elapsed between Figures 10.10B and 10.10C.

Some extratropical cyclones continue deepening even after occlusion has begun. Typically, the upper-air low center is not directly over the surface low center while deepening is occurring, but it may be fairly close. These systems develop a closed, vertically stacked (or nearly so) circulation that is troposphere-deep and typically move much more slowly than at earlier stages, at around 30 km per hr (19 mph), and may even stall. Occasionally, the track of the surface low becomes erratic, with the low sometimes moving toward the west as the system becomes "cut-off" or detached from the westerly flow in the middle to upper troposphere. This circulation tends to draw the occluded front around the low center into a configuration sometimes described as a *bent-back occlusion.* The central pressure of the low may now be 980 mb or lower, with perhaps a half dozen or more closely spaced closed isobars, and surface winds that can exceed 75 km per hr (45 mph). Such intense cyclones commonly occur over the North Atlantic or North Pacific Oceans. Although still present, the surface warm sector is detached from the cyclone center and the occluded front extends as the warm sector is pushed aloft.

Cold, warm, and occluded fronts intersect at the point of occlusion, or **triple point**, where conditions can favor development of a new cyclone, sometimes called a *secondary cyclone.* (Note the similar appearance of the triple point with its fronts in Figure 10.10D and the incipient low with its fronts in Figure 10.10A). At this stage, the cloud pattern typically becomes a spiraling swirl with enhanced bands associated with the fronts. The spiraling cloud band may circle the center of an intense low (central pressure of 960 mb or lower) several times.

Eventually, the cyclone weakens as its central pressure rises, the horizontal pressure gradient weakens, and winds diminish; this process is called **cyclolysis** or *filling.* The cyclone can weaken during any of the stages described above if its upper-air support decreases to the point that horizontal divergence aloft becomes less than horizontal convergence near the surface. Then the surface pressure at the low center rises, and the surrounding horizontal pressure gradient weakens. As the central pressure rises, the cyclone loses its identity in the sea-level pressure field and is marked only by an area of cloudy skies and drizzle. Such weakening is inevitable once the system becomes vertically stacked, but the filling process may not begin for many hours and may proceed slowly, allowing an intense circulation to persist for days.

The cyclone life cycle described above may occur over many days, or it may be completed in a day or so. If upper-air support is less favorable, the storm may spend a longer time in any one of the early stages, even weaken temporarily, and still become fully occluded. Sometimes cyclones develop with meager upper-air support (weak divergence aloft) and are poorly defined. At other times, widespread cloudiness and precipitation are linked to an upper-air or surface trough with no closed cyclonic circulation at the surface. When upper-level conditions are ideal, the entire life cycle from incipient cyclone to bent-back occlusion can occur in less than a day and a half.

CONVEYOR BELT MODEL

The Norwegian cyclone model provides a reasonably good description of the characteristics and life cycle of a typical extratropical cyclone. Improved understanding of upper-air circulation in the years since the Norwegian model was first developed led to a new three-dimensional cyclone model, combining horizontal and vertical air motions. This **conveyor belt model** depicts the circulation within a mature extratropical cyclone in terms of three broad interacting air streams, referred to as *conveyor belts.* Just as mechanical conveyor belts transport goods (or even people) from one location to another, atmospheric conveyor belts transport air with certain properties from one location to another and are named for the type of air they transport. The three conveyor belts in extratropical cyclones are (1) warm and humid, (2) cold, and (3) dry. A schematic drawing of this model is shown in Figure 10.12

In the conveyor belt model, a warm air stream originates in the cyclone's warm sector and follows the warm side of the cold front near the Earth's surface, south and east of the storm center. Typically, this warm and humid air stream ascends slightly as it flows northward in the cyclone's warm sector at low levels. The warm conveyor belt then ascends more rapidly as it glides northward along the sloping warm frontal surface, which serves as the boundary between the warm air and the dome of cooler air to the north of the surface warm front. As the ascending air expands and cools adiabatically, water vapor condenses forming clouds, and rain or snow develops. Latent heat released during condensation adds to the buoyancy of air in the warm conveyor belt.

Recall that at this mature stage in the cyclone's life cycle, winds aloft blow more or less from the southwest, with the surface cyclone located between the upper-level trough and ridge. The warm conveyor belt thus turns from a southerly to a southwesterly, or even westerly, direction as it ascends over the warm front and follows the upper-level flow. This air stream therefore helps explain the broad region of stratiform cloudiness and steady precipitation north of the warm front.

While the warm conveyor belt is gliding up and over the warm frontal surface, winds in the lower troposphere in the region northeast of the surface low pressure system blow from the east or northeast and are relatively cold. This flow forms the cold conveyor belt and, at low levels, is located just to the cold (north) side of the warm front. Like the warm conveyor belt, this air stream ascends as it progresses toward the west. Warm frontal precipitation falls through the cold air stream, increasing its relative humidity to saturation so that clouds and precipitation develop in the cold conveyor belt. As the cold conveyor belt ascends, it comes under the influence of mid- and upper-level tropospheric winds and turns clockwise, following the southwesterly or westerly flow aloft. The ascending saturated cold conveyor belt produces the *comma cloud* to the northwest of the cyclone center.

The third conveyor belt is the dry air stream. Whereas air ascends at the cyclone center and along associated fronts, air descends just west of the cold front. Descending air originates aloft in the upper troposphere and lower stratosphere upstream of the upper-level trough and is very dry, especially compared to the air in the warm conveyor belt. As the dry conveyor belt descends, one branch turns southward behind the cold front, causing clearing skies. The other branch of the dry air stream first descends as it moves northward toward the surface low center, and then ascends as it comes under the influence of the winds aloft east of the upper-level trough. As the dry air stream moves toward the surface low center, it forms the *dry slot* that separates the head and tail of the comma cloud and is especially prominent in the later phases of the cyclone's life cycle (Figure 10.11).

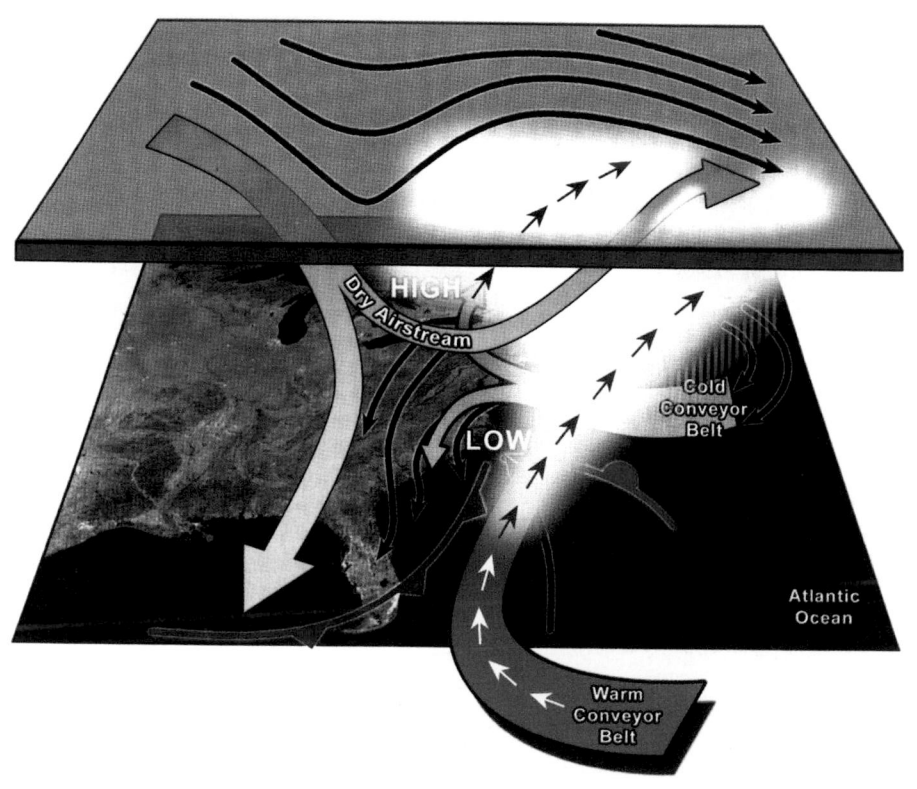

FIGURE 10.12
Schematic representation of the three-dimensional circulation within an intense coastal extratropical cyclone showing warm, cold, and dry air streams. [From P.J. Kocin and L.W. Uccellini, *Northeast Snowstorms.* Boston, MA: American Meteorological Society, 2004, p. 129]

CYCLONE WEATHER

As an illustration of typical extratropical cyclone weather, consider a winter wave cyclone that developed a well-defined warm sector as it moved into the Upper Midwest. Although the storm is still intensifying at map time, its circulation, clouds, and precipitation already affect weather over a broad region. The typical diameter of such a cyclone is between 1000 and 2000 km (about 600 to 1200 mi). Figure 10.13 is a schematic representation of (A) the surface winds, (B) air temperatures, (C) cloud shield and precipitation, and (D) the air pressure tendency pattern associated with this mature low.

Ideally, based on surface weather, a mature wave cyclone can be divided into four sectors about the low center. The lowest air temperatures occur to the northwest of the storm center, where continental polar or Arctic air flow southward and eastward. Because of the relatively steep air pressure gradient typically found on the west side of the surface low, winds are strong, making the air feel colder than the actual temperature (wind-chill effect). Stratiform clouds and non-convective precipitation in this northwest sector are associated with the head of the comma cloud; precipitation may be either rain or snow depending upon air temperature. The cold front is south of the low center; it is accompanied by a relatively narrow band of ascending air, cumuliform clouds, and convective-type showers and thunderstorms.

Sinking air and generally clear skies characterize much of the southwest sector of the storm system. The mildest air is in the southeast (warm) sector of the cyclone, where south and southeast winds advect maritime tropical air northward from the Gulf of Mexico. Skies are generally partly cloudy, dewpoints are high, and scattered convective showers are possible, triggered by afternoon solar heating. To the north and northeast of the Low center is an extensive zone of overrunning as maritime tropical air surges over the wedge of cold air maintained by east and northeast winds at the surface. In the northeast sector, skies are obscured by an extensive shield of stratiform clouds, and the non-convective precipitation is steady and substantial.

Weather conditions in the various sectors of a mature wave cyclone are consistent with our earlier

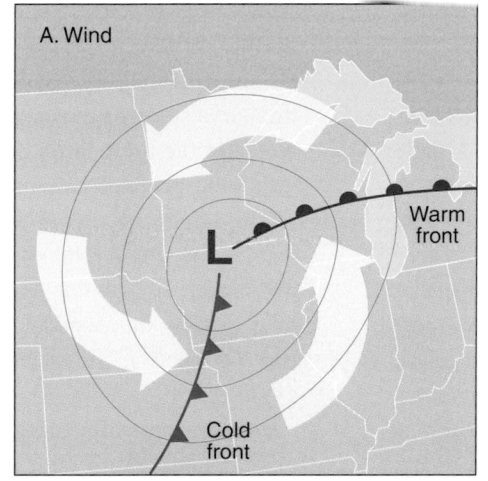

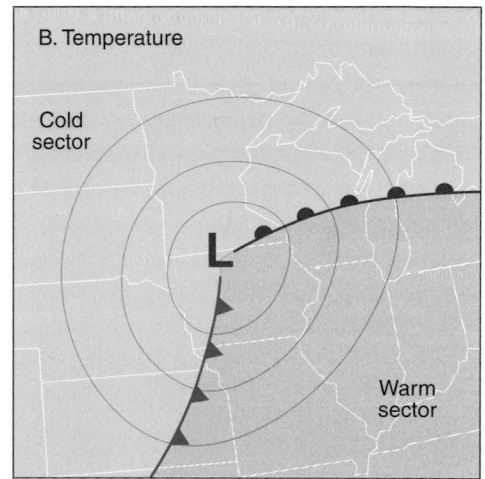

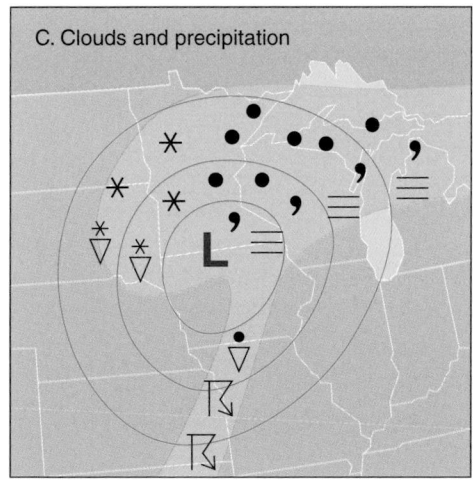

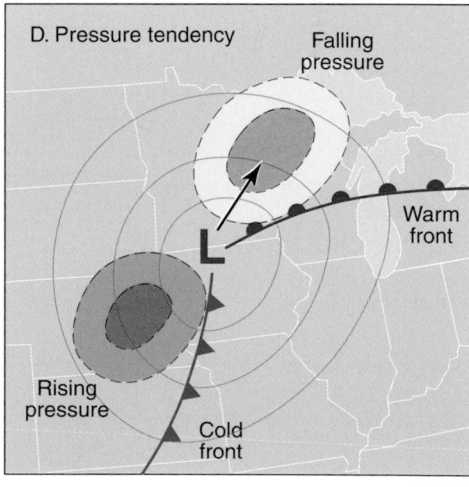

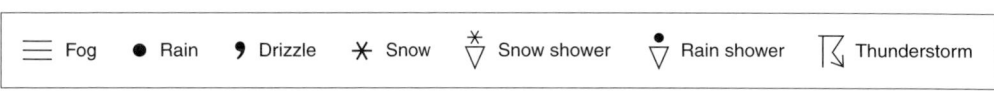

FIGURE 10.13

A mature extratropical cyclone centered on the border between Minnesota and Iowa, showing typical patterns of (A) surface winds, (B) surface air temperatures, (C) areas of cloudiness and precipitation, and (D) areas of greatest positive and negative pressure tendency with the low moving northeastward.

description of cold and warm frontal weather. Although the flow in and around extratropical cyclones is mostly horizontal, uplift of air along frontal surfaces is responsible for the cloud and precipitation pattern illustrated in Figure 10.13.

PRINCIPAL CYCLONE TRACKS

The track of an extratropical cyclone depends on the direction and strength of the upper-level westerlies in which the storm system is embedded. The Low center tends to move in the direction of the wind blowing directly above the surface Low in the mid-troposphere (i.e., at or near the 500-mb level). As a general rule, the Low center advances at about half the speed of the 500-mb winds. Keep in mind, however, that the upper-level horizontal circulation pattern (including the steering winds) also changes with time and the cyclone's path varies accordingly.

Principal cyclone tracks across the lower 48 states of the United States are plotted in Figure 10.14. Note how cyclones tend to converge toward the northeast; their ultimate destination usually being the semi-permanent Icelandic low of the North Atlantic Ocean or Western Europe. Although many extratropical cyclones appear to originate just east of the Rocky Mountains, in reality most form over the Pacific Ocean, near the Aleutians or in the Gulf of Alaska. As a Low travels through mountainous terrain, it may temporarily lose its identity, but redevelops on the Great Plains just east of the Front Range of the Rockies (Figure 10.15). For more on what happens, see

this chapter's first *For Further Exploration*, The Case of the Missing Storm.

Similarities in tracking have earned some cyclone paths characteristic nicknames including the panhandle hook, Alberta clipper, and the nor'easter. The *panhandle hook* develops to the lee of the central Rockies, travels southeastward over the Oklahoma and Texas Panhandles, before turning almost due north over the Southern Plains in a hook-like path, hence the name. The *Alberta clipper* develops to the lee of the Canadian Rockies in Alberta and travels rapidly from west to east across southern Canada or the U.S. northern tier states. The center of a *nor'easter* tracks toward the northeast along the U.S. East Coast. Specific tracks of nor'easters are plotted as "Texas" and "East Gulf" in Figure 10.14. A nor'easter usually deepens off the North Carolina coast and then tracks up the coast in a southwest to northeast direction. It may be confusing to think of a storm as moving toward the northeast, while the winds in the northeast sector of the storm blow from the northeast. In effect, two motions occur simultaneously: (1) the forward movement of the Low center along the coast and (2) the counterclockwise circulation of surface winds about the Low center. For the most part, the circulation of a Low is independent of its track, much as the spin of a Frisbee® is independent of its trajectory. Of all the extratropical cyclones to affect the U.S. coastal zone, the nor'easter generally receives the most attention because of its potential for producing strong onshore winds, heavy precipitation and coastal erosion, and the large number of

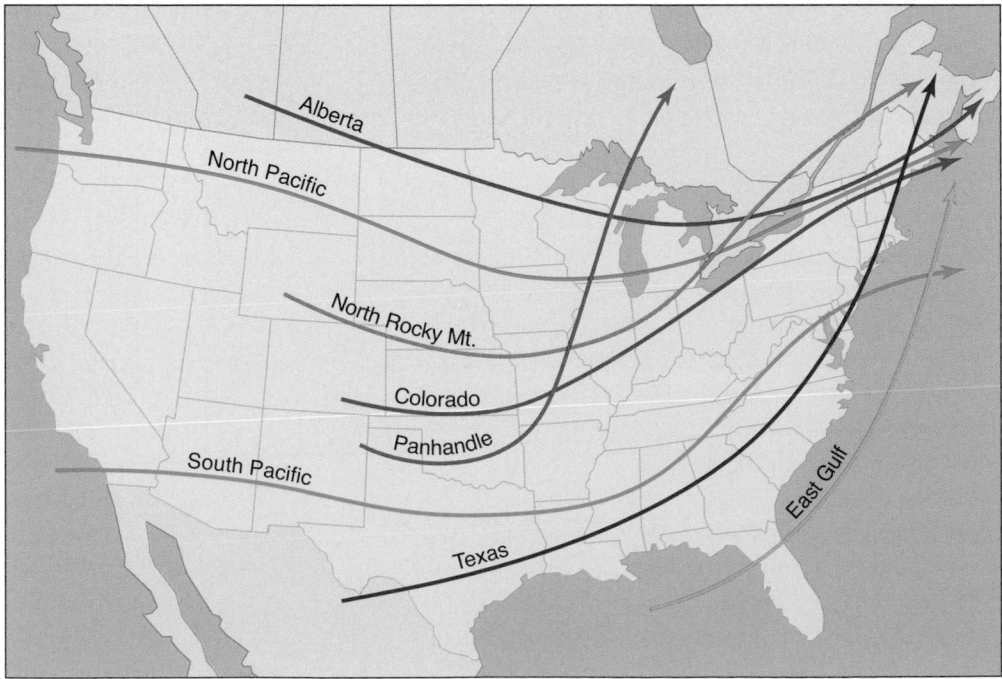

FIGURE 10.14
Principal extratropical cyclone tracks across the lower 48-states of the United States.

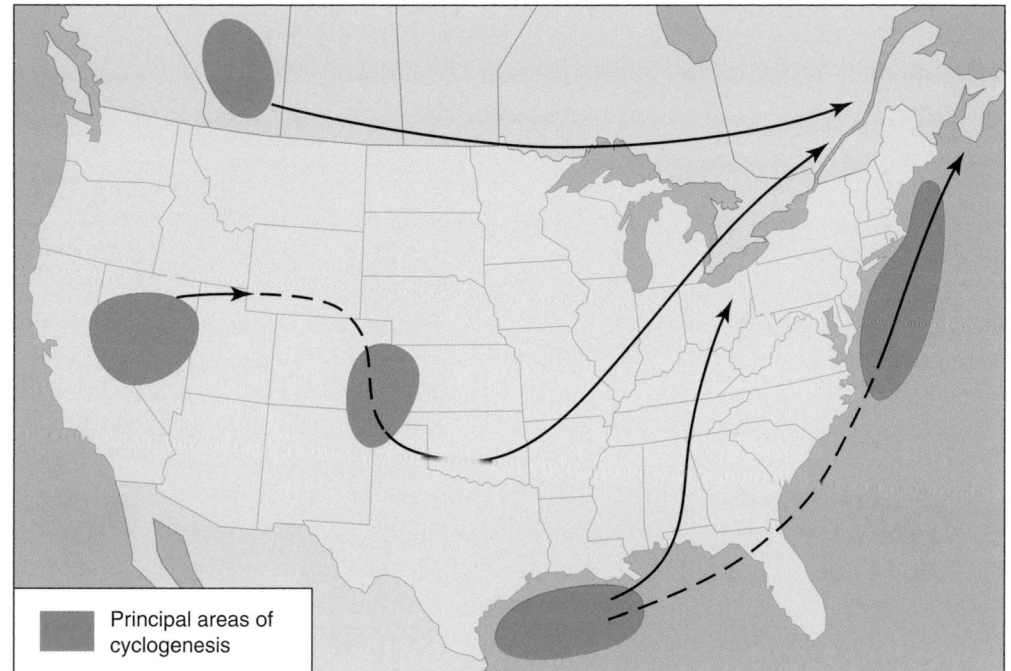

FIGURE 10.15
Principal areas of cyclogenesis and associated cyclone tracks. Note that the dashed portions of the tracks represent areas where a Low may temporarily loose its identity prior to restrengthening.

Principal areas of cyclogenesis

people likely to be impacted. For more on the nor'easter, refer to this chapter's second *For Further Exploration*.

Generally, extratropical cyclones that form in the south yield more precipitation than those that develop in the north because the southerly flow in the warm sector of southern cyclones is better positioned to draw moisture-rich maritime tropical air from source regions such as the Gulf of Mexico. For this reason, Alberta cyclones typically yield only light amounts of rain or snow, whereas Colorado- and Gulf-track storms often produce heavy accumulations of rain or snow. The forward speed of the Low also affects the total precipitation received at a particular location. For places in the path of a cyclone, a fast-moving system may yield rain or snow for only a few hours whereas a slower moving storm may precipitate for 12 hours or longer.

In Chapter 9, we examined how the circulation in the mid- and upper-troposphere supplies *upper-air support* (horizontal divergence) for extratropical cyclones. Just as the planetary-scale circulation undergoes seasonal changes, so too do cyclones and cyclone tracks. In summer, when the mean position of the polar front and jet stream is across southern Canada, few well-organized cyclones occur in the United States, and the Alberta storm track shifts northward across Canada. In winter, however, when the mean position of the polar front and jet stream shifts southward, cyclogenesis is more frequent in the United States. Alberta-track cyclones are most common because they occur year-round, whereas storms with more

southerly tracks develop primarily in winter. In fact, one sign of the beginning of winter circulation patterns is the appearance of Colorado lows.

COLD SIDE/WARM SIDE

An extratropical cyclone has a cold side and a warm side. In Figure 10.13, the coldest air is northwest of the low center, where surface winds are blowing from the northwest, while the warmest air is to the southeast of the low center, where surface winds are blowing from the south. Hence, as an extratropical cyclone moves eastward across the continent, the weather to the north or left (cold) side of the storm track can differ considerably from the weather to the south or right (warm) side of the storm track.

Consider, for example, the two storm tracks plotted in Figure 10.16. A winter cyclone develops over eastern Colorado. As the storm matures, it moves northeastward toward the Great Lakes region following either track A or track B, depending on the direction of the steering winds aloft. In both cases, the center of the cyclone passes within 150 km (95 mi) of Chicago. With track A, the storm moves west and then north of Chicago whereas track B takes the storm to the south and then east of the city. The storm's influence on Chicago's weather depends on its track, as summarized in Table 10.3.

If the storm follows track A, Chicago residents experience the warm side of the storm. As the storm approaches, stratiform clouds thicken and lower, with

FIGURE 10.16
A winter cyclone develops over eastern Colorado and tracks northeastward toward the western Great Lakes. Track A takes the storm to the west and north of Chicago so that the city is on the warm side of the system. Track B takes the storm to the south and east of Chicago so that the city is on the cold side of the system.

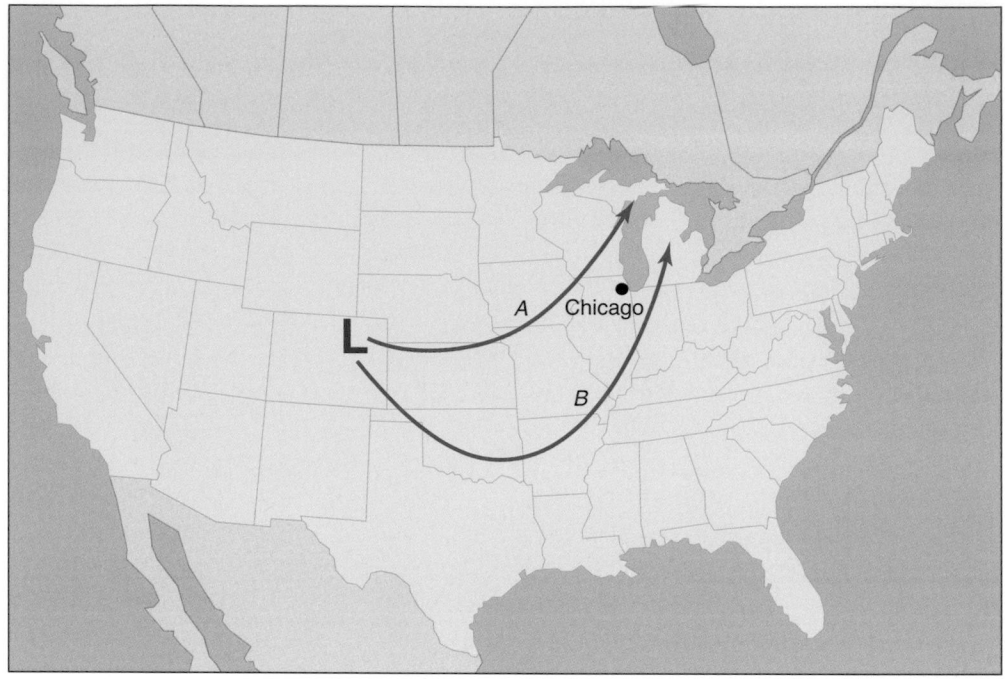

steady rain—or perhaps snow briefly at the onset—giving way to drizzle and fog after 12 to 24 hrs, signaling the arrival of the warm front. As the warm front passes over the city, skies partially clear and winds shift abruptly from east to southeast and then to the south, advecting warm and humid (*mT*) air at the surface. At this time, the warm sector is over Chicago as the Low center passes to the north. Clearing is short-lived, however, as convective clouds develop, accompanied by scattered showers and thunderstorms, heralding the arrival of colder air behind the cold front. The surface cold front passes through the city and then on toward the east and northeast. Winds *veer* (turn clockwise with time), blowing initially from the southwest, then west, and finally northwest. Stratocumulus clouds follow the frontal passage and then skies clear again as the air temperature and dewpoint fall.

In contrast, if the storm takes track B, Chicago residents experience the cold side of the storm and no frontal passages. With the approach of the storm, lowering and thickening stratiform clouds accompany gusty east and northeast winds that drive steady snow or rain (depending on air temperatures) for 12 hrs or longer. As the storm passes to the south of the city, winds gradually *back* (turn counterclockwise with time) to a northerly direction, precipitation tapers off to snow flurries or showers, and air temperatures drop. Finally, winds shift to northwest,

TABLE 10.3
General Sequence of Weather Conditions at Chicago as Winter Cyclone Tracks West of the City (*Track A*) or East of the City (*Track B*)

	Track A						*Track B*			
Wind direction	E	SE	S	SW	W	NW	E	NE	N	NW
Frontal passage	-	WF	-	CF	-	-	-	-	-	-
Advection	-	Wm	Wm	Cd	Cd	Cd	-	-	Cd	Cd
Air pressure tendency	F	F	F	R	R	R	F	F	R	R
			Time	⇒				Time	⇒	

WF = Warm front, CF = Cold front, Wm = Warm air advection, Cd = Cold air advection
F = Pressure falling, R = Pressure rising

the sky begins to clear, and air temperatures continue to fall in response to strong cold air advection as the Low tracks toward the northeast away from Chicago.

In summary, if you are located on the warm side of an extratropical cyclone's track, the wind direction veers with time and a cold front follows a warm front. However, if you are on the cold side of the storm, the wind direction backs with time without the passage of fronts. In winter in a northern location, substantial snowfall is much more likely on the cold side than on the warm side of the cyclone's track, with the axis of heaviest snowfall running parallel about 240 km (150 mi) to the northwest of the storm track.

WINTER STORMS

Extratropical cyclones develop along the polar front throughout the year. However, these storms become more intense and pose a greater threat to middle and high latitudes from late autumn, through winter, and then into early spring as the polar front shifts equatorward and the temperature contrast across the front increases. Extratropical cyclones are among the most dangerous aspects of winter weather in many regions of the United States and Canada. Except for the Gulf Coast and some coastal sections of southern California, the weather of few places across the continent escape the influence of winter storms.

A **winter storm** is an extratropical cyclone that produces some combination of frozen or freezing precipitation, including snow, ice pellets (sleet) and/or freezing rain (Chapter 7). In the more intense winter storms, heavy snow combined with low air temperatures and strong winds can produce potentially life-threatening conditions (Figure 10.17). Strong winds cause blowing and drifting snow, sometimes leading to a blizzard. During the American Civil War, the term "blizzard" referred to a heavy volley of musketry. In the spring of 1870, a writer for a newspaper

in Estherville, IA, was apparently the first to describe a severe snowstorm accompanied by strong winds as a blizzard. Over the following decade, use of the term became widespread. Today, the National Weather Service defines a **blizzard** as a severe storm characterized by high winds and reduced visibility due to falling or blowing snow. A blizzard requires sustained winds of 56 km per hr (35 mph) or greater and sufficient falling, blowing or drifting snow to reduce the horizontal visibility to less than 400 m (0.25 mi). Blizzard conditions may persist long after the passage of a Low as strengthening winds pick up loose snow from the ground. A *whiteout* occurs when visibility is reduced to the point that it is nearly impossible for a person to distinguish the sky from the ground—everything is white. Driving is impossible and people caught in a whiteout become disoriented and lost. (Recall the Case-in-Point in Chapter 1.)

Ice storms form a coating of glaze on cold surfaces that makes for treacherous walking and driving conditions and can bring down tree limbs that sever power lines causing widespread electrical power outages. The cold wave that often follows on the heels of a winter storm exacerbates the situation by increasing the potential for frost bite and hypothermia for people who venture outdoors without adequate clothing.

The necessary ingredients for development of a winter storm include cold air, a moisture supply,

FIGURE 10.17
Heavy snowfall can seriously disrupt all forms of transportation including driving. [Utah Department of Transportation]

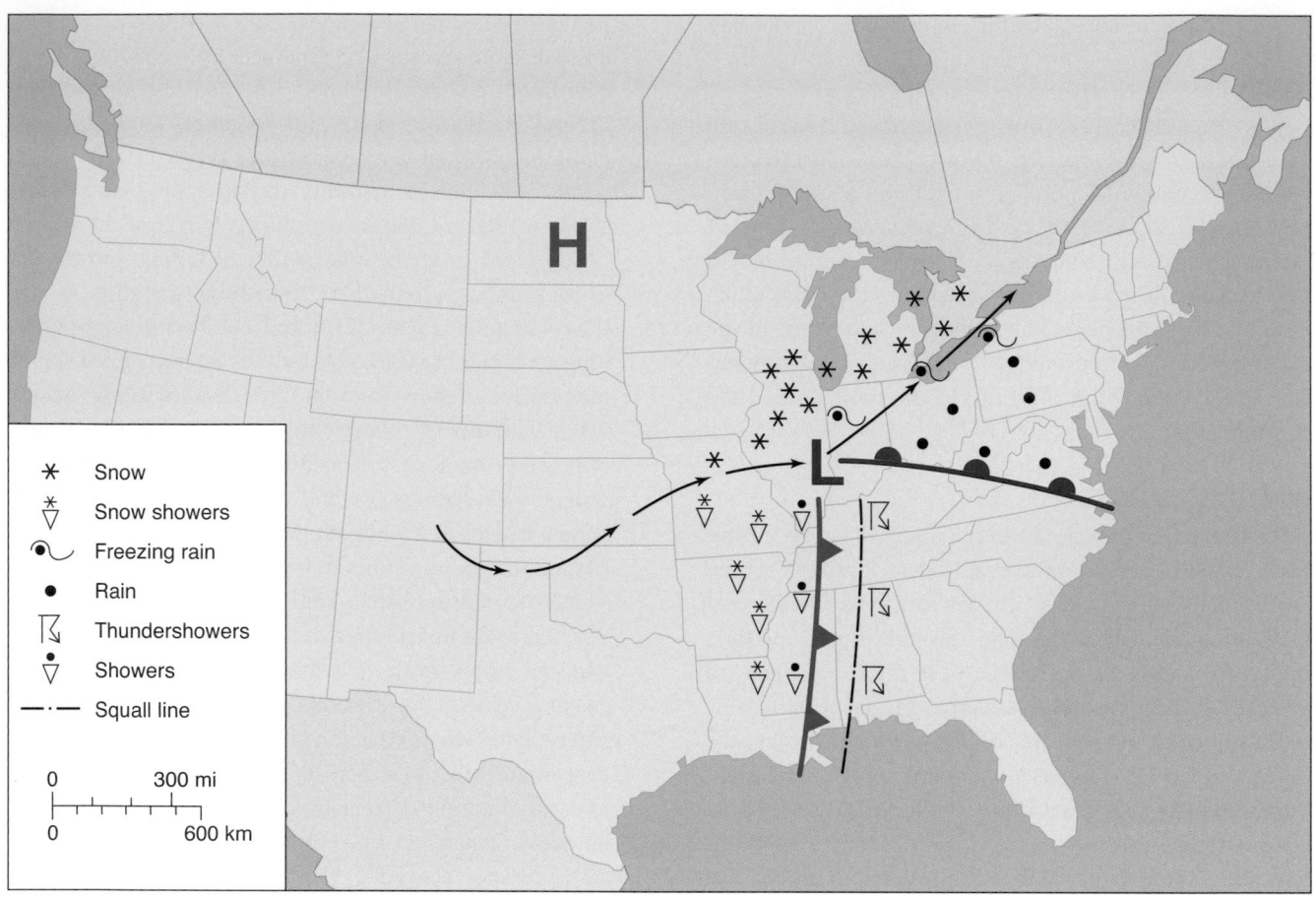

FIGURE 10.18

A schematic representation of a surface weather map showing the various forms of precipitation associated with a mature Colorado-track winter storm system.

and mechanisms that provide uplift. Cold air is needed to ensure that the precipitation falls as snow or another frozen form of precipitation. Typically, a sprawling cold High is located to the north of the track of the Low and helps convey cold air into the Low (Figure 10.18). For a major snowstorm, warm and humid air is conveyed northward into the storm system. Panhandle hook Lows typically produce more snow across the Plains and Midwest as abundant Gulf moisture is available, whereas Alberta Clippers have little available moisture, resulting in relatively light accumulations of snow. For a winter storm tracking to the northeast, the heaviest snow tends to accumulate to the west and north of the track of the Low center, on the cold side of the storm.

COLD- AND WARM-CORE SYSTEMS

An occluded extratropical cyclone is a cold-core system; that is, the lowest temperatures occur within the column of air above the low-pressure center. In vertical cross section, constant pressure (isobaric) surfaces dip downward toward the center of an occluded **cold-core cyclone** (concave upward) (Figure 10.19). Furthermore, the depth of the low increases with altitude implying that a cyclonic circulation prevails throughout the troposphere and is most intense at high altitudes. The thickness of an air layer defined above and below by pressure surfaces is directly proportional to the mean temperature of that bounded air layer. The requirement that the thickness (mean temperature) be lowest at the center of the low produces the characteristic isobaric pattern. This structure is consistent with the occluded cyclone shown in Figure 10.10D.

On the other hand, the lowest air temperatures in a non-occluded wave cyclone occur northwest of the system's center and the highest temperatures are to the southeast. Figure 10.20 is a vertical cross-section through the low from northwest to southeast. Thickness (mean temperature) arguments indicate that until occlusion the low center aloft is not located above the low center near the surface, but rather is displaced to the cold side of the storm,

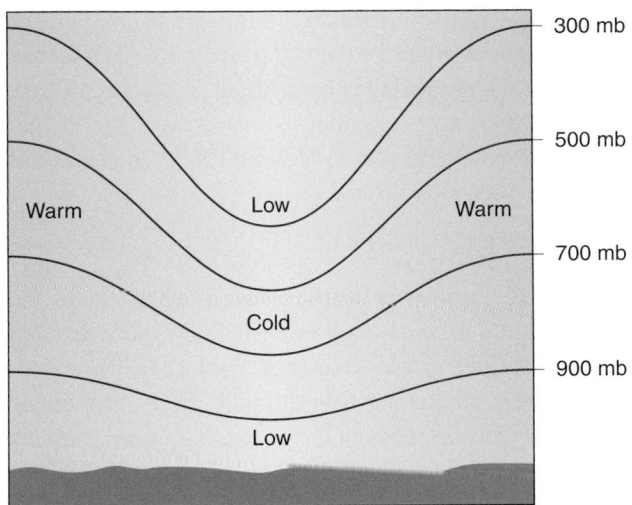

FIGURE 10.19
West-to-east oriented vertical cross-section through an occluded cold-core extratropical cyclone showing isobaric (constant pressure) surfaces. The vertical scale is greatly exaggerated.

implying that the system tilts with altitude. This structure is consistent with Figure 10.10B, which shows the upper-level trough lagging behind the surface cyclone.

A different type of low that sometimes is plotted on a surface weather map has characteristics markedly different from those of cold-core or tilted lows. Cyclones of this type are stationary, have no fronts, and usually are associated with fair weather. In response to intense solar heating of the ground, they form over a broad expanse of arid or semiarid land, including the interior of Mexico and the southwestern United States.

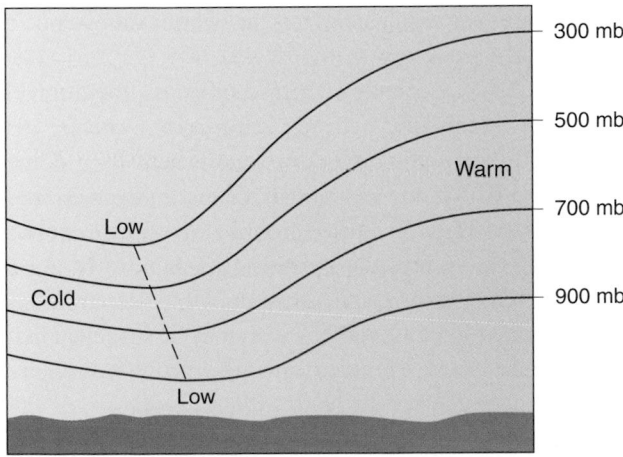

FIGURE 10.20
West-to-east oriented vertical cross-section through a non-occluded wave cyclone from northwest (cold) to southeast (warm) showing isobaric (constant pressure) surfaces. The vertical scale is greatly exaggerated.

The hot surface heats the overlying air, lowering the density of the air column over an area broad enough for a synoptic-scale low to appear on a surface weather map. This **warm-core cyclone** (or *thermal low*) is very shallow, and its circulation weakens rapidly with altitude, that is, away from the immediate heat source (the ground). The surface counterclockwise circulation frequently reverses at some altitude, and an anticyclone overlies the thermal low at middle and upper levels of the troposphere. Few clouds and little precipitation develop near the center of the thermal low that often overlies the lower Colorado River Valley in summer. To the east, a flow of moist air on southeast winds produces afternoon showers and thunderstorms over the mountains of Arizona and New Mexico as part of the Southwest Monsoon (Chapter 9). In addition, tropical cyclones (e.g., tropical storm, hurricanes) are warm-core lows that have no associated fronts (Chapter 12).

In this section, we have focused on the life cycle, principal tracks, and general weather conditions associated with extratropical cyclones. In the daily march of weather in middle latitudes, anticyclones follow cyclones.

Anticyclones

As noted earlier, the circulation in an extratropical cyclone favors convergence of contrasting air masses and development of fronts, clouds, and precipitation. In an **anticyclone**, by contrast, subsiding air and diverging surface winds favor formation of a uniform air mass, no fronts, and generally fair skies (Figure 10.21). Air modifies as it moves away from a center of high pressure so that a semi-permanent anticyclone can be the source of different types of air masses. Like cyclones, anticyclones have either cold or warm cores.

ARCTIC AND POLAR HIGHS

A **cold-core anticyclone** is actually a dome of continental polar (*cP*) or Arctic (*A*) air and, depending on the specific type of air mass, is labeled either a **polar high** or an **Arctic high**. Cold anticyclones are products of extreme radiational cooling over the often snow-covered continental interior of North America well north of the polar front. They are shallow systems in which the clockwise circulation weakens with altitude and frequently reverses direction aloft. Hence, a cold trough typically overlies a cold anticyclone.

Cold-core anticyclones exert the highest surface pressures in winter, when the associated air mass is coldest and air is most dense. The air is extremely stable

FIGURE 10.21
A slow-moving anticyclone is responsible for bright sunny skies with just a few scattered clouds and light winds.

in these systems, and soundings indicate a temperature inversion in the lowest kilometer or two, associated with strong subsidence and adiabatic heating above the inversion. Massive Arctic or polar anticyclones with very high central pressures tend to remain stationary over their source region. The *Siberian high*, for example, is centered near Lake Baikal and from late November to early March its average sea level pressure is greater than 1030 mb. Lobes of cold air (smaller cold highs) often break away from Arctic or polar highs. In North America, cold air masses move out of source regions in Alaska and northwest Canada and slide southeasterly across the Prairie Provinces of southern Canada into the United States east of the Rockies. Cold anticyclones interact with the circulation of extratropical cyclones, helping to maintain and strengthen the temperature contrast across the cyclone's cold front. Hence, clearing skies and sharply lower temperatures usually follow winter storms.

On some occasions in winter, a particularly strong Arctic high brings a surge of bitterly cold air that sweeps as far south as Florida, and rarely can even traverse the Gulf of Mexico into Central America. The resulting subfreezing temperatures can spell disaster for citrus growers and other agricultural interests. In the 1980s, for example, three exceptional cold waves invaded the Florida peninsula. For three days, beginning on Christmas Day 1983, an Arctic air mass gripped Florida, dropping temperatures well below freezing through much of the state. Citrus trees covering 93,000 hectares (230,000 acres) were either killed or damaged for a total loss in excess of $1 billion. Again, on 21 January 1985, a cold wave of even greater severity than the one in 1983 further damaged surviving citrus trees. Subfreezing temperatures were reported as far south as Miami on the morning of 22 January. Losses

from this double blow reduced citrus-producing acreage by almost 90% in Lake County, formerly Florida's second largest citrus-producing county. Some growers replanted, and the groves were beginning to recover when yet another deep freeze over Christmas weekend 1989 ruined much of central Florida's citrus crop.

WARM HIGHS

A **warm-core anticyclone** forms south of the polar front and consists of extensive areas of subsiding warm, dry air. Like cold-core cyclones, warm-core anticyclones strengthen with altitude. They are massive systems with a circulation extending from Earth's surface up to the tropical tropopause. The semipermanent subtropical anticyclones, such as the *Bermuda-Azores high*, are examples of warm-core systems (Chapter 9). While extensions of these subtropical highs may stretch across North America, other warm-core anticyclones may develop over the interior of North America, especially in summer.

A cold high coincides with a shallow mass of cold, dense air and produces high surface air pressures compared to surrounding areas. How does a warm anticyclone produce high surface pressures? Whereas the central surface air pressure may be similar in both cold and warm anticyclones, an important difference is found in their vertical structure and the density contrast between warm and cold air. The column of cold dense air in a cold-core anticyclone is much shallower than the column of less dense warm air in a warm-core anticyclone. The total mass of air over the center of a warm-core anticyclone, related to a higher tropopause, compared to the surrounding atmosphere, is responsible for the warm anticyclone's high surface pressure.

Cold-core anticyclones modify as they travel and may eventually become warm-core systems. As noted earlier, a cold-core anticyclone is actually a dome of relatively cold air, and as that air mass traverses land that is snow free, it is heated from below and moderates considerably, its pressure decreasing significantly. As a cold-core High drifts southeastward over the coterminous United States, air mass modification may be sufficient that the pressure system eventually merges with the warm-core subtropical High over the North Atlantic Ocean.

ANTICYCLONE WEATHER

An anticyclone is a fair-weather system because diverging surface winds induce subsidence of air over a broad area. Because subsiding air is compressionally warmed, the relative humidity drops and clouds usually dissipate or fail to develop. Although anticyclones are fair-weather systems, they can produce weather and climate extremes. The lowest air temperatures of the winter are usually associated with an Arctic high, whereas drought and excessive heat in summer are associated with a warm anticyclone that stalls. Sinking air also becomes more stable and a temperature inversion forms as we saw with the trade wind inversion (Chapter 9). Greater stability along with weak winds suppresses mixing, leading to reduced air quality under a slow moving or stalled anticyclone.

The horizontal air pressure gradient is weak over a broad region about the center of an anticyclone so that prevailing winds are light or the air is calm. Clear skies and light winds favor intense radiational cooling, perhaps resulting in dew, frost, or fog as nighttime air temperatures fall to the dewpoint or frost point. Away from the broad central region of an anticyclone, the horizontal air pressure gradient strengthens and so does the wind. With stronger winds, significant advection can occur. Typically, well to the east of the center of a Northern Hemisphere high, northwest winds advect cold air southeastward, whereas to the west of the high center, southeast winds advect warm air northwestward. Air mass advection helps to increase the temperature contrast across the low pressure trough that separates highs, thereby favoring *frontogenesis*.

An understanding of the circulation in an anticyclone enables us to anticipate the sequence of weather as a high moves into and out of a middle latitude location. Consider what happens in winter as a cold anticyclone slides southeastward out of southern Canada and into the northeastern United States. Depending on the anticyclone's forward speed, it may take several days to a week to play out.

Ahead of the anticyclone, strong northwest winds bring a surge of cold continental polar or Arctic air. Strong winds and falling temperatures produce low wind-chill temperatures. To the lee of the Great Lakes, heavy lake-effect snow showers break out (Chapter 8). Even hundreds of kilometers downwind of the lakes, instability showers bring light accumulations of snow (e.g., over the hills of West Virginia and western Pennsylvania). However, as the center of the anticyclone drifts closer, winds slacken, skies clear, and radiational cooling produces very low surface temperatures at night. Under these conditions, air temperatures dip to their lowest readings, especially where the ground is snow-covered. Then, as the anticyclone drifts away toward the southeast, winds again strengthen, but this time blowing from the south, and warm air advection begins. The first sign of warm air advection is the appearance of high, thin cirrus clouds in the western sky.

At times in summer, a Canadian high-pressure system causes the same advection patterns as in winter except that the temperature contrast between air masses is considerably less. Air advected ahead of the high on northwesterly winds may be not much cooler than the air advected behind the high on southerly winds because of the extensive sunlit daytime hours. Often, the most noticeable difference between northerly and southerly winds at middle latitudes in summer is a contrast in humidity. Air advected ahead of the high is often less humid with lower dewpoints, and therefore is more comfortable than the air advected behind the high.

At times in summer, however, an anticyclone becomes established east of the Rocky Mountains and may stall there for weeks. Such anticyclones are warm-core systems with a deep circulation that is not readily displaced. Aloft, over the high is a warm ridge that may become cut-off from the prevailing westerly flow. With a *block* in place (Chapter 9), subsiding air associated with the anticyclone suppresses cloudiness and precipitation. Persistence of this pattern causes unusually high temperatures and drought-producing conditions.

The pattern of air mass advection in an anticyclone also applies to an upper-level ridge. Cold air advection usually occurs ahead (to the east) of a ridge, while warm air advection occurs behind (to the west of) a ridge. The circulation in an anticyclone (or ridge) does not occur in isolation from the circulation in a cyclone (or trough). The atmosphere is a continuous fluid, with anticyclones following cyclones and cyclones following anticyclones. Northwest winds develop ahead of the high and on the back (west) side of a retreating low. Winds are caused by horizontal air pressure gradients that develop between migrating anticyclones and cyclones.

Local and Regional Circulation Systems

Air masses, fronts, cyclones, and anticyclones are synoptic-scale systems that dominate the weather outside of the tropics. Many other weather systems operating at smaller spatial and temporal scales also contribute to the mix of weather and climate of middle and high latitudes. In some cases, synoptic-scale systems set the boundary conditions that make smaller scale circulation systems possible. In this section, we open our discussion of mesoscale and microscale systems with sea (or lake) and land breezes, mountain and valley breezes, chinook winds, Santa Ana winds, katabatic winds, and desert winds.

SEA (OR LAKE) BREEZE AND LAND BREEZE

For people living near the ocean or a large lake, sea or lake breezes bring welcome respite from the oppressive heat of a summer afternoon. On warm days, if the synoptic-scale air pressure gradient is relatively weak, a cool wind sweeps inland from over the sea or large lake. Depending on the source, this refreshing wind is called either a **sea breeze** or a **lake breeze**. Both breezes owe their existence to differential heating of land and water surfaces.

When land and water are exposed to the same intensity of solar radiation, the land surface warms more than the water surface (Chapter 4). The relatively warm land heats the overlying air, thereby lowering air density. Compared to the land, the water surface is relatively cool, as is the air overlying the water. A local horizontal air pressure gradient develops between land and water, with the higher pressure over the water surface and lower pressure over the land (Figure 10.22A). In response to this horizontal air pressure gradient, cool air sweeps inland. Aloft, continuity requires a return flow of air directed from land to water, with air rising over the relatively warm land and sinking over the relatively cool water.

Sea (or lake) breezes are shallow circulation systems, generally confined to the lowest 1000 m (3300 ft) of the troposphere. Typically, the breeze begins near the shoreline several hours after sunrise and gradually expands both inland and out over the body of water, attaining maximum strength by middle afternoon. The inland extent of the breeze varies from only a few hundred meters to tens of kilometers, depending in part on local topography.

Near sunset, the sea (or lake) breeze dies down. By late evening, surface winds begin to blow offshore as a **land breeze** due to a reversal in heat differential between land and water. At night, radiational cooling chills the land surface more than the water surface. Air over the land surface thus becomes cooler (denser) than air over the water surface. A horizontal gradient in air density gives rise to a horizontal air pressure gradient with higher pressure over the land and lower pressure over the sea (or lake). A cool offshore breeze develops, along with a return airflow aloft; air sinks over the relatively cool land and rises over the relatively warm water (Figure 10.22B). A land breeze attains maximum strength around sunrise but is generally weaker than a sea (or lake) breeze.

Earth's rotation typically does not significantly influence the direction of sea (or lake) and land breezes because the duration of these wind regimes is too brief and distances too small. In some places, however, the Coriolis

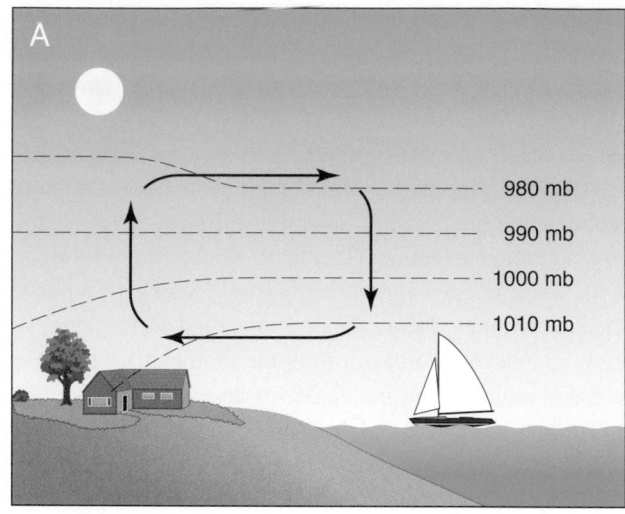

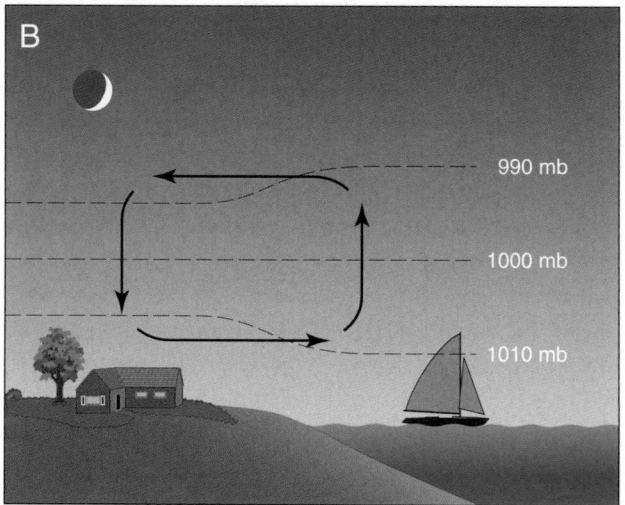

FIGURE 10.22
Vertical cross-sections showing isobaric (constant pressure) surfaces and circulation in a (A) sea (or lake) breeze and (B) land breeze. A sea (or lake) breeze blows onshore during the day whereas a land breeze blows offshore at night.

Effect is responsible for a gradual shift in the direction of a sea breeze through the course of a day. Furthermore, in some localities, uplift produced along a sea-breeze front spurs the development of convective clouds and perhaps thunderstorms (Chapter 11).

MOUNTAIN BREEZE AND VALLEY BREEZE

In summer, a localized circulation system may develop in wide, deep mountain valleys that are sunlit during the day (Figure 10.23). After the winter snows have melted, bare valley walls strongly absorb solar radiation and sensible heating raises the temperature of air in contact with the valley walls. Air adjacent to the

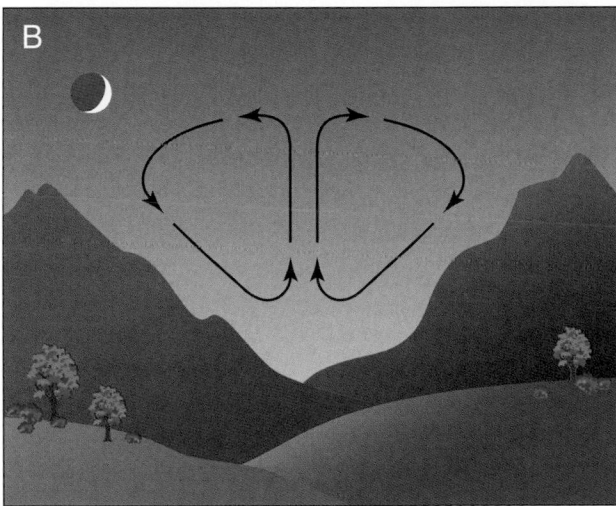

FIGURE 10.23
Schematic representation of the circulation in (A) valley and (B) mountain breezes.

valley wall becomes warmer and less dense than air at the same altitude out over the valley floor. The cooler, denser air over the valley sinks as air adjacent to the valley walls blows upslope as a **valley breeze**. The ascending valley breeze expands and cools and may trigger development of cumulus clouds near the summit.

A valley breeze is best developed between late morning and sunset. By midnight, the circulation reverses direction and persists until about sunrise. Under clear skies, nocturnal radiational cooling chills the valley walls and the air in contact with the walls also cools. Now air adjacent to the valley walls is colder and denser than the air at the same altitude above the valley floor. Air over the valley

is forced to ascend as the cold, gusty **mountain breeze** blows downslope. Cold, dense air accumulates in the valley bottom where additional radiational cooling may lead to formation of fog or low stratus clouds.

Mountain and valley breezes are most common during fair weather and when synoptic-scale winds are light. Hence, these localized winds typically occur when mountainous regions are under the influence of a slow-moving anticyclone.

CHINOOK WIND

A **chinook wind** is a relatively warm and dry downslope wind usually blowing from the west or southwest. It develops when air aloft (in the mid-troposphere above ridge tops) is adiabatically compressed as it descends the leeward (eastern) slopes of the Rocky Mountains in the U.S. and Canada. (The name comes from the Chinook Indians who lived along the lower Columbia River valley and coastal Washington and Oregon.) For every 1000 m of descent, the air temperature rises 9.8 Celsius degrees (the *dry adiabatic lapse rate*) so that air flowing down the slopes of high mountain ranges can undergo considerable warming. As this unsaturated air warms, its relative humidity decreases. During a chinook, mountain-wave clouds form the *chinook arch* along the north-south crest and eastern slopes of the Rockies in Montana and Alberta. The chinook arch corresponds to the location where clouds that formed on the windward slopes vaporize on the leeward slopes.

Typically, a chinook develops when strong synoptic-scale winds force a layer of stable air in the lower troposphere to ascend the windward slopes of a mountain range. When the air layer crosses over the ridge crest to the leeward slopes, its stability causes it to descend to its original altitude. The larger-scale circulation causes further descent of the air. For example, chinook winds may be directed down the leeward slopes of the Rocky Mountain Front Range by strong west winds associated with cyclones and anticyclones centered over the Great Plains well east of the mountains. In another case, downslope winds may be associated with the circulation produced by a large High centered over the Great Basin (in Utah) and a Low tracking across the Canadian Rockies.

At the onset of a chinook, surface air temperatures often climb abruptly tens of degrees in response to the arrival of compressionally warmed air. On 6 January 1966, at Pincher Creek, Alberta, a chinook sent the temperature soaring 21 Celsius degrees (38 Fahrenheit degrees) in only 4 minutes! The sudden spring-like warmth may just as quickly give way to bitter cold as the chinook winds

FIGURE 10.24
Boulder, CO, situated in the foothills of the Rockies to its west, experiences some particularly strong and destructive downslope winds. [Photo by Hustvedt/License: Creative Commons Attribution ShareAlike 3.0]

abate. For example, along the foot of the Rockies, a shift of synoptic-scale winds from west to north brings an abrupt end to the chinook, the return of polar or Arctic air, and plunging temperatures.

According to Native American tradition, chinook means *snow eater*, referring to the catastrophic effect of the warm, dry wind on a snow cover. As noted in Chapter 6, air ascending the windward slopes of a mountain range loses much of its water vapor to condensation and deposition (cloud formation) and precipitation. As air descends the leeward slopes and is compressionally warmed, the relative humidity drops dramatically. Because the chinook is both warm and very dry, a snow cover sublimates and melts rapidly. It is not unusual for a foot of snow to disappear in this way in only a few hours.

Chinook winds are strong and gusty and, locally, may reach destructive speeds, especially along the foothills of the Front Range of the Rocky Mountains. At Boulder, CO (Figure 10.24), in the foothills just northwest of Denver, violent downslope winds, sometimes gusting to 160 km per hr (100 mph) or higher, unroof buildings and topple power poles. On average, the community sustains about $1 million in property damage each year because of these destructive winds.

Researchers do not agree on the precise mechanism that triggers destructive chinook winds, but according to a prevalent view, strong downslope winds are linked to the disturbance of the planetary-scale westerlies as they pass over the Rockies. Recall from Chapter 7 that a prominent north-south oriented mountain range such as the Rockies deflects a westerly air flow into a *standing wave*, a stationary pattern of crests and troughs extending downwind of the mountain range that gives rise to *mountain-wave clouds*. A chinook is actually a segment of the wave that dips down the leeward slopes. Apparently, in a violent chinook, very energetic turbulent eddies that are generated aloft are transported downward into the foothills so that surface winds become very strong and gusty.

Chinook-type winds are not restricted to the leeward slopes of the Rockies. A more general name for a dry compressionally warmed downslope wind is *foehn* (or *föhn*). Such winds blow into the Alpine valleys of Austria, Germany, and Switzerland. The same type of wind is drawn down the leeward slopes of the Argentine Andes, where it is known as the *zonda*.

SANTA ANA WIND

The **Santa Ana** is a hot, dry wind usually blowing from the northeast or east that impacts portions of southern California from fall into early winter (October-March). Named for the Santa Ana Canyon or the Santa Ana Mountains southeast of Los Angeles, the wind originates in the Great Basin and the high Mojave Desert where the air is cool and very dry. Gravity initiates downslope flow of this relatively dense air. In addition, a strong air pressure gradient associated with a high pressure system over the Great Basin drives winds toward the southwest from the desert plateaus of Utah and Nevada, around the Sierra Nevada Mountains, and as far west as coastal southern California (Figure 10.25). Adiabatic compression produces hot, dry winds. Along the southern California coast, the year's highest temperatures usually occur during a Santa Ana episode. Winds are also strong and gusty, accelerating as air flow is constricted by river valleys and mountain passes (Figure 10.26). Santa Ana winds sometimes gust to 130 to 145 km per hr (80 to 90 mph) and almost always cause some property damage. During 30 November-1 December 2011, the most powerful Santa Ana winds in a decade peaked at 97 km per hr (60 mph) in Los Angeles and up to 225 km per hr (140 mph) in the nearby mountains. Trees were knocked down, trucks overturned, and hundreds of thousands of customers were without electric power.

Santa Ana winds desiccate vegetation and contribute to outbreaks of forest and brush fires. For more on the Santa Ana wind and the wildfire hazard in southern California, refer to this chapter's third *For Further Exploration*, Santa Ana Winds and California Fire Weather.

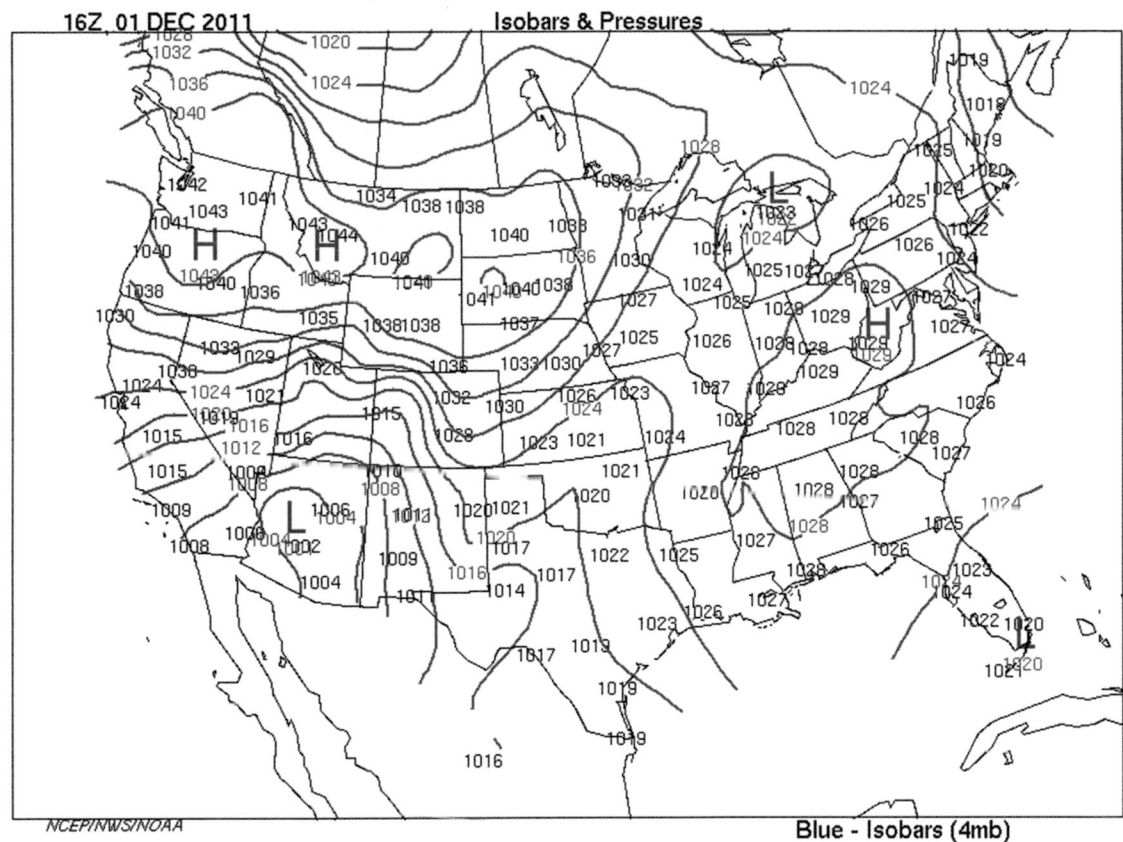

FIGURE 10.25
The surface weather pattern on 1 December 2011 that developed the most powerful Santa Ana winds in a decade over southern California. Solid lines are isobars. The strong air pressure gradient between high pressure on the northern portion of the Great Basin and low pressure over Arizona drove the Santa Ana winds.

FIGURE 10.26
Typical paths followed by Santa Ana winds in southern California. Winds converge and accelerate in mountain passes and river valleys. Adiabatic compression of the air as it descends produces hot, dry winds. [Background physiographic map from NOAA National Geophysical Data Center]

KATABATIC WIND

A shallow layer of cold, dense air flows downhill under the influence of gravity. This **katabatic wind** usually originates in winter over an extensive snow-covered plateau or other highland. Although adiabatic compression warms the descending air to some extent, the air is so cold to begin with and blows over a snow- and ice-covered surface that katabatic winds are still very cold when they reach the lowlands. Among the best known katabatic winds are the mistral and bora occurring in winter. The **mistral** descends from the snow-capped Alps down the Rhone River Valley of France and into the Gulf of Lyons along the Mediterranean coast. The **bora** originates in the high plateau region of Croatia and cascades onto the narrow Dalmatian coastal plain along the Adriatic Sea. The most extreme katabatic winds blow off the glacial plateaus of Greenland and Antarctica.

Katabatic winds typically are weak with speeds averaging less than 10 km per hr (6 mph). But in some places, such as inlets of the coastal mountain ranges of

British Columbia and Alaska, katabatic winds are forced to blow through narrow valleys, and this constricted flow sometimes accelerates the wind to potentially destructive speeds. Steep slopes also accelerate katabatic flow. Along the edge of the Greenland and Antarctic ice sheets, for example, katabatic winds frequently top 100 km per hr (62 mph). The bora also may produce gusts of 50 to 100 km per hr (31 to 62 mph).

DESERT WINDS

Deserts are windy places primarily because of intense solar heating of the ground. In deserts, most of the absorbed radiation goes into sensible heating because water is scarce, vegetation is sparse, and relatively little heat is used for evapotranspiration. In some spots, the midday temperature of the ground surface can exceed 55 °C (131 °F). Such a hot surface generates a *superadiabatic lapse rate* in the lowest air layer (that is, a lapse rate greater than 9.8 Celsius degrees per 1000 m). A superadiabatic lapse rate means great instability, vigorous convection, and gusty surface winds. But with negligible water vapor, few clouds form. The strength and gustiness of the wind vary with the intensity of solar radiation so that wind speeds and gusts usually peak in the early afternoon and during the warmest months.

A **dust devil**, a whirling mass of dust-laden air, is a common sight over flat, dry terrain (Figure 10.27). Dust devils develop on sunny days in response to local variations in surface characteristics (i.e., albedo, moisture supply, topography) that give rise to localized hot spots. Air over a hot spot is heated and forced to rise by cooler surface winds that converge toward the hot spot. Shear in the horizontal wind causes the column of rising hot air to spin about a nearly vertical axis. The source of wind shear may be nearby obstacles that disturb the horizontal wind or the overturning of air induced by extreme instability. In the process, dust is lifted off the ground and whirled about, making the system visible. Unlike a tornado, a dust devil is not linked to a cloud.

Dust devils are microscale circulation systems. The most common ones are small whirls less than 1.0 m (about 3 ft) across that typically last less than a minute. On rare occasions, a dust devil exceeds 100 m (330 ft) in diameter and whirls about for 20 minutes or longer. Such intense dust devils may be visible to altitudes topping 900 m (3000 ft), but the invisible portion of the rising air column may reach 4500 m (15,000 ft).

Most dust devils are too weak to cause serious property damage. The larger ones, however, are known to produce winds in excess of 75 km per hr (45 mph) and may

FIGURE 10.27
A large dust devil. [NASA]

cause some damage. According to the National Weather Service (NWS), every year in New Mexico, several large dust devils cause substantial damage to mobile homes, travel trailers, and buildings under construction. In the spring of 1991, a powerful dust devil that fortuitously passed over anemometers at NOAA's NWS Forecast Office at Albuquerque produced a wind gust of 113 km per hr (70 mph).

Recent research revealed that dust devils account for about 25% of the mineral dust conveyed into the atmosphere, some three times the amount of dust lifted by large dust storms that sweep across the deserts perhaps a few times each month. According to studies conducted at the University of Michigan, surprisingly strong electrical fields develop within dust devils and help the system lift dust off the ground. Collisions between dust particles in the turbulent dust devil induce electrical charges that tend to be negative on lighter particles and positive for heavier particles. The resulting separation of charged particles helps vortices lift dust off the ground. Research by Jacquelin Koch, an atmospheric scientist at the University of Michigan, suggests that a large dust devil having a base diameter of 100 m (328 ft) can lift about 15 metric tons of dust during its 30-minute life cycle. The flux of dust into the atmosphere by dust devils may have implications for climate.

Thunderstorms or migrating cyclones produce larger scale winds in deserts. Surface winds associated with these weather systems can give rise to dust storms or sandstorms, the difference between the two hinging on the size range of the loose surface sediments lifted by the wind. Dust consists of very small particles (diameters less than 0.06 mm or 0.002 in.) that can be carried by winds to great altitudes. Sand, typically covering only a small fraction of desert terrain, consists of larger particles (diameters of 0.06 to 2.0 mm or 0.002 to 0.08 in.) that are transported by the wind to a few meters above the ground. Dust storms and sandstorms can be hazardous, abruptly reducing visibility to perhaps only a few meters, contributing to vehicular accidents on highways.

The strong, gusty downdraft of a thunderstorm generates one of the most spectacular dust storms, known as a **haboob** (Figure 10.28). In a desert, rain falling from thunderstorm clouds often evaporates completely in the dry air beneath the cloud base and does not reach the ground. The cooling by evaporation counteracts some of the warming by compression as the air subsides. A thunderstorm downdraft, therefore, exits the cloud base, striking the ground as a surge of cool, gusty air that lifts dust off the ground. The dusty mass rolls along the ground as a huge ominous black cloud that severely restricts visibility. A haboob may be more than 100 km (60 mi) wide and may reach altitudes of several thousand meters. These dust storms are most common in the northern and central Sudan especially near Khartoum. They also occur in the American southwest deserts.

HEAT BURST

An abrupt rise in surface air temperature can occur via mechanisms not involving the arrival of a chinook wind. A **heat burst** can occur if convective rain showers fall from clouds having a high base and relatively dry air below. Just such a situation developed across south-central and southeast Wisconsin on 8 June 2011. A line of rain showers moving through the area generated localized heat bursts, accompanied by winds gusting to 65-90 km per hr (40-55 mph) along with a significant rise in temperature. With a cloud base of 3000 to 3700 m (10,000 to 12,000 ft), much of the rain evaporated prior to reaching the ground. Evaporation of the rain cooled the air below

FIGURE 10.28
A haboob approaching a neighborhood in Phoenix, AZ, on 17 July 2011. [Photo by Joseph Welker]

cloud base and the rain-cooled air accelerated toward the ground but then began warming due to adiabatic compression. Once the compressionally warmed air reached the ground, the air temperature rose significantly. For example, the temperature rose from 76 °F to 92 °F at Milton, from 78°F to 89 °F at Whitewater, and 75 °F to 82 °F at Sullivan, WI. Heat bursts are more common over the Great Plains where cloud bases tend to be higher and temperatures have been known to rise from the 70s and 80s °F to the 100 to 105 °F range.

Conclusions

We have now examined the features of the principal weather makers of middle latitudes: air masses, fronts, cyclones, and anticyclones. We have seen how these synoptic-scale weather systems interact with one another and how they are linked to the planetary-scale circulation described in Chapter 9. Also recall from our discussion in Chapter 4 that synoptic-scale weather systems play an important role in poleward heat transport.

We also covered the characteristics of selected local and regional weather systems. Domination of smaller scale weather systems by the larger scale atmospheric circulation is apparent; that is, the planetary- and synoptic-scale patterns set boundary conditions for any smaller scale circulation system. In some cases, synoptic-scale winds reinforce mesoscale winds, as when regional winds blow in the same direction as a sea breeze. In other cases, synoptic-scale winds overwhelm mesoscale winds, as when northerly winds sweep Arctic air along the eastern edge of the Rocky Mountains eliminating the possibility of chinook winds. We continue our examination of weather systems in the next chapter with a look at mesoscale and microscale systems, including thunderstorms and tornadoes.

Basic Understandings

- Air masses are classified on the basis of temperature and humidity, characteristics that are acquired in their source region. The four basic air mass types are cold and dry, cold and humid, warm and dry, and warm and humid.
- Air masses form over land (continental) and ocean (maritime), and at high latitudes (polar) and low latitudes (tropical). Arctic air is distinguished from polar air by its bitter cold.
- Air masses modify as they travel from one place to another. The degree of modification depends on air mass stability and the nature of the surface over which the air travels, specifically whether the surface is warmer or colder than the air mass.

- A front is a narrow zone of transition between air masses that contrast in temperature and/or humidity. The four types of fronts are:
  1. stationary,
  2. warm,
  3. cold, and
  4. occluded.

- A stationary front exhibits essentially no lateral movement; it transitions to a cold front if it begins moving such that colder (more dense) air displaces warmer (less dense) air. A stationary front transitions to a warm front if it begins moving such that the warm (less dense) air advances while the cold (more dense) air retreats.

- An occluded front forms late in the life cycle of a wave cyclone, as the system moves into colder air. The cold front sweeps around the center of the Low, forming the occluded front as it overtakes and merges with the warm front, forcing the warm air aloft. The distinction between cold and warm occluded fronts depends on the temperature (density) contrast between the air behind the cold front and the air ahead of the warm front.

- Weather associated with stationary or warm fronts typically consists of a broad stratiform cloud and precipitation shield that may extend hundreds of kilometers on the cold side of the surface front. Weather along or ahead of a cold front usually consists of a narrow band of cumuliform clouds and brief rain or snow showers, or thunderstorms.

- As an extratropical cyclone progresses through its life cycle, it is supported and steered by the upper-level circulation toward the east and northeast. An extratropical cyclone typically begins as a wave along the polar front and deepens as the surface air pressure continues to drop. Winds strengthen and frontal weather develops. The cyclone finally occludes as the faster-moving cold front plows into the slower-moving warm front and the upper-level trough becomes vertically stacked over the surface cyclone.

- The life cycle of an extratropical cyclone can vary significantly in duration. A typical cyclone requires about one day in each of the stages leading up to maximum intensity. However, when conditions are optimal, a cyclone can progress from birth to peak intensity in 36 hrs or less. In some cases, the cyclone never attains occlusion, instead spending most of its life span as a weak wave cyclone rippling along a stationary front.

- The conveyor belt model represents the three-dimensional structure of a mature extratropical cyclone in terms of three air streams: warm and humid, cold, and dry.

- The track followed by a cyclone is key to the type of weather experienced at a given locality. At locations on the cold side of the storm track, winds back with time (turn counterclockwise) and there are no frontal passages. At places on the warm side of the storm track, winds veer (turn clockwise) with time and a cold front follows a warm front.

- Across eastern North America, Alberta-track cyclones are most frequent, but Colorado and coastal storm tracks are responsible for heavier precipitation, especially in winter.

- A warm-core cyclone (thermal low) is stationary, has no fronts, and is associated with hot dry weather. These relatively shallow systems typically develop in the American southwest desert.

- A cold-core anticyclone is a shallow system that coincides with a dome of continental polar or Arctic air. Cold air advection occurs ahead of the pressure center of an advancing High and warm air advection occurs behind the high pressure center.

- A warm-core anticyclone, such as a semipermanent subtropical high, extends from Earth's surface to the tropopause, and produces a broad area of dry air warmed by compression associated with subsidence.

- When synoptic-scale winds are weak, localized horizontal air pressure gradients can develop above neighboring land and sea (or lake) surfaces. In response, winds blow onshore during the day as a sea or lake breeze, and offshore at night as a land breeze.

- Mountain and valley breezes develop in summer in deep, wide sunlit mountain valleys. A valley breeze is an upslope wind that forms during the day, and a mountain breeze is a downslope wind that forms at night.

- A chinook wind consists of compressionally warmed, stable air that is forced down the leeward slopes of a mountain range by the circulation about cyclones or anticyclones.

- A shallow layer of cold, dense air flows downhill under the influence of gravity. This katabatic wind usually originates in winter over an extensive snow-covered plateau or other highland.

- Intense solar heating of desert terrain produces a steep temperature lapse rate in the lowest air layer. Dust devils may develop in such unstable air. In addition, strong winds associated with thunderstorms or migrating cyclones may cause dust storms or sandstorms.

Enduring Ideas

- Stationary, warm, cold, and occluded fronts are narrow zones of transition between contrasting air masses. Properties that define a front include marked differences in temperature and/or humidity, wind shift, convergence, and a trough in the pressure pattern over short distances. Clouds and precipitation can develop along fronts when there is a significant density contrast between air masses and an adequate supply of water vapor.
- The Norwegian cyclone model describes an extratropical (cold-core) cyclone's progression from a wave along the polar front to an occluded system. The conveyor belt model illustrates air flow within a mature cyclone. Weather is typically much different on the warm and cold sides of an extratropical cyclone. A cyclone center is generally steered by winds near the 500-mb level.
- A continental polar or Arctic air mass is typically associated with a cold-core anticyclone, a shallow system that weakens with height. By contrast, a warm-core anticyclone's circulation strengthens with height. The semipermanent subtropical highs are examples of warm-core anticyclones.
- Weather systems operating at smaller spatial and temporal scales than synoptic-scale cyclones and anticyclones also contribute to the weather and climate of middle and high latitudes. Included in these mesoscale and microscale systems are sea (or lake) and land breezes, chinook winds, desert winds, mountain and valley breezes, and katabatic winds.

Key Terms

air mass
continental tropical air (*cT*)
maritime tropical air (*mT*)
maritime polar air (*mP*)
continental polar air (*cP*)
Arctic air (*A*)
air mass modification
Pacific air
front
stationary front
overrunning
warm front
frontal fog
cold front
squall line
back-door cold front
occluded front

frontogenesis
frontolysis
extratropical cyclone
Norwegian cyclone model
cyclogenesis
comma cloud
dry slot
triple point
cyclolysis
conveyor belt model
winter storm
blizzard
cold-core cyclone
warm-core cyclone
anticyclone
cold-core anticyclone
polar high

Arctic high
warm-core anticyclone
sea breeze
lake breeze
land breeze
valley breeze
mountain breeze
chinook wind
Santa Ana wind
katabatic wind
mistral
bora
dust devil
haboob
heat burst

Review

1. The average temperature contrast between winter and summer is greater for continental polar air than for maritime tropical air. Explain why.
2. Identify the various processes that could be involved in the modification of air masses.
3. Over the Great Plains and Midwest, Pacific air is relatively mild and dry. How does Pacific air acquire these characteristics if it originated over the Pacific Ocean?
4. The cloud and precipitation pattern associated with a cold front is usually narrower than the cloud and precipitation pattern associated with a warm front. Explain why.
5. Under what conditions in the upper troposphere is cyclogenesis most likely to take place?
6. Describe the general weather conditions in the southeast sector of a typical mature extratropical cyclone.
7. Along the path of a mature extratropical cyclone, where is snowfall most likely?
8. What is the relationship between an Arctic High and an Arctic air mass?
9. Why is a Chinook wind both warm and dry?
10. What is the principal force that drives a katabatic wind?

Critical Thinking

1. In winter, continental polar air modifies more rapidly when traveling over bare ground than snow-covered ground. Explain why.
2. How does air mass stability affect the type of weather that occurs along or ahead of a front?
3. Describe the changes that take place in cloud cover as a warm front approaches your locality.
4. In winter, Colorado-track *Lows* typically bring heavier snowfall to the Great Lakes than do Alberta-track *Lows*. Explain why.
5. The leading edge of a sea breeze resembles a miniature cold front. What type of weather might develop along a sea breeze front?
6. Colder weather usually follows in the wake of an intense winter cyclone. Explain why.
7. Why is a surface temperature inversion associated with an Arctic high?
8. What is the significance of a *triple point* located along the coast of North Carolina?
9. Why is a *dry slot* dry?
10. What are some of the differences between a cold-core cyclone and a warm-core cyclone?

The Case of the Missing Storm

A cyclone that sweeps ashore along the west coast of North America seems to disappear from the sea-level pressure pattern as it tracks inland over the mountainous West. A few days later, a low develops to the east of the Rocky Mountain Front Range, typically on the plains of Alberta or eastern Colorado. What actually happened to the storm as it crossed the mountains?

An extratropical cyclone tracks eastward onshore from the Pacific Ocean. Visualize the cyclone as a huge cylinder of air spinning about a vertical axis in a counterclockwise direction (Figure 1). The cylinder has Earth's surface as its lower boundary and the tropopause as its upper boundary. Because the tropopause remains at essentially the same altitude over the region, the column is forced to shrink in height as the cylinder moves up the windward slopes of a mountain range. The mass of air in the cylinder widens as its vertical dimension decreases and the spin of the cylinder slows; that is, the storm's circulation weakens. As the cylinder of air then descends the leeward slopes of the mountain range, it stretches vertically and contracts horizontally, and the cyclonic spin strengthens. Weakening cyclonic circulation upslope followed by strengthening cyclonic circulation downslope account for the seeming disappearance and reappearance of the storm as it traverses mountainous terrain.

The changes in the storm's circulation are analogous to what happens to an ice skater performing a spin. The skater changes her spin rate by extending or drawing in her arms. When she extends her arms (analogous to

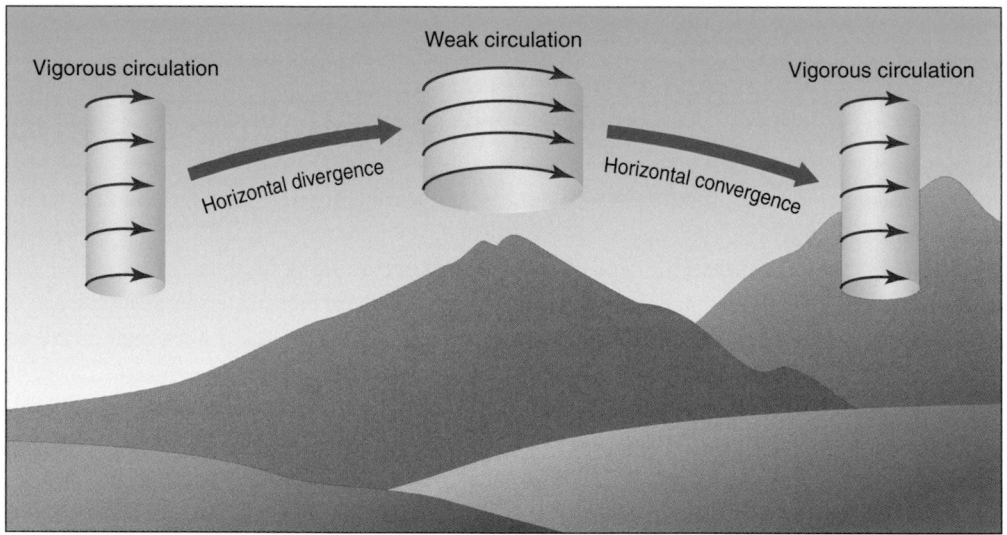

FIGURE 1
The cyclonic spin of a cylinder of air weakens as it ascends the windward slopes of a mountain range and strengthens as it descends the leeward slopes of the mountain range.

horizontal divergence and the widening of our cylinder), her spin rate slows. When the skater brings her arms close to her body (analogous to horizontal convergence and the contracting of our cylinder), she spins faster. Both the spinning skater and the cyclone passing over the mountains conserve angular momentum. Simply put, *conservation of angular momentum* means that a change in the radius of a rotating mass is balanced by a change in its rotational speed. An increase in radius (horizontal divergence) due to shrinking of the vertical column is accompanied by a reduction in rotational rate whereas a decrease in radius (horizontal convergence) due to stretching of the column is accompanied by an increase in rotational rate.

Nor'easters

An intense extratropical cyclone that tracks along the East Coast of North America, a *nor'easter* is named for the direction from which its powerful onshore winds blow (Figure 1). Most frequent from October through April when seasonal contrasts in air mass characteristics (i.e., temperature and humidity) are greatest, these storm systems also track to the northeast. If centered just offshore, they bring strong onshore winds on their northern and northwestern flanks (blowing from the east and northeast) that can cause considerable flooding, coastal erosion, and property damage.

Although rarely attaining the strength of even a weak East Coast hurricane, a nor'easter can pack a powerful punch because it impacts a much greater swath of coastline. While a hurricane impacts only about 100 to 150 km (60 to 90 mi) of coastline, a nor'easter often tracks parallel to the coast with onshore winds that can sweep over more than 1500 km (900 mi) of coastline. Additionally, the diameter of an average nor'easter is about three times that of an average hurricane. Both hurricanes and nor'easters, when either has strong onshore winds and low air pressure, can produce a great dome of seawater that inundates the coast, known as a *storm surge*. In an intense nor'easter, strong winds can build sea waves to heights of 10 m (33 ft) superimposed on a storm surge of up to 5 m (16 ft). Because winds in the slow-moving system blow in the same onshore direction for a longer period, a less intense but slow-moving nor'easter

FIGURE 1
Color-enhanced infrared satellite image of a nor'easter on 29 October 2011. This storm brought early snowfall to many areas of the Northeast. [NASA Goddard Space Flight Center, data from NOAA GOES]

can cause more damage than a more intense, fast-moving nor'easter. Persistent onshore winds increase the likelihood of a destructive storm surge and high waves. These effects are amplified at locations experiencing high tide.

Unlike hurricanes, nor'easters may originate as relatively weak Lows over land, forming along the boundary (a *front*) between air masses that contrast in temperature and humidity. Like hurricanes, they often intensify over warm ocean waters. A low-level temperature contrast between land and sea, abundant moisture (supplied by the Gulf of Mexico or Atlantic Ocean), areas of rising air, and a pool of Arctic air (an *Arctic* high-pressure system) to the north or northwest are essential to the formation and intensity of a nor'easter along or near the coast. The subtropical jet stream over the southeast U.S. and the polar jet stream to the north work in tandem to induce and sustain regions of rising air throughout the troposphere. For a nor'easter to produce copious amounts of snow, the polar front jet stream should be in a north-south oriented "blocking pattern", which not only channels cold air into the coastal storm but also slows the forward speed of the system, thereby favoring its intensification. The heaviest snow usually falls in one or more bands located 160 to 320 km (100 to 200 mi) to the northwest of the storm center. These conditions most often come together during winters characterized by the warm phase of the El Niño-Southern Oscillation and the negative phase of the North Atlantic Oscillation/Arctic Oscillation (Chapter 9).

Energized by latent heat acquired from evaporation of relatively warm ocean water, some nor'easters develop very rapidly. By convention, a rapidly intensifying extratropical cyclone is labeled a *bomb* if its central pressure drops at least 24 mb in 24 hrs. Few cyclones actually meet this criterion and, of those that do, most develop over warm surface ocean currents such as the Gulf Stream or the Kuroshio Current. In early January 1989, an extreme bomb hit the East Coast. With a central air pressure of 996 mb, the incipient cyclone was first identified off Cape Hatteras, NC, at about 7:00 p.m. (EST) on 4 January. Twenty-four hours later, the storm was centered about 700 km (435 mi) south of Newfoundland with a central pressure of 936 mb. The storm had intensified (pressure decreased) by 60 mb in 24 hrs, 2.5 times the criterion for a bomb.

For its widespread impact on society, one of the most noteworthy nor'easters of the 20th century occurred on 12-15 March 1993 when a major storm system tracked from the Gulf of Mexico up the Eastern Seaboard. Although the storm system's size and minimum central pressure (960 mb) were not particularly unusual, its track along the densely populated East Coast had a huge effect on humanity. The storm system drew a tremendous amount of moisture evaporated from the warm ocean waters just off the coast into its circulation while the cold air inland helped the storm produce heavy snow over the highly populated areas to the north and west of its track from Alabama to Maine.

Snow fell along the Gulf Coast with 7.6 cm (3 in.) reported at Mobile, AL, and up to 13 cm (5 in.) measured in the Florida Panhandle, the greatest single snowfall in the state's history. The 33 cm (13 in.) that blanketed Birmingham, AL, not only set a new 24-hr snowfall record for any month, but also set records for maximum snow depth, greatest snow fall in a single storm, and maximum snowfall for a single month. Total storm snowfall was tremendous in the Appalachians, including 142 cm (56 in.) at Mt. LeConte, TN, and 127 cm (50 in.) at Mount Mitchell, NC. Just about every official weather station in West Virginia set a new 24-hr snowfall record with 76 cm (30 in.) reported at Beckley, WV. Further to the north, snow totaled 64 cm (25 in.) at Pittsburgh, PA, 69 cm (27 in.) at Albany, NY, and 109 cm (43 in.) at Syracuse, NY. Winds along the coast gusting in excess of hurricane force whipped heavy snow into huge drifts; Boston, MA, recorded gusts to 131 km per hr (81 mph). For the first time in history, snow forced the closing of every major East Coast airport. Thousands of people were isolated in the Appalachian Mountains, more than 3 million customers were without electricity, hundreds of roofs collapsed under the weight of accumulated snow, and about 200 homes on North Carolina's Outer Banks were severely damaged by winds and flooding.

In addition to property damage caused by heavy snow and high winds, a *squall line* (an elongated group of strong to severe thunderstorms) associated with the storm produced 27 tornadoes in Florida while a 3 m (10 ft) storm surge in the Gulf of Mexico flooded the Apalachicola area. All told, this nor'easter claimed about 270 lives, more than three times the combined death toll of Hurricanes Hugo and Andrew, and caused damage estimated at $7 billion (in 2002 dollars), the costliest extratropical cyclone in U.S. history.

FIGURE 2
MODIS image taken from NASA's Terra Satellite of the mid-Atlantic snowcover on 11 February 2010. [NASA image by Jeff Schmaltz, MODIS Rapid Response Team at NASA GSFC]

While the impact of the March 1993 nor'easter remains unmatched, overall the winter of 2009-2010 was historic for many areas of the mid-Atlantic. Four nor'easters each brought widespread 25-50 cm (10-20 in.) swaths of snow, with much greater amounts locally. Two of the storms arrived back-to-back (4-7 February and 9-11 February), shutting down school systems for over a week and Federal government offices for 4 successive days. Seasonal snowfall records were shattered in Washington, DC, Baltimore, MD, and Philadelphia, PA. Baltimore received a total of 196 cm (77 in.), a far cry from the average seasonal total of 46 cm (18.2 in.). Figure 2 is a Moderate Resolution Imaging Spectroradiometer (MODIS) display of the blanket of snow covering the mid-Atlantic on 11 February 2010.

Focusing on Northeast snowstorms' potential disruption of transportation systems and economic effects on the rest of the nation, in 2006 NOAA introduced a scale to

Category	NESIS Value	Description
1	1 - 2.499	Notable
2	2.5 - 3.99	Significant
3	4 - 5.99	Major
4	6 - 9.99	Crippling
5	10.0+	Extreme

FIGURE 3
The Northeast Snowfall Impact Scale (NESIS). [Image adapted from table at http://www.ncdc.noaa.gov/snow-and-ice/nesis.php]

quickly categorize the impact of major snowstorms (including nor'easters) on the Northeast. The *Northeast Snowfall Impact Scale (NESIS)* (Figure 3) rates snowstorms on the amount of snowfall, the area affected, and the population in the path of the storm, and allows for assessment of snowstorm impact in days instead of weeks. Snowstorms are assigned to one of five categories: 1 (notable), 2 (significant), 3 (major), 4 (crippling), or 5 (extreme), the last for storms producing heavy snowfall over large geographical areas that encompass major metropolitans. For example, the 12-15 March 1993 nor'easter rated 5 on the NESIS (Figure 4), and the three of the winter 2009-2010 nor'easters were each assigned category 3. Development of the scale was the joint effort of Louis Uccellini, director of NOAA's National Centers for Environmental Prediction (NCEP) and Paul Kocin, a meteorologist at NCEP. The NESIS joins the Saffir-Simpson Hurricane Wind Scale (Chapter 12) and the Enhanced Fujita scale for tornadoes (Chapter 11) as tools for better informing the public on weather extremes.

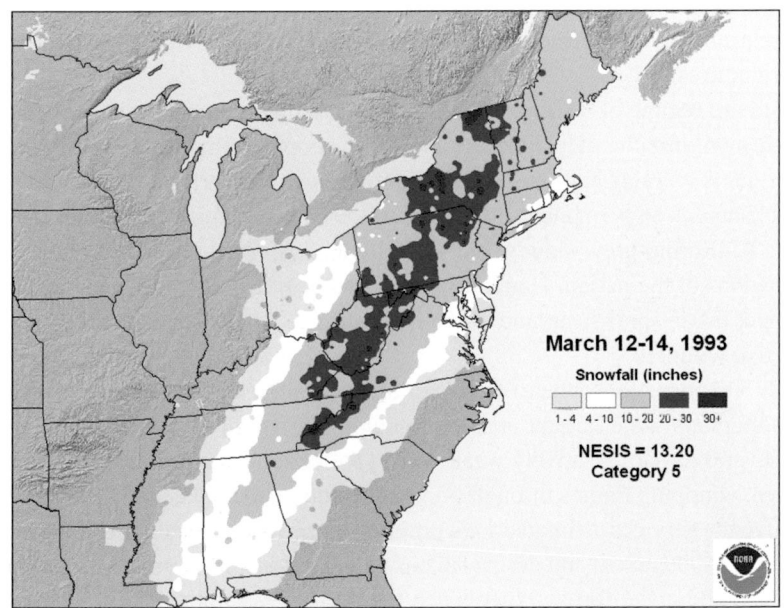

March 12-14, 1993

Snowfall (inches)

1 - 4   4 - 10   10 - 20   20 - 30   30+

NESIS = 13.20
Category 5

FIGURE 4
Gauge-based snowfall totals from the March 1993 nor-easter. This storm was a NESIS Category 5. [NOAA]

For Further Exploration

Santa Ana Winds and California Fire Weather

Every autumn, *Santa Ana winds* descend the mountain slopes of interior southern California and sweep over the coastal plain to the sea. These hot dry winds blow over a landscape parched by the long dry summer. Santa Ana winds further desiccate the vegetation, making southern California particularly vulnerable to wildfires. Also contributing to the fire danger is a shrub community known as *chaparral* (Figure 1) that dominates the hill slopes. The tissues of these plants contain oils that readily ignite; in fact, chaparral shrubs practically explode when exposed to flame. Periodic wildfires, whipped by Santa Ana winds, roar down the canyons of southern California. One of the earliest accounts of such fires was in October 1542 when Juan Rodriguez Cabrillo (ca. 1499-1543), whose ship lay anchored off the Los Angeles Basin, observed hot desert winds blowing and fires burning.

FIGURE 1
Chaparral surround a California lake. [U.S. Department of the Interior/Bureau of Reclamation/Mid-Pacific Region/ Lake Berryessa Recreation Division, Administrative Office]

In southern California, winter rains (from November through April) usually follow the summer dry season. Falling on the nutrient-rich ashes of burned vegetation, winter rains spur resprouting of shrubs. Furthermore, fire followed by rain also triggers germination of seeds that have been dormant in the soil. Renewal of vegetative cover helps re-stabilize hillsides denuded by wildfires. However, if heavy rains fall before vegetation has a chance to stabilize burnt-over slopes, soil and debris are washed from the bare ground and flushed into streambeds. Rainwater also percolates into the loosely consolidated upper soil layer. When this layer becomes saturated, it flows down steep hillsides as a river of mud, covering roadways and inundating homes. These mudflows are the greatest post-fire problem. Furthermore, even a single winter of below-average rainfall significantly elevates the potential for autumnal wildfires.

Native Americans adapted to this climate/fire regime of southern California by building temporary shelters that they could readily move out of harm's way should a wildfire threaten. Furthermore, these people routinely set small fires (*controlled burns*) that improved the habitat for game animals and thus made hunting easier. The arrival of European settlers in southern California marked the establishment of permanent dwellings and efforts to suppress wildfires. In subsequent centuries the population of southern California grew slowly, but since the end of World War II, southern California has been one of the fastest growing regions of the nation. Hundreds of thousands of homes were constructed in the chaparral-covered foothills. A combination of urban sprawl, summer drought, fire-prone vegetation, and autumnal Santa Ana winds set the stage for highly destructive wildfires.

Such a destructive event erupted on 27 October 1993 when fifteen major fires broke out on hillsides from Ventura County south to San Diego County. Firefighters were largely unable to contain the wildfires until Santa Ana winds finally died down three days later. After several days of relatively weak winds, the regional air pressure gradient again strengthened and Santa Ana winds resumed, whipping flames through previously untouched areas of Malibu and Topanga. In the aftermath, state and federal emergency services estimated total property damage at more than $1 billion. Fires scorched more than 80,000 hectares (almost 200,000 acres) and destroyed more than 1200 structures.

About ten years later, in October 2003, southern California experienced the most devastating wildfires in the state's history. Fourteen major fires broke out affecting Ventura, Los Angeles, San Bernardino, and San Diego counties (Figure 2). The fires claimed 24 lives, destroyed 3710 homes, and burned over 303,542 hectares (750,043 acres). Thousands of residents fled their homes and headed for evacuation centers. At the peak of the fire siege, more than 14,000 firefighters from federal, state, and local agencies battled the flames from the ground and air. Once again Santa Ana winds fanned the flames that consumed the considerable fuel (dead, dying, and diseased vegetation) that had accumulated at the wildland/urban interface.

Massive wildfires erupted from Malibu, CA to San Diego County in October and November of 2007. These 23 fires took 10 lives, injured 292 people, burned 211,407 hectares (522,398 acres), and destroyed more than 3290 structures. In southern California, 7 counties were declared major disaster areas.

What can be done to reduce the Santa Ana fire hazard? Nothing can be done to alter the normal sequence of summer drought, autumnal Santa Ana winds, and variable winter rains. People can change where they chose to live, however. Some people argue that zoning ordinances should be enacted and enforced that prohibit construction of houses in locales subject to natural hazards such as wildfires and mudflows. Alternatively, ordinances could require that dwellings be constructed of fireproof materials (particularly roofs and outside walls). Furthermore, chaparral vegetation could be cleared within a designated perimeter of buildings. At present, many people permit shrubby vegetation to grow up to the edge of their homes for privacy and a sense of living with nature.

If wildfire is suppressed, chaparral shrubs grow larger and denser and annually add more leaves and twigs to the ground litter layer. Because of the buildup of fuel, the chaparral, once ignited, burns even more intensely. Periodic setting of controlled burns would reduce the frequency and severity of wildfires. With such a strategy, people can work with and not against the forces of nature.

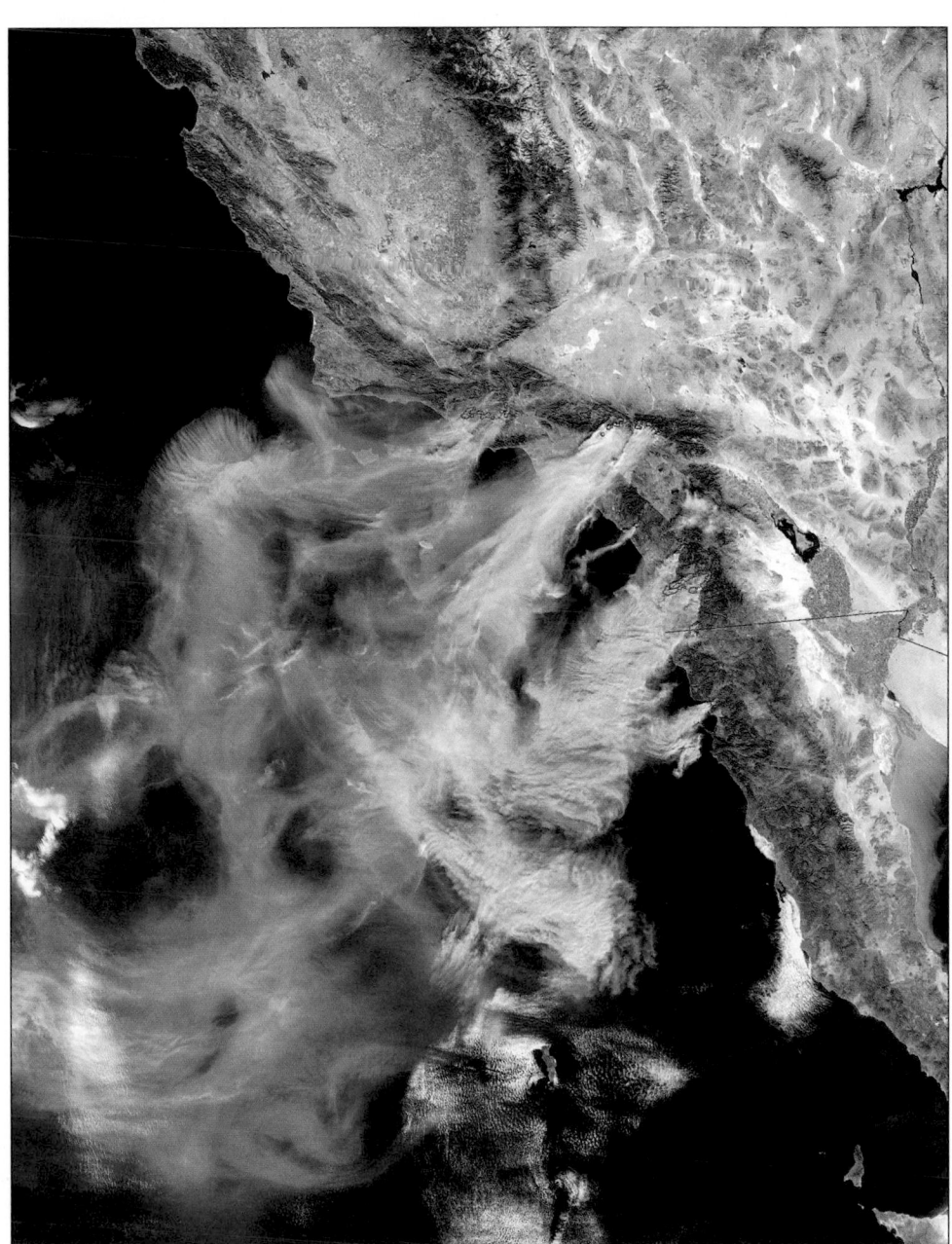

FIGURE 2
Several massive wildfires were raging across southern California over the weekend of October 25, 2003. Red denotes active fire locations at that time. Whipped by the hot, dry Santa Ana winds that blow toward the coast from interior deserts, at least one fire grew 10,000 acres in just 6 hours. The Moderate Resolution Imaging Spectroradiometer (MODIS) on the Terra satellite captured this image of the fires and clouds of smoke spread over the region on October 26, 2003. [NASA]

Twisted guard rail wrapped around a tree next to the Wrangler Distribution Center in Hackleburg, Alabama, which was destroyed by an EF-5 tornado in Marion County on April 27, 2011. [Photo by Christopher Mardorf / FE]

THUNDERSTORMS & TORNADOES

Chapter Highlights

Case-in-Point
  Super Tornado Outbreak of 2011
Thunderstorm Life Cycle
  Towering Cumulus Stage
  Mature Stage
  Dissipating Stage
Thunderstorm Classification
Where and When
Severe Thunderstorms
Thunderstorm Hazards
  Lightning
  Downbursts
  Derechos
  Flash Floods
  Hail
Tornadoes
Tornado Characteristics
Where and When
Tornado Hazards and the EF-Scale
The Tornado-Thunderstorm Connection
Monitoring Tornadic Thunderstorms
Conclusions
Basic Understandings/Enduring Ideas
Key Terms/Review/Critical Thinking
For Further Exploration
  Lightning Safety
  Hail Suppression
  Tornado Look-Alikes

Learning Objectives

Identify the distinguishing characteristics of each of the three stages in the life cycle of a thunderstorm cell.

Explain the role of atmospheric stability in the development of a thunderstorm cell.

Identify the various factors that govern thunderstorm frequency across the coterminous United States.

Explain the thunderstorm frequency maximum in Florida.

List the characteristics of a severe thunderstorm.

Sketch the synoptic-scale weather pattern that favors development of severe thunderstorms.

Distinguish between a microburst and a macroburst.

Explain why a microburst can be hazardous to aviation.

Identify the characteristics of a thunderstorm that is most likely to produce flooding rains.

Explain why urban areas are particularly vulnerable to flash flooding.

Describe the origins and characteristics of hail.

Distinguish between a tornado and a funnel cloud.

Identify the principal force operating in a tornado.

List the characteristics that are responsible for the relatively high frequency of tornadoes in tornado alley.

Explain why severe thunderstorms and tornadoes are most likely to occur in spring.

Describe the linkage between severe thunderstorm cells and tornadoes.

What conditions in the atmosphere favor development of severe convective weather systems?

Case-in-Point

Super Tornado Outbreak of 2011

In the United States, 2011 was a record-breaking year for nature's most violent storms—tornadoes (Table 11.1). With 1690 tornadoes confirmed nationwide, 2011 was second only to 2004 since record keeping began in 1950. The spring of 2011 was the most active of any three-month period (April, May, and June) on record with 1150 confirmed tornadoes. Of them, April had the most tornadoes, including two tornado outbreaks on 14-16 April and 25-28 April.

The most active and destructive day came on 27 April 2011 (Figure 11.1). Detecting atmospheric conditions favorable for an extreme tornado outbreak, the National Centers for Environmental Prediction's *Storm Prediction Center (SPC)* advised residents that the southern U.S. was at a high risk for severe weather. A deep upper level low pressure system was moving through the nation's mid-section. Ahead of the associated cold front, warm humid air with dewpoints in the mid and

TABLE 11.1
2011 U.S. Tornado Data Compiled by SPC*

Record	Data	Ranking
States Reporting Tornadoes	48	1st (tied with 1989)
Total Annual Tornadoes	1690	2nd
Greatest Monthly Total	758	1st
Greatest Daily Total	200 (27 April)	1st
Tornado Deaths	550	4th
Tornado Injuries	5400	2nd
Deadliest Tornado	Joplin, MO (158)	7th
Tornadoes Rated (E)F4-5	22	4th (tied with 1953)
Tornadoes Rated (E)F-5	6	2nd
Estimated Property/Crop Losses	10 billion USD	1st

*Comparisons are made for the period 1950-2011. Annual/single tornado fatality records are extended back to the 1925 (Tri-State Tornado Outbreak).

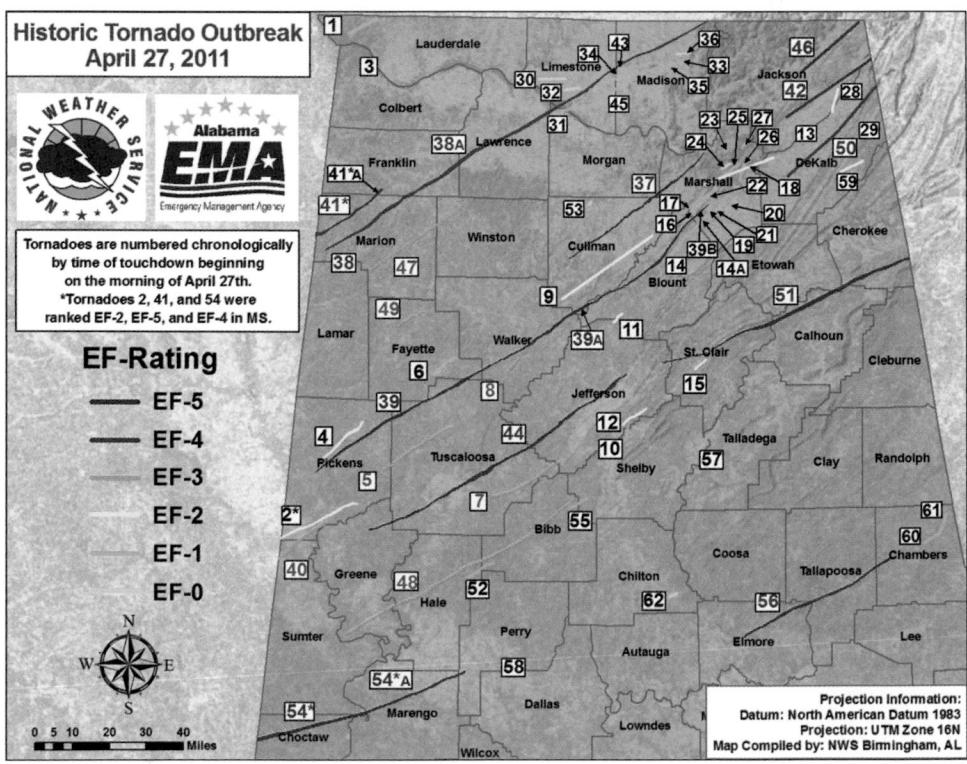

FIGURE 11.1
Tornado damage swaths in Alabama on 27 April 2011. [NOAA/NWS and Alabama EMA]

upper 60s °F flowed northward. As the day progressed, surface winds backed to the south-southeast while winds at the 850-mb level increased to 50-55 knots and became more southerly. This vertical wind shear (winds changing direction and/or speed with altitude) is a key ingredient for development of severe weather. Throughout the day, the atmospheric moisture content increased and wind shear strengthened, making the atmosphere increasingly unstable. Convective currents surged into the upper troposphere, and even into the lower stratosphere. Massive supercell thunderstorms developed explosively and spawned violent tornadoes.

FIGURE 11.2
An EF-4 tornado near Arab, AL, on 27 April 2011. [Photo courtesy of Charles Whisenant, The Arab Tribune]

A total of 200 tornadoes touched down, four of which rated EF-5 (the highest rating on the *Enhanced Fujita Scale*) indicating peak winds around 339 km per hr (210 mph), and eight of which rated EF-4 (Figure 11.2). When one of the EF-5 tornadoes passed through Smithville, MS, it swept newly constructed brick homes from their foundations, debarked trees, and removed sections of an asphalt road. In Rainsville, AL, an EF-5 tornado picked up an 800-lb steel safe anchored to the foundation, ripped the door from its hinges, and threw it 183 m (600 ft).

The tornado that hit Hackleburg, AL, was the first EF-5 to strike Alabama in more than a decade. According to the American Red Cross, 75% of Hackleburg was destroyed. Among the buildings it leveled was the Wrangler Jean Distribution Center (Figure 11.3); residents of a nearby town reported seeing jeans falling from the sky. With the longest track distance of any tornado in the outbreak, over 210 km (130 mi), this one tornado killed 72 people, all in Alabama, which also made it the deadliest tornado ever to strike the state. Later that same day, a large and destructive EF-4 tornado struck Tuscaloosa, AL, killing more than 40 people, before continuing into the suburbs of Birmingham, AL, where it flattened entire neighborhoods.

This day became the deadliest tornado day in the U.S. since the 1925 "Tri-State" tornadoes. More than 300 people lost their lives with 235 of these fatalities in Alabama. In addition, 2000 people were injured. Alabama declared a state of emergency.

All told, in the 25-28 April 2011 outbreak, 342 tornadoes were confirmed by the National Weather Service in 21 states from Texas to New York and 321 people lost their lives. This event is now known as the 2011 Super Outbreak and is the largest tornado outbreak recorded and the fourth deadliest in U.S. history, since modern records began in 1950. This outbreak was also one of the costliest natural disasters in U.S. history, totaling over $10 billion (in 2011 dollars).

FIGURE 11.3
Debris removal at the Wrangler Distribution Center in Hackleburg, AL on 19 July 2011. [Photo by Christopher Mardorf/FEMA]

Most of us are familiar with typical thunderstorm weather—the blackening sky and abrupt freshening of wind, followed by bursts of torrential rain, flashes of lightning, and rumbles of thunder. Often the storm's cool breezes and rains bring welcome relief on a hot, muggy summer afternoon. For a farmer whose crops are wilting under the searing summer Sun, the rains may be an economic lifesaver. Some thunderstorms become violent, however, and wreak havoc. Lightning starts fires, strong winds uproot trees and damage buildings, torrential rains cause flash flooding, and some thunderstorms produce destructive hail or tornadoes.

Although less than 1% of all thunderstorms spawn tornadoes, the possibility of a tornado is the principal reason why many people fear thunderstorms. Tornadoes are the most intense of all weather systems. They are associated with certain severe thunderstorms and can take lives and cause many injuries and considerable property damage. This chapter covers thunderstorms and tornadoes, their characteristics, life cycles, geographical and seasonal distribution, and associated hazards.

Thunderstorm Life Cycle

A **thunderstorm** is a storm accompanied by lightning and thunder. It affects a relatively small area and is short-lived. A thunderstorm is the product of vigorous convection that extends high into the troposphere, sometimes reaching the tropopause or even higher. Upward surging air currents are made visible by billowing cauliflower-shaped cumuliform clouds. A thunderstorm consists of one or more convection cells, each of which progresses through a life cycle that is divided into three stages: towering cumulus, mature, and dissipating (Figure 11.4).

TOWERING CUMULUS STAGE

If conditions in the atmosphere are favorable, cumulus clouds build both vertically and laterally, as shown in Figure 11.5. This is the initial or **towering cumulus stage** of thunderstorm formation. Over a period of perhaps 10 to 15 minutes, cumulus cloud tops surge upward to altitudes of 8000 to 10,000 m (26,000 to 33,000 ft). At the same time, neighboring cumulus clouds merge so that by the end of the towering cumulus stage, the storm's lateral dimension may be 10 to 15 km (6 to 9 mi).

Cumulus clouds are products of convection operating within the atmosphere (Chapter 4). Based on mode of origin, convection is either free or forced. Intense solar heating of Earth's surface triggers *free convection*, but usually free convection is not sufficiently energetic to generate a thunderstorm. Frontal or orographic lifting or converging surface winds strengthen convection, a process known as *forced convection*. Most thunderstorms are products of forced convection.

Visualize the ascending branch of a convection current as a continuous stream of parcels of warm, unsaturated air. The rising parcels expand and initially cool at the dry adiabatic lapse rate (9.8 Celsius degrees

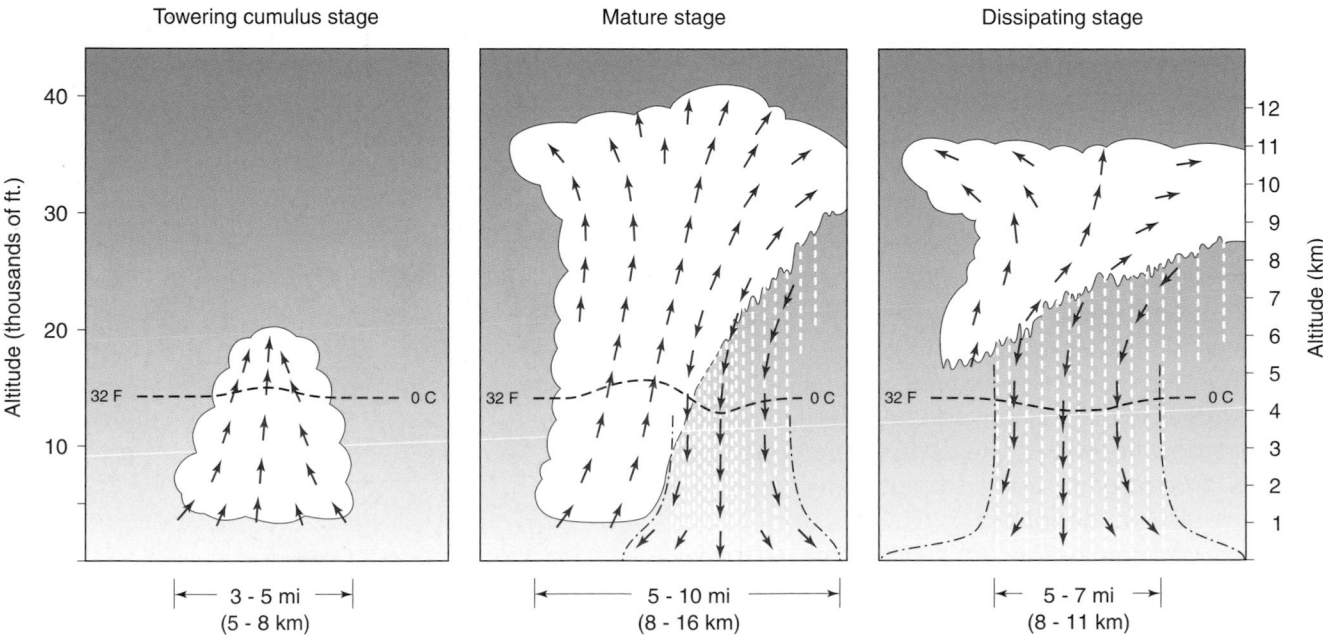

FIGURE 11.4
The life cycle of a thunderstorm cell consists of towering cumulus, mature, and dissipating stages.

FIGURE 11.5
Towering cumulus clouds may signal the initial stage in the development of a thunderstorm.

per 1000 m) until they reach the *convective condensation level (CCL)*, where the relative humidity reaches 100%, water vapor begins to condense, and cumulus clouds form. The more humid the air is to begin with, the less expansional cooling needed for air parcels to achieve saturation and the lower is the base of cumulus clouds. The bases of cumulus clouds usually are lower in Florida, where the near-surface relative humidity is higher, than in New Mexico, where the relative humidity is lower.

Latent heat released during condensation adds to the buoyancy of the saturated (cloudy) air parcels, and they surge upward while now cooling at the moist adiabatic lapse rate (averaging 6 Celsius degrees per 1000 m). Air parcels continue to ascend as long as they are warmer (less dense) than the surrounding air, that is, as long as the ambient air is unstable (Chapter 6). Some of the saturated parcels surge upward at the cloud top and evaporate in the relatively dry air above the cloud, increasing the water vapor concentration of that air. Because air above the cloud is now more humid, subsequent parcels are able to ascend higher before evaporating. As this process is repeated, the cumulus cloud billows upward. A cumuliform cloud that shows significant vertical growth and resembles a huge cauliflower is known as a **cumulus congestus cloud**. If vertical growth continues, a cumulus congestus cloud builds into a **cumulonimbus cloud**, a thunderstorm cloud with characteristic anvil top, producing precipitation, lightning, and thunder.

During the initial stage of the thunderstorm life cycle, saturated air streams upward throughout the cell as an **updraft**. The updraft is strong enough to keep water droplets and ice crystals suspended in the upper reaches of the cloud. For this reason, precipitation does not occur during the towering cumulus stage.

MATURE STAGE

By convention, the towering cumulus stage ends and the **mature stage** begins when precipitation first reaches Earth's surface. Typically, this stage lasts about 10 to 20 minutes. The cumulative weight of growing water droplets and ice crystals eventually becomes so great that they can no longer be supported by the updraft. Rain, ice pellets, and snow descend through the cloud and drag the adjacent air downward, creating a strong **downdraft** alongside the updraft. At the same time, unsaturated air at the edge of the cloud is drawn into the cloud, a process known as *entrainment*. (Actually, entrainment occurs throughout the life cycle of a thunderstorm cell.) Entrained air mixes with the cloudy (saturated) air, causing some of the water droplets and ice crystals to vaporize. The consequent evaporative cooling weakens the buoyant uplift and strengthens the downdraft.

The downdraft exits the base of the cloud and spreads out along Earth's surface, reaching well in advance of the parent thunderstorm cell, as a mass of cool, gusty air. The air from the downdraft is relatively cool because of evaporative cooling beneath the cloud base that offsets to some extent the compressional warming of the downdraft. At the surface, the arc-shaped leading edge of downdraft air resembles a miniature cold front and is called a **gust front** (Figure 11.6). Uplift along the advancing gust front sometimes produces additional cumuliform clouds that may evolve into secondary thunderstorm cells tens of kilometers ahead of the parent cell.

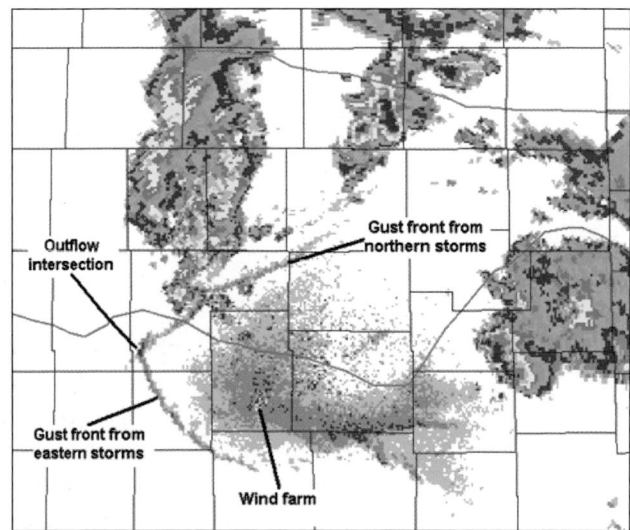

FIGURE 11.6
This radar image segment shows gust fronts, also known as outflow boundaries ahead of both the northern and eastern storms. [Courtesy of NOAA/NWS/Storm Prediction Center]

FIGURE 11.7
Roll cloud over a beach in Uruguay. [Photo by Daniela Mirner Eberl/License: Creative Commons Attribution ShareAlike 3.0]

Ominous-appearing low clouds are sometimes associated with thunderstorm gust fronts. A **roll cloud** (Figure 11.7) is an elongated, tube-shaped cloud that appears to rotate slowly about its horizontal axis. The roll cloud occurs behind the gust front and beneath, but detached from, the cumulonimbus cloud. How this cloud forms is not fully understood. Although appearances might suggest otherwise, roll clouds are seldom accompanied by severe weather. This is not the case for shelf clouds. A **shelf cloud** (Figure 11.8), also called an *arcus cloud*, is a low, elongated cloud that is wedge-shaped with a flat base. This cloud appears at the edge of a gust front and beneath and attached to the cumulonimbus cloud. A shelf cloud is thought to develop from the uplift of stable warm and humid air along the gust front. Damaging surface winds may occur under a shelf cloud, and sometimes this cloud is associated with a severe thunderstorm.

A thunderstorm cell attains maximum intensity during its mature stage. Rain is heaviest (especially during the first 5 minutes or so), lightning is most frequent, and hail, strong surface winds, and even tornadoes may develop. Cloud tops can build to altitudes in excess of 18,000 m (about 60,000 ft). Strong winds at such great altitudes distort the cloud top into an anvil shape (Figure 11.9). The flat top of the anvil indicates that convection currents have reached the extremely stable air of the tropopause. Only in severe thunderstorms will convection currents overshoot this altitude, causing clouds to billow into the lower stratosphere before collapsing back into the troposphere. Temperatures within the upper portion of the cloud are so low that the anvil is composed exclusively of ice crystals, giving it a fibrous appearance.

Viewed from space by satellite sensors, clusters of mature thunderstorm cells appear as bright white blotches (Figure 11.10). The brightness of these clusters on visible satellite images is due to the high albedo of the cloud tops. Much of the solar radiation that penetrates cumulonimbus clouds is scattered and ultimately absorbed, so that sunlight is weakened considerably when it emerges at the cloud base. For this reason, from our perspective on Earth's surface, the daytime sky darkens with the approach of a thunderstorm. White blotches on infrared satellite images are indicative of the relatively cold anvil tops of cumulonimbus clouds.

FIGURE 11.8
Shelf cloud at the base of a cumulonimbus cloud. Shelf clouds are often associated with damaging surface winds and a severe thunderstorm. [Photo by Jim Kurdzo/University of Oklahoma]

FIGURE 11.9
When the upward billowing cumulonimbus cloud reaches the tropopause, it spreads out forming a flat anvil top.

DISSIPATING STAGE

As precipitation spreads throughout the thunderstorm cell, so does the downdraft, heralding the cell's demise. During the **dissipating stage**, subsiding air replaces the updraft throughout the cloud, effectively cutting off the supply of moisture delivered by the updraft. Adiabatic compression warms the subsiding air, the relative

FIGURE 11.10
In this visible satellite image from 27 April 2011, clusters of intense thunderstorm cells appear as bright white blotches over portions of Mississippi, Alabama, and Tennessee. [NASA with data from GOES]

humidity below the cloud base decreases, precipitation tapers off and ends, and remaining convective clouds gradually vaporize.

Thunderstorm Classification

Thunderstorms are **mesoscale convective systems (MCS)** and are classified based on the number, organization, and intensity of their constituent cells. Thunderstorms occur as single cells, multi-cellular clusters, and supercells (Figure 11.11).

A *single-cell thunderstorm* is usually a relatively weak weather system that appears to pop up randomly within a warm, humid air mass. In reality, a single-cell thunderstorm is not a random phenomenon and almost always develops along some boundary within an air mass. The boundary, for example, may be the leading edge of the rain-cooled outflow from a distant thunderstorm cell. Or it may be the edge of a bubble of cool, relatively dense air that is the remnant of another thunderstorm cell that dissipated. As noted earlier, outside of the tropics, solar heating of Earth's surface usually is not sufficiently strong to generate a thunderstorm cell via convection alone. Frontal or orographic lifting or converging surface winds are needed to strengthen vertical motion.

Typically, a thunderstorm cell completes its life cycle in 30 minutes or so, but sometimes lightning, thunder, and bursts of heavy rain persist for many hours. This is because most thunderstorms are multi-cellular; that is, a thunderstorm usually consists of more than one cell. Each cell may be at a different stage of its life cycle, with new cells forming and old cells dissipating continually. A succession of many cells is thus responsible for a prolonged period of thunderstorm weather. Although a locality may be in the direct path of a distant, intense thunderstorm cell, the relatively brief life

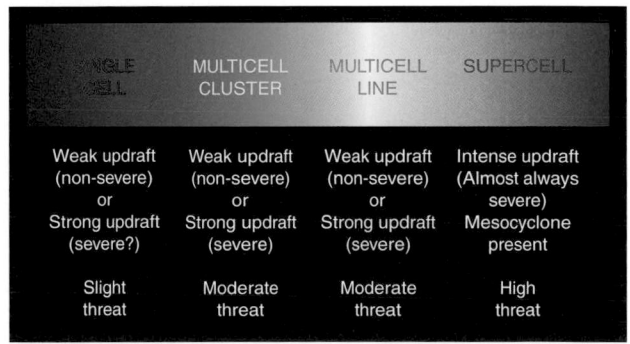

SINGLE CELL	MULTICELL CLUSTER	MULTICELL LINE	SUPERCELL
Weak updraft (non-severe) or Strong updraft (severe?)	Weak updraft (non-severe) or Strong updraft (severe)	Weak updraft (non-severe) or Strong updraft (severe)	Intense updraft (Almost always severe) Mesocyclone present
Slight threat	Moderate threat	Moderate threat	High threat

FIGURE 11.11
Classification of thunderstorms and the likelihood of severe weather. [Adapted from NOAA, *Basic Spotters' Field Guide*]

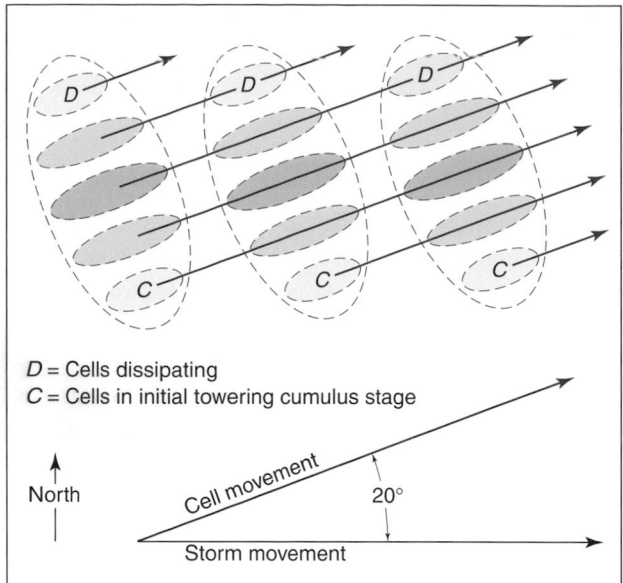

FIGURE 11.12
In this idealized situation viewed from above, the component cells of a multicellular thunderstorm travel at about 20 degrees to the eastward moving thunderstorm. As they travel toward the northeast, the individual cells progress through their life cycle.

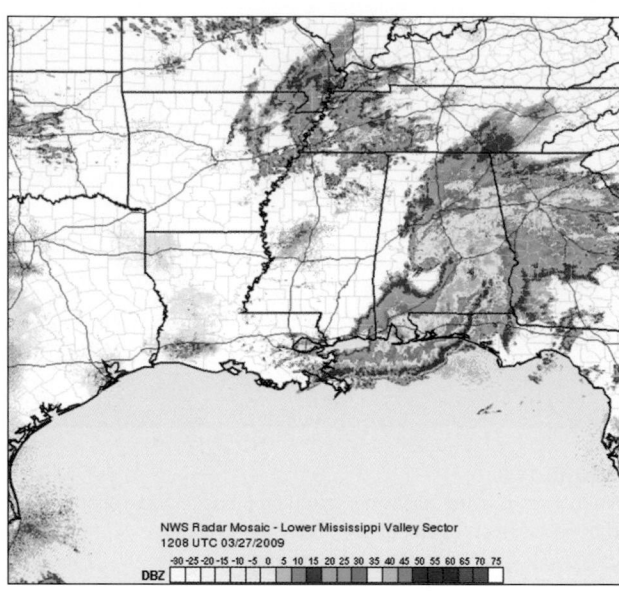

FIGURE 11.13
Radar image showing a squall line, an elongated cluster of thunderstorm cells. The squall line begins in southeast Alabama, and stretches across the Gulf of Mexico to southeast Louisiana. [NOAA/NWS]

span of an individual cell means that severe weather may dissipate before reaching that locality. The multi-cellular nature of most thunderstorms complicates the motion of the weather system because a thunderstorm may track at an angle to the paths of its constituent cells. For example, a thunderstorm may track from west to east, while its component cells head generally toward the east-northeast (Figure 11.12). In terms of organization, two types of multi-cellular thunderstorms are the squall line and the mesoscale convective complex. Cells in either of these systems can produce severe weather, which is most likely to occur near the interface between the updraft and the downdraft of a mature cell.

A **squall line** is an elongated cluster of thunderstorm cells that is accompanied by a continuous gust front at the line's leading edge (Figure 11.13). A squall line is most likely to develop in the warm southeast sector of a mature extratropical cyclone ahead of and parallel to the cold front. (A more general name for a thunderstorm formed by uplift along cold or warm fronts is *frontal thunderstorm*.) Squall line thunderstorm cells are usually more intense than single-cell thunderstorms because of the contribution of the synoptic-scale circulation to vertical motion. That is, the circulation within the associated low pressure system strengthens the updraft. As the squall-line gust front surges forward, warm, humid air is lifted and forced into the updraft located at the leading edge of

the squall line. The heaviest rain or hail typically occurs just behind (to the west of) this updraft. Light to moderate precipitation is produced by the older cells situated behind the mature cells at the leading edge.

A **mesoscale convective complex (MCC)** is a nearly circular cluster of many interacting thunderstorm cells covering an area that may be a thousand times larger than that of a single isolated thunderstorm cell. In fact, it is not unusual for a single MCC to cover an area equal to that of the state of Iowa (Figure 11.14). MCCs are primarily warm-season (March through September) phenomena that generally develop at night over the eastern two-thirds of the United States, where more than 50 may be expected in a single season.

An MCC is not associated with a front and usually develops during weak synoptic-scale flow, often near an upper-level ridge of high pressure and on the cool side of a stationary front. A low-level jet feeds warm humid air into the developing MCC while a low pressure system forms at mid-levels of the troposphere. Rising temperatures at low levels and radiational cooling at upper levels destabilizes the troposphere. New cells form while old cells dissipate continually within an MCC, so the life expectancy of the system is at least 6 hrs and often 12 to 24 hrs. The longevity and typically slow movement (15 to 30 km per hr, or 9 to 18 mph) of an MCC mean that rainfall is widespread and substantial. MCCs account for

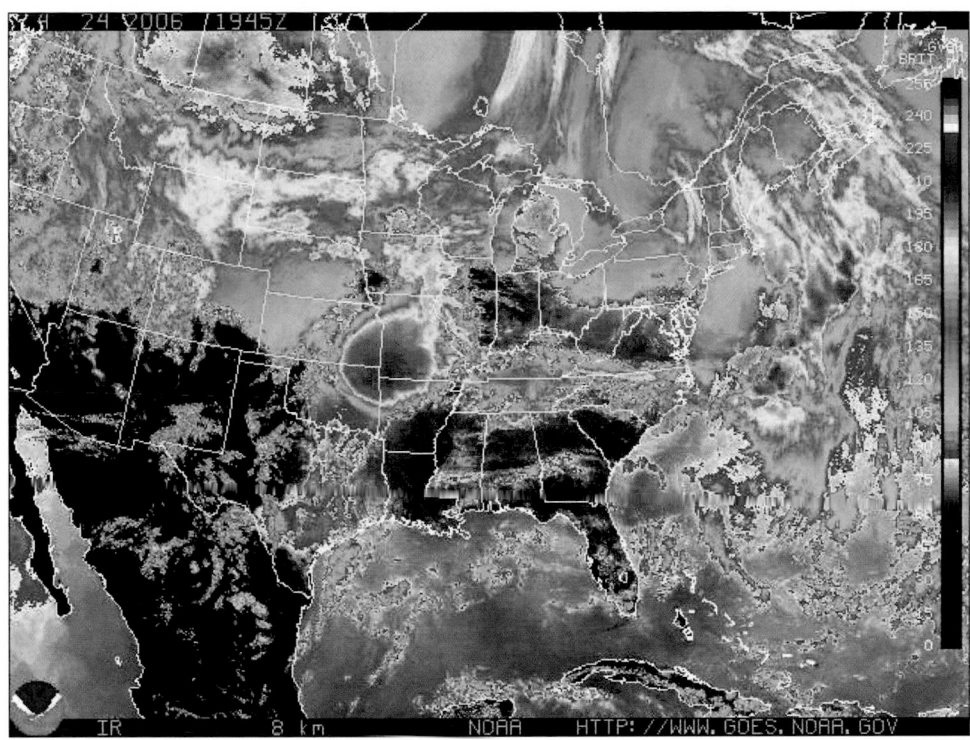

FIGURE 11.14
Infrared satellite image showing a MCC over the south-central U.S., centered on the tri-state border area of Kansas, Missouri, and Oklahoma. [NOAA]

perhaps 80% of growing season rainfall across the Great Plains and Midwest. An MCC also has the potential to produce severe weather, including weak tornadoes, moderate-sized hail, and flash flooding.

A **supercell thunderstorm** is a relatively long-lived, large, and intense system. It consists of a single cell with an exceptionally strong updraft, in some cases estimated at 240 to 280 km per hr (150 to 175 mph). A distinguishing feature of a supercell is the tendency for the updraft to develop a rotational circulation that may evolve into a tornado. Because of their association with strong to violent tornadoes, supercell thunderstorms are covered in more detail later in this chapter.

Where and When

Atmospheric conditions that favor thunderstorm development vary with latitude, season, and time of day. Key to understanding the spatial and temporal variability of thunderstorms is the conditions required for their formation: humid air in the low- to mid-troposphere, atmospheric instability, and a source of uplift.

Most thunderstorms develop within a mass of warm, humid air, that is, maritime tropical (*mT*) air, when that air mass is destabilized by uplift. Usually, maritime tropical air is conditionally stable and becomes unstable (buoyant) only when lifted to the convective condensation level. Recall from Chapter 6 that *conditional stability* means that the ambient air is stable for unsaturated (clear) air parcels but unstable for saturated (cloudy) air parcels. The more humid the air, the less the ascent (expansional cooling) needed for destabilization. Most thunderstorms develop when maritime tropical air is lifted along fronts, up mountain slopes, or via converging surface winds. Furthermore, cold air advection aloft and/or warm air advection at the surface enhances the potential instability of *mT* air. Either of these processes increases the air temperature lapse rate thereby reducing ambient air stability.

Solar heating drives atmospheric convection, so it is not surprising that thunderstorms tend to be most frequent when and where solar radiation is most intense, that is, during the warmest hours of the day. However, there are many exceptions to this rule. Mesoscale convective complexes and squall lines develop both night and day. In the Missouri River Valley and adjacent portions of the upper Mississippi River Valley, even single-cell thunderstorms are more frequent at night than during the day. One possible explanation for the nocturnal thunderstorm maximum in the upper Missouri/Mississippi Valleys is based on the role of a *low-level jet stream* of maritime tropical air that flows from the Gulf of Mexico northward up the Mississippi River Valley. This jet stream strengthens at night and causes warm air advection at low levels that destabilizes the air and spurs the buildup of cumuliform clouds.

Thunderstorm frequency is usually expressed as the number of thunderstorm days per year, where a *thunderstorm day* is defined as a day when thunder is heard. This conventional method of expressing thunderstorm frequency likely underestimates the actual number of thunderstorms, particularly if more than one line of thunderstorms passes over a weather station on the same day. With this limitation in mind, thunderstorms occur with greatest frequency over the continental interiors of tropical latitudes. The steamy Amazon Basin of Brazil, the Congo Basin of equatorial Africa, and the islands of Indonesia have the highest frequency of thunderstorms in the world, experiencing at least 100 thunderstorm days per year. Because the surfaces of large bodies of water do not warm as much as land surfaces in response to the same intensity of solar radiation, thunderstorms are less frequent over adjacent bodies of water.

In the subtropics and tropics, intense solar heating may combine with converging surface winds to trigger thunderstorm development. As noted in Chapter 9, this combination characterizes the intertropical convergence zone (ITCZ), a discontinuous band of thunderstorms more or less paralleling the equator that moves north and south seasonally with the Sun.

In North America, thunderstorm frequency generally increases from north to south with the highest frequency over central Florida (Figure 11.15). Some interior localities of that state average 100 thunderstorm days per year. The Florida thunderstorm maximum near Orlando is due to convergence of sea breezes that develop along both the east and west coasts of the Florida peninsula (Figure 11.16). Sea breeze convergence over the interior induces ascent of maritime tropical air and formation of cumulonimbus clouds.

Portions of the Rocky Mountain Front Range rank second to interior Florida in thunderstorm frequency over North America. On average, more than 60 thunderstorm days occur per year in a band from southeastern Wyoming southward through central Colorado and into north central New Mexico. This high thunderstorm frequency is linked to topographically related differences in heating.

Mountain slopes facing the Sun absorb direct solar radiation and become relatively warm. The warm slopes, in turn, heat the air in immediate contact with the slopes, and that air rises. At the same time, the air at the same altitude, but located to the east of the mountains out over the relatively flat terrain of the western Great Plains, is much cooler. As warm air rises over the mountain slopes, it is replaced by cooler air sweeping westward from over the Plains. Updrafts over the mountain slopes produce cumuliform clouds that often evolve into thunderstorm cells, particularly during the warmest hours of the day. Thunderstorm development is enhanced whenever the synoptic-scale air pressure pattern favors winds blowing from the east over the western Great Plains.

To this point, our discussion has focused on conditions conducive to thunderstorm development. Under other conditions, convection and ascent of air is inhibited and thunderstorms do not form. This is the case when air masses reside or travel over relatively cold surfaces and are thereby stabilized. For example, snow-

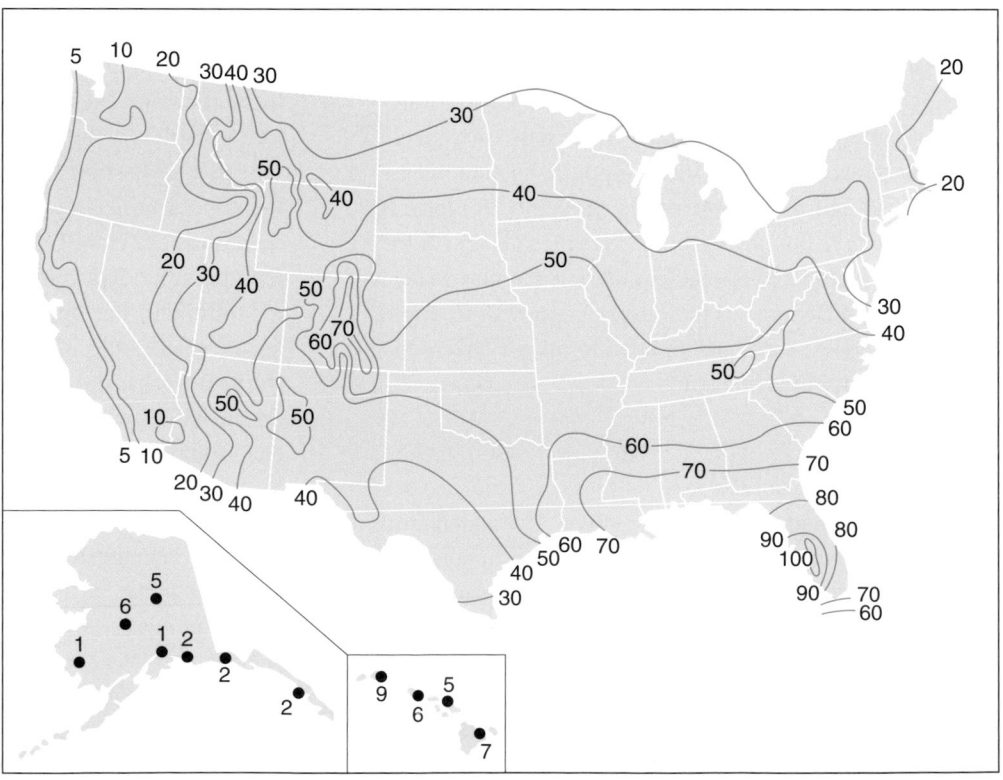

FIGURE 11.15
Average annual number of thunderstorm days in the United States.

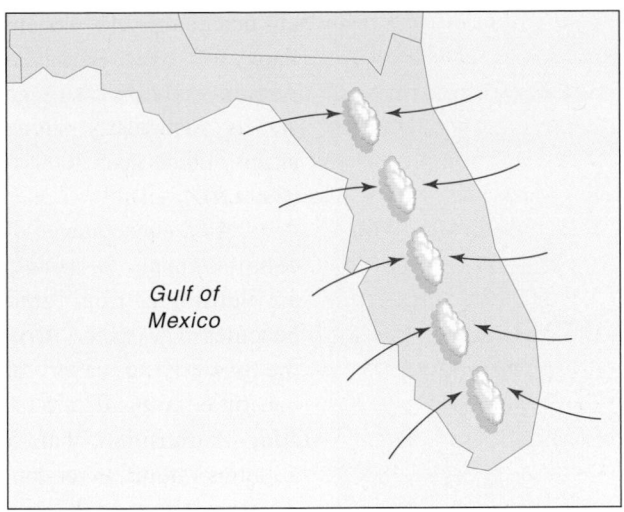

FIGURE 11.16
The relatively high frequency of thunderstorms over the interior of the Florida peninsula is linked to the convergence of sea breezes blowing inland from both the Gulf and East Coasts.

covered ground cools and stabilizes the overlying air. Convection is suppressed, so thunderstorms are relatively rare at middle and high latitudes in winter. Nonetheless, thunderstorms can be associated with winter extratropical cyclones, usually developing ahead of the surface warm front as lifting over the front destabilizes the near surface air. In that case, *thundersnow* can be locally heavy.

Thunderstorms are unusual over coastal areas that are downwind from relatively cold ocean waters. For example, thunderstorms are infrequent along coastal California, where prevailing winds are onshore and a shallow layer of maritime polar air often flows inland from off the relatively cold California Current. Cool *mP* air at low levels suppresses deep convection and thunderstorm development. Thus, the average annual number of thunderstorm days is only 2 at San Francisco, 6 at Los Angeles, and 5 at San Diego.

Thunderstorms are also relatively rare in Hawaii, primarily because of the trade wind inversion (Chapter 9). At Hilo and Honolulu, the average annual number of thunderstorm days is under 10. The trade wind inversion, at an average altitude of about 2000 m (6600 ft), restricts vertical development of towering cumulus clouds. Cumulus clouds normally do not attain the altitude needed to develop into cumulonimbus clouds.

Severe Thunderstorms

According to the National Weather Service (NWS) official criterion, a thunderstorm qualifies as severe if the storm system has surface winds stronger than 50 knots (93 km per hr or 58 mph) and/or produces hailstones 1.0 in. (2.5 cm) or larger in diameter (penny-size). **Severe thunderstorms** may also produce flash floods or tornadoes.

As a general rule, the greater the altitude of the top of a thunderstorm, the more likely it is that the system will produce severe weather. Why do some thunderstorm cells surge to great altitudes and trigger severe weather, whereas others do not? The most important factor appears to be the magnitude of the vertical wind shear. Vertical **wind shear** is the change in horizontal wind speed or direction with increasing altitude. Weak vertical wind shear (little change in wind with altitude) favors short-lived updrafts, low cloud tops, and weak thunderstorms, whereas strong vertical wind shear favors vigorous updrafts, great vertical cloud development, and severe thunderstorms.

With weak vertical wind shear, the flow of warm humid air into the cell is relatively weak. The downdraft's outflow pushes the gust front well ahead of the system, eventually cutting off the supply of warm humid air entirely. Also, precipitation particles fall through (against) the updraft, thereby weakening it. With increasing vertical wind shear, however, the gust front cannot advance as far from the cell and the inflow of warm humid air is sustained for a longer period. Furthermore, most precipitation falls alongside rather than through (and against) the tilted updraft so that the updraft maintains its strength and the thunderstorm surges to great altitudes. On the other hand, if the vertical wind shear were too strong, a developing cumulonimbus cloud would be torn apart.

In the United States and Canada, most severe thunderstorms break out over the Great Plains and are associated with mature extratropical cyclones. Severe thunderstorm cells usually form as part of a squall line within the cyclone's warm sector, ahead of and parallel to a fast-moving and well-defined cold front. Severe cells can produce large hail, heavy rain, and downbursts (discussed later in this chapter). Most tornadoes that do form are weak, although some squall line cells evolve into supercells that can spawn strong to violent tornadoes.

A synoptic situation that favors severe thunderstorm development is shown schematically in Figure 11.17. A mature cyclone is centered over western Kansas with a well-defined cold front trailing across the Oklahoma and Texas panhandles into West Texas and a warm front that stretches southeastward from the low center. The polar front jet stream and a low-level (*mT*) jet,

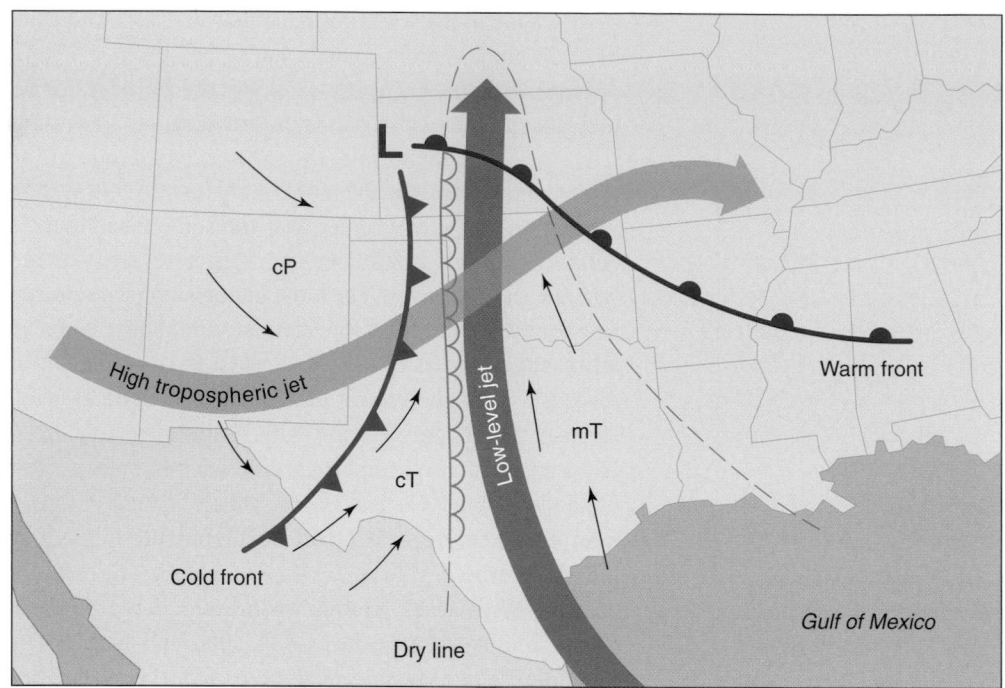

FIGURE 11.17
A synoptic weather pattern that favors development of severe thunderstorms.

which are also plotted on the map, cross to the southeast of the cyclone center. The most intense thunderstorm cells are likely to develop near the intersection of these two jets. The *subtropical jet stream* (Chapter 9), although not plotted on Figure 11.17, is sometimes also present at high levels in the troposphere, blowing from west to east over the warm sector of the cyclone. In that event, intense squall lines are likely to form between the polar front and subtropical jet streams.

The polar front jet stream produces strong vertical wind shear that maintains a vigorous updraft and favors great vertical development of thunderstorm cells. In addition, the jet contributes to a stratification of air that increases the potential instability of the troposphere. As described in Chapter 9, a jet streak induces both horizontal divergence and horizontal convergence of air in the upper troposphere. Recall that diverging horizontal winds trigger ascent of air and cyclone development under the left-front quadrant of a jet streak. Meanwhile, air converges in the right-front quadrant of a jet streak, causing weak subsidence of air over the warm sector of the cyclone. The subsiding air is compressionally warmed and its relative humidity decreases. The subsiding, warming air is prevented from reaching Earth's surface by a shallow layer of maritime tropical air that surges northward as a low-level jet from the Gulf of Mexico. The jet of maritime tropical

air occurs on the western flank of the Bermuda-Azores subtropical high and is particularly strong at an altitude near 3000 m (9800 ft).

As a consequence of compressional warming, air subsiding from aloft becomes warmer than the underlying layer of maritime tropical air. A zone of transition, that is, a temperature inversion, develops between the two layers of air (Figure 11.18). An air layer characterized by a temperature inversion is extremely stable, so the two air layers do not mix and convection is confined to the surface layer of *mT* air. In this circumstance, the inversion is known as a **capping inversion**. For as long as this stratification persists, the contrast between air layers increases; that is, subsiding air becomes drier while the underlying *mT* air becomes more humid. The potential for severe weather continues to grow; all that is needed is a trigger that causes updrafts to penetrate the capping inversion.

The necessary upward impetus may be supplied by a combination of the intense solar heating of mid-afternoon and the lifting of air caused by an approaching cold front or jet streak. Updrafts eventually break through the capping inversion, and cumulus clouds billow upward at explosive speeds that may exceed 100 km per hr (62 mph). Such updrafts can even penetrate the tropopause and surge into the lower stratosphere. A severe thunderstorm is born.

With the synoptic pattern shown in Figure 11.17, air is quite dry ahead of the cold front but west of the warm, humid tongue (enclosed by a dashed line). For this reason, the western boundary of the *mT* air mass is called the **dryline**. West of the dryline and east of the cold front, hot and dry continental tropical (*cT*) air flows eastward and downhill from the southwest deserts and Mexican Plateau. Because of the density contrast associated with differences in atmospheric water vapor content, a dryline brings about uplift in a manner similar to a cold front. A dryline can extend hundreds of kilometers from Texas

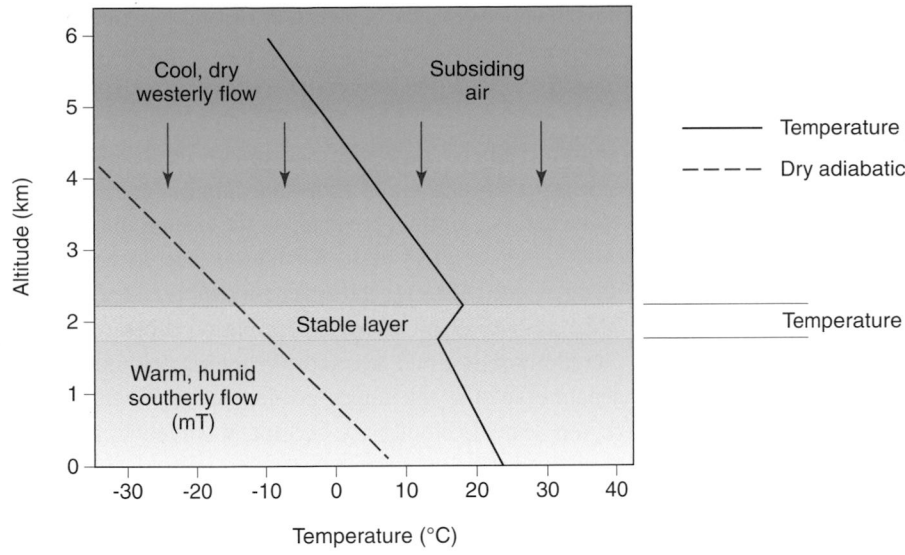

FIGURE 11.18
A temperature sounding that favors the development of severe thunderstorm cells. A capping temperature inversion separates subsiding dry air aloft from warm, humid air near the surface.

northward onto the western Plains and is often the site of squall line and severe thunderstorm activity. Typically, the dryline advances eastward during the day and retreats westward at night.

Sometimes, ominous pouch-like **mammatus clouds** appear at the base of the spreading anvil top of a cumulonimbus cloud (Figure 11.19). (They are also associated with a variety of other cloud types including cirrus, altocumulus, and stratocumulus.) Contrary to popular belief, mammatus clouds do not always indicate a severe thunderstorm. In fact, in the mountainous western United States, they often are associated with relatively weak convective showers. Their unusual appearance is attributed to blobs of cold, cloudy air that descend from the anvil into the unsaturated (clear) air beneath the anvil. Ice crystals at the margin of a blob sublimate, causing further

FIGURE 11.19
Pouch-like mammatus clouds occur on the underside of a thunderstorm anvil and sometimes indicate a severe storm system.

cooling of the blob. This cooling offsets compressional warming of the descending air so that the blob sinks, bulging downward into the dry air below. Ultimately, the size of the bulge is limited by the increasing rate of sublimation of the constituent ice crystals.

Thunderstorm Hazards

Thunderstorm hazards include lightning, downbursts, flash floods, hail, and tornadoes. Because tornadoes are an especially severe weather hazard, they are covered in a later section of this chapter.

LIGHTNING

A convective rain or snow shower is a thunderstorm if accompanied by lightning, which also produces thunder. **Lightning** (Figure 11.20) is a brilliant flash of light caused by an electrical discharge within a cumulonimbus cloud (the most common case) or between the cloud and Earth's surface (the most dangerous case). Lightning flashes also travel horizontally through the sides of thunderclouds and vertically through the top of the thunderstorm cloud (called a *gigantic jet*) some 80 km into space. Worldwide, 100 flashes of lightning occur each second. Lightning is a weather phenomenon that can be directly hazardous to human life. Over a recent 30-year period (1981-2010), lightning killed an average of 55 people in the United States annually. Lightning's death toll may seem surprisingly high, perhaps because fatal lightning strikes are typically isolated events that seldom make headlines. A single, fatal lightning strike is not as newsworthy as a single disastrous tornado that may take many lives. However, so many lightning strikes

FIGURE 11.20
Lightning is a brilliant flash of light associated with an electrical discharge between clouds and Earth's surface, within a cloud, or between clouds. [NOAA]

occur daily that fatalities and injuries add up. For tips on lightning safety, refer to this chapter's first *For Further Exploration*.

Lightning not only kills and injures people, it also ignites forest and brush fires. In the Rocky Mountain region, for example, lightning is the most common cause of forest fires, starting more than 9000 each year. A person might assume that heavy rain associated with a thunderstorm would quickly quench a lightning-ignited wildfire. In the western basins, however, the base of a cumulonimbus cloud is usually so far above the ground (perhaps 2400 m or 8000 ft), and the air below the cloud so dry, that much if not all the rain evaporates before reaching a wildfire.

Lightning is very costly to electrical utilities, each year causing tens of millions of dollars in damage to equipment (power lines, transformers) and in service restoration expenses. A useful tool in utilities' efforts to cope with the lightning hazard is the **lightning detection network (LDN)**, a system that provides real-time information on the location and severity of lightning strikes. Each monitoring station in the network consists of direction-finding equipment that can sense electromagnetic fields associated with cloud-to-ground lightning. A computer plots the locations of lightning strikes on a regional map. An electrical utility that subscribes to a LDN service accesses lightning data through a computer link. Such information enables the utility to mobilize repair crews more efficiently and speeds up restoration of disrupted service.

What causes lightning? The potential for an electrical discharge exists whenever a charge difference develops between two objects. A normally neutral object becomes negatively charged when it gains electrons (negatively charged subatomic particles) and positively charged when it loses electrons. That is, the object is *ionized*. When large differences in electrical charge develop within a cloud, between clouds, or between a cloud and the ground, the stage is set for lightning.

On a clear day, Earth's surface is negatively charged with respect to the upper atmosphere (ionosphere). This fair-weather charge distribution changes as a cumulonimbus cloud develops. Within the cloud, charges separate so that the upper portion and a much smaller region near the cloud base become positively charged. In between, a disk-shaped zone of negative charge forms that is a few hundred meters thick and several kilometers in diameter. At the same time, the region of negative charge in the developing cumulonimbus induces a positive charge on the ground directly under the cloud.

Air is a very good electrical insulator, so that as a thunderstorm cell forms and electrical charges separate and build an electric field in the thundercloud, a potential develops for lightning. Recent discoveries that high-energy X-rays and gamma rays often accompany lightning suggest a mechanism (known as *runaway breakdown*) for initiating the lightning process that involves delivery of high-speed electrons by cosmic rays striking the atmosphere. Cosmic rays also collide with other electrons (in the thundercloud) causing them to speed up. About the time the thunderstorm cell enters its mature stage, the electrical field has strengthened to the point that the electrical resistance of air breaks down and electrons flow as a lightning flash, thereby neutralizing the electrical charges

What causes electrical charges to separate within cumulonimbus clouds is not well understood, but field studies and simulations offer a promising explanation focusing on graupel and the convective circulation within cumulonimbus clouds. Graupel (German for *soft hail*) consists of millimeter- to centimeter-sized ice pellets formed when supercooled water droplets collide and freeze on impact. Within the cloud, as graupel descend and rub against smaller ice crystals in their path, opposite charges develop on the graupel and the ice crystals. For collisions that take place at cloud temperatures below about −15 °C (5 °F), graupel particles become negatively charged while ice crystals acquire a positive charge. Vigorous updrafts separate the particles, carrying the smaller positively charged ice crystals to the upper portion of the cloud, whereas the larger negatively charged graupel concentrate mostly in the lower portion of the cloud. This mechanism

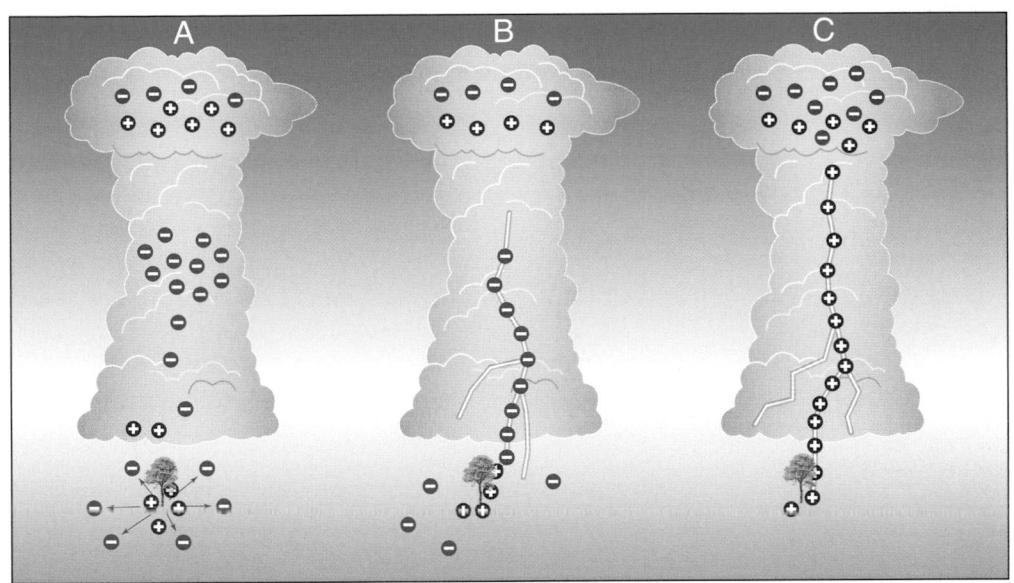

FIGURE 11.21
Steps in cloud-to-ground lightning discharge. (A) The clouds's large negative charge repels the negative charge on the ground leaving a positive charge. (B) A stepped leader of negatively charged electrons begins zigzagging toward the ground. As the stepped leader nears the ground it attracts a streamer of opposite charge, usually through the highest point in the area below the stepped leader. (C) When the leader and streamer connect, a large electrical current begins flowing to the ground. Contact also sends a wave of positive charge zipping up, creating the light we see - the return stroke. It travels upward at nearly 50,000 km (31,000 mi) per second. [Adapted from *The AMS Weather Book*, © 2009 by the American Meteorological Society]

explains the positive charge of the upper portion of the cloud and the negative charge of the disk-shaped zone of the lower cloud region.

What accounts for the positive charge near the cloud base? Typically, temperatures near the cloud base are higher than −15 °C (5 °F). At such temperatures, collisions between graupel and ice crystals induce a positive charge on the graupel and a negative charge on ice crystals. Ice crystals are conveyed upward in the updraft, leaving heavier graupel to accumulate near the cloud base, giving that region a positive charge.

Lightning strikes between a cloud and the ground (Figure 11.21) pose the greatest hazard for people, although these represent only about 20% of all lightning flashes, with in-cloud lightning accounting for nearly 80% of the flashes. Using high-speed photography (some video cameras capable of shooting 4 million frames per second), scientists found that a lightning flash involves a regular sequence of events. Initially, streams of electrons surge from the cloud base toward the ground in discrete steps, each about 20 to 100 m (65 to 330 ft) long. These *stepped leaders* describe a faintly visible branching path and produce a narrow ionized channel. When a branch of the stepped leaders comes within about 100 m (330 ft) of the ground, it is met by a positively charged streamer

ascending from the ground. A luminescent return stroke forms as an ascending electric current when the positive and negative electric charges recombine. The *return stroke* follows the path of least resistance and often emanates from tall, pointed structures such as a metal flagpole or tower. Now an ionized channel, only a few centimeters in diameter, links Earth's surface to the cloud. Electrons flow, neutralization occurs, and the channel is illuminated.

Following this initial electrical discharge, subsequent surges of electrons from the cloud, called *dart leaders*, follow the same conducting path. Each dart leader is met by a return stroke (from the ground), and the conducting path is again illuminated. Typically, a single lightning flash consists of two to four dart leaders plus return strokes. Sometimes, a dart leader is met by a return stroke that forges a new conducting path from the ground. The result is a forked lightning flash that strikes the ground at more than one place.

The sequence just described is the most common occurrence of cloud-to-ground lightning. In less than 10% of cases, a positively charged leader emanates from the cloud and initiates a lightning flash. Much more rarely, positive or negative stepped leaders propagate upward from the ground and meet a return stroke surging downward from a cloud. This ground-to-cloud lightning

TABLE 11.2 Types of Lightning	
Ball lightning	A reddish, luminous sphere, about 30 cm in diameter, that moves rapidly along a solid surface or appears to float in air; may be accompanied by a hissing sound.
Bead lightning	Resembles a series of beads on a string; actually a normal zigzag lightning stroke or bolt, segments of which are viewed end-on, giving the impression of high intensity at a series of points along the lightning channel.
Heat lightning	Light reflected by clouds or within clouds from lightning in thunderstorms that are too far away for thunder to be heard.
Ribbon lightning	A lightning stroke that appears to spread horizontally into a ribbon of parallel luminous streaks when a strong wind is blowing at some angle to the observer's line of sight.
Sheet lightning	Clouds illuminated by cloud-to-cloud lightning so that they appear bright white; clouds block the view of the lightning stroke.
Streak lightning	Discharge between cloud and ground that is concentrated in a single, relatively straight channel.

usually is initiated from mountaintops or tall structures such as antenna towers.

Electricity flows at an astonishing rate of nearly 50,000 km (31,000 mi) per second; hence, the entire lightning sequence takes place in less than two-tenths of a second. The human eye has difficulty separating the individual flashes of light that constitute a single lightning flash, so that we perceive a lightning flash as a flickering light. The visible lightning flash takes on many distinctive forms (Table 11.2).

Where there is lightning, there is thunder, although sometimes we see distant lightning but cannot hear the thunder. Lightning heats the air along the narrow conducting path to temperatures that may exceed 25,000 °C (45,000 °F). Such intense heating occurs so rapidly that air density cannot respond, at least initially. The rapid rise in air temperature is accompanied by a tremendous increase in air pressure locally that generates a shock wave. The shock wave propagates outward, producing sound waves that we hear as the rumble of **thunder**. The first thunder heard is generated from the nearest part of the lightning flash. Subsequent sound waves reach us from portions of the flash that are progressively further away; hence, the rumble of thunder. Thunderstorm cells that are more than 20 km (12 mi) away are too distant for thunder to be heard although lightning can be seen. This is called *heat lightning*, a name that is somewhat misleading because

people may assume that heat causes the lightning when in fact it occurs in the same way as all lightning.

Light travels about a million times faster than sound so that we see lightning almost instantaneously, but we hear thunder later. The closer we are to a thunderstorm cell, the shorter is the time interval between the lightning flash and thunder. As a rule, thunder takes about 3 seconds to travel 1 km (or 5 seconds to travel 1 mi). If you must wait 9 seconds between lightning flash and thunderclap, the lightning is about 3 km (1.8 mi) away. This is the *flash-to-bang* method of determining the distance to a thunderstorm. By noting the time difference between the flash of light and thunder of successive lightning discharges, you can deduce whether the thunderstorm cell is approaching or moving away, or even predict when it is dangerously close. As a general guideline, if the time between flash and bang is 30 seconds or less, you are well advised to take precautions to avoid the lightning.

DOWNBURSTS

Severe, and sometimes not so severe, thunderstorms can produce a **downburst**, an exceptionally strong downdraft that, upon striking Earth's surface, diverges horizontally as a surge of potentially destructive winds. Downbursts occur with or without rain; if without rain, the downburst usually occurs under a curtain of *virga* (Chapter 7). T. Theodore Fujita (1920-1998) of the University of Chicago is credited with discovering downbursts, and he coined the term. Fujita's discovery stemmed from his airborne survey of property damage near Beckley, WV, shortly after the Super Tornado Outbreak of 3-4 April 1974. He documented storm debris spread over the countryside in a starburst pattern—distinct from the swirling pattern that is typical of tornado damage. Downbursts blow down trees, flatten crops, and wreck buildings. Based on size of the affected area, Fujita classified a downburst as either a macroburst or a microburst.

A **macroburst** cuts a swath of destruction over a distance of more than 4 km (2.5 mi) with surface winds that may top 210 km per hr (130 mph). The leading edge of a macroburst may be marked by a gust front, and the system lasts up to 30 minutes. A **microburst** is smaller and shorter-lived than a macroburst. By convention, its path of destruction is 4 km (2.5 mi) or less, top surface winds may be as high as 270 km per hr (170 mph), and its life expectancy is less than 10 minutes.

Microbursts are particularly dangerous at airports, where they can play havoc with aircraft, especially on takeoff or landing. Microbursts trigger *wind shear*, a change in wind speed or direction with distance, which can disturb the lift forces acting on the wings of an aircraft—perhaps causing an abrupt change in its altitude (Figure 11.22). An aircraft that flies into a microburst first encounters a strong headwind and then a strong tailwind. In a matter of seconds, the aircraft's speed relative to air can drop by more than 160 km per hr (100 mph). If this happens during takeoff or landing, the aircraft's air speed may drop below the minimum required for flight. Fujita underscored the microburst hazard for aircraft when he

demonstrated that microbursts probably contributed to two commercial aircraft accidents in 1975. Apparently, in both cases, the pilots unwittingly flew their jet planes through the center of a microburst, one on takeoff and the other while attempting to land. Both planes abruptly lost altitude and crashed.

In response to the microburst hazard to aviation, the Federal Aviation Administration (FAA) now requires airlines to install an approved microburst detection system on their aircraft. The most effective systems employ sensors that can detect wind shear using the same principle as Doppler radar (Chapter 7). The goal is to provide pilots with at least 20 to 40 seconds advance warning of microburst-induced wind shear. Such advance warning is considered sufficient for pilots to take evasive action.

Pilots also rely on the airport based Terminal Doppler Weather Radar (TDWR) system operating at 48 U.S. and Puerto Rico airports, which detects microbursts during precipitation events by measuring the velocity of precipitation particles. However, TDWR has limitations in dry climates as it cannot consistently detect microbursts when precipitation does not reach the ground and a significant amount of ground clutter is present. In 2007 the FAA began testing a light detection and ranging (LIDAR) system at Las Vegas International Airport. LIDAR transmits pulses of infrared light, and can determine the velocity of dust particles. The LIDAR system at Las Vegas International Airport detected 91% of dry microburst events. Based on results from this study, the FAA determined that LIDAR is a viable dry wind shear detection system and should be considered at airports prone to dry wind shear events and significant ground clutter. The LIDAR was procured for Las Vegas

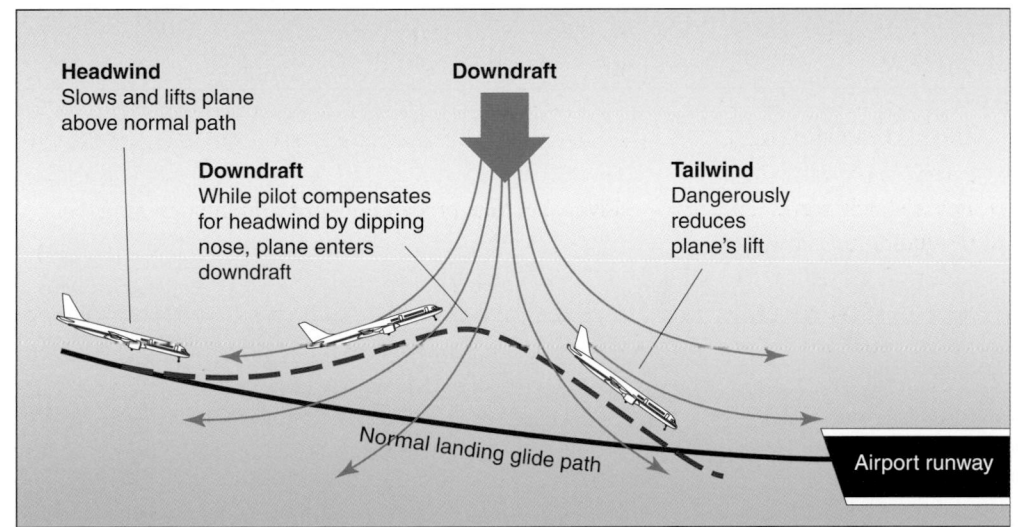

Headwind
Slows and lifts plane above normal path

Downdraft

Downdraft
While pilot compensates for headwind by dipping nose, plane enters downdraft

Tailwind
Dangerously reduces plane's lift

Normal landing glide path

Airport runway

FIGURE 11.22
A microburst causes a landing aircraft to depart from its normal glide path. [Adapted from Michael Shibao, National Center for Atmospheric Research]

and is currently combined with the Integrated Terminal Weather System which will use TDWR and LIDAR to generate wind shear alerts at the control tower.

DERECHO

A squall line or mesoscale convective complex may produce a family of straight-line downburst winds that impacts a swath that by definition is at least 400 km (240 mi) long with wind gusts of at least 93 km per hr (58 mph). At any location along the path of the system, the period of strongest winds typically lasts 10 to 20 minutes or so. This convectively induced windstorm is known as a **derecho** (derived from the Spanish meaning *straight ahead* or *direct* as contrasted with the Spanish word *tornar*, meaning *to turn*, and thought to be the origin of the word tornado). Gustavus Hinrichs (1836-1923), a physics professor at the University of Iowa, coined the term derecho (pronounced deh-RAY-cho) in 1888 and applied the term to any convectively-induced straight-line windstorm.

Derechos are associated with rapidly moving bands of showers and thunderstorms that exhibit a curved or bowed shape. Bow-shaped systems are called **bow echoes**, a name coined by T. Theodore Fujita based on his study of the damage swath produced by derechos that impacted northern Wisconsin on 4 July 1977. Derechos are typically associated with either a long-lived bow echo or a series of bow echoes and generally track from northwest to southeast (Figure 11.23).

In the November 2005 issue of the *Bulletin of the American Meteorological Society*, Walter S. Ashley of Northern Illinois University, and Thomas L. Mote of the University of Georgia, published a derecho climatology for the U.S. based on observations from 1986 to 2003. During that 18-year period, 377 derechos were reported (an average of 21 per year) in seasonal corridors mostly over the eastern two-thirds of the nation. Derecho frequency was highest over the southern Great Plains, extending from northeast Oklahoma southeast toward the southern Mississippi Valley, and from the upper Midwest into the Ohio Valley. A derecho is a warm season phenomenon with most (about 75%) occurring from May through August. Derechos are responsible for deaths, injuries, and property damage. Some are as destructive as tornadoes or U.S. landfalling hurricanes. In the 18-year period examined by Ashley and Mote, fatalities totaled 155 with about half involving boating and motor vehicles. More than 2600 people were injured. In addition, a derecho's strong winds have caused much forest blow down.

On 4 July 1977, a derecho, consisting of 25 individual downbursts, struck several counties in northern Wisconsin, felling trees and buildings along a path 268 km (166 mi) long and 27 km (17 mi) wide. One person was killed, 35 were injured, and damage to buildings and timber totaled in the millions of dollars. Based on the extent of damage, surface winds may have been as high as 250 km per hr (155 mph). Twenty-two years later, on 4 July 1999, a derecho struck Minnesota's Boundary Waters Canoe Area Wilderness. In about 20 minutes, winds gusting as high as 150 km per hr (93 mph) blew down more than 20 million trees along a path measuring 60 km (37 mi) long and 20 km (12 mi) wide. Fortunately, there were no fatalities.

The morning of 11 July 2011, a line of severe thunderstorm cells swept from west to east over Chicago. According to weather radar, the cells produced a bow echo accompanied by straight-line winds with maximum gusts generally in the range of 113 to 129 km per hr (70 to 80 mph). These derecho winds caused considerable damage, ripping roofs off buildings, uprooting large

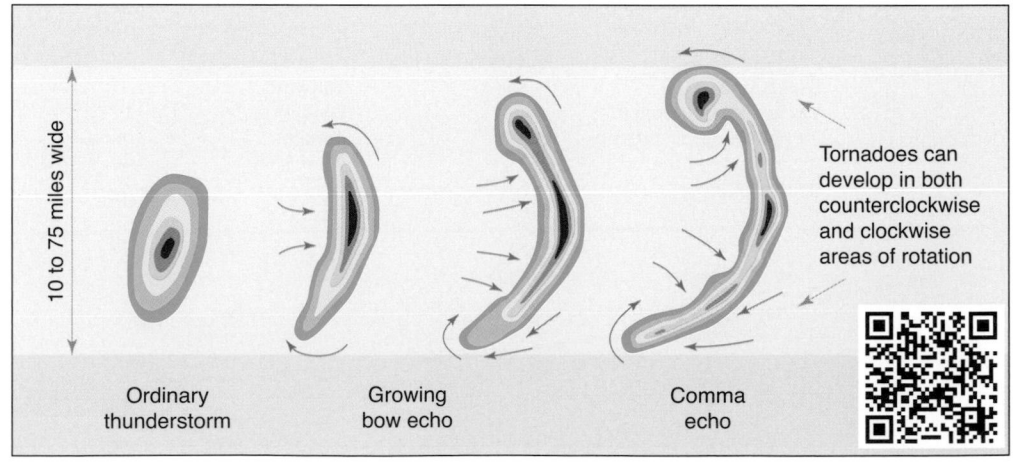

FIGURE 11.23
Weather radar images document the development of a bow echo over a period of several minutes to a few hours as the system progresses from left to right beginning as an ordinary thunderstorm. Rainfall intensity is color coded from the lightest (green) to the heaviest (red). Blue arrows represent wind relative to the system. Downbursts, derechos, or weak tornadoes may be associated with a bow echo.

10 to 75 miles wide

Ordinary thunderstorm

Growing bow echo

Comma echo

Tornadoes can develop in both counterclockwise and clockwise areas of rotation

trees, and downing power lines. Chicago's electric power utility (ComEd) reported that at one time more than 800,000 of their customers had lost power. The storm system that blasted Chicago was part of a rapidly moving complex of derechos that produced a nearly continuous swath of wind damage over a distance of 2260 km (1400 mi) in sections of 17 states in 30 hrs (Figure 11.24). The system formed in western Nebraska on the mid-afternoon of 10 July 2011, tracked generally eastward reaching Chicago just after 8 a.m. the next day, and by sunset was dissipating over eastern Maryland. Along the path, severe weather reports numbered more than 770 (mostly strong winds).

FIGURE 11.24
Trees and power lines down on cars and homes in Vinton, IA, due to the 10 July 2011 derecho. [NOAA/NWS Quad Cities IA/IL]

FLASH FLOODS

A **flash flood** is a short-term, localized, and often unexpected rise in stream level above bankfull, usually because of torrential rain falling over a relatively small geographical area. Typically, the stream level rises and falls within 6 hrs of the rain event. Figure 11.25 is an example of a hydrograph showing dramatic changes in gauge level (stage) and discharge in response to an episode of heavy rain. Excessive rainfall may occur when a succession of thunderstorm cells, parts of a squall line or MCC, matures

over the same area. Alternatively, a stationary or slow-moving intense thunderstorm cell may produce flooding rains. A thunderstorm is stationary or slow-moving when the system is embedded in weak steering winds aloft and/or maintained by a persistent flow of humid air up the slopes of a mountain range.

Atmospheric conditions that favor flash floods differ somewhat from those that give rise to other types of severe weather (e.g., hail and tornadoes). Thunderstorms

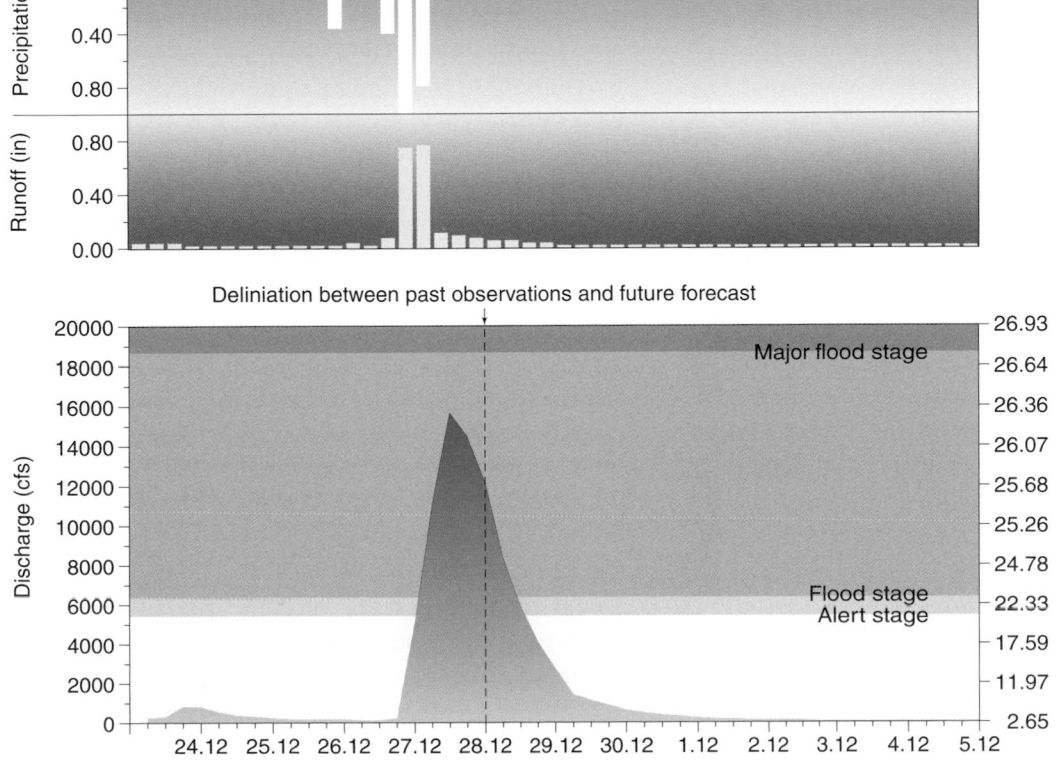

FIGURE 11.25
A hydrograph showing changes in water gauge level (stage) and discharge in response to a heavy rain event. On the top graph, precipitation and runoff are shown in 6-hr intervals. The total rainfall was about 2.2 inches. The vertical line on the bottom graph separates observed discharge (to the left of the line) and forecasted discharge (to the right), calculated every 6 hrs. [Adapted from Scipio, Kansas on Pottawatomie Creek hydrograph, NOAA/NWS/ Missouri Basin, Pleasant Hill River Forecast Center]

producing flash floods are more common at night and form in an atmosphere with weak vertical wind shear and abundant moisture through great depths. Flash flooding is most likely in an atmosphere that is *precipitation efficient* with high values of precipitable water and relative humidity, reducing the amount of precipitation that vaporizes. In addition, temperatures of the cloud base are above freezing, favoring the collision-coalescence process that leads to exceptionally heavy rainfall.

A combination of topography and favorable atmospheric conditions contributed to the flash flood that claimed 139 lives (many of them campers) and caused $35.5 million in property damage in the Big Thompson Canyon of Colorado on 31 July 1976. Big Thompson Canyon is located in the Colorado Front Range about 80 km (50 mi) northwest of Denver along a stream that flows eastward from Rocky Mountain National Park. A persistent flow of humid air from the east up the mountain slopes triggered development of a succession of thunderstorm cells and heavy rainfall. With weak winds aloft, thunderstorm cells remained nearly stationary over essentially the same geographical area for at least 4 hrs. Rainfall in the headwaters of the Big Thompson River was estimated at 25 to 30 cm (10 to 12 in.) with perhaps 20 cm (8 in.) falling in only 2 hrs. Runoff cascaded down the steep mountain slopes and into the river that winds along the narrow canyon floor. The river level rose abruptly and overflowed its banks as a flash flood. At one place along the river, the discharge (volume of water flowing per second) was more than 200 times greater than the long-term average. A wall of water almost 6 m (20 ft) high destroyed 418 houses and washed away 197 motor vehicles.

Some 21 years later, on 28 July 1997, Colorado residents were reminded of the Big Thompson Canyon disaster when a thunderstorm complex deluged nearby Fort Collins with record rainfall. Between 5:30 and 11:00 p.m. (MDT) up to 25 cm (10 in.) of rain drenched the southwest side of Fort Collins. A flash flood on Spring Creek, just downstream from where rainfall was heaviest, claimed five lives. Floodwaters also caused property damage (estimated at more than $200 million) over a large part of the city; especially hard-hit were buildings on the Colorado State University campus. Flood plain management measures, initiated well before the flood, particularly along Spring Creek, likely prevented greater loss of life and residential property damage. Management strategies included moving residents out of the floodway and constructing retention basins along the floodplain that could accommodate excess water during a major rainfall event.

FIGURE 11.26
Road sign in the Colorado Rockies warns visitors to climb to higher ground in the event of a flash flood.

A combination of topography and favorable atmospheric conditions was responsible for the extreme rainfall of 28 July 1997 just as at Big Thompson Canyon. Fort Collins is located just east of the Rocky Mountain Front Range, about 105 km (65 mi) north of Denver. A persistent westward flow of unusually humid air over eastern Colorado toward higher terrain plus weak steering winds aloft set the stage for a succession of slow moving thunderstorm cells over the Front Range, dumping heavy rain for hours over essentially the same geographical area.

Flash flooding is especially hazardous in mountainous terrain, and motorists and campers are well advised to head for higher ground in the event of a flash flood warning (Figure 11.26). But even where the topography is relatively flat, a prolonged period of heavy rain (more than 7.6 mm, or 0.3 in., per hr) can greatly exceed the infiltration capacity of the ground. Simply put, the ground becomes saturated and can absorb no additional rainwater. The excess water runs off to creeks, streams, rivers, or sewers, or collects in other low-lying areas. If the drainage system cannot accommodate the sudden input of huge quantities of water, flooding is the consequence.

Because of their design and composition, urban areas are prone to flash floods during intense downpours. Concrete and asphalt surfaces of a city are virtually impervious to water, so elaborate storm sewer systems are required to transport runoff to nearby natural drainageways. Storm sewer systems have a limited capacity for water, however, and may be unable to handle the excess runoff produced during a torrential rainfall. Water collects in underpasses, at intersections, in dips in the roadway, and in other low-lying areas (Figure 11.27). Sometimes water levels rise so quickly in these areas that motorists

FIGURE 11.27
Extreme flooding in Nashville, TN, in May 2010. [David Fine/FEMA]

are trapped in their vehicles. In many cases, motorists misjudge the depth of the water and do not realize the power of even shallow water in motion. They unwittingly drive into water that sweeps their vehicle downstream and puts their lives in extreme danger.

Severe thunderstorms are not the only cause of flash flooding. Breaching of a dam or levee, or the sudden release of water during breakup of a river ice jam, can also cause an abrupt rise in water level. In any event, when in flood-prone areas, people are well advised to heed the precautions listed in Table 11.3.

HAIL

Hail is frozen precipitation in the form of balls or lumps of ice more than 5 mm (0.2 in.) in diameter, called *hailstones* (Figure 11.28). The size of hailstones is usually reported to the National Weather Service by comparison to the size of a common object (Table 11.4). Hailstones range in size from a pea to the size of an orange or even larger. Hail is described as *severe* if its diameter is equal to or greater than 25.4 mm (1 in.), which is about the size of a quarter. *Extreme* hail has a diameter equal to or greater than 51 mm (2 in.) and accounts for less than 10% of all hail reports. In the United States, the heaviest and largest diameter hailstone fell at Vivian, SD, on 23 July 2010. It weighed about 879 g (1.94 lb) and measured 47.3 cm (18.62 in.) in circumference and 20.3 cm (8 in.) in diameter, about the size of a cantaloupe. Vivian resident Les Scott discovered the hailstone when it crashed through his deck during a severe thunderstorm. He commented that the area was littered with large hailstones.

Hail almost always falls from cumulonimbus clouds characterized by strong updrafts, great vertical

<table>
<tr><td align="center">**TABLE 11.3**
Some Flash Flood Safety Tips[a]</td></tr>
</table>

- Avoid driving on flooded roadways or bridges. Floodwaters only 0.5 m (1.5 ft) deep can carry away most automobiles. Also, flooded roads may be undermined.
- If your vehicle stalls in high water, abandon it and seek high ground.
- On foot, never attempt to cross a stream if the water level is above your knees.
- Keep children away from drainage ditches and culverts.
- Find out the elevation of your property and the flood history of your community. If there is a risk of flooding, prepare an evacuation plan.
- Exercise caution when hiking or camping in remote areas. Avoid camping in mountain valleys or near dry streambeds. Take along a battery-powered or hand crank radio, or smartphone to monitor changing weather conditions.

[a]Based on NOAA recommendations.

development, and an abundance of supercooled water droplets. A hailstone develops when an ice pellet is transported vertically through portions of a cumulonimbus cloud containing varying concentrations of supercooled water droplets. The ice pellet may descend slowly through the entire cloud, or it may follow a more complex pattern of ascent and descent as it is caught alternately in updrafts and downdrafts. In the process, the ice pellet grows by accretion (addition) of freezing water droplets.

FIGURE 11.28
Hailstones are lumps of ice that fall from intense thunderstorm cells.

TABLE 11.4
National Weather Service Hail Conversion Chart

Common objects	Diameter in cm (in.)
Penny	1.9 cm (0.75 in.)
Nickel/Mothball	2.2 cm (0.88 in.)
Quarter	2.54 cm (1.0 in.)
Half Dollar	3.2 cm (1.25 in.)
Walnut/Ping Pong Ball	3.8 cm (1.50 in.)
Golf Ball	4.4 cm (1.75 in.)
Hen Egg	5.1 cm (2.00 in.)
Tennis Ball	6.4 cm (2.50 in.)
Baseball	7.0 cm (2.75 in.)
Tea Cup	7.6 cm (3.00 in.)
Grapefruit	10.2 cm (4.00 in.)
Softball	13.7 cm (5.40 in.)

In general, the stronger the updraft, the greater is the size of the ice pellet. For example, an updraft of about 60 km per hr (37 mph) is required to produce a hailstone 1.9 cm (0.75 in.) in diameter whereas an updraft of roughly 160 km per hr (100 mph) is needed for a 7.6 cm (3.0 in.) diameter hailstone. When it grows too large and heavy to be supported by updrafts, the ice pellet descends and falls out of the cloud base. If the ice does not melt completely during its journey through the above-freezing air beneath cloud base, the ice pellet reaches Earth's surface as a hailstone.

When an ice pellet enters a portion of the cloud containing a relatively high concentration of supercooled water droplets, water collects on the ice pellet as a liquid film, which freezes slowly to form a transparent layer, or *glaze*. When the ice pellet travels through a portion of the cloud where the concentration of supercooled water droplets is relatively low, droplets freeze immediately on contact with the ice pellet. As droplets freeze, many tiny air bubbles are trapped within the ice, producing an opaque whitish layer of granular ice, or *rime*. Hence, a hailstone is composed of alternating layers of clear (glaze) and opaque (rime) ice, which, in cross section, resembles the internal structure of an onion. Scientists have counted as many as 25 layers in a single large hailstone.

A column of hail falling from a thunderstorm cell is called a *hailshaft* and the accumulation of hail on the ground in a long, narrow strip is known as a **hailstreak** (or *hail swath*). A typical hailstreak may be 2 km (1.2 mi) wide and 10 km (6.2 mi) in length, and a single large thunderstorm may produce several hailstreaks. Based on analysis of numerous hailstorms and hailstreaks, scientists at the Illinois State Water Survey devised a model to simulate hailstreak development (Figure 11.29). Hail forms in the upper portion of a cumulonimbus cloud. After several minutes of hail formation, the updraft weakens, allowing the hail to descend within the cloud; about 4 minutes later, the first hailstones reach the ground. In the ensuing 10 minutes, as the thunderstorm continues its lateral motion, the entire volume of hail is deposited along the ground as a hailstreak.

On rare occasions, the fall of hail is so great that snowplows must be called out to clear highways. This happened in Milwaukee, WI on 4 September 1988, when a heavy fall of pea-sized hail formed drifts to 46 cm

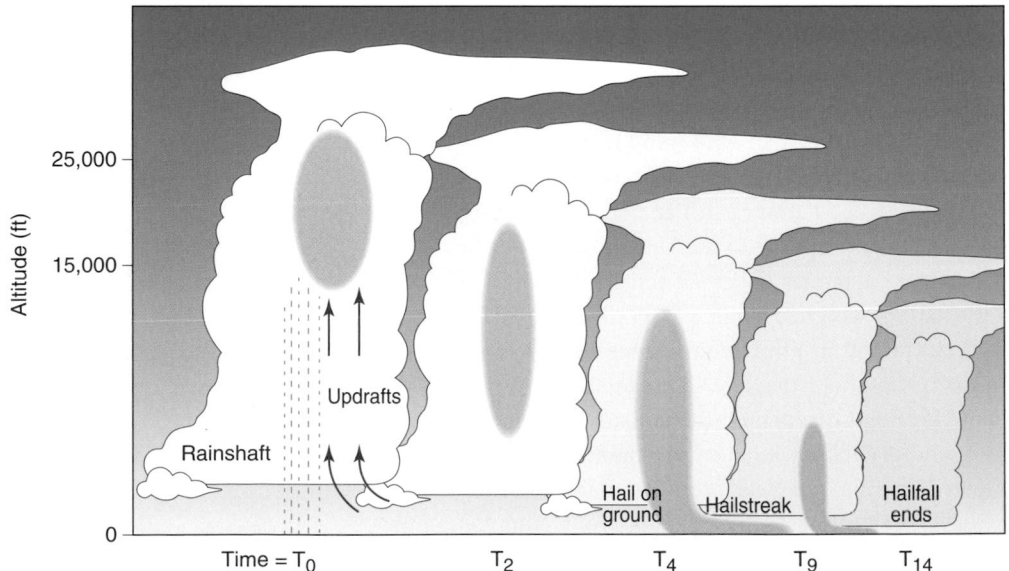

FIGURE 11.29
Model of hailstreak development. [Adapted from S.A. Changnon and J.L. Ivens, *Hail in Illinois, Public Information Brochure* 13, Illinois State Water Survey, 1987, p. 5.]

(18 in.) in some north side neighborhoods. On 6 August 1980, at Orient, IA drifts of hail were reported to be 1.8 m (6 ft) deep. On the afternoon of 13 June 1984, hailstones as large as golf balls fell for up to 1.5 hrs in the western suburbs of Denver, CO producing an accumulation of 25 cm (10 in.), with drifts greater than a meter.

Large hailstones can smash windows and dent automobiles (Figure 11.30), but the most costly damage is to crops. In the U.S. each year, hail causes an average $1 billion in damage, primarily to crops, livestock, and roofs. Hail usually falls during the growing season and in only minutes can wipe out the fruits of a farmer's year of labor. On 11 July 1990, softball-sized hailstones in Denver caused $625 million in property damage, mostly to automobiles and roofs. On 5 May 1995, the most costly hailstorm in U.S. history struck the Fort Worth, TX, area. Golf ball- to baseball-size hailstones injured scores of people who were caught out in the open during the storm. Total property damage was estimated at more than $2 billion. Traditionally, farmers cope with the hail hazard by purchasing insurance. Illinois farmers, for example, lead the United States in crop-hail insurance, purchasing an average annual liability coverage that tops $600 million. Refer to this chapter's second *For Further Exploration* for information on efforts to suppress hail formation.

According to the Storm Prediction Center (SPC) in Norman, OK, the annual number of severe hail reports in the contiguous United States increased exponentially from less than 350 in 1955 to about 9415 in 2011 (Figure 11.31), with even higher numbers of reports in recent years. The upward trend likely can be attributed to greater public awareness of the hail hazard, proliferation of cell phones

FIGURE 11.30
Car heavily damaged by hail on the New Mexico Tech campus on 5 October 2004. [Photo by Fred M. Phillips, New Mexico Tech, Socorro, NM]

making it easier for the public to file reports, improved capabilities of weather radar (especially the *WSR-88D*), and the National Weather Service's more aggressive verification of hail reports.

In the contiguous U.S., severe hail is most likely in a belt stretching northward from south-central Texas to the Valley of the Red River of the North in North Dakota and Minnesota, roughly corresponding to *tornado alley* (described later in this chapter). This locale of large hail is consistent with the region's relatively high frequency of supercell thunderstorms. Other areas of enhanced hail activity include a band from northeastern Texas eastward to North Carolina and a portion of northeast Colorado.

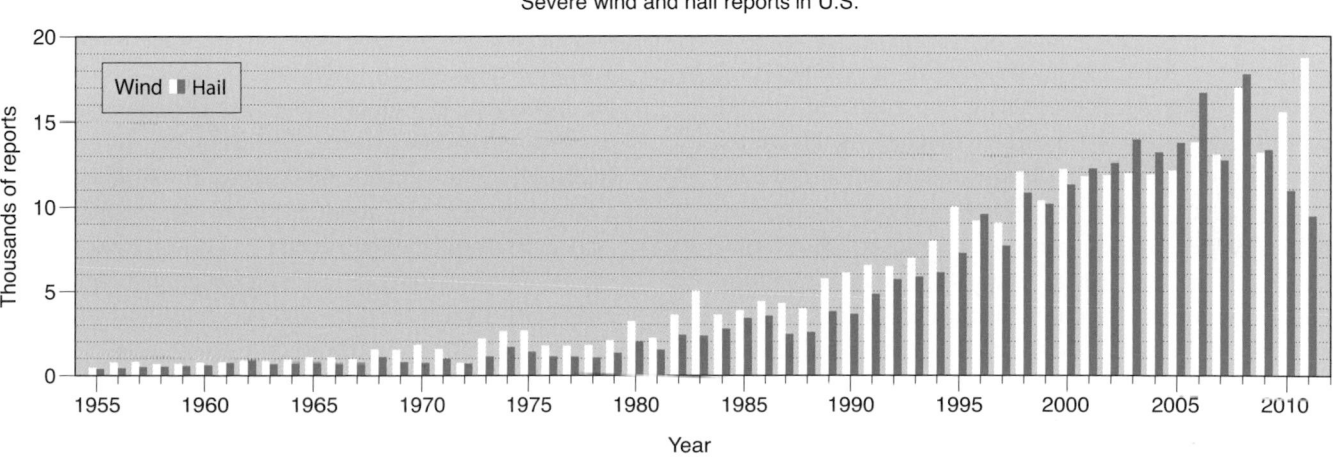

FIGURE 11.31
Annual reports of severe hail and wind occurrences in the contiguous United States have trended rapidly upward since 1955. [NOAA, NCEP, Storm Prediction Center]

The principal site of hail activity shifts seasonally from the southeast in winter to Oklahoma by April and then to eastern Colorado and northern Minnesota by July.

Tornadoes

Of the roughly 10,000 severe thunderstorms that occur in the United States in an average year, about 10% produce tornadoes. A **tornado** is a violently rotating column of air in contact with the ground that is usually produced by a thunderstorm. The system often (but not always) is made visible by water droplets formed by condensation and/or by dust and debris drawn into the tornado (Figure 11.32).

Tornadoes are the most violent of all weather systems. Fortunately, most are small and short-lived and often strike sparsely populated regions. Occasionally, however, a major tornado outbreak causes incredible devastation, death, and injury. As noted in this chapter's Case-in-Point, on 25-28 April 2011, a record super tornado outbreak claimed 321 lives. The National Weather Service confirmed 342 tornadoes in 21 states, from Texas to New York. Over a 16-hr period on 3-4 April 1974, 148 tornadoes struck 13 states in the east central United States, the most widespread and costly outbreak of tornadoes in the nation's history. This super outbreak left 315 people dead, 6142 people injured, and caused property damage in excess of $600 million. In early May 1999, an outbreak of very intense tornadoes hit portions of Oklahoma, Kansas, Texas, and Tennessee. Oklahoma City, OK, was hardest hit; 55 people lost their lives and property damage topped $1.1 billion.

Tornado Characteristics

The most striking characteristic of a tornado is a localized lowering of cloud base into a tapered rotating column composed of tiny water droplets. If this vortex remains aloft it is called a **funnel cloud**, but if it extends to the ground it is classified as a tornado. Although often funnel-shaped, in fact, tornadoes take a variety of forms, ranging from cylindrical masses of roughly uniform dimensions in cross-section, to long and slender, ropelike pendants. A funnel cloud forms in response to the steep air pressure gradient directed from the tornado's outer edge toward its center. Humid air expands and cools as it is drawn inward toward the center of the system. Cooling of air to below its dewpoint causes water vapor to condense into cloud droplets. However, if the air is exceptionally dry, a funnel cloud may not form, and the

FIGURE 11.32
A tornado is a small-scale weather system that has the potential of taking lives and causing injuries and property damage. [Photo by Joe Willey, Plainview, TX]

tornado may be made visible only by a whirl of dust and debris that is lifted off the ground.

A funnel cloud may not reach the ground. If there is no corresponding whirl of dust or debris on the ground, the system is reported as a *funnel aloft* and the system may or may not develop into a tornado. In some cases, a whirl of dust or debris appears on the ground prior to the appearance of a funnel cloud. Furthermore, the actual tornadic circulation covers a much wider area than is suggested by the funnel cloud. Typically, the diameter of a funnel cloud is only about one-tenth the diameter of the associated tornadic circulation.

A weak tornado's path on the ground typically is less than 1.6 km (1 mi) long and 100 m (330 ft) wide, and the system has a life expectancy of only a few minutes. Wind speeds are less than 180 km per hr (110 mph). While weak tornadoes account for about three-quarters of all tornadoes, they were responsible for only about 7% of all tornado fatalities from 1999-2008. At the other extreme, a violent tornado can carve out a path of destruction more than 160 km (100 mi) long and 1.0 km (3000 ft) wide, and the lifetime of the system may be from 10 minutes to more than 2 hrs. One of the broadest tornadoes on record struck Hallam, NE on 22 May 2004; its diameter was about 4 km (2.5 mi). Based on indirect measurements, wind speeds in violent tornadoes range up to 500 km per hr (300 mph).

The deadliest tornado in North American history was the Tri-State tornado of 18 March 1925. Traveling at a maximum forward speed of 118 km per hr (73 mph), the system persisted for 3.5 hrs and produced a 353-km (219-mi) path of devastation from southeastern Missouri through the southern tip of Illinois and into southwest Indiana. The tornado caused 695 fatalities and 2000 injuries, and 11,000 people were made homeless. (Actually, more than one tornado may have contributed to the Tri-State disaster.) That same day, seven other tornadoes claimed an additional 97 lives across Kentucky and Tennessee.

Most tornadoes are spawned by and travel with intense thunderstorm cells. Tornadoes and their parent cells usually track from southwest to northeast, but any direction is possible. Tornado trajectories are often erratic, with many tornadoes causing a hopscotch pattern of destruction as they alternately touch down and lift off the ground. Tornadoes have been known to move in circles and even to describe figure eights. Average forward speed is around 48 km per hr (30 mph), although there are reports of tornadoes racing along at speeds approaching 120 km per hr (75 mph) (e.g., the 1925 Tri-State tornado).

An exceptionally great horizontal air pressure gradient is responsible for a tornado's vigorous circulation. The air pressure drop over a horizontal distance of only 100 m (330 ft) may equal the normal air pressure drop between sea level and an altitude of 1000 m (3300 ft), that is, a reduction of about 10%. The continually changing direction of the inward-directed pressure gradient force produces the centripetal force that maintains the rotation of the air column. Most

Northern Hemisphere tornadoes rotate in a counterclockwise direction (viewed from above); only about 5% rotate clockwise. While the Coriolis Effect has a negligible influence at the small spatial scale represented by a funnel cloud, the counterclockwise bias is inherited from the larger parent thunderstorm.

Where and When

The central United States is one of only a few places in the world where synoptic weather conditions and terrain are ideal for tornado development; interior Australia is another. Although tornadoes have been reported in all 50 states and throughout southern Canada, most occur in **tornado alley**, a north-south corridor stretching from eastern Texas and the Texas Panhandle northward through Oklahoma, Kansas, Nebraska, and into southeastern South Dakota. Kansas and Oklahoma have the highest annual incidence of significant tornadoes (EF-2 or higher) per unit area, whereas local tornado frequency maxima occur from central Iowa eastward into central Indiana, and along the Gulf Coast. In terms of average annual number of tornadoes per 10,000 square miles by state, Kansas is the highest with 14 for the period 2001-2010 (Figure 11.33).

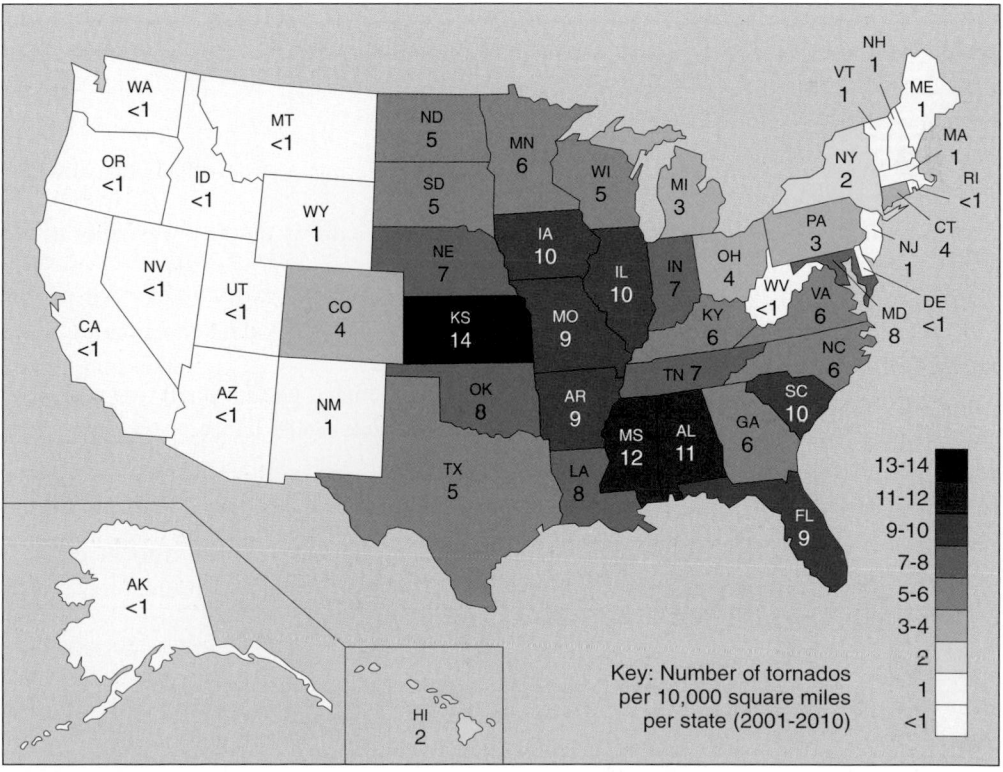

FIGURE 11.33
Average annual number of tornadoes per 10,000 square miles by state, 2001-2010. [NOAA/NWS/SPC]

Concentration of tornadoes over the central Plains is sometimes taken to imply that tornadoes do not occur in regions of great topographical relief. Actually, strong to violent tornadoes are largely unaffected by terrain; there are numerous examples of such tornadoes traversing rugged and even mountainous topography. On the other hand, weak tornadoes are more likely over flat than rough terrain.

Each year, the United States can anticipate nearly 1300 tornadoes. A national record was set in 2004 with 1817 tornadoes reported but fatalities totaled only 36. In 2011, 1690 tornadoes were confirmed, the second most tornadoes in a single year. The number of tornado deaths in 2011, however, amounted to 550, the 4th most on record since 1950. To be prepared in the event that a tornado threatens your area, study the recommendations in Table 11.5. Slightly more than half of all tornadoes develop during the warmest hours of the day (10 a.m. to 6 p.m., local time), and almost three-quarters of tornadoes in the United States occur from March to July. The months of peak tornado activity are May and June when atmospheric conditions are optimal for deep convection and the severe thunderstorms that spawn tornadoes. As of this writing, the record for the greatest number of tornadoes in a month was set in April 2011 with 758 reported; the previous monthly record was 543 in May 2003.

One factor that contributes to the spring peak in tornado frequency is the relative instability of the lower atmosphere at that time of year. During the transition from winter to summer, daylight lengthens and solar radiation incident on Earth's surface becomes more intense. Heat is transported from the relatively warm ground into the cold troposphere, but it takes time for the temperature of the entire troposphere to adjust to heating from below. The upper troposphere, in fact, usually retains its winter-like cold well into spring. The steep air temperature lapse rate (warm at low levels and cold aloft) favors vigorous convection and severe thunderstorm development.

Another factor that contributes to the spring tornado maximum is the greater likelihood that favorable synoptic weather conditions will occur at that time of year. Recall from earlier in this chapter that severe thunderstorms typically develop in the warm southeast sector of a deep extratropical cyclone. Such cyclones achieve their greatest intensity when sharp north-south temperature contrasts develop across the continent, that is, in spring when the polar front is well defined.

TABLE 11.5
Some Tornado Safety Tips[a]

- Seek shelter in a tornado (storm) cellar, an underground excavation, or a steel-framed or substantial reinforced concrete building.

- Avoid auditoriums, gymnasiums, supermarkets, or other structures that have wide, free-span roofs.

- In an office building or school, go to an interior hallway on the lowest floor or a designated shelter area. Lie flat on the floor with your head covered.

- In a high-rise, go to the center of the building (an elevator shaft or staircase). Stay away from glass.

- At home, go to the basement. If there is no basement, go to a small room (closet, bathroom, or interior hallway) in the center of the house on the lowest floor. Seek shelter under a mattress or a sturdy piece of furniture.

- Stay clear of all windows and outside walls. Protect yourself from flying debris.

- In open country, never try to outrun a tornado in a motor vehicle. Tornadoes often move too fast and their paths are too erratic to avoid even if you drive at right angles to the apparent track of the storm. Instead stop and seek shelter indoors, and if this is not possible, lie flat in a ravine, dry creek bed, or open ditch.

- Do not seek shelter in a motor vehicle or under a highway overpass.

- Get out of mobile homes (manufactured housing).

---

[a]Based on NOAA recommendations.

From late winter through spring, tornado occurrences progress northward. In effect, the center of maximum tornado frequency follows the Sun, as do the polar front jet stream, the principal storm tracks, and northward incursions of maritime tropical air. By late February, maximum tornado frequency, on average, is along the central Gulf States. In April, the maximum frequency shifts to the southeast Atlantic States; in May and June, the highest tornado incidence is usually over the southern Plains, and by early summer, it moves into the northern Plains, the Prairie Provinces east of the Rockies, and the Great Lakes region. From late summer through autumn, tornado occurrence shifts southward. During spring, most tornadoes travel from southwest to northeast, steered by southwesterly winds in the middle and upper troposphere, but in summer and into fall, many tornadoes travel from northwest toward the southeast as the steering winds shift.

Your chances of experiencing a tornado are very slim. Less than 1% of all thunderstorms produce tornadoes, and even in the most tornado-prone regions of North America, a tornado is likely to strike a given locale only once every 250 years. There are, of course, exceptions to the rule. Tornadoes hit the Oklahoma City area 147 times from 1890 to 2011, according to the NWS Forecast Office in Norman, OK.

Tornado Hazards and the EF-Scale

Tornadoes threaten people and property because of extremely high winds, a strong updraft, subsidiary vortices, and an abrupt drop in air pressure. Winds that may reach hundreds of kilometers per hour blow down trees, power poles, buildings, and other structures. Flying debris causes much of the death and injury associated with tornadoes. Broken glass, splintered lumber, and even vehicles become lethal projectiles. In violent tornadoes, the updraft near the center of the storm's funnel may top 160 km per hr (100 mph) and is sometimes strong enough to lift a railroad car off its tracks or a house off its foundation.

Some tornadoes consist of two or more *subsidiary vortices* that orbit each other or about a common center within a massive tornado. These **multi-vortex tornadoes** are usually the most destructive of all tornadoes (Figure 11.34). A particularly destructive multi-vortex tornado struck Joplin, MO, on Sunday, 22 May 2011. Part of a larger outbreak, the tornado touched down at 5:34 p.m. CDT on a path that would take it eastward through the southern part of the city (Figure 11.35). The tornado then gradually weakened as it continued eastward across Interstate Highway 44 and into more rural areas finally lifting off east of Diamond at 6:12 p.m. CDT. The total track length measured at least 35.6 km (22.1 mi). The system achieved a maximum width of more than a mile and was rated as high as EF-5. The death toll was 158 and property damage totaled $2.8 billion (in 2011 dollars) (Figure 11.36).

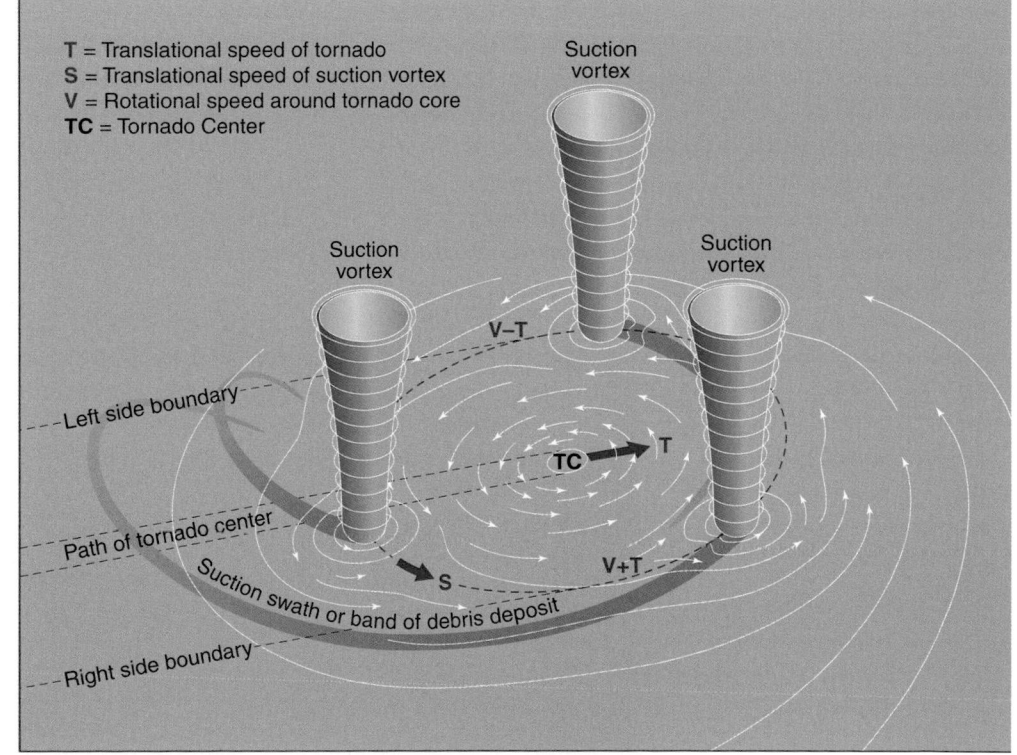

FIGURE 11.34
Model of tornado with multiple subsidiary vortices. [Above color drawing adapted from original "Fig. 8. Model of Multi-Suction Tornado" in Proposed Mechanism of Suction Spots Accompanied by Tornadoes, by T. Theodore Fujita, SMRP Research Paper 102, reprinted from preprint Seventh Conference on Severe Local Storms, Oct. 5-7, 1971, Kansas City, MO, by the Satellite & Mesometeorology Research Project, University of Chicago, p. 11, courtesy the Tetsuya "Ted" Fujita Collection, Southwest Collection/Special Collections Library, Texas Tech University]

FIGURE 11.35

Path of the Joplin, MO, tornado of 22 May 2011. Numbers along the path indicate ratings on the Enhanced Fujita tornado intensity scale. The tornado formed just west of Joplin and traveled eastward, reaching EF-5 intensity. The tornado then turned southeastward and gradually weakened. [NOAA]

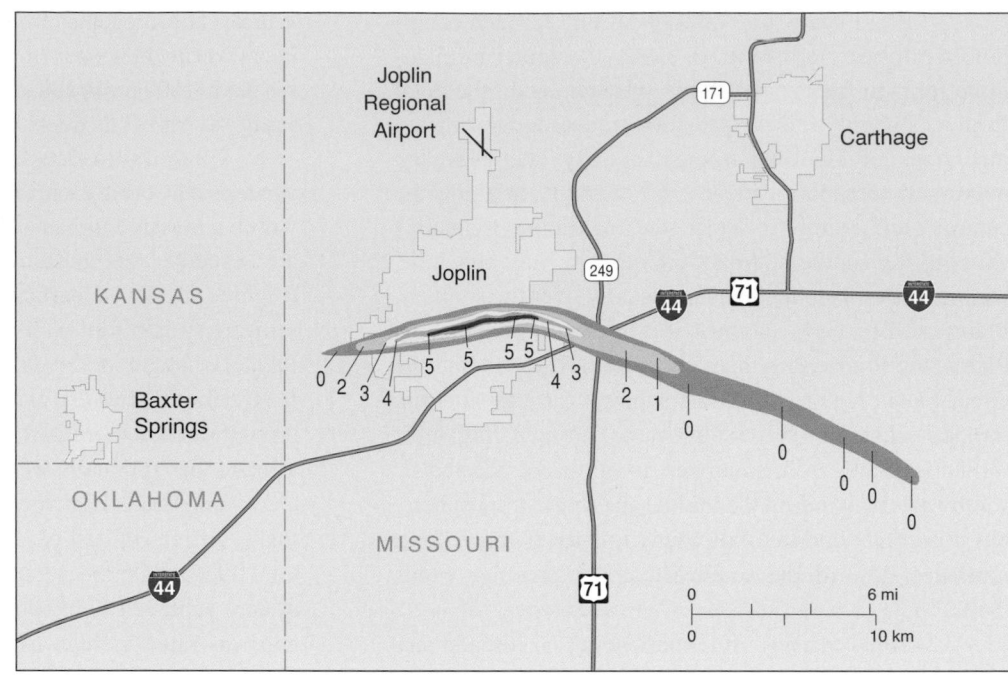

It was once widely believed that a tornado caused buildings to explode, presumably because the air pressure within the building could not adjust rapidly enough to the abrupt pressure drop associated with the tornado. In the event of a tornado sighting, people were advised to open windows to help equalize the internal and external air pressure. In fact, most buildings have sufficient air leaks that a potentially explosive pressure differential never develops. Hence, people are now advised to stay away from windows, closing them only if time permits. Structural damage to a building is caused either by winds slamming debris against the walls or by very strong currents of air that stream over the roof, causing it to lift. (The same lifting happens as air flows over the curved upper surface of an airplane wing.) As the roof is lifted, the walls collapse or are blown in, and the building disintegrates.

In 1971, T.T. Fujita devised a six-point intensity scale for rating tornado strength and damage to structures. The *F-scale*, ranging from F-0 to F-5, was based on rotational wind speeds estimated from property damage (not measured). To more closely align estimated wind speeds with associated storm damage, the F-scale was revised, becoming operational in February 2007 as the **Enhanced F-scale**, ranging from EF-0 to EF-5 (Table 11.6). An EF-0 tornado (with estimated wind gusts to 137 km per hr or 85 mph) produces minor damage, snapping tree branches and perhaps breaking some windows. EF-1 and EF-2 tornadoes can cause

FIGURE 11.36

Tornado damage in Joplin, MO. [Steve Zumwalt/FEMA]

TABLE 11.6
The Enhanced Fujita Tornado Intensity Scale

EF-Scale	Damage	3 Second Wind Gust km/hr	mph
0	light	105-137	65-85
1	moderate	138-178	86-110
2	considerable	179-218	111-135
3	severe	219-266	136-165
4	devastating	267-322	166-200
5	incredible	> 322	> 200

moderate to considerable property damage and even take lives. EF-1 tornadoes (with estimated wind gusts up to 178 km per hr or 110 mph) can down trees and shift mobile homes off their foundations, and an EF-2 tornado (with estimated wind gusts up to 218 km per hr or 135 mph) can rip roofs off frame houses, demolish mobile homes, and uproot large trees. An EF-3 tornado (with estimated wind gusts to 266 km per hr or 165 mph) can partially destroy even well constructed buildings and lift motor vehicles off the ground. At the violent end of the EF-scale, destruction is described as "devastating" to "incredible," with the potential for many fatalities. An EF-4 tornado (with estimated wind gusts up to 322 km per hr or 200 mph) can level sturdy buildings and other structures and toss automobiles about like toys. In an EF-5 tornado, sturdy frame houses are lifted and transported some distance before disintegrating from estimated wind gusts over 322 km per hr (200 mph) and devastating crash landings.

Fortunately, EF-5 tornadoes are rare. Of the nearly 1300 tornadoes that strike the United States in an average year, perhaps only one will be rated EF-5. Since 1950 in the contiguous U.S., only 52 tornadoes rated F-5 or EF-5. When one does occur, however, the impact can be catastrophic. The Xenia, OH, tornado, part of the Super Tornado Outbreak of 3-4 April 1974, rated F-5 over a portion of its 51-km (32-mi) path; that tornado claimed 34 lives. In about 1 minute, an F-5 tornado leveled the village of Barneveld, WI, on 8 June 1984; 100 homes were totally destroyed and 9 lives were lost.

On 4 May 2007 at 9:45 p.m., an EF-5 tornado struck the town of Greensburg, KS, a community of about 1400 in the south-central part of the state. According to the Federal Emergency Management Agency's *Tornado Damage Report*, the tornado created a damage path 2.7 km (1.7 mi) across and destroyed or severely damaged most structures. The tornado caused the deaths of 10 people, but the death toll likely would have been much greater were it not for the tornado warning issued by the National Weather Service 20 minutes prior to the arrival of the tornado.

According to NOAA's National Climatic Data Center, about 77% of tornadoes in the U.S. are rated weak (EF-0 or EF-1) and about 95% are below EF-3 intensity. Only 0.1% of all tornadoes reach EF-5 status. The few violent systems are responsible for the majority of all fatalities, however. Between 2001 and 2010, the average annual number of fatalities from tornadoes was 56. In an average year tornadoes injure about 1000 people.

The Tornado-Thunderstorm Connection

Most strong to violent tornadoes are spawned by a highly organized convective system known as a *supercell thunderstorm* (or just *supercell*). Supercells are very energetic with updraft speeds sometimes in excess of 240 km per hr (150 mph); they can last for several hours and produce more than one tornado. Schematic views of a tornadic supercell are shown from the side in Figure 11.37A and from above in Figure 11.37B.

Tornado development in a supercell begins with an interaction between the updraft and the larger-scale horizontal wind. In a tornadic supercell, the horizontal wind exhibits strong vertical shear in both speed and direction. That is, the wind strengthens and veers (turns clockwise) with altitude, from south or southeast at the surface to southwest or west aloft. This shear in wind speed causes the air to rotate about a horizontal axis in a rolling motion. When this rotation interacts with the updraft, the tube of rotating air is tilted from a horizontal to nearly vertical orientation. The shear in the horizontal wind direction also adds to the rotation of air about a vertical axis. Consequently, the updraft spins in a counterclockwise direction (viewed from above), forming a **mesocyclone**, 3 to 10 km (2 to 6 mi) across. Initially, the mesocyclone develops in the middle troposphere and stretches vertically, but in only about 10% of all cases does a mesocyclone evolve into a tornado.

A **wall cloud** (or *inflow cloud*), a roughly circular lowered portion of the rain-free base of a thunderstorm, often accompanies a mesocyclone. A wall cloud is typically about 3 km (2 mi) in diameter and forms in the region of strongest updraft, usually to the rear (south or southwest) of

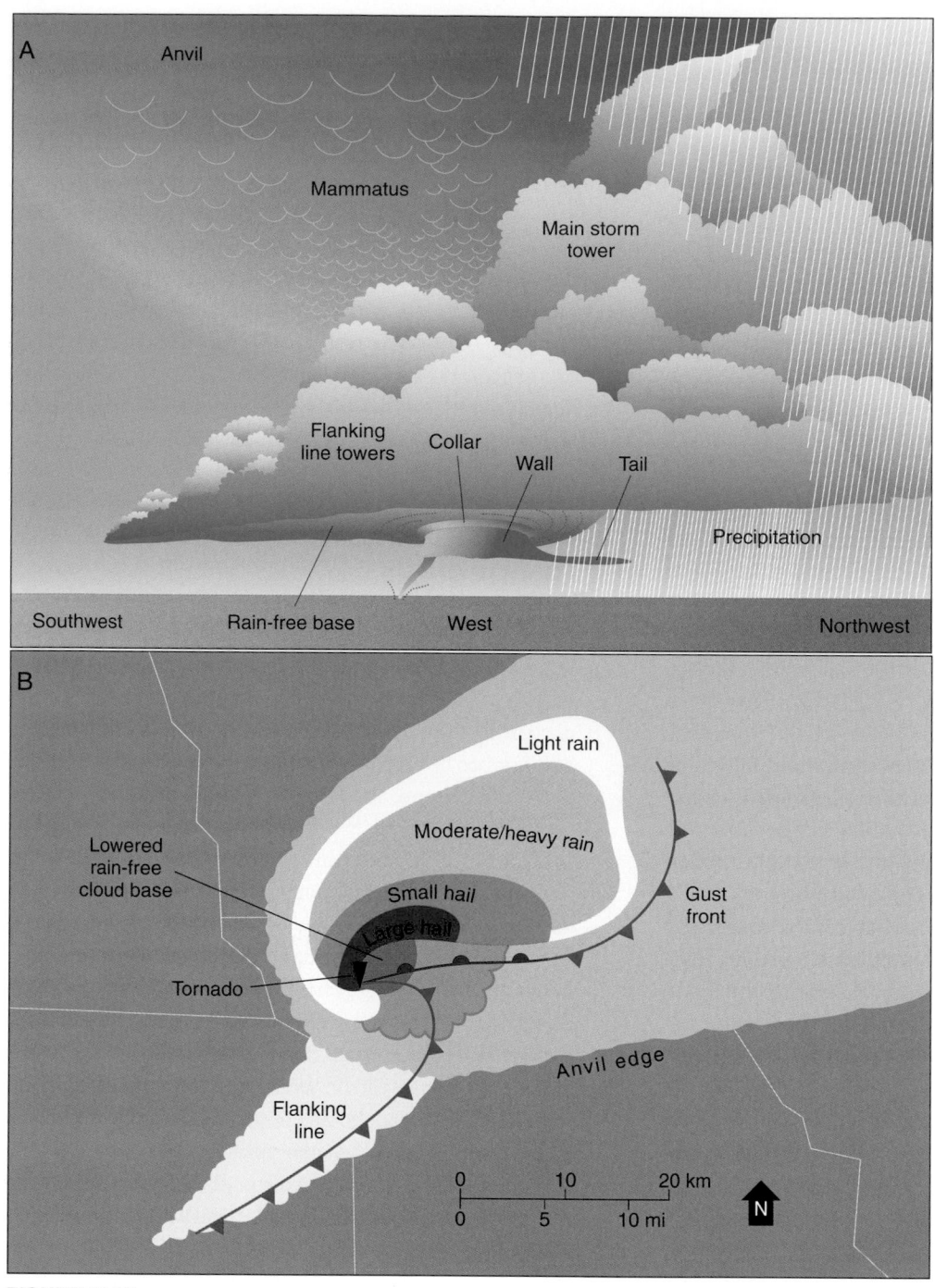

FIGURE 11.37
(A) Schematic view of a tornadic supercell thunderstorm. (B) Plan view of a tornadic supercell thunderstorm. [Adapted from NOAA, *Advanced Spotter's Field Guide*]

the main shaft of precipitation. The updraft draws humid, rain-cooled air into the system, and as that air ascends, expands and cools, water vapor begins to condense at an altitude lower than main cloud base. Hence, the base of the thunderstorm descends as a wall cloud (Figure 11.38A). As foreboding as they may appear, most wall clouds do not produce tornadoes. Tornadic wall clouds typically

are relatively long-lived, preceding the tornado by perhaps 10 to 20 minutes. Rotation of the wall cloud is obvious and persistent, becoming violent just prior to appearance of the tornado. Also, surface winds blowing in toward a tornadic wall cloud are strong (in the range of 40 to 55 km per hr, or 25 to 35 mph).

A mesocyclone circulation is most intense at an altitude of about 6100 m (20,000 ft), and from there it builds both upward and downward. Meanwhile, the updraft strengthens as more air converges toward the base of the supercell. The updraft sometimes becomes so strong that it overshoots the top of the thunderstorm and produces a dome-like cloud bulge on top of the spreading anvil cloud. In a tornadic supercell, the mesocyclone narrows and builds downward toward the ground. Relatively low air pressure in the mesocyclone causes water vapor to condense so that a funnel cloud forms (Figure 11.38B). As the spinning column of air narrows, its circulation strengthens, sometimes to extremely high speeds. (This increase in wind speed is analogous to what happens when an ice skater performs a spin: As the skater pulls her arms closer to her body, her spin rate increases.) A tornado typically appears near the updraft and toward the rear of a supercell (Figure 11.38C). Hence, in some instances, a shaft of heavy rain falling between the tornado and an observer makes visual detection of an approaching tornado difficult.

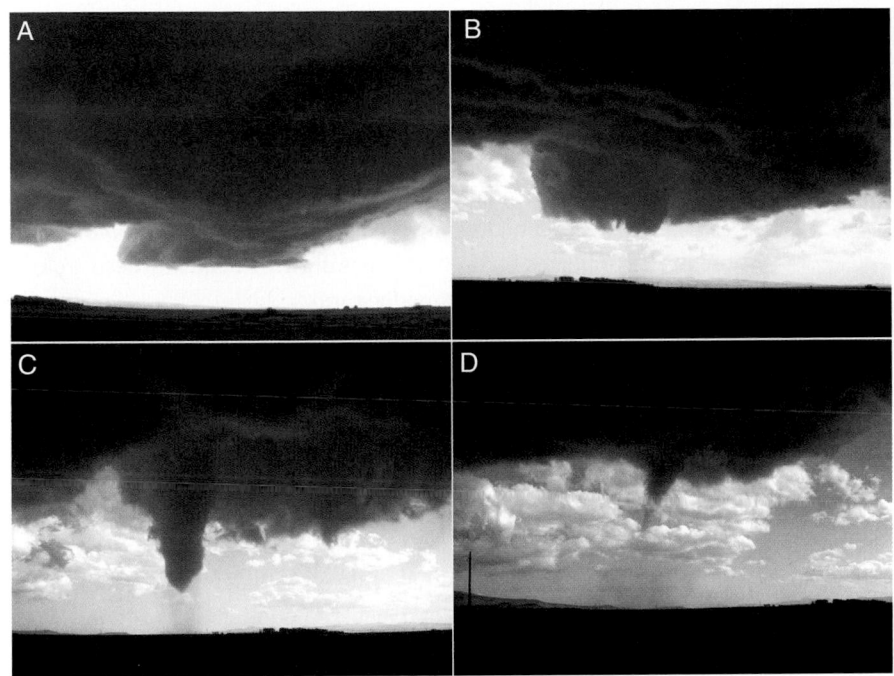

FIGURE 11.38
Tornado development in the Arbon Valley, looking Southwest from the Pocatello, Idaho Airport. (A) Wall cloud descends; (B) Funnel cloud develops from wall cloud; (C) Tornado forms; (D) Tornado dissipates. [NOAA/NWS Forecast Office, Pocatello]

layer. For example, very weak tornadoes can spin off a thunderstorm gust front. Recall that a *gust front* is the leading edge of a downdraft after it strikes the ground and begins spreading out ahead of a thunderstorm cell. These short-lived tornadoes are sometimes called *gustnadoes*. Another example of a non-supercell tornado that begins as swirling winds at the Earth's surface is a *landspout*. An updraft builds and pulls the swirling air upward. Vertical stretching causes the horizontal rotation to strengthen as the updraft eventually reaches cloud base. The landspout is made visible by ground debris that is pulled into the circulation. A similar phenomenon, called a waterspout, can occur over water. It is made visible by spray. We have more to say on waterspouts in this chapter's third *For Further Exploration, Tornado Look-Alikes.*

At about the same time that the tornadic circulation begins to descend towards Earth's surface, a downdraft develops near the rear edge of the supercell. The downdraft strikes the Earth's surface within minutes of the tornado touchdown and begins to wrap around the tornado and mesocyclone. Eventually, the downdraft completely surrounds the tornado and mesocyclone and cuts off the inflow of warm, humid air. Consequently, the tornado weakens. As the tornado dissipates, its funnel shrinks, tilts and develops a ropelike appearance (Figure 11.38D).

Although supercells typically produce the most devastating tornadoes, potentially destructive tornadoes can develop in multi-cellular clusters. Squall-line tornadoes are most common in cells located just north of a break in the line or in the southernmost (*anchor*) cell. Perhaps 90% of all North American tornadoes are spawned by thunderstorms that are associated with mature extratropical cyclones. Most other tornadoes are products of convective instability triggered by hurricanes. In fact, many hurricanes that make landfall along the southeastern United States coast are accompanied by tornadoes. Tornadoes often develop on the northeast flank of an Atlantic hurricane, after the system has turned toward the north and northeast (Chapter 12).

In some cases relatively weak tornadoes develop without any pre-existing mid-level mesocyclone. Instead, the initial rotation develops in the atmospheric boundary

Monitoring Tornadic Thunderstorms

Scientific understanding of tornadoes and their genesis is somewhat tentative because direct monitoring of tornadoes generally is not feasible; traditional weather instruments are widely spaced and not sufficiently durable to withstand tornadic winds. Between 1981 and 1983, storm chasers from the University of Oklahoma tried to deploy a 180-kg (400-lb) tornado-resistant instrument package in the path of tornadoes without much success. Today, storm chasers rely primarily on digital cameras, balloon-borne instruments that monitor atmospheric conditions surrounding severe storms, and portable Doppler radar that can resolve the circulation within supercells.

In 2009 and 2010, the *Verification of the Origins of Rotation in Tornadoes Experiment 2 (VORTEX2)* was conducted to intensely investigate tornado formation. NOAA and NSF supported this largest tornado research project in history. About 100 scientists from around the world examined tornadic supercells using various instrumentation, including portable Doppler weather radar (Figure 11.39). Data collected is enhancing understanding of how, when, and why tornadoes form, and is enabling improvement to weather prediction models.

FIGURE 11.39
Truck-mounted portable Doppler weather radar unit (*Doppler On Wheels*). VORTEX2 researchers investigated this Wyoming tornado in 2009. [Herb Stein/CSWR]

In communities where severe weather is a major concern, volunteer *storm spotters*, trained by National Weather Service personnel, monitor atmospheric conditions for signs of severe thunderstorms. Spotters provide a valuable service that complements weather radar, satellites, and networks of weather instruments and lightning detectors. But even trained storm spotters must be cautious of phenomena that may resemble tornadoes or funnel clouds. For more on this topic, refer to this chapter's third *For Further Exploration*.

As noted in Chapter 7, weather radar is an indispensable tool for monitoring severe weather systems. Operating in the reflectivity mode, weather radar cannot detect a tornado directly, but in some cases when the parent mesocyclone is present, a hook-shaped echo appears on the radar screen on the southeast side of a severe thunderstorm cell (Figure 11.40A). A **hook echo** is produced by rainfall as it is drawn around the mesocyclone within a severe thunderstorm.

Weather radar operating in the velocity (Doppler) mode monitors the circulation within a weather system rather than just the intensity, location, and movement of areas of precipitation (Figure 11.40B). Doppler radar sees inside severe thunderstorms and detects mesocyclones, developing tornadoes, gust fronts, and strong wind shear associated with microbursts. Mesocyclone signatures are sometimes identified up to 30 minutes prior to formation of a tornado and at distances up to 230 km (140 mi) from the radar site. Doppler radar can monitor the mesocyclone as it evolves and before it descends to Earth's surface to become a tornado. A tornadic circulation within several tens of kilometers of the radar may show up on the radar image as a *tornado vortex signature (TVS)*, a small region of rapidly changing wind direction within a mesocyclone. Doppler radar provides the public with more advance warning of severe weather than was possible with reflectivity-only radars. Advance warning of up to 10 minutes or more (compared to less than 5 minutes with the conventional radars in use in the late 1980s) is feasible and has saved many lives. Phased array weather radar, the

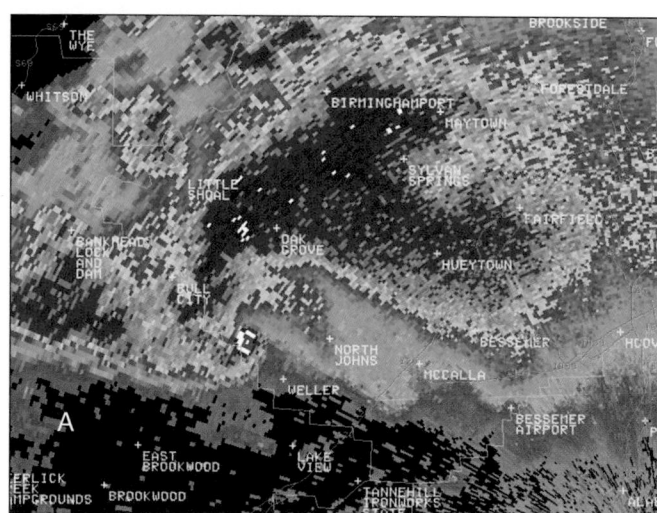
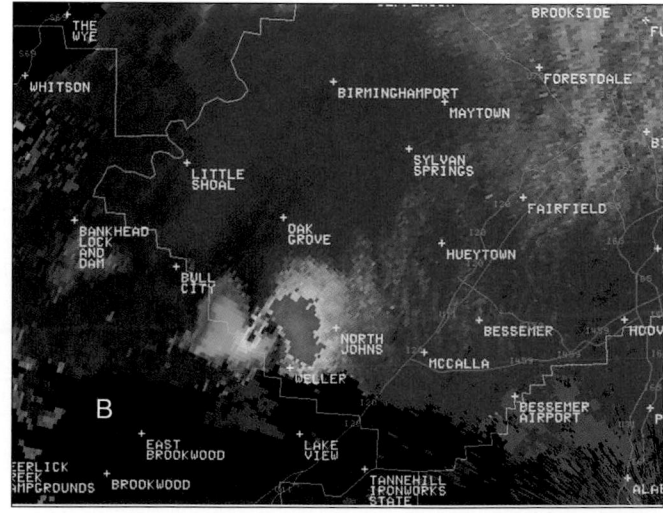

FIGURE 11.40
On 27 April 2011, an EF-4 tornado devastated areas of Tuscaloosa and Birmingham, AL. (A) is the weather radar reflectivity image showing the hook-shaped echo associated with this tornado. In (B), the corresponding radar velocity image shows a concentrated area of counterclockwise circulation in the area of the hook echo. [NOAA/NWS Birmingham, AL]

Tornado reports in U.S.

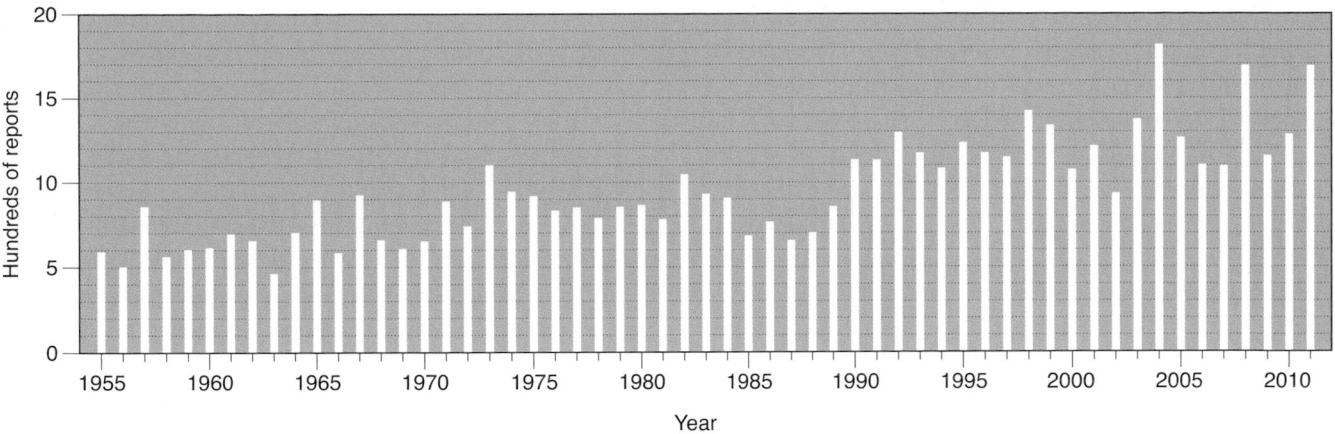

FIGURE 11.41
Annual number of reports of tornadoes in the United States for the period 1955-2011. [NOAA/NWS/Storm Prediction Center]

likely long term future of radar (Chapter 7), has the potential of extending warning lead times to 18 - 22 minutes.

Although the advent of Doppler weather radar has greatly improved the ability of meteorologists to forecast tornadoes, storm spotters and visual surveillance of thunderstorms are still needed for several reasons:

1. Not all tornadoes are associated with mesocyclones and these tornadoes are particularly difficult to forecast.
2. Most mesocyclone signatures and some tornado vortex signatures are not associated with tornadoes.
3. Many tornado-producing thunderstorms do not produce distinctive radar signatures and some storms give false signatures.
4. The resolution of radar imagery is not fine enough to detect some severe weather phenomena.

A technique assisting with improved tornado identification is the national implementation of dual-polarization weather radar, described in detail in Chapter 7. Dual-polarization radar can detect tornado debris, confirming the existence of a tornado on the ground, and lengthen tornado warning lead times for areas in the storm's path.

According to scientists at NOAA's Storm Prediction Center, the annual number of tornado reports has trended upward, more than doubling since the mid-1950s (Figure 11.41). (On the other hand, the number of tornadoes rated F-2/EF-2 or higher declined from about 200 per year in the mid-1950s to about 125 per year today.) Is the upward trend in tornado reports the consequence of a major shift in atmospheric conditions—perhaps involving climate change? Probably not. More likely, the reason for the upward trend is a combination of non-meteorological factors. Today, more people live in rural areas and observe storms meaning that storm damage that would have gone unseen in the past is now more likely to be reported. Furthermore, people today tend to have greater weather awareness and access to better and more convenient communications systems (e.g., cell phones) to report sightings of tornadoes. The spread of storm spotter networks and the advent of Doppler radar are also major contributing factors.

Conclusions

A thunderstorm is the product of convection currents that surge to great altitudes within the troposphere. As such, thunderstorms channel excess heat at Earth's surface into the atmosphere via a combination of sensible and latent heating. Thunderstorms occur most frequently in the warmest regions of Earth, especially over the continental interiors of tropical latitudes. Most thunderstorms consist of more than one cell, each of which completes its life cycle (i.e., towering cumulus, mature, dissipating stages) within an hour. A thunderstorm cell reaches its maximum intensity during its mature stage when it produces lightning, heavy rain, and strong gusty surface winds.

Although less than 1% of all thunderstorms produce tornadoes, in some areas of North America, the possibility of a tornado is a principal reason why people are fearful of thunderstorms. A tornado is a short-lived, small-scale weather system that usually is spawned by a severe thunderstorm. Tornado development requires a special combination of atmospheric conditions and terrain, so tornadoes are most frequent in spring over the central United States. Hazards of tornadoes are extremely high winds, a powerful updraft, subsidiary vortices, and an abrupt drop in air pressure. The intensity of a tornado is rated on the EF-scale from EF-0 to EF-5 based on wind speed estimated from property damage. The most intense tornadoes develop in the strong updraft of supercell thunderstorms. Our discussion of weather systems continues in the next chapter with a focus on tropical cyclones (e.g., tropical storms and hurricanes).

Basic Understandings

- A thunderstorm is a mesoscale weather system produced by strong convection currents that surge to great altitudes in the troposphere and on occasion may push into the lower stratosphere. The life cycle of a thunderstorm cell consists of a three-stage sequence: towering cumulus, mature, and dissipating.

- During the towering cumulus stage, cumulus clouds build vertically and laterally, updrafts characterize the entire system, and there is no precipitation. The mature stage begins when precipitation reaches Earth's surface. At that stage, updrafts occur alongside downdrafts and the system attains peak intensity. During the dissipating stage, subsiding air spreads through the entire cell and clouds vaporize.

- A thunderstorm usually consists of more than one cell, each of which may be at a different stage in its life cycle. A cluster of thunderstorm cells may track in a different direction than the individual constituent cells.

- Most thunderstorms develop in maritime tropical air as a consequence of uplift (1) along fronts, (2) on mountain slopes, (3) via convergence of surface winds, or (4) through intense solar heating of Earth's surface.

- Thunderstorms develop along boundaries within a maritime tropical air mass, as a squall line along or ahead of a cold front, or in a mesoscale convective complex (MCC). MCCs account for a substantial portion of growing season rainfall over the Great Plains and Midwest.

- Worldwide, thunderstorms are most common over the continental interiors of tropical latitudes. In North America, thunderstorm days are most

frequent in central Florida where converging sea breezes induce uplift of maritime tropical air, while thunderstorms are also frequent on the eastern slopes of the central and southern Rockies, associated with heating of the mountain slopes and orographic lifting. Convection is inhibited and thunderstorms are unlikely to develop in air masses that reside or travel over relatively cold surfaces and are thereby stabilized.

- Severe thunderstorm cells typically form as part of a squall line ahead of a fast-moving, well-defined cold front associated with a mature extratropical cyclone. The polar front jet stream causes dry air to subside over a surface layer of maritime tropical air. This produces a layering of air that sets the stage for explosive convection and development of severe thunderstorm cells.

- Lightning is a brilliant flash of light produced by an electrical discharge within a cloud, between clouds, or between clouds and the ground. Cloud-to-ground lightning consists of a very rapid sequence of events involving stepped leaders, bright return strokes, and dart leaders.

- What causes electrical charge separation within a cumulonimbus cloud is not well understood, but collisions between graupel and ice crystals, plus updrafts and downdrafts within the cloud likely play important roles.

- Some thunderstorm cells produce downbursts, which are intense downdrafts that spread out (diverge) at Earth's surface as potentially destructive winds. Based on size, a downburst is classified as either a macroburst or microburst. A squall line or mesoscale convective complex sometimes produces a family of straight-line downburst winds, known as a derecho.

- A flash flood is a short-term, localized, and often unexpected rise in stream level causing the stream to flow over its banks. Flash flooding is especially a hazard in mountainous terrain, where steep slopes channel excess runoff into narrow stream and river valleys, as well as in urban areas, where impervious surfaces cause excess runoff to collect in low-lying areas.

- Hail develops in intense thunderstorm cells characterized by strong updrafts, great vertical development, and an abundant supply of supercooled water droplets.

- A tornado is a small mass of air that whirls rapidly about a nearly vertical axis and is usually made visible by condensed water vapor (funnel cloud), and dust and debris drawn into the system.

- An exceptionally steep horizontal air pressure gradient between the tornado center and outer edge is the force ultimately responsible for the violence of a tornado.
- Most tornadoes occur in spring within a corridor stretching from Texas northward to southeastern South Dakota, from central Iowa eastward to central Indiana, and along the Gulf Coast states.
- Synoptic weather conditions that favor the outbreak of tornadoes progress northward (with the Sun), from the Gulf Coast in early spring to southern Canada by early summer.
- When a tornado strikes, very high winds, a strong updraft, subsidiary vortices, and an abrupt air pres-

sure drop are responsible for considerable property damage. The Enhanced Fujita Scale rates the intensity of tornadoes based on winds estimated from structural damage. Most tornadoes cause light or moderate damage (EF-0 or EF-1), but most fatalities are due to rare violent tornadoes that cause devastating or incredible damage (EF-4 or EF-5).
- Most intense tornadoes develop out of a mesocyclone that forms in the strong updraft of a supercell thunderstorm. A rotating mesocyclone circulation is generated by the interaction between a thunderstorm updraft and strong vertical shear in the horizontal wind.
- Doppler radar can monitor a tornado as it evolves from a mesocyclone and before it descends to Earth's surface.

Enduring Ideas

- Thunderstorms can bring beneficial rains, but can also cause potentially destructive lightning, downbursts, flash floods, hail, and tornadoes. An individual thunderstorm cell's life cycle consists of the towering cumulus, mature, and dissipating stages.
- Atmospheric conditions needed for thunderstorm formation include humid air in the low- to mid-troposphere, instability, and a source of uplift. Additionally, strong vertical wind shear and a capping inversion favor severe thunderstorm development. Thunderstorms are classified based on the number, organization, and intensity of their constituent cells.
- Less than 1% of thunderstorms are severe enough to spawn tornadoes. About 10% of severe thunderstorms produce a tornado, a whirling mass of air about a vertical axis, made visible by an associated funnel cloud and dust and debris drawn into the system. Tornadoes are classified on the EF-scale based on wind estimates from structural damage. The majority of tornadoes occur in spring in "tornado alley" and along the Gulf Coast, but have been reported in all 50 states.
- Most intense tornadoes are associated with a mesocyclone that forms in a supercell thunderstorm's powerful updraft. Doppler radar can monitor a tornado as it evolves from a mesocyclone and before it descends to Earth's surface.

Key Terms

thunderstorm
towering cumulus stage
cumulus congestus cloud
cumulonimbus cloud
updraft
mature stage (thunderstorm cell)
downdraft
entrainment
gust front
roll cloud
shelf cloud
dissipating stage (thunderstorm cell)
mesoscale convective systems
  (MCS)
squall line

mesoscale convective complex
  (MCC)
supercell thunderstorm
severe thunderstorm
wind shear
capping inversion
dryline
mammatus clouds
lightning
lightning detection network (LDN)
graupel
thunder
downburst
macroburst
microburst

light detection and ranging (LIDAR)
derecho
flash flood
hail
hailstreak
tornado
funnel cloud
tornado alley
multi-vortex tornadoes
enhanced F-scale
mesocyclone
wall cloud
hook echo

Review

1. Briefly describe the characteristics of the mature stage of the life cycle of a thunderstorm cell.
2. Distinguish between a roll cloud and a shelf cloud.
3. What causes the anvil top of a cumulonimbus cloud?
4. Distinguish between two types of multi-cellular thunderstorms.
5. Describe how and why thunderstorm frequency varies with Earth's surface characteristics.
6. Why are severe thunderstorms and tornadoes more likely in spring than winter?
7. Describe the *flash-to-bang* method of determining the distance to an approaching thunderstorm cell.
8. In the coterminous U.S., the number of reports of severe hail has increased exponentially since 1955. Explain why.
9. Describe the location and significance of tornado alley.
10. What is the basis of the EF-scale?

Critical Thinking

1. Although most thunderstorm cells complete their life cycle in 30 minutes or so, thunderstorm weather may persist at a specific locality for 6-12 hrs or longer. Explain why.
2. What causes the Florida thunderstorm maximum?
3. Describe the role of vertical wind shear in the development of a severe thunderstorm.
4. Why are microbursts a hazard to aviation?
5. Flash flooding is most likely in an atmosphere that is precipitation efficient. What is meant by precipitation efficient?
6. What forces govern the wind speed and direction in a tornado?
7. What is the relationship between a mesocyclone and a tornado?
8. Describe the formation of a wall cloud and a mammatus cloud.
9. What combination of atmospheric conditions favors development of a supercell thunderstorm?
10. What causes a radar hook echo and what might it signify?

For Further Exploration

Lightning Safety

In the U.S. on average each year, more people are killed by lightning (Figure 1) than by tornadoes, hurricanes, or strong winds. Compared to all other thunderstorm-related hazards, only floods are responsible for more average annual fatalities. According to NOAA's National Weather Service, the odds of being struck by lightning in one's lifetime is 1:10,000 with most victims surviving. The odds of being struck and killed by lightning is 1:600,000. Although the danger of lightning cannot be ignored, some simple precautions will minimize the risk of death or injury from lightning. No place is absolutely safe, but the risk can be significantly reduced in even the most lightning-prone area of the nation, south and central Florida, where an estimated 10 lightning flashes strike every square kilometer of land annually. In their comprehensive investigation of lightning mortality in the U.S., Walker S. Ashley and Christopher W. Gilson of Northern Illinois University, reported that from 1959 to 2006, the Miami, Fort Lauderdale, Miami Beach, FL, area ranked first among 25 metropolitan areas with a total of 107 lightning fatalities. (The Ashley-Gilson report appears in the October 2009 issue of the *Bulletin of the American Meteorological Society*.)

Even as a thunderstorm approaches, some people believe that they are safe as long as it is not raining. They continue doing what they have been doing, even if it is an inherently dangerous activity during a thunderstorm (e.g., getting in that one last hole of golf or inning of baseball). Unfortunately, lightning can strike well outside the area of rainfall. In fact, "bolts from the blue" have been reported as far as 16 km (10 mi) beyond the main shaft of thunderstorm precipitation. And in arid or semi-arid regions, thunderstorm precipitation may completely vaporize in the dry air below cloud base. Such thunderstorms still produce lightning.

When a thunderstorm threatens, use the *flash-to-bang* method to determine the distance to the thunderstorm and whether it is approaching you (described elsewhere in this chapter). If the time between flash and bang is 30 seconds or less, the wisest strategy is to seek shelter in a house or other building, avoiding contact with conductors of electricity that provide pathways for lightning. These include pipes (do not shower), stoves (do not cook), and wires (do not use the computer or land-line telephone). Electrical appliances pose no hazard if grounded, but why tempt fate by assuming they are properly wired? Also, it is a good idea to wait 30 minutes following the last thunderclap before resuming outdoor activities as new thunderstorms cells can still develop behind the main line of thunderstorms.

FIGURE 1
Night-time lightning. [Photo by Jim Kurdzo/University of Oklahoma]

Some confusion surrounds the safety of motor vehicles during a lightning storm. Contrary to popular belief, the rubber tires of a motor vehicle provide occupants with little or no protection from lightning. When lightning strikes a motor vehicle, the electrical current is conducted over the metal surface of the vehicle, through the steel frame to the tires, and then to the ground. Without direct contact with the metal frame, passengers usually escape injury. All things considered, a person is much safer inside a motor vehicle (not a convertible, the open back of a pickup truck, or a golf cart) than outdoors during a thunderstorm.

What about the lightning threat to aircraft? Pilots steer clear of thunderstorm cells not only because of lightning but also because of the associated strong wind shear and the possibility of hail. All large aircraft (and many small ones) carry so-called static discharge wicks on various trailing edges of the craft that are designed to bleed electrical charges back into the atmosphere. This prevents a build-up of electrical charge and lessens the chance of a lightning flash between clouds and aircraft.

If caught out in the open and a building or motor vehicle is not accessible when a thunderstorm approaches, find shelter under a cliff, in a cave, or in a low area, such as a ravine, a valley, or even a roadside ditch (not subject to flooding). A group of people in the open should spread out, staying several meters apart. Avoid (1) tall, isolated structures, such as trees, telephone poles, and flagpoles, (2) metallic objects, such as wire fencing, rails, wire clotheslines, bicycles, and golf clubs, (3) high areas, such as the tops of mountains, hills, and roofs, and (4) bodies of water such as swimming pools and lakes. Stay off riding lawn mowers and farm tractors (unless equipped with a metal cab). While isolated trees in open spaces are hazardous, a thick grove of small trees surrounded by taller trees may offer safe haven.

If outdoors and your hair stands on end or your skin tingles, lightning may be about to strike nearby. Immediately crouch down with only the balls of your feet touching the ground, placing your hands on your knees, and bend forward. In this way, you present the smallest possible target for a return stroke. Do not lie flat on the ground because if lightning does strike, the electrical charge running along the ground can pass through you. Such a charge traveling through your body between your head and feet while you lie on the ground will deliver much more electrical energy to you than if the charge runs along the ground between your feet as you crouch. Furthermore, unless they are wet, shoes will reduce the charge conducted to your body should a current move along the ground.

Contrary to popular belief, lightning can strike the same structure more than once (obviously as long as the structure was not destroyed). The top of the Empire State Building in New York City typically is struck more than 20 times a year, and on one occasion was struck 15 times in only 15 minutes.

About 90% of people *struck* by lightning survive, and two-thirds of people struck by lightning recover fully. Most survivors are jolted by a nearby lightning flash and are not hit directly. Such near misses, however, can produce severe burns because of the intense heat generated by lightning. If an electrical current from a lightning strike passes through the body, however, it can disrupt a special system in the heart that generates rhythmical electrical impulses that control the regular contractions of the heart muscle. If the disruption is severe, heart muscles are unable to work together and cardiac output declines. Blood flow can slow to the point that body tissues are damaged by an inadequate supply of oxygen and nutrients, a condition known as *circulatory shock*. However, the body possesses a number of mechanisms that attempt to return cardiac output to normal.

Emergency medical personnel should be summoned immediately for a lightning-strike victim. Meanwhile, cardiopulmonary resuscitation (CPR), mouth-to-mouth resuscitation, or a defibrillator may revive the victim. Covering the victim with a blanket or additional clothing will reduce heat loss from the body. It is a myth that victims of lightning strikes carry an electrical charge; they should be attended to without delay. Treatment can aid bodily processes that counteract shock. Hence, if circulatory shock is not severe, the victim usually recovers.

With severe electrical disruption of the heart, circulatory shock can reach a critical stage whereby shock breeds more shock. That is, reduced cardiac output causes the circulatory system (including the heart) to deteriorate, which in turn, causes a further reduction in blood flow. This condition, known as *progressive shock*, can trigger a vicious cycle of cardiovascular deterioration. Without immediate medical intervention, progressive shock can rapidly reach a critical stage where the person cannot be saved.

Hail Suppression[1]

Humankind's efforts to suppress hail have deep historical roots. In 14th century Europe, church bells were rung and cannons fired in the belief that the attendant noise would ward off hail. A period of particularly intense hail suppression activity took place in the grape-growing regions of Austria, France, and Italy during the late 19th century. M. Albert Stiger, a wine grower and burgomaster of Windisch-Feistritz, Austria, designed and built a funnel shaped hail suppression cannon. Stiger believed that the smoke particles in the cannon fire would inhibit hailstone development. Amazingly, in experimental firings in 1896-97 at Windisch-Feistritz, Stiger reported no hail, although severe hail damage occurred in neighboring areas.

Word of Stiger's apparent success spread throughout the vineyard regions of Europe, and hail cannons soon became commonplace (Figure 1). There were so many cannons that accidental shooting of people became a serious problem in some localities. After Stiger's much heralded success was not duplicated elsewhere, interest in hail cannons rapidly waned, and by 1905 this early attempt at weather modification had ended.

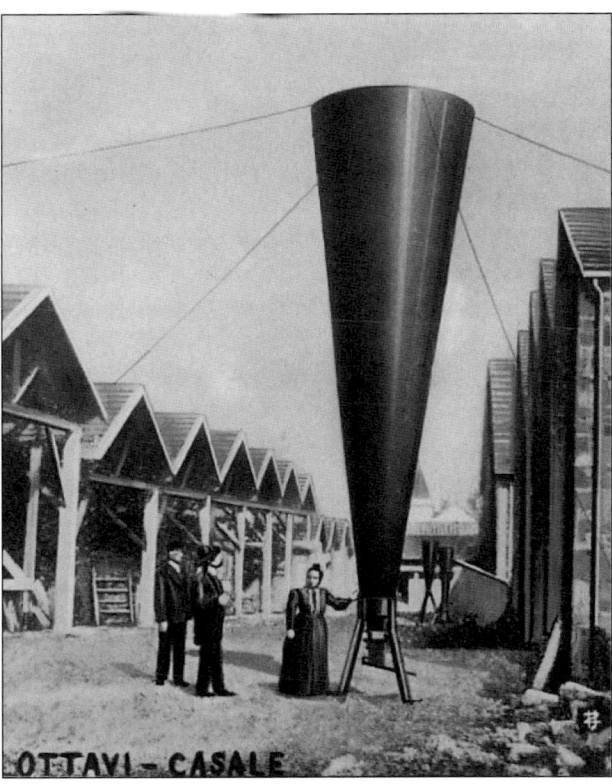

FIGURE 1
An unusually large hail suppression cannon used in Casale, Italy in the late 19th century. [Morgan, G.M., 1973. "A general description of the hail problem in the Po Valley of northern Italy," *J. Appl. Meteorol.,* 12:338-353.]

The modern era of hail suppression experimentation began after World War II. Although founded on a much better, albeit incomplete, understanding of cloud physics, the new techniques shared some similarities with earlier efforts. For example, until the practice was outlawed in the early 1970s, farmers in Italy routinely fired explosive rockets into

1 Much of this discussion is based on S.A. Changnon, Jr., and J.L. Ivens, "History Repeated: The Forgotten Hail Cannons of Europe," *Bulletin of the American Meteorological Society* 62(1981):368-375.

threatening clouds in an attempt to shatter developing hailstones. In the former Soviet Union, scientists fired silver iodide (AgI) crystals, a cloud-seeding agent, into cumulonimbus clouds. They hypothesized that AgI crystals would stimulate the formation of large numbers of small ice pellets (graupel), which would melt long before reaching the ground, instead of the normal development of small numbers of larger ice pellets that reach the ground as hailstones.

U.S. agricultural losses to hail total hundreds of millions of dollars every year. It is not surprising then that U.S. scientists set out to test the Soviet hypothesis of hail suppression and to learn more about hail-producing thunderstorms. To these ends, the *National Hail Research Experiment* was launched over northeastern Colorado in 1972. Although much was learned about hailstorms, three years of seeding potential hail-producing thunderstorms failed to confirm the Soviet hypothesis. Some scientists, however, questioned the experimental design and whether it was a viable test of the Soviet technique. In an interesting parallel with events in 19th century Europe, declining public confidence in the effectiveness of modern hail-suppression efforts brought an end to federal funding of hail suppression research in 1979.

Tornado Look-Alikes

Experienced sky watchers are well aware that a number of atmospheric phenomena resemble funnel clouds or tornadoes. These include waterspouts, virga, scud clouds, and dust devils. Even distant smoke plumes are sometimes mistaken for tornadoes.

A *waterspout* is a tornado-like system that occurs over the ocean or over a large inland lake (Figure 1). It is so named because it consists of a whirling mass of spray that appears to stream out of the base of its parent cloud, either cumulus congestus or cumulonimbus. A waterspout is usually

FIGURE 1
Waterspout in the Peace River on 15 July 2005. [Photo by the Punta Gorda, FL, Police department]

considerably less energetic, smaller, and shorter-lived than a tornado. The rare intense waterspout may well be a tornado that formed over land and then traveled out over a body of water, while on occasion a waterspout is counted as a tornado upon moving ashore. In any event, boaters should steer clear of waterspouts.

Virga or scud clouds are sometimes mistaken for a tornado or funnel cloud because their occurrence might coincidently exhibit a cylindrical or funnel-shaped profile. *Virga* is a curtain of rain or snow that vaporizes prior to reaching the ground (Chapter 7). *Scud clouds* (also called *fractus clouds*) are low, ragged stratiform or cumuliform cloud elements that are not attached to a thunderstorm cloud (Figure 2). Virga and scud clouds lack organized rotation about a vertical or near-vertical axis, clearly distinguishing them from tornadoes or funnel clouds. In arid or semi-arid areas, or wherever the soil dries out, intense solar heating often gives rise to a swirling mass of dust, known as a *dust devil*. A dust devil (Chapter 10) resembles a tornado, but forms near the ground, is not attached to any clouds, and causes little if any property damage.

FIGURE 2
Scud clouds over Olney, TX. [NOAA/National Severe Storms Laboratory (NSSL) photo by Roger Edwards]

Galveston, Texas, September 16, 2008—debris piled up by Hurricane Ike's storm surge. More debris is scattered throughout the area. [Photo courtesy of Jocelyn Augustino/FEMA]

TROPICAL WEATHER SYSTEMS

Chapter Highlights

Case-in-Point
Lessons Learned from Hurricane Katrina
Weather in the Tropics
Hurricane Characteristics
Where and When
Hurricane Life Cycle
Hurricane Hazards
Inland Flooding
Wind
Storm Surge
Saffir-Simpson Hurricane Wind Scale
Trends in Hurricane Frequency
Hurricane Threat to the Southeast United
  States
Barrier Islands
Evacuation
Long Range Forecasting of Atlantic
  Hurricanes
Hurricane Modification
Conclusions
Basic Understandings/Enduring Ideas
Key Terms/Review/Critical Thinking
For Further Exploration
Polar Lows with Hurricane Characteristics
Naming Hurricanes
Variability in Atlantic Hurricane Activity

Learning Objectives

Identify the general ways the weather of
  the tropics differs from that in middle
  latitudes.
Describe the thermal characteristics of a
  tropical cyclone.
Compare and contrast a hurricane with a
  typical extratropical cyclone.
Identify the atmospheric and oceanic
  conditions required for hurricane formation.
Locate the major hurricane breeding grounds
  worldwide.
Explain why hurricanes rarely strike the west
  coast of North America.
Explain why hurricanes are most common in
  late summer and autumn.
Summarize the life cycle of an Atlantic basin
  hurricane.
Identify the changes that take place in a
  tropical cyclone as it tracks from sea to
  land.
Identify and describe the hazards of
  hurricanes and tropical storms.
Explain the purpose of the Saffir-Simpson
  Hurricane Wind Scale.
Explain why the coastal southeast U.S. is
  particularly vulnerable to a destructive
  hurricane.
Summarize the basis for long-range
  forecasting of Atlantic hurricanes.

*What conditions are required for the development of tropical
cyclones?*

Lessons Learned from Hurricane Katrina

On Sunday evening, 28 August 2005, Hurricane Katrina's outer rain bands began sweeping ashore along the north central Gulf Coast. Within 24 hrs, Katrina's ferocious winds, storm surge, and heavy rains would devastate the Gulf Coast of Louisiana and Mississippi, obliterating coastal communities and devastating New Orleans. According to the National Weather Service, Katrina, with a minimum central air pressure of 920 mb, was the third most intense hurricane (after the "Labor Day" hurricane of 1935 and Hurricane Camille in 1969) to make landfall in the U.S. since reliable records began in 1851. Katrina directly claimed about 1200 lives, and more than 600 fatalities were indirectly linked to Katrina. In terms of economic loss, Katrina ranks as the most destructive hurricane to strike the U.S.

FIGURE 12.1

Satellite image of Katrina making landfall as a category 3 hurricane on the Louisiana coast on 29 August 2005. Katrina would be the most destructive hurricane in U.S. history. [NOAA]

Katrina began in the southeastern Bahamas on 23 August 2005 as the twelfth tropical depression of the season; the next day the system strengthened to a tropical storm and was assigned its name. Katrina intensified as it slowly moved northwestward and then westward through the Bahamas. On Thursday, 25 August, Katrina made landfall at 6:30 a.m. EDT, between Hallandale Beach and North Miami Beach in south Florida as a category 1 hurricane on the Saffir-Simpson Hurricane Wind Scale, producing maximum sustained winds of 130 km per hr (80 mph) and heavy rain (more than 12.5 cm or 5 in. in southeast Florida). Tracking almost due west and fueled by the warm waters of the Gulf of Mexico, Katrina quickly intensified while gradually turning toward the northwest and then north. By the afternoon of 26 August, Katrina was a major hurricane. As a precaution, public safety officials issued mandatory evacuation orders for the people of New Orleans, LA, Gulfport, MS, and sections of Mobile, AL. Earlier, personnel on platforms and oil rigs in the Gulf had been evacuated.

By the morning of 28 August, Katrina was a category 5 hurricane with maximum sustained surface winds of 282 km per hr (175 mph) and a minimum central pressure of 902 mb, the sixth lowest on record for an Atlantic hurricane. Katrina weakened to a strong category 3 system prior to making landfall at 7:10 a.m. on Monday, 29 August, in southern Plaquemines Parish, just south of the town of Buras, LA, at the mouth of the Mississippi River (Figure 12.1). Maximum sustained winds were 202 km per hr (125 mph) east of the storm center with hurricane force winds extending 195 km (120 mi) from its center, driving a 6 to 9 m (20 to 30 ft) storm surge that reached well inland (to land elevations of 30 ft) along the Louisiana and Mississippi Gulf Coast and as far east as Mobile, AL, flooding parts of the city. Several hours later, after passing over Breton Sound and Lake Borgne, the eye of the storm made another landfall at the mouth of the Pearl River on the Louisiana-Mississippi border. By then, Katrina had weakened somewhat but was still a category 3. At 11 p.m. EDT, the north/northeast moving system was centered near Columbus, MS, downgraded to tropical storm status. Along much of Katrina's path, total rainfall exceeded 20 to 25 cm (8 to 10 in.).

Topography made New Orleans particularly vulnerable to Hurricane Katrina. The city occupies a bowl between the Mississippi River and Lake Pontchartrain. Much of the bowl, home to more than 1 million residents and businesses, is 1.8 m (6 ft) below sea level. The people of New Orleans depend on earthen levees and concrete floodwalls to keep the water out, and pumps to remove accumulating waters, including those from flooding. However, according to the U.S. Army Corps of Engineers, the levees were designed to withstand a hurricane no stronger than a category 3. The combination of Katrina's strong winds, storm surge, and heavy rainfall caused numerous breaches in the levee system, the pumps failed, and up to 80% of New Orleans was flooded to depths that in many places reached 6 m (20 ft).

Although warned in advance of the approaching potentially catastrophic storm, thousands of people either were unable or chose not to evacuate New Orleans. As weather conditions deteriorated and floodwaters began to rise, many people fled to the Louisiana Superdome, the city's shelter of last resort. On 29 August, winds gusting to more than 160 km per hr (100 mph) ripped off part of the roof of the Superdome and knocked out electricity. An estimated 30,000 people spent many days in overcrowded, deplorable conditions without adequate food, water, sanitary facilities, or air conditioning. Other residents of New Orleans fled the rapidly rising waters, climbing to the upper floors, attics or roofs of their homes where they awaited rescue by boat or helicopter. The damage to New Orleans was so extensive that recovery is still ongoing.

Making matters worse, less than a month after Katrina devastated portions of Louisiana, Mississippi, and Alabama, another potentially catastrophic hurricane again threatened the U.S. Gulf Coast. Over the Gulf of Mexico, Hurricane Rita intensified to a category 5 with maximum sustained winds of 282 km per hr (175 mph) and a minimum central pressure of 897 mb, the fourth most intense hurricane since records began in the mid 19th century. Rita weakened to a category 3 before making landfall on 24 September 2005 on the extreme southwest coast of Louisiana near Sabine Pass, TX. Although New Orleans was spared a direct hit by Rita, the heavy rains and storm surge associated with the approaching storm caused new breaks in the recently repaired levee system. Flood waters up to 1.5 m (5 ft) deep spread over parts of the city.

The devastating impact of Hurricane Katrina and the scare caused by Hurricane Rita greatly increased the awareness of persons in the New Orleans area and made possible a strict evacuation plan for future threats. These plans were put in practice in late August 2008, when Hurricane Gustav intensified to a category 3 system in the Gulf of Mexico (after making landfall in Cuba), and moved directly toward the Louisiana coastline. Three million people fled portions of the Gulf Coast, especially southern Louisiana, and Major Ray Nagin ordered a mandatory evacuation of New Orleans. Officials used buses and contraflow lane reversal on major highways to expedite the evacuation, set up shelters away from the city (instead of using the Superdome), and Gov. Bobby Jindal ordered a pre-storm state of emergency, activating several thousand members of the Louisiana National Guard. Gustav just spared New Orleans, making landfall on 1 September 2008 about 110 km (70 mi) to the southwest of the city in Cocodrie, LA, as a category 2 hurricane. Gustav's storm surge overtopped levees and floodwalls in a few parts of New Orleans, but flooding was minor compared to that caused by Katrina. Gustav spawned 41 tornadoes (11 in Louisiana) and caused heavy rains and moderate flooding in Louisiana and Arkansas. Baton Rouge, LA, was especially hard hit by Gustav's strong winds, which knocked down trees and power lines, and effectively shut down the city for several days. Total U.S. damage from Gustav was estimated at $4.5 billion (in year 2011 dollars). (Unless otherwise noted, all damage amounts mentioned in this chapter have been standardized to 2011 dollars).

This chapter focuses on tropical weather systems, primarily tropical cyclones. **Tropical cyclones** are synoptic-scale low-pressure systems that originate over the tropical ocean and include tropical depressions, tropical storms, and hurricanes (typhoons). At maturity, a tropical cyclone is one of the most intense and destructive storms on Earth. We describe the characteristics, geographical and seasonal distribution, life cycle, and hazards of tropical cyclones. We also consider trends in hurricane frequency, the hurricane threat to the southeastern United States, efforts to forecast hurricanes based on atmospheric conditions at lower latitudes around the globe, and unsuccessful experiments to modify hurricanes. We set the stage by summarizing some of the basic characteristics of weather in the tropics.

Weather in the Tropics

The tropics, the belt between the Tropics of Cancer and Capricorn (from 23.5 degrees N to 23.5 degrees S), exhibits relatively little seasonal variation in temperature because of year-round high noontime solar altitude and nearly uniform day length (Chapter 3). In fact, through much of the tropics, the diurnal (daily) temperature variation is greater than the range in monthly mean temperatures through the course of a year.

In the tropics, broad expanses of air are uniformly warm and humid so that fronts and frontal weather are absent. Intense solar radiation heats Earth's surface driving deep convection and isolated afternoon thunderstorms that produce brief bursts of heavy rain. For reasons presented in Chapters 4 and 11, these thunderstorms are more frequent over continents and large islands than the adjacent ocean. Sometimes thunderstorm cells are aligned as narrow bands known as *tropical non-squall clusters*. On other occasions, a band of more intense cells forms, similar to a squall line of middle latitudes. In these so-called *tropical squall clusters*, winds can be strong and gusty and rainfall may persist for several hours and can be very heavy especially at the onset. The ITCZ (intertropical convergence zone) stimulates thunderstorm activity and its north-south seasonal shifts are responsible for the seasonal variations in precipitation in portions of the tropics (Chapter 9). The ITCZ follows the Sun, moving northward in the Northern Hemisphere spring and southward in the Northern Hemisphere autumn. The rainy season is summer (high Sun) and the dry season is winter (low Sun).

Little variation in the thermal and moisture properties of tropical air means that there is little horizontal variation in surface air pressure. Hence, isobaric maps are of little value in routine weather analysis and forecasting in the tropics. Instead, meteorologists rely upon streamline analysis. A **streamline** is a line drawn on a map that is everywhere parallel to the wind direction, thereby portraying graphically the horizontal flow of air. Streamline analysis can be used to identify regions of divergence and convergence such as associated with an easterly wave. As described later in this chapter, an *easterly wave* is an important weather-maker in the tropics and sometimes is the precursor of a tropical cyclone.

Lengthy episodes of tranquil warm weather are typical of the tropics, but occasionally tropical cyclones develop that bring torrential rains and strong winds. The most feared and potentially most damaging of these tropical cyclones is the hurricane.

Hurricane Characteristics

Hurricane is likely derived from *Haracan*, the name of the storm god of the Taino people who inhabited Caribbean islands at the time of Spanish exploration of the New World. A **hurricane** is an intense cyclone that originates over tropical ocean waters, usually in late summer or early fall (when sea-surface temperatures are highest), and has a maximum sustained wind speed of at least 119 km per hr (74 mph). By convention in the United States, *sustained wind speed* is a one-minute average measured at the standard anemometer height of 10 m (33 ft).

A convenient way to describe a hurricane is to contrast it with the extratropical cyclone examined in Chapter 10. A hurricane develops in a uniformly warm and humid air mass, so the system has no associated fronts or frontal weather (Figure 12.2). Isobars plotted on a surface weather map form a pattern of closely spaced nearly concentric circles about the system's low pressure center. Typically, the central pressure at sea level is considerably lower and the horizontal air pressure gradient much greater in a hurricane than in an extratropical cyclone. Reconnaissance aircraft extrapolated a surface air pressure of 882 mb (26.05 in.) at the center of Hurricane Wilma while the storm was over the northwest Caribbean Sea early on 19 October 2005; this is the lowest sea-level air pressure on record in the Western Hemisphere. A hurricane is usually a much smaller system, averaging a third the diameter of a typical extratropical cyclone. Rarely do hurricane-force winds extend much more than 120 km (75 mi) beyond the system's center. Structurally, a mature hurricane is a warm-core low that weakens rapidly with altitude, especially above 3000 m (9800 ft). In the upper troposphere, at altitudes above about 12,000 m (40,000 ft),

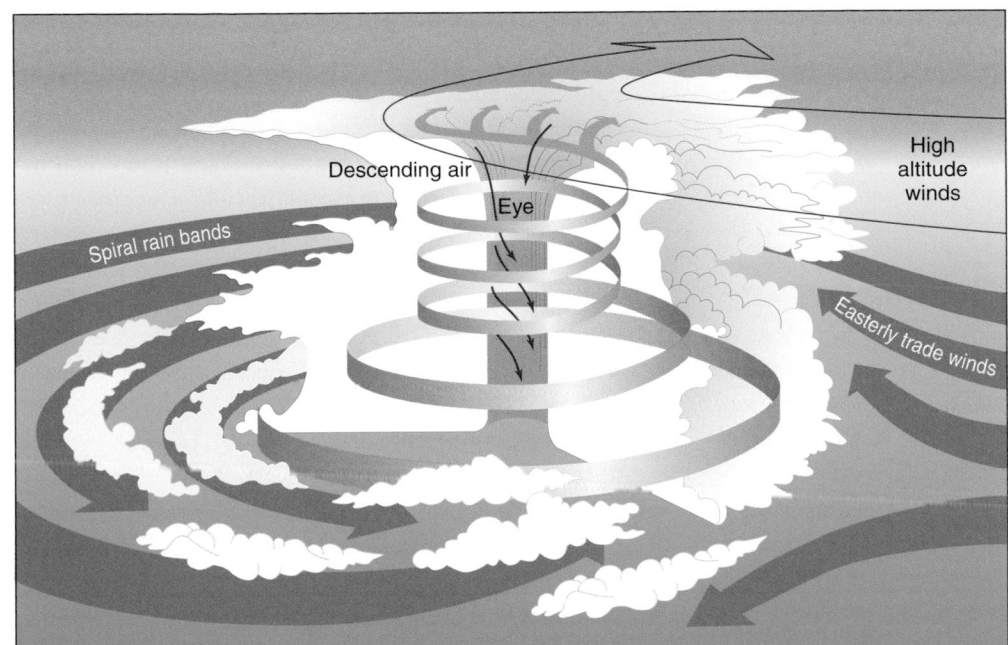

FIGURE 12.2
Schematic diagram of the internal structure of a hurricane. Note that the vertical dimension is greatly exaggerated. [From NOAA, Hurricane. Washington, DC: Superintendent of Documents, 1977]

the circulation usually becomes anticyclonic (clockwise when viewed from above in the Northern Hemisphere). Recall that an extratropical cyclone is a cold-core system whose circulation strengthens with altitude.

At the center of a hurricane is an area of almost cloudless skies, subsiding air, and light winds (less than 25 km per hr or 16 mph), called the **eye** of the storm (visible in Figure 12.1). The eye generally ranges from 10 to 65 km (6 to 40 mi) across, shrinking in diameter as the hurricane intensifies and winds strengthen. At a hurricane's typical rate of forward motion, the eye may take up to an hour to pass over a given locality. People are sometimes lulled into thinking the storm has ended when skies clear and winds abruptly slacken following a hurricane's initial blow. They may be experiencing passage of the hurricane's eye; heavy rains and ferocious winds will soon resume but blow from the opposite direction.

Bordering the eye of a mature hurricane is the **eye wall**, a ring of thunderstorm (cumulonimbus) clouds that produce heavy rains and very strong winds (Figure 12.3). The most dangerous and potentially most destructive part of a hurricane is the portion of the eye wall on the side of the advancing system where the wind blows in the same direction as the storm's forward motion. On that side, hurricane winds combine with the storm's forward motion producing the system's strongest surface winds. In the Northern Hemisphere, this dangerous semicircle of high winds and high ocean waves is on the right side of the hurricane when facing in the direction of the system's forward movement. Cloud bands, producing heavy convective showers and hurricane-force winds, spiral inward toward the eye wall. Meanwhile, at high altitudes cirrus or cirrostratus clouds spiral outward from the system.

Intense, long-lived hurricanes often undergo an *eye wall replacement cycle*. Over a period of a day or so, the main eye wall collapses and the storm weakens. Then an outer eye wall contracts and takes the place of the original eye wall. The hurricane then intensifies.

FIGURE 12.3
The eye wall of Hurricane Katrina taken on 28 August 2005, as seen from a NOAA P-3 hurricane hunter aircraft while the storm was over the Gulf of Mexico. [NOAA]

Whereas hurricanes differ substantially from extratropical cyclones, they share some characteristics with intense storms that sometimes develop over the ocean at high latitudes. For more on these systems, refer to this chapter's first *For Further Exploration, Polar Lows with Hurricane Characteristics.*

Where and When

Three conditions are necessary for a tropical cyclone to form: relatively high sea-surface temperatures, adequate Coriolis Effect, and weak winds aloft. In addition, relatively humid air in the mid-troposphere (near 4900 m or 16,000 ft) favors tropical cyclone development. To a large extent, these requirements dictate where and when tropical cyclones form.

Tropical cyclones require a sea-surface temperature (SST) of at least 26.5 °C (80 °F) through an ocean depth of at least 45 m (150 ft). Such exceptionally warm ocean water sustains the system's circulation by the latent heat released when water vapor, evaporated from the ocean surface, is conveyed upward and condenses within the storm system. Recall from Chapter 6 that temperature largely governs the rate of evaporation of water, so the higher the SST, the greater is the supply of latent heat for the storm system. Furthermore, the spray from breaking ocean waves readily evaporates, adding to the supply of water vapor that condenses and releases latent heat in the developing tropical cyclone.

As a tropical cyclone makes landfall or moves over colder water, however, it loses its warm-water energy source and weakens. The strong winds of a tropical cyclone stir up surface ocean waters and can bring about a drop in sea-surface temperatures. Cyclonic winds induce *Ekman transport* and divergence of surface waters under the storm system, bringing cold water to the surface (Chapter 9). Until the normal thermal structure of the ocean is restored, lower than usual SST can inhibit development of subsequent tropical cyclones over the same region of the ocean. In addition, atmospheric scientists discovered that changes in SST associated with warm- and cold-core ocean rings influence tropical cyclone development. A **ring** is a large turbulent eddy that forms when a meander in a current forms a loop that pinches off, separating from the main ocean surface current such as the Gulf Stream. Rings can be monitored remotely via sensors onboard Earth-orbiting satellites. Ocean water to the north of the Gulf Stream is relatively cold with temperatures averaging less than 10 °C (50 °F). Rings that form on the north side of the Gulf Stream have entrained relatively warm water

from the Sargasso Sea whereas rings that form on the south side of the Gulf Stream have entrained relatively cold water. In the Northern Hemisphere, *warm-core rings* rotate in a clockwise direction, have diameters to 200 km (125 mi), and consist of a pool of warm water that may reach a depth of 1500 m (4900 ft). *Cold-core rings* rotate in a counterclockwise direction, are larger (diameters to 300 km or 185 mi) and their associated pool of cold water may reach the ocean floor (to a depth of more than 4000 m or 13,000 ft). Tropical cyclones may intensify over warm-core rings but weaken over cold-core rings. For example, the Loop Current in the Gulf of Mexico is a source of warm-core rings that may intensify tropical cyclones passing overhead.

The second condition required for tropical cyclone development is a significant Coriolis Effect; that is, the influence of Earth's rotation must be sufficiently strong to initiate a cyclonic circulation. As noted in Chapter 8, the Coriolis Effect is zero at the equator and increases as latitude increases. With very rare exceptions, tropical cyclones do not form within 5 degrees of the equator (a distance of about 480 km or 300 mi). One exception to this rule was Typhoon Vamei which formed near Singapore on 27 December 2001. Vamei was centered near 1.5 degrees N and its circulation extended into both hemispheres producing maximum sustained surface winds of 140 km per hr (87 mph).

The first two conditions that favor tropical cyclone formation (i.e., high SST and sufficient Coriolis Effect) are present only over certain portions of the ocean. The main ocean breeding grounds for tropical cyclones along with average storm trajectories are plotted in Figure 12.4. Most hurricanes form in the 8- to 20-degree latitude belts. Major hurricane breeding grounds are: (1) the tropical North Atlantic west of Africa (including the Caribbean Sea and Gulf of Mexico), (2) the North Pacific Ocean west of Mexico, (3) the western tropical North Pacific and China Sea, where a hurricane is called a *typhoon* (from the Cantonese, *tai-fung*, meaning great wind), (4) the South Indian Ocean east of Madagascar, (5) the North Indian Ocean (including the Arabian Sea and the Bay of Bengal), and (6) the South Pacific Ocean from the east coast of Australia eastward to about 140 degrees W. In the Indian Ocean and near Australia, hurricanes are simply called *cyclones.*

The requirement of high sea-surface temperature explains the seasonal occurrence of tropical cyclones (Figure 12.5). For reasons presented in Chapter 4, the temperature of surface ocean waters lags the regular seasonal variations in incoming solar radiation. Sea-

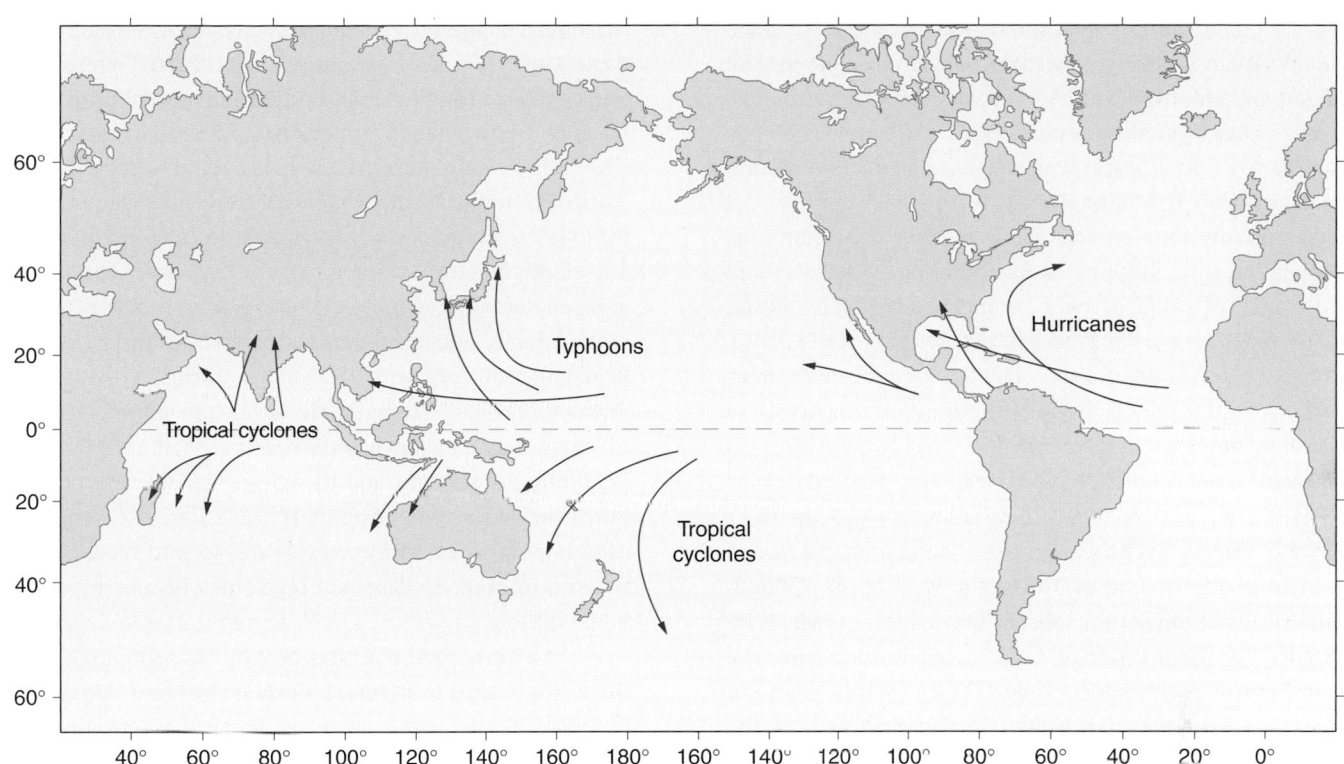

FIGURE 12.4
Tropical cyclone breeding grounds are located only over certain regions of the world ocean. Arrows indicate average hurricane trajectories.

surface temperatures reach a seasonal maximum about 6 to 8 weeks after the date of most intense incoming solar radiation. Most Atlantic hurricanes develop when surface waters are warmest, that is, in late summer and early autumn; the official hurricane season runs from 1 June to 30 November, with the peak hurricane threat for the U.S. coastline between mid-August and late October.

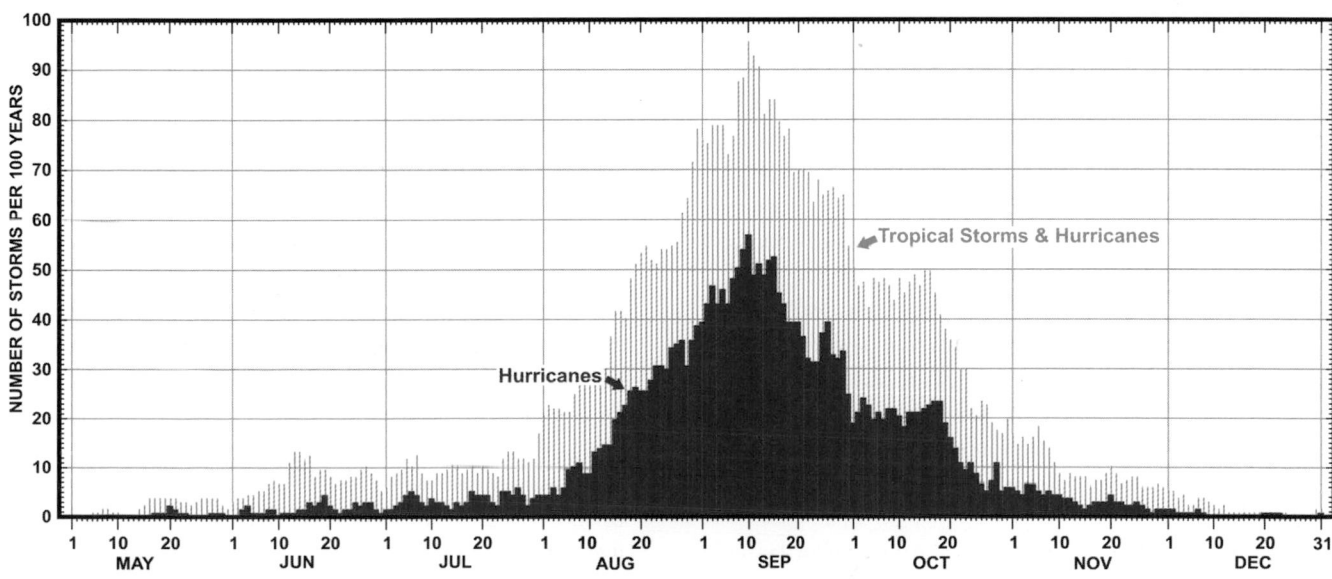

FIGURE 12.5
The frequency of Atlantic basin tropical storms and hurricanes by date (number of storms per 100 years). The date of peak frequency occurs about 2.5 months following the summer solstice. [NOAA]

The third condition for tropical cyclone development is relatively weak winds aloft (in the middle and upper troposphere) over oceanic breeding grounds. Weak winds aloft allow a cluster of cumulonimbus clouds to organize over tropical seas, the first step in the evolution of a hurricane. By contrast, strong west-to-east winds aloft shear off the tops of westward tracking thunderstorms, preventing the systems from building vertically and organizing. During El Niño, winds are unusually strong at altitudes near 12,000 m (40,000 ft) over the Atlantic hurricane breeding ground. Hence, Atlantic hurricanes usually are infrequent during El Niño. For example, the Atlantic hurricane season of 1997 occurred during an intense El Niño and produced relatively few hurricanes (3), only one of which made landfall on the U.S. coast.

Strong vertical *wind shear* is the principal reason hurricanes rarely form off the east or west coasts of South America (although Caribbean hurricanes occasionally impact the north coast of Venezuela). Furthermore, the intertropical convergence zone (ITCZ) usually is absent over the Southern Hemisphere ocean and without the ITCZ, there is no source of synoptic-scale vorticity (spin) or convergence needed for tropical cyclone development. Nonetheless, on 28 March 2004, a rare hurricane that formed in the South Atlantic Ocean made landfall as a category 1 system on the coast of Brazil about 800 km (500 mi) south of Rio de Janeiro. Hurricane Catarina, as it was later called, claimed at least two lives, destroyed 500 homes, and left 1500 people homeless.

According to NOAA's National Climatic Data Center *International Best Track Archive for Climate Stewardship (IBTrACS)*, for the 30-year period 1980-2009, the global annual average of named tropical cyclones is 86.5. Slightly more than half of those storms (45.4) developed into a hurricane or typhoon (category 1-3), and of those, 21.9 reached major storm status (category 4-5). The western Pacific Ocean, with its vast expanse of warm surface waters, is the most active area for tropical cyclones, with an average of 26.3 systems each season (1980-2009). More than half of those systems, 15.9, develop into a typhoon (category 1-3), and of those, 8.0 reach major storm status (category 4-5). Some of these typhoons can become very intense. A typhoon with maximum sustained winds of 240 km per hr (150 mph) or stronger is called a *supertyphoon*. Typically, four supertyphoons develop in the western Pacific annually.

Hurricanes spawned over the tropical Atlantic, Caribbean Sea, and Gulf of Mexico pose the most serious threat to coastal North America. According to the National Hurricane Center, based on the period 1966-2010, a seasonal average 11.5 named tropical storms formed over these waters. Of these systems, on average, 6.2 intensified into hurricanes and of these 2.4 became major hurricanes. On average, two hurricanes strike the U.S. coast each year. The 2005 Atlantic hurricane season set a new record with 27 named tropical cyclones (and one unnamed subtropical cyclone); the previous record was 21 set in 1933. In 2005, 15 Atlantic tropical cyclones became hurricanes (a record) and 4 major hurricanes struck the U.S. (also a record). The 1995 Atlantic hurricane season was the third most active in recorded history with 19 tropical storms, 11 of which attained hurricane strength. However, the annual number of hurricanes may have little bearing on the number of landfalling hurricanes and their impact. Whereas only 4 hurricanes occurred during the 1992 Atlantic season, one of them, Hurricane Andrew, was the second most costly in terms of property damage in U.S. history, amounting to $44.8 billion.

Hurricanes have hit every Gulf and Atlantic Coast State from Texas to Maine. Florida is the most hurricane-prone of all the states, with 66 hurricanes crossing its coastline between 1900 and 2011. In the same period, Texas was second with 41 hurricanes, while Louisiana had 31, and North Carolina had 30.

The primary breeding ground for hurricanes in the Atlantic shifts east and then west with the seasons. Early in the hurricane season (May and June), hurricanes form mostly over the Gulf of Mexico and the western Caribbean. By July, the main area of hurricane development begins to shift eastward across the tropical North Atlantic. By mid-September, most hurricanes develop in a belt stretching from the Lesser Antilles (in the eastern Caribbean Sea) eastward to south of the Cape Verde Islands (off Africa's West Coast). After mid-September, hurricanes again originate mostly over the Gulf of Mexico and the western Caribbean.

According to the National Hurricane Center, the Pacific Ocean off Mexico and Central America ranks second to the western North Pacific in average annual number of tropical cyclones (15.1 per year between 1966 and 2010); the majority of these (8.4) develop into hurricanes. The Pacific Coast of the United States is rarely a target of hurricanes, although one or two tropical cyclones typically make landfall on the Mexican Pacific Coast each year. Prevailing winds (northeast trades) are directed offshore and usually steer tropical cyclones that form west of Central America away from the coast. Also, the southward flowing cold California Current plus upwelling just off the southern California (and Baja California) coast produce sea-surface temperatures that normally are too low to sustain hurricanes that travel toward the northeast. However, during unusual atmospheric/oceanic circulation regimes, hurricanes have struck coastal Southern California and even traveled over the Desert Southwest. For example, on 9-10 September 1976, Hurricane Kathleen crossed Baja California and tracked near Yuma, AZ, bringing torrential rains and flooding to the Imperial and Lower Colorado River Valleys. Another example was Hurricane Nora in September 1997. Nora made landfall on 25 September near Punta Eugenia on the Baja California coast with maximum sustained winds of 137 km per hr (85 mph). The system weakened to a tropical storm and crossed into the United States near the California/Arizona border. Rainfall up to 7.5 cm (3 in.) was common in the Desert Southwest and some mountain locations in Arizona reported almost 30 cm (12 in.). For comparison, Yuma, AZ, normally receives 7.64 cm (3.01 in.) of precipitation annually.

The Hawaiian Islands are sometimes threatened by tropical cyclones that develop over the central tropical Pacific or track westward into that region from the Pacific hurricane breeding grounds west of Mexico. Fortunately, in an average year, only 3 to 4 tropical cyclones affect the central Pacific and since 1957 only 4 hurricanes have impacted the Islands. The most recent one, Iniki in September 1992, devastated the island of Kauai. Iniki tracked directly over the island with estimated maximum sustained winds of 185 km per hr (115 mph) and gusts to 258 km per hr (160 mph). A storm surge of 1.5 m (5 ft) topped by wind-driven waves caused extensive flooding. Seven people lost their lives. Total property damage was estimated at $3.0 billion, making this the most costly natural disaster in the history of the State of Hawaii.

Hurricane Life Cycle

The first sign that a hurricane may be forming is the appearance of an organized cluster of cumulonimbus clouds over tropical seas. This region of convective activity is labeled a **tropical disturbance** if a center of low pressure is detected at the surface. Chances are that the tropical disturbance was triggered by the ITCZ, a trough in the westerlies that intruded into the tropics from middle latitudes, a West African disturbance line (similar to a squall line), or a wave (or ripple) in the easterly trade winds (an *easterly wave*).

The majority of tropical cyclones that threaten North America develop out of convective cloud clusters associated with easterly waves traveling westward off the West African coast. In fact, easterly waves are precursors

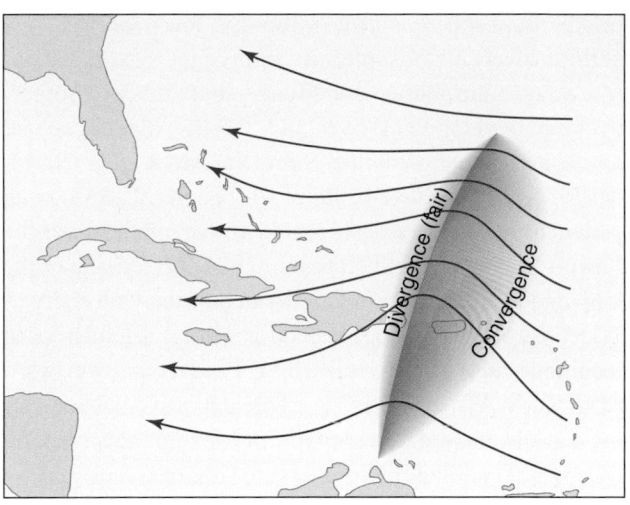

FIGURE 12.6
Schematic drawing of an easterly wave moving westward across the western tropical Atlantic. An easterly wave may play an important role in triggering the development of a tropical cyclone.

of perhaps 65% of all named tropical cyclones in the Atlantic basin. An **easterly wave** is a ripple in the tropical easterlies (trade winds) featuring a weak trough of low pressure (Figure 12.6). The system likely owes its origin to the disturbance of the prevailing easterlies by the Ethiopian Highlands of East Africa. Most common from August through October, an easterly wave propagates from east to west across the tropical Atlantic generally more slowly than the flow in which it is embedded. Some waves pass over Central America and into the eastern Pacific where they can trigger tropical cyclone development.

Over the ocean, to the west of the trough line within an easterly wave is an area of horizontal divergence in the lower troposphere, a shallow moist layer, and fair weather. To the east of the trough line is an area of intense lower-tropospheric convergence, a deep moist layer, and heavy rain showers. Convergence on the east side of an easterly wave helps to organize convective activity into a developing tropical cyclone. Storm systems originating in this way over the Atlantic are referred to as *Cape Verde-type hurricanes*.

Only a small percentage of convective cloud clusters that appear in the tropical Atlantic actually evolve into full-blown hurricanes for the following reasons: (1) Subsidence of air on the eastern flank of the Bermuda-Azores anticyclone and the associated *trade-wind inversion* inhibit deep convection (Chapter 9). (2) Vertical wind shear (change of horizontal wind with altitude) over the tropical Atlantic is usually too great for hurricane formation; strong horizontal winds aloft disrupt deep convection. (3) The middle troposphere is usually too dry; low vapor pressure at these levels inhibits intensification of the system.

Atmospheric conditions that inhibit tropical cyclone development over the North Atlantic appear to be associated with the **Saharan Air Layer (SAL)**, described in the Case-in-Point of Chapter 2. SAL is an elevated mass of dry dusty stable air originating over the Sahara Desert of North Africa. In 2004, J.P. Dunion of the University of Miami and C.S. Velden of the University of Wisconsin-Madison reported on a special satellite-based technique they used to track SAL's movement westward from North Africa across the tropical North Atlantic, over the Caribbean, and into the Gulf of Mexico. This tracking technique enabled researchers to observe interactions between the SAL, easterly waves, and tropical cyclones. In 2006, A.T. Evan of the University of Wisconsin-Madison and colleagues followed up on the 2004 report, demonstrating a strong link between satellite-measured mean dust coverage and tropical cyclone activity over the North Atlantic.

Over its desert source region, the Saharan Air Layer is well mixed up to an altitude of almost 5500 m (18,000 ft)—near the 500-mb level. However, as SAL advances westward over the ocean, it is underlain by a layer of cool, humid air (the *marine layer*) with a depth of about 900 to 1800 m (2900 to 5900 ft). An elevated temperature inversion separating the warm SAL above from the cool marine layer below reinforces the pre-existing trade wind inversion. In the SAL's interior, this inversion coupled with strong vertical shear in the prevailing easterly winds suppresses the deep convection required for tropical cyclone development. Furthermore, as a tropical storm nears the Saharan Air Layer, its circulation draws in dry stable air producing downdrafts. Hence, when SAL engulfs a tropical cyclone (or an easterly wave), convection diminishes and the system weakens even though sea-surface temperatures are above the threshold for tropical cyclone formation. On the other hand, a tropical cyclone that has weakened as a result of an encounter with SAL may quickly regain strength after the system moves beyond the influence of the SAL.

SAL is most frequent from late spring through early fall, overlapping much of the Atlantic hurricane season. SAL typically covers an area of the Atlantic that is somewhat larger than the 48 contiguous United States and maintains its Saharan characteristics as it travels many thousands of kilometers westward over the Atlantic. Suppression of convection by SAL may explain why tropical cyclones are less common in the North Atlantic than in the eastern North Pacific.

If atmospheric/oceanic conditions favor hurricane development and if those conditions persist, the surface air pressure falls and a cyclonic circulation develops. Water vapor condenses within the storm, releasing latent heat of vaporization, and the heated air is buoyed upward. Expansional cooling of the ascending air triggers more condensation, release of even more latent heat, and an additional increase in buoyancy. Rising temperatures in the core of the storm, coupled with anticyclonic outflow of air aloft, cause a sharp drop in surface air pressure, which in turn, induces more rapid convergence of humid air at the surface. The consequent uplift surrounding the developing *eye* leads to additional condensation and release of latent heat.

Through this process, a tropical disturbance intensifies and its winds strengthen. When maximum sustained wind speeds reach 37 km per hr (23 mph) or higher, the developing system is called a **tropical depression**. When maximum sustained wind speeds reach at least 63 km per hr (39 mph), the system is classified as a **tropical storm** and assigned a name, such as Ana or Bill. For more on naming hurricanes, refer to this chapter's second *For Further Exploration*. Once maximum sustained winds reach 119 km per hr (74 mph) or higher, the storm is officially designated a hurricane. As a hurricane weakens and decays, the system is downgraded by reversing this classification scheme.

Hurricanes that form over the Atlantic near the Cape Verde Islands usually drift slowly westward with the trade winds (along the southern flank of the Bermuda-Azores subtropical high) across the tropical North Atlantic and into the Caribbean. At this stage in the storm's trajectory, it is not unusual for the system to travel at a mere 10 to 20 km per hr (6 to 12 mph) and take a week to cross the Atlantic. Once over the western Atlantic, however, the forward speed of the storm usually increases and the storm begins curving northward steered by the circulation on the western flank of the subtropical high, and then northeastward as the system comes under the influence of the middle latitude westerlies. Precisely where this curvature takes place determines whether the hurricane enters the Gulf of Mexico (perhaps then tracking up the lower Mississippi River Valley or over the Southeastern States), moves up the Eastern Seaboard, or curves northward into the Atlantic.

Upon reaching about 30 degrees N, an Atlantic hurricane may begin to acquire extratropical characteristics as colder air circulates into the system and fronts develop. From then on, the storm resembles an extratropical cyclone and completes its life cycle, usually over the North Atlantic.

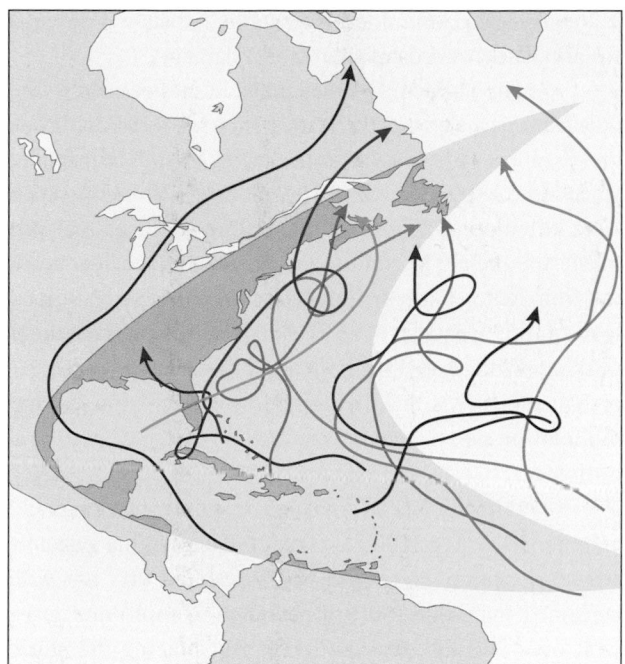

FIGURE 12.7
Tropical cyclone trajectories are often erratic, as shown by these samples. As indicated by the shaded area, however, Atlantic tropical cyclones initially drift westward and curve toward the north and northeast when they reach the western Atlantic. [From NOAA, Hurricane. Washington, DC: Superintendent of Documents, 1977]

Many hurricanes, however, depart significantly from the track just described. Some of the hurricane tracks plotted in Figure 12.7, for example, are very erratic. A hurricane can describe a complete circle or reverse direction. Since the middle of the 20th century, seven named tropical cyclones moved westward from the Atlantic to the Pacific, while two storms moved eastward from the Pacific to the Atlantic. In November 1999, Hurricane Lenny (only the fifth Atlantic basin storm in November on record to achieve category 3 or 4 status) tracked toward the east and northeast across the Caribbean Sea—opposite the direction of motion of most Atlantic basin hurricanes. According to the National Hurricane Center, this unusual track caused unprecedented wave and storm surge damage to westward facing harbors on Caribbean islands. During the 20th century only three other hurricanes tracked mostly east or northeastward through the Caribbean.

Some hurricanes, fueled by warm Gulf Stream waters, maintain tropical characteristics far north along the Atlantic Coast. The eye of Hurricane Hazel, for example, was still discernible when the system passed over Toronto, ON in October 1954, after making landfall about 1100 km (700 mi) to the south near Cape Fear, NC. The remnants of Hazel moved northward into Ontario and

became one of the most memorable storms in Canadian history. Winds gusted to 120 km per hr (75 mph) and up to 18.3 cm (7.2 in.) of rain fell. Eighty people died, mostly from flooding in the Toronto area.

New England, situated more than 25 degrees of latitude north of the usual Atlantic hurricane breeding grounds, has been the target of many full-blown hurricanes. The most deadly of these was the unnamed hurricane of 21 September 1938 (category 3) that raced through New England and into Canada at an estimated 80 km per hr (50 mph). A storm surge ravaged the New England coast, with a 4-m (16-ft) wall of water sweeping into downtown Providence, RI. Winds gusting over 200 km per hr (125 mph) severely damaged forests, and torrential rains caused flash flooding of rivers and streams. Fatalities were estimated at 600 and property damage totaled $400 million ($6.1 billion in year 2011 dollars). A more recent New England hurricane was Bob (category 2) in August 1991. Hurricane Bob moved swiftly up the East Coast from its origin in the northern Bahamas. The storm center passed just east of Cape Hatteras, NC and then remained offshore until striking Rhode Island's south coast. The eye then passed just east of Boston, MA and Portland, ME. Top wind speed was 185 km per hr (115 mph), the death toll was 18, and property damage totaled $2.0 billion.

Hurricane Hazards

Hazards of hurricanes are (1) heavy rains and inland flooding, (2) strong winds, (3) tornadoes, and (4) storm surge. According to the National Hurricane Center, freshwater flooding was responsible for almost 60% of the 600 U.S. deaths attributed to tropical cyclones or their remnants during the period from 1970-1999. In those three decades, far more people (351) died from inland flooding than from coastal storm surge flooding (only 6 deaths). For example, of the 56 people who perished during Hurricane Floyd (1999), 50 drowned due to inland flooding. On the other hand, the majority of the approximately 1200 fatalities directly attributed to Hurricane Katrina in August 2005, were caused by storm surge. The storm surge remains the most serious potential impact of a landfalling hurricane along coasts and is the primary reason why people are evacuated from low-lying coastal areas and barrier islands should a hurricane threaten.

INLAND FLOODING

Hurricanes and tropical storms produce very heavy rainfall with amounts typically in the range of 13 to 25 cm (5 to 10 in.). Even if the system tracks well

FIGURE 12.8
This roadside marker in central Virginia commemorates the fatalities and property damage caused by the torrential rains associated with the remnants of Hurricane Camille in August 1969.

inland, heavy rains often persist and may trigger costly flooding (Figure 12.8).

In June 1972, rains from the remnants of Hurricane Agnes accounted for much of the storm's $11 billion in property damage. Agnes tracked up the Eastern Seaboard from southwest of the Florida Keys to North Carolina and then northward into Pennsylvania (Figure 12.9). The storm brought beneficial rains to the Southeast, which had been experiencing a dry spell, but heavy rains in the mid-Atlantic region were anything but beneficial. During the week prior to Agnes' arrival, frontal showers and thunderstorms produced soaking rains from Virginia to New England. Rainfall averaged 3 to 8 cm (1 to 3 in.) and locally topped 15 cm (6 in.). Agnes' torrential rains falling on already saturated soils produced runoff that triggered many record-breaking river crests and devastating floods.

Heaviest rains from Agnes fell in a band about 350 km (215 mi) wide from western North Carolina northeastward into central New York State. In this hilly terrain, rainfall totals were quite variable but generally ranged from 20 to 30 cm (8 to 12 in.) during the period 18-25 June. Some localities reported total rainfall in excess of 46 cm (18 in.) with more than 25 cm (10 in.) falling in only 24 hours. Flooding was particularly severe in central Pennsylvania where rising waters forced more than 250,000 people from their homes. The Susquehanna River crested at 4 to 5.5 m (13 to 18 ft) above flood stage. Of the 122 fatalities attributed to Agnes, 50 were

in Pennsylvania, and flood damage in that state accounted for almost two-thirds of total storm damage.

The impact of Agnes pales in comparison to the devastation caused by Hurricane Mitch in 1998. Mitch was the deadliest hurricane to strike the Western Hemisphere since the "Great Hurricane of 1780," and was estimated to have killed up to 22,000 people on the Caribbean islands of Martinique, St. Eustatius, and Barbados. The final death toll from Mitch likely will never be known, but the official loss of life was listed at 11,000 with thousands more missing. More than 3 million people were made homeless or had their lives and livelihoods severely affected and the total property damage was estimated as at least $6.9 billion.

Mitch began as a weak tropical depression on 22 October 1998. By 2100 UTC on 26 October, the system's sea-level central pressure had dropped to 905 mb with sustained winds of 290 km per hr (180 mph) and gusts well over 320 km per hr (200 mph), placing the storm in the category of most powerful hurricanes (category 5). Mitch remained at this high level of intensity for 33 consecutive hours. After threatening Jamaica, Mitch tracked westward and by 2100 UTC on 27 October, was centered about 95 km (60 mi) north of Trujilo on the north coast of Honduras. Over the next two days, Mitch slowly drifted southward and finally made landfall, devastating offshore islands and coastal areas of Honduras. Mitch

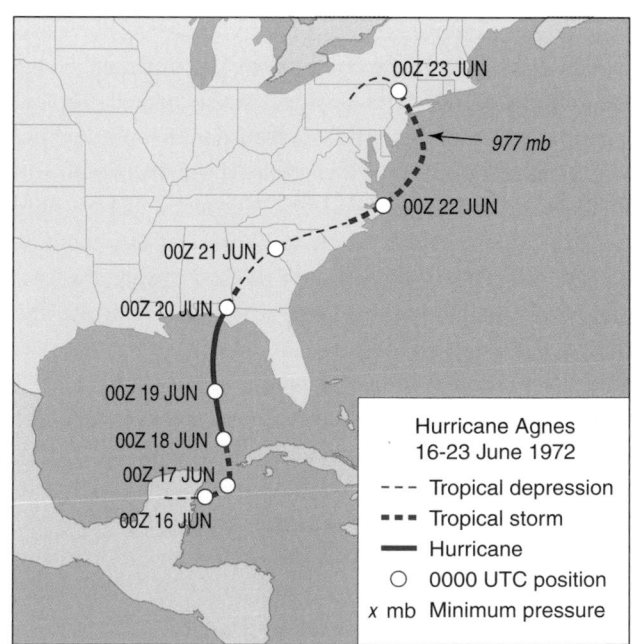

FIGURE 12.9
The track of Hurricane Agnes in June 1972. This relatively weak system brought flooding rains from Virginia north to New England. [NOAA data]

then drifted westward through the mountainous interior of Honduras, finally reaching the border with Guatemala on 31 October. Although winds slowly abated, orographically induced rainfall was torrential, falling at the rate of 30 to 60 cm (12 to 24 in.) per day in many mountainous regions and totaling in some places as much as 190 cm (75 in.) for the duration of the storm. Flooding and mudflows swept away entire villages and their inhabitants, virtually destroyed the infrastructure of Honduras, and devastated parts of Nicaragua, Guatemala, Belize, and El Salvador.

Alberto and Allison are other examples of how devastating rains can be produced by the remnants of a tropical cyclone. The first tropical storm of the 1994 season, Alberto never achieved hurricane status, yet its rains took a terrible toll in a band from southeastern Alabama to central Georgia. Alberto developed in the western Caribbean Sea and strengthened to a tropical storm on 1 July. After drifting slowly west/northwestward, Alberto turned northward, and on the morning of 3 July, came ashore near Fort Walton Beach, FL with peak sustained wind speeds near 97 km per hr (60 mph). The storm then weakened rapidly as it drifted across southeastern Alabama and into central Georgia.

Alberto likely would have received little notoriety except that its moisture-rich remnants slowly meandered over central Georgia and eastern Alabama for three days. Many area weather stations reported 3-day (3-5 July) rainfall totals in excess of 25 cm (10 in.). Americus, GA set an all-time state record for 24-hr rainfall with 53.6 cm (21.1 in.). (This record stood until 20 July 1997 when 82.6 cm (32.52 in.) fell at Dauphin Island Sea Lab, GA.) Alberto's torrential rains caused rivers to rise to record high crests, and flooding was extensive. The Ocmulgee River at Macon, GA broke its old record by 1.7 m (5.5 ft) when it crested at more than 5.2 m (17 ft) above flood stage. Most of the storm's 32 fatalities were victims of floodwaters. Property damage was estimated at $759 million.

With maximum sustained winds of 97 km per hr (60 mph), Allison was a minimal tropical storm before moving ashore at the eastern end of Galveston Island, TX on 5 June 2001 with sustained winds of 80 km per hr (50 mph). Yet this weak system produced near-record rainfall and massive flooding from eastern Texas across the Gulf States and along the mid-Atlantic coast. Allison ranks as the most deadly and costly tropical storm ever to strike the U.S. mainland.

Because of weak steering winds aloft, Allison followed an erratic path from 4-9 June, describing a meandering counterclockwise loop over eastern Texas,

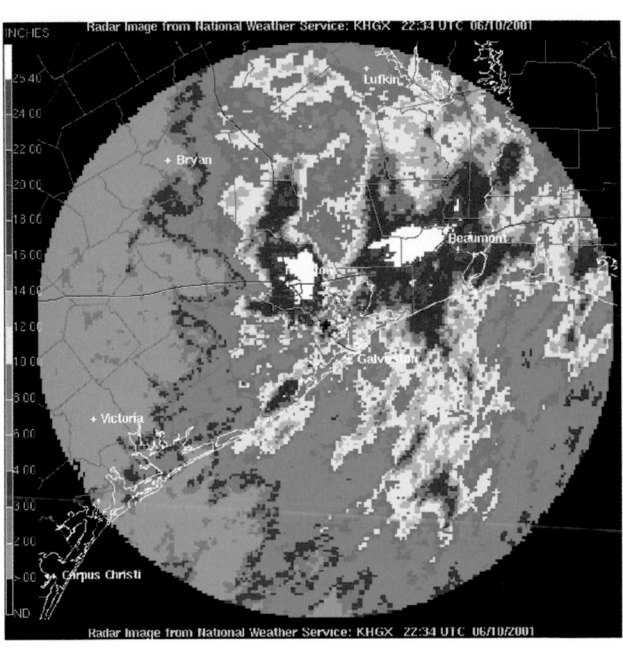

FIGURE 12.10
Radar-determined cumulative rainfall over southeast Texas (Houston and Galveston near center) produced by the remnants of tropical storm Allison for the period ending on 10 June 2001.

weakening to a tropical depression on the 7th, and twice soaking the Houston metropolitan area (Figure 12.10). During this period, rainfall totals in the Houston area included 94 cm (36.99 in.) at Port of Houston and 90.1 cm (35.67 in.) at Greens Bayou. The system's remnant circulation then drifted southward into the northwestern Gulf of Mexico where it intensified into a *subtropical cyclone* (a weather system having characteristics of both tropical and extratropical cyclones). On the 11th, Allison moved inland over Louisiana where torrential rains totaled 75.8 cm (29.86 in.) at Thibodaux and 70 cm (27.55 in.) at Salt Point. The system slowly drifted over southeastern Mississippi and then moved northeastward to eastern North Carolina where it again stalled on 14 June. Tallahassee, FL recorded a 24-hr record rainfall of 25.7 cm (10.13 in.) from the 11th to the 12th. On 17 June, Allison tracked toward the northeast off the mid-Atlantic coast and eventually dissipated on the 19th over the North Atlantic southeast of Nova Scotia. All told, Allison took the lives of 41 people (including 23 in Texas and 8 in Florida) and caused at least $6.4 billion in property damage.

In August 2008, Tropical Storm Fay produced copious amounts of rainfall as it made a record 4 landfalls over Florida and spent 7 days over the state. Rainfall totals in some parts of the state were greater than 64 cm (25 in.), leading to extensive flooding. U.S. damage totals exceeded $585 million.

Irene was one of the costliest hurricanes to impact the northeast U.S. It was the ninth named tropical cyclone and the first hurricane of the 2011 Atlantic hurricane season. Irene evolved from a well-defined tropical wave east of the Lesser Antilles. On 20 August, the developing system was named Tropical Storm Irene and later the same day made landfall on St. Croix. As the system continued to slowly intensify, Irene made landfall early on 21 August on Puerto Rico as a category 1 hurricane, followed by four landfalls in the Bahamas, reaching its maximum intensity as a category 3 with winds peaking at 195 km per hr (120 mph). Irene then curved northward, passed Grand Bahama, weakening to a category 1 before making landfall at Cape Lookout just west of the Outer Banks of North Carolina at 7:30 a.m. EDT on 27 August (Figure 12.11). Maximum sustained winds were 140 km per hr (85 mph). Irene gradually weakened as it hugged the coastline tracking north-northeast from southeast Virginia to the Atlantic and then made landfall again as a tropical storm on the southeast New Jersey shore on 28 August. Irene's final landfall was near New York City, NY. From there the system began developing extratropical characteristics as it turned northeastward and accelerated into New England, crossing the Vermont/New Hampshire border, then tracking into eastern Quebec and Newfoundland.

Irene was responsible for loss of life estimated at 56 in the Caribbean and U.S. and considerable property damage mostly due to coastal and inland flooding and severe beach erosion. Total damage in the Caribbean was estimated at up to $3.1 billion and in the U.S. estimated near $15.6 billion. Storm surge heights included 1.4 m (4.56 ft) at Sewells Point, VA, and 1.4 m (4.50 ft) at the Battery in New York City. Rainfall was very heavy, with storm totals generally ranging from 13 to 38 cm (5 to 15 in.) from eastern North Carolina into parts of New England (Figure 12.12). These rains falling on already saturated soils resulted in considerable runoff and record flooding from the northern mid-Atlantic to New England (Figure 12.13). On 27-28 August, millions of customers were without electricity because of strong winds bringing down tree limbs and power lines. Heavy rains coupled with hilly terrain caused heavy flooding in south and central Vermont that cut off many towns. On 29 August, Otter Creek in Rutland, VT, reached 2.8 m (9.21 ft) above flood stage, breaking the previous record stage of 1.7 m (5.45 ft) following the 1938 New England hurricane. In New York City, the flood threat closed the subway network and the city's three major airports, cancelling thousands of flights. Along the coast more than 300,000 people were evacuated from low-lying areas.

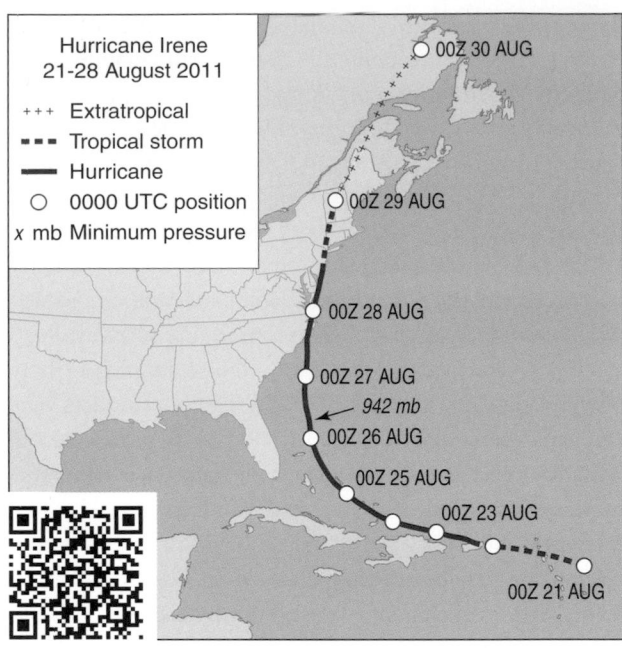

FIGURE 12.11
The track of Hurricane Irene in August 2011. [NOAA data]

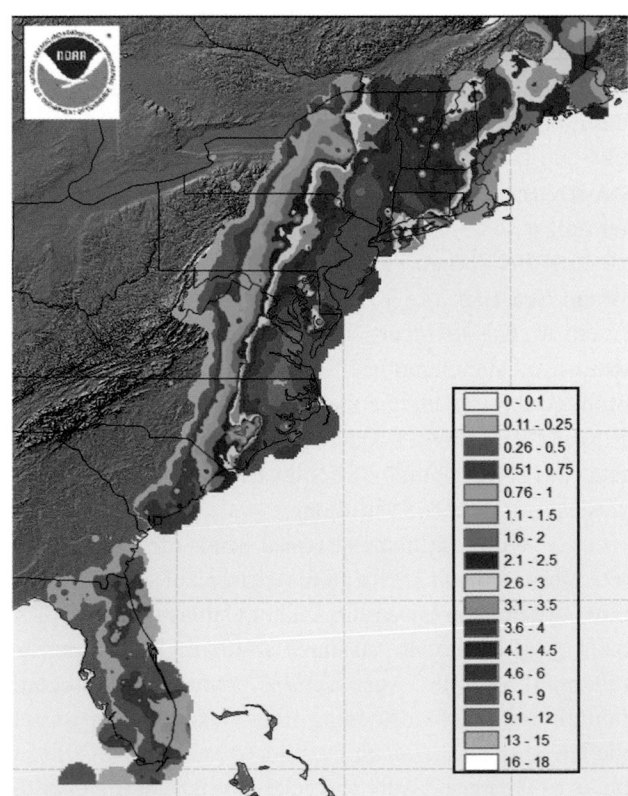

FIGURE 12.12
Hurricane Irene preliminary rainfall data (in inches) from 25 - 29 August 2011. [Map created on 1 September 2011 by Michael Kruk, NOAA/NCDC]

FIGURE 12.13
Severe urban flooding on 28 August 2011 in Lincoln Park, NJ.
[Photo by Christopher Mardorf/FEMA]

WIND

From 1970 to 1999, about 12% of tropical cyclone fatalities were wind-related. Winds pushing on the outside walls of buildings exert a pressure that increases dramatically as winds strengthen. *Wind pressure*, the force per unit area caused by air in motion, increases with the square of the wind speed. Hence, tripling the wind speed increases the wind pressure on an external wall by a factor of nine ($3^2 = 9$). Furthermore, debris transported by the wind and hurled against structures adds to the damage potential of strong winds.

Meteorologists had long assumed that strong gusts were responsible for the most serious hurricane wind-related property damage. Following detailed analysis of the aftermath of Hurricane Andrew, however, T.T. Fujita argued that small but powerful whirlwinds embedded in the hurricane circulation were responsible for the most severe property damage. Whirlwinds combined with the larger-scale hurricane circulation to produce winds estimated at 320 km per hr (200 mph). Based on their mode of origin, Fujita described these whirlwinds as *spin-up vortices*. Apparently, small eddies spin off the eye wall and are stretched vertically by powerful convection currents. Stretching accelerates the eddy circulation.

Hurricane winds diminish rapidly once the system makes landfall, so that most wind damage is confined to within about 200 km (125 mi) of the coastline. In fact, the database of hurricane tracks maintained by the National Hurricane Center shows no hurricane centered more than about 325 km (200 mi) inland in the United States. Two factors account for the abrupt drop in wind speed once a hurricane makes landfall. A hurricane over land is no longer in contact with its energy source, warm ocean water. In addition, the frictional resistance offered by the rougher land surface slows the wind and shifts the wind direction toward the low-pressure center of the system (Chapter 8). This wind shift causes the storm to begin to fill; that is, its central pressure rises, the horizontal air pressure gradient weakens, and winds slacken.

Although wind speed decreases once a hurricane makes landfall, the system may produce tornadoes (Chapter 11). As a hurricane moves over land its surface winds weaken rapidly while winds aloft remain strong. The wind shear created between horizontal winds at the surface and aloft contributes to tornado development. Usually only a few tornadoes occur with a hurricane, but in 1967, Hurricane Beulah reportedly spawned as many as 115 tornadoes across southern Texas. About the same number of tornadoes accompanied Hurricane Ivan as it tracked northward along the Eastern Seaboard in 2004. Tornadoes are most likely after the hurricane enters the westerly steering current and curves toward the north and northeast; they form mostly to the northeast of the storm center, often outside the region of hurricane-force winds.

STORM SURGE

A **storm surge** is a dome of ocean water, perhaps 80 to 160 km (50 to 100 mi) wide that sweeps over the coastline near the hurricane's landfall (Figure 12.14). A storm surge is caused by a combination of strong winds and low barometric pressure and is most likely on the side of the approaching hurricane where winds are blowing toward the shore. Wind-driven waves with heights of at least 1.5 to 10 m (5 to 33 ft) top the dome of sea water and, armed with floating debris, are responsible for much of the structural damage and beach erosion attributed to a storm surge. A storm surge can wash over barrier islands, wash out roads and railway beds, and demolish marinas, piers, cottages, and other coastal structures. In extreme cases, entire communities are washed away.

Prior to 1970, storm surges were responsible for the majority of hurricane-related fatalities. In fact, in October 1972, the American Meteorological Society adopted a policy statement noting that 90% of all hurricane-related fatalities near the U.S. coast were attributed to the combination of storm surge and accompanying high waves. Subsequently, a better-informed population, early storm warnings, and evacuation of people from vulnerable coastal areas spurred a sharp decline in storm surge related deaths. As pointed out earlier in this chapter, during the 30-year period from 1970 to 1999, the majority of hurricane-related deaths in the United States were from drowning in freshwater floods while a total of only 6 deaths were due to storm surges. Then came Katrina in 2005 and the death and destruction caused by its storm surge. With the recent upsurge in the frequency of major hurricanes (discussed later in this chapter), coastal residents are advised to remain vigilant should a hurricane threaten, for history recounts many catastrophic storm surges.

A storm surge accounted for much of the considerable loss of life in a hurricane that struck the southeast United States on 27 August 1893, one of six hurricanes to strike the U.S. mainland that year (a record number that was equaled in 1916 and 1985). The unnamed hurricane came ashore at night just north of Savannah, GA with little warning. U.S. Weather Bureau meteorologists

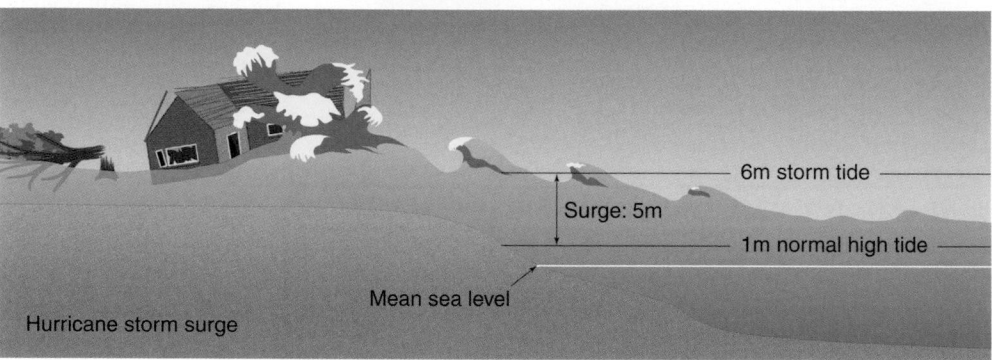

FIGURE 12.14

Onshore winds and low air pressure associated with a hurricane cause a dome of water to move shoreward, causing considerable coastal erosion and flooding. Adding to the destructive potential of a storm surge are wind-driven waves and floating debris.

predicted that the hurricane, first detected at Nassau in the Bahamas the previous day, would make landfall somewhere along the East Coast south of New York. The actual track placed the system's dangerous semicircle, the band of strongest onshore winds, over the Sea Islands of Beaufort County, SC. Most of the inhabitants of the Sea Islands were freed slaves, their children and grandchildren. They were primarily farmers who raised a highly valued variety of cotton. No telegraph lines connected the islands to the mainland so there was no warning of the hurricane's approach. With no means of evacuation, residents were completely at the mercy of 175 to 210 km per hr (110 to 130 mph) winds and a 3.7 to 4.6 m (12 to 15 ft) storm surge that swept over the islands. Some 20,000 to 30,000 people were left homeless, their crops were destroyed, and the death toll was estimated at 2000 (and possibly as high as 2500). The Sea Islands disaster came at a time of national recession and prior to the time of organized federal or state natural disaster assistance. Intervention by the Red Cross was all that saved the survivors from almost certain famine.

In the most deadly natural disaster in U.S. history more than 8000 people perished, mostly by drowning,

when a hurricane storm surge of over 4.6 m (15 ft) devastated Galveston, TX, on 8 September 1900. Ocean waters flooded the entire city situated on a barrier island, where the greatest elevation was only 2.7 m (8 ft). Some 3600 homes were destroyed; a dam of wreckage that built up six blocks inland from the beach probably spared the business district total destruction. To protect the city from future storm surges, the U.S. Army Corps of Engineers elevated the city an average of 3 m (10 ft) and erected a massive concrete seawall along much of the city's Gulf shore.

On 13 September 2008, Hurricane Ike made landfall along the north end of Galveston Island, TX. Ike was a strong category 2 hurricane at landfall, but its unusually broad wind field caused a storm surge typical of a more intense hurricane. The National Hurricane Center estimated that Galveston Island experienced a 3 to 4.6 m (10 to 15 ft) storm surge, which caused tremendous structural damage, but the flooding would likely have been far worse without the seawall. Ike's highest storm surge (estimate at 4.6 to 6.1 m or 15 to 20 ft) occurred just east of Galveston Island on the Bolivar Peninsula of Texas (Figure 12.15), where most areas were inundated by 3 m

FIGURE 12.15
An aerial view of the damage Hurricane Ike inflicted upon Gilchrist, TX, a community on the Bolivar Peninsula. [Jocelyn Augustino/ FEMA]

(10 ft) of water. Ike is the fourth costliest U.S. hurricane with a preliminary damage estimate of $20 billion, after Hurricanes Katrina (2005), Andrew (1992), and Wilma (2005).

Hurricane Camille, an exceptionally intense system with peak winds to 300 km per hr (186 mph), produced a maximum storm surge of 7.3 m (24.3 ft) at Pass Christian, MS on 17 August 1969. The total death toll from the storm was 256, about half from coastal flooding. The surge destroyed or badly damaged more than 18,000 homes and 700 businesses. Scientists had a difficult time determining the maximum storm surge produced by Hurricane Katrina in August 2005 because the storm destroyed many of the local tide gauges, but it appears likely that Katrina's storm surge was higher than that of Camille. As detailed in this chapter's Case-in-Point, Katrina's storm surge contributed to breaches in the New Orleans levee system, leading to catastrophic flooding of the city (Figure 12.16). In 1970, one of the most disastrous storm surges ever flooded the Bay of Bengal coast of East Pakistan (now Bangladesh). A storm surge of nearly 7 m (23 ft) spread over the vast low-lying coastal plain of the Ganges Delta, claiming an estimated 300,000 lives, mostly by drowning, because there was no high ground for people to seek refuge.

Strong onshore winds plus relatively low air pressure are responsible for storm surge. Relatively low air pressure in a hurricane causes sea level to rise (about 0.5 m for every 50 mb drop in pressure or about 1 ft for every 1 in. of mercury drop in pressure). The height of a storm surge is the difference between the observed rise in local sea level and the regular astronomical tide. In general, a storm surge of 1 to 2 m (3 to 6.5 ft) can be expected with a weak hurricane, whereas the storm surge accompanying a violent hurricane may top 5 m (16.4 ft). According to NOAA researchers, storm surge height depends on many factors including the intensity and speed of the hurricane, direction of hurricane motion relative to the shoreline, slope of the seafloor along the shore, configuration of the shoreline, and the height of the astronomical tide. The greatest potential for a highly destructive storm surge occurs with strong onshore winds, along a shallow sloping shoreline, during high tide, and in densely populated areas lacking coastal buffers such as barrier islands, coral reefs, marshes, or swamps. Typically, the rise in sea level in a storm surge lasts for 4 to 8 hours but it may take days for floodwaters to recede.

A special numerical model developed in 1979 by NOAA predicts the location and height of a storm surge with a high degree of accuracy. Weather forecasters report

FIGURE 12.16
An aerial view of downtown New Orleans, LA, taken from a U.S. Navy helicopter on 31 August 2005, two days after Katrina made landfall. The Superdome can be seen near the center of the image. [U.S. Navy]

considerable success with the **SLOSH (Sea, Lake, and Overland Surges from Hurricanes)** model. Using output from SLOSH and consulting local topographic maps, forecasters can identify areas most likely to be inundated by floodwaters when a hurricane threatens. Hypothetical hurricanes are simulated using various combinations of storm intensity, forward speed, and expected location of landfall to predict the maximum high water levels for various SLOSH basins along the coast. Public safety officials use this information to develop evacuation plans. SLOSH model coverage includes the entire East and Gulf Coasts, as well as parts of Hawaii, Guam, Puerto Rico, and the Virgin Islands.

SAFFIR-SIMPSON HURRICANE WIND SCALE

In the early 1970s, H.S. Saffir (1917-2007), a consulting engineer, and R.H. Simpson, former director of the National Hurricane Center, designed a rating system for hurricanes known today as the **Saffir-Simpson Hurricane Wind Scale** (Table 12.1). Hurricanes are rated from category 1 up to category 5 as indicated by the maximum wind speed. The scale, first included in hurricane advisories in 1975, provides an estimate of potential coastal flooding and property damage from a hurricane landfall. Wind speed is the primary determining factor for a hurricane's rating on the Saffir-Simpson Scale as storm surge heights are highly dependent on bathymetry, topography, and hurricane size, forward speed, and angle to the shoreline and other factors in the region of landfall. Each intensity category specifies a range of maximum sustained wind speed.

Property damage potential rises rapidly with a hurricane's ranking on the Saffir-Simpson Scale. In fact, destruction from a category 4 or 5 hurricane can be 100 to 300 times greater than that caused by a category 1 hurricane. A category 1 (weak) hurricane generally causes no real damage to buildings but may affect unanchored mobile homes, shrubbery, and trees, and cause minor damage to piers. With a category 2 (moderate) hurricane, we can expect some damage to roofing materials, doors, and windows of buildings. Damage to trees, mobile homes, and piers can be considerable and small craft in unprotected anchorages may break free of their moorings. A category 3 (strong) hurricane can destroy mobile homes, cause some structural damage to small residences and utility buildings, and blow down large trees. Flooding is likely along the coast and some evacuation may be necessary. With a category 4 (very strong) hurricane, the roof structure of small residences may fail. Beaches experience major erosion and considerable damage is done to structures near the shore. The threat of coastal flooding calls for evacuation of low-lying residential areas. In a category 5 (devastating) hurricane, we can expect roofs to be ripped off of homes and industrial buildings. In some instances, buildings experience complete failure and are swept from the foundation. Major coastal flooding is likely, requiring massive evacuation of residential areas.

Of the 185 hurricanes that struck the U.S. Atlantic or Gulf coasts between 1901 and 2010, 69 (37.3%) were classified as major; that is, they rated category 3 or higher on the Saffir-Simpson Hurricane Wind Scale. The 25 major hurricanes that made landfall along the Gulf or Atlantic coast between 1949 and 1990 accounted for three-quarters of all property damage from all landfalling tropical storms and hurricanes during the same period. Tables 12.2 and 12.3 list the most intense, and the 10 deadliest hurricanes (including offshore losses), respectively, to strike the United States since 1900.

TABLE 12.1 Saffir-Simpson Hurricane Wind Scale		
Category	Wind speed km per hr (mph)	Damage potential
1	119-153 (74-95)	*Very dangerous winds will produce some damage.*
2	154-177 (96-110)	*Extremely dangerous winds will cause extensive damage.*
3	178-208 (111-129)	*Devastating damage will occur.*
4	209-251 (130-156)	*Catastrophic damage will occur.*
5	252 or higher (157 or higher)	*Catastrophic damage will occur.*

TABLE 12.2
Most Intense U.S. Hurricanes at Landfall (based on lowest central pressure), 1900-2011[a]

Rank	Hurricane	Year	Category	Pressure (mb)
1	"Labor Day," Florida Keys	1935	5	892
2	Camille (MS, LA, VA)	1969	5	909
3	Katrina (LA, MS, AL)	2005	3	920
4	Andrew (FL, LA)	1992	5	922
5	FL Keys, south TX	1919	4	927
6	FL (SE/Lake Okeechobee)	1928	4	929
7	Donna (FL, eastern US)	1960	4	930
7	FL (Miami)/MS/AL/Pensacola	1926	4	930
9	Carla (TX)	1961	4	931
10	S TX	1916	4	932
10	Hugo (SC)	1989	4	934
12	TX (Galveston)	1900	4	936
13	Rita (NE TX, W LA)	2005	3	937

[a]From NOAA, National Hurricane Center

TABLE 12.3
The Ten Deadliest U.S. Hurricanes, 1900-2011[a]

Rank	Hurricane	Year	Category	Deaths
1	TX (Galveston)	1900	4	8000[+]
2	FL (SE/Lake Okeechobee)	1928	4	2500-3000
3	Katrina (LA, MS, AL)	2005	3	1200[++]
4	FL Keys, south TX	1919	4	600[&]
5	"New England"	1938	3[*]	600[&]
6	Audrey, SW LA and N TX	1957	4	>416
7	"Labor Day," Florida Keys	1935	5	408
8	NE United States	1944	3[*]	390[#]
9	FL (Miami)/MS/AL/Pensacola	1926	4	372
10	Grand Isle, LA	1909	4	350

[a]From NOAA, National Hurricane Center
[+] Could be as high as 12,000
[++] Could be higher than 1800 if indirect Katrina-related deaths are included
[&] Total including offshore losses is 600
[*] Moving at more than 48 km per hr (30 mph)
[#] Some 326 of these lost on ships at sea

FIGURE 12.17
Memorial to those lives lost in the hurricane that struck the Florida Keys in 1935. As of 2011, this was the most intense landfalling hurricane since 1900. [Photo by J.A. Brey]

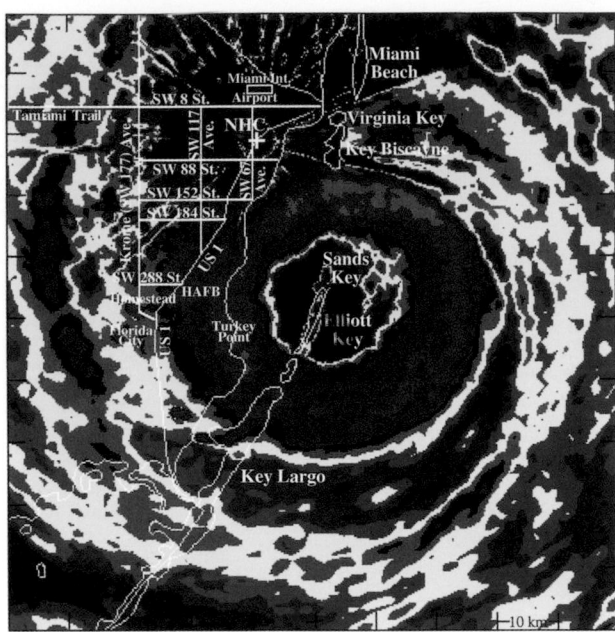

FIGURE 12.18
Radar reflectivity image of the spiral rain bands associated with Andrew as it came ashore as a category 5 hurricane south of Miami Beach, FL on 24 August 1992. [NOAA]

Figure 12.17 displays a memorial constructed in the Florida Keys in remembrance of the 1935 hurricane. This is the most intense (in terms of central pressure at landfall) and one of the deadliest hurricanes to make landfall in the United States.

As of the 2011 hurricane season, the last Atlantic basin hurricane to make landfall as a category 5 was Andrew (Figure 12.18). In the three hours it took Hurricane Andrew to cross extreme southern Florida on the morning of 24 August 1992, its peak sustained winds of 266 km per hr (165 mph), with much stronger gusts, contributed to the deaths of 15 people and made 180,000 homeless. To make matters worse, Andrew continued to track west and northwestward across the Gulf of Mexico and, two days later, struck the Louisiana coast where

it claimed four more lives. With total property damage of $42.5 billion, Andrew at the time was the costliest hurricane in U.S. history.

Prior Atlantic hurricanes to make landfall as category 5 storms were Gilbert in September 1988 and Camille in August 1969. Gilbert originated as a tropical wave off the West African coast on 3 September 1988. The wave followed a steady west-northwest track and seven days later strengthened to a hurricane while south of Puerto Rico. Gilbert then passed over the length of Jamaica on the 12th, intensified over the northwest Caribbean with the minimum central pressure falling to a then record low 888 mb, and on 14 September made landfall (as category 5) near Cozumel on Mexico's Yucatan Peninsula. At landfall, maximum sustained winds were estimated at 275 km per hr (171 mph). Gilbert then tracked northwestward over the Gulf of Mexico and made its final landfall (as category 3) on the coast of Mexico about 200 km (125 mi) south of the Texas border. Gilbert brought much death and destruction to both Mexico and Jamaica. The death toll was 202 in Mexico and 45 in Jamaica, and combined property damage in the two countries was at least $7.8 billion.

Trends in Hurricane Frequency

In recent years, the frequency of intense hurricanes has increased. In 2005, researchers at the Georgia Institute of Technology and the National Center for Atmospheric Research (NCAR) reported that the number of category 4 and 5 hurricanes worldwide nearly doubled over the 35-year period, 1970-2004 (Figure 12.19). This upward trend occurred even though the overall number of hurricanes worldwide has declined since the 1990s. In the 1970s, the number of category 4 and 5 hurricanes globally averaged about 10 per year, but since 1990, that number has increased to about 18 per year. Much of the increased frequency of intense hurricanes occurred in the North Pacific, Southwest Pacific, and Indian Ocean with a slightly smaller increase in the North Atlantic.

The only region of the globe where tropical cyclone activity has increased recently is the North Atlantic (Figure 12.20). According to the National Hurricane Center, from 1995-2010 the average annual number of named tropical cyclones (tropical storms and hurricanes) is 14.8 compared to 9.3 during the period 1970-1994. Since 1995, the Atlantic averages 7.9 hurricanes per season, 3.8 of which are major hurricanes. In the 25 years prior to 1995, the seasonal average for hurricanes was 5, with 1.5 rated major. For more about the variability of Atlantic hurricane activity, see this chapter's third *For Further Exploration*.

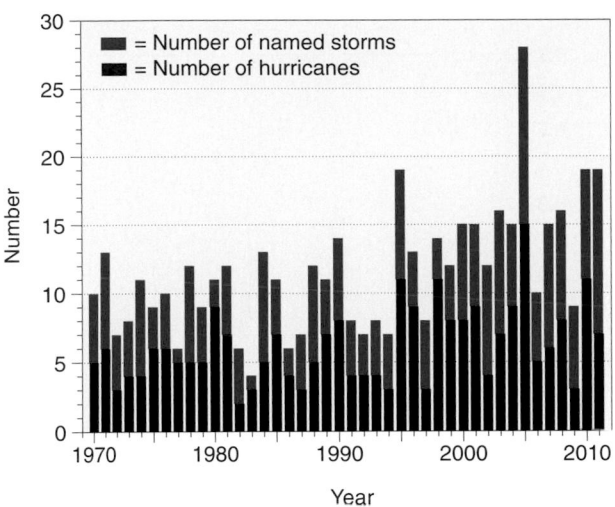

FIGURE 12.20
Number of Atlantic tropical storms and hurricanes, 1970-2011. [Data from the NOAA Atlantic Oceanographic and Meteorological Laboratory/Hurricane Research Division HURDAT]

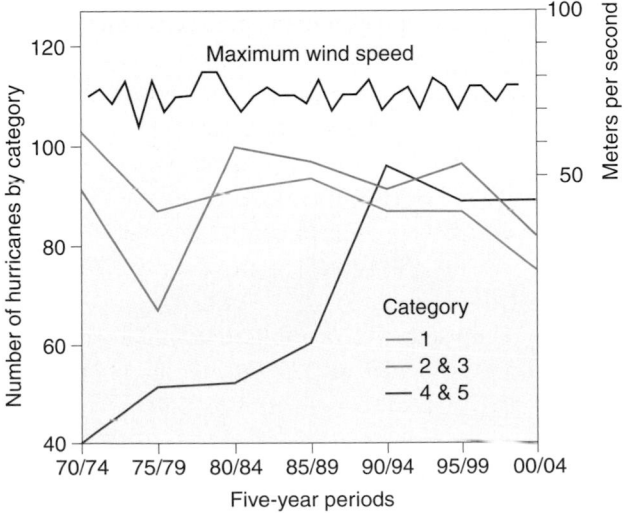

FIGURE 12.19
Recent upward trend in the number of intense (category 4 and 5) hurricanes. [Modified after Peter Webster, Georgia Institute of Technology]

Why the recent increase in Atlantic hurricanes? Several factors contribute. Sea-surface temperatures (SST) have been higher in the tropical Atlantic. In the atmosphere, an amplified ridge, favorable for tropical cyclones, has been in place over the central and eastern North Atlantic. Vertical wind shear has weakened in the deep tropics over the central North Atlantic. In addition, the African easterly jet stream has been more favorable for tropical cyclone development.

Multidecadal changes in North Atlantic SST appear to be related to alternating weakening and strengthening of the ocean thermohaline circulation. North Atlantic SST rise and fall over a 60-70 year quasi-periodic cycle known as the *Atlantic Multidecadal Oscillation (AMO)*. Recall from Chapter 4 that the ocean thermohaline circulation transports heat energy through the ocean basins and is linked to surface ocean currents such as the Gulf Stream. When the thermohaline circulation weakens, SST fall, and tropical cyclones are less frequent. Relatively cool AMO years and reduced tropical cyclone activity prevailed in the Atlantic from 1903 to 1925 and again from 1971 to 1994. When the circulation strengthens, SST rise, and tropical cyclones are more frequent. Relatively warm AMO years and increased tropical cyclone activity prevailed in the Atlantic from 1926 to 1970, and we are currently in a warm phase that began in 1994 (Figure 12.21).

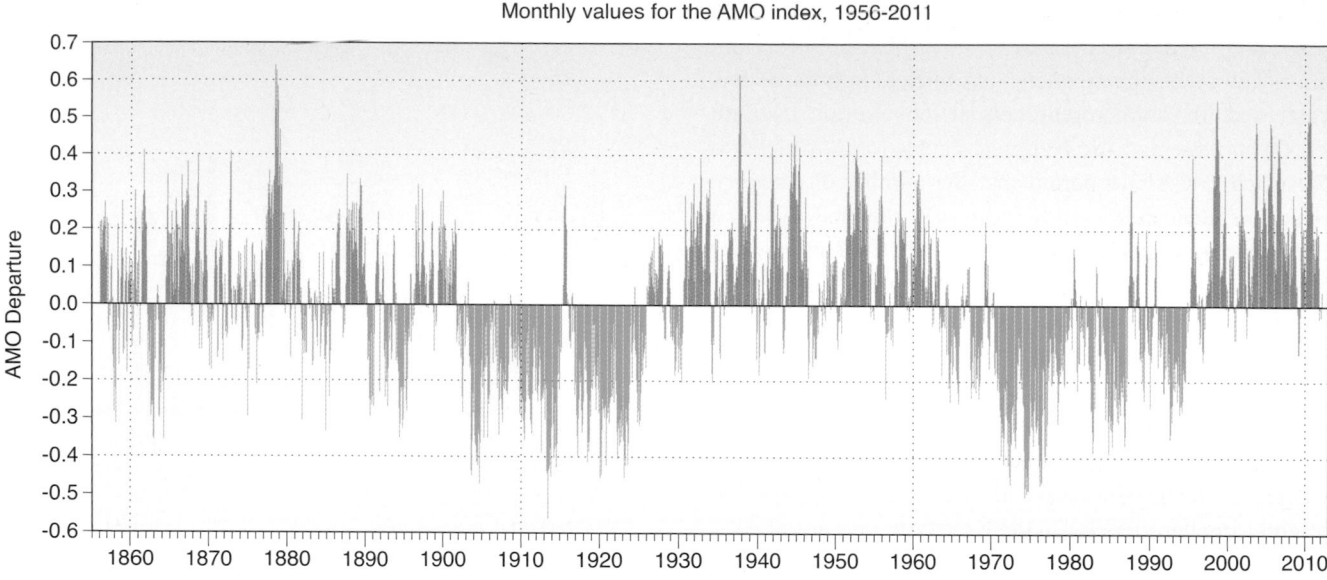

FIGURE 12.21
Atlantic Multidecadal Oscillation time series, 1856–2011. [Source data from NOAA Earth Systems Research Laboratory]

Hurricane Threat to the Southeast United States

From 1981 to 2000, only 10 major hurricanes (category 3 or higher) made landfall on the U.S. mainland. This was similar to the 1971 through 1990 period when only 9 major hurricanes made landfall on the U.S. coast. From 10 August 1980 to 17 August 1983, no hurricanes struck the United States. The infrequency of major hurricanes during the 1970s and 1980s lulled many coastal residents of the southeast United States into a false sense of security and encouraged development and population growth in areas that could be devastated by a major hurricane. In spite of Hugo in 1989 and Andrew in 1992, population growth continued unabated. Development of the U.S. coastal zone is now proceeding at a pace 2 to 3 times the national average. More and more resort hotels, high-rise condominiums, and expensive homes are being constructed perilously close to the shoreline and even among coastal sand dunes (Figure 12.22).

In 2011, the National Oceanic and Atmospheric Administration (NOAA) reported U.S. Census Bureau data showing that coastal watershed counties were home to about 52% of all Americans. Some 161 million people resided in these counties—some 50.9 million more than in 1970—and the population is expected to swell by 14.9 million by 2020. By convention, a county is described as coastal if it is on a coast (i.e., Atlantic, Gulf, Pacific, Great Lakes) or at least 15% of the county's land area is in a coastal watershed. The 650 coastal counties (including the District of Columbia and excluding Alaska coastal areas) represent 17% of the nation's land area and have about five times the average population density of inland counties.

About 45 million permanent residents inhabit hurricane-prone portions of the nation's coastline. Population growth is most rapid from Texas through Virginia with Florida leading the nation in both population growth

FIGURE 12.22
The relative infrequency of hurricanes and tropical storms in the 1970s and 1980s lulled some residents of coastal areas into a false sense of security and inspired construction within meters of high-tide level. [Courtesy of Joseph Schlueter]

(by percent) and hurricane potential. The population of coastal counties along the Gulf of Mexico increased by 150% between 1960 and 2008. Until the recent upturn in hurricane frequency, most Atlantic and Gulf Coast residents never experienced the full impact of a major hurricane. Some of them may have weathered with relative ease a weak hurricane or the fringes of a strong system. But such an experience may lull them into complacency so that they are less likely to prepare adequately should a major hurricane threaten. Compounding the problem of the growth of the resident population in the Southeast is the arrival of holiday, weekend, and seasonal visitors at seaside resorts. The human population in some of these locales swells ten- to one-hundred-fold during vacation periods. Many of these resorts are located in low-lying coastal areas or on exposed beaches that are subject to rapid inundation by a storm surge.

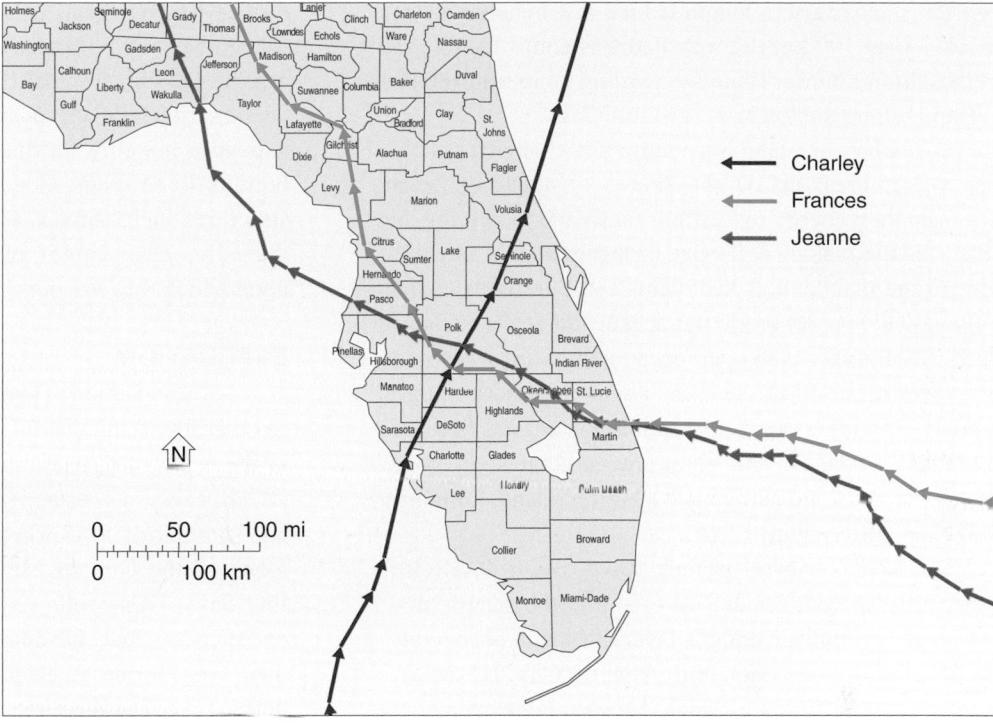

FIGURE 12.23
Tracks of three of the four hurricanes that struck Florida in 2004. Hurricane Ivan (not shown) made landfall on the western Florida panhandle (off map). [National Hurricane Center]

Public safety officials are concerned by the post-1995 trend toward a greater number of tropical cyclones in the Atlantic basin. The Atlantic hurricane seasons of 2004, 2005, and 2008 were particularly active with considerable loss of life and property damage caused by landfalling hurricanes. During the 2004 Atlantic hurricane season, 15 named tropical cyclones formed; 9 intensified into hurricanes and of these, 6 became major hurricanes. Eight tropical cyclones formed during August—a new record for the month. Four hurricanes (Charley, Frances, Ivan, and Jeanne) struck Florida, with three crossing the same area of central Florida (Figure 12.23). The total death toll across the western Atlantic, Caribbean, and Gulf of Mexico was more than 3100, the second greatest number of tropical cyclone fatalities in 30 years. While many of the deaths occurred in Haiti due to flooding from Hurricane Jeanne, the direct death toll in the U.S. was 60 and property damage (from 5 landfalling hurricanes) reached $52.4 billion. As noted earlier, these losses were greatly exceeded in 2005.

The 2008 Atlantic hurricane season included 16 tropical cyclones, 8 of which intensified into hurricanes, and of these, 5 became major hurricanes. It was notable for being the only year on record in which major hurricanes were present in all 6 months of the season. According to NOAA's National Climatic Data Center, properly damage was estimated upwards of $56.4 billion mostly from Hurricane Ike.

The 2010 season was also well above average, including 19 tropical cyclones, 12 hurricanes, and 5 major hurricanes, but fortunately there were no U.S. landfalling hurricanes. The 2011 season featured 19 tropical cyclones and 7 hurricanes, 4 of which intensified into major hurricanes. One landfalling hurricane (Irene) and one landfalling tropical storm (Lee) produced more than $20 billion in property damage, mostly due to extensive inland flooding from Irene.

BARRIER ISLANDS

The hurricane danger is particularly acute for people living on or visiting the nearly 300 barrier islands that fringe portions of the Atlantic and Gulf coasts, from Maine to Texas. A **barrier island** is an elongated, narrow accumulation of sand oriented parallel to the coast and separated from the mainland by a lagoon, estuary, or bay.

Barrier islands vary in length from a few hundred meters to more than 100 km (62 mi). Padre Island is the longest of the nation's barrier islands extending more than 180 km (112 mi) along the lower Texas Gulf Coast.

A barrier island is a continually changing system. Sea waves breaking on the shores of a barrier island dissipate their energy by shifting sands and modifying the shape of the island. A barrier island gradually migrates toward the mainland as sediments settle in the land-side lagoon while waves erode the ocean side of the island. A barrier island faces the open ocean and absorbs the brunt of powerful storm-driven sea waves and storm surges thereby providing some protection for coastal beaches, estuaries, wetlands, and shoreline structures. In some cases, a storm surge will breach a barrier island, forming a new tidal inlet (Figure 12.24).

Many barrier islands have been developed, especially for cottages and resorts, and the sand is now temporarily stabilized under a layer of asphalt or concrete. Some coastal cities, including Atlantic City, NJ, Miami Beach, FL, and Virginia Beach, VA, are built entirely on

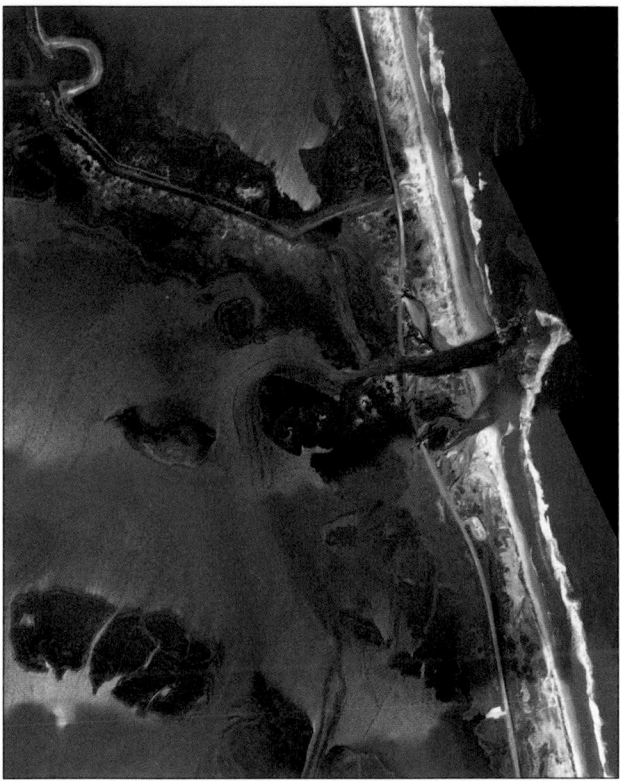

FIGURE 12.24
Aerial photograph of a section of Hatteras Island where Hurricane Irene caused a large breech. Damage to Highway 12 (shown as the gray north-south line on the western side of the island), which extends along North Carolina's Outer Banks, stranded about 2500 people. [Image courtesy of the NOAA Hurricane Irene Project]

barrier islands. Such exposed locations are particularly vulnerable to the ravages of high winds and storm surge associated with tropical cyclones. On developed barrier islands, much of the energy of storm waves would be expended not only shifting sands but also in demolishing buildings and roads. Conflict is inevitable between rigid structures such as roads and buildings and the inherently dynamic (changeable) platforms (i.e., barrier islands) upon which they are built.

EVACUATION

Evacuation of people from barrier islands, as well as other low-lying coastal areas, is the traditional strategy in the event of a major storm threat (Figure 12.25). The effectiveness of coastal evacuation plans was tested in the late summer of 1985 when Hurricane Elena (category 3) menaced the Gulf of Mexico coast (Figure 12.26). For four days, Elena followed an erratic path over the Gulf of Mexico, first heading toward southern Louisiana, then the Florida panhandle, and later central Florida before reversing direction and finally coming ashore near Biloxi, MS on 2 September. Nearly a million people from Sarasota, FL to New Orleans, LA were forced to leave coastal communities and flee to inland shelters. Some returned home only to evacuate again as Elena changed course. Although property damage was considerable ($2.6 billion) because of extensive flooding and winds that exceeded 160 km per hr (100 mph), only 4 fatalities were attributed to Elena and none in the area of hurricane landfall. Timely evacuation of residents of low-lying coastal areas also saved many lives when category 4 Hugo (September 1989) and category 5 Andrew (August 1992) threatened.

The potential downside of evacuation was illustrated in September 1999 when Floyd, an unusually massive hurricane, threatened much of the Eastern Seaboard. More than 2 million residents of the coastal area from South Florida to South Carolina took to the roads and fled inland. In many areas the result was gridlock—too many vehicles on too few highways. As it turned out, Floyd spared most of the evacuated area and made landfall near Cape Fear, NC on 16 September 1999 as a category 2 system. The gradually weakening storm crossed eastern North Carolina and Virginia and then tracked northeastward along the New Jersey coast and became an extratropical cyclone over New England. Torrential rainfall, totaling as much as 38 to 50 cm (15 to 20 in.) over portions of eastern North Carolina and Virginia caused extensive inland flooding, 56 deaths, and property damage of about $6.1 billion. Floyd's track along

FIGURE 12.25
In hurricane-prone areas, evacuation routes are marked by special road signs. In some hurricane-prone coastal communities, public safety officials are rethinking the traditional evacuation strategy that clogs highways with too many escaping motor vehicles. Instead, they are considering "vertical" evacuation, that is, evacuation to the middle and upper floors of sturdy public buildings.

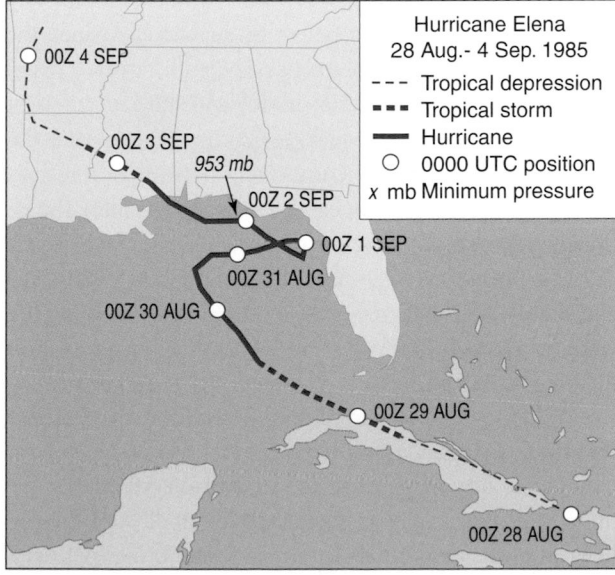

FIGURE 12.26
The erratic track of Hurricane Elena from 28 August through 4 September 1985. The hurricane's position is shown every 24 hours. [NOAA data]

the East Coast required the posting of hurricane warnings from South Florida to Massachusetts.

Successful evacuation of coastal communities hinges on sufficient advance warning of a hurricane's approach, but hurricanes are notorious for sudden changes in direction, forward speed, and intensity. (In retrospect, about 25% of evacuations were followed by landfalls. Most of the rest were merited on the basis of potential loss of life and property.) A hurricane's erratic behavior is especially troublesome for people living in isolated localities (e.g., a barrier island linked to the mainland by a single bridge) and congested cities where highway systems have not kept pace with population growth. In such places the time required for evacuation may be lengthy. The *Federal Emergency Management Agency (FEMA)* estimates evacuation times during the peak tourist season as up to 50-60 hrs for New Orleans, LA, Ocean City, MD, and Fort Myers, FL, up to 30-39 hrs for the Florida Keys, the Outer Banks of North Carolina, Cape May County, NJ, and Atlantic City, NJ, and up to 20-29 hrs for Long Island, NY and Galveston, TX.

As coastal communities continue to grow, the time required for evacuation of their population lengthens. Evacuation must begin earlier when a tropical cyclone is farther away and greater uncertainty surrounds its likely track. Such uncertainty necessitates a broader zone of evacuation that translates into greater economic losses associated with evacuation (e.g., closed businesses). As a general rule, the cost of evacuation amounts to about $1 million per mile of coastline. Cognizant of the problem of lengthy evacuation times, some coastal communities are considering the option of *vertical evacuation*, moving people to shelters on the floors of well-constructed public buildings, which are well above the storm surge, but below floor areas impacted by the highest hurricane winds, for example, in a high rise building. Vertical evacuation may be a viable alternative to sending people fleeing on congested roads and highways for large urban areas where evacuation options are limited.

Other strategies adopted or advocated by public safety and other government officials to minimize the loss of life and property to hurricanes are (1) stringent building codes, (2) preservation of mangrove swamps, and (3) elimination of federal floodplain insurance. Building codes may call for new homes to be more wind resistant by requiring bolts that anchor the floor to the foundation and steel brackets that attach the roof to the walls. In addition, in many areas, all new buildings must be elevated above the once-in-a-century flood level. Some buildings are constructed so that the lower floor

FIGURE 12.27
This home, located on North Carolina's Outer Banks, is designed so that the first floor will give way to storm-surge floodwaters. The second and third floor living area of the home is supported by wooden beams driven deeply into the sand.

gives way to flood waters, thus reducing damage to main structural supports (Figure 12.27). Unfortunately, many structures were built prior to adoption of these tough codes, and stringent building codes are not always enforced.

Mangrove swamps along tropical coastlines are the lines of natural defense against hurricanes and other ocean storms. Their extensive root systems anchor soil and vegetation and dissipate the energy of a storm surge. In too many cases, however, shoreline development has destroyed these swamps, often because the vegetation obstructed the ocean view for residents of seaside dwellings. Preservation of mangrove swamps and other natural buffers, such as coastal sand dunes, has received renewed emphasis in hurricane-prone areas.

A controversial strategy is to eliminate federal flood insurance for flood-prone coastal areas as has been done already on undeveloped portions of barrier islands. Some people argue that by providing policies at low cost, federal flood insurance programs actually encourage development in flood-prone areas. Furthermore, this insurance enables homeowners to rebuild structures destroyed by flooding in the same hazardous location.

Long-Range Forecasting of Atlantic Hurricanes

Since the early 1980s, Dr. William M. Gray and his colleagues (currently Dr. Phil Klotzbach) at Colorado State University's Tropical Meteorology Project have issued long-range forecasts of Atlantic basin hurricane activity for the upcoming hurricane season. Forecasts specify the number of expected tropical cyclones, hurricanes, and major hurricanes, along with integrated hurricane metrics such as Accumulated Cyclone Energy, described below, and Net Tropical Cyclone (NTC) activity. The forecast also includes the probability that at least one major (category 3 or higher) hurricane will make landfall on the U.S. coast somewhere from Texas to Maine, as well as in the Caribbean. The initial extended range forecast is issued in early April, with subsequent updated hurricane forecasts made public early in the months of June and August.

The basis of Gray's original hurricane forecasting technique was his discovery of an apparent linkage between Atlantic basin hurricane activity and El Nino Southern Oscillation (ENSO). Gray also factored into his forecasting scheme the stratospheric *quasi-biennial oscillation (QBO)* that also appeared related to Atlantic basin hurricane activity. The QBO is a reversal of the winds in the lower portion of the tropical stratosphere (from west to east and then reverse) over a period that averages about 26 months. In the early 1990s, the scheme was updated to include the discovery of a relationship between the frequency of major hurricanes in the tropical Atlantic and rainfall in West Africa.

Gray's forecasting scheme gave reasonably skillful results (predicted values nearer to observed values than to the long-term average) until around 1995, when the relationships between the QBO and African rainfall and Atlantic basin hurricanes began to degenerate. Since the late 1990s, new relationships have been discovered using the NCEP/NCAR Reanalysis products which have allowed for improved skill in recent years. In recent years, the forecast team has been incorporating dynamical forecasts of ENSO from the European Centre for Medium Range Weather Forecasts into their statistical schemes. Forecast schemes developed over the recent period from 1982-2010 and then tested over the earlier part of the 20th century were employed to make sure that those relationships remained fairly stable over the past 100+ years.

Forecasters also comb the historical database for appropriate analogs, that is, global atmosphere/ocean conditions that are similar to current conditions for the initial forecast. Analogs may provide indications as to what the following hurricane season may bring. All three seasonal forecasts (April, June, August) utilize a similar approach but select different predictors. In general, forecasts have increased skill as the peak of the hurricane season approaches. During the peak months of the hurricane season from August-October, the Tropical Meteorology Project also issues two-week forecasts which are based on the phase and amplitude of the *Madden-Julian Oscillation (MJO)*, a relationship between large-scale atmospheric circulation and deep convection in the tropics. The MJO is characterized by large, eastward-moving areas of enhanced and suppressed tropical rainfall in the Indian and Pacific Oceans.

NOAA has also issued a Seasonal Outlook for Atlantic basin hurricane activity since 1998. This outlook, first released in May and then updated in August, gives probabilities of overall seasonal activity compared to normal, as well as the likely ranges of named tropical cyclones, hurricanes, major hurricanes, and Accumulated Cyclone Energy (ACE). ACE is a method of describing the activity of an individual tropical cyclone or an entire season. An individual cyclone's ACE is a sum of the squares of wind speed every 6 hrs; the seasonal ACE totals the values of all individual cyclones. The outlook is based on forecasts of climate variables that strongly influence seasonal activity, as well as an analysis of past seasons similar to the current one.

Hurricane Modification

Spurred by six destructive hurricanes that lashed the U.S. East Coast during the mid-1950s, atmospheric scientists stepped up efforts to learn more about hurricanes and to devise hurricane modification techniques. To these ends, the U.S. government funded *Project STORMFURY* beginning in 1961. Until the project's termination in 1983, staff scientists advanced our understanding of the formation, structure, and dynamics of tropical cyclones, improved hurricane forecasting, and upgraded hurricane reconnaissance methods. The original goal of hurricane modification, however, was never realized.

The working hypothesis of Project STORMFURY was that seeding hurricanes with silver iodide (AgI) crystals would reduce wind strength. The argument went as follows: Seeding the band of convective clouds just beyond the *eye wall* would trigger additional latent heat release (as supercooled water droplets converted to ice crystals) and enhance convection. This artificially invigorated convection would then dominate convection in the eye wall, and a new eye wall would form at a greater distance from the storm center. Thereby, the band of strongest winds would be displaced farther out from the eye center and the hurricane's circulation would weaken (just as a skater's spin rate slows when she extends her arms).

The 1960s and 1970s presented few opportunities for hurricane seeding experiments primarily because few hurricanes met the selection criteria. Systems selected for seeding had to be mature, relatively intense, expected to remain far from land for at least 24 hrs following seeding, and have a well-developed eye wall. Some success was reported, however, in reducing winds (between 10% and 30%) in four hurricanes that were seeded. Later, these apparent successes were dismissed when new data called into question the original Project STORMFURY hypothesis. For one, it was discovered that convective clouds in hurricanes contain too little supercooled water for seeding to be effective. For another, monitoring of unmodified hurricanes found that changes in eye-wall diameter occur as part of the system's natural evolution. It is likely, then, that STORMFURY's apparent success at hurricane modification was by chance rather than by seeding.

Conclusions

Hurricanes are intense, long-lived, tropical cyclones that develop over tropical ocean waters but can track into middle latitudes. The requirements of relatively high sea-surface temperature, significant Coriolis Effect, and weak vertical wind shear restrict hurricane formation to certain ocean regions of the globe. In the North Atlantic Ocean basin, these systems initially drift westward in the trade winds and then turn north and northeastward as they are caught up in the prevailing westerlies of middle latitudes. Torrential rains, inland flooding, strong winds, and storm surge associated with a hurricane can take many lives and cause considerable property damage. To relate possible damage to wind speeds, a hurricane is assigned a rating from 1 to 5 on the Saffir-Simpson Hurricane Wind Scale.

With this chapter, we have completed our description of the genesis, life cycle, and characteristics of the major atmospheric circulation systems that affect the weather of middle latitudes. In the next chapter, we examine the techniques and challenges of weather forecasting.

Basic Understandings

- Weather in the tropics exhibits very little seasonal temperature variation because of year-round high noontime solar altitudes and nearly uniform daylight lengths. Rainfall is primarily the result of convective activity driven by intense solar heating. Principal weather systems that bring rainfall to the tropics are the ITCZ, easterly waves, and tropical cyclones.

- A hurricane is a violent tropical cyclone with a maximum sustained wind speed of 119 km per hr (74 mph) or higher. The system originates over tropical ocean waters, usually in late summer or early fall. The warm-core low pressure system develops in a warm and humid air mass, has no fronts, and on average is about one-third the diameter of an extratropical cyclone.

- At the center of a hurricane is an area of almost cloudless skies, subsiding air, and light winds, called the eye of the storm. The most dangerous and potentially most destructive part of a hurricane is the portion of the surrounding eye wall on the side of the advancing system where the wind blows in the same direction as the storm's forward motion.

- For a tropical cyclone to develop, sea-surface temperatures must be 26.5 °C (80 °F) or higher through a depth of about 45 m (150 ft), the Coriolis Effect must be sufficient (latitude of more than 5 degrees north or south of the equator) to initiate a cyclonic circulation, and the vertical wind shear must be relatively weak. The first two requirements mean that tropical cyclone development is limited to certain regions of the tropical ocean. Weak winds aloft allow a cluster of cumulonimbus clouds to organize over tropical seas, the first stage in the evolution of a hurricane.

- The annual number of hurricanes may have little bearing on the number of hurricanes that make landfall and their impact.

- Most major hurricanes that make landfall in North America originate as convective cloud clusters associated with easterly waves. An easterly wave is a westward-propagating ripple in the trade winds featuring a weak trough of low pressure and areas of diverging and converging winds.

- Atmospheric conditions that inhibit tropical cyclone development over the Atlantic appear to be associated with the Saharan Air Layer, an elevated mass of dry dusty stable air that originates over the Sahara Desert and moves westward over the ocean.

- If atmospheric and sea-surface conditions are favorable, convective cloud clusters evolve into a tropical disturbance that can intensify into a tropical depression, tropical storm, and finally a hurricane.

- The track of a hurricane can be erratic, although in the Atlantic there tends to be a westward drift at low latitudes and a more rapid northeastward movement at middle latitudes, following the circulation of the Bermuda-Azores subtropical high. Some Atlantic hurricanes, fueled by warm Gulf Stream waters, maintain tropical characteristics far north along the Eastern Seaboard.

- Hazards of hurricanes include heavy rains, inland flooding, strong winds, tornadoes, and storm surge. The storm surge is the most serious potential impact of a landfalling hurricane on coasts and is the principal reason why people are evacuated from low-lying coastal regions.

- Once a hurricane makes landfall, it quickly loses its warm-water energy source and experiences the frictional resistance offered by the greater surface roughness of land versus water. Its circulation weakens rapidly; hence, most wind damage is confined to within about 200 km (125 mi) of the coastline. Torrential rains often continue well inland, however, and can cause severe flooding. Freshwater flooding is the major cause of hurricane-related deaths.

- Although winds weaken once a hurricane makes landfall, the system may produce tornadoes. Tornadoes form mostly to the northeast of the storm center and often outside the region of hurricane-force winds.

- A storm surge is a dome of ocean water topped by high waves that is driven onshore by storm winds. The greatest potential for coastal flooding and beach erosion occurs along a shallow sloping shoreline when a storm surge coincides with high tide.

- Based on sustained wind speed, a hurricane is rated from category 1 (weak) to category 5 (devastating) on the Saffir-Simpson Hurricane Wind Scale.

- In recent years, the frequency of intense hurricanes (categories 4 and 5) has increased. Also, since 1995, the number of tropical cyclones in the Atlantic is greater likely because of higher sea-surface temperatures, weaker vertical wind shear, and more favorable atmospheric circulation patterns.

- Rapid population growth in the coastal zone coupled with the lull in intense hurricane activity during the 1970s and 1980s has heightened the hurricane hazard in the southeastern United States. The hurricane danger is particularly acute for people living on or visiting the barrier islands that fringe portions of the Atlantic and Gulf coasts.

- Attempts to modify hurricanes through cloud seeding have been unsuccessful. However, from such efforts has come a greater understanding of the dynamics of hurricanes.

Enduring Ideas

- The main weather systems in the tropics include the ITCZ, easterly waves, and tropical cyclones. A tropical cyclone is a warm-core low pressure system that develops in a warm and humid air mass, has no fronts, and is about one-third the size of an extratropical cyclone. Its formation requires high sea-surface temperatures, a sufficient Coriolis Effect, and weak winds aloft. Tropical cyclones generally move westward at low latitudes, steered by the trade winds, and then northeastward at middle latitudes.
- A tropical cyclone weakens after landfall as it loses its warm-water energy source (latent and sensible heating) and encounters the frictional resistance of land. Hazards associated with tropical cyclones include heavy rains and inland flooding, strong winds, tornadoes, and storm surge. These hazards are especially acute for those living in coastal areas of the Southeastern U.S., particularly on barrier islands.
- A tropical cyclone is classified as a hurricane when its maximum sustained wind speed reaches 119 km per hr (74 mph). The Saffir-Simpson Hurricane Wind Scale rates storms from 1 to 5. The frequency of category 4 and 5 hurricanes globally has increased in recent years and overall tropical cyclone numbers in the North Atlantic have increased since 1995.

Key Terms

tropical cyclones	tropical disturbance	SLOSH
streamline	easterly waves	Saffir-Simpson Hurricane Wind
hurricane	Saharan Air Layer (SAL)	Scale
eye (of the storm)	tropical depression	barrier island
eye wall	tropical storm	
ring	storm surge	

Review

1. In general terms, compare and contrast the weather in the tropics with the weather at middle latitudes.
2. Contrast the characteristics of a hurricane with those of a typical extratropical cyclone.
3. Describe the typical weather in the eye of a hurricane.
4. What three conditions are required for a tropical cyclone to form?
5. Why does a hurricane weaken when the system tracks from ocean to land?
6. Why is the U.S. Pacific coast rarely the target of hurricanes?
7. What distinguishes a tropical depression from a tropical storm?
8. Identify and describe the hazards of hurricanes.
9. Which sector of a landfalling hurricane is responsible for the greatest storm surge?
10. What is the basis for the Saffir-Simpson Hurricane Wind Scale?

Critical Thinking

1. Why are Atlantic hurricanes most likely to form in late summer and early autumn—much later than the time of peak incoming solar radiation?
2. Describe how the Saharan Air Layer (SAL) influences Atlantic tropical cyclones.
3. How is it possible for a tropical storm to cause considerable inland flooding?
4. Why are tornadoes associated with hurricanes?
5. Describe the recent trend in hurricane frequency in the Atlantic basin. What factors may be responsible for that trend?
6. What factors contribute to the rapid growth in human population of hurricane-prone regions of the southeast United States?
7. What role is played by barrier islands in the hurricane threat to the U.S. Atlantic and Gulf coasts?
8. What are some of the problems associated with evacuating large numbers of people from low-lying coastal areas when a major hurricane threatens?
9. What are some alternatives to evacuating people from low-lying coastal areas when a hurricane approaches?
10. Comment on efforts to modify hurricanes.

Polar Lows with Hurricane Characteristics

Some intense storms that form over the ocean off the coasts of Alaska and Norway have characteristics in common with hurricanes and tropical storms. For this reason, these polar lows are sometimes called *Arctic hurricanes* (Figure 1).

Polar lows are similar to tropical hurricanes in several respects. Both are relatively small-scale cyclones that are highly symmetric and feature a relatively warm core. Most clouds are convective with a ring of cumulonimbus clouds surrounding a calm eye, which is sometimes clear, and the strongest winds are associated with those cumulonimbus clouds. Both systems are triggered by other pre-existing atmospheric disturbances and both derive their energy from the ocean surface.

Arctic hurricanes differ from tropical hurricanes in the following ways: Arctic hurricanes are shallower systems that develop in cold air north of the polar front and evolve to full maturity in only 12 to 24 hrs. On the other hand, tropical hurricanes form in warm, humid air and may require 3 to 7 days to attain full strength. In addition, Arctic hurricanes travel nearly twice as fast as tropical hurricanes. Sustained wind speeds are somewhat higher in a tropical hurricane than in an Arctic hurricane. Maximum sustained wind speeds are about 110 km per hr (68 mph) in an Arctic hurricane and, by convention, must be 119 km per hr (74 mph) or higher in a tropical hurricane.

FIGURE 1
Measurements obtained by the Moderate Resolution Imaging Spectrometer (MODIS) on NASA's Aqua satellite show a polar low in the Arctic Ocean. [Jeff Schmaltz/NASA Visible Earth]

In a tropical hurricane, the flux of heat from the sea-surface to the atmosphere is the principal source of energy that sustains the system. Most of this energy flux consists of latent heating (that is, evaporation of water at the sea-surface and subsequent condensation and release of latent heat within the storm system). Much less important is sensible heating (that is, conduction and convection of heat from the sea-surface to the atmosphere). Although the total heat flux from sea to air is roughly the same for both Arctic and tropical hurricanes, much larger fluxes of sensible heat are produced by Arctic hurricanes because of the much greater air-sea temperature contrast in the Arctic versus the tropics. Furthermore, lower SST in the Arctic means less latent heating than in the tropics.

As noted elsewhere in this chapter, a tropical hurricane is triggered by the ITCZ, a trough in the westerlies, or an easterly wave. The trigger for an Arctic hurricane may be a polar trough or jet streak. Development is most likely along a front that marks the transition zone between modified and unmodified arctic air. Arctic air is modified (becoming milder and more humid) by virtue of its trajectory over ocean water.

Naming Hurricanes

When sustained winds in a tropical depression reach 63 km per hr (39 mph), the intensifying system is designated a *tropical storm* and assigned a name. This practice is intended to improve communication between forecasters and the general public especially because tropical storms or hurricanes are relatively long-lived weather systems and more than one may occur in the same ocean basin at the same time.

Originally, hurricanes were identified by their latitude and longitude and later by letters of the alphabet. Late in the 19th century, an Australian forecaster, Clement Wragge (1852-1922), was apparently the first to apply women's names to tropical cyclones; he also named them for politicians he disliked. During World War II, U.S. Army Air Corps and Navy meteorologists informally applied women's names to tropical cyclones. From 1950 to 1952, tropical cyclones in the North Atlantic were labeled using the phonetic alphabet (e.g., Able, Baker, etc.). In 1953, U.S. Weather Bureau meteorologists began using an alphabetical sequence of female names for tropical storms and hurricanes in the Atlantic basin. Women's names were first used for tropical cyclones near Hawaii in 1959 and for the rest of the northeast Pacific basin beginning the following year. Starting in 1978, the list consisted of alternating male and female names in alphabetical order, in English, Spanish, and French. The same practice began in the Atlantic basin in 1979 (Table 1). Six lists of names are repeated every 6 years. In the event that more than 21 named tropical cyclones occur in a single season, the National Hurricane Center uses the Greek alphabet, beginning with Alpha. This happened during the 2005 Atlantic hurricane season.

TABLE 1
Tropical Cyclone Names in the Atlantic Basin

2012	*2013*	*2014*	*2015*	*2016*	*2017*
Alberto	Andrea	Arthur	Ana	Alex	Arlene
Beryl	Barry	Bertha	Bill	Bonnie	Bret
Chris	Chantal	Cristobal	Claudette	Colin	Cindy
Debby	Dorian	Dolly	Danny	Danielle	Don
Ernesto	Erin	Edouard	Erika	Earl	Emily
Florence	Fernand	Fay	Fred	Fiona	Franklin
Gordon	Gabrielle	Gonzalo	Grace	Gaston	Gert
Helene	Humberto	Hanna	Henri	Hermine	Harvey
Issac	Ingrid	Isaias	Ida	Ian	Irma
Joyce	Jerry	Josephine	Joaquin	Julia	Jose
Kirk	Karen	Kyle	Kate	Karl	Katia
Leslie	Lorenzo	Laura	Larry	Lisa	Lee
Michael	Melissa	Marco	Mindy	Matthew	Maria
Nadine	Nestor	Nana	Nicholas	Nicole	Nate
Oscar	Olga	Omar	Odette	Otto	Ophelia
Patty	Pablo	Paulette	Peter	Paula	Philippe
Rafael	Rebekah	Rene	Rose	Richard	Rina
Sandy	Sebastien	Sally	Sam	Shary	Sean
Tony	Tanya	Teddy	Teresa	Tobias	Tammy
Valerie	Van	Vicky	Victor	Virginie	Vince
William	Wendy	Wilfred	Wanda	Walter	Whitney

Source: NOAA, National Hurricane Center

The names of very destructive hurricanes have been retired from the list. The following is a list of retired names for the Atlantic Ocean, Caribbean Sea, and Gulf of Mexico since 1965: Betsy, 1965; Inez, 1966; Beulah, 1967; Edna, 1968; Camille, 1969; Celia, 1970; Agnes, 1972; Carmen and Fifi, 1974; Eloise, 1975; Anita, 1977; David and Frederic, 1979; Allen, 1980; Alicia, 1983; Elena and Gloria, 1985; Gilbert and Joan, 1988; Hugo, 1989; Diana and Klaus, 1990; Bob, 1991; Andrew, 1992; Luis, Marilyn, Opal and Roxanne, 1995; Cesar, Fran, and Hortense, 1996; Georges and Mitch, 1998; Floyd and Lenny, 1999; Keith, 2000; Allison, Iris, and Michelle, 2001; Isidore and Lili, 2002; Fabian, Isabel, and Juan, 2003; Charley, Frances, Ivan, and Jeanne, 2004; Dennis, Katrina, Rita, Stan, and Wilma, 2005; Dean, Felix, and Noel, 2007; Gustav, Ike, and Paloma, 2008; Igor and Tomas, 2010 (Figure 1); and Irene, 2011.

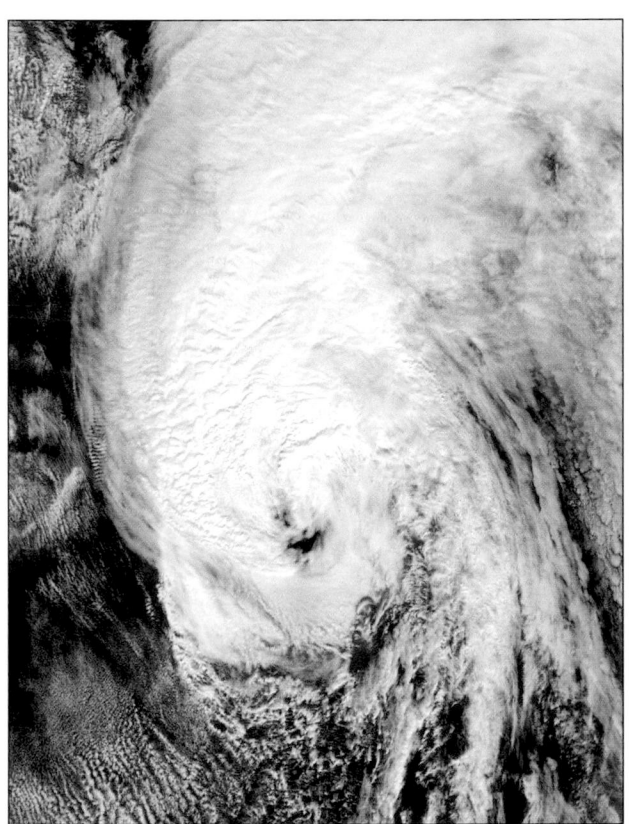 

FIGURE 1
Retired hurricanes from the 2010 Atlantic Hurricane season include Igor (left) and Tomas (right). Igor was memorable for rapid intensification into category 4 status on the Saffir-Simpson Hurricane Wind Scale while over tropical seas and for being the most destructive tropical cyclone to strike Newfoundland (as a category 1 hurricane). The image above shows Hurricane Igor about to make landfall on the southeast coast of Newfoundland. Igor caused severe flooding and road washout in Newfoundland, temporarily isolating about 150 communities. Tomas impacted Barbados and the Windward Islands before reaching peak strength as a category 2 hurricane in the Caribbean. The image above shows Hurricane Tomas about to move through the Windward Passage just west of Haiti. Heavy rains caused flooding and mudslides in Haiti, adding to the devestation caused by the powerful earthquake earlier that year. [NASA Goddard Space Flight Center]

In the northwest Pacific Basin, women's names were used for tropical cyclones from 1945 to 1979. Then, alternating male and female names were used until 1 January 2000. On that date, the World Meteorological Organization's Typhoon Committee adopted a unique list of Asian names, most of which are not personal names but rather the names of flowers, animals, birds, and trees. Furthermore, the list of names is organized by the contributing nation (in alphabetical order). The Tokyo Typhoon Centre of the Japanese Meteorological Agency is responsible for assigning names from this list to developing tropical storms.

Separate lists of names are maintained for the southwest Indian Ocean (west of 90 degrees E), western Australia (90 degrees to 125 degrees E), northern Australia (125 degrees to 137 degrees E), and eastern Australia (137 degrees to 160 degrees E). Lists of names are also available for the regions near Fiji and Papua New Guinea. Tropical cyclones in the north Indian Ocean are not assigned names.

Variability in Atlantic Hurricane Activity

Climatologists are wrestling with the question of whether global climate change has or will affect the frequency and intensity of hurricanes. Following the lull in hurricane activity during the 1970s and 1980s, the frequency of major Atlantic hurricanes (category 3 or higher on the Saffir-Simpson Hurricane Wind Scale) increased substantially. From 1971-94, the annual average of major hurricanes was 1.5, but from 1995-2010, that annual average increased to 3.8. Is this change within the normal variability of hurricane frequency or is it the consequence of global climate change?

The detailed reliable instrument-based record of Atlantic hurricanes is limited to the past 130 years or so. Geoscientists have employed many methods to extend this record as far back in time as possible. An obvious advantage of a lengthy record is possible insight on the variability of hurricane frequency and an historical perspective on the current active phase.

As we saw elsewhere in this chapter, two keys to hurricane formation are sea-surface temperatures (SST) and vertical wind shear. Even when SST are well above average, hurricane frequency depends more on strength of wind shear than SST. In 2007, K. Halimeda Kilbourne, a NOAA paleoclimatologist, and colleagues reconstructed the wind shear record back to 1730. Exposing coral growth rings to a UV radiation source reveals the luminescence of growth rings. Luminescence, in turn, is a measure of the amount of organic materials washed by thunderstorms from the land to the sea. Strong wind shear inhibits the development of thunderstorms and hurricanes. Hence, the luminescence of coral growth rings served as a proxy for Atlantic hurricane frequency.

Kilbourne and colleagues concluded that large variations in hurricane frequency are the norm in the Atlantic. From 1730 to 2005, the annual average of major hurricanes was 3.25, somewhat less than it was during the recent active period. Researchers identified at least 6 intervals since 1730 when hurricane activity was comparable in frequency to the present time.

Jeffrey P. Donnelly of Woods Hole Oceanographic Institution and colleagues demonstrated a relationship between Atlantic hurricane frequency and El Niño. Analysis of a 5000-year record of sedimentation from a lake in Ecuador and a lagoon in eastern Puerto Rico identify episodes of strong and frequent El Niño. Strong wind shear associated with El Niño favored a lower annual average frequency of Atlantic hurricanes.

In the March/April 2007 issue of *American Scientist*, Kam-biu Liu of Louisiana State University reported on his reconstruction of the frequency of major hurricanes along the Gulf Coast over a time frame of thousands of years. His working hypothesis was that the waves and storm surge associated with an intense hurricane would wash sand into coastal lakes located behind sandy beaches or dunes. In vertical cross-section, a sharply bounded sand layer situated between layers of fine mud served as a signature of hurricane impact. Radiocarbon or other radiometric dating technique was used to establish a hurricane chronology.

Liu demonstrated that along the Gulf Coast, major hurricane activity varied on time scales of centuries to millennia. At four coastal sites between Louisiana and Florida, catastrophic hurricanes (category 4 or 5) occurred 10 to 12 times every 3800 years or once every 300-350 years. Hurricane activity along the Gulf Coast was relatively low from 5000 to 3800 years before present and again during the past 1000 years when an average of only one catastrophic hurricane occurred per millennium. From 3800 to 1000 years ago, hurricane activity was relatively high with a catastrophic hurricane striking as frequently as once every two centuries.

Lightning strikes during a football game at Lane Stadium/Worsham Field in Blacksburg, Virginia. Weather forecasts can be critical for ensuring public safety during outdoor events. [Courtesy NOAA]

WEATHER ANALYSIS & FORECASTING

Chapter Highlights

Case-in-Point
   Evolution of Tornado Forecasting
International Cooperation
Acquisition of Weather Data
   Surface Weather Observations
   Upper-Air Weather Observations
Weather Data Assimilation, Depiction and
   Analysis
   Surface Weather Maps
   Upper-Air Weather Maps
Weather Prediction
   Numerical Weather Forecasting
   Forecasting Tropical Cyclones
   Forecasting for Aviation
   Forecasting Severe Storms
   River and Flood Forecasting
   Marine Forecasting
   Space Weather Forecasting
   Forecast Skill
   Long-Range Forecasting
   Single-Station Forecasting
   Private Sector Forecasting
Communication and Dissemination
   Weather-Ready Nation
Conclusions
Basic Understandings/Enduring Ideas
Key Terms/Review/Critical Thinking
For Further Exploration
   Marine Weather Statements
   Chaos and the Limits to Forecasting

Learning Objectives

Explain why international exchange of weather observations is essential for weather forecasting.

Describe how weather data are obtained from Earth's surface and upper atmosphere.

Identify the sources of weather data from the ocean surface.

Explain why it is essential that weather observations worldwide adhere to a standard time system (i.e., UTC).

Describe how warm air advection and cold air advection affect the height of an isobaric surface (e.g., the 800-mb level).

Describe the complementary roles of meteorologists and numerical models in weather data analysis and forecasting.

Describe the various divisions of NOAA's National Centers for Environmental Prediction.

Explain how and why weather forecasting skill changes with the lengthening of the forecast period.

Define and identify the purpose of ensemble forecasting and model comparison.

Describe how meteorologists prepare long-range (e.g., seasonal) weather forecasts.

Describe the role played by teleconnections in seasonal weather forecasting.

Distinguish between weather watches and warnings.

Summarize how to make reasonably accurate single-station weather forecasts.

Identify the fundamental objective of NOAA's *Weather Ready Nation* program.

What are processes involved in making a scientific forecast of the weather?

Evolution of Tornado Forecasting

In the late 19th century John P. Finley of the U.S. Army Signal Corps pioneered tornado forecasting. Finley enlisted the aid of volunteer observers in the Midwest to help him document the occurrence and tracks of tornadoes. By mid-1884, about 950 "tornado reporters" were gathering observational data. Finley also consulted records of tornado sightings by weather networks operated by the U.S. Army Medical Department, the Smithsonian Institution, the Army Corps of Engineers, and the Army Signal Corps. Finley's criteria for a tornado was a funnel cloud observed by a credible witness or indications of a violently rotating wind based on the pattern of property damage. From his analysis of these data plus his personal observations in tornado-prone areas of the country, Finley developed a list of rules for forecasting tornadoes and began issuing the first tornado predictions on an experimental basis on 10 March 1884.

Finley's experiment in tornado forecasting lasted only two years and his forecasts were not issued to the public. In 1886, the Army Signal Corps banned the use of the word "tornado" in weather forecasts, although a special warning could be included if violent storms were possible. This action stemmed from the fear that a tornado forecast would cause the general public to panic, resulting in more harm than the storm system itself. Six decades would pass before "tornado" was included in an official U.S. Weather Bureau forecast.

In the late 1940s and early 1950s, U.S. Air Force Air Weather Service meteorologists Major Ernest J. Fawbush and Captain Robert C. Miller developed a method for forecasting tornadoes but issued those forecasts primarily for military installations. Fawbush and Miller's interest in tornado forecasting stemmed from their experience with a tornado that struck Tinker Air Force Base in Oklahoma City, OK on 20 March 1948, injuring 8 people and causing more than $10 million ($95 million in year 2011 dollars) in damage to aircraft. The tornado struck without warning while Miller was on duty. In response, the commanding general of the Oklahoma City Air Materiel Area almost immediately directed Tinker's meteorologists to study the possibility of tornado forecasting. After analyzing atmospheric conditions preceding the 20 March tornado as well as other tornadoes and finding similarities, Fawbush and Miller developed a list of six atmospheric conditions that together precede a tornado outbreak. On 25 March, only five days after the earlier tornado, those same atmospheric conditions came together again in the vicinity of Oklahoma City prompting Fawbush and Miller to issue their first tornado forecast. Base personnel enacted a newly drawn-up tornado safety plan (e.g., moving aircraft to hangers, diverting approaching aircraft, and seeking shelter). The tornado developed along a squall line and followed a track similar to the earlier tornado through Tinker Air Force Base. This time, no one was injured although damage to planes amounted to about $6 million ($56 million in year 2011 dollars).

The Fawbush-Miller tornado forecasting method proved so successful that it was extended to other areas near military bases. In spite of this success, the U.S. Weather Bureau remained reluctant to issue tornado forecasts. Francis W. Reichelderfer (1895-1983), chief of the U.S. Weather Bureau (USWB) from 1938 until 1963, believed that tornado forecasting was not sufficiently precise for distribution to the general public. However, on 6 May 1949, Fawbush and Miller correctly forecast a tornado that struck Amarillo, TX, killing 7 and injuring 82. Shortly afterward, Reichelderfer authorized USWB forecasters to include the possibility of severe or destructive local storms if warranted in forecasts for public distribution. Eventually, the U.S. Weather Bureau followed the lead of the Air Weather Service, forming the Severe Local Storm Warning Center (now the Storm Prediction Center in Norman, OK). The ban on the use of the word "tornado" in weather forecasts for civilians was finally abolished on 17 March 1952, when meteorologists at the new center issued the first tornado watch.

Most people readily recall occasions when an erroneous weather forecast upset their plans. It may have been an unexpected thundershower that brought an abrupt end to a softball game, or a raging blizzard that appeared instead of the anticipated clearing skies, or the promised spring-like weekend that turned out to be chilly and dreary. People seem to remember missed weather forecasts all too clearly and overlook the fact that most forecasts are on target.

Weather forecasts, especially for the next day or so, are generally quite accurate and forecasters achieve some *skill* (or accuracy) with forecasts out to 7 days. Nonetheless, most people remember when the weather pulled unpleasant, and in some cases, hazardous surprises on them. In many situations, these surprises were not from the lack of a weather forecast, watch, or warning. The fact is changes in weather or the potential for change, including the more dramatic and hazardous changes, are routinely predicted to a high degree of accuracy. Unfortunately, forecasts might not reach as far into the future as we would like or need. The quality of weather forecasts, especially short-term forecasts, has made them very valuable planning tools. Simply stated, people who know what the weather is and what it is likely to be, benefit in many ways.

Of course, not all weather events are predicted nor are they currently predicted with the accuracy one would like beyond a few days. Our present understanding of the workings of the atmosphere and our ability to monitor it are neither perfect nor complete. Also, weather forecasts are not always prepared to the detail we might desire for our particular situation. Weather prediction will never be perfect because forecasters will always be working with incomplete information on the initial state of the atmosphere and some scientific questions are yet to be answered. However, when viewed with the objectivity of statistical analysis, short-range weather forecasting is surprisingly accurate. The U.S. **National Weather Service (NWS)**, an agency of the **National Oceanic and Atmospheric Administration (NOAA)**, issues 24-hr weather forecasts of temperature that are accurate to within 5 Fahrenheit degrees of error about 85% of the time. The popular notion that weather forecasting is not very accurate stems from the simple fact that a missed forecast is more memorable because of the inconvenience a person might have experienced as a result.

How are weather forecasts made? What are the limits of forecast accuracy? On the basis of what you learn in this course, how can you make your own weather forecasts? These are the principal questions addressed in this chapter.

International Cooperation

The atmosphere is a continuous fluid that envelops the globe, so that weather observation, analysis, and forecasting require international cooperation. The American naval officer Matthew Fontaine Maury (1806-1873) was an early leader in international efforts to coordinate meteorology at sea. Maury helped convene the first International Meteorological Conference held in Brussels in August 1853. At a subsequent meeting in Vienna in September 1873, an agreement was drafted to form the *International Meteorological Organization (IMO)*. The IMO was founded in 1878 by representatives of the newly formed weather services of a dozen sea-faring nations, including the United States. In 1951, the IMO became the **World Meteorological Organization (WMO)**, an agency of the United Nations. Today, the WMO, headquartered in Geneva, Switzerland, coordinates the efforts of 188 nations and territories in a standardized global weather-monitoring program known as *World Weather Watch (WWW)*. WWW combines standardized observing systems, communications facilities, and data analysis/forecasting centers operated by member nations and territories to make available meteorological information internationally.

At standard observation times, the state of the atmosphere is monitored worldwide. This *Global Observing System* (Figure 13.1) consists of 6 geostationary and 3 polar-orbiting satellites, complemented by experimental and back-up satellites, along with around 11,000 land stations—about 4000 of which contribute data to regional synoptic weather networks. Additional observational data are gathered by some 4000 ships at sea, more than 3000 reconnaissance and commercial aircraft (*Aircraft Communications Addressing and Reporting System*), weather radar, about 1300 radiosonde stations, and 1200 drifting and 200 moored buoys. The *Global Data-Processing System (GDPS)* makes available meteorological analyses and various forecast products to WMO member nations via the *Global Telecommunications System (GTS)*. Observational data are transmitted to the three World Meteorological Centers (located near Washington, DC, Moscow, Russia, and Melbourne, Australia) where maps and charts representing the current state of the atmosphere are prepared. From human and computer analysis of this information, weather forecasts are created. Maps, charts, and forecasts are then sent to regional and national meteorological centers where weather information and forecasts are generated and interpreted for each center's area of responsibility and distributed to local weather service forecast offices and then to the public.

FIGURE 13.1
The Global Observing System. NMS refers to National Meteorological Centers. [World Meteorological Organization]

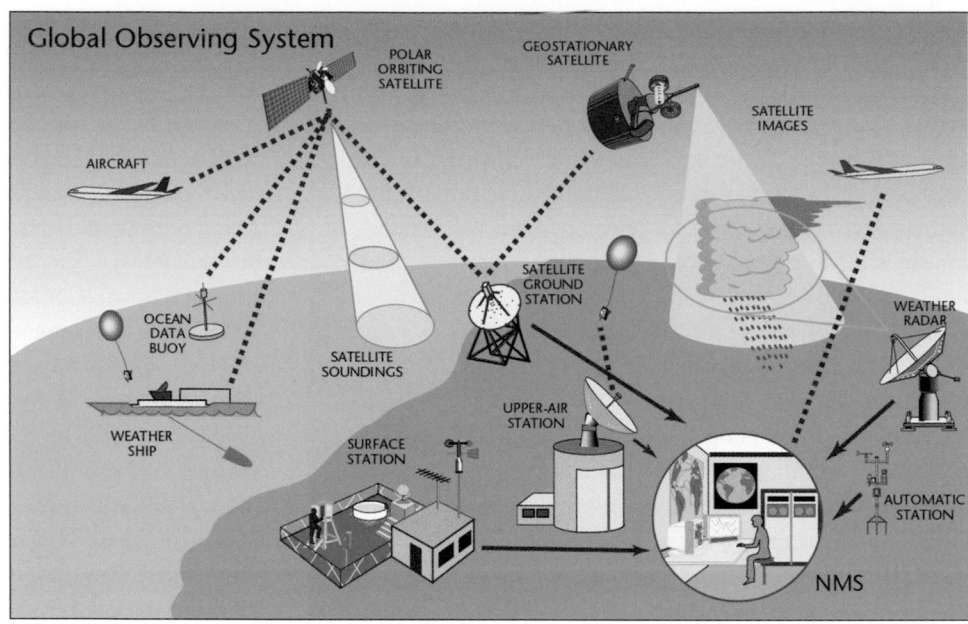

Timely and accurate weather forecasts for the United States are the responsibility of the **National Centers for Environmental Prediction (NCEP)**, part of NOAA's National Weather Service. Originally established as the former National Meteorological Center in 1958, NCEP was headquartered in Camp Springs, MD, until 2012, when it moved to a state-of-the-art facility collocated at the University of Maryland, just outside Washington, DC. NCEP's goal is to "protect life and property, as well as mitigate economic loss, by providing accurate forecasts and forecast guidance products to weather service field offices." NCEP meteorologists produce short-term national forecasts as well as longer-term weather and climate outlooks. At present, meteorologists issue weather forecasts out to 7 days and climate predictions for two weeks to a year or so. Nine centers comprise NCEP: Central Operations, Environmental Modeling Center, Hydrometeorological Prediction Center, Ocean Prediction Center, Climate Prediction Center (all located at NCEP Headquarters), Aviation Weather Center (Kansas City, MO), Storm Prediction Center (Norman, OK), National Hurricane Center (Miami, FL), and the Space Weather Prediction Center (Boulder, CO). We have more on NCEP later in this chapter.

Weather forecasting entails (1) acquisition of present weather data, (2) graphical depiction of the state of the atmosphere by plotting those data on maps and charts, (3) analysis of data and maps, (4) prediction of the future state of the atmosphere, and (5) dissemination of weather information and forecasts to the public.

Acquisition of Weather Data

Since invention of the first weather instruments in the 17th century, weather observation has undergone considerable refinement (Chapter 2). Denser monitoring networks, more sophisticated calibrated instruments and communications systems, and better trained weather observers have produced an increasingly detailed, reliable, and representative record of weather and climate.

SURFACE WEATHER OBSERVATIONS

About 2000 stations across the United States routinely monitor surface weather. These stations may be operated by any of the following: (1) National Weather Service personnel, (2) the staff of other government agencies, including the *Federal Aviation Administration (FAA)*, or (3) private citizens or businesses in cooperation with the NWS. At sea and on the Great Lakes, more than 1200 ships also voluntarily gather weather data.

The NWS also maintains networks of automated weather stations in locations where manned observations are not feasible. As of this writing, the **National Data Buoy Center (NDBC)** operates 254 such stations, including 109 Meteorological and Oceanographic Weather Buoys (Figure 13.2), 51 Coastal Meteorological Automated Network (CMAN) Stations at lighthouses, fishing piers, and offshore oil platforms, 39 Tsunami Warning Buoys, and 55 buoys included in the Tropical Atmosphere Ocean Array (TAO) which support El Niño – Southern Oscillation (ENSO) forecasting. NDBC stations provide data on storm intensity and track and are considered an

FIGURE 13.2
This NDBC buoy, located off the coast of Oregon, has a 3 m (9.8 ft) hull diameter. The anemometers are 5 m (16.4 ft) above the water line and the temperature and humidity sensors are in the two cylinders below the anemometers. The two rectangular panels are solar panels. The ocean wave, air pressure, and water temperature sensors are inside the hull. (The grill encompassing the deck is to keep sea lions from climbing onboard.) [NOAA/NDBC]

essential part of the hurricane warning system. Satellites relay data from the buoy network to users.

Weather stations gather data for (1) preparation of weather maps and forecasts, (2) exchange with other nations, and (3) use by aviation and marine interests. Stations report cloud height and sky cover, wind speed and direction, visibility, precipitation type and amount, air temperature, dewpoint, and air pressure. Also, airport weather stations report altimeter settings for guidance to aviation.

Weather observations must be taken at the same time everywhere so as to accurately represent the state of the atmosphere. To this end, weather observers adhere to **Coordinated Universal Time (UTC)** (or *Universel Temps Coordinné*), previously known as *Greenwich Mean Time (GMT)*, the time along the prime meridian, 0 degree longitude. (The prime meridian passes through the Old Royal Observatory, Greenwich, England.) For reference, at 0600 UTC, it is midnight Central Standard Time (CST)

in Chicago, IL and 10 p.m. Pacific Standard Time (PST) in San Francisco, CA. See Chapter 1 for more on time keeping.

During the 1990s, weather-observing facilities of the National Weather Service underwent extensive restructuring and modernization with the goal of upgrading the quality and reliability of weather observation and forecasting. NWS Weather Forecast Offices were consolidated, and are now located at 122 sites across the United States (Figure 13.3), each with a designated area of forecast and warning responsibility. The National Weather Service replaced old weather radars with Doppler radars, currently at 122 sites nationwide. The FAA and the U.S. Department of Defense (DOD) operate another 38 Doppler units. As noted in Chapters 7 and 11, Doppler radar offers significant (possibly lifesaving) advantages over the old weather radar in providing more advance warning of the development and approach of severe weather systems. The recent installation of dual-polarization radar provides even more detailed information about severe weather and is expected to increase tornado warning lead times.

Automated weather stations replaced the old manual system of hourly observations. A joint effort of the NWS, FAA, and DOD, the **Automated Surface Observing System (ASOS)** consists of modern sensors, computers, and fully automated communications ports (Figure 2.11). As of this writing, a network of 947 ASOS units operates continuously (updating observations every minute, 24-hrs a day), feeding observational data to NWS Forecast Offices and local airport control towers. Using FAA ground-to-air radio, ASOS transmits computer-generated voice observations directly to aircraft flying near airports. ASOS reports a variety of weather elements: temperature (ambient, dewpoint), pressure

FIGURE 13.3
Exterior view of a National Weather Service Weather Forecast Office. [Courtesy of NOAA/NWS Forecast Office New York]

(sea-level, altimeter setting), wind (direction, speed), precipitation accumulation, visibility (to at least 16 km or 10 mi), obstructions to vision (e.g., fog, haze), present weather, and sky condition (cloud height and amount of cloud cover) up to 3660 m (12,000 ft). In addition, ASOS detects significant changes in weather (e.g., wind shift, rapid pressure changes, beginning and ending times for precipitation). Similar to ASOS is the *Automated Weather Observation System (AWOS)*, located at many small airports. As of this writing, 163 FAA-owned and 1149 non-Federal AWOS units are operating in the U.S. In addition, FAA operates 50 *Automated Weather Sensor System (AWSS)* units, also similar to ASOS.

In addition to the numerous land-based weather stations that provide information of potential use for weather forecasting and aviation, nearly 10,000 cooperative weather stations are scattered across the United States (Figure 13.4). These stations are cooperative in that people volunteer their time and labor to make observations and the NWS provides instruments and data management. The principal function of the **NWS Cooperative Observer Network** is to record daily precipitation and

temperatures for hydrologic, agricultural, and climatic purposes. Observers report 24-hr precipitation totals and maximum/minimum temperatures based on observations made daily by 8 a.m., local time; some observers also report river levels. Traditionally, observers telephoned reports to the local NWS Weather Forecast Office but a new program that automates cooperative stations enables the observer to enter data into a computer that formats and transmits data to computer workstations in the NWS Advanced Weather Interactive Processing System (AWIPS) (described later).

UPPER-AIR WEATHER OBSERVATIONS

As noted in Chapter 2, meteorologists still rely primarily on radiosondes to monitor conditions in the troposphere and lower stratosphere (Figure 13.5). A **radiosonde** is a radio-equipped instrument package carried aloft by a balloon that transmits to a ground station temperature, pressure, and humidity profiles (soundings) to a maximum altitude of about 30,000 m (100,000 ft). In addition, winds at various levels in the atmosphere are computed by tracking the balloon's drift using a radio

FIGURE 13.4
A meteorologist checks the calibration of a rain gauge at a NWS cooperative weather station. These stations provide data that are useful for hydrologic, agricultural, and climatic purposes.

FIGURE 13.5
A radiosonde instrument package carried aloft by a balloon measures vertical profiles of temperature, pressure, and humidity to a maximum altitude of about 30,000 m (100,000 ft).

direction-finding antenna (a *rawinsonde* observation). Worldwide, readings are taken twice each day, at 0000 UTC and 1200 UTC.

As of this writing, the National Weather Service is implementing the *Radiosonde Replacement System (RRS)* to upgrade the current generation of radiosondes. The RRS consists of a global positioning system (GPS) tracking antenna, GPS radiosondes that operate at a radio frequency of 1680 MHz, plus a NT-based computer workstation. The new system provides more detailed and accurate upper-air data. Readings are available at 1-second intervals, corresponding to altitude levels about 5 m (16 ft) apart.

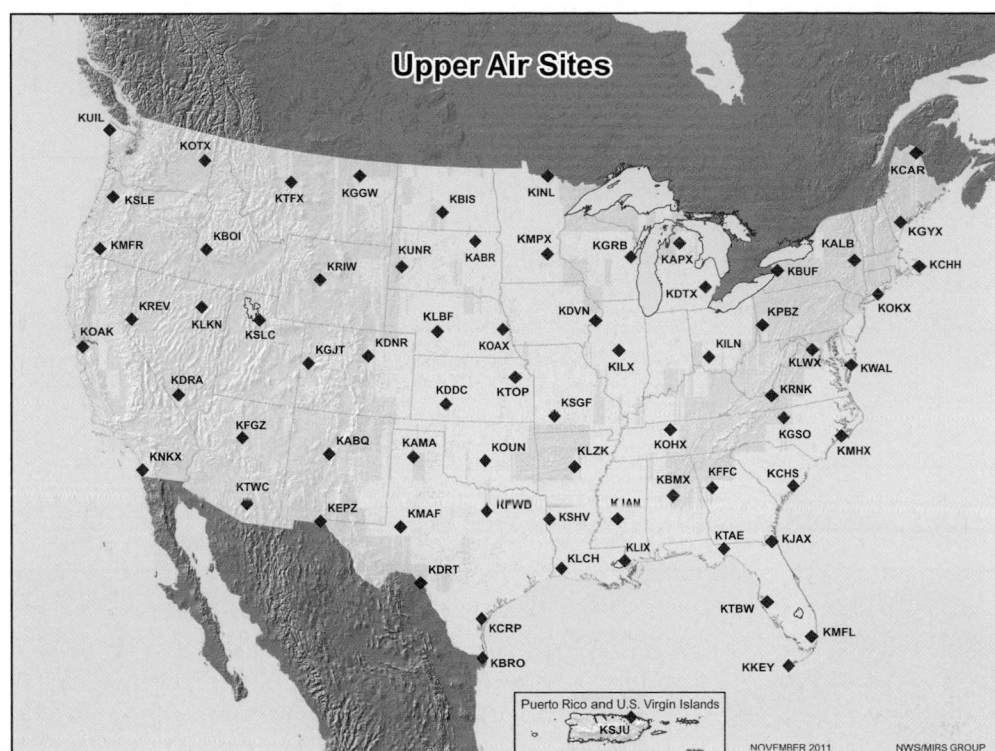

FIGURE 13.6
Locations of radiosonde observation (RAOB) stations in the continental U.S. and Puerto Rico. [NOAA]

The U.S. operates 92 radiosonde observation (RAOB) stations, 69 of which are located in the continental U.S., 13 in Alaska, 2 in Hawaii, and one each in Guam and Puerto Rico (Figure 13.6). These sites use a balloon designed to obtain data from the surface to altitudes above 30 km (18.6 mi) or more. The balloons reach 30 km at least 60% of the time. Occasionally, meteorological rockets are employed to reach much higher altitudes (up to 100 km or 62 mi) but data from these probes are primarily for research purposes. In some urban areas, low-level soundings (up to 3000 m or 9800 ft) monitor atmospheric conditions to assess air pollution potential. Satellites, weather radar, aircraft, wind profilers, and *dropwindsondes* (similar to a radiosonde but dropped from an aircraft) also supply upper-air weather data. Satellite-derived observational data currently account for 95% of all data assimilated by NCEP numerical models.

Weather Data Assimilation, Depiction and Analysis

Weather data collected from surface and upper-air stations in the network are transmitted to national weather centers where these data are subjected to quality control prior to being displayed on maps for weather analysis or used in numerical weather prediction models. Accurate observational data must be obtained rapidly from as many locations as possible. Computers at the *Hydrometeorological Prediction Center (HPC)*, one of the National Centers for Environmental Prediction (NCEP), use special symbols for plotting weather observations on synoptic and hemispheric maps. The weather reported by each observation station is depicted on a map by following a conventional **station model**. The station model in Figure 13.7 shows symbols for surface weather conditions. By international agreement, the same station model format and symbols are used throughout the world, thereby overcoming language barriers and permitting easy interpretation of weather data by meteorologists anywhere.

Weather systems are three-dimensional so that both surface and upper-air weather maps are needed to represent the state of the atmosphere. A very different approach is used for the two types of maps, however. Surface weather data are plotted on a constant-altitude surface (usually sea-level), whereas upper-air weather data are plotted on constant-pressure (isobaric) surfaces.

FIGURE 13.7
U.S. Surface weather station model.

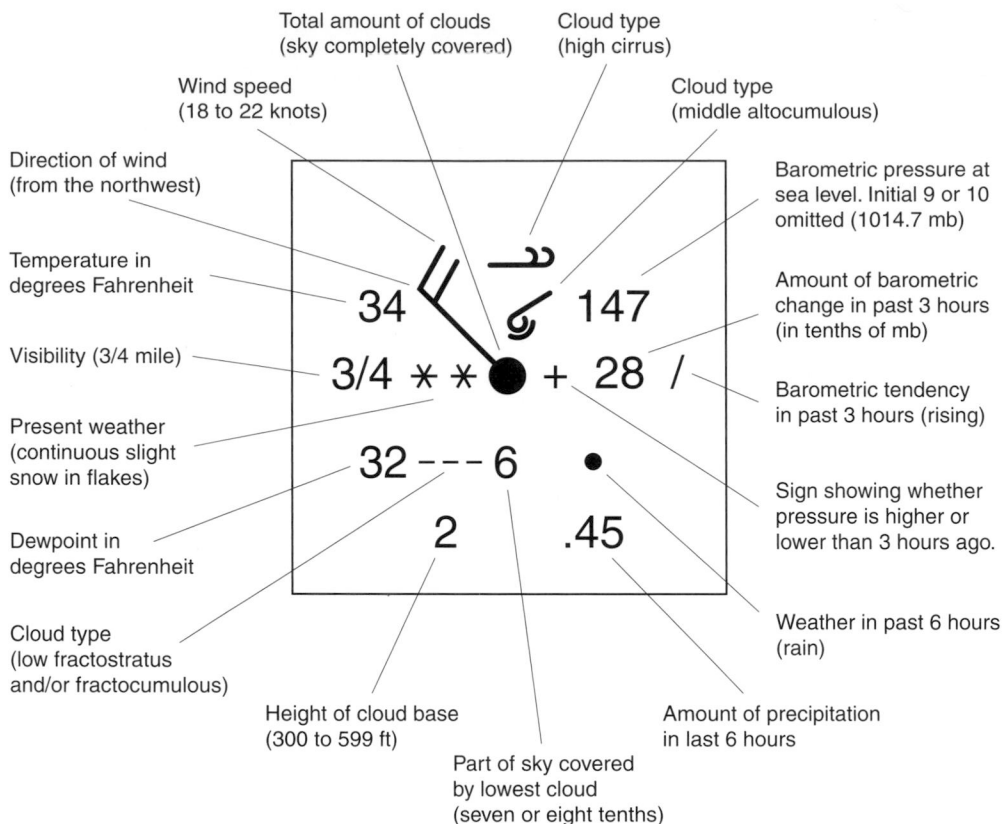

SURFACE WEATHER MAPS

It is standard practice for meteorologists to adjust surface air pressure readings to sea level (Chapter 5), thereby eliminating the influence of station elevation on air pressure readings. This procedure enables meteorologists to compare surface air pressure readings at weather stations that are located at different elevations above sea level. The adjusted air pressure readings are plotted on surface weather maps. Meteorologists (or their computers) draw **isobars**, lines of equal air pressure, at 4-mb intervals, a procedure that requires interpolation between stations. An isobaric analysis reveals the location of such features as anticyclones (*highs*) and cyclones (*lows*), troughs and ridges, as well as horizontal air pressure gradients (Figure 13.8).

On surface weather maps, a cyclone center, where air pressure is lowest locally, is indicated by the symbol *L* or *Low*. Closely spaced isobars surrounding the storm center indicate a relatively steep horizontal air pressure gradient and strong surface winds. Fronts originate at the storm center and typically appear in pressure troughs revealed as bends in isobars. Because we can infer surface wind direction from the isobar pattern, bending of isobars at fronts indicates that the wind changes direction as we cross a front. As noted

in Chapter 10, a wind shift is one of the characteristics associated with a frontal passage. A relatively weak horizontal air pressure gradient (widely spaced isobars) and weak winds or calm air generally occur over a broad area about the center of an anticyclone (mapped as *H* or *High*).

Surface synoptic weather maps are drawn every 3 hrs for North America and at 6-hr intervals for the Northern Hemisphere. Special maps and charts are also constructed that summarize a variety of surface weather elements including, for example, maximum and minimum temperatures for a 24-hr period, precipitation amounts for 6 and 24 hrs, and observed snow cover. An animated national composite radar reflectivity loop is also updated every 30 minutes or less.

UPPER-AIR WEATHER MAPS

Upper-air weather data acquired mostly by radiosondes are plotted on constant-pressure surfaces following a modified station model. Applying basic laws of atmospheric physics, meteorologists compute the altitudes corresponding to these pressure values. For example, meteorologists determine the altitude of the 500-mb surface, that is, the altitude where the air pressure is about one-half its average sea-level

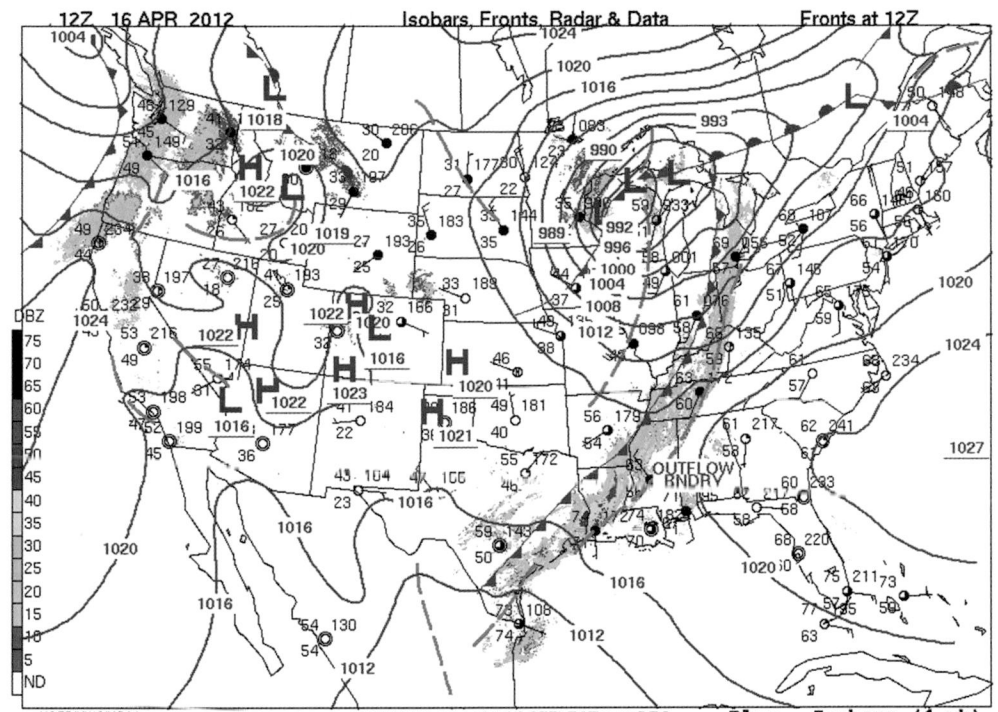

value. By plotting altitudes of the 500-mb level as monitored simultaneously by all radiosonde stations, meteorologists construct a map representing the topography of the 500-mb surface (Figure 13.9). The 500-mb map is particularly useful for weather

forecasting because the atmospheric circulation at that level closely approximates the upper-level steering winds and the trough and ridge patterns responsible for the development and movement of surface weather systems such as extratropical cyclones.

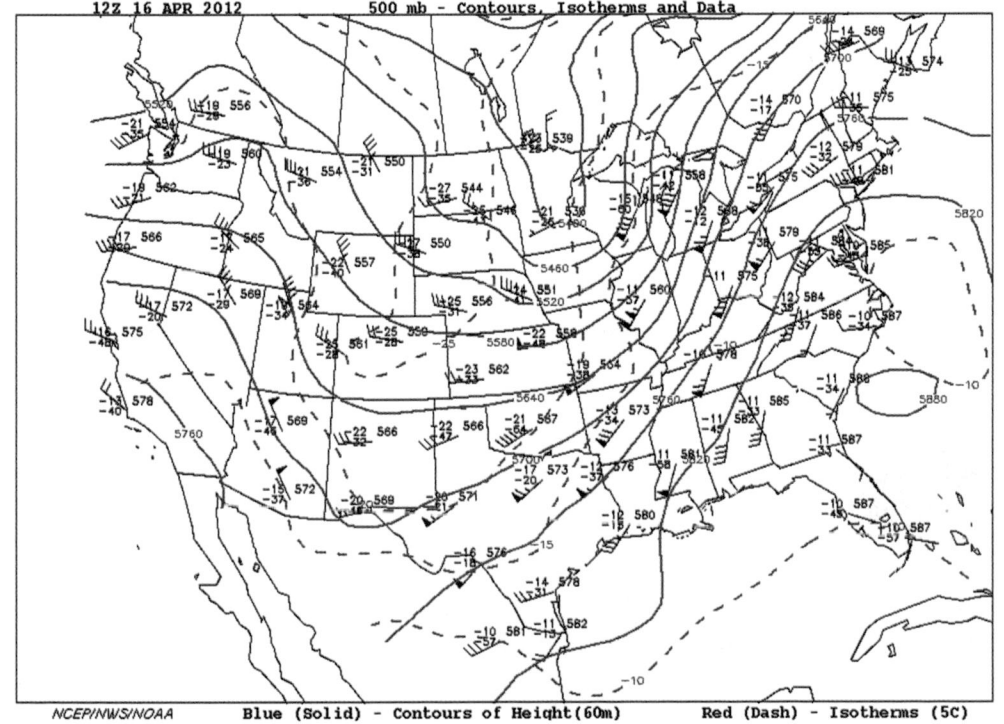

Meteorologists (or computer analysis programs) draw height contour lines where the 500-mb level is at the same altitude, interpolating between weather stations. By convention, height contours are labeled in meters above sea level and the contour interval (difference between successive contours) is 60 m. Dashed lines, also plotted on the map, are *isotherms*, lines of equal temperature, labeled in degrees Celsius. Arrows and barbs represent wind direction and speed at the 500-mb level.

The altitude of a pressure surface (such as the 500-mb level) varies from one place to another, primarily because of differences in mean temperature of the air below that pressure surface. Air pressure drops with altitude more rapidly in cold air than in warm air. Raising the temperature of air reduces its density so that greater vertical distances are required for warm air to exhibit the same drop in pressure as cold air. This means, for example, that the 500-mb level is at a lower altitude where the air below is relatively cold, but at a higher altitude where the air below is relatively warm. Hence, height contours plotted on an isobaric surface (e.g., the 500-mb chart) generally show a gradual slope downward from the warm tropics to the colder polar latitudes.

Upper-air observations indicate that horizontal winds generally parallel the height contour lines at the 500-mb level. Where contours are closely spaced (a relatively steep height gradient), winds are strong, and where contours are far apart (a relatively weak height gradient), winds are light. Why are winds associated with a gradient in the height of a pressure surface? A height gradient develops in response to a horizontal gradient in air temperature, and with a horizontal air temperature gradient, there are also horizontal gradients in air density and air pressure. As discussed in Chapter 8, a horizontal air pressure gradient generates wind. The 500-mb surface is so far above the *atmospheric boundary layer* that the planetary- and synoptic-scale winds are essentially *geostrophic* where contours are straight, and *gradient* where contours are curved. Large-scale winds blowing at the 500-mb level are the product of interactions among a horizontal height gradient, the Coriolis Effect, and the centripetal force.

On upper-air weather maps, contours exhibit both cyclonic (counterclockwise) and anticyclonic (clockwise) curvature. These reveal the *troughs* and *ridges*, respectively, in the prevailing westerlies described in Chapter 9. Contours sometimes define a series of nearly concentric circles, perhaps indicating a cutoff or blocking circulation pattern. At the center of a ridge, the air column is relatively warm and contour heights are

high, so we label the ridge with an *H*. An upper-air ridge often is linked to a warm-core anticyclone at the surface. In contrast, near the center of a trough, the air column is relatively cold and contour heights are low, so the trough is labeled with an *L*. An upper-air trough may be linked to a cold-core extratropical cyclone at the surface. Warm-core cyclones (thermal lows) and cold-core anticyclones (polar and arctic highs) are too shallow to appear on 500-mb maps. Interestingly, a ridge may be found on upper-air charts above a warm-core cyclone such as a thermal low (Chapter 10) or hurricane (Chapter 12). Similarly, a trough may occur above a shallow cold-core high.

Winds blowing across regional isotherms produce cold or warm air advection (Chapter 4). **Cold air advection** takes place where winds blow from colder localities toward warmer localities. **Warm air advection** occurs where winds blow in the opposite direction. Warm air advection causes the 500-mb surface to rise, whereas cold air advection lowers the 500-mb surface. Cold air advection between the surface and 500-mb thus deepens troughs and weakens ridges, whereas warm air advection strengthens ridges and weakens troughs. When cold air advects into troughs at the same time that warm air advects into ridges, the circulation becomes more meridional. On the other hand, if warm air advects into troughs while cold air advects into ridges, the circulation becomes more zonal. As described in Chapter 9, shifts in the upper-air circulation pattern between meridional flow and zonal flow have important implications for north/south air mass exchange and the development and track of extratropical cyclones.

NCEP's Hydrometeorological Prediction Center issues 500-mb maps twice each day based on upper-air observations at 0000 UTC and 1200 UTC. One set of maps covers North America and another covers the Northern Hemisphere. Although we have focused our discussion of upper-air weather maps on the 500-mb level, similar analyses are routinely constructed twice daily for the 850-, 700-, 300-, 250-, 200-, and 100-mb levels. The 300-, 250-, and 200-mb charts are useful in locating the polar front jet stream. Lines, called *isotachs*, are drawn on these charts connecting points having the same wind speed and they highlight the regions of strongest upper tropospheric winds, which represent the location of the jet stream. The altitude of the jet stream depends on the mean temperature of the underlying air column and is lower in winter than in summer. Hence, for jet stream analysis, meteorologists rely upon the 300-mb chart during the coldest time of year and the 200-mb chart during the warmest time of year. The display of the position of the jet stream on these upper

FIGURE 13.10
AWIPS (Advanced Weather Interactive Processing System) workstation at a National Weather Service Forecast Office. AWIPS enables meteorologists to overlay and interpret a variety of graphical meteorological data. [Photo by R.S. Weinbeck]

tropospheric charts helps meteorologists locate regions of divergence and convergence that provide upper-air support for the development of surface weather systems.

The deluge of real-time weather information (e.g., weather maps, satellite images, soundings, composite radar maps) has spurred development of computerized data management systems. An example is **AWIPS (Advanced Weather Interactive Processing System)**, a technologically advanced interactive computer system that helps forecasters analyze huge amounts of data.

AWIPS plays a critical role in the ability of U.S. forecasters to make weather predictions that can save lives and safeguard property. It is a complex network of systems that ingest and integrate meteorological, hydrological, satellite, and radar data. Meteorologists then display and overlay images, graphics, and other data to prepare and issue more accurate and timely forecasts and warnings (Figure 13.10). Forecasters at more than 130 weather forecast offices and river forecast centers across the nation utilize the capabilities of AWIPS to make increasingly accurate weather, water, and climate predictions. AWIPS II, a newer, updated version, began its roll-out in 2011.

Weather Prediction

Meteorologists at the Hydrometeorological Prediction Center analyze satellite images and weather observations plotted on surface and upper-air weather maps and computer model output. From these analyses, they prepare general short- to medium-range weather forecasts for the United States. They issue short-term 12-, 24-, 36-, and 48-hr forecasts plus medium-range 3- to 7-day extended outlooks each day. Based upon the guidance provided by these forecasts plus the output of regional numerical models, NWS Weather Forecast Offices prepare detailed short-term forecasts for their region of responsibility. In addition, the *Climate Prediction Center* generates 6-10 day, 8-14 day, monthly (30-day), seasonal (90-day), and multi-seasonal outlooks.

Weather forecasting is an extremely challenging endeavor, primarily because it involves many variables and a vast quantity of weather data. For these reasons, computerized numerical models of the atmosphere have been developed to assist forecasters.

NUMERICAL WEATHER FORECASTING

The fundamentals of scientific weather forecasting were proposed in the early 20th century, first in 1904 by the Norwegian scientist Vilhelm Bjerknes (1862-1951), who provided the first schematic outline for numerical weather prediction and later by English physicist Lewis F. Richardson (1881-1953) who envisioned a large amphitheatre of clerks who would use desk calculators to compute a forecast. However, the absence of sufficiently rapid computing capabilities made these early ideas unfeasible. In the early 1950s, John von Neumann (1903-1957) and Jule Charney (1917-1981) pioneered the use of electronic computers for weather forecasting at the Institute for Advanced Study at Princeton University. A primitive computer, less powerful than today's tablet computer, successfully forecast the horizontal air pressure pattern at 5000 m (16,400 ft) altitude over North America 24 hrs in advance. By 1955, computers were routinely generating weather forecasts from surface and upper-air weather observations.

A computer is programmed with a numerical model of the atmosphere, that is, a model consisting of computational versions of the mathematical equations that relate winds, temperature, pressure, and water vapor concentration. Beginning with present (real-time) weather data, numerical models predict the values of various atmospheric properties on a uniform grid of points on pressure surfaces for some time in the future—say a few minutes from now. With these predicted conditions as a new starting point, another forecast is then computed for the subsequent few minutes. The computer repeats this procedure again and again until weather maps are generated for the next 12, 24, 36, and 48 hrs (or longer). In this iterative process, more than a trillion computations are performed each second on a vast array of observational

FIGURE 13.11
IBM supercomputers used for climate and weather forecasts. [NOAA]

data; hence the need for a high-speed computer (supercomputer) that can accommodate huge quantities of data. (In fact, meteorology pioneered the application of supercomputers.) As of this writing, NCEP's latest high performance IBM supercomputers each process 71 trillion calculations per second at maximum performance and ingest more than 2 billion global observations daily (Figure 13.11).

Some computerized numerical models of the atmosphere are designed to operate over different spatial scales depending on the forecast range. For medium-range forecasts (up to 10 days), observational data are fed into the computer from all over the globe, because within that forecast range a weather system may travel thousands of kilometers. On the other hand, for short-range forecasts (up to 3 days), the model utilizes data drawn from a more restricted region of the globe. Compared to a global model, a regional model offers the advantage of greater resolution of data over a smaller area of interest. Thus, tomorrow's forecast of weather across the nation is based upon data obtained from all over the North American continent and surrounding ocean, but the run of forecasts for the next week requires data from around the globe.

NCEP computers are programmed with several different numerical models of the Earth-atmosphere system which differ in spatial scale, resolution, and forecast period. The *North American Mesoscale (NAM)* model slices the troposphere into 60 layers or levels with 12 km (7.4 mi) between data points and generates forecasts every 6 hours out to 84 hrs (3.5 days). In October 2011, the NAM model underwent major upgrades including changing the basic framework from the *Weather Research and Forecast Model (WRF)* to the *NOAA Environmental Modelling System (NEMS)*.

Four higher resolution subdomains were also added to the main 12 km domain. The subdomains cover the U.S. (4 km resolution), Alaska (6 km resolution), Hawaii (3 km resolution), and Puerto Rico (3 km resolution). These run out to 60 hours each. There is also a new grid (1.33-1.5 km resolution) that runs out to 36 hours to support fire weather forecasting. The *High Resolution Window (HRW)* model, also known as *Nested Window Run (NWR)*, contains images from the WRF versions of two additional models. The WRF is run 4 times a day and produces forecast graphics at 3-hr increments out to 2 days. The *Rapid Update Cycle (RUC)* model features 50 levels with a horizontal resolution of 13 km (7.5 mi). RUC provides short-range, hourly numerical weather guidance for general forecasting, as well as for aviation and severe weather forecasting (at a frequency of every hour out to 18 hours).

The *Rapid Refresh (RR)* model, which includes all of North America in its domain, is planned to replace the RUC in 2012. The *Global Forecast System (GFS)* consists of one 64-level model operating at different resolutions for early (higher resolution) and later (lower resolution) forecast periods. The T574 version of the GFS model has a horizontal data interval of about 27 km (16.8 mi) and provides forecasts out to 192 hrs (8 days). From there, the T190 version of the GFS model, with a horizontal resolution of about 70 km (43 mi), generates forecasts out to 384 hrs (16 days). The GFS model is run four times per day.

Meteorologists routinely employ at least two techniques to optimize the skill of weather forecasts based on numerical models: (1) **ensemble forecasting** and (2) **model comparison**. Traditionally, forecasts are made using the output from a single run of the "best" model until it loses its skill (produces unlikely results), typically out to about 6 to 8 days. Ensemble forecasts can be skillful up to 15 days. Instead of using just one model run, ensemble forecasts are created by conducting many runs of one model, or even many different models, where each run is based on slightly different sets of initial conditions. An average of the different forecasts is created. This average, or ensemble-mean, forecast improves upon any individual member of the ensemble because meaningful parts of the forecast tend to be reflected in the mean, while random errors, which may be positive or negative in sign, tend to be reduced through averaging. NCEP uses two major ensemble forecast models in daily operation, the *Global Ensemble Forecast System (GEFS)* and the *Short Range Ensemble Forecast (SREF)*. GEFS provides global forecasts every 6 hrs out to 16 days. SREF is a regional model that provides numerical weather guidance for general forecasting as well as for aviation forecasting. It generates North American forecasts every 3 hrs out to 87 hrs (3.6 days). Improvements in one- to three-day forecasts for the continental U.S. have been observed.

The *North American Ensemble Forecast System (NAEFS)* is run by the Meteorological Service of Canada (MSC) and the U.S. NWS, with additional participation from the National Meteorological Service of Mexico (NMSM). It combines NCEP's GEFS with Canada's version of GEFS into one joint model that provides North American forecasts every 6 hrs out to 384 hrs (16 days). There are plans to expand the NAEFS with the addition of the U.S. Navy Fleet Numerical Meteorology and Oceanography Center (FNMOC) ensemble. The *High-Resolution Ensemble Forecast (HREF)* is used for forecasting high-impact weather, such as heavy precipitation events. Forecasts are generated hourly out to 48 hrs for the continental U.S. and Alaska at a 5 km resolution.

In the second technique of model comparison, a comparison is made among forecasts made by several different models (i.e., NCEP models and models operated by other national weather services). If they all agree, then the forecast is issued with a relatively high level of confidence. Using either technique, if forecasts are inconsistent, any one forecast is considered unreliable.

Not all weather forecasts are the primary responsibility of NCEP's Hydrometeorological Prediction Center. Special forecast centers are responsible for predicting tropical cyclones (hurricanes and tropical storms), severe weather, river flow and flooding, weather for mariners, and providing guidance information for domestic and international aviation. Also, scientists at the Space Weather Prediction Center at Boulder, CO predict the "weather" of Earth's near-space environment for satellite and space shuttle operations.

FORECASTING TROPICAL CYCLONES

In 1873, the U.S. Army Signal Corps (then in charge of the nation's system of weather observation and forecasting) began gathering weather reports from Havana and Santiago, Cuba to help detect tropical cyclones in the Caribbean. A hurricane was plotted for the first time on a surface weather map on 28 September 1874 (centered offshore near Savannah, GA). In 1890, with the transfer of weather observation and forecasting responsibilities from military to civilian hands (the U.S. Weather Bureau in the Department of Agriculture), primary emphasis was on observing and forecasting weather over the continent with little attention paid to tropical cyclones. Although six

land-falling hurricanes claimed more than 4000 U.S. lives in 1893, the primary impetus for expanding the number of weather stations in the Caribbean was the Spanish-American War of 1898 and President William McKinley's (1843-1901) fear of what a hurricane near Cuba could do to the U.S. fleet. A U.S. Weather Bureau office was even established in Havana.

In subsequent decades, technological advances in communications and remote sensing greatly benefited the monitoring and understanding of tropical cyclones. Invention of the radio made possible ship-to-shore reports on the state of the atmosphere and seas near tropical cyclones; the first such report was filed in August 1909. By the late 1930s, the upper-air was regularly monitored providing insights on the relationship between tropical cyclones and steering winds aloft. Beginning in the late 1950s, weather radar at coastal stations observed rain bands and other features of an approaching hurricane. Since the 1960s, sensors onboard weather satellites have proved valuable in tracking and monitoring changes in tropical cyclones as they progress through their life cycle. The importance of satellites in data gathering is underscored by the fact that there is only one upper-air station for 10 million square km (4 million square mi) of the tropical and subtropical Atlantic. Furthermore, in recent decades, coastal and offshore buoys provide information on sea-surface conditions.

Specially instrumented reconnaissance aircraft complement satellite, radar, and buoy surveillance of tropical cyclones when they threaten any land in the Atlantic, Caribbean, Gulf of Mexico, Hawaii, or California. Reconnaissance flights into tropical cyclones began during World War II and originally involved aircraft from both the U.S. Navy and Army Air Forces (later the U.S. Air Force). Today, the 53rd Weather Reconnaissance Squadron, a U.S. Air Force reserve unit, conducts most hurricane reconnaissance from Keesler Air Force Base in Biloxi, MS. In addition, NOAA supports special aircraft designed for hurricane research and reconnaissance. Based at MacDill Air Force Base in Tampa, FL, these flights gather data for input into computer models at the Hydrometeorological Prediction Center with products sent on to the National Hurricane Center.

Reconnaissance aircraft fly directly through and over the storm to determine the location of the eye, measure wind speeds, and since 1997 obtain soundings by deploying global positioning system (GPS) dropwindsondes. A *dropwindsonde* (similar to a radiosonde except that the instrument package is dropped from an aircraft) provides data from flight-level (usually about 3000 m or 10,000 ft) down to Earth's surface. Based on several hundred GPS dropwindsonde soundings, meteorologists determined that surface wind speeds in the hurricane eyewall average about 90% of what they are at flight-level. Today, National Hurricane Center meteorologists typically use a factor of 90% to estimate a hurricane's maximum surface winds from flight-level measurements. Also, by extrapolating from air pressure readings at flight-level, meteorologists estimate the sea-level air pressure at the storm center.

During the 1940s, responsibility for forecasting Atlantic hurricanes was split among U.S. Weather Bureau offices in New Orleans, Miami, Washington, DC, Boston, and San Juan. In 1967, the U.S. Weather Bureau office in Miami was officially designated as the **National Hurricane Center (NHC)**. In 1984, the NHC was separated from the Miami NWS Weather Forecast Office. Today, responsibility for forecasting tropical cyclones that could impact the U.S. is divided between the National Hurricane Center on the campus of Florida International University in Miami, FL and the Central Pacific Hurricane Center in Honolulu, HI. The NHC is responsible for issuing statements covering tropical storms and hurricanes in the North Atlantic basin (including the Gulf of Mexico and the Caribbean Sea) and the eastern Pacific basin (to 140 degrees W). When a tropical cyclone develops in the central Pacific (west to the international dateline), the Honolulu NWS Forecast Office activates the Central Pacific Hurricane Center that is responsible for issuing tropical cyclone watches, warnings, and advisories.

The NHC has responsibility for producing and coordinating analysis and forecasting of tropical cyclones for 24 nations in the Americas, Caribbean, and for the waters of the North Atlantic, Caribbean Sea, Gulf of Mexico, and the eastern North Pacific. The *Tropical Analysis and Forecast Branch* of the NHC provides year-round marine weather analysis and forecasts for the tropical and subtropical waters of the eastern North and South Pacific and the North Atlantic basin.

At the NHC, meteorologists watch for development of tropical cyclones over the Atlantic basin (including the Caribbean and Gulf of Mexico) primarily from 1 June through 30 November, plus the eastern Pacific from 15 May through 30 November. The Center operates the SLOSH model for prediction of storm surges, prepares and distributes hurricane watches and warnings for the public, conducts research on hurricane forecasting techniques, and sponsors public awareness programs. Local NWS Weather Forecast Offices transmit information from the NHC as advisories, warnings, or special weather statements to people in their areas of responsibility. The

goal is to provide at least 12 hrs of daylight warning for coastal residents so that they can prepare for a hurricane and evacuate if necessary.

The principal challenge for NHC forecasters is predicting the track and intensity of a hurricane. Such forecasts are issued every 6 hrs and, until 2001, covered periods up to 72 hrs in the future. In 2001, the track forecast period was extended to 96 hrs (4 days) and 120 hrs (5 days). The basis for hurricane track forecasts is a blend of climatology (records of tracks of similar hurricanes in the past), special numerical models, and the experience of the forecaster (Figure 13.12). How skillful are these track forecasts? Figure 13.13 shows trends in the error of Atlantic hurricane track forecasts (in nautical miles) for forecast periods of 72 hrs, 48 hrs, and 24 hrs in advance from 1990 to 2011, and for 120 hrs and 96 hrs from 2001 to 2011. Trend lines indicate that hurricane track errors were cut in half in 15 years. Further reduction in track error likely will come with better numerical models and higher quality observational data that are fed into the models. Research directed at improving track forecast models focuses primarily on better simulations of the oceanic energy source, interactions of tropical cyclones with the surrounding atmosphere, and the storm system's internal dynamics.

While track forecasts have improved, numerical models have not done well in predicting changes in the intensity of tropical cyclones—especially rapidly intensifying systems. With rapid intensification, a category 1 or 2 hurricane can become a category 4 or 5 system in less than 24 hours. As a result, NHC intensity errors have remained relatively flat at all forecast periods, increasing as the forecast period lengthens (Figure 13.14).

If the hurricane track forecast is on target, the SLOSH model can accurately predict the location and height of the storm surge (a dome of water pushed ashore by storm winds) along coastal areas (Chapter 12). By consulting local topographic maps, public safety officials use SLOSH predictions to identify areas likely to be inundated by floodwaters and order evacuation of residents accordingly.

Since the 1983 hurricane season, the NHC has included a probability forecast as part of its advisory statements. A probability forecast states the percent chance that the center of a hurricane or tropical storm will pass within 105 km (65 mi) of any one of 46 designated Gulf and East Coast communities from Brownsville, TX to Eastport, ME. The first probability forecast is usually issued 72 hrs in advance of the storm's anticipated landfall. At that time, by convention, the probability of a hurricane landfall is set no higher than 10% for any community.

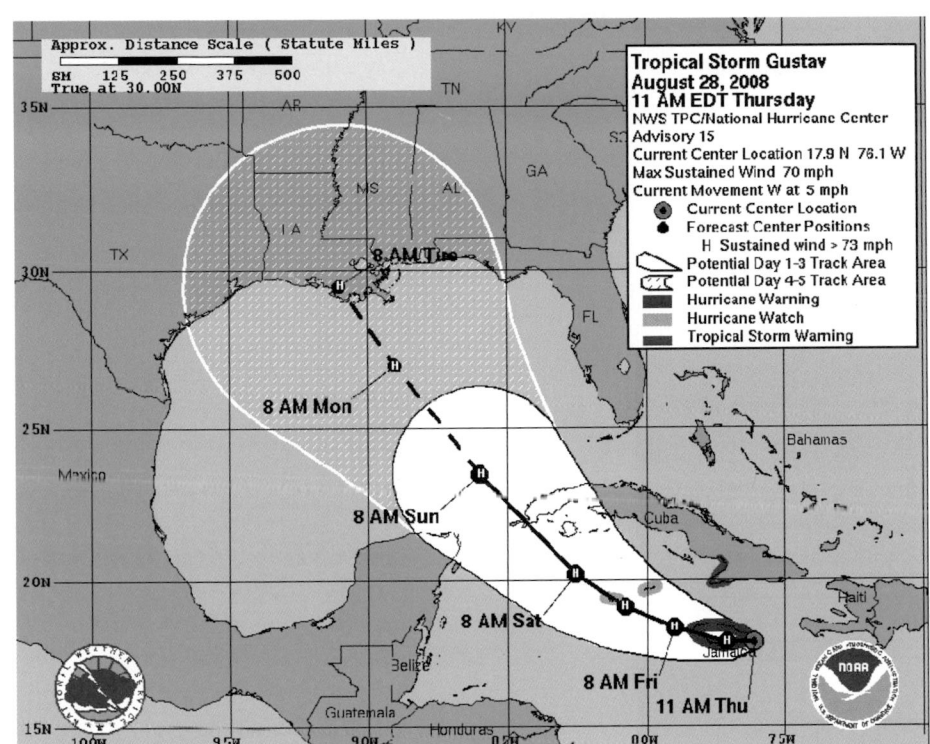

FIGURE 13.12
Track forecast for Hurricane Gustav issued at 11 a.m. EDT on 28 August 2008. The white area (the "cone of uncertainty") encompasses possible variations in the track based on output of many numerical models. [NOAA, National Hurricane Center]

Probabilities increase to 13% to 18% at 48 hrs, 20% to 25% at 36 hrs, 35% to 50% at 24 hrs, and 60% to 80% at 12 hrs, prior to the storm's expected landfall.

A *hurricane watch* is issued for a coastal area when sustained winds of 74 mph (119 km per hr) or higher are possible within the next 36 hrs; a *hurricane warning* means that hurricane conditions are expected in 24 hrs or less. (Watches and warnings are also issued for tropical storms.) The uncertainty in hurricane track forecasts and the considerable threat posed by a hurricane requires forecasters to over-warn the public. That is, hurricane watches and warnings are issued for a stretch of coastline that is much longer than the diameter of the approaching hurricane. Over-warning is costly for people who prepare for a hurricane that never shows up. Millions of dollars are spent on preparations, such as boarding up windows and

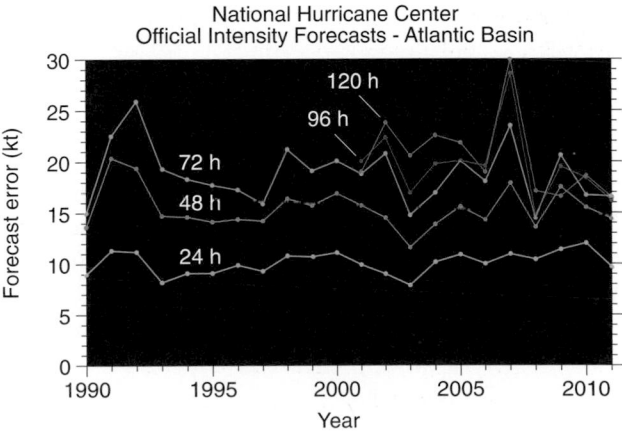

FIGURE 13.13
Atlantic hurricane track forecast error (in nautical miles) for forecast periods of 72, 48, and 24 hours (1990-2011) and for forecast periods of 120 and 96 hours (2001-2011). [NOAA, NHC]

FIGURE 13.14
Atlantic hurricane basin intensity forecast errors (in nautical miles per hour) have remained fairly flat at all forecast periods. With lengthening forecast period, the error increases. [NOAA, NHC]

FIGURE 13.15
During the 20th century, tropical cyclone fatalities in the United States generally trended downward. This trend was interrupted in the first decade of the 21st century due to 1200 fatalities directly caused by Hurricane Katrina in 2005. Property damage (normalized to 2010 dollars) has trended upward since the 1980s. [NOAA]

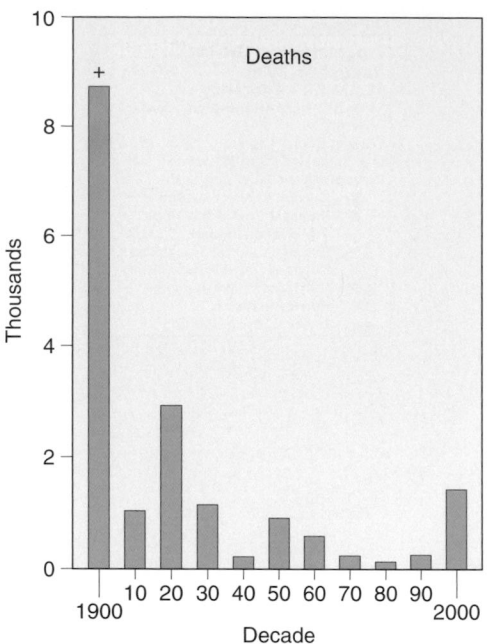

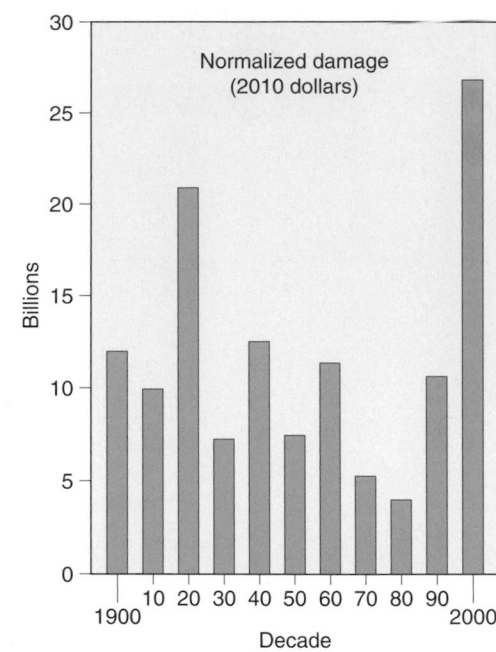

closing businesses, and additional millions of dollars are lost due to disruption of commerce and industry. The cost of evacuation can also be considerable. Nonetheless, these costs are outweighed by the benefits of being prepared. The lull in Atlantic hurricane activity during the 1970s and 1980s coupled with more skillful hurricane forecasting and more effective public response programs was responsible for a dramatic decline in hurricane related fatalities during the 20th century (Figure 13.15). This all changed in August 2005 when Hurricane Katrina directly claimed 1200 lives along the Louisiana and Mississippi Gulf Coast. Also, during the 20th century, the dollar cost of property damage skyrocketed, presumably because of greater coastal development.

FORECASTING FOR AVIATION

As of 1 October 1995, the National Severe Storms Forecast Center (NSSFC) in Kansas City, MO became the *Aviation Weather Center (AWC)* and the *Storm Prediction Center (SPC)*. The AWC remains in Kansas City where it shares facilities with the National Weather Service Training Center (Figure 13.16). The SPC moved to Norman, OK in 1997 and is co-located with the *National Severe Storms Laboratory (NSSL)*, a center for severe weather research.

Meteorologists at the Aviation Weather Center issue warnings, forecasts, and analysis of weather conditions for aviation interests. These functions support the Federal Aviation Administration (FAA) Air Traffic Control responsibility to safely and efficiently manage the nation's airspace. According to NOAA's National

Weather Service, weather is a factor in more than one-quarter to one-half of all aviation accidents each year. In addition, weather-related aviation delays result in an estimated economic loss of $1 billion per year. AWC meteorologists issue forecasts for weather conditions that may impact domestic and international airspace during the subsequent 24 hours, analyze weather hazards for aircraft in flight, and generate warnings that are made available to the aviation community immediately. AWC forecasts complement airport terminal and enroute forecasts provided by local NWS Weather Forecast Offices. In addition to the public safety aspect, efficient

FIGURE 13.16
The National Weather Service Training Center co-located with the Aviation Weather Center at Kansas City, MO. [Photo by R.S. Weinbeck]

The AWC provides a number of products and services. These include the following:

Area Forecasts—weather conditions for six geographical areas issued three times daily.

SIGMET Advisories—issued for severe turbulence, severe icing, intense sand or dust storms or volcanic dust clouds. These advisories are provided to pilots, dispatchers, and air traffic controllers, apply to all aircraft flying at or below 45,000 ft, and are valid for up to four hours.

AIRMET Advisories—issued for less severe weather hazards.

Collaborative Convective Forecast Product—issued every two hours and provides two-, four-, and six-hour forecasts of convective activity.

Convective SIGMETs—inflight advisories for thunderstorms that are especially hazardous for aviation; issued for three geographical areas each hour.

Low Level Significant Weather Charts—depict areas of low cloud ceilings and visibilities, turbulence, and freezing levels; cover the 48 contiguous states and southern Canada and extend from the surface to 24,000 ft.

use of aviation forecasts has proven economically beneficial to airlines through fuel savings.

FORECASTING SEVERE STORMS

Meteorologists at the Storm Prediction Center (SPC) monitor atmospheric conditions for potential development of severe local storms, including severe thunderstorms and tornadoes over the contiguous U.S. The SPC also monitors heavy rain, blizzards and fire weather and issues guidance products for these hazards. Products are used by NWS Weather Forecast Offices, emergency managers, broadcast meteorologists, aviation interests, and other groups.

Several times daily, the SPC issues *convective outlooks* identifying areas around the nation expected to experience both severe and non-severe thunderstorms over the following 1 to 3 days. Outlooks specify areas of severe thunderstorm risk (slight, moderate, or high) depending on the coverage and intensity of thunderstorms expected in a region (Figure 13.17). Forecasters also issue outlooks for tornado, severe hail, and severe wind probabilities. With the passage of time as the threat of severe weather becomes better defined, SPC meteorologists issue a *mesoscale discussion*, which describes the evolving severe weather threat (including non-thunderstorm hazards such as heavy snow). If severe thunderstorm development is

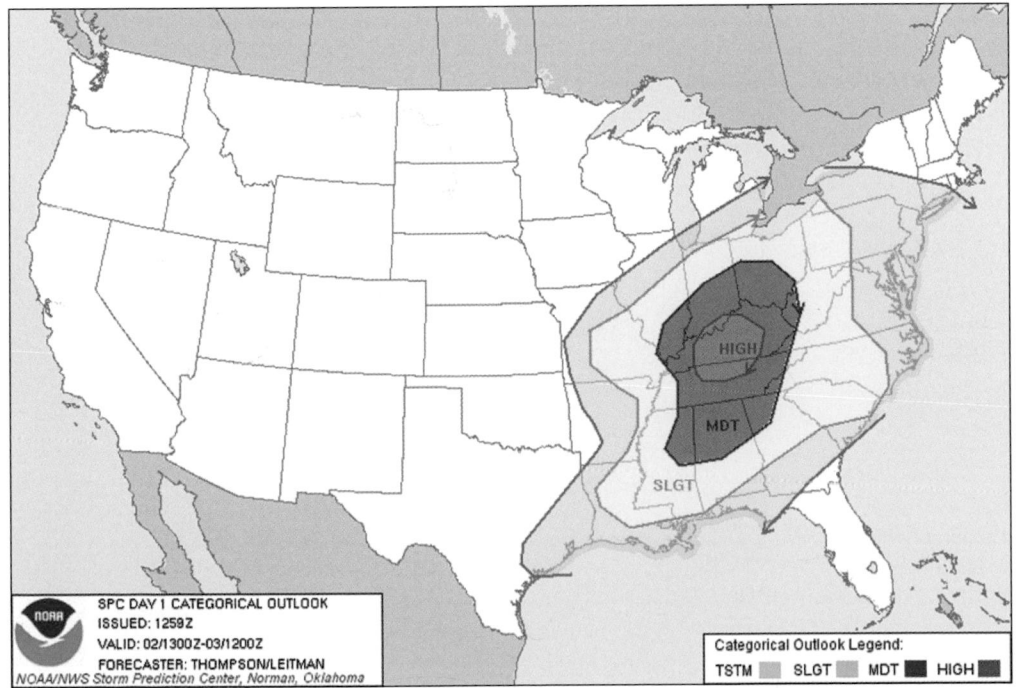

FIGURE 13.17
The SPC Day 1 convective outlook issued at 1259Z on 2 March 2012 shows areas of slight (yellow), moderate (red), and high (purple) severe thunderstorm risk. The shaded green region is a general area that could experience thunderstorms. [NOAA/NWS/Storm Prediction Center]

imminent or likely to occur in the next several hours, the SPC issues watches for severe thunderstorms or tornadoes as appropriate. Watches alert the public to the threat for severe weather and activate storm spotter networks. Watches essentially fine-tune forecast areas identified in the convective outlook. A typical watch encompasses about 65,000 square km (25,000 square mi) and is valid for 4 to 6 hrs. Beginning in 2005, watches apply to geographic areas defined by counties. Warnings are issued by the local NWS Weather Forecast Office for the areas expected to be more directly affected.

Statistics on verification of severe weather watches issued by the Storm Prediction Center demonstrate significant improvement from 1970 through 2010. As shown in Figure 13.18, the percent of watches that actually contained severe weather trended upward as did the percent of cases of severe weather in a watch area.

RIVER AND FLOOD FORECASTING

The U.S. Army Signal Corps began the first river and flood forecast service in 1872. A disastrous flood on the Kansas River in 1903 prompted Congress to make the river and flood forecast service a special division of the U.S. Weather Bureau (now the National Weather Service). By 1913, the service had expanded to 483 river gauging stations (many operated by the U.S. Geological Survey). In 1946, the Weather Bureau established the nation's first *River Forecast Centers (RFCs)* at Cincinnati, OH and Kansas City, MO. Currently, 13 RFCs serve the nation

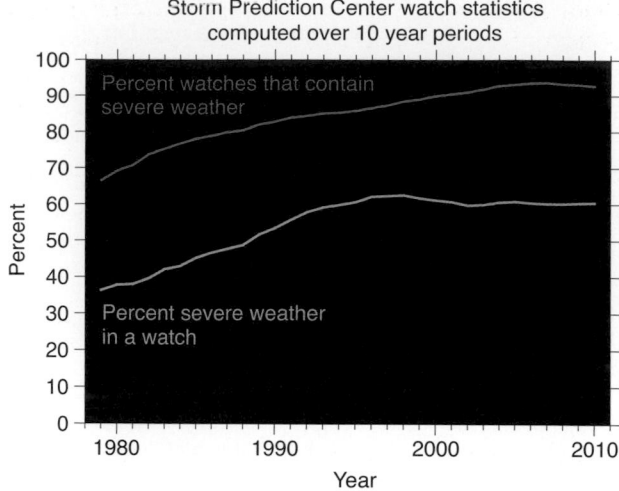

FIGURE 13.18

Upward trend in the skill of severe weather watches issued over the period 1970-2010. [NOAA/NWS/Storm Prediction Center]

with the primary goal of minimizing the loss of life and property damage caused by floods by providing the public with timely river and flood forecasts (Figure 13.19).

At RFCs, staff hydrologists and hydro-meteorologists develop river, reservoir, and flood forecasts using special numerical models that simulate the response of river and stream flow to rainfall and snowmelt. The objective is to predict the volume of water that runs off the land surface into a river or stream, the discharge at specific locations along major rivers or small streams near urban

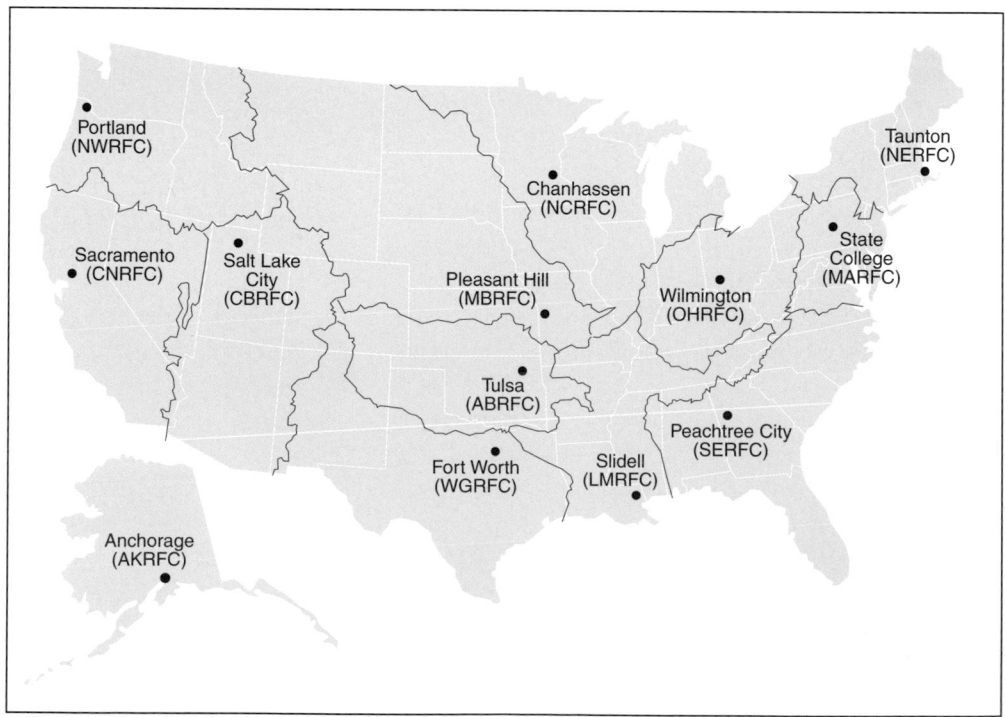

FIGURE 13.19
Locations of the River Forecast Centers (RFCs) and their regions of responsibility. [NOAA]

areas, and variations in river discharge and stage (height of the river) as the water volume moves downstream. In late winter and early spring, RFC staff prepares spring snowmelt outlooks to identify the potential for spring flooding, assuming a normal rate of snowmelt. Specialized services include low flow forecasts during dry periods, ice advisories for navigable rivers in winter, and reservoir inflow forecasts. Forecasts are forwarded to personnel at local NWS Weather Forecast Offices who in turn issue forecasts and outlooks for water resource managers, emergency management officials, and the public. Flood forecasts specify the expected height of the flood crest, the date and time when the river is expected to flood, and the date and time when the river is likely to return to its banks.

MARINE FORECASTING

Meteorologists and oceanographers at NCEP's *Ocean Prediction Center* in Camp Springs, MD issue forecasts, warnings, and other guidance information for a diverse user community including mariners, fisheries, and recreational boaters. The goal is to "ensure the safety of life and property at sea, manage marine natural resources and protect the environment." Operations benefit marine safety, navigation, fisheries, and the health of coastal ecosystems. Analysis and forecasting cover such conditions as wind, waves, fog, and ice accretion. In addition, the Ocean Prediction Center works with the National Weather Service Forecast Offices, the Federal Emergency Management Agency, and the U.S. Coast Guard Search and Rescue in responding to emergency situations. For more on marine weather analysis and forecasting, refer to this chapter's first *For Further Exploration*.

SPACE WEATHER FORECASTING

Scientists at NCEP's *Space Weather Prediction Center (SWPC)* in Boulder, CO observe and forecast solar and geophysical events that impact a variety of technological systems, such as GPS and the national power grid. Scientists monitor phenomena such as the aurora, solar wind, and progression of the solar cycle. SWPC uses scales, which have numbered levels from 1 to 5 indicating severity, to communicate the possible effects of current and future space weather conditions on systems such as satellite operations and radio communications. The scales also describe potential biological hazards, for example increased radiation exposure to airline passengers at high latitudes. Refer to the Chapter 2 *For Further Exploration*, *Space Weather Prediction*, to learn more about space weather phenomena and the role of the SWPC.

FORECAST SKILL

Just how accurate are today's computer-guided weather forecasts? This question can be answered by comparing the skill of modern weather forecasting with that of predictions based on *persistence* (forecasting no change in present weather) or *climatology* (forecasts derived from past weather records). On this basis, weather forecasting skill declines rapidly for periods longer than 48 hrs and is minimal beyond 10 days.

Several factors contribute to the drop in forecasting skill as the forecast period lengthens. Computerized forecasts are only as accurate as the input (observational) data and predictive equations (numerical models) allow them to be. Errors are introduced by (1) missing or inaccurate observational data, (2) failure of weather stations to detect all mesoscale and microscale circulation systems, and (3) imprecise equations in numerical models that include assumptions and first approximations. Unfortunately, the adverse effects of errors in initial conditions grow cumulatively as the forecast period lengthens. In today's numerical models of the atmosphere, the impact of even small errors doubles about every 2 to 2.5 days over the forecast period. When computerized numerical models are run out to 10 days, some elements of daily forecasts are useful for some localities to about 6 or 7 days. Beyond that, chaos, the term for sensitivity to initial conditions, sets in. For more on chaos and the limits to weather forecasting, see this chapter's second *For Further Exploration*.

The accuracy of numerical weather forecasting, particularly in the 1- to 5-day range, has shown slow but steady improvement over the past few decades, and prospects for continued progress are good. This optimistic outlook stems from: (1) a better understanding of atmospheric processes that makes possible the development of more realistic numerical models of the Earth-atmosphere system, (2) larger and faster computers, (3) more reliable and sophisticated observational tools, including Doppler radar and remote sensing by satellite, and (4) denser weather observational networks worldwide.

Although computerized numerical models have revolutionized weather forecasting, computers are never likely to replace human weather forecasters. The products of computerized numerical models function as guidance materials for the forecaster. He or she must understand the basic atmospheric processes, and the characteristic errors and biases of the various numerical models and weather-observing systems. As much as numerical weather prediction has advanced in recent years, the best

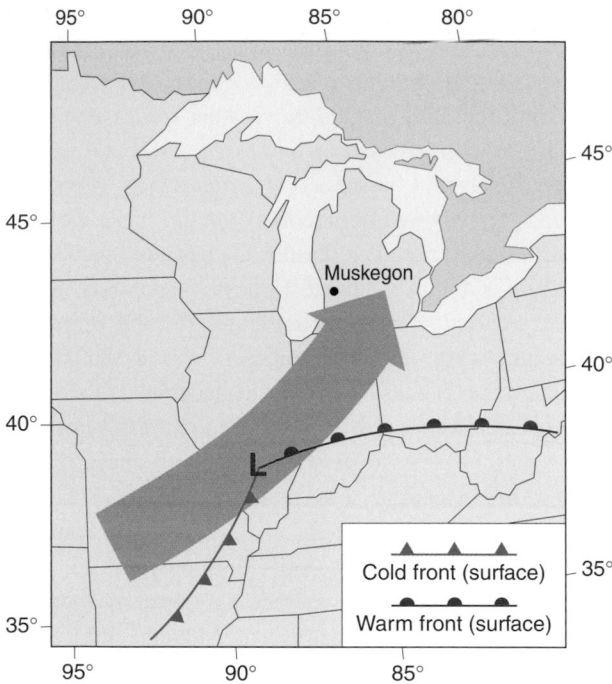

FIGURE 13.20
An early winter cyclone tracks northeastward through the Midwest, causing heavy snowfall at Muskegon, MI.

forecasters still rely heavily on personal experience and intuition, tempered by their knowledge of how the Earth-atmosphere system works. A good forecaster does not start with the model output in preparing a forecast, but rather he or she begins with previous and current observations, forming a view of what the atmosphere has been doing and ideas on what is likely to happen in the future. Furthermore, special local or regional conditions may call for significant modification of a computer-generated weather forecast. Forecasters must analyze and interpret computerized predictions and adapt those forecasts as necessary to regional and local circumstances. Consider an example.

Suppose that an intense early winter extratropical cyclone tracks northeastward through the Midwest, across central Illinois, and into eastern Lower Michigan, as shown in Figure 13.20. Muskegon, MI is on the cold, snowy side of the storm's path, so residents experience snow and strong northeast winds. After the storm passes, the wind shifts to the north and then northwest, and cold air advection begins. A computerized numerical model might predict clearing skies for Muskegon, but local conditions dictate a different result. Strong winds from the northwest advect cold, dry air across the relatively warm waters of Lake Michigan, giving rise to localized *lake-effect*

snows on the lake's downwind eastern shore (Chapter 8). Because of this local effect, northwest winds may bring more snow to Muskegon after the storm has passed than northeast winds did when the storm was nearby.

LONG-RANGE FORECASTING

Meteorologists at NCEP's *Climate Prediction Center* prepare 30-day (monthly), 90-day (seasonal), and multi-seasonal generalized climate outlooks that map areas where the probability of total precipitation and temperature is expected to depart from normal, that is, long-term averages (Figure 13.21). Outlooks are issued from two weeks to 13 months in advance for the contiguous U.S., Hawaii, and other Pacific islands.

A promising area of research in long-range weather forecasting relies on identification of teleconnections. A **teleconnection** is a linkage between changes in atmospheric circulation occurring in widely separated regions of the globe, often many thousands of kilometers apart. For example, Jerome Namias (1910-1997) and his colleagues at the Scripps Institution of Oceanography argued that a coupling between ocean and atmosphere triggered the 1988 Midwestern drought. Anomalously cold surface water in the central and eastern tropical Pacific (associated with La Niña) plus unusually warm surface water to the north (near Hawaii) interacted with the atmospheric circulation to produce a long-wave pattern that was more meridional than usual. As described in Chapter 9, this circulation pattern featured a warm anticyclone that persisted over the central contiguous United States through much of the spring and summer of 1988. Severe drought was the consequence.

Identification of teleconnections associated with El Niño and La Niña is the primary basis for extension of long-range forecasting to multi-seasons. As described in detail in Chapter 9, meteorologists report success in predicting the onset of El Niño a year or so in advance and have identified linkages between El Niño and weather anomalies in middle latitudes—especially during winter.

Successfully forecasting the prevailing circulation pattern at the 700-mb level is the first step in predicting the probability of monthly temperature and precipitation departures from normal. The present circulation pattern is extrapolated into the future, although an effort is also made, based on historical data, to identify features of the present pattern that are most likely to persist. The prevailing westerly flow at the 700-mb level permits identification of areas of strong warm and cold air advection as well as the principal storm tracks.

Monthly & Seasonal Color Outlook Maps
Issued 19 April 2012

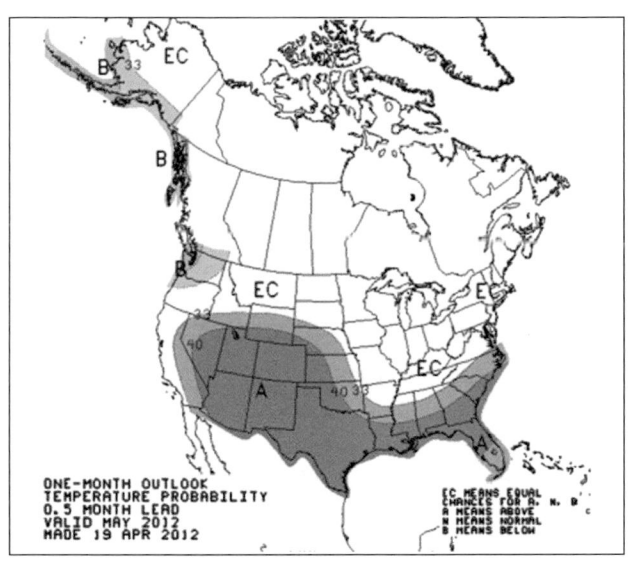

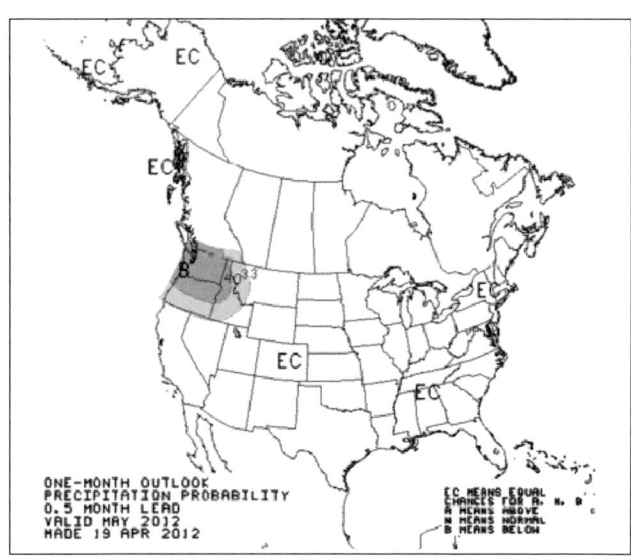

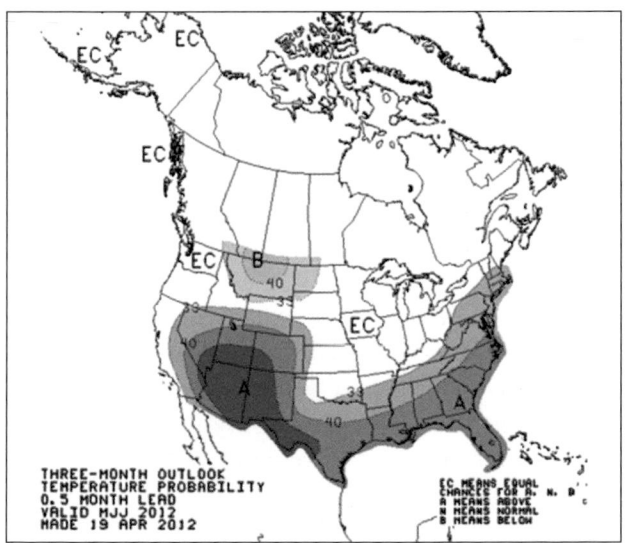

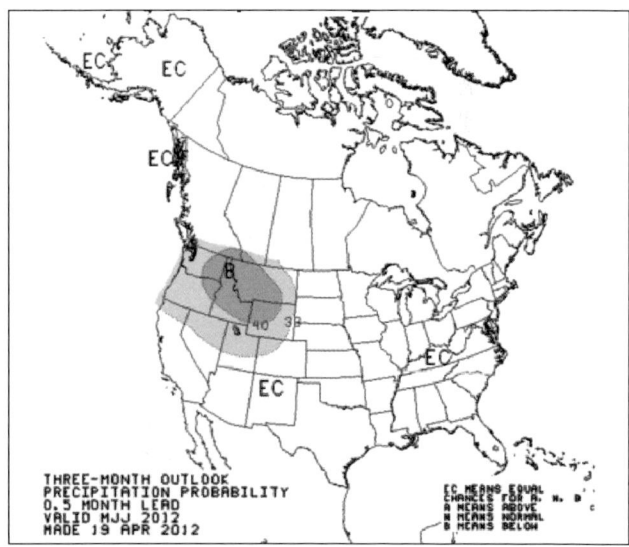

FIGURE 13.21

Sample monthly and seasonal outlook maps of the probability of total precipitation and temperature departing from normals, i.e., long-term averages. *A* indicates a probability of higher than normal values whereas *B* indicates a probability of lower than normal values. EC (white area) denotes equal chance and is therefore unpredictable. [NOAA/NWS/Climate Prediction Center]

Predictions of surface temperature and precipitation anomalies are derived from this analysis.

A somewhat different approach is taken for 90-day outlooks (issued once each month near mid-month). Forecasters rely more on long-term trends and recurring events, attempting to isolate persistent circulation features from prior months and seasons. A suite of 40 NCEP *Climate Forecasting System* (CFS) runs yield a mean prediction. Model runs depend heavily on tropical Pacific sea-surface temperatures (SST) specified by a coupled ocean-atmosphere model. The model-generated outlook accounts primarily for the influence of El Niño and La Niña, and other low frequency phenomena influenced by tropical SST forcing (Chapter 9).

In January 1995, meteorologists at the Climate Prediction Center began issuing 15-month (multi-seasonal) outlooks for regions of expected positive and negative anomalies in temperature and precipitation. Each month 13 forecasts are issued, each one covering a 3-month period. The first forecast covers the three months beginning two weeks from the date of the forecast's release. Each subsequent 3-month forecast overlaps the previous one by two months.

SINGLE-STATION FORECASTING

Even without access to an official NWS forecast, short-term weather prediction based on weather observations at one location, known as **single-station forecasting**, may be derived from the principles of weather behavior examined in this course. Because such forecasts are based on rules applied at only one location, they tend to be generalized and tentative, with complications often cropping up as local conditions are modified by changes elsewhere. Table 13.1 is a sample list of rules of thumb applicable to middle latitudes. You may wish to add to this list. Other approaches base forecasts on fair-weather bias, persistence, or climatology.

Analysis of records of past weather events may aid single-station weather forecasting. Such records reveal a **fair-weather bias**. That is, fair-weather days outnumber stormy days almost everywhere. In fact, if we boldly predict that all days will be fair, we probably will be correct more than half the time. The only merit of such an exercise is to establish a baseline for evaluating the skill of more sophisticated weather forecasting techniques. That is, we would expect traditional forecasting methods to score higher than forecasting methods based solely on fair-weather bias.

Another characteristic behavior of weather is **persistence**, that is, the tendency for the same weather

---

TABLE 13.1
Some Rules of Thumb for Single-Station Weather Forecasting

- At night, air temperatures are lower if the sky is clear than if the sky is cloud-covered.
- Clear skies, light winds, and snow-covered ground favor extreme nocturnal radiational cooling and very low air temperatures by sunrise.
- Steadily falling air pressure often signals the approach of stormy weather whereas rising air pressure suggests that continued fair weather or improving weather is in the offing.
- Appearance of cirrus, cirrostratus, and altostratus clouds, in that order, indicates overrunning ahead of a warm front and the likelihood of precipitation.
- A counterclockwise wind shift from northeast to north to northwest (called *backing*) is usually accompanied by clearing skies and cold air advection.
- A clockwise wind shift from east to southeast to south (called *veering*) is often accompanied by clearing skies and warm air advection.
- A wind shift from northwest to west to southwest is usually accompanied by warm air advection and increasing clouds.
- If radiation fog lifts by late morning, a fair afternoon is likely.
- With west to northwest winds, a steady or rising barometer, and scattered cumulus clouds, fair weather is likely to persist.
- Towering cumulus clouds by mid-morning forewarn of afternoon thunderstorms.

---

episode to persist for some period of time. For example, if the weather has been cold and stormy for several days, the weather may well continue that way for many more days. Weather records show, however, that at middle latitudes one weather type typically gives way to another weather type very abruptly, usually in a day or less. Weather forecasts based on persistence alone are therefore prone to error.

A third approach is based on **climatology**; that is, we prepare a weather forecast for a particular day based on the type of weather that occurred on the same day in years past. Suppose, for example, that the climate record of your area indicates that it has rained on 9 August only 12 times in the past 100 years. Accordingly, you predict the probability of rain next 9 August to be only 12%, and confidently plan a picnic for that day. The problem with the climatology approach is that, statistics aside, there is no guarantee that it will not rain on the next 9 August.

PRIVATE SECTOR FORECASTING

Our discussion thus far has focused mainly on the role of the federal government in weather forecasting. In addition, many private sector meteorologists analyze weather maps and other guidance materials supplied by the National Centers for Environmental Prediction (NCEP). Many television and radio stations and some newspapers employ their own meteorologists or contract with private forecast services. Private meteorologists tailor their forecasts to the specific needs of their commercial, agricultural, or industrial clients. For example, a private weather forecaster retained by an appliance store chain might alert store executives to a pending heat wave so that stores might be stocked with an adequate supply of fans and air conditioners. Another private forecaster might advise an electric power utility of expected summer temperatures so the energy supplier can better anticipate customer demand for air conditioning. A private forecaster might provide frost and freeze warning service for orchards or cranberry bogs. In this way, private sector forecasters supplement the efforts of government weather forecasters.

Communication and Dissemination

Weather maps, charts, and forecasts issued by the National Centers for Environmental Prediction are transmitted to local NWS Weather Forecast Offices to guide staff meteorologists in preparing specific forecasts for their areas of responsibility. Weather information is then distributed to public safety officials and the general public via a variety of communications systems.

The National Weather Service issues outlooks, watches, warnings, and advisories when hazardous weather threatens. An *outlook* is provided for people who require considerable advance notice so that they might adequately prepare for a specific weather event. For example, the outlook for flooding from spring snowmelt in the Upper Midwest may be issued many weeks in advance.

When hazardous weather is occurring or appears possible or probable, the National Weather Service issues event-specific watches, warnings, and advisories. These statements cover severe local storms, winter storms, floods, hurricanes, and non-precipitation hazards. A **weather watch**, such as a severe thunderstorm watch or winter storm watch, is indicated when hazardous weather is possible based on current or anticipated atmospheric conditions. The actual occurrence or timing of hazardous weather can be uncertain. People in the designated area need not interrupt their normal activities except to remain alert for threatening weather and to keep the television, radio, computer, or smartphone on for further information. A **weather warning** (e.g., tornado warning, heavy snow warning) is issued when hazardous weather is taking place somewhere in the region or is imminent. People are advised to take all necessary safety precautions. *Advisories* refer to anticipated weather hazards that are less serious than those covered by a warning. Examples include a winter weather advisory, heavy surf advisory, or heat advisory.

The local NWS Weather Forecast Office issues a *tornado warning* only after detection of a thunderstorm that is known or likely to include a tornado. Indication of a tornadic circulation is based on Doppler radar and/or reports of funnel clouds or cloud-base rotation by specially trained storm spotters. A warning covers a much smaller area than a watch, usually all or part of a single county, and specifies the location of the tornado, its anticipated path, and the time when the tornado is expected in the warning area. Tornado warnings usually are valid for 30 to 60 minutes.

Winter storm warnings may specify heavy snow, blizzard conditions, or an ice storm. Usually, a *winter storm warning for heavy snow* is issued if snowfall is expected to total at least 4 to 6 in. (10 to 15 cm) in less than 12 hrs, or at least 6 to 8 in. (15 to 20 cm) in less than 24 hrs; exact thresholds depend upon the locale. A *blizzard warning* means that falling or blowing snow is expected to be accompanied by sustained winds (or frequent gusts) of 35 mph (56 km per hr) or higher, reducing visibility to less than 1300 ft (400 m). Conditions are expected to persist for at least 3 hrs. A *severe blizzard* produces winds stronger than about 45 mph (73 km per hr), visibility near zero because of blowing snow, air temperature below 10 °F (−12 °C), and dangerously low wind-chills. An *ice storm warning* indicates that potentially dangerous accumulations of freezing rain or sleet are expected on the ground and other exposed surfaces.

To alert the public to the flash flood hazard, personnel at the local NWS Weather Forecast Offices issue watches, warnings, and advisories. A *flash flood watch* indicates that flash flooding is possible within or close to the designated watch area based on current or anticipated weather conditions. Residents of the watch area are advised to prepare to take action in the event that a flash flood warning is issued or flooding is observed. A *flash flood warning* is issued when a dangerously rapid rise in river level is imminent in the warning area or is currently occurring due to heavy rain (or dam failure). The public is advised to take appropriate action immediately. An urban and small stream flood advisory alerts the public to potential flooding that may cause inconvenience but is not life threatening for those living in the affected area. These advisories are issued when heavy rain is anticipated, which could flood streets and other low-lying urban areas, or if small streams are expected to reach bank full.

Because weather is changeable, weather observations and guidance information such as weather maps, forecasts, watches, and warnings must be communicated as rapidly as possible both nationally and internationally. For this reason, the World Meteorological Organization and weather services of its member nations maintain elaborate communications networks consisting of a variety of systems. Weather information is relayed by satellite, radio, Internet, and facsimile systems (which reproduce maps, charts, and satellite images). Modernization of U.S. National Weather Service facilities, including the Automated Surface Observing System (ASOS) and the Advanced Weather Interactive Processing System (AWIPS), is making the weather communications network more efficient and reliable.

Through television (*The Weather Channel* and other channels), Internet, radio, newspaper, weather apps for laptops, tablets, and mobile devices, and NOAA Weather Radio, the public has numerous ways to access weather reports and forecasts. As technology continues to evolve, forecasts will improve, communication will increase, and the public will be better prepared.

WEATHER-READY NATION

According to NOAA, over a recent 31-year period ending in 2011, the U.S. has experienced a total of 114 weather-related disasters in which overall damages reached or exceeded $1 billion dollars. Included within that list are Hurricane Irene and the 2011 tornado season which caused over $7.3 billion and $20 billion (2011 dollars) in damages, respectively. Tragically, these disasters have also resulted in many fatalities. More than 500 people lost their lives during the 2011 tornado season, fourth on the all-time list. The devastating impacts of extreme events can be reduced though through improved preparedness, the goal of NOAA's *Weather-Ready Nation*. This comprehensive initiative is about building community resilience as communities across the country become increasingly vulnerable to severe weather events, such as intense heat waves, droughts, flooding, hurricanes, and solar storms that threaten electrical and communication systems.

The goals of the program are to improve forecasts and communication of risk to local authorities, develop mobile-ready emergency response teams, continue to develop better technology, such as dual-polarization radar, and strengthen partnerships to enhance community preparedness. In the end, emergency managers, first responders, government officials, and the public will be better informed to make faster and smarter decisions to save lives.

Conclusions

Weather forecasting is a complex and challenging science that depends on the efficient interplay of weather observation, data analysis by meteorologists and computers, and rapid communication systems. Meteorologists have achieved a high level of skill for short-range weather forecasting. Further improvement is expected with denser surface and upper-air observational networks, more precise numerical models of the atmosphere, larger and faster computers, and more sophisticated techniques of remote sensing by satellite. If these advances are to be realized, however, continued international cooperation is essential, for the atmosphere is a continuous fluid with no political boundaries.

Basic Understandings

- The fluid nature of the atmosphere means that international cooperation is required in gathering and interpreting surface and upper-air weather data. To this end, the World Meteorological Organization (WMO) coordinates an international effort of weather observation, analysis, and forecasting.
- Timely and accurate weather forecasts for the United States are the responsibility of the National Centers for Environmental Prediction (NCEP), part of NOAA's National Weather Service. NCEP's goal is to "protect life and property, as well as mitigate economic loss, by providing accurate forecasts and forecast guidance products to weather service field offices." Nine centers comprise NCEP, each responsible for forecasting specific weather or other environmental phenomena.

- Weather forecasting entails acquisition of present weather data, graphical depiction of the state of the atmosphere on weather maps and charts, computer-aided analysis of data and maps, prediction of the future state of the atmosphere, and dissemination of weather information and predictions to the public.

- Surface weather is monitored at land stations and by automated weather stations and ships at sea. Rawinsondes (profiling temperature, pressure, humidity, and wind), radar, aircraft, and satellites monitor the atmosphere above the Earth's surface. Modernization of the National Weather Service during the 1990s upgraded NWS Weather Forecast Offices, and installed the Automated Surface Observing System (ASOS) and Doppler radar.

- Using standard station models, surface weather observations are plotted on constant altitude (sea-level) maps whereas upper-air weather observations are plotted on maps of constant pressure (isobaric) surfaces (e.g., 500-mb map). The altitude of a specific pressure surface varies from one place to another primarily because of differences in mean temperature of the air below that pressure surface. For example, the 500-mb surface is at a lower altitude where the air below is relatively cold, but at a higher altitude where the air below is relatively warm.

- The NWS Advanced Weather Interactive Processing System (AWIPS) consists of computer workstations that ingest and organize ASOS, satellite, and radar data plus analysis and guidance products from the National Centers for Environmental Prediction. AWIPS enables meteorologists to display, process, and overlay images, graphics, and other data.

- With numerical weather forecasting, a computer is programmed with a mathematical model of the Earth-atmosphere system that relates winds, temperature, pressure, and humidity. Current weather observational data are used to initialize the model and the model then predicts a future state of the atmosphere. With the predicted conditions serving as a new starting point, another forecast is computed for a subsequent period of time. Repeated iterations yield weather forecasts for the next 12, 24, 36, and 48 hrs or beyond.

- Special NWS forecast centers are responsible for predicting tropical cyclones (hurricanes and tropical storms), severe weather (e.g., tornadoes), weather for domestic and international aviation, river flow and flooding, space weather, and weather for mariners.

- Although the skill of short- and medium-range weather forecasting has improved steadily in recent decades, forecasting skill declines rapidly for periods longer than 48 hrs and is minimal for periods beyond 10 days. Errors are introduced by missing or inaccurate observational data, failure of weather stations to detect all micro-scale and mesoscale weather systems, and equations in numerical models that include imprecise assumptions and first approximations. These errors in initial state grow with the forecast period.

- NCEP's Climate Prediction Center prepares 30-day, 90-day, and multi-seasonal generalized climate outlooks that identify the probability of departures from normal (long-term average) of temperature and total precipitation. Teleconnections, linkages between changes in atmospheric circulation occurring in widely separated regions of the globe, are particularly useful in long-range weather forecasting. Examples are winter weather anomalies at middle latitudes during El Niño or La Niña episodes.

- Single-station weather forecasting is based upon principles of meteorology, fair-weather bias, climatology, or persistence of weather episodes.

- NOAA's Weather Ready Nation is a comprehensive initiative that will build community resilience in response to increasing vulnerability to severe weather events.

Enduring Ideas

- The NOAA/NWS/National Centers for Environmental Prediction are charged with monitoring and prediction of weather in the United States. A network of land and sea-based stations monitors surface weather conditions, which are plotted on a constant altitude (sea-level) map for analysis. Radiosondes monitor weather variables above Earth's surface while being tracked for wind information; these conditions are plotted on constant pressure upper-air maps. Surface and upper-air data are ingested into sophisticated computer forecast models, which then predict future states of the atmosphere. Meteorologists use model output as guidance in making their forecasts. Forecasting skill generally declines rapidly for periods longer than 2 days.
- Forecasters at nine specialized NCEP facilities predict phenomena such as tropical cyclones (National Hurricane Center), severe weather (Storm Prediction Center), and climate (Climate Prediction Center). The Climate Prediction Center uses teleconnections to produce seasonal forecasts of temperature and total precipitation as a probability of departure from normal values.

Key Terms

National Weather Service (NWS)

National Oceanic and Atmospheric Administration (NOAA)

World Meteorological Organization (WMO)

National Centers for Environmental Prediction (NCEP)

National Data Buoy Center (NDBC)

Coordinated Universal Time (UTC)

Automated Surface Observing System (ASOS)

NWS Cooperative Observer Network

radiosonde

station model

isobars

cold air advection

warm air advection

Advanced Weather Interactive Processing System (AWIPS)

ensemble forecasting

model comparison

National Hurricane Center (NHC)

teleconnection

single-station forecasting

fair-weather bias

persistence

climatology

weather watch

weather warning

Review

1. What role might selective memory play in our personal assessment of the skill of weather forecasts?
2. Weather observation and forecasting require international cooperation. Why?
3. List the main steps involved in the preparation of a weather forecast.
4. What are the sources of surface weather information on land and at sea?
5. Why must weather observations be taken at the same time everywhere?
6. What is the chief source of observational data that are used to plot an upper-air weather map?
7. Surface air pressure readings are adjusted to sea level. Explain why.
8. Explain why the height of the 500-mb surface generally slopes downward from the tropics to the polar region.
9. Why does weather forecast skill decline as the forecast period lengthens?
10. Explain why computer models are never likely to replace human weather forecasters.

Critical Thinking

1. A cold-core anticyclone (polar high or arctic high) does not appear on a 500-mb weather map. Explain why.
2. Describe how and why the height of the polar front jet stream changes between winter and summer.
3. Describe the role of numerical models in scientific weather forecasting.
4. Weather observation stations are much more closely spaced over the continents than the ocean. Speculate on how this difference might influence a meteorologist's ability to represent accurately the state of the atmosphere.
5. Why are special forecast centers (e.g., Storm Prediction Center) needed?
6. Describe the advection pattern that would cause a meridional flow pattern at the 500-mb level to become more zonal.
7. How are teleconnections used in long-range forecasting?
8. What are some of the factors that RFC hydrometeorologists must take into account when forecasting floods?
9. What might account for the decline in Atlantic hurricane track error since 1990?
10. Distinguish between a weather watch and a weather warning.

Marine Weather Statements

Weather systems moving across the ocean can produce life-threatening conditions not only for mariners at sea (Figure 1) but also for residents of coastal communities. Hence, NOAA's National Weather Service (NWS) has a facility that monitors the weather, prepares weather forecasts, and issues warnings for marine and coastal interests. The NWS area of responsibility includes coastal and open waters of the Atlantic and Pacific Oceans, the Gulf of Mexico, and the Great Lakes. Data used in preparation of these forecasts are obtained from a variety of sources, including ships, buoys, and satellites.

Coastal or near shore forecasts are intended for those mariners staying in coastal waters that are roughly within 37 km (20 nautical mi) of the Atlantic coast and within 110 km (60 nautical mi) of the Pacific coast. Offshore forecasts are for those mariners operating farther from the coast, typically a day or more from safe harbor, or from about 37 to 463 km (20 to 250 nautical mi) offshore of the Atlantic coast and 110 to 463 km (60 to 250 nautical mi) offshore of the Pacific coast. High seas forecasts are mainly geared for large ocean-going vessels operating more than 463 km (250 nautical mi) out to sea. In addition to forecasts, various marine-related advisories, watches and warnings are issued to the public. These pertain to a variety of severe weather types as well as unusual water, wave, and current conditions that could affect life and property.

The Ocean Prediction Center (OPC), formerly known as the Marine Prediction Center (MPC), is the component of the National Centers for Environmental Prediction (NCEP) that issues marine forecasts (out to 5 days) for coastal and offshore waters as well as the high seas of the western North Atlantic (west of 35 degrees W) and much of the North Pacific poleward of 30 degrees N to 66 degrees N. This NCEP center also issues marine warnings for conditions not involving tropical weather systems, which is the responsibility of the National Hurricane Center (discussed elsewhere in this chapter).

FIGURE 1
NOAA ship *Ronald H. Brown.* [NOAA]

Watches, warnings, and advisories can be issued by the National Weather Service for several types of coastal events, as described below.

Coastal flood watch — Issued to alert coastal residents of the possibility of the inundation of land areas along the coast within the next 12 to 36 hrs.

Coastal flood warning — Issued to warn residents of coastal areas that land areas along the coast will be inundated by sea water above the typical tide action.

Heavy surf advisory — Issued to inform the public that high ocean surf might pose a threat to life or property. The criteria for such advisories depend upon the locale, but typically these include minimum wave heights of 2.4 to 3.7 m (8 to 12 ft) with periods on the order of 10 seconds. The heavy surf is typically produced by large ocean swells associated with a distant storm system over the ocean, supplemented at times by astronomical high tides.

Tsunami watch/warning — Issued by the National Weather Service to either alert residents in coastal regions along the Pacific Ocean that an impending tsunami (seismic sea wave) may cause damage to low lying regions. The type of bulletin is based on the magnitude and the location of the source underwater geological event.

Small-craft advisory — Issued to advise mariners of sustained (exceeding two hours) weather and/or sea conditions, either present or forecast, potentially hazardous to small boats. These conditions generally include winds of 18-33 knots (21-38 mph) and/or dangerous wave conditions. Small-craft advisories may be issued also for hazardous sea conditions or lower wind speeds that may affect small craft operations. Advisories can be issued up to 12 hrs prior to the onset of adverse conditions. The small-craft advisory signals for this condition consist of one triangular red pennant by day, and a red lantern over a white lantern by night.

Small-craft warning — Issued as a warning for marine interests of impending winds up to 28 knots (32 mph); used mostly in coastal or inland waters.

Gale warning — A storm warning for marine interests of impending winds associated with an extratropical cyclone with speeds ranging from 34 to 47 knots (39 to 54 mph) within a 24-hr period. The warning signals for this condition are two triangular red pennants by day, and a white lantern over a red lantern by night.

Storm (also known as whole-gale) warning — A warning for marine interests of impending winds associated with an extratropical cyclone that are greater than 48 knots (55 mph). The warning signals for this condition are one square red flag with black center by day, and two red lanterns by night.

Hurricane warning — A warning that hurricane conditions with sustained winds of 85 knots (74 mph) or higher are expected within the specified coastal area. The warning signals for this condition are two square red flags with black centers by day, and one white lantern between two red lanterns by night.

Special marine warning — A warning issued for marine interests of potentially hazardous over-water events of relatively short duration, usually up to 2 hrs. Typically, these warnings are issued for strong to severe thunderstorms that may include strong winds, frequent lightning, heavy rain or waterspouts.

Marine weather information, including forecasts and warnings, are transmitted to mariners and other interested parties by a variety of methods. The U.S. Coast Guard transmits weather maps to ships at sea by HF Radiofax and forecasts by either voice (HF, VHF or MF radio) or text transmission (NAVTEX). Coastal and near shore forecasts (typically within about 40 km (25 mi) of shore) can be obtained from the NOAA Weather Radio network of stations (Chapter 1). Graphics and text are also available on the Internet through the National Weather Service.

Chaos and the Limits to Forecasting

Can the fluttering of a butterfly's wings in Brazil eventually spawn a tornado in Texas? This question, originally attributed to the late atmospheric scientist Edward N. Lorenz (1917-2008) of the Massachusetts Institute of Technology, provides the key to understanding the limits to weather forecasting and why modern science can fly a person to the Moon but cannot accurately predict the weather more than 10 days from now. In the early 1960s, Lorenz coined the term *butterfly effect* in reference to the fact that future states of certain systems (e.g., the Earth-atmosphere system) are highly sensitive to initial conditions. Even the slightest differences in initial conditions can rapidly magnify with time making long-range forecasting of weather events untenable.

A system consists of a number of components that function and interact in an orderly manner. The human respiratory and digestive systems are familiar examples. A *dynamical system* is one whose evolution from an initial state can be described by one or more mathematical equations. The Earth-atmosphere is a dynamical system and, as noted elsewhere in this chapter, meteorologists have built computerized numerical models that attempt to predict its future behavior.

As the forecast period lengthens beyond a few days or so, however, weather forecasts derived from numerical models quickly go awry. The problem is not so much the accuracy of the component equations as it is the imprecise description of the initial state of the system. That description is based on weather observations from a network of widely spaced stations utilizing instruments whose measurements are not always reliable or representative. The system is so sensitive that even slight changes in initial conditions can give rise to totally different forecasts. Hence, meteorologists cannot accurately forecast the weather more than 10 days from now. On the other hand, scientists know much more precisely the initial conditions and governing physical laws that enabled them to send a manned rocket to the Moon.

Researchers with NOAA's National Weather Service have experimented with two techniques to detect the butterfly effect: (1) *ensemble forecasting* and (2) *model comparison*. In the first, a numerical model generates several forecasts, each based on slightly different sets of initial conditions (Figure 1). If the forecasts are consistent, then the forecast is considered reliable. In the second technique, forecasts generated by many different models for the same period are compared. If they all agree, then the forecast is issued with relatively high confidence. If, however, through either technique, forecasts are inconsistent, then any one forecast is considered unreliable.

According to *chaos theory*, internal instabilities cause complex behavior in a system. Because its initial conditions are imperfectly known, the Earth-atmosphere is a chaotic dynamical system. The term chaotic should not be construed as implying random behavior, however. Certain physical laws govern the system so that there are boundaries within which chaos takes place; that is, the chaos has a certain underlying order or structure. For example, in the Northern Hemisphere, winds about a cyclone will always blow in a counterclockwise direction regardless of the storm's track.

A dynamical system whose initial conditions are known with precision may be drawn toward some stable state known as an *attractor*. Hence, an attractor imposes predictability. In a chaotic dynamical system, on the other hand, there is some range of possible future states that is encompassed by so-called *strange attractors*. Strange attractors contain the system's preferred behaviors and hence, if identified, can be used to predict the future state of the system.

Not surprisingly, some scientists are searching for strange attractors, that is, the structure in complex climate data. Others are searching for multiple attractors. The latter effort is inspired by the fact that the atmosphere shifts between a finite number of circulation patterns. This may imply that the system follows one attractor for a period of time and then shifts to another.

The dimensions of a strange attractor indicate the complexity of a system. In effect, the number of dimensions provides modelers with an estimate of the number of variables needed to describe the behavior of the system. From studies to date, the Earth-atmosphere system appears to have a relatively high dimensional strange attractor. Chaos theory thus raises the question as to whether there are internal constraints on the ultimate predictability of weather. It also prompts the question as to whether the recent global warming trend might be merely an expression of the natural variability of the Earth-atmosphere system.

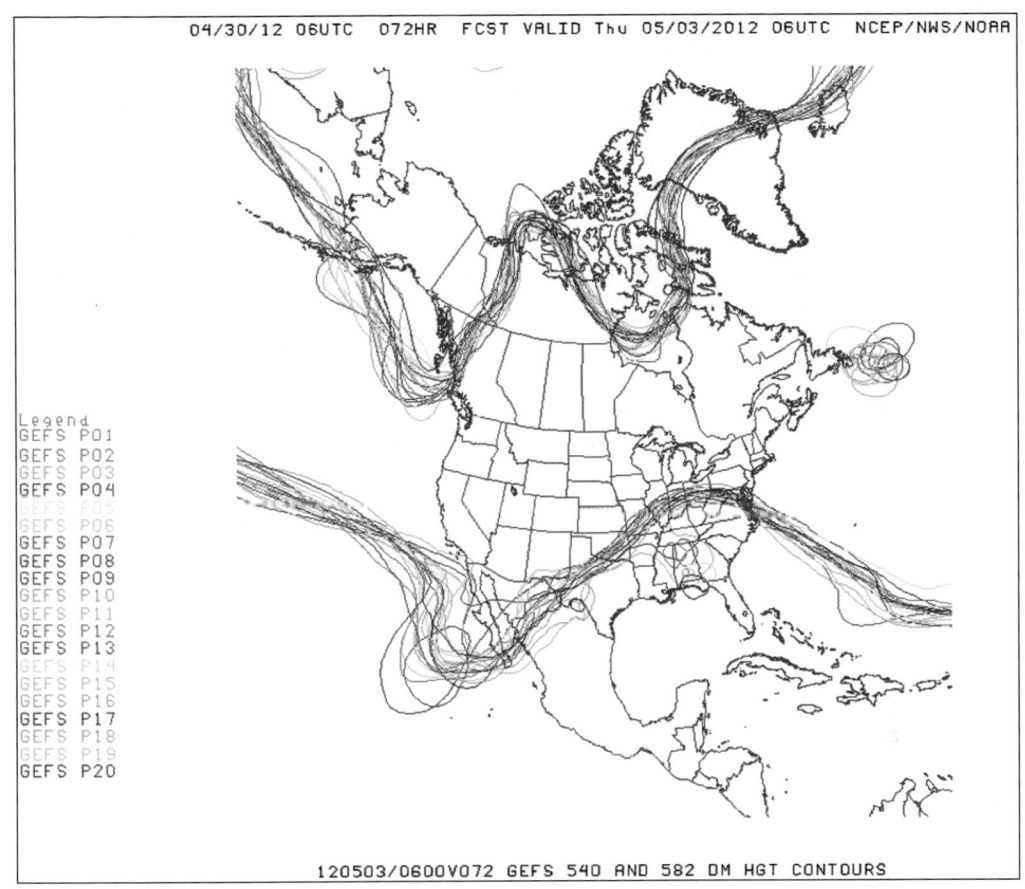

FIGURE 1

A 72-hr ensemble forecast of the positions of the 540 (northern areas) and 582 (southern) height contours on a 500-mb map. These forecasts are based on 20 slight permutations of the NOAA/NWS/NCEP Global Ensemble Forecasting System (GEFS). [NOAA/NWS/NCEP]

A colorful view of airglow layers at Earth's horizon is featured in this image photographed by a crewmember on the Space Shuttle Endeavour while docked with the International Space Station. Airglow is a photochemical luminescence arising from chemical reactions in the upper atmosphere over middle and low altitudes. [Courtesy NASA]

LIGHT & SOUND IN THE ATMOSPHERE

Chapter Highlights

Case-in-Point
  Fata Morgana
Atmospheric Optics
  Visible Light and Color Perception
  Red Sun, White Clouds, and Blue Sky
  Halo
  Rainbow
  Corona
  Glory
  Mirage
  Sunrise-Sunset and Twinkling Stars
  Twilight
Atmospheric Acoustics
  Sound Waves
  Thunder
  Sonic Boom
  Aeolian Sounds
Conclusions
Basic Understandings/Enduring Ideas
Key Terms/Review/Critical Thinking
For Further Exploration
  Blue Haze

Learning Objectives

Distinguish among scattering, refraction, and diffraction of light.
Describe the roles of rods and cones in the sensitivity of the human eye to light.
Explain how reflected visible electromagnetic radiation is responsible for the color of an object.
Explain why the daytime clear sky is blue.
Distinguish between Rayleigh scattering and Mie scattering in the atmosphere.
Describe the formation of a halo around the Sun (or Moon) and its meteorological relevance.
Identify the types of clouds most likely to produce a halo around the Sun or Moon.
Contrast the optics of a rainbow with that of a glory.
Explain why a rainbow appears to an observer in the sky opposite the Sun.
Explain how the occurrence of a rainbow may be a harbinger of weather to come.
Identify two atmospheric optical phenomena that involve diffraction of light.
Describe atmospheric conditions that can lead to formation of a mirage.
Explain the cause and significance of twilight.
Distinguish between light and sound waves.
Identify the factors that influence sound propagation within the atmosphere.
Explain how and why sound waves undergo refraction.
Identify some of the consequences of atmospheric refraction of sound waves.

What is responsible for various optical phenomena observed in the atmosphere and how does sound propagate through the atmosphere?

Case-in-Point

Fata Morgana

Fata Morgana is an example where meteorology provides a scientific explanation for myth and legend. Fata Morgana is an optical phenomenon in which images of distant objects, such as houses and bluffs, are distorted vertically so that they resemble castles or walls with spires. According to tradition, the mirage was named after Morgan le Fay reputed to be the half-sister of King Arthur. Legend has it that she was a fairy enchantress who could change her shape, a skill she may have learned from Merlin the Magician. Morgan le Fay was said to live in a submerged castle which she could elevate to above the sea surface. Sailors who were drawn to the castle thinking it offered safe harbor were lured to their deaths on rocky shores.

The Fata Morgana phenomenon is a **mirage**, an optical illusion caused by refraction (bending) of light rays as they travel through the atmosphere. It occurs where alternating layers of cold and warm air occur near the ground or over a water surface. Fata Morgana is most commonly observed on the horizon after sunrise following a clear calm night of extreme radiational cooling and the development of low-level temperature inversions. It is a complex mirage with multiple distortions in which several images of an object are superimposed. While one image may be upright, others can be inverted and as air layers move up and down, the images undergo rapid change.

In addition to their occurrence in coastal areas, Fata Morgana is relatively common in high mountain valleys in winter (e.g., Colorado's San Luis Valley) and over the frozen Arctic Ocean. It can also be observed in Antarctica (Figure 14.1).

FIGURE 14.1
Photo taken on 1 October 2004 looking south across the Ross Ice Shelf from McMurdo Station, Ross Island, Antarctica, toward White Island and Black Island. [Copyright Alan Robock]

A mirage, such as Fata Morgana, is one of many types of optical phenomena that occur in the atmosphere and involve visible light. Others include the blue of the daytime sky, halos, rainbows, and twilight. These are subjects of the science of *atmospheric optics*. This chapter focuses on the cause and meteorological significance of various atmospheric optical phenomena. We then examine the role of the atmosphere in the propagation of sound plus sounds of meteorological origin.

Atmospheric Optics

As the Sun's rays travel through the atmosphere, they can be scattered, reflected or refracted by cloud droplets or ice crystals, or by raindrops, and even air itself. The consequence is a variety of optical phenomena, including halos, rainbows, coronae, and glories. In these cases, refraction (turning or bending) of light occurs when solar radiation travels from one transparent medium into another (i.e., from air into water or ice) with varying speeds of transmission. Solar rays are also refracted as they pass through the atmosphere because of variations in air density. This type of refraction gives rise to mirages. In addition, scattering of sunlight by particles within the atmosphere is responsible for the blue of the daytime sky, the white of clouds, a red horizon at sunset, and twilight. In observing various atmospheric optical phenomena, extreme caution should be exercised so as not to look directly at the Sun, except when it is near the horizon, because the intense solar rays can cause irreparable damage to your eyes.

VISIBLE LIGHT AND COLOR PERCEPTION

Sunlight, which is visible light, represents only a small portion of the electromagnetic spectrum. Recall from Chapter 3 the relationship between color and wavelength in the visible portion of the electromagnetic spectrum. The visible portion of the spectrum bounds the region of maximum intensity of light emitted from the Sun, an object with a radiating temperature of approximately 6000 K. In fact, visible light that is received from the Sun represents approximately 45% of the total solar irradiance (rate of energy incident on a unit surface area). In 1666, Sir Issac Newton (1643-1727) discovered that when sunlight (white light) passed through a glass prism and onto a screen, the sunlight separated into a spectrum of sensations to which we have termed colors. Thus, Newton demonstrated that white sunlight is polychromatic, that is, composed of several individual primary colors.

The human eye is light sensitive, capable of detecting electromagnetic radiation having wavelengths between 0.38 and 0.76 micrometers (the normal photopic range), with the greatest sensitivity at approximately 0.55 micrometers. Light activates rod and cone cells in the retina of the eye. Rods, which are concentrated in the peripheral portions of the retina, respond to radiation of all visible wavelengths. Therefore, rods distinguish between light and dark and are responsible for scotopic (lower illumination) vision. Cones, which are concentrated in the central portion of the retina, respond to radiation at specific wavelengths in the visible range. Therefore, cones distinguish colors and are responsible for photopic (brighter illumination) vision.

Color is a sensation produced through the excitation of the cones in the retina by radiation at discrete wavelengths within the visible region of the electromagnetic spectrum. The normal human eye (no color blindness) perceives color. The primary colors of the spectrum are monochromatic, extending over very limited specific wavelength regions of the visible spectrum. White light contains light of all colors striking and stimulating the cone detectors in the eye with nearly uniform intensity. Black light (ultraviolet radiation) contains no light in the visible portion of the electromagnetic spectrum and hence, does not stimulate any of the detectors of the retina.

We see some objects because their temperature is relatively high so that they emit visible radiation. For these objects, such as the Sun or a glowing piece of hot metal, the perceived color is visible electromagnetic radiation with wavelength bands surrounding the wavelength of maximum emission from the object, as described by Wien's displacement law (Chapter 3). The temperature of most objects on Earth, such as humans, vegetation, and soil, is such that they do not emit visible radiation. Their perceived color is that of the visible wavelength bands reflected from the object; other visible wavelengths are absorbed by the object. For example, something green, such as a plant leaf, reflects only radiation in the green portion of the visible electromagnetic spectrum, while the radiation from all other visible wavelengths that falls upon the leaf is absorbed by the leaf. Moonlight is reflected sunlight and under most atmospheric conditions appears white.

RED SUN, WHITE CLOUDS, AND BLUE SKY

Preferential scattering of sunlight is responsible for the blue color of the daytime sky in directions other than directly at the Sun (Figure 14.2). **Scattering** occurs when tiny particles in the atmosphere interact with light waves and send those light waves in different directions. If the radius of the scattering particles is much smaller than the wavelength of the scattered light, then the amount

FIGURE 14.2
Scattering of sunlight by small air molecules is responsible for the blue of the daytime sky. Scattering of sunlight by the relatively larger ice crystals and water droplets composing clouds is responsible for the white of clouds.

FIGURE 14.3
The setting Sun turns the horizon red. [Photo by A. A. Nugnes]

of scattering varies with wavelength. This happens, for example, when visible light is scattered by the gas molecules (mostly nitrogen and oxygen) composing air. In a now classic experiment performed in 1881, the English physicist and mathematician Lord Rayleigh (1842-1919) demonstrated that this type of scattering is inversely proportional to the fourth power of the wavelength. Hence, violet light, at the short-wavelength end of the visible spectrum, is scattered much more efficiently than red light, at the long-wavelength end of the visible spectrum. This optical effect is known as **Rayleigh scattering**.

As sunlight travels through the atmosphere, component colors are selectively scattered out of the solar beam: violet is scattered more than blue, blue more than green, green more than yellow, and so forth. Dependence of scattering on wavelength predicts that scattered sunlight should be mostly violet. But the daytime sky appears blue rather than violet. One reason is the greater sensitivity of the human eye to blue light rather than violet light, so that the sky appears bluer than it really is. (The cones in the eye register the dominantly violet/blue/green illumination as blue.) Another factor contributing to the blue of the daytime sky is dilution of violet light by all the other scattered colors; although the other colors are scattered less than violet, they tend to wash violet into blue.

We can now understand why the setting Sun turns the horizon red on a clear evening (Figure 14.3). When the Sun is on the horizon, the path of the solar beam through the atmosphere is about 40 times longer than when the Sun is directly overhead. Consequently, at sunset (or sunrise), the interaction (e.g., scattering) between incoming solar radiation and the atmosphere's component gas molecules is

considerable. Arriving at an observer's sea level location, the horizontal solar beam has about 23% of its initial red light but only 0.000006% of its initial violet light. Just before sunrise or just after sunset (during twilight), the clear sky horizon turns red. On the Moon, which has a highly rarefied atmosphere, Apollo astronauts found that the Sun appears as a white disk in a black sky and stars were visible even in bright sunshine. Since scattering of sunlight is inconsequential in the extremely thin lunar atmosphere, the lack of diffuse sunlight means that dark shadows are cast by lunar boulders (Figure 14.4).

The result is different when the particles doing the scattering are much larger than the wavelength of light. In such instances, scattering is not wavelength dependent; that is, visible radiation is scattered more or less equally at all wavelengths. The particles composing clouds (i.e., water droplets and/or ice crystals) are sufficiently large to scatter sunlight in this way. For this reason, sunlit clouds when seen from above or the side have the same color as the Sun, that is, they are white during the day and red or pink at sunrise and sunset.

Particles that are about the same size as the wavelength of light are responsible for **Mie scattering**, named for the German physicist Gustav Mie (1868-1957). In 1908, Mie demonstrated that particles having diameters slightly less than 1 micrometer scatter the most light for their size (scatter most efficiently). Aerosols (solid and liquid particles suspended in the atmosphere) between about 1 and 2 micrometers in diameter scatter red light more efficiently than violet light. Sulfurous aerosols injected into the stratosphere by violent volcanic eruptions are in this size range and are responsible for the enhanced

FIGURE 14.4
In this scene of the surface of the Moon, the sky appears black and the boulder casts sharp black shadows. [NASA]

red and orange sunsets that may persist for more than a year following such a volcanic eruption. For example, vivid sunsets were reported worldwide following the eruption of Krakatoa in 1883 and Mount Pinatubo in 1991. Most non-volcanic aerosols have diameters of 0.02 to 0.5 micrometer and scatter sunlight with equal efficiency at all wavelengths. Hence, an abundance of these aerosols turn the sky a hazy white. In this chapter's *For Further Exploration* we describe the reason for the blue haze that often overhangs heavily forested areas.

HALO

A **halo** is a whitish (sometimes slightly colored) ring of light surrounding the Sun or the Moon (Figure 14.5). It forms when the tiny ice crystals that compose high, thin clouds, such as cirrus or cirrostratus, refract the Sun's rays. **Refraction** is the bending of light as it passes from one transparent medium (such as air) into another transparent medium (such as ice or water). The light rays bend because the speed of light is greater in air than in ice or water. Refraction occurs whenever light rays strike the interface between two different transparent media at an angle other than 90 degrees (Figure 14.6). Rays of light striking water or ice surfaces at 90 degrees are not refracted.

Ice crystals composing clouds occur as hexagonal (six-sided) plates, columns, or dendrites (stars). Most halos form when light is refracted upon entering and

exiting either plates or columns. A light ray entering an ice crystal through one columnar side and exiting through another columnar side is refracted by at least 22 degrees (Figure 14.7). If ice crystals are relatively small and

FIGURE 14.5
A halo about the Sun is caused by refraction of sunlight by tiny ice crystals in high thin clouds. The innermost halo in this image is a luminous circle having a radius of 22 degrees about the Sun. The rare, outermost halo is a circle with a radius of about 46 degrees. [Photo by Jon Oldroyd, taken with a fisheye lens at Halley 5, Brunt Ice Shelf, Coats Land, Antarctica]

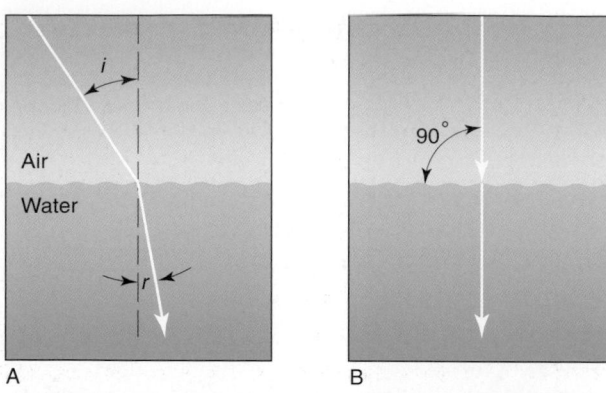

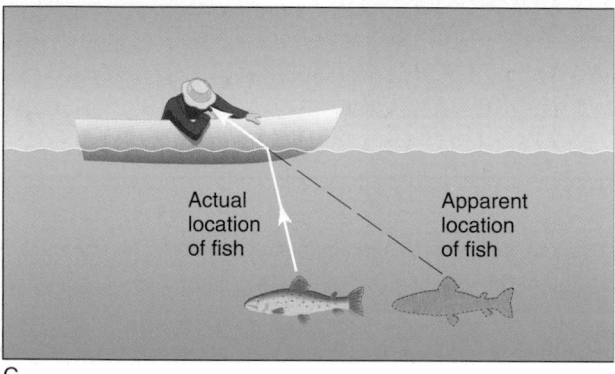

FIGURE 14.6
Light rays may be refracted (bent) as they travel from one transparent medium into another as in (A) where a light ray is refracted as it travels from air to water. The speed of light is less in water than in air, so the light ray is bent toward a line drawn perpendicular to the water surface, and angle *r* is less than angle *i*. (B) Light rays that enter the water at a 90 degree angle are not refracted. (C) Refracted light can be deceptive. The fish appears to be farther away from the boat than it really is because of refraction and because human perception assumes that light always travels in a straight line.

FIGURE 14.7
A ray of sunlight is refracted as it passes through a hexagonal ice crystal from one side (top side in the drawing) to another side (bottom side in the drawing). The angle between the two sides of the ice crystal is 60 degrees. This type of refraction produces a 22-degree halo centered on the Sun or Moon.

randomly oriented, most rays are refracted at 22 degrees so that light is focused in a luminous circle having a radius of about 22 degrees. Suppose, for example, that you see a halo surrounding the Moon. With one eye closed, visualize two straight lines: one extending from your eye to the Moon's center, and the other from your eye to any point on the halo. The acute angle formed at your eye between the two imaginary lines is 22 degrees. For reference, the halo's radius appears to be the same as the width of this book when held at arm's length.

A less commonly observed halo has a radius of about 46 degrees about the Sun (or Moon). In this case, light is refracted by columnar ice crystals with diameters in the range of 15 to 25 micrometers. Light rays travel through an ice crystal from one columnar side to the top end or from side to base rather than from side to side (Figure 14.8).

The Zuni Tribe of New Mexico interprets a halo to be a harbinger of stormy weather: "When the Sun is in his house, it will rain soon." In this weather proverb, the phrase "in his house" refers to the Sun surrounded by a halo. The appearance of cirrus or cirrostratus clouds may signal the beginning of overrunning as warm humid air is lifted by the circulation associated with an approaching low pressure system (cyclone). However, the appearance of a halo around the Sun (or Moon) is no guarantee that precipitation will follow. Cirrus or cirrostratus clouds may spread as much as 1000 km (620 mi) ahead of the storm center, and the storm could change direction or even weaken and die out prior to reaching your location.

Sometimes light is concentrated near and on the 22 degree halo horizontally in one or both sides of the Sun as two brilliant spots, known as **sundogs** because they appear to follow the Sun (Mercury's chariot) around the sky, also called *parhelia* (from the Greek, meaning "beside the Sun") (Figure 14.9). Cirrus clouds composed of relatively large plate-like ice crystals are responsible for sundogs. Because of air resistance, the top and bottom surfaces of these crystals remain nearly horizontal as they settle through the atmosphere to the Earth's surface. The rectangular sides of the ice crystals are vertically oriented

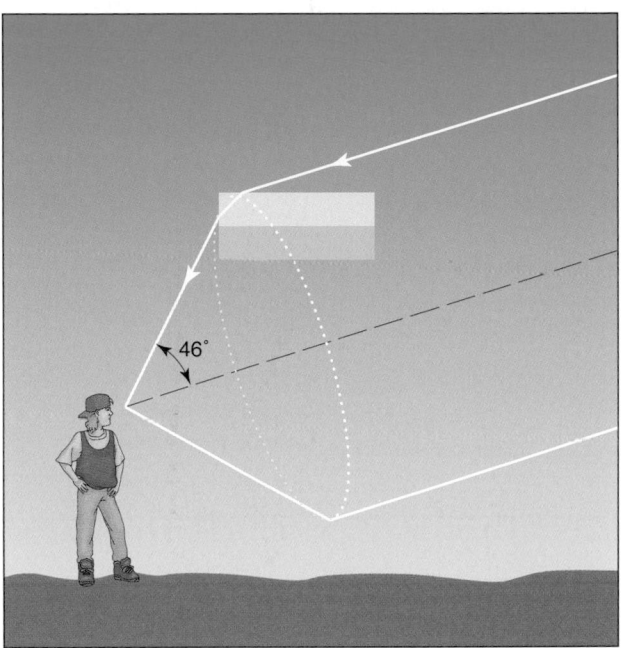

FIGURE 14.8
In a less common situation than shown in Figure 14.7, a ray of sunlight is refracted from side to top or base, or vice versa, as it passes through a columnar ice crystal. The angle between the side and top or base is 90 degrees. This type of refraction produces a 46-degree halo centered on the Sun or Moon.

so that light is refracted either to the right or left so that sundogs always appear at the same altitude as the Sun.

An ice crystal acts like a glass prism, dispersing sunlight into its component colors. The more energetic short-wavelength end of the visible spectrum (violet light) is refracted the most, so that the violet portion of a sundog is farthest from the Sun. The less energetic long-wavelength end of the visible spectrum (red light) is refracted the least, so the red portion of a sundog is closest to the Sun. Because of the optics involved, halos and sundogs are brightest at low solar altitudes and when ice clouds are relatively thin.

Because ice crystals refract sunlight, you may wonder why halos are only weakly colored at best, unlike rainbows for example. In fact, ice crystals disperse light into its component colors, but because the size and shape of ice crystals vary more than the size and shape of raindrops, the colors produced by an assemblage of ice crystals tend to overlap one another rather than form discrete bands. In halos, colors therefore wash out, although occasionally a reddish tinge is visible on the inside of a halo (as in Figure 14.5).

A bright light shaft sometimes appears as a vertical extension above or below the setting or rising Sun.

FIGURE 14.9
Bright colored spots appearing on either side of the Sun and at the same elevation, called parhelia or sundogs, are caused by refraction of sunlight by plate-like ice crystals falling through the atmosphere. [© 2005 Roger Edwards / Insojourn]

This shaft, called a *sun pillar* (Figure 14.10), may have a length from 5 to 20 degrees above the Sun, generally tapering to a point. These pillars are usually not colored, except when the Sun is at or below the horizon, when they become reddish. Sun pillars are produced by reflection of sunlight off the bottom or top faces of horizontally oriented ice crystals; generally, large sized plates and prism crystals (capped columns) are involved.

FIGURE 14.10
Sun pillar - most cases require ice crystals falling from a cloud. [NOAA Photo Library]

FIGURE 14.11
A rainbow is a circular arc of color caused by refraction and internal reflection of sunlight by falling raindrops. Refraction disperses visible light into its component colors. [U.S. Department of the Interior/Bureau of Land Reclamation Mid-Pacific Region]

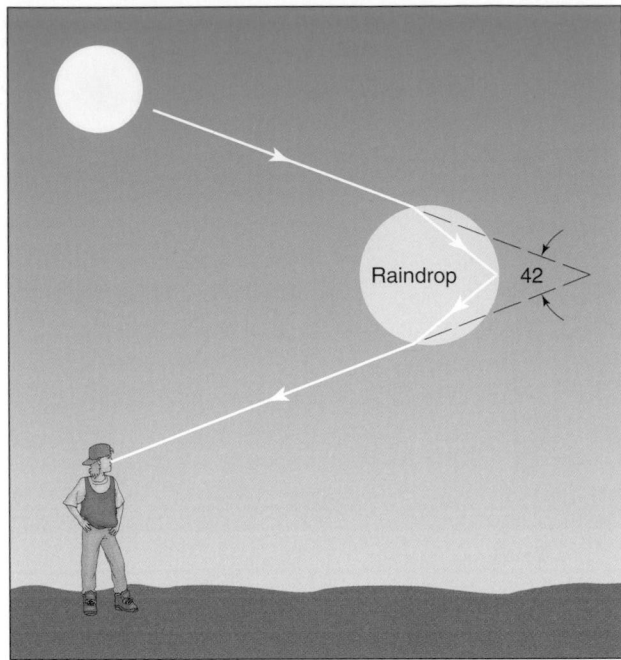

FIGURE 14.12
With a primary rainbow, a solar ray is refracted upon entering a raindrop, reflected internally, and refracted again upon exiting the raindrop.

RAINBOW

A **rainbow** is an arc of concentric colored bands, caused by a combination of refraction and reflection of sunlight (or rarely of moonlight) due to raindrops (Figure 14.11). In the early 1600s, René Descartes (1596-1650) conducted experiments that explained the optics of the rainbow. Sunlight striking a shaft of falling raindrops is refracted twice and internally reflected by each drop of rain. As shown in Figure 14.12, a solar ray is refracted as it enters a raindrop; then the ray is reflected by the inside back of the drop before being refracted again as it exits the drop. Each person sees his or her own rainbow (actually two, one from each eye).

Because of reflection, a rainbow appears to an observer whose back is to the Sun and is facing a distant rain shower (Figure 14.13). A rainbow never forms when the sky is completely cloud covered; the Sun must be shining on the raindrops. Because of geometric considerations, the Sun can be no higher than 42 degrees above the local

horizon in order for someone on the ground to see a rainbow. If the solar altitude is greater than 42 degrees, the cone-shaped surface on which the rainbow could form is below ground level. Hence, a rainbow is more likely during a morning or evening shower than a shower at mid-day. You can create your own rainbow at any time on a sunny day by directing the spray from a garden hose so that you look down and observe the spray with the Sun at your back. At middle latitudes, weather systems usually progress from west to east so that appearance of a rainbow to the east in the early evening usually signals improving weather. Rain showers are moving away toward the east and clearing skies are approaching from the west, where the Sun is setting. Conversely, a morning rainbow to the west possibly signals approaching rain.

Similar to a glass prism, raindrop refraction disperses sunlight into its component colors forming the concentric bands of color of a *primary rainbow*. From outer to innermost band, the colors are red, orange, yellow, green, blue, and violet. In many cases, a much dimmer, *secondary rainbow* appears about 8 degrees beyond the primary rainbow (visible in Figure 14.11). Double reflection within raindrops produces the secondary rainbow with the order of colors reverse that of the primary rainbow (Figure 14.14).

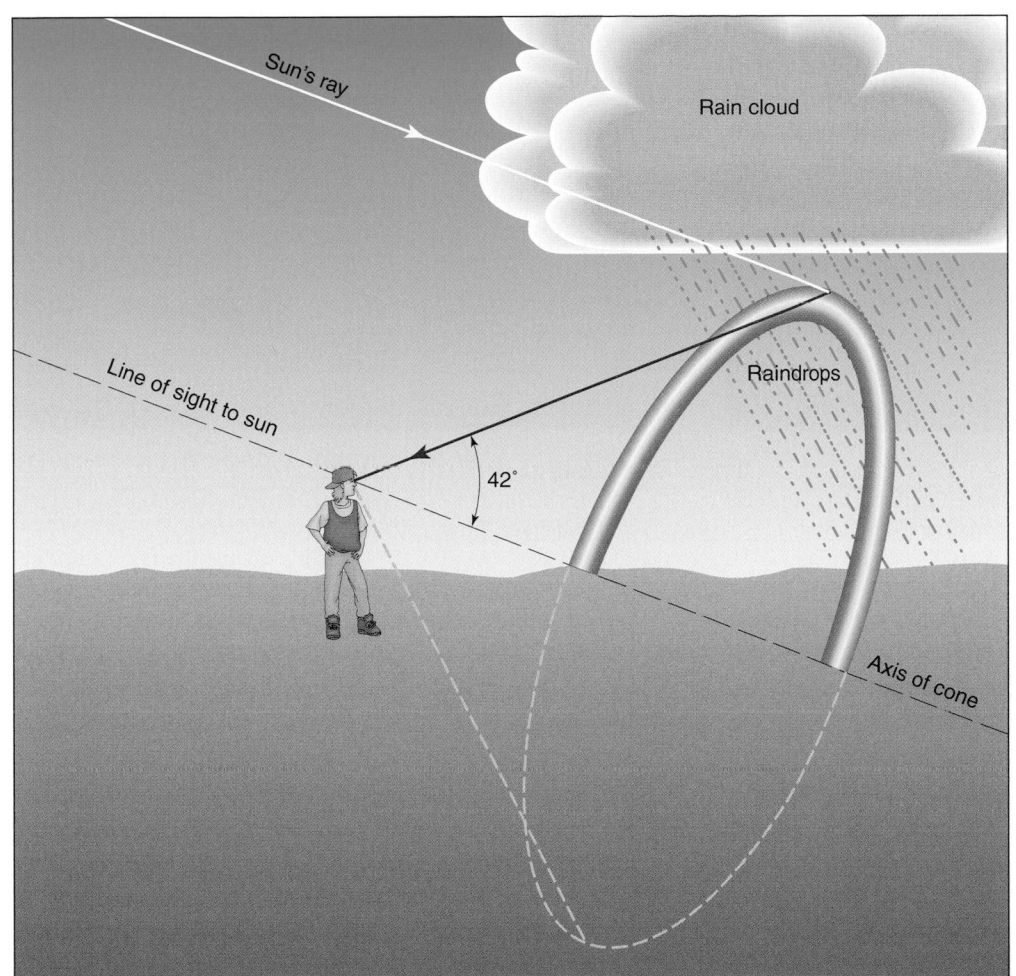

FIGURE 14.13
A rainbow appears to an observer who has his or her back to the Sun and faces a distant rain shower.

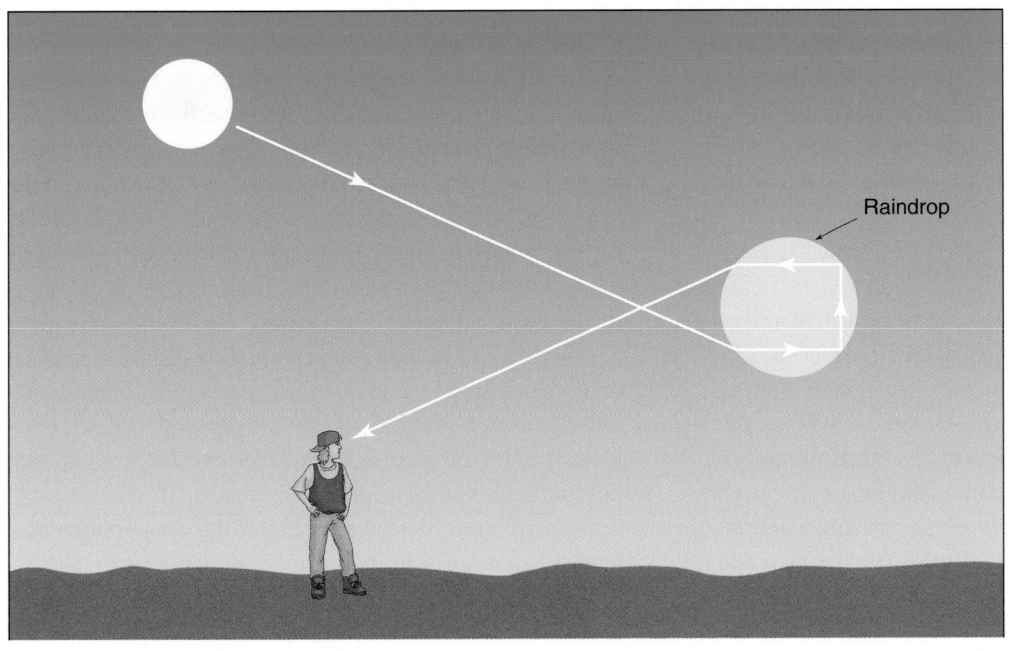

FIGURE 14.14
Refraction of solar rays by raindrops plus double reflection within raindrops produces a dimmer secondary rainbow just above the primary rainbow. A secondary rainbow is visible in Figure 14.11.

FIGURE 14.15
A fog bow observed in San Francisco. [Photo by Mila Zinkova/ License: Creative Commons Attribution ShareAlike 2.5]

FIGURE 14.16
Iridescent clouds exhibit brilliant spots or borders of colors, usually red and green. [Photo by E.J. Hopkins]

A *fog bow* (Figure 14.15) forms in a similar fashion to a rainbow, but is broader and appears almost white, with very faint red on the outermost band to very faint blue and violet on the innermost band. A fog bow is caused by diffraction of reflected light by the very small cloud droplets composing fog. The wavelengths of visible light interfere with one another and the broad color bands overlap, giving the fog bow its washed out appearance.

CORONA

At times, a series of alternating light and dark concentric rings appears to closely surround the Moon or, less often, the Sun when one looks at these light sources through a thin cloud veil. This optical phenomenon is known as a **corona** (not to be confused with the region of hot rarefied and ionized gases that surrounds the Sun). Typically, the radius of a corona is only a few degrees, far smaller than a halo. It is caused by diffraction of light around similarly sized water droplets that compose a thin, translucent veil of altocumulus, altocumulus lenticularis, or stratocumulus clouds (Chapter 7). Often only one ring is actually observed. Smaller sized cloud droplets usually produce the best corona.

Diffraction refers to the slight bending of a light wave as it moves along the boundary of an object such as a water droplet, with production of an interference pattern as light waves bend behind the obstruction. Where the crests of one light wave coincide with the crests of another wave, interference is constructive and a larger wave results, with a bright band due to greater illumination. On the other hand, where the crests of one wave coincide with the troughs of another wave, the interference is destructive, and the waves cancel each other leading to a dark band. If cloud droplets are uniform in size, a corona will be colored with blue-violet on the inside and red on the outside of each ring. Dependency of diffraction on wavelength is responsible for this separation of color: the longer wavelength red light is diffracted more than the shorter wavelength blue-violet light.

Diffraction also is involved in **iridescent clouds**. These are thin clouds with nearly uniformly sized cloud droplets (usually altocumulus, cirrostratus, or cirrocumulus) having bright spots, bands, or borders of delicate colors, usually red and green (Figure 14.16). Iridescent clouds typically appear up to about 30 degrees from the Sun. Iridescent clouds are essentially corona fragments.

GLORY

Prior to the age of aircraft travel, about the only way a person could view a glory was from the top of a lofty mountain peak. To see a glory, an observer must be in bright sunshine above a warm cloud or fog layer, and the Sun must be situated so as to cast the observer's shadow on the clouds below. The observer then sees a **glory** as concentric rings of color centered about the shadow of his or her head. Although less distinct, the colors of a glory are the same as in a primary rainbow, with the innermost band being violet and the outermost band being red. Today, aircraft pilots and observant passengers often see a glory around the shadow of their aircraft on a cloud deck below (Figure 14.17).

A glory depends on much the same optics as a primary rainbow with two important differences, that is, the size of the particles doing the reflecting and refracting, and the direction of reflected and refracted light. Whereas

FIGURE 14.17
A glory appears centered on the shadow of an aircraft on a deck of clouds below. [Photo by E.J. Hopkins]

rainbows occur when sunlight strikes a mass of falling raindrops, glories are the consequence of sunlight interacting with a mass of much smaller suspended water droplets of nearly uniform size (radii less than about 25 micrometers) that compose a cloud. In both cases, the Sun's rays undergo refraction upon entering the droplet, followed by a single internal reflection, and refraction upon exiting the droplet. In a glory, sunlight is refracted and reflected directly back toward the Sun due to the small droplet size. A small spherical cloud droplet diffracts light rays ever so slightly toward the droplet, so that light rays incident on a cloud droplet parallel those returning from the cloud droplet (Figure 14.18). The diffraction patterns produce bands that may be slightly colored.

The special optics of a glory explains why it appears about the shadow of the observer who is situated in the direct path of both the incident solar rays and the returning (refracted and reflected) solar rays. This also

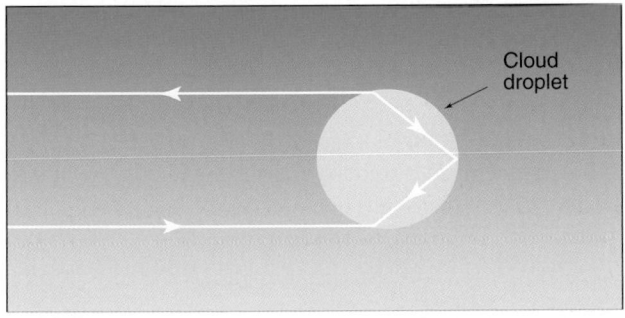

FIGURE 14.18
In the special optics that produce a glory, both the incident and returning solar rays are diffracted slightly toward the surface of the cloud droplet. Consequently, the incident and returning rays follow almost parallel paths.

explains an observation that must have fascinated ancient mountain mystics. Suppose that you and a friend are standing side by side high on a mountain slope at a site favorable for viewing a glory (cloud deck or fog bank below and the Sun behind your shoulder). On the cloud deck below, you see your own shadow next to that of your friend, but a glory appears about your head and not about your friend's head. Lest you presume that you have been singled out, note that your friend has essentially the same observation, for your friend sees a glory only about his or her head. The fact is that each observer is in a position to view only one glory.

MIRAGE

Appearances can be deceiving, especially in the case of mirages. A **mirage** is an optical phenomenon in which an image of some distant object appears to be displaced from its normal view. Distant buildings or hills can appear higher or lower than they really are. A nonexistent pool of water may suddenly appear on the highway ahead, or a sailboat viewed from shore may appear upside down. Mirages are caused by the refraction of light rays within the lower atmosphere.

Light travels at different speeds through different transparent substances. Light changes speed, for example, as it travels from air into an ice crystal. For reasons discussed earlier in this chapter, a light ray that strikes an interface at other than a 90 degree angle is refracted at the interface between two different transparent media. The speed of light also varies within a single medium if the density of that medium is not uniform. Hence, light rays bend as they pass through a substance of varying density, such as the atmosphere.

If the density of the atmosphere were uniform throughout, then light rays would always travel along straight paths at constant speed, and no refraction would occur. As described in Chapter 5, however, air density varies with changes in temperature and pressure. Because our concern here is with optical phenomena that occur within the lower troposphere and involve relatively short viewing distances, we need not be concerned with the influence of horizontal air pressure gradients on air density. For the same reason, we also ignore the effects of horizontal temperature gradients on air density. But we cannot ignore the effect of vertical temperature profiles on the change of air density with altitude, which occur at every location.

As a rule, light rays traveling through the atmosphere are refracted such that denser air is on the inside (concave side) of the bend, and less dense air is on

FIGURE 14.19
Inferior mirage in the Mohave Desert. The "lake" that appears on the ground is not real. [Photo by Mila Zinkova in Primm, Nevada on 4 April 2007]

the outside (convex side) of the bend. Air density almost always decreases with altitude so that light reflected from a distant object follows a downward curved path to the observer. Because human perception is based on the assumption that light travels in a straight path, to the observer the object appears to be higher than it really is.

If the air temperature decreases with altitude at less than the usual rate, or if there is a *temperature inversion* (an increase of temperature with altitude), air density decreases with altitude at a faster rate than normal. Hence, light rays reflected from a distant object bend more sharply than usual before reaching the viewer and objects appear higher that normal. This is called a *superior mirage*. On the other hand, if the lowest air layer features a greater than usual temperature lapse rate, rays are refracted less than normal and objects appear lower than we usually see them. This is called an *inferior mirage* (Figure 14.19).

These are a few examples of the many possible types of mirages. As the vertical air temperature profile becomes more complex, so too do the types of mirages that can appear. Recall our discussion of Fata Morgana in this chapter's Case-in-Point. Fata Morgana involves the simultaneous occurrence of superior and inferior mirages. The familiar oasis mirage in a desert is an inverted image of the sky seen below the horizon. The bluish colored mirage is caused by the light from the sky that arrives at the observer's eye by a curved ray, not the reflection of the skylight from a non-existent water surface. All mirages are displacements or distortions of something real.

SUNRISE-SUNSET AND TWINKLING STARS

Because the atmosphere refracts sunlight, the image of the setting or rising Sun that we see is slightly higher in the sky than it would be without an atmosphere. One implication of this involves the length of daylight on the equinox. As noted in Chapter 3, on the equinox, day and night are assumed to be of equal length (12 hrs) everywhere (except right at the poles). Actually, days are slightly longer than nights on the equinox. At Washington, DC, for example, day length on the vernal equinox of 20 March 2012 was 12 hrs and 10 minutes. The *length of daylight* is defined as the period between sunrise and sunset, and sunrise and sunset occur when the upper edge of the solar disk is just visible on the unobstructed level horizon. Because the Sun appears higher than it actually is at both sunrise and sunset, there is slightly more than 12 hrs between sunrise and sunset on the equinox.

The shape of the Sun or Moon near the horizon appears distorted from the usual circular disk, especially when strong atmospheric stratification is present. Rays from the lower rim of the disk are lifted more than the rays from the upper rim, thereby flattening the perceived solar or lunar disk (Figure 14.20). However, atmospheric refraction is not responsible for the larger apparent size of the Sun or Moon near the horizon than at higher altitude angles. This enlargement effect is an optical illusion, caused by how we judge the apparent size based upon the presence of reference landmarks near the horizon.

As light from stars and other celestial bodies passes through the atmosphere, fluctuations in air density (mostly due to temperature differences) cause rapid changes in brightness. This explains the apparent twinkling of stars, essentially single points of light, at

FIGURE 14.20
Sunset at Mackinaw City, MI. The image shows both Lakes Huron (near) and Michigan (far). [© 2007 Roger Edwards / Insojourn]

night. Light rays reaching the observer from a star may be momentarily turned away from the observer's line of sight by turbulent air motions, causing an abrupt change in the apparent luminosity of the star. This phenomenon, known as **scintillation**, is especially noticeable on cold, clear nights, when rapid, small-scale fluctuations in air density alter the path of starlight through the atmosphere. The planets, with larger diameters, have a more constant luminosity because their light is not as affected by scintillation.

TWILIGHT

Scattering of sunlight is also responsible for **twilight**, the period following sunset or before sunrise when the sky is illuminated. In this discussion, assume that the sky is not cloud covered. Multiple scattering of sunlight by constituents of the upper atmosphere illuminates the sky. Because of geometrical considerations involved with the path of the apparent Sun with respect to the local horizon, the length of twilight varies with latitude and time of year. The annual range increases with latitude so that in polar regions, for example, the length of twilight ranges from 24-hrs to none at all. In the tropics, on the other hand, the duration of twilight is relatively short. At middle latitudes, twilight length is longer near the solstices, but shorter at the equinoxes. Twilight is divided into three sequential stages based on the level of illumination: civil twilight, nautical twilight, and astronomical twilight.

Illumination during *civil twilight* is just adequate for outdoor activities without the need for artificial lighting. It covers the period between sunrise/sunset and when the center of the Sun's disk is 6 degrees below the horizon. On a clear day, a faint purple glow may appear over a large region of the western sky during the latter portion of civil twilight in the evening. During *nautical twilight*, light is sufficient to distinguish the outlines of distant objects on the ground. While the ocean horizon is visible at the beginning of nautical twilight during the evening, it is not at the end when the center of the Sun's disk has sunk to 12 degrees below the horizon. *Astronomical twilight* occurs when the center of the Sun's disk is between 12 and 18 degrees below the horizon. Sixth magnitude stars are visible directly overhead.

At the beginning of evening twilight, observant sky watchers may see crepuscular rays or the green flash. **Crepuscular rays** appear as beams of sunlight radiating from the Sun. They consist of alternating light and dark bands (solar rays and shadows) that diverge in a fanlike pattern from the vicinity of the solar disk near twilight (Figure 14.21). The rays are visible because of scattering.

FIGURE 14.21
Crepuscular rays consist of alternating light and dark bands that appear to diverge fanlike from the Sun's position during twilight. [© 2003 Roger Edwards / Insojourn]

Actually, solar rays are parallel and only appear to diverge because of the perspective of the observer. Crepuscular rays can also occur at other times during the day as the Sun's rays pass through holes in clouds or between clouds in a hazy sky.

The **green flash** is a thin green rim that appears briefly at the upper edge of the Sun at sunrise or sunset and is best seen on a distant horizon when the atmosphere is very clear (Figure 14.22). The green flash is primarily the consequence of atmospheric refraction and scattering of light from a low Sun. Refraction is most pronounced when the Sun is on the horizon and, as noted earlier, blue/violet light at the short wavelength end of the visible spectrum is refracted (bent) more than red light at the long wavelength end. This explains the red rim at the bottom of the low Sun and should produce a blue/violet rim at the top of the solar

FIGURE 14.22
Green flash (at upper edge of Sun) in San Francisco, CA. [Photo by Mila Zinkova/License: Creative Commons Attribution ShareAlike 2.5]

disc. However, Rayleigh scattering removes most of the blue/violet light leaving behind a sliver of green.

The sliver of green produced by atmospheric refraction alone is too small to be seen by the naked eye. With an ideal temperature profile (e.g., superadiabatic lapse rate near the surface), the atmosphere acts like a lens to magnify the green rim to visible proportions. The green flash is actually a mirage that appears as an island of green light floating above the setting or rising Sun. At middle latitudes, the Sun rises and sets so quickly that the green flash lasts only about 1 second (hence the "flash" designation). But at polar latitudes, the Sun rises and sets more slowly and the green flash can lasts upward of 30 minutes.

Atmospheric Acoustics

Vibrations of an object generate sound waves that propagate outward in all directions. The sensation of hearing is the stimulation of auditory nerves in the ear by vibrations, that is, sound waves. Sound waves enter the ear and cause the eardrum and inner ear to vibrate. In response, auditory nerves send impulses to the brain. Unlike electromagnetic waves that can travel through a vacuum, sound waves require a transmitting medium such as air or water. Without this medium, all would be silent. Our focus in this section is sound transmission in the atmosphere and sounds produced by atmospheric phenomena.

SOUND WAVES

A **sound wave** is a compressional wave, consisting of alternate compressions and rarefactions

of air. Transmission of sound energy is via alternating increases and decreases in air pressure produced by waves that radiate outward in all directions from a source, much like the waves generated by a stone tossed into a quiet pool of water. *Wave frequency* is the number of oscillations per second, measured in hertz (Hz). With increasing frequency, the sound increases in pitch. The audible range for most humans is between 20 and 20,000 Hz. Also, the attenuation of sound waves is directly proportional to the square of the frequency. Hence, low frequency sound waves can propagate greater distances than high frequency sound waves. In fact, audible sound waves can travel thousands of kilometers through the atmosphere. The 1883 violent eruption of Krakatoa, a volcano in the Sunda Strait between Java and Sumatra in Indonesia, was heard as far away as Rodriguez Island, 4653 km (2885 mi) across the Indian Ocean.

The intensity (loudness) of sound depends on the pressure exerted by sound waves and is measured in *decibels (dB)*. Decibel levels corresponding to the sound of a variety of devices and activities are listed in Table 14.1. Any sound over 85 dB is potentially harmful for humans, and over 120 dB is painful.

Wind and air temperature affect the speed of sound waves traveling through the atmosphere. The speed of a sound wave is the sum of the speed of sound through the medium (air) plus the motion of the medium (wind). Hence, sounds are heard better downwind of the

TABLE 14.1 Sound Decibel (dB) Levels[a]	
Soft whisper, quiet library	30
Rustling leaves	40
Rainfall, refrigerator	50
Normal conversation, air conditioner	60
City or freeway traffic, sewing machine	70
Hair dryer, alarm clock	80
Power lawn mower, motorcycle	90
Garbage truck, snowmobile	100
Shouting in one's ear, dance club, race car	110
Jet plane taking off, car stereo on full volume, band practice	120
Live rock music, jack hammer	130
Firecracker, nearby gunshot blast, jet engine	140

[a]Source: U.S. Congress Select Committee on Children, Youth, and Families

source than upwind. In addition, atmospheric turbulence (that tends to increase with increasing wind speed) helps to dissipate the energy of sound waves. Sound waves propagate by collisions of gas molecules and the average kinetic-molecular activity of gases increases with rising temperature. Hence, sound waves travel faster in warm air than cold air. At sea-level air pressure and an air temperature of 0 °C (32 °F), the speed of sound is 332 m (1088 ft) per second. Raising the air temperature to 20 °C (68 °F) increases the speed of sound to 343 m (1125 ft) per second.

Dependence of the speed of sound on wind and temperature is the operating principle of the *sonic anemometer*, an instrument that emits sound waves to measure wind speed (Figure 8.25). Sound waves are transmitted in opposite directions across known paths. The speed of propagation of those sound waves equals the vector sum of the speed of sound in air (at the measured temperature) plus the wind speed. Hence, the wind speed can be calculated by comparing the actual speed of sound to the speed based on temperature alone. As of this writing, the sonic anemometer has replaced the cup anemometer as the standard wind speed sensor at over 90% of NOAA's National Weather Service observing systems (e.g., ASOS).

A change in air temperature with distance (*temperature gradient*) alters the speed of sound, which can lead to the bending or refraction of sound waves. Like light waves, sound waves are refracted away from regions where they travel fastest. Only in air layers that are isothermal (uniform temperature) do sound waves propagate in straight lines. When the air temperature decreases with altitude (the usual condition in the troposphere), sound waves follow a path that is concave upward; that is, sound is deflected upward (away from the warmer air at the surface). When the air temperature increases with altitude (a *temperature inversion*), the path of sound waves curves downward; that is, sound is deflected downward toward the colder air at the surface. This explains why in winter human voices can sometimes be heard over horizontal distances of several kilometers within a mass of arctic air. In an arctic air mass, over a broad area winds are light or the air is calm (no turbulent dissipation) and the coldest air is at the surface. Sound waves originating near the ground follow a path convex to the sky as they are refracted by the warmer air aloft back to the ground some distance away. On the other hand, elevated observers can hear sounds traveling up from the ground better than those hearing sounds traveling downward.

Refraction of sound waves explains the formation of an **acoustic shadow**, an area where sound is not heard even though the area is relatively close to the source of the sound. For example, news reports from American Civil War battlefields described situations when the sounds of cannon fire were not heard at positions near the battle whereas more distant observers could readily hear the sounds of battle. Charles D. Ross of Longwood College, VA estimates that acoustic shadows and blockage of sound waves by hills affected the outcome of 10 Civil War battles. At Gettysburg, PA, a Confederate commander could not hear an artillery barrage that was meant to signal his troops to attack Union forces. Those Union troops then defeated another Confederate force.

Fresh snow on the ground absorbs sound waves thereby lowering the ambient noise level over the landscape. Even a snow depth of as little as a few centimeters has a significant sound-absorbing effect. The explanation is the air trapped between the individual snowflakes that attenuate vibrations. Sound waves produced by multiple sources can also undergo interference. Constructive interference causes louder sounds whereas destructive interference results in quiet zones.

THUNDER

Where there's lightning, there's thunder, although sometimes we see distant lightning but do not hear the thunder. **Thunder** is the sharp clap or rumbling sound heard following a lightning flash (Chapter 11). Lightning heats the air along the narrow conducting path to temperatures that may top 25,000 °C (45,000 °F). The rapid rise in air temperature is accompanied by a tremendous increase in air pressure locally that generates a shock wave. The shock wave propagates outward, producing sound waves that are heard as thunder. The initial thunder clap we hear is generated by the nearest part of the lightning flash. Subsequent sound waves reach us from portions of the flash that are progressively further away; hence, the rumble of thunder.

Thunderstorm cells that are more than 20 km (12 mi) away are too distant for thunder to be audible, mainly due to the sound's refraction upward, although lightning is observed. This is called *heat lightning*. Light waves are also refracted but much less than sound waves so that lightning is seen but thunder is not heard.

In the atmosphere, light travels about a million times faster than sound so that we see lightning almost instantaneously, but we hear thunder later. The closer we are to a thunderstorm cell, the shorter is the time interval between the lightning flash and thunder. As a rule, thunder takes about 3 seconds to travel 1 km (or 1 second to travel 1000 ft and 5 seconds to travel 1 mi). If you must wait 9 seconds between lightning flash and thunderclap, the lightning is about 3 km (1.8 mi) away. This is the *flash-to-bang* method of determining the distance to a thunderstorm. By noting the time difference between lightning flash and thunderclap of successive lightning discharges, you can deduce whether the thunderstorm is approaching or moving away, or even predict when it is dangerously close. As a general guideline, if the time between flash and bang is 30 seconds or less, you are well advised to take precautions to avoid the lightning.

SONIC BOOM

A **sonic boom** can be loud enough to break windows. They are caused by aircraft traveling at speeds that exceed the speed of sound, known as *Mach 1*. The Mach scale is named for the Austrian physicist Ernst Mach (1838-1916) who was an expert on sound propagation. An aircraft moving at Mach 2 is traveling at twice the speed of sound.

Noise from an aircraft moving at subsonic speeds propagates in all directions faster than the forward speed of the aircraft. When a plane accelerates past the "sound barrier" to supersonic speeds, the plane moves faster than the leading edge of the pressure wave generated by the aircraft. A narrow conical zone of compressed air is produced in the form of a shock wave through which the pressure changes greatly over a short distance. The shock wave propagates at the same speed as the aircraft. Listeners in the area affected by the sudden increase in pressure will hear a sonic boom, similar to a thunderclap, but the pilot does not hear it.

AEOLIAN SOUNDS

Winds blowing over obstacles such as roofs, telephone or power transmission wires produce humming, singing or whistling sounds called **aeolian sound**. Turbulent eddies that form immediately downwind of the obstacle are responsible for the sounds. Thinner wires produce smaller eddies and a higher pitch than thicker wires. During the cold of winter, wires contract and become taut. The increased tension is transmitted to the utility poles that act as sounding boards (similar to a piano sounding board) and the volume of humming, singing or whistling increases. Also, the pitch increases with higher wind speeds.

Other sounds of meteorological origin include the rattle of ice pellets (sleet), squeaking of snow, clatter of hail, patter of rain, and the rustle of leaves. Many people describe the sound of a nearby tornado as similar to the roar of a passing freight train. The squeaking sound heard when one walks on a snow surface at temperatures below −10 °C (14 °F) is due to the crushing of ice crystals. Ice crystals at warmer temperatures are covered by a liquid-like surface layer, which allows crystals to slide past each other when pressed together. As temperatures lower, the thickness of the layer decreases, until at about -10 °C (14 °F), it essentially disappears. Pressed ice crystals then no longer slip; they fracture.

Conclusions

We have seen that interactions of sunlight (or moonlight) with clouds or rainfall produce a variety of optical phenomena including halos, rainbows, coronae, and glories. These interactions involve reflection, refraction, and/or diffraction. In these cases, refraction takes place as light rays change speed as they pass from one transparent medium into another (e.g., air to water or air to ice). In addition, because of variations in air density within the atmosphere, light rays are refracted and give rise to mirages and the green flash. Scattering of sunlight by air molecules is wavelength-dependent and responsible for the blue of the daytime sky whereas scattering of sunlight by cloud particles (droplets and ice crystals) is uniform across wavelengths and accounts for the white of clouds.

We have also seen that sound propagation through the atmosphere is influenced by wind and temperature. Sound waves can be refracted by the atmosphere and heard great distances from the source. Furthermore, atmospheric phenomena produce a variety of sounds.

Basic Understandings

- Rayleigh scattering of sunlight by the molecules composing air accounts for the blue of the daytime sky. If the radius of the scattering particles is much smaller than the wavelength of the scattered light, then the amount of scattering is inversely proportional to the fourth power of the wavelength of the light being scattered. The human eye's greater sensitivity to blue light than violet light plus the dilution of violet light by the other scattered colors also play a role in the blue of the sky.
- Scattering is not wavelength dependent when light is scattered by particles having a radius that is much greater than the wavelength of the radiation being scattered. This is the case for clouds and explains why clouds appear white

during the day and pink or red near sunset or sunrise. Mie scattering applies to particles that have diameters about the same as the wavelength of visible light. Scattering of visible light is most efficient for particles having a diameter of 1 micrometer.

- A halo is a whitish (or slightly colorized) ring of light surrounding the Sun or Moon, formed when the ice crystals in high thin clouds refract the Sun's rays. Refraction is the bending of light as it passes from one transparent medium into another. Cirrus clouds composed of relatively large plate-like ice crystals are responsible for sundogs, brilliant spots of light located on either side of the Sun.

- A rainbow is an arc of concentric colored bands, caused by a combination of refraction and internal reflection of sunlight by raindrops. In many cases a much dimmer secondary rainbow appears about 8 degrees beyond the primary rainbow. Double reflection within raindrops produces the secondary rainbow with the order of colors reverse that of the primary rainbow.

- A fog bow forms in a similar fashion to a rainbow, but appears almost white due to diffraction of light by the small cloud droplets composing fog.

- A corona consists of alternating light and dark rings that encase the Moon (or less often the Sun) caused by diffraction of light around water droplets that compose a thin, translucent veil of altocumulus or stratocumulus clouds. Diffraction is the slight bending of a light wave as it moves along the boundary of an object such as a water droplet, producing an interference pattern.

- A glory consists of concentric rings of color that encase a shadow and is caused by interactions between sunlight and a uniform mass of water droplets composing a warm cloud. The optics of a glory differ from that of a rainbow in that spherical cloud droplets diffract light rays ever so slightly toward the droplet, so that light rays incident on a cloud droplet parallel those returning from the cloud droplet.

- A mirage is an optical phenomenon in which an image of some distant object appears to be displaced from its normal view. It is caused by refraction of light as it travels through the atmosphere. Changes in air density that accompany variations in the vertical temperature profile are responsible for the refraction of light.

- As light from stars passes through the atmosphere, fluctuations in air density (mostly due to temperature differences) cause rapid changes in brightness to the point sources of light. This explains the apparent twinkling of stars at night.

- Twilight is a period after sunset or before sunrise when the sky is still subjected to scattered sunlight by the upper atmosphere. Based on decreasing level of illumination, and the angular position of the Sun below the horizon, twilight is designated civil, nautical, and astronomical.

- Possible phenomena near twilight include crepuscular rays and the green flash. Crepuscular rays consist of alternating light and dark bands (solar rays and shadows) that appear to diverge in a fanlike pattern from the Sun's position at about twilight. The green flash is a thin green rim that appears briefly at the upper edge of the Sun at sunrise or sunset and is primarily the consequence of atmospheric refraction and scattering of light from a low Sun.

- A sound wave is a compressional wave, consisting of alternate compressions and rarefactions of air. Transmission of sound energy occurs via alternating increasing and decreasing pressure changes that emanate outward from a source.

- Wind and air temperature affect the speed of sound waves traveling through the atmosphere. The speed of a sound wave is the sum of the speed of sound through the medium (air) plus the motion of the medium (wind). Hence, sound is heard more clearly downwind of the source than upwind. Sound waves propagate by collisions of gas molecules and the average kinetic-molecular activity of gases increases with rising temperature. Hence, sound waves travel faster in warm air than cold air.

- A change in air temperature with distance alters the speed of sound that can cause refraction of sound waves. Sound waves are refracted away from regions where they travel fastest. When the air temperature decreases with altitude (the usual condition in the troposphere), sound waves follow a path that is concave upward; that is, sound is deflected upward (away from the warmer air at the surface). When the air temperature increases with altitude (a temperature inversion), the path of sound waves curves downward; that is, sound is deflected downward toward the colder air at the surface.

- Thunder is the sharp clap of sound emitted during a lightning flash. Lightning heats the air along the narrow conducting path to temperatures that may top 25,000 °C (45,000 °F). The rapid rise in air temperature is accompanied by a tremendous increase in air pressure locally that generates a shock wave that propagates outward, producing sound waves that are heard as thunder. A rumble is heard when sound starting at essentially the same time along a lightning channel (or from different lightning strikes) travels different distances to the listener.

- Winds blowing over obstacles such as roofs, telephone or power transmission wires produce humming, singing or whistling sounds called aeolian sound. Turbulent eddies that form immediately downwind of the obstacle are responsible for these sounds. Such aeolian sounds also occur in the free atmosphere from turbulence.

Enduring Ideas

- Scattering, reflection, refraction, and/or diffraction of sunlight by cloud droplets, ice crystals, and raindrops produce a variety of optical phenomena, such as halos, rainbows, coronae, and glories. Mirages can occur when sunlight is refracted as it passes through the atmosphere. Scattering of solar rays by atmospheric particles, including molecules, causes the blue of the sky, the white of clouds, a red horizon at sunset, and twilight.

- The type of scattering that occurs is dependent upon the size of the scattering particles relative to the wavelengh of light. For example, if the radius of the scattering particles (e.g. gas molecules composing air) is much smaller than the wavelength of the scattered light, then the amount of scattering varies with wavelength. Violet and blue are scattered more efficiently than red light; this accounts for the blue of the daytime sky.

- Atmospheric refraction causes the images of the setting and rising Sun to be higher than if an atmosphere were not present. For this reason, the length of daylight at an equinox is slightly more than 12 hrs.

- When an object vibrates, it generates sound waves that propagate outward. Electromagnetic waves can travel through a vacuum, but sound waves need a transmitting medium such as air or water. Sound propagation through the atmosphere is influenced by wind, temperature, and atmospheric refraction. Weather phenomena produce a variety of sounds, such as thunder and aeolian sounds.

Key Terms

Fata Morgana	rainbow	crepuscular rays
mirage	fog bow	green flash
scattering	corona (optical)	sound wave
Rayleigh scattering	diffraction	acoustic shadow
Mie scattering	iridescent clouds	thunder
halo	glory	sonic boom
refraction	scintillation	aeolian sound
sundogs	twilight	

Review

1. What two processes explain the color of an object?
2. Why is the color of the clear daytime sky blue?
3. Near sunset clouds may be red or pink. Explain why.
4. Why are light rays refracted as they travel from air into raindrops and ice crystals?
5. A halo around the Moon may be a precursor of stormy weather. Explain why.
6. Why does a rainbow never form when the sky is completely cloud covered?
7. Provide two examples of atmospheric optical phenomena that involve diffraction.
8. Under what conditions might a glory be visible from an aircraft window?
9. Explain why the shape of the Sun near the horizon appears distorted from the usual circular disk.
10. Distinguish between sound waves and electromagnetic waves.

Critical Thinking

1. On a cloud-free evening, the setting Sun turns the horizon red. Explain why.
2. On the Moon, the sky is black rather than blue. Why?
3. In the year or so following a violent volcanic eruption, sunsets tend to be more red and orange than usual. How is this explained by Mie scattering?
4. Explain why sundogs and a halo about the Sun can appear in the sky at the same time.
5. Identify the various atmospheric optical effects that combine to produce a green flash.
6. How does an acoustic shadow form?
7. Under what condition will sound waves propagate in straight lines? Explain.
8. Why is there a lag time between a lightning flash and thunder?
9. Identify the various factors that influence sound propagation in the atmosphere.
10. Why is the daytime sky blue rather than violet?

Blue Haze

Many heavily forested areas are often shrouded in a blue haze. In fact, blue haze is such a usual feature of some regions that its presence is reflected in their names. The Blue Ridge Mountains of Virginia, the Blue Mountains of Australia, and the Blue Hills outside Boston are examples. Also, the Cherokee Indians' name for the Great Smoky Mountains (Figure 1) of Tennessee was the "Place of Blue Smoke." Studies suggest that blue haze plays a major role in the generation of tropospheric ozone, an important component of photochemical smog that plagues the air quality of many metropolitan areas. These findings have raised questions regarding the composition and production of blue haze as well as the most appropriate strategy to reduce tropospheric ozone (O_3).

Scientists have established that blue haze is a complex and variable mixture of gaseous hydrocarbons (volatile organic compounds) emitted by vegetation. One major class of these volatile hydrocarbons is essential oils, compounds that are responsible for the characteristic aromas of certain plant foliage. Familiar examples of plants that produce essential oils include peppermint, sage, basil, and pine and eucalyptus trees. Essential oils serve to protect plants from herbivores because these compounds have toxic properties.

Air quality research indicates that isoprene is often the most common hydrocarbon component of blue haze. Isoprene is a breakdown product of a compound in the synthetic pathway that leads to the production of essential oils. Many, but not all, plants emit isoprene, and isoprene emissions are strongly influenced by leaf temperature. Over short periods of time, isoprene emissions from leaves of the most temperature-sensitive species can increase tenfold with a rise in air temperature of 10 Celsius degrees (18 Fahrenheit degrees). Hence, the highest levels of atmospheric isoprene occur

FIGURE 1
Great Smoky Mountains National Park. [National Park Service]

on hot summer days when winds are light. Furthermore, isoprene readily reacts with nitrogen oxides in the presence of sunlight to form ozone.

The role of blue haze in the generation of ozone prompts the question as to what fraction of all atmospheric hydrocarbons are biogenic (natural origin) and what fraction can be attributed to anthropogenic sources (e.g., motor vehicles, power plants, and industry). What fraction of emissions is of local or distant origin? The answers to these questions are important because federal law mandates that individual states reduce tropospheric ozone levels. Ozone abatement efforts thus far have focused primarily on reduction of anthropogenic sources, but if biogenic sources comprise a major component of atmospheric hydrocarbons, ozone abatement strategies would need to be reevaluated.

Monitoring of hydrocarbon levels has brought some surprises. In heavily wooded Atlanta, GA, for example, a study found that 83% of all hydrocarbon emissions were biogenic with isoprene accounting for most of the natural volatile organic compounds. This and similar findings from other urban areas indicate that biogenic emissions of hydrocarbons can override efforts to reduce anthropogenic sources. Furthermore, short of cutting down millions of trees, nothing can be done to reduce biogenic hydrocarbons.

In addition to hydrocarbons, the other main ingredient in the generation of tropospheric ozone is nitrogen oxides, most of which have anthropogenic (industrial, transportation) sources. Given the relatively high levels of biogenic hydrocarbons and our inability to control their emissions, many air quality specialists argue for a shift in ozone-abatement strategies from a reduction in anthropogenic hydrocarbons to a reduction in anthropogenic nitrogen oxides.

Coral reefs are threatened by ocean acidification and other effects of global climate change. [Courtesy NOAA]

CLIMATE & CLIMATE CHANGE

Chapter Highlights

Case-in-Point
  Global Warming and Sea Level Rise
Earth's Climate System
  The Climatic Norm
  Climatic Anomalies
Climate Boundary Conditions
Global Climate Patterns
  Temperature
  Precipitation
  Climate Classification
The Climate Record
  Geologic Time
  Past Two Million Years
  Instrument-Based Climate Trends
Causes of Climate Variability and Change
  Solar Variability
  Earth's Orbit
  Volcanoes
  Earth's Surface Properties
  Human Activity
  Anthropogenic versus Natural Forcing
   of Climate
The Climate Future
  Global Climate Models
  Search for Cycles and Analogs
  Enhanced Greenhouse Effect and Global
   Warming
Potential Impacts of Global Climate Change
  Shrinking Glaciers and Rising Sea Level
  Arctic Environment
  Other Impacts
Conclusions
Basic Understandings/Enduring Ideas
Key Terms/Review/Critical Thinking
For Further Exploration
  Sources of Proxy Climate Data
  Lessons of the Climate Past

Learning Objectives

Distinguish between weather and climate.
Explain why the climate of a location is
    described in terms of average weather
    plus extremes in weather.
Provide a definition for climatic norm.
Identify the forcing mechanisms and agents
    that operate in Earth's climate system.
Describe the general global patterns of
    surface air temperature and precipitation.
Explain why precipitation is distinctly
    seasonal in certain regions of the world.
List some of the methods used to reconstruct
    past climates.
Explain how plate tectonics likely contributed
    to long-term climate change in the
    geological past.
Sketch the principal features of the global
    climate record over the past 2 million years.
Explain the significance of polar amplification
    for future climate change.
Present the significance of the Little Ice Age.
Explain the geographic non-uniformity of
    climate change.
Describe the climatic significance of the
    Maunder minimum.
Explain how the Milankovitch cycles impact
    large-scale fluctuations in Earth's glacial
    ice cover.
Present the properties of volcanic eruptions
    most likely to impact global climate.
Explain how changes in the properties of
    Earth's surface can influence climate.
Describe how feedback in Earth's climate
    system affects climate change.
Explain how human activity enhances the
    greenhouse effect.

How and why does climate change?

Case-in-Point

Global Warming and Sea Level Rise

The consensus of scientific opinion is that the present global warming trend is largely anthropogenic in origin. Human activity, principally the combustion of fossil fuels, is responsible for the build-up of carbon dioxide and other infrared-absorbing gases in the atmosphere, enhancing Earth's greenhouse effect. One of the most pervasive effects of global warming is the slow rise in mean sea level (msl) due to thermal expansion of seawater and the melting of land-based glaciers and ice sheets. Rising sea level threatens to submerge islands and inundate low-lying coastal areas, forcing the relocation of human populations.

Even if the international community agrees to sharp cuts in greenhouse gas emissions within this century, the *tipping point* may already have been reached for many small island nations. Simply put, the amount of anthropogenic CO_2 already emitted into the atmosphere ensures a magnitude of warming that will cause a disastrous rise in sea level in some localities. Removal of CO_2 from the atmosphere via natural processes is insufficient to head off this warming. In the most conservative scenario, the 2007 *IPCC Fourth Assessment Report* estimates that msl will be 0.2 to 0.6 m higher than now by the year 2100. However, this estimate is based primarily on thermal expansion of seawater and ignores the melting of ice sheets. At the other extreme, W.T. Pfeffer and colleagues of the Institute of Arctic and Alpine Research at the University of Colorado at Boulder include estimates of the contributions of glacier melt in their projection of a sea level rise of 0.8 to 2 m by 2100.

An example of an island nation that is particularly vulnerable to rising sea level is the Maldives, a chain of 1200 small flat islands near the middle of the Indian Ocean having a population of about 360,000 (Figure 15.1). Another example is Kiribati, consisting of three island groups located in the Pacific Ocean about half way between Hawaii and Fiji. Kiribati is home to about 100,000 people. The islands of both nations are 2 m or less above msl.

Adaptation to higher sea level is not a viable option for the people of the Maldives, Kiribati, and other low lying island nations; migration may be their only alternative. Mohamed Nasheed, president of the Maldives from November 2008 to February 2012, proposed to raise funds to help purchase land abroad for a new homeland for his people. Relocation would take place during the present century. Anote Tong, president of Kiribati since 2003, has a somewhat different plan for his threatened island nation. He is asking for help from nearby Australia and New Zealand to train Kiribati's younger people in skilled professions so that they might find jobs and new homes abroad.

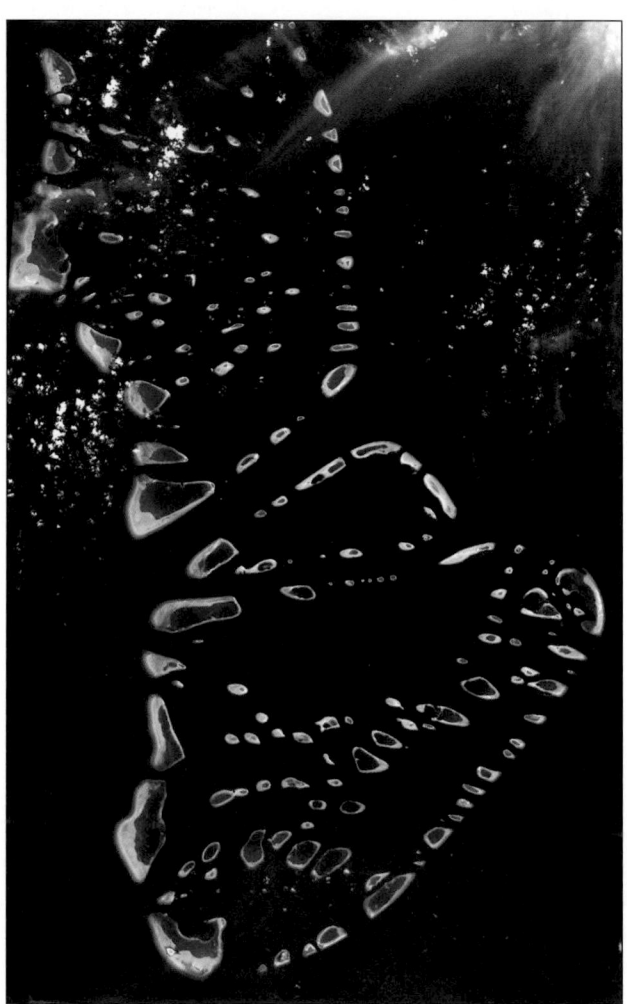

FIGURE 15.1
A small portion of the 1200 flat islands composing the Maldives. [NASA/GSFC/METI/ERSDAC/JAROS, and U.S./Japan ASTER Science Team]

This chapter covers some of the basics of Earth's climate system, climate variability, and climate change. We examine the instrument-based and reconstructed climate record with the objective of learning more about how climate changes through time. These lessons of the climate past are useful in establishing a perspective on the present climate and how climate might change in the future. Two of the most obvious lessons of the climate past are that climate changes over a broad range of temporal scales, from years to decades to centuries to millennia, and many forces working together are responsible for climate change.

Variability in incoming solar radiation, regular changes in Earth's orbit, volcanic eruptions, changes in Earth's surface and atmospheric properties, and human activities are among the many factors that contribute to climate change. With the continued build-up of atmospheric carbon dioxide and other greenhouse gases, global warming appears likely to continue at least through the present century. Scientists rely on global climate models to predict the magnitude of potential warming arising from an enhanced greenhouse effect. The impacts of global climate change on all sectors of society are likely to be significant.

Earth's Climate System

Climate is defined as the weather of some locality averaged over some time period plus extremes in weather. Climate must be specified for a particular place and time interval because, like weather, climate varies both spatially and temporally. Thus, for example, the climate of Minneapolis is different from that of Miami, and winters in Minneapolis were somewhat milder in recent decades than they were a half century or more ago. Extremes in weather are important aspects of the climate record because what has happened in the past could happen again. Hence, daily weather reports usually include the highest and lowest temperatures ever recorded for the date and climatic summaries typically identify such extremes as the coldest, warmest, driest, wettest, snowiest, or cloudiest month or year on record.

Climate is usually described quantitatively in terms of normals, means, and extremes of a variety of weather elements including temperature, precipitation, and wind. Climate summaries are available in tabular form for climatic divisions of each state and major cities along with a narrative description of local or regional climate. The U.S. National Weather Service is responsible for gathering the basic weather data used in the nation's climatological summaries. Data are processed, archived, and made available for users via NOAA's **National Climatic Data Center (NCDC)** in Asheville, NC.

THE CLIMATIC NORM

Traditionally, the **climatic norm** or normal is equated to the average value of some climatic element such as temperature or snowfall. By international convention, climatic norms are computed from averages of weather elements compiled over a standard 30-year period. In 1935, delegates to the International Meteorological Conference at Warsaw, Poland selected the 30-year period of 1901-1930 as the initial reference for calculating normals, with subsequent updating at the close of each decade. Current climate summaries are based on weather records from 1981 to 2010. Average July rainfall, for example, is the simple average of the total rainfall during each of 30 consecutive Julys from 1981 through 2010. The 30-year period is adjusted every 10 years by dropping the earliest decade and adding the latest one. In the United States, 30-year averages are computed for temperature, precipitation, and air pressure only. Averages of other climatic elements such as wind speed and cloudiness are derived from the entire period of record at a particular weather station. Extremes such as the highest temperature or driest month are also drawn from the entire period of record.

People find that averages and extremes in the climate record constitute a useful guide to future expectations. However, the use of climate norms has some limitations. *Normal* may be taken to imply that climate is static when, in fact, climate is inherently variable with time. Furthermore, normal may imply a Gaussian (bell-shaped) probability distribution, although many climate elements are non-Gaussian. For our purposes, we can think of the climatic norm of some locality as encompassing the total variability in the climate record, that is, averages plus extremes. This implies, for example, that an exceptionally cold winter may not be "abnormal" because its mean temperature may fall within the expected range of variability of winter temperature.

Many people assume that the mean value of some climatic element is the same as the median (middle value); that is, 50% of all cases are above the mean and 50% of all cases are below the mean. This is a reasonable assumption for air temperature, which approximates a simple Gaussian-type probability distribution. Hence, we might expect about half the Januarys to be warmer and half the Januarys to be colder than the 30-year mean January temperature. On the other hand, the distribution of some climatic elements, such as precipitation, is non-Gaussian, and the mean value is not likely to be the same as the median value. For example, in a dry climate subject to infrequent deluges of rain during the summer, considerably fewer than half the Julys are wetter than the mean and many more than half of Julys are drier than the mean. For many purposes, the median value of precipitation is a more useful description of climate than the mean value.

For some applications, selection of a 30-year period for averaging weather data may be inappropriate because climate varies over a broad range of time scales and can change significantly in periods much shorter than 30 years. For some purposes, a 30-year period provides a shortsighted view of the climate record. Compared with the long-term climate record, the current norm, for example, was unusually mild over much of the United States.

CLIMATIC ANOMALIES

Climatologists find it useful to compare the average weather of a specific week, month, or year with past climate records. Such comparisons carried out over a broad geographical region show that departures from long-term climatic averages, called **climatic anomalies**, do not occur with the same sign or magnitude everywhere. For example, as shown in Figure 15.2A, the average temperature during December 2011 was above the long-term average (*positive anomalies*) through much of the eastern two-thirds of the lower U.S., but below the long-term average (*negative anomalies*) in much of the western one-third. Furthermore, the magnitude of the anomaly, positive or negative, varied from one place to another.

The geographical patterns of precipitation anomalies are typically more complex than tempera-

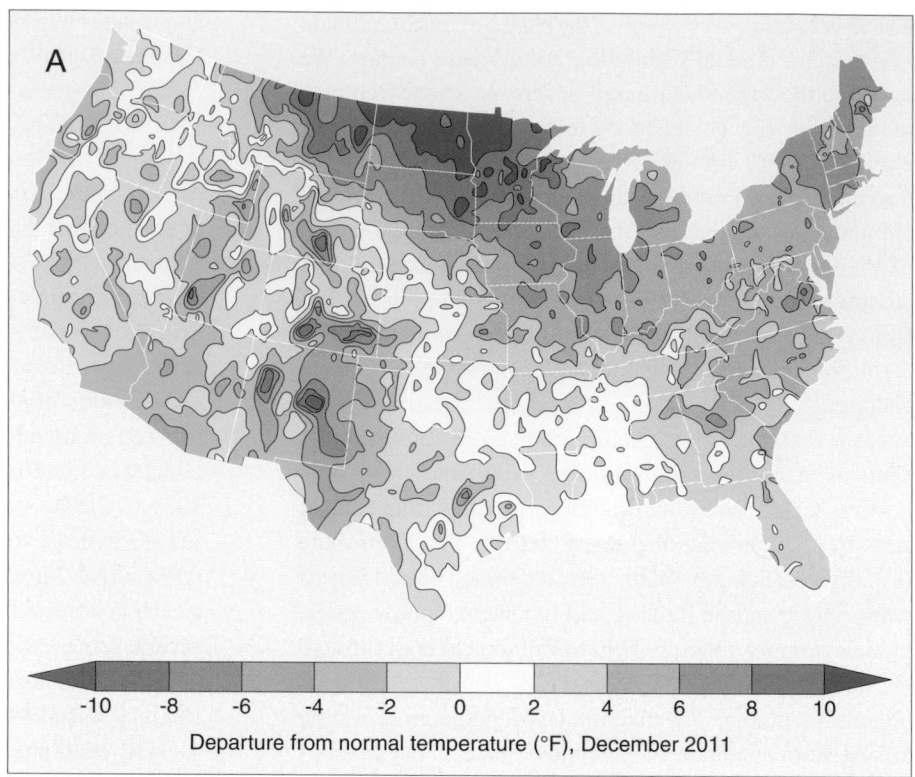

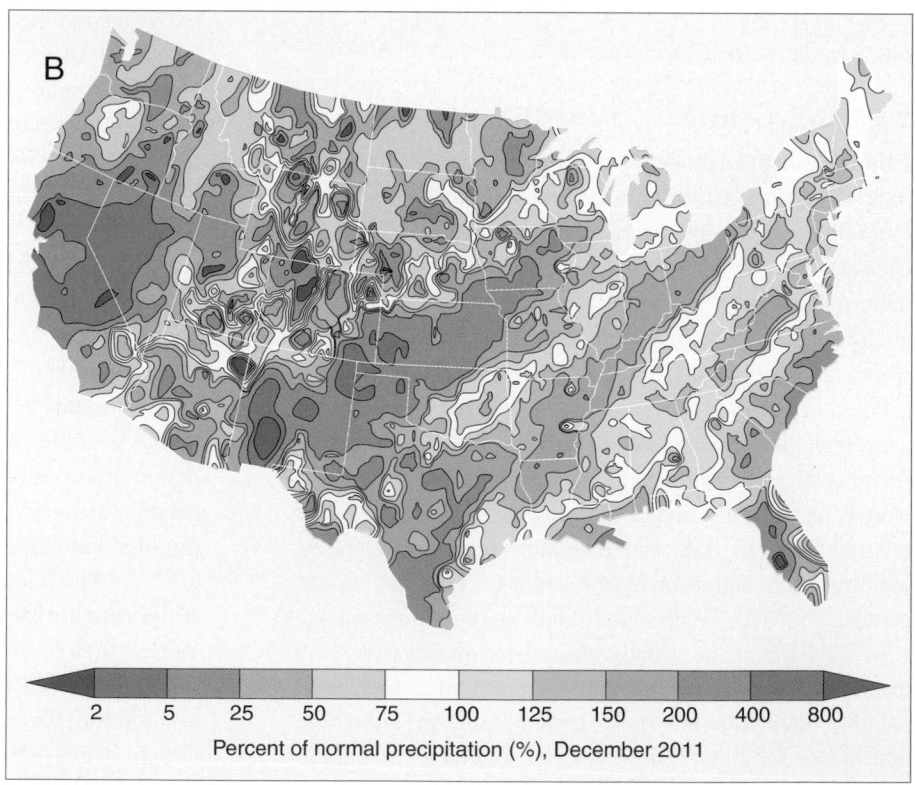

FIGURE 15.2

(A) Departure of average temperature from the long-term average (1981-2010) in Fahrenheit degrees for December 2011. (B) Percent of long-term average (1981-2010) precipitation (rain plus melted snow) for December 2011. [NOAA, Climate Prediction Center]

ture anomalies (Figure 15.2B). This is due to greater spatial differences in precipitation arising from the variability of storm tracks and the almost random distribution of convective showers. For these reasons, in spring and summer in middle latitudes, even adjoining counties may experience opposite rainfall anomalies (one with above-average rainfall and the other below-average rainfall). From an agricultural perspective, the geographic non-uniformity of climatic anomalies may be advantageous in that some compensation is implied. That is, poor growing weather and consequent low crop yields in one area may be compensated to some extent by better growing weather and increased crop yields elsewhere. **Agroclimatic compensation** generally applies to crops such as corn, soybeans, and other grains that are grown over broad geographical areas.

At middle and high latitudes, the geographic non-uniformity of climatic anomalies is linked to the prevailing westerly wave pattern that ultimately governs cold and warm air advection, cyclogenesis, and storm tracks (Chapter 9). Hence, the pattern of the westerlies determines the location of weather extremes such as drought or areas of exceptionally low temperature. In view of the number of westerly waves that typically encircle the hemisphere, a single weather extreme never occurs over an area as large as the United States; that is, severe cold or drought never grips the entire nation at the same time.

Geographic non-uniformity also applies to trends in climate. Hence, the trend in the average annual temperature of the Northern Hemisphere is not necessarily representative of all localities within the hemisphere. During the same period, some places experience cooling whereas other places experience warming regardless of the direction of the overall hemispheric (or global) temperature trend. Not only is it misleading to assume that the direction of large-scale climatic trends applies to all localities, but also it is erroneous to assume that the magnitude of climatic trends is the same everywhere. A small change in average hemispheric temperature typically translates into a much greater change in some areas and little or no change in other areas.

Climate Boundary Conditions

The climate of a locale or the entire planet is constrained by physical principles associated with the conservation of energy and mass (primarily water). The globe is a mosaic of many different climate types including, for example, the hot wet tropics, warm and cold deserts, temperate regions, and polar ice caps. This section summarizes the various boundary conditions operating in Earth's climate system.

Many factors working together shape the climate of any locality. Boundary conditions are imposed on climate due to latitude, elevation, topography, proximity to large bodies of water, Earth's surface characteristics, atmospheric composition, long-term average atmospheric circulation, and prevailing ocean circulation. On time scales extending from millions to hundreds of millions of years, all climate boundary conditions are variable. Continents have drifted to different latitudes, ocean basins have opened and closed, and mountain ranges have risen and eroded away—all with implications for climate. On shorter time scales (e.g., the range of human existence), for all practical purposes, the first four climate boundary conditions are fixed and exert regular and predictable influences on climate.

Seasonal changes in incoming solar radiation (solar altitude and length of daylight) vary with latitude and Earth's surface temperature responds to those regular variations. Air temperature drops with increasing elevation and determines whether precipitation falls as rain or snow. Topography can affect the distribution of clouds and precipitation so that the windward slopes of high mountain barriers (facing the oncoming wind) usually are wetter than the leeward slopes (facing downwind). The relatively great thermal inertia of large bodies of water (especially the ocean) moderates the temperature of downwind localities, reducing the temperature contrast between summer and winter (Chapter 4). Earth's surface characteristics (e.g., ocean versus land, vegetative cover, semi-permanent snow and ice cover) influence the amount of incident solar radiation that is absorbed and converted to heat and how that heat is used (e.g., raising air temperature, evaporating water, melting snow and ice).

Atmospheric circulation encompasses the combined influence of weather systems operating at all spatial and temporal scales ranging from sea breezes to the prevailing wind belts that encircle the planet. Although strongly influenced by the other boundary conditions, atmospheric circulation is considerably less regular and less predictable than the others. This variability is especially evident in weather systems such as thunderstorms and hurricanes that are smaller than the planetary scale. Planetary-scale circulation systems (e.g., prevailing winds, subtropical anticyclones), exert a more systematic influence on climate, determining for example, where precipitation is seasonal and the location of the major biomes, such as subtropical deserts.

The ocean is a major player in Earth's climate system operating on time scales of days to millennia and spatial scales from local to global. The ocean influences radiational heating and cooling of the planet. Covering

FIGURE 15.3
The ocean is a major player in Earth's climate system in terms of storage and exchange of heat, water, and greenhouse gases.

about 71% of Earth's surface, the ocean is a primary control of how much solar radiation is absorbed (converted to heat) at Earth's surface. Also, the ocean is the main source of atmospheric water vapor and is a major regulator of the atmospheric concentration of the greenhouse gas carbon dioxide (Figure 15.3).

On an annual average, the ocean absorbs about 92% of the solar radiation striking its surface; the balance is reflected. Most of this absorption takes place within about 200 m (650 ft) of the ocean surface with the depth of penetration of sunlight limited by the amount of suspended particles and discoloration caused by dissolved substances. On the other hand, at high latitudes highly reflective multi-year pack ice greatly reduces the amount of solar radiation absorbed by the ocean. The snow-covered surface of sea ice absorbs only about 15% of incident solar radiation and reflects away the rest. At present, multi-year pack ice covers about 7% of the ocean surface with greater coverage in the Arctic Ocean than the Southern Ocean (mostly in Antarctica's Weddell Sea). The Arctic is an ocean surrounded by continents whereas the Antarctic is a continent roughly centered on the pole and surrounded by ocean. Without an Antarctic continent, there would be considerably more sea-ice coverage in the Southern Ocean than is now the case.

The ocean influences the planetary energy budget not only by radiational heating and cooling, but also by contributing to the non-radiative latent and sensible heat fluxes at the air-sea interface. Heat is transferred from Earth's surface to the atmosphere via latent heating (vaporization of water at the surface followed by cloud formation in the atmosphere) and sensible heating (conduction plus convection). On a global average annual basis, about ten times more heat is transferred from the ocean surface to the atmosphere via latent heating than sensible heating.

The ocean and atmosphere are closely coupled. This coupling is most apparent in the ocean's surface waters where temperatures and wind-driven currents respond to variations in atmospheric conditions within hours to days. On the other hand, the deeper basin-scale thermohaline circulation responds more sluggishly to changes in atmospheric conditions, taking decades to centuries or longer to fully adjust. In turn, ocean currents strongly influence climate. Cold surface currents, such as the California Current, are heat sinks; they chill and stabilize the overlying air, increasing the frequency of sea fogs and reducing the likelihood of thunderstorms. Relatively warm surface currents, such as the Gulf Stream, are sources of heat and moisture for the overlying air, destabilizing the lower troposphere and energizing storm systems. Ocean surface currents and thermohaline circulation transport heat from the tropics to higher latitudes (Chapter 4).

Global Climate Patterns

How the various boundary conditions interact to shape the climates of the continents will become clearer as we summarize the basic patterns of climate on Earth. From a global perspective, climate exhibits some regular patterns. As we examine these patterns, keep in mind that they may be significantly modified by local and regional climate boundary conditions.

TEMPERATURE

Ignoring the influence of mountainous terrain on air temperature, mean annual isotherms roughly parallel latitude circles, underscoring the influence of solar radiation (solar altitude and day length) on climate. The latitude of highest mean annual surface temperature, the **heat equator**, is located about 10 degrees north of the geographical equator. Mean annual isotherms are symmetrical with respect to the heat equator, decreasing in magnitude toward the poles.

The heat equator is in the Northern Hemisphere because overall that hemisphere is warmer than the Southern Hemisphere for several reasons. For one, the polar regions of the two hemispheres have different radiational characteristics so that the Arctic is warmer than the Antarctic. Most of the Antarctic continent is ice and snow covered, so the surface has a very high albedo for solar radiation and is the site of intense radiational cooling, especially during the long polar night. By contrast, the Northern Hemisphere polar region is mostly ocean. Although the Arctic Ocean is usually ice covered, patches of open water develop in summer and lower the overall surface albedo. A second factor contributing to the relative warmth of the Northern Hemisphere is the greater fraction of land in that hemisphere's tropical latitudes. Because land surfaces warm more than water surfaces in response to the same incoming solar radiation, tropical latitudes are warmer in the Northern Hemisphere than in the Southern Hemisphere. A third contributing factor is ocean circulation which transports more warm water to the Northern Hemisphere than to the Southern Hemisphere.

Systematic patterns also appear when we consider the worldwide distribution of mean January temperature (Figure 15.4) and mean July temperature (Figure 15.5); January and July are usually the coldest/warmest months of the year in the respective hemisphere. Neglecting the influence of topography on air temperature, isotherms tend to parallel latitude circles. However, monthly isotherms exhibit some notable north-to-south bends, primarily because of land/sea contrasts and the influence of ocean currents. As the year progresses from January to July and to January again, isotherms in both hemispheres follow the Sun, shifting north and south in tandem. The latitudinal (north-south) shift in isotherms is greater over the continents than over the ocean; that is, the annual range in air temperature is greater over land than over the sea. Furthermore, the north-south mean temperature gradient is greater in the winter hemisphere than in the summer hemisphere, the result of greater north-south differences in incident solar radiation in which higher latitudes experience little or no daylight during fall and winter, while the tropics continue to receive a relatively large influx of solar radiation. A steeper temperature gradient means a more vigorous circulation and stormier weather in the winter hemisphere.

PRECIPITATION

The global pattern of mean annual precipitation (rain plus melted snow) exhibits considerable spatial variability (Figure 15.6). Some of this variability can be attributed to topography and the distribution of land and sea, but the planetary-scale circulation is also important. The intertropical convergence zone (ITCZ), subtropical anticyclones, and prevailing wind belts impose a roughly zonal signature on precipitation distribution. In addition, regular shifts in these circulation features through the year are responsible for the seasonality of precipitation that characterizes the climate of many localities.

In tropical latitudes, convective activity associated with intense solar heating and the trade wind convergence triggers abundant rainfall year-round. In the adjacent belt poleward to about 20 degrees latitude, rainfall depends on seasonal shifts of the ITCZ and the subtropical anticyclones. Shift of the ITCZ toward the pole causes summer rains, whereas shift of the subtropical highs toward the equator brings a dry winter. This climatic zone includes the belt of tropical monsoon winds, described in Chapter 9.

Poleward of this belt, from about 20 to 35 degrees N and S, subtropical anticyclones, centered over the ocean basins, dominate the climate all year. Subsiding dry air on the anticyclones' eastern flanks is responsible for Earth's major subtropical deserts (e.g., the Sahara). On the other hand, unstable humid air on the western flanks of subtropical anticyclones causes relatively moist conditions (e.g., over the southeastern U.S.). Between about 35 and 40 degrees latitude, the prevailing westerlies and subtropical anticyclones govern precipitation. Typically, on the western side of continents, winter cyclones migrating with the westerlies bring moist weather, but in summer, westerlies shift poleward and the area lies under the dry eastern flank of a subtropical anticyclone. Hence, summers are dry. At the same latitudes, but on the eastern side of continents, the climate is dominated by westerlies in winter and the moist airflow on the western flank of a subtropical anticyclone in summer. Thus, rainfall is triggered by cyclonic activity in winter and by convection in summer, resulting in little seasonal variability in monthly precipitation totals.

Precipitation generally declines poleward of about 40 degrees latitude as lower temperatures reduce the amount of precipitable water (Chapter 6). Although precipitation is generally not seasonal in the continental interior, the tendency is for more precipitation in summer. The summer precipitation maximum is due to higher air temperatures, greater precipitable water, and more vigorous convection at that time of year.

Our description of the global pattern of annual precipitation is somewhat idealistic and requires some qualification. Land/sea distribution and topography complicate the generally zonal distribution of precipitation. While more rain falls over the ocean than over the continents, as pointed out in Chapter 6, more precipitation falls on land than evaporates from it whereas less precipitation falls on the ocean than evaporates from it. Mountain belts induce wet windward slopes and extensive leeward rain shadows. Furthermore, annual precipitation totals fail to convey some other important aspects of precipitation, including the average daily rainfall and the season-to-season and year-to-year reliability of precipitation. As a rule, rainfall is most reliable in maritime climates, less reliable in continental localities, and least reliable in arid regions. However, drought is possible anywhere, even in maritime climates.

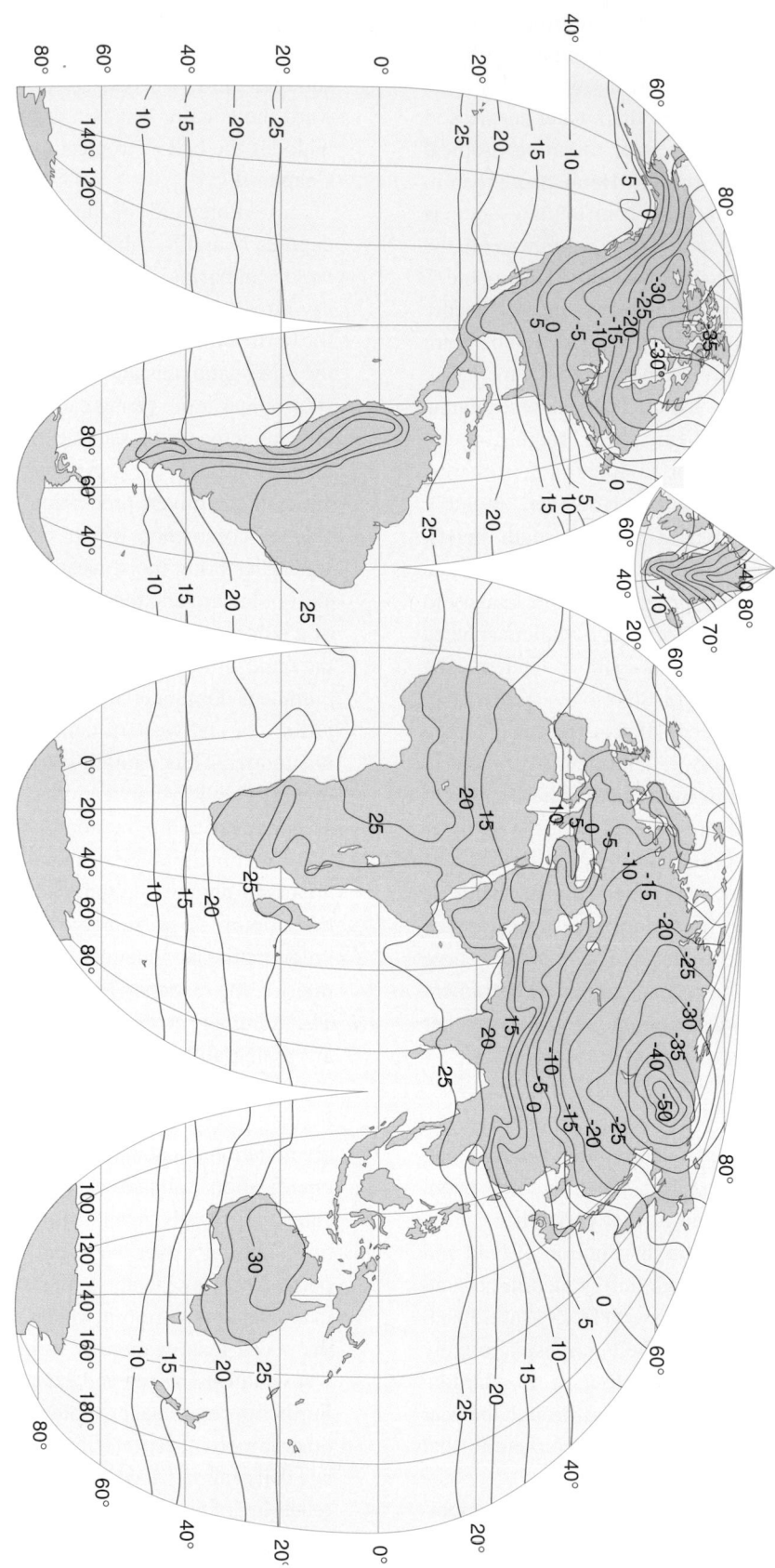

FIGURE 15.4
Mean surface air temperature for January in degrees Celsius.

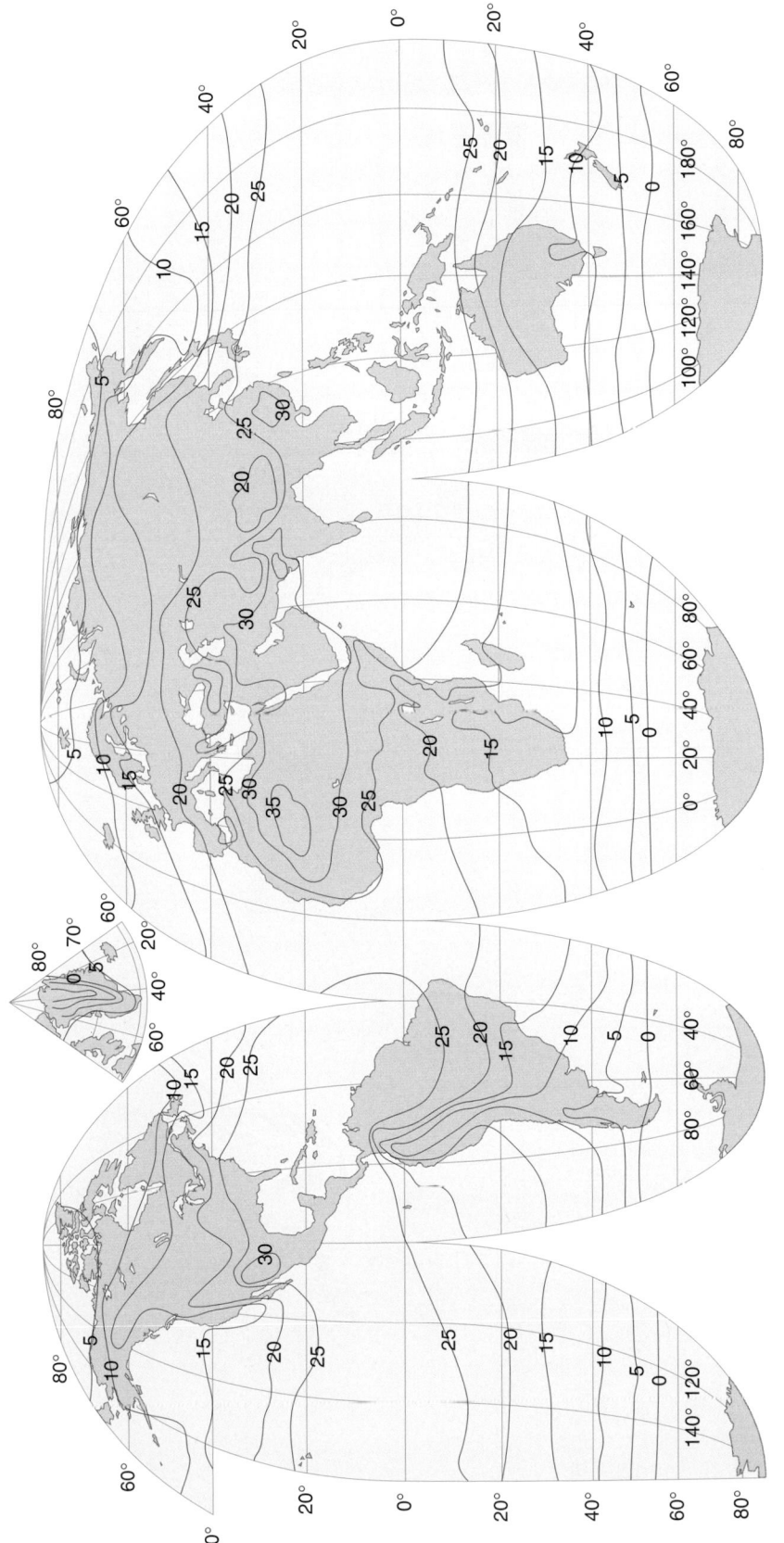

FIGURE 15.5
Mean surface air temperature for July in degrees Celsius.

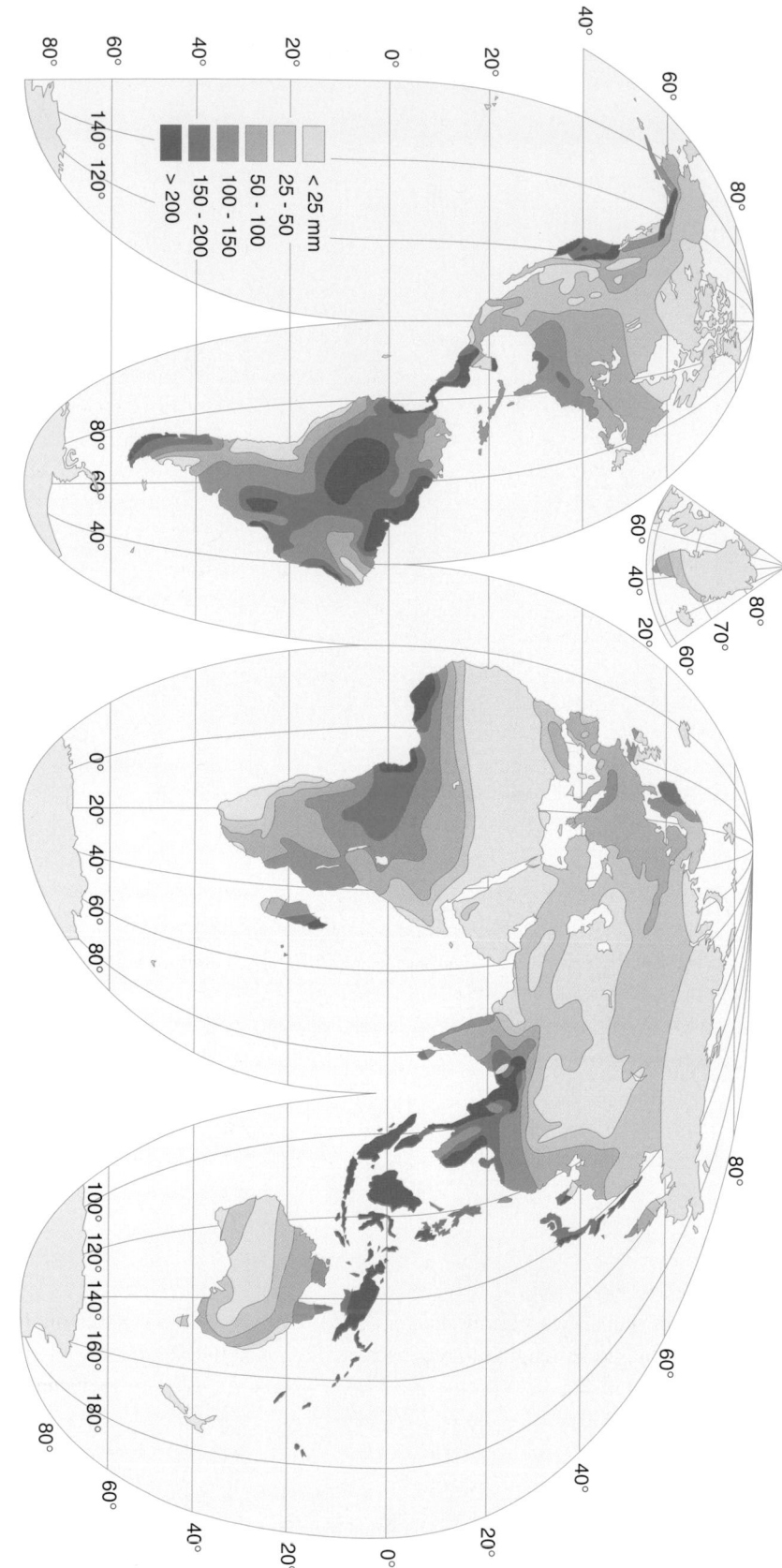

FIGURE 15.6
Mean annual precipitation (rain plus melted snow) in millimeters (mm).

CLIMATE CLASSIFICATION

For more than a century, climatologists have attempted to organize Earth's myriad of climate types by devising classification schemes that group together climates having common characteristics. Classification schemes typically group climates according to the meteorological basis of climate, or the environmental effects of climate. The first is a genetic classification that asks why climate types occur where they do. The second is an empirical classification that infers the type of climate from such environmental indicators as the distribution of indigenous vegetation. In addition, the advent of computers and databases has made possible numerical climate classification schemes that utilize sophisticated statistical techniques. Appendix III provides information on one of the more popular climate classification schemes.

The Climate Record

Climate varies not only from one place to another but also with time. Examining the record of the climate past reveals some basic understandings regarding the nature of climate behavior and provides a useful perspective on current and future climate. In most places, however, the reliable instrument-based record of past weather and climate is limited to not much more than the past century or so. For information on earlier fluctuations in climate, scientists rely on reconstructions of climate based on historical documents (e.g., personal diaries, ship's logs) and longer-term climate-sensitive geological and biological evidence such as fossil plants and animals, pollen profiles, tree growth rings, glacial ice cores, and deep-sea sediment cores. (Refer to this chapter's first *For Further Exploration* for information on how past climatic data are derived from some of these non-instrument sources.) This section summarizes major features of Earth's climate record from geologic time through to the present.

GEOLOGIC TIME

Throughout most of the 4.5 billion years that constitute geologic time, which is divided based on large-scale geological events into eons, eras, periods, and epochs (Figure 15.7), global climate is challenging to reconstruct. Lengthy gaps in proxy climate records, problems dating specific events, and limitations in correlating events in widely separated locations make describing early climate difficult. Furthermore, the movement of tectonic plates complicates climate reconstruction that spans hundreds of millions of years. Nonetheless, the available evidence supports some general conclusions regarding the climate over geologic time.

According to the theory of **plate tectonics**, the solid outer skin of the planet is divided into a dozen massive rigid plates (and many smaller ones) that are slowly driven (typically less than 20 cm per year) across the face of the globe moving continents and opening and closing ocean basins. Mountain building and most volcanic activity occur along plate boundaries. This model originated around 1910 when the German meteorologist Alfred Wegener (1880-1930) and the American geologist Frank B. Taylor (1860-1939) independently proposed **continental drift**, the hypothesis that the continents move over the surface of the planet. Wegener proposed that the supercontinent *Pangaea* split into *Laurasia* (encompassing present day North America and Eurasia) and *Gondwanaland* (encompassing South America, Africa, India, Australia, and Antarctica). He based his idea on the observed close fit between continental margins, similarities in fossil plants and animals, and the continuity of rock formations and mountain ranges between continents on either side of the Atlantic. The scientific community was slow in accepting the theory of plate tectonics until the 1960s when the assemblage of supporting geological evidence became convincing.

Through the span of human existence, topography and the geographical distribution of the ocean and continents can be considered as essentially fixed boundary conditions in Earth's climate system. This was not the case over the vast expanse of geologic time. Plate tectonics has probably operated on the planet for at least 2 billion years and explains such seemingly anomalous finds as glacial deposits in the Sahara Desert, fossil tropical plants in Greenland, and fossil coral in Wisconsin (Figure 15.8). These discoveries reflect climate conditions millions of years ago when landmasses were situated at different latitudes than they are today (Figure 15.9). Plate tectonics also affected global climate by altering the course of heat-transporting surface ocean currents and the thermohaline circulation.

About 570 million years ago as the Proterozoic eon neared its end, just before the Phaerozoic eon, the planet experienced extreme climate fluctuations. Along Namibia's Skeleton Coast, in southwest Africa on the Atlantic, are layers of rock that formed in tropical seas amid layers of glacial deposits, indicating abrupt changes in climate between heat and cold. According to the *snowball Earth hypothesis*, during as many as four cold episodes, each lasting 10 million years, the continents were encased in glacial ice and the ocean was frozen to a depth of more than 1000 m (3300 ft). At the close of each cold episode, temperatures rose rapidly, and within only a few centuries, all the ice melted.

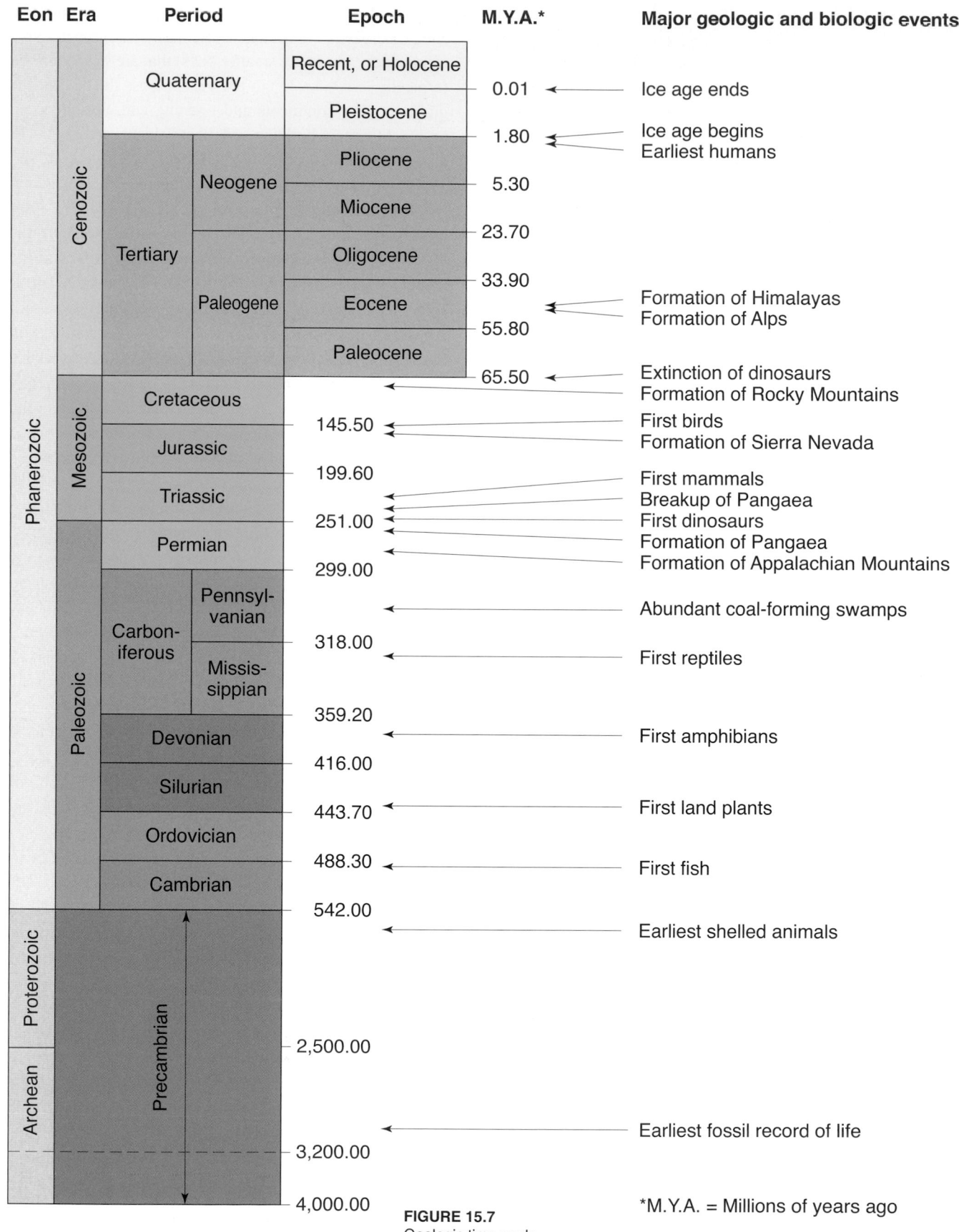

FIGURE 15.7
Geologic time scale.

*M.Y.A. = Millions of years ago

FIGURE 15.8
This bedrock exposed in northeastern Wisconsin contains fossil coral that dates from nearly 400 million years ago. Based on the environmental requirements of modern coral, scientists conclude that 400 million years ago, Wisconsin's climate was tropical marine. Plate tectonics can explain such a drastic change between ancient and modern conditions.

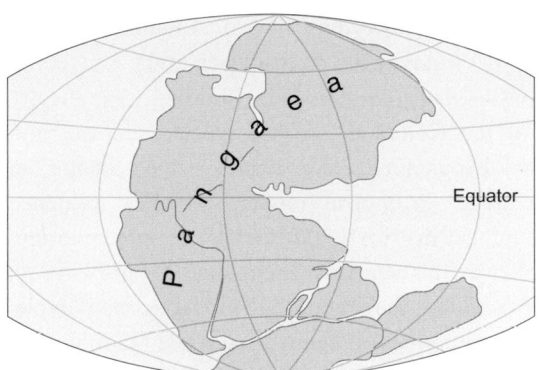

Permian - 260 m.y.a.

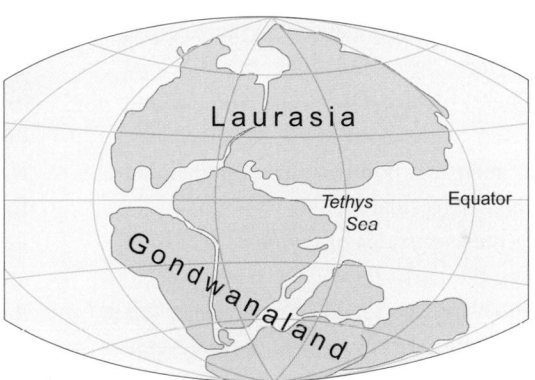

Triassic - 210 m.y.a

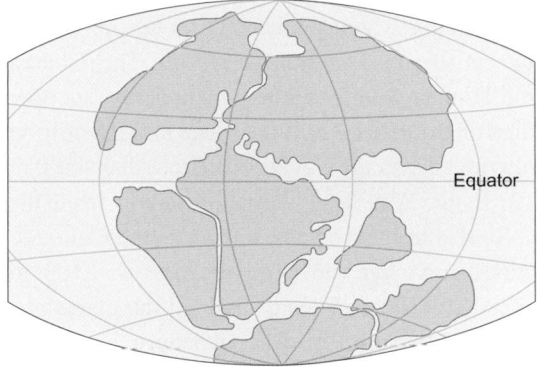

Jurassic - 150 m.y.a.

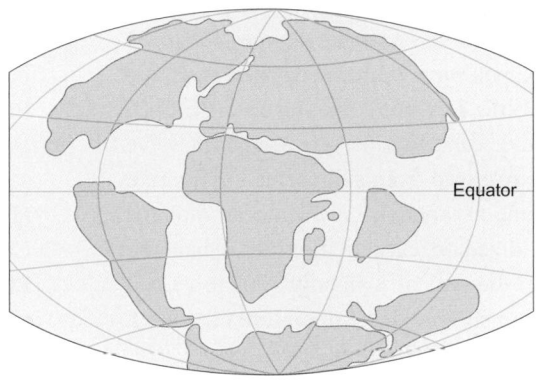

Cretaceous - 66 m.y.a.

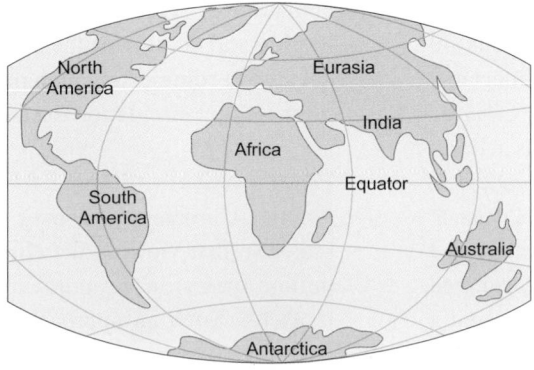

Present day

FIGURE 15.9
About 200 million years ago, the super-continent Pangaea began to split apart into separate continents that slowly drifted apart. Ocean basins opened and eventually the continents reached their present positions. Continental drift, a manifestation of plate tectonics, is responsible for climate changes operating over hundreds of millions of years. [U.S. Geological Survey]

The Mesozoic era, 251 million to 65.5 million years ago, was characterized by a generally warm Earth free of large glacial ice sheets. Between the Triassic and Jurassic periods, the first and second of the Mesozoic era, the global mean temperature rose 3 to 4 Celsius degrees (5.5 to 7 Fahrenheit degrees), contributing to a major extinction or displacement of many animal and plant species. At peak warming during the Cretaceous period, the last of this era, the global mean temperature was 6 to 8 Celsius degrees (11 to 14 Fahrenheit degrees) higher than now. Subtropical plants and animals lived as far north as 60 degrees N, the present polar region, and dinosaurs roamed what is now the North Slope of Alaska. A major mass extinction at the end of the Cretaceous period, about 65.5 million years ago, was likely caused by an asteroid impact on Earth's surface, throwing huge quantities of dust into the atmosphere, blocking sunlight, and causing cooling that contributed to the demise of the dinosaurs. Without competition, mammal populations exploded and diversified as the atmosphere cleared.

With the warmest episodes of the past 80 million years occurring during the early Paleogene period (45 to 65.5 million years ago), and the glacial climate intervals of the Pleistocene Ice Age beginning 1.8 million years ago, drastic climate fluctuations have characterized the present Cenozoic era. About 56 million years ago, near the transition between the Paleocene and Eocene epochs, deep ocean temperatures rose 6 to 7 Celsius degrees (11 to 13 Fahrenheit degrees), making a warm planet even warmer. This **Paleocene-Eocene Thermal Maximum (PETM)**, spanning 170 thousand years, was likely caused by massive amounts of methane released from submarine gas hydrate deposits. The methane escaped from the ocean into the atmosphere and oxidized to carbon dioxide, enhancing the greenhouse effect dramatically. In only 1000 to 10,000 years, the global mean surface air temperature climbed 5 to 10 Celsius degrees (9 to 18 Fahrenheit degrees).

Deep-sea sediment cores from the Southern Ocean indicate a significant increase in CO_2 levels during the middle Eocene epoch, after the PETM about 40 million years ago, that enhanced the greenhouse effect and interrupted a cooling trend. The *Middle Eocene Climatic Optimum (MECO)* was among the hottest intervals in Earth history. The MECO may be linked to tectonic processes. Following the break-up of the supercontinent Pangaea, the tectonic plate carrying the Indian subcontinent moved northward until, about 50 million years ago, it slammed into Asia. Prior to this collision, its approach triggered a million years of volcanic activity along Asia's southern border. Before this lava erupted, it incorporated carbon

and other elements from melted sea floor sediments. This increased atmospheric CO_2 to more than 1000 ppmv, raising temperatures worldwide. When the plates finally collided (forming the Himalayans), the supply of carbonate sediments was cut off. At the same time, weathering and erosion of rock exposed on the Indian subcontinent acted as a sink for atmospheric CO_2, decreasing its concentration to 300 ppmv by 30 million years ago. The consequent weakening of the greenhouse effect was responsible for long-term cooling.

Earth's climate not only became cooler but also drier and more variable, setting the stage for the Pleistocene Ice Age, an epoch of numerous major glacial advances and recessions that began 1.8 million years ago and ended 10,500 years ago. According to W.F. Ruddiman of the University of Virginia and J.E. Kutzbach of the University of Wisconsin-Madison, mountain building, specifically the rise of the Colorado Plateau, Tibetan Plateau, and Himalayan Mountains, may explain this change in Earth's climate. Prominent mountain ranges influence the geographical distribution of clouds and precipitation, and can alter the planetary-scale circulation. Furthermore, mountain building may alter the global carbon cycle. Enhanced weathering of bedrock exposed in mountain ranges sequesters more atmospheric carbon dioxide in sediments thereby weakening the greenhouse effect.

Although mountain building began about 40 million years ago, about half of the total uplift took place between 10 and 5 million years ago. The Tibetan Plateau and Himalayan Mountains of southern Asia now cover an area of more than 2 million square km (0.8 million square mi) with an average elevation of more than 4500 m (14,700 ft). In the American West, the region from the California Sierras to the Rockies, known as the Colorado Plateau, has an average elevation of 1500 to 2500 m (5000 to 8200 ft). These plateaus diverted the planetary-scale westerlies into a more meridional pattern, increasing the north-south exchange of air masses and altering the climate over a broad region of the globe. Also, seasonal heating and cooling of the plateaus cause low pressure to develop in summer and high pressure in winter, enhancing the monsoon circulation over southern Asia (Chapter 9).

PAST TWO MILLION YEARS

Climate varies over a broad spectrum of time scales so that viewing the climate record of the past two million years in progressively narrower time frames is useful. Such an approach resolves the oscillations of climate into more detailed fluctuations especially over the recent past. Over the past two million years, plate tectonics was not a major factor in climate change. For practical purposes, mountain ranges, continents, and ocean basins were essentially as they are today. Compared to the climate that prevailed through most of geologic time, the climate of the last two million years was unusual in favoring the development of huge glacial ice sheets (although evidence also exists of ice ages earlier in geologic time). During much of Earth's history, the average global temperature may have been 10 Celsius degrees (18 Fahrenheit degrees) higher than it was over the past two million years.

During the Pleistocene Ice Age, the climate shifted numerous times between glacial climates and interglacial climates. A **glacial climate** favors the thickening and expansion of glaciers whereas an **interglacial climate** favors the thinning and retreat of existing glaciers or no glaciers at all. During major glacial climatic episodes of the Pleistocene, the Laurentide ice sheet developed over central Canada and spread westward to the Rocky Mountains, eastward to the Atlantic Ocean, and southward over the northern tier states of the United States (Figure 15.10). At the same time, mountain glaciers in the Rockies coalesced into the Cordilleran ice sheet, a relatively thin ice sheet covered the Arctic Archipelago, and an ice sheet much smaller than the Laurentide developed over northwest Europe including the British Isles and Scandinavia. The vast quantity of water locked up in these ice sheets caused sea level to drop by 113 to 135 m (370 to 443 ft), exposing portions of the continental shelf, including a land bridge linking Siberia and North America. The Laurentide and European ice sheets thinned and retreated, and may even have disappeared entirely, during relatively mild interglacial climatic episodes, which typically lasted about 10,000 years. Throughout these interglacials, glacial ice persisted over most of Antarctica and Greenland, as it still does today.

During glacial climatic episodes, temperatures were lower than they are today but the magnitude of cooling was not the same everywhere. A variety of geologic evidence indicates that during the Pleistocene, temperature fluctuations between major glacial and interglacial climatic episodes typically amounted to as much as 5 Celsius degrees (9 Fahrenheit degrees) in the tropics, 6 to 8 Celsius degrees (11 to 14 Fahrenheit degrees) at middle latitudes, and 10 Celsius degrees (18 Fahrenheit degrees) or more at high latitudes. An increase in the magnitude of a temperature change with increasing latitude is known as **polar amplification**, indicating that polar areas are subject to greater changes in climate.

Oxygen isotope analysis of deep-sea sediment cores reveals numerous fluctuations in glacial ice volume between major glacial and interglacial climatic episodes over the past 600,000 years (Figure 15.11A). Shifting focus to the past 160,000 years, resolution of the climate record improves. The temperature curve in Figure 15.11B is based on analysis of an ice core extracted from the Antarctic ice sheet at Vostok. A relatively mild interglacial episode, referred to as the *Eemian*, began about 127,000 years ago and persisted for about 7000 years. In some localities, temperatures may have been 1 to 2 Celsius degrees (2 to 4 Fahrenheit degrees) higher than during the warmest portion of the present interglacial. The Eemian interglacial was followed by numerous fluctuations between glacial and interglacial climatic episodes. The last major glacial climatic episode began about 27,000 years ago and reached its peak about 20,000 to 18,000 years ago when glacial ice cover over North America was about as extensive as it had ever been.

Ice cores from both Greenland and Antarctica reveal an approximately 100,000 year Ice Age cycle consisting of cold glacial climatic episodes sandwiched between mild interglacial climatic episodes. Perhaps 16 of these long-term cycles operated over the span of the Pleistocene Epoch. As noted later in this chapter, evidence from deep-sea sediment cores indicates that regular variations in Earth-Sun geometry (the Milankovitch cycles) drive this approximately 100,000-year glacial/interglacial cycle.

Greenland and Antarctic ice-core records correlate well both in terms of magnitude of temperature

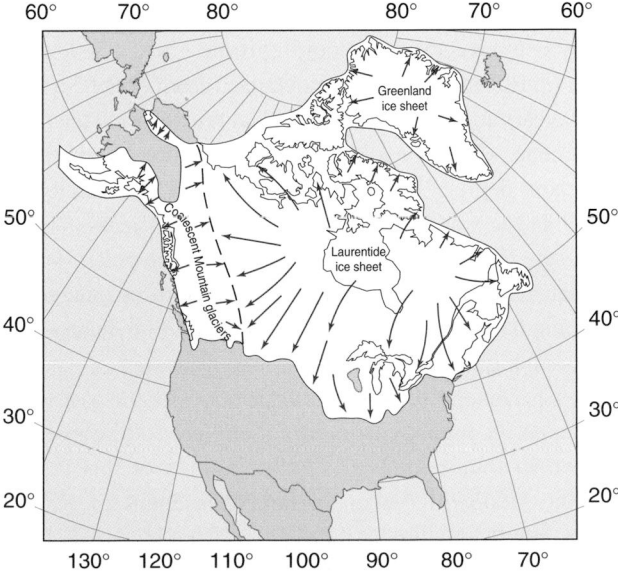

FIGURE 15.10
The extent of glacial ice cover over North America about 20,000 to 18,000 years ago, the time of the last glacial maximum.

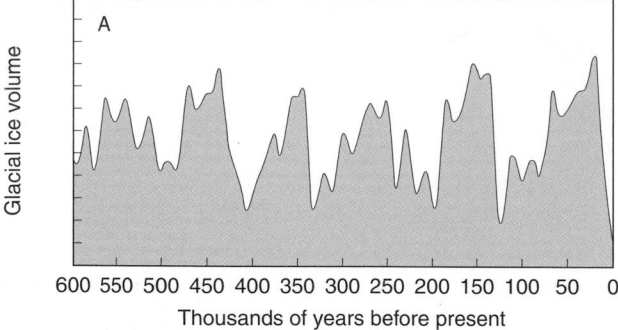

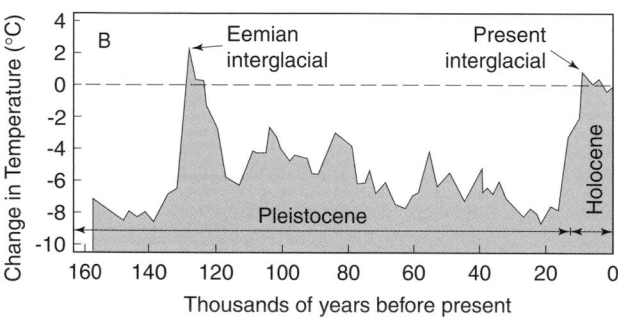

FIGURE 15.11
(A) Variation in global glacial ice volume from the present back to about 600 million years ago based on analysis of oxygen isotope ratio in deep-sea sediment cores. (B) Temperature variation over the past 160,000 years based on oxygen isotope analysis of an ice core extracted from the Antarctic ice sheet at Vostok and expressed as a departure in Celsius degrees from the 1900 mean global temperature. [Compiled by R.S. Bradley and J.A. Eddy from J. Jousel et al., *Nature* 329(1987):403-408 and reported in *EarthQuest* 5, No. 1 (1991)]

change and the timing of events suggesting that the 100,000-year Ice Age cycles were globally synchronous. However, comparison of the Greenland and Antarctic ice core data over the most recent Ice Age cycle (i.e., from about 142,000 years ago to 10,500 years ago) reveals marked differences between the Southern and Northern Hemispheres. Whereas the Antarctic record is reasonably smooth and "calm," the Greenland record shows numerous abrupt and drastic flip-flops between glacial and interglacial climatic episodes. Temperatures fluctuated by as much as 7 Celsius degrees (12.6 Fahrenheit degrees) over periods of decades or less (in some cases in only 3 years.) These abrupt temperature changes, having two basic periods of 2000 to 3000 years and 7000 to 12,000 years, occurred during the final (Wisconsinan) stage of the Pleistocene but not during the subsequent Holocene epoch.

The most likely explanation for these short-term changes in temperature is the alternate weakening and strengthening of the ocean's meridional overturning circulation (Chapter 4). This may explain, for example, the occurrence of the relatively cool episode from 11,000 to 10,000 years ago, known as the **Younger Dryas** (named for the polar wildflower *Dryas octopetala* that reappeared in portions of Europe at the time). The return of glacial climatic conditions triggered short-lived re-advances of remnant ice sheets in North America, Scotland, and Scandinavia.

The Younger Dryas began abruptly when glacial ice lobes disrupted drainage patterns, diverting meltwater from the Mississippi River into the St. Lawrence River and North Atlantic. With this input of fresh water, North Atlantic surface waters became less saline and eventually were not sufficiently dense to sink. This weakened the meridional overturning circulation (MOC), which in turn diminished the warm water flowing into the central and northern North Atlantic, causing a marked cooling of the surrounding lands. The Younger Dryas ended just as abruptly as it began when the input of fresh water into the North Atlantic decreased and the MOC strengthened.

Glacial ice finally withdrew from the Great Lakes region of North America about 10,500 years ago ushering in the present interglacial, the **Holocene epoch**. Although the Laurentide ice sheet was melting and disappeared almost entirely 5500 years ago, the Holocene was an epoch of spatially and temporally variable temperature and precipitation. Cores extracted from the Greenland ice sheet and sediment cores taken from the bottom of the North Atlantic Ocean reveal that the overall post-glacial warming trend was interrupted by abrupt millennial-scale fluctuations in climate. Post-glacial warming during the Holocene gave way to a cold episode about 8200 years ago; significant cooling also occurred from about 3100 to 2400 years ago. On the other hand, at times during the mid-Holocene, mean annual global temperature was perhaps 1 Celsius degree (2 Fahrenheit degrees) higher than it was in 1900, the warmest in more than 110,000 years, that is, since the Eemian interglacial. According to a pollen-based climate reconstruction, July mean temperatures some 6000 years ago were about 2 Celsius degrees (3.6 Fahrenheit degrees) higher than now over most of Europe.

Generalized temperature curves for the past 1300 years, derived mostly from proxy climate data sources, is shown in Figure 15.12. Notable features of this record are the **Medieval Warm Period**, from about CE 950 to 1250, and the cooling that followed, from about CE 1400 to 1900, a period now known as the **Little Ice Age**. The Medieval Warm Period and the Little Ice Age were not episodes of sustained warming and cooling, respectively. On the contrary, sediment and glacial ice core records plus

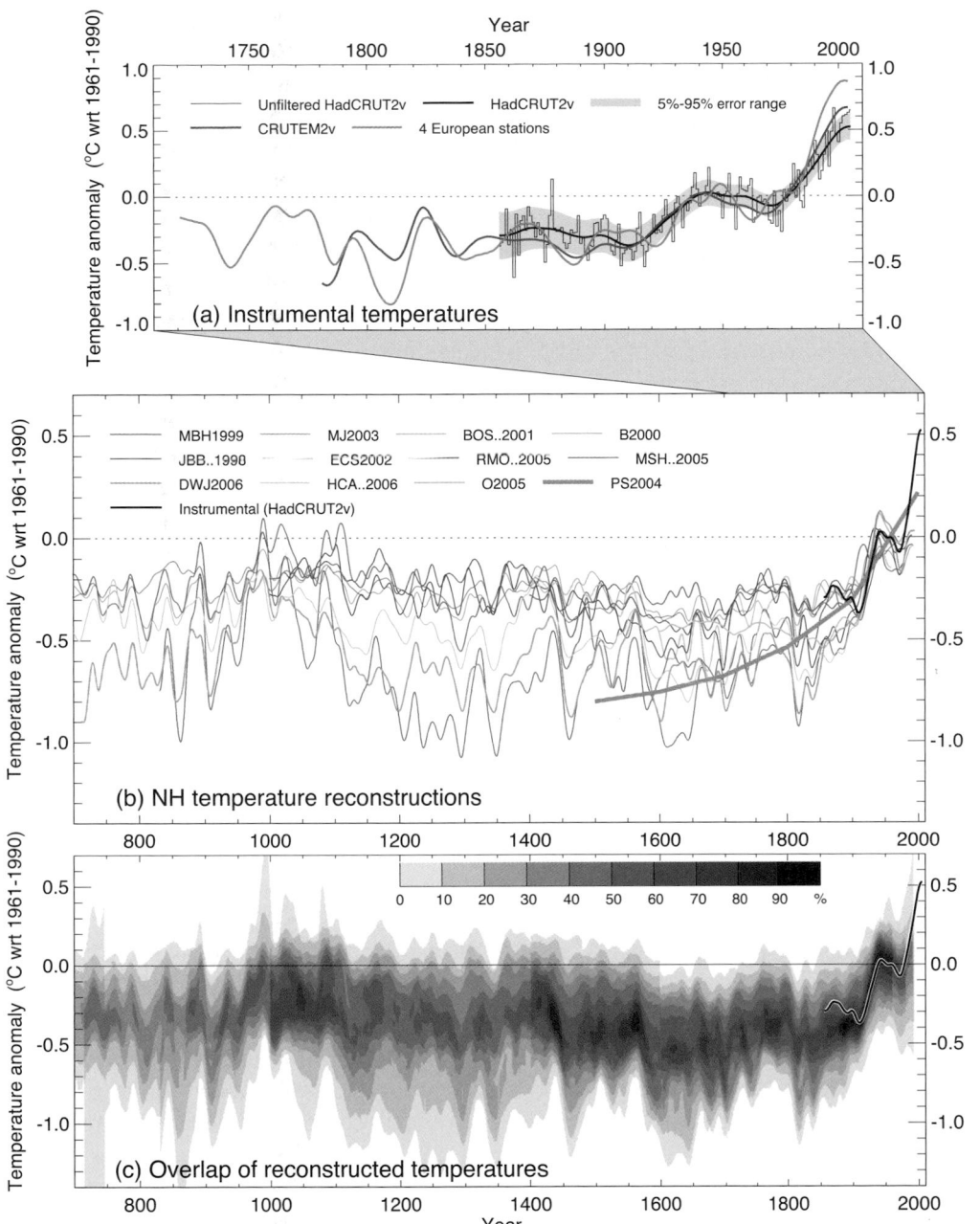

FIGURE 15.12
Records of Northern Hemisphere temperature variation during the last 1300 years. These reconstructions use multiple climate proxy records and an instrumental temperature record. All series have been smoothed with a Gaussian-weighted filter to remove fluctuations on time scales less than 30 years. All temperatures represent departures from the 1961 to 1990 mean (in Celsius degrees). [*Climate Change 2007: The Physical Science Basis*, Working Group I Contribution to the Fourth Assessment Report of the Report of the Intergovernmental Panel on Climate Change, Figure 6.10, Cambridge University Press, U.K., 2007]

historical documents point to decadal-scale fluctuations in temperature and precipitation. During much of the Medieval Warm Period, global temperature averaged about 0.5 Celsius degree (0.9 Fahrenheit degree) higher than in 1900. The first Norse settlements were established along the southern coast of Greenland and vineyards thrived in the British Isles. Independent lines of evidence confirm that the Little Ice Age was a relatively cool period in many regions with mean annual global temperatures perhaps 0.5 Celsius degree (0.9 Fahrenheit degrees) lower than it was in 1900. Sea-ice cover expanded, mountain glaciers advanced, growing seasons shortened, and erratic harvests

caused much hardship for many people including the end of the Norse settlements in Greenland. Note the dramatic warming trend underway during the 20th century and first part of the 21st century.

INSTRUMENT-BASED CLIMATE TRENDS

Invention of weather instruments, establishment of weather observing networks around the world, plus standardized methods of observation and record-keeping made the climate record much more detailed and dependable. The most reliable temperature records date from the late 1800s with the birth of national weather services, including those of the U.S. and Canada, along with the predecessor to today's *World Meteorological Organization (WMO)*. Examination of temperature trends over the past 120 years or so is instructive as to climate variability and climate change.

In the late 1990s, climatologist Michael Mann and his colleagues at Pennsylvania State University combined a dozen Northern Hemisphere temperature records covering the past 1000 years. Most of the combined records were derived from proxy climate data sources such as tree growth rings, whereas the most recent segment of the record also included instrument-based measurements. Because of its shape, the temperature curve was referred to as a "hockey stick." For most of the 1000-year period, the temperature trended gently downward (the handle) and then sharply upward into the 21st century (the blade). From this, Mann and his colleagues concluded that the final decades of the 20th century were likely warmer than any prior comparable period in the past 1000 years. This conclusion is consistent with an ice-core derived temperature record extracted from a glacier at 7163 m (23,500 ft) in the Himalayan Mountains. Based on its review of research conducted by Mann and colleagues, in 2006 a National Research Council (NRC) panel generally concurred with the "hockey stick" model although expressing less confidence in the early part of the record.

In Chapter 3 of the text, Figure 3.39 plots the 1880 to 2010 instrument-derived record of variations in (1) global (land plus sea-surface) mean annual temperature, (2) global mean sea-surface temperature, and (3) global mean land surface temperature. In all three cases, the temperature is expressed as a departure (in Celsius degrees and Fahrenheit degrees) from the long-term century (1901-2010) average. For reasons given in Chapter 4, these temperature series, assembled by NOAA's *National Climatic Data Center (NCDC)*, indicate greater year-to-year variability over land than ocean. The trend in global mean temperature is initially downward and then back upward from 1880 until about 1940, downward or steady from 1940 to about 1970, and upward again through the 1990s and early 2000s. The overall temperature fluctuation is mostly ±0.5 Celsius degree (±0.9 Fahrenheit degree) about the century average. Note that this temperature record for the globe as a whole is not representative (in direction or magnitude) of all locations worldwide; that is, the trend was amplified or reversed, or both, in specific regions of the world.

In 2011, NOAA reported that 2010 was the 34th consecutive year in which global temperatures were above the 1901-2000 average. Combined global land and ocean annual surface temperatures in 2010 tied with 2005 as the warmest, at 0.62 ± 0.07 Celsius degrees above the 20th century average, since the instrumental record began in 1880. As shown in Figure 15.12, most of the warming of the 20th century took place from 1956 to 2000 (0.65 ± 0.15 Celsius degrees). NASA data shows that the climate has warmed by about 0.2 Celsius degrees from the late 1970s to 2010. Climate reconstructions (mostly in the Northern Hemisphere) indicate that the 20th century was the warmest in 1000 years. For example, the ice core record extracted from a glacier at 7163 m (23,500 ft) in the Himalayan Mountains revealed that the 1990s and the last half of the 20th century were the warmest of any equivalent periods in the past 1000 years.

A general consensus in the scientific community holds that an overall global-scale warming trend has prevailed since the end of the Little Ice Age. The most reasonable scientific explanation for the observed warming trend is the steady build-up of carbon dioxide (CO_2) and other infrared-absorbing gases in the atmosphere (primarily because of combustion of fossil fuels) and the consequent enhancement of the natural greenhouse effect (Chapter 3). Continuation of the upward trend in the concentration of greenhouse gases inevitably will lead to global warming throughout this century and perhaps beyond. For a summary of the lessons of the climate record, see this chapter's second *For Further Exploration*.

Causes of Climate Variability and Change

There is no single, simple explanation for why climate changes. The complex spectrum of climate variability and change is a response to the interactions of many agents and forcing mechanisms both internal and external to the Earth's climate system (Figure 15.13). In addition, positive and negative feedback can play important roles in the magnitude of climate change and variability.

One way to organize our thinking on the many possible causes of climate change is to match a possible cause (forcing agent or mechanism) with a specific climatic oscillation (or response), based on similar periods of oscillation (Figure 15.14). With plate tectonics, for example, atmospheric and oceanic circulation patterns change in response to continental drift, the opening and closing of ocean basins, and mountain building. Hence, plate tectonics explains long-term climate changes operating over hundreds of millions of years. As well, systematic changes in Earth's orbit about the Sun affect

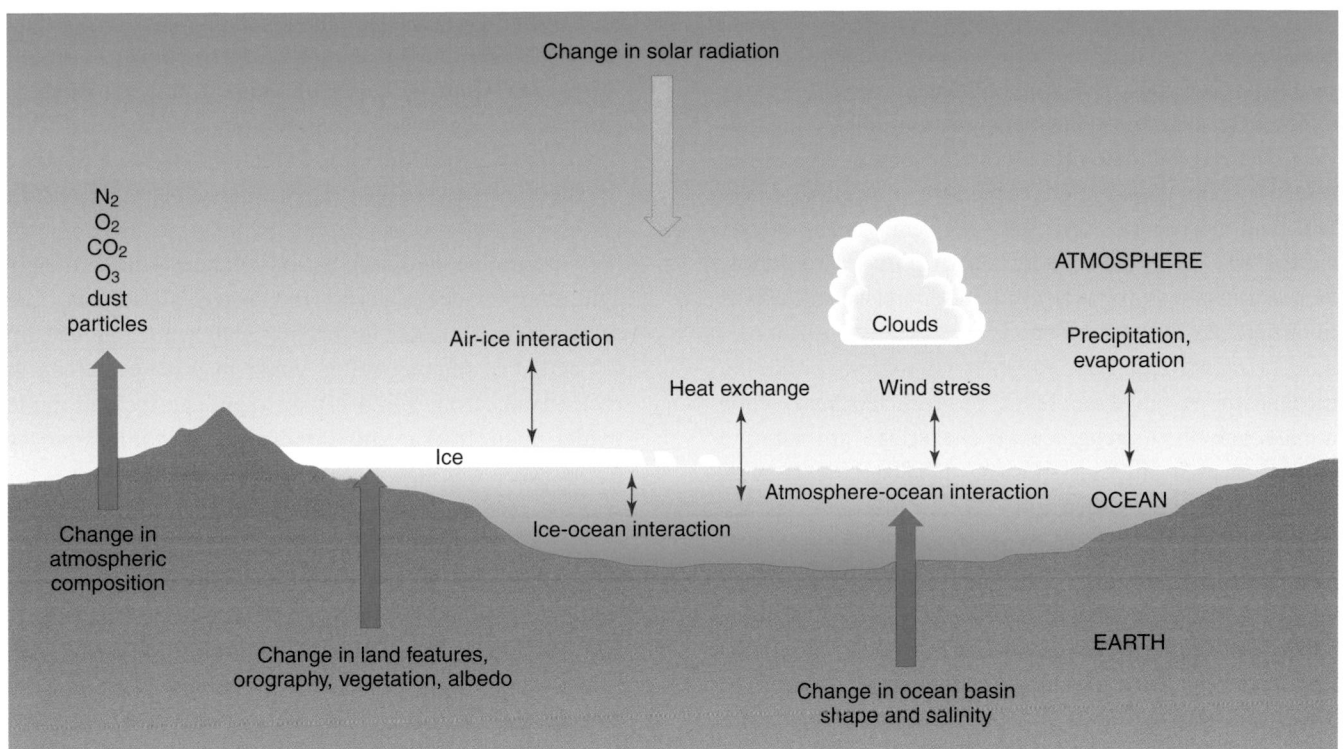

FIGURE 15.13
The complex spectrum of climate variability and climate change is a response to the interactions of many forcing mechanisms that operate both internal and external to the Earth-atmosphere-ocean system. [Modified from W.L. Gates and Y. Mintz, *Understanding Climatic Change: A Program for Action*, 1975, National Academy of Sciences, National Academy Press, Washington, DC]

the latitudinal and seasonal distribution of incoming solar radiation, accounting for climate shifts of 10,000 to 100,000 years. Variations in the number of sunspots and energy from the Sun may be associated with climate fluctuations of decades to centuries. Explosive volcanic eruptions, El Niño, or La Niña may account for interannual climate variability. But matching a forcing mechanism with a climate response based on a similar periodicity is no guarantee of a real physical relationship.

Another way to think about the possible causes of climate change is in terms of the global energy budget.

With **global radiative equilibrium**, energy entering the Earth-atmosphere system (i.e., absorbed solar radiation) ultimately must equal energy leaving the system (i.e., infrared radiation emitted to space). Any change in energy input, thus affecting energy output, would shift Earth's planetary system to a new equilibrium and alter the planet's climate. In Chapter 9, we described the influence of ENSO on inter-annual climate variability. In this section, we summarize the many other factors that may contribute to climate change and variability over a broad spectrum of time scales.

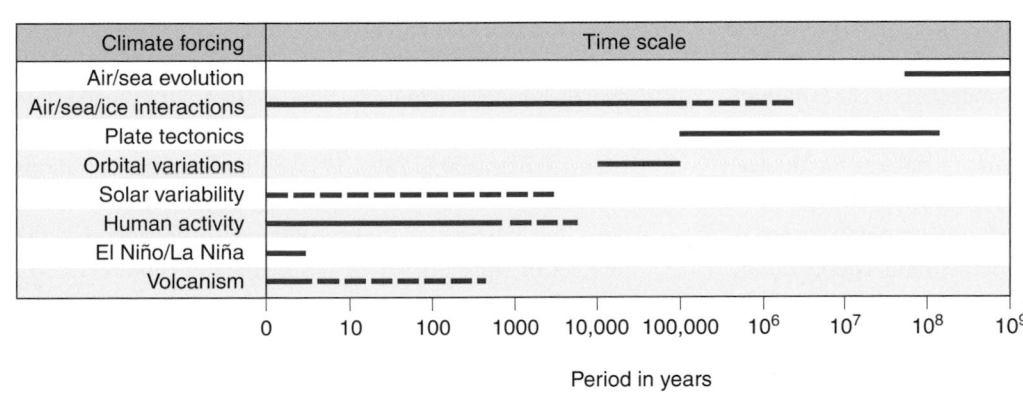

FIGURE 15.14
The various causes of climate change operate over a range of time scales.

In discussing the possible causes of climate change, we must take into account positive and negative feedback processes. In Earth's climate system, **feedback** refers to the sequence of interactions among climate controls (or variables) that determines how the system responds to disturbance (or perturbation) of boundary conditions. Many feedback mechanisms or processes operate in the climate system and can be positive or negative. *Positive feedback* amplifies whereas *negative feedback* diminishes the effects of changes in climate forcing.

Feedback that enhances an initial warming mechanism is an example of positive feedback while a feedback that strengthens (i.e., further promotes) a cooling mechanism is also positive. Feedback which tends to reduce the process that caused it is negative. Feedback that works counter to a warming process is negative, as is a feedback that reduces the effect of a cooling mechanism so that less cooling takes place.

In summary, feedback which strengthens the original forcing mechanism, regardless of whether in a positive or negative direction, is a positive feedback. On the other hand, feedback which weakens the original forcing mechanism, regardless of whether it is in a positive or negative direction, is negative. For realistic predictions of the climate future, climate models must account for feedback in Earth's climate system. When the climate is altered, feedback will either amplify (positive feedback) or dampen (negative feedback) that change. While climate forcing agents and mechanisms drive climate change, feedback controls the magnitude of climate change.

SOLAR VARIABILITY

Fluctuations in the Sun's energy output, sunspots, or regular variations in Earth's orbital parameters are external factors in Earth's climate system that can alter Earth's climate. Satellite measurements since the 1980s confirmed that the Sun's total energy output at all wavelengths (total *solar irradiance*) is not constant. Numerical global climate models predict that a 1% change in the Sun's energy output could significantly alter the mean temperature of Earth's planetary system.

According to standard models of stellar evolution, stars such as the Sun gradually become brighter (increasing luminosity) during their lifetime. Changes in nuclear fusion reactions within the Sun's interior are responsible for this steady increase in solar energy output (Chapter 3). On this basis, 4 billion years ago the Sun's energy output was about 70% of what it is today. With the same level of greenhouse gases in the atmosphere as today, Earth's average surface air temperature would have been about 25 Celsius degrees (45 Fahrenheit degrees) lower than now, so that water on Earth's surface would have been frozen and would remain frozen until 1 or 2 billion years ago. As first proposed by astronomers Carl Sagen (1934-1996) and George Mullen in 1972, this creates a paradox for there is evidence that liquid water flowed on Earth's surface from very early in the planet's history. Some of the oldest rocks on Earth, dating to about 4 billion years ago, show evidence of flowing water. These rocks are sedimentary and contain pebbles that were rounded by the action of running water. Other features preserved in rocks dating from 3 to 4 billion years ago include ripple marks, mud cracks, and fossil algae, all indicative of liquid water.

Higher than expected temperatures at Earth's surface is explained by a greenhouse effect that was stronger than at present. Early on, Earth's atmosphere contained very little free oxygen (O_2). In the presence of oxygen, methane (CH_4) breaks down to carbon dioxide (CO_2), so that we can assume that, in the absence of oxygen, the atmospheric concentration of methane, a very efficient greenhouse gas, was much higher than at present. Furthermore, volcanic activity elevated the atmosphere's concentration of carbon dioxide. Hence, warming due to a stronger greenhouse effect partially offset the climatic impact of a Sun that was 30% fainter than now so Earth's surface was warm enough for water to exist in the liquid phase.

Changes in solar energy output are related to the number of sunspots. A **sunspot** is an irregularly shaped dark blotch on the face of the Sun, typically thousands of kilometers across (Figure 15.15). Sunspots develop where an intense magnetic field suppresses the flow of gases transporting heat energy from the Sun's interior. A sunspot appears dark because its temperature is 400 to 1800 Celsius degrees (720 to 3240 Fahrenheit degrees) lower than the temperature of the surrounding surface of the Sun, the *photosphere*. At least as early as BCE 28, Chinese astronomers observed sunspots with the unaided eye by viewing the Sun's reflection on the surface of a quiet pond. Galileo Galilei (1564-1642) is credited with being among the first to study sunspots telescopically in 1610, and thereafter sunspots became objects of considerable scientific interest. In 1843, the German astronomer Samuel Heinrich Schwabe (1789-1875) reported regular variations in sunspot activity. A sunspot typically lasts only a few days, but the rate of sunspot generation is such that the number of sunspots varies systematically (Figure 15.16). Successive sunspot maxima or minima averages about 11 years with a range of 10 to 12 years. Also, an approximate 22-year oscillation (*double sunspot cycle*) characterizes

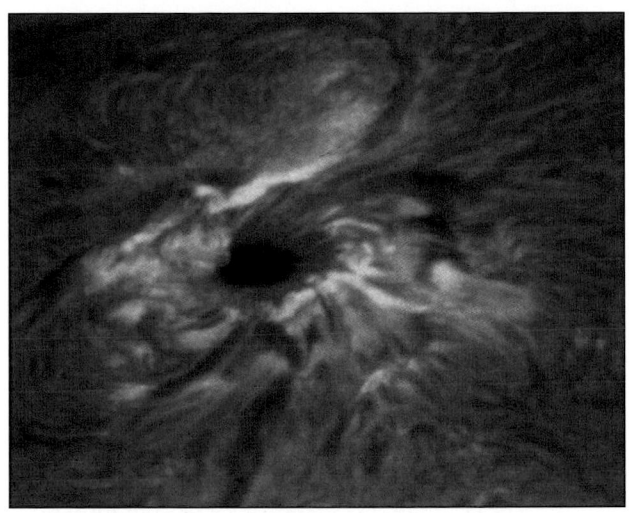

FIGURE 15.15
A sunspot is a dark blotch that appears on the sun's photosphere, typically thousands of kilometers across, that develops where an intense magnetic field suppresses the flow of gases transporting heat from the Sun's interior. [Courtesy of Donat G. Wentzel, University of Maryland and the National Optical Astronomical Observatories]

the strong magnetic field associated with sunspots. For example, the sunspot number reached a minimum in 1996, a maximum in 2001, and a minimum in late 2008. The next solar maximum is expected in 2013.

Satellite monitoring reveals that the Sun's energy output varies directly with sunspot number. When the Sun is slightly brighter, it has more sunspots whereas when it is slightly dimmer, sunspots are fewer in number. A brighter Sun is associated with more sunspots because of

a concurrent increase in bright areas, known as *faculae*, which appear near sunspots on the photosphere. When faculae dominate sunspots, the Sun brightens. More sunspots may contribute to a warmer global climate here on Earth and fewer sunspots may contribute to a colder global climate. However, the total solar energy output varies less than 0.1% through an 11-year sunspot cycle, much less than that expected to cause climate change. Also, most of the variation is in the ultraviolet (UV) portion of the solar spectrum.

In the late 1880s, the German astronomer F.W. Gustav Spörer (1822-1895) and the English solar astronomer E. Walter Maunder (1851-1928) reported that sunspot activity was greatly diminished during the 55-year period from 1645 to 1700, which coincided with a cold episode in Europe. This period of greatly reduced solar activity is now referred to as the **Maunder minimum**. Until the American astronomer John A. Eddy (1931-2009) reinvestigated the phenomenon, the scientific community largely ignored the Maunder minimum. Eddy pointed out that the Maunder minimum plus a prior 90-year period of reduced sunspot number, called the *Spörer minimum* (1460 to 1550), occurred about the same time as relatively cold phases of the Little Ice Age in Western Europe. Furthermore, the Medieval Warm Period was an interval of heightened sunspot activity from 1100 to 1250.

The connections between sunspots and climate fluctuations are not universally accepted in the scientific community. Cold episodes occurred in Europe just prior to and after the Maunder minimum, and these relatively cool conditions did not persist through the Maunder minimum,

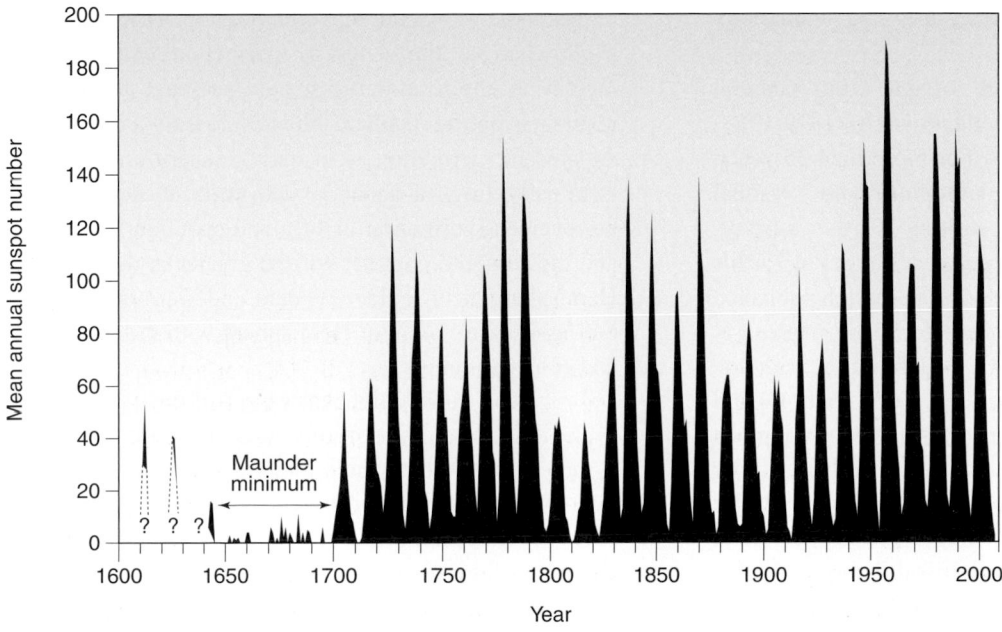

FIGURE 15.16
Variation in mean annual sunspot number since early in the 17th century. [National Geophysical Data Center]

nor were they global events. Furthermore, the variation in the solar radiation output during an 11-year sunspot cycle may be too weak to significantly alter Earth's climate. Still, mechanisms operating within the atmosphere could amplify changes in total solar output, which could make the slight brightening and dimming of the Sun an important player in Earth's climate system.

EARTH'S ORBIT

In 1842, the French mathematician Joseph Alphonse Adhémar (1797-1862) proposed that regular variations in the shape (eccentricity) of Earth's orbit about the Sun could explain the climate changes that paced the glacial fluctuations of the Pleistocene Ice Age. Subsequently, Adhémar's ideas were expanded upon by James Croll and later by Milutin Milankovitch.

Over twenty years later, James Croll (1821-1890), a self-educated Scottish scientist, attributed the climate changes responsible for the Ice Age to fluctuations in incoming solar radiation that accompany regular changes in Earth's orbital parameters (eccentricity of the orbit, tilt of the rotational axis, and precession of the axis). On this basis, Croll worked out an Ice Age chronology, arguing that less incoming solar radiation in winter favored greater accumulations of snow and the more extensive snow cover would further chill the atmosphere (*positive feedback*), eventually culminating in glaciation. Croll predicted multiple glaciations, an idea that was later confirmed through field work by the American geologist Thomas C. Chamberlain (1843-1928). However, by the close of the 19th century, Croll's theory fell into disfavor when European geologists discovered a serious discrepancy between his astronomically-based Ice Age chronology and field evidence.

In the second decade of the 20th century, Serbian astronomer Milutin Milankovitch (1879-1958) revived interest in Croll's work. For more than 25 years, Milankovitch calculated the latitudinal and seasonal variations in solar radiation striking Earth's surface, which arise from the long-term regular changes in Earth's three orbital parameters. In 1938, Milankovitch published radiation curves for latitudes from 55 to 65 degrees N for the past 600,000 years. His collaboration with the German climatologist Wladimir Köppen (1846-1940) and meteorologist Alfred Wegener (1880-1930) convinced Milankovitch that reduced solar radiation during summer (and not winter as Croll had proposed) at high latitudes was key to initiating glaciation. In the 1930s and 1940s, support for Milankovitch's astronomical theory of the Ice Age was considerable, especially in Europe. However, in

a dramatic turnabout, by the mid-1950s Milankovitch's views were rejected by most geologists. It was not until the mid-1970s that independent corroborative evidence from deep-sea sediment cores firmly established the Milankovitch cycles as the pacemaker of the major climatic fluctuations of the Pleistocene epoch.

Milankovitch cycles are regular variations in the eccentricity of Earth's orbit about the Sun, and the tilt and precession of Earth's rotational axis (Figure 15.17). These changes in Earth-Sun geometry are caused by gravitational influences exerted on Earth by other large planets, the Moon, and the Sun. Combined, Milankovitch cycles drive climate fluctuations operating over tens of thousands to hundreds of thousands of years (Figure 15.18).

The eccentricity (shape) of Earth's elliptical orbit about the Sun shifts from relatively high (elongated oval) to low (nearly circular) during cycles of approximately 100,000 and 400,000 years. Variation in orbital eccentricity alters the distance between Earth and Sun at *perihelion*, when Earth is closest to the Sun, and *aphelion*, when Earth is farthest from the Sun. When Earth's orbit over a year is highly elliptical, Earth intercepts less solar radiation than when the orbit is less elliptical.

The obliquity (tilt) of Earth's spin axis shifts from 22.1 degrees to 24.5 degrees from the perpendicular to the plane of its orbit and back to 22.1 degrees over a period of 41,000 years. As the axial tilt increases, winters become colder and summers become warmer in both hemispheres. Earth's axial tilt has been decreasing for about 10,000 years and will continue to do so for the next 10,000 years. (Presently, the tilt is 23.5 degrees.)

Over a period of 23,000 years, Earth's spin axis circles like the wobble of a spinning top. This precession cycle changes the dates of perihelion and aphelion, increasing the summer-to-winter seasonal temperature contrast in one hemisphere while decreasing it in the other. At present, perihelion is in early January and aphelion is in early July. In about 10,000 years, those dates will be reversed (perihelion in July and aphelion in January) and the seasonal contrast will be greater in the Northern Hemisphere, with colder winters and warmer summers, but less in the Southern Hemisphere with milder winters and cooler summers.

The most significant aspect of the Milankovitch cycles is not the amount of solar energy the Earth's planetary system receives, but where and when it arrives, creating long-term variations at different latitudes throughout the seasons. However, the amplitude of glacial/interglacial climate cycles is not explained exclusively by the relatively small variations in solar radiation associated with the

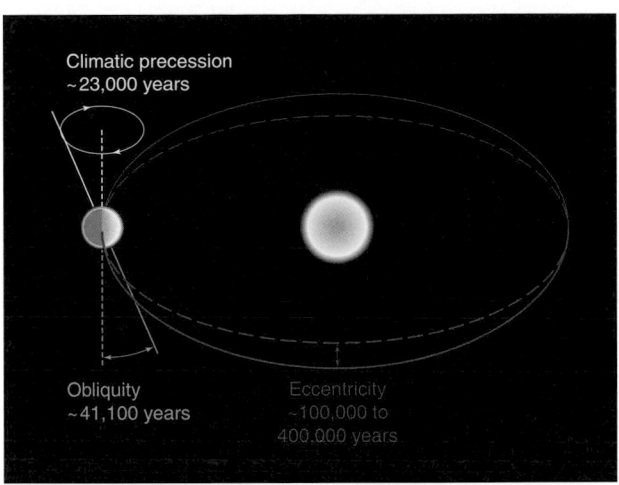

FIGURE 15.17
Milankovitch cycles likely explain the large-scale fluctuations of glacial ice cover during the Pleistocene Epoch. Note that diagram greatly exaggerates changes in Earth-Sun geometry.

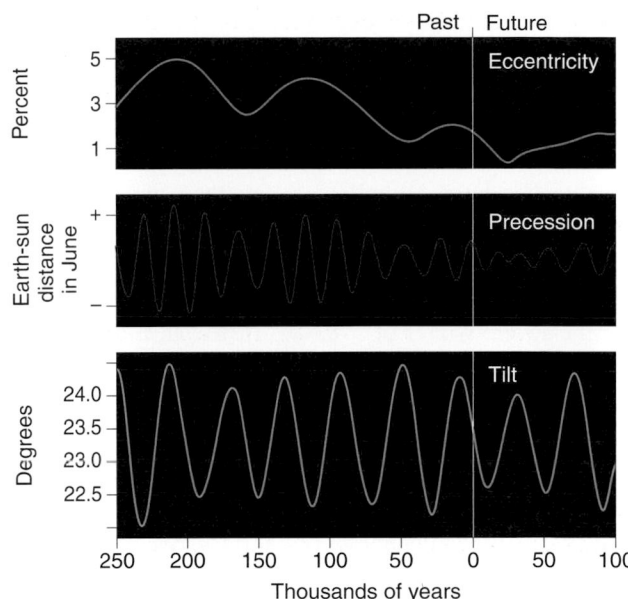

FIGURE 15.18
Past and future variations in the Milankovitch cycles.

Milankovitch cycles. Feedback processes operate within Earth's climate system amplifying the orbital variations with a major player being the greenhouse gas carbon dioxide. Glacial ice cores from Antarctica, spanning the past 650,000 years, reveal that the concentration of atmospheric CO_2 varies inversely with changes in the volume of glacial ice. The concentration is higher (260 to 280 ppmv) during interglacial episodes, and lower (200 ppmv) during glacial episodes. During interglacial episodes, rising levels of CO_2 adds to the warming caused by orbital variations, whereas during glacial episodes, falling levels of CO_2 contributes to the cooling caused by orbital variations.

Based on the three orbital cycles, Milankovitch developed a numerical model that calculated incoming solar radiation and the corresponding surface temperature, by latitude for the 600,000 years prior to 1800. He proposed that glacial climatic episodes began when Earth-Sun geometry favored an extended period of increased solar radiation in winter and decreased solar radiation in summer at 65 degrees N. More intense winter radiation in eastern Canada translated into somewhat higher temperatures, with higher humidity, and more snowfall. If coupled with weaker solar radiation in the summer, some of the winter's snow cover, especially north of 60 degrees N, would survive the summer. A succession of many such cool summers would favor formation of a glacier which, once formed, would reflect away incoming solar radiation (*positive feedback*). This process, according to Milankovitch, was the origin of the Laurentide ice sheet. At other times, Earth-Sun geometry favored enhanced solar radiation in summer at 65 degrees N, triggering

an interglacial climatic episode and shrinkage of the Laurentide ice sheet.

Meanwhile, Willard F. Libby (1908-1980) of the University of Chicago developed the radiocarbon method of age-determination for the remains of long-dead organisms. By 1951, when geologists applied the new dating technique to organic materials associated with the last major glacial advance, they discovered that the astronomical chronology did not match the radiocarbon chronology. However, the radiocarbon chronology was derived mostly from terrestrial sources and some geoscientists believed a more reliable chronology could be obtained from deep-sea sediment cores.

Because the rate of sedimentation in the open ocean is extremely slow, ranging from 1 to 3 mm per century, a relatively short undisturbed core contains information on climate fluctuations through much of the Pleistocene epoch with a resolution of 10,000 years. Reconstructing past changes in climate from analysis of deep-sea sediment cores, using new age-dating techniques (based on reversals in Earth's magnetic field) and oxygen isotope analysis, an international team of scientists analyzed two cores extracted from the bottom of the Indian Ocean that went back 450,000 years. In December 1976, they announced that major changes in climate occurred at essentially the same frequency as variations in Earth-Sun geometry (eccentricity, obliquity, and precession) even describing the cycles as the "pacemakers of the ice ages." Today, Milankovitch's astronomical theory is widely accepted as the explanation for the major climate fluctuations of the Pleistocene epoch.

VOLCANOES

The idea that volcanic eruptions influence climate has been around for more than two centuries. Benjamin Franklin (1706-1790) proposed that hazy skies in Europe associated with the eruption of Iceland's Laki fissure zone during the summer of 1783 was responsible for the severe winter of 1783-84. However, Franklin was unaware that additional larger eruptions had also occurred in eastern Asia that same year, which added to the aerosol loading of the atmosphere. The climatic impact of the violent eruption of the Peruvian volcano Huaynaputina in 1600 likely contributed to one of Russia's worst famines. The unusually cool summer of 1816 followed the violent eruption of Tambora, an Indonesian volcano, in the spring of 1815. Several relatively cold years occurred on the heels of the 1883 eruption of Krakatau, also an Indonesian volcano. Is the relationship between volcanic eruptions and climate change real or coincidental?

Only explosive volcanic eruptions rich in sulfur dioxide (SO_2) are likely to affect global or hemispheric climate and then only for a few years. A violent volcanic eruption can send sulfur dioxide high into the stratosphere where it combines with water vapor to form tiny droplets of sulfuric acid (H_2SO_4) and sulfate particles, collectively called **sulfurous aerosols**. The small size of sulfurous aerosols (averaging about 0.1 micrometer in diameter) coupled with the absence of precipitation in the stratosphere, allow sulfurous aerosols to remain suspended in the stratosphere for many months to perhaps a year or longer before they cycle out of the atmosphere to Earth's surface. Successive volcanic eruptions have produced a sulfurous aerosol veil in the stratosphere at altitudes of about 15 to 25 km (9 to 16 mi). Sulfur dioxide emissions from clusters of volcanic eruptions that temporarily thicken the stratospheric aerosol veil influence the climate.

Research conducted after the 1991 eruption of Mount Pinatubo (Figure 15.19) provided insights on the relationship between sulfurous aerosols of volcanic origin and large-scale climate fluctuations. On 15-16 June 1991, Mount Pinatubo, located on Luzon Island in the Philippines, erupted violently injecting an estimated 20 megatons of sulfur dioxide into the stratosphere. This was the most massive stratospheric volcanic aerosol cloud of the 20th century. By altering both incoming and outgoing radiation, sulfurous aerosols affected temperatures in the stratosphere and troposphere, atmospheric circulation, and surface air temperatures around the globe.

FIGURE 15.19
The June 1991 explosive eruption of Mount Pinatubo (in the Philippines) was rich in sulfur dioxide. The resulting sulfurous aerosol veil in the stratosphere was responsible for cooling at Earth's surface, temporarily interrupting the post-1970s global warming trend. [U.S. Geological Survey photo by Dave Harlow]

Sulfurous aerosols absorb both incoming solar radiation and outgoing infrared radiation, which warms the lower stratosphere, especially in the tropics. Sulfurous aerosols also reflect solar radiation to space. NASA scientists reported that in the months following the Mount Pinatubo eruption, satellite sensors measured a 3.8% increase in solar radiation reflected to space. Also, in the presence of chlorine, sulfurous aerosols destroy ozone (O_3), especially at high latitudes, allowing more solar UV radiation to reach Earth's surface. (This is essentially the same mechanism responsible for the Antarctic ozone hole described in Chapter 3.) In the two years following the Pinatubo eruption, ozone levels at middle latitudes declined by about 5% to 8%. The combination of stratospheric warming at low latitudes and ozone depletion at high latitudes strengthened the circumpolar vortex. Associated with this large-scale circulation change was a non-uniform change in surface temperatures; that is, some places cooled while other places warmed.

Through its impacts on the flux of radiation and attendant changes in global atmospheric circulation patterns, the Pinatubo eruption was likely responsible for the cool summer of 1992 over continental areas of the Northern Hemisphere (with surface temperatures up to 2.0 Celsius degrees or 3.6 Fahrenheit degrees lower than the long-term average) (Figure 15.20). The Pinatubo eruption was also implicated in the temperature

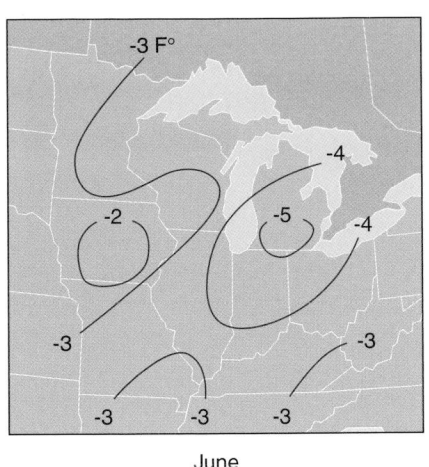 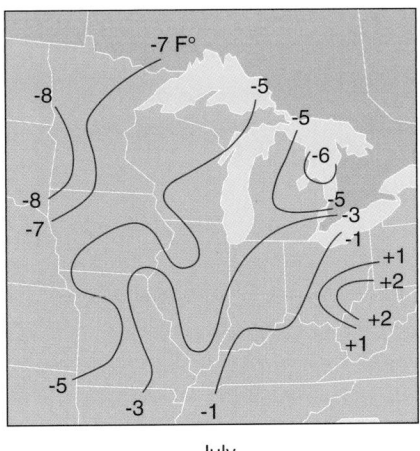 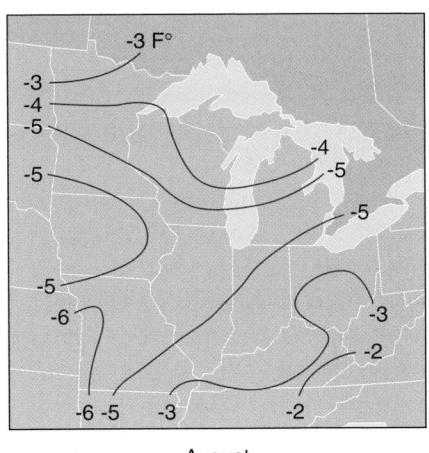

June   July   August

FIGURE 15.20
Temperature anomalies in Fahrenheit degrees in the Midwest during June, July, and August 1992. [From W.M. Wendland and J. Dennison, *Weather and Climate Impacts in the Midwest*, Midwestern Climate Center, Illinois State Water Survey, Champaign, IL]

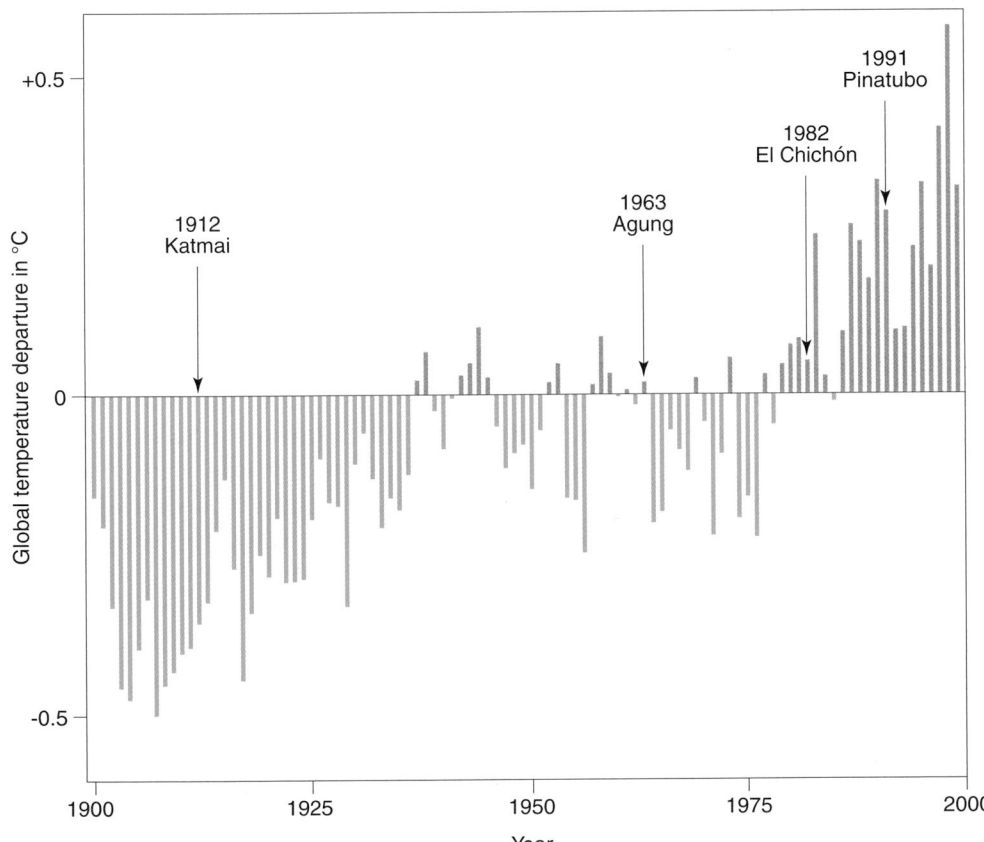

FIGURE 15.21
Large-scale cooling often followed massive volcanic eruptions that emitted sulfur dioxide (SO_2) into the stratosphere. The greatest volcano-induced temperature change on the hemispheric or global scale is usually about 1 Celsius degree or less. At smaler spatial scales, the cooling may be several degrees greater in magnitude during a particular season of the year. Further complicating the temperature response is the fact that often more than one large-scale SO_2-enriched volcanic eruption occurs at the same time in different parts of the world. These factors help account for the geographic non-uniformity of climate variability.

anomalies of the winters of 1991-92 and 1992-93, featuring higher than average temperatures over most of North America, Europe, and Siberia and lower than average temperatures over Alaska, Greenland, the Middle East, and China.

A violent sulfur-rich volcanic eruption is unlikely to lower the mean hemispheric or global surface temperature by more than about 1.0 Celsius degree (1.8 Fahrenheit degrees) although the magnitude of local and regional temperature change may be greater (Figure 15.21). The 1963 eruption of Agung on the island of Bali lowered the mean temperature of the Northern Hemisphere an estimated 0.3 Celsius degree (0.5 Fahrenheit degree) for a year or two. The violent eruption of the Mexican volcano

El Chichón in March-April 1982 may have produced hemispheric cooling of about 0.2 Celsius degree (0.4 Fahrenheit degree). Cooling associated with the Mount Pinatubo eruption temporarily interrupted the post-1970s global warming trend; from 1991 to 1992, the global mean annual temperature dropped 0.4 Celsius degree (0.7 Fahrenheit degree).

EARTH'S SURFACE PROPERTIES

Earth's surface, which is mostly ocean water, is the principal absorber of solar radiation. Any change in the physical properties of Earth's water or land surfaces or in the relative distribution of ocean, land, and ice may affect Earth's radiation budget and climate.

Changes in mean regional snow cover may contribute to climate variability because an extensive snow cover has a refrigerating effect on the atmosphere. Fresh-fallen snow typically reflects 80% or more of incident solar radiation, thereby substantially reducing the amount of solar heating and lowering the daily maximum air temperature (Figure 15.22). Snow is also an excellent emitter of infrared radiation, efficiently radiating heat to space (especially on nights when the sky is clear), and lowering the daily minimum air temperature. Because of this radiational feedback, an extensive snow cover tends to be self-sustaining.

Persistence of a snow cover may be further enhanced by extratropical cyclones that tend to track along the periphery of a regional snow cover, where horizontal air temperature gradients are relatively great. This places the snow-covered region on the cold, snowy side of migrating storms, thereby adding to the snow cover and reinforcing the chill (Chapter 10). An unusually widespread winter snow cover favors persistence of an episode of cold weather. On the other hand, less than the usual extent of winter snow cover raises average air temperatures.

Whereas changes in regional snow cover might impact climate over the short-term (seasonal), changes in Earth's sea ice or glacial ice cover are likely to have longer-lasting effects on climate. Sea ice (formed from the freezing of seawater) covers an average area of about 25 million km^2 (9.6 million mi^2), about the area of the North American continent. Terrestrial ice sheets, ice caps, and mountain glaciers cover a total area of about 15 million km^2 (5.8 million mi^2), roughly 10% of the land area of the planet. Ice (especially snow-covered ice) reflects much more incident solar radiation than either the ocean or snow-free land so that any change in glacial or sea ice cover would affect climate.

FIGURE 15.22
Snow covered ground lowers both the daily maximum and minimum air temperature.

Changes in ocean circulation and sea-surface temperatures (SST) contribute to large-scale climate change and climate variability. As described in detail in Chapter 9, changes in SST patterns accompanying El Niño and La Niña significantly influence inter-annual climate variability not only in the tropical Pacific but through teleconnections around the globe. Ocean circulation includes warm and cold surface currents and the deep-ocean thermohaline circulation that transports heat energy throughout the world. Regular changes in the strength of this circulation may explain millennial-scale (1400- to 1500-year) climate cycles over the past 10,000 years. A strong thermohaline circulation brings a relatively mild climate (for the latitude) to Western Europe whereas a weakening of the thermohaline circulation triggers cooling. Such climate shifts can be abrupt, occurring in a decade or less.

HUMAN ACTIVITY

Many human activities affect climate over broad ranges of spatial and temporal scales. Humans modify the landscape (e.g., urbanization, clear-cutting of forests) thereby altering radiational properties of Earth's surface. For reasons presented in Chapter 4, cities are slightly warmer than the surrounding countryside (the *urban heat island effect*). Combustion of fossil fuels (i.e., coal, oil, and natural gas) alters concentrations of certain key gaseous and aerosol components of the atmosphere. Of these human impacts on Earth's climate system, the burning of fossil fuels is most likely influencing climate on a global scale.

Many scientists and public policy makers are concerned about the impact on global climate by the steadily rising concentrations of atmospheric carbon dioxide (CO_2) and other infrared-absorbing gases due to human activity. Higher levels of these gases enhance the greenhouse effect, contributing to warming on a global scale. This is the *Callendar effect* described in Chapter 3.

Recall from Chapter 3 that systematic monitoring of atmospheric carbon dioxide levels began in 1957 at NOAA's Mauna Loa Observatory in Hawaii under the direction of Charles D. Keeling (1928-2005) of Scripps Institution of Oceanography. Also since 1957, atmospheric CO_2 has been monitored at the South Pole station of the U.S. Antarctic Program and that record closely parallels the one at Mauna Loa. The Mauna Loa record (the *Keeling curve*) shows a sustained increase in average annual atmospheric carbon dioxide concentration from about 316 ppmv (parts per million by volume) in 1959 to 389 ppmv near the end of 2011 (Figure 3.36).

The anthropogenic contribution to the buildup of atmospheric CO_2 may have begun thousands of years ago with land clearing for agriculture and settlement. By the middle of the 19th century, growing dependency on coal burning associated with the beginnings of the Industrial Revolution triggered a more rapid rise in atmospheric CO_2 concentration. Carbon dioxide is a byproduct of the burning of coal and other fossil fuels. The concentration of atmospheric CO_2 is now about 35% higher than it was in the pre-industrial era. Fossil fuel combustion accounts for roughly 75% of the increase in atmospheric carbon dioxide while deforestation (and other land clearing) is likely responsible for the balance. With continued growth in fossil fuel combustion, the atmospheric carbon dioxide concentration could top 550 ppmv (double the pre-industrial level) by the end of the present century.

The ocean is a major reservoir in the global carbon cycle and as such plays an important role in governing the amount of carbon dioxide in the atmosphere. From the beginning of the Industrial Revolution, scientists were able to estimate the amount of CO_2 released to the atmosphere by human activity. But in measuring the actual amount of carbon dioxide in the atmosphere, about half was missing. The "missing" carbon is distributed between sinks in the ocean and on land. According to IPCC estimates in 2007, the ocean takes up 56.2% of the carbon dioxide of anthropogenic origin (via photosynthesis and cold surface waters absorbing CO_2 and sinking) while terrestrial biomass is a sink for 13.7%.

Besides carbon dioxide, rising levels of other infrared-absorbing gases associated with human activity (e.g., methane, nitrous oxide, halocarbons, and tropospheric ozone) enhance the greenhouse effect (Chapter 3). Although occurring in extremely low concentrations (typically measured in parts per billion), methane and nitrous oxide are very efficient absorbers of infrared radiation. In response to international efforts to protect the stratospheric ozone shield, the atmospheric concentration of halocarbons began leveling off in the early 1990s.

Aerosols are tiny solid and liquid particles suspended in the atmosphere. Atmospheric aerosols that are byproducts of human activity apparently have the opposite effect on temperatures at Earth's surface as greenhouse gases; that is, they cause cooling rather than warming. Aerosols vary in size, shape, and chemical composition. Larger aerosols have short residence times and tend to settle out of the atmosphere quickly whereas smaller ones may remain suspended for many days to weeks and can be transported thousands of kilometers by the wind, possibly impacting large-scale climate. Perhaps 90% of anthropogenic aerosols are byproducts of fossil fuel burning in the Northern Hemisphere.

Sulfur oxides emitted from power plant smokestacks and boiler vent pipes combine with water vapor in the air to produce tiny droplets of sulfuric acid and sulfate particles. These sulfurous aerosols raise the atmosphere's albedo directly by reflecting sunlight to space and indirectly by spurring cloud development. Aerosols of anthropogenic origin function as cloud condensation nuclei (CCN) that favor the formation of more numerous and smaller cloud droplets and brighter clouds that reflect more solar radiation to space. Greater reflectivity cools the lower atmosphere. Sulfurous aerosols in the troposphere have a shorter-term impact on climate than carbon dioxide and other greenhouse gases. Rain and snow wash sulfurous aerosols from the atmosphere so that the residence time of these substances in the atmosphere is typically only a few days. On the other hand, a CO_2 molecule resides in the atmosphere for a much longer period before being cycled out by natural processes operating as part of the global carbon cycle. Approximately one-fifth of the carbon we release now will still be in the atmosphere after 200 years; after several thousand years about one-twentieth will remain.

Prior to the late 1980s, aerosol cooling in the Earth-atmosphere system partially offset enhanced greenhouse warming. Consequently, scientists may have underestimated the magnitude of potential warming from the build up of greenhouse gases. Aerosol masking began to decline in the 1990s, perhaps explaining why the global warming trend was greater than expected over the past few decades.

The idea that anthropogenic changes in land use and land cover contribute to climate change has deep historical roots. During much of the 18th and 19th centuries, natural scientists argued whether deforestation and cultivation practices were changing the climate of new settlements in America. Today, much of the research on the climatic impacts of changes in land use and land cover focuses on deforestation in the tropics (e.g., the Amazon Basin). But as pointed out in a 2005 NASA news item, human development has transformed an estimated one-third to one-half of Earth's land surfaces. The possible implications of this transformation on the climate system are far reaching.

In 2001, Gordon Bonan of the National Center for Atmospheric Research (NCAR) found that at middle latitudes, clearing of forests for agriculture and reforestation of abandoned farm land caused regional climate change. Bonan compared the *diurnal temperature range (DTR)* at cropland sites in the Midwest with forested sites in the Northeast. He selected 65 climate stations that were not near cities or water bodies and examined temperature records for 1986-1995. The DTR was lower in the Midwest than the Northeast primarily because the albedo of cropland was higher than forested land. Daytime heating was less in the Midwest than Northeast even though the Northeast had more cloudiness. Midwest cooling was greater in late spring and summer and less in fall after the crops were harvested. In expanding his study to a 100-year record of U.S. climate, Bonan found that prior to 1940 when the percent of cultivated land was similar in both the Midwest and the Northeast, the difference in DTR between the two regions was less than in recent decades.

ANTHROPOGENIC VERSUS NATURAL FORCING OF CLIMATE

The diagram in Figure 15.23, provided by the IPCC, summarizes the contributions of the various climatic forcing agents or mechanisms (excluding volcanic aerosols) in 2005. These are global average radiative forcings in units of watts per square meter (W/m^2) grouped by anthropogenic and natural sources. Positive forcings (e.g., greenhouse gases) cause the climate to become warmer whereas negative forcings (e.g., aerosols) cause the climate to become cooler. Black error bars represent the level of certainty of each forcing agent or mechanism; that is, the probability that values lie within the error bar is 90%. Also included is the typical spatial scale and assessed level of scientific understanding (LOSU) for each forcing agent or mechanism.

FIGURE 15.23
Estimates of global average radiative forcing (RF) in 2005 for greenhouse gases and other important climate change agents and mechanism, along with typical geographical extent (spatial scale) of the forcing and level of scientific understanding (LOSU). [From *IPCC, Climate Change 2007: The Physical Science Basis*]

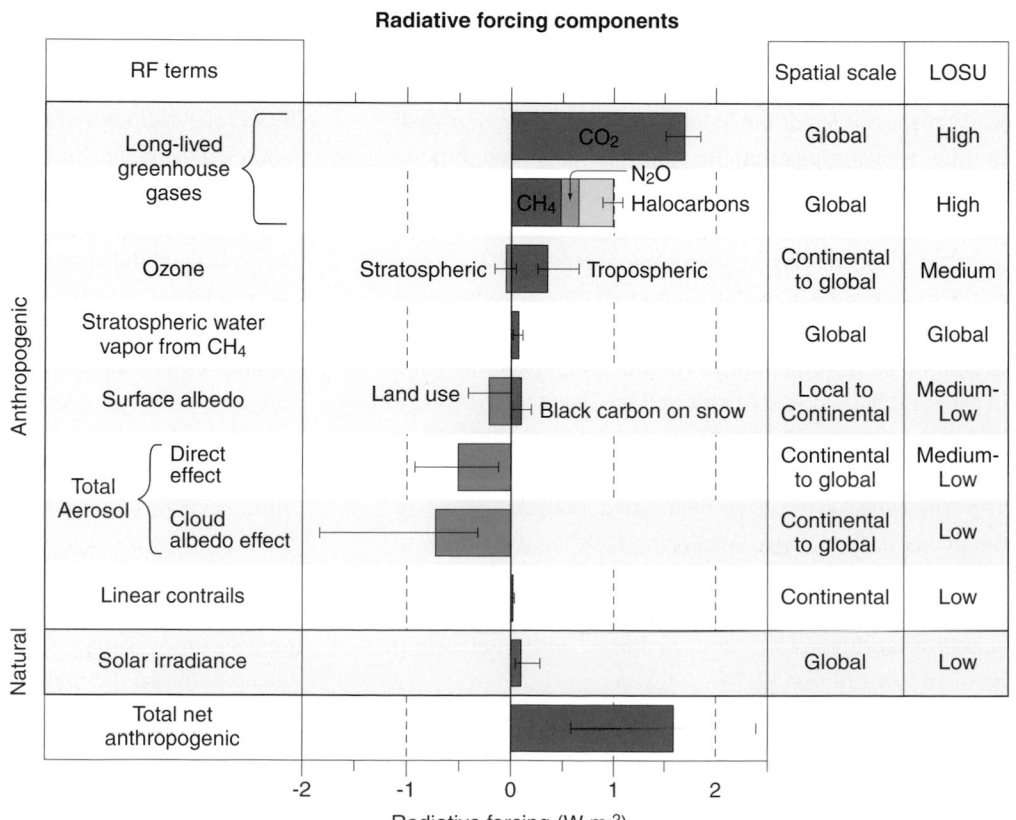

According to the diagram, the net radiative forcing is anthropogenic in origin. Specifically, the build-up of the greenhouse gases CO_2, CH_4, N_2O, halocarbons, and O_3 is primarily responsible for a net positive radiative forcing. This results in global warming as evidenced in the observational record.

The Climate Future

What does the climate future hold? Atmospheric scientists attempt to answer this question primarily by relying on global climate models that run on supercomputers.

GLOBAL CLIMATE MODELS

A **global climate model (GCM)** is a simulation of Earth's climate system. One type of global climate model consists of mathematical equations that describe the physical interactions among the various components of the climate system, that is, the atmosphere, ocean, land, ice-cover, and biosphere. A global climate model differs from numerical models used for forecasting weather in that it predicts broad regions of expected positive and negative temperature and precipitation anomalies (departures from long-term averages) and the mean location of circulation features such as jet streams and principal storm tracks over much longer time scales.

Global climate models are used to predict the potential climatic impacts of rising levels of atmospheric carbon dioxide (or other greenhouse gases). Using current boundary conditions, a global climate model simulates the present climate. Then, holding constant all other variables in the model, the concentration of atmospheric CO_2 (or other greenhouse gas) is elevated. Two different approaches are taken in adding CO_2. In a *transient run*, CO_2 is slowly added to the model and the effects are evaluated from moment to moment whereas in an *equilibrium run*, CO_2 is added all at once and the model is run until it achieves a new equilibrium. By comparing the new climate state with the present climate, scientists deduce the impact of an enhanced greenhouse effect on patterns of temperature and precipitation. Using boundary conditions derived from proxy climate data sources, global climate models have also been used to predict climates that prevailed in the geologic past such as the last glacial maximum (20,000 to 18,000 years ago).

Most modelers argue that global climate models are in need of considerable refinement. Today's models may not adequately simulate the role of small-scale weather systems (e.g., thunderstorms) or accurately portray local and regional conditions and may miss important feedback processes. A major uncertainty is the net feedback of clouds. Clouds cause both cooling (by reflecting sunlight to space) and warming (by absorbing and emitting to Earth's surface outgoing infrared radiation). According to one view, the cooling effect prevails with an increase in low cloud cover whereas the warming effect prevails with an increase in high cloud cover.

Problems with global climate models stem in part from the limited spatial resolution of the models. Today's models partition the global atmosphere into a three-dimensional grid of boxes with each box typically having an area of 250 km^2 (155 mi^2) and a thickness of 1 km (0.6 mi). Limited spatial resolution in climate models is caused by limited computational speed. Although today's supercomputers can perform trillions of operations per second, the complexity of the climate system means that simulation of climate change over a century requires months of computing time. Much greater resolution by global climate models will come with development of faster supercomputers.

SEARCH FOR CYCLES AND ANALOGS

An alternate approach to predicting future climate is empirical in nature and seeks to identify the various factors that may have contributed to past fluctuations in climate and to extrapolate their influence into the future. Atmospheric scientists have probed the instrument-based and reconstructed climate record in search of regular cycles that might be extended into the future, and analogs that might provide clues as to how the climate in specific regions responds to global-scale climate change. One formidable challenge in this search is to separate signal from noise. Time series of some climatic elements are so variable (noisy) that detection of any cycles or trends (the signal) requires close scrutiny of the climate record. Use of computers programmed with sophisticated statistical routines has greatly facilitated the search for regular climatic rhythms and trends. The motivation behind this effort is obvious: Identification of any statistically significant periodicities or trends in the climate record would be a powerful tool in climate forecasting.

Few of the quasi-regular oscillations that appear in the climate record have much practical value for climate forecasting, at least over the next century. Cycles established as significant in a rigorous statistical sense are the familiar annual and diurnal radiation/temperature cycles and a less familiar quasibiennial cycle (about every two years) in various climatic elements. The first merely means that summers are warmer than winters and days are warmer than nights. Examples of the quasibiennial cycle include an approximate two-year fluctuation in Midwestern rainfall, and an approximately two-year cycle in the strength of the trade winds over the western Pacific and eastern Indian Oceans. Trends may be visible in the climate record, but unless a trend is demonstrated to be part of a statistically significant cycle, there is no guarantee that the trend will not end abruptly or reverse direction at any time.

Although the climate record yields much useful information on how climate behaves through time, the search for realistic analogs of future global warming has been unsuccessful. Proposed analogs include relatively warm episodes of the mid Holocene Epoch and the Eemian interglacial. But those analogs are inappropriate because the mid Holocene and Eemian warming episodes primarily affected seasonal temperatures with only a slight rise in global mean temperature. Furthermore, boundary conditions were different. During the mid Holocene, sea level was lower, ice sheets were more extensive, and the seasonal and latitudinal receipt of solar radiation (due to different dates of perihelion and aphelion) were not the same as they are now or will be over the next several centuries. Although the level of atmospheric CO_2 trended upward during the mid Holocene, the rate of increase (about 0.5 ppmv per century) was much less than at present (more than 60 ppmv per century).

ENHANCED GREENHOUSE EFFECT AND GLOBAL WARMING

In the long run, during the next 10,000 to 100,000 years, the Milankovitch Earth-Sun orbital cycles favor a return to Ice Age conditions. Over the next few decades, if all other boundary conditions remain fixed, rising concentrations of atmospheric CO_2 and other greenhouse gases will cause global warming to persist well beyond this century. The magnitude of warming will depend on continued emissions of greenhouse gases.

In 1979, President Jimmy Carter asked the National Academy of Sciences (NAS) to report on the potential impacts of the upward trend in atmospheric concentration of carbon dioxide. Jule G. Charney (1917-1981) of the Massachusetts Institute of Technology (MIT) led the NAS investigation team that designed the now classic experiment in which numerical models of Earth's climate system had their CO_2 concentration doubled while holding all other variables constant.

Addition of CO_2 makes the atmosphere more opaque for outgoing infrared radiation, warming the lower atmosphere and cooling the upper atmosphere. Applying

basic radiation laws, Charney and his colleagues found that doubling the CO_2 concentration would reduce the net radiative flux (from Earth to Space) at the tropopause by a global average of about 4 watts per square meter (W/m^2). How much warmer would Earth's surface become as a consequence of this enhanced greenhouse effect? According to the Stefan-Boltzmann law, the radiation emitted by an object is directly proportional to the fourth power of the object's absolute temperature. Following a doubling of atmospheric CO_2, Earth would have to radiate an additional 4 W/m^2 to space, brought about by a global warming of 1.2 Celsius degrees (or 0.3 Celsius degrees per W/m^2) to regain radiative equilibrium.

Charney's experiment did not include the effects of feedback processes. Recall that forcing agents and mechanisms drive climate change while positive and negative feedbacks determine the magnitude. Accounting for feedbacks, the 1979 NAS study would predict global warming ranging from 2 to 3.5 Celsius degrees (3.6 to 6.3 Fahrenheit degrees). A more recent IPCC report (AR4) estimates that the magnitude of warming with feedbacks incorporated as 3 Celsius degrees (5.4 Fahrenheit degrees) with an uncertainty range of 2 to 4.5 Celsius degrees (3.6 to 8.1 Fahrenheit degrees). According to the *2007 IPCC Assessment Report*, climate models predict that over the subsequent 20 years, the global mean annual temperature will increase at an average rate of about 0.2 Celsius degrees per decade. Again depending on the future greenhouse gas emission scenario, over the present century, climate models project that the globally averaged surface temperature will rise by 1.8 to 4.0 Celsius degrees (3.2 to 7.2 Fahrenheit degrees). Recall, however, that climate change is geographically non-uniform (in both magnitude and direction) so that this projected rise in global mean annual temperature is not necessarily representative of what might happen everywhere. For example, *polar amplification* suggests that global warming will be greater at higher latitudes. In any event, enhancement of the greenhouse effect could cause a climate change that would be greater in magnitude than any previous climate change over the past 10,000 years.

Once in the atmosphere, the concentration of a greenhouse gas depends on how the rates of emission of the gas into the atmosphere compare to the rates whereby other processes remove it from the atmosphere. These interactions determine the **lifetime of a gas** in the atmosphere, which is defined as the time it takes for the gas to be reduced to 37% of its original amount. Even if greenhouse gas emissions were to stabilize at present levels, global warming would likely continue well beyond the 21st century.

Models of global warming based on an enhanced greenhouse effect assume that all other boundary conditions remain constant. Whether that actually happens is not known. Comparing the post-1957 trend in atmospheric carbon dioxide to the trend in mean annual global temperature strongly suggests that recent climate has been shaped by many interacting factors. The rapid rise in atmospheric CO_2 concentration was not accompanied by a consistent rise in global mean temperature over the same period. Recall, for example, that sulfurous aerosols from the June 1991 eruption of Mount Pinatubo apparently were responsible for significant global-scale cooling the following year. Also, El Niño and La Niña influence inter-annual climate variability in some areas of the globe (Chapter 9). Furthermore, analysis of the composition of tiny air bubbles in cores extracted from the Greenland and Antarctic ice sheets indicate that during the Pleistocene Ice Age, atmospheric CO_2 varied between about 260 and 280 ppmv. Although CO_2 was consistently higher during milder interglacial climatic episodes than during colder glacial climatic episodes, fluctuations in carbon dioxide levels lag reconstructed variations in temperature. For example, at the beginning of the last major glacial climatic episode, the decline in CO_2 concentration significantly lagged cooling in the Antarctic. Hence, fluctuations in atmospheric CO_2 may have been a response to large-scale climate oscillations rather than a cause of those oscillations.

Further complicating efforts to predict the climate future are indications that the global radiation budget currently is not in equilibrium. In 2005, scientists reported that the planet was absorbing about 0.85 W/m^2 more energy from the Sun than it is emitting as infrared radiation to space; the imbalance is approximately 0.06% of the total incoming solar radiation at the top of the atmosphere. This conclusion is supported by measurements made by sensors onboard Earth-orbiting satellites and instrumented buoys indicating an increase in the heat content of the ocean over the previous decade. The imbalance represents the delay in the response of Earth's climate (i.e., surface temperature) to some agent or forcing mechanism (i.e., build up of greenhouse gases). The delay, in turn, is the consequence of considerable thermal inertia in Earth's climate system, mostly due to the ocean. On this basis, global climate models predict a yet unrealized additional warming of 0.6 Celsius degree (1.1 Fahrenheit degrees) with no change in atmospheric composition as Earth shifts to a new state of global radiative equilibrium.

Potential Impacts of Global Climate Change

When it comes to climate variability and climate change, the primary focus is on temperature. The prospect of continuation of the current global warming trend usually takes center stage in discussions regarding the climate future and possible societal impacts. After all, this anthropogenic influence on Earth's climate system is most directly experienced by people. But just as important as temperature trends (if not more so) are concurrent changes in the global water cycle, including precipitation amounts and patterns. For example, changes in the frequency and/or persistence of drought or episodes of exceptionally heavy rainfall are likely to seriously impact agricultural systems, urban infrastructure, and hydroelectric facilities.

Global climate models predict that continuation of the current warming trend will be accompanied by an overall increase in precipitation. But climate change is geographically non-uniform in direction and magnitude so that rainfall is expected to increase in some areas and decrease in other areas. Already the instrument-based record is indicating that some regions are experiencing a greater frequency of extreme rainfall events and more days with rainfall. In 2008, P. Groisman and R. Knight of NOAA's National Climatic Data Center reported on their analysis of rainfall trends over the previous 40 years at 4000 stations across the coterminous U.S. They found a significant increase in the mean duration of prolonged dry episodes during the warm season. Over the eastern U.S., the return period (or recurrence interval) for a dry episode lasting one month decreased from every 15 years to 6-7 years.

While global-scale trends in climate do not impact all regions in the same way, if global warming persists, the consequences for society are likely to be extensive. Rising sea level is one of the most serious potential consequences because a third of the human population lives within 91 m (300 ft) of sea level

SHRINKING GLACIERS AND RISING SEA LEVEL

Waxing and waning of glaciers on land plus ocean temperature fluctuations are two factors that govern **eustasy**, the global variation in sea level brought about by a change in the volume of water occupying the ocean basins. Persistence of the current global warming trend appears likely to cause sea level to rise in response to melting of land-based polar ice sheets and mountain glaciers, coupled with thermal expansion of seawater. When glacial ice leaves the land and enters the ocean, it begins to float, displacing a quantity of seawater of equivalent weight, and causing an immediate rise in sea level.

How has *mean sea level (msl)* responded to the present global warming trend? For most of the 20th century, coastal tide gauges were the principal source of data on sea level change. Care must be taken to exclude (or adjust for) tide gauge records that are influenced by geological processes (i.e., tectonic uplift, subsidence, or post-glacial rebound). Based on adjusted tide gauge records, msl from 1870 to 1993 is estimated to have risen at an average rate of 1.7 mm per year (Figure 15.24A). Beginning in 1993, microwave altimeters on Earth-orbiting satellites have provided more precise measurements of the rise in msl (Figure 15.24B). From 1993 to 2009, msl increased an estimated 3.3 ± 0.4 mm per year on average, showing that sea level rise is accelerating. In total, mean sea level is estimated to have risen about 180 mm (7.1 in.) during the 20th century.

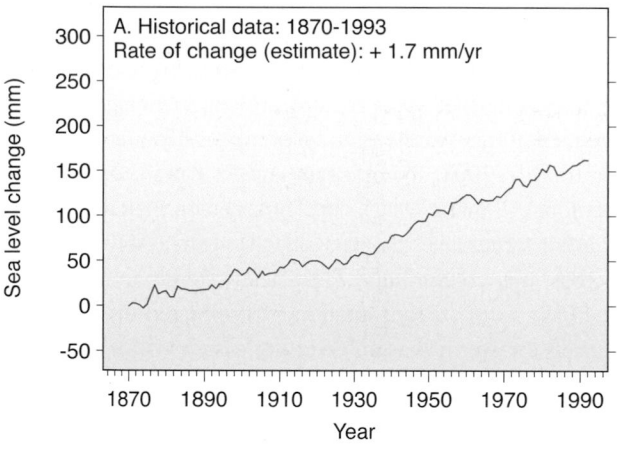

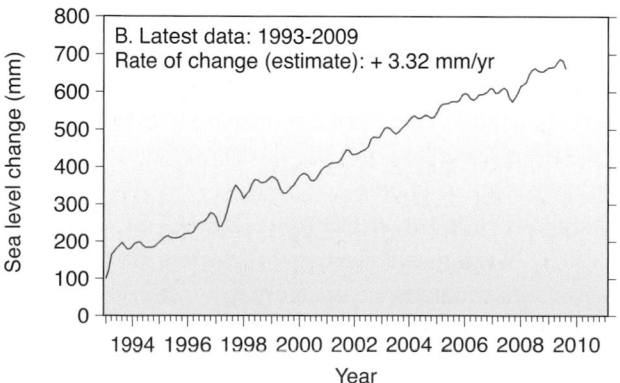

FIGURE 15.24
Upward trend in mean sea level as indicated by (A) coastal tide gauge records, 1870-1993, and (B) measurements by microwave altimeters flown onboard Earth-orbiting satellites from 1993 to 2009. [From CSIRO (A) and CLS/Cnes/Legos (B)]

How much of the recent rise in msl was due to melting of glacial ice and how much was due to thermal expansion of warming ocean waters is not known. Most mountain glaciers have been shrinking since the mid 20th century, portions of the Greenland and West Antarctic ice sheets recently have shown signs of accelerating mass loss, and portions of the ocean are warming.

Amplification of the warming trend at higher latitudes would threaten the ice sheets of Antarctica and Greenland. About 90% of the planet's glacial ice blankets Antarctica and melting could cause a considerable rise in sea level. How likely is this to happen? Two ice sheets cover Antarctica, separated by the Transantarctic Mountains. The West Antarctic ice sheet sits on a former ocean bottom and is mostly below sea level whereas the East Antarctic ice sheet is situated on a continent and is well above sea level. While geological evidence suggests that the East Antarctic ice sheet has been stable for the past 30 million years and remains fairly stable today, the West Antarctic sheet has undergone episodes of rapid disintegration and may have completely melted at least once in the past 600,000 years.

The behavior of ice streams is an important consideration in predicting how the Antarctic and Greenland ice sheets influence sea level. An **ice stream** is a zone of relatively rapidly flowing ice within an ice sheet (Figure 15.25). As part of the global water cycle, excess water on land flows via rivers and streams to the ocean. Similarly, most of the ice (perhaps 90%) that flows from Antarctica and Greenland to the surrounding ocean occurs via ice streams and *outlet glaciers* (ice streams bounded by mountains). Similar in dimension to the Byrd Glacier shown in Figure 15.25, the Rutford Ice Stream in West Antarctica is 150 km (93 mi) long, 25 km (16 mi) wide, up to 3 km (2 mi) thick, and moves at an average speed of 1.0 m per day. Many of Antarctica's ice streams feed ice shelves, floating ice attached to land and fringing about 44% of the coast.

Although the average speed of ice streams is greater than that of the surrounding ice, ice stream speed can be highly variable depending on the frictional resistance of the terrain under the ice and the temperature of the ice. Recently, scientists confirmed the existence of large lakes underlying portions of some ice streams. The *subglacial lake* under Russia's Vostok Station in East Antarctica is about the size of Lake Ontario. (Lake Vostok is about 4000 m (13,100 ft) under the surface of the ice.) The ice in the water and the lake surface (water-ice interface) offers essentially no frictional resistance to the moving ice plus the ice is warmed by the water thereby

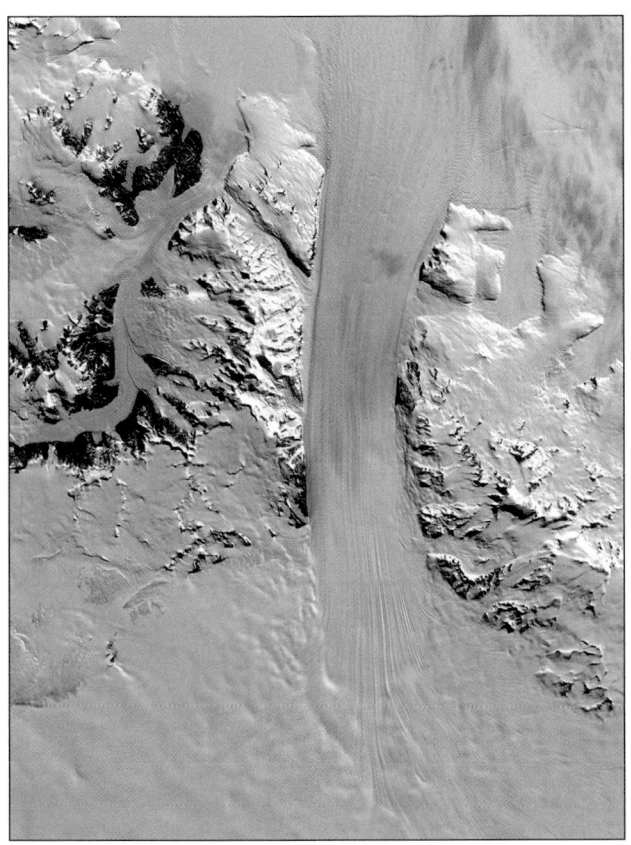

FIGURE 15.25
Located near McMurdo Station, the principal U.S. Antarctic Research Base, the Byrd Glacier plunges through a deep, 15-mile-wide valley in the Transantarctic Mountains to create a 100-mile-long, rock-floored ice stream. [NASA image by Jesse Allen, made from the Landsat Image Mosaic of Antarctica (LIMA).]

reducing its viscosity. Hence, the flow of ice streams accelerates over subglacial lakes. Along the way, the ice stream erodes sediment and deposits it as a wedge where the ice stream enters the ocean. This wedge of sediment acts like a speed bump slowing the advancing ice.

More than 30 years ago, recognition of the relative instability of the West Antarctic ice sheet prompted speculation among some scientists that ice streams flowing from the interior of the glacier to the Ross and Ronne ice shelves might cause a total collapse of the ice sheet in a few centuries or less. Such a catastrophic event would greatly accelerate the rate of sea level rise. (Complete disintegration of the West Antarctic ice sheet would raise sea level by 3.2 m or 10.5 ft. on average.)

In 2001, concerns about the possible disintegration of the West Antarctic ice sheet were alleviated with the discovery that new snowfall was keeping pace with the loss of ice from bergs breaking off the Ross Ice Shelf. In early 2002, scientists at the California Institute of Tech-

nology and the University of California-Santa Cruz reported that based on satellite measurements of the flow of the Ross ice streams, the West Antarctic ice sheet appears to be thickening. Meanwhile, the region of the ice sheet that feeds the Thwaites and Pine Island glaciers is thinning. These glaciers transport ice directly into the ocean (rather than adding to an ice shelf) and are responsible for about 10% of the average annual rise in sea level.

More recently, Eric Rignot of NASA's Jet Propulsion Laboratory in Pasadena, CA and colleagues used satellite imagery to measure the thickness of ice fringing the coast of Antarctica. These data along with estimates of ice stream discharge and snowfall enabled them to determine the annual loss of ice from Antarctica. They concluded that the ice loss from Antarctica (mostly West Antarctica) was about 75% greater in 2006 than in 1996. Nonetheless, the consensus of scientific opinion today is that the West Antarctic ice sheet is unlikely to catastrophically accelerate the current rise in sea level. According to some experts, long-term gradual shrinkage of the West Antarctic ice could raise sea level at a rate of about 1.0 m (3.3 ft) per 500 years.

Melting (and the rate of sea level rise) could accelerate if the Antarctic ice sheets begin to feel the effects of global warming. Over most of Antarctica, the mean air temperature has fluctuated very little over the past 50 years. The Antarctic Peninsula is the only part of the continent that has shown significant warming with summer mean temperatures rising more than 2 Celsius degrees (3.6 Fahrenheit degrees) over the past half century. This warming has been accompanied by breakup of ice shelves along the coast.

NASA research results released in 2000 concluded that while the central interior of the Greenland ice sheet showed no sign of thinning, about 70% of the margin was thinning substantially based on observations made between 1993 and 1999. The maximum melting rate at the margin was about 1.0 m per year. An estimated 50 km³ (12 mi³) of Greenland's ice melts each year, enough to raise sea level by 0.13 mm annually. Scientists at the University of Colorado reported that during the summer of 2002, surface melting on the Greenland ice sheet encompassed an area of about 695,000 km² (265,000 mi²)—about 9% greater than observed during any summer since monitoring by satellite began 24 years previously. The duration and extent of melting at elevations above 2000 m set a new record in 2007. Researchers at the Goddard Space Flight Center monitored snow melt on the ice sheet surface using a satellite microwave sensor and computed a snow melt index by multiplying the area

FIGURE 15.26
(A) A meltwater lake on the surface of the Greenland ice sheet, one of thousands that form each summer. [Photo by Ian Joughin, University of Washington Polar Science Center] (B) A large ice canyon in the Greenland ice sheet excavated over many years by a meltwater stream. [Photo by Sarah Das, Woods Hole Oceanographic Institution]

of snowmelt by the duration of melting. During 2007, the melting index above 2000-m elevation was about 2.5 times greater than the annual average from 1988 to 2006. Currently, it appears that winter snowfall is insufficient to offset the summer melt so that overall the Greenland ice sheet is shrinking.

In 2002, scientists discovered that meltwater lakes that form in summer on the surface of the Greenland ice sheet (known as *supraglacial lakes*) can force open a crevasse allowing its waters to drain catastrophically to the base of the glacier. This phenomenon was documented by researchers from the Woods Hole Oceanographic Institution and the University of Washington in July 2006 (Figure 15.26). In only 90 minutes, a supraglacial lake covering 5.7 km^2 (2.2 mi^2) and containing 11.6 billion gallons of water drained through a crevasse, plunging some 980 m (3215 ft) to the base of the glacier. Writing

in the 14 November 2008 issue of *Science*, glaciologist Richard Alley of Pennsylvania State University and colleagues proposed that the lubricating effect of this water is not as important as the heat it delivers to the base of the ice sheet in accelerating glacier flow. Should the atmospheric warming trend lengthen the melt season and create meltwater lakes further inland on the Greenland ice sheet, the downward cascading meltwater could thaw areas where the glacier is now frozen to the ground. This is likely to significantly accelerate glacier flow and perhaps destabilize the Greenland ice sheet. Complete melting of the Greenland ice sheet would raise mean sea level by an estimated 7.3 m (24 ft).

Alley and colleagues also point to another factor that influences the flow of outlet glaciers to the sea. At the glacier/ocean interface, outlet glaciers move faster when they do not encounter obstacles such as ice shelves and grounded ice. Most ice shelves surround Antarctica but grounded ice, bedrock highs, and coastal landforms obstruct the flow of Greenland outlet glaciers. Melting of ice shelves or grounded ice would accelerate the flow of outlet glaciers.

Roger G. Barry of the *National Snow and Ice Data Center (NSIDC)* of the Cooperative Institute for Research in Environmental Sciences at the University of Colorado, Boulder, reports that the rate of melting of most of the world's mountain glaciers accelerated after the mid-1900s and especially since the mid-1970s (Figure 15.27). Some mountain glaciers have disappeared entirely. According to the U.S. Geological Survey, in 2010 only 25 glaciers larger than 10 hectares (25 acres) remain of the 150 glaciers that existed in Montana's Glacier National Park a century ago. Barry estimates that runoff from melting mountain glaciers contributes about 0.4 mm to the annual rise in msl.

A major concern is the shrinkage of mountain glaciers and ice fields whose seasonal runoff is the principal source of fresh water for people, their crops and livestock. In India, for example, 500 million people depend on runoff from the glaciers of the Himalayas for their fresh water supply. Although many glaciers are shrinking, it is doubtful that the glaciers of the Himalayas will soon disappear. The largest glaciers are at high elevations, fed by copious monsoon moisture. Of more immediate concern are the many smaller glaciers and ice fields at lower elevations; they are much more vulnerable to warming and monsoon failure.

According to the 2007 *IPCC Fourth Assessment Report*, thermal expansion of warming seawater will be a greater contributor to mean sea level rise than melting

FIGURE 15.27
The present global warming trend has caused mountain glaciers to shrink with much of the meltwater eventually draining into the ocean. The dramatic recession of Grinnell Glacier in Montana's Glacier National Park is shown in photographs taken in 1938, 1981, 1998, and 2006. [Photos courtesy of Glacier National Park Archives and the USGS]

of land-based glaciers during the 21st century. Climate models predict that global warming will cause a rise in mean sea level in the range of 0.2 to 0.6 m (8 to 24 in.) during this century. Thermal expansion of ocean waters would account for more than 60% of the rise with the balance due to melting glaciers. At least over the short-term (the coming few decades), the behavior of ice sheets is the largest unknown regarding the magnitude of sea level rise.

Some scientists view the IPCC estimates as too conservative, underestimating the impact of melting glaciers and ice sheets. For example, W.T. Pfeffer and colleagues at the Institute of Arctic and Alpine Research at the University of Colorado, Boulder, include estimates of glacier melt in their projected msl rise of 0.8 to 2.0 m (2.6 to 6.7 ft) by 2100. Other climate models predict sea level rise to be in the range of 0.3 to 1.8 m (1 to 5.9 ft) by 2100.

Higher mean sea level would accelerate coastal erosion by wave action, inundate wetlands, estuaries and some islands, and make low-lying coastal plains more vulnerable to storm surges. Recall from this chapter's Case-in-Point, the threat posed by rising sea level to inhabitants of the Maldives in the Indian Ocean and Kiribati in the Pacific Ocean. People may have to abandon these island nations as sea level rises. Globally, a 50-cm (20-in.) rise in sea level would double the number of people at risk from storm surges from about 45 million at present to more than 90 million, not counting any additional population growth in the coastal zone.

Rising sea level would disrupt coastal ecosystems, ruin agricultural lands, and could threaten historical, cultural, and recreational resources. In some coastal areas, higher sea level is likely to exacerbate saltwater intrusion into groundwater. According to a 1997 report by the U.S. Office of Science and Technology Policy, a 50-cm (20-in.) rise in sea level would result in a substantial loss of coastal land, especially along the U.S. Gulf and southern Atlantic coasts. Particularly vulnerable is South Florida where one-third of the Everglades is less than 30 cm (12 in.) above sea level.

While higher temperatures would mean higher sea level, the level of North America's Great Lakes is likely to fall. Higher summer temperatures coupled with less winter ice cover on the Great Lakes are likely to translate into greater evaporation. And less winter snowfall would reduce spring runoff. Depending on the model used, forecasts call for a drop in mean water level of Lake Michigan of up to 2 m (6.5 ft) by the year 2070.

ARCTIC ENVIRONMENT

A major concern associated with the current global warming trend is shrinkage of Arctic sea-ice cover. Although melting of floating sea ice does not raise sea level, it can alter climate significantly. Shrinkage of Arctic sea ice is likely to trigger an **ice-albedo feedback** mechanism that would accelerate melting of sea ice and amplify warming (Figure 15.28). Sea ice insulates the overlying air from warmer seawater and reflects much more incident solar radiation than ocean water. The albedo of snow-covered sea ice is about 85% whereas ice-free Arctic Ocean water has an average albedo of only about 7%. As sea ice cover shrinks, the greater area of

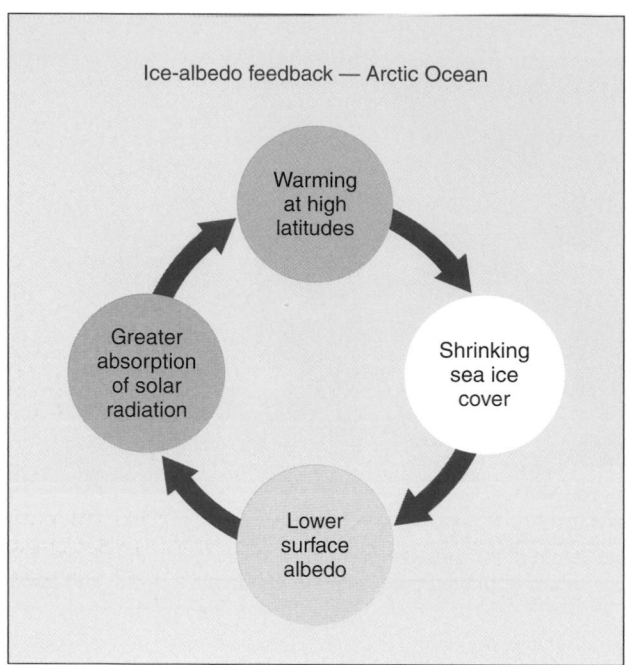

FIGURE 15.28
Positive ice-albedo feedback in the Arctic is likely to accelerate warming of surface waters and shrinkage of sea ice cover.

ice-free ocean waters absorbs more solar radiation, sea-surface temperatures rise, and more ice melts. Furthermore, warmer water slows the formation of ice in autumn. This positive feedback could rapidly reduce Arctic sea ice cover and greatly alter the flux of heat energy and moisture between the ocean and atmosphere with possible ramifications for global climate.

Less sea ice cover on the Arctic Ocean is likely to increase the humidity of the overlying air leading to more cloudiness. Clouds cause both cooling (by reflecting sunlight to space) and warming (by absorbing and emitting to Earth's surface outgoing infrared radiation). During the long dark polar winter, clouds would have a warming effect. In summer, the impact of a greater cloud cover depends on the altitude of the clouds. Cooling would prevail with an increase in low cloud cover whereas warming would likely accompany an increase in high cloud cover.

What is the trend in Arctic sea ice? Arctic sea ice cover varies seasonally (Figure 15.29) and exhibits some long term trends. A variety of sources provides information on the extent of Arctic sea ice cover since the early part of the 20th century. Ship and aircraft observations in-

FIGURE 15.29
Total area covered by sea ice (not including the open water within the pack ice) in the Northern Hemisphere polar region on 7 March 2011 (top image) and 9 September 2011 (bottom). March and September are the months of maximum and minimum ice extent respectively over a 12-month period. Average ice conditions are estimated using passive microwave sensors on Earth-orbiting satellites. [Image courtesy of NASA's Scientific Visualization Studio, Goddard Space Flight Center]

March 2011

September 2011

dicate that the multi-year Arctic ice cover remained essentially constant in all seasons through the first half of the 20th century. But beginning in the 1950s, observations by ships and aircraft detected shrinkage in the summer minimum extent of ice while the winter maximum remained nearly constant. By the mid-1970s, surveillance by satellites and submarines plus ice-core measurements found a decline in the winter maximum extent of ice as well.

After 2000, the rate of reduction of Arctic sea ice cover accelerated. From analysis of satellite data, NSIDC scientists reported that the extent of Arctic sea ice in 2002 was the lowest in the satellite record, likely the lowest since the early 1950s, and perhaps the lowest in several centuries. In September 2002, sea ice covered about 6.0 million km^2 (2.3 million mi^2) compared to the long-term average of about 6.3 million km^2 (2.4 million mi^2). According to the IPCC, from 1979 through 2006, the extent of Arctic sea ice decreased by 2.7 ± 0.6 percent per decade in the annual average and 7.4 ± 2.4 percent per decade for the end-of-summer. Between 1981 and 2000, average ice thickness decreased by about 1.13 m (3.7 ft) or 22%.

In September 2011, the average monthly extent of Arctic sea ice cover decreased to 4.61 million square km (1.78 million square mi), the second lowest average sea-ice extent since satellite monitoring of the polar region began in 1979 (Figure 15.30). This allowed temporary opening of the Northwest Passage shipping lanes, which connect the Atlantic and Pacific Oceans through the Arctic Ocean (Figure 15.31).

In September 2007, the summer monthly average extent of sea-ice cover reached a record minimum of 4.28 million square km (1.65 million square mi), which was 38% below the long-term average and 23% below the previous record low. This sharp decline in Arctic sea ice cover was attributed to a combination of factors: the persistence of the long-term trend in ice thinning and shrinkage, unusually strong summer winds that conveyed large amounts of ice out of the Arctic basin and created a large area of open water and thin ice in the Arctic Ocean, less than the usual summer cloud cover that allowed more solar radiation to reach the ocean surface, and the ice-albedo feedback that accelerated the warming and melting during the summer.

The shrinkage of Arctic sea ice may be a direct

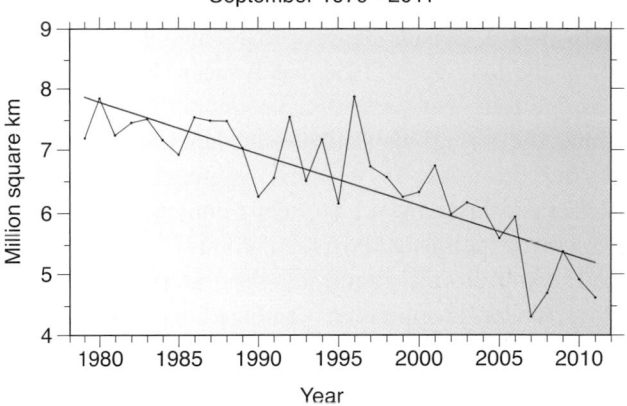

FIGURE 15.30

Arctic sea ice has been shrinking at least since the time when satellite monitoring began in 1979. The record lowest September sea ice extent was in 2007, followed by 2011. [Adapted from NOAA's National Snow and Ice Data Center, Boulder, CO]

consequence of higher air temperatures or indirectly the result of changes in ocean circulation (i.e., greater input of warmer Atlantic water under the Arctic sea ice). Some scientists argue that these are natural variations in the Arctic climate associated with the *Arctic Oscillation (AO)* and that the ice-cover will return to normal after the AO shifts phase. However the Arctic sea ice cover has declined to record low levels over the past decade while the phase of the AO has not been consistently positive. Instead, several other sea level air pressure patterns have, perhaps coincidentally, generated winds that exported sea ice out of the Arctic. It is possible the wind more readily transports thinner ice. Furthermore, research shows that less cold water flowing into the Arctic basin remains below the halocline, and has not obviously contributed to sea ice shrinkage.

Mean annual air temperatures in the Arctic region have climbed about 0.5 Celsius degree (0.9

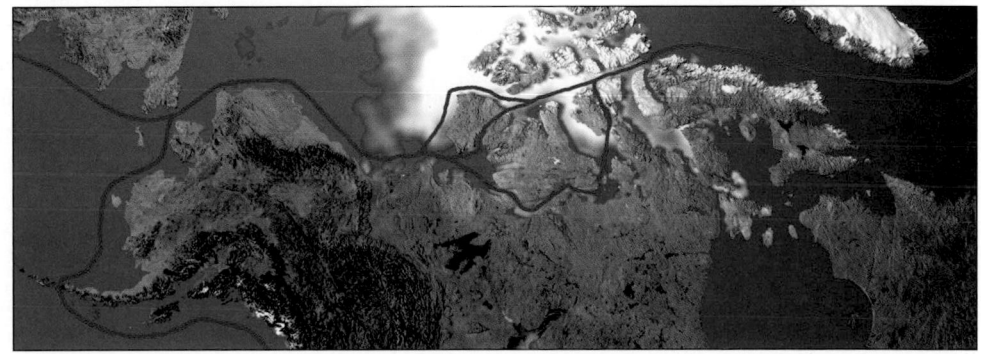

FIGURE 15.31

Northwest Passage sea routes. [NASA]

Fahrenheit degree) per decade over the past thirty years. In 2007, mean annual temperatures in the Arctic were more than 3 Celsius degrees (5.4 Fahrenheit degrees) above the long-term (1951-1980) mean. Proxy climate data indicate that present temperatures (especially in winter and spring) may have reached their highest levels in four centuries. In response, mountain glaciers in Alaska are shrinking at historically unprecedented rates, *permafrost* (permanently frozen ground) is beginning to thaw, and freshwater runoff into the ocean has increased. More winter precipitation combined with accelerated flow of groundwater (due to the thawing of permafrost) is likely responsible for a 7% increase in the discharge of six major Eurasian rivers into the Arctic Ocean since the 1930s.

OTHER IMPACTS

Whereas regional response to global climate change is uncertain, scientists and policy makers are confident that localities most vulnerable to water resource problems are places where the quality and quantity of fresh water are already a problem, that is, in semi-arid and arid regions. In the U.S., water supply problems are most likely in the drainage basins of the Missouri, Arkansas, Rio Grande, and lower Colorado Rivers. Water shortages in the Middle East and Africa may heighten political tensions, especially in nations that depend on water supplies that originate outside their borders.

Higher temperatures and more frequent drought may severely affect food production in certain regions of the world. Developed nations have the capacity to adapt their agricultural systems to climate change. While overall U.S. food production is expected to climb throughout the 21st century, there will be some losers. For example, greater demand for irrigation waters may further stress important reservoirs of groundwater (e.g., the High Plains aquifer). On the other hand, the agricultural systems of developing nations are much more vulnerable to the adverse effects of warming and are most likely to experience a significant decline in food production.

Some scientists speculated that the record high Atlantic hurricane activity in 2005 was a consequence of global warming. The prospect of continued warming and higher sea-surface temperatures prompted some scientists to predict an overall upturn in the number and intensity of tropical cyclones (i.e., hurricanes and tropical storms) through the remainder of this century. Recall from Chapter 12 that tropical cyclones ultimately derive their energy from evaporation of seawater. However, relatively high sea-surface temperature is only one of several factors required for tropical cyclone formation. In fact, some climate models predict that stronger winds aloft will accompany a warmer climate producing wind shear that would inhibit tropical cyclone development. Recent studies suggest that while the total number of tropical cyclones has not increased, the number of major hurricanes appears to be increasing.

Conclusions

The interaction of many factors is responsible for the inherent variability of climate. Although we can isolate specific boundary conditions that are internal or external to Earth's climate system, our understanding of how those boundary conditions interact is far from complete. This state of the art limits the ability of atmospheric scientists to forecast the climatic future. Continued research on climate is needed. It is reasonable to assume that physical laws govern climatic change; that is, climate variability and climate change are not arbitrary, random events. As scientists more fully comprehend the laws regulating climate variability and change, their ability to predict the climate future will improve, aided by faster supercomputers and more realistic global climate models. Meanwhile, trends in climate must be monitored closely, especially in view of the strong dependence of our food and water supplies and the effect of our energy demand on climate. In spite of, or perhaps because of, the scientific uncertainty, some experts argue that actions should be taken now to reduce anthropogenic contributions to climate change.

Basic Understandings

- Climate is defined as the weather of some locality averaged over some time period plus extremes in weather. The climatic norm (or normal) at some place encompasses the total variability of the climate record, that is, averages plus extremes.
- By international convention, climatic norms are computed from averages of weather elements compiled over a standard 30-year period. Current climate summaries are based on weather records covering the period from 1981 to 2010.
- Departures from long-term climatic averages, called anomalies, do not occur with the same sign or magnitude everywhere. At middle and high latitudes, the geographic non-uniformity of climatic anomalies is linked to the prevailing

westerly wave pattern that ultimately governs cold and warm air advection, cyclogenesis, and storm tracks. Large-scale trends in climate are also geographically non-uniform.

- Many factors working together shape the climate of any locality. Boundary conditions are imposed on climate due to latitude, elevation, topography, proximity to large bodies of water, Earth's surface characteristics, long-term average atmospheric circulation, and prevailing ocean circulation.

- The ocean is a major player in Earth's climate system. The ocean influences radiational heating and cooling, is the main source of water vapor, and a major regulator of atmospheric carbon dioxide (CO_2). On a global average annual basis, about ten times more heat is transferred from the ocean surface to the atmosphere via latent heating than sensible heating.

- Coupling of the atmosphere and ocean is most apparent in the ocean's surface waters where temperatures and wind-driven currents respond to variations in atmospheric conditions within hours to days. On the other hand, the deeper basin-scale thermohaline circulation responds more sluggishly to changes in atmospheric conditions, taking decades to centuries or longer to fully adjust.

- Mean annual isotherms roughly parallel latitude circles, underscoring the influence of solar radiation and solar altitude on climate. Through the course of a year, mean monthly isotherms in both hemispheres follow the Sun, shifting north and south in tandem.

- The global pattern of mean annual precipitation (rain plus melted snowfall) exhibits considerable spatial variability in response to topography, distribution of land and sea, and the planetary-scale circulation.

- For information on climates prior to the era of reliable instrument-based records, scientists rely on climatic inferences drawn from historical documents and climate-sensitive geological/biological evidence such as fossil plants and animals, pollen, tree growth rings, glacial ice cores, and deep-sea sediment cores.

- Plate tectonics complicates climate reconstruction of periods spanning hundreds of millions of years. In the context of geologic time, topography and the geographical distribution of continents and the ocean are variable controls of climate.

- By 40 million years ago, global climate began shifting toward colder, drier, and more variable conditions. Scientists have implicated tectonic forces and the building of the Colorado Plateau, Tibetan Plateau, and Himalayan Mountains as the principal causes of this climate change.

- Cooling after 40 million years ago culminated in the Pleistocene Ice Age that began about 1.8 million years ago. During the Pleistocene, the climate shifted numerous times between glacial climatic episodes (favoring expansion of glaciers) and interglacial climatic episodes (favoring shrinkage of glaciers or no glaciers at all).

- Climate change during the Pleistocene epoch was geographically non-uniform in magnitude; cooling was maximum at high latitudes and minimum in the tropics. This latitudinal variation in the magnitude of temperature change is known as polar amplification.

- The last major glacial climatic episode began about 27,000 years ago and reached its peak about 20,000 to 18,000 years ago when the glacial ice cover over North America was about as extensive as it had ever been.

- Notable post-glacial climatic episodes included the Medieval Warm Period (from about CE 950 to 1250), and the Little Ice Age (from about 1400 to 1900).

- The instrument-based record of global mean temperature indicates a gradual warming trend since 1880, interrupted by cooling at middle and high latitudes of the Northern Hemisphere from about 1940 to 1970.

- Matching some forcing mechanism with a climate response based on similar periods of oscillation is no guarantee of a real physical relationship. Factors that could alter the global radiative equilibrium and change Earth's climate include fluctuations in solar energy output, changes in Earth's orbital parameters, volcanic eruptions, changes in Earth's surface properties, and certain human activities.

- Changes in the Sun's total energy output at all wavelengths apparently are related to sunspot activity. Solar output varies directly and minutely with sunspot number; that is, a slightly brighter Sun has more sunspots (because of a concurrent

- increase in faculae, bright areas), and a slightly dimmer Sun exhibits fewer sunspots.

- Variations in the seasonal and latitudinal receipt of solar energy available to planet Earth associated with the Milankovitch cycles apparently drive climate oscillations operating over tens of thousands to hundreds of thousands of years. They were likely responsible for the major advances and recessions of the Laurentide ice sheet over North America during the Pleistocene epoch. Cycles consist of regular variations in precession and tilt of Earth's rotational axis and the eccentricity of its orbit about the Sun. These same cycles show up in deep-sea sediment cores that date from the Pleistocene Ice Age.

- Only violent volcanic eruptions rich in sulfur dioxide are likely to impact global or hemispheric-scale climate by affecting the transparency of the atmosphere to incoming solar radiation and outgoing infrared radiation from Earth's surface. Such an eruption has a non-uniform effect on surface air temperature but is unlikely to lower the mean hemispheric temperature by more than about 1.0 Celsius degree (1.8 Fahrenheit degrees).

- Earth's surface, a large fraction of which is covered by ocean water, is the prime absorber of solar radiation so that any change in the physical properties of the water or land surfaces or in the relative distribution of ocean and land may impact the global radiation balance and climate.

- Human activity may impact global-scale climate by increasing the concentration of greenhouse gases (causing warming) or sulfurous aerosols (causing cooling). The upward trend in atmospheric carbon dioxide since the beginning of the Industrial Revolution is primarily due to burning of fossil fuels and to a lesser extent the clearing of vegetation.

- A global climate model consists of mathematical equations that describe the physical laws that govern the interactions among the various components of Earth's climate system and feedback in the climate system. They are used to predict broad regions of expected positive and negative temperature and precipitation anomalies.

- The climate record does not appear to contain reliable cycles that can be used to forecast the climate over a period of several centuries. Also, no satisfactory analogs exist in the climate record that would allow scientists to predict accurately the regional response to global-scale climate change.

- Current global climate models predict that significant global warming will accompany a doubling of atmospheric CO_2 (possible by the close of this century if assumptions concerning future emission rates of this gas turn out to be correct). Warming may be greater in magnitude than during any prior climate change of the past 10,000 years.

- Global climate models predict that continuation of the current warming trend will be accompanied by an overall increase in precipitation. However, significant variations will occur geographically with some regions wetter and other regions drier.

- Projected increases in global temperature would likely cause sea level to rise in response to melting of glaciers and expansion of seawater. Higher sea level would accelerate coastal erosion, inundate wetlands, estuaries and some islands, and make low-lying coastal plains more vulnerable to storm surges. On the other hand, warming is likely to cause levels of the Great Lakes to fall in response to increased evaporation and reduced winter snow pack.

- Melting of floating sea ice can alter climate in a significant way by greatly reducing the surface albedo. Beginning in the 1950s, observations by ships and aircraft detected shrinkage in the summer minimum extent of Arctic sea ice while the winter maximum remained nearly constant. By the mid-1970s, satellite surveillance, submarines, and ice-cores were finding a decline in the winter maximum as well. Shrinkage of Arctic sea ice raises concerns about a possible ice-albedo feedback mechanism that would accelerate melting of sea ice and amplify warming.

- Higher global temperatures may translate into more severe drought in some locations and floods in other places. Unfortunately, global climate models do not agree on the locations of precipitation extremes. Arid and semi-arid localities already experiencing water supply problems are most vulnerable to the adverse effects of global climate change.

Enduring Ideas

- Climate is the state of the atmosphere (weather) of some locality averaged over some time period (usually 30 years) plus extremes in weather over the entire period of record. Many boundary conditions are imposed on climate due to latitude, elevation, topography, proximity to large bodies of water, Earth's surface properties, and long-term average atmospheric and ocean circulations.
- Interactions among these variables can result in positive and negative feedback.
- For the climate record prior to the instrument era, scientists rely on interpretation of proxy climatic data sources such as tree growth rings and glacial ice cores. The inherent temporal variability of climate is one of the most important lessons of the reconstructed and instrument-based climate record.
- Many agents and forcing mechanisms, operating over a broad range of temporal scales, are responsible for climate change and variability. Climate change is geographically non-uniform in both sign and magnitude.
- Human activity (primarily burning of fossil fuels) is changing global-scale climate chiefly by increasing the atmospheric concentrations of greenhouse gases (causing warming) or sulfurous aerosols (causing cooling).
- Global climate models predict continued warming through at least the present century and an overall increase in precipitation but with significant geographical variations (some regions wetter and other regions drier). Consequences of warming are predicted to include rising sea level and shrinkage of the Arctic ice cover enhancing high latitude warming through an ice-albedo feedback mechanism.

Key Terms

climate
National Climatic Data Center (NCDC)
climatic norm
climatic anomalies
agroclimatic compensation
heat equator
geologic time
plate tectonics
continental drift

Paleocene-Eocene Thermal Maximum (PETM)
glacial climate
interglacial climate
polar amplification
Younger Dryas
Holocene epoch
Medieval Warm Period
Little Ice Age
global radiative equilibrium feedback

sunspot
Maunder minimum
Milankovitch cycles
sulfurous aerosols
aerosols
global climate model (GCM)
lifetime of a gas
eustasy
ice stream
ice-albedo feedback

Review

1. The climatic norm (or normal) is often equated to an average value of some climatic element. Provide a more complete definition of the climatic norm.
2. Is a 30-year period likely to include the full range of climate variability? Explain your response.
3. Explain how climatic anomalies are computed.
4. Identify several boundary conditions that operate in Earth's climate system.
5. How does plate tectonics complicate efforts to reconstruct climate over periods of tens of millions to hundreds of millions of years?
6. Distinguish between a glacial climate and an interglacial climate.
7. What is meant by polar amplification? What are some possible implications of polar amplification should global warming continue well into the future?
8. What is the criterion for establishing that climate change is abrupt?
9. Explain how solar energy output varies with sunspot number.
10. What are the characteristics of a volcanic eruption that is most likely to impact global climate?

Critical Thinking

1. Why are precipitation anomaly patterns more complex than temperature anomaly patterns?
2. Demonstrate that the ocean is a major player in Earth's climate system.
3. How do regular season-to-season changes in the meridional temperature gradient affect the circulation of the westerlies in mid latitudes?
4. In the context of human existence, some climatic boundary conditions are essentially fixed whereas others are variable. Provide examples of both types of boundary conditions.
5. Identify the various factors that may affect the integrity of the long-term instrument-based air temperature record.
6. Explain why the Milankovitch astronomical theory of Ice Ages was not widely accepted by the scientific community until after the mid 1970s.
7. Past climatic episodes differ from present climatic episodes in frequency rather than type of episode. Elaborate on the meaning and significance of this statement.
8. Why is it difficult to establish the likely impact of global warming on the frequency of Atlantic tropical cyclones (tropical storms and hurricanes)?
9. How do aerosols of anthropogenic origin mask the effects of carbon dioxide on Earth's surface air temperature?
10. Even if greenhouse gas emissions were to stabilize at present levels, the global warming trend is likely to persist well beyond the 21st century. Explain why.

For Further Exploration

Sources of Proxy Climate Data

For times and places where no instrument-derived record of climate exists, past climate information may be inferred from various sensors that substitute for actual weather instruments. These sensors of climate are called *proxy climate data sources*. They include, for example, historical documents, tree growth rings, pollen profiles, deep-sea sediment cores, and glacial ice cores. No one type of proxy alone is sufficient to enable scientists to reconstruct broad scale patterns of climate.

Under cautious scrutiny, certain historical documents archived in libraries and museums can yield a wealth of information on past climates. Personal diaries, almanacs, old newspapers, and mariner's log books may yield qualitative and some quantitative references to weather and climate. Other types of documents refer indirectly to weather and climate but can be useful nonetheless. Records of success of grain harvests, quality of wine, or phenological events (such as dates of blooming of plants in spring) provide indirect indications of growing season weather. For example, researchers at the University of Bern reconstructed summer weather/climate patterns for parts of Switzerland based on records of grape harvests dating as far back as the late 15th century. The growth of grapevines and the ripening of the fruit strongly respond to average April through August temperatures.

Caution must be exercised in inferring climate information from historical documents because many factors, in addition to weather and climate, usually influence such records. For example, aside from growing season weather, harvest dates in vineyards are affected by fluctuations in the wine market. Usually, the authors of such documents had no intention of chronicling the weather or climate and it is also important to bear in mind that people have long applied ingenuity to moderate the impact of climate—particularly extremes in climate. Hence, climate information derived from written records of human activity is not always reliable and corroborating data from other independent sources are necessary to support these climatic inferences.

Analysis of variations in the thickness and density of annual growth rings of certain trees can yield detailed information on past climates (Figure 1). The study of tree growth rings for climate data is known as *dendroclimatology*. Andrew E. Douglass (1867-1962), a solar astronomer, pioneered this work in the American Southwest between 1894 and 1901 while at Lowell Observatory, a private non-profit research institution in Flagstaff, AZ. In 1937, he founded the Laboratory of Tree-Ring Research at the University of Arizona. Today, his successors at the Laboratory reconstruct past climates using computers programmed with special statistical techniques.

At the onset of the growing season in spring, plant tissue located immediately beneath tree bark produces relatively large thin-walled wood cells, which give the wood a relatively light appearance. Wood cells produced in summer, however, are thick-walled, giving the wood a darker appearance. A year's growth of spring wood plus summer wood constitutes an annual growth ring, so counting the number of growth rings gives the age of the tree in years. Because the width of growth rings normally decreases as the tree ages, widths are usually expressed in terms of a *tree-growth index*, defined as the ratio of the actual tree growth-ring width to the width expected based on the tree's age. The index is relatively low in stressful growing seasons and high in favorable growing seasons.

Only trees living in subpolar terrestrial regions are useful in dendroclimatic research and primarily record conditions during the warm season. (Tropical species do not have well-defined growth rings.) Trees growing near the limits of their range are the most sensitive to climate variability so that their growth rings are the most reliable sensors of climate. A simple hollow drill is used to extract cores from living trees or cut timber. Usually cores are taken from many trees at one site, and tree-ring indexes are averaged. Typically, tree ring

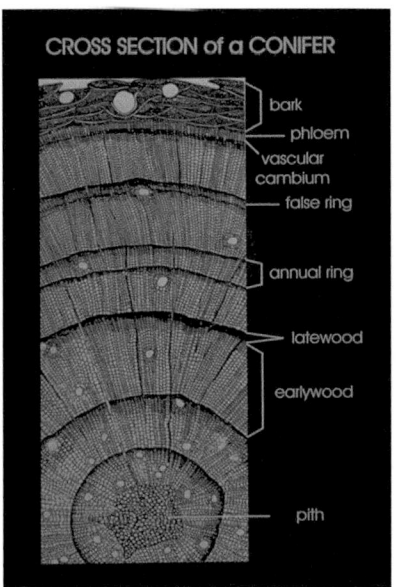

FIGURE 1

This cross-section through a conifer tree species shows annual growth rings whose thickness and density may reveal information on the climate of past growing seasons. A potential source of error is the presence of a "false" ring, an apparent sequence of latewood, earlywood, and then latewood within the bounds of an actual tree growth ring. Stressful conditions during the growing season may be responsible for a false ring. [NOAA Paleoclimatology Program]

chronologies date back some 500 to 700 years but range up to 11,000 years in a few cases. In Western and Southwestern North America, the primary locale for dendroclimatic research, the ponderosa pine, Douglas fir, or the exceptionally long-lived bristlecone pine are sampled. Some of the longest tree ring records have been obtained from several thousand year-old bristlecone pine, found in the White Mountains of eastern California (Figure 2). By assiduous matching of tree growth-ring records from living trees with those from timbers in prehistoric dwellings, detailed tree-ring chronologies are extended back in time thousands of years. This matching technique is known as *cross-dating*.

Although other environmental factors (e.g., soil type) can be important, the thickness and density of tree growth rings are especially sensitive to moisture stress and have been used to reconstruct lengthy drought chronologies. For example, Connie A. Woodhouse, a climatologist at the University of Arizona, and her colleagues found tree growth ring evidence that a 60-year *megadrought* impacted the Colorado Plateau during the 12th century. Analysis of tree growth rings enabled these scientists to extend the region's drought chronology back to CE 762. They relied on data from both living trees and cross-dating of ring patterns in tree trunks scattered throughout the upper Colorado River drainage basin, where dry conditions preserved them.

FIGURE 2
Bristlecone pines in the White Mountains of California. [Photo by Jonathan Pilcher, Palaeoecology Centre, Queen's University, Belfast; NOAA/NCDC/ Paleoclimatology Program]

Ponds, peat bogs, marshes, and swamps are favorable sites for the accumulation and preservation of wind-borne pollen. *Pollen* is the tiny dust-like fertilizing component of a seed plant that is dispersed by the wind. Mixing with other sediments (clay, silt and organic particles), pollen grains settle and accumulate in low-lying depositional areas. Upward of 20,000 pollen grains may be mixed in a single cubic centimeter of pond mud. Assuming that the pollen is the product of nearby vegetation and that climate largely governs vegetation types, climate may be inferred from pollen. When climate changes, vegetation and pollen types also change. Thus changes in the abundance of pollen of different species at various depths within accumulated sediment may reflect variations in climate.

Scientists use a corer to extract a sediment column (core), then separate pollen from its host sediments and reconstruct the sequence of past changes in vegetation. From the climate requirements of the reconstructed vegetation (based on modern species distribution and modern climate), scientists decipher the sequence of past climatic episodes.

Pollen is a valuable source of information on the vegetation and climate of the late Pleistocene Ice Age and subsequent Holocene Epoch, especially over the past 15,000 years. Using sophisticated statistical techniques to calibrate climate and pollen, scientists have reconstructed remarkably detailed quantitative climate data. For example, a pollen record (profile) from Kirchner Marsh near Minneapolis, MN, yielded a record of variations in July mean temperature and annual precipitation back to 12,000 years ago. Unfortunately, relatively few sites favor the accumulation and preservation of continuous long-term pollen/climate records. In North America, most such records come from a few geographical areas, including western mountain valleys, the Great Lakes region, and interior New England.

Cores extracted from sediments that blanket the ocean floor yield a continuous record of sedimentation dating back many hundreds of thousands of years and in some places, millions of years (Figure 3). Much of what we know about the climate of the Pleistocene Ice Age is based on analysis of the shell and skeletal remains of microscopic marine organisms found in deep-sea sediment cores. Identification of the environmental requirements of these organisms, plus oxygen isotope analysis of their remains, enables scientists to distinguish between relatively cold and warm climatic episodes of the past.

With *oxygen isotope analysis*, scientists use a special property of water to reconstruct large-scale climate fluctuations of the Pleistocene Ice Age. A water molecule (H_2O) is composed of one of two stable isotopes of oxygen, ^{16}O or ^{18}O. (*Isotopes* consist of atoms that are chemically identical, with the same number of protons but different number of neutrons in the nucleus.) In the Earth system, the lighter isotope (^{16}O) is much more abundant than the heavier

A   B   C

FIGURE 3
Sediment cores extracted from beneath the ocean floor provide valuable information on the geologic and climatic past. (A) A hollow pipe lined with plastic tubing and coupled to a weight at the top is lowered over the side of a ship. When within about 8 m of the bottom, the corer free-falls into the sediment while a piston suctions sediment into the tube. The coring device is recovered and the sediment core is removed and split lengthwise for analysis. (B) USGS crew on the research vessel *G.K. Gilbert* collect a 20-ft sediment core, using an electric coring system and hydraulic crane. [USGS] (C) Core racks holding a total of 72,000 m of sediment cores at the Deep-Sea Sample Repository of Lamont-Doherty Earth Observatory in Palisades, NY. [Courtesy of the Lamont-Doherty Earth Observatory]

isotope (^{18}O); only one ^{18}O exists for every thousand or so ^{16}O. Nonetheless, small but significant variations occur in the ratio of light oxygen to heavy oxygen circulating in the global water cycle, which have important implications for past fluctuations in glacial ice volume.

On average, water molecules containing the lighter ^{16}O isotope move slightly faster than water molecules containing the heavier ^{18}O isotope and therefore evaporate more readily so water molecules that evaporated are enriched with light oxygen, compared to heavy oxygen. The amount of ^{16}O compared to ^{18}O is greater in cloud particles and precipitation versus liquid water on Earth's surface. When it rains or snows, the ^{16}O returns to the ocean, replenishing the ocean's supply of light oxygen and maintaining a relatively constant average ratio of light to heavy oxygen. However, geographical variations in the oxygen isotope ratio of seawater arise because of differences in precipitation amounts and evaporation rates. Seawater has more ^{18}O at subtropical latitudes, where evaporation exceeds precipitation, and less in middle latitudes, where rainfall is greater.

During a climatic episode that favors the formation or growth of a glacier, snow that accumulates on land converts to ice. Heavy water molecules condense and precipitate slightly more readily than light water molecules. Moisture plumes moving from the tropics to high latitudes lose heavy oxygen along the way, so snow falling at high latitudes has less ^{18}O than rain falling in the tropics. The result is that growing glacial ice sheets sequester more and more light oxygen, while ocean water has less and less. With a shift to an interglacial climate, ice sheets shrink and meltwater rich in ^{16}O drains back into the ocean, increasing the ratio of ^{16}O to ^{18}O.

Organic sediments on the ocean floor record fluctuations in the oxygen isotope ratio of seawater. Marine organisms, such as foraminifera living in the sunlit surface waters, build their shells from calcium carbonate ($CaCO_3$) that is dissolved in seawater. Shells formed during warmer interglacial climatic episodes contain more light oxygen than those formed during colder glacial climatic episodes. When these organisms die, their shells settle to the ocean floor and mix with other marine sediments. With specially outfitted deep-sea drilling ships, scientists extract cores from an undisturbed sequence of ocean bottom sediments. In the laboratory, the core is split open, and shells are extracted and

FIGURE 4
Removal of ice core from ice coring barrel during drilling on the Antarctic ice sheet. [Courtesy USGS]

analyzed for their oxygen isotope ratio. The youngest sediments are at the top of the core and the oldest sediments at the bottom. Variations in oxygen isotope ratio document changes in the planet's glacial ice volume and show past changes in climate. The proportion of light to heavy oxygen in ocean water decreases with increasing glacial ice volume.

Oxygen isotope analysis of deep-sea sediment cores indicates that the Pleistocene Ice Age (1.8 million to 10,500 years ago) was punctuated by numerous abrupt changes between glacial and interglacial climatic episodes. Oxygen isotope analysis has also been applied to ice layers within cores extracted from the Greenland and Antarctic ice sheets. These analyses confirm the abrupt change behavior of climate dating back hundreds of thousands of years.

To better understand Earth's climate system and climate change, scientists are collecting and analyzing climate data from the Pleistocene Ice Age, hoping to predict future climate. Ice cores (Figures 4 and 5) extracted from the Antarctic and Greenland ice sheets are important sources of information on climate change, as well as the chemical composition of air during the Pleistocene Ice Age.

In 1988, Soviet and French scientists reported on their analysis of a 2200 m (7200 ft) ice core extracted at Vostok station on the East Antarctic ice sheet. The ice core spanned 160,000 years. Oxygen isotope analysis yielded a temperature record, and chemical analysis of trapped air bubbles revealed trends in the greenhouse gases carbon dioxide and methane. During the summers of 1991-1993, two independent scientific teams, one American and the other European, drilled into the thickest portion of the Greenland ice sheet. The two drill sites were located within 30 km (19 mi) of each other, about 650 km (404 mi) north of the Arctic Circle. Both cores were about 3000 m (9840 ft) in length and spanned a time interval of roughly 200,000 years. In the mid-1990s, drilling at Vostok recovered a 3100 m (10,170 ft) ice core spanning the past 425,000 years. In 2004, the *European Project for Ice Coring in Antarctica (EPICA)* extracted an ice core from East Antarctica representing a time interval of 740,000 years. By 2008, the recovered ice core encompassed about 800,000 years.

To derive air temperature from glacial ice cores, scientists use either *oxygen isotope analysis* (described above) or analysis of the ratio of deuterium to hydrogen in ice. The water molecule is composed of two different isotopes of hydrogen: ^1H and ^2H. ^1H consists of one proton and no neutrons whereas ^2H (also called deuterium or D) has one proton and one neutron. Isotopic ratios are

FIGURE 5
This 6-m (20-ft) long ice core was extracted from the Greenland ice sheet, and is a source of information on past variations in climate and atmospheric composition. [Photo by Mark Twickler, University of New Hampshire; NOAA/NCDC/Paleoclimatology Project]

compared to that of *standard mean ocean water (SMOW)*. Compared to SMOW, glacial ice cores contain slightly less of the heavier oxygen or deuterium isotopes. As the temperature falls, there is comparatively less and less ^{18}O or D.

In addition to reconstructing temperature, glacial ice cores are also analyzed for changes in the composition of the atmosphere, especially greenhouse gases. The present level of atmospheric carbon dioxide is about 27% higher than the highest levels detected in air bubbles trapped in glacial ice cores dating back 650,000 years. In addition, the reconstructed temperature record closely parallels the concentration of the greenhouse gases carbon dioxide and methane. The question remains, however, whether greenhouse gas concentration drives climate or climate drives greenhouse gas concentration, or some combination of the two.

Lessons of the Climate Past

What does the climate record tell us about the behavior of climate through time? The following lessons of the climate past are useful in assessing prospects for the climate future and the possible impacts of climate change.

Climate is inherently variable over a broad spectrum of time scales ranging from years to decades, to centuries, to millennia. Variability and change are endemic characteristics of climate. The question for the future is not whether the climate will change but in which direction climate will change and by how much.

Climate variability and change are geographically non-uniform in both sign (warming or cooling) and magnitude. Some areas may experience warming while other areas experience cooling over the same period. Also, some areas may become wetter (with potentially disastrous flooding) while other areas experience drying and more frequent drought.

Global- and hemispheric-scale trends in climate are not necessarily duplicated at particular locations, although the magnitude of temperature change tends to amplify with increasing latitude (polar amplification). For example, more rapid warming takes place at high latitudes than low latitudes. Partially for this reason, global climate change is not likely to have the same environmental or societal impact everywhere.

Climate change may consist of a long-term trend in various elements of climate (e.g., mean temperature or average precipitation) and/or a change in the frequency of extreme weather events (e.g., drought, excessive cold). Climate encompasses mean values plus extremes in climatic elements. A trend toward warmer or cooler, wetter or drier conditions may or may not be accompanied by a change in frequency of weather extremes. On the other hand, a climatic regime featuring relatively little change in mean temperature or precipitation through time could be accompanied by an increase or decrease in frequency of weather extremes.

Climate change tends to be more abrupt than gradual. If the time of transition between climatic episodes (e.g., glacial climate versus interglacial climate) was much shorter than the duration of the individual episodes, then the transition would be considered relatively abrupt. Analysis of cores extracted from the Greenland ice sheet indicates that millennial-scale cold and warm climatic episodes were punctuated by abrupt change over periods as brief as a single decade or even several years. The abrupt-change nature of climate would test the resilience of society to respond effectively to climate change.

Few cyclical variations can be discerned from the long-term climate record. Regular cycles include diurnal and seasonal changes in incoming solar radiation (the forcing) and temperature (the response). This means simply that days are usually warmer than nights and summers are warmer than winters. Quasi-regular variations in climate include El Niño (occurring about every 3 to 7 years), Holocene millennial-scale fluctuations identified in glacial ice cores, and the major glacial-interglacial climate shifts of the Pleistocene Ice Age unlocked from analysis of deep-sea sediment cores and operating over tens of thousands to hundreds of thousands of years (i.e., the Milankovitch cycles).

Climate change impacts society. History recounts numerous instances when climate change significantly impacted society. Consider some examples. Researchers have implicated prolonged drought in the decline and fall of Middle Eastern empires around BCE 2000, the Harappan civilization of the Indus Valley of India about BCE 1700, the Mycenaean civilization of Greece in BCE 1200, and the Anasazi abandonment of Mesa Verde (southwestern Colorado) around CE 1300. The cooling trend that heralded the Little Ice Age was likely a major factor in the disappearance of Norse settlements in Greenland about CE 1400. Although modern societies may be more capable of dealing with climate change than early peoples, a rapid and significant change in climate likely would seriously impact all sectors of modern society.

CONVERSION FACTORS

	Multiply	*By*	*To obtain*
LENGTH	inches (in.)	2.54	centimeters (cm)
	centimeters	0.3937	inches
	feet (ft)	0.3048	meters (m)
	meters	3.281	feet
	statute miles (mi)	1.6093	kilometers (km)
	statute miles (mi)	0.869	nautical miles
	kilometers	0.6214	statute miles
	kilometers	0.54	nautical miles
	nautical miles	1.852	kilometers
	kilometers	3281	feet
	feet	0.0003048	kilometers
SPEED	miles per hour (mph)	1.6093	kilometers per hour (kph)
	miles per hour	0.869	knots (kts)
	miles per hour	0.447	meters per second (m/s)
	knots	1.1508	miles per hour
	knots	0.5144	meters per second
	kilometers per hour	0.6214	miles per hour
	kilometers per hour	0.540	knots
	meters per second	1.944	knots
	meters per second	2.237	miles per hour
WEIGHTS AND MASS	ounces (oz)	28.35	grams (g)
	grams	0.0353	ounces
	pounds (lb)	0.4536	kilograms (kg)
	kilograms	2.205	pounds
	tons	0.9072	metric tons
	metric tons	1.102	tons
LIQUID	fluid ounces (fl oz)	0.0296	liters (L)
	gallons (gal)	3.785	liters
	liters	0.2642	gallons
	liters	33.814	fluid ounces

AREA			
	acres (A)	0.4047	hectares (ha)
	square yards (yd^2)	0.8361	square meters (m^2)
	square miles (mi^2)	2.590	square kilometers (km^2)
	hectares	0.010	square kilometers
	hectares	2.471	acres

PRESSURE/ FORCE			
	pounds force (lb)	4.448	newtons (N)
	newtons	0.2248	pounds force
	millimeters of mercury at 0 ºC	133.32	pascals (Pa; N per m^2)
	pounds per square inch (psi)	6.895	kilopascals (kPa; 1000 pascals)
	pascals	0.0075	millimeters of mercury at 0 ºC
	kilopascals	0.1450	pounds per square inch
	bars	1000	millibars (mb)
	bars	100,000	pascals
	bars	0.9869	atmospheres (atm)

ENERGY			
	joules (J)	0.2389	calories (cal)
	kilocalories (kcal)	1000	calories
	joules	1.0	watt-seconds (W-sec)
	kilojoules (kJ)	1000	joules
	calories	0.00397	Btu (British thermal units)
	Btu	252	calories

POWER			
	joules per second	1.0	watts (W)
	kilowatts (kW)	1000	watts
	megawatts (MW)	1000	kilowatts
	kilocalories per minute	69.78	watts
	watts	0.00134	horsepower (hp)
	kilowatts	56.87	Btu per minute
	Btu per minute	0.0236	horsepower

MILESTONES IN THE HISTORY OF ATMOSPHERIC SCIENCE

ca. 525 B.C.E.	Greek philosopher Anaximenes of Miletus proposed that winds, clouds, rain and hail are formed by thickening of air, the primary substance.
ca. 500 B.C.E.	Parmenides classified world climates by latitude as torrid, temperate, or frigid.
ca. 400 B.C.E.	Rainfall measured in India using the first known rain gauge.
350-340 B.C.E.	Aristotle produced his *Meteorologica*, the first work on the atmospheric sciences.
330 B.C.E.	Hippocrates wrote *Air, Waters, and Places,* a treatise on climate and medicine.
ca. 135 B.C.E.	The Chinese writer Han Ying first notes the hexagonal shape of snowflakes in *Moral discourses illustrating the Han text of the "Book of Songs."*
ca. 28 B.C.E.	Chinese astronomers observed sunspots with the unaided eye.
61 C.E.	Seneca complained of the air pollution of Rome.
1287	First snow gauges in China.
1304	Theodoric of Friebourg (1250-1310) wrote *De iride* (On the Rainbow) describing his experiments with bulbs filled with water.
1442	Rain gauges used in Korea.
1450	Cardinal Nicholas of Cusa (1400-1464) constructed the first balance hygrometer.
1591	English mathematician Thomas Harriot (1560-1621) notes that snowflakes have six sides.
1592	Galileo Galilei (1564-1642) invented the thermoscope, forerunner of the thermometer.
1610	Galileo Galilei was among the first to study sunspots telescopically.
1621	Dutch physicist Willebrord Snell (1590-1626) discovered the law of refraction of light.
1637	French philosopher René Descartes (1596-1650) published *Discours de la méthode,* which includes his ideas about the rainbow and cloud formation.

1638	Galileo Galilei determined that air has weight.
1639	The first mention of a rain gauge in European literature.
1641	Ferdinand II of Tuscany (1610-1670) constructs a thermometer containing liquid.
1643	Italian mathematician Evangelista Torricelli (1608-1647) invented the mercury barometer.
1644	Rev. John Campanius (1601-1683) recorded the first weather observations in America, near the present site of Wilmington, DE.
1648	Blaise Pascal (1623-1662) and Florin Périer ascended the Puy-de-Dôme in France to demonstrate that air pressure decreases with increasing altitude.
1660	German scientist Otto von Guericke (1602-1686) noted that a severe storm is followed by a sudden drop in barometric pressure.
1666	Sir Isaac Newton (1643-1727) used a glass prism to demonstrate that sunlight is made up of a spectrum of colors.
ca. 1670	Mercury was used in thermometers for the first time.
1683	English astronomer Edmund Halley (1656-1742) published the first comprehensive map of winds along with a partial explanation of the trade winds.
1686	Edmund Halley proposed an explanation for monsoons.
1687	Isaac Newton developed his three laws of motion.
1687	French physician Guillaume Amontons (1663-1705) invented a hygrometer.
1709	German physicist Gabriel Daniel Fahrenheit (1686-1736) constructed a thermometer with alcohol as the working fluid.
1714	Gabriel Daniel Fahrenheit introduced the Fahrenheit temperature scale.
1730	French entomologist and physicist René Réamur (1683-1757) introduced a thermometer using a water/alcohol mixture.
1735	German explorer Johann G. Gmelin (1709-1755) discovered permafrost in Siberia.
1735	George Hadley (1685-1768) proposed the Hadley cell circulation.
1742	Swedish astronomer Anders Celsius (1701-1744) introduced the Celsius temperature scale.
1743	Jean-Pierre Christin (1683-1755) inverted the fixed points on the Celsius temperature scale, producing the scale used today.
1743	Benjamin Franklin (1706-1790) deduced the progressive movement of a storm along the U.S. East Coast.

1747	Benjamin Franklin began a series of experiments on electricity that led to the use of the lightning rod.
1749	Benjamin Franklin attached a lightning rod to his house in Philadelphia.
1749	In Glasgow, Scotland, Alexander Wilson (1714-1786) used thermometers attached to a kite to obtain the first temperature profile of the lower atmosphere (up to perhaps 60 m or 200 ft).
1752	Benjamin Franklin performed his famous kite experiment demonstrating the electrical nature of lightning.
1760	Scottish chemist Joseph Black (1728-1799) formulated the concept of specific heat.
1766	English chemist Henry Cavendish (1731-1810) discovered hydrogen.
1772	Scottish medical student Daniel Rutherford (1749-1819) was first to publish the discovery of nitrogen.
1774	Carl Scheele (1742-1786) and Joseph Priestly (1733-1804) discovered oxygen.
1778	Thomas Jefferson (1743-1826) and James Madison (1751-1836) took the nation's first simultaneous weather observations.
1781	Systematic weather observations began at New Haven, CT.
1783	Swiss naturalist Horace de Saussure (1740-1799) demonstrated supercooling of water.
1800	Astronomer Sir William Herschel (1738-1822) discovered energy transfer in the infrared (IR).
1802-03	British pharmacist and amateur meteorologist Luke Howard (1772-1864) developed his classification of cloud types.
1802	French physicist J. L. Gay-Lussac (1778-1850) published the *gas law*; British chemist John Dalton (1766-1844) defined *relative humidity*.
1804	J. L. Gay-Lussac and Jean Biot (1774-1862) conducted the first manned balloon exploration of the atmosphere, reaching a maximum altitude of 7000 m (23,000 ft).
1805	British admiral Sir Francis Beaufort (1774-1857) proposed a scale of winds (now known as the *Beaufort scale*).
1811	Italian chemist Amedeo Avogadro (1776-1856) first stated Avogadro's law.
1814	U.S. Surgeon General James Tilton (1745-1822) directed the Army Medical Corps to begin a diary of weather conditions at army posts.
1819	German meteorologist Heinrich Wilhelm Brandes (1777-1834) drew the first weather map, depicting a storm over the English Channel.
1820	English chemist John Frederic Daniell (1790-1845) invented the dew-point hygrometer.
1821	Swiss geologist Ignatz Venetz (1788-1857) proposed that glaciers formerly occurred throughout Europe.

1824	French mathematician Jean Baptiste Joseph Fourier (1768-1830) studied the heating of enclosed spaces such as greenhouses.
1825	E.F. August developed the psychrometer.
1835	French mathematician Gaspard Gustave de Coriolis (1792-1843) demonstrated quantitatively how Earth's rotation affects large-scale motion in the atmosphere or ocean.
1837	Swiss naturalist Louis Agassiz (1807-1873) used the term *Eiszeit* (Ice Age) for the first time; he championed the glacial theory during the mid 1800s.
1837	American meteorologist James P. Espy (1785-1860) published the first U.S. weather map.
1838	British Navy adopted the Beaufort scale for estimating wind speed from the state of the sea.
1840	Swiss palenontologist Louis Agassiz (1807-1873) publishes his *Etudes des glaciers* (*Studies on Glaciers*).
1842	Austrian physicist Johann Christian Doppler (1803-1853) first explained the Doppler effect.
1842	French mathematician Joseph Alphonse Adhémar (1797-1862) proposed that regular variations in Earth's orbit explained climate fluctuations during the Ice Age.
1843	German astronomer Samuel Heinrich Schwabe (1789-1875) discovered the sunspot cycle.
1844	Aneroid barometer invented.
1850	French chemist Jean Barrel (1819-1884) confirmed the existence of supercooled cloud droplets in the free atmosphere.
1853	American meteorologist James Coffin (1806-1873) suggested that there are three distinct wind zones in the Northern Hemisphere.
1853	First International Meteorological Conference, held in Brussels.
1856	American meteorologist William Ferrel (1817-1891) proposed a model of the general circulation of the atmosphere consisting of three cells.
1857	American physicist Lorin Blodget (1823-1901) published *Climatology of the United States*.
1857	Dutch meteorologist Christopher Buys-Ballot (1817-1890) developed a rule that relates wind direction to the location of a low pressure center (*Buys Ballot's law*).
1859	John Tyndall (1820-1893), an Irish physical scientist, conducted the first experiments on the radiative properties of gases and established the experimental basis for the greenhouse effect.
1862	English meteorologist James Glaisher (1809-1903) and Henry Coxwell (1819-1900) set the manned balloon altitude record of 9000 m (29,500 ft) over Wolverhampton, England.
1864	Scottish scientist James Croll (1821-1890) furthered study on the astronomical theory of the Ice Age.

1869	In Cincinnati, OH, Cleveland Abbe (1838-1916) prepared the first regular weather maps for part of the United States.
1870	President Ulysses S. Grant (1822-1885) signed legislation establishing a national weather service operated by the U.S. Army Signal Corps.
1871	Cleveland Abbe appointed chief meteorologist of the new national weather service.
1874	Based on field studies, American geologist Thomas Chrowder Chamberlin (1843-1928) suggested that there were several Ice Ages separated by nonglacial epochs.
1874	Hurricane plotted for the first time on a surface weather map (offshore near Savannah, GA).
1878	The International Meteorological Organization (IMO) was founded.
1881	English Lord Rayleigh (1842-1919) demonstrated that scattering of visible light by gas molecules is inversely proportional to the fourth power of the wavelength.
1884	Austrian physicist Ludwig Boltzmann (1844-1906) derived the Stefan-Boltzmann law.
1884	International time zones were adopted at the International Meridian Conference in Washington, DC.
1884	John P. Finley (1854-1943) of the U.S. Army Signal Corps issued the first experimental tornado forecasts.
1885	First reported sighting of noctilucent clouds.
1886	U.S. Army Signal Corps banned the use of "tornado" in weather forecasts.
1888	Gustavus Hinrichs (1836-1923), an Iowa weather researcher, first used the word *derecho* for a straight-line windstorm.
1891	Luis Carranza, President of the Lima Geographical Society, describes a countercurrent flowing north to south from Paita to Pacasmayo and called El Niño by the local fishermen.
1891	The national weather service was transferred from military to civilian control in a new Weather Bureau within the U.S. Department of Agriculture.
1893	British astronomer E. Walter Maunder (1851-1928) identified a period of low solar activity between 1645 and 1715 (now known as the *Maunder minimum*).
1894	German physicist Wilhelm Wien (1864-1928) developed his *displacement law of radiation*.
1894	First sounding of the atmosphere using a self-recording thermometer attached to a kite at Blue Hill Observatory, Milton, MA.
1896	Svante August Arrhenius (1859-1927), a Swedish electrochemist, proposed that increasing levels of atmospheric carbon dioxide could alter Earth's heat budget.
1900	Russian-born German climatologist Wladimir Peter Köppen (1846-1940) first published his original climate classification scheme.

1902	French meteorologist Léon Philippe Teisserenc de Bort (1855-1913) identified and named the troposphere and stratosphere as layers of the atmosphere.
1904	Vilhelm Bjerknes (1862-1951) laid the foundations of synoptic meteorology at the Bergen School of Meteorology in Norway.
1905	Swedish physicist V. Walfrid Ekman (1874-1954) first described mathematically the coupling of surface winds and surface ocean waters (the *Ekman spiral*).
1910	German meteorologist Alfred Wegener (1880-1930) and American geologist Frank B. Taylor (1860-1939) independently proposed the concept of continental drift.
1916	Pyranometer for measuring global radiation developed.
1917	Vilhelm Bjerknes formulated polar front theory.
1920s	Serbian astronomer Milutin Milankovitch (1879-1958) revived the theory of long-term cyclic climate change based on regular changes in Earth-Sun geometry (*Milankovitch cycles*); he calculated latitudinal and seasonal changes in incoming solar radiation.
1920	Theory of atmospheric front was developed by Vilhem Bjerknes, Holver Solberg (1895-1974), and Jacob Bjerknes (1897-1975).
1924	Sir Gilbert Walker (1868-1958) discovered the Southern Oscillation.
1927	G. Stüve (1888-1935) introduces a type of thermodynamic diagram (the *Stüve diagram*).
ca. 1928	First radiosonde developed.
1929	American physicist Robert H. Goddard (1882-1945) conducted the first rocket probe of the atmosphere.
1933	Swedish meteorologist Tor Bergeron (1891-1977) published his paper on *Physics of Clouds and Precipitation*.
1935	Radar invented.
1935	Selection of the 30-year period 1901-1930 as the basis for calculating climatic normals, at the International Meteorological Conference in Warsaw, Poland.
1937	Carl-Gustav Rossby (1898-1957) introduced techniques for forecasting upper-level westerly waves.
1937	Andrew E. Douglass (1867-1962), a solar astronomer, founded the Laboratory of Tree-Ring Research at the University of Arizona.
1937	First official U.S. Weather Bureau radiosonde launch, at East Boston, MA.
1938	The Callendar Effect was articulated by the British engineer Guy Stewart Callendar (1898-1964).
1939	Work by Walter Findeisen supported Bergeron's theory of precipitation development in cold clouds (later known as the *Bergeron-Findeisen process*).

1941	Radar applied to weather systems for the first time; used to track a thunderstorm on the south coast of England.
1944	American meteorologist Hurd Curtis Willett (1903-1992) produced atmospheric cross sections showing the jet stream.
1946	American chemists Vincent J. Schaefer (1906-1993) and Irving Langmuir (1881-1957) performed the first cloud-seeding experiments.
1946	Hungarian-born American mathematician John von Neumann (1903-1957) and colleagues began mathematical modeling of the atmosphere.
1946	U.S. Weather Bureau established the first River Forecast Centers at Cincinnati, OH and Kansas City, MO.
1947	American chemist Willard F. Libby (1908-1980) invented the radiocarbon dating technique, a valuable tool in reconstructing late-glacial climates.
1947	First photographs of the Earth's cloud cover obtained by a V2 rocket at altitudes of 110 to 165 km (70 to 100 mi).
1948	E.J. Fawbush (1932-2009) and Robert C. Miller (1920-1998) of the U.S. Air Force Air Weather Service developed and applied a method for forecasting tornadoes.
1950	A team led by John von Neumann produced the first computer-generated weather forecasts using the ENIAC computer.
1951	The World Meteorological Organization (WMO) commenced operation.
1951	The American geographer and climatologist Arthur N. Strahler (1918-2002) introduced a climate classification scheme based on air masses.
1952	U.S. Weather Bureau lifted the ban on the use of the word "tornado" in civilian weather forecasts.
1955	Beginning of era when electronic computers were routinely generating weather forecasts from surface and upper-air weather observations.
1956	The British government attempted to reduce smog by limiting smoke emissions.
1957	Atmospheric CO_2 measurements began at Mauna Loa and Antarctica.
1959	A temperature-humidity index first introduced by the U.S. Weather Bureau.
1959	Satellite surveillance of Earth's climate system began with the U.S. launch of *Explorer I*. An onboard radiometer provided the first measurements of Earth's radiation budget from space.
1959	WSR-57 radar installed at National Hurricane Center in Miami, FL.
1960	The United States successfully orbited the first weather satellite, TIROS-I.

1961	American meteorologist Edward N. Lorenz (1917-2008) observed that computer predictions of weather are highly sensitive to small differences in initial conditions, an aspect of chaos theory.
1963	Edward N. Lorenz applied chaos theory to meteorology.
1964	U.S. government introduced the first Clean Air Act to set national ambient air quality standards.
1965	British climatologist Hubert H. Lamb (1913-1997) was the first to describe the Medieval Warm Period as an episode of mild winters and warm, dry summers in western Europe from C.E. 1100 to 1200.
1966	Jacob Bjerknes (1897-1975) demonstrated a relationship between El Niño and the Southern Oscillation.
1967	American inventor Verner E. Suomi (1915-1995) processed the first geostationary satellite image.
1970s	American engineer Herbert Saffir (1917-2007) and meteorologist Robert Simpson (1912-) developed their hurricane intensity scale (the *Saffir-Simpson scale*).
1971	Japanese-born American meteorologist Tetsuya T. Fujita (1920-1998) with Allen Pearson (1925-) introduced their tornado intensity scale (*Fujita* or *Fujita-Pearson scale*).
1972	Earth Resources Technology Satellite 1 (later called Landsat-1) was launched.
1974	The Global Atmospheric Research Program (GARP) conducted the GARP Atlantic Tropical Experiment (GATE) to improve understanding of the tropical atmosphere.
1974	Mexican chemist M. Molina (1943-) and American chemist Frank S. Rowland (1927-2012) warned of the CFC threat to the stratospheric ozone layer.
1974	Tetsuya T. Fujita discovered downbursts and coined the term.
1975	First GOES (Geostationary Operational Environmental Satellite) was launched.
1975	The European Space Agency was formed.
1975	The Saffir-Simpson Hurricane Intensity Scale became operational.
1976	Milankovitch cycles were discovered in deep-sea sediment cores.
1977	The first Meteosat weather satellite was launched by the European Space Agency.
1979	Robert G. Steadman developed the heat index (apparent temperature index).
1979	NOAA scientists developed the SLOSH model that predicts the location and height of a storm surge.
1985	The British Antarctic Survey first reported a drastic decline in stratospheric ozone over Antarctica during the Southern Hemisphere spring.
1985	The Vienna Convention for the Protection of the Ozone Layer was convened.
1985	The Tropical Ocean-Global Atmosphere Program (TOGA) began.

1987	International adoption of the *Montreal Protocol on Substances That Deplete the Ozone Layer*.
1987	*The Joint Global Ocean Flux Study (JGOFS)* initiated to study the fluxes of carbon in the ocean.
1988	The Global Energy and Water Cycle Experiment (GEWEX) launched to examine the global hydrologic cycle.
1988	Formation of the Intergovernmental Panel on Climate Change (IPCC) by the World Meteorological Organization and the UN Environmental Programme.
1990	First Doppler radar (WSR-88D) was introduced into meteorological service.
1992	*UN Framework Convention on Climate Change* was ratified.
1992	NOAA's wind profiler network was up and running.
1994	NWS and EPA introduced the Ultraviolet (UV) Index.
1997	The Kyoto Protocol addressed reductions in greenhouse gas emissions.
1997	Launch of the TRMM satellite for remote sensing of precipitation between 40 degrees S and 40 degrees N.
1999	Landsat-7 was launched.
2001	NWS implemented a major revision of the wind-chill index.
2004	New Ultraviolet (UV) Index introduced.
2007	Enhanced F-Scale (EF-Scale) became operational for rating tornadoes.
2007	The *IPCC Fourth Assessment Report* concluded that global warming since the mid 20th century very likely caused by human activity.
2009	Through VORTEX2 (Verification of the Origins of Rotation in Tornadoes EXperiment 2) a tornado in Goshen County, WY, on 5 June was the most documented through instrumentation.
2010	The Saffir-Simpson Hurricane Wind Scale become operational, focusing on wind speed and excluding storm surge ranges, flooding impact and central pressure.
2011	Deployment begins for WSR-88D Dual Polarization Weather Radar.

CLIMATE CLASSIFICATION

Tropical Humid Climates
Dry Climates
Subtropical Climates
Snow Forest Climates
Polar Climates
Highland Climates

One of the most widely used climate classification systems was designed by German climatologist and plant geographer Wladimir Köppen (1846-1940) and was subsequently modified by his students R. Geiger and W. Pohl. Recognizing that indigenous vegetation is a natural indicator of regional climate, Köppen and his students looked for patterns in annual and monthly mean temperature and precipitation, which closely correspond to the limits of vegetative communities, thereby revealing broad-scale climatic boundaries throughout the world. Records of annual and monthly mean temperature and precipitation are sufficiently long and reliable in many parts of the world that they serve as a good first approximation of climate. Since its introduction in the early 1900s, Köppen's climate classification has undergone numerous and substantial revisions by Köppen himself and other climatologists, and has had a variety of applications.

As shown in the Table, the Köppen climate classification system identifies six main climate groups, four based on temperature, one on precipitation, and one applies to mountainous regions. Köppen's scheme uses letters to symbolize major climatic groups: Tropical humid (A), Dry (B), Subtropical (C), Snow forest (D), Polar (E), and Highland (H). Additional letters further differentiate climate types.

Tropical Humid Climates

Tropical humid climates (A) constitute a discontinuous belt straddling the equator and extending poleward to

---

TABLE
Köppen-Based Climate Classification

Tropical humid (A)
    Af tropical wet
    Am tropical monsoon
    Aw tropical wet-and-dry

Dry (B)
    BS steppe or semiarid (BSh, BSk)
    BW arid or desert (BWh, BWk)
    BWn foggy desert

Subtropical (C)
    Cs subtropical dry summer (Csa, Csb)
    Cw subtropical dry winter
    Cf subtropical humid (Cfa, Cfb, Cfc)

Snow forest (D)
    Dw dry winter (Dwa, Dwb, Dwc, Dwd)
    Df year-round precipitation (Dfa, Dfb, Dfc, Dfd)
    Ds dry summer

Polar (E)
    Et tundra
    Ef ice cap

Highland (H)

near the Tropic of Cancer in the Northern Hemisphere and the Tropic of Capricorn in the Southern Hemisphere. Mean monthly temperatures are high and exhibit little variability throughout the year. The mean temperature of the coolest month is no lower than 18 °C (64 °F), so there is no frost. The temperature contrast between the warmest and coolest month is typically less than 10 Celsius degrees (18 Fahrenheit degrees). In fact, the diurnal (day-to-night) temperature range generally exceeds the annual temperature range. This monotonous air temperature regime is the consequence of consistently intense incoming solar radiation associated with a high maximum solar altitude and little variation in the period of daylight throughout the year.

Although tropical humid climate types are not readily distinguishable on the basis of temperature, important differences occur in precipitation regime. Tropical humid climates are subdivided into tropical wet (Af), tropical monsoon (Am), and tropical wet-and-dry (Aw). Although these climate types generally feature abundant annual rainfall, more than 100 cm (40 in.) on average, their rainy seasons differ in length and, for both Am and Aw, there is a pronounced dry season and wet season. In tropical wet climates, the yearly average rainfall of 175 to 250 cm (70 to 100 in.) supports the world's most lush vegetation. Tropical rainforests occupy the Amazon Basin of Brazil (Figure 1), the Congo Basin of Africa, the islands of Micronesia, and American Samoa (Figure 2). For the most part, rainfall is distributed uniformly throughout the year, although some areas experience a brief (one or two month) dry season. Rainfall occurs as heavy downpours in frequent thunderstorms triggered by local convection and the intertropical convergence zone (ITCZ). Convection is largely controlled by solar radiation and rainfall typically peaks in midafternoon, the warmest time of day. Because the water vapor concentration is very high, even the slightest cooling at night leads to saturated air and the formation of dew or radiation fog, giving these regions a sultry, steamy appearance.

Tropical monsoon (Am) climates feature a seasonal rainfall regime with extremely heavy rainfall during several months and a lengthy dry season. The principal control for these climates involves seasonal shifts in wind from land to sea, typified by the Asian monsoon. During the low-Sun season, high air pressure over the Asian continent causes dry air to flow southward into parts of Southeast Asia and India. During the high-Sun season, low air pressure covers the Tibetan Plateau and the winds reverse direction, advecting moisture inland from over the Indian Ocean. Local convection, orographic

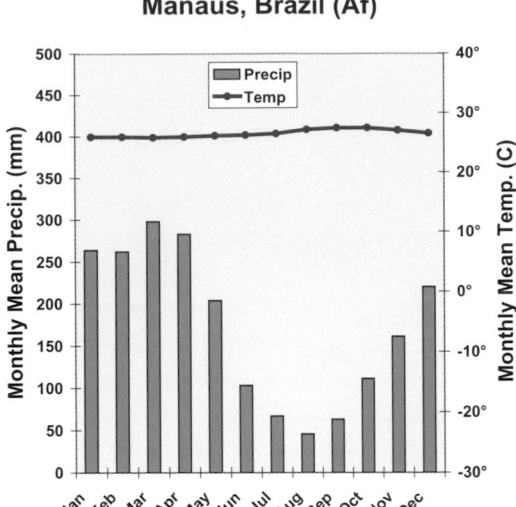

FIGURE 1
Monthly mean precipitation and temperature in Manaus, Brazil.

lifting, and shifts of the ITCZ combine to deluge the land with torrential rains. Am climates also occur in western Africa and northeastern Brazil.

For the most part, tropical wet-and-dry climates (Aw) border tropical wet climates (Af) and are transitional to subtropical dry climates in a poleward direction. Aw climates support the savanna, tropical grasslands with scattered deciduous trees. Summers are wet and winters are dry, with the dry season lengthening poleward. This marked seasonality of rainfall is linked to shifts of the intertropical convergence zone (ITCZ) and semipermanent subtropical anticyclones, which follow the seasonal excursions of the Sun. In summer (*high-Sun* season),

FIGURE 2
NOAA's Fagatele Bay National Marine Sanctuary in American Samoa, an example of an Af climate. [NOAA Photo Library]

surges of the ITCZ trigger convective rainfall while in winter (*low-Sun* season), the dry eastern flank of the subtropical anticyclones dominates the weather.

The annual mean temperature in Aw climates is only slightly lower, and the seasonal temperature range is only slightly greater, than in the tropical wet climates (Af). The diurnal temperature range varies seasonally, however. In summer, frequent cloudy skies and high humidity suppress the diurnal temperature range by reducing both solar heating during the day and radiational cooling at night. In winter, on the other hand, persistent fair skies have the opposite effect on radiational heating and cooling, which increases the diurnal temperature range. Cloudy, rainy summers plus dry winters also mean that the year's highest temperatures typically occur toward the close of the dry season in late spring.

Dry Climates

Dry climates (B) characterize those regions where average annual potential evaporation exceeds average annual precipitation. *Potential evaporation* is the quantity of water that would vaporize into the atmosphere from a surface of fresh water during long-term average weather conditions. Air temperature largely governs the rate of evaporation so it is not possible to specify a maximum rainfall amount as the criterion for dry climates. Rainfall is not only limited in B climates but also highly variable and unreliable. As a general rule, the lower the mean annual rainfall, the greater is its variability from one year to the next.

Earth's dry climates encompass a larger land area than any other single climate grouping. Perhaps 30% of the planet's land surface, stretching from the tropics into midlatitudes, experience a moisture deficit of varying degree. These are the climates of the world's deserts and steppes, where vegetation is sparse and equipped with special adaptations that permit survival under conditions of severe moisture stress. Based on the degree of dryness, we distinguish between two dry climate types: steppe or semiarid (BS) and arid or desert (BW). Steppe or semiarid climates are transitional between more humid climates and arid or desert climates. Mean annual temperature is latitude dependent, as is the range in variation of mean monthly temperatures through the year. Hence, a distinction is made between warm dry climates of tropical latitudes (BSh and BWh) and cold, dry climates of higher latitudes (BSk and BWk).

Dryness is the consequence of subtropical anticyclones, cold surface ocean currents, or the rain shadow effect of high mountain ranges. Subsiding stable

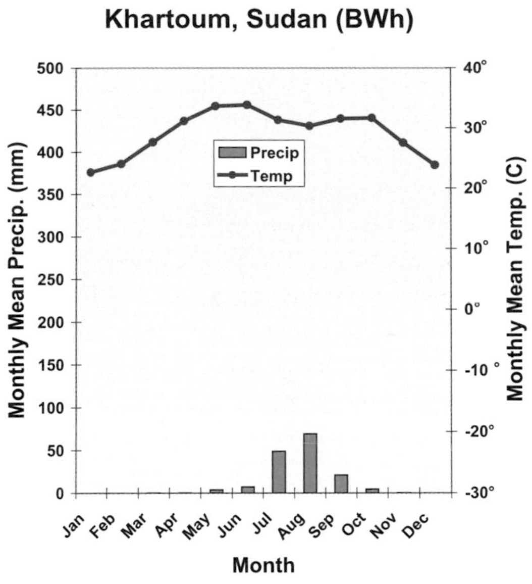

FIGURE 3
Monthly mean precipitation and temperature in Khartoum, Sudan.

air on the eastern flanks of subtropical anticyclones gives rise to tropical dry climates, designated as BSh and BWh (Figure 3). These huge semipermanent pressure systems, centered over the ocean basins, dominate the weather year-round near the Tropics of Cancer and Capricorn. Consequently, dry climates characterize North Africa eastward to northwest India, the southwestern United States (Figure 4) and northern Mexico, coastal Chile and Peru, southwest Africa, and much of the interior of Australia.

Although persistent and abundant sunshine is generally the rule in dry tropical climates, there are some important exceptions. Where cold ocean waters border a coastal desert, a shallow layer of stable marine air drifts inland. The desert air thus features high relative humidity, persistent low stratus clouds and fog, and considerable dew formation. Examples are the Atacama Desert of Peru and Chile, the Namib Desert of southwest Africa, and portions of the coastal Sonoran Desert of Baja California and stretches of the coastal Sahara Desert of northwest Africa. These anomalous foggy desert climates are designated BWn.

Cold, dry climates of higher latitudes (BWk and BSk) are situated in the rain shadows of great mountain ranges. They occur primarily in the Northern Hemisphere, to the lee of the Sierra Nevada and Cascade ranges in North America and the Himalayan chain in Asia. Because these dry climates are at higher latitudes than their tropical counterparts, mean annual temperatures

FIGURE 4
Desert near Yuma, AZ, an example of a BWh climate. [Photo by Jason West]

are lower and the seasonal temperature contrast is greater. Anticyclones dominate winter, bringing cold and dry conditions, whereas summers are hot and generally dry. Scattered convective showers, mostly in summer, produce relatively meager precipitation.

Subtropical Climates

Subtropical climates (C) are located just poleward of the Tropics of Cancer and Capricorn and are dominated by seasonal shifts of subtropical anticyclones. There are three basic climate types: subtropical dry summer (or *Mediterranean*) (Cs), subtropical dry winter (Cw), and subtropical humid (Cf), which receive precipitation throughout the year.

Mediterranean climates occur on the western side of continents between about 30 and 45 degrees latitude. In North America, mountain ranges confine this climate to a narrow coastal strip of California. Elsewhere, Cs climates rim the Mediterranean Sea and occur in portions of extreme southern Australia. Summers are dry because at that time of year Cs regions are under the influence of stable subsiding air on the eastern flanks of the semi-permanent subtropical highs. Equatorward shift of subtropical highs in autumn allows extratropical cyclones to migrate inland, bringing moderate winter rainfall. Mean annual precipitation varies greatly-ranging from 30 to 300 cm (12 to 80 in.) with the wettest winter month typically receiving at least three times the precipitation of the driest summer month.

Although Mediterranean climates exhibit a pronounced seasonality in precipitation (dry summers and wet winters), the temperature regime is quite variable.

In coastal areas, cool onshore breezes prevail, lowering the mean annual temperature and reducing seasonal temperature contrasts. Well inland, away from the ocean's moderating influence, summers are considerably warmer; hence, inland mean annual temperatures are higher and seasonal temperature contrasts are greater than in coastal Cs localities. Climatic records of coastal San Francisco and inland Sacramento, CA illustrate the contrast in temperature regime within Cs regions. Although the two cities are separated by only about 145 km (90 mi.), the climate of Sacramento is much more continental (much warmer summers and somewhat cooler winters) than that of San Francisco (Figure 5). The warm climate subtype is designated Csa and the cooler subtype is Csb.

Subtropical dry winter climates (Cw) are transitional between Aw and BS climates and located in South America and Africa between 20 and 30 degrees S. Cw climates also occur between the Aw and H climates of the Himalayas and Tibetan plateau and between the BS and Cfa climates of Southeast and East Asia (Figure 6). Northward shift of the subtropical high pressure systems is responsible for the dry winter in South America and Africa. The narrowness of the two continents between 20 and 30 degrees S means a relatively strong maritime influence exists and dictates against extreme dryness. In spring, subtropical highs shift southward and rains return. In Asia, winter dryness is caused by winds radiating outward from the massive cold Siberian high. As the continent warms in spring, the Siberian high weakens and eventually is replaced by low pressure. Moist winds then flow inland bringing summer rains. Mean annual precipitation in Cw climates is in the range of 75 to 150 cm (30 to 60 in.).

Subtropical humid climates (Cf) occur on the eastern side of continents between about 25 and 40

FIGURE 5
San Francisco, CA, an example of a Csb climate. [NOAA Photo Library]

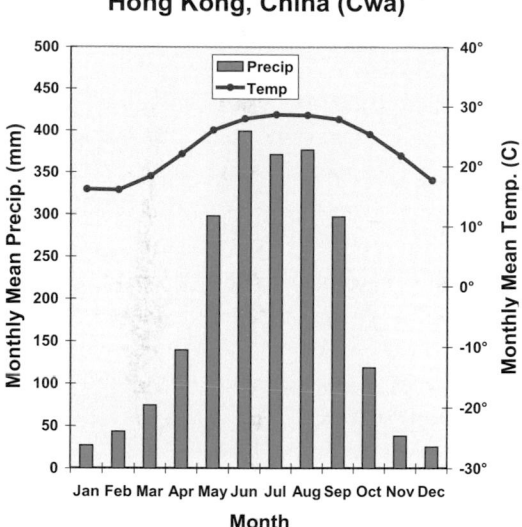

FIGURE 6

Monthly mean precipitation and temperature in Hong Kong, China.

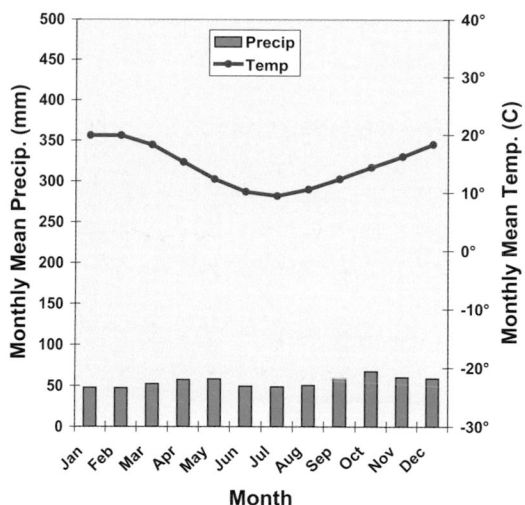

FIGURE 7

Monthly mean precipitation and temperature in Melbourne, Australia

degrees latitude (and even more poleward where the maritime influence is strong). Cfa climates are the most important of the Cf climate subtypes in terms of land area and number of people impacted. Cfa climates are situated primarily in the southeastern United States, a portion of southeastern South America, eastern China, southern Japan, on the extreme southeastern coast of South Africa, and along much of the east coast of Australia. These climates feature abundant precipitation (75 to 200 cm, or 30 to 80 in., on average annually), which is distributed throughout the year. In summer, Cfa regions are dominated by a flow of sultry maritime tropical air on the western flanks of the subtropical anticyclones. Consequently, summers are hot and humid with frequent thunderstorms, which can produce brief periods of substantial rainfall. Hurricanes and tropical storms contribute significant rainfall (up to 15% to 20% of the annual total) to some North American and Asian Cfa regions, especially from summer through autumn. In winter, after the subtropical highs shift toward the equator, Cfa regions come under the influence of migrating extratropical cyclones and anticyclones.

In Cfa localities, summers are hot and winters are mild. Mean temperatures of the warmest month are typically in the range of 24 to 27 °C (75 to 81 °F). Average temperatures for the coolest months typically range from 4 to 13 °C (39 to 55 °F). Subfreezing temperatures and snowfalls are infrequent.

A strong maritime influence is responsible for the cool summers and mild winters of Cfb climates. These climates occur over much of Northwest Europe, New Zealand, and portions of southeastern South America, southern Africa, and Australia (Figure 7). The coldest subtype, Cfc, is relegated to coastal areas of southern Alaska, Norway, and the southern half of Iceland. Cfb and Cfc climates are relatively humid with mean annual precipitation ranging between 100 and 200 cm (40 and 80 in.).

Snow Forest Climates

Snow forest climates (D) occur in the interior and to the leeward sides of large continents. The name emphasizes the link between biogeography and the Köppen climate classification system. These climates feature cold snowy winters (except for the Dw subtype in which the winter is dry) and occur only in the Northern Hemisphere. Snow forest climates are subdivided according to seasonal precipitation regimes with Df climates experiencing year-round precipitation whereas Dw climates have a dry winter. D climates with dry summers (Ds) are rare and small in extent. Additional distinction is made between warmer subtypes (Dwa, Dfa, Dwb, and Dfb) and colder subtypes (Dwc, Dfc, Dwd, Dfd).

The warmer subtypes, sometimes termed *temperate continental*, have warm summers (mean

FIGURE 8
Autumn colors on Clopper Lake in Seneca Creek State Park, Maryland, an example of a Dfa climate. [NOAA Photo Library]

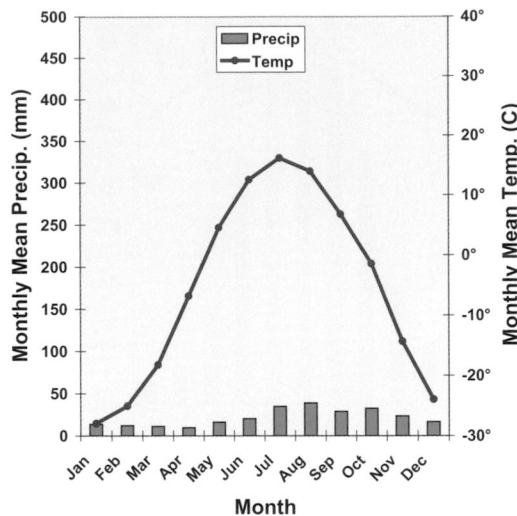

FIGURE 9
Monthly mean precipitation and temperature in Yellowknife, Canada.

temperature of the warmest month greater than 22 °C or 71 °F) and cold winters. They are located in Eurasia, the northeastern third of the United States, southern Canada, and extreme eastern Asia. Continentality increases inland with maximum temperature contrasts between the coldest and warmest months as great as 25 to 35 Celsius degrees (45 to 63 Fahrenheit degrees). The southerly Dfa climates (Figure 8) have cool winters and warm summers and the more northerly Dfb climates have cold winters and mild summers. The freeze-free period varies in length from 7 months in the south to only 3 months in the north. The weather in these regions is very changeable and dynamic because these areas are swept by extratropical cyclones and anticyclones and by surges of contrasting air masses. Polar front cyclones dominate winter, bringing episodes of light to moderate frontal precipitation. These storms are followed by incursions of dry polar and arctic air masses. In summer, cyclones are weak and infrequent as the principal storm track shifts poleward. Summer rainfall is mostly convective, and locally amounts can be very heavy in severe thunderstorms and mesoscale convective complexes (MCCs). Although precipitation is distributed rather uniformly throughout the year, most places experience a summer maximum.

In northern portions, winter snowfall becomes an important factor in the climate. Mean annual snowfall and the persistence of a snow cover increase northward. Because of its high albedo for solar radiation and its efficient emission of infrared, a snow cover chills and stabilizes the overlying air. For these reasons, a snow cover tends to be self-sustaining; once established in early winter, an extensive snow cover tends to persist.

Moving poleward, summers get colder and winters are very cold. These so-called *boreal climates* (Dfc, Dfd, Dwc, Dwd) occur only in the Northern Hemisphere as an east-west band between 50 to 55 degrees N and 65 degrees N (Figure 9). It is a region of extreme continentality and very low mean annual temperature. Summers are short and cool, and winters are long and bitterly cold. Because midsummer freezes are possible, the growing season is precariously short. Both continental polar (cP) and arctic (A) air masses originate here, and this area is the site of an extensive coniferous (boreal) forest. In summer, the mean position of the leading edge of arctic air (the arctic front) is located along the northern border of the boreal forest. In winter, the mean position of the arctic front is situated along the southern border of the boreal forest.

Weak cyclonic activity occurs throughout the year and yields meager annual precipitation (typically less than 50 cm, or 20 in.). Convective activity is rare. A summer precipitation maximum is due to the winter dominance of cold, dry air masses. Snow cover persists throughout the winter and the range in mean temperature between winter and summer is among the greatest in the world.

Polar Climates

Polar climates (E) occur poleward of the Arctic and Antarctic circles. These boundaries correspond roughly to localities where the mean temperature for the warmest

McMurdo, Antarctica (Ef)

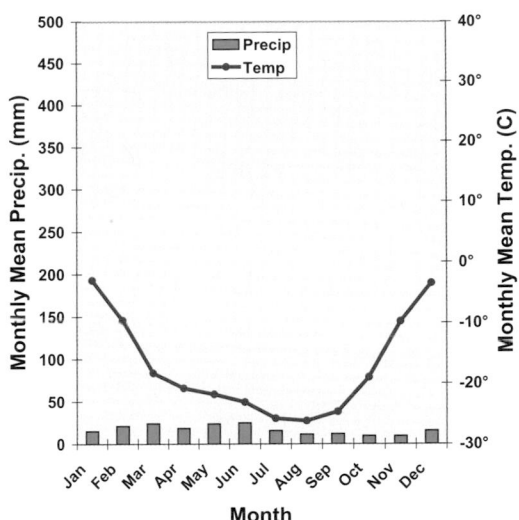

FIGURE 10
Monthly mean precipitation and temperature in McMurdo, Antarctica.

month is 10 °C (50 °F). These limits also approximate the tree line, the poleward limit of tree growth. Poleward are tundra and the Greenland and Antarctic ice sheets. A distinction is made between tundra (Et) and ice cap (Ef) climates, with the dividing criterion being 0 °C (32 °F) for the mean temperature of the warmest month. Vegetation is sparse in Et regions. It is almost nonexistent in Ef areas (Figures 10 and 11).

Polar climates are characterized by extreme cold and slight precipitation, which falls mostly in the form of snow (less than 25 cm, or 10 in., melted, per year). Greenland and Antarctica could be considered

FIGURE 11
Penguins explore snow dunes in Antarctica, an example of a Ef climate. [NOAA Photo Library]

deserts for lack of significant precipitation, despite the presence of large ice sheets. Although summers are cold, the winters are so extremely cold that polar climates feature a marked seasonal temperature contrast. Mean annual temperatures are the lowest of any place in the world.

Highland Climates

Highland Climates (H) encompass a wide variety of climate types that characterize mountainous terrain (Figure 12). Altitude, latitude, and exposure are among the factors that shape a complexity of climate types. For example, temperature decreases rapidly with increasing altitude and windward slopes tend to be wetter than leeward slopes. Climate-ecological zones are telescoped in mountainous terrain. In ascending several thousand meters of altitude, the same bioclimatic zones that are experienced over several thousand kilometers of latitude are encountered. As a general rule, every 300 m (980 ft) of elevation corresponds roughly to a northward advance of 500 km (310 mi).

FIGURE 12
Different climate-ecological zones can be found in mountainous terrain, such as in this view of Mount Rainier, WA, from Indian Henrys Hunting Grounds. [Photo by Daniel Keebler]

GLOSSARY

A

absolute humidity—The mass of water vapor per unit volume of air containing the water vapor and usually expressed as grams of water vapor per cubic meter of humid air.

absolute instability—Property of an ambient air layer that enhances vertical motion for both saturated (cloudy) and unsaturated (clear) air parcels; the *temperature* of the ambient air is dropping more rapidly with altitude than the *dry adiabatic lapse rate.*

absolute stability—Property of an ambient air layer that suppresses vertical motion for both saturated (cloudy) and unsaturated (clear) air parcels. Occurs when the *temperature* in the ambient air layer drops more slowly with altitude than the *moist adiabatic lapse rate*, the temperature does not change with altitude in the ambient air layer (isothermal), or the temperature increases with altitude in the ambient air layer (temperature inversion).

absolute zero—The theoretical *temperature,* 0 K (−273.15 °C or −459.57 °F), at which a body does not emit *electromagnetic radiation* and all molecular motion ceases (usually some subatomic activity takes place).

absorption—The energy conversion process whereby a portion of the radiation incident on an object is converted to *heat* energy. Absorption is a function of the surface and composition of the absorbing material and the *wavelength* of the incident radiation.

acid—Hydrogen-containing compound that releases positively charged hydrogen ions (H^+) when dissolved in water.

acid deposition—The delivery of acidic particles, usually sulfuric acid and nitric acid, to Earth's surface.

acid rain—Rain having a pH lower than 5.6, representing the pH of natural rain water saturated with carbon dioxide.

acoustic shadow—An area where sound is not heard due to the refraction of sound waves.

adiabatic process—A thermodynamic process in which no *heat* is exchanged between a mass and its environment. Within the *atmosphere, expansional cooling* and *compressional warming* of unsaturated air parcels are essentially adiabatic processes, that is, no heat is exchanged between the air parcels and the surrounding air.

Advanced Weather Interactive Processing System (AWIPS)—A computerized workstation that receives and organizes *Automated Surface Observing System* data and other analysis and guidance products from the *National Centers for Environmental Prediction* and enables *National Weather Service* meteorologists to display, process, and overlay images, graphics, and other data.

advection fog—Ground-level *cloud* generated by the cooling of a mild, humid *air mass* (to its *dewpoint*) as it travels over a relatively cool surface.

aeolian sound—Humming, singing, or whistling sounds produced by winds blowing over obstacles.

aerosols—Tiny liquid or solid particles of various compositions that occur suspended in the *atmosphere*.

aerovane—A wind sensor designed to measure wind speed and direction.

agroclimatic compensation—Poor growing *weather* and decreased crop yields in one area are offset to some extent by better growing weather and increased crop yields in other areas.

air density—Mass of molecules per unit volume of air, about 1.275 kg per cubic meter at sea-level temperature of 0 °C and pressure of 1000 millibars.

air mass—A huge volume of air covering thousands of square kilometers that has relatively uniform *temperature* and water vapor concentration (*humidity*) horizontally. The specific characteristics of an air mass depend on the type of surface over which the air mass forms (its source region) and travels.

air mass advection—Movement of an *air mass* from one place to another, one air mass replacing another air mass having different *temperature* and/or *humidity* characteristics.

air mass modification—Changes in the *temperature*, *humidity*, and/or stability of an *air mass* as it travels away from its source region. Occurs primarily by exchange of *heat*, moisture, or both with the surface over which the air mass travels, *radiational heating* or *radiational cooling*, or adiabatic heating or cooling associated with large-scale vertical air motion.

air parcel—An imaginary volume of air useful for visualizing atmospheric processes, containing a fixed number of molecules.

air pollutant—A gas or *aerosol* that occurs at a concentration that threatens the wellbeing of living organisms (especially humans) or disrupts the orderly functioning of the environment.

air pressure—The cumulative force exerted on any surface by the molecules composing air; usually expressed as the weight of a column of air per unit surface area.

air pressure gradient—Change in *air pressure* with distance.

air pressure tendency—Change in *air pressure* with time; on a surface *weather* map, the air pressure change over the prior 3 hrs.

albedo—The fraction or percent of radiation striking a surface (or interface) that is reflected by that surface (or interface); usually applied to the reflectivity of an object to visible radiation.

alkaline substance—When dissolved in water, it releases negatively charged hydroxyl ions (OH⁻). Also known as a base.

altimeter—Any device used to determine altitude such as an *aneroid barometer* calibrated to read in altitude or elevation.

altocumulus clouds (Ac)—Middle level *clouds* occurring as roll-like patches or puffs forming waves or parallel bands. Altocumulus are distinguished from *cirrocumulus* by the larger size of the cloud patches and by sharper edges indicating the presence of supercooled water droplets rather than ice crystals.

altocumulus lenticularis clouds—Lens-shaped *altocumulus*; *mountain-wave clouds* generated by the disturbance of horizontal airflow by a prominent mountain range.

altostratus clouds (As)—Middle level *clouds* occurring as uniformly gray or bluish white layers totally or partially covering the sky. The Sun is typically only dimly visible through altostratus. These clouds are composed of supercooled water droplets and ice crystals.

aneroid barometer—A portable instrument that utilizes a flexible partially-evacuated metal chamber and spring to measure *air pressure*; may be used as an *altimeter*.

Antarctic Circle—Poleward of this latitude (66 degrees 33 minutes S) there are 24 hrs of sunlight on the austral summer *solstice* and 24 hrs of darkness on the austral winter solstice.

Antarctic ozone hole—A large area of significant stratospheric ozone depletion over the Antarctic continent that typically develops annually between late August and early October, and generally ends in mid-November. Ozone thinning is attributed to the action of chlorine (Cl) liberated from a group of chemicals known as chlorofluorocarbons (CFCs).

anticyclone—A dome of air that exerts relatively high surface *air pressure* compared with surrounding air; same as a *High*. Viewed from above, surface *winds* in an anticyclone blow clockwise and outward in the Northern Hemisphere but counterclockwise and outward in the Southern Hemisphere.

aphelion—The time of the year when the Earth is farthest (152 million km or 94 million mi) from the Sun (currently about 4 July).

Arctic air (A)—Exceptionally cold and dry *air masses* that form primarily in winter over the Arctic Basin, Greenland, and the northern interior of North America north of about 60 degrees N.

Arctic Circle—Poleward of this latitude (66 degrees 33 minutes N) there are 24 hrs of sunlight at the boreal summer *solstice* and 24 hrs of darkness at the boreal winter solstice.

Arctic high—A *cold-core anticyclone* originating in a source region for *Arctic air*; the product of extreme *radiational cooling* over the *snow*-covered continental interior of North America well north of the *polar front*.

Arctic Oscillation (AO)—A sesaw variation in sea level air pressure between the Arctic and the portion of the middle latitudes centered from 37 to 45 degrees N. Changes in the surface air pressue gradient alter the strength of the bands of winds aloft (the polar vortex) blowing in a mostly zonal pattern counterclockwise around the Arctic.

asteroids—Large, celestrial bodies of rock many kilometers across.

atmosphere—A thin envelope of gasses (also containing suspended solid and liquid particles and *clouds*) that encircles the planet.

atmospheric boundary layer—The atmospheric zone to which frictional resistance (*eddy viscosity*) is essentially confined; on average the zone extends from Earth's surface to an altitude of about 1000 m (3300 ft).

atmospheric rivers (AR)—Relatively narrow bands of concentrated water vapor transported in the lower atmosphere, and responsible for most horizontal air flow of water vapor outside the tropics.

atmospheric stability—Property of ambient air that either enhances (unstable) or suppresses (stable) vertical motion of air parcels; depends on the vertical *temperature* profile or *sounding* of the ambient air and whether air parcels are saturated (cloudy), with cooling or warming at the *moist adiabatic lapse rate*, or unsaturated (clear), with cooling or warming at the *dry adiabatic lapse rate*. See *stable air layer* and *unstable air layer*.

atmospheric windows—*Wavelength* bands within which atmospheric gases absorb little or no *electromagnetic radiation*.

aurora—A luminous phenomenon in the night sky of overlapping curtains of greenish-white light, sometimes fringed with pink, appearing in the *ionosphere* of high latitudes. Light is emitted by atoms and molecules excited by beams of electrons from complex interactions between *solar wind* and Earth's *atmosphere*. Known as the aurora borealis (northern lights) in the Northern Hemisphere and the aurora australis (southern lights) in the Southern Hemisphere.

Automated Surface Observing System (ASOS)—Meteorological sensors, computers, and fully automated communication ports that record and transmit atmospheric conditions automatically. A network of ASOS units feeds observational data to *National Weather Service* Forecast Offices and local airport control towers 24-hrs per day.

Avogadro's law—Equal volumes of gases measured under the same temperature and pressure contain the same number of molecules.

B

back-door cold front—A *cold front* that propagates southward or southwestward along the North Atlantic coast east of the Appalachian Mountains. These fronts occur most frequently in the summer and fall, and usher in Canadian *continental polar air* or Atlantic *maritime polar air*.

barograph—A recording instrument that provides a continuous trace of *air pressure* with time and helps determine *air pressure tendency*.

barometer—An instrument used to measure *air pressure* and monitor its changes.

barrier island—A long, narrow strip of sand parallel to the coast and separated from the mainland by a lagoon or backwater bay. Since a barrier island faces the open ocean, it absorbs the brunt of *tropical cyclone*, and other sea storm-driven waves and *storm surge*, providing some protection for coastal beaches, wetlands, and shoreline structures.

Beaufort scale—A scale of *wind* speed based originally on visual assessment of the effects of wind on seas and later extended to describe the effects of wind on land-based flexible objects such as trees. The scale ranges from 0 for calm conditions to 12 for *hurricane*-strength winds.

Bergeron-Findeisen process—*Precipitation* formation in *cold cloud*s whereby ice crystals grow at the expense of supercooled water droplets in response to differences in *vapor pressure* relative to water and ice surfaces; also known as the ice-crystal process.

blackbody—A hypothetical "body" that absorbs all *electromagnetic radiation* that is incident on it at every *wavelength* and emits all radiation at every wavelength; no radiation is reflected or transmitted. The "body" must be large compared to the wavelength of incident radiation.

blizzard—A severe storm characterized by high winds and reduced visibility due to falling or blowing snow.

blocking system—A *cyclone* or *anticyclone* cutoff from the main westerly airflow that blocks the usual west-to-east progression of *weather* systems. A blocking system may be responsible for weather extremes, such as drought or flooding *rains* or excessive *heat* and cold.

bora—A cold *katabatic wind* that flows downslope and onto the coastal plain of the Adriatic Sea.

Bowen ratio—For a moist surface, the ratio of *heat* energy used for *sensible heating* (*conduction* and *convection*) to the heat energy used for *latent heating* (*evaporation* of water or *sublimation* of snow). The Bowen ratio varies from one locality to another depending on the amount of surface moisture, ranging from 0.1 for the ocean surface to about 5.0 for deserts; negative values are also possible.

bow echos—Rapidly moving band of showers and thunderstorms that exhibit a curved or bowed shape, associated with *derechos*.

Boyle's Law—When temperature is constant, the pressure and density of an ideal gas are directly proportional.

British thermal unit (Btu)—The quantity of *heat* needed to raise the *temperature* of one pound of water one Fahrenheit degree (technically, from 62 to 63 °F).

C

Callendar effect—The theory that global climate change can be brought about by enhancement of the greenhouse effect by increased levels of atmospheric carbon dioxide from anthropogenic sources.

calorie (cal)—The amount of *heat* needed to raise the *temperature* of one gram of water one Celsius degree (technically, from 14.5 to 15.5 °C).

capping inversion—An elevated temperature inversion layer that caps a convective boundary layer, keeping the convective elements from rising higher into the atmosphere.

centripetal force—An inward-directed force that acts on an object moving in a curved path, confining the object to the curved path; the result of other forces.

Charles' Law—With constant pressure, the absolute temperature (in kelvins) of an ideal gas is inversely proportional to the density of the gas.

chinook wind—Air that is adiabatically compressed as it is drawn down the leeward slope of a mountain range. As a consequence, the air is warm and dry, causing rapid temperature rises and reduction in snow cover.

chromosphere—Portion of the Sun above the *photosphere*; consists of transparent ionized hydrogen and helium at 4000 to 40,000 °C (7200 to 72,000 °F).

circumpolar vortex—The planetary-scale circulation regime that surrounds the cold pool of air in the polar regions.

cirriform cloud—A thin wispy or fibrous cloud composed of ice crystals.

cirrocumulus clouds (Cc)—High *clouds* that exhibit a wavelike pattern of small, white, rounded patches rarely covering the entire sky. Cirrocumulus are composed of ice crystals.

cirrostratus clouds (Cs)—High, thin, layered *clouds* composed of ice crystals that form a thin white veil over the sky. Cirrostratus are nearly transparent, so the Sun or Moon readily shines through them.

cirrus clouds (Ci)—High, thin, nearly transparent *clouds* composed of ice crystals and occurring as delicate silky strands, which are streaks of falling ice particles blown laterally by strong *winds*.

climate—*Weather* of some locality averaged over some time period as well as extremes in weather behavior.

climatic anomalies—Departures from long-term climatic averages.

climatic norm—The average value of some climatic element such as temperature or snowfall.

climatology—Study of *climate*, its controls and spatial and temporal variability.

cloud—A visible suspension of minute water droplets and/or ice crystals in the *atmosphere* above the Earth's surface. Clouds differ from *fog* only in that the latter is, by definition, in contact with Earth's surface. Clouds form in the free atmosphere primarily as a result of *condensation* or *deposition* of water vapor in ascending air that nears *saturation*.

cloud condensation nuclei (CCN)—Tiny solid and liquid particles that promote the *condensation* of water vapor at *temperatures* both above and below the freezing point of water, may include *hygroscopic nuclei*.

cloud streets—*Clouds* aligned in rows due to strong vertical shear in horizontal *wind* speed or direction; these rows often extend for hundreds of kilometers.

cold air advection—Flow of air across regional isotherms from relatively cool localities to relatively warm localities.

cold cloud—*Cloud* composed of ice crystals or supercooled water droplets or a mixture of both which have *temperatures* below 0 °C (32 °F).

cold front—*Front* that moves in such a way that relatively cold (more dense) air advances and replaces relatively warm (less dense) air.

cold-core anticyclone—Shallow high-pressure system that coincides with a dome of *continental polar air* or *arctic air*; labeled as either a *polar high* or an *arctic high*. The high's shallow anticyclonic circulation weakens rapidly with altitude and often reverses to a cyclonic circulation in the middle and upper *troposphere*.

cold-core cyclone—An occluded low-pressure system in which the lowest *temperatures* occur throughout the column of air above the low-pressure center. Isobaric surfaces dip downward above the low center and the depth of the low increases with altitude, implying that a cyclonic circulation prevails throughout the *troposphere* and is most intense at high altitudes.

collision-coalescence process—Growth of *cloud* droplets into raindrops within a *warm cloud*; droplets merge upon impact. This process takes place in a cloud made up of droplets of different sizes; larger droplets with higher *terminal velocity* overtake, and then collide and coalesce with smaller droplets in their paths.

comma cloud—The pattern of cloudiness associated with a mature *extratropical cyclone*; the head of the comma stretches from the low center to the northwest and its tail follows along the *cold front*. This pattern reflects the strengthening of the system's circulation.

comet—A relatively small celestial body composed of meteoric dust and ice that travels in a highly elliptical orbit around the Sun.

compressional warming—A *temperature* rise that accompanies a pressure increase on and resultant compression of a gas (or mixture of gases). The work of compressing air is converted to *heat*, raising the temperature of the gas. Air parcels that descend within the *atmosphere* undergo compressional warming in response to increasing *air pressure*.

condensation—The process that produces a net gain of liquid water mass at the interface between liquid water and air; more water changes phase from vapor to liquid than liquid to vapor. The phase change from vapor to liquid releases the *latent heat of vaporization*.

conditional stability—Property of an ambient air layer that suppresses vertical motion for unsaturated (clear) air parcels and enhances vertical motion for saturated (cloudy) air parcels. The *sounding* in a conditionally stable layer of air lies between the *dry adiabatic lapse rate* and the *moist adiabatic lapse rate.*

conduction (of heat)—Transfer of kinetic energy of atoms or molecules via collisions between the neighboring atoms or molecules.

continental climate—An inland climate that experiences greater contrast between summer and winter temperatures due to a large landmass's thermal inertia.

continental drift—Slow movement of continents, which are parts of gigantic tectonic plates, across the face of the globe.

continental polar air (cP)—Relatively dry *air masses* that develop over the northern interior of North America; these air masses are very cold in winter and relatively mild in summer.

continental tropical air (cT)—Warm, dry *air masses* that form over the subtropical deserts of Mexico and the southwestern United States primarily in summer.

contrails—Bright white streamers of ice crystals that form in the exhaust of jet aircraft; short for *con*densation *trails.*

convection—The transport of *heat* within a fluid via motions of the fluid itself; generally occurs only in liquids and gases. Convection is much more important than *conduction* in transporting heat vertically within the *troposphere.*

convective condensation level (CCL)—The altitude at which *condensation* begins to occur through *convection*; coincides with the base of *cumuliform clouds,* typically at altitudes between 1000 and 2000 m (3600 and 6000 ft).

converging winds—*Winds* that blow toward a central point of a column of air.

conveyor belt model—Three-dimensional depiction of a mature *extratropical cyclone* and its corresponding *fronts* in terms of three interacting airstreams, often referred to as conveyor belts. This model developed out of a better understanding of upper-air circulation in the years since the *Norwegian cyclone model.*

Coordinated Universal Time (UTC)—Formerly known as Greenwich Mean Time, the time along the prime meridian, zero degree longitude.

coral bleaching—The loss of color by coral polyps and their subsequent death caused by unusually high sea-surface temperature.

Coriolis Effect—An apparent deflective force arising from the rotation of the Earth on its axis; affects principally synoptic-scale and planetary-scale *winds.* Winds are deflected to the right of their initial direction in the Northern Hemisphere and to the left in the Southern Hemisphere. Magnitude depends on latitude and speed of the moving object.

corona (optical)—A series of alternating light and dark colored rings that surround the Moon or Sun; due to *diffraction* of light by spherical *cloud* droplets.

corona (solar)—See *Solar corona.*

crepuscular rays—Alternating light and dark bands (solar rays and shadows) that appear to diverge in a fanlike pattern from the solar disk. They can be observed at the beginning of *twilight* or during the day as the Sun's rays pass through holes in *clouds* or between clouds in a hazy sky.

cumuliform clouds—*Clouds* that exhibit significant vertical development; often produced by updrafts in *convection* currents.

cumulonimbus clouds (Cb)—*Thunderstorm clouds* that form as a consequence of deep *convection* in the *atmosphere*; have tops that sometimes reach altitudes of 20,000 m (60,000 ft) or higher. The upper portions of these clouds are anvil-shaped and composed of mostly ice crystals, the middle portions are composed of supercooled water droplets or a mixture of supercooled water droplets and ice crystals, and the lower portions are often composed of ordinary water droplets.

cumulus clouds (Cu)—Vertically developed *clouds*, resembling ball of cotton, that form as a consequence of the updraft in *convection* currents. Cumulus are normally fair-*weather* clouds and consist of mostly water droplets, which may be supercooled.

cumulus congestus cloud—Upward-building convective *clouds*, resembling cauliflower, with vertical development between those of *cumulus* and *cumulonimbus*. Cumulus congestus can occur when the ambient air in the middle to upper *troposphere* is unstable for saturated air. These clouds consist of mostly water droplets, which are frequently supercooled.

cup anemometer—An instrument used to monitor *wind* speed. Wind rotates the 3 or 4 open hemispheric cups, which spin horizontally on a vertical shaft, and that motion is calibrated in wind speed.

curvature effect—The effect a curved water surface has on the ability of water molecules to vaporize from the surface. The smaller the radius of a droplet, the greater the concentration of surrounding water vapor that is necessary for the droplet to grow. Due to differences in the bond strength of the water molecules on the droplet surface, water molecules more readily vaporize from a small droplet than a large droplet, explaining why the *saturation vapor pressure* over a small droplet is greater than over a large droplet at the same *temperature*. Tiny droplets cannot form or exist without a supersaturated environment.

cyclogenesis—The birth and development of a *cyclone*; for an *extratropical cyclone*, cyclogenesis usually takes place along the *polar front* directly under an area of strong horizontal *divergence* in the upper *troposphere*.

cyclolysis—Process whereby a *cyclone* weakens; usually takes place when the central pressure rises and winds slacken. Upper-air support for the cyclone diminishes to the point where horizontal *divergence* aloft becomes weaker than horizontal *convergence* near the surface.

cyclone—A *weather* system characterized by relatively low surface *air pressure* compared with the surrounding air; same as a *Low*. Viewed from above, surface *winds* blow counterclockwise and inward in the Northern Hemisphere but clockwise and inward in the Southern Hemisphere.

D

Dalton's law—A scientific law that states that the total pressure exerted by a mixture of gases is equal to the sum of the partial pressures of each constituent gas. Each gas species in the mixture acts independently of the other molecules.

deposition—Phase change of water from vapor to ice (without first becoming liquid) that produces a net gain of ice mass at the interface between ice and air; more water changes phase from vapor to ice than ice to vapor.

derecho—A family of straight-line *downburst winds* in excess of 94 km per hr (58 mph) that impacts a path up to hundreds of kilometers long; may be produced by a *squall line* or a *mesoscale convective complex*.

dew—Tiny water droplets formed by *condensation* of water vapor on a relatively cold surface. For water vapor to condense as dew on the surface of an object, the *temperature* of that surface must cool below the *dewpoint*.

dewpoint—*Temperature* to which air must be cooled (at constant *air pressure* and constant water vapor content) to achieve *saturation* of the air relative to liquid water (if at or above 0 °C or 32 °F).

dewpoint hygrometer—An instrument that measures the *dewpoint* of the air. In one design, air passes over the surface of a metallic mirror that is cooled electronically. An electronic sensor continually monitors the *temperature* of the mirror at the same time that a beam of *infrared radiation* is pointed at the mirror. When a thin *condensation* film forms on the mirror, the *reflectivity* of the infrared beam changes, and the mirror temperature is automatically recorded as the dewpoint.

diffraction—The slight bending of a light wave as it moves along a boundary of an object such as a water droplet. As light waves bend, they interfere with each other either constructively, producing a larger wave, or destructively, canceling each other out. When interference of light waves is constructive, a ring of bright light is seen, and when interference is destructive, darkness is perceived.

dissipating stage—The final phase in the life cycle of a *thunderstorm* cell; features downdrafts throughout the system, the tapering off and ending of *precipitation*, and vaporization of *clouds*.

distillation—The purification of water through phase changes (e.g., *evaporation* followed by *condensation*). When water vaporizes, all suspended and dissolved substances are left behind.

diverging winds—*Winds* that blow away from a central point in a column of air.

doldrums—An east-west equatorial belt of light and variable surface *winds* where the *trade winds* of the two hemispheres converge.

Doppler effect—A shift in the frequency (or phase) of an electromagnetic or sound wave due to the relative movement of the source or the observer.

Doppler radar—*Weather radar* that determines the velocity of targets (*precipitation*, dust particles) moving directly toward or away from the radar unit based on the difference in frequency (or phase) between the outgoing and returning radar signal.

downburst—A strong and potentially destructive *thunderstorm* downdraft; depending on size, classified as either a *microburst* or a *macroburst*.

downdraft—A strong, downward-flowing air current within a thunderstorm, usually associated with precipitation.

downwelling—Where Ekman transport moves surface waters toward the coast, piling them and thus forcing them to sink.

drizzle—A form of liquid *precipitation* consisting of water droplets having diameters between 0.2 and 0.5 mm (0.01 and 0.02 in.); falls very slowly from low *stratus* clouds.

dry adiabatic lapse rate—The adiabatic expansional cooling of ascending, clear (unsaturated) air parcels at the rate of about 9.8 Celsius degrees per 1000 m of uplift (or 5.5 Fahrenheit degrees per 1000 ft). If clear (unsaturated) air parcels descend in the *atmosphere*, the parcels warm adiabatically at the same rate.

dry slot—Dry air descending behind a cold front and drawn into the center of a cyclone.

dryline—A boundary between *continental tropical air* and *maritime tropical air* in the southeast sector of a mature *extratropical cyclone*; likely site for *squall line* and *severe thunderstorm* development.

dust devil—A swirling mass of dust triggered by intense solar heating of dry surface areas. The most common dust devil is less than 1 m (about 3 ft) in diameter, lasts less than a minute, and is too weak to cause serious property damage.

E

Earth-atmosphere system—The Earth's surface (i.e., continents, ocean, ice sheets) plus the *atmosphere*.

easterly wave—A ripple in the tropical easterlies (*trade winds*) featuring a weak trough of low pressure that propagates from east to west. Lower-tropospheric convergence on the east side of the wave can help organize convective activity into a *tropical disturbance*.

eddy viscosity—Fluid *friction* (*viscosity*) arising from eddies (irregular whirls) within a fluid such as air or water.

Ekman spiral—A spiral in water motion in the top 100 m (330 ft) or so of the ocean produced by the directional change and decreasing horizontal motion of successively lower layers of water. Due to a balance between the *Coriolis Effect* and frictional drag, the surface ocean layer moves at an angle of up to 45 degrees to the right or left (depending on the hemisphere) of the surface *wind* direction. Due to the same balance of forces, the layer immediately below the surface ocean layer moves slower and at an angle to the motion of the layer above. This spiraling effect is responsible for the *Ekman transport*.

Ekman transport—The net horizontal movement of water in the top 100 m (330 ft) or so of the ocean induced by the coupling of surface *winds* with ocean surface waters. The *Ekman spiral* causes this net transport to be about 90 degrees to the right of the surface wind direction in the Northern Hemisphere and to the left of the surface wind direction in the Southern Hemisphere.

El Niño—An anomalous warming of surface ocean waters in the eastern tropical Pacific accompanied by suppression of *upwelling* off the coasts of Peru and Ecuador.

electromagnetic radiation—Energy in the form of waves that have both electrical and magnetic properties and travel through gases, liquids, and solids, and occur even in a vacuum. All objects emit all forms of electromagnetic radiation, although each object emits its peak radiation at a certain *wavelength* within the *electromagnetic spectrum*.

electromagnetic spectrum—Range of *electromagnetic radiation* types arranged by *wavelength* or by *wave frequency* or both. Although the electromagnetic spectrum is continuous, different names are assigned to different segments. These segments extend from the longest wavelength (lowest frequency) radio waves through microwaves, *infrared radiation*, *visible radiation*, *ultraviolet radiation*, X-rays, to the shortest wavelength (highest frequency) gamma radiation.

electronic hygrometer—Measures the amount of water vapor in the air based on changes in the electrical resistance of certain chemicals as they adsorb water vapor from the air. Variations in electrical resistance are calibrated in terms of *relative humidity* or *dewpoint*. An electronic hygrometer is flown aboard *radiosondes*.

emissivity—A measure of how close a radiating object approximates a *blackbody*, a perfect radiator. The emissivity of a blackbody is 1.0 and less than 1.0 for all other objects.

energy—The capacity for doing work.

Enhanced F-scale—A revised version of T.T. Fujita's tornado intensity scale that more closely aligns estimated wind speed with associated storm damage; ranges from EF-0 to EF-5.

ensemble forecasting—A numerical model which generates several forecasts each based on slightly different sets of initial conditions.

ENSO (El Niño/Southern Oscillation)—The relationship between *El Niño* and the *Southern Oscillation*. An El Niño episode begins when the weakening of the surface *air pressure gradient* between the western and central tropical Pacific heralds the slackening of the *trade winds*.

equinoxes—The first days of spring and autumn when day and night are of approximately equal length at all latitudes (except the poles) and the noon Sun is directly over the equator (*solar altitude* of 90 degrees).

eustasy—The global variation in sea level brought about by a change in the volume of water occupying the ocean basins.

evaporation—The process which produces a net loss of water vapor mass at the interface between liquid water and air; more water changes phase from liquid to vapor than vapor to liquid. The phase change from liquid to vapor requires the *latent heat of vaporization* and occurs at a *temperature* below the boiling temperature of water.

evapotranspiration—Direct *evaporation* from Earth's surface plus the release of water vapor by vegetation (*transpiration*).

expansional cooling—A *temperature* drop that accompanies a pressure reduction on and resultant expansion of a gas (or mixture of gases). The work of expansion requires energy, which is drawn from the internal energy of the gas, reducing its temperature. Air parcels that ascend within the *atmosphere* undergo expansional cooling in response to falling *air pressure*; expansional cooling is the principal means of *cloud* formation in the *atmosphere*.

extratropical cyclone—A *synoptic-scale* low pressure system that occurs in midlatitudes, often forming along the *polar front*. This low, characterized by *fronts* and a *comma cloud* pattern, becomes a *cold-core cyclone* especially in the later stage of its life cycle.

eye—An area of almost cloudless skies, subsiding air, and light *winds* (less than 25 km per hr or 16 mph) at the center of a *hurricane*. The eye diameter ranges from 10 to 65 km (6 to 40 mi) across and typically shrinks as the system strengthens.

eye wall—A circle of *cumulonimbus* clouds that surrounds the *eye* of a mature *hurricane* and produces heavy *rains* and very strong *winds*. The most dangerous and potentially most destructive part of a hurricane is the portion of the eye wall where the wind blows in the same direction as the hurricane's forward motion.

F

fair-weather bias—The observation that fair-*weather* days outnumber stormy days almost everywhere.

Fata Morgana—An optical phenomenon in which images of distant objects, such as houses and bluffs, are distorted vertically so that they resemble castles or walls with spires.

feedback—In Earth's climate system, the sequence of interactions among climate controls that determines how the system responds to disturbances of boundary conditions.

flash flood—A short-term, localized, and often unexpected rise in river or stream level causing flooding. Usually occurs in response to torrential *rain* falling over a small geographical area.

fog—A visibility-restricting suspension of tiny water droplets or ice crystals in an air layer next to Earth's surface; *stratus clouds* in contact with the ground.

force—A push or pull on an object computed as mass times acceleration.

freezing rain—Supercooled raindrops that at least partially freeze on contact with cold surfaces.

friction—The resistance an object encounters as it comes into contact with other objects; the friction of fluid flow is known as *viscosity.*

front—A narrow zone of transition between *air masses* of contrasting *air density,* that is, air masses of different *temperature, humidity,* or both. Fronts are classified as stationary, warm, cold, and occluded.

frontal fog—*Fog* formed when *precipitation* falling from relatively warm air aloft into a wedge of relatively cool air near the Earth's surface evaporates and raises the *vapor pressure* in the cool air to saturation; occurs either just ahead of a *warm front* or just behind a *cold front.*

frontal uplift—Ascent of air along the surface of a boundary between air masses of contrasting *air density* and often accompanied by cloud development.

frontogenesis—The formation or strengthening of a *front.*

frontolysis—Weakening of a *front.*

frost—Ice crystals that are formed by *deposition* of water vapor on a relatively cold surface. For water vapor to deposit as frost on the surface of an object, the *temperature* of that surface must cool below the *frost point.*

frost point—The *temperature* to which air must be cooled at constant *air pressure* to achieve saturation at or below 0 °C (32 °F).

F-scale (Fujita scale)—A six-category *tornado* intensity scale developed by T. Theodore Fujita that rates tornadoes from F0 to F5 on the basis of rotational *wind* speed estimated from property damage. Categories are weak (F0, F1), strong (F2, F3), and violent (F4, F5). See *Enhanced F-scale.*

funnel cloud—A tornadic circulation extending below *cloud* base but not reaching the ground; made visible by a cone-shaped cloud.

G

gas law—The relationship between the *variables of state* of a gas (or mixture of gases) stating that the pressure exerted by a gas is directly proportional to the product of its density and *temperature.*

Geostationary Operational Environmental Satellite (GOES)—A geostationary satellite that revolves around Earth at the same rate and direction as the planet so it remains positioned over the same spot. Operated by NOAA, it is designed to remotely monitor *weather* systems.

geostrophic wind—A hypothetical, unaccelerated horizontal *wind* that flows along a straight path parallel to *isobars* or height contours above the *atmospheric boundary layer*; results from a balance between the horizontal *pressure gradient force* and the *Coriolis Effect.*

glacial climate—Long-term average conditions favorable to the initiation and growth of glacial ice.

global climate model (GCM)—A *numerical model* that describes the physical interactions among the various components of the *climate* system: the *atmosphere*, ocean, land, ice-cover, and biosphere. It predicts broad regions of expected positive and negative *temperature* and *precipitation anomalies*, and the mean location of circulation features such as jet streams and principal storm tracks.

global radiative equilibrium—The balance between net incoming solar radiation and *infrared radiation* emitted to space by the *Earth-atmosphere system*.

global warming potential (GWP) index—An index that compares the contributions of various gases to the *greenhouse effect*.

global water budget—Balance sheet for the inputs and outputs of water to and from the various global water reservoirs. The global water budget indicates an annual net gain of water mass on land and an annual net loss of water mass from the ocean; the excess water mass on land flows to the sea and balances the budget.

global water cycle—The ceaseless movement of water among the atmospheric, oceanic, terrestrial, and biospheric reservoirs on a planetary scale; assumes an essentially fixed amount of water in the *Earth-atmosphere system*.

glory—Concentric rings of color about the shadow of an observer's head that appear on top of a *warm cloud* situated below the observer. A glory is caused by much the same optics as a primary *rainbow*, two important differences being the size of the reflecting and refracting particles, and the direction of reflected and refracted light. A glory is the consequence of sunlight interacting with a mass of much smaller cloud droplets that compose a *warm cloud*.

gradient wind—A hypothetical horizontal *wind* that blows parallel to curved *isobars* or height contours, above the *atmospheric boundary layer*. The gradient wind differs from the *geostrophic wind* in that the path of the gradient wind is curved.

graupel—See *snow pellets*.

gravity—The force that accelerates air downward to the Earth's surface. It is the net result of gravitation, the force of attraction between Earth and all other objects, and the *centripetal force* arising from Earth's rotation on its axis.

green flash—A brilliant, thin, green rim that occasionally appears on the upper limb of the Sun as it rises or sets (at *twilight*). A green flash is primarily the consequence of atmospheric *refraction* and *scattering* of light from a low Sun; it is best seen on a distant horizon when the *atmosphere* is clear.

greenhouse effect—Heating of Earth's surface and lower atmosphere as a consequence of differences in atmospheric transparency to *electromagnetic radiation*. The *atmosphere* is nearly transparent to incoming solar radiation, but much less transparent to outgoing *infrared radiation*. Terrestrial infrared radiation is absorbed and radiated primarily by water vapor and, to a lesser extent, by carbon dioxide and other trace gases, thereby slowing the loss of *heat* to space from the *Earth-atmosphere system*.

greenhouse gases—Those gases that absorb and emit appreciable *infrared radiation* and contribute to the *greenhouse effect* in the *Earth-atmosphere system*. The main greenhouse gas is water vapor, others are carbon dioxide, ozone, methane and nitrous oxide.

ground clutter—A pattern of *radar echoes* produced by *reflection* of radar signals by fixed objects such as buildings on Earth's surface.

gust front—Leading edge of a mass of relatively cool gusty air that flows out of the base of a *thunderstorm cloud* (downdraft) and spreads along the ground well in advance of the parent thunderstorm cell; a mesoscale *cold front*. Uplift along the gust front may produce additional *cumulus clouds* that may evolve into secondary thunderstorm cells tens of kilometers ahead of the parent cell.

H

haboob—A dust- or sandstorm caused by the strong, gusty downdraft of a desert *thunderstorm*. The mass of dust or sand rolls along the ground as a huge ominous dark *cloud* that may be more than 100 km (60 mi) wide and may reach altitudes of several thousand meters.

Hadley cell—Thermally-driven air circulation in tropical and subtropical latitudes of both hemispheres resembling a huge convective cell with rising air near the equator in the *intertropical convergence zone* and sinking air in the *subtropical anticyclones*. Equatorward blowing surface winds and poleward directed upper-level *winds* complete the circulation.

hail—*Precipitation* in the form of jagged to nearly spherical chunks of ice often characterized by internal concentric layering. Hail is associated with *thunderstorm* cells that have strong updrafts, an abundant supply of supercooled water droplets, and great vertical *cloud* development.

hailstreak—Accumulation of *hail* in a long narrow path along the ground typically around 2 km (1.2 mi) wide and 10 km (6.2 mi) long.

hair hygrometer—An instrument designed to monitor *relative humidity* by measuring the changes in the length of human hair that accompany *humidity* variations.

halo—A whitish ring of light surrounding the Sun or Moon formed when the tiny ice crystals that compose high, thin *clouds* (such as *cirrus* or *cirrostratus*) refract the Sun's rays.

heat—A form of energy transferred between systems, or components of a system, in response to differences in *temperature*. Heat energy is always transferred from a warmer system to a colder system.

heat burst—An abrupt rise in surface air temperature that can occur if convective rain showers fall from clouds having a high base and relatively dry air below.

heat conductivity—The ratio of the rate of heat transport across an area to a temperature gradient.

heat equator—The latitude (about 10 degrees N) of highest mean annual surface air *temperature*. The mean position of the *intertropical convergence zone* approximately corresponds to the heat equator.

heterosphere—The *atmosphere* above 80 km (50 mi) and the *homosphere* where gases are stratified; concentrations of the heavier gases decrease more rapidly with altitude than concentrations of the lighter gases.

Holocene Epoch—The interval since the end of the Pleistocene glaciation (approximately 10,500 years ago) that represents the present interglacial. This epoch has been characterized by spatially and temporally variable *temperature* and *precipitation*.

homosphere—The *atmosphere* up to 80 km (50 mi) where the relative proportions of principal gases, such as nitrogen and oxygen, are constant.

hook echo—A distinctive hook-shaped reflectivity pattern in a *radar echo* that often indicates the presence of a *severe thunderstorm* cell and perhaps tornadic circulation. This pattern is produced by rainfall being drawn around the *mesocyclone*.

horse latitudes—Areas of persistent light *winds* or calm air between about 30 and 35 degrees N and S under *subtropical anticyclones*.

hot-wire anemometer—An instrument that measures *wind* speed based on the rate of *heat* loss to air flowing passed a heated wire. The rate of heat loss from the wire to the air increases as wind speed increases.

humidity—A general term referring to any one of many ways of describing the amount or concentration of water vapor in the air.

hurricane—An intense *tropical cyclone* that originates over tropical ocean basins; called a typhoon in the western Pacific Ocean. Maximum near-surface sustained *winds* are 119 km per hr (74 mph) or higher.

hydrostatic equilibrium—Balance between the *atmosphere's* vertical *pressure gradient force* and the equal, but oppositely directed force of *gravity*.

hygrograph—An instrument that records a continuous trace of *relative humidity* variations with time. The most common hygrograph is a *hair hygrometer* designed to move a pen on a clock-driven drum.

hygrometer—An instrument that measures the amount of water vapor in the air. See *dewpoint hygrometer*, *hair hygrometer*, and *electronic hygrometer*.

hygroscopic nuclei—A special category of *cloud condensation nuclei* that have a special chemical affinity for water molecules, so that *condensation* may take place on these nuclei at *relative humidities* under 100%.

hypothesis—A proposed explanation for some observation or phenomenon which is tested through the *scientific method*.

I

ice pellets—*Precipitation* consisting of frozen raindrops 5 mm (0.2 in.) or less in diameter that bounce on impact with the ground; also called sleet.

ice stream—A zone of relatively rapidly flowing ice within a glacial ice sheet.

ice-albedo feedback—Warming that causes a reduction in Arctic sea-ice cover is enhanced by the lower albedo of the open ocean waters.

ice-forming nuclei (IN)—Tiny particles that promote the formation of ice crystals at *temperatures* well below freezing.

ideal gas—A gas which obeys *Boyle's law* and *Charles' law*.

infrared radiation (IR)—*Electromagnetic radiation* at *wavelengths* ranging from 0.8 micrometer (near-infrared) to about 0.1 mm (far infrared). Infrared radiation has wavelengths shorter than *microwaves* and longer than visible red light; most objects in the *Earth-atmosphere system* have their peak emission in the infrared.

infrared radiometer—An instrument that measures the intensity of infrared radiation emitted by the surface of some object, such as clouds or the ocean.

infrared satellite image—Picture or image processed from radiometers onboard a satellite that sense thermal (or infrared) radiation (typically, from *wavelengths* of approximately 8 to 12 micrometers) emitted by Earth and *cloud* surfaces of the *Earth-atmosphere system*. Infrared radiation signals are routinely calibrated to give the surface *temperature* of objects in the sensor's field of view.

interglacial climate—Climate episode that favors the thinning or retreat of existing glaciers or no glaciers at all.

Intergovernmental Panel on Climate Change (IPCC)—Entity established by WMO and UNEP to assess the technical, scientific, and socio-economic information relevant for the understanding of climate change and its potential impacts.

internal energy—A measure of the molecular activity of a system or the summation of total energies of all molecules in a specific mass.

intertropical convergence zone (ITCZ)—Discontinuous low-pressure belt of *thunderstorms* paralleling the equator and marking the convergence of the Northern and Southern Hemisphere surface *trade winds*.

inverse square law—Intensity of radiation emitted by a point source (e.g., the Sun) decreases as the inverse square of distance traveled.

ion—An atomic molecule that caries an electric charge.

ionosphere—Region of the upper *atmosphere* above 80 km (50 mi) that contains a relatively high concentration of *ions* (electrically charged particles). The ionosphere is located primarily in the *thermosphere*.

iridescent cloud—A *cloud* (usually *altocumulus*, *cirrostratus*, or *cirrocumulus*) having bright spots, bands, or borders of color, usually red or green. Iridescent clouds are produced by *diffraction* and typically appear up to about 30 degrees from the Sun.

isobar—A line plotted on a map joining locations reporting the same *air pressure*. Drawing an isobar on a map of sea-level air pressures requires interpolation between reporting *weather* stations.

isotherms—Lines drawn on a map through localities having the same air *temperature*.

J

jet streak—An area of accelerated air flow within a jet stream; the *wind* may strengthen by as much as an additional 100 km per hr (62 mph). Jet streaks occur where surface horizontal *temperature gradients* are particularly steep and play an important role in the generation and maintenance of synoptic-scale *cyclones*. The strongest jet streaks develop during winter in the *polar front jet stream* along the East Coasts of North America and Asia.

K

katabatic wind—Shallow layer of cold, dense air that flows downhill under the influence of *gravity*.

L

La Niña—A period of particularly strong *trade winds* and unusually low sea-surface *temperatures* (SST) in the central and eastern tropical Pacific; opposite of *El Niño*. The strong trade winds induce exceptionally vigorous *upwelling* in the eastern tropical Pacific, which causes unusually low SST.

lake breeze—A relatively cool surface *wind* directed from a large lake toward land in response to differential heating between land and lake; develops during daylight hours.

land breeze—A relatively cool surface *wind* directed from land to sea or land to lake in response to differential cooling between land and a body of water; develops at night.

latent heat—The quantity of *heat* involved in the phase changes of water.

latent heat of fusion—*Heat* released to the environment when water changes phase from liquid to solid; releases 80 *calories* per gram.

latent heat of vaporization—*Heat* required to change the phase of water from liquid to vapor; requires 540 to 600 *calories* per gram, depending on the *temperature* of the water.

latent heating—Transport of *heat* from one place to another within the *atmosphere* as a consequence of phase changes of water. Heat is supplied for melting, *evaporation*, and *sublimation* of water at the Earth's surface, and heat is released during freezing, *condensation*, and *deposition* within the atmosphere.

law of energy conservation— Energy is neither created nor destroyed but can change from one form to another; also known as the first law of thermodynamics.

lee-wave clouds—Lens-shaped (lenticular) *clouds* that form in the crests of a standing wave downwind from a prominent mountain range.

lifetime of a gas—The time is takes for the gas in the atmosphere to be reduced to 37% of its original amount.

lifting condensation level (LCL)—The altitude to which air must be lifted so that *expansional cooling* leads to *condensation* (or *deposition*) and *cloud* development; corresponds to the base of clouds.

lightning—A brilliant flash of light produced by an electrical discharge in response to the buildup of an electrical potential between a *cloud* and Earth's surface, between different clouds, or between different portions of the same cloud.

lightning detection network (LDN)—System that provides real-time information on the location and severity of *lightning* strokes.

Little Ice Age—The interval of the *Holocene* from about 1450 to 1850 when average global *temperatures* were lower, and alpine glaciers increased in size and advanced down mountain valleys. The Little Ice Age followed the *Medieval Warm Period*.

M

macroburst—A *downburst* that affects a path longer than 4.0 km (2.5 mi), has maximum surface *winds* that may top 210 km per hr (130 mph), and has a life expectancy of up to 30 minutes. The leading edge of a macroburst may be marked by a *gust front*.

magnetosphere—Region of the upper *atmosphere* encompassed by the Earth's magnetic field and deflected by *solar wind* into a teardrop-shaped cavity.

mammatus clouds—*Clouds* that form on the underside of a *thunderstorm* anvil and exhibit pouchlike, downward protuberances; may indicate turbulent air.

maritime climate—The climate immediately downwind of an ocean, or large lake, that experiences less contrast between winter and summer temperatures due to the greater thermal inertia of water.

maritime polar air (mP)—Cool, humid *air masses* that form over the cold ocean waters of the North Pacific and North Atlantic, especially north of 40 degrees N. Along the West Coast, maritime polar air contributes to heavy winter *rains* and *snows* and persistent summer coastal *fogs*.

maritime tropical air (mT)—Warm, humid *air masses* that form over tropical and subtropical ocean waters (e.g., the Gulf of Mexico). Maritime tropical air is responsible for oppressive summer *heat* and *humidity* east of the Rocky Mountains.

mature stage—The middle and most intense phase in the life cycle of a *thunderstorm* cell; begins when *precipitation* reaches Earth's surface and is characterized by both updrafts and downdrafts. *Rain* is heaviest, *lightning* is most frequent, and *hail*, strong surface *winds*, and *tornadoes* may develop during this stage, which typically lasts about 10 to 20 minutes.

Maunder minimum—A 70-year period from 1645 to 1715 when *sunspots* were rare.

Medieval Warm Period—A relatively mild episode of the *Holocene* between about C.E. 950 and 1250.

mercury barometer—A mercury-filled tube used to measure *air pressure*; the standard barometric instrument, which features great precision.

meridional flow pattern—Flow of the *planetary-scale* westerlies in a series of deep troughs and sharp ridges, exhibiting considerable amplitude. In this pattern, cold *air masses* surge southward and warm air masses stream northward, leading to strong *temperature gradients*.

mesocyclone—A vertical column of cyclonically rotating air 3 to 10 km (2 to 6 mi) across that develops in the updraft of a *severe thunderstorm* cell; an early stage in the development of a *tornado*. A mesocyclone forms a tornado about 10% of the time.

mesopause—Narrow zone of transition between the *mesosphere* below and the *thermosphere* above; the top of the mesosphere. Has an average altitude of 80 km (50 mi) and features the lowest average *temperature* in the *atmosphere*.

mesoscale convective complex (MCC)—A nearly circular organized cluster of many interacting *thunderstorm* cells covering an area of many thousands of square kilometers. MCCs occur mostly in the warm-season over the eastern two-thirds of the United States and develop at night during weak synoptic-scale flow. Lasting from 6 to 12 hrs, they are driven by a flow of warm, humid air at low levels and *radiational cooling* at upper levels, which work together to destabilize the *troposphere*.

mesoscale convective system (MCS)—A regional weather system such as a thunderstorm.

mesoscale systems—*Weather* phenomena that may influence the weather in only a portion of a large city or county; includes *thunderstorms*, *sea breezes*, and *lake breezes*. These systems have dimensions of 1 to 100 km (1 to 60 mi) and last from hours to a day or so.

mesosphere—The atmospheric layer between the *stratosphere* and the *thermosphere*; *temperature* falls with increasing altitude. Located between average altitudes of 50 to 80 km (31 to 50 mi).

meteorology—The scientific study of the *atmosphere*, atmospheric processes, and the life cycle of *weather* systems.

microburst—A *downburst* that affects a path on the ground that is 4.0 km (2.5 mi) or shorter, has maximum surface *winds* as high as 270 km per hr (170 mph), and has a life expectancy of less than 10 minutes.

microscale systems— *Weather* phenomena that represent the smallest spatial subdivision of atmospheric circulation, such as a weak *tornado*. These systems have dimensions of 1 m to 1 km (3 ft to 1 mi) and last from seconds to an hour or so.

midlatitude westerlies—Prevailing planetary-scale *winds* in the middle and upper *troposphere* between about 30 and 60 degrees of latitude, blowing, on average from the southwest in the Northern Hemisphere and from the northwest in the Southern Hemisphere out of the poleward flanks of the *subtropical anticyclones*.

Mie scattering—The optical effect whereby light is scattered equally at all *wavelengths*. It is produced by spherical particles having about the same diameter as the wavelength of *visible radiation;* responsible for a milky white sky when large numbers of *aerosols* are present.

Milankovich cycles—Systematic changes in three elements of Earth-Sun geometry: precession of the *solstices* and *equinoxes*, tilt of Earth's rotational axis, and orbital eccentricity; affect the seasonal and latitudinal distribution of incoming solar radiation and influence climatic fluctuations over tens to hundreds of thousands of years.

mirage—An optical phenomenon that makes an object appear to be displaced from its true position. It is caused by *refraction* of light rays due to the change in *air density* with altitude within the lower *atmosphere*. When an object appears higher that it actually is, it is called a superior mirage. When an object appears lower that it actually is, it is called an inferior mirage.

mist—Very thin *fog* in which visibility is greater than 1.0 km (0.62 mi). Mist is also known as light *drizzle*.

mistral—A *katabatic wind* that descends from the snow-capped Alps down the Rhone River Valley of France and into the Gulf of Lyons along the Mediterranean coast.

mixing ratio—Mass of water vapor per mass of the remaining dry air and usually expressed as so many grams of water vapor per kilogram of dry air.

model comparison—To optmize the accuracy of weather forecasts, a comparison is made among forecasts from several different numberical models. If they all agree, then the forecast is issued with a relatively high level of confidence.

moist adiabatic lapse rate—A variable rate of cooling applicable to saturated (cloudy) air parcels that are ascending within the *atmosphere*. This rate is less than the *dry adiabatic lapse rate* because some of the *expansional cooling* is compensated by *latent heat* released during *condensation* or *deposition* of water vapor.

molecular viscosity—Fluid *friction* (*viscosity*) arising from the random motions and interactions of molecules composing a fluid such as air or water.

monsoon circulation—Seasonal reversals in prevailing *winds* that cause wet summers and relatively dry winters. The most vigorous monsoon circulation occurs over Africa and southern Asia.

mountain breeze—A shallow, gusty downslope flow of cool air that develops at night in some mountain valleys in response to differential heating between the air adjacent to the valley wall and air at the same altitude out over the valley floor.

mountain-wave clouds—Stationary lenticular (lens-shaped) *clouds* situated over and downwind of a prominent mountain range; formed by the disturbance of the large-scale horizontal *winds* by the mountain range.

multi-vortex tornadoes—A tornado consisting of two or more subsidiary vortices that orbit each other or about a common center within a massive tornado.

N

nacreous clouds—Rarely seen *clouds* with a soft, pearly luster that form in the upper *stratosphere*; may be composed of ice crystals or supercooled water droplets. Nacreous clouds are *cirrus* and *altocumulus lenticularis* that form on sulfuric acid nuclei. Also called mother-of-pearl clouds or polar stratospheric clouds.

National Centers for Environmental Prediction (NCEP)—Centers responsible for the interpretation of weather maps and charts, and the generation of forecasts and other guidance products for the nation and for exchange with other nations. Part of *National Oceanic and Atmospheric Administration (NOAA)/National Weather Service.*

National Climatic Data Center (NCDC)—An agency, located in Asheville, NC, of the *National Oceanographic and Atmospheric Administration* that archives climatic data of the United States. Part of *NOAA.*

National Data Buoy Center (NDBC)—Operates a network of automated weather stations attached to moored buoys in offshore locations and at lighthouses, fishing piers and offshore oil platforms. Part of *NOAA/National Weather Service.*

National Hurricane Center (NHC)— Part of the *National Centers for Environmental Prediction* located in south Florida that is responsible for forecasting *tropical cyclones* in the Atlantic Basin, Gulf of Mexico, and the eastern tropical Pacific Ocean.

National Oceanic and Atmospheric Administration (NOAA)—The administrative unit within the U.S. Department of Commerce that oversees the *National Weather Service.*

National Weather Service (NWS)—The agency of the *National Oceanic and Atmospheric Administration* responsible for *weather* data acquisition, data analysis, weather forecasting and dissemination, and *storm watches* and *warnings.*

nebula—In astronomy, an immense rotating interstellar cloud of dust, ice and gases.

neutral air layer—An ambient air layer in which an ascending or descending air parcel always has the same *temperature* (and *air density*) as its surroundings.

Newton's first law of motion—An object at rest or in straight-line, unaccelerated motion remains that way unless acted upon by a net external force.

Newton's second law of motion—A net force is required to cause a unit mass of a substance to accelerate; force equals mass times acceleration.

nimbostratus clouds (Ns)—Low, gray, layered *clouds* that resemble *stratus* but are thicker, appear darker gray, have a less uniform base, and yield more substantial *precipitation*. Nimbostratus are composed of mostly water droplets.

NOAA Weather Radio—Low power, VHF-high-band FM radio transmitters operated by the *National Oceanic and Atmospheric Administration* that broadcast continuous *weather* information (e.g., regional conditions, local forecasts, marine warnings) directly from *National Weather Service* Forecast Offices 24 hrs per day.

noctilucent clouds—Wavy, thin *clouds* resembling *cirrus*, but usually bluish-white or silvery; best seen at high latitudes just before sunrise or just after sunset. These rare clouds occur in the upper *mesosphere* at altitudes above about 80 km (50 mi) and may be composed of ice deposited on meteoric dust.

Norwegian cyclone model—An approximation of the structure and life cycle of an extratropical low-pressure system based mostly on surface observations, first proposed during World War I by researchers at the Norwegian School of Meteorology at Bergen.

North Atlantic Oscillation—The seasaw variation in sea level air pressure between the Azores and Iceland. The air pressure gradient between the Azores high and Icelandic low governs the strength and direction of westerly winds, the middle latitude jet stream, and storm tracks across the North Atlantic.

nuclei—Tiny solid or liquid particles suspended in the *atmosphere* that provide surfaces on which water vapor condenses into droplets or deposits into ice crystals; essential for *cloud* formation.

number density—Number of molecules of a gas (or mixture of gases) per unit volume.

NWS Cooperative Observer Network—Consists of more than 10,400 *weather* stations across the United States that record daily *precipitation* and maximum/minimum *temperatures* for hydrologic, agricultural, and climatic purposes. These manned stations complement observations from the *Automated Surface Observing System.*

O

occluded front—A *front* formed late in the life cycle of an extratropical *cyclone*; its behavior depends upon the characteristics of air behind the *cold front* and ahead of the *warm front*. Also known as an occlusion.

orographic lifting—The forced rising of air up the slopes of a hill or mountain. Air that is forced to ascend the slopes facing the oncoming *wind* (windward slopes) expands and cools, which increases its *relative humidity*. With sufficient cooling, *clouds* and *precipitation* develop.

outgassing—Release of gasses to the *atmosphere* from hot, molten rock during volcanic activity and from impact of meteorites on the rocky surface of the planet; the origins of most atmospheric gases.

overrunning—The process whereby less dense air flows up and over denser air; occurs along a *warm front* and may occur along a *stationary front*. Ascending air cools by expansion, leading to the formation of *clouds* and perhaps *precipitation* over a widespread area.

P

Pacific air—Term used to describe cool, humid *maritime polar air* swept inland from the Pacific Ocean that undergoes *air mass modification* over the Rocky Mountains, emerging milder and drier to the east of the mountains. During a *zonal flow pattern*, Pacific air floods the eastern two-thirds of the United States and southern Canada, causing mild and generally dry *weather*.

Pacific Decadal Oscillation (PDO)—Long lived variation in climate over the North Pacific and North America sea surface temperatures between the north central Pacific (north of 20 degrees N) and the west coast of North America.

Paleocene-Eocene Thermal Maximum (PETM)—A geologically brief interval of widespread warming associated with a massive buildup of greenhouse gases; about 55 million years ago.

perihelion—The time of the year when the Earth's orbital path brings it closest (147 million km or 91 million mi) to the Sun (at present, about 3 January).

persistence—Tendency for *weather* episodes to continue for some period of time.

pH scale—A measure of the range of acidity and alkalinity of different substances, extending from 0 to 14. A pH of 7 is neutral. Acids have values less than 7 and alkaline substances greater than 7.

photosphere—The intensely bright portion of the Sun visible to the unaided eye; the several-hundred-kilometer-thick layer is what we perceive as the surface of the Sun. Features such as *sunspots* and *faculae* are observed on the photosphere.

photosynthesis—The process whereby plants use sunlight, water, and carbon dioxide to manufacture their food and generate oxygen as a byproduct.

planetary albedo—The fraction (or percent) of incident solar radiation that is scattered and reflected back into space by the *Earth-atmosphere system*; measurements by satellite sensors indicate a planetary albedo of about 30%.

planetary-scale systems—*Weather* phenomena operating at the largest spatial scale of atmospheric circulation; includes the global *wind* belts and *semipermanent pressure systems*. These systems have dimensions of 10,000 to 40,000 km (6000 to 24,000 mi) and exhibit patterns that persist from weeks to months.

plate tectonics—Concept that the outer 100 km (60 mi) of solid Earth is divided into a dozen or more gigantic rigid plates that move relative to one another slowly across the surface of the planet. The drift of these plates moves continents and opens and closes ocean basins over the vast expanse of geologic time; mountain building and most volcanic activity occur at plate boundaries.

polar amplification—The tendency for a major *temperature* change to increase in magnitude with latitude.

polar front—Narrow transition zone where the relatively mild *midlatitude westerlies* meet and override the relatively cold polar easterlies. When the *temperature gradient* across the *front* is steep, the front is well defined and is a potential site for development of *extratropical cyclones*.

polar front jet stream—A corridor of strong westerlies in the upper *troposphere* between the midlatitude *tropopause* and the polar tropopause and directly over the *polar front*.

polar high—A *cold-core anticyclone* originating in a source region for *continental polar air*; this shallow system is the product of intense *radiational cooling* over the *snow*-covered continental interior of North America, well north of the *polar front*.

Polar-orbiting Operation Environmental Satellite (POES)—A satellite whose orbital plane passes near the North and South poles. Earth rotates under the plane of the satellite.

polar stratospheric clouds (PSCs)—Clouds composed of water ice, nitric acid and sulfuric acid which form during the Antarctic winter when extreme radiational cooling causes stratospheric temperatures to drop below −88 °C (−126 °F).

poleward heat transport—Flow of *heat* from tropical to middle and high latitudes in response to latitudinal imbalances in net *radiational heating* and *radiational cooling*. Poleward heat transport is accomplished primarily by *air mass* exchange, but also by storms and ocean circulation.

precipitable water—The depth of water that would be produced if all the water vapor in a vertical column of air, extending from Earth's surface to the top of the *troposphere,* were condensed.

precipitation—Water in solid, liquid, or freezing form that falls to Earth's surface from *clouds*; can be in the form of *rain, drizzle, snow, ice pellets, hail* or *freezing rain.*

pressure gradient force—A force operating in the *atmosphere* that accelerates air parcels away from regions of high *air pressure* directly across *isobars* toward regions of low air pressure in response to an *air pressure gradient.*

pressure systems—Individual synoptic-scale features of atmospheric circulation; commonly denoted as highs (or *anticyclones*) or lows (or *cyclones*), less frequently ridges or troughs.

psychrometer—An instrument used to measure the amount of water vapor in the air. It consists of two identical liquid-in-glass *thermometers* mounted side by side with the bulb of one thermometer wrapped in a muslin wick. To take a reading, the wick-covered bulb is first soaked in distilled water and then the instrument is ventilated, either by being whirled about (*sling psychrometer*) or with a small fan (*aspirated psychrometer*). The dry bulb thermometer measures the actual air *temperature.* Water vaporizes from the muslin wick into the air streaming past the wet-bulb thermometer and evaporative cooling lowers the *wet-bulb temperature* to a steady value. By referring to special psychrometric tables, the *relative humidity* or the *dewpoint* can be determined from the difference between the dry-bulb temperature and wet-bulb temperature (*wet-bulb depression*).

pyranometer—The standard instrument for measuring solar radiation incident on a horizontal surface; calibrates the *temperature* response of a special sensor in units of radiation flux, such as watts per square meter.

R

radar—An acronym for *radio detection and ranging*. In *meteorology*, a remote sensing tool which broadcasts and receives microwave signals back from targets for the purpose of determining the location, height, movement and intensity of precipitation areas.

radar echo—*Microwave* signals emitted by *weather radar* that are scattered or reflected by *rain* or *snow* back to a receiver where they are electronically processed and displayed on a cathode ray tube.

radiation fog—Ground-level *cloud* formed by nocturnal *radiational cooling* of a humid air layer near the ground so that its *relative humidity* approaches 100%; sometimes called *ground fog,*

radiational cooling—The drop in *temperature* of an object or a surface accomplished whenever the object or surface undergoes a net loss of *heat* due to a greater rate of emission of *electromagnetic radiation* than *absorption.*

radiational heating—The rise in *temperature* of an object or a surface accomplished whenever the object or surface undergoes a net gain of *heat* due to a greater rate of *absorption* of *electromagnetic radiation* than emission.

radiosonde—A small balloon-borne instrument package equipped with a radio transmitter that takes altitude readings (*soundings*) of *temperature, air pressure,* and *humidity* in the *atmosphere.*

rain—Form of *precipitation* consisting of liquid water drops having diameters generally between 0.5 and 6.0 mm (0.02 and 0.2 in.) that falls mostly from *nimbostratus* and *cumulonimbus* clouds.

rain gauge—A device for collecting and measuring rainfall (or melted snowfall); a standard rain gauge consists of a cylindrical container equipped with a cone-shaped funnel at the top.

rain shadow—A region situated downwind (often hundreds of kilometers) of a prominent mountain range and characterized by descending air and, as a consequence, a relatively dry *climate*.

rainbow—An arc of concentric colored bands formed by *refraction* and internal *reflection* of sunlight by raindrops. To see a rainbow, an observer must be looking at a distant *rain* shower with the Sun at his/her back.

Rayleigh scattering—The optical effect where light at the short-*wavelength* end of the visible *electromagnetic spectrum* is scattered much more efficiently than light at the long-wavelength end of the visible spectrum. This wavelength-dependent *scattering* is caused by spherical scattering particles having diameters much smaller than the wavelength of scattered radiation. Rayleigh scattering is responsible for the blue of the daytime sky.

reflection—The process whereby a portion of the radiation that strikes the interface between two different media is redirected (backscattered). Reflection is a special case of *scattering*.

refraction—The bending of a light ray as it passes from one transparent medium into another (from air to water, for example). Bending is due to the differing speeds of light in the two media.

relative humidity—Compares the amount of water vapor in the air with the amount of water vapor in the same air at *saturation*. Relative humidity is a measure of how close air is to *saturation* at a specific *temperature*, always expressed as a percentage. It can be computed from either the ratio of the *vapor pressure* to the *saturation vapor pressure* or the ratio of the *mixing ratio* to the *saturation mixing ratio*.

remote sensing—The technology of measuring or acquiring data and information about an object or phenomenon with a device that is not in physical contact with it.

ring—A large turbulent eddy that forms when a meander in an ocean current forms a loop which pinches off, separating from the main current.

roll cloud—A low, cylindrically-shaped and elongated *cloud* occurring behind a *gust front*; associated with but detached from a *cumulonimbus* cloud. The cloud appears to rotate slowly about its horizontal axis.

Rossby waves—Series of long-*wavelength* troughs and ridges that characterize the planetary-scale westerlies (above the 500-mb level) as they encircle the globe; also called *long waves*. Typically, between 2 and 5 waves encircle the hemisphere at one time.

S

Saffir-Simpson Hurricane Intensity Scale—A five-category *hurricane* intensity scale developed by H. S. Saffir and R. H. Simpson that rates hurricanes from 1 to 5 on the basis of maximum sustained *wind* speed. Categories are 1 (weak), 2 (moderate), 3 (strong), 4 (very strong), and 5 (devastating).

Saharan Air Layer (SAL)—An elevated mass of dry stable air originating over the Sahara Desert.

Santa Ana wind—A hot, dry *chinook wind* that blows from the desert plateaus of Utah and Nevada across the Sierra Nevada and downslope toward coastal southern California. This *wind* desiccates vegetation and contributes to outbreaks of forest and brush fires.

saturation absolute humidity—The value of *absolute humidity* when air is saturated with respect to water vapor.

saturation mixing ratio—The value of the *mixing ratio* when air is saturated with respect to water vapor; varies directly with *temperature*, and to a lesser extent upon *air pressure*.

saturation specific humidity—The value of the *specific humidity* when air is saturated with respect to water vapor; varies directly with *temperature*.

saturation vapor pressure—The value of the *vapor pressure* when air is saturated with respect to water vapor; varies directly with *temperature*.

scattering—The process by which *aerosols* and molecules disperse radiation in all directions. Scattering is a function of the *wavelength* of the incident radiation and the size, surface, and composition of the scattering aerosol or molecule, where size is defined relative to the wavelength. No energy transformation results from scattering.

scientific method—A systematic form of inquiry that involves observation, speculation, and the formulation and testing of hypotheses.

scientific model—An approximate representation or simulation of a real system that omits all but the most essential variables of the system.

scientific theory—A *hypothesis* that has stood the test of time and is generally accepted by the scientific community.

scintillation—Generic term for rapid variations in apparent position, brightness, or color of a distant luminous object viewed through the atmosphere, caused by changing refraction of light passing along an atmospheric path with rapidly changing density.

sea breeze—A relatively cool mesoscale surface *wind* directed from the ocean toward land in response to differential heating of land and sea; develops during daylight hours.

second law of thermodynamics—All systems tend toward a state of disorder. A gradient in a system, such as a *temperature gradient*, signals order in the system. As a system tends towards disorder, gradients are eliminated.

semi-permanent pressure systems—Persistent areas of high and low *air pressure* that are *planetary-scale systems*. They undergo some important seasonal changes in location and surface air pressures and include *subtropical anticyclones*, the *intertropical convergence zone*, *subpolar lows*, and *polar highs*.

sensible heating—Transport of *heat* from one location or object to another via *conduction*, *convection* or both.

severe thunderstorm—*Thunderstorms* accompanied by locally damaging surface *winds*, frequent *lightning*, or large *hail*. The official *National Weather Service* criterion for the severe thunderstorm designation includes any one or a combination of the following: hailstones with diameters 1 in. (2.5 cm) or larger, *tornadoes* or *funnel clouds*, surface winds stronger than 58 mph (93 km per hr).

shelf cloud—A low, wedge-shaped, and elongated *cloud* that occurs along a *gust front*; associated with and attached to a *cumulonimbus cloud*. Damaging surface *winds* may occur under a shelf cloud, and this cloud may be associated with a *severe thunderstorm*. Also known as an arcus cloud.

short waves—Relatively small short-*wavelength* ripples (troughs and ridges) superimposed on *Rossby waves* in the planetary-scale westerlies; they propagate with the airflow in the middle and upper *troposphere*. Typically, a dozen or more short waves encircle the hemisphere at one time.

single-station forecasts—*Weather* forecasts based on observations at one location.

sleet—See *ice pellets.*

SLOSH (Sea, Lake, & Overland Surges from Hurricanes)—A numerical model developed in 1979 by NOAA to predict the location and height of a storm surge in a coastal area.

snow—A frozen form of *precipitation* consisting of an assemblage of ice crystals in the form of flakes. Although snowflakes vary in shape and size, the constituent ice crystals are hexagonal (six-sided) and may consist of needles, dendrites, plates, or columns.

snow grains—Frozen form of *precipitation* consisting of flat particles of opaque, white ice having diameters less than 1 mm (0.04 in.); originates in the same way as *drizzle* except that particles freeze prior to reaching the ground. Also known as granular snow.

snow pellets—Frozen form of *precipitation* consisting of soft spherical (or sometimes conical) particles of opaque, white ice having diameters of 2 to 5 mm (0.08 to 0.2 in.). They often break up when striking a hard surface and differ from *snow grains* in being softer and larger. Formally called soft hail or graupel.

solar altitude—The angle of the Sun 90 degrees or less above the horizon that influences the intensity of solar radiation striking Earth's surface. At a maximum possible solar altitude of 90 degrees, the solar rays are most intense; the intensity declines with decreasing solar altitude (towards 0 degrees).

solar constant—The flux of solar radiational energy falling on a surface that is positioned at the outer edge of the *atmosphere* and oriented perpendicular to the solar beam when Earth is at its average distance from the Sun.

solar corona—Outermost portion of the solar atmosphere (above the *chromosphere*) that is a region of extremely hot (1 to 4 million °C or 1.8 to 7.2 million °F), highly rarefied gases extending millions of kilometers into space, to the outer limits of the solar system. The *solar wind* originates in the solar corona.

solar wind—A stream of super-hot electrically charged subatomic particles (mainly protons and electrons) flowing into space from the Sun. The solar wind originates in the *solar corona.*

solstice—A time during the year when the Sun is at its maximum poleward location relative to the Earth (23 degrees, 30 minutes, North or South); the first days of astronomical summer and winter.

sonic anemometer—An instrument sensor that emits sound waves to determine *wind* speed.

sonic boom—A shock wave caused by an aircraft flying at supersonic speeds, heard as a loud boom.

sound wave—A compressional wave consisting of alternate compressions and rarefactions of air.

Southern Oscillation—Opposing swings of surface *air pressure* between the western and central tropical Pacific Ocean. When air pressure is low at Darwin (Australia) it is high at Tahiti (a south Pacific island) and when air pressure is high at Darwin it is low at Tahiti.

Southern Oscillation Index (SOI)—An index calculated from the monthly or seasonal fluctuations in the air pressure between Darwin, Australia and Tahiti. Positive values indicate La Niña conditions and negative values El Niño conditions.

Southwest Monsoon—Also called the North American Monsoon System (NAMS); circulation brings summer rainfall to the American Southwest from moisture sources in the Gulf of California and the Gulf of Mexico.

specific heat—The amount of *heat* required to raise the *temperature* of 1 gram of a substance by 1 Celsius degree.

specific humidity—The ratio of the mass of water vapor (in grams) to the mass (in kilograms) of air containing the water vapor; that is, the combined mass of dry air plus water vapor.

split flow pattern—Wave pattern in the *planetary-scale circulation* regime where westerlies to the north have a wave configuration that differs from that of westerlies to the south.

squall line—An elongated cluster of intense *thunderstorm* cells that is accompanied by a continuous *gust front* at the cluster's leading edge. A squall line is located parallel to and up to 300 km (180 mi) ahead of a fast-moving well-defined *cold front*.

stable air layer—An ambient air layer characterized by a vertical *temperature* profile such that air parcels return to their original altitudes following any upward or downward displacement. An ascending air parcel becomes cooler (denser) than the ambient air and a descending air parcel becomes warmer (less dense) than the ambient air, returning in both cases to the original altitude.

standard atmosphere—A model that represents the state of the *atmosphere* averaged for all latitudes and seasons. It features a fixed sea-level air *temperature*, *air pressure*, and *air density* plus fixed vertical profiles of *temperature* and *air pressure*.

station model—A specified pattern for entering, on a weather map, the meteorological symbols that represent the state of the weather at a particular observing station.

stationary front—A *front* that exhibits essentially no lateral movement; *winds* blow parallel but in opposite directions on either side of the *front*.

steam fog—The general name for *fog* produced when extremely cold, dry air comes in contact with a relatively warm (unfrozen) water surface; has the appearance of rising streamers. Also known as Arctic sea smoke.

Stefan-Boltzmann law—A radiation law that states that the total energy radiated by a *blackbody* at all *wavelengths* is directly proportional to the fourth power of the absolute *temperature* (in kelvins) of the body. For example, the Sun's energy output per square meter is about 190,000 times that of the *Earth-atmosphere system*.

storm surge—A dome of water perhaps 80 to 160 km (50 to 100 mi) wide that sweeps over the coastline near the landfall of a tropical or extratropical *cyclone*, often causing considerable coastal erosion and flooding. The dome of water is caused primarily by strong onshore *winds* and, to a lesser extent, low *air pressure* associated with the storm system.

stratiform clouds—Layered *clouds*, such as *altostratus*, often produced by *overrunning*.

stratocumulus clouds (Sc)—Low *clouds* occurring as large, irregular puffs or rolls arranged in a layer. Stratocumulus are composed of mostly water droplets.

stratopause—Transition zone between the *stratosphere* below and the *mesosphere* above.

stratosphere—The *atmosphere's* thermal subdivision, situated between the *troposphere* and *mesosphere*, at an average altitude of 50 km (31 mi) and the primary site of ozone formation. Air *temperature* in the lower portion of the stratosphere is constant then increases with altitude to the *stratopause*.

stratospheric ozone shield—Ozone in the *stratosphere* that shields organisms at the Earth's surface from exposure to potentially lethal intensities of solar *ultraviolet radiation*.

stratus clouds (St)—Low *clouds* that occur as a uniform gray layer stretching from horizon to horizon. They may produce *drizzle*, and where they intersect the ground, they are classified as *fog*. Stratus are composed of water droplets.

streamline—A line that graphically portrays the flow of *wind* and can be used to identify regions of divergence and convergence.

Stüve thermodynamic diagram—A graphical representation of *soundings* and *adiabatic processes* in the *atmosphere*.

sublimation—The process which produces a net loss of water mass at the interface between ice and air; more water changes phase from ice to vapor than vapor to ice. The phase change of water from ice to vapor (without first becoming liquid) requires the addition of the *latent heat of melting* plus the *latent heat of vaporization*.

sub-polar gyres—Roughly circular surface ocean current systems that occur at high latitudes of the Northern Hemisphere.

subpolar lows—High-latitude, semipermanent *cyclones* marking the convergence of planetary-scale surface southwesterlies of midlatitudes with surface northeasterlies of polar latitudes in the Northern Hemisphere or midlatitude northwesterlies and polar southeasterlies in the Southern Hemisphere. The Icelandic low and Aleutian low are Northern Hemisphere examples.

subtropical anticyclones—Semipermanent high-pressure systems centered over subtropical latitudes (on average, near 30 degrees N and S) of the Atlantic, Pacific, and Indian Oceans. These warm-core systems extend from the ocean surface up to the *tropopause*.

subtropical gyres—Large-scale roughly circular surface ocean current systems, centered near 30 degrees latitude in the North and South Atlantic, the North and South Pacific, and the Indian Ocean.

subtropical jet stream—A zone of relatively strong *winds* aloft situated between the tropical *tropopause* and the midlatitude tropopause, on the poleward side of the *Hadley cell*.

sulfurous aerosols—Tiny droplets of sulfuric acid (H_2SO_4) and sulfate particles formed when volcanic eruptions send sulfur dioxide into the stratosphere to combine with water vapor.

sundog—A colored luminous spot at the same altitude as the Sun produced by *refraction* of light by ice crystals. In some cases, sundogs are part of a *halo* surrounding the Sun. Also called parhelia.

sunspot—Relatively dark, cool area that develops on the surface of the Sun's *photosphere* where an intense magnetic field suppresses the flow of gases transporting *heat* from the Sun's interior.

supercell thunderstorm—A relatively long-lived, large and intense *thunderstorm* cell characterized by an exceptionally strong updraft sometimes in excess of 240 km per hr (150 mph); may produce a *tornado*.

supercooling—The reduction of temperature of any liquid below the melting point of the substance's solid phase; that is, cooling beyond its normal freezing point while remaining a liquid.

synoptic-scale systems—*Weather* phenomena operating at the continental or oceanic spatial scale; includes migrating *cyclones* and *anticyclones*, *hurricanes*, *air masses* and *fronts*. These systems have dimensions of 100 to 10,000 km (60 to 6000 mi) and last from days to a week or so.

system—An entity having components that function and interact with one another in an orderly and predictable manner that can be described by fundamental physical principles or natural laws.

T

teleconnection—A linkage between changes in atmospheric circulation occurring in widely separated regions of the globe, often many thousands of kilometers apart.

temperature—A measure of the average *kinetic energy* of the individual atoms or molecules composing a substance.

temperature gradient—*Temperature* change with distance.

terminal velocity—Constant downward-directed speed of a particle within a fluid due to a balance between *gravity* (acting downward) and fluid resistance (directed upward).

thermal inertia—Resistance to a change in *temperature*.

thermograph—A recording instrument that provides a continuous trace of *temperature* variations with time.

thermohaline circulation Subsurface movement of ocean water masses caused by density differences arising from differences in temperature and salinity.

thermometer—An instrument used for measuring *temperature* by incorporating a thermal sensor that utilizes the variation of the physical properties of substances according to their thermal states. Thermal sensors include liquid-in-glass, deformation, and electronic thermometers, along with radiometers that sense *electromagnetic radiation*.

thermosphere—The outermost thermal subdivision of the *atmosphere* above the *mesopause* in which the air *temperature* is isothermal in the lower reaches and then increases with altitude. The thermosphere starts at an average altitude of 80 km (50 mi) and includes most or all of the *ionosphere*.

thunder—Sound accompanying *lightning*; intense heating of air by a lightning discharge is accompanied by a tremendous local increase in air pressure that generates a shock wave. The shock wave propagates outward producing sound waves heard as thunder.

thunderstorm—A *mesoscale system* produced by strong *convection* current that surge to great altitudes within the *troposphere*, sometimes reaching the *tropopause* or higher. It consists of *cumulonimbus* accompanied by *lightning* and *thunder* and, often, locally heavy rainfall (or snowfall) and gusty surface *winds*. A thunderstorm consists of one or more convective cells, each of which progresses through the life cycle of *towering cumulus stage*, *mature stage*, and *dissipating stage*.

tipping-bucket rain gauge—A recording *rain gauge* that collects rainfall in increments of 0.01 in. by containers that alternately fill, tip and empty.

tornado—A small mass of air in contact with the ground that whirls rapidly about an almost vertical axis; made visible by water droplets formed by *condensation* and by dust and debris sucked into the system.

tornado alley—Region of maximum *tornado* frequency in North America; a corridor stretching from eastern Texas and the Texas panhandle northward into Oklahoma, Kansas and Nebraska, and into southeastern South Dakota.

towering cumulus stage—Initial stage in the life cycle of a *thunderstorm* cell when *cumulus* clouds build both vertically and laterally over a period of about 10 to 15 minutes. The updraft throughout the cell is sufficiently strong to keep water droplets and ice crystals suspended in the *cloud*; therefore *precipitation* does not occur during this stage.

trade wind inversion—An elevated *stable air layer* that occurs on the eastern flank of *subtropical anticyclones* in the vicinity of the *trade winds*; a persistent and climatically significant feature. Formed when the subsiding, compressionally warmed air in a subtropical anticyclone encounters the marine air layer, a layer of cool, humid, and stable air formed where sea-surface *temperatures* are relatively low. A temperature inversion develops at the altitude where air subsiding from above meets the top of the marine air layer.

trade winds—Prevailing planetary-scale surface *winds* in tropical latitudes blowing from the northeast in the Northern Hemisphere and from the southeast in the Southern Hemisphere out of the equatorward flanks of the *subtropical anticyclones*.

transpiration—Process by which water that is taken up by plant roots escapes as vapor through tiny leaf pores. Measurements of direct *evaporation* from Earth's surface plus transpiration are usually combined as *evapotranspiration*.

triple point—The point of occlusion in an *extratropical cyclone* where *cold front*, *warm front*, and *occluded front* all come together. Conditions at this location are favorable for the formation of a new secondary extratropical *cyclone*.

Tropic of Cancer—A *solstice* position of the Sun with a latitude of 23 degrees 27 minutes N. On 21 June, the Sun's noon rays are vertical (*solar altitude* of 90 degrees) at this latitude.

Tropic of Capricorn—A *solstice* position of the Sun with a latitude of 23 degrees, 27 minutes S. On 21 December, the Sun's noon rays are vertical (*solar altitude* of 90 degrees) at this latitude.

tropical cyclone—A *synoptic-scale system* of low pressure that originates over the tropical ocean; includes *tropical depression*, *tropical storm*, *hurricane*, and *typhoon*. Example of a *warm-core cyclone*.

tropical depression—A *tropical cyclone* with sustained *wind* speeds of at least 37 km per hr (23 mph) but less than 63 km per hr (39 mph); an early stage in the development of a *hurricane*.

tropical disturbance—A region of convective activity over tropical seas with a detectable center of low *air pressure* at the surface; the initial stage in the development of a *hurricane*. A tropical disturbance is typically triggered by the *intertropical convergence zone*, by a trough in the westerlies intruding into the tropics, or by a wave in the easterly *trade winds* (an *easterly wave*).

tropical storm—A *tropical cyclone* having sustained *wind* speeds of 63 to 118 km per hr (39 to 73 mph); a storm at pre-*hurricane* stage. When a tropical cyclone reaches tropical storm strength, it is assigned a name.

tropopause—Zone of transition between the *troposphere* below and the *stratosphere* above.

troposphere—Lowest thermal subdivision of the *atmosphere* in which air *temperature* normally drops with altitude; the site of most *weather*. Located between Earth's surface and the *tropopause* at an average altitude ranging from 6 km (3.7 mi) at the poles to about 20 km (12 mi) at the equator.

turbulence—A state of fluid flow characterized by irregular (eddy) motion.

twilight—A period after sunset or before sunrise when the sky is illuminated by sunlight scattered by constituents of the upper *atmosphere*. Twilight is divided into the stages of civil twilight, nautical twilight, and astronomical twilight.

U

unstable air layer—An ambient air layer characterized by a vertical *temperature* profile such that air parcels accelerate upward or downward and away from their original altitudes. An ascending air parcel remains warmer (less dense) than the ambient air and continues to ascend and a descending air parcel remains cooler (denser) than the ambient air and continues to descend.

updraft—Upward-moving current of air.

upslope fog—*Fog* formed as a consequence of the *expansional cooling* of humid air that is forced to ascend a mountain slope.

upwelling—The upward movement of cold, nutrient-rich water from depths of 200 to 1000 m (about 650 to 3300 ft) toward the ocean surface and associated with the offshore flow of near-surface waters.

urban heat island—An area of higher air *temperatures* in a city setting compared to the air temperatures of the suburban and rural surroundings; shows up as an island in the pattern of isotherms on a surface map.

V

valley breeze—Shallow, upslope flow of relatively warm air that develops during daylight hours within *snow*-free mountain valleys in response to differential heating between the air adjacent to the valley wall and air at the same altitude out over the valley floor.

vapor pressure—The portion of the total *air pressure* exerted by the water vapor component of air; increases as the content of water vapor in air increases.

variables of state—Descriptors of the physical state of a thermodynamic system; for a gas (and mixtures of gas) these include *temperature*, pressure and density. The variables of state are interrelated through the *gas law*.

virga—Streaks of water and ice particles falling from a *cloud* that vaporize before reaching Earth's surface.

viscosity—*Friction* within fluids such as air and water.

visible radiation (light)—*Electromagnetic radiation* that is perceptible to the human eye with *wavelengths* ranging from about 0.40 (violet) to 0.70 (red) micrometers. Visible radiation has wavelengths longer than *infrared radiation* but shorter than *ultraviolet radiation*.

visible satellite image—Image processed from radiometers onboard a satellite that sense visible solar radiation reflected or back-scattered from surfaces in the *Earth-atmosphere system*.

W

Walker Circulation—Direct zonal tropical circulation, thermally driven, in which air rises over the warm western Pacific Ocean and sinks over the cool eastern Pacific.

wall cloud—A roughly circular lowered portion of the *rain*-free base of a *cumulonimbus cloud* about 3 km (2 mi) in diameter associated with a humid updraft. A wall cloud forms in the region of strongest *thunderstorm* updraft and often accompanies a mesocyclone.

warm air advection—The flow of air across regional isotherms from a relatively warm locality to a relatively cool locality.

warm cloud—*Cloud* composed of liquid water droplets having a *temperatures* above 0 °C (32 °F).

warm front—A *front* that moves in such a way that the cold (more dense) air retreats, allowing relatively warm (less dense) air to advance. May be associated with a broad band of cloudiness and light to moderate *precipitation*.

warm-core anticyclone—High-pressure system occupying a thick column of subsiding warm, dry air. Isobaric surfaces bulge upward above the high center and the anticyclonic circulation is most intense at high altitudes. *Subtropical anticyclones* are examples.

warm-core cyclone—A surface, synoptic-scale stationary low-pressure system that develops as a consequence of intense solar heating of a large, relatively dry geographical area; same as a thermal low. The thermal low's shallow cyclonic circulation weakens rapidly with altitude and often reverses to an anticyclonic circulation in the middle and upper *troposphere*.

water vapor imagery—Processed from a satellite sensor which monitors infrared radiation emitted by clouds and water vapor in the Earth's atmosphere.

water vapor plume—Extensive stream of water vapor that often originates over tropical seas and is transported horizontally, and is easily identified on *water vapor satellite imagery*. A typical water vapor plume can be several hundred kilometers wide and thousands of kilometers long. Plumes supply moisture to *hurricanes*, *thunderstorm* clusters, and winter storms.

water vapor satellite imagery—Picture or image processed from a satellite radiometer that senses *infrared radiation* at those *wavelengths* (typically near 6.7 micrometers) emitted by *clouds* and water vapor in the *atmosphere*; this imagery displays flow patterns in the middle *troposphere* near the top of the bulk of atmospheric water.

wave frequency—Number of crests or troughs of a wave that pass a given point in a specified period of time, usually 1 second.

wavelength—The distance between successive wave crests (or equivalently, wave troughs).

weather—The state of the *atmosphere* at some place and time described in terms of such variables as *temperature*, *humidity*, cloudiness, *precipitation*, and *wind* speed and direction.

weather radar—An adaptation of radar for meteorological purposes. The *scattering* of microwave radiation (*wavelengths* of a few millimeters to several centimeters) by raindrops, snowflakes, or hailstones is used for locating and tracking the movement of areas of *precipitation* and monitoring the circulation within small-scale *weather* systems such as *thunderstorms*.

weather warning—Issued by the NWS, a warning, such as a tornado warning or heavy snow warning, indicates when hazardous weather is present within that region or imminent. People are advised to take all necessary safety precautions.

weather watch—Issued by the NWS, a watch, such as a severe thunderstorm watch or winter storm watch, indicates when hazardous weather is possible based on current or anticipated atmospheric conditions. People are advised to be more alert but not interrupt their schedules.

weighing-bucket rain gauge—A recording *rain gauge* that is calibrated so that the weight of cumulative rainfall is recorded directly in terms of water depth (in millimeters or inches).

wet-bulb depression—On a *psychrometer*, the difference between the dry-bulb *temperature* and the *wet-bulb temperature*; used to determine *relative humidity*.

wet-bulb temperature—The *temperature* an air parcel would have if cooled adiabatically to saturation at constant *air pressure* by *evaporation* of water into it; measured using a *psychrometer* in which the wick-covered bulb is first soaked in distilled water and the instrument is ventilated to promote evaporation.

Wien's displacement law—A radiation law whereby the *wavelength* of maximum emission by a *blackbody* is inversely proportional to its absolute *temperature* (in kelvins). For example, hot objects (such as the Sun) emit peak radiation at relatively short wavelengths, whereas cold objects (such as the *Earth-atmosphere system*) emit peak radiation at longer wavelengths.

wind—Air in motion measured relative to the Earth's rotating surface.

wind shear—Change in horizontal *wind* speed or direction with increasing distance.

wind vane—An instrument used to monitor *wind* direction that consists of a free-swinging shaft with a vertical plate at one end and a counterweight (arrowhead) at the other end. The counterweight always points into the wind.

windsock—An instrument used to monitor *wind* direction that consists of a cone-shaped cloth bag opened at both ends. The larger end is held open by a metal ring attached to a pole; air enters the larger opening, rotating the sock and stretching it downwind.

winter storm—An extratropical cyclone that produces any combination of frozen or freezing precipitation, including snow, ice pellets or freezing rain.

World Meteorological Organization (WMO)—The agency of the United Nations that coordinates *weather* data collection and analysis by 185 member nations and territories; based in Geneva, Switzerland.

Y

Younger Dryas—A relatively cool climatic episode from 11,000 to 10,000 years ago that triggered short-lived re-advances of remnant ice sheets in North America, Scotland, and Scandinavia.

Z

zonal flow pattern—Flow of the *planetary-scale* westerlies almost directly from west to east, exhibiting little amplitude. In this pattern, the north-south exchange of *air masses* is minimal.

Index

A

Abbe, Cleveland, 22, 40
absolute humidity, 172-173
absolute instability, 184
absolute stability, 184-185
absolute temperature, 66
absolute zero, 109
absorption, of radiation, 77
absorptivity, 76
acceleration, 249
acclimatization, 159
Accumulated Cyclone Energy (ACE), 429
acid, 227, 228
acid deposition, 227-228
acid rain, 227
Acid Rain and Related Programs 2007 Progress Report, 228
Acid Rain Program, 228
acoustic shadow, 485
Adams, NY, 269
Adhémar, J.A., 514
adiabatic process, 154-155, 182
Advanced Weather Interactive Processing System (AWIPS), 42, 449
advection, 125-126, 150, 448
advection fog, 218
advective cooling, 218
aeolian sound, 486
aerosols, 28-29, 35
        and climate change, 519
        sulfurous, 516-518
aerovane, 264
Africa, sub-Saharan drought, 278-279
African Monsoon Multidisciplinary Analysis (AMMA), 279
Agata, Siberia, 145
agricultural drought, 317
agroclimatic compensation, 497
air
        ambient, 36
        composition of, 34-35
        density of, 144-145
        heat conductivity of, 113
        number density of, 34, 146
        pollution, 35-36, 198
        pressure of, 7-9, 143
        quality of, 198-201
        saturated, 173-175, 182-183
        stability of, 184-185
        water vapor in, 172-175
air mass, 10, 149, 323-325
        advection of, 218
        arctic, 323-324
        continental polar, 323, 324
        continental tropical, 323, 324
        exchange of, 122
        marine air layer, 28, 284
        maritime polar, 323, 324, 325
        maritime tropical, 323, 324
        modification of, 150, 324-325
        North American types, 323-324
        source region, 10, 323-324
        stability of, 324, 325
air mass modification, 324-325
air parcel, 152
air pollutant, 35-36
air pollution, 35-36
        and air stability, 198-201
        episode, 198
air pressure, 12, 143
        conversion factors, 145
        definition of, 7-9, 143
        effect of convergence on, 151
        effect of divergence on, 151
        effect of humidity on, 150-151
        effect of temperature on, 149-150
        global pattern of, 281-282
        gradients in, 249-250
        horizontal variations in, 148-151
        human responses to, 158-160
        measurement of, 143-145
        reduction to sea level, 148
        tendency of, 145
        versus water pressure, 161
        units of, 145
        variation with altitude, 145-148
air pressure gradient, 249-250
air quality standards, 36
Airborne Gamma Radiation Snow Survey Program, 236
Aircraft Communications Addressing and Reporting System, 441
albedo, 76, 77, 83
        cloud top, 77
        feedback (Arctic), 527-530
        ocean, 83-84
        planetary, 84, 85
Alberta clipper, 336
Aleutian low, 283, 312, 336
alkaline substance, 227
Alley, Richard, 526
altimeter, 148, 162
        microwave, 163
        pressure, 162
        radio (or electronic), 163
altimetry, 148, 162
altitude
        indicated, 162
        true, 162
altocumulus cloud, 211
        lenticularis, 213-214
altostratus cloud, 210, 211

ambient air quality standards, 36
Anasazi, 540
anchor cells, 391
anemometer, 162, 164
aneroid barometer, 144-145
anomalous propagation, 230
Antarctic Circle, 72, 73
Antarctic circumpolar vortex, 283
Antarctic ice sheet, 524-525
Antarctic ozone hole, 37, 80-81, 216
anthropogenic influence on climate, 520-521
anticyclone, 151, 256-257, 341-343
    and air quality, 198-199
    arctic, 341-342
    circulation in, 256-260
    cold-core, 341-342
    cutoff, 291-292
    polar, 341-342
    Siberian, 323, 342
    subtropical, 282
    surface winds in, 9
    warm-core, 342
    weather of, 342-343
anvil cloud top, 366-367
aphelion, 70, 71
apparent temperature, 194-195
apparent temperature index, 194-195
Arctic air (A), 323-324
Arctic Circle, 72, 73
Arctic high, 341-342
Arctic hurricanes, 433
Arctic Oscillation (AO), 311-312, 529
Arctic ozone, 81-82
Arctic sea ice cover, 527-530
Arctic sea smoke, 218-219
arcus cloud, 366
Argentia, Newfoundland, 216
argon, origin of, 32-33
Aristotle, 37
Army Medical Corps, 40
Army Signal Corps, 40
Arthasastra, 233
Ashley, W.S., 378
aspirated psychrometer, 179
asteroids, 32
Astoria, OR, 188, 189
astronomical twilight, 483
Atlantic Multidecadal Oscillation (AMO), 423-424
atmosphere, 31-36
    acoustics, 484-486
    composition of, 34-35
    evolution of, 31-36
    homogeneous, 146
    interaction with solar radiation, 76-78
    models of, 38-39
    monitoring of, 39-46
    optics, 473-484
    stability of, 183-187, 198-201
    standard, 146-148
    temperature profile of, 46-47
atmospheric acoustics, 484-486
atmospheric boundary layer, 254, 448

atmospheric fixation, 34
atmospheric instability, 183-187
atmospheric models, 38-39
atmospheric optics, 473-484
atmospheric river, 166-167, 171, 181-182
atmospheric stability, 183-187
    and air quality, 198-201
atmospheric windows, 86-87
attainment area, 36
attractor, 468
aurora, 47-49
    australis, 47
    borealis, 47
auroral ovals, 48, 49
autoconvective lapse rate, 146-147
Automated Surface Observing System (ASOS), 41, 443-444
Automated Weather Observation System (AWOS), 444
Automated Weather Sensor System (AWSS), 444
Aviation Weather Center (AWC), 454, 455
Avagadro, Amedeo, 150
Avogadro's law, 150
Azizia, Libya, 106

B

back (wind), 338
back-door cold front, 329
ball lightning, 376
Bangladesh, East Pakistan, storm surge, 419
banner cloud, 214
Barneveld, WI, 1984 tornado, 389
barograph, 145, 149
barometer, 143-145
    aneroid, 144-145
    piezoelectric, 144
    mercury, 143-144
Barrel, Jean, 208
barrier islands, 425-426
Barry, Roger G., 526
bead lightning, 376
Beaufort, Sir Francis, 262
Beaufort scale, 262-263
bent-back occlusion, 333
Bergeron, Tor, 221, 331
Bergeron-Findeisen process, 221-222
Bermuda-Azores high, 278, 282, 283, 310, 342, 372
Big Thompson Canyon, Colorado, flash flood of 1976, 380
billow clouds, 215
biological fixation, 34
Biot, Jean, 43
Bjerknes, Jacob, 297, 331
Bjerknes, Vilhelm, 331, 449
Black, Joseph, 31, 115
blackbody, 66, 93
    radiation laws, 66-68
blizzard, 339
Blizzard of '88, 2-3
blizzards of 2010, 3
blocking systems, 291, 292
blue haze, 490
Blue Hill Observatory, 43, 490
Blue Mountains, Australia, 490

blue of the sky, 474
Blue Ridge Mountains of Virginia, 490
bomb (cyclone), 355-356
Bonan, Gordon, 520
bora, 347
Boulder, CO, chinook winds, 346
boundary layer (skin), 137
Boundary Waters Canoe Area Wilderness, MN, 378
bow echo, 378-379
Bowen, I.S., 120
Bowen ratio, 120-121
Boyle, Robert, 152
Boyle's Law, 152
Brandes, Heinrich Wilhelm, 22
breeze, see specific types (lake, land, mountain, sea, and valley)
British Antarctic Survey, 80
British thermal unit (Btu), 110
Bruintjes, Roelof, 243
butterfly effect, 468
Buys-Ballot, Christopher H.D., 258

C

Cabrillo, Juan Rodriquez, 358
Cahokia, 62
California
        fire weather, 358-359
        hurricane threat, 440-441
        Santa Ana winds, 346-347
        windmill farms, 274-275
Callendar, G.S., 87
Callendar effect, 87-88
Calorie, 110
Campanius, John, 40
Canadian Network for Sampling Precipitation, 227
Cape Disappointment, WA, 216
Cape Verde-type hurricanes, 411
Capping inversion, 372, 373
Carbon dioxide, 33
        and climatic change, 87-90
        and the greenhouse effect, 86-88
Carter, President Jimmy, 522
Cary Institute of Ecosystem Studies, 227
Castelli, Benedetto, 233
Cavendish, Henry, 34
Celsius, Anders, 109
Celsius temperature scale, 109
Cenozoic Era, 504
Central Pacific Hurricane Center, 452
centripetal force, 250-251
CFCs, see chlorofluorocarbons
Challenger Deep, 161
Chamberlain, T.C., 514
Chankillo, 62-63
chaos theory, 468
chapparral, 358
Charles, Jacques, 152
Charles' law, 152
Charleston, SC, 41, 285-286
Charney, Jule, 449, 522
Chicago, heat wave of 1995, 193
Children's Blizzard, 2-3

chinook arch, 345
Chinook Indians, 345
chinook winds, 345-346
chlorofluorocarbons (CFCs), 79-80
Chomolungma, 142
chromosphere, 69
circulatory shock, 398
circumpolar vortex, 81
cirriform cloud, 208
cirrocumulus cloud, 210
cirrostratus cloud, 210
cirrus cloud, 210, 327, 328
civil twilight, 483
Clemens, S.L., 31
climate
        analogs, 521-522
        anomalies, 496
        boundary conditions, 497-498
        classification of, 503, 555
        continental, 116-117
        cycles, 521-522
        definition of, 4-5, 495
        dry, 557
        future, 521-523
        of geologic time, 503-506
        glacial, 507
        global patterns of, 498-502
        highland, 561
        interglacial, 507
        lessons of the past, 540
        maritime, 116-117
        Mediterranean, 558
        monsoon, 278, 287-290
        and the ocean, 497-498
        of the past two million years, 506-509
        polar, 560-561
        reconstruction of, 535-539
        record of, 90-91, 503-510
        snow forest, 559-560
        and solar variability, 512-514
        subtropical, 558-559
        system, 495
        tropical humid, 555-556
climate change, 510-521
        and aerosols, 519
        and carbon dioxide, 87-90
        causes of, 510-521
        and Earth's orbit, 514-516
        and Earth's surface, 518, 520
        and greenhouse gases, 522-523
        impacts of, 523-530
        and solar variability, 512-514
        and volcanoes, 516-518
Climate Prediction Center (CPC), 449, 458
climatic anomalies, 496-497
climatic norm, 495
climatology, 5, 457, 460-461
cloud, 18-19, 182
        albedo of, 77
        altocumulus, 211
        altocumulus lenticularis, 213-214
        altostratus, 210, 211

arcus, 366
banner, 214
billow, 215
canals in, 222-223
castellanus, 210
cirriform, 208
cirrocumulus, 210
cirrostratus, 210
cirrus, 210, 327, 328
classification of, 208-216
cold, 209
comma, 332, 333, 334
congestus, 210
cover, 207
cumuliform, 18, 209, 212-213
cumulonimbus, 18, 19, 120, 211, 328, 365
cumulus, 18, 19, 120, 212-213
cumulus congestus, 212-213, 365
development of, 206-208
fair-weather cumulus, 120, 212
formation of, 206-208
fractus, 210
funnel, 384
and the greenhouse effect, 87
high, 209-210
holes, in, 222-223
incus, 210
inflow, 389-390
iridescent, 480
lee-wave, 214-215
lenticularis, 213-214
low, 211-212
mammatus, 373
middle, 210-211
by mixing, 241
mountain-wave, 214-215, 346
mother-of-pearl, 216
nacreous, 216
nimbostratus, 211
noctilucent, 215-216
nuclei of, 206-208
pileus, 210
polar stratospheric, 81, 216
roll, 366
rotor, 215
scud, 401
seeding of, 222-223, 242-244
shelf, 366
stratiform, 18, 209
stratocumulus, 211
stratus, 211
uncinus, 210
unusual, 213-216
vertical development of, 212-213
wall, 389-390
warm, 209
see also specific types
cloud condensation nuclei (CCN), 207
cloud seeding, 222-223, 242-244
cloud streets, 213, 325
coastal upwelling, 301
cold air advection, 125-126, 150, 448

cold air drainage, 217
cold cloud, 209
cold-cloud precipitation, 221-222
cold-core anticyclone, 283, 341-342
cold-core cyclone, 340-341
cold-core rings, 408
cold front, 10, 11, 328-329
cold occlusion, 329-330
collision-coalescence process, 220-221
collision efficiency, 221
color perception, 473
Colorado low, 336-337
Colorado Plateau, 506
Columbus, Christopher, 283
comet, 32
comma cloud, 332, 333, 334
compressional warming, 152-155
conceptual model, 38
condensation, 114, 115, 170
condensation trails, 204-205, 241
conditional stability, 184, 369
conduction, of heat, 112-115
conservation of angular momentum, 354
conservation of energy, 64, 68, 76, 122, 153
continental climate, 116-117
continental drift, 221, 503, 505
continental polar air (cP), 323, 324
continental tropical air (cT), 323, 324
continuity, 260-261
contrails, 204-205, 241
controlled burns, 358
convection, 112-114, 187-188
    and cloud development, 187-188
    forced, 364
    free, 364
convective condensation level (CCL), 212, 365
convective outlook, 455
convergence,
    in anticyclones, 260
    caused by friction, 261
    in cyclones, 260
    and the jet stream, 295-296
    in westerly waves, 295-296
conveyor-belt model (cyclone), 333-334
Cook, K., 279
cooling degree-day, 135-136
Coordinated Universal Time (UTC), 443
coral bleaching, 303
coral reef, 28-29
Coriolis, Gaspard-Gustave de, 252
Coriolis Effect, 251-253, 256, 266, 280, 287, 294, 299-300
    and hurricanes, 408
corona (optical), 480
corona (solar), 69
coronal hole, 58
coronal mass ejection, 58
Coxwell, Henry, 43
crepuscular rays, 483
Croll, J., 514
cross-dating, 536
Crutzen, P.J., 80
cumuliform cloud, 18, 209, 212-213

cumulonimbus cloud, 365
    see also thunderstorm
cumulus cloud, 212
cumulus congestus cloud, 212, 365
cup anemometer, 262, 264
curvature effect, 206
cutoff high, 291, 292
cutoff low, 291, 292, 333
cyclogenesis, 331, 337
cyclolysis, 333
cyclone, 151, 256-257, 322
    Aleutian, 283, 312, 336
    bomb, 355-356
    cold-core, 340-341
    conveyor-belt model, 333-334
    cutoff, 291, 292
    extratropical, 322, 331-341, 354
    heat transport by, 122
    versus hurricanes, 406-407
    Icelandic, 283
    and the jet stream, 295-296
    life cycle of, 331-333
    Norwegian model, 322, 331, 333
    principal tracks of, 336-339
    secondary, 333
    surface winds in, 257-260
    thermal, 285, 289, 341
    tropical, 28-29, 303, 406, 451-454
    and upper-air support, 295-296
    warm-core, 285, 289, 341
    warm sector of, 332
    weather associated with, 335-336

D

Dalton, John, 143
Dalton's law of partial pressures, 143, 172
dart leader, 375
Darwin, Australia, 297
daylight, length of, 73-74, 75
Daylight Savings Time, 7
Death Valley, CA, 106-107
decibels (dB), 484
deep-sea sediment cores, 537-538
deepening, 332
Defoe, Daniel, 322
deforestation, 279
degree-days, 135-136
dendroclimatology, 535
deposition, 114, 115, 171
    nuclei, 207
derecho, 378-379
Descartes, René, 478
desert winds, 348-349
deserts, 282
desertification, 279
dew, 176
dewpoint, 12, 175-176, 177
dewpoint hygrometer, 178, 179
diffraction, 480
dissipating stage (thunderstorm), 364, 367
distillation, 171

diurnal temperature range (DTR), 204-205
divergence
    in anticyclones, 260
    caused by friction, 261
    in cyclones, 260
    and the jet stream, 295-296
    in westerly waves, 295-296
Dobson units, 78
doldrums, 283
Donnelly, J.P., 437
Donora, Pennsylvania, air pollution, 198
Doppler effect, 17, 230-232
Doppler, Johann Christian, 230
Doppler radar, 17, 230-232, 272, 443
double sunspot cycle, 512-513
Douglass, Andrew E., 535
downburst, 376-378
downwelling, 300
drainage basin, 172
drizzle, 224
dropwindsonde, 45, 445, 451-452
drought, 125, 317-318, 343
    agricultural, 317
    hydrologic, 317
    meteorological, 317
dry adiabat, 154
dry adiabatic lapse rate, 154
dry deposition, 227
dry-bulb temperature, 179, 180
dry ice, 242
dry slot, 333, 334
Dryas octopetala, 508
dryline, 372-373
dual polarization weather radar, 232
dust devil, 348-349, 401
dynamical model, 306
dynamical system, 468

E

ear popping, 159-160
easterly wave, 411
Eddy, J.A., 513
eddy viscosity, 253
Edmund Fitzgerald, 248
Eemian Interglacial, 507, 508
Einstein, A., 68
Ekman spiral, 299-300
Ekman transport, 299-300, 408
Ekman, V. Walfred, 299
El Chichón, 1982 eruption of, 517-518
El Niño, 296, 297-298, 302-304
    frequency, 309-310
    historical perspective, 297-298
    of 1982-83, 298, 319
    of 1997-98, 303, 308, 319
electromagnetic radiation, 64-66
    see also radiation; specific types
electromagnetic spectrum, 64-66
electronic altimeter, 163
electronic hygrometer, 178-179
emissivity, 92

empirical model, 306
energy, 64
energy conservation, law of, 64, 68, 76, 122, 153
Enhanced F-scale, 388-389
ensemble forecasting, 450, 468, 469
ENSO, 298, 306-309
ENSO Alert System, 309
ENSO Observing System, 306
entrainment, 365
Environmental Science Services Administration (ESSA), 41
equatorial upwelling, 300
equilibrium run, 521
equinoxes, 72
Espy, James P., 22
Eustachian tube, 159-160
eustasy, 523
evacuation from hurricanes, 426-427
evaporation, 114, 115, 170
      measurement of, 197
evaporation pan, 197
evapotranspiration, 171
Everest, Sir George, 142
excess death rate, 193
expansional cooling, 152-155, 182
Explorer VII, 46
extratropical cyclones, 322, 331-341
    and cyclolysis, 333
    and hurricanes, 406-407
    and upper-air support, 331, 337
extreme hail, 381
eye (of a hurricane), 407
eye wall, 407
    replacement cycle, 407

F

FAA, 204, 377, 442, 454
F-scale, 388
faculae, 68
fair-weather bias, 460
Fahrenheit, Gabriel D., 109
Fahrenheit temperature scale, 109
Fata Morgana, 472
Fawbush, Ernest J., 440
Fawbush-Miller tornado forecasting method, 440
Federal Aviation Administration (FAA), 204, 377, 442, 454
Federal Emergency Management Agency (FEMA),
    and hurricane evacuation, 427
feedback, 512, 515, 522
fetch, 269
filling, 333
Findeisen, Walter, 221
Finley, John P., 440
fire weather (California), 358-359
first law of thermodynamics, 64, 153
fire prediction, 455
flash flood, 289, 379-381
    safety, 381
flash-to-bang (thunderstorm), 376, 397
flat plate solar collector, 101
floods, Midwest in 1993, 292-293
florida

freeze in, 342
    rainmaking, 243-244
    thunderstorm frequency, 370
foehn, 346
fog, 18, 19, 216-219
fog bow, 480
fog drip, 171
force, 143, 249
forced convection, 364
forecast skill, 457
forecasting, *see* weather forecasting
Fort Collins, Colorado, flash flood of 1997, 380
Fram, 299
Franklin, Benjamin, 322, 516
free convection, 364
freezing, 114, 115
freezing drizzle, 225
freezing nuclei, 207
freezing rain, 225-226
friction, 253-254
    effect on surface winds, 257-260
front, 10-11, 325-330
    cold, 10, 11, 328-329
    and fog, 328
    gust, 365, 391
    occluded, 329-330
    polar, 283
    stationary, 10, 11, 326-327
    uplift along, 10-11, 188, 325-330
    warm, 10, 11, 327-328
    weather produced by, 10-11, 325-330
frontal fog, 328
frontal thunderstorm, 328, 368
frontal uplift, 10-11, 188, 325-330
frontogenesis, 330, 343
frontolysis, 330
frost, 176
frost point, 12, 176
Fujita, T.T., 376-377, 378, 388, 417
funnel aloft, 384
funnel cloud, 384

G

Galilei, Galileo, 110, 143, 512
Galveston, Texas, hurricane in 1900, 418
gamma radiation, 64, 65
gas constant, 152
gas law, 151-152
gas thermoscope, 110
Gay Lussac, J.L., 43
General Electric Research Laboratory, 242
geologic time, 503
geologic time scale, 504
Geophysical Institute, Bergen, 322, 331
Geostationary Operational Environmental Satellite (GOES), 13, 14
geostationary satellite, 13, 14
geostrophic wind, 255-256, 448
Ghezzi, Iván, 63
glacial climate, 507
Glacier National Park, 527
Glaisher, James, 43

glaze, 225
global climate models (GCM), 521
  equilibrium run, 521
  transient run, 521
Global Data Processing System (GDPS), 441
Global Forecast System (GFS), 450
global mean annual temperature, 91
global mean land surface temperature, 91
global mean sea surface temperature, 91
Global Observing System, 441, 442
global positioning system (GPS), 16, 24-25, 163, 445
global radiative equilibrium, 68, 117, 511
Global Telecommunications System (GTS), 441
global warming, 90-91, 522-523
  and mountain glaciers, 90
  and polar ice sheets, 90
  and potential effects on agriculture, 90
  and sea level, 90, 494
global warming potential (GWP), index, 89
global water budget, 171-172
global water cycle, 33, 168-172
glory, 480-481
Goddard, Robert H., 46
Gomez, Nuno, 161
Gondwanaland, 503
GPS, 24-25
gradient wind, 256-257, 448
Grant, Ulysses S., 40
granules, 68
graphical model, 38
graupel, 374
gravitation, 254
gravity, 143, 219, 254
Great Lakes, 83
  and climate change, 527
  and lake-effect snow, 269-271
Great Oxidation Event, 33-34
Great Salt Lake, lake-effect snows, 271
green flash, 483-484
greenhouse effect, 32, 66, 85-86, 117, 204
  and carbon dioxide, 38-39
  and clouds, 87
  and global warming, 90-91, 522-523
  and water vapor, 86
greenhouse gases, 85-87
Greenland
  ice sheet, 525-526
Greenland Ranch, CA, 106
Greensburg, KS, 2007 tornado, 389
Greenwich Mean Time (GMT), 6, 443
Grimaldi, Francesco, 252
ground clutter, 230
gust front, 365, 391
gustnadoes, 391
gyres (ocean), 286, 287

H

Habeler, Peter, 142
haboob, 349
Hackleburg, AL, 363
Hadley, G., 284

Hadley cell, 283-284, 295
hail, 226-227, 381
  cannon, 399-400
  extreme, 381
  severe, 381
  suppression of, 399-400
  swath, 382
hailshaft, 382
hailstones, 226-227
hailstreak, 382
hair hygrometer, 178
Hallam, NE tornado, 285
Halley, Edmund, 287
halo, 475-477
Hawaii
  drought in, 303
  hurricanes, 411
  Mauna Loa Observatory, 87
  orographic precipitation, 188, 284-285
  thunderstorm frequency, 284-285, 371
  and trade wind inversion, 284-285
Hawaiian high, 282
heat
  conduction of, 113
  definition of, 108
  imbalances in, 117-123
  latent, 114
  specific, 115-116
  transfer of, 112-115, 117-121, 121-123
  units of, 110
heat burst, 349-350
heat conductivity, 113
heat equator, 283, 498-499
heat index, 194-195
heat island (urban), 126-128
heat lightning, 376, 485
heat transfer, 112-115, 117-121, 121-123
heat units, 110
heating degree-day, 135-136
heatstroke, 194-195
heliostats, 101
Henry, Joseph, 22, 40
heterogeneous nucleation, 208
heterosphere, 34
high, *see* anticyclone
high-altitude pulmonary edema, 158
high clouds, 209-210
high fog, 219
High Resolution Window (HRW), 450
Hillary, Sir Edmund, 142
Himalayan Mountains, 142, 506
Hinrichs, Gustavus, 378
Hobbs, P.V., 222
Hoerling, M., 303
Holloway, L., 98
Holocene epoch, 508
homogeneous nucleation, 208
homosphere, 34
hook echo, 392
Hopkins, E.J., 381
horse latitudes, 282
hot-wire anemometer, 264

Howard, Luke, 208
Huaynaputina volcano, 516
human comfort, 193-196
humidity, 72-175
    absolute, 172-173
    effect on air pressure, 150-151
    and human comfort, 193-196
    instruments, 178-181
    relative, 175-176
    specific, 172
    see also water vapor
hurricane, 406-429
    Agnes, 414
    Andrew, 410, 417, 419, 422
    arctic, 433
    Beulah, 417
    Camille, 414, 419
    Cape Verde type, 411
    Catarina, 410
    characteristics of, 406-408
    Charley, 425
    Elena, 426, 427
    and El Niño, 410
    evacuation from, 426-427
    and extratropical cyclones, 406-407
    eye of, 407
    Floyd, 426-427
    Frances, 425
    frequency, 423, 437
    Gilbert, 422
    geographical distribution, 408-411
    Gustav, 405
    hazards of, 413-422
    Hazel, 413
    Hugo, 421
    Igor, 435
    Ike, 418-419
    Iniki, 411
    Irene, 416
    Ivan, 417
    Jeanne, 425
    Kathleen, 411
    Katrina, 404-405, 418, 419
    life cycle of, 411-413
    long-range forecasting of, 428-429
    Mitch, 414-415
    modification, 429
    naming of, 434-436
    New England, 413
    Nora, 411
    Rita, 405
    seasonality of, 409, 410
    spin-up vortices in, 417
    storm surge, 355
    threat to the Southeast United States, 424-425
    Tomas, 435
    and tornadoes, 417
    warning, 453
    watch, 453
    Wilma, 406, 419
hydrologic drought, 317
Hydrometeorological Prediction Center (HPC), 445, 449

hydrostatic equilibrium, 255
hygrograph, 178
hygrometer, 178
hygroscopic nuclei, 208
hyperthermia, 195
hypothermia, 2, 137
hypothesis, 37
hypoxia, 158

I

ice-albedo feedback (Arctic), 527-530
ice cores, 538-539
ice-crystal process, 221-222
ice embryo, 208
ice fog, 216
ice-forming nuclei, 207
ice pellets, 225
ice stream, 524
Icelandic low, 283, 310
ideal gas, 152
ideal gas law, 152
indicated altitude, 162
indications, 2
inertial oscillation, 256
inferior mirage, 482
infiltration component, 171-172
inflow cloud, 389-390
infrared atmospheric windows, 86-87
infrared radiation, 64, 66, 85-90
infrared radiometer, 92
infrared satellite image, 14, 15
inland flooding, 413-416, 417
instrument-based temperature records, 509-510
instrument shelter, 41, 106, 111
interglacial climate, 507
Intergovernmental Panel on Climate Change (IPCC), 89, 494
internal energy, 108, 153
International Best Track Archive for Climate
    Stewardship (IBTrACS), 410
International Geophysical Year (IGY), 297
International Meridian Conference, 6
International Meteorological Conference of 1935, 495
International Meteorological Organization (IMO), 441
international time zones, 6-7
intertropical convergence zone (ITCZ), 283, 285, 406
        and monsoon circulation, 278
inverse square law, 67-68
ion, 47
ionosphere, 24, 47-49
    radio transmission in, 55-56
IPCC Fourth Assessment Report 2007, 494, 522, 526
iridescent clouds, 480
isobar, 148, 151, 249-250
isoprene, 490
isotachs, 295-296, 448
isotherms, 126, 448
ITCZ, *see* intertropical convergence zone

J

Jason 1, 307

Jason 2, 307
Jefferson, Thomas, 40
jet streak, 294, 295-296, 331, 372
jet stream, 294-295, 448-449
  and cyclones, 295-296
  and the polar front, 294
  seasonal shifts of, 294-295
  streak, 294
  subtropical, 295, 372
  tropical easterly, 295
Joplin, MO, tornado of 2011, 387-388
joule, 110
Jupiter, 32

K

katabatic wind, 347-348
Katmai, 517
Kautila, 233
Keeling, Charles D., 87
Keeling curve, 87, 88
Keesler Air Force Base, MS, 451
Kelvin, Lord, 109
Kelvin scale, 109
Kelvin-Helmholz waves, 215
Kepler, Johannes, 224
Kilbourne, K.H., 437
kinetic molecular theory, 152
kiribati, 494
kites, in weather observation, 43-44
Knight, Nancy, 224
Knox, J.A., 256
Koch, Jacqueline, 348-349
Kocin, P., 357
Köppen, Wladimir, 514, 555
Krakatoa, 215, 484, 516
Kutzbach, J.E., 506

L

La Niña, 296, 298, 304-305, 309-310
Laboratory of Tree-Ring Research, 535
lake breeze, 12, 344
lake-effect snow, 12, 269-271, 458
Laki, Iceland, eruption, 516
land breeze, 270, 344
land spout, 391
Langmuir, Irving, 242
lapse rate
  dry adiabatic, 154
  moist adiabatic, 154-155
Laskin, David, 3
latent heat, 114
latent heat of fusion, 119
latent heat of vaporization, 119
latent heating, 118-119
Laurasia, 503
Laurentide ice sheet, 507
law of energy conservation, 64, 68, 76, 122, 153
lee-wave clouds, 214-215
length of daylight, 73-74, 75, 482
Libby, W.F., 515

lifetime of a gas, 522-523
lifting condensation level (LCL), 188
lifting processes, 167, 187-189
Lightfoot, Gordon, 248
lightning, 373-376
  and circulatory shock, 398
  and progressive shock, 398
  return stroke, 375
  safety, 397-398
  stepped leaders, 375
  and thunder, 376
lightning detection and ranging (LIDAR), 377-378
lightning detection network (LDN), 374
Likens, G.E., 227
Little Ice Age, 91, 508-509, 513
Liu, Kam-biu, 437
local solar noon, 69
long waves (in westerlies), 290-294
Loomis, Elias, 22
Lord Rayleigh, 474
Lorenz, Edward N., 468
Lovell, Joseph, 40
low, *see* cyclone
low clouds, 211-212
low-level jet stream, 369, 372
Lysimeter, 197

M

MacDill Air Force Base, FL, 451
Mach 1, 486
Mach, Ernst, 486
macroburst, 377
Madison, James, 40
magnetosphere, 48
Maldives, 494
mammatus clouds, 373
mangrove swamps, 428
Mann, Michael, 510
Marianas Trench, 161
marine air layer, 28, 284, 412
maritime climate, 116-117
maritime polar air (mP), 323, 324, 325
maritime tropical air (mT), 323, 324
Mars, 67-68
  atmosphere of, 35, 52-54
  volcanic activity on, 52
  water on, 52-53
Mars Exploration Rovers (MERS), 53
Martian atmosphere, 52
Martian climate, 54
mature stage (thunderstorm), 365-366
Mauna Loa Observatory, Hawaii, 87
Maunder, E.W., 513
Maunder minimum, 513
Maury, Matthew Fontaine, 441
Maximum-Minimum Temperature System (MMTS), 110-111
maximum temperature, 12
McKinley, President William, 451
McSorley, Ernest, 248
mechanical turbulence, 254
Medieval Warm Period, 508-509

Mediterranean climate, 558
melting, 114-115
mercury barometer, 143-145
meridional flow pattern, 290-291
meridional overturning circulation (MOC), 123, 508
mesocyclone, 389
mesopause, 46, 47
mesoscale convective complex (MCC), 368-369
mesoscale convective systems (MCS), 367
mesoscale systems, 265
mesosphere, 46, 47
Mesozoic Era, 506
Messner, Reinhold, 142
meteorograph, 43
Meteorologica, 37
meteorological drought, 318
Meteorological Services of Canada (MSC), 137
Meteorological Society of the Palatinate, 22
meteorology, 4
methane, 80, 86, 88, 216
METROMEX, 208
Metropolitan Meteorological Experiment (METROMEX), 208
Mexican monsoon, 289
microburst, 377
microscale systems, 265
microwave, 64-66
middle clouds, 210-211
Middle Eocene Climatic Optimum (MECO), 33, 506
midlatitude westerlies, 282, 283
Mie, Gustav, 474
Mie scattering, 77, 474-475
Milankovitch cycles, 514-516
Milankovitch, Milutin, 514
Miller, Robert C., 440
Millibar, 145
Milwaukee, WI (dewpoint), 176
Milwaukee, WI, hail storm of 1988, 382-383
minimum temperature, 12
mirages, 472, 481-482
mist, 216
Mistake Island, Maine, 216
mistral, 347
mixing depth, 198
mixing ratio, 172
      at saturation, 174
model comparison, 450, 451, 468
modern phase, 34-35
moist adiabat, 154
moist adiabatic lapse rate, 154-155
molecular diffusion, 198
molecular viscosity, 253
Molina, M.J., 80
monsoon active phase, 288
monsoon circulation, 278, 287-290
monsoon dormant phase, 288
monsoon high, 289
Montreal Protocol on Substances that Deplete the Ozone Layer, 82-83
Moon, 68
Moose Peak Lighthouse, ME, 216
Mote, T.L., 378
mother-of-pearl clouds, 216
Mount Everest, 142

Mount Pinatubo, 1991 eruption of, 516
Mount Weather Observatory, 43
mountain breeze, 344-345
mountain sickness, 158-159, 346
mountain-wave clouds, 214-215
mudslides, 167
Mullen, George, 512
multi-vortex tornadoes, 387

N

Nabta, 62, 63
nacreous clouds, 216
Nansen, Fridtjof, 299
NAO Index, 310-311
Nasheed, Mohamed, 494
National Atmospheric Deposition Program, 227
National Centers for Environmental Prediction (NCEP), 442
National Climatic Data Center (NCDC), 495
National Data Buoy Center (NDBC), 442-443
National Hail Research Experiment, 400
National Hurricane Center (NHC), 452
National Oceanic and Atmospheric Administration (NOAA), 5, 41, 441
      Environmental Modeling System (NEMS), 450
      Space Weather Prediction Center (SWPC), 457
National Research Council (NRC), 510
National Severe Storms Laboratory (NSSL), 454
National Solar Radiation Data Base (NSRDB), 100
National Weather Service (NWS), 441
      Cooperative Observer Network, 41-42, 444
      Weather Forecast Offices, 443
nautical twilight, 483
Navstar, 24
nebula, 31-32
Nested Window Run (NWR), 450
neutral air layer, 185
neutral conditions (tropical Pacific), 301-302
New Delhi, India, 288
New England, hurricane threat to, 413
New Orleans, Hurricane Katrina, 404-405
Newell, Reginald, 166
Newton, Isaac, Sr., 473
Newton's laws of motion
      first law, 219, 251, 255
      second law, 249
Nicholas, of Cusa (Cardinal), 178
Nielsen-Gammon, 305
nimbostratus cloud, 211
nitrogen, 34
nitrous oxide, 80, 89
noctilucent clouds, 215-216
NOAA, *see* National Oceanic and Atmospheric Administration
nonattainment area, 36
Nor'easter, 2, 322, 336-337, 355-357
Norgay, Tenzing, 142
North African dust, 28, 29
North American Ensemble Forecast System (NAEFS), 451
North American Mesoscale Model (NAM), 450
North American Monsoon System (NAMS), 289-290
North Atlantic Oscillation (NAO), 310-311, 355
Northeast Snowfall Impact Scale (NESIS), 356-357
northern lights, *see* aurora

Norwegian cyclone model, 322, 331, 333
NWS, *see* National Weather Service
number density, 34
numerical model, 38
numerical weather forecasting, 449-451

O

occluded front, 329-330, 332
occlusion, 329-330, 332
      bent-back, 333
      cold, 329-330
      warm, 330
ocean, albedo of, 83-84
ocean circulation, heat transport by, 122-123
ocean effect snow, 271
ocean gyres, 286-287
Ocean Prediction Center, 457, 466-467
ocean surface currents, 286-287
Opportunity, 53
optics (atmospheric), 473-484
orographic lifting, 167, 188, 189, 325
orographic precipitation, 167
outflow boundary, 365
outgassing, 32-33, 52
outlet glaciers, 524
outlook (weather), 461
overgrazing, 279
overrunning, 327
oxygen
      isotope analysis, 507, 536
      in the atmosphere, 31-35
ozone
      in the stratosphere, 78-83
ozone shield, 34

P

pH scale, 227-228
Pacific air, 290, 325
Pacific Decadal Oscillation (PDO), 312
Pacific Marine Environmental Laboratory (PMEL), 306
Paleocene-Eocene Thermal Maximum (PETM), 506
Palmer Drought Severity Index, 318
pan evaporimeter, 197
Pangaea, 503
panhandle hook, 336
Parent, Robert, 46
parhelia, 476-477
Pascal, Blaise, 146
Pascal (Pa), 145
Pathfinder mission to Mars, 53
Périer, Florin, 146
perihelion, 70, 71
persistence, 457
Peruvian fishery, 302-303
phased array weather radar, 45-46
Pfeffer, W.T., 494, 527
Phoenix, AZ, 196, 349
Phoenix Mars Lander, 53
photolysis, 89
photochemical smog, 36

photosphere, 68, 512
photosynthesis, 33-34
photovoltaic cell (PV cell), 101-102
physical model, 38
Piezoelectric barometer, 144
Pineapple Express, 167
planetary albedo, 84, 85
planetary-scale circulation, 265, 280-287
      features of, 281-287
          idealized circulation pattern, 280-281
          seasonal shifts, 285-286
plate tectonics, 503, 505
Pleistocene Ice Age, 33, 506-508
Point Reyes, CA, 216
polar amplification, 89, 507, 522
polar front, 283
      and the jet stream, 294
polar front jet stream, 294
polar highs, 283
Polar-orbiting Operational Environmental Satellite (POES), 13, 14
polar stratospheric clouds (PSCs), 81, 216
Polaris, 72
poleward heat transport, 121-123, 284
pollen, 536
pollen-based climate reconstruction, 536
potential evapotranspiration, 197
power tower system, 101
precipitable water, 176-178, 245
precipitation, 171, 219-236
      forms of, 223-227
      global pattern, 499, 502
      measurement of, 12, 233-236
      orographic, 167
      processes, 219-223
      and radar, 235
      remote sensing of, 235-236
precipitation efficient, 380
pressure, *see* air pressure; vapor pressure
pressure altimeter, 162
pressure gradient force, 250
pressure systems, 7-10
Priestly, Joseph, 31
primary air pollutant, 36
primary air quality standards, 36
primary rainbow, 478
prime meridian, 6
primeval phase, 31-34
private sector forecasting, 461
progressive shock, 398
Project STORMFURY, 429
proxy climate data sources, 535-539
psychrometer, 179
      aspirated, 179
      sling, 179
psychrometric table, 180
Puy-de-Dôme, France, 146
pyranometer, 91-92

Q

quiet zone, 55

R

radar, 45-46, 229-233
    clear-air mode, 229
    dual polarization, 232
    echo, 229
    phased array, 233
    precipitation measurement, 233-236
    reflectivity mode, 229-230
    velocity (Doppler) mode, 230-233
radiation
    electromagnetic, 64-66, 112
    fog, 216-218
    laws of, 66-68
    measurement of, 91-92
    see also specific types
radiational cooling, 112
    and fog development, 216-218
radiational heating, 112
radiational temperature inversion, 200
radio waves
    and the ionosphere, 55-56
radiosonde, 42, 44, 45, 444-445
radiosonde observation stations, (RAOB), 45
Radiosonde Replacement System (RRS), 42, 445
radome, 45, 229, 230
rain, 223-224
    acid, 227-228
    freezing, 225-226
    gauge, 44, 233-235, 444
    shadow, 188, 189
rainbow, 478-479
rain shadow, 188, 189
rainmaking, 242-244
Rapid Refresh (RR) model, 450
Rapid Update Cycle (RUC) model, 450
rawinsonde, 44, 445
Rayleigh scattering, 77, 474
reflection, 77
refraction, 472, 475
Reichelderfer, Francis W., 440
relative humidity, 12, 175, 182
    temperature dependence of, 175
remote sensing, 45-46
return stroke, 375
ribbon lightning, 376
Riccioli, Giovanni, 252
Richardson, L.F., 449
ridge, 448
Rignot, Eric, 525
ring, 408
River Forecast Centers (RFCs), 456-457
rocketsondes, 46
roll cloud, 366
Ross, C.D., 485
Rossby, Carl-Gustav, 290
Rossby waves, 290, 295
rotor cloud, 215
Rowland, F.S., 80
Ruddiman, W.F., 506
runaway breakdown, 374
runoff, 171-172

Rutherford, D., 31

S

Saffir, H.S., 420
Saffir-Simpson Hurricane Wind Scale, 420
Sagen, Carl, 512
Saharan Air Layer (SAL), 28-29, 412
Saharan Desert, 28-29
Sahel, African drought, 278-279
Saint Louis, MO, 116-117, 208
SAME (Specific Area Message Encoding), 6
San Diego, CA, 285-286
San Francisco,
    climate of, 116-117
    fog, 218
San Joaquin Valley, CA, 218
Santa Ana wind, 346-347, 358-359
satellite
    geostationary, 13, 14
    polar-orbiting, 13, 14
saturated air, 173-175, 182-183
saturation absolute humidity, 173
saturation, achieving, 182-183
saturation mixing ratio, 173
    variation with temperature, 174
saturation specific humidity, 173
saturation vapor pressure, 173
    variation with temperature, 174
Saussure, Horace de, 208
scattering, of radiation, 76-77, 473-474
scatterometer, 265
Schaefer, V.J., 208, 242
Schaefer point, 208
Schwabe, S.H., 512
scientific method, 37-38
scientific theory, 37
scintillation, 483
scud clouds, 401
sea breeze, 12, 344
Sea Islands, SC, hurricane, 418
Sea, Lake and Overland Surges from Hurricanes
    (SLOSH), 419-420, 452
sea level
    fluctuations of, 302, 308, 494, 523-527
seasons, reasons for, 70-74
second law of thermodynamics, 112
secondary air pollutant, 36
secondary ambient air quality standards, 36
secondary cyclone, 333
secondary rainbow, 478
semi-permanent pressure systems, 281-281
sensible heating, 114, 119-120
severe thunderstorm, 371-373
shadow rule, 98-99
sheet lightning, 376
shelf cloud, 366
shore-parallel snow bands, 270-271
short wave, 295
Siberian Express, 324
Siberian high, 323, 342
Sierra Cooperative Pilot Project (SCPP), 243

silver iodide, 242
Simpson, R.H., 420
single-cell thunderstorm, 367
single-station forecasting, 460-461
Skeleton Coast, Namibia, 503
sky cover, 13
sleet, *see* ice pellets
sling psychrometer, 179
SLOSH (Sea, Lake and Overland Surges from Hurricanes), 419-420, 452
snow, 224-225
 and air temperature, 245
 albedo, 125
 bay-effect, 271
 grains, 225
 lake-effect, 269-271
 measurement of, 233-235
 meltwater equivalent, 235
 ocean effect, 271
 pellets, 225
snow bands
 shore-parallel, 270-271
 wind-parallel, 270-271
snow burst, 269
snow fence, 254
snow grains, 225
snow pellets, 225
Snowball Earth hypothesis, 503
Sojourner, 53
solar altitude, 69
solar cell, 101-103
 thin-film, 103
solar collector, 100-101
solar constant, 74-76
solar corona, 69
solar flare, 49-57
solar power, 100-103
solar pyranometer, 91-92
solar radiation
 absorptivity, 76
 budget of, 84-85
 interactions with the atmosphere, 76-78
 interactions with the Earth's surface, 83-85
 and the seasons, 70-74
solar wind, 48
solstice, 71, 72-73
solstitium, 73
sonic anemometer, 264, 485
sonic boom, 486
sound waves, 484-485
sounding, 184-185
southern lights, *see* aurora
Southern Oscillation, 297
Southern Oscillation Index (SOI), 297
Southwest Monsoon, 289-290
space weather prediction, 57-59
Space Weather Prediction Center, 57, 457
Spanish-American War of 1898, 451
Special Sensor Microwave Images (SSM/I), 181
Specific Area Message Encoding (SAME), 6
specific heat, 115-116
specific humidity, 172

speed convergence, 260-261
speed divergence, 260-261
speed of light, 65
spin-up vortices, 417
Spirit, 53
split flow pattern, 291
Spörer minimum, 513
Sputnik I, 46
squall line, 328, 356, 368
stable air layer, 184
stability
  absolute, 184-185
  of air masses, 324, 325
  conditional, 184
standard atmosphere, 146-148, 162
standing wave, 346
station model, 445, 446
stationary front, 10, 11, 326-327
Steadman, R.G., 194
steam fog, 218-219, 269
Stefan-Boltzmann law, 67-68
stepped leaders, 375
Stiger, M. Albert, 399
Stonehenge, 62
Storm Prediction Center (SPC), 440, 454
storm spotters, 392
storm surge, 355, 417-420
stratiform cloud, 18, 209
stratocumulus cloud, 211
stratopause, 46, 47
stratosphere, 46, 47
 ozone layer in, 34
stratospheric ozone shield, 34, 78-83, 96
stratus cloud, 211
streak lightning, 376
streamline, 406
Stüve thermodynamic diagram, 185-187
sublimation, 114, 115, 171
subglacial lakes, 524
sub-polar gyres, 286-287
subpolar lows, 283
Sub-Saharan Africa, drought, 278-279
sub-satellite point, 13,14
subsidence temperature inversion, 199
subtropical anticyclone, 282
subtropical gyres, 286, 287
subtropical jet stream, 295, 372
sudden ionospheric disturbances (SIDs), 56
sulfurous aerosols, 516-518
sun pillar, 477
sun protection factor (SPF), 99
sunbathing, hazards of, 96-99
sundogs, 476-477
sunspots, 57, 68, 512-514
Suomi, Verner E., 46
supercell thunderstorm, 369, 389
supercooled water, 173, 208
supercooling, 208
supergranules, 68
superior mirage, 482
supertyphoon, 410
Super Tornado Outbreak of 2011, 362-363

supraglacial lakes, 526
surface tension, 223
surface weather maps, 446
surface weather observations, 39-42, 442-444
sustained wind speed, 406
synoptic scale systems, 265, 323-343
system, 38

T

Tahiti, 297
Tai-fung, 408
Taino people, 406
Tambora, 516
TAO/TRITON, 306, 307
Taylor, Frank B., 503
teleconnection, 303, 458
temperature
    absolute, 66
    annual cycle of, 124
    apparent, 194-195
    definition of, 108
    dewpoint, 12, 175-176, 177
    dry-bulb, 179, 180
    effect on air pressure, 149-150
    frost point, 12
    gradients in, 112
    inversion, 184, 199-201, 482
    maximum daily, 12
    measurement of, 110-112
    minimum daily, 12
    profile of the atmosphere, 46-47
    radiational controls of, 124, 125
    scales, 108-109
    wet-bulb, 179, 180
    wind chill equivalent, 137-139
temperature inversion, 184, 199-201, 482
temperature scales, 108-109
Terminal Doppler Weather Radar (TDWR), 377
terminal velocity, 219-220
thermal inertia, 116, 287
thermal low, 285, 289, 341
thermal turbulence, 254
thermocline, 301
thermohaline circulation, 123
thermograph, 111
thermometer, 110-112
    response time, 111
thermosphere, 46, 47, 148
thin-film solar cell, 103
Thomson, William (Lord Kelvin), 109
thunder, 376, 485-486
thundersnow, 371
thunderstorm, 11, 12, 328, 364-384
    charge separation in, 374-376
    classification, 367-369
    days, 369
    dissipating stage, 367
    and downbursts, 376-378
    downdraft, 365
    and flash-to-bang method, 376
    and Florida, 370-371
    frontal, 328, 368
    geographical distribution of, 369-371
    hail in, 226-227
    hazards of, 373-384
    and lightning, 373-376
    life cycle, 364-367
    mature stage, 365-366
    multicellular, 367-369
    severe, 371-373
    single-cell, 367
    stages in life cycle of, 364-367
    supercell, 369, 389
    and tornadoes, 389-391
    towering cumulus stage, 364-365
thunderstorm day, 369-370
Tibetan Plateau, 506
Tilton, James, 40
time zones, 6-7, 73
time keeping, 6-7
Tinker Air Force Base, OK, 440
tipping point, 494
tipping-bucket rain gauge, 234
TIROS-I (Television and Infrared Observation Satellite), 46
TOGA (Tropical Ocean Global Atmosphere) program, 306
Tong, Anote, 494
TOPEX/Poseidon, 306
tornado, 384-393
    alley, 385-386
    characteristics of, 384-385
    geographical distribution, 385-386
    and hurricanes, 417
    look-alikes, 401
    multi-vortex, 387
    safety during, 386
    with subsidiary vortices, 387
    thunderstorm, connection with, 389-391
Tornado Alley, 385-386
Tornado Vortex Chamber, 38, 39
Torricelli, Evangelista, 143
Total Ozone Mapping Spectrometer (TOMS), 81
towering cumulus stage (thunderstorm), 364-365
trade wind inversion, 284-285
trade winds, 282
transmissivity, 76
transpiration, 118, 170-171
Travis, David J., 204
tree growth index, 535
tree growth rings, 535
trilateration, 24
triple point, 333
Tri-State tornadoes of 1925, 363, 385
TRMM Microwave Imager (TMI), 236, 307
TRMM (Tropical Rainfall Measuring Mission), 236, 307
Tropic of Cancer, 72
Tropic of Capricorn, 73
tropical cyclones, 28-29, 303, 406, 451-454
tropical depression, 412
tropical disturbance, 411
tropical non-squall cluster, 406
Tropical Ocean Global Atmosphere (TOGA) program, 306
Tropical Rainfall Measuring Mission (TRMM), 236, 301
tropical squall cluster, 406

tropical storm, 412
  Alberto, 415
  Allison, 415
  Fay, 415
tropics, weather in, 406
tropopause, 46, 47, 284
troposphere, 46, 47
trough, 448
true altitude, 162
Tucson, AZ, 289-290
turbulence, 215, 254
Twain, Mark, 30-31
twilight, 483
twinkling stars, 482-483
typhoon, 408
Typhoon Tip, 145
Typhoon Vamei, 409

U

Uccellini, L., 357
UFOs (Unidentified Flying Objects), 213
ultraviolet radiation (UV), 64, 65, 78-83
  hazards of, 96-98
UN Environmental Programme, 97
UV Index, 97-98
UV radiation, 34
UVA, 96
UVB, 79
UVC, 96
Universel Temps Coordinné (UTC), 443
updraft, 365
upper-air support, 331, 337
upper-air weather maps, 446-449
upper-air weather observations, 42-45, 444-445
upslope fog, 219
upwelling, 300
urban heat island effect, 126-128
U.S. Army Signal Corps, 40, 440, 451, 456
U.S. Department of Defense, 443
U.S. Environmental Protection Agency (EPA), 89

V

valley breeze, 344-345
vapor pressure, 172
  at saturation, 174
variables of state, 151
Venus, 35, 67-68
Verification of the Origins of Rotation in Tornadoes
  Experiment 2 (VORTEX 2), 391, 392
vertical evacuation, 427
vertically integrated water vapor, 181
Viking spacecraft, 52
virga, 222, 230, 401
viscosity, 253-254
visible light, 64, 65
visible radiation, 64, 65
  see also solar radiation
visible satellite image, 14, 15
Vizy, E., 279
Von Neumann, John, 449

Vonnegut, Bernard, 242

W

Walker, Sir Gilbert, 297, 310
Walker circulation, 302
wall cloud, 389-390
warm air advection, 125-126,150, 448
warm cloud, 209
warm-cloud precipitation, 220-221
warm-core anticyclone, 342
warm-core cyclone, 341
warm-core rings, 408
warm front, 10, 11, 327-328
warm occlusion, 330
water
  change in phase of, 114, 115, 170-171
  cycling of, 33, 168-171
  global budget of, 171-172
  origin of, 32, 168
  phase changes of, 114-115
  precipitable, 176-178, 245
  specific heat of, 115-116
  supercooled, 173, 208
water vapor
  monitoring, 178-182
  plume, 181
  satellite imagery, 14, 16
water vapor satellite imagery, 14, 16, 181-182
watershed, 172
waterspout, 401
wave cyclone, 332
wave frequency, 64-65
wavelength, of radiation, 64, 65
weather, 4-5
  analysis, 22
  definition, 4
  maps, 8
  persistence, 457
  radar, 16-17, 45-46, 229-233
  risk, 133-134
  satellites, 13-16
  warning, 13
  watch, 13
  see also weather forecasting; weather maps
The Weather Channel, 5
weather instruments, see specific type
weather maps,
  historical perspective, 22-23
weather radar, 16-17, 45-46, 229-233
  dual polarization, 232
  phased array, 233
  reflectivity mode, 229-230
  velocity (Doppler) mode, 230-232
weather radio, 5-6
Weather-Ready Nation, 462
Weather Reconnaissance Squadron (53rd), 451
Weather Research and Forecast Model (WRF), 450
weather risk, 133-134
Weather Satellite Imagery, 13-16
weather systems, 7-12
weather warning, 461

weather watch, 461
Wegener, Alfred, 221, 503, 514
weighing-bucket rain gauge, 234
weight, 143
West African monsoon, 278
wet deposition, 227
white dew, 176
whiteout, 339
Wien's displacement law, 66-67, 92, 473
Wilson, Alexander, 43
wind
       continuity of, 260-261
       forces governing, 249-255
       geostrophic, 255-256
       gradient, 256-257
       and human comfort, 137-139
       measurement of, 12-13
       monitoring, 261-265
       pressure, 417
       profiler, 272
       shear, 213
wind power, 273-275
wind shear, 371, 410
wind vane, 262
wind-chill, 137-138
wind-chill equivalent temperature (WET), 137-138
wind-chill index, 137-138
windmill farms, 274-275
windsock, 262
winter storms, 339-340
Wisconsin, 503, 505
World Meteorological Organization (WMO), 441, 510
World Weather Watch (WWW), 441
Wragge, Clement, 434
WSR-88D radar, 45, 230-232

X

X rays, 64, 65

Y

Yakima, WA, 188, 189
Younger Dryas, 508

Z

Zhu, Yong, 166
zonal flow pattern, 290, 291
zonda, 346